FOOD CHEMICALS CODEX

SECOND EDITION

FOOD CHEMICALS CODEX

SECOND EDITION

Prepared by the
Committee on Specifications,
Food Chemicals Codex,
of the
Committee on Food Protection
National Research Council

National Academy of Sciences
Washington, D. C.
1972

NOTICE

The study reported herein was undertaken under the aegis of the National Research Council with the approval of its Governing Board. Such approval indicated that the Board considered the problem of national significance and that the resources of NRC were particularly suitable to the conduct of the project.

The members of the committee were selected for their individual scholarly competence and judgment with due consideration for the balance and breadth of disciplines. Responsibility for all aspects of this report rests with the study committee, to whom sincere appreciation is expressed.

Although the reports of committees are not submitted for approval to the Academy membership or to the Council, each is reviewed according to procedures established and monitored by the Academy's Report Review Committee. The report is distributed only after satisfactory completion of this review process.

Acknowledgment. This study was supported in part by U. S. Food and Drug Administration Research Grant No. FD 00213, administered by the Department of Health, Education, and Welfare, U. S. Public Health Service.

Compliance with Federal Statutes. The fact that an article appears in this Food Chemicals Codex does not exempt it from compliance with requirements of Acts of Congress or with regulations and rulings issued by agencies of the United States Government under authority of these Acts.

Revisions of the federal requirements that affect the Codex standards will be included in Codex Supplements as promptly as practicable.

ISBN 0-309-01949-4

Library of Congress Catalog Card No. 66-61563

Printed in the United States of America

Contents

Organization of the Food Chemicals Codex vii

Preface to the Second Edition ix

Preface to the First Edition xiv

General Provisions 1

Monographs 9

General Tests and Apparatus 861

Solutions and Indicators 983

Former and Current Titles of Food Chemicals Codex Monographs 1009

Food Chemicals Codex Substances Classified by Functional Use in Foods 1011

Index 1021

Organization of the Food Chemicals Codex 1966–1972

National Academy of Sciences—National Research Council

PHILIP HANDLER, *President* (1969–   )
FREDERICK SEITZ, *President* (until 1969)

DIVISION OF BIOLOGY AND AGRICULTURE

D. S. FARNER, *Chairman* (1969–   )
A. G. NORMAN, *Chairman* (until 1969)

R. B. STEVENS, *Executive Secretary*

FOOD AND NUTRITION BOARD

D. MARK HEGSTED, *Chairman* (1968–1972)
GRACE A. GOLDSMITH, *Chairman* (until 1968)

PAUL E. JOHNSON, *Executive Secretary* (1971–   )
LEROY VORIS, *Executive Secretary* (until 1971)

COMMITTEE ON FOOD PROTECTION

L. J. FILER, *Chairman* (1971–1972)
W. J. DARBY, *Chairman* (until 1971)

LORNE A. CAMPBELL, *Staff Officer* (1971–   )
PAUL E. JOHNSON, *Executive Secretary* (until 1971)

FOOD CHEMICALS CODEX

DURWARD F. DODGEN, *Director* (1969–   )
Consultant (1966–1969)

PAUL E. JOHNSON, *Director* (1966–1969)

JUSTIN L. POWERS, *Consultant*

ADVISORY PANEL ON THE FOOD CHEMICALS CODEX (1966–1970)

R. BLACKWELL SMITH, JR.†
Chairman
Richmond, Va.

EDWARD G. FELDMANN
Washington, D. C.

LLOYD C. MILLER
San Diego, Calif.

HENRY FISCHBACH
Washington, D. C.

R. H. PHILBECK
Washington, D. C.

H. C. SPENCER
Midland, Mich.

† Deceased

COMMITTEE ON SPECIFICATIONS

HENRY FISCHBACH
Chairman
Washington, D. C.

C. J. KRISTER
Vice Chairman
Wilmington, Del.

JOHN D. BRANDNER
Wilmington, Del.

ROBERT S. BRYANT (1971–)
Westport, Conn.

DOUGLAS G. CHAPMAN (1970–)
Ottawa, Ontario, Canada

B. F. DAUBERT (1966–1971)
White Plains, N. Y.

WILLIAM S. ENGLE (1966–1970)
Charleston, W. Va.

EDWARD C. FEARNS
Union Beach, N. J.

ANTHONY FILANDRO (1969–)
Brooklyn, N. Y.

J. P. FLETCHER (1970–)
S. Charleston, W. Va.

CHESTER L. FRENCH
St. Louis, Mo.

THOMAS C. GRENFELL (1966–1970)
New York, N. Y.

RICHARD L. HALL (1966–1968)
Cockeysville, Md.

WILLIAM H. HUNT (1966–1971)
St. Louis, Mo.

JOHN C. KIRSCHMAN (1971–)
White Plains, N. Y.

FRED A. MORECOMBE (1968–)
North Chicago, Ill.

L. L. RAMSEY
Washington, D. C.

ANDREW SCHMITZ, JR. (1970–)
New York, N. Y.

HOWARD C. SPENCER (1966–1971)
Midland, Mich.

ADVISORS TO THE COMMITTEE ON SPECIFICATIONS

W. STANLEY CLABAUGH (1966–1970)
Port Republic, Md.

MAURICE F. LOUCKS
Midland, Mich.

FRED A. MORECOMBE (1966–1968)
North Chicago, Ill.

R. O. READ (1970–)
Ottawa, Ontario, Canada

Preface to the Second Edition

The historical events and developmental work leading to the publication of the First Edition of the Food Chemicals Codex in October, 1966, under the direction of Dr. Justin L. Powers, are summarized in the *Preface to the First Edition* reprinted herein on pages xiv–xvi.

Organization. Since its inception, the Food Chemicals Codex project has been under the administrative supervision of the Committee on Food Protection of the Food and Nutrition Board. An Advisory Panel was formed in 1961, during the first year of the project, to set general policy and to develop guidelines to be followed in the preparation of the First Edition, and a Committee on Specifications of the Advisory Panel was organized as the working group to develop the monographs and general test procedures. Both of these groups continued to operate until 1970, when the Advisory Panel, having served its function in guiding the project through the First Edition, was released. The Committee on Specifications then came under the direct supervision of the Committee on Food Protection.

Scope. The scope of the Second Edition was broadened slightly over that of the First Edition. Substances included in the First Edition were limited largely to chemicals added directly to foods to perform some desired function, whereas for the Second Edition many substances not added directly to foods but which come into contact with foods, such as food processing aids (e.g., extraction solvents, filter media), have been included. Furthermore, a number of substances that are not considered to be "chemical additives" in the conventional sense have been included in the new edition; examples are the modified food starches, the masticatory substances used in chewing gum base, and pectin.

All of the 512 monographs from the First Edition, plus the 45 monographs added via the four supplements to the First Edition, have been incorporated in the Second Edition, except for five monographs: calcium cyclamate, cyclohexylsulfamic acid, diethylpyrocarbonate, nordihydroguaiaretic acid, and sodium cyclamate. The addition of 87 completely new monographs not previously published brings to 639 the total number of monographs in the Second Edition.

Sources of Specifications. The sources of specifications used in the development of new monographs for the Second Edition were essentially the same as those employed for the First Edition, except that sources from other compendia have been utilized to a very minor extent, and greater reliance has been placed upon cooperation with individual suppliers of food-grade chemicals and with industrial trade organizations (see *Acknowledgments*).

Design and **Mechanism of Compilation.** The design of the Second Edition is the same as that of the First Edition. The mechanism of compilation followed for the Second Edition differs from that of the First Edition in that neither the provisional specifications nor the loose-leaf specifications were published. However, all proposed specifications were reviewed by the manufacturers or suppliers of the substances, and all monographs prepared from the specifications were approved by the Committee on Specifications before final adoption. It should be emphasized, however, that the Codex specifications are under continuing scrutiny and that many revisions that could not be implemented in time for publication in the Second Edition will be made via supplemental revisions.

In addition to the new monographs in the Second Edition, specifications and test procedures in many of the individual monographs from the First Edition have been revised. These changes, deletions, or additions in the specifications have been made, where possible hazard was involved, with the approval of the Subcommittee on Toxicology of the Committee on Food Protection.

Limits of Impurities. With one exception, the policy regarding limits of impurities that was followed for the First Edition has also been observed in preparing the Second Edition. The policy followed in developing specifications for the First Edition is quoted below:

> It will be the policy of the Food Chemicals Codex to set maximum limits for trace impurities wherever they are deemed to be important for a particular chemical, and they shall be set at levels consistent with safety and good manufacturing practice. The maximum limits for heavy metals shall be 40 parts per million, for lead 10 parts per million, and for arsenic 3 parts per million, except in instances where higher levels cannot be avoided [under conditions of good manufacturing practice]. Where a heavy metals limit of 10 parts per million can be established, a separate limit for lead need not be specified.
>
> Flavoring agents used in foods at levels of 0.01 percent or less require only a heavy metals limit of 40 parts per million, and separate arsenic and lead tolerances may be safely omitted from [specifications for] these substances.
>
> Maximum limits for other inorganic traces impurities [e.g., fluoride, mercury, selenium] will be included in any monographs where safety or manufacturing experience indicate their desirability.

The policy in connection with limits of impurities in flavoring agents was modified for the Second Edition. Long experience had shown that the heavy metals test is always negative when a vast class of flavoring agents is tested, specifically those agents that (a) are organic liquids, (b) are purified by distillation, (c) are immiscible with water, and (d) do not dissolve inorganic substances. In such cases, the heavy metals limits do not contribute to safety or to good manufacturing practice, especially when the flavoring agents are used in foods at levels of 0.01 percent or lower. Consequently, it was decided that limits for arsenic, heavy metals, and lead need not be included in the specifications for flavoring agents used in foods at levels of 0.01 percent or lower, provided that they meet the above criteria.

Nomenclature. The titles of a number of monographs have been changed in going from the First to the Second Edition. A listing of former and current titles is provided on page 1009. One title, "Poloxamer," which is applied to two substances belonging to a series of polyols, was coined specifically for use in this Codex. This name was developed with the cooperation and assistance of the U. S. Adopted Names (USAN) Council. It is expected that additional nonproprietary names to be used as titles of Codex monographs will be developed in this manner for other substances hitherto known only by their tradenames or by long, unwieldy chemical names.

Test Procedures. Almost all of the test procedures employed in the First Edition have been retained for use in the Second Edition, and several new procedures have been added. Many of the older procedures have been substantially revised. With few exceptions, the methods employed in this Codex are considered to be adequate for their intended use in determining compliance of the substances to the specifications.

The limitations of the conventional heavy metals test procedure, however, are well known, and an extensive two-year collaborative study was conducted in an effort to improve the accuracy and precision of the method. The study, while inconclusive in other respects, indicated that the reproducibility of the method is approximately 7 parts per million at the 95 percent confidence level for substances containing 10 parts per million of heavy metals. While it is recognized that the test procedure seldom gives a truly accurate indication of the actual heavy metals content, it does demonstrate that the test substance is not grossly contaminated with heavy metals. In this manner, the test does serve a useful function in partially defining the purity of food-grade chemicals.

Nevertheless, a continuing effort is being made to improve the test procedure and the methods of sample preparation to the greatest extent practicable. Any such improvements will be announced by supplemental revisions to this Codex. Where it is found, with certain substances, that the method cannot be improved to give a reasonably accurate indication of the heavy metals content, consideration will be given to the use of other procedures (e.g., atomic absorption spectrophotometry) that will determine individual elements, for which limits will be specified in lieu of a general heavy metals limit.

Legal Status. The First Edition of the Food Chemicals Codex was given quasi-legal recognition by means of a letter of endorsement from the Commissioner of Foods and Drugs, which was reprinted in the book. At that time (April, 1966), the Commissioner stated that "The

FDA will regard the specifications in the Food Chemicals Codex as defining an 'appropriate food grade' within the meaning of Sec. 121.101(b)(3) and Sec. 121.1000(a)(2) of the food additive regulations," although such endorsement could not be construed to exempt substances from compliance with requirements of Acts of Congress or with regulations and rulings issued by the FDA under authority of such Acts.

Subsequently, the Codex was given official recognition by the U. S. Food and Drug Administration when the "Definitions and Procedural and Interpretative Regulations," relating to the eligibility of substances for classification as generally recognized as safe (GRAS), were revised and published in the Federal Register on June 25, 1971 (Vol. 36, No. 123, page 12093). Reference is made in Sec. 121.3(d) of the regulation as follows:

> Any substance used in food must be of food-grade quality. The Commissioner regards the applicable specifications in the current edition of "Food Chemicals Codex" as establishing food grade unless he has by Federal Register promulgation established other specifications.

Food Chemicals Codex specifications have also been adopted, under certain conditions, by the Food and Drug Directorate of Canada (January 29, 1970) and by the Food Additives and Contaminants Committee, Ministry of Agriculture, Fisheries and Food, of Great Britain (1968).

Acknowledgments. The Committee on Food Protection and the Codex Committee on Specifications wish to express appreciation for encouragement and support by the U. S. Food and Drug Administration, whose Research Grant No. FD 00213, administered by the Department of Health, Education, and Welfare, U. S. Public Health Service, has made possible the development of the specifications published in this Codex.

Many individuals associated with food processors and chemical manufacturers have contributed to this project by supplying information relating to specifications and analytical procedures for food-grade chemicals and by reviewing proposed specifications prior to their publication. This assistance is gratefully acknowledged. Thanks is also extended to the many industrial firms, associations, and foundations that contributed grants during the first five years of the project, leading to the publication of the First Edition.

The following organizations assisted in developing or reviewing specifications for a number of new substances included in this edition: American Spice Trade Association, Corn Refiners Association, Essential Oil Association, Flavor and Extract Manufacturers Association, and National Association of Chewing Gum Manufacturers. The cooperation of these organizations is greatly appreciated.

Portions of many of the monographs and most of the general tests included in the First Edition were adapted from, and used with permission granted by the parent organizations of, the following publications: United States Pharmacopeia, Sixteenth Revision; National Formulary, Eleventh Edition; Reagent Chemicals—A.C.S. Specifications 1960; Official and Tentative Methods of the American Oil Chemists Society; Essential Oil Association of USA Specifications, Infrared Spectra, and Revisions; and Specifications for Flavoring Materials, Flavoring and Extract Manufacturers Association. In addition, the following A.S.T.M. procedures (many of which were included in the First Edition) have been adapted for use in the Second Edition, with permission granted by the American Society for Testing Materials: B 214-66, D 1078-58, D 1347-56, D 1416, D 1417, D 1824, D 1439, D 1493-58T, E 28-67, E 77-58, and EL-62.

Future Revisions. Following the mechanism used to make revisions in or additions to the specifications in the First Edition, the Second Edition will also be kept up-to-date by the issuance of annual supplements, which will be sent to all holders of the book at no charge. It is expected that a Third Edition of the Codex will be published in 1978.

The Committee on Specifications would welcome any constructive criticisms and suggestions regarding the specifications and analytical procedures incorporated in this Codex. Reports of errors, suggestions for revisions, and any other comments should be addressed to: Food Chemicals Codex, National Academy of Sciences, 2101 Constitution Avenue, N. W., Washington, D. C. 20418.

May, 1972

Preface to the First Edition

The need for a compilation of standards for food-grade chemicals has been recognized for quite some time, but it was not until 1958, soon after the enactment of the Food Additives Amendment, that any positive action was taken to compile such a compendium. Although the federal Food and Drug Administration (FDA) had by regulations and informal statements defined in general terms quality requirements for food chemicals generally recognized as safe (GRAS), these requirements were not designed to be sufficiently specific to serve as release, procurement, and acceptance specifications by primary chemical manufacturers and food processors. Since complete specifications and quality-control procedures required by the FDA in food-additive petitions for chemicals not included in the GRAS lists were not published in the official regulations, their use for general guidance was restricted. It was therefore incumbent upon food processors to provide detailed procurement specifications when ordering food-additive chemicals from primary manufacturers or distributors. This system may have functioned satisfactorily in most instances, but it was generally believed that the availability of a book of standards designed especially for food-additive chemicals would be more convenient and would promote greater uniformity of quality and thus provide added assurance of safety.

For these and other reasons, the Food Protection Committee of the National Academy of Sciences-National Research Council received requests in 1958 from its Industry Liaison Panel and other sources to undertake a project designed to produce a Food Chemicals Codex comparable in many respects to the United States Pharmacopeia (U.S.P.) and the National Formulary (N.F.).

In response to these requests, advice was sought from special committees composed of representatives of industry, government agencies, and others experienced in the operation of the U.S.P. and the N.F. It was the consensus of these groups that there was a definite need for a Food Chemicals Codex and that the Food Protection Committee was a suitable agency to assume responsibility for the project.

This first edition of the Food Chemicals Codex, parts of which were published in loose-leaf form between 1963 and 1966, is the result of an effort by the Food Protection Committee started in 1961 to provide objective quality standards for food-grade chemicals. The aim of the Codex is to define a substantial number of food-grade chemicals in terms of minimum identity and purity specifications based on the elements of safety and good manufacturing practice. It is believed that this objective has been achieved. As indicated in a letter written by Dr. James L. Goddard, Commissioner of Food and Drugs, the Food Chemicals Codex specifications have received endorsement by the federal Food and Drug Administration as constituting adequate minimum requirements of purity for chemicals permitted for intentional and purposeful use in food for man. With this official endorsement, it is expected that the Codex standards will be utilized by food processors as procurement and acceptance specifications and by primary manufacturers of food-grade chemicals as release specifications.

Scope. The scope of this first edition of the Codex is limited to substances amenable to chemical characterization or biological standardization which are added directly to food to perform some desired function. Such substances were selected from food additives generally recognized as safe, those approved by prior sanctions, and those for which special use tolerances have been established by FDA regulations.

Sources of Specifications. Specifications and analytical procedures required for the Codex have been adapted from compendia devoted to standards for chemicals, from original scientific literature sources, and from data supplied by chemical manufacturers and food processors. In some instances where procedures required laboratory study, the facilities of commercial consulting laboratories have been utilized, but often the necessary work has been done in industry laboratories as a service to the Codex project.

Design. Specifications and procedures for their determination are presented in the form of monographs, which constitute the major portion of the Codex. Other sections cover subjects such as general provisions designed to interpret the relative significance that should be attached to the different types of specifications, and general tests and solutions frequently referred to in the monographs.

Mechanism of Compilation. In general, provisional specifications, based on information obtained from reliable sources, were prepared in the office of the director of the project and then circulated for review to selected members of committees and panels associated with the Food Protection Committee and the Codex, and to all manufacturers who submitted data on their products. Suggestions and recommendations for revisions received from these sources resulted in revisions prior to the publication between 1963 and 1966 of a loose-leaf edition of the Codex in ten parts, which was made generally available upon a subscription basis. Finally, the loose-leaf edition was further revised, the pages collated in appropriate sequence, and the material published in its present form.

Future Revisions. If the Food Chemicals Codex is to function effectively as an authoritative book of standards for food-grade chemicals, provision for its continuous revision under appropriate sponsorship and supervision is highly essential. It is a source of gratification to those who have made the publication of this first edition possible that such provision has been made. The Governing Board of the National Academy of Sciences has approved a plan for continuing the sponsorship of the Codex for a second five-year period under the administrative supervision of the Food Protection Committee of the Food and Nutrition Board. The approved plan provides for the issuance and distribution of interim revision supplements whenever necessary and the publication of a second, completely revised edition of the Codex in 1971.

Assistance and Support. During the course of the development and compilation of the Food Chemicals Codex, cooperation was received from many sources.

The many constructive suggestions offered by the members of the Food Protection Committee, its subcommittees, and its Liaison Panel have been most helpful.

In devising suitable specifications for flavoring agents, the assistance of the Scientific Section of the Essential Oil Association of the USA and Scientific Research Committee of the Flavoring Extract Manufacturers' Association has been particularly notable, and appreciation is expressed to these two groups for their valuable contributions.

Many individuals associated with food processors and primary manufacturers of chemicals have contributed greatly to the project by furnishing information and advice relating to specifications and analytical procedures for food-grade chemicals and by reviewing provisional specifications prior to their publication.

The Food Protection Committee and those directly responsible for the compilation of the Food Chemicals Codex wish to express appreciation for encouragement and support by the Public Health Service whose Research Grant No. EF-00222 from the *Division of Environmental Engineering and Food Protection* has made possible the compilation and publication of the Food Chemicals Codex. Comparable appreciation should also be recorded for the contribution of supplementary grants in support of the Codex project by industry, and by associations and foundations.

June, 1966

General Provisions

Applying to Standards, Tests, and Assays of the Food Chemicals Codex

TITLE

The title of this book, including supplements thereto issued separately, is the *Food Chemicals Codex*, Second Edition. It may be abbreviated to F. C. C. II.

When manufacturers of F. C. C. substances wish to indicate on their labels that the substances conform to F. C. C. specifications, the designation "Food Chemicals Codex Grade," or "F. C. C. Grade," or simply "F. C. C." (implying the concurrent edition of the F. C. C.) may be used.

Where the term "Codex" is used without further qualification in the text of this book, it applies to the *Food Chemicals Codex*, Second Edition.

CODEX STANDARDS

The specifications for identity and purity described in the *Food Chemicals Codex* are designed to serve for substances of a quality level sufficiently high to ensure their safety under usual conditions of intentional use in foods or in food processing. These specifications generally represent acceptable levels of quality and purity of food-grade substances available in the United States. The main titles of substances in Codex monographs are in most instances the common or usual names. The standards apply equally to substances bearing the main titles, synonyms listed under the main titles, and names derived by transposition of definitive words in main titles.

The assays and tests described constitute methods upon which the standards of the *Food Chemicals Codex* depend. The analyst is not prevented, however, from applying alternative methods if he is satisfied that the procedures he uses will produce results of equal accuracy. In the event of doubt or disagreement concerning a substance purported to comply with the quality requirements of this Codex, only the methods described herein are applicable and authoritative.

ATOMIC WEIGHTS AND CHEMICAL FORMULAS

Computation of molecular weights and volumetric and gravimetric factors stated in tests and assays are based upon the 1969 Revision of the International Atomic Weights.

Molecular and structural formulas and molecular weights immediately following monograph titles are included for the purpose of information and are not to be considered an indication of the purity of the compound. Molecular formulas given in specifications, tests, and assays, however, denote the pure chemical entity.

ASSAYS AND TESTS

Analytical Samples. In the description of assays and tests, the approximate quantity of the analytical sample to be used is usually indicated. The quantity actually used, however, should not deviate by more than 10 percent from that stated.

Some substances are directed to be dried before a sample is taken for an assay or test. When a *Loss on drying* or *Water* test is given in the monograph, the undried substance may be used and the results calculated on the dried basis, provided that any moisture or other volatile matter in the undried sample does not interfere with the specified assay and test procedures.

The word "accurately," used in connection with gravimetric or volumetric measurements, means that the operation should be carried out within the limits of error prescribed under *Volumetric Apparatus*, page 976, or under *Weights and Balances*, page 980. The same significance also applies to the term "exactly" or expressions such as "100.0 ml." or "50.0 mg."

The word "transfer," when used in describing assays and tests, means that the procedure should be carried out quantitatively.

Apparatus. With the exception of volumetric flasks and other exact measuring or weighing devices, directions to use a definite size or type container or other laboratory apparatus are intended only as recommendations, unless otherwise specified.

Where an instrument for physical measurement, such as a thermometer, spectrophotometer, gas chromatograph, etc., is designated by its distinctive name or tradename in a test or assay, a similar instrument of equivalent or greater sensitivity or accuracy may be employed.

Blank Tests. Where a blank determination is specified in a test or assay, it is to be conducted by using the same quantities of the same reagents and by the same procedure repeated in every detail except that the substance being tested is omitted.

A *residual blank titration* may be stipulated in assays and tests involving a back titration in which a volume of a volumetric solution larger than is required to react with the sample is added, and the excess of this solution is then titrated with a second volumetric solution. Where a *residual blank titration* is specified or where the procedure involves such a titration, a blank is run as directed in the preceding paragraph. The volume of the titrant consumed in the back titration is then subtracted from the volume required for the blank. The difference between the two, equivalent to the actual volume consumed by the sample, is the corrected volume of the volumetric solution to be used in calculating the quantity of the substance being determined.

Constant Weight. A direction that a substance is to be "dried to constant weight" means that the drying should be continued until two consecutive weighings differ by not more than 0.5 mg. per gram of sample taken, the second weighing to follow an additional hour of drying.

The direction "ignite to constant weight" means that the ignition should be continued at 800° ± 25°, unless otherwise specified, until two consecutive weighings do not differ by more than 0.5 mg. per gram of sample taken, the second weighing to follow an additional 15-minute ignition period.

Desiccators and Desiccants. The expression "in a desiccator" means the use of a tightly closed container of appropriate design in which a low moisture content can be maintained by means of a suitable desiccant. Preferred desiccants include anhydrous calcium chloride, magnesium perchlorate, phosphorus pentoxide, and silica gel.

Identification. The tests described under this heading in monographs are designed for application to substances taken from labeled containers and are provided only as an aid to substantiate identification. These tests, regardless of their specificity, are not necessarily sufficient to establish proof of identity, but failure of a substance taken from a labeled container to meet the requirements of a prescribed identification test is an indication that it will not conform to the specifications in the monograph.

Indicators. The quantity of an indicator solution used should be 0.2 ml. (approximately 3 drops) unless otherwise directed in an assay or test.

Loss on drying and Water. In general, a limit test, to be determined by the *Karl Fischer Titrimetric Method*, is provided under the heading *Water* for compounds containing water of crystallization or adsorbed water.

Limit tests under the heading *Loss on drying*, determined by other methods, are designed for compounds in which the loss on drying may not necessarily be attributable to water.

Negligible. The term "negligible," as used in some *Residue on ignition* specifications, indicates a quantity not exceeding 0.5 mg.

Odorless. This term, when used in describing a substance, applies to the examination, after exposure to air for 15 minutes, of about 25 grams of the substance that has been transferred quickly from the original container to an open evaporating dish of about 100-ml. capacity. If the package contains 25 grams or less, the entire contents should be examined.

Reagents. Specifications for reagents are not included in the *Food Chemicals Codex.* Unless otherwise specified, reagents required in tests and assays should conform to the specifications of the current editions of the "United States Pharmacopeia," the "National Formulary," or "Reagent Chemicals–American Chemical Society Specifications." Reagents not covered by any of these specifications should be of the best grade available and should be examined for interfering impurities.

Reference Standards. Some instrumental and chromatographic tests and assays specify the use of a reference standard. Where a reference standard is designated as "U.S.P." it may be obtained from the United States Pharmacopeia, 12601 Twinbrook Parkway, Rockville, Md. 20852. Where identified as "N. F.," the reference standard is available from the American Pharmaceutical Association, 2215 Constitution Avenue, N. W., Washington, D. C. 20037. Reference standards bearing the abbreviation F. C. C. are supplied by the *Food Chemicals Codex*, 2101 Constitution Avenue, N. W., Washington, D. C. 20418.

Significant Figures. Where tolerance limits are expressed numerically, the values are considered to be significant to the number of digits indicated. Thus, "not less than 99.0 percent" means 99.0 but not 99.00. Values should be rounded off to the nearest indicated digit according to the commonly used practice of rejecting or increasing numbers less than or greater than 5. For example, a requirement of not less than 96.0 percent would be met by a result of 95.96 but not by a result of 95.94. When the digit to be dropped is exactly 5, the value should be rounded off to the closest even digit. Thus, both 1.4755 and 1.4765 would be rounded off to 1.476. When a range is stated, the upper and lower limits are inclusive so that the range consists of the two values themselves, properly rounded off, and all intermediate values between them. Where limits are given as parts per million, with the percentage value in parentheses, the percentage value will indicate the number of significant digits.

Solutions. All solutions, unless otherwise specified in an individual monograph, are to be prepared with distilled or deionized water conforming to the U. S. P. requirements for *Purified Water.*

The concentration of solutions designated by expressions such as "1 in 10" or "10 percent" means that *1 part by volume* of a liquid or *1 part by weight* of a solid is to be dissolved in a volume of the diluent or solvent sufficient to make the finished solution 10 parts by volume. Directions for the preparation of test solutions and colorimetric solutions designated by the abbreviations T.S. and C.S., respectively, are provided on pages 985 and 983.

Where volumetric solutions of definite concentrations are directed to be used in a test or assay, standardized solutions of other normalities or molarities may be employed if allowance is made for the factor and if the error of measurement is not known to be increased thereby.

Specific Gravity. Numerical values for specific gravity, unless otherwise noted, refer to the ratio of the weight of a substance in air at 25° to that of an equal volume of water at the same temperature. Specific gravity may be determined by any reliable method, unless otherwise specified.

Time Limits. Unless otherwise specified, 5 minutes is to be allowed for a reaction to take place in conducting limit tests for trace impurities such as chloride, iron, etc.

Expressions such as "exactly 5 minutes" mean that the stated period should be accurately timed.

Temperatures. Unless otherwise specified, temperatures are expressed in centigrade (Celsius) degrees and all measurements are to be made at 25° unless otherwise directed.

Test Solutions. See *Solutions.*

Tolerances. The minimum purity tolerances specified for *Food Chemicals Codex* items have been established with the expectation that the substances to which they apply will be used as food additives, ingredients, or food processing aids. These tolerance limits should neither bar the use of lots of articles which more nearly approach 100 percent purity nor should they constitute a basis for a claim that such lots exceed the quality prescribed by the *Food Chemicals Codex.*

When a maximum tolerance is not given, the assay should show the equivalent of not more than 100.5 percent.

Trace Impurities. Tests for inherent trace impurities are provided to limit such substances to levels consistent with good manufacturing practice and that are safe and otherwise unobjectionable under conditions in which the food additive or ingredient is customarily employed. In instances where both a heavy metals and a lead limit are specified in a monograph and the former is found to be 10 parts per million or less, the lead content need not be determined.

It is obviously impossible to provide limits and tests in each monograph for the detection of all possible unusual or unexpected impurities, the presence of which would be inconsistent with good manufacturing practice. The limits and tests provided are those considered to be necessary according to currently recognized methods of manufacture. If other methods of manufacture or other than the usual raw materials are used, or if other possible impurities may be present, additional tests may be required and should be applied, if necessary, by the vendor or user to demonstrate that the substance is suitable for its intended application in foods or in food processing.

Microbiological criteria, which are not within the scope of this Food Chemicals Codex, should be applied as necessary to ensure that the substance is not contaminated with pathogenic or other objectionable organisms and that the substance is otherwise suitable for its intended use.

Vacuum. The unqualified use of the term "in vacuum" means a pressure at least as low as that obtainable by an efficient aspirating water pump (not higher than 20 mm. of mercury).

Weights and Measures. The metric system of weights and measures is used in most specifications, assays, and tests in this Food Chemicals Codex. The units and abbreviations commonly employed are: m. = meter; cm. = centimeter; mm. = millimeter; μ = micron (0.001 mm.); mμ = millimicron (nanometer); kg. = kilogram; mg. = milligram; mcg. = microgram; l. = liter; ml. = milliliter; μl. = microliter. Other abbreviations are: amp. = ampere; ma. = milliampere; v. = volt; mv. = millivolt.

GENERAL SPECIFICATIONS

Certain specifications in the monographs of the *Food Chemicals Codex* are not amenable to precise description and accurate determination within narrow limiting ranges. Because of the subjective or general nature of these specifications, good judgment, based upon experience, must be used in interpreting and attaching significance to them. Specifications which are most likely to cause doubt are discussed in the subsequent paragraphs.

Description. The material given under this heading in monographs includes a description of physical characteristics such as color, odor, taste, form, etc., and information on stability under certain conditions of exposure to air and light. Statements in this section may also cover approximate indications of properties such as solubility (see below) in various solvents, pH, melting point, and boiling point, with numerical values modified by "about," "approximately," "usually," and other comparable nonspecific terms. Descriptions are provided as general information and are not intended to be interpreted as rigidly as measurable characteristics described in tests and assays.

Solubility. Statements included in the *Specifications* section in monographs under headings such as *Solubility in alcohol* express exact requirements and constitute quality specifications.

Statements relating to solubility given in monographs under the heading *Description*, however, are intended as information regarding approximate solubilities only and are not to be considered as Codex quality requirements. Such statements are considered to be of minor significance as a means of identification or determination of purity. For those purposes, dependence must be placed upon other tests provided in the monographs.

Approximate solubilities are indicated by the following descriptive terms:

Descriptive Term	*Parts of Solvent Required for 1 Part of Solute*
Very soluble	Less than 1
Freely soluble	From 1 to 10
Soluble	From 10 to 30
Sparingly soluble	From 30 to 100
Slightly soluble	From 100 to 1000
Very slightly soluble	From 1000 to 10,000
Practically insoluble or insoluble	More than 10,000

Soluble substances, when brought into solution, may show slight physical impurities, such as fragments of filter paper, fibers, and dust particles, unless excluded by definite tests or other requirements in the individual monograph; however, significant amounts of black specks, metallic chips, glass fragments, or other insoluble matter are not permitted.

Functional Use in Foods. A statement of functional classification is provided in each monograph as useful information to indicate the principal applications of the substance in foods or in food processing. The statement is not intended to limit in any way the choice or use of the substance or to indicate that it has no other utility.

Packaging and Storage. Statements in monographs relating to packaging are advisory in character and are intended only as general information to emphasize instances where deterioration may be accelerated under adverse packaging and storage conditions such as exposure to air, light, or extremes of temperature, or where safety hazards are involved.

The definitions employed in designating preferred types of containers and storage temperatures are adapted with minor variations from those appearing in the current editions of the U. S. P. and the N. F.

Containers. The container is the device which holds the substance and which is or may be in direct contact with it. The immediate container is that which is in direct contact with the substance at all times. The closure is a part of the container.

The container should not interact physically or chemically with the material that it holds so as to alter its strength, quality, or purity, and the food (additive) contact surface of the container should comply with the food additive regulations promulgated under the F. D. & C. Act*.

Light-Resistant Container. A light-resistant container is designed to prevent deterioration of the contents beyond the prescribed limits of strength, quality, or purity, under the ordinary or customary conditions of handling, shipment, storage, and sale. A colorless container may be made light-resistant by enclosing it in an opaque carton or wrapper.

* Or with applicable regulations promulgated by the responsible government agency in other countries in which Codex specifications are recognized.

Well-Closed Container. A well-closed container protects the contents from extraneous solids and from loss of the chemical under the ordinary or customary conditions of handling, shipment, storage, and sale.

Tight Container. A tight container protects the contents from contamination by extraneous liquids, solids or vapors, from loss of the chemical, and from efflorescence, deliquescence, or evaporation under the ordinary or customary conditions of handling, shipment, storage, and sale, and is capable of tight reclosure.

ADDED SUBSTANCES

Unless otherwise specified in an individual monograph, suitable anticaking agents, antioxidants, preservatives, stabilizers, etc., may be added to a Food Chemicals Codex substance to enhance its stability or utility, provided that the added agent (a) is permitted for use in foods by the U.S. Food and Drug Administration*; (b) is of appropriate food-grade quality and meets the requirements of the Food Chemicals Codex, if listed herein; (c) is used in an amount not to exceed the minimum required to impart its intended technical effect; (d) will not result in concentrations exceeding permitted levels in any food as a consequence of permitted use of the Food Chemicals Codex substance or substances; and (e) does not interfere with the assays and tests prescribed for determining compliance of the substance with the Food Chemicals Codex requirements.

* Or by the responsible government agency in other countries in which Codex specifications are recognized.

Monographs

ACACIA

Gum Arabic

DESCRIPTION

A dried gummy exudation obtained from the stems and branches of *Acacia senegal* (L.) Willd. or of related species of *Acacia* (Fam. *Leguminosae*). Unground acacia occurs as white or yellowish white spheroidal tears of varying size or in angular fragments. It is also available commercially in the form of white to yellowish white flakes, granules, or powder. One gram dissolves in 2 ml. of water forming a solution which flows readily and is acid to litmus. It is insoluble in alcohol. A 1 in 10 solution is slightly levorotatory.

IDENTIFICATION

To 10 ml. of a cold 1 in 50 solution of acacia add 0.2 ml. of diluted lead subacetate T.S. A flocculent, or curdy, white precipitate is formed immediately.

SPECIFICATIONS

Limits of Impurities

Arsenic (as As). Not more than 3 parts per million (0.0003 percent).
Ash (Total). Not more than 4 percent.
Ash (Acid-insoluble). Not more than 0.5 percent.
Heavy metals (as Pb). Not more than 40 parts per million (0.004 percent).
Insoluble matter. Not more than 1 percent.
Lead. Not more than 10 parts per million (0.001 percent).
Loss on drying. Not more than 15 percent.
Starch or dextrin. Passes test.
Tannin-bearing gums. Passes test.

TESTS

Arsenic. A *Sample Solution* prepared as directed for organic compounds meets the requirements of the *Arsenic Test*, page 865.

Ash (Total). Determine as directed in the general method, page 868.

Ash (Acid-insoluble). Determine as directed in the general method, page 869.

Heavy metals. Prepare and test a 500-mg. sample as directed in *Method II* under the *Heavy Metals Test*, page 920, using 20 mcg. of lead ion (Pb) in the control (*Solution A*).

Insoluble matter. Dissolve a 5-gram sample in about 100 ml. of water contained in a 250-ml. Erlenmeyer flask, add 10 ml. of diluted hydrochloric acid T.S., and boil gently for 15 minutes. Filter the hot solution by suction through a tared filtering crucible, wash thoroughly with hot water, dry at 105° for 2 hours, and weigh.

Lead. A *Sample Solution* prepared as directed for organic compounds meets the requirements of the *Lead Limit Test*, page 929, using 10 mcg. of lead ion (Pb) in the control.

Loss on drying, page 931. Dry at 105° for 5 hours. Unground samples should be powdered to pass through a No. 40 sieve and mixed well before weighing.

Starch or dextrin. Boil a 1 in 50 solution, cool, and add a few drops of iodine T.S. No bluish or reddish color is produced.

Tannin-bearing gums. To 10 ml. of a 1 in 50 solution add about 0.1 ml. of ferric chloride T.S. No blackish coloration or blackish precipitate is formed.

Packaging and storage. Store in well-closed containers.

Functional use in foods. Stabilizer; thickener; emulsifier.

ACETALDEHYDE

Ethanal

CH_3CHO

C_2H_4O Mol. wt. 44.05

DESCRIPTION

A flammable, colorless liquid with a characteristic odor. It is miscible with water and with alcohol and various other organic solvents. It boils at about 21°. *Caution: Acetaldehyde is very flammable. Do not use where it may be ignited.*

IDENTIFICATION

A. To 1 ml. of the sample add sodium bisulfite T.S. dropwise. A white, crystalline precipitate is produced which is soluble in water but insoluble in an excess of the reagent.

B. To 5 ml. of hot alkaline cupric tartrate T.S. add a few drops of the sample. A copious yellow to red precipitate of cuprous oxide is formed.

SPECIFICATIONS

Assay. Not less than 99.0 percent, by weight, of C_2H_4O.

Specific gravity. Between 0.804 and 0.811 at 0°/20°.

Limits of Impurities

Acidity (as acetic acid). Not more than 0.1 percent.

Heavy metals (as Pb). Not more than 10 parts per million (0.001 percent).

Nonvolatile residue. Not more than 60 parts per million (0.006 percent).

TESTS

Assay

Triethanolamine, 0.5 N. Dilute 65 ml. (74 grams) of 98 percent triethanolamine with water to make 1000.0 ml., and mix.

Hydroxylamine Hydrochloride Reagent. Dissolve 35 grams of hydroxylamine hydrochloride in 150 ml. of water, dilute to 1000.0 ml. with methanol, and mix. To 500 ml. of this solution, add 6 ml. of bromophenol blue T.S., and titrate with *0.5 N Triethanolamine* until the solution appears greenish blue by transmitted light.

Pressure Bottles. Heat-resistant pressure bottles (approx. 500 ml. capacity), provided with rubber gaskets and caps capable of being securely fastened. Immediately before capping, purge the bottles for 2 minutes with a gentle stream of nitrogen.

Procedure. Transfer 65.0 ml. of *Hydroxylamine Hydrochloride Reagent* and 50.0 ml. of *0.5 N Triethanolamine* into a pressure bottle, and add about 600 mg. of the sample contained in a tared, sealed glass ampule and accurately weighed. Add several pieces of 8-mm. glass rod, cap the bottle, and shake vigorously to break the ampule. Allow to stand at room temperature for 30 minutes, swirling occasionally, then cool slightly with tap water and uncap carefully to prevent any loss of reaction material. Allow the mixture to return to room temperature, and titrate with 0.5 *N* sulfuric acid to the same end-point obtained with a blank treated with the same quantities of the same reagents and in the same manner, approaching the end-point dropwise until the colors match by transmitted light. Calculate the difference between the volume of 0.5 *N* sulfuric acid required for the blank and that required for the sample. Each ml. of 0.5 *N* sulfuric acid is equivalent to 22.03 mg. of C_2H_4O.

Specific gravity. Determine at 0° ± 0.05° by means of a hydrometer calibrated to give the apparent specific gravity at 0°/20° and capable of being read to the nearest 0.0005 unit (see page 5).

Acidity. Mix a 3.9-ml. (3-gram) sample with 10 ml. of water, previously cooled to about 5°, add phenolphthalein T.S., and titrate with 0.1 *N* alcoholic potassium hydroxide to a pink color that persists for 15 seconds. Not more than 0.5 ml. is required.

Heavy metals. Evaporate a 2.6-ml. (2-gram) sample to dryness with 10 mg. of sodium carbonate, heat gently to volatilize any organic matter, and dissolve the residue in 25 ml. of water. This solution meets the requirements of the *Heavy Metals Test,* page 920, using 20 mcg. of lead ion (Pb) in the control (*Solution A*).

Nonvolatile residue. Evaporate 100 ml. (about 80 grams) in a tared platinum dish on a steam bath, and dry at 105° for 1 hour. The weight of the residue does not exceed 5 mg.

Packaging and storage. Store in tight containers in a cold place, preferably below 15°.

Functional use in foods. Flavoring agent.

ACETANISOLE

p-Methoxyacetophenone

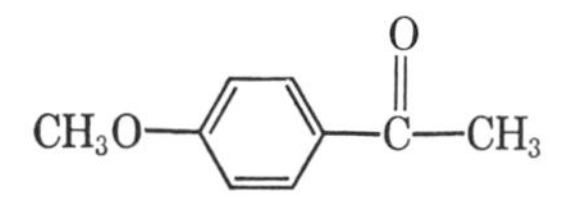

$C_9H_{10}O_2$ Mol. wt. 150.18

DESCRIPTION

A colorless to pale yellow, fused solid with an odor suggestive of hawthorn. It is soluble in most fixed oils and in propylene glycol. It is insoluble in glycerin and in mineral oil.

SPECIFICATIONS

Assay. Not less than 98.0 percent of $C_9H_{10}O_2$.

Solidification point. Not lower than 36°.

Solubility in alcohol. Passes test.

Limits of Impurities

Arsenic (as As). Not more than 3 parts per million (0.0003 percent).

Chlorinated compounds. Passes test.

Heavy metals (as Pb). Not more than 40 parts per million (0.004 percent).

Lead. Not more than 10 parts per million (0.001 percent).

TESTS

Assay. Weigh accurately about 1.2 grams, and proceed as directed under *Aldehydes and Ketones-Hydroxylamine Method,* page 894, using 75.09 as the equivalence factor (*E*) in the calculation.

Solidification point. Determine as directed in the general method, page 954.

Solubility in alcohol. Proceed as directed in the general method, page 899. One ml. dissolves in 5 ml. of 50 percent alcohol.

Arsenic. A *Sample Solution* prepared as directed for organic compounds meets the requirements of the *Arsenic Test*, page 865.

Chlorinated compounds. Proceed as directed in the general method, page 895.

Heavy metals. Prepare and test a 500-mg. sample as directed in *Method II* under the *Heavy Metals Test*, page 920, using 20 mcg. of lead ion (Pb) in the control (*Solution A*).

Lead. A *Sample Solution* prepared as directed for organic compounds meets the requirements of the *Lead Limit Test*, page 929, using 10 mcg. of lead ion (Pb) in the control.

Packaging and storage. Store in full, tight, preferably glass, aluminum, tin-lined, or other suitably lined containers in a cool place protected from light.

Functional use in foods. Flavoring agent.

ACETIC ACID, GLACIAL

CH_3COOH

$C_2H_4O_2$ Mol. wt. 60.05

DESCRIPTION

A clear, colorless liquid having a pungent, characteristic odor and, when well diluted with water, an acid taste. It boils at about 118° and has a specific gravity of about 1.049. It is miscible with water, with alcohol, and with glycerin. A 1 in 3 solution gives positive tests for *Acetate*, page 925.

SPECIFICATIONS

Assay. Not less than 99.5 percent, by weight, of $C_2H_4O_2$.

Solidification point. Not lower than 15.6°.

Limits of Impurities

Arsenic (as As). Not more than 3 parts per million (0.0003 percent).

Heavy metals (as Pb). Not more than 10 parts per million (0.001 percent).

Nonvolatile residue. Not more than 0.005 percent.

Readily oxidizable substances. Passes test.

TESTS

Assay. Measure about 2 ml. into a tared, glass-stoppered flask, and weigh accurately. Add 40 ml. of water, then add phenolphthalein T.S., and titrate with 1 *N* sodium hydroxide. Each ml. of 1 *N* sodium hydroxide is equivalent to 60.05 mg. of $C_2H_4O_2$.

Solidification point. Determine as directed in the general procedure, page 954.

Arsenic. A solution of 1 gram in 10 ml. of water meets the requirements of the *Arsenic Test*, page 865.

Heavy metals. To the residue obtained in the test for *Nonvolatile residue* add 8 ml. of 0.1 *N* hydrochloric acid, warm gently until solution is complete, and dilute to 100 ml. with water. A 10-ml. portion of this solution diluted to 25 ml. meets the requirements of the *Heavy Metals Test*, page 920, using 20 mcg. of lead ion (Pb) in the control (*Solution A*).

Nonvolatile residue. Evaporate 19 ml. (20 grams), accurately measured, in a tared dish on a steam bath, and dry at 105° for 1 hour.

Readily oxidizable substances. Dilute 2 ml. in a glass-stoppered container with 10 ml. of water, and add 0.1 ml. of 0.1 *N* potassium permanganate. The pink color is not changed to brown within 2 hours.

Packaging and storage. Store in tight containers.

Functional use in foods. Acidifier; flavoring agent.

ACETOIN

Acetyl Methyl Carbinol; Dimethylketol; 3-Hydroxy-2-butanone

$CH_3CH(OH)COCH_3$

$C_4H_8O_2$ Mol. wt. 88.11

DESCRIPTION

Acetoin is usually prepared by fermentation, or by partial reduction of diacetyl. It is a colorless, or pale yellow liquid (monomer), having a characteristic buttery odor, or a white crystalline powder (dimer). The solid form is converted to the liquid form by melting, or dissolving. It is miscible with alcohol, with propylene glycol, and with water. It is practically insoluble in vegetable oils.

SPECIFICATIONS

Assay. Not less than 96.0 percent of $C_4H_8O_2$.

Refractive index. Between 1.417 and 1.420 at 20°

Specific gravity. Between 1.005 and 1.019.

TESTS

Assay. Determine the percent of acetoin by gas-liquid chromatography using an instrument containing a thermal conductivity detector. Prepare a 1.5-meter × 6.35-mm. column consisting of 20 percent Carbowax 20 M on acid washed, 60/80-mesh Chromosorb W, or other components capable of separating diacetyl, water, and acetoin. Observe the following operating conditions during the determination: *Sample*, 2 microliters; *Injector*, about 195°; *Column*, about 130°; *Detector*, about 230°; and *Helium flow rate*, about 35 ml. per minute. The approximate retention time for diacetyl is 2¼ minutes, for water 3 minutes, and for acetoin 12 minutes. The area of the acetoin peak is not less than 96.0 percent of the total area of all peaks.

Refractive index, page 945. Determine with an Abbé or other refractometer of equal or greater accuracy.

Specific gravity. Determine by any reliable method (see page 5).

Packaging and storage. Store in full, tight, preferably glass, aluminum, or tin-lined, light-resistant containers in a cool place.

Functional use in foods. Flavoring agent.

ACETONE

2-Propanone; Dimethyl Ketone

CH_3COCH_3

C_3H_6O Mol. wt. 58.08

DESCRIPTION

A clear, colorless, volatile liquid having a characteristic odor. It is miscible with water, with alcohol, with ether, with chloroform, and with most volatile oils. Its refractive index is about 1.356. *Caution: Acetone is highly flammable.*

IDENTIFICATION

Mix 0.1 ml. of the sample with 10 ml. of water, add 5 ml. of sodium hydroxide T.S., warm, and add 5 ml. of iodine T.S. A yellow precipitate of iodoform is produced.

SPECIFICATIONS

Assay. Not less than 99.5 percent of C_3H_6O, by weight.

Distillation range. Within a range of 1°, including 56.1°.

Solubility in water. Passes test.

Limits of Impurities

Acidity (as acetic acid). Not more than 20 parts per million (0.002 percent).

Aldehydes (as formaldehyde). Not more than 20 parts per million (0.002 percent).

Alkalinity (as NH_3). Not more than 10 parts per million (0.001 percent).

Heavy metals (as Pb). Not more than 1 part per million (0.0001 percent).
Methanol. Not more than 0.05 percent.
Nonvolatile residue. Not more than 10 parts per million (0.001 percent).
Phenols. Passes test.
Substances reducing permanganate. Passes test.
Water. Not more than 0.5 percent.

TESTS

Assay. Its specific gravity, determined by any reliable method (see page 5), is not greater than 0.7880 at 25°/25° (equivalent to 0.7930 at 20°/20).

Distillation range. Proceed as directed in the general method, page 890.

Solubility in water. Mix 38 ml. (about 30 grams) of the sample with an equal volume of carbon dioxide-free water. The solution remains clear for at least 30 minutes.

Acidity. Mix 38 ml. (about 30 grams) of the sample with an equal volume of carbon dioxide-free water, add 0.1 ml. of phenolphthalein T.S., and titrate with 0.1 *N* sodium hydroxide. Not more than 0.1 ml. is required to produce a pink color.

Aldehydes. Dilute 2.5 ml. (about 2 grams) of the sample with 7.5 ml. of water. Prepare a standard solution containing 40 mcg. of formaldehyde in 10 ml. of water. To each solution add 0.15 ml. of a 5 percent solution of 5,5-dimethyl-1,3-cyclohexanedione in alcohol, and evaporate on a steam bath until the acetone is volatilized. Dilute to 10 ml. with water, and cool quickly in an ice bath while stirring vigorously. Any turbidity in the sample solution does not exceed that produced in the standard.

Alkalinity. Add 1 drop of methyl red T.S. to 25 ml. of water, add 0.1 *N* sulfuric acid until a red color just appears, then add 23 ml. (about 18 grams) of the sample, and mix. Not more than 0.1 ml. of 0.1 *N* sulfuric acid is required to restore the red color.

Heavy metals. Evaporate 25 ml. (about 20 grams) of the sample to dryness on a steam bath in a glass evaporating dish. Cool, add 2 ml. of hydrochloric acid, and slowly evaporate to dryness again on the steam bath. Moisten the residue with 1 drop of hydrochloric acid, add 10 ml. of hot water, and digest for 2 minutes. Cool, and dilute to 25 ml. with water. This solution meets the requirements of the *Heavy Metals Test*, page 920, using 20 mcg. of lead ion (Pb) in the control (*Solution A*).

Methanol. Dilute 10 ml. of the sample to 100 ml. with water. Prepare a standard solution in water containing 40 mcg. of methanol in each ml. To 1 ml. of each solution add 0.2 ml. of 10 percent phosphoric acid and 0.25 ml. of potassium permanganate solution (1 in 20).

Allow to stand for 15 minutes, then add 0.3 ml. of sodium bisulfite solution (1 in 10), and shake until colorless. Slowly add 5 ml. of ice-cold 80 percent sulfuric acid, keeping the mixture cold during the addition. Add 0.1 ml. of chromotropic acid solution (1 in 100), mix, and digest on a steam bath for 20 minutes. Any violet color produced in the sample solution does not exceed that produced in the standard.

Nonvolatile residue. Evaporate 125 ml. (about 100 grams) of the sample to dryness in a tared dish on a steam bath, dry the residue at 105° for 30 minutes, cool, and weigh.

Phenols. Evaporate 3 ml. of the sample to dryness at 60°. To the residue add 3 drops of a solution of 100 mg. of sodium nitrite in 5 ml. of sulfuric acid, allow to stand for about 3 minutes, and then carefully add 3 ml. of 2 *N* sodium hydroxide. No color is produced.

Substances reducing permanganate. Transfer 10 ml. of the sample into a glass-stoppered cylinder, add 0.05 ml. of 0.1 *N* potassium permanganate, mix, and allow to stand for 15 minutes. The pink color is not entirely discharged.

Water. Determine by the *Karl Fischer Titrimetric Method*, page 977, using freshly distilled pyridine instead of methanol as the solvent.

Packaging and storage. Store in tight containers, remote from fire.

Functional use in foods. Extraction solvent.

ACETONE PEROXIDES

DESCRIPTION

A mixture of monomeric and linear dimeric acetone peroxides (mainly 2,2-hydroperoxypropane), with minor proportions of higher polymers, usually mixed with an edible carrier such as cornstarch. The cornstarch mixture is a fine, white, free-flowing powder having a sharp, acrid odor similar to that of hydrogen peroxide when the container is first opened.

Caution: Acetone peroxides are strong oxidizing agents. Exposure to the skin and eyes should be avoided.

SPECIFICATIONS

Assay. It yields an amount of hydrogen peroxide equivalent to not less than 16.0 percent of acetone peroxides.

Limits of Impurities

Arsenic (as As). Not more than 3 parts per million (0.0003 percent).

Heavy metals (as Pb). Not more than 10 parts per million (0.001 percent).

TESTS

Assay. Transfer about 200 mg., accurately weighed, into a 250-ml. beaker, add 50 ml. of dilute sulfuric acid (1 in 10), allow to stand for at least 3 minutes, stirring occasionally, and titrate with 0.1 *N* potassium permanganate to a light pink color that persists for at least 20 seconds. Calculate the total peroxides, *P*, as grams of hydrogen peroxide equivalents per 100 grams of the sample, by the formula $(V)(N)(0.017)(100)/W$, in which *V* and *N* are the volume and exact normality, respectively, of the potassium permanganate, 0.017 is the milliequivalent weight of H_2O_2, and *W* is the weight, in grams, of the sample taken. Multiply the value *P* so obtained by 1.6 to convert to percent of acetone peroxides.

Sample Solution for the Determination of Arsenic and Heavy Metals. Mix 10 grams with 100 ml. of dilute sulfuric acid (1 in 10), allow to stand for 5 minutes, stirring occasionally, and filter. Heat the filtrate on a steam bath for 15 minutes, then boil for 1 minute, cool, and dilute to 100 ml. with water.

Arsenic. A 10-ml. portion of the *Sample Solution* meets the requirements of the *Arsenic Test,* page 865.

Heavy Metals. Dilute 20 ml. of the *Sample Solution* to 25 ml. with water. This solution meets the requirements of the *Heavy Metals Test,* page 920, using 20 mcg. of lead ion (Pb) in the control (*Solution A*).

Packaging and storage. Store in a cool, dry place, preferably below 24°.

Functional use in foods. Bleaching agent; maturing agent; dough conditioner.

ACETOPHENONE

Methyl Phenyl Ketone

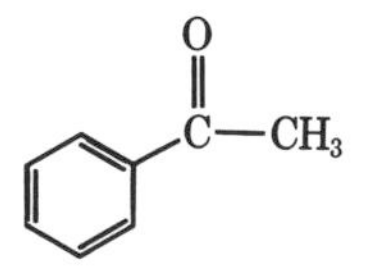

C_8H_8O Mol. wt. 120.15

DESCRIPTION

A practically colorless liquid, at temperatures above 20°, having a very sweet, pungent odor. It is very soluble in propylene glycol and in most fixed oils; soluble in alcohol, in chloroform, and in ether; slightly soluble in water and in mineral oil; and insoluble in glycerin.

SPECIFICATIONS

Assay. Not less than 98.0 percent of C_8H_8O.

Refractive index. Between 1.533 and 1.535 at 20°, in supercooled liquid form.

Solidification point. Not lower than 19°.

Solubility in alcohol. Passes test.

Specific gravity. Between 1.025 and 1.028.

Limits of Impurities

Chlorinated compounds. Passes test.

TESTS

Assay. Weigh accurately about 1 gram, and proceed as directed under *Aldehydes and Ketones—Hydroxylamine Method*, page 894, using 60.08 as the equivalence factor (E) in the calculation.

Refractive index, page 945. Determine with an Abbé or other refractometer of equal or greater accuracy.

Solidification point. Determine as directed in the general method, page 954.

Solubility in alcohol. Proceed as directed in the general method, page 899. One ml. dissolves in 5 ml. of 50 percent alcohol.

Specific gravity. Determine by any reliable method (see page 5).

Chlorinated compounds. Determine as directed in the general method, page 895.

Packaging and storage. Store in full, tight, preferably glass, aluminum, tin-lined, or iron containers in a cool place protected from light.

Functional use in foods. Flavoring agent.

ACETYLATED MONOGLYCERIDES

Acetylated Mono- and Diglycerides

DESCRIPTION

Acetylated monoglycerides consist of partial or complete esters of glycerin with a mixture of acetic acid and edible fat-forming fatty acids. They may be manufactured by the interesterification of edible fats with triacetin and glycerin in the presence of catalytic agents, followed by molecular distillation, or by the direct acetylation of edible monoglycerides with acetic anhydride without the use of catalyst or molecular distillation. They vary in consistency from clear, thin liquids to solids and are from white to pale yellow in color. They may have an acetic acid odor, but are practically bland in taste. They are insoluble in water, but are soluble in alcohol, in acetone, and in other organic solvents, the extent of solubility depending upon the degree of esterification and the melting range.

SPECIFICATIONS

Reichert-Meissl value. Between 75 and 150.

The following specifications should conform to the representations of the vendor: **Free glycerin, Iodine value,** and **Saponification value.**

Limits of Impurities

Acid value. Not more than 6.

Arsenic (as As). Not more than 3 parts per million (0.0003 percent).

Heavy metals (as Pb). Not more than 10 parts per million (0.001 percent).

TESTS

Reichert-Meissl value. Determine as directed in the general method, page 912.

Free glycerin. Determine as directed in the general method, page 904.

Iodine value. Determine by the *Wijs Method,* page 906.

Saponification value. Determine as directed in the general method, page 914.

Acid value. Determine as directed under *Method II* in the general procedure, page 902.

Arsenic. A *Sample Solution* prepared as directed for organic compounds meets the requirements of the *Arsenic Test,* page 865.

Heavy metals. Prepare and test a 2-gram sample as directed in *Method II* under the *Heavy Metals Test,* page 920, using 20 mcg. of lead ion (Pb) in the control (*Solution A*).

Packaging and storage. Store in well-closed containers.

Functional use in foods. Emulsifier; coating agent; texture-modifying agent; solvent; lubricant.

ADIPIC ACID

Hexanedioic Acid; 1,4-Butanedicarboxylic Acid

$HOOC(CH_2)_4COOH$

$C_6H_{10}O_4$ Mol. wt. 146.14

DESCRIPTION

White crystals or crystalline powder. It is soluble in acetone, freely soluble in alcohol, and slightly soluble in water. It is not hygroscopic.

SPECIFICATIONS

Assay. Not less than 99.6 percent and not more than the equivalent of 101.0 percent of $C_6H_{10}O_4$.

Melting range. Between 151.5° and 154°.

Limits of Impurities

Arsenic (as As). Not more than 3 parts per million (0.0003 percent).

Heavy metals (as Pb). Not more than 10 parts per million (0.001 percent).

Residue on ignition. Not more than 0.002 percent.

Water. Not more than 0.2 percent.

TESTS

Assay. Mix about 1.5 grams, accurately weighed, with 75 ml. of recently boiled and cooled water in a 250-ml. glass-stoppered Erlenmeyer flask, add phenolphthalein T.S., and titrate with 0.5 *N* sodium hydroxide to the first appearance of a faint pink end-point which persists for at least 30 seconds, shaking the flask as the end-point is approached. Each ml. of 0.5 *N* sodium hydroxide is equivalent to 36.54 mg. of $C_6H_{10}O_4$.

Melting range. Determine as directed in the general procedure, page 931.

Arsenic. A *Sample Solution* prepared as directed for organic compounds meets the requirements of the *Arsenic Test*, page 865.

Heavy metals. Prepare and test a 2-gram sample as directed in *Method II* under the *Heavy Metals Test*, page 920, using 20 mcg. of lead ion (Pb) in the control (*Solution A*).

Residue on ignition. Transfer 100.0 grams to a tared 125-ml. platinum dish which has been previously cleaned by fusing with 5 grams of potassium pyrosulfate or bisulfate, followed by boiling in diluted sulfuric acid T.S. and rinsing with water. Melt the sample completely over a gas burner, then ignite the melt with the burner. After ignition starts, lower or remove the flame in order to prevent the sample from boiling and to keep it burning slowly until it is completely carbonized. Ignite at 850° in a muffle furnace for 30 minutes or until the carbon is completely removed, cool, and weigh.

Water. Determine by the *Karl Fischer Titrimetric Method,* page 977.

Packaging and storage. Store in well-closed containers.

Functional use in foods. Buffer; neutralizing agent.

AGAR

DESCRIPTION

A dried hydrophylic, colloidal polygalactoside extracted from *Gelidium cartilagineum* (L.) Gaillon (Fam. *Gelidiaceae*), *Gracilaria confervoides* (L.) Greville (Fam. *Sphaerococcaceae*), and related red algae (Class *Rhodophyceae*). It is commercially available in bundles consisting of thin, membranous agglutinated strips, or in cut, flaked, granulated, or powdered forms. It is white to pale yellow in color and is either odorless, or has a slight characteristic odor, and a mucilaginous taste. Agar is insoluble in cold water, but soluble in boiling water.

IDENTIFICATION

A. Place a few fragments of unground agar or a small amount of the powder on a slide, add a few drops of water, and examine microscopically. The agar appears granular and somewhat filamentous. A few fragments of the spicules of sponges and a few frustules of diatoms may be present.

B. Boil 1 gram with 65 ml. of water for 10 minutes with continuous stirring, and adjust to a concentration of 1.5 percent, by weight, with hot water. A clear liquid is obtained which congeals between 32° and 39° to form a firm, resilient gel which does not liquefy below 85°.

SPECIFICATIONS

Water Absorption. Passes test.

Limits of Impurities

Arsenic (as As). Not more than 3 parts per million (0.0003 percent).

Ash (Total). Not more than 6.5 percent on the dried basis.

Ash (Acid-insoluble). Not more than 0.5 percent on the dried basis.

Gelatin. Passes test.

Heavy metals (as Pb). Not more than 40 parts per million (0.004 percent).

Insoluble matter. Not more than 1 percent.

Lead. Not more than 10 parts per million (0.001 percent).

Loss on drying. Not more than 20 percent.

Starch. Passes test.

TESTS

Water absorption. Place 5 grams in a 100-ml. graduated cylinder, fill to the mark with water, mix and allow to stand at about 25° for 24 hours. Pour the contents of the cylinder through moistened glass wool, allowing the water to drain into another 100-ml. graduated cylinder. Not more than 75 ml. of water is obtained.

Arsenic. A *Sample Solution* prepared as directed for organic compounds meets the requirements of the *Arsenic Test*, page 865.

Ash (**Total**). Determine as directed in the general method, page 868.

Ash (**Acid-insoluble**). Determine as directed in the general method, page 869.

Gelatin. Dissolve about 1 gram in 100 ml. of boiling water, and allow to cool to about 50°. To 5 ml. of the solution add 5 ml. of trinitrophenol T.S. No turbidity appears within 10 minutes.

Heavy metals. Prepare and test a 500-mg. sample as directed in *Method II* under the *Heavy Metals Test*, page 920, using 20 mcg. of lead ion (Pb) in the control (*Solution A*).

Insoluble matter. To 7.5 grams add sufficient water to make 500 grams, boil for 15 minutes, and readjust to the original weight. To 100 grams of the mixture add hot water to make 200 ml., heat almost to boiling, filter while hot through a tared filtering crucible, rinse the container with several portions of hot water, and pass the rinsings through the crucible. Dry the crucible and its contents at 105° to constant weight, cool, and weigh. The weight of the residue does not exceed 15 mg.

Lead. A *Sample Solution* prepared as directed for organic compounds meets the requirements of the *Lead Limit Test*, page 929, using 10 mcg. of lead ion (Pb) in the control.

Loss on drying, page 931. Dry at 105° for 5 hours. Cut unground agar into pieces from 2 to 5 mm. square before drying.

Starch. Boil 100 mg. in 100 ml. of water, cool, and add a few drops of iodine T.S. No blue color is produced.

Packaging and storage. Store in well-closed containers.

Functional use in foods. Stabilizer; emulsifier; thickener.

DL-ALANINE

DL-2-Aminopropanoic Acid

$CH_3CH(NH_2)COOH$

$C_3H_7NO_2$ Mol. wt. 89.09

DESCRIPTION

A white, odorless, crystalline powder, having a sweetish taste. It is freely soluble in water, but sparingly soluble in alcohol. It is optically inactive. The pH of a 1 in 20 solution is between 5.5 and 7.0. It melts with decomposition at about 198°.

IDENTIFICATION

A. Heat 5 ml. of a 1 in 1000 solution with 1 ml. of triketohydrindene hydrate T.S. for 3 minutes. A violet color is produced.

B. Dissolve 200 mg. in 10 ml. of water, add 100 mg. of potassium permanganate, and heat to boiling. The odor of acetaldehyde is detected.

SPECIFICATIONS

Assay. Not less than 98.5 percent and not more than the equivalent of 102.0 percent of $C_3H_7NO_2$, calculated on the dried basis.

Limits of Impurities

Arsenic (as As). Not more than 3 parts per million (0.0003 percent).

Heavy metals (as Pb). Not more than 20 parts per million (0.002 percent).

Lead. Not more than 10 parts per million (0.001 percent).

Loss on drying. Not more than 0.3 percent.

Residue on ignition. Not more than 0.2 percent.

TESTS

Assay. Dissolve about 200 mg., accurately weighed, in 3 ml. of formic acid and 50 ml. of glacial acetic acid, add 2 drops of crystal violet T.S., and titrate with 0.1 *N* perchloric acid to a bluish green end-point. Perform a blank determination (see page 2), and make any necessary correction. Each ml. of 0.1 *N* perchloric acid is equivalent to 8.909 mg. of $C_3H_7NO_2$.

Arsenic. A *Sample Solution* prepared as directed for organic compounds meets the requirements of the *Arsenic Test*, page 865.

Heavy metals. Prepare and test a 1-gram sample as directed in *Method II* under the *Heavy Metals Test*, page 920, using 20 mcg. of lead ion (Pb) in the control (*Solution A*).

Lead. A *Sample Solution* prepared as directed for organic compounds meets the requirements of the *Lead Limit Test*, page 929, using 10 mcg. of lead ion (Pb) in the control.

Loss on drying, page 931. Dry at 105° for 3 hours.

Residue on ignition, page 945. Ignite 1 gram as directed in the general method.

Packaging and storage. Store in well-closed, light-resistant containers.

Functional use in foods. Nutrient; dietary supplement.

L-ALANINE

L-2-Aminopropanoic Acid

$CH_3CH(NH_2)COOH$

$C_3H_7NO_2$ Mol. wt. 89.09

DESCRIPTION

A white, odorless, crystalline powder, having a sweetish taste. It is freely soluble in water, sparingly soluble in alcohol, and insoluble in ether. The pH of a 1 in 20 solution is between 5.5 and 7.0.

IDENTIFICATION

A. Heat 5 ml. of a 1 in 1000 solution with 1 ml. of triketohydrindene hydrate T.S. for 3 minutes. A violet color is produced.

B. Dissolve 200 mg. in 10 ml. of water, add 100 mg. of potassium permanganate, and heat to boiling. The odor of acetaldehyde is detected.

SPECIFICATIONS

Assay. Not less than 98.5 percent and not more than the equivalent of 102.0 percent of $C_3H_7NO_2$, calculated on the dried basis.

Specific rotation, $[\alpha]_D^{20°}$. Between +13.5° and +15.5°, on the dried basis.

Limits of Impurities

Arsenic (as As). Not more than 3 parts per million (0.0003 percent).

Heavy metals (as Pb). Not more than 20 parts per million (0.002 percent).

Lead. Not more than 10 parts per million (0.001 percent).

Loss on drying. Not more than 0.3 percent.

Residue on ignition. Not more than 0.2 percent.

TESTS

Assay. Dissolve about 200 mg., accurately weighed, in 3 ml. of formic acid and 50 ml. of glacial acetic acid, add 2 drops of crystal violet T.S., and titrate with 0.1 *N* perchloric acid to a bluish green end-point. Perform a blank determination (see page 2) and make any necessary correction. Each ml. of 0.1 *N* perchloric acid is equivalent to 8.909 mg. of $C_3H_7NO_2$.

Specific rotation, page 939. Determine in a solution containing 10 grams of a previously dried sample in sufficient 6 *N* hydrochloric acid to make 100 ml.

Arsenic. A *Sample Solution* prepared as directed for organic compounds meets the requirements of the *Arsenic Test,* page 865.

Heavy metals. Prepare and test a 1-gram sample as directed in *Method II* under the *Heavy Metals Test,* page 920, using 20 mcg. of lead ion (Pb) in the control (*Solution A*).

Lead. A *Sample Solution* prepared as directed for organic compounds meets the requirements of the *Lead Limit Test,* page 929, using 10 mcg. of lead ion (Pb) in the control.

Loss on drying, page 931. Dry at 105° for 3 hours.

Residue on ignition, page 945. Ignite 1 gram as directed in the general method.

Packaging and storage. Store in well-closed, light-resistant containers.

Functional use in foods. Nutrient; dietary supplement.

ALGINIC ACID

$(C_6H_8O_6)_n$

Equiv. wt., *Calculated,* 176.13
Equiv. wt., *Actual* (Avg.), 200.00

DESCRIPTION

Alginic acid is a hydrophilic colloidal carbohydrate extracted by the use of dilute alkali from various species of brown seaweeds (*Phaeophyceae*). It may be described chemically as a linear glycuronoglycan consisting mainly of β-(1 → 4) linked D-mannuronic and L-guluronic acid units in the pyranose ring form. It occurs as a white to yellowish-white, fibrous powder. It is odorless and tasteless. Alginic acid is insoluble in water, readily soluble in alkaline solutions, and insoluble in organic solvents. The pH of a 3 in 100 suspension in water is between 2.0 and 3.4.

IDENTIFICATION

A. To 5 ml. of a 1 in 150 solution in 0.1 *N* sodium hydroxide add 1 ml. of calcium chloride T.S. A voluminous gelatinous precipitate is formed.

B. To 5 ml. of the solution prepared for *Identification test A* add 1 ml. of diluted sulfuric acid T.S. A heavy gelatinous precipitate is formed.

C. To about 5 mg., contained in a test tube, add 5 ml. of water, 1 ml. of a freshly prepared 1 in 100 solution of naphthoresorcinol in ethanol, and 5 ml. of hydrochloric acid. Heat the mixture to boiling, boil gently for about 3 minutes, and then cool to about 15°. Transfer the contents of the test tube to a 30-ml. separator with the aid of 5 ml. of water and extract with 15 ml. of isopropyl ether. Perform a blank test (see page 2). The isopropyl ether extract from the sample exhibits a deeper purplish hue than that from the blank.

SPECIFICATIONS

Assay. It yields, on the dried basis, not less than 20.0 percent and not more than 23.0 percent of carbon dioxide (CO_2) corresponding to between 91.0 and 104.5 percent of alginic acid (Equiv. wt. 200.00).

Limits of Impurities

Arsenic (as As). Not more than 3 parts per million (0.0003 percent).
Ash. Not more than 4 percent.
Heavy metals (as Pb). Not more than 40 parts per million (0.004 percent).
Insoluble matter. Not more than 0.2 percent.
Lead. Not more than 10 parts per million (0.001 percent).
Loss on drying. Not more than 15 percent.

TESTS

Assay. Proceed as directed under *Alginates Assay*, page 863. Each ml. of 0.25 *N* sodium hydroxide consumed in the assay is equivalent to 25 mg. of alginic acid (Equiv. wt. 200.00).

Arsenic. A *Sample Solution* prepared as directed for organic compounds meets the requirements of the *Arsenic Test*, page 865.

Ash. Weigh accurately about 3 grams, previously dried at 105° for 4 hours, in a tared crucible, and incinerate at a low temperature, not exceeding a dull red heat until free from carbon. Cool the crucible and its contents in a desiccator, weigh, and determine the weight of the ash.

Heavy metals. Prepare and test a 500-mg. sample as directed in *Method II* under the *Heavy Metals Test*, page 920, but use a platinum crucible for the ignition. Any color does not exceed that produced in a control (*Solution A*) containing 20 mcg. of lead ion (Pb).

Insoluble matter. Transfer about 500 mg., accurately weighed, into a 600-ml. beaker, add 7.5 ml. of 0.1 *N* sodium hydroxide, and dilute the mixture to 200 ml. Cover the beaker, heat to boiling, and boil gently for 1 hour with frequent stirring. Filter while hot through a tared Gooch crucible provided with an asbestos mat, wash thoroughly with hot water, dry at 105° for 1 hour, cool, and weigh.

Lead. A *Sample Solution* prepared as directed for organic compounds meets the requirements of the *Lead Limit Test*, page 929, using 10 mcg. of lead ion (Pb) in the control.

Loss on drying, page 931. Dry at 105° for 4 hours.

Packaging and storage. Store in well-closed containers.

Functional use in foods. Stabilizer; thickener; emulsifier.

ALLYL CYCLOHEXANEPROPIONATE

Allyl-3-cyclohexanepropionate

C_6H_{11}—$CH_2CH_2COOCH_2CH{=}CH_2$

$C_{12}H_{20}O_2$ Mol. wt. 196.29

DESCRIPTION

A colorless liquid having a pineapple-like odor. It is practically insoluble in water and in glycerin; miscible with alcohol, with chloroform, and with ether.

SPECIFICATIONS

Assay. Not less than 98.0 percent of $C_{12}H_{20}O_2$.

Refractive index. Between 1.457 and 1.462 at 20°.

Solubility in alcohol. Passes test.

Specific gravity. Between 0.945 and 0.950.

Limits of Impurities

Acid value. Not more than 5.0.

Heavy metals (as Pb). Not more than 40 parts per million (0.004 percent).

TESTS

Assay. Weigh accurately about 1.2 grams, and proceed as directed under *Ester Determination*, page 896, using 98.15 as the equivalence factor (*e*) in the calculation.

Refractive index, page 945. Determine with an Abbé or other refractometer of equal or greater accuracy.

Solubility in alcohol. Proceed as directed in the general method, page 899. One ml. dissolves in 4 ml. of 80 percent alcohol.

Specific gravity. Determine by any reliable method (see page 5).

Acid value. Determine as directed in the general method, page 893.

Heavy metals. Prepare and test a 500 mg. sample as directed in *Method II* under the *Heavy Metals Test*, page 920, using 20 mcg. of lead ion (Pb) in the control (*Solution A*).

Packaging and storage. Store in tight, light-resistant containers.

Functional use in foods. Flavoring agent.

ALLYL HEXANOATE

Allyl Caproate

$CH_3(CH_2)_4COOCH_2CH{=}CH_2$

$C_9H_{16}O_2$ Mol. wt. 156.23

DESCRIPTION

A colorless to light yellow liquid having a strong pineapple-like odor. It is miscible with alcohol, with most fixed oils, and with mineral oil. It is practically insoluble in propylene glycol.

SPECIFICATIONS

Assay. Not less than 98.0 percent of $C_9H_{16}O_2$.

Refractive index. Between 1.422 and 1.426 at 20°.

Solubility in alcohol. Passes test.

Specific gravity. Between 0.884 and 0.890.

Limits of Impurities

Acid value. Not more than 1.0.

TESTS

Assay. Weigh accurately about 1 gram and proceed as directed under *Ester Determination*, page 896, using 78.12 as the equivalence factor (*e*) in the calculation.

Refractive index, page 945. Determine with an Abbé or other refractometer of equal or greater accuracy.

Solubility in alcohol. Proceed as directed in the general method, page 899. One ml. dissolves in from 3.5 to 5 ml. of 70 percent alcohol.

Specific gravity. Determine by any reliable method (see page 5).

Acid value. Determine as directed in the general method, page 893.

Packaging and storage. Store in full, tight, preferably glass or aluminum containers in a cool place protected from light.

Functional use in foods. Flavoring agent.

ALLYL α-IONONE

Allyl Ionone

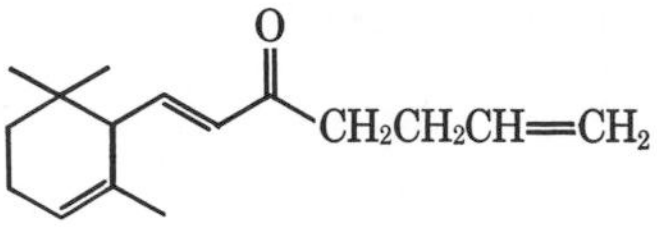

$CH_2CH_2CH{=}CH_2$

$C_{16}H_{24}O$ Mol. wt. 232.37

DESCRIPTION

A yellow liquid having a strong, fruity, pineapple-like odor. It is soluble in alcohol, but insoluble in water.

SPECIFICATIONS

Assay. Not less than 88.0 percent of $C_{16}H_{24}O$.

Refractive index. Between 1.503 and 1.507 at 20°.

Solubility in alcohol. Passes test.

Specific gravity. Between 0.928 and 0.935.

Limits of Impurities

Arsenic (as As). Not more than 3 parts per million (0.0003 percent).

Heavy metals (as Pb). Not more than 40 parts per million (0.004 percent).

Lead. Not more than 10 parts per million (0.001 percent).

TESTS

Assay. Weigh accurately about 1.5 grams, and proceed as directed for ketones under *Aldehydes and Ketones-Hydroxylamine Method,* page 894, using 116.18 as the equivalence factor (*e*) in the calculation.

Refractive index, page 945. Determine with an Abbé or other refractometer of equal or greater accuracy.

Solubility in alcohol. Proceed as directed in the general method, page 899. One ml. dissolves in 8 ml. of 70 percent alcohol to form a clear solution.

Specific gravity. Determine by any reliable method (see page 5).

Arsenic. A *Sample Solution* prepared as directed for organic compounds meets the requirements of the *Arsenic Test,* page 865.

Heavy metals. Prepare and test a 500-mg. sample as directed in *Method II* under the *Heavy Metals Test,* page 920, using 20 mcg. of lead ion (Pb) in the control (*Solution A*).

Lead. A *Sample Solution* prepared as directed for organic compounds meets the requirements of the *Lead Limit Test,* page 929, using 10 mcg. of lead ion (Pb) in the control.

Packaging and storage. Store in full, tight, preferably glass, tin-lined or other suitably lined containers in a cool place protected from light.

Functional use in foods. Flavoring agent.

ALLYL ISOTHIOCYANATE

Mustard Oil, Volatile

$CH_2{=}CH{-}CH_2{-}N{=}C{=}S$

C_3H_5NCS Mol. wt. 99.15

DESCRIPTION

The oil obtained by maceration with water and subsequent distillation of the dried ripe seed (free from fixed oil) of *Brassica nigra* (L.) Koch, or of *Brassica juncea* (L.) Czerniaew (Fam. *Cruciferae*), or prepared synthetically. It is a colorless or pale yellow, strongly refractive liquid, having a very pungent, irritating odor and an acrid taste. It is miscible with alcohol, with ether, and with carbon disulfide. It is optically inactive.

Caution: Exercise great care in smelling allyl isothiocyanate. It should be tasted only when highly diluted.

IDENTIFICATION

To a 3-ml. sample, gradually add 3 ml. of sulfuric acid, keeping the mixture cool, then cautiously agitate the liquid. The mixture evolves sulfur dioxide and retains a light yellow color, but loses the pungent odor of the oil.

SPECIFICATIONS

Assay. Not less than 93.0 percent of C_3H_5NCS.

Distillation range. Between 148° and 154°.

Refractive index. Between 1.527 and 1.531 at 20°.

Specific gravity. Between 1.013 and 1.020.

Limits of Impurities

Phenols. Passes test.

TESTS

Assay. Transfer about 4 ml., accurately weighed, into a 100-ml. volumetric flask, and add sufficient alcohol to make 100.0 ml. Pipet 5 ml. of this solution into a 100-ml. volumetric flask, and add 50.0 ml. of 0.1 *N* silver nitrate and 5 ml. of ammonia T.S. Connect the flask to a reflux condenser, heat it on a water bath for 1 hour, and allow the liquid to cool to room temperature. Disconnect the flask from the condenser, add sufficient water to make the mixture measure 100.0 ml., mix well, and filter through a dry filter, rejecting the first 10 ml. of filtrate. To 50 ml. of the subsequent filtrate, accurately measured, add about 5 ml. of nitric acid and 2 ml. of ferric ammonium sulfate T.S., and titrate the excess silver nitrate with 0.1 *N* ammonium thiocyanate. Perform a blank determination, using 5 ml. of alcohol in place of the sample solution, and make any necessary corrections. Each ml. of 0.1 *N* silver nitrate is equivalent to 4.958 mg. of C_3H_5NCS.

Distillation range. Proceed as directed under *Distillation Range*, page 890. The sample distills completely within the specified range.

Refractive index, page 945. Determine with an Abbé or other refractometer of equal or greater accuracy.

Specific gravity. Determine by any reliable method (see page 5).

Phenols. Dilute 1 ml. of sample with 5 ml. of alcohol, and add 1 drop of ferric chloride T.S. A blue color is not produced immediately.

Packaging and storage. Store in full, tight containers in a cool place protected from light.

Labeling. Label allyl isothiocyanate to indicate whether it was made synthetically or distilled from one of the plants mentioned in the *Description*.

Functional use in foods. Flavoring agent.

ALMOND OIL, BITTER, FFPA

Bitter Almond Oil Free from Prussic Acid

DESCRIPTION

A volatile oil obtained from *Prunus amygdalus* Batsch var. *amara* (De Candolle) Focke (Fam. *Rosaceae*), apricot kernel (*Prunus armeniaca* L.), and other fruit kernels containing amygdalin. It is prepared by steam distillation of a water-macerated, powdered and pressed cake which has been specially treated and redistilled to remove hydrocyanic acid. It is a colorless to slightly yellow liquid, having a strong almond-like aroma and a slightly astringent, mild taste. It is soluble in most fixed oils and in propylene glycol, and it is slightly soluble in mineral oil. It is insoluble in glycerin.

SPECIFICATIONS

Assay. Not less than 95.0 percent of aldehydes, calculated as benzaldehyde (C_7H_6O).

Acid value. Not more than 8.0.

Angular rotation. Optically inactive, or not more than $\pm 0.15°$.

Refractive index. Between 1.541 and 1.546 at 20°.

Solubility in alcohol. Passes test.

Specific gravity. Between 1.040 and 1.050.

Limits of Impurities

Arsenic (as As). Not more than 3 parts per million (0.0003 percent).

Chlorinated compounds. Passes test.

Heavy metals (as Pb). Not more than 40 parts per million (0.004 percent).

Hydrocyanic acid. Passes test (about 0.15 percent).

Lead. Not more than 10 parts per million (0.001 percent).

TESTS

Assay. Weigh accurately about 1 ml., and proceed as directed under *Aldehydes*, page 894, using 53.05 as the equivalence factor (e) in the calculation.

Acid value. Determine as directed in the general method, page 893.

Angular rotation. Determine in a 100-mm. tube as directed under *Optical Rotation*, page 939.

Refractive index, page 945. Determine with an Abbé or other refractometer of equal or greater accuracy.

Solubility in alcohol. Proceed as directed in the general method, page 899. One ml. dissolves to form a clear solution in 2 ml. of 70 percent alcohol.

Specific gravity. Determine by any reliable method (see page 5).

Arsenic. A *Sample Solution* prepared as directed for organic compounds meets the requirements of the *Arsenic Test*, page 865.

Chlorinated compounds. Proceed as directed in the general method, page 895.

Heavy metals. Prepare and test a 500-mg. sample as directed in *Method II* under the *Heavy Metals Test*, page 920, using 20 mcg. of lead ion (Pb) in the control (*Solution A*).

Hydrocyanic acid. To a 1-ml. sample in a test tube, add 1 ml. of water, 5 drops of a 1 in 10 sodium hydroxide solution, and 5 drops of a 1 in 10 ferrous sulfate solution. Shake thoroughly and acidify with 0.5 N hydrochloric acid. No blue precipitate or color forms.

Lead. A *Sample Solution* prepared as directed for organic compounds meets the requirements of the *Lead Limit Test*, page 929, using 10 mcg. of lead ion (Pb) in the control.

Packaging and storage. Store in full, tight, glass, aluminum, tin-lined, or other suitably lined containers in a cool place protected from light.

Functional use in foods. Flavoring agent.

ALUMINUM AMMONIUM SULFATE

Ammonium Alum

$AlNH_4(SO_4)_2 . 12H_2O$ Mol. wt. 453.32

DESCRIPTION

Large, colorless crystals, white granules, or a powder. It is odorless and has a sweetish, strongly astringent taste. One gram dissolves in 7 ml. of water at 25° and in about 0.3 ml. of boiling water. It is insoluble in alcohol, and is freely, but slowly soluble in glycerin. Its solutions are acid to litmus. A 1 in 20 solution gives positive tests for *Aluminum* and for *Ammonium*, page 926, and for *Sulfate*, page 928.

SPECIFICATIONS

Assay. Not less than 99.5 percent of $AlNH_4(SO_4)_2 . 12H_2O$.

Limits of Impurities

Alkalies and alkaline earths. Passes test.
Arsenic (as As). Not more than 3 parts per million (0.0003 percent).
Fluoride. Not more than 30 parts per million (0.003 percent).
Heavy metals (as Pb). Not more than 20 parts per million (0.002 percent).
Lead. Not more than 10 parts per million (0.001 percent).
Selenium. Not more than 30 parts per million (0.003 percent).

TESTS

Assay. Weigh accurately about 1 gram, dissolve it in 50 ml. of water, add 50.0 ml. of 0.05 *M* disodium ethylenediaminetetraacetate, and boil gently for 5 minutes. Cool, and add in the order given and with continuous stirring 20 ml. of pH 4.5 buffer solution (77.1 grams of ammonium acetate and 57 ml. of glacial acetic acid in 1000 ml. of solution), 50 ml. of alcohol, and 2 ml. of dithizone T.S. Titrate with 0.05 *M* zinc sulfate to a bright rose-pink color, and perform a blank determination (see page 2). Each ml. of 0.05 *M* disodium ethylenediaminetetraacetate is equivalent to 22.67 mg. of $AlNH_4(SO_4)_2 . 12H_2O$.

Alkalies and alkaline earths. Completely precipitate the aluminum from a boiling solution of 1 gram of the sample in 100 ml. of water by the addition of enough ammonia T.S. to render the solution distinctly alkaline to methyl red T.S., and filter. Evaporate the filtrate to dryness, and ignite. The weight of the residue does not exceed 5 mg.

Arsenic. A solution of 1 gram in 35 ml. of water meets the requirements of the *Arsenic Test*, page 865.

Fluoride

Lime Suspension. Carefully slake about 56 grams of low-fluorine calcium oxide (about 2 parts per million F) with 250 ml. of water, and add 250 ml. of 60 percent perchloric acid slowly and with stirring. Add a few glass beads, and boil to copious fumes of perchloric acid, then cool, add 200 ml. of water, and boil again. Repeat the dilution and boiling once more, cool, dilute considerably, and filter through a fritted glass filter, if precipitated silicon dioxide appears. Pour the clear solution, with stirring, into 1000 ml. of sodium hydroxide solution (1 in 10), allow the precipitate to settle, and siphon off the supernatant liquid. Remove the sodium salts from the precipitate by washing 5 times in large centrifuge bottles, shaking the mass thoroughly each time. Finally, shake the precipitate into a suspension and dilute to 2000 ml. Store in paraffin-lined bottles and shake well before use. (*Note:* 100 ml. of this suspension should give no appreciable fluoride blank when evaporated, distilled, and titrated as directed in the *Fluoride Limit Test*, page 917.)

Procedure. Assemble the distilling apparatus as described in the *Fluoride Limit Test*, page 917, and add to the distilling flask 1.67 grams of the sample, accurately weighed, and 25 ml. of dilute sulfuric acid (1 in 2). Distil until the temperature reaches 160°, then maintain at 160° to 165° by adding water from the funnel, collecting 300 ml. of distillate. Oxidize the distillate by the cautious addition of 2 or 3 ml. of fluorine-free 30 percent hydrogen peroxide (to remove sulfites), allow to stand for a few minutes, and evaporate in a platinum dish with an excess of *Lime Suspension*. Ignite briefly at 600°, then cool and wet the ash with about 10 ml. of water. Cover the dish with a watch glass, and cautiously introduce under cover just sufficient 60 percent perchloric acid to dissolve the ash. Add the contents of the dish through the dropping funnel of a freshly prepared distilling apparatus (the distilling flask should contain a few glass beads), using a total of 20 ml. of the perchloric acid for dissolving the ash and transferring the solution. Add 10 ml. of water and a few drops of silver perchlorate solution (1 in 2) through the dropping funnel, and continue as directed in the *Fluoride Limit Test*, page 917, beginning with "Distil until the temperature reaches 135°. . ."

Heavy metals. Dissolve 1 gram in 20 ml. of water, add a few drops of diluted hydrochloric acid T.S., and evaporate to dryness in a porcelain dish. Treat the residue with 20 ml. of water, and add 50 mg. of hydroxylamine hydrochloride. Heat on a steam bath for 10 minutes, cool, and dilute to 25 ml. with water. This solution meets the requirements of the *Heavy Metals Test*, page 920, using 20 mcg. of lead ion (Pb) and 50 mg. of hydroxylamine hydrochloride in the control (*Solution A*).

Lead. A solution of 1 gram in 10 ml. of water meets the requirements of the *Lead Limit Test*, page 929, using 10 mcg. of lead ion (Pb) in the control.

Selenium. A solution of 2 grams in 40 ml. of dilute hydrochloric acid (1 in 2) meets the requirements of the *Selenium Limit Test*, page 953.

Packaging and storage. Store in well-closed containers.

Functional use in foods. Buffer; neutralizing agent.

ALUMINUM POTASSIUM SULFATE

Potassium Alum

$AlK(SO_4)_2 . 12H_2O$ Mol. wt. 474.38

DESCRIPTION

Large, transparent crystals or crystalline fragments, or a white crystalline powder. It is odorless and has a sweetish, astringent taste. One gram dissolves in 7.5 ml. of water at 25° and in about 0.3 ml. of boiling water. It is insoluble in alcohol, but is freely soluble in glycerin. Its solutions are acid to litmus. A 1 in 20 solution gives positive tests for *Aluminum*, page 925, for *Potassium*, page 928, and for *Sulfate*, page 928.

SPECIFICATIONS

Assay. Not less than 99.5 percent of $AlK(SO_4)_2 . 12H_2O$.

Limits of Impurities

Ammonium salts. Passes test.

Arsenic (as As). Not more than 3 parts per million (0.0003 percent).

Fluoride. Not more than 30 parts per million (0.003 percent).

Heavy metals (as Pb). Not more than 20 parts per million (0.002 percent).

Lead. Not more than 10 parts per million (0.001 percent).

Selenium. Not more than 30 parts per million (0.003 percent).

TESTS

Assay. Weigh accurately about 1 gram, dissolve it in 50 ml. of water, add 50.0 ml. of 0.05 *M* disodium ethylenediaminetetraacetate, and boil gently for 5 minutes. Cool, and add in the order given and with continuous stirring 20 ml. of pH 4.5 buffer solution (77.1 grams of ammonium acetate and 57 ml. of glacial acetic acid in 1000 ml. of solution), 50 ml. of alcohol, and 2 ml. of dithizone T.S. Titrate with 0.05 *M* zinc sulfate to a bright rose-pink color, and perform a blank determination (see page 2). Each ml. of 0.05 *M* disodium ethylenediaminetetraacetate is equivalent to 23.72 mg. of $AlK(SO_4)_2 . 12H_2O$.

Ammonium salts. Heat 1 gram with 10 ml. of sodium hydroxide T.S. on a steam bath for 1 minute. The odor of ammonia is not perceptible.

Arsenic. A solution of 1 gram in 35 ml. of water meets the requirements of the *Arsenic Test*, page 865.

Heavy metals. Dissolve 1 gram in 20 ml. of water, add a few drops of diluted hydrochloric acid T.S., and evaporate to dryness in a porcelain dish. Treat the residue with 20 ml. of water, and add 50 mg. of hydroxylamine hydrochloride. Heat on a steam bath for 10 minutes, cool, and dilute to 25 ml. with water. This solution meets the requirements of the *Heavy Metals Test*, page 920, using 20 mcg. of lead ion (Pb) and 50 mg. of hydroxylamine hydrochloride in the control (*Solution A*).

Fluoride. Determine as directed in the test for *Fluoride* under *Aluminum Ammonium Sulfate*, page 34.

Lead. A solution of 1 gram in 10 ml. of water meets the requirements of the *Lead Limit Test*, page 929, using 10 mcg. of lead ion (Pb) in the control.

Selenium. A solution of 2 grams in 40 ml. of dilute hydrochloric acid (1 in 2) meets the requirements of the *Selenium Limit Test*, page 953.

Packaging and storage. Store in well-closed containers.

Functional use in foods. Buffer; neutralizing, firming agent.

ALUMINUM SODIUM SULFATE

Soda Alum; Sodium Alum

$AlNa(SO_4)_2$ Mol. wt. 242.09

DESCRIPTION

Aluminum sodium sulfate is anhydrous or may contain up to 12 molecules of water of hydration. It occurs as colorless crystals, or white granules or powder. It is odorless and has a saline, astringent taste. The anhydrous form is slowly soluble in water. The dodecahydrate is freely soluble in water, and it effloresces in air. Both forms are insoluble in alcohol. It responds to the flame test for *Sodium*, page 928, and gives positive tests for *Aluminum*, page 925, and for *Sulfate*, page 928.

SPECIFICATIONS

Assay. *Anhydrous form*, not less than 96.5 percent of $AlNa(SO_4)_2$ after drying; *dodecahydrate*, not less than 99.5 percent of $AlNa(SO_4)_2$ after drying.

Loss on drying. *Anhydrous form*, not more than 10 percent; *dodecahydrate*, not more than 47.2 percent.

Neutralizing value. *Anhydrous form*, between 103 and 107.

Limits of Impurities

Ammonium salts. Passes test.

Arsenic (as As). Not more than 3 parts per million (0.0003 percent).

Fluoride. Not more than 30 parts per million (0.003 percent).

Heavy metals (as Pb). Not more than 20 parts per million (0.002 percent).

Lead. Not more than 10 parts per million (0.001 percent).

Selenium. Not more than 30 parts per million (0.003 percent).

TESTS

Assay. Weigh accurately about 500 mg. of a sample previously dried as directed in the test for *Loss on drying*, moisten with 1 ml. of acetic acid, and dissolve it in 50 ml. of water, warming gently on a steam bath until solution is complete. Cool, neutralize with ammonia T.S., add 50.0 ml. of 0.05 *M* disodium ethylenediaminetetraacetate, and boil gently for 5 minutes. Cool, and add in the order given and with continuous stirring 20 ml. of pH 4.5 buffer solution (77.1 grams of ammonium acetate and 57 ml. of glacial acetic acid in 1000 ml. of solution), 50 ml. of alcohol, and 2 ml. of dithizone T.S. Titrate with 0.05 *M* zinc sulfate to a bright rose-pink color, and perform a blank determination (see page 2). Each ml. of 0.05 *M* disodium ethylenediaminetetraacetate is equivalent to 12.10 mg. of $AlNa(SO_4)_2$.

Loss on drying, page 931. *Anhydrous form:* dry at 200° for 16 hours. *Dodecahydrate:* dry first at 50–55° for 1 hour, then at 200° for 16 hours.

Neutralizing value. Weigh accurately 500 mg. of the anhydrous form into a 200 ml. Erlenmeyer flask, add 30 ml. of water and 4 drops of phenolphthalein T.S., and boil until the sample dissolves. Add 13.0 ml. of 0.5 *N* sodium hydroxide, boil for a few seconds, and titrate with 0.5 *N* hydrochloric acid to the disappearance of the pink color, adding the acid dropwise and agitating vigorously after each addition. Calculate the neutralizing value, as parts of $NaHCO_3$ equivalent to 100 parts of the sample, by the formula $8.4V$, in which V is the volume, in ml., of 0.5 *N* sodium hydroxide consumed by the sample.

Ammonium salts. Heat 1 gram with 10 ml. of sodium hydroxide T.S. on a steam bath for 1 minute. The odor of ammonia is not perceptible.

Arsenic. A solution of 1 gram in 35 ml. of water meets the requirements of the *Arsenic Test,* page 865.

Fluoride. Determine as directed in the test for *Fluoride* under *Aluminum Ammonium Sulfate,* page 34.

Heavy metals. Dissolve 1 gram in 20 ml. of water, add a few drops of diluted hydrochloric acid T.S., and evaporate to dryness in a porcelain dish. Treat the residue with 20 ml. of water, and add 50 mg. of hydroxylamine hydrochloride. Heat on a steam bath for 10 minutes, cool, and dilute to 25 ml. with water. This solution meets the requirements of the *Heavy Metals Test,* page 920, using 20 mcg. of lead ion (Pb) and 50 mg. of hydroxylamine hydrochloride in the control (*Solution A*).

Lead. A solution of 1 gram in 10 ml. of water meets the requirements of the *Lead Limit Test,* page 929, using 10 mcg. of lead ion (Pb) in the control.

Selenium. A solution of 2 grams in 40 ml. of dilute hydrochloric acid (1 in 2) meets the requirements of the *Selenium Limit Test,* page 953.

Packaging and storage. Store in tight containers.

Functional use in foods. Buffer; neutralizing, firming agent.

ALUMINUM SULFATE

$Al_2(SO_4)_3 . xH_2O$ Mol. wt. (anhydrous) 342.15

DESCRIPTION

Aluminum sulfate is anhydrous or contains 18 molecules of water of crystallization. Due to efflorescence, the hydrate may have a composition approximating the formula $Al_2(SO_4)_3 . 14H_2O$. It occurs as a white powder, as shining plates, or as crystalline fragments. It is odorless and has a sweet taste, becoming mildly astringent. One gram of the hydrate dissolves in about 2 ml. of water. The anhydrous product approaches the same solubility, but the rate of solution is so slow that it initially appears to be relatively insoluble. The pH of a 1 in 20 solution is 2.9 or above. A 1 in 10 solution gives positive tests for *Aluminum*, page 925, and for *Sulfate*, page 928.

SPECIFICATIONS

Assay. $Al_2(SO_4)_3$ (anhydrous), not less than 99.5 percent of $Al_2(SO_4)_3$, calculated on the ignited basis; $Al_2(SO_4)_3 . 18H_2O$ (hydrate), not less than 99.5 percent and not more than the equivalent of 114.0 percent of $Al_2(SO_4)_3 . 18H_2O$, corresponding to not more than approximately 101.7 percent of $Al_2(SO_4)_3 . 14H_2O$.

Limits of Impurities

Alkalies and alkaline earths. Passes test (about 0.4 percent).

Ammonium salts. Passes test.

Arsenic (as As). Not more than 3 parts per million (0.0003 percent).

Heavy metals (as Pb). Not more than 40 parts per million (0.004 percent).

Lead. Not more than 10 parts per million (0.001 percent).

Loss on ignition. $Al_2(SO_4)_3$ (anhydrous), not more than 5 percent. [*Note:* This specification does not apply to $Al_2(SO_4)_3 . 18H_2O$.]

Selenium. Not more than 30 parts per million (0.003 percent).

TESTS

Assay. Weigh accurately an amount of sample equivalent to about 4 grams of $Al_2(SO_4)_3$, transfer into a 250-ml. volumetric flask, dissolve in water, dilute to volume with water, and mix. Pipet 10 ml. of this solution into a 250-ml. beaker, add 25.0 ml. of 0.05 *M* disodium ethylenediaminetetraacetate, and boil gently for 5 minutes. Cool, and add in the order given and with continuous stirring 20 ml. of pH 4.5 buffer solution (77.1 grams of ammonium acetate and 57 ml. of glacial acetic acid in 1000 ml. of solution), 50 ml. of alcohol, and 2 ml. of dithizone T.S. Titrate with 0.05 *M* zinc sulfate until the color changes from green-violet to rose-pink, and perform a blank determination (see page 2), substituting 10 ml. of water for the sample.

Each ml. of 0.05 *M* disodium ethylenediaminetetraacetate is equivalent to 8.554 mg. of $Al_2(SO_4)_3$ or to 16.67 mg. of $Al_2(SO_4)_3 . 18H_2O$.

Alkalies and alkaline earths. To a boiling solution of 2 grams in 150 ml. of water add a few drops of methyl red T.S., and then add ammonia T.S. until the color of the solution just changes to a distinct yellow. Add hot water to restore the original volume, and filter while hot. Evaporate 75 ml. of the filtrate to dryness, and ignite to constant weight. Not more than 4 mg. of residue remains.

Ammonium salts. Heat 1 gram with 10 ml. of sodium hydroxide T.S. on a steam bath for 1 minute. The odor of ammonia is not perceptible.

Arsenic. A solution of 1 gram in 35 ml. of water meets the requirements of the *Arsenic Test,* page 865.

Heavy metals. Dissolve 500 mg. in 20 ml. of water, add a few drops of diluted hydrochloric acid T.S., and evaporate to dryness in a porcelain dish. Treat the residue with 20 ml. of water, and add 50 mg. of hydroxylamine hydrochloride. Heat on a steam bath for 10 minutes, cool, and dilute to 25 ml. with water. This solution meets the requirements of the *Heavy Metals Test,* page 920, using 20 mcg. of lead ion (Pb) and 50 mg. of hydroxylamine hydrochloride in the control (*Solution A*).

Lead. A solution of 1 gram in 10 ml. of water meets the requirements of the *Lead Limit Test,* page 929, using 10 mcg. of lead ion (Pb) in the control.

Loss on ignition. Weigh accurately about 2 grams of $Al_2(SO_4)_3$ (anhydrous), and ignite, preferably in a muffle furnace, at about 500° for 3 hours. [*Note:* This test does not apply to $Al_2(SO_4)_3 . 18H_2O$.]

Selenium. A solution of 2 grams in 40 ml. of dilute hydrochloric acid (1 in 2) meets the requirements of the *Selenium Limit Test,* page 953.

Packaging and storage. Store in well-closed containers.

Functional use in foods. Firming agent.

ALUMINUM SULFATE SOLUTION

Alum Liquor

DESCRIPTION

Aluminum sulfate solution is usually available as a solution containing approximately 50 percent of $Al_2(SO_4)_3 . 18H_2O$. It is a clear to slightly turbid, colorless to slightly colored, liquid having a mildly astringent taste. It is acid to litmus, and a 1 in 20 solution [based on $Al_2(SO_4)_3 . 18H_2O$] has a pH between 2.5 and 3.5. Diluted 1 in 5, it gives positive tests for *Aluminum,* page 925, and for *Sulfate,* page 928.

SPECIFICATIONS

Assay. Not less than 90.0 percent and not more than 110.0 percent, by weight, of the labeled amount of aluminum sulfate, expressed as $Al_2(SO_4)_3$, $Al_2(SO_4)_3 . 18H_2O$, or Al_2O_3.

Limits of Impurities

Alkalies and alkaline earths. Passes test.

Ammonium salts. Passes test.

Arsenic (as As). Not more than 3 parts per million (0.0003 percent), calculated on the $Al_2(SO_4)_3 . 18H_2O$ determined in the *Assay.*

Heavy metals (as Pb). Not more than 40 parts per million (0.004 percent), calculated on the $Al_2(SO_4)_3 . 18H_2O$ determined in the *Assay.*

Lead. Not more than 10 parts per million (0.001 percent), calculated on the $Al_2(SO_4)_3 . 18H_2O$ determined in the *Assay.*

Selenium. Not more than 30 parts per million (0.003 percent), calculated on the $Al_2(SO_4)_3 . 18H_2O$ determined in the *Assay.*

TESTS

Assay. Transfer an accurately weighed amount of the solution, equivalent to about 7.5 grams of $Al_2(SO_4)_3 . 18H_2O$, into a 250-ml. volumetric flask, dilute to volume with water, and mix. Pipet 10 ml. of this solution into a 250-ml. beaker, add 25.0 ml. of 0.05 *M* disodium ethylenediaminetetraacetate, and boil gently for 5 minutes. Cool, and add in the order given and with continuous stirring 20 ml. of pH 4.5 buffer solution (77.1 grams of ammonium acetate and 57 ml. of glacial acetic acid in 1000 ml. of solution), 50 ml. of alcohol, and 2 ml. of dithizone T.S. Titrate with 0.05 *M* zinc sulfate until the color changes from green-violet to rose-pink, and perform a blank determination (see page 2). Each ml. of 0.05 *M* disodium ethylenediaminetetraacetate is equivalent to 8.554 mg. of $Al_2(SO_4)_3$, 16.67 mg. of $Al_2(SO_4)_3 . 18H_2O$, or 2.549 mg. of Al_2O_3.

Alkalies and alkaline earths. Add a few drops of methyl red T.S. to an accurately weighed amount of the solution equivalent to 2 grams of $Al_2(SO_4)_3 . 18H_2O$, dilute to about 150 ml. with water, heated to boiling, and then add ammonia T.S. until the color of the solution just changes to a distinct yellow. Add hot water to restore the original volume, and filter while hot. Evaporate 75 ml. of the filtrate to dryness, and ignite to constant weight. Not more than 4 mg. of residue remains.

Ammonium salts. Heat an accurately weighed amount of the solution equivalent to 1 gram of $Al_2(SO_4)_3 . 18H_2O$ with 10 ml. of sodium hydroxide T.S. on a steam bath for 1 minute. The odor of ammonia is not perceptible.

Arsenic. Dilute an accurately weighed amount of the solution equivalent to 1 gram of $Al_2(SO_4)_3 . 18H_2O$ to 35 ml. with water. The resulting solution meets the requirements of the *Arsenic Test*, page 865.

Heavy metals. Dilute an accurately weighed amount of the solution equivalent to 500 mg. of $Al_2(SO_4)_3.18H_2O$ to 25 ml. with water, add a few drops of diluted hydrochloric acid T.S., and evaporate to dryness in a porcelain dish. Treat the residue with 20 ml. of water, and add 50 mg. of hydroxylamine hydrochloride. Heat on a steam bath for 10 minutes, cool, and dilute to 25 ml. with water. This solution meets the requirements of the *Heavy Metals Test*, page 920, using 20 mcg. of lead ion (Pb) in the control (*Solution A*).

Lead. Dilute an accurately weighed amount of the solution equivalent to 1 gram of $Al_2(SO_4)_3.18H_2O$ to 10 ml. with water. The resulting solution meets the requirements of the *Lead Limit Test*, page 929, using 10 mcg. of lead ion (Pb) in the control.

Selenium. Dilute an accurately weighed amount of the solution equivalent to 2 grams of $Al_2(SO_4)_3.18H_2O$ to 40 ml. with diluted hydrochloric acid T.S. The resulting solution meets the requirements of the *Selenium Limit Test*, page 953.

Packaging and storage. Store in tight containers.

Functional use in foods. Firming agent.

AMBRETTE SEED OIL

Ambrette Seed Liquid

DESCRIPTION

The volatile oil obtained by steam distillation from the partially dried and crushed seeds of the plant *Abelmoschus moschatus* Moench, syn. *Hibiscus Abelmoschus* L. (Fam. *Malvaceae*). It is refined by solvent extraction to remove fatty acids, or precipitation of the fatty acid salts. It is a clear yellow to amber liquid, having the strong musky odor of ambrettolide. It is soluble in most fixed oils and in mineral oil, often with cloudiness. It is relatively insoluble in glycerin and in propylene glycol.

SPECIFICATIONS

Acid value. Not more than 3.0.

Angular rotation. Between $-2.5°$ and $+3°$.

Refractive index. Between 1.468 and 1.485 at 20°.

Saponification value. Between 140 and 200.

Specific gravity. Between 0.898 and 0.920.

TESTS

Acid value. Determine as directed in the general method, page 893.

Angular rotation. Determine in a 100-mm. tube as directed under *Optical Rotation*, page 939.

Refractive index, page 945. Determine with an Abbé or other refractometer of equal or greater accuracy.

Saponification value. Determine as directed in the general method, page 896, using about 1 gram, accurately weighed.

Specific gravity. Determine by any reliable method (see page 5).

Packaging and storage. Store in full, preferably glass, aluminum, tin-lined, or other suitably lined containers in a cool place protected from light.

Functional use in foods. Flavoring agent.

AMMONIUM ALGINATE

Algin

$(C_6H_7O_6NH_4)_n$

Equiv. wt., *Calculated,* 193.16
Equiv. wt., *Actual* (Avg.), 217.00

DESCRIPTION

The ammonium salt of alginic acid (see *Alginic Acid,* page 26) occurs as a white to yellowish, fibrous or granular powder. It dissolves in water to form a viscous, colloidal solution. It is insoluble in alcohol and in hydro-alcoholic solutions in which the alcohol content is greater than about 30 percent by weight. It is insoluble in chloroform, in ether, and in acids having a pH lower than about 3.

IDENTIFICATION

A. To 5 ml. of a 1 in 100 solution add 1 ml. of calcium chloride T.S. A voluminous, gelatinous precipitate is formed.

B. To 10 ml. of a 1 in 100 solution add 1 ml. of diluted sulfuric acid T.S. A heavy gelatinous precipitate is formed.

C. Ammonium alginate meets the requirements of *Identification Test C* under *Alginic Acid,* page 26.

D. To about 1 gram of ammonium alginate contained in a test tube add 5 ml. of sodium hydroxide T.S. and shake the mixture briefly. The odor of ammonia is evolved.

SPECIFICATIONS

Assay. It yields, on the dried basis, not less than 18.0 percent and not more than 21.0 percent of carbon dioxide (CO_2) corresponding to between 88.7 and 103.6 percent of ammonium alginate (Equiv. wt. 217.00).

Limits of Impurities

Arsenic (as As). Not more than 3 parts per million (0.0003 percent).

Ash. Not more than 4 percent.
Heavy metals (as Pb). Not more than 40 parts per million (0.004 percent).
Insoluble matter. Not more than 0.2 percent.
Lead. Not more than 10 parts per million (0.001 percent).
Loss on drying. Not more than 15.0 percent.

TESTS

Assay. Proceed as directed under *Alginates Assay*, page 863. Each ml. of 0.25 *N* sodium hydroxide consumed in the assay is equivaent to 27.12 mg. of ammonium alginate (Equiv. wt. 217.00).

Arsenic. A *Sample Solution* prepared as directed for organic compounds meets the requirements of the *Arsenic Test*, page 865.

Ash. Determine as directed under *Ash* in the monograph on *Alginic Acid*, page 26.

Heavy metals. Prepare and test a 500-mg. sample as directed in *Method II* under the *Heavy Metals Test*, page 920, but use a platinum crucible for the ignition. Any color does not exceed that produced in a control (*Solution A*) containing 20 mcg. of lead ion (Pb).

Insoluble matter. Determine as directed under *Insoluble matter* in the monograph on *Alginic Acid*, page 26.

Lead. A *Sample Solution* prepared as directed for organic compounds meets the requirements of the *Lead Limit Test*, page 929, using 10 mcg. of lead ion (Pb) in the control.

Loss on drying, page 931. Dry at 105° for 4 hours.

Packaging and storage. Store in well-closed containers.
Functional use in foods. Stabilizer; thickener; emulsifier.

AMMONIUM BICARBONATE

NH_4HCO_3 Mol. wt. 79.06

DESCRIPTION

White crystals or a crystalline powder having a slight odor of ammonia. At a temperature of 60° or above it volatilizes rapidly, dissociating into ammonia, carbon dioxide, and water, but at room temperature it is quite stable. One gram dissolves in about 6 ml. of water. It is insoluble in alcohol. It gives positive tests for *Ammonium*, page 926, and for *Bicarbonate*, page 926.

SPECIFICATIONS

Assay. Not less than 99.0 percent of NH_4HCO_3.

Limits of Impurities

Arsenic (as As). Not more than 3 parts per million (0.0003 percent).

Chloride. Not more than 30 parts per million (0.003 percent).

Heavy metals (as Pb). Not more than 10 parts per million (0.001 percent).

Nonvolatile residue. Not more than 0.05 percent.

Sulfur compounds. Not more than 70 parts per million (0.007 percent).

TESTS

Assay. Weigh accurately about 3 grams, dissolve it in 40 ml. of water, add methyl orange T.S., and titrate with 1 *N* sulfuric acid. Each ml. of 1 *N* sulfuric acid is equivalent to 79.06 mg. of NH_4HCO_3.

Arsenic. A solution of 1 gram in 10 ml. of water meets the requirements of the *Arsenic Test*, page 865.

Chloride, page 879. Any turbidity produced by a 500-mg. sample does not exceed that shown in a control containing 15 mcg. of chloride ion (Cl).

Heavy metals. Dissolve the residue from the test for *Nonvolatile residue* in 1 ml. of diluted hydrochloric acid T.S., evaporate to dryness, and dissolve the residue in 50 ml. of water. A 25-ml. portion of this solution meets the requirements of the *Heavy Metals Test*, page 920, using 20 mcg. of lead ion (Pb) in the control (*Solution A*).

Nonvolatile residue. Transfer 4 grams into a tared dish, add 10 ml. of water, evaporate on a steam bath, and then dry at 105°. The weight of the residue does not exceed 2 mg. Retain the residue for the *Heavy Metals Test.*

Sulfur compounds. Dissolve 4 grams in 40 ml. of water, add about 10 mg. of sodium carbonate and 1 ml. of 30 per cent hydrogen peroxide, and evaporate the solution to dryness on a steam bath. Treat the residue as directed in the *Sulfate Limit Test*, page 879. Any turbidity produced does not exceed that shown by 280 mcg. of sulfate ion (SO_4).

Packaging and storage. Store in well-closed containers.

Functional use in foods. Alkali; leavening agent.

AMMONIUM CARBONATE

DESCRIPTION

Ammonium carbonate consists of ammonium bicarbonate (NH_4HCO_3) and ammonium carbamate ($NH_2.COONH_4$) in varying proportions. It occurs as a white powder or hard, white or translucent masses. Its solutions are alkaline to litmus. On exposure to air, it becomes opaque, and is finally converted into porous lumps or a white powder of ammonium bicarbonate due to the loss of ammonia and carbon dioxide. One gram dissolves slowly in about 4 ml. of water.

IDENTIFICATION

When heated, it volatilizes without charring and the vapor is alkaline to moistened litmus paper. A 1 in 20 solution effervesces upon the addition of an acid.

SPECIFICATIONS

Assay. Not less than 30.0 percent and not more than 34.0 percent of NH_3.

Limits of Impurities

Arsenic (as As). Not more than 3 parts per million (0.0003 percent).
Chloride. Not more than 30 parts per million (0.003 percent).
Heavy metals (as Pb). Not more than 10 parts per million (0.001 percent).
Nonvolatile residue. Not more than 0.05 percent.
Sulfur compounds. Not more than 50 parts per million (0.005 percent).

TESTS

Assay. Place about 10 ml. of water in a weighing bottle, tare the bottle and its contents, add about 2 grams of ammonium carbonate, and weigh accurately. Transfer the contents of the bottle to a 250-ml. flask, and slowly add, with mixing, 50.0 ml. of 1 *N* sulfuric acid, allowing for the release of carbon dioxide. When solution has been effected, wash down the sides of the flask with a few ml. of water, add methyl orange T.S., and titrate the excess acid with 1 *N* sodium hydroxide. Each ml. of 1 *N* sulfuric acid is equivalent to 17.03 mg. of NH_3.

Arsenic. A solution of 1 gram in 10 ml. of diluted hydrochloric acid T.S. meets the requirements of the *Arsenic Test*, page 865.

Chloride. Dissolve 500 mg. in 10 ml. of hot water, add about 5 mg. of sodium carbonate, and evaporate to dryness on a steam bath. Treat the residue as directed in the *Chloride Limit Test*, page 879. Any turbidity produced does not exceed that shown in a control containing 15 mcg. of chloride ion (Cl).

Heavy metals. Dissolve the residue from the test for *Nonvolatile residue* in 1 ml. of diluted hydrochloric acid T.S., and evaporate to dryness. Dissolve the residue in water to make 50 ml. A 25-ml. portion of this solution meets the requirements of the *Heavy Metals Test*, page 920, using 20 mcg. of lead ion (Pb) in the control (*Solution A*).

Nonvolatile residue. Transfer 4 grams into a tared dish, add 10 ml. of water, evaporate on a steam bath, and then dry for 1 hour at 105°. The weight of the residue does not exceed 2 mg. Retain the residue for the *Heavy Metals Test.*

Sulfur compounds. Dissolve 4 grams in 40 ml. of water, add about 10 mg. of sodium carbonate and 1 ml. of 30 percent hydrogen peroxide, and evaporate the solution to dryness on a steam bath. Treat the residue as directed in the *Sulfate Limit Test*, page 879. Any turbidity produced does not exceed that shown in a control containing 200 mcg. of sulfate (SO_4).

Packaging and storage. Store in tight, light-resistant containers, preferably at a temperature not exceeding 30°.

Functional use in foods. Miscellaneous and general purpose; buffer; neutralizing agent.

AMMONIUM CHLORIDE

NH_4Cl Mol. wt. 53.49

DESCRIPTION

Colorless crystals, or a white, fine or coarse, crystalline powder. It has a cool, saline taste, and is somewhat hygroscopic. One gram dissolves in 2.6 ml. of water at 25°, in 1.4 ml. of boiling water, in about 100 ml. of alcohol, and in about 8 ml. of glycerin. The pH of a 1 in 20 solution is between 4.5 and 6.0. A 1 in 10 solution gives positive tests for *Ammonium*, page 926, and for *Chloride*, page 926.

SPECIFICATIONS

Assay. Not less than 99.0 percent of NH_4Cl after drying.

Limits of Impurities

Arsenic (as As). Not more than 3 parts per million (0.0003 percent).

Heavy metals (as Pb). Not more than 10 parts per million (0.001 percent).

Loss on drying. Not more than 0.5 percent.

TESTS

Assay. Dry about 200 mg. over silica gel for 4 hours, weigh accurately, and dissolve it in about 40 ml. of water in a glass-stoppered flask. Add, while agitating, 3 ml. of nitric acid, 5 ml. of nitrobenzene, 50.0 ml. of 0.1 *N* silver nitrate, shake vigorously, then add 2 ml. of ferric ammonium sulfate T.S., and titrate the excess silver nitrate with 0.1 *N* ammonium thiocyanate. Each ml. of 0.1 *N* silver nitrate is equivalent to 5.349 mg. of NH_4Cl.

Arsenic. A solution of 1 gram in 35 ml. of water meets the requirements of the *Arsenic Test*, page 865.

Heavy metals. A solution of 2 grams in 25 ml. of water meets the requirements of the *Heavy Metals Test*, page 920, using 20 mcg. of lead ion (Pb) in the control (*Solution A*).

Loss on drying, page 931. Dry over silica gel for 4 hours.

Packaging and storage. Store in tight containers.

Functional use in foods. Yeast food; dough conditioner.

AMMONIUM HYDROXIDE

Strong Ammonia Solution; Stronger Ammonia Water

NH_4OH Mol. wt. 35.05

DESCRIPTION

A clear, colorless solution of NH_3, having an exceedingly pungent, characteristic odor. Upon exposure to air it loses ammonia rapidly. Its specific gravity is about 0.90. Dense, white fumes are produced when a glass rod wet with hydrochloric acid is held near the surface of the liquid.

SPECIFICATIONS

Assay. Not less than 27.0 percent and not more than 30.0 percent, by weight, of NH_3.

Limits of Impurities

Arsenic (as As). Not more than 3 parts per million (0.0003 percent).

Heavy metals (as Pb). Not more than 5 parts per million (0.0005 percent).

Nonvolatile residue. Not more than 0.02 percent.

Readily oxidizable substances. Passes test.

TESTS

Assay. Tare accurately a 125-ml. glass-stoppered Erlenmeyer flask containing 35.0 ml. of 1 *N* sulfuric acid. Partially fill a 10-ml. graduated pipet from near the bottom of a sample, previously cooled in the original sample bottle to 10° or lower. (Do not use vacuum for drawing up the sample.) Wipe off any liquid adhering to the outside of the pipet, and discard the first ml. Hold the pipet just above the surface of the acid, and transfer 2 ml. into the flask, leaving at least 1 ml. in the pipet. Stopper the flask, mix, and weigh again to obtain the weight of the sample. Add methyl red T.S., and titrate the excess acid with 1 *N* sodium hydroxide. Each ml. of 1 *N* sulfuric acid is equivalent to 17.03 mg. of NH_3.

Arsenic. Evaporate 11 ml. (10-gram sample) to about 2 ml. on a steam bath, dilute to 50 ml. with water, and mix. A 5-ml. portion of this solution meets the requirements of the *Arsenic Test*, page 865.

Heavy metals. Transfer 22 ml. (20-gram sample) to a beaker, add about 5 mg. of sodium chloride, evaporate to dryness on a steam bath, and dissolve the residue in 2 ml. of diluted acetic acid T.S. and sufficient water to make 50 ml. A 10-ml. portion of this solution, diluted to 25 ml. with water, meets the requirements of the *Heavy Metals Test*, page 920, using 20 mcg. of lead ion (Pb) in the control (*Solution A*).

Nonvolatile residue. Evaporate 11 ml. (10-gram sample) in a tared platinum or porcelain dish to dryness, dry at 105° for 1 hour, cool, and weigh.

Readily oxidizable substances. Dilute 4 ml. with 6 ml. of water, and add a slight excess of diluted sulfuric acid T.S. and 0.1 ml. of 0.1 *N* potassium permanganate. The pink color does not completely disappear within 10 minutes.

Packaging and storage. Store in tight containers, preferably at a temperature not exceeding 25°.

Functional use in foods. Alkali.

AMMONIUM PHOSPHATE, DIBASIC

Diammonium Phosphate

$(NH_4)_2HPO_4$ Mol. wt. 132.06

DESCRIPTION

White, odorless crystals, crystalline powder, or granules having a cooling, saline taste. It is freely soluble in water. The pH of a 1 in 100 solution is between 7.6 and 8.2. A 1 in 20 solution gives positive tests for *Ammonium*, page 926, and for *Phosphate*, page 928.

SPECIFICATIONS

Assay. Not less than 96.0 percent and not more than 102.0 percent of $(NH_4)_2HPO_4$.

Limits of Impurities

Arsenic (as As). Not more than 3 parts per million (0.0003 percent).
Fluoride. Not more than 10 parts per million (0.001 percent).
Heavy metals (as Pb). Not more than 10 parts per million (0.001 percent).

TESTS

Assay. Dissolve about 600 mg., accurately weighed, in 40 ml. of water, and titrate to a pH of 4.6 with 0.1 *N* sulfuric acid. Each ml. of 0.1 *N* sulfuric acid is equivalent to 13.21 mg. of $(NH_4)_2HPO_4$.

Arsenic. A solution of 1 gram in 35 ml. of water meets the requirements of the *Arsenic Test*, page 865.

Fluoride. Proceed as directed in the *Fluoride Limit Test*, page 917.

Heavy metals. A solution of 2 grams in 25 ml. of water meets the requirements of the *Heavy Metals Test*, page 920, using 20 mcg. of lead ion (Pb) in the control (*Solution A*). (*Note:* Use glacial acetic acid in making the pH adjustment.)

Packaging and storage. Store in tight containers.

Functional use in foods. Buffer; dough conditioner; leavening agent; yeast food.

AMMONIUM PHOSPHATE, MONOBASIC

Monoammonium Phosphate

$NH_4H_2PO_4$ Mol. wt. 115.03

DESCRIPTION

White, odorless crystals, crystalline powder, or granules. It is freely soluble in water. The pH of a 1 in 100 solution is between 4.3 and 5.0. A 1 in 20 solution gives positive tests for *Ammonium*, page 926, and for *Phosphate*, page 928.

SPECIFICATIONS

Assay. Not less than 96.0 percent and not more than 102.0 percent of $NH_4H_2PO_4$.

Limits of Impurities

Arsenic (as As). Not more than 3 parts per million (0.0003 percent).

Fluoride. Not more than 10 parts per million (0.001 percent).

Heavy metals (as Pb). Not more than 10 parts per million (0.001 percent).

TESTS

Assay. Dissolve about 500 mg., accurately weighed, in 50 ml. of water, and titrate to a pH of 8.0 with 0.1 *N* sodium hydroxide. Each ml. of 0.1 *N* sodium hydroxide is equivalent to 11.50 mg. of $NH_4H_2PO_4$.

Arsenic. A solution of 1 gram in 35 ml. of water meets the requirements of the *Arsenic Test*, page 865.

Fluoride. Proceed as directed in the *Fluoride Limit Test*, page 917.

Heavy metals. A solution of 2 grams in 25 ml. of water meets the requirements of the *Heavy Metals Test,* page 920, using 20 mcg. of lead ion (Pb) in the control (*Solution A*).

Packaging and storage. Store in tight containers.

Functional use in foods. Buffer; dough conditioner; leavening agent; yeast food.

AMMONIUM SACCHARIN

1,2-Benzisothiazolin-3-one 1,1-Dioxide Ammonium Salt

$C_7H_8N_2O_3S$ Mol. wt. 200.21

DESCRIPTION

White crystals or a white crystalline powder. It is freely soluble in water. The pH of a 1 in 3 solution is between 5 and 6. It is about 500 times as sweet as sucrose in dilute solutions.

IDENTIFICATION

A. Dissolve about 100 mg. in 5 ml. of sodium hydroxide solution (1 in 20), evaporate to dryness, and gently fuse the residue over a small flame until it no longer evolves ammonia. After the residue has cooled dissolve it in 20 ml. of water, neutralize the solution with diluted hydrochloric acid T.S., and filter. The addition of a drop of ferric chloride T.S. to the filtrate produces a violet color.

B. Mix 20 mg. with 40 mg. of resorcinol, cautiously add 10 drops of sulfuric acid, and heat the mixture in a liquid bath at 200° for 3 minutes. After cooling, add 10 ml. of water and an excess of sodium hydroxide T.S. A fluorescent green liquid results.

C. A 1 in 10 solution gives positive tests for *Ammonium,* page 926.

D. To 10 ml. of a 1 in 10 solution add 1 ml. of hydrochloric acid. A crystalline precipitate of saccharin is formed. Wash the precipitate well with cold water and dry at 105° for 2 hours. It melts between 226° and 230° (*Class Ia,* page 931).

SPECIFICATIONS

Assay. Not less than 98.0 percent and not more than the equivalent of 101.0 percent of $C_7H_8N_2O_3S$.

Limits of Impurities

Arsenic (as As). Not more than 3 parts per million (0.0003 percent).

Benzoate and salicylate. Passes test.

Heavy metals (as Pb). Not more than 10 parts per million (0.001 percent).

Readily carbonizable substances. Passes test.
Selenium. Not more than 30 parts per million (0.003 percent).
Water. Not more than 0.3 percent.

TESTS

Assay. Weigh accurately about 500 mg., and transfer it quantitatively to a separator with the aid of 10 ml. of water. Add 2 ml. of diluted hydrochloric acid T.S., and extract the precipitated saccharin first with 30 ml., then with five 20-ml. portions of a solvent composed of 9 volumes of chloroform and 1 volume of alcohol. Filter each extract through a small filter paper moistened with the solvent mixture, and evaporate the combined filtrates on a steam bath to dryness with the aid of a current of air. Dissolve the residue in 75 ml. of hot water, cool, add phenolphthalein T.S., and titrate with 0.1 *N* sodium hydroxide. Perform a blank determination, and make any necessary correction (see page 2). Each ml. of 0.1 *N* sodium hydroxide is equivalent to 20.02 mg. of $C_7H_8N_2O_3S$.

Arsenic. A *Sample Solution* prepared as directed for organic compounds meets the requirements of the *Arsenic Test*, page 865.

Benzoate and salicylate. To 10 ml. of a 1 in 20 solution previously acidified with 5 drops of acetic acid, add 3 drops of ferric chloride T.S. No precipitate or violet color appears.

Heavy metals. Prepare and test a 2-gram sample as directed in *Method II* under the *Heavy Metals Test*, page 920, using 20 mcg. of lead ion (Pb) in the control (*Solution A*).

Readily carbonizable substances, page 943. Dissolve 200 mg. in 5 ml. of sulfuric acid T.S., and keep at a temperature of 48° to 50° for 10 minutes. The color is no darker than *Matching Fluid A*.

Selenium. Prepare and test a 2-gram sample as directed in the *Selenium Limit Test*, page 953.

Water. Determine by the *Karl Fischer Titrimetric Method*, page 977.

Packaging and storage. Store in well-closed containers.
Functional use in foods. Nonnutritive sweetener.

AMMONIUM SULFATE

$(NH_4)_2SO_4$ Mol. wt. 132.14

DESCRIPTION

Colorless or white, odorless crystals or granules which decompose at temperatures above 280°. One gram is soluble in about 1.5 ml. of water and is insoluble in alcohol. The pH of a 0.1 *M* solution is about 5.5. It gives positive tests for *Ammonium*, page 926, and for *Sulfate*, page 928.

SPECIFICATIONS

Assay. Not less than 99.0 percent of $(NH_4)_2SO_4$.

Limits of Impurities

Arsenic (as As). Not more than 3 parts per million (0.0003 percent).
Heavy metals (as Pb). Not more than 10 parts per million (0.001 percent).
Residue on ignition. Not more than 0.25 percent.
Selenium. Not more than 30 parts per million (0.003 percent).

TESTS

Assay. Transfer about 2 grams, accurately weighed, into a 250-ml. flask and dissolve it in 100 ml. of water. To the solution add 40 ml. of a mixture of equal volumes of formaldehyde and water, previously neutralized to phenolphthalein T.S. with 1 *N* sodium hydroxide. Mix, allow to stand for 30 minutes, and titrate the mixture with 1 *N* sodium hydroxide to a pink end-point that persists for 5 minutes. Each ml. of 1 *N* sodium hydroxide is equivalent to 66.07 mg. of $(NH_4)_2SO_4$.

Arsenic. A solution of 1 gram in 10 ml. of water meets the requirements of the *Arsenic Test*, page 865.

Heavy metals. A solution of 2 grams in 25 ml. of water meets the requirements of the *Heavy Metals Test*, page 920, using 20 mcg. of lead ion (Pb) in the control (*Solution A*).

Residue on ignition, page 945. Ignite 1 gram as directed in the general method.

Selenium. A solution of 2 grams in 40 ml. of dilute hydrochloric acid (1 in 2) meets the requirements of the *Selenium Limit Test*, page 953.

Packaging and storage. Store in well-closed containers.

Functional use in foods. Miscellaneous or general purpose; dough conditioner; yeast food.

α-AMYLCINNAMALDEHYDE

Amylcinnamaldehyde

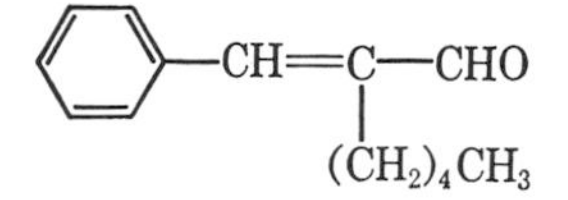

$C_{14}H_{18}O$ Mol. wt. 202.30

DESCRIPTION

A yellow liquid, having a strong floral odor which becomes suggestive of jasmin on dilution. It is soluble in most fixed oils and in mineral oil. It is insoluble in glycerin and in propylene glycol.

SPECIFICATIONS

Assay. Not less than 97.0 percent of $C_{14}H_{18}O$.

Refractive index. Between 1.554 and 1.559 at 20°.

Solubility in alcohol. Passes test.

Specific gravity. Between 0.963 and 0.968.

Limits of Impurities

Acid value. Not more than 5.0.

Chlorinated Compounds. Passes test.

TESTS

Assay. Weigh accurately about 1.5 grams, and proceed as directed under *Aldehydes and Ketones-Hydroxylamine Method*, page 894, using 101.2 as the equivalence factor (E) in the calculation. Allow the sample and the blank to stand at room temperature for one-half hour before titrating.

Refractive index, page 945. Determine with an Abbé or other refractometer of equal or greater accuracy.

Solubility in alcohol. Proceed as directed in the general method, page 899. One ml. dissolves in 4.5 ml. of 80 percent alcohol.

Specific gravity. Determine by any reliable method (see page 5).

Acid value. Determine as directed in the general method, page 893.

Chlorinated compounds. Proceed as directed in the general method, page 895.

Packaging and storage. Store in full, tight, preferably glass, aluminum, steel, or tin-lined containers in a cool place protected from light. It is very susceptible to oxidation by air. It can not be stored unless protected by a suitable antioxidant.

Functional use in foods. Flavoring agent.

AMYRIS OIL

West Indian Sandalwood Oil

DESCRIPTION

The volatile oil obtained by steam distillation from the wood of *Amyris balsamifera* L. (Fam. *Rutaceae*). It is a clear, pale yellow viscous liquid having a distinct odor suggestive of sandalwood. It is soluble in most fixed oils, and usually in mineral oil. It is soluble in an equal volume of propylene glycol, the solution often becoming opalescent on further dilution. It is practically insoluble in glycerin.

SPECIFICATIONS

Acid value. Not more than 3.0.

Angular rotation. Between +10° and +53°.

Ester value. Not more than 7.

Ester value after acetylation. Between 115 and 165.

Refractive index. Between 1.503 and 1.512 at 20°.

Solubility in alcohol. Passes test.

Specific gravity. Between 0.943 and 0.976.

Limits of Impurities

Arsenic (as As). Not more than 3 parts per million (0.0003 percent).

Heavy metals (as Pb). Not more than 40 parts per million (0.004 percent).

Lead. Not more than 10 parts per million (0.001 percent).

TESTS

Acid value. Determine as directed in the general method, page 893.

Angular rotation. Determine in a 100-mm. tube as directed under *Optical Rotation*, page 939.

Ester value. Determine as directed in the general method, page 897, using about 5 grams, accurately weighed.

Ester value after acetylation. Proceed as directed under *Total Alcohols*, page 893, using about 2 grams of the dried acetylated oil, accurately weighed. Reflux for a period of 2 hours. Calculate the *Ester Value after Acetylation* by the formula $A \times 28.05/B$, in which A is the number of ml. of 0.5 N alcoholic potassium hydroxide consumed in the saponification, and B is the weight, in grams, of the acetylated oil used in the test.

Refractive index, page 945. Determine with an Abbé or other refractometer of equal or greater accuracy.

Solubility in alcohol. Proceed as directed in the general method, page 899. One ml. dissolves in 3 ml. of 80 percent alcohol, often with opalescence.

Specific gravity. Determine by any reliable method (see page 5).

Arsenic. A *Sample Solution* prepared as directed for organic compounds meets the requirements of the *Arsenic Test*, page 865.

Heavy metals. Prepare and test a 500-mg. sample as directed in *Method II* under the *Heavy Metals Test*, page 920, using 20 mcg. of lead ion (Pb) in the control (*Solution A*).

Lead. A *Sample Solution* prepared as directed for organic compounds meets the requirements of the *Lead Limit Test*, page 929, using 10 mcg. of lead ion (Pb) in the control.

Packaging and storage. Store in full, tight, preferably aluminum, glass, or tin-lined containers in a cool place protected from light.

Functional use in foods. Flavoring agent.

ANETHOLE

p-Propenylanisole

$$CH_3O-C_6H_4-CH{=}CHCH_3$$

$C_{10}H_{12}O$ Mol. wt. 148.21

DESCRIPTION

Anethole is obtained from anise oil and other sources, or it is prepared synthetically. It is a colorless or faintly yellow liquid at or above 23°. It has a sweet taste and a characteristic anise-like odor. It is affected by light. It is slightly soluble in water, and is miscible with chloroform and with ether. Its alcohol solutions are neutral to litmus.

SPECIFICATIONS

Angular rotation. Between −0.15° and +0.15°.

Distillation range. Between 231° and 237°.

Refractive index. Between 1.557 and 1.561 at 20°.

Solidification point. Not below 20°.

Solubility in alcohol. Passes test.

Specific gravity. Between 0.983 and 0.988.

Limits of Impurities

Aldehydes and ketones. Passes test.

Arsenic (as As). Not more than 3 parts per million (0.0003 percent).

Heavy metals (as Pb). Not more than 40 parts per million (0.004 percent).

Lead. Not more than 10 parts per million (0.001 percent).

Phenols. Passes test.

TESTS

Angular rotation. Determine in a 100-mm. tube as directed under *Optical Rotation*, page 939.

Distillation range, page 890. Distil a 100-ml. sample as directed in the general method.

Refractive index, page 945. Determine with an Abbé or other refractometer of equal or greater accuracy.

Solidification point. Determine as directed in the general method, page 954.

Solubility in alcohol. Proceed as directed in the general method, page 899. One ml. dissolves in 2 ml. of alcohol.

Specific gravity. Determine by any reliable method (see page 5).

Aldehydes and ketones. Shake 10 ml. with 50 ml. of a saturated solution of sodium bisulfite in a glass-stoppered, graduated cylinder, and allow the mixture to stand for six hours. The volume of the sample does not diminish appreciably, and no crystalline deposit separates.

Arsenic. A *Sample Solution* prepared as directed for organic compounds meets the requirements of the *Arsenic Test,* page 865.

Heavy metals. Prepare and test a 500-mg. sample as directed in *Method II* under the *Heavy Metals Test,* page 920, using 20 mcg. of lead ion (Pb) in the control (*Solution A*).

Lead. A *Sample Solution* prepared as directed for organic compounds meets the requirements of the *Lead Limit Test,* page 929, using 10 mcg. of lead ion (Pb) in the control.

Phenols. Shake 1 ml. with 20 ml. of water, and allow the liquids to separate. Filter the water layer through a filter paper previously moistened with water, and to 10 ml. of the filtrate add 3 drops of ferric chloride T.S. No purplish color is produced.

Packaging and storage. Store in full, tight containers in a cool place protected from light.

Functional use in foods. Flavoring agent.

ANGELICA ROOT OIL

DESCRIPTION

Angelica root oil is obtained by steam distillation of the dried slender rootlets of *Angelica archangelica* L., which grows primarily in Europe. It is a pale yellow to deep amber liquid having a warm pungent odor and bitter-sweet taste. It is soluble in most fixed oils, slightly soluble in mineral oil, but relatively insoluble in glycerin and in propylene glycol.

SPECIFICATIONS

Acid value. Not more than 7.0.

Angular rotation. Optically inactive, or not more than +46.0°.

Ester value. Between 10 and 65.

Refractive index. Between 1.473 and 1.487 at 20°.

Solubility in alcohol. Passes test.

Specific gravity. Between 0.850 and 0.880.

TESTS

Acid value. Determine as directed in the general method, page 893.

Angular rotation. Determine in a 100-mm. tube as directed under *Optical Rotation*, page 939.

Ester value. Determine as directed in the general method, page 897, using about 5 grams, accurately weighed.

Refractive index, page 945. Determine with an Abbé or other refractometer of equal or greater accuracy.

Solubility in alcohol. Proceed as directed in the general method, page 899. One ml. dissolves in 1 ml. of 90 percent alcohol, often with turbidity, and remains in solution on further addition of alcohol to a total of 10 ml.

Specific gravity. Determine by any reliable method (see page 5).

Packaging and storage. Store in full, tight, preferably dark glass bottles, or aluminum, or tin-lined containers in a cool place protected from light. The oils increase in specific gravity and viscosity on storage.

Functional use in foods. Flavoring agent.

ANGELICA SEED OIL

DESCRIPTION

Angelica Seed Oil is obtained by steam distillation of the fresh seeds of *Angelica archangelica* L., grown primarily in Europe. It is a light yellow liquid having a sweeter and more delicate aroma than the root oil. It is soluble in most fixed oils, slightly soluble in mineral oil, but relatively insoluble in glycerin and in propylene glycol.

SPECIFICATIONS

Acid value. Not more than 3.0.

Angular rotation. Between +4° and +16°.

Ester value. Between 14.0 and 32.0.

Refractive index. Between 1.480 and 1.488 at 20°.

Solubility in alcohol. Passes test.

Specific gravity. Between 0.853 and 0.876.

TESTS

Acid value. Determine as directed in the general method, page 893.

Angular rotation. Determine in a 100-mm. tube as directed under *Optical Rotation*, page 939.

Ester value. Determine as directed in the general method, page 897, using about 5 grams, accurately weighed.

Refractive index, page 945. Determine with an Abbé or other refractometer of equal or greater accuracy.

Solubility in alcohol. Proceed as directed in the general method, page 899. One ml. dissolves in 4 ml. of 50 percent alcohol, often with considerable turbidity, and remains in solution on further addition of alcohol to a total of 10 ml.

Specific gravity. Determine by any reliable method (see page 5).

Packaging and storage. Store in full, tight, preferably dark glass bottles, or aluminum, or tin-lined containers in a cool place protected from light.

Functional use in foods. Flavoring agent.

ANISE OIL

DESCRIPTION

Anise Oil is obtained by steam distillation of the dried, ripe fruit of *Pimpinella anisum* L. (Fam. *Umbelliferae*), or *Illicium verum* Hooker filius (Fam. *Magnoliaceae*). It is a colorless to pale yellow, strongly refractive liquid, having the characteristic odor and taste of anise.

Note: If solid material has separated, carefully warm the anise oil until it is completely liquefied, and mix before using it.

SPECIFICATIONS

Angular rotation. Between $-2°$ and $+1°$.

Refractive index. Between 1.553 and 1.560 at 20°.

Solidification point. Not lower than 15°.

Solubility in alcohol. Passes test.

Specific gravity. Between 0.978 and 0.988.

Limits of Impurities

Arsenic (as As). Not more than 3 parts per million (0.0003 percent).

Heavy metals (as Pb). Not more than 40 parts per million (0.004 percent).

Lead. Not more than 10 parts per million (0.001 percent).

Phenols. Passes test.

TESTS

Angular rotation. Determine in a 100-mm. tube as directed under *Optical Rotation*, page 939.

Refractive index, page 945. Determine with an Abbé or other refractometer of equal or greater accuracy.

Solidification point. Determine as directed in the general method, page 954.

Solubility in alcohol. Proceed as directed in the general method, page 899. One ml. dissolves in 3 ml. of 90 percent alcohol.

Specific gravity. Determine by any reliable method (see page 5).

Arsenic. A *Sample Solution* prepared as directed for organic compounds meets the requirements of the *Arsenic Test*, page 865.

Heavy metals. Prepare and test a 500-mg. sample as directed in *Method II* under the *Heavy Metals Test*, page 920, using 20 mcg. of lead ion (Pb) in the control (*Solution A*).

Lead. A *Sample Solution* prepared as directed for organic compounds meets the requirements of the *Lead Limit Test*, page 929, using 10 mcg. of lead ion (Pb) in the control.

Phenols. Prepare a 1 in 3 solution of recently distilled anise oil in 90 percent alcohol. It is neutral to moistened litmus paper, and the mixture develops no blue or brownish color upon the addition of 1 drop of ferric chloride T.S. to 5 ml. of the solution.

Packaging and storage. Store in full, tight containers. Avoid exposure to excessive heat.

Functional use in foods. Flavoring agent.

ANISOLE

Methylphenyl Ether

C_6H_5—OCH_3

C_7H_8O Mol. wt. 108.14

DESCRIPTION

A colorless liquid having a phenolic, anise-like odor. It is soluble in alcohol and in ether, but insoluble in water.

SPECIFICATIONS

Distillation range. Not more than 2°.

Refractive index. Between 1.515 and 1.518 at 20°.

Specific gravity. Between 0.990 and 0.993.

Limits of Impurities

Arsenic (as As). Not more than 3 parts per million (0.0003 percent).

Heavy metals (as Pb). Not more than 40 parts per million (0.004 percent).

Lead. Not more than 10 parts per million (0.001 percent).

Phenols. Passes test.

TESTS

Distillation range, page 890. Distil 100 ml. The difference between the temperatures observed when 1 ml. and 95 ml. have distilled does not exceed 2°. The boiling point of pure anisole at 760 mm. pressure is 155.5°.

Refractive index, page 945. Determine with an Abbé or other refractometer of equal or greater accuracy.

Specific gravity. Determine by any reliable method (see page 5).

Arsenic. A *Sample Solution* prepared as directed for organic compounds meets the requirements of the *Arsenic Test,* page 865.

Heavy metals. Prepare and test a 500-mg. sample as directed in *Method II* under the *Heavy Metals Test,* page 920, using 20 mcg. of lead ion (Pb) in the control (*Solution A*).

Lead. A *Sample Solution* prepared as directed for organic compounds meets the requirements of the *Lead Limit Test,* page 929, using 10 mcg. of lead ion (Pb) in the control.

Phenols. Shake 1 ml. with about 20 ml. of water, allow the layers to separate, collect the water layer in a test tube, and add to it a few drops of ferric chloride T.S. No greenish, bluish, or purplish color is produced.

Packaging and storage. Store in full, tight, preferably glass, tin-lined or other suitably lined containers in a cool place protected from light.

Functional use in foods. Flavoring agent.

ANISYL ACETATE

p-Methoxybenzyl Acetate

$CH_3O—C_6H_4—CH_2OOCCH_3$

$C_{10}H_{12}O_3$ Mol. wt. 180.20

DESCRIPTION

A colorless to slightly yellow liquid, having a floral, fruity odor. It is soluble in most fixed oils and in mineral oil, but it is insoluble in glycerin and in propylene glycol.

SPECIFICATIONS

Assay. Not less than 97.0 percent of $C_{10}H_{12}O_3$.

Refractive index. Between 1.511 and 1.516 at 20°.

Solubility in alcohol. Passes test.

Specific gravity. Between 1.104 and 1.107.

Limits of Impurities

Acid value. Not more than 1.0.

TESTS

Assay. Weigh accurately about 1.5 grams, and proceed as directed under *Ester Determination*, page 896, using 90.11 as the equivalence factor (*e*) in the calculation.

Refractive index, page 945. Determine with an Abbé or other refractometer of equal or greater accuracy.

Solubility in alcohol. Proceed as directed in the general method, page 899. One ml. dissolves in 6 ml. of 60 percent alcohol and remains in solution on dilution to 10 ml.

Specific gravity. Determine by any reliable method (see page 5).

Acid value. Determine as directed in the general method, page 893.

Packaging and storage. Store in full, tight, preferably glass, aluminum, stainless steel, or other suitably lined containers in a cool place protected from light.

Functional use in foods. Flavoring agent.

ANISYL ALCOHOL

Anisic Alcohol; *p*-Methoxybenzyl Alcohol

CH_3O—(benzene ring)—CH_2OH

$C_8H_{10}O_2$ Mol. wt. 138.17

DESCRIPTION

Anisyl alcohol, a constituent of vanilla pods, may be prepared by reduction from anisic aldehyde. It is a colorless to slightly yellow liquid with a floral odor. It is soluble in most fixed oils, and is sparingly soluble in glycerin. It is insoluble in mineral oil.

SPECIFICATIONS

Assay. Not less than 97.0 percent of $C_8H_{10}O_2$.

Refractive index. Between 1.543 and 1.545 at 20°.

Solidification point. Not less than 23.5°.

Solubility in alcohol. Passes test.

Specific gravity. Between 1.110 and 1.115.

Limits of Impurities

Acid value. Not more than 1.0.

Aldehydes. Not more than 1.0 percent, calculated as anisic aldehyde ($C_8H_8O_2$).

TESTS

Assay. Proceed as directed under *Total Alcohols*, page 893. Weigh accurately about 1.5 grams of the acetylated alcohol for the saponification, and use 69.09 as the equivalence factor (*e*) in the calculation.

Refractive index, page 945. Determine with an Abbé or other refractometer of equal or greater accuracy.

Solidification point. Proceed as directed in the general method, page 954.

Solubility in alcohol. Proceed as directed in the general method, page 899. One ml. dissolves in 1 ml. of 50 percent alcohol and remains in solution on dilution to 10 ml.

Specific gravity. Determine by any reliable method (see page 5).

Acid value. Determine as directed in the general method, page 893.

Aldehydes. Weigh accurately about 5 grams, and proceed as directed under *Aldehydes and Ketones-Hydroxylamine Method*, page 894, using 68.07 as the equivalence factor (*E*) in the calculation for anisic aldehyde ($C_8H_8O_2$). Allow the mixture to stand at room temperature for 30 minutes before titrating.

Packaging and storage. Store in full, tight, preferably glass, or suitably lined containers in a cool place protected from light. *Do not use aluminum containers.*

Functional use in foods. Flavoring agent.

L-ARGININE

L-1-Amino-4-guanidovaleric Acid

```
NHCH2CH2CH2CHCOOH
|             |
C=NH          NH2
|
NH2
```

$C_6H_{14}N_4O_2$ Mol. wt. 174.20

DESCRIPTION

White crystals or a white crystalline powder. It is soluble in water, but insoluble in ether and sparingly soluble in alcohol. It is strongly alkaline, and its water solutions absorb carbon dioxide from the air.

IDENTIFICATION

To 5 ml. of a 1 in 1000 solution add 1 ml. of triketohydrindene hydrate T.S. A reddish purple color appears.

SPECIFICATIONS

Assay. Not less than 98.0 percent and not more than the equivalent of 102.0 percent of $C_6H_{14}N_4O_2$, calculated on the dried basis.

Specific rotation, $[\alpha]^{20°}_{D}$. Between +25.0° and +27.9°, on the dried basis.

Limits of Impurities

Arsenic (as As). Not more than 3 parts per million (0.0003 percent).

Heavy metals (as Pb). Not more than 20 parts per million (0.002 percent).

Lead. Not more than 10 parts per million (0.001 percent).

Loss on drying. Not more than 1 percent.

Residue on ignition. Not more than 0.2 percent.

TESTS

Assay. Dissolve about 200 mg., accurately weighed, in 3 ml. of formic acid and 50 ml. of glacial acetic acid, add 2 drops of crystal violet T.S., and titrate with 0.1 *N* perchloric acid to a green end-point or until the blue color disappears completely. Each ml. of 0.1 *N* perchloric acid is equivalent to 8.710 mg. of $C_6H_{14}N_4O_2$.

Specific rotation, page 939. Determine in a solution containing 8 grams of a previously dried sample in sufficient 6 *N* hydrochloric acid to make 100 ml.

Arsenic. A *Sample Solution* prepared as directed for organic compounds meets the requirements of the *Arsenic Test,* page 865.

Heavy metals. Prepare and test a 1-gram sample as directed in *Method II* under the *Heavy Metals Test,* page 920, using 20 mcg. of lead ion (Pb) in the control (*Solution A*).

Lead. A *Sample Solution* prepared as directed for organic compounds meets the requirements of the *Lead Limit Test,* page 929, using 10 mcg. of lead ion (Pb) in the control.

Loss on drying, page 931. Dry at 80° for 3 hours.

Residue on ignition, page 945. Ignite 1 gram as directed in the general method.

Packaging and storage. Store in well-closed, light-resistant containers.

Functional use in foods. Nutrient; dietary supplement.

L-ARGININE MONOHYDROCHLORIDE

L-1-Amino-4-guanidovaleric Acid Monohydrochloride

```
NHCH2CH2CH2CHCOOH
|              |
CH=NH          NH2
|
NH2.HCl
```

$C_6H_{14}N_4O_2.HCl$ Mol. wt. 210.67

DESCRIPTION

A white or nearly white, practically odorless, crystalline powder. It is soluble in water, slightly soluble in hot alcohol, and insoluble in ether. It melts with decomposition at about 235°.

IDENTIFICATION

A. Heat 5 ml. of a 1 in 1000 solution with 1 ml. of triketohydrindene hydrate T.S. A reddish purple color is produced.

B. A 1 in 1000 solution gives positive tests for *Chloride*, page 926.

SPECIFICATIONS

Assay. Not less than 98.0 percent and not more than the equivalent of 102.0 percent of $C_6H_{14}N_4O_2.HCl$, calculated on the dried basis.

Specific rotation, $[\alpha]_D^{20°}$. Between +20.5° and +23.0°, on the dried basis.

Limits of Impurities

Arsenic (as As). Not more than 3 parts per million (0.0003 percent).

Heavy metals (as Pb). Not more than 20 parts per million (0.002 percent).

Lead. Not more than 10 parts per million (0.001 percent).

Loss on drying. Not more than 0.3 percent.

Residue on ignition. Not more than 0.1 percent.

TESTS

Assay. Transfer about 200 mg., accurately weighed, into a 250-ml. flask, and dissolve in 3 ml. of formic acid and 50 ml. of glacial acetic acid. Add 10 ml. of mercuric acetate T.S. and 2 drops of crystal violet T.S., and titrate with 0.1 *N* perchloric acid to the first appearance of a pure green color or until the blue color disappears completely. Each ml. of 0.1 *N* perchloric acid is equivalent to 10.53 mg. of $C_6H_{14}N_4O_2.HCl$.

Specific rotation, page 939. Determine in a solution containing 8 grams of a previously dried sample in sufficient 6 *N* hydrochloric acid to make 100 ml.

Arsenic. A *Sample Solution* prepared as directed for organic compounds meets the requirements of the *Arsenic Test*, page 865.

Heavy metals. Prepare and test a 1-gram sample as directed in *Method II* under the *Heavy Metals Test,* page 920, using 20 mcg. of lead ion (Pb) in the control (*Solution A*).

Lead. A *Sample Solution* prepared as directed for organic compounds meets the requirements of the *Lead Limit Test,* page 929, using 10 mcg. of lead ion (Pb) in the control.

Loss on drying, page 931. Dry at 105° for 3 hours.

Residue on ignition, page 945. Ignite 1 gram as directed in the general method.

Packaging and storage. Store in well-closed, light-resistant containers.

Functional use in foods. Nutrient; dietary supplement.

ASCORBIC ACID

Vitamin C; L-Ascorbic Acid

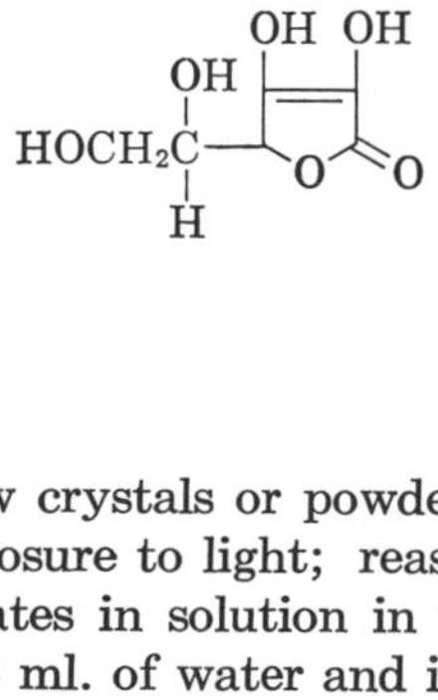

OH OH
OH
HOCH$_2$C— O O
H

$C_6H_8O_6$ Mol. wt. 176.13

DESCRIPTION

White or slightly yellow crystals or powder, melting at about 190°. Gradually darkens on exposure to light; reasonably stable in air when dry, but rapidly deteriorates in solution in the presence of air. One gram is soluble in about 3 ml. of water and in about 30 ml. of alcohol. It is insoluble in chloroform, in ether, and in benzene.

IDENTIFICATION

A. A 1 in 50 solution slowly reduces alkaline cupric tartrate T.S. at 25°, but more readily upon heating.

B. To 2 ml. of a 1 in 50 solution of the sample add 4 drops of methylene blue T.S., and warm to 40°. The deep blue color is practically discharged within 3 minutes.

C. Dissolve 15 mg. of the sample in 15 ml. of a 1 in 20 solution of trichloroacetic acid, add about 200 mg. of activated charcoal, shake vigorously for 1 minute, and filter through a small fluted filter, returning the filtrate, if necessary, until clear. To 5 ml. of the filtrate add 1 drop of pyrrole, agitate gently until dissolved, and then heat in a water bath at 50°. A blue color develops.

SPECIFICATIONS

Assay. Not less than 99.0 percent of $C_6H_8O_6$.

Specific rotation, $[\alpha]_D^{25°}$. Between +20.5° and +21.5°.

Limits of Impurities

Arsenic (as As). Not more than 3 parts per million (0.0003 percent).

Heavy metals (as Pb). Not more than 20 parts per million (0.002 percent).

Lead. Not more than 10 parts per million (0.001 percent).

Residue on ignition. Not more than 0.1 percent.

TESTS

Assay. Dissolve about 400 mg., accurately weighed, in a mixture of 100 ml. of water, recently boiled and cooled, and 25 ml. of diluted sulfuric acid T.S. Titrate the solution immediately with 0.1 *N* iodine, adding starch T.S. near the end-point. Each ml. of 0.1 *N* iodine is equivalent to 8.806 mg. of $C_6H_8O_6$.

Specific rotation, page 939. Determine in a solution containing 1 gram in 10 ml. of water.

Arsenic. A *Sample Solution* prepared as directed for organic compounds meets the requirements of the *Arsenic Test,* page 865.

Heavy metals. A solution of 1 gram in 25 ml. of water meets the requirements of the *Heavy Metals Test,* page 920, using 20 mcg. of lead ion (Pb) in the control (*Solution A*).

Lead. A *Sample Solution* prepared as directed for organic compounds meets the requirements of the *Lead Limit Test,* page 929, using 10 mcg. of lead ion (Pb) in the control.

Residue on ignition, page 945. Ignite 2 grams as directed in the general method.

Packaging and storage. Store in tight, light-resistant containers.

Functional use in foods. Preservative; antioxidant; nutrient; dietary supplement.

ASCORBYL PALMITATE

Palmitoyl L-Ascorbic Acid

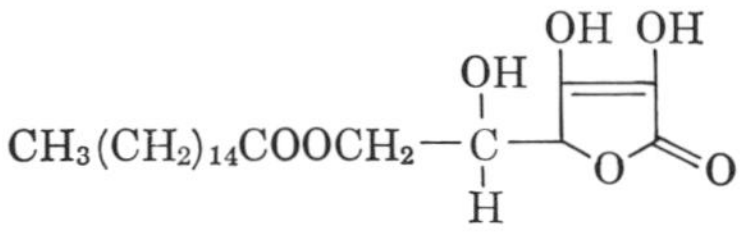

$C_{22}H_{38}O_7$ Mol. wt. 414.54

DESCRIPTION

A white or yellowish white powder having a characteristic citrus-like odor. It is very slightly soluble in water and in vegetable oils. One gram dissolves in about 4.5 ml. of alcohol. A 1 in 10 solution in alcohol decolorizes a 1 in 1000 solution of 2,6-dichlorophenol-indophenol sodium.

SPECIFICATIONS

Assay. Not less than 95.0 percent of $C_{22}H_{38}O_7$ calculated on the dried basis.

Melting range. Between 107° and 117°.

Specific rotation, $[\alpha]_D^{25°}$. Between +21° and +24° calculated on the dried basis.

Limits of Impurities

Arsenic (as As). Not more than 3 parts per million (0.0003 percent).

Heavy metals (as Pb). Not more than 10 parts per million (0.001 percent).

Loss on drying. Not more than 2 percent.

Residue on ignition. Not more than 0.1 percent.

TESTS

Assay. Dissolve about 300 mg., accurately weighed, in 50 ml. of alcohol in a 250-ml. Erlenmeyer flask, add 30 ml. of water, and immediately titrate with 0.1 *N* iodine to a yellow color which persists for at least 30 seconds. Each ml. of 0.1 *N* iodine is equivalent to 20.73 mg. of $C_{22}H_{38}O_7$.

Melting range. Determine as directed in *Procedure for Class Ia,* page 931.

Specific rotation, page 939. Determine in a solution containing 1 gram in 10 ml. of methanol.

Arsenic. A *Sample Solution* prepared as directed for organic compounds meets the requirements of the *Arsenic Test,* page 865.

Heavy metals. Prepare and test a 2-gram sample as directed in *Method II* under the *Heavy Metals Test,* page 920, using 20 mcg. of lead ion (Pb) in the control (*Solution A*).

Loss on drying, page 931. Dry in a vacuum oven at 56° to 60° for 1 hour.

Residue on ignition. Ignite 2 grams as directed in the general method, page 945.

Packaging and storage. Store in tight containers, preferably in a cool, dry place.

Functional use in foods. Antioxidant.

DL-ASPARTIC ACID

DL-Aminosuccinic Acid

$HOOCCH_2CH(NH_2)COOH$

$C_4H_7NO_4$ Mol. wt. 133.10

DESCRIPTION

Colorless or white, odorless crystals having an acid taste. It is slightly soluble in water, but insoluble in alcohol and in ether. It is optically inactive and melts with decomposition at about 280°.

IDENTIFICATION

To 5 ml. of a 1 in 1000 solution add 1 ml. of triketohydrindene hydrate T.S. A bluish purple color is produced.

SPECIFICATIONS

Assay. Not less than 98.0 percent and not more than the equivalent of 102.0 percent of $C_4H_7NO_4$, calculated on the dried basis.

Limits of Impurities

Arsenic (as As). Not more than 3 parts per million (0.0003 percent).

Heavy metals (as Pb). Not more than 20 parts per million (0.002 percent).

Lead. Not more than 10 parts per million (0.001 percent).

Loss on drying. Not more than 0.3 percent.

Residue on ignition. Not more than 0.1 percent.

TESTS

Assay. Dissolve about 250 mg., accurately weighed, in 100 ml. of recently boiled and cooled water, add phenolphthalein T.S., and titrate with 0.1 *N* sodium hydroxide to the first appearance of a faint pink color that persists for at least 30 seconds. Each ml. of 0.1 *N* sodium hydroxide is equivalent to 13.31 mg. of $C_4H_7NO_4$.

Arsenic. A *Sample Solution* prepared as directed for organic compounds meets the requirements of the *Arsenic Test*, page 865.

Heavy metals. Prepare and test a 1-gram sample as directed in *Method II* under the *Heavy Metals Test,* page 920, using 20 mcg. of lead ion (Pb) in the control (*Solution A*).

Lead. A *Sample Solution* prepared as directed for organic compounds meets the requirements of the *Lead Limit Test,* page 929, using 10 mcg. of lead ion (Pb) in the control.

Loss on drying, page 931. Dry at 105° for 3 hours.

Residue on ignition, page 945. Ignite 1 gram as directed in the general method.

Packaging and storage. Store in well-closed, light-resistant containers.

Functional use in foods. Nutrient; dietary supplement.

AZODICARBONAMIDE

$$H_2N-\overset{\displaystyle O}{\overset{\|}{C}}-N=N-\overset{\displaystyle O}{\overset{\|}{C}}-NH_2$$

$C_2H_4N_4O_2$ Mol. wt. 116.08

DESCRIPTION

A yellow to orange-red, odorless, crystalline powder. It is practically insoluble in water and in most organic solvents. It is slightly soluble in dimethyl sulfoxide. It melts above 180° with decomposition.

SPECIFICATIONS

Assay. Not less than 98.6 percent of $C_2H_4N_4O_2$ after drying.

Nitrogen. Between 47.2 percent and 48.7 percent.

pH of a 2 percent suspension. Not less than 5.0.

Limits of Impurities

Arsenic (as As). Not more than 3 parts per million (0.0003 percent).

Heavy metals (as Pb). Not more than 30 parts per million (0.003 percent).

Lead. Not more than 10 parts per million (0.001 percent).

Loss on drying. Not more than 0.5 percent.

Residue on ignition. Not more than 0.15 percent.

TESTS

Assay. Transfer about 150 mg., previously dried in a vacuum oven at 50° for 2 hours and accurately weighed, into a standard taper glass-stoppered 250-ml. iodine flask. Wet the sample thoroughly with about 3 ml. of water by trituration with the flattened end of a stirring rod. Add 7 grams of potassium iodide dissolved in 50 ml. of water followed by 5 ml. of hydrochloric acid. Stopper the flask, place it in a water bath at about 60°, and swirl it occasionally for 5 minutes. Remove the flask from the bath, allow it to stand in a dark place another 5 minutes, and then titrate with 0.1 *N* sodium thiosulfate using starch T.S. as the indicator. Perform a blank determination (see page 2) and make any necessary correction. Each ml. of 0.1 *N* sodium thiosulfate is equivalent to 5.804 mg. of $C_2H_4N_4O_2$.

Nitrogen. Transfer about 50 mg. into a 100-ml. Kjeldahl flask, add 3 ml. of hydriodic acid, and digest the mixture with gentle heating for 1.25 hours, adding sufficient water, when necessary, to maintain the original volume. Increase the heat at the end of the digestion period and continue heating until the volume is reduced by about one-half. Cool to room temperature, add 1.5 grams of potassium sulfate, 3 ml. of water, and 4.5 ml. of sulfuric acid, and heat until iodine fumes are no longer evolved. Allow the mixture to cool, wash down the sides of the flask with water, heat until charring occurs, and again cool to room temperature. To the charred material add 40 mg. of mercuric oxide, heat until the color of the solution is pale yellow, then cool, wash down the sides of the flask with a few ml. of water, and digest the mixture for 3 additional hours. Cool the digest, add 20 ml. of ammonia-free water, 16 ml. of a 50 percent sodium hydroxide solution, and 5 ml. of a 44 percent sodium thiosulfate solution. Immediately connect the flask to a distillation apparatus as directed under *Nitrogen Determination*, page 937, and distil, collecting the distillate in 10 ml. of a 4 percent boric acid solution. Add a few drops of methyl red-methylene blue T.S. to the distillate and titrate with 0.05 *N* sulfuric acid. Perform a blank determination (see page 2). Each ml. of 0.05 *N* sulfuric acid is equivalent to 0.7004 mg. of N.

pH of a 2 percent suspension. Add 2 grams to 100 ml. of water, agitate the mixture with a power stirrer for 5 minutes, and determine the pH of the resulting suspension potentiometrically.

Arsenic. A *Sample Solution* prepared as directed for organic compounds meets the requirements of the *Arsenic Test*, page 865.

Heavy metals. Prepare and test a 670-mg. sample as directed in *Method II* under the *Heavy Metals Test*, page 920, using 20 mcg. of lead ion (Pb) in the control (*Solution A*).

Lead. A *Sample Solution* prepared as directed for organic compounds meets the requirements of the *Lead Limit Test*, page 929, using 10 mcg. of lead ion (Pb) in the control.

Loss on drying, page 931. Dry in a vacuum oven at 50° for 2 hours.

Residue on ignition. Ignite 1.5 grams as directed in the general method, page 945.

Packaging and storage. Store in well-closed, light-resistant containers.

Functional use in foods. Maturing agent for flour.

BALSAM PERU OIL

DESCRIPTION

The oil obtained by extraction or distillation from Peruvian Balsam obtained from *Myroxylon pereirae* Royle Klotzsche (Fam. *Leguminosae*). It is a yellow to pale brown, slightly viscous liquid, having a sweet balsamic odor. Occasionally crystals may separate from the liquid. It is soluble in most fixed oils, and soluble, with turbidity, in mineral oil. It is partly soluble in propylene glycol, but it is practically insoluble in glycerin.

SPECIFICATIONS

Acid value. Between 30 and 60.

Angular rotation. Between $-1°$ and $+2°$.

Ester value. Between 200 and 225.

Solubility in alcohol. Passes test.

Refractive index. Between 1.567 and 1.579 at 20°.

Specific gravity. Between 1.095 and 1.110.

Limits of Impurities

Arsenic (as As). Not more than 3 parts per million (0.0003 percent).

Heavy metals (as Pb). Not more than 40 parts per million (0.004 percent).

Lead. Not more than 10 parts per million (0.001 percent).

TESTS

Acid value. Determine as directed in the general method, page 893.

Angular rotation. Determine in a 100-mm. tube as directed under *Optical Rotation*, page 939.

Ester value. Proceed as directed in the general method, page 897, using about 1 gram, accurately weighed.

Solubility in alcohol. Proceed as directed in the general method, page 899. One ml. dissolves in 0.5 ml. of 90 percent alcohol, and remains in solution upon dilution to 10 ml.

Refractive index, page 945. Determine with an Abbé or other refractometer of equal or greater accuracy.

Specific gravity. Determine by any reliable method (see page 5).

Arsenic. A *Sample Solution* prepared as directed for organic compounds meets the requirements of the *Arsenic Test*, page 865.

Heavy metals. Prepare and test a 500-mg. sample as directed in *Method II* under the *Heavy Metals Test*, page 920, using 20 mcg. of lead ion (Pb) in the control (*Solution A*).

Lead. A *Sample Solution* prepared as directed for organic compounds meets the requirements of the *Lead Limit Test,* page 929, using 10 mcg. of lead ion (Pb) in the control.

Packaging and storage. Store in full, tight, preferably glass, tin-lined, or aluminum containers in a cool place protected from light.

Functional use in foods. Flavoring agent.

BASIL OIL

DESCRIPTION

Basil oil is obtained by steam distillation of the flowering tops or the entire plant of *Ocimum basilicum* L., grown on Réunion Island and on the Islands of the Comoros. It may be distinguished from other types, such as European sweet basil oil, by its camphoraceous odor and physicochemical constants. It is a light yellow liquid with a spicy odor. It is soluble in most fixed oils, and, with turbidity, in mineral oil. One ml. is soluble in 20 ml. of propylene glycol with slight haziness, but it is insoluble in glycerin.

SPECIFICATIONS

Acid value. Not more than 1.0.

Angular rotation. Between 0° and +2°.

Ester value after acetylation. Between 25 and 45.

Refractive index. Between 1.512 and 1.519 at 20°.

Saponification value. Between 4 and 10.

Solubility in alcohol. Passes test.

Specific gravity. Between 0.952 and 0.973.

TESTS

Acid value. Determine as directed in the general method, page 893.

Angular rotation. Determine in a 100-mm. tube as directed under *Optical Rotation,* page 939.

Ester value after acetylation. Proceed as directed under *Linalool Determination,* page 897, using 2.5 grams of the dry acetylated oil, accurately weighed, for the saponification. Calculate the *Ester value after acetylation* by the formula: $(a) \times (28.05)/(b)$, in which (a) is the number of ml. of 0.5 *N* alcoholic potassium hydroxide consumed in the saponification, and (b) is the weight of the acetylated oil, in grams, used in the test.

Refractive index, page 945. Determine with an Abbé or other refractometer of equal or greater accuracy.

Saponification value. Determine as directed in the general method, page 896, using about 5 grams of sample, accurately weighed.

Solubility in alcohol. Proceed as directed in the general method, page 899. One ml. dissolves in 4 ml. of 80 percent alcohol.

Specific gravity. Determine by any reliable method (see page 5).

Packaging and storage. Store in full, tight, preferably glass, or aluminum containers in a cool place protected from light.

Functional use in foods. Flavoring agent.

BAY OIL

Myrcia Oil

DESCRIPTION

The volatile oil distilled from the leaves of *Pimenta acris* Kostel. It occurs as a yellow or brownish yellow liquid with a pleasant aromatic odor, and a pungent, spicy taste. It is soluble in alcohol and in glacial acetic acid. Its solutions in alcohol are acid to litmus.

IDENTIFICATION

A. Mix a sample with an equal volume of a concentrated solution of sodium hydroxide. A semi-solid mass forms.

B. Dissolve 2 drops in 4 ml. of alcohol and add 1 drop of ferric chloride T.S. A light green color is produced. If the same test is made with 1 drop of dilute ferric chloride (1 volume of T.S. with 4 volumes of water), a yellow color is produced which soon disappears.

C. Shake 1 ml. with 20 ml. of hot water and filter. The filtrate gives not more than a slight acid reaction with litmus, and on the addition of 1 drop of ferric chloride T.S., yields only a transient grayish green, not a blue or purple color.

SPECIFICATIONS

Assay. Not less than 50 percent and not more than 65 percent, by volume, of phenols.

Angular rotation. Levorotatory, but not more than $-3°$.

Refractive index. Between 1.507 and 1.516 at 20°.

Solubility in alcohol. Passes test.

Specific gravity. Between 0.950 and 0.990.

TESTS

Assay. Proceed as directed under *Phenols*, page 898.

Angular rotation. Determine in a 100-mm. tube as directed under *Optical Rotation*, page 939.

Refractive index, page 945. Determine with an Abbé or other refractometer of equal or greater accuracy.

Solubility in alcohol. Proceed as directed in the general method, page 899. One ml. dissolves in 1 ml. of alcohol to form a clear or only slightly turbid solution.

Specific gravity. Determine by any reliable method (see page 5).

Packaging and storage. Store in full, tight containers in a cool place protected from light.

Functional use in foods. Flavoring agent.

BEESWAX, WHITE

White Wax

DESCRIPTION

The bleached, purified wax from the honeycomb of the bee, *Apis mellifera* L. (Fam. *Apidae*). It is a yellowish white solid, somewhat translucent in thin layers, having a faint characteristic odor, free from rancidity. Its specific gravity is about 0.95. White beeswax is insoluble in water and sparingly soluble in cold alcohol. Boiling alcohol dissolves cerotic acid and part of the myricin, which are constituents of the wax. It is completely soluble in chloroform, in ether, and in fixed and volatile oils. It is partly soluble in cold benzene and in cold carbon disulfide, but completely soluble in these liquids at temperatures of 30° or above.

SPECIFICATIONS

Acid value. Between 17 and 24.

Ester value. Between 72 and 79.

Melting range. Between 62° and 65°.

Limits of Impurities

Arsenic (as As). Not more than 3 parts per million (0.0003 percent).

Carnauba wax. Passes test.

Fats, Japan wax, rosin, and soap. Passes test.

Heavy metals (as Pb). Not more than 40 parts per million (0.004 percent).

Lead. Not more than 10 parts per million (0.001 percent).

Saponification cloud test. Passes test.

TESTS

Acid value, page 902. Warm about 3 grams, accurately weighed, in a 200-ml. flask with 25 ml. of absolute alcohol, previously neutralized to phenolphthalein with potassium hydroxide, until the sample is melted. Shake the mixture, add 1 ml. of phenolphthalein T.S., and titrate the warm solution with 0.5 *N* alcoholic potassium hydroxide to a permanent, faint pink color.

Ester value. To the solution resulting from the determination of *Acid value*, add 25.0 ml. of 0.5 *N* alcoholic potassium hydroxide and 50 ml. of alcohol, heat the mixture under a reflux condenser for 4 hours, and titrate the excess alkali with 0.5 *N* hydrochloric acid. Perform a residual blank titration, and calculate the *Ester value* as the number of mg. of potassium hydroxide required for each gram of the sample taken for the test.

Melting range. Determine as directed for *Class II* substances in the general procedure, page 931.

Arsenic. A *Sample Solution* prepared as directed for organic compounds meets the requirements of the *Arsenic Test*, page 865.

Carnauba wax. Place 100 mg. in a test tube and add 20 ml. of *n*-butanol. Immerse the test tube in boiling water and shake the mixture gently until solution is complete. Transfer the test tube into a beaker of water at 60° and allow it to cool to room temperature. A loose mass of fine, needle-like crystals separate from a clear mother liquor. Under the microscope the crystals appear as loose needles or stellate clusters and no amorphous masses are observed (*absence of carnauba wax*).

Fats, Japan wax, rosin, and soap. Boil 1 gram for 30 minutes with 35 ml. of a 1 in 7 solution of sodium hydroxide, maintaining the volume by the occasional addition of water, and cool the mixture. The wax separates and the liquid remains clear. Filter the cold mixture and acidify the filtrate with hydrochloric acid. No precipitate is formed.

Heavy metals. Prepare and test a 500-mg. sample as directed in *Method II* under the *Heavy Metals Test*, page 920, using 20 mcg. of lead ion (Pb) in the control (*Solution A*).

Lead. A *Sample Solution* prepared as directed for organic compounds meets the requirements of the *Lead Limit Test*, page 929, using 10 mcg. of lead ion (Pb) in the control.

Saponification cloud test

Saponifying Solution. Dissolve 40 grams of potassium hydroxide in about 900 ml. of aldehyde-free alcohol maintained at a temperature of 15° until solution is complete, then warm to room temperature and add sufficient aldehyde-free alcohol to make 1000 ml.

Procedure. Transfer 3.00 grams into a round-bottom, 100-ml. boiling flask provided with a ground-glass joint, add 30 ml. of the *Saponifying Solution*, attach a reflux condenser to the flask, and heat the mixture gently on a steam bath for 2 hours. At the end of this period, remove the reflux condenser, insert a thermometer into the solution, and place the flask in a water bath at a temperature of 80°. Rotate the flask while both the bath and the solution cool to 65°. The solution shows no cloudiness or globule formation before this temperature is reached.

Packaging and storage. Store in well-closed containers.

Functional use in foods. Candy glaze and polish; miscellaneous and general purpose; flavoring agent.

BEESWAX, YELLOW

Yellow Wax

DESCRIPTION

The purified wax from the honeycomb of the bee, *Apis mellifera* L. (Fam. *Apidae*). It is a yellowish to grayish brown solid having an agreeable, honey-like odor. It is somewhat brittle when cold, and presents a dull, granular, noncrystalline fracture when broken. It becomes pliable at a temperature of about 35°. Its specific gravity is about 0.95. Yellow beeswax is insoluble in water and sparingly soluble in cold alcohol. Boiling alcohol dissolves cerotic acid and part of the myricin which are constituents of the wax. It is completely soluble in chloroform, in ether, and in fixed and volatile oils. It is partly soluble in cold benzene and in cold carbon disulfide, but completely soluble in these solvents at temperatures of 30° or above.

SPECIFICATIONS

Acid value. Between 18 and 24.

Melting range. Between 62° and 65°.

Ester value. Between 72 and 77.

Limits of Impurities

Arsenic (as As). Not more than 3 parts per million (0.0003 percent).

Carnauba wax. Passes test.

Fats, Japan wax, rosin, and soap. Passes test.

Heavy metals (as Pb). Not more than 40 parts per million (0.004 percent).

Lead. Not more than 10 parts per million (0.001 percent).

Saponification cloud test. Passes test.

TESTS

For the determination of *Acid value, Ester value, Melting range, Arsenic, Carnauba wax, Fats, Japan wax, rosin and soap, Heavy metals, Lead,* and *Saponification cloud test,* proceed as directed in the monograph on *White Beeswax,* page 75.

Packaging and storage. Store in well-closed containers.

Functional use in foods. Candy glaze and polish; miscellaneous and general purpose; flavoring agent.

BENZALDEHYDE

Benzoic Aldehyde

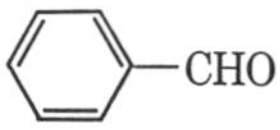

C_7H_6O Mol. wt. 106.12

DESCRIPTION

Benzaldehyde occurs as a constituent of oils of bitter almond, peach, and apricot kernel. It is usually prepared synthetically. It is a colorless liquid having an odor resembling that of bitter almond oil, and a burning taste. It is affected by light and it oxidizes in air to benzoic acid. It is soluble in about 350 volumes of water, and is miscible with alcohol, ether, fixed oils, or volatile oils.

SPECIFICATIONS

Assay. Not less than 98.0 percent of C_7H_6O.

Refractive index. Between 1.544 and 1.547 at 20°.

Specific gravity. Between 1.041 and 1.046.

Limits of Impurities

Arsenic (as As). Not more than 3 parts per million (0.0003 percent).
Chlorinated compounds. Passes test.
Heavy metals (as Pb). Not more than 40 parts per million (0.004 percent).
Hydrocyanic acid. Passes test (about 0.15 percent).
Lead. Not more than 10 parts per million (0.001 percent).
Solubility in sodium bisulfite. Passes test.

TESTS

Assay. Weigh accurately about 1 gram, and proceed as directed under *Aldehydes*, page 894, using 53.06 as the equivalence factor (E) in the calculation.

Refractive index, page 945. Determine with an Abbé or other refractometer of equal or greater accuracy.

Specific gravity. Determine by any reliable method (see page 5).

Arsenic. A *Sample Solution* prepared as directed for organic compounds meets the requirements of the *Arsenic Test*, page 865.

Chlorinated compounds. Determine as directed in the general method, page 895.

Heavy metals. Prepare and test a 500-mg. sample as directed in *Method II* under the *Heavy Metals Test*, page 920, using 20 mcg. of lead ion (Pb) in the control (*Solution A*).

Hydrocyanic acid. Shake 0.5 ml. with 5 ml. of water, add 0.5 ml. of sodium hydroxide T.S. and 0.1 ml. of ferrous sulfate T.S., and warm the mixture gently. Upon the addition of a slight excess of hydrochloric acid, no greenish blue color or blue precipitate is produced within 15 minutes.

Lead. A *Sample Solution* prepared as directed for organic compounds meets the requirements of the *Lead Limit Test*, page 929, using 10 mcg. of lead ion (Pb) in the control.

Solubility in sodium bisulfite. Transfer 5 ml. to a 100 ml. Cassia flask. Add 70 ml. of a 10 percent, by weight, solution of sodium metabisulfite. Shake the mixture vigorously for 15 minutes at room temperature. Add sufficient bisulfite solution to bring the liquid into the graduated neck of the flask. No oil should separate, and the solution should be clear, or not more than slightly cloudy.

Packaging and storage. Store in tight, preferably full, light-resistant containers. Avoid heat and light.

Functional use in foods. Flavoring agent.

BENZOIC ACID

$C_6H_5.COOH$

$C_7H_6O_2$ Mol. wt. 122.12

DESCRIPTION

White crystals, scales, or needles. It is odorless or has a slightly benzoin- or benzaldehyde-like odor. It begins to sublime at about 100° and is volatile with steam. One gram is soluble in 275 ml. of water at 25°, in 20 ml. of boiling water, in 3 ml. of alcohol, in 5 ml. of chloroform, and in 3 ml. of ether. It is soluble in fixed and in volatile oils and is sparingly soluble in solvent hexane.

IDENTIFICATION

Dissolve 1 gram in a mixture of 20 ml. of water and 1 ml. of sodium hydroxide T.S., filter the solution, and add about 1 ml. of ferric chloride T.S. A buff colored precipitate is produced.

SPECIFICATIONS

Assay. Not less than 99.5 percent of $C_7H_6O_2$ after drying.

Solidification point. Between 121° and 123°.

Limits of Impurities

Arsenic (as As). Not more than 3 parts per million (0.0003 percent).

Heavy metals (as Pb). Not more than 10 parts per million (0.001 percent).

Readily carbonizable substances. Passes test.

Readily oxidizable substances. Passes test.

Residue on ignition. Not more than 0.05 percent.

TESTS

Assay. Dissolve about 500 mg., previously dried over silica gel for 3 hours and accurately weighed, in 25 ml. of 50 percent alcohol previously neutralized with 0.1 *N* sodium hydroxide, add phenolphthalein T.S., and titrate with 0.1 *N* sodium hydroxide. Each ml. of 0.1 *N* sodium hydroxide is equivalent to 12.21 mg. of $C_7H_6O_2$.

Solidification point. Determine as directed in the general method, page 954.

Arsenic. A *Sample Solution* prepared as directed for organic compounds meets the requirements of the *Arsenic Test*, page 865.

Heavy metals. Volatilize 2 grams over a low flame. To the residue add 2 ml. of nitric acid and about 10 mg. of sodium carbonate, and evaporate to dryness on a steam bath. Dissolve the residue in a mixture of 1 ml. of diluted acetic acid T.S. and 24 ml. of water. This solution meets the requirements of the *Heavy Metals Test*, page 920, using 20 mcg. of lead ion (Pb) in the control (*Solution A*).

Readily carbonizable substances, page 943. Dissolve 500 mg. in 5 ml. of sulfuric acid T.S. The color is no darker than *Matching Fluid Q*.

Readily oxidizable substances. To a mixture of 100 ml. of water and 1.5 ml. of sulfuric acid heated to 100°, add dropwise 0.1 *N* potassium permanganate until a pink color persists for 30 seconds. Dissolve 1.0 gram of the benzoic acid in the hot solution and titrate with 0.1 *N* potassium permanganate to a pink color that persists for 15 seconds. The volume of 0.1 *N* potassium permanganate consumed does not exceed 0.5 ml.

Residue on ignition, page 945. Ignite 2 grams as directed in the general method.

Packaging and storage. Store in well-closed containers.

Functional use in foods. Preservative; antimicrobial agent.

BENZOPHENONE

Diphenyl Ketone; Benzoylbenzene

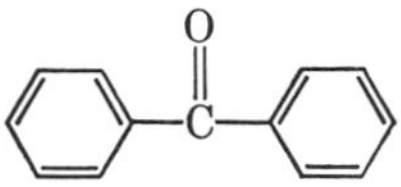

$C_{13}H_{10}O$ Mol. wt. 182.22

DESCRIPTION

A white, rhombic, crystalline, or flaky solid, having a delicate, persistent rose-like odor. It is soluble in most fixed oils and in mineral oil. It is slightly soluble in propylene glycol, but it is insoluble in glycerin.

SPECIFICATIONS

Solidification point. Not lower than 47°.

Solubility in alcohol. Passes test.

Limits of Impurities

Arsenic (as As). Not more than 3 parts per million (0.0003 percent).

Chlorinated compounds. Passes test.

Heavy metals (as Pb). Not more than 40 parts per million (0.004 percent).

Lead. Not more than 10 parts per million (0.001 percent).

TESTS

Solidification point. Determine as directed in the general method, page 954.

Solubility in alcohol. Proceed as directed in the general method, page 899. One gram dissolves in 10 ml. of 80 percent alcohol.

Arsenic. A *Sample Solution* prepared as directed for organic compounds meets the requirements of the *Arsenic Test*, page 865.

Chlorinated compounds. Proceed as directed in the general method, page 899, after liquefying the sample for testing.

Heavy metals. Prepare and test a 500-mg. sample as directed in *Method II* under the *Heavy Metals Test*, page 920, using 20 mcg. of lead ion (Pb) in the control (*Solution A*).

Lead. A *Sample Solution* prepared as directed for organic compounds meets the requirements of the *Lead Limit Test*, page 929, using 10 mcg. of lead ion (Pb) in the control.

Packaging and storage. Store in fiber drums, or suitably lined cans in a cool, dry place.

Functional use in foods. Flavoring agent.

BENZOYL PEROXIDE

Benzoyl Superoxide

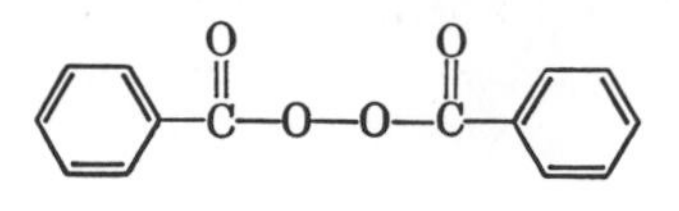

$C_{14}H_{10}O_4$ Mol. wt. 242.23

DESCRIPTION

A colorless, crystalline solid having a faint odor of benzaldehyde. It is insoluble in water, slightly soluble in alcohol, and soluble in benzene, chloroform, and ether. One gram dissolves in 40 ml. of carbon disulfide. It melts between 103° and 106° with decomposition.

Caution: Benzoyl peroxide, especially in the dry form, is a dangerous, highly reactive, oxidizing material and has been known to explode spontaneously. Observe safety precautions printed on the label of the container.

IDENTIFICATION

To 500 mg. of the sample add 50 ml. of 0.5 *N* alcoholic potassium hydroxide, heat gradually to boiling, and continue boiling for 15 minutes. Cool, dilute to 200 ml. with water, and make the solution strongly acid with 0.5 *N* hydrochloric acid. Extract with ether, dry the extract with anhydrous sodium sulfate, and then evaporate to dryness on a steam bath. The residue of benzoic acid so obtained melts between 121° and 123° (see page 931).

SPECIFICATIONS

Assay. Not less than 96.0 percent of $C_{14}H_{10}O_4$.

Limits of Impurities

Arsenic (as As). Not more than 3 parts per million (0.0003 percent).

Heavy metals (as Pb). Not more than 40 parts per million (0.004 percent).

Lead. Not more than 10 parts per million (0.001 percent).

TESTS

Assay. Dissolve about 250 mg., accurately weighed, in 15 ml. of acetone in a 100-ml. glass-stoppered bottle, and add 3 ml. of potassium iodide solution (1 in 2). Swirl for 1 minute, then immediately titrate with 0.1 *N* sodium thiosulfate (without the addition of starch T.S.). Each ml. of 0.1 *N* sodium thiosulfate is equivalent to 12.11 mg. of $C_{14}H_{10}O_4$.

Arsenic. Mix 1 gram with 10 ml. of sodium hydroxide T.S., slowly evaporate to dryness on a steam bath, and cool. A *Sample Solution*, prepared as directed for organic compounds from the residue obtained, meets the requirements of the *Arsenic Test*, page 865.

Heavy metals. Mix 500 mg. with 5 ml. of sodium hydroxide T.S., slowly evaporate to dryness on a steam bath, cool, and dissolve the residue in 25 ml. of water. This solution meets the requirements of the *Heavy Metals Test*, page 920, using 20 mcg. of lead ion (Pb) and 5 ml. of sodium hydroxide T.S. in the control (*Solution A*).

Lead. Mix 1 gram with 10 ml. of sodium hydroxide T.S., slowly evaporate to dryness on a steam bath, and cool. A *Sample Solution*, prepared as directed for organic compounds from the residue so obtained, meets the requirements of the *Lead Limit Test*, page 929, using 10 mcg. of lead ion (Pb) in the control.

Packaging and storage. Store in the original container and observe the safety precautions printed on the label.

Functional use in foods. Bleaching agent.

BENZYL ACETATE

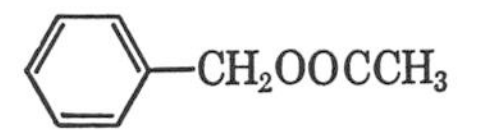

$C_9H_{10}O_2$ Mol. wt. 150.18

DESCRIPTION

A colorless liquid having a characteristic floral odor. It is soluble in most fixed oils, in mineral oil, and in propylene glycol, but it is insoluble in glycerin and in water.

SPECIFICATIONS

Assay. Not less than 98.0 percent of $C_9H_{10}O_2$.

Refractive index. Between 1.501 and 1.504 at 20°.

Solubility in alcohol. Passes test.

Specific gravity. Between 1.052 and 1.056.

Limits of Impurities

Acid value. Not more than 1.0.

Chlorinated compounds. Passes test.

TESTS

Assay. Weigh accurately about 900 mg., and proceed as directed under *Ester Determination*, page 896, using 75.10 as the equivalence factor (*e*) in the calculation. The results should be corrected for the acid value.

Refractive index, page 945. Determine with an Abbé or other refractometer of equal or greater accuracy.

Solubility in alcohol. Proceed as directed in the general method, page 899. One ml. dissolves in 5 ml. of 60 percent alcohol.

Specific gravity. Determine by any reliable method (see page 5).

Acid value. Determine as directed in the general method, page 893.

Chlorinated compounds. Proceed as directed in the general method, page 895.

Packaging and storage. Store in full, tight, preferably glass, aluminum, or tin-lined containers in a cool place protected from light.

Functional use in foods. Flavoring agent.

BENZYL ALCOHOL

Phenyl Carbinol

C_6H_5—CH_2OH

C_7H_8O Mol. wt. 108.14

DESCRIPTION

A colorless liquid with a faint, aromatic odor and a sharp burning taste. It boils without decomposition at about 206° and is neutral to litmus. One gram dissolves in about 30 ml. of water. One volume dissolves in 1.5 volumes of 50 percent alcohol. It is miscible with alcohol, with ether, and with chloroform.

IDENTIFICATION

Add 2 or 3 drops to 5 ml. of a 1 in 20 solution of potassium permanganate, and acidify with diluted sulfuric acid T.S. The odor of benzaldehyde is noticeable.

SPECIFICATIONS

Distillation range. Not less than 95.0 percent distils between 202.5° and 206.5°.

Refractive index. Between 1.538 and 1.540 at 20°.

Specific gravity. Between 1.042 and 1.047.

Limits of Impurities

Aldehydes. Not more than 0.2 percent.

Arsenic (as As). Not more than 3 parts per million (0.0003 percent).

Chlorinated compounds. Passes test.

Heavy metals (as Pb). Not more than 10 parts per million (0.001 percent).

Residue on ignition. Not more than 0.005 percent.

TESTS

Distillation range. Proceed as directed in the general procedure, page 890.

Refractive index, page 945. Determine with an Abbé or other refractometer of equal or greater accuracy.

Specific gravity. Determine by any reliable method (see page 5).

Aldehydes. Transfer 5.0 ml. of the sample into a 250-ml. Erlenmeyer flask, add 75.0 ml. of *Hydroxylamine Hydrochloride Solution* to this flask and to a second flask to serve as the blank, and proceed as directed under *Aldehydes,* page 894, using 53.06 as the equivalence factor (E) in the calculation.

Arsenic. A *Sample Solution* prepared as directed for organic compounds meets the requirements of the *Arsenic Test,* page 865.

Chlorinated compounds. Determine as directed in the general method, page 895.

Heavy metals. Prepare and test a 2-gram sample as directed in *Method II* under the *Heavy Metals Test,* page 920, using 20 mcg. of lead ion (Pb) in the control (*Solution A*).

Residue on ignition, page 945. Use a 20-ml. sample and determine as directed in *Method II* of the general procedure.

Packaging and storage. Store in tight containers.

Functional use in foods. Flavoring agent.

BENZYL BENZOATE

C_6H_5—COOCH$_2$—C_6H_5

$C_{14}H_{12}O_2$ Mol. wt. 212.25

DESCRIPTION

A clear, colorless, oily liquid having a slight aromatic odor and a sharp, burning, taste. It is practically insoluble in water and in glycerin. It is miscible with alcohol, with chloroform, and with ether.

IDENTIFICATION

Evaporate the solution obtained in the *Assay* to a volume of about 15 ml., filter, and place 5 ml. of the clear filtrate into each of two test tubes. Render the contents of the first tube faintly acid with diluted hydrochloric acid T.S., and add a few drops of ferric chloride T.S. A salmon-colored precipitate is formed. To the contents of the second tube add 2 ml. of diluted hydrochloric acid, mix, and collect the precipitate on a filter. Wash with ten 1-ml. portions of water, and dry at 60° in a vacuum oven. The benzoic acid so obtained melts between 121° and 123° (see page 931).

SPECIFICATIONS

Assay. Not less than 99.0 percent of $C_{14}H_{12}O_2$.

Refractive index. Between 1.568 and 1.570 at 20°.

Solidification point. Not lower than 18°.

Specific gravity. Between 1.116 and 1.120.

Limits of Impurities

Acid value. Not more than 1.0.

Chlorinated compounds. Passes test.

TESTS

Assay. Weigh accurately about 1 gram, and proceed as directed under *Ester Determination*, page 896, using 106.1 as the equivalence factor (e) in the calculation.

Refractive index, page 945. Determine with an Abbé or other refractometer of equal or greater accuracy.

Solidification point. Determine as directed in the general method, page 954.

Specific gravity. Determine by any reliable method (see page 5).

Acid value. Determine as directed in the general method, page 893.

Chlorinated compounds. Proceed as directed in the general method, page 895.

Packaging and storage. Store in full, tight containers in a cool place protected from light.

Functional use in foods. Flavoring agent.

BENZYL BUTYRATE

Benzyl *n*-Butyrate

C_6H_5—$CH_2OOCCH_2CH_2CH_3$

$C_{11}H_{14}O_2$ Mol. wt. 178.23

DESCRIPTION

A colorless liquid having a fruity odor suggestive of plum. It is soluble in most fixed oils and in mineral oil. It is practically insoluble in propylene glycol, and it is insoluble in glycerin.

SPECIFICATIONS

Assay. Not less than 98.0 percent of $C_{11}H_{14}O_2$.

Refractive index. Between 1.492 and 1.496 at 20°.

Solubility in alcohol. Passes test.

Specific gravity. Between 1.006 and 1.009.

Limits of Impurities

Acid value. Not more than 1.0.

TESTS

Assay. Weigh accurately about 1 gram, and proceed as directed under *Ester Determination*, page 896, using 89.10 as the equivalence factor (*e*) in the calculation.

Refractive index, page 945. Determine with an Abbé or other refractometer of equal or greater accuracy.

Solubility in alcohol. Proceed as directed in the general method, page 899. One ml. dissolves in 2 ml. of 80 percent alcohol.

Specific gravity. Determine by any reliable method (see page 5).

Acid value. Determine as directed in the general method, page 893.

Packaging and storage. Store in full, tight, preferably glass, tin or other suitably lined containers in a cool place protected from light.

Functional use in foods. Flavoring agent.

BENZYL CINNAMATE

—CH=CHCOOCH$_2$—

$C_{16}H_{14}O_2$ Mol. wt. 238.29

DESCRIPTION

A white to pale yellow solid having a sweet, balsamic odor. It is soluble in most fixed oils and sparingly soluble in mineral oil. It is insoluble in glycerin and in propylene glycol.

SPECIFICATIONS

Assay. Not less than 99.0 percent of $C_{16}H_{14}O_2$.

Solidification point. Between 33.0° and 34.5°.

Solubility in alcohol. Passes test.

Limits of Impurities

Acid value. Not more than 1.0.

Arsenic (as As). Not more than 3 parts per million (0.0003 percent).

Chlorinated compounds. Passes test.

Heavy metals (as Pb). Not more than 40 parts per million (0.004 percent).

Lead. Not more than 10 parts per million (0.001 percent).

TESTS

Assay. Weigh accurately about 1.6 grams, and proceed as directed under *Ester Determination*, page 896, using 119.1 as the equivalence factor (*e*) in the calculation.

Solidification point. Determine as directed in the general method, page 954.

Solubility in alcohol. Proceed as directed in the general method, page 899. One ml. dissolves in 8 ml. of 90 percent alcohol.

Acid value. Determine as directed in the general method, page 893.

Arsenic. A *Sample Solution* prepared as directed for organic compounds meets the requirements of the *Arsenic Test*, page 865.

Chlorinated compounds. Proceed as directed in the general method, page 895.

Heavy metals. Prepare and test a 500-mg. sample as directed in *Method II* under the *Heavy Metals Test*, page 920, using 20 mcg. of lead ion (Pb) in the control (*Solution A*).

Lead. A *Sample Solution* prepared as directed for organic compounds meets the requirements of the *Lead Limit Test*, page 929, using 10 mcg. of lead ion (Pb) in the control.

Packaging and storage. Store in full, tight, preferably glass, aluminum, or tin-lined containers in a cool place protected from light.

Functional use in foods. Flavoring agent.

BENZYL PHENYLACETATE

C_6H_5—CH_2COOCH_2—C_6H_5

$C_{15}H_{14}O_2$ Mol. wt. 226.28

DESCRIPTION

A colorless liquid having a sweet, floral odor with a honey undertone. It is miscible with alcohol, with chloroform, and with ether.

SPECIFICATIONS

Assay. Not less than 98.0 percent of $C_{15}H_{14}O_2$.

Refractive index. Between 1.553 and 1.558 at 20°.

Solubility in alcohol. Passes test.

Specific gravity. Between 1.095 and 1.099.

Limits of Impurities

Acid value. Not more than 1.0.

Arsenic (as As). Not more than 3 parts per million (0.0003 percent).

Heavy metals (as Pb). Not more than 40 parts per million (0.004 percent).

Lead. Not more than 10 parts per million (0.001 percent).

TESTS

Assay. Weigh accurately about 1.5 grams, and proceed as directed under *Ester Determination*, page 896, using 113.14 as the equivalence factor (*e*) in the calculation.

Refractive index, page 945. Determine with an Abbé or other refractometer of equal or greater accuracy.

Solubility in alcohol. Proceed as directed in the general method, page 899. One ml. dissolves in 3 ml. of 90 percent alcohol to form a clear solution.

Specific gravity. Determine by any reliable method (see page 5).

Acid value. Determine as directed in the general method, page 893.

Arsenic. A *Sample Solution* prepared as directed for organic compounds meets the requirements of the *Arsenic Test*, page 865.

Heavy metals. Prepare and test a 500-mg. sample as directed in *Method II* under the *Heavy Metals Test*, page 920, using 20 mcg. of lead ion (Pb) in the control (*Solution A*).

Lead. A *Sample Solution* prepared as directed for organic compounds meets the requirements of the *Lead Limit Test*, page 929, using 10 mcg. of lead ion (Pb) in the control.

Packaging and storage. Store in full, tight, preferably glass, tin-lined, or other suitably lined containers in a cool place protected from light.

Functional use in foods. Flavoring agent.

BENZYL PROPIONATE

C_6H_5—$CH_2OOCCH_2CH_3$

$C_{10}H_{12}O_2$ Mol. wt. 164.20

DESCRIPTION

A colorless liquid with a sweet, fruity odor. It is soluble in most fixed oils, sparingly soluble in propylene glycol, and slightly soluble in mineral oil. It is insoluble in glycerin.

SPECIFICATIONS

Assay. Not less than 98.0 percent of $C_{10}H_{12}O_2$.

Refractive index. Between 1.496 and 1.500 at 20°.

Solubility in alcohol. Passes test.

Specific gravity. Between 1.028 and 1.032.

Limits of Impurities

Acid value. Not more than 1.0.

TESTS

Assay. Weigh accurately about 1 gram, and proceed as directed under *Ester Determination*, page 896, using 82.10 as the equivalence factor (*e*) in the calculation.

Refractive index, page 945. Determine with an Abbé or other refractometer of equal or greater accuracy.

Solubility in alcohol. Proceed as directed in the general method, page 899. One ml. dissolves in 3 ml. of 70 percent alcohol, and remains in solution on dilution to 10 ml.

Specific gravity. Determine by any reliable method (see page 5).

Acid value. Determine as directed in the general method, page 893.

Packaging and storage. Store in full, tight, preferably glass, tin-lined or other suitably lined containers in a cool place protected from light.

Functional use in foods. Flavoring agent.

BENZYL SALICYLATE

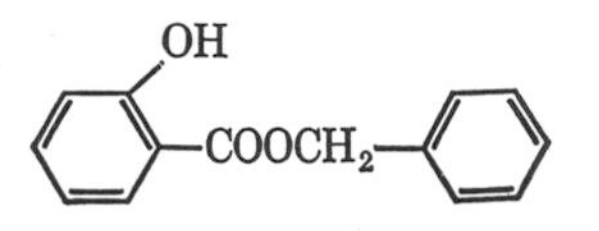

$C_{14}H_{12}O_3$ Mol. wt. 228.25

DESCRIPTION

An almost colorless liquid having a faint sweet odor. It is soluble in most fixed oils and in mineral oil, but it is practically insoluble in propylene glycol. It is insoluble in glycerin.

SPECIFICATIONS

Assay. Not less than 98.0 percent of $C_{14}H_{12}O_3$.

Refractive index. Between 1.579 and 1.582 at 20°.

Solidification point. Not less than 23.5°.

Solubility in alcohol. Passes test.

Specific gravity. Between 1.176 and 1.180.

Limits of Impurities

Acid value. Not more than 1.0.

TESTS

Assay. Weigh accurately about 1.4 grams, and proceed as directed under *Ester Determination*, page 896, using 114.1 as the equivalence factor (*e*) in the calculation. Modify the procedure by refluxing for 2 hours and using phenol red T.S. as indicator.

Refractive index, page 945. Determine with an Abbé or other refractometer of equal or greater accuracy.

Solidification point. Determine as directed in the general method, page 954.

Solubility in alcohol. Proceed as directed in the general method, page 899. One ml. dissolves in 9 ml. of 90 percent alcohol.

Specific gravity. Determine by any reliable method (see page 5).

Acid value. Determine as directed in the general method, page 893, using phenol red T.S. as indicator.

Packaging and storage. Store in full, tight, preferably glass, aluminum, or tin-lined containers in a cool place protected from light.

Functional use in foods. Flavoring agent.

BERGAMOT OIL, EXPRESSED

DESCRIPTION

A volatile oil obtained by expression, without the aid of heat, from the fresh peel of the fruit of *Citrus bergamia* Risso et Poiteau (Fam. *Rutaceae*). It is a green to yellowish green or yellowish brown liquid having a fragrant, sweet-fruity odor. It is miscible with alcohol and with glacial acetic acid. It is soluble in most fixed oils, but is insoluble in glycerin and in propylene glycol.

SPECIFICATIONS

Assay. Not less than 36.0 percent of esters, calculated as linalyl acetate ($C_{12}H_{20}O_2$).

Angular rotation. Between +8° and +24°.

Refractive index. Between 1.465 and 1.468 at 20°.

Residue on evaporation. Not more than 6.0 percent.

Solubility in alcohol. Passes test.

Specific gravity. Between 0.875 and 0.880.

Ultraviolet absorbance. Not less than 0.32.

Limits of Impurities

Arsenic (as As). Not more than 3 parts per million (0.0003 percent).

Heavy metals (as Pb). Not more than 40 parts per million (0.004 percent).

Lead. Not more than 10 parts per million (0.001 percent).

TESTS

Assay. Weigh accurately about 2 grams, and proceed as directed under *Ester Determination*, page 896, but heat the mixture for 30 minutes on the steam bath. Use 98.15 as the equivalence factor (*e*) in the calculation.

Angular rotation. Determine in a 100-mm. tube as directed under *Optical Rotation*, page 939.

Refractive index, page 945. Determine with an Abbé or other refractometer of equal or greater accuracy.

Residue on evaporation. Proceed as directed in the general method, page 899, heating for 5 hours.

Solubility in alcohol. Proceed as directed in the general method, page 899. One ml. dissolves in 2 ml. of 90 percent alcohol.

Specific gravity. Determine by any reliable method (see page 5).

Ultraviolet absorbance. Proceed as directed under *Ultraviolet Absorbance of Citrus Oils*, page 900, using about 50 mg. of sample, accurately weighed. The absorbance maximum occurs at 315 ± 3 mμ.

Arsenic. A *Sample Solution* prepared as directed for organic compounds meets the requirements of the *Arsenic Test*, page 865.

Heavy metals. Prepare and test a 500-mg. sample as directed in *Method II* under the *Heavy Metals Test*, page 920, using 20 mcg. of lead ion (Pb) in the control (*Solution A*).

Lead. A *Sample Solution* prepared as directed for organic compounds meets the requirements of the *Lead Limit Test*, page 929, using 10 mcg. of lead ion (Pb) in the control.

Packaging and storage. Store in full, tight, preferably glass, aluminum, tin-lined or other suitably lined containers in a cool place protected from light.

Functional use in foods. Flavoring agent.

BHA

Butylated Hydroxyanisole

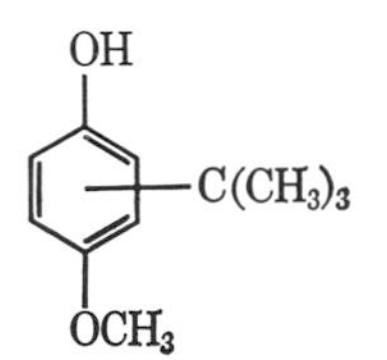

$C_{11}H_{16}O_2$ Mol. wt. 180.25

DESCRIPTION

BHA is predominately 3-*tert*-butyl-4-hydroxyanisole (3-BHA), with varying amounts of 2-*tert*-butyl-4-hydroxyanisole (2-BHA). It occurs as a white or slightly yellow, waxy solid having a faint characteristic odor. It is insoluble in water, but is freely soluble in alcohol and in propylene glycol. It melts between 48° and 63°.

IDENTIFICATION

To 5 ml. of a 1 in 10,000 solution of the sample in 72 percent alcohol add 2 ml. of sodium borate T.S. and 1 ml. of a 1 in 10,000 solution of 2,6-dichloroquinonechlorimide in absolute alcohol, and mix. A blue color develops.

SPECIFICATIONS

Assay. Not less than 98.5 percent of $C_{11}H_{16}O_2$.

Limits of Impurities

Arsenic (as As). Not more than 3 parts per million (0.0003 percent).

Heavy metals (as Pb). Not more than 10 parts per million (0.001 percent).

Residue on ignition. Not more than 0.01 percent.

TESTS

Assay

Standard Reference Curve. Weigh accurately 900.0 mg., 950.0 mg., and 1.0000 gram of F.C.C. 3-*tert*-Butyl-4-hydroxyanisole Reference Standard, representing 90.0, 95.0, and 100.0 percent of $C_{11}H_{16}O_2$, respectively, into separate 10-ml. volumetric flasks. Dissolve each standard in carbon disulfide, dilute to volume with carbon disulfide, and mix. Measure the infrared absorption spectrum of each solution from 10.5 μ to 12.5 μ with a suitable double beam infrared spectrophotometer, using 0.15 mm. sample cells, a 1.3 cm. rock salt plate in the reference beam, 2 $\times$ slits, and normal scanning speed. Draw a background line on the spectrogram from 11.2 μ to 12.0 μ, and determine the net absorbance of each solution at 11.42 μ by subtracting the background absorbance at this wavelength from the total absorption of each solution. Plot the calculated net absorbances against the percent of $C_{11}H_{16}O_2$ in each solution.

Assay Preparation. Transfer 1.0000 gram of the sample, accurately weighed, into a 10-ml. volumetric flask, dissolve it in carbon disulfide, dilute to volume with carbon disulfide, and mix. Measure the infrared absorption spectrum of this solution using the same conditions described above, determine the net absorbance of the sample at 11.42 μ, and obtain the apparent percentage of $C_{11}H_{16}O_2$ by means of the *Standard Reference Curve.*

Calculation. Calculate the true, total percentage of $C_{11}H_{16}O_2$ in the sample of butylated hydroxyanisole taken by the formula [(Apparent % BHA) + 0.16(100 − % 3-BHA)], in which *Apparent % BHA* is the percentage found by the procedure described under *Assay Preparation,* 0.16 is a factor to correct for the decreased absorbance of the 2-isomer (which absorbs only 84 percent as strongly as does the 3-isomer at 11.42 μ), and *% 3-BHA* is the percentage of 3-*tert*-butyl-4-hydroxy anisole in the sample obtained by the procedure described below.

% 3-BHA. Accurately weigh 1.0000 gram of the sample, previously melted in a water bath and thoroughly mixed, transfer it into a 10-ml. volumetric flask, dilute to volume with carbon disulfide, and mix. Measure the infrared absorption spectrum of this solution from 10 μ to 12 μ, using a 0.4 mm. sample cell and a 1.3 cm. rock salt plate in the reference beam. Subtract the background absorbance at 10.40 μ from the absorbance at 10.75 μ and at 10.95 μ. Using these net absorbance values, calculate the absorbance ratio $A_{10.75}/A_{10.95}$ by dividing the net absorbance found at 10.75 μ by that found at 10.95 μ. (The exact position of these absorption bands may vary, depending upon the instrument. If a recording instrument is used, the position of maximum absorbance on the spectrogram should be taken. With a non-recording instrument, the exact wavelength and slit settings should be determined.) Determine the percentage of 3-*tert*-butyl-4-hydroxy-

anisole in the sample taken by means of a calibration curve obtained as follows: Prepare three 10.0-ml. solutions in carbon disulfide containing, respectively, the following quantities of the specified F.C.C. Reference Standards: (1) 1.0000 gram of 3-*tert*-butyl-4-hydroxyanisole, (2) 900.0 mg. of 3-*tert*-butyl-4-hydroxyanisole and 100.0 mg. of 2-*tert*-butyl-4-hydroxyanisole, and (3) 800.0 mg. of 3-*tert*-butyl-4-hydroxyanisole and 200.0 mg. of 2-*tert*-butyl-4-hydroxyanisole. Measure the infrared absorption spectrum of each solution under the same conditions employed for the sample, and plot the calculated absorbance ratios against the corresponding percentages of 3-*tert*-butyl-4-hydroxyanisole.

Arsenic. A *Sample Solution* prepared as directed for organic compounds meets the requirements of the *Arsenic Test*, page 865.

Heavy metals. Prepare and test a 2-gram sample as directed in *Method II* under the *Heavy Metals Test*, page 920, using 20 mcg. of lead ion (Pb) in the control (*Solution A*).

Residue on ignition. Ignite 10 grams as directed in the general method page 945.

Packaging and storage. Store in well-closed containers.

Functional use in foods. Antioxidant.

BHT

Butylated Hydroxytoluene; 2,6-Di-*tert*-butyl-*p*-cresol

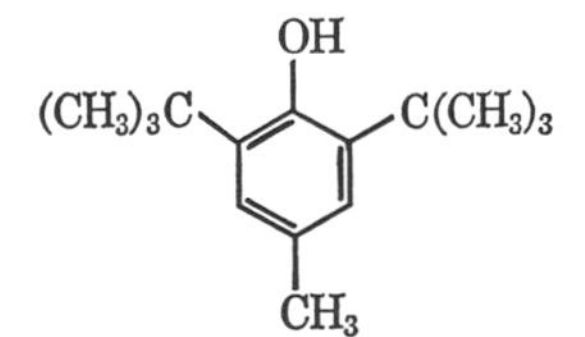

$C_{15}H_{24}O$ Mol. wt. 220.36

DESCRIPTION

A white crystalline solid having a faint characteristic odor. It is insoluble in water and in propylene glycol, but is freely soluble in alcohol.

IDENTIFICATION

To 10 ml. of a 1 in 100,000 solution of the sample in methanol add 10 ml. of water, 2 ml. of sodium nitrite solution (3 in 1000), and 5 ml. of dianisidine solution (200 mg. of 3,3′-dimethoxybenzidine dissolved in a mixture of 40 ml. of methanol and 60 ml. of 1 *N* hydrochloric acid). An orange-red color develops within 3 minutes. Add 5 ml. of chloroform and shake. The chloroform layer exhibits a purple or magenta color which fades when exposed to light.

SPECIFICATIONS

Assay. Not less than 99.0 weight percent of $C_{15}H_{24}O$.

Limits of Impurities

Arsenic (as As). Not more than 3 parts per million (0.0003 percent).
Heavy metals (as Pb). Not more than 10 parts per million (0.001 percent).
Residue on ignition. Not more than 0.002 percent.

TESTS

Assay. Its solidification point (see page 954) is not lower than 69.2°, indicating a purity of not less than 99.0 percent, of $C_{15}H_{24}O$.

Arsenic. A *Sample Solution* prepared as directed for organic compounds meets the requirements of the *Arsenic Test*, page 865.

Heavy metals. Prepare and test a 2-gram sample as directed in *Method II* under the *Heavy Metals Test*, page 920, using 20 mcg. of lead ion (Pb) in the control (*Solution A*).

Residue on ignition. Transfer a 50-gram sample into a tared crucible, ignite until thoroughly charred, and cool. Moisten the ash with 1 ml. of sulfuric acid, and complete the ignition by heating for 15-minute periods at 800° ± 25° to constant weight.

Packaging and storage. Store in well-closed containers.

Functional use in foods. Antioxidant.

BIOTIN

cis-Hexahydro-2-oxo-1H-thieno [3,4]-imidazole-4-valeric Acid

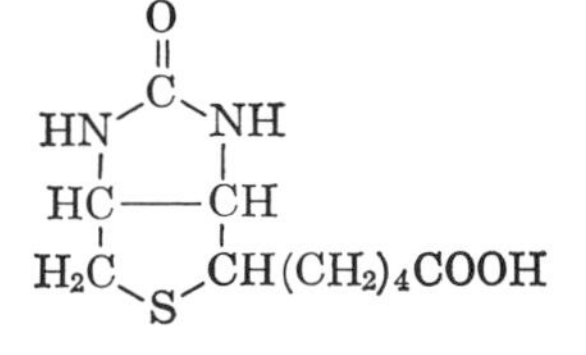

$C_{10}H_{16}N_2O_3S$ Mol. wt. 244.31

DESCRIPTION

A practically white, crystalline powder. It is stable to air and heat.

One gram dissolves in about 5000 ml. of water at 25° and in about 1300 ml. of alcohol; it is more soluble in hot water and in dilute alkali, and is insoluble in other common organic solvents. A saturated solution in warm water decolorizes bromine T.S., added dropwise.

SPECIFICATIONS

Assay. Not less than 97.5 percent of $C_{10}H_{16}N_2O_3S$.

Melting range. Between 229° and 232°.

Specific rotation, $[\alpha]_D^{25°}$. Between +89° and +93°.

Limits of Impurities

Arsenic (as As). Not more than 3 parts per million (0.0003 percent).

Heavy metals (as Pb). Not more than 10 parts per million (0.001 percent).

TESTS

Assay. Mix about 500 mg., accurately weighed, with 100 ml. of water, add phenolphthalein T.S., and titrate the suspension slowly with 0.1 *N* sodium hydroxide to a pink color, stirring continuously. Each ml. of 0.1 *N* sodium hydroxide is equivalent to 24.43 mg. of $C_{10}H_{16}$-N_2O_3S.

Melting range. Determine as directed in the general procedure, page 931.

Specific rotation, page 939. Determine in a solution in 0.1 *N* sodium hydroxide containing 500 mg. in each 25 ml.

Arsenic. A *Sample Solution* prepared as directed for organic compounds meets the requirements of the *Arsenic Test*, page 865.

Heavy metals. Prepare and test a 2-gram sample as directed in *Method II* under the *Heavy Metals Test*, page 920, using 20 mcg. of lead ion (Pb) in the control (*Solution A*).

Packaging and storage. Store in tight containers.

Functional use in foods. Nutrient; dietary supplement.

BIRCH TAR OIL, RECTIFIED

DESCRIPTION

The pyroligneous oil obtained by dry distillation of the bark and the wood of *Betula pendula* Roth and related species of *Betula* (Fam. *Betulaceae*) and rectified by steam distillation. It is a clear, dark brown liquid having a strong leather-like odor. It is soluble in most fixed oils, but it is insoluble in glycerin, in mineral oil, and in propylene glycol.

SPECIFICATIONS

Solubility in alcohol. Passes test.

Specific gravity. Between 0.886 and 0.950.

Limits of Impurities

Arsenic (as As). Not more than 3 parts per million (0.0003 percent).

Heavy metals (as Pb). Not more than 40 parts per million (0.004 percent).

Lead. Not more than 10 parts per million (0.001 percent).

TESTS

Solubility in alcohol. Proceed as directed in the general method, page 899. One ml. dissolves in 3 ml. of absolute alcohol.

Specific gravity. Determine by any reliable method (see page 5).

Arsenic. A *Sample Solution* prepared as directed for organic compounds meets the requirements of the *Arsenic Test*, page 865.

Heavy metals. Prepare and test a 500-mg. sample as directed in *Method II* under the *Heavy Metals Test*, page 920, using 20 mcg. of lead ion (Pb) in the control (*Solution A*).

Lead. A *Sample Solution* prepared as directed for organic compounds meets the requirements of the *Lead Limit Test*, page 929, using 10 mcg. of lead ion (Pb) in the control.

Packaging and storage. Store in full, tight preferably glass containers in a cool place protected from light.

Functional use in foods. Flavoring agent.

BOIS DE ROSE OIL

DESCRIPTION

The volatile oil obtained by steam distillation from the chipped wood of *Aniba rosaeodora* var. *amazonica* Ducke, (Fam. *Lauraceae*). The oils from the coastal region of Brazil and the Amazon valley tend to differ in odor and in linalool content from that produced in the Loreto province of Peru. The oil is a colorless to pale yellow liquid, having a slightly camphoraceous, pleasant floral odor. It is soluble in most fixed oils and in propylene glycol. It is soluble in mineral oil, occasionally with turbidity, but only slightly soluble in glycerin.

SPECIFICATIONS

Assay. Not less than 82.0 percent and not more than 92.0 percent of total alcohols, calculated as linalool ($C_{10}H_{18}O$).

Angular rotation. Between $-4°$ and $+6°$.

Distillation range. Not less than 70 percent distils between 195° and 205°.

Refractive index. Between 1.462 and 1.470 at 20°.
Solubility in alcohol. Passes test.
Specific gravity. Between 0.868 and 0.889.

TESTS

Assay. Proceed as directed under *Linalool Determination*, page 897, using about 1.2 grams of the acetylated oil, accurately weighed.

Angular rotation. Determine in a 100-mm. tube as directed under *Optical Rotation*, page 939.

Distillation range. Proceed as directed in the general method, page 890, using 50 ml. of the sample, previously dried over anhydrous sodium sulfate, and employing a 125-ml. flask.

Refractive index, page 945. Determine with an Abbé or other refractometer of equal or greater accuracy.

Solubility in alcohol. Proceed as directed in the general method, page 899. One ml. dissolves in 6 ml. of 60 percent alcohol.

Specific gravity. Determine by any reliable method (see page 5).

Packaging and storage. Store in full, tight, preferably glass, aluminum, tin-lined, or other suitably lined containers in a cool place protected from light.
Functional use in foods. Flavoring agent.

BORNYL ACETATE

Levo-Bornyl Acetate

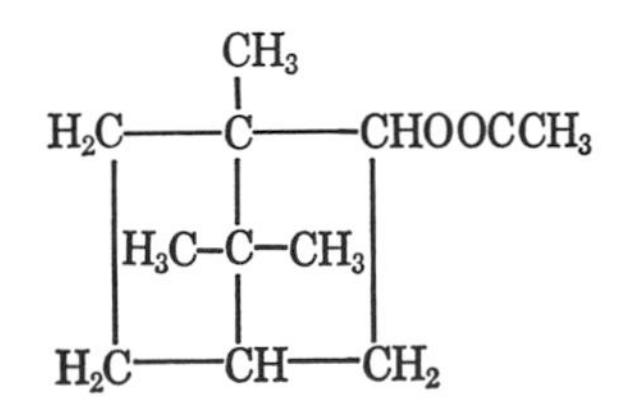

$C_{12}H_{20}O_2$ Mol. wt. 196.29

DESCRIPTION

Bornyl acetate may be obtained from various pine needle oils, or it may be prepared by acetylation of borneol. It is a colorless liquid, a semicrystalline mass, or a white crystalline solid, having a strong piney odor. It is soluble in most fixed oils and in mineral oil. It is practically insoluble in glycerin and in propylene glycol.

SPECIFICATIONS

Assay. Not less than 98.0 percent of $C_{12}H_{20}O_2$.

Angular rotation. Between −39.5° and −45.0°.

Refractive index. Between 1.462 and 1.466 at 20°.

Solidification point. Not less than 25°.

Solubility in alcohol. Passes test.

Specific gravity. Between 0.981 and 0.985.

Limits of Impurities

Acid value. Not more than 1.0.

TESTS

Assay. Weigh accurately about 1 gram, and proceed as directed under *Ester Determination*, page 896, using 98.14 as the equivalence factor (*e*) in the calculation.

Angular rotation. Determine in a 100-mm. tube as directed under *Optical Rotation*, page 939.

Refractive index, page 945. Determine with an Abbé or other refractometer of equal or greater accuracy.

Solidification point. Determine as directed in the general method, page 954.

Solubility in alcohol. Proceed as directed in the general method, page 899. One ml. dissolves in 3 ml. of 70 percent alcohol, and remains in solution on dilution to 10 ml.

Specific gravity. Determine by any reliable method (see page 5).

Acid value. Determine as directed in the general method, page 893.

Packaging and storage. Store in full, tight, preferably glass, tin-lined, or aluminum containers in a cool place protected from light.

Functional use in foods. Flavoring agent.

BROMINATED VEGETABLE OIL

DESCRIPTION

Brominated vegetable oil is a bromine addition product of vegetable oil or oils. It is a pale yellow to dark brown, viscous, oily liquid having a bland or fruity odor and a bland taste. It is insoluble in water, but is soluble in alcohol, in chloroform, in ether, in hexane, and in fixed oils.

IDENTIFICATION

Mix about 0.2 ml. of the sample with 1 gram of anhydrous sodium carbonate in a suitable crucible, cover the mixture with an additional

1 gram of sodium carbonate, compact the mixture by gentle tapping, and heat the crucible rapidly and strongly for 10 minutes. Cool the crucible and its contents, dissolve the residue in 20 ml. of hot water, and filter. To the filtrate add diluted nitric acid T.S. until effervescence ceases, then add 1 ml. of silver nitrate T.S. A curdy, yellowish precipitate, which is insoluble in nitric acid but soluble in an excess of stronger ammonia water, is formed.

SPECIFICATIONS

Specific gravity. Within the range specified by the vendor.

Limits of Impurities

Arsenic (as As). Not more than 3 parts per million (0.0003 percent).

Free bromine. Passes test

Free fatty acids (as oleic). Not more than 2.5 percent.

Heavy metals (as Pb). Not more than 10 parts per million (0.001 percent).

Iodine value. Not more than 16.

TESTS

Specific gravity. Determine as directed in the general procedure, page 915, at the temperature specified by the vendor.

Arsenic. A *Sample Solution* prepared as directed for organic compounds meets the requirements of the *Arsenic Test*, page 865.

Free bromine. Dissolve 1 gram in 20 ml. of acetone, add 1 gram of sodium iodide, and allow to stand in a stoppered flask in the dark for 30 minutes, with occasional shaking. Add 25 ml. of water and 1 ml. of starch T.S. No blue color is produced.

Free fatty acids. Determine as directed in the general procedure, page 903, using 28.2 as the equivalence factor (*e*) in the calculation for oleic acid. Titrate with the appropriate normality of sodium hydroxide solution, shaking vigorously, to the first permanent pink color of the same intensity as that of the neutralized alcohol, or, if the color of the sample interferes, titrate to a pH of 8.5, determined with a suitable instrument.

Heavy metals. Prepare and test a 2-gram sample as directed in *Method II* under the *Heavy Metals Test*, page 920, using 20 mcg. of lead ion (Pb) in the control (*Solution A*).

Iodine value. Determine by the *Wijs Method*, page 906.

Packaging and storage. Store in well-closed containers.

Functional use in foods. Flavoring adjunct; beverage stabilizer.

BUTADIENE-STYRENE 75/25 RUBBER

SBR 2006 Type Latex; SBR 1027 Type Solid

DESCRIPTION

A synthetic liquid latex (SBR 2006 Type) or solid rubber (SBR 1027 Type) produced by the emulsion copolymerization of butadiene and styrene, using fatty acid soaps (free from chick-edema factor) as emulsifiers, a persulfate catalyst, and a suitable shortstop. The latex, which has a pH between 9.5 and 11.0 and a solids content of 26–42 percent, is coagulated with or without other food-grade ingredients in a heated kettle, the coagulated mass is squeezed to drain off serums, and the coagulum is washed with hot water (with or without alkali) and rinsed with water until the batch is neutral. Finally, the coagulum is dried to remove residual volatiles. When butadiene-styrene rubber is purchased in the latex form, it must be washed by the preceding or an equivalent procedure. The solid form is supplied by the manufacturer (prewashed by an equivalent procedure) either in slab or a free-flowing uniform crumb. The crumb form may contain a suitable food-grade partitioning agent.

Note: The following *Specifications* apply to SBR solids.

SPECIFICATIONS

Bound styrene. Between 22.0 percent and 26.0 percent.

Limits of Impurities

Arsenic (as As). Not more than 3 parts per million (0.0003 percent).
Heavy metals (as Pb). Not more than 40 parts per million (0.004 percent).
Lead. Not more than 3 parts per million (0.0003 percent).
Quinones. Not more than 20 parts per million (0.002 percent).
Residual Styrene. Not more than 10 parts per million (0.001 percent).

TESTS

Bound styrene. Determine as directed in the general method, page 873.

Arsenic. Prepare a *Sample Solution* as directed in the general method under *Chewing Gum Base,* page 877. This solution meets the requirements of the *Arsenic Test,* page 865.

Heavy metals. Prepare and test a 500-mg. sample as directed in *Method II* under the *Heavy Metals Test,* page 920, using 20 mcg. of lead ion (Pb) in the control (*Solution A*).

Lead. Prepare a *Sample Solution* as directed in the general method under *Chewing Gum Base,* page 878. This solution meets the requirements of the *Lead Limit Test,* page 929, using 10 mcg. of lead ion (Pb) in the control.

Quinones. Determine as directed in the general method, page 876.

Residual styrene. Determine as directed in the general method, page 877.

Packaging and storage. Store in well-closed containers.

Functional use in foods. Masticatory substance in chewing gum base.

BUTADIENE-STYRENE 50/50 RUBBER

SBR 2000 Type Latex; SBR 1028 Type Solid

DESCRIPTION

A synthetic liquid latex (SBR 2000 Type) or solid rubber (SBR 1028 Type) produced by the emulsion copolymerization of butadiene and styrene, using rosin acid soaps or fatty acid soaps (free from chick-edema factor) as emulsifiers, a persulfate catalyst, and a suitable short-stop. The latex, which has a pH between 10.0 and 11.5 and a solids content of 41–63 percent, is coagulated with or without other food-grade ingredients in a heated kettle, the coagulated mass is squeezed to drain off serums, and the coagulum is washed with hot water (with or without alkali) and rinsed with water until the batch is neutral. Finally, the coagulum is dried to remove residual volatiles. When butadiene-styrene rubber is purchased in the latex form, it must be washed by the preceding or an equivalent procedure. The solid form is supplied by the manufacturer (prewashed by an equivalent procedure) either in a slab or a free-flowing uniform crumb. The crumb form may contain a suitable food-grade partitioning agent.

Note: The following *Specifications* apply to SBR solids.

SPECIFICATIONS

Bound styrene. Between 45.0 percent and 50.0 percent.

Limits of Impurities

Arsenic (as As). Not more than 3 parts per million (0.0003 percent).

Heavy metals (as Pb). Not more than 40 parts per million (0.004 percent).

Lead. Not more than 3 parts per million (0.0003 percent).

Quinones. Not more than 20 parts per million (0.002 percent).

Residual styrene. Not more than 10 parts per million (0.001 percent).

TESTS

Bound styrene. Determine as directed in the general method, page 873.

Arsenic. Prepare a *Sample Solution* as directed in the general method under *Chewing Gum Base*, page 877. This solution meets the requirements of the *Arsenic Test*, page 865.

Heavy metals. Prepare and test a 500-mg. sample as directed in *Method II* under the *Heavy Metals Test*, page 920, using 20 mcg. of lead ion (Pb) in the control (*Solution A*).

Lead. Prepare a *Sample Solution* as directed in the general method under *Chewing Gum Base*, page 878. This solution meets the requirements of the *Lead Limit Test*, page 929, using 10 mcg. of lead ion (Pb) in the control.

Quinones. Determine as directed in the general method, page 876.

Residual styrene. Determine as directed in the general method, page 877.

Packaging and storage. Store in well-closed containers.

Functional use in foods. Masticatory substance in chewing gum base.

2-BUTANONE

Methyl Ethyl Ketone

$CH_3COCH_2CH_3$

C_4H_8O Mol. wt. 72.11

DESCRIPTION

A clear, colorless, mobile liquid having a characteristic sharp penetrating odor. It is miscible in all proportions with alcohol and with ether. One ml. dissolves in about 4 ml. of water.

SPECIFICATIONS

Assay. Not less than 99.5 percent of C_4H_8O.

Distillation range. Within a 1.5° range. The boiling point of 2-butanone at 760 mm. pressure is 79.6°.

Specific gravity. Between 0.801 and 0.803.

Limits of Impurities

Acidity (as acetic acid). Not more than 0.003 percent.

Arsenic (as As). Not more than 3 parts per million (0.0003 percent).

Heavy metals (as Pb). Not more than 10 parts per million (0.001 percent).

Residue on evaporation. Not more than 2 mg. per 100 ml.

Water. Not more than 0.2 percent.

TESTS

Assay. Determine the purity of 2-butanone by gas-liquid chromatography (see page 886), using an instrument containing a thermal conductivity detector. Prepare a 4-meter × 6-mm. column consisting of a blend of equal quantities of 20 percent Carbowax 20M on acid-washed, 60/80-mesh Chromosorb W, and 20 percent tetrahydroxyethyl ethylenediamine on 30/60-mesh Chromosorb P, or use other suitable column materials capable of separating 2-butanone and the impurities whose retention times are listed below. Observe the following operating conditions during the determination: *Sample,* 10 μl.; *Column temperature,* about 80°; *Helium flow rate,* 30 to 32 ml. per minute; and *Detector voltage,* 8.0. The approximate retention times, in minutes, are as follows: acetone 7; ethyl acetate, 9; 2-butanone, 11; *t*-butanol, 14; methanol, 15; ethanol, 19; 2-butanol, 30; and propanol, 36.

Measure the area under each peak of the chromatogram so obtained, calculate the area percent of each impurity, and record the sum of the impurities as *A*. (*Note:* In this procedure, area percent and weight percent may be assumed to be identical.) Calculate the percent of C_4H_8O in the sample by the formula, $100.0 - (A + B + C)$, in which *B* is the percent of water determined under *Water,* and *C* is the acidity (as acetic acid) as determined under *Acidity.*

Distillation range. Distil 100 ml. as directed in the general method, page 890.

Specific gravity. Determine by any reliable method (see page 5).

Acidity. Transfer 75 ml. (60 grams) into a 250-ml. Erlenmeyer flask, add phenolphthalein T.S., and titrate with 0.02 *N* alcoholic potassium hydroxide to a pink end-point that persists for at least 15 seconds. Each ml. of 0.02 *N* alcoholic potassium hydroxide is equivalent to 1.20 mg. of $C_2H_4O_2$.

Arsenic. A *Sample Solution* prepared as directed for organic compounds meets the requirements of the *Arsenic Test,* page 865.

Heavy metals. Prepare and test a 2-gram sample as directed in *Method II* under the *Heavy Metals Test,* page 920, using 20 mcg. of lead ion (Pb) in the control (*Solution A*).

Residue on evaporation, page 899. Evaporate 100 ml. to dryness in a tared platinum dish, on a steam bath, and dry at 105° for 30 minutes. Cool in a desiccator and weigh. The weight of the residue does not exceed 2 mg.

Water. Determine by the *Karl Fischer Titrimetric Method,* page 977, using freshly distilled pyridine instead of methanol as the solvent.

Packaging and storage. Store in tight containers.

Functional use in foods. Flavoring agent.

BUTYL ACETATE

n-Butyl Acetate

$CH_3COOCH_2CH_2CH_2CH_3$

$C_6H_{12}O_2$ Mol. wt. 116.16

DESCRIPTION

A clear, colorless, mobile liquid having a strong, characteristic, fruity odor. One ml. dissolves in about 145 ml. of water. It is miscible in all proportions with alcohol, with ether, and with propylene glycol.

SPECIFICATIONS

Assay. Not less than 98.0 percent of $C_6H_{12}O_2$.

Distillation range. Between 120° and 128°.

Refractive index. Between 1.393 and 1.395 at 20°.

Specific gravity. Between 0.876 and 0.880.

Limits of Impurities

Acidity (as acetic acid). Not more than 0.01 percent.

Nonvolatile residue. Not more than 60 parts per million (0.006 percent).

TESTS

Assay. Transfer about 1.5 grams of the sample, accurately weighed, into a 250-ml. Erlenmeyer flask, add 25.0 ml. of 1 *N* potassium hydroxide and 25 ml. of anhydrous isopropanol, swirl to effect complete solution, and allow to stand at room temperature for 30 minutes. Add about 1 ml. of phenolphthalein T.S. to the mixture, and titrate with 0.5 *N* sulfuric acid to the disappearance of the pink color. Perform a residual blank determination (see page 2). Each ml. of 0.5 *N* sulfuric acid is equivalent to 58.08 mg. of $C_6H_{12}O_2$.

Distillation range. Distil 100 ml. as directed in the general method, page 890.

Refractive index, page 945. Determine with an Abbé or other refractometer of equal or greater accuracy.

Specific gravity. Determine by any reliable method (see page 5).

Acidity. Transfer 69.0 ml. (60 grams) into a 250-ml. Erlenmeyer flask, add phenolphthalein T.S., and titrate with 0.1 *N* alcoholic potassium hydroxide to a pink end-point that persists for at least 15 seconds. Not more than 1.0 ml. is required.

Nonvolatile residue. Evaporate 95 ml. (about 83.5 grams) to dryness in a tared platinum dish on a steam bath, and dry at 105° for 30 minutes. Cool in a desiccator and weigh. The weight of the residue does not exceed 5 mg.

Packaging and storage. Store in tight containers and avoid excessive temperatures.

Functional use in foods. Flavoring agent.

BUTYL ALCOHOL

1-Butanol

$CH_3(CH_2)_2CH_2OH$

$C_4H_{10}O$ Mol. wt. 74.12

DESCRIPTION

A clear, colorless, mobile liquid having a characteristic, penetrating vinous odor. One ml. dissolves in about 15 ml. of water. It is miscible in all proportions with alcohol, with ether, and with many other organic solvents.

SPECIFICATIONS

Distillation range. Within a 1.5° range.

Specific gravity. Between 0.807 and 0.809.

Limits of Impurities

Acidity (as acetic acid). Not more than 0.005 percent.
Aldehydes. Passes test.
Butyl ether. Not more than 0.2 percent.
Residue on evaporation. Not more than 4 mg. per 100 ml.
Water. Not more than 0.1 percent.

TESTS

Distillation range. Distil 100 ml. as directed in the general method, page 890. The difference between the temperature observed at the beginning and the end of the distillation does not exceed 1.5°. The boiling point of pure butyl alcohol at 760 mm. of mercury pressure is 117.7°.

Specific gravity. Determine by any reliable method (see page 5).

Acidity. Transfer 74 ml. (60 grams) into a 250-ml. Erlenmeyer flask, add phenolphthalein T.S., and titrate with 0.02 *N* alcoholic potassium hydroxide to a pink end-point that persists for at least 15 seconds. Each ml. of 0.02 *N* alcoholic potassium hydroxide is equivalent to 1.20 mg. of $C_2H_4O_2$.

Aldehydes. Transfer 10 ml. of ammoniacal silver nitrate T.S. into a 20 × 150-mm. test tube, add 10 ml. of the sample, mix thoroughly, and allow the mixture to stand in a dark place for 30 minutes. No color is produced, but a slight precipitate may form at the interface of the two layers.

Butyl ether. Determine the percent of butyl ether in the sample by gas-liquid chromatography using an instrument containing a thermal conductivity detector. Prepare a 2-meter × 6-mm. column consisting of 25 percent, by weight, β,β'-thiodipropionitrile on 30/40-mesh Chromosorb P and operated at a constant temperature of about 85°. Establish a helium flow rate of about 75 ml. per minute, adjust the inlet pressure to about 30 psi, and inject a sample of about 10 microliters. The recorder should be operated in the 0.1 mv. range, with the detector voltage set at about 8, depending upon the particular instrument used. Under the conditions described, the butyl ether is eluted in about 6 minutes. The area of the butyl ether peak is not more than 0.2 percent of the total area of all peaks.

Residue on evaporation, page 899. Evaporate 100 ml. to dryness in a tared platinum dish, on a steam bath, and dry at 105° for 30 minutes. Cool in a desiccator and weigh. The weight of the residue does not exceed 4 mg.

Water. Determine by the *Karl Fischer Titrimetric Method*, page 977.

Packaging and storage. Store in tight containers and avoid excessive temperatures.

Functional use in foods. Flavoring agent.

BUTYLATED HYDROXYMETHYLPHENOL

4-Hydroxymethyl-2,6-di-*tert*-butylphenol

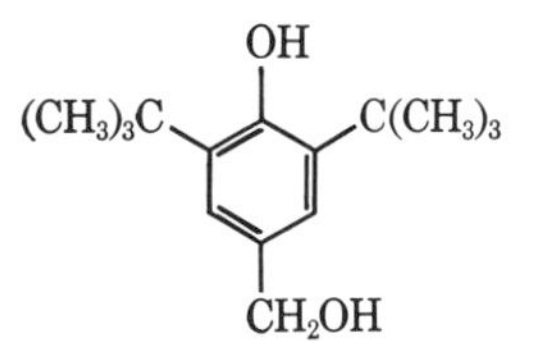

$C_{15}H_{24}O_2$ Mol. wt. 236.36

DESCRIPTION

A nearly white crystalline solid having a faint characteristic odor. It is insoluble in water and in propylene glycol, but is freely soluble in alcohol.

SPECIFICATIONS

Assay. Not less than 98.0 percent of $C_{15}H_{24}O_2$.

Limits of Impurities

Arsenic (as As). Not more than 3 parts per million (0.0003 percent).

Heavy metals (as Pb). Not more than 10 parts per million (0.001 percent).

TESTS

Assay. Its solidification point (see page 954) is not lower than 140°, indicating a purity of not less than 98.0 percent, by weight, of $C_{15}H_{24}O_2$.

Arsenic. A *Sample Solution* prepared as directed for organic compounds meets the requirements of the *Arsenic Test*, page 865.

Heavy metals. Prepare and test a 2-gram sample as directed in *Method II* under the *Heavy Metals Test*, page 920, using 20 mcg. of lead ion (Pb) in the control (*Solution A*).

Packaging and storage. Store in well-closed containers.

Functional use in foods. Antioxidant.

BUTYL BUTYRATE

n-Butyl *n*-Butyrate

$CH_3CH_2CH_2COOC_4H_9$

$C_8H_{16}O_2$ Mol. wt. 144.21

DESCRIPTION

A colorless liquid having a fruity odor which upon dilution is suggestive of pineapple. It is miscible with alcohol, with ether, and with most vegetable oils. It is slightly soluble in water and in propylene glycol. One part is soluble in 3 parts of 70 percent alcohol.

SPECIFICATIONS

Assay. Not less than 98.0 percent of $C_8H_{16}O_2$.

Refractive index. Between 1.405 and 1.407 at 20°.

Specific gravity. Between 0.867 and 0.871.

Limits of Impurities

Acid value. Not more than 1.0.

TESTS

Assay. Weigh accurately about 1 gram, and proceed as directed under *Ester Determination*, page 896, using 72.10 as the equivalence factor (*e*) in the calculation.

Refractive index, page 945. Determine with an Abbé or other refractometer of equal or greater accuracy.

Specific gravity. Determine by any reliable method (see page 5).

Acid value. Determine as directed in the general method, page 893.

Packaging and storage. Store in full, tight, preferably glass, aluminum, or tin-lined containers in a cool place protected from light.

Functional use in foods. Flavoring agent.

BUTYL BUTYRYLLACTATE

Butyryllactic Acid, Butyl Ester; Lactic Acid, Butyl Ester, Butyrate

$$\begin{array}{r} CH_3CHCOOC_4H_9 \\ | \\ CH_3CH_2CH_2COO \end{array}$$

$C_{11}H_{20}O_4$ Mol. wt. 216.28

DESCRIPTION

A colorless liquid with a mild buttery odor. It is miscible with alcohol and most fixed oils, soluble in propylene glycol, and insoluble in glycerin and in water.

SPECIFICATIONS

Assay. Not less than 95.0 percent of $C_{11}H_{20}O_4$.

Specific gravity. Between 0.970 and 0.974.

Solubility in alcohol. Passes test.

Refractive index. Between 1.420 and 1.423 at 20°.

Limits of Impurities

Acid value. Not more than 1.0.

TESTS

Assay. Weigh accurately about 1 gram, and proceed as directed under *Ester Determination*, page 896, using 54.07 as the equivalence factor (e) in the calculation.

Specific gravity. Determine by any reliable method (see page 5).

Solubility in alcohol. Proceed as directed in the general method, page 899. One ml. dissolves in 3 ml. of 70 percent alcohol.

Refractive index, page 945. Determine with an Abbé or other refractometer of equal or greater accuracy.

Acid value. Determine as directed in the general method, page 893.

Packaging and storage. Store in full, tight, glass or aluminum, or suitably lined containers in a cool place protected from light.

Functional use in foods. Flavoring agent.

1,3-BUTYLENE GLYCOL

$CH_2OHCH_2CHOHCH_3$

$C_4H_{10}O_2$ Mol. wt. 90.12

DESCRIPTION

A clear, colorless, hygroscopic, viscous liquid having a slight, characteristic taste. It is practically odorless. It is miscible with water, with acetone, and with ether in all proportions, but it is immiscible with fixed oils. It dissolves most essential oils and synthetic flavoring substances.

SPECIFICATIONS

Assay. Not less than 99.0 percent of $C_4H_{10}O_2$.

Distillation range. Between 200° and 215°.

Specific gravity. Between 1.004 and 1.006 at 20°.

Limits of Impurities

Arsenic (as As). Not more than 3 parts per million (0.0003 percent).

Heavy metals (as Pb). Not more than 10 parts per million (0.001 percent).

TESTS

Assay. Prepare an acetylating reagent, within one week of use, by mixing 3.4 ml. of water and 130 ml. of acetic anhydride with 1000 ml. of anhydrous pyridine. Pipet 20 ml. of this reagent into a 250-ml. iodine flask, and add about 1 gram of the sample, accurately weighed. Attach a dry reflux condenser to the flask, and reflux for 1 hour. Allow the flask to cool to room temperature, then rinse the condenser with 50 ml. of chilled (10°) carbon dioxide-free water, allowing the water to drain into the flask. Stopper the flask, cool to below 20°, add phenolphthalein T.S., and titrate with 0.5 *N* sodium hydroxide, swirling the contents of the flask continuously during the titration. Perform a blank determination (see page 2). Each ml. of 0.5 *N* sodium hydroxide is equivalent to 2.253 mg. of $C_4H_{10}O_2$.

Distillation range. Proceed as directed in the general method, page 890.

Specific gravity. Determine by any reliable method (see page 5).

Arsenic. A *Sample Solution* prepared as directed for organic compounds meets the requirements of the *Arsenic Test*, page 865.

Heavy metals. Prepare and test a 2-gram sample as directed in *Method II* under the *Heavy Metals Test*, page 920, using 20 mcg. of lead ion (Pb) in the control (*Solution A*).

Packaging and storage. Store in well-closed containers.

Functional use in foods. Solvent for flavoring agents.

BUTYRALDEHYDE

Butyl Aldehyde

$CH_3(CH_2)_2CHO$

C_4H_8O Mol. wt. 72.11

DESCRIPTION

A clear, colorless, mobile liquid having a characteristic sharp, pungent odor. One ml. dissolves in about 15 ml. of water. It is miscible in all proportions with alcohol and with ether.

SPECIFICATIONS

Assay. Not less than 98.0 percent of C_4H_8O.

Distillation range. Between 72° and 80° (first 95 percent).

Specific gravity. Between 0.797 and 0.802.

Limits of Impurities

Acidity (as butyric acid). Not more than 0.75 percent.

Para-butyraldehyde. Not more than 2.5 percent of $C_6H_{12}O$.

Water. Not more than 0.5 percent.

TESTS

Assay. Transfer 65 ml. of 0.5 *N* hydroxylamine hydrochloride and 50.0 ml. of 0.5 *N* triethanolamine into a suitable heat-resistant pressure bottle provided with a tight closure that can be fastened securely. Replace the air in the bottle by passing a gentle stream of nitrogen for two minutes through a glass tube positioned so that the end is just above the surface of the liquid. To the mixture in the pressure bottle add about 900 mg. of the sample contained in a sealed ampul and accurately weighed. Introduce several pieces of 8-mm. glass rod, cap the bottle, and shake vigorously to break the ampul. Allow the bottle to stand at room temperature, swirling occasionally, for 60 minutes. Cool, if necessary, uncap the bottle, and titrate with 0.5 *N* sulfuric acid to a greenish blue end-point. Perform a residual blank titration (see *Blank Tests*, page 2). Each ml. of 0.5 *N* sulfuric acid is equivalent to 36.06 mg. of C_4H_8O.

Distillation range. Distil 100 ml. as directed in the general method, page 890.

Specific gravity. Determine by any reliable method (see page 5).

Acidity. Transfer 54 ml. (44 grams) into a 250-ml. Erlenmeyer flask through which a gentle stream of nitrogen has been passed for 2 minutes. Add phenolphthalein T.S. and titrate with 0.1 *N* alcoholic potassium hydroxide in an atmosphere of nitrogen to a pink end-point that persists for at least 15 seconds. Each ml. of 0.1 *N* alcoholic potassium hydroxide is equivalent to 8.81 mg. of $C_4H_8O_2$.

Para-butyraldehyde. Transfer about 30 ml. of a freshly prepared 1 in 5 solution of sodium bisulfite into a Babcock bottle having a neck graduated into 8 units of 0.2 ml. and each of which is further subdivided into 10 parts, representing a volume of 0.02 ml. each. To the sodium bisulfite solution add 5.0 ml. of the sample, stopper the bottle, immerse in an ice-water bath, and shake vigorously until heat is no longer evolved. Remove the bottle from the bath, shake it at room temperature for about 5 minutes, then fill to the highest graduation with sodium bisulfite solution, mix, and centrifuge for 5 minutes. Read the volume of the oil layer in ml. and calculate the percent of para-butyraldehyde by the formula: [(ml. of oil layer + 0.05) × 92.0]/5 × specific gravity of the butyraldehyde, in which 0.05 is a solubility factor and 92.0 relates to the specific gravity of para-butyraldehyde.

Water. Determine by the *Karl Fischer Titrimetric Method*, page 977, using freshly distilled pyridine instead of methanol as the solvent.

Packaging and storage. Store in tight containers

Functional use in foods. Flavoring agent.

BUTYRIC ACID

$CH_3(CH_2)_2COOH$

$C_4H_8O_2$ Mol. wt. 88.11

DESCRIPTION

A clear, colorless liquid having a strong, penetrating odor similar to that of rancid butter. It is miscible with alcohol, propylene glycol, most fixed oils, water, and alcohol.

SPECIFICATIONS

Assay. Not less than 99.0 percent and not more than the equivalent of 101.0 percent of $C_4H_8O_2$.

Refractive index. Between 1.397 and 1.399 at 20°.

Specific gravity. Between 0.957 and 0.961.

Limits of Impurities

Arsenic (as As). Not more than 3 parts per million (0.0003 percent).

Heavy metals (as Pb). Not more than 40 parts per million (0.004 percent).

Lead. Not more than 10 parts per million (0.001 percent).

Reducing substances. Passes test.

TESTS

Assay. Transfer about 1.5 grams, accurately weighed, into a 250-ml. Erlenmeyer flask containing about 75 ml. of water, add phenolphthalein T.S., and titrate with 0.5 *N* sodium hydroxide. Each ml. of 0.5 *N* sodium hydroxide is equivalent to 44.06 mg. of $C_4H_8O_2$.

Refractive index, page 945. Determine at 20° using an Abbé or other refractometer of equal or greater accuracy.

Specific gravity. Determine by any reliable method (see page 5).

Arsenic. A *Sample Solution* prepared as directed for organic compounds meets the requirements of the *Arsenic Test,* page 865.

Heavy metals. Prepare and test a 500-mg. sample as directed in *Method II* under the *Heavy Metals Test,* page 920, using 20 mcg. of lead ion (Pb) in the control (*Solution A*).

Lead. A *Sample Solution* prepared as directed for organic compounds meets the requirements of the *Lead Limit Test,* page 929, using 10 mcg. of lead ion (Pb) in the control.

Reducing substances. Dilute 2 ml. in a glass-stoppered flask with 50 ml. of water and 5 ml. of sulfuric acid, shaking the flask during the addition. While the solution is still warm, titrate with 0.1 *N* potassium permanganate. Not more than 1 ml. is required to produce a pink color that persists for 30 minutes.

Packaging and storage. Store in tight containers.

Functional use in foods. Flavoring agent.

CAFFEINE

1,3,7-Trimethylxanthine

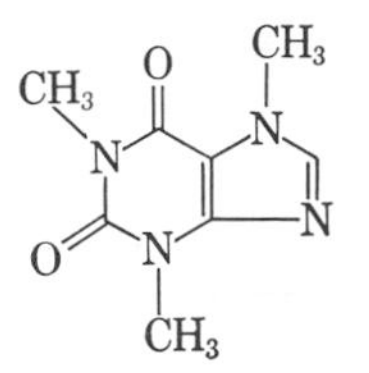

$C_8H_{10}N_4O_2$ Mol. wt. 194.19

DESCRIPTION

Caffeine is anhydrous or contains one molecule of water of hydration. A white powder, or white, glistening needles, usually matted together. It is odorless and has a bitter taste. Its solutions are neutral to litmus. The hydrate is efflorescent in air. One gram of hydrated caffeine is soluble in about 50 ml. of water, in 75 ml. of alcohol, in about 6 ml. of chloroform, and in 600 ml. of ether.

IDENTIFICATION

A. Dissolve about 5 mg. in 1 ml. of hydrochloric acid in a porcelain dish, add 50 mg. of potassium chlorate, and evaporate on a steam bath to dryness. Invert the dish over a vessel containing a few drops of ammonia T.S. The residue acquires a purple color, which disappears upon the addition of a solution of a fixed alkali.

B. To a saturated solution of caffeine add tannic acid T.S. A precipitate, which is soluble in an excess of the reagent, is formed.

C. To 5 ml. of a saturated solution of caffeine add 5 drops of iodine T.S. No precipitate is formed. Then add 3 drops of diluted hydrochloric acid T.S. A red-brown precipitate, which redissolves when a slight excess of sodium hydroxide T.S. is added, is formed.

SPECIFICATIONS

Assay. Not less than 98.5 percent and not more than the equivalent of 101.0 percent of $C_8H_{10}N_4O_2$, calculated on the anhydrous basis.

Melting range. Between 235° and 237.5°.

Water. Anhydrous caffeine, not more than 0.5 percent; hydrous caffeine, not more than 8.5 percent.

Limits of Impurities

Arsenic (as As). Not more than 3 parts per million (0.0003 percent).

Heavy metals (as Pb). Not more than 20 parts per million (0.002 percent).

Lead. Not more than 10 parts per million (0.001 percent).

Readily carbonizable substances. Passes test.

Residue on ignition. Not more than 0.1 percent.

TESTS

Assay. Dissolve about 800 mg., accurately weighed, of finely powdered caffeine, with warming, in a mixture of 80 ml. of acetic anhydride and 180 ml. of benzene. Cool, and titrate with 0.1 *N* perchloric acid, determining the end-point potentiometrically. Each ml. of 0.1 *N* perchloric acid is equivalent to 19.42 mg. of $C_8H_{10}N_4O_2$.

Melting range. Dry at 80° for 4 hours and then determine as directed in the general procedure, page 931.

Water. Determine the water content by drying at 80° for 4 hours (page 931) or by the *Karl Fischer Titrimetric Method*, page 977.

Arsenic. A *Sample Solution* prepared as directed for organic compounds meets the requirements of the *Arsenic Test*, page 865.

Heavy metals. A solution of 500 mg. in 2.5 ml. of hydrochloric acid and 23 ml. of water meets the requirements of the *Heavy Metals Test*, page 920, using 10 mcg. of lead ion (Pb) in the control (*Solution A*).

Lead. A *Sample Solution* prepared as directed for organic compounds meets the requirements of the *Lead Limit Test*, page 929, using 10 mcg. of lead ion (Pb) in the control.

Readily carbonizable substances, page 943. Dissolve 500 mg. in 5 ml. of sulfuric acid T.S. The color is no darker than *Matching Fluid D*.

Residue on ignition, page 945. Ignite 2 grams as directed in the general method.

Packaging and storage. Store hydrous caffeine in tight containers and anhydrous caffeine in well-closed containers.

Labeling. Label caffeine to indicate whether it is anhydrous or hydrous.

Functional use in foods. Central stimulant in cola-type beverages.

CALCIUM ACETATE

$Ca(C_2H_3O_2)_2$ Mol. wt. 158.17

DESCRIPTION

A fine, white, bulky, odorless powder. It is freely soluble in water and is slightly soluble in alcohol. A 1 in 10 solution gives positive tests for *Calcium*, page 926, and for *Acetate*, page 925.

SPECIFICATIONS

Assay. Not less than 99.0 percent of $Ca(C_2H_3O_2)_2$, calculated on the anhydrous basis.

Limits of Impurities

Arsenic (as As). Not more than 3 parts per million (0.0003 percent).

Chloride. Not more than 500 parts per million (0.05 percent).
Heavy metals (as Pb). Not more than 25 parts per million (0.0025 percent).
Lead. Not more than 10 parts per million (0.001 percent).
Sulfate. Not more than 0.1 percent.
Water. Not more than 7 percent.

TESTS

Assay. Dissolve about 300 mg., accurately weighed, in 150 ml. of water containing 2 ml. of diluted hydrochloric acid T.S. While stirring, preferably with a magnetic stirrer, add about 30 ml. of 0.05 *M* disodium ethylenediaminetetraacetate from a 50-ml. buret, then add 15 ml. of sodium hydroxide T.S. and 300 mg. of hydroxy naphthol blue indicator, and continue the titration to a blue end-point. Each ml. of 0.05 *M* disodium ethylenediaminetetraacetate is equivalent to 7.909 mg. of $Ca(C_2H_3O_2)_2$.

Arsenic. A *Sample Solution* prepared as directed for organic compounds meets the requirements of the *Arsenic Test*, page 865.

Chloride, page 879. Any turbidity produced by a 40-mg. sample does not exceed that shown in a control containing 20 mcg. of chloride ion (Cl).

Heavy metals. A solution of 1.2 grams in 25 ml. of water meets the requirements of the *Heavy Metals Test*, page 920, using 20 mcg. of lead ion (Pb) and 400 mg. of the sample in the control (*Solution A*).

Lead. A *Sample Solution* prepared as directed for organic compounds meets the requirements of the *Lead Limit Test*, page 929, using 10 mcg. of lead ion (Pb) in the control.

Sulfate, page 879. Any turbidity produced by a 200-mg. sample does not exceed that shown in a control containing 200 mcg. of sulfate (SO_4).

Water. Determine by the *Karl Fischer Titrimetric Method*, page 977.

Packaging and storage. Store in well-closed containers.
Functional use in foods. Sequestrant.

CALCIUM ALGINATE

Algin

$[(C_6H_7O_6)_2Ca]_n$

Equiv. wt., *Calculated*, 195.16
Equiv. wt., *Actual* (Avg.), 219.00

DESCRIPTION

The calcium salt of alginic acid (see *Alginic Acid*, page 26), occurs as a white to yellowish, fibrous or granular powder. It is nearly odorless and tasteless. It is insoluble in water, but is soluble in alkaline solutions or in solutions of substances which combine with the calcium. It is insoluble in organic solvents.

IDENTIFICATION

A. To 5 ml. of a 1 in 100 solution in 0.1 *N* sodium hydroxide add 1 ml. of calcium chloride T.S. A voluminous, gelatinous precipitate is formed.

B. To 10 ml. of the solution prepared for *Test A* add 1 ml. of diluted sulfuric acid T.S. A heavy gelatinous precipitate is formed.

C. Calcium alginate meets the requirements of *Identification Test C* under *Alginic Acid*, page 26.

D. Extract the *Ash* from calcium alginate with diluted hydrochloric acid T.S. and filter. The filtrate gives positive tests for *Calcium*, page 926.

SPECIFICATIONS

Assay. It yields, on the dried basis, not less than 18.0 percent and not more than 21.0 percent of carbon dioxide (CO_2) corresponding to between 89.6 and 104.5 percent of calcium alginate (Equiv. wt. 219.00).

Ash. Between 12 and 18 percent on the dried basis.

Limits of Impurities

Arsenic (as As). Not more than 3 parts per million (0.0003 percent).

Heavy metals (as Pb). Not more than 40 parts per million (0.004 percent).

Insoluble matter. Not more than 0.2 percent.

Lead. Not more than 10 parts per million (0.001 percent).

Loss on drying. Not more than 15 percent.

TESTS

Assay. Proceed as directed under *Alginates Assay*, page 863. Each ml. of 0.25 *N* sodium hydroxide consumed in the assay is equivalent to 27.38 mg. of calcium alginate (Equiv. wt. 219.00).

Ash. Determine as directed under *Ash* in the monograph on *Alginic Acid*, page 26.

Arsenic. A *Sample Solution* prepared as directed for organic compounds meets the requirements of the *Arsenic Test*, page 865.

Heavy metals. Prepare and test a 500-mg. sample as directed in *Method II* under the *Heavy Metals Test*, page 920, but use a platinum crucible for the ignition. Any color does not exceed that produced in a control (*Solution A*) containing 20 mcg. of lead ion (Pb).

Insoluble matter. Determine as directed under *Insoluble matter* in the monograph on *Alginic Acid*, page 26, but use 15 ml. of 0.1 *N* sodium hydroxide in the solvent.

Lead. A *Sample Solution* prepared as directed for organic compounds meets the requirements of the *Lead Limit Test,* page 929, using 10 mcg. of lead ion (Pb) in the control.

Loss on drying, page 931. Dry at 105° for 4 hours.

Packaging and storage. Store in well-closed containers.

Functional use in foods. Stabilizer; thickener; emulsifier.

CALCIUM ASCORBATE

$C_{12}H_{14}CaO_{12}.2H_2O$ Mol. wt. 426.35

DESCRIPTION

A white to slightly yellow, odorless, crystalline powder. It is soluble in water, slightly soluble in alcohol, and insoluble in ether. The pH of a 1 in 10 solution is between 6.8 and 7.4. A 1 in 10 solution gives positive tests for *Calcium,* page 926, and it decolorizes dichlorophenol-indophenol T.S.

SPECIFICATIONS

Assay. Not less than 98.0 percent of $C_{12}H_{14}CaO_{12}.2H_2O$.

Specific rotation, $[\alpha]_D^{25°}$. Between +95° and +97°.

Limits of Impurities

Arsenic (as As). Not more than 3 parts per million (0.0003 percent).

Fluoride. Not more than 10 parts per million (0.001 percent).

Heavy metals (as Pb). Not more than 10 parts per million (0.001 percent).

Oxalate. Passes test.

TESTS

Assay. Dissolve about 300 mg., accurately weighed, in 50 ml. of water in a 250-ml. Erlenmeyer flask, and immediately titrate with 0.1 *N* iodine to a pale yellow color which persists for at least 30 seconds. Each ml. of 0.1 *N* iodine is equivalent to 10.66 mg. of $C_{12}H_{14}CaO_{12} \cdot 2H_2O$.

Specific rotation, page 939. Determine in a solution containing 1 gram in each 20 ml.

Arsenic. A *Sample Solution* prepared as directed for organic compounds meets the requirements of the *Arsenic Test,* page 865.

Fluoride. Proceed as directed in the *Fluoride Limit Test,* page 917.

Heavy metals. A solution of 2 grams in 25 ml. of water meets the requirements of the *Heavy Metals Test,* page 920, using 20 mcg. of lead ion (Pb) in the control (*Solution A*).

Oxalate. To a solution of 1 gram in 10 ml. of water add 2 drops of glacial acetic acid and 5 ml. of a 1 in 10 solution of calcium acetate. The solution remains clear after standing for 5 minutes.

Packaging and storage. Store in tight containers, preferably in a cool, dry place.

Functional use in foods. Antioxidant.

CALCIUM BROMATE

$Ca(BrO_3)_2.H_2O$ Mol. wt. 313.90

DESCRIPTION

A white crystalline powder. It is very soluble in water.

IDENTIFICATION

A. A 1 in 20 solution in diluted hydrochloric acid T.S. imparts a transient yellowish red color to a nonluminous flame.

B. To a 1 in 20 solution add sulfurous acid dropwise. A yellow color is produced which disappears upon the addition of an excess of sulfurous acid.

SPECIFICATIONS

Assay. Not less than 99.8 percent of $Ca(BrO_3)_2.H_2O$.

Limits of Impurities

Arsenic (as As). Not more than 3 parts per million (0.0003 percent).

Heavy metals (as Pb). Not more than 40 parts per million (0.004 percent).

Lead. Not more than 10 parts per million (0.001 percent).

TESTS

Assay. Dissolve about 900 mg., accurately weighed, in 50 ml. of water in a 250-ml. glass-stoppered Erlenmeyer flask. Add 3 grams of potassium iodide, followed by 3 ml. of hydrochloric acid. Allow the mixture to stand for 5 minutes, add 100 ml. of cold water, and titrate the liberated iodine with 0.1 *N* sodium thiosulfate, adding starch T.S. near the end-point. Perform a blank determination (see page 2). Each ml. of 0.1 *N* sodium thiosulfate is equivalent to 26.16 mg. of $Ca(BrO_3)_2.H_2O$.

Arsenic. Dissolve 1 gram in a mixture of 5 ml. of hydrochloric acid and 5 ml. of water, and evaporate the solution until crystals appear. Cool, dissolve the residue in water, and dilute to 35 ml. This solution meets the requirements of the *Arsenic Test*, page 865.

Sample Solution for the Determination of Heavy Metals and Lead. Dissolve 2 grams in 10 ml. of water, add

10 ml. of hydrochloric acid, and evaporate to dryness on a steam bath. Dissolve the residue in 5 ml. of hydrochloric acid, again evaporate to dryness, and then dissolve the residue in 40 ml. of water.

Heavy metals. A 10-ml. portion of the *Sample Solution*, diluted to 25 ml. with water, meets the requirements of the *Heavy Metals Test*, page 920, using 20 mcg. of lead ion (Pb) in the control (*Solution A*).

Lead. A 20-ml. portion of the *Sample Solution* meets the requirements of the *Lead Limit Test*, page 929, using 10 mcg. of lead ion (Pb) in the control.

Packaging and storage. Store in well-closed containers.

Functional use in foods. Maturing agent; dough conditioner.

CALCIUM CARBONATE

Precipitated Calcium Carbonate

$CaCO_3$ Mol. wt. 100.09

DESCRIPTION

A fine, white microcrystalline powder. It is odorless and tasteless, and is stable in air. It is practically insoluble in water and in alcohol. The presence of any ammonium salt or carbon dioxide increases its solubility in water, but the presence of any alkali hydroxide reduces the solubility. It dissolves with effervescence in diluted acetic acid T.S., in diluted hydrochloric acid T.S., and in diluted nitric acid T.S., and the resulting solutions, after boiling, give positive tests for *Calcium*, page 926.

SPECIFICATIONS

Assay. Not less than 98.0 percent of $CaCO_3$ after drying.

Limits of Impurities

Acid-insoluble substances. Not more than 0.2 percent.

Arsenic (as As). Not more than 3 parts per million (0.0003 percent).

Fluoride. Not more than 40 parts per million (0.004 percent).

Heavy metals (as Pb). Not more than 30 parts per million (0.003 percent).

Lead. Not more than 10 parts per million (0.001 percent).

Loss on drying. Not more than 2 percent.

Magnesium and alkali salts. Not more than 1 percent.

TESTS

Assay. Transfer about 200 mg., previously dried at 200° for 4 hours and accurately weighed, into a 400-ml. beaker, add 10 ml. of water, and swirl to form a slurry. Cover the beaker with a watch glass, and introduce 2 ml. of diluted hydrochloric acid T.S. from a pipet inserted between the lip of the beaker and the edge of the watch glass. Swirl the contents of the beaker to dissolve the sample. Wash down the sides of the beaker, the outer surface of the pipet, and the watch glass, and dilute to about 100 ml. with water. While stirring, preferably with a magnetic stirrer, add about 30 ml. of 0.05 *M* disodium ethylenediaminetetraacetate from a 50-ml. buret, then add 15 ml. of sodium hydroxide T.S. and 300 mg. of hydroxy naphthol blue indicator, and continue the titration to a blue end-point. Each ml. of 0.05 *M* disodium ethylenediaminetetraacetate is equivalent to 5.004 mg. of $CaCO_3$.

Acid-insoluble substances. Suspend 5 grams in 25 ml. of water, cautiously add with agitation 25 ml. of dilute hydrochloric acid (1 in 2), then add water to make a volume of about 200 ml. Heat the solution to boiling, cover, digest on a steam bath 1 hour, cool, and filter. Wash the precipitate with water until the last washing shows no chloride with silver nitrate T.S., and then ignite it. The weight of the residue does not exceed 10 mg.

Arsenic. A solution of 1 gram in 10 ml. of diluted hydrochloric acid T.S. meets the requirements of the *Arsenic Test*, page 865.

Fluoride. Determine as directed in *Method III* under the *Fluoride Limit Test*, page 919.

Sample Solution for the Determination of Heavy Metals and Lead. Cautiously dissolve 5 grams in 25 ml. of dilute hydrochloric acid (1 in 2) and evaporate to dryness on a steam bath. Dissolve the residue in about 15 ml. of water and dilute to 25 ml. (1 ml. = 200 mg.).

Heavy metals. Neutralize 3.3 ml. (667 mg.) of the *Sample Solution* with sodium hydroxide T.S., using phenolphthalein as the indicator, and dilute to 25 ml. This solution meets the requirements of the *Heavy Metals Test*, page 920, using 20 mcg. of lead ion (Pb) in the control (*Solution A*).

Lead. A 5-ml. portion of the *Sample Solution* meets the requirements of the *Lead Limit Test*, page 929, using 10 mcg. of lead ion (Pb) in the control.

Loss on drying, page 931. Dry at 200° for 4 hours.

Magnesium and alkali salts. Mix 1 gram with 40 ml. of water, carefully add 5 ml. of hydrochloric acid, mix and boil for 1 minute. Rapidly add 40 ml. of oxalic acid T.S., and stir vigorously until precipitation is well established. Immediately add 2 drops of methyl red T.S., then add ammonia T.S., dropwise, until the mixture is just alkaline, and cool. Transfer the mixture to a 100-ml. cylinder, dilute with water to 100 ml., let it stand for 4 hours or overnight, then decant the clear, supernatant liquid through a dry filter paper. To 50 ml. of the clear filtrate in a platinum dish add 0.5 ml. of sulfuric acid, and evaporate the mixture on a steam bath to a small volume. Carefully evaporate the remaining liquid to dryness over a free flame, and continue heating until the ammonium salts have been completely decomposed and volatilized. Finally, ignite the residue to constant weight. The weight of the residue does not exceed 5 mg.

Packaging and storage. Store in well-closed containers.

Functional use in foods. Alkali; nutrient supplement; dough conditioner; firming agent; yeast food.

CALCIUM CHLORIDE

$CaCl_2 . 2H_2O$ Mol. wt. 147.02

DESCRIPTION

White, hard, odorless fragments or granules. It is deliquescent. One gram dissolves in 1.2 ml. of water at 25°, in 0.7 ml. of boiling water, in 10 ml. of alcohol at 25°, and in 2 ml. of boiling alcohol. The pH of a 1 in 20 solution is between 4.5 and 8.5. A 1 in 10 solution gives positive tests for *Calcium*, page 926, and for *Chloride*, page 926.

SPECIFICATIONS

Assay. Not less than 99.0 percent and not more than the equivalent of 107.0 percent of $CaCl_2 . 2H_2O$.

Limits of Impurities

Arsenic (as As). Not more than 3 parts per million (0.0003 percent).
Fluoride. Not more than 40 parts per million (0.004 percent).
Heavy metals (as Pb). Not more than 20 parts per million (0.002 percent).
Lead. Not more than 10 parts per million (0.001 percent).
Magnesium and alkali salts. Not more than 1 percent.

TESTS

Assay. Transfer about 1.5 grams, accurately weighed, into a 250-ml. volumetric flask, dissolve it in a mixture of 100 ml. of water and 5 ml. of diluted hydrochloric acid T.S., dilute to volume with water, and mix. Transfer 50.0 ml. of this solution into a suitable container, and add 50 ml. of water. While stirring, preferably with a magnetic stirrer, add about 30 ml. of 0.05 *M* disodium ethylenediaminetetraacetate from a 50-ml. buret, then add 15 ml. of sodium hydroxide T.S. and 300 mg. of hydroxy naphthol blue indicator, and continue the titration to a blue end-point. Each ml. of 0.05 *M* disodium ethylenediaminetetraacetate is equivalent to 7.351 mg. of $CaCl_2 . 2H_2O$.

Arsenic. A solution of 1 gram in 10 ml. of water meets the requirements of the *Arsenic Test*, page 865.

Fluoride. Determine as directed in *Method III* under the *Fluoride Limit Test*, page 919.

Heavy metals. Dissolve 1 gram in 2 ml. of diluted acetic acid T.S., and add water to make 25 ml. This solution meets the requirements of the *Heavy Metals Test*, page 920, using 20 mcg. of lead ion (Pb) in the control (*Solution A*).

Lead. A solution of 1 gram in 20 ml. of water meets the requirements of the *Lead Limit Test*, page 929, using 10 mcg. of lead ion (Pb) in the control.

Magnesium and alkali salts. Dissolve 1 gram in about 50 ml. of water, add 500 mg. of ammonium chloride, mix and boil for 1 minute. Rapidly add 40 ml. of oxalic acid T.S., and stir vigorously until precipitation is well established. Immediately add 2 drops of methyl red T.S., then add ammonia T.S., dropwise, until the mixture is just alkaline, and cool. Transfer the mixture to a 100-ml. cylinder, dilute with water to 100 ml., let it stand for 4 hours or overnight, then decant the clear, supernatant liquid through a dry filter paper. To 50 ml. of the clear filtrate in a platinum dish add 0.5 ml. of sulfuric acid, and evaporate the mixture on a steam bath to a small volume. Carefully evaporate the remaining liquid to dryness over a free flame, and continue heating until the ammonium salts have been completely decomposed and volatilized. Finally, ignite the residue to constant weight. The weight of the residue does not exceed 5 mg.

Packaging and storage. Store in tight containers.

Functional use in foods. Miscellaneous and general purpose; sequestrant; firming agent.

CALCIUM CHLORIDE, ANHYDROUS

$CaCl_2$ Mol. wt. 110.99

DESCRIPTION

White, hard, odorless fragments or granules. It is deliquescent. One gram dissolves in 1.5 ml. of water at 25°, in 0.7 ml. of boiling water, in 8 ml. of alcohol at 25°, and in 1.6 ml. of boiling alcohol. A 1 in 10 solution gives positive tests for *Calcium*, page 926, and for *Chloride*, page 926.

SPECIFICATIONS

Assay. Not less than 93.0 percent of $CaCl_2$.

Limits of Impurities

Arsenic (as As). Not more than 3 parts per million (0.0003 percent).

Fluoride. Not more than 40 parts per million (0.004 percent).

Heavy metals (as Pb). Not more than 20 parts per million (0.002 percent).

Lead. Not more than 10 parts per million (0.001 percent).

Magnesium and alkali salts. Not more than 5 percent.

TESTS

Assay. Transfer about 1 gram, accurately weighed, into a 250-ml. volumetric flask, dissolve it in a mixture of 100 ml. of water and 5 ml. of diluted hydrochloric acid T.S., dilute to volume with water, and mix. Transfer 50.0 ml. of this solution into a suitable container, and add 50 ml. of water. While stirring, preferably with a magnetic stirrer, add about 30 ml. of 0.05 *M* disodium ethylenediaminetetraacetate from a 50-ml. buret, then add 15 ml. of sodium hydroxide T.S. and 300 mg. of hydroxy naphthol blue indicator, and continue the titration to a blue end-point. Each ml. of 0.05 *M* disodium ethylenediaminetetraacetate is equivalent to 5.550 mg. of $CaCl_2$.

Arsenic. A solution of 1 gram in 10 ml. of water meets the requirements of the *Arsenic Test*, page 865.

Fluoride. Determine as directed in *Method III* under the *Fluoride Limit Test*, page 919.

Heavy metals. Dissolve 1 gram in 2 ml. of diluted acetic acid T.S., and add water to make 25 ml. This solution meets the requirements of the *Heavy Metals Test*, page 920, using 20 mcg. of lead ion (Pb) in the control (*Solution A*).

Lead. A solution of 1 gram in 20 ml. of water meets the requirements of the *Lead Limit Test*, page 929, using 10 mcg. of lead ion (Pb) in the control.

Magnesium and alkali salts. Dissolve 1 gram in about 50 ml. of water, add 500 mg. of ammonium chloride, mix, and boil for about 1 minute. Rapidly add 40 ml. of oxalic acid T.S., and stir vigorously until precipitation is well established. Immediately add 2 drops of methyl red T.S., then add ammonia T.S., dropwise, until the mixture is just alkaline, and cool. Transfer the mixture into a 100-ml. cylinder, dilute with water to 100 ml., let it stand for 4 hours or overnight, and then decant the clear, supernatant liquid through a dry filter paper. To 50 ml. of the clear filtrate in a platinum dish add 0.5 ml. of sulfuric acid, and evaporate the mixture on a steam bath to a small volume. Carefully evaporate the remaining liquid to dryness over a free flame, and continue heating until the ammonium salts have been completely decomposed and volatilized. Finally, ignite the residue to constant weight. The weight of the residue does not exceed 25 mg.

Packaging and storage. Store in tight containers.

Functional use in foods. Miscellaneous and general purpose; sequestrant; firming agent.

CALCIUM CITRATE

$Ca_3(C_6H_5O_7)_2.4H_2O$ Mol. wt. 570.51

DESCRIPTION

A fine, white, odorless powder. It is very slightly soluble in water and insoluble in alcohol.

IDENTIFICATION

A. Dissolve 500 mg. in 10 ml. of water and 2.5 ml. of diluted nitric acid T.S., add 1 ml. of mercuric sulfate T.S., heat to boiling and then add potassium permanganate T.S. A white precipitate is formed.

B. Ignite 500 mg. completely at as low a temperature as possible, cool, and dissolve the residue in 10 ml. of water and 1 ml. of glacial acetic acid. Filter and add 10 ml. of ammonium oxalate T.S. to the filtrate. A voluminous white precipitate appears which is soluble in hydrochloric acid.

SPECIFICATIONS

Assay. Not less than 97.5 percent of $Ca_3(C_6H_5O_7)_2$ after drying.

Loss on drying. Between 10 and 13.3 percent.

Limits of Impurities

Arsenic (as As). Not more than 3 parts per million (0.0003 percent).

Fluoride. Not more than 30 parts per million (0.003 percent).

Heavy metals (as Pb). Not more than 20 parts per million (0.002 percent).

Lead. Not more than 10 parts per million (0.001 percent).

TESTS

Assay. Dissolve about 350 mg., previously dried at 150° to constant weight and accurately weighed, in a mixture of 10 ml. of water and 2 ml. of diluted hydrochloric acid T.S., and dilute to about 100 ml. with water. While stirring, preferably with a magnetic stirrer, add about 30 ml. of 0.05 *M* disodium ethylenediaminetetraacetate from a 50-ml. buret, then add 15 ml. of sodium hydroxide T.S. and 300 mg. of hydroxy naphthol blue indicator, and continue the titration to a blue end-point. Each ml. of 0.05 *M* disodium ethylenediaminetetraacetate is equivalent to 8.303 mg. of $Ca_3(C_6H_5O_7)_2$.

Loss on drying, page 931. Dry to constant weight at 150°.

Arsenic. A solution of 1 gram in 5 ml. of diluted hydrochloric acid T.S. meets the requirements of the *Arsenic Test*, page 865.

Fluoride. Weigh accurately 1.67 grams, and proceed as directed in the *Fluoride Limit Test*, page 917.

Heavy metals. Dissolve 1 gram in 20 ml. of water and 2 ml. of hydrochloric acid, add 1.5 ml. of stronger ammonia T.S., and dilute to 25 ml. with water. This solution meets the requirements of the *Heavy Metals Test*, page 920, using 20 mcg. of lead ion (Pb) in the control (*Solution A*).

Lead. Dissolve 1 gram in 10 ml. of water and 1 ml. of hydrochloric acid. This solution meets the requirements of the *Lead Limit Test*, page 929, using 10 mcg. of lead ion (Pb) in the control.

Packaging and storage. Store in well-closed containers.

Functional use in foods. Sequestrant; buffer; firming agent

CALCIUM DISODIUM EDTA

Calcium Disodium Ethylenediaminetetraacetate; Calcium Disodium (Ethylenedinitrilo)tetraacetate; Calcium Disodium Edetate

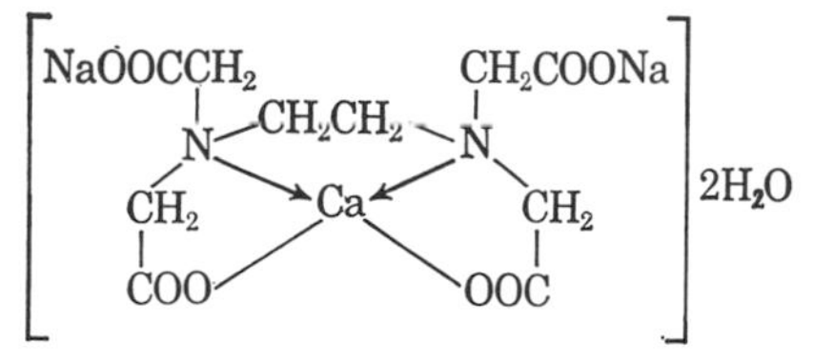

$C_{10}H_{12}CaN_2Na_2O_8 . 2H_2O$ Mol. wt. 410.31

DESCRIPTION

White, odorless crystalline granules or a white to off-white powder. It is slightly hygroscopic, has a faint saline taste, and is stable in air. It is freely soluble in water.

IDENTIFICATION

A. A 1 in 20 solution responds to the oxalate test for *Calcium*, page 926, and to the flame test for *Sodium*, page 928.

B. The infrared absorption spectrum of a liquid petrolatum dispersion of the sample exhibits maxima only at the same wavelengths as that of a similar preparation of U.S.P. Calcium Disodium Edetate Reference Standard.

C. To 5 ml. of water in a test tube add 2 drops of ammonium thiocyanate T.S. and 2 drops of ferric chloride T.S. To the deep red solution so obtained add about 50 mg. of the sample, and mix. The deep red color disappears.

SPECIFICATIONS

Assay. Not less than 97.0 percent and not more than the equivalent of 102.0 percent of $C_{10}H_{12}CaN_2Na_2O_8$, calculated on the anhydrous basis.

pH of a 1 in 100 solution. Between 6.5 and 7.5.

Water. Not more than 13 percent.

Limits of Impurities

Arsenic (as As). Not more than 3 parts per million (0.0003 percent).

Heavy metals (as Pb). Not more than 20 parts per million (0.002 percent).

Lead. Not more than 10 parts per million (0.001 percent).

Magnesium-chelating substances. Passes test.

TESTS

Assay. Transfer from 1.2 to 1.5 grams, accurately weighed, into a 250-ml. beaker, dissolve in about 50 ml. of water, and add 0.5 ml. of xylenol orange T.S. and 10 ml. of buffer solution (350 ml. of glacial acetic acid and 17 grams of anhydrous sodium acetate diluted to 1000 ml. with water). Titrate with 0.1 *M* thorium nitrate to a sharp color change from yellow to red. Each ml. of 0.1 *M* thorium nitrate is equivalent to 37.43 mg. of $C_{10}H_{12}CaN_2Na_2O_8$.

pH of a 1 in 100 solution. Determine by the *Potentiometric Method*, page 941.

Water. Determine by the *Karl Fischer Titrimetric Method*, page 977.

Arsenic. Prepare a *Sample Solution* as directed for organic compounds on page 867, but use 70 percent perchloric acid instead of 30 percent hydrogen peroxide in the decomposition of the sample. The resulting solution meets the requirements of the *Arsenic Test*, page 865.

Heavy metals. Prepare and test a 1-gram sample as directed in *Method II* under the *Heavy Metals Test*, page 920, using 20 mcg. of lead ion (Pb) in the control (*Solution A*).

Lead. Prepare a *Sample Solution* as directed for organic compounds on page 930, but use 70 percent perchloric acid instead of 30 percent hydrogen peroxide in the decomposition of the sample. The resulting solution meets the requirements of the *Lead Limit Test*, page 929.

Magnesium-chelating substances. Transfer a 1-gram sample, accurately weighed, to a small beaker, and dissolve it in 5 ml. of water. Add 5 ml. of a buffer solution prepared by dissolving 67.5 grams of ammonium chloride in 200 ml. of water, adding 570 ml. of stronger ammonia T.S., and diluting with water to 1000 ml. Then to the buffered solution add 5 drops of eriochrome black T.S., and titrate with 0.1 *M* magnesium acetate to the appearance of a deep wine-red color. Not more than 2.0 ml. is required.

Packaging and storage. Store in well-closed containers.

Functional use in foods. Preservative; sequestrant.

CALCIUM GLUCONATE

$[CH_2OH(CHOH)_4.COO]_2Ca$

$C_{12}H_{22}CaO_{14}$ Mol. wt. 430.38

DESCRIPTION

White, crystalline granules or powder. It is odorless, tasteless, and stable in air. Its solutions are neutral to litmus. One gram dissolves slowly in about 30 ml. of water at 25° and in about 5 ml. of boiling water. It is insoluble in alcohol and in many other organic solvents.

IDENTIFICATION

A. A 1 in 50 solution gives positive tests for *Calcium*, page 926.

B. Place 500 mg. in a test tube and dissolve it in 5 ml. of water by warming. To the warm solution add about 0.7 ml. of glacial acetic acid and 1 ml. of freshly distilled phenylhydrazine, heat on a steam bath for 30 minutes, and allow to cool. Induce crystallization by scratching the inner surface of the tube with a glass rod. Crystals of gluconic acid phenylhydrazide form.

SPECIFICATIONS

Assay. Not less than 98.0 percent and not more than the equivalent of 102.0 percent of $C_{12}H_{22}CaO_{14}$ after drying.

Limits of Impurities

Arsenic (as As). Not more than 3 parts per million (0.0003 percent).

Heavy metals (as Pb). Not more than 20 parts per million (0.002 percent).

Lead. Not more than 10 parts per million (0.001 percent).

Loss on drying. Not more than 3 percent.

Sucrose and reducing sugars. Passes test.

TESTS

Assay. Dissolve about 800 mg., previously dried at 105° for 16 hours and accurately weighed, in 100 ml. of water containing 2 ml. of diluted hydrochloric acid T.S. While stirring, preferably with a magnetic stirrer, add about 30 ml. of 0.05 *M* disodium ethylenediaminetetraacetate from a 50-ml. buret, then add 15 ml. of sodium hydroxide T.S. and 300 mg. of hydroxy naphthol blue indicator, and continue the titration to a blue end-point. Each ml. of 0.05 *M* disodium ethylenediaminetetraacetate is equivalent to 21.52 mg. of $C_{12}H_{22}CaO_{14}$.

Arsenic. A *Sample Solution* prepared as directed for organic compounds meets the requirements of the *Arsenic Test*, page 865, substituting nitric acid for hydrogen peroxide in the wet digestion of the sample.

Heavy metals. Mix a 1-gram sample with 4 ml. of 1 *N* hydrochloric acid, dilute to 25 ml. with water, warm gently until dissolved, and cool. This solution meets the requirements of the *Heavy Metals Test,* page 920, using 20 mcg. of lead ion (Pb) in the control (*Solution A*).

Lead. A *Sample Solution* prepared as directed for organic compounds meets the requirements of the *Lead Limit Test,* page 929, using 10 mcg. of lead ion (Pb) in the control.

Loss on drying, page 931. Dry at 105° for 16 hours.

Sucrose and reducing sugars. Dissolve 500 mg. in 10 ml. of hot water, add 2 ml. of diluted hydrochloric acid T.S., boil for about 2 minutes, and cool. Add 5 ml. of sodium carbonate T.S., allow to stand for 5 minutes, dilute with water to 20 ml., and filter. Add 5 ml. of the clear filtrate to about 2 ml. of alkaline cupric tartrate T.S., and boil for 1 minute. No red precipitate is formed immediately.

Packaging and storage. Store in well-closed containers.

Functional use in foods. Miscellaneous and general purpose; buffer; firming agent; sequestrant.

CALCIUM GLYCEROPHOSPHATE

$C_3H_7CaO_6P$ Mol. wt. 210.14

DESCRIPTION

A fine, white, odorless, almost tasteless powder. It is somewhat hygroscopic. One gram dissolves in about 50 ml. of water at 25°. It is more soluble in water at a lower temperature, and citric acid increases its solubility in water. It is insoluble in alcohol.

IDENTIFICATION

A. A saturated solution gives positive tests for *Calcium,* page 926.

B. Heat a mixture of 100 mg. of the sample with 500 mg. of potassium bisulfate. Pungent vapors of acrolein are evolved.

SPECIFICATIONS

Assay. Not less than 98.0 percent of $C_3H_7CaO_6P$ after drying.

Limits of Impurities

Alkalinity. Passes test.

Arsenic (as As). Not more than 3 parts per million (0.0003 percent).

Heavy metals (as Pb). Not more than 40 parts per million (0.004 percent).

Lead. Not more than 10 parts per million (0.001 percent).

Loss on drying. Not more than 12 percent.

TESTS

Assay. Weigh accurately about 2 grams, previously dried at 150° for 4 hours, and dissolve in 100 ml. of water and 5 ml. of diluted hydrochloric acid T.S. Transfer the solution to a 250-ml. volumetric flask, dilute to volume with water, and mix well. Pipet 50.0 ml. of this solution into a suitable container, and add 50 ml. of water. While stirring, preferably with a magnetic stirrer, add about 30 ml. of 0.05 *M* disodium ethylenediaminetetraacetate from a 50-ml. buret, then add 15 ml. of sodium hydroxide T.S. and 300 mg. of hydroxy naphthol blue indicator, and continue the titration to a blue end-point. Each ml. of 0.05 *M* disodium ethylenediaminetetraacetate is equivalent to 10.51 mg. of $C_3H_7CaO_6P$.

Alkalinity. A solution of 1 gram in 60 ml. of water requires not more than 1.5 ml. of 0.1 *N* sulfuric acid for neutralization, using 3 drops of phenolphthalein T.S. as indicator.

Arsenic. A *Sample Solution* prepared as directed for organic compounds meets the requirements of the *Arsenic Test*, page 865.

Heavy metals. Dissolve 500 mg. in 3 ml. of diluted acetic acid T.S., and dilute to 25 ml. with water. This solution meets the requirements of the *Heavy Metals Test*, page 920, using 20 mcg. of lead ion (Pb) in the control (*Solution A*).

Lead. A *Sample Solution* prepared as directed for organic compounds meets the requirements of the *Lead Limit Test*, page 929, using 10 mcg. of lead ion (Pb) in the control.

Loss on drying, page 931. Dry at 150° for 4 hours.

Packaging and storage. Store in tight containers.

Functional use in foods. Nutrient; dietary supplement.

CALCIUM HYDROXIDE

Slaked Lime

$Ca(OH)_2$ Mol. wt. 74.09

DESCRIPTION

A white powder, possessing an alkaline, slightly bitter taste. One gram dissolves in 630 ml. of water at 25°, and in 1300 ml. of boiling water. It is soluble in glycerin and in a saturated solution of sucrose, but is insoluble in alcohol.

IDENTIFICATION

A. When mixed with from 3 to 4 times its weight of water, it forms a smooth magma. The clear, supernatant liquid from the magma is alkaline to litmus.

B. Mix 1 gram with 20 ml. of water, and add sufficient acetic acid to effect solution. The resulting solution gives positive tests for *Calcium*, page 926.

SPECIFICATIONS

Assay. Not less than 95.0 percent of $Ca(OH)_2$.

Limits of Impurities

Acid-insoluble substances. Not more than 0.5 percent.
Arsenic (as As). Not more than 3 parts per million (0.0003 percent).
Carbonate. Passes test.
Fluoride. Not more than 50 parts per million (0.005 percent).
Heavy metals (as Pb). Not more than 40 parts per million (0.004 percent).
Lead. Not more than 10 parts per million (0.001 percent).
Magnesium and alkali salts. Not more than 4.8 percent.

TESTS

Assay. Weigh accurately about 1.5 grams, transfer to a beaker, and gradually add 30 ml. of diluted hydrochloric acid T.S. When solution is complete, transfer it to a 500-ml. volumetric flask, rinse the beaker thoroughly, adding the rinsings to the flask, dilute with water to volume, and mix. Transfer 50.0 ml. of this solution into a suitable container, and add 50 ml. of water. While stirring, preferably with a magnetic stirrer, add about 30 ml. of 0.05 *M* disodium ethylenediaminetetraacetate from a 50-ml. buret, then add 15 ml. of sodium hydroxide T.S. and 300 mg. of hydroxy naphthol blue indicator, and continue the titration to a blue end-point. Each ml. of 0.05 *M* disodium ethylenediaminetetraacetate is equivalent to 3.705 mg. of $Ca(OH)_2$.

Acid-insoluble substances. Dissolve 2 grams in 30 ml. of dilute hydrochloric acid (1 in 3), and heat to boiling. Filter the mixture, wash the residue with hot water, and ignite. The weight of the residue does not exceed 10 mg.

Arsenic. A solution of 1 gram in 15 ml. of diluted hydrochloric acid T.S. meets the requirements of the *Arsenic Test*, page 865.

Carbonate. Mix 2 grams of the sample with 50 ml. of water, and add an excess of diluted hydrochloric acid T.S. No more than a slight effervescence is produced.

Fluoride. Weigh accurately 1.0 gram, and proceed as directed in the *Fluoride Limit Test*, page 917.

Heavy metals. Dissolve 500 mg. in 10 ml. of diluted hydrochloric acid T.S., and evaporate to dryness on a steam bath. Dissolve the residue in 25 ml. of water, and filter. The filtrate meets the requirements of the *Heavy Metals Test*, page 920, using 20 mcg. of lead ion (Pb) in the control (*Solution A*).

Lead. A solution of 1 gram in 15 ml. of diluted hydrochloric acid T.S. meets the requirements of the *Lead Limit Test*, page 929, using 10 mcg. of lead ion (Pb) in the control.

Magnesium and alkali salts. Dissolve 500 mg. in a mixture of 30 ml. of water and 10 ml. of diluted hydrochloric acid T.S., and boil for 1 minute. Rapidly add 40 ml. of oxalic acid T.S., and stir vigorously until precipitation is well established. Immediately add 2 drops of methyl red T.S., then add ammonia T.S., dropwise, until the mixture is just alkaline, and cool. Transfer the mixture to a 100-ml. cylinder, dilute with water to 100 ml., let it stand for 4 hours or overnight, then decant the clear, supernatant liquid through a dry filter paper. To 50 ml. of the clear filtrate in a platinum dish add 0.5 ml. of sulfuric acid, and evaporate the mixture on a steam bath to a small volume. Carefully evaporate the remaining liquid to dryness over a free flame, and continue heating until the ammonium salts have been completely decomposed and volatilized. Finally, ignite the residue to constant weight. The weight of the residue does not exceed 12 mg.

Packaging and storage. Store in tight containers.

Functional use in foods. Miscellaneous and general purpose; buffer; neutralizing agent; firming agent.

CALCIUM IODATE

$Ca(IO_3)_2.H_2O$ Mol. wt. 407.90

DESCRIPTION

A white powder. It is odorless or has a slight odor. It is slightly soluble in water, and is insoluble in alcohol.

IDENTIFICATION

To 5 ml. of a saturated solution of the sample add 1 drop of starch T.S. and a few drops of 20 percent hypophosphorous acid. A transient blue color appears.

SPECIFICATIONS

Assay. Not less than 99.0 percent and not more than the equivalent of 101.0 percent of $Ca(IO_3)_2.H_2O$.

Limits of Impurities

Arsenic (as As). Not more than 3 parts per million (0.0003 percent).

Heavy metals (as Pb). Not more than 10 parts per million (0.001 percent).

TESTS

Assay. Weigh accurately about 600 mg., dissolve it in 10 ml. of 70 percent perchloric acid and 10 ml. of water, heating gently if necessary, and dilute with water to 250.0 ml. Transfer 50.0 ml. to a 250-ml.

glass-stoppered Erlenmeyer flask, add 1 ml. of 70 percent perchloric acid and 5 grams of potassium iodide, stopper the flask, and swirl briefly. Let stand for 5 minutes, then titrate with 0.1 *N* sodium thiosulfate, adding starch T.S. just before the end-point is reached. Each ml. of 0.1 *N* sodium thiosulfate is equivalent to 33.98 mg. of $Ca(IO_3)_2.H_2O$.

Arsenic. Mix 3 ml. of hydrochloric acid with a 1-gram sample, evaporate to dryness on an asbestos board on a hot plate, and cool. Add 5 ml. of hydrochloric acid and again evaporate to dryness. Dissolve the residue in 15 ml. of water, heat nearly to boiling, and add just enough hydrazine sulfate to discharge any yellow color. Cool, and dilute to 35 ml. with water. This solution meets the requirements of the *Arsenic Test*, page 865.

Heavy metals. Mix 5 ml. of hydrochloric acid with a 2-gram sample, evaporate to dryness on an asbestos board on a hot plate, and cool. Add 5 ml. of hydrochloric acid and again evaporate to dryness. Dissolve the residue in 15 ml. of water, heat nearly to boiling, and add just enough hydrazine sulfate to discharge any yellow color. Cool, and dilute to 25 ml. with water. This solution meets the requirements of the *Heavy Metals Test*, page 920, using 20 mcg. of lead ion (Pb) in the control (*Solution A*).

Packaging and storage. Store in well-closed containers.

Functional use in foods. Maturing agent; dough conditioner.

CALCIUM LACTATE

$[CH_3CH(OH)COO]_2Ca.xH_2O$

$C_6H_{10}CaO_6.xH_2O$ Mol. wt. (anhydrous) 218.22

DESCRIPTION

White to cream colored, almost odorless, crystalline powder or granules, containing up to 5 molecules of water of crystallization. The pentahydrate is somewhat efflorescent and at 120° becomes anhydrous. It is soluble in water and practically insoluble in alcohol. A 1 in 20 solution gives positive tests for *Calcium*, page 926, and for *Lactate*, page 927.

SPECIFICATIONS

Assay. Not less than 98.0 percent and not more than 101.0 percent of $C_6H_{10}CaO_6$, after drying.

Loss on drying. *Pentahydrate:* between 24 percent and 30 percent; *trihydrate:* between 15 percent and 20 percent; *monohydrate:* between 5 percent and 8 percent; *dried form:* not more than 3 percent.

Limits of Impurities

Acidity. Passes test (about 0.45 percent, as lactic acid).

Arsenic (as As). Not more than 3 parts per million (0.0003 percent).

Fluoride. Not more than 15 parts per million (0.0015 percent).

Heavy metals (as Pb). Not more than 20 parts per million (0.002 percent).

Lead. Not more than 10 parts per million (0.001 percent).

Magnesium and alkali salts. Not more than 1 percent.

Volatile fatty acids. Passes test.

TESTS

Assay. Dissolve an accurately weighed amount of the sample, equivalent to about 350 mg. of $C_6H_{10}CaO_6$, in 150 ml. of water containing 2 ml. of diluted hydrochloric acid T.S. While stirring, preferably with a magnetic stirrer, add about 30 ml. of 0.05 *M* disodium ethylenediaminetetraacetate from a 50-ml. buret, then add 15 ml. of sodium hydroxide T.S. and 300 mg. of hydroxy naphthol blue indicator, and continue the titration to a blue end-point. Each ml. of 0.05 *M* disodium ethylenediaminetetraacetate is equivalent to 10.91 mg. of $C_6H_{10}CaO_6$.

Loss on drying, page 931. Distribute a sample of about 1.5 grams evenly in a suitable weighing dish to a depth not exceeding 3 mm., and dry at 120° for 4 hours.

Acidity. Dissolve 1 gram in 20 ml. of water, add 3 drops of phenolphthalein T.S., and titrate with 0.1 *N* sodium hydroxide. Not more than 0.5 ml. is required.

Arsenic. A *Sample Solution* prepared as directed for organic compounds meets the requirements of the *Arsenic Test*, page 865.

Fluoride. Proceed as directed in the *Fluoride Limit Test*, page 917, using *Method I* (3.3-gram sample) or *Method III* (1.0-gram sample).

Heavy metals. A solution of 1 gram in 25 ml. of water meets the requirements of the *Heavy Metals Test*, page 920, using 20 mcg. of lead ion (Pb) in the control (*Solution A*).

Lead. Dissolve 1 gram in 3 ml. of dilute nitric acid (1 in 2), boil for 1 minute, cool, and dilute to 20 ml. with water. This solution meets the requirements of the *Lead Limit Test*, page 929, using 10 mcg. of lead ion (Pb) in the control.

Magnesium and alkali salts. Mix 1 gram with 40 ml. of water, carefully add 1 ml. of hydrochloric acid, boil for 1 minute, and add rapidly 40 ml. of oxalic acid T.S. Add immediately to the warm mixture 2 drops of methyl red T.S., then add ammonia T.S., dropwise, from a buret until the mixture is just alkaline, and cool to room temperature. Transfer the mixture into a 100-ml. graduate, dilute with water to 100 ml., mix, and allow to stand for 4 hours or overnight. Decant the clear, supernatant liquid through a dry filter paper, transfer 50 ml. of the clear filtrate to a tared platinum dish, and add 0.5 ml. of sulfuric acid. Evaporate to a small volume on a steam bath, then carefully heat over a free flame to dryness, and continue heating to complete decomposition and volatilization of the ammonium salts. Finally, ignite the residue to constant weight. The weight of the residue does not exceed 5 mg.

Volatile fatty acids. Stir about 500 mg. of the sample with 1 ml. of sulfuric acid, and warm. The mixture does not emit an odor of volatile fatty acids.

Packaging and storage. Store in tight containers.

Functional use in foods. Buffer; dough conditioner; yeast food.

CALCIUM LACTOBIONATE

Calcium 4-(β,D-Galactosido)-D-gluconate

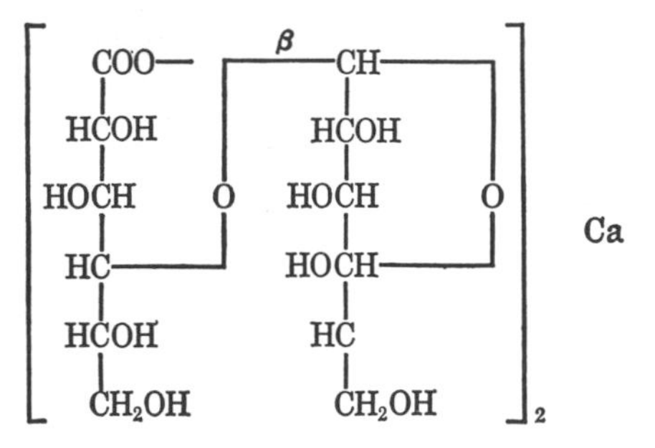

$C_{24}H_{42}CaO_{24}$ Mol. wt. 754.67

DESCRIPTION

A white, odorless, free-flowing powder. It is freely soluble in water but is insoluble in alcohol and in ether. It has a bland taste, and it readily forms double salts, such as the chloride, bromide, and gluconate. It decomposes at about 120°. The pH of a 1 in 10 solution is between 6.5 and 7.5. It gives positive tests for *Calcium*, page 926.

SPECIFICATIONS

Calcium content. Not less than 4.5 percent and not more than 5.0 percent of Ca.

Specific rotation, $[\alpha]_D^{20°}$. Between +23° and +25°.

Limits of Impurities

Arsenic (as As). Not more than 3 parts per million (0.0003 percent).

Bromide. Passes test.

Heavy metals (as Pb). Not more than 20 parts per million (0.002 percent).
Lead. Not more than 10 parts per million (0.001 percent).
Loss on drying. Not more than 8 percent.
Reducing substances. Not more than 5 percent.
Sulfate. Not more than 0.7 percent.

TESTS

Calcium content. Weigh accurately about 1.5 grams, and dissolve it in 100 ml. of water containing 2 ml. of diluted hydrochloric acid T.S. While stirring, preferably with a magnetic stirrer, add about 30 ml. of 0.05 *M* disodium ethylenediaminetetraacetate from a 50-ml. buret, then add 15 ml. of sodium hydroxide T.S. and 300 mg. of hydroxy naphthol blue indicator, and continue the titration to a blue end-point. Each ml. of 0.05 *M* disodium ethylenediaminetetraacetate is equivalent to 2.004 mg. of Ca.

Specific rotation, page 939. Determine in a solution containing 500 mg., calculated on the anhydrous basis, in each 10 ml.

Arsenic. A *Sample Solution* prepared as directed for organic compounds meets the requirements of the *Arsenic Test,* page 865.

Bromide. To 20 ml. of a 1 in 10 solution of the sample add 1 ml. of silver nitrate T.S. No yellowish white precipitate is formed.

Heavy metals. Prepare and test a 1-gram sample as directed in *Method II* under the *Heavy Metals Test,* page 920, using 20 mcg. of lead ion (Pb) in the control (*Solution A*).

Lead. A *Sample Solution* prepared as directed for organic compounds meets the requirements of the *Lead Limit Test,* page 929, using 10 mcg. of lead ion (Pb) in the control.

Loss on drying, page 931. Dry at 105° for 8 hours.

Reducing substances. Dissolve 1.0 gram in 35 ml. of water in a 400-ml. beaker, and mix. Add 50 ml. of alkaline cupric tartrate T.S., cover the beaker with a watch glass, heat the mixture at such a rate that it comes to a boil in about 4 minutes, and boil for exactly 2 minutes. Collect the precipitated cuprous oxide in a tared Gooch crucible previously washed with hot water, alcohol, and ether, and dried at 100° for 30 minutes. Thoroughly wash the collected cuprous oxide on the filter with hot water, then with 10 ml. of alcohol, and finally with 10 ml. of ether, and dry at 100° for 30 minutes. The weight of the cuprous oxide does not exceed 76 mg.

Sulfate. Transfer about 25 grams, accurately weighed, into a 600-ml. beaker, dissolve it in 200 ml. of water, adjust the solution to a pH between 4.5 and 6.5 with diluted hydrochloric acid T.S., and filter, if necessary. Heat the filtrate or clear solution to just below the boiling point, then add 10 ml. of barium chloride T.S., stirring vigorously, boil gently for 5 minutes, and allow to stand for at least 2 hours, or preferably overnight. Collect the precipitate of barium sulfate on a tared Gooch crucible, wash until free from chloride, dry, and ignite at 600° to constant weight. The weight of barium sulfate so obtained, multiplied by 0.412, represents the weight of SO_4 in the sample taken.

Packaging and storage. Store in well-closed containers.

Functional use in foods. Firming agent in dry pudding mixes.

CALCIUM OXIDE

Lime

CaO Mol. wt. 56.08

DESCRIPTION

Hard, white or grayish white masses or granules, or a white to grayish white powder. It is odorless. One gram dissolves in about 840 ml. of water at 25°, and in about 1740 ml. of boiling water. It is soluble in glycerin, but is insoluble in alcohol.

IDENTIFICATION

Slake 1 gram with 20 ml. of water, and add acetic acid until the sample is dissolved. The resulting solution gives positive tests for *Calcium*, page 926.

SPECIFICATIONS

Assay. Not less than 95.0 percent of CaO after ignition.

Limits of Impurities

Acid-insoluble substances. Not more than 1 percent.

Alkalies or magnesium. Not more than 3.6 percent.

Arsenic (as As). Not more than 3 parts per million (0.0003 percent).

Fluoride. Not more than 50 parts per million (0.005 percent).

Heavy metals (as Pb). Not more than 40 parts per million (0.004 percent).

Lead. Not more than 10 parts per million (0.001 percent).

Loss on ignition. Not more than 10 percent.

TESTS

Assay. Ignite about 1 gram to constant weight, and dissolve the ignited sample, accurately weighed, in 20 ml. of diluted hydrochloric acid T.S. Cool the solution, dilute with water to 500.0 ml., and mix. Pipet 50.0 ml. of this solution into a suitable container, and add 50 ml. of water. While stirring, preferably with a magnetic stirrer, add about 30 ml. of 0.05 *M* disodium ethylenediaminetetraacetate from a 50-ml. buret, then add 15 ml. of sodium hydroxide T.S. and 300 mg.

of hydroxy naphthol blue indicator, and continue the titration to a blue end-point. Each ml. of 0.05 *M* disodium ethylenediaminetetra-acetate is equivalent to 2.804 mg. of CaO.

Acid-insoluble substances. Slake a 5-gram sample, then mix it with 100 ml. of water and sufficient hydrochloric acid, added dropwise, to effect solution. Boil the solution, cool, add hydrochloric acid, if necessary, to make the solution distinctly acid, and filter through a tared crucible. Wash the residue with water until free of chlorides, dry at 105° for 1 hour, cool, and weigh.

Alkalies or magnesium. Dissolve 500 mg. in 30 ml. of water and 15 ml. of diluted hydrochloric acid T.S. Heat the solution and boil for 1 minute. Add rapidly 40 ml. of oxalic acid T.S., and stir vigorously. Add 2 drops of methyl red T.S., and neutralize the solution with ammonia T.S. to precipitate the calcium completely. Heat the mixture on a steam bath for 1 hour, cool, dilute to 100 ml. with water, mix well, and filter. To 50 ml. of the filtrate add 0.5 ml. of sulfuric acid, then evaporate to dryness and ignite to constant weight in a tared platinum crucible.

Arsenic. A solution of 1 gram in 15 ml. of diluted hydrochloric acid T.S. meets the requirements of the *Arsenic Test*, page 865.

Fluoride. Weigh accurately 1.0 gram, and proceed as directed in the *Fluoride Limit Test*, page 917.

Heavy metals. Mix 2 grams with 25 ml. of water, cautiously add 7 ml. of hydrochloric acid, followed by 3 ml. of nitric acid, and evaporate to dryness on a steam bath Dissolve the residue in 1 ml. of diluted hydrochloric acid T.S. and 25 ml. of hot water, filter, wash with a few ml. of water, and dilute the filtrate to 100 ml. with water. A 25-ml. portion of this solution meets the requirements of the *Heavy Metals Test*, page 920, using 20 mcg. of lead ion (Pb) in the control (*Solution A*).

Lead. A solution of 1 gram in 15 ml. of diluted hydrochloric acid T.S. meets the requirements of the *Lead Limit Test*, page 929, using 10 mcg. of lead ion (Pb) in the control.

Loss on ignition. Ignite 1 gram to constant weight in a tared platinum crucible with a blast lamp.

Packaging and storage. Store in tight containers.

Functional use in foods. Alkali; nutrient; dietary supplement; dough conditioner; yeast food.

CALCIUM PANTOTHENATE

Dextro Calcium Pantothenate

$[HOCH_2C(CH_3)_2CH(OH)CONH(CH_2)_2COO]_2Ca$

$C_{18}H_{32}CaN_2O_{10}$ Mol. wt. 476.54

DESCRIPTION

The calcium salt of the dextrorotatory isomer of pantothenic acid occurs as a slightly hygroscopic, white powder. It is odorless and has a bitter taste. It is stable in air. One gram dissolves in about 3 ml. of water. It is soluble in glycerin, but is practically insoluble in alcohol, in chloroform, and in ether.

IDENTIFICATION

A. A 1 in 20 solution gives positive tests for *Calcium*, page 926.

B. The infrared absorption spectrum of a potassium bromide dispersion of the sample, previously dried at 105° for 3 hours, exhibits maxima only at the same wavelengths as that of a similar preparation of U.S.P. Calcium Pantothenate Reference Standard.

C. Boil 50 mg. in 5 ml. of 1 *N* sodium hydroxide for 1 minute, cool, and add 5 ml. of 1 *N* hydrochloric acid and 2 drops of ferric chloride T.S. A strong yellow color is produced.

SPECIFICATIONS

Assay. Not less than 90.0 percent and not more than the equivalent of 110.0 percent of dextrorotatory calcium pantothenate ($C_{18}H_{32}CaN_2O_{10}$), calculated on the dried basis.

Calcium content. Not less than 8.2 percent and not more than 8.6 percent of Ca after drying.

Specific rotation, $[\alpha]_D^{25°}$. Between +25° and +27.5°.

Limits of Impurities

Alkalinity. Passes test.

Alkaloids. Passes test.

Heavy metals (as Pb). Not more than 20 parts per million (0.002 percent).

Loss on drying. Not more than 5 percent.

TESTS

Assay. Proceed as directed under *Calcium Pantothenate Assay*, page 869.

Calcium content. Weigh accurately about 950 mg., previously dried at 105° for 3 hours, and dissolve it in 100 ml. of water containing 2 ml. of diluted hydrochloric acid T.S. While stirring, preferably with a magnetic stirrer, add about 30 ml. of 0.05 *M* disodium ethylenediaminetetraacetate from a 50-ml. buret, then add 15 ml. of sodium hydroxide T.S. and 300 mg. of hydroxy naphthol blue indicator, and continue the titration to a blue end-point. Each ml. of 0.05 *M* disodium ethylenediaminetetraacetate is equivalent to 2.004 mg. of Ca.

Specific rotation, page 939. Determine in a solution containing 500 mg., calculated on the dried basis, in each 10 ml.

Alkalinity. Dissolve 1 gram in 15 ml. of recently boiled and cooled water in a small flask. As soon as solution is complete, add 1.0 ml. of 0.1 *N* hydrochloric acid, then add 0.05 ml. of phenolphthalein T.S., and mix. No pink color is produced within 5 seconds.

Alkaloids. Dissolve 200 mg. in 5 ml. of water, and add 1 ml. of diluted hydrochloric acid T.S., and 2 drops of mercuric-potassium iodide T.S. No turbidity is produced in 1 minute.

Heavy metals. A solution of 1 gram in 25 ml. of water meets the requirements of the *Heavy Metals Test,* page 920, using 20 mcg. of lead ion (Pb) in the control (*Solution A*).

Loss on drying, page 931. Dry at 105° for 3 hours.

Packaging and storage. Store in tight containers.

Functional use in foods. Nutrient; dietary supplement.

CALCIUM PANTOTHENATE, RACEMIC

$C_{18}H_{32}CaN_2O_{10}$ Mol. wt. 476.54

DESCRIPTION

A mixture of the calcium salts of the dextrorotatory and levorotatory isomers of pantothenic acid. It occurs as a white, slightly hygroscopic powder. It is odorless, has a bitter taste, and is stable in air. Its solutions are neutral or alkaline to litmus. It is optically inactive. It is freely soluble in water. It is soluble in glycerin, and is practically insoluble in alcohol, in chloroform, and in ether.

Note: The physiological activity of racemic calcium pantothenate is approximately one-half that of d-calcium pantothenate.

IDENTIFICATION

A. A 1 in 20 solution gives positive tests for *Calcium,* page 926.

B. The infrared absorption spectrum of a potassium bromide dispersion of the sample, previously dried at 105° for 3 hours, exhibits maxima only at the same wavelengths as that of a similar preparation of U.S.P. Calcium Pantothenate Reference Standard.

C. Boil 50 mg. in 5 ml. of 1 *N* sodium hydroxide for 1 minute, cool, and add 5 ml. of 1 *N* hydrochloric acid and 2 drops of ferric chloride T.S. A strong yellow color is produced.

SPECIFICATIONS

Assay. Not less than 42.5 percent of dextrorotatory calcium pantothenate ($C_{18}H_{32}CaN_2O_{10}$), calculated on the dried basis.

Calcium content. Not less than 8.2 percent and not more than 8.6 percent of Ca after drying.

Alkalinity. Passes test.

Limits of Impurities

Alkaloids. Passes test.

Heavy metals (as Pb). Not more than 20 parts per million (0.002 percent).

Loss on drying. Not more than 5 percent.

TESTS

Assay. Proceed as directed under *Calcium Pantothenate Assay*, page 869.

Calcium content. Weigh accurately about 950 mg., previously dried at 105° for 3 hours, and dissolve it in 100 ml. of water containing 2 ml. of diluted hydrochloric acid T.S. While stirring, preferably with a magnetic stirrer, add about 30 ml. of 0.05 *M* disodium ethylenediaminetetraacetate from a 50-ml. buret, then add 15 ml. of sodium hydroxide T.S. and 300 mg. of hydroxy naphthol blue indicator, and continue the titration to a blue end-point. Each ml. of 0.05 *M* disodium ethylenediaminetetraacetate is equivalent to 2.004 mg. of Ca.

Alkalinity. Dissolve 1 gram in 15 ml. of recently boiled and cooled water in a small flask. As soon as solution is complete, add 1.6 ml. of 0.1 *N* hydrochloric acid, then add 0.05 ml. of phenolphthalein T.S., and mix. No pink color is produced within 5 seconds.

Alkaloids. Dissolve 200 mg. in 5 ml. of water, and add 1 ml. of diluted hydrochloric acid T.S. and 2 drops of mercuric-potassium iodide T.S. No turbidity is produced in 1 minute.

Heavy metals. A solution of 1 gram in 25 ml. of water meets the requirements of the *Heavy Metals Test*, page 920, using 20 mcg. of lead ion (Pb) in the control (*Solution A*).

Loss on drying, page 931. Dry at 105° for 3 hours.

Packaging and storage. Store in tight containers.

Functional use in foods. Nutrient; dietary supplement.

CALCIUM PANTOTHENATE, RACEMIC—CALCIUM CHLORIDE COMPLEX

Calcium Chloride Double Salt of *dl*-Calcium Pantothenate

$C_{18}H_{32}CaN_2O_{10}.CaCl_2$ Mol. wt. 587.53

DESCRIPTION

A chemical complex composed of approximately equimolecular quantities of racemic calcium pantothenate and calcium chloride. It occurs as a white, odorless, free-flowing, fine powder having a bitter taste. It is freely soluble in water, but insoluble in alcohol. Its solutions in water are alkaline to litmus.

IDENTIFICATION

A. A 1 in 20 solution gives positive tests for *Calcium*, page 926.

B. Dissolve 50 mg. in 5 ml. of 1 *N* sodium hydroxide and filter. To the filtrate add 1 drop of cupric sulfate T.S. A deep blue color develops.

C. Stir 1.0 gram of a dried sample with 15 ml. of dimethylformamide for 5 minutes. Centrifuge the mixture, then transfer 2.0 ml. of the clear supernatant liquid to a weighing dish, evaporate it under vacuum on a steam bath, and dry the residue in an oven at 105° for 1 hour. The weight of the residue, composed of uncombined racemic calcium pantothenate and calcium chloride, in grams, multiplied by 750 equals the percent of uncomplexed material in the sample. It does not exceed 10.0 percent of the weight of the sample.

SPECIFICATIONS

Assay. Not less than the equivalent of 37.0 percent of dextro calcium pantothenate, calculated on the dried basis.

Calcium content. Between 12.4 percent and 13.6 percent of Ca after drying.

Chloride. Between 10.5 percent and 12.1 percent of Cl after drying.

Loss on drying. Not more than 5 percent.

Limits of Impurities

Arsenic (as As). Not more than 3 parts per million (0.0003 percent).

Heavy metals (as Pb). Not more than 20 parts per million (0.002 percent).

Lead. Not more than 10 parts per million (0.001 percent).

TESTS

Assay. Proceed as directed under *Calcium Pantothenate Assay*, page 869.

Calcium content. Proceed as directed for *Calcium content* under *Calcium Pantothenate*, page 140.

Chloride. Transfer about 1 gram, previously dried in vacuum for 1 hour and accurately weighed, into a 250-ml. beaker, and add sufficient water to make 100 ml. Equip a pH meter with glass and silver electrodes, and set it on the "+ millivolt" scale. Insert the electrodes and a motor driven glass stirring rod into the sample beaker. Add 1–2 drops of methyl orange T.S. Stir and add, dropwise, 10 percent nitric acid until a pink color is obtained, then add 10 ml. excess. Titrate the solution with 0.1 *N* silver nitrate to a reading of +1.0 millivolt on the pH meter. Each ml. of 0.1 *N* silver nitrate is equivalent to 3.545 mg. of Cl.

Loss on drying, page 931. Dry in vacuum at 100° for 1 hour.

Arsenic. A solution of 1 gram in 25 ml. of water meets the requirements of the *Arsenic Test*, page 865.

Heavy metals. A solution of 1 gram in 25 ml. of water meets the requirements of the *Heavy Metals Test*, page 920, using 20 mcg. of lead ion (Pb) in the control (*Solution A*).

Lead. A solution of 1 gram in 25 ml. of water meets the requirements of the *Lead Limit Test*, page 929, using 10 mcg. of lead ion (Pb) in the control.

Packaging and storage. Store in tight containers.

Functional use in foods. Nutrient; dietary supplement.

CALCIUM PEROXIDE

Calcium Dioxide; Calcium Superoxide

CaO_2 Mol. wt. 72.08

DESCRIPTION

A white or yellowish, odorless, almost tasteless powder or granular material. It decomposes in moist air. It is practically insoluble in water. It dissolves in acids, forming hydrogen peroxide. A 1 in 100 aqueous slurry has a pH of about 12.

IDENTIFICATION

Cautiously dissolve 250 mg. in 5 ml. of glacial acetic acid, and add a few drops of a saturated solution of potassium iodide. Iodine is liberated. Add 20 ml. of water and sufficient sodium thiosulfate T.S. to remove the iodine color. The resulting solution gives positive tests for *Calcium*, page 926.

SPECIFICATIONS

Assay. Not less than 60.0 percent of CaO_2.

Limits of Impurities

Arsenic (as As). Not more than 3 parts per million (0.0003 percent).

Fluoride. Not more than 50 parts per million (0.005 percent).

Heavy metals (as Pb). Not more than 40 parts per million (0.004 percent).

Lead. Not more than 10 parts per million (0.001 percent).

TESTS

Assay. Transfer about 3.6 grams, accurately weighed, into a 250-ml. beaker, cautiously add 50 ml. of glacial acetic acid, and mix thoroughly. Add immediately 1 ml. of a saturated solution of potassium iodide, and stir for exactly 1 minute. Discharge the brownish iodine color with 0.1 *N* sodium thiosulfate, added from a buret, then add several drops of the titrant in excess and continue stirring for 45 seconds. Add 100 ml. of water, back titrate the excess sodium thiosulfate with 0.1 *N* iodine to a light brown color, then add 0.1 *N* sodium thiosulfate, dropwise, until the brownish color disappears. Subtract the volume of 0.1 *N* iodine required from the total volume of 0.1 *N* sodium thiosulfate required to obtain the corrected volume of 0.1 *N* sodium thiosulfate, each ml. of which is equivalent to 36.04 mg. of CaO_2.

Sample Solution for the Determination of Arsenic, Heavy **Metals, and Lead.** Weigh accurately 4.0 grams of the sample into a 250-ml. beaker, cautiously add 50 ml. of nitric acid, and evaporate just to dryness on a steam bath. Add 20 ml. of nitric acid, repeat the evaporation, cool, and dissolve the residue in sufficient water, containing 4 drops of nitric acid, to make 40.0 ml.

Arsenic. A 10-ml. portion of the *Sample Solution* meets the requirements of the *Arsenic Test*, page 865.

Fluoride. Weigh accurately 1.0 gram, and proceed as directed in the *Fluoride Limit Test*, page 917.

Heavy metals. A 10-ml. portion of the *Sample Solution* meets the requirements of the *Heavy Metals Test*, page 920, using 40 mcg. of lead ion (Pb) in the control (*Solution A*), and adjusting the solutions to a pH of 2.0, instead of between 3.0 and 4.0.

Lead. A 10-ml. portion of the *Sample Solution* meets the requirements of the *Lead Limit Test*, page 929, using 10 mcg. of lead ion (Pb) in the control.

Packaging and storage. Store in tight containers, and avoid contact with readily oxidizable materials. Observe safety precautions printed on the label of the original container.

Functional use in foods. Dough conditioner; oxidizing agent.

CALCIUM PHOSPHATE, DIBASIC

Dicalcium Phosphate

$CaHPO_4 . 2H_2O$ Mol. wt. 172.09

DESCRIPTION

Dibasic calcium phosphate is anhydrous or contains two molecules of water of hydration. It occurs as a white, odorless, tasteless powder which is stable in air. It is practically insoluble in water, but is readily soluble in dilute hydrochloric and nitric acids. It is insoluble in alcohol.

IDENTIFICATION

A. Dissolve about 100 mg. by warming with a mixture of 5 ml. of diluted hydrochloric acid T.S. and 5 ml. of water, add 2.5 ml. of ammonia T.S., dropwise, with shaking, and then add 5 ml. of ammonium oxalate T.S. A white precipitate is formed.

B. To 10 ml. of a warm solution (1 in 100) in a slight excess of nitric acid add 10 ml. of ammonium molybdate T.S. A yellow precipitate of ammonium phosphomolybdate is formed.

SPECIFICATIONS

Assay. $CaHPO_4$ (anhydrous), not less than 39.0 percent and not more than 42.0 percent of CaO; $CaHPO_4 . 2H_2O$ (dihydrate), not less than 31.9 percent and not more than 33.5 percent of CaO.

Loss on ignition. $CaHPO_4$ (anhydrous), between 7.0 and 8.5 percent; $CaHPO_4 . 2H_2O$ (dihydrate), between 24.5 and 26.5 percent.

Limits of Impurities

Arsenic (as As). Not more than 3 parts per million (0.0003 percent).

Fluoride. Not more than 50 parts per million (0.005 percent).

Heavy metals (as Pb). Not more than 30 parts per million (0.003 percent).

Lead. Not more than 5 parts per million (0.0005 percent).

TESTS

Assay. Weigh accurately a portion of the sample equivalent to about 325 mg. of $CaHPO_4$, dissolved in 10 ml. of diluted hydrochloric acid T.S., add about 120 ml. of water and a few drops of methyl orange T.S., and boil for 5 minutes, keeping the volume and pH of the solution constant during the boiling period by adding hydrochloric acid or water, if necessary. Add 2 drops of methyl red T.S. and 30 ml. of ammonium oxalate T.S., then add dropwise, with constant stirring, a mixture of equal volumes of ammonia T.S. and water until the pink color of the indicator just disappears. Digest on a steam bath for 30 minutes, cool to room temperature, allow the precipitate to settle, and filter the supernatant liquid through an asbestos mat in a Gooch crucible, using gentle suction. Wash the precipitate in the beaker with about 30 ml. of cold (below 20°) wash solution, prepared by diluting 10 ml. of ammonium oxalate T.S. to 1000 ml. Allow the precipitate to settle, and pour the supernatant liquid through the filter. Repeat this washing by decantation three more times. Using the wash solution, transfer the precipitate as completely as possible to the filter. Finally, wash the beaker and the filter with two 10-ml. portions of cold (below 20°) water. Place the Gooch crucible in the beaker, and add 100 ml. of water and 50 ml. of cold dilute sulfuric acid (1 in 6). Add from a buret 35 ml. of 0.1 *N* potassium permanganate, and stir until the color disappears. Heat to about 70°, and complete the titration with 0.1 *N* potassium permanganate. Each ml. of 0.1 *N* potassium permanganate is equivalent to 2.80 mg. of CaO.

Loss on ignition. Weigh accurately about 1 gram, and ignite, preferably in a muffle furnace, at 800° to 825° to constant weight.

Arsenic. A solution of 1 gram in 5 ml. of diluted hydrochloric acid T.S. meets the requirements of the *Arsenic Test*, page 865.

Fluoride. Weigh accurately 1.0 gram, and proceed as directed in the *Fluoride Limit Test*, page 917.

Heavy metals. Warm 1.33 grams with 5 ml. of diluted hydrochloric acid T.S. until no more dissolves, dilute to 50 ml. with water, and filter. A 25-ml. portion of the filtrate meets the requirements of the *Heavy Metals Test*, page 920, using 20 mcg. of lead ion (Pb) in the control (*Solution A*).

Lead. A solution of 250 mg. in 5 ml. of diluted hydrochloric acid T.S. meets the requirements of the *Lead Limit Test*, page 929, using 1.25 mcg. of lead ion (Pb) in the control.

Packaging and storage. Store in well-closed containers.

Labeling. Label to indicate whether it is anhydrous or the dihydrate.

Functional use in foods. Dough conditioner; nutrient supplement; yeast food.

CALCIUM PHOSPHATE, MONOBASIC

Monocalcium Phosphate; Calcium Biphosphate; Acid Calcium Phosphate

$Ca(H_2PO_4)_2$ Mol. wt. 234.05

DESCRIPTION

Monobasic calcium phosphate is anhydrous or contains one molecule of water of hydration, but, due to its deliquescent nature, more than the calculated amount of water may be present. It occurs as white crystals or granules, or as a granular powder. It is sparingly soluble in water and is insoluble in alcohol.

IDENTIFICATION

A. Dissolve 100 mg. by warming in a mixture of 2 ml. of diluted hydrochloric acid T.S. and 8 ml. of water, and add 5 ml. of ammonium oxalate T.S. A white precipitate forms.

B. To a warm solution of the sample in a slight excess of nitric acid add ammonium molybdate T.S. A yellow precipitate forms.

SPECIFICATIONS

Assay. $Ca(H_2PO_4)_2$ (anhydrous), not less than 23.5 percent and not more than 25.6 percent of CaO; $Ca(H_2PO_4)_2.H_2O$ (monohydrate), not less than 22.2 percent and not more than 24.7 percent of CaO.

Loss on drying. $Ca(H_2PO_4)_2.H_2O$ (monohydrate), not more than 0.6 percent.

Loss on ignition. $Ca(H_2PO_4)_2$ (anhydrous), between 14.0 and 15.5 percent.

Neutralizing value. Not less than 80.

Limits of Impurities

Arsenic (as As). Not more than 3 parts per million (0.0003 percent).

Fluoride. Not more than 25 parts per million (0.0025 percent).

Heavy metals (as Pb). Not more than 30 parts per million (0.003 percent).

Lead. Not more than 5 parts per million (0.0005 percent).

TESTS

Assay. Weigh accurately a portion of the sample equivalent to about 475 mg. of $Ca(H_2PO_4)_2$, dissolve it in 10 ml. of diluted hydrochloric acid T.S., add a few drops of methyl orange T.S., and boil for 5 minutes, keeping the volume and pH of the solution constant during the boiling period by adding hydrochloric acid or water, if necessary. Add 2 drops of methyl red T.S. and 30 ml. of ammonium oxalate T.S., then add dropwise, with constant stirring, a mixture of equal volumes of ammonia T.S. and water until the pink color of the indicator just disappears. Digest on a steam bath for 30 minutes, cool to room temperature, allow the precipitate to settle, and filter the supernatant liquid through an asbestos mat in a Gooch crucible, using gentle suction. Wash the precipitate in the beaker with about 30 ml. of cold (below 20°) wash solution, prepared by diluting 10 ml. of ammonium oxalate T.S. to 1000 ml. Allow the precipitate to settle, and pour the supernatant liquid through the filter. Repeat this washing by decantation three more times. Using the wash solution, transfer the precipitate as completely as possible to the filter. Finally, wash the beaker and the filter with two 10-ml. portions of cold (below 20°) water. Place the Gooch crucible in the beaker, and add 100 ml. of water and 50 ml. of cold dilute sulfuric acid (1 in 6). Add from a buret 35 ml. of 0.1 *N* potassium permanganate, and stir until the color disappears. Heat to about 70°, and complete the titration with 0.1 *N* potassium permanganate. Each ml. of 0.1 *N* potassium permanganate is equivalent to 2.804 mg. of CaO.

Loss on drying, page 931. Dry $Ca(H_2PO_4)_2.H_2O$ (monohydrate) at 60° for 3 hours.

Loss on ignition. Weigh accurately about 1 gram of $Ca(H_2PO_4)_2$ (anhydrous), and ignite, preferably in a muffle furnace, at 800° for 30 minutes.

Neutralizing value. Transfer 840 mg., accurately weighed, into a 375-ml. casserole, add 24 ml. of cold water, and stir for a few seconds. For the monohydrate, add 90.0 ml. of 0.1 *N* sodium hydroxide, and bring the suspension to a boil in exactly 2 minutes; for the anhydrous product, add 100.0 ml. of 0.1 *N* sodium hydroxide, and stir intermittently for about 5 minutes before heating to a boil. Boil for 1 minute. While the solution is still boiling hot, add 1 drop of phenolphthalein T.S., and titrate the excess alkali with 0.2 *N* hydrochloric acid until the pink color just disappears. Calculate the neutralizing value, as parts of $NaHCO_3$ equivalent to 100 parts of the sample, by the formula $V - 2v$, in which V is the volume, in ml., of 0.1 *N* sodium hydroxide added, and v the volume, in ml., of 0.2 *N* hydrochloric acid consumed in the titration.

Arsenic. A solution of 1 gram in 5 ml. of diluted hydrochloric acid T.S. meets the requirements of the *Arsenic Test*, page 865.

Fluoride. Weigh accurately 2.0 grams, and proceed as directed in the *Fluoride Limit Test*, page 917.

Heavy metals. Warm 1.33 grams with 5 ml. of diluted hydrochloric acid T.S. until no more dissolves, dilute to 50 ml. with water, and filter. A 25-ml. portion of the filtrate meets the requirements of the *Heavy Metals Test*, page 920, using 20 mcg. of lead ion (Pb) in the control (*Solution A*).

Lead. A solution of 250 mg. in 5 ml. of diluted hydrochloric acid T.S. meets the requirements of the *Lead Limit Test*, page 929, using 1.25 mcg. of lead ion (Pb) in the control.

Packaging and storage. Store in well-closed containers.

Functional use in foods. Buffer; dough conditioner; firming agent; leavening agent; nutrient; dietary supplement; yeast food; sequestrant.

CALCIUM PHOSPHATE, TRIBASIC

Tricalcium Phosphate; Precipitated Calcium Phosphate

DESCRIPTION

Tribasic calcium phosphate consists of a variable mixture of calcium phosphates having the approximate composition of $10CaO.3P_2O_5.H_2O$. It occurs as a white, odorless, tasteless powder which is stable in air.

It is insoluble in alcohol and almost insoluble in water, but it dissolves readily in dilute hydrochloric and nitric acids.

IDENTIFICATION

A. To a warm solution of the sample in a slight excess of nitric acid add ammonium molybdate T.S. A yellow precipitate forms.

B. Dissolve about 100 mg. by warming with 5 ml. of diluted hydrochloric acid T.S. and 5 ml. of water, add 1 ml. of ammonia T.S., dropwise, with shaking, and then add 5 ml. of ammonium oxalate T.S. A white precipitate forms.

SPECIFICATIONS

Assay. Not less than the equivalent of 90.0 percent of $Ca_3(PO_4)_2$, calculated on the ignited basis.

Titration value. Passes test.

Loss on ignition. Not more than 10 percent.

Limits of Impurities

Arsenic (as As). Not more than 3 parts per million (0.0003 percent).

Fluoride. Not more than 50 parts per million (0.005 percent).

Heavy metals (as Pb). Not more than 30 parts per million (0.003 percent).

Lead. Not more than 5 parts per million (0.0005 percent).

TESTS

Assay. Weigh accurately about 200 mg., and dissolve it in a mixture of 25 ml. of water and 10 ml. of diluted nitric acid T.S. Filter, if necessary, wash any precipitate, add sufficient ammonia T.S. to the filtrate to produce a slight precipitate, then dissolve the precipitate by the addition of 1 ml. of diluted nitric acid T.S. Adjust the temperature to about 50°, add 75 ml. of ammonium molybdate T.S., and maintain the temperature at about 50° for 30 minutes, stirring occasionally. Wash the precipitate once or twice with water by decantation, using from 30 to 40 ml. each time. Transfer the precipitate to a filter, and wash with potassium nitrate solution (1 in 100) until the last washing is not acid to litmus paper. Transfer the precipitate and filter to the precipitation vessel, add 40.0 ml. of 1 *N* sodium hydroxide, agitate until the precipitate is dissolved, add 3 drops of phenolphthalein T.S., and then titrate the excess alkali with 1 *N* sulfuric acid. Each ml. of 1 *N* sodium hydroxide corresponds to 6.743 mg. of $Ca_3(PO_4)_2$.

Titration value. Weigh accurately about 2 grams, and dissolve, by warming, in 50.0 ml. of 1 *N* hydrochloric acid. Cool, add 1 or 2 drops of methyl orange T.S., and slowly titrate the excess of 1 *N* hydrochloric acid with 1 *N* sodium hydroxide to a yellow color, vigorously shaking the mixture during the titration. Not less than 12.5 ml. and not more than 13.8 ml. of 1 *N* hydrochloric acid is consumed for each gram of salt, calculated on the ignited basis.

Loss on ignition. Weigh accurately about 1 gram, and ignite, preferably in a muffle furnace, at 800° to 825° to constant weight.

Arsenic. A solution of 1 gram in 25 ml. of diluted hydrochloric acid T.S. meets the requirements of the *Arsenic Test*, page 865.

Fluoride. Weigh accurately 1.0 gram, and proceed as directed in the *Fluoride Limit Test*, page 917.

Heavy metals. Warm 1.33 grams with 7 ml. of diluted hydrochloric acid T.S. until no more dissolves, dilute to 50 ml. with water, and filter. A 25-ml. portion of the filtrate meets the requirements of the *Heavy Metals Test*, page 920, using 20 mcg. of lead ion (Pb) in the control (*Solution A*). (*Note:* Filter the mixture after pH adjustment.)

Lead. A solution of 250 mg. in 5 ml. of diluted hydrochloric acid T.S. meets the requirements of the *Lead Limit Test*, page 929, using 1.25 mcg. of lead ion (Pb) in the control.

Packaging and storage. Store in well-closed containers.

Functional use in foods. Anticaking agent; buffer; nutrient; dietary supplement.

CALCIUM PROPIONATE

$(CH_3CH_2COO)_2Ca$

$C_6H_{10}CaO_4$ Mol. wt. 186.22

DESCRIPTION

White crystals or crystalline solid, possessing not more than a faint odor of propionic acid. One gram dissolves in about 3 ml. of water. The pH of a 1 in 10 solution is between 8 and 10.

IDENTIFICATION

A. A 1 in 20 solution gives positive tests for *Calcium*, page 926.

B. Upon ignition at a relatively low temperature, it yields an alkaline residue which effervesces with acids.

C. Warm a small sample with sulfuric acid. Propionic acid, recognizable by its odor, is evolved.

SPECIFICATIONS

Assay. Not less than 98.0 percent of $C_6H_{10}CaO_4$, calculated on the anhydrous basis.

Limits of Impurities

Arsenic (as As). Not more than 3 parts per million (0.0003 percent).

Fluoride. Not more than 30 parts per million (0.003 percent).

Heavy metals (as Pb). Not more than 10 parts per million (0.001 percent).

Insoluble substances. Not more than 0.2 percent.
Magnesium (as MgO). Passes test (about 0.4 percent).
Water. Not more than 5 percent.

TESTS

Assay. Dissolve about 400 mg., accurately weighed, in 100 ml. of water. While stirring, preferably with a magnetic stirrer, add about 30 ml. of 0.05 *M* disodium ethylenediaminetetraacetate from a 50-ml. buret, then add 15 ml. of sodium hydroxide T.S. and 300 mg. of hydroxy naphthol blue indicator, and continue the titration to a blue end-point. Each ml. of 0.05 *M* disodium ethylenediaminetetraacetate is equivalent to 9.311 mg. of $C_6H_{10}CaO_4$.

Arsenic. A *Sample Solution* prepared as directed for organic compounds meets the requirements of the *Arsenic Test*, page 865.

Fluoride. Proceed as directed in the *Fluoride Limit Test*, page 917, using *Method I* (1.67-gram sample) or *Method III* (1.0-gram sample).

Heavy metals. A solution of 2 grams in 25 ml. of water meets the requirements of the *Heavy Metals Test*, page 920, using 20 mcg. of lead ion (Pb) in the control (*Solution A*).

Insoluble substances. Dissolve 10 grams in 100 ml. of hot water, filter through a tared filtering crucible, wash the insoluble residue with hot water, and dry at 105° to constant weight.

Magnesium. Place 400.0 mg. of the sample, 5 ml. of diluted hydrochloric acid T.S., and about 10 ml. of water in a small beaker, and dissolve the sample by heating on a hot plate. Evaporate the solution to a volume of about 2 ml., and cool. Transfer the residual liquid into a 100-ml. volumetric flask, dilute to volume with water, and mix. Dilute 7.5 ml. of this solution to 20 ml. with water, add 2 ml. of sodium hydroxide T.S. and 0.05 ml. of a 1 in 1000 solution of Titan yellow (Clayton yellow), mix, allow to stand for 10 minutes, and shake. Any color does not exceed that produced by 1.0 ml. of *Magnesium Standard Solution* (50 mcg. Mg ion) in the same volume of a control containing 2.5 ml. of the sample solution (10-mg. sample) and the quantities of the reagents used in the test.

Water. Determine by the *Karl Fischer Titrimetric Method*, page 977.

Packaging and storage. Store in tight containers.

Functional use in foods. Preservative; mold and rope inhibitor.

CALCIUM PYROPHOSPHATE

$Ca_2P_2O_7$ Mol. wt. 254.10

DESCRIPTION

A fine, white, odorless and tasteless powder. It is insoluble in water, but is soluble in dilute hydrochloric and nitric acids.

IDENTIFICATION

A. Dissolve about 100 mg. by warming with a mixture of 5 ml. of diluted hydrochloric acid T.S. and 5 ml. of water, add 2.5 ml. of ammonia T.S., dropwise, with shaking, then add 5 ml. of ammonium oxalate T.S. A white precipitate is formed.

B. To 1 ml. of a 1 in 100 solution in diluted hydrochloric acid T.S. add a few drops of silver nitrate T.S. A white precipitate is formed which is soluble in diluted nitric acid T.S.

SPECIFICATIONS

Assay. Not less than 96.0 percent of $Ca_2P_2O_7$.

Limits of Impurities

Arsenic (as As). Not more than 3 parts per million (0.0003 percent).

Fluoride. Not more than 50 parts per million (0.005 percent).

Heavy metals (as Pb). Not more than 30 parts per million (0.003 percent).

Lead. Not more than 5 parts per million (0.0005 percent).

Loss on ignition. Not more than 1 percent.

TESTS

Assay. Dissolve about 300 mg., accurately weighed, in 10 ml. of diluted hydrochloric acid T.S., add about 120 ml. of water and a few drops of methyl orange T.S., and boil for 30 minutes, keeping the volume and pH of the solution constant during the boiling period by adding hydrochloric acid or water, if necessary. Add 2 drops of methyl red T.S. and 30 ml. of ammonium oxalate T.S., then add dropwise, with constant stirring, a mixture of equal volumes of ammonia T.S. and water until the pink color of the indicator just disappears. Digest on a steam bath for 30 minutes, cool to room temperature, allow the precipitate to settle, and filter the supernatant liquid through an asbestos mat in a Gooch crucible, using gentle suction. Wash the precipitate in the beaker with about 30 ml. of cold (below 20°) wash solution, prepared by diluting 10 ml. of ammonium oxalate T.S. to 1000 ml. Allow the precipitate to settle, and pour the supernatant liquid through the filter. Repeat this washing by decantation three more times. Using the wash solution, transfer the precipitate as completely as possible to the filter. Finally, wash the beaker and the filter with two 10-ml. portions of cold (below 20°) water. Place the Gooch crucible in the beaker, and add 100 ml.

of water and 50 ml. of cold dilute sulfuric acid (1 in 6). Add from a buret 35 ml. of 0.1 *N* potassium permanganate, and stir until the color disappears. Heat to about 70°, and complete the titration with 0.1 *N* potassium permanganate. Each ml. of 0.1 *N* potassium permanganate is equivalent to 6.35 mg. of $Ca_2P_2O_7$.

Arsenic. A solution of 1 gram in 5 ml. of diluted hydrochloric acid T.S. meets the requirements of the *Arsenic Test*, page 865.

Fluoride. Weigh accurately 1.0 gram, and proceed as directed in the *Fluoride Limit Test*, page 917.

Heavy metals. Warm 1.33 grams with 7 ml. of diluted hydrochloric acid T.S. until no more dissolves, dilute to 50 ml. with water, and filter. A 25-ml. portion of the filtrate meets the requirements of the *Heavy Metals Test*, page 920, using 20 mcg. of lead ion (Pb) in the control (*Solution A*).

Lead. A solution of 250 mg. in 5 ml. of diluted hydrochloric acid T.S. meets the requirements of the *Lead Limit Test*, page 929, using 1.25 mcg. of lead ion (Pb) in the control.

Loss on ignition. Weigh accurately about 1 gram, and ignite, preferably in a muffle furnace, at 800° to 825° for 30 minutes.

Packaging and storage. Store in well-closed containers.

Functional use in foods. Buffer; neutralizing agent; nutrient; dietary supplement.

CALCIUM SACCHARIN

1,2-Benzisothiazolin-3-one 1,1-Dioxide Calcium Salt

$C_{14}H_8CaN_2O_6S_2 . 3^1/_2 H_2O$ Mol. wt. 467.49

DESCRIPTION

White crystals or a white, crystalline powder. It is odorless or has a faint, aromatic odor and an intensely sweet taste even in dilute solutions. In dilute solutions it is about 500 times as sweet as sucrose. One gram is soluble in 1.5 ml. of water.

IDENTIFICATION

A. Dissolve about 100 mg. in 5 ml. of sodium hydroxide solution (1 in 20), evaporate to dryness, and gently fuse the residue over a small flame until it no longer evolves ammonia. After the residue has cooled, dissolve it in 20 ml. of water, neutralize the solution with diluted hydrochloric acid T.S., and filter. The addition of a drop of ferric chloride T.S. to the filtrate produces a violet color.

B. Mix 20 mg. with 40 mg. of resorcinol, add 10 drops of sulfuric acid, and heat the mixture in a liquid bath at 200° for 3 minutes. After cooling, add 10 ml. of water and an excess of sodium hydroxide T.S. A fluorescent green liquid results.

C. A 1 in 10 solution gives positive tests for *Calcium*, page 926.

D. To 10 ml. of a 1 in 10 solution add 1 ml. of hydrochloric acid. A crystalline precipitate of saccharin is formed. Wash the precipitate well with cold water and dry at 105° for 2 hours. It melts between 226° and 230° (*Class Ia*, page 931).

SPECIFICATIONS

Assay. Not less than 95.0 percent of $C_{14}H_8CaN_2O_6S_2$, calculated on the anhydrous basis.

Water. Between 3.0 percent and 15.0 percent.

Limits of Impurities

Arsenic (as As). Not more than 3 parts per million (0.0003 percent).

Benzoate and salicylate. Passes test.

Heavy metals (as Pb). Not more than 10 parts per million (0.001 percent).

Readily carbonizable substances. Passes test.

Selenium. Not more than 30 parts per million (0.003 percent).

TESTS

Assay. Weigh accurately about 500 mg. and transfer it quantitatively to a separator with the aid of 10 ml. of water. Add 2 ml. of diluted hydrochloric acid T.S., and extract the precipitated saccharin first with 30 ml., then with five 20-ml. portions, of a solvent composed of 9 volumes of chloroform and 1 volume of alcohol. Filter each extract through a small filter paper moistened with the solvent mixture, and evaporate the combined filtrates on a steam bath to dryness with the aid of a current of air. Dissolve the residue in 75 ml. of hot water, cool, add phenolphthalein T.S., and titrate with 0.1 *N* sodium hydroxide. Perform a blank determination, and make any necessary correction (see page 2). Each ml. of 0.1 *N* sodium hydroxide is equivalent to 20.22 mg. of $C_{14}H_8CaN_2O_6S_2$.

Water. Determine by the *Karl Fischer Titrimetric Method*, page 977.

Arsenic. A *Sample Solution* prepared as directed for organic compounds meets the requirements of the *Arsenic Test*, page 865.

Benzoate and salicylate. To 10 ml. of a 1 in 20 solution previously acidified with 5 drops of acetic acid, add 3 drops of ferric chloride T.S. No precipitate or violet color appears.

Heavy metals. Prepare and test a 2-gram sample as directed in *Method II* under the *Heavy Metals Test*, page 920, using 20 mcg. of lead ion (Pb) in the control (*Solution A*).

Readily carbonizable substances, page 943. Dissolve 200 mg. in 5 ml. of sulfuric acid T.S., and keep at a temperature of 48° to 50° for 10 minutes. The color is no darker than *Matching Fluid A*.

Selenium. Prepare and test a 2-gram sample as directed in the *Selenium Limit Test*, page 953.

Packaging and storage. Store in well-closed containers.

Functional use in foods. Nonnutritive sweetener.

CALCIUM SILICATE

DESCRIPTION

A hydrous or anhydrous silicate with varying proportions of CaO and SiO_2. It occurs as a white to off-white free-flowing powder which remains so after absorbing relatively large amounts of water or other liquids. It is insoluble in water, but forms a gel with mineral acids. The pH of a 1 in 20 aqueous slurry is between 8.4 and 10.2.

IDENTIFICATION

A. Mix about 500 mg. with 10 ml. of dilute hydrochloric acid T.S., filter, and neutralize the filtrate to litmus paper with ammonia T.S. The neutralized filtrate gives positive tests for *Calcium*, page 926.

B. Prepare a bead by fusing a few crystals of sodium ammonium phosphate on a platinum loop in the flame of a Bunsen burner. Place the hot, transparent bead in contact with a sample, and again fuse. Silica floats about in the bead, producing, upon cooling, an opaque bead with a web-like structure.

SPECIFICATIONS

Calcium oxide and silicon dioxide. Not less than the percentages stated or within the range claimed by the vendor.

Loss on drying or loss on ignition. Not more than the percentages stated or within the range claimed by the vendor.

Limits of Impurities

Arsenic (as As). Not more than 3 parts per million (0.0003 percent).

Fluoride. Not more than 50 parts per million (0.005 percent).

Heavy metals (as Pb). Not more than 40 parts per million (0.004 percent).

Lead. Not more than 10 parts per million (0.001 percent).

TESTS

Assay for silicon dioxide. Transfer about 400 mg. of the sample, accurately weighed, into a beaker, add 5 ml. of water and 10 ml. of perchloric acid, and heat until dense white fumes of perchloric acid are evolved. Cover the beaker with a watch-glass, and continue to heat for 15 minutes longer. Allow to cool, add 30 ml. of water, filter, and wash the precipitate with 200 ml. of hot water. Retain the combined filtrate and washings for use in the *Assay for calcium oxide.* Transfer the filter paper and its contents to a platinum crucible, heat slowly to dryness, and then heat sufficiently to char the filter paper. After cooling, add a few drops of sulfuric acid, and then ignite at about 1300° to constant weight. Moisten the residue with 5 drops of sulfuric acid, add 15 ml. of hydrofluoric acid, heat cautiously on a hot plate until all of the acid is driven off, and ignite to constant weight at a temperature not lower than 1000°. Cool in a desiccator and weigh. The loss in weight is equivalent to the SiO_2 in the sample taken.

Assay for calcium oxide. To the combined filtrate and washings retained in the *Assay for silicon dioxide,* add, while stirring, about 30 ml. of 0.05 *M* disodium ethylenediaminetetraacetate from a 50-ml. buret, then add 15 ml. of sodium hydroxide T.S. and 300 mg. of hydroxy naphthol blue indicator, and continue the titration to a blue end-point. Each ml. of 0.05 *M* disodium ethylenediaminetetraacetate is equivalent to 2.804 mg. of CaO.

Loss on drying, page 931. Dry at 105° for 2 hours.

Loss on ignition. Transfer about 1 gram, previously dried at 105° for 2 hours and accurately weighed, into a suitable tared crucible, and ignite at 900° to constant weight.

> **Sample Solution for the Determination of Arsenic, Heavy Metals, and Lead.** Transfer 10.0 grams of the sample into a 250-ml. beaker, add 50 ml. of 0.5 *N* hydrochloric acid, cover with a watch glass, and heat slowly to boiling. Boil gently for 15 minutes, cool, and let the undissolved material settle. Decant the supernatant liquid through Whatman No. 4, or equivalent, filter paper into a 100-ml. volumetric flask, retaining as much as possible of the insoluble material in the beaker. Wash the slurry and beaker with three 10-ml. portions of hot water, decanting each washing through the filter paper into the flask. Finally, wash the filter paper with 15 ml. of hot water, cool the filtrate to room temperature, dilute to volume with water and mix.

Arsenic. A 10-ml. portion of the *Sample Solution* meets the requirements of the *Arsenic Test,* page 865.

Fluoride. Weigh accurately 1.0 gram and proceed as directed in the *Fluoride Limit Test,* page 917.

Heavy metals. A 5-ml. portion of the *Sample Solution* meets the requirements of the *Heavy Metals Test,* page 920, using 20 mcg. of lead ion (Pb) in the control (*Solution A*).

Lead. A 10-ml. portion of the *Sample Solution* meets the requirements of the *Lead Limit Test,* page 929, using 10 mcg. of lead ion (Pb) in the control.

Packaging and storage. Store in well-closed containers.

Functional use in foods. Anticaking agent.

CALCIUM STEARATE

DESCRIPTION

Calcium stearate is a compound of calcium with variable proportions of stearic and palmitic acids. It occurs as a fine, white to yellowish white, bulky powder, having a slight, characteristic odor. It is unctuous, and is free from grittiness. It is insoluble in water, in alcohol, and in ether. It conforms to the regulations of the federal Food and Drug Administration pertaining to specifications for salts of fatty acids and fatty acids derived from edible fat sources.

IDENTIFICATION

A. Heat 1 gram with a mixture of 25 ml. of water and 5 ml. of hydrochloric acid. Fatty acids are liberated, floating as an oily layer on the surface of the liquid. The water layer gives positive tests for *Calcium*, page 926.

B. Mix 25 grams of the sample with 200 ml. of hot water, then add 60 ml. of diluted sulfuric acid T.S., and heat the mixture, with frequent stirring, until the fatty acids separate cleanly as a transparent layer. Wash the fatty acids with boiling water until free from sulfate, collect them in a small beaker, and warm on a steam bath until the water has separated and the fatty acids are clear. Allow the acids to cool, pour off the water layer, then melt the acids, filter into a dry beaker, and dry at 105° for 20 minutes. The solidification point of the fatty acids so obtained is not below 54° (see page 931).

SPECIFICATIONS

Assay. Not less than the equivalent of 9.0 percent and not more than the equivalent of 10.5 percent of CaO.

Limits of Impurities

Arsenic (as As). Not more than 3 parts per million (0.0003 percent).

Free fatty acid (as stearic acid). Not more than 3.0 percent.

Heavy metals (as Pb). Not more than 10 parts per million (0.001 percent).

Loss on drying. Not more than 4 percent.

TESTS

Assay. Boil about 1.2 grams, accurately weighed, with 50 ml. of 0.1 *N* hydrochloric acid for 10 minutes, or until the fatty acid layer is clear, adding water if necessary to maintain the original volume. Cool, filter, and wash the filter and flask thoroughly with water until the last washing is not acid to litmus. Neutralize the filtrate to litmus with sodium hydroxide T.S. While stirring, preferably with a magnetic stirrer, add about 30 ml. of 0.05 *M* disodium ethylenediaminetetraacetate from a 50-ml. buret, then add 15 ml. of sodium hydroxide T.S. and 300 mg. of hydroxy naphthol blue indicator, and continue the titration to a blue end-point. Each ml. of 0.05 *M* disodium ethylenediaminetetraacetate is equivalent to 2.804 mg. of CaO.

Arsenic. Mix 1 gram of the sample with 10 ml. of hydrochloric acid and 8 drops of bromine T.S., and heat on a steam bath until a transparent layer of melted fatty acid forms. Add 50 ml. of water, boil down to about 25 ml., and filter while hot. Cool, neutralize with a 1 in 2 solution of sodium hydroxide, and dilute to 35 ml. with water. This solution meets the requirements of the *Arsenic Test*, page 865.

Free fatty acid. Transfer 2 grams of the sample, accurately weighed, into a dry 125-ml. Erlenmeyer flask containing 50 ml. of acetone, fit an air-cooled reflux condenser onto the neck of the flask, boil the mixture on a steam bath for 10 minutes, and cool. Filter through two layers of Whatman No. 42, or equivalent, filter paper, and wash the flask, residue, and filter with 50 ml. of acetone. Add phenolphthalein T.S. and 5 ml. of water to the filtrate, and titrate with 0.1 *N* sodium hydroxide. Perform a blank determination, using 100 ml. of acetone and 5 ml. of water (see page 2). Each ml. of 0.1 *N* sodium hydroxide is equivalent to 28.45 mg. of stearic acid ($C_{18}H_{36}O_2$).

Heavy metals, page 920. Place 2.5 grams of the sample in a porcelain dish, place 500 mg. of the sample in a second dish for the control, and to each add 5 ml. of a 1 in 4 solution of magnesium nitrate in alcohol. Cover the dishes with 3-inch short stem funnels so that the stems are straight up. Heat for 30 minutes on a hot plate at the low setting, then heat for 30 minutes at the medium setting, and cool. Remove the funnels, add 20 mcg. of lead ion (Pb) to the control, and heat each dish over an Argand burner until most of the carbon is burned off. Cool, add 10 ml. of nitric acid, and transfer the solutions into 250-ml. beakers. Add 5 ml. of 70 percent perchloric acid, evaporate to dryness, then add 2 ml. of hydrochloric acid to the residues, and wash down the inside of the beakers with water. Evaporate carefully to dryness again, swirling near the dry point to avoid spattering. Repeat the hydrochloric acid treatment, then cool, and dissolve the residues in about 10 ml. of water. To each solution add 1 drop of phenolphthalein T.S. and sufficient sodium hydroxide T.S. until the solutions just turn pink, and then add diluted hydrochloric acid T.S. until the solutions become colorless. Add 1 ml. of diluted acetic acid T.S. and a small amount of charcoal to each solution, and filter through Whatman No. 2, or equivalent, filter paper into 50-ml. Nessler tubes. Wash with water, dilute to 40 ml., and add 10 ml. of hydrogen sulfide T.S. to each tube. The color in the solution of the sample does not exceed that produced in the control.

Loss on drying, page 931. Dry at 105° to constant weight, using 2-hour increments of heating.

Packaging and storage. Store in well-closed containers.

Functional use in foods. Anticaking agent; binder; emulsifier.

CALCIUM STEAROYL-2-LACTYLATE

DESCRIPTION

A mixture of calcium salts of stearoyl lactic acid, with minor proportions of other salts of related acids. It occurs as a cream colored powder having a mild, caramel-like odor. It is slightly soluble in hot water. It conforms to the regulations of the federal Food and Drug Administration pertaining to specifications for fats or fatty acids derived from edible sources.

IDENTIFICATION

Calcium stearoyl-2-lactylate responds to the tests for *Identification* under *Calcium Stearate*, page 158.

SPECIFICATIONS

Acid value. Between 50 and 86.

Calcium content. Between 4.2 percent and 5.2 percent.

Ester value. Between 125 and 164.

Total lactic acid. Between 32 percent and 38 percent.

Limits of Impurities

Arsenic (as As). Not more than 3 parts per million (0.0003 percent).

Heavy metals (as Pb). Not more than 10 parts per million (0.001 percent).

TESTS

Acid value. Transfer about 1 gram, accurately weighed, to a 125-ml. volumetric flask, add 25 ml. of alcohol, previously neutralized to phenolphthalein T.S., and heat on a hot plate until the sample is dissolved. Cool, add 5 drops of phenolphthalein T.S., and titrate rapidly with 0.1 N sodium hydroxide to the first pink color that persists for at least 30 seconds. Calculate the acid value by the formula $56.1V \times N/W$, in which V is the volume, in ml., and N is the normality, respectively, of the sodium hydroxide solution, and W is the weight, in grams, of the sample taken. Retain the neutralized solution for the determination of *Ester value.*

Calcium content

Stock lanthanum solution. Transfer 5.86 grams of lanthanum oxide, La_2O_3, into a 100-ml. volumetric flask, wet with a few ml. of water, slowly add 25 ml. of hydrochloric acid, and swirl until the material is completely dissolved. Dilute to volume with water, and mix.

Stock calcium solution. Use a solution containing 0.5 mg. of Ca in each ml. (500 parts per million Ca). The solution may be obtained commercially or prepared as follows: transfer 124.8 mg. of calcium carbonate, $CaCO_3$, previously dried at 200° for 4 hours, into a 100-ml. volumetric flask, carefully dissolve in 2 ml. of diluted hydrochloric acid T.S., dilute to volume with water, and mix.

Standard preparations. Transfer 10.0 ml. of the *Stock lanthanum solution* into each of three 50-ml. volumetric flasks. Using a microliter syringe, transfer 0.20 ml. of the *Stock calcium solution* into the first flask, 0.40 ml. into the second flask, and 0.50 ml. into the third flask. Dilute each flask to volume with water, and mix. The flasks contain 2.0, 4.0, and 5.0 mcg. of Ca per ml., respectively. Prepare these solutions fresh daily.

Sample preparation. Transfer about 250 mg. of the sample, accurately weighed, into a 30-ml. beaker, dissolve with heating in 10 ml. of alcohol, and quantitatively transfer the solution into a 25-ml. volumetric flask. Wash the beaker with two 5-ml. portions of alcohol, adding the washings to the flask, dilute to volume with alcohol, and mix. Transfer 5.0 ml. of the *Stock lanthanum solution* to a second 25-ml. volumetric flask. Using a microliter syringe, transfer 0.25 ml. of the alcoholic solution of the sample to the second flask, dilute to volume with water, and mix.

Procedure. Concomitantly determine the absorbance of each *Standard preparation* and of the *Sample preparation* at 4227 Å, with a suitable atomic absorption spectrophotometer, following the operating parameters as recommended by the manufacturer of the instrument. Plot the absorbance of the *Standard preparations* vs. concentration of Ca, in mcg. per ml., and from the curve so obtained determine the concentration, C, in mcg. per ml., of Ca in the *Sample preparation.* Calculate the quantity, in mg., of Ca in the sample taken by the formula $2.5C$.

Ester value. To the neutralized solution retained in the test for *Acid value* add 10.0 ml. of alcoholic potassium hydroxide solution prepared by dissolving 11.2 grams of potassium hydroxide in 250 ml. of alcohol and diluting with 25 ml. of water. Add 5 drops of phenolphthalein T.S., connect a suitable condenser, and reflux for 2 hours. Cool, add 5 additional drops of phenolphthalein T.S., and titrate the excess alkali with 0.1 N sulfuric acid. Perform a blank determination using 10.0 ml. of the alcoholic potassium hydroxide solution. Calculate the ester value by the formula $56.1(B - S)N/W$, in which $B - S$ represents the difference between the volumes of 0.1 N sulfuric acid required for the blank and the sample, respectively, N is the normality of the sulfuric acid, and W is the weight, in grams, of the sample taken.

Total Lactic Acid

Standard curve. Dissolve 1.067 grams of lithium lactate in sufficient water to make 1000.0 ml. Transfer 10.0 ml. of this solution into a 100-ml. volumetric flask, dilute to volume with water, and mix. Transfer 1.0, 2.0, 4.0, 6.0, and 8.0 ml. of the diluted standard solution into separate 100-ml. volumetric flasks, dilute each flask to volume with water, and mix. These standards represent 1, 2, 4, 6, and 8 mcg. of lactic acid per ml., respectively. Transfer 1.0 ml. of each solution into separate test tubes, and continue as directed in the *Procedure,* beginning with "Add 1 drop of cupric sulfate T.S. . . . " After color development and reading the absorbance values, construct a *Standard curve* by plotting absorbance versus mcg. of lactic acid.

Test preparation. Transfer about 200 mg. of the sample, accurately weighed, into a 125-ml. Erlenmeyer flask, add 10 ml. of 0.5 *N* alcoholic potassium hydroxide and 10 ml. of water, attach an air condenser, and reflux gently for 45 minutes. Wash the sides of the flask and the condenser with about 40 ml. of water, and heat on a steam bath until no odor of alcohol remains. Add 6 ml. of dilute sulfuric acid (1 in 2), heat until the fatty acids are melted, then cool to about 60°, and add 25 ml. of petroleum ether. Swirl the mixture gently, and transfer quantitatively to a separator. Collect the water layer in a 100-ml. volumetric flask, and wash the petroleum ether layer with two 20-ml. portions of water, adding the washings to the volumetric flask. Dilute to volume with water, and mix. Transfer 1.0 ml. of this solution into a second 100-ml. volumetric flask, dilute to volume with water, and mix.

Procedure. Transfer 1.0 ml. of the *Test preparation* into a test tube, and transfer 1.0 ml. of water to a second test tube to serve as the blank. Treat each tube as follows: Add 1 drop of cupric sulfate T.S., swirl gently, and then add rapidly from a buret 9.0 ml. of sulfuric acid. Loosely stopper the tube, and heat in a water bath at 90° for exactly 5 minutes. Cool immediately to below 20° in an ice bath for 5 minutes, add 3 drops of *p*-phenylphenol T.S., shake immediately, and heat in a water bath at 30° for 30 minutes, shaking the tube twice during this time to disperse the reagent. Heat the tube in a water bath at 90° for exactly 90 seconds, and then cool immediately to room temperature in an ice water bath. Determine the absorbance of the solution in a 1-cm. cell, at 570 mμ, with a suitable spectrophotometer, using the blank to set the instrument. Obtain the weight, in mcg., of lactic acid in the portion of the *Test preparation* taken for the *Procedure* by means of the *Standard curve.*

Arsenic. A *Sample Solution* prepared as directed for organic compounds meets the requirements of the *Arsenic Test*, page 865.

Heavy metals. Prepare and test a 2-gram sample as directed in *Method II* under the *Heavy Metals Test*, page 920, using 20 mcg. of lead ion (Pb) in the control (*Solution A*).

Packaging and storage. Store in tight containers in a cool dry place.

Functional use in foods. Dough conditioner; stabilizer; whipping agent.

CALCIUM SULFATE

$CaSO_4.xH_2O$ Mol. wt. (anhydrous) 136.14

DESCRIPTION

Calcium sulfate is anhydrous or contains two molecules of water of hydration. It occurs as a fine, white to slightly yellow-white, odorless powder.

IDENTIFICATION

Dissolve about 200 mg. by warming with a mixture of 4 ml. of diluted hydrochloric acid T.S. and 16 ml. of water. A white precipitate forms when 5 ml. of ammonium oxalate T.S. is added to 10 ml. of the solution. Upon the addition of barium chloride T.S. to the remaining 10 ml., a white precipitate forms which is insoluble in hydrochloric and nitric acids.

SPECIFICATIONS

Assay. Not less than 99.0 percent of $CaSO_4$, calculated on the dried basis.

Loss on drying. $CaSO_4$ (anhydrous), not more than 1.5 percent; $CaSO_4.2H_2O$ (dihydrate), between 19 and 23 percent.

Limits of Impurities

Arsenic (as As). Not more than 3 parts per million (0.0003 percent).

Fluoride. Not more than 30 parts per million (0.003 percent).

Heavy metals (as Pb). Not more than 10 parts per million (0.001 percent).

Selenium. Not more than 30 parts per million (0.003 percent).

TESTS

Assay. Dissolve about 350 mg., accurately weighed, in 100 ml. of water and 4 ml. of diluted hydrochloric acid T.S. While stirring, preferably with a magnetic stirrer, add about 30 ml. of 0.05 *M* disodium ethylenediaminetetraacetate from a 50-ml. buret, then add 15 ml. of sodium hydroxide T.S. and 300 mg. of hydroxy naphthol blue indicator, and continue the titration to a blue end-point. Each ml. of 0.05 *M* disodium ethylenediaminetetraacetate is equivalent to 6.807 mg. of $CaSO_4$.

Loss on drying, page 931. Dry at 250° to constant weight.

Arsenic. Mix 1 gram with 10 ml. of water, add 12 ml. of diluted hydrochloric acid T.S., and heat to boiling to dissolve the sample. Cool, filter, and dilute the filtrate to 35 ml. with water. This solution meets the requirements of the *Arsenic Test*, page 865.

Fluoride. Weigh accurately 1.67 grams, and proceed as directed in the *Fluoride Limit Test*, page 917.

Heavy metals. Mix 2 grams with 20 ml. of water, add 25 ml. of diluted hydrochloric acid T.S., and heat to boiling to dissolve the sample. Cool, and add ammonium hydroxide to a pH of 7. Filter, evaporate to a volume of about 25 ml., and refilter if necessary to obtain a clear solution. This solution meets the requirements of the *Heavy Metals Test*, page 920, using 20 mcg. of lead ion (Pb) in the control (*Solution A*).

Selenium. A solution of 2 grams in 40 ml. of dilute hydrochloric acid (1 in 2) meets the requirements of the *Selenium Limit Test*, page 953.

Packaging and storage. Store in well-closed containers.

Functional use in foods. Nutrient; dietary supplement; yeast food; dough conditioner; firming agent; sequestrant.

CANANGA OIL

DESCRIPTION

The oil obtained by distillation from the flowers of the tree *Cananga odorata* Hook f. et Thoms., (Fam. *Anonaceae*). It is a light to deep yellow liquid, having a harsh floral odor, suggestive of ylang ylang. It is soluble in most fixed oils and in mineral oil, but it is practically insoluble in glycerin and in propylene glycol.

SPECIFICATIONS

Angular rotation. Between $-15°$ and $-30°$.

Refractive index. Between 1.495 and 1.505 at 20°.

Saponification value. Between 10 and 40.

Solubility in alcohol. Passes test.

Specific gravity. Between 0.904 and 0.920.

TESTS

Angular rotation. Determine in a 100-mm. tube as directed under *Optical Rotation*, page 939.

Refractive index, page 945. Determine with an Abbé or other refractometer of equal or greater accuracy.

Saponification value. Determine as directed in the general method, page 896, using about 5 grams, accurately weighed.

Solubility in alcohol. Proceed as directed in the general method, page 899. One ml. dissolves in 0.5 ml. of 95 percent alcohol, usually becoming cloudy on further dilution.

Specific gravity. Determine by any reliable method (see page 5).

Packaging and storage. Store in full, tight, preferably glass, aluminum, tin-lined, or other suitably lined containers in a cool place protected from light.

Functional use in foods. Flavoring agent.

CANDELILLA WAX

DESCRIPTION

A purified wax obtained from the leaves of the candelilla plant, *Euphorbia antisyphilitica*. It is a hard, yellowish-brown, opaque to translucent wax. Its specific gravity is about 0.983. It is soluble in chloroform and in toluene, but is insoluble in water.

SPECIFICATIONS

Acid value. Between 12 and 22.

Melting range. Between 68.5° and 72.5°.

Saponification value. Between 43 and 65.

Limits of Impurities

Arsenic (as As). Not more than 3 parts per million (0.0003 percent).

Heavy metals (as Pb). Not more than 40 parts per million (0.004 percent).

Lead. Not more than 3 parts per million (0.0003 percent).

TESTS

Acid value. Determine as directed under *Method I* in the general procedure, page 902.

Melting range. Determine as directed for *Class II* substances in the general procedure, page 931.

Saponification value. Determine as directed in the general procedure, page 914.

Arsenic. Prepare a *Sample Solution* as directed in the general method under *Chewing Gum Base*, page 877. This solution meets the requirements of the *Arsenic Test*, page 865.

Heavy metals. Prepare and test a 500-mg. sample as directed in *Method II* under the *Heavy Metals Test*, page 920, using 20 mcg. of lead ion (Pb) in the control (*Solution A*).

Lead. Prepare a *Sample Solution* as directed in the general method under *Chewing Gum Base*, page 878. This solution meets the requirements of the *Lead Limit Test*, page 929, using 10 mcg. of lead ion (Pb) in the control.

Packaging and storage. Store in well-closed containers.

Functional use in foods. Masticatory substance in chewing gum base; surface-finishing agent.

CARAWAY OIL

DESCRIPTION

A volatile oil distilled from the dried, ripe fruit of *Carum carvi* L. (Fam. *Umbelliferae*). It is a colorless to pale yellow liquid, having the characteristic odor and taste of caraway.

SPECIFICATIONS

Assay. Not less than 50 percent, by volume, of ketones as carvone.

Angular rotation. Between +70° and +80°.

Refractive index. Between 1.484 and 1.488 at 20°.

Solubility in alcohol. Passes test.

Specific gravity. Between 0.900 and 0.910.

Limits of Impurities

Arsenic (as As). Not more than 3 parts per million (0.0003 percent).

Heavy metals (as Pb). Not more than 40 parts per million (0.004 percent).

Lead. Not more than 10 parts per million (0.001 percent).

TESTS

Assay. Proceed as directed under *Aldehydes and Ketones-Neutral Sulfite Method,* page 895.

Angular rotation. Determine in a 100-mm. tube as directed under *Optical Rotation,* page 939.

Refractive index, page 945. Determine with an Abbé or other refractometer of equal or greater accuracy.

Solubility in alcohol. Proceed as directed in the general method, page 899. One ml. dissolves in 8 ml. of 80 percent alcohol.

Specific gravity. Determine by any reliable method (see page 5).

Arsenic. A *Sample Solution* prepared as directed for organic compounds meets the requirements of the *Arsenic Test,* page 865.

Heavy metals. Prepare and test a 500-mg. sample as directed in *Method II* under the *Heavy Metals Test,* page 920, using 20 mcg. of lead ion (Pb) in the control (*Solution A*).

Lead. A *Sample Solution* prepared as directed for organic compounds meets the requirements of the *Lead Limit Test,* page 929, using 10 mcg. of lead ion (Pb) in the control.

Packaging and storage. Store in full, tight containers in a cool place protected from light.

Functional use in foods. Flavoring agent.

CARBON, ACTIVATED

Activated Charcoal; Decolorizing Carbon; Active Carbon

DESCRIPTION

A solid, porous, carbonaceous material prepared by carbonizing and activating organic substances. The raw materials, which include sawdust, peat, lignite, coal, cellulose residues, coconut shells, petroleum coke, etc., may be carbonized and activated at a high temperature with or without the addition of inorganic salts in a stream of activating gases such as steam or carbon dioxide. Alternatively, carbonaceous matter may be treated with a chemical activating agent such as phosphoric acid or zinc chloride and the mixture carbonized at an elevated temperature, followed by removal of the chemical activating agent by water washing. Activated carbon occurs as a black, tasteless substance, varying in particle size from coarse granules to a fine powder, depending upon its intended use. It is insoluble in water and in organic solvents.

IDENTIFICATION

A. Place about 3 grams of powdered sample in a glass-stoppered Erlenmeyer flask containing 10 ml. of dilute hydrochloric acid (5 percent), boil for 30 seconds, and cool to room temperature. Add 100 ml. of iodine T.S., stopper, and shake vigorously for 30 seconds. Filter through Whatman No. 12 filter paper, or equivalent, discarding the first portion of filtrate. Compare 50 ml. of the subsequent filtrate with a reference solution prepared by diluting 10 ml. of iodine T.S. to 50 ml. with water, but not treated with carbon. The color of the carbon-treated iodine solution is no darker than that of the reference solution, indicating the adsorptivity of the sample.

B. Ignite a portion of the sample in air. Carbon monoxide and carbon dioxide are produced, and an ash remains.

SPECIFICATIONS

Limits of Impurities

Arsenic (as As). Not more than 3 parts per million (0.0003 percent).

Cyanogen compounds. Passes test.

Heavy metals (as Pb). Not more than 40 parts per million (0.004 percent).

Higher aromatic hydrocarbons. Passes test.

Lead. Not more than 10 parts per million (0.001 percent).

Water extractables. Not more than 4 percent.

The following additional *Specifications* should conform to the representations of the vendor: **Loss on drying** and **Residue on ignition.**

TESTS

Arsenic. A 20-ml. portion of the filtrate obtained in the test for *Water extractables,* diluted to 35 ml. with water, meets the requirements of the *Arsenic Test,* page 865.

Cyanogen compounds. Mix 5 grams of the sample with 50 ml. of water and 2 grams of tartaric acid, and distil the mixture, collecting 25 ml. of distillate below the surface of a mixture of 2 ml. of sodium hydroxide T.S. and 10 ml. of water contained in a small flask placed in an ice bath. Dilute the distillate to 50 ml. with water, and mix. Add 12 drops of ferrous sulfate T.S. to 25 ml. of the diluted distillate, heat almost to boiling, cool, and add 1 ml. of hydrochloric acid. No blue color is produced.

Heavy metals. A 10-ml. portion of the filtrate obtained in the test for *Water extractables* meets the requirements of the *Heavy Metals Test,* page 920, using 20 mcg. of lead ion (Pb) in the control (*Solution A*).

Higher aromatic hydrocarbons. Extract 1 gram of the sample with 12 ml. of cyclohexane in a continuous-extraction apparatus for 2 hours. The extract shows no more color and not more fluorescence than does a solution of 100 mcg. of quinine sulfate in 1000 ml. of 0.1 *N* sulfuric acid when observed in ultraviolet light.

Lead. A 20-ml. portion of the filtrate obtained in the test for *Water extractables* meets the requirements of the *Lead Limit Test,* page 929, using 10 mcg. of lead ion (Pb) in the control.

Water extractables. Transfer 5.00 grams of the sample into a 250-ml. flask provided with a reflux condenser and a Bunsen valve. Add 100 ml. of water and several glass beads, and reflux for 1 hour. Cool slightly, and filter through Whatman No. 12 or equivalent filter paper, discarding the first 10 ml. of filtrate. Cool the subsequent filtrate to room temperature, and pipet 25.0 ml. into a tared crystallization dish. (*Note:* Retain the remainder of the filtrate for the *Arsenic, Heavy metals,* and *Lead* tests.) Evaporate the filtrate in the dish to incipient dryness on a hot plate, never allowing the solution to boil. Dry for 1 hour at 100° in a vacuum oven, cool, and weigh.

Loss on drying, page 931. Dry at 120° for 4 hours.

Residue on ignition. Ignite 500 mg. as directed in the general method, page 945.

Packaging and storage. Store in well-closed containers.

Functional use in foods. Decolorizing agent; taste- and odor-removing agent; purification agent in food processing.

CARDAMOM OIL

DESCRIPTION

The volatile oil distilled from the seed of *Elettaria cardamomum* (L.) Maton (Fam. *Zingiberaceae*). It is a colorless or very pale yellow liquid with the aromatic, penetrating, and somewhat camphoraceous odor of cardamom, and a pungent, strongly aromatic taste. It is affected by light. It is miscible with alcohol.

SPECIFICATIONS

Angular rotation. Between +22° and +44°.

Refractive index. Between 1.463 and 1.466 at 20°.

Solubility in alcohol. Passes test.

Specific gravity. Between 0.917 and 0.947.

Limits of Impurities

Arsenic (as As). Not more than 3 parts per million (0.0003 percent).

Heavy metals (as Pb). Not more than 40 parts per million (0.004 percent).

Lead. Not more than 10 parts per million (0.001 percent).

TESTS

Angular rotation. Determine in a 100-mm. tube as directed under *Optical Rotation*, page 939.

Refractive index, page 945. Determine with an Abbé or other refractometer of equal or greater accuracy.

Solubility in alcohol. Proceed as directed in the general method, page 899. One ml. dissolves in 5 ml. of 70 percent alcohol.

Specific gravity. Determine by any reliable method (see page 5).

Arsenic. A *Sample Solution* prepared as directed for organic compounds meets the requirements of the *Arsenic Test*, page 865.

Heavy metals. Prepare and test a 500-mg. sample as directed in *Method II* under the *Heavy Metals Test*, page 920, using 20 mcg. of lead ion (Pb) in the control (*Solution A*).

Lead. A *Sample Solution* prepared as directed for organic compounds meets the requirements of the *Lead Limit Test*, page 929, using 10 mcg. of lead ion (Pb) in the control.

Packaging and storage. Store in full, tight containers in a cool place protected from light.

Functional use in foods. Flavoring agent.

CARNAUBA WAX

DESCRIPTION

A purified wax obtained from the leaf buds and leaves of the Brazilian wax palm, *Copernicia cereferia* (Arruda) Mart. It is hard and brittle, has a resinous fracture, and ranges in color from light brown to pale yellow. Its specific gravity is about 0.997. It is partially soluble in boiling alcohol, soluble in chloroform and in ether, but insoluble in water.

SPECIFICATIONS

Acid value. Between 2 and 10.

Ester value. Between 75 and 85.

Melting range. Between 82° and 86°.

Unsaponifiable matter. Between 50 and 55 percent.

Limits of Impurities

Arsenic (as As). Not more than 3 parts per million (0.0003 percent).

Heavy metals (as Pb). Not more than 40 parts per million (0.004 percent).

Lead. Not more than 10 parts per million (0.001 percent).

TESTS

Acid value. Determine as directed under *Method I* in the general procedure, page 902.

Ester value. Weigh accurately about 5 grams, and determine the *Saponification value* as directed in the general procedure, page 914. Subtract the *Acid Value* from the *Saponification value* to obtain the *Ester value.*

Melting range, page 931. Determine as directed in *Procedure for Class II.*

Unsaponifiable matter. Determine as directed in the general method, page 915.

Arsenic. A *Sample Solution* prepared as directed for organic compounds meets the requirements of the *Arsenic Test,* page 865.

Heavy metals. Prepare and test a 500-mg. sample as directed in *Method II* under the *Heavy Metals Test,* page 920, using 20 mcg. of lead ion (Pb) in the control (*Solution A*).

Lead. A *Sample Solution* prepared as directed for organic compounds meets the requirements of the *Lead Limit Test,* page 929, using 10 mcg. of lead ion (Pb) in the control.

Packaging and storage. Store in well-closed containers.

Functional use in foods. Candy glaze and polish.

β-CAROTENE

$$\text{(structural formula: } \beta\text{-carotene; two trimethylcyclohexenyl rings joined by } -CH{=}CHC(CH_3){=}CHCH{=}CHC(CH_3){=}CHCH{=}CHCH{=}C(CH_3)CH{=}CHCH{=}C(CH_3)CH{=}CH-)$$

$C_{40}H_{56}$ Mol. wt. 536.89

DESCRIPTION

Red crystals or crystalline powder. It is insoluble in water and in acids and alkalies, but is soluble in carbon disulfide, in benzene, and in chloroform. It is sparingly soluble in ether, in solvent hexane, and in vegetable oils, and is practically insoluble in methanol and in ethanol.

IDENTIFICATION

A. Determine the absorbance of *Sample Solution B* (prepared for the *Assay*) at 455 mμ and at 483 mμ. The ratio A_{455}/A_{483} is between 1.14 and 1.18.

B. Determine the absorbance of *Sample Solution B* at 455 mμ and that of *Sample Solution A* at 340 mμ. The ratio $A_{455} \times 10/A_{340}$ is not lower than 15.

SPECIFICATIONS

Assay. Not less than 96.0 percent and not more than the equivalent of 101.0 percent of $C_{40}H_{56}$.

Melting range. Between 176° and 182°, with decomposition.

Solution in chloroform. Passes test.

Limits of Impurities

Arsenic (as As). Not more than 3 parts per million (0.0003 percent).

Heavy metals (as Pb). Not more than 10 parts per million (0.001 percent).

Residue on ignition. Not more than 0.2 percent.

TESTS

Assay

Note: Carry out all work in low-actinic glassware and in subdued light.

Sample Solution A. Transfer about 50 mg., accurately weighed, into a 100-ml. volumetric flask, dissolve in 10 ml. of acid-free chloroform, immediately dilute to volume with cyclohexane, and mix. Pipet 5 ml. of this solution into a second 100-ml. volumetric flask, dilute to volume with cyclohexane, and mix.

Sample Solution B. Pipet 5 ml. of *Sample Solution A* into a 50-ml volumetric flask, dilute to volume with cyclohexane, and mix.

Procedure. Determine the absorbance of *Sample Solution B* in a 1-cm. cell at the wavelength of maximum absorption at about 455 mμ, with a suitable spectrophotometer, using cyclohexane as the blank. Calculate the quantity, in mg., of $C_{40}H_{56}$ in the sample taken by the formula 20,000*A*/250, in which *A* is the absorbance of the solution, and 250 is the absorptivity of pure β-carotene.

Melting range, page 931. Determine as directed under *Procedure for Class Ia.*

Solution in chloroform. A 1 in 100 solution in chloroform is complete and clear.

Arsenic. A *Sample Solution* prepared as directed for organic compounds meets the requirements of the *Arsenic Test,* page 865.

Heavy metals. Prepare and test a 2-gram sample as directed in *Method II* under the *Heavy Metals Test,* page 920, using 20 mcg. of lead ion (Pb) in the control (*Solution A*).

Residue on ignition. Ignite 2 grams as directed in the general method, page 945.

Packaging and storage. Store in tight, light-resistant containers.

Functional use in foods. Nutrient; dietary supplement; color.

CARRAGEENAN

DESCRIPTION

A hydrocolloid consisting mainly of a sulfated polysaccharide the dominant hexose units of which are galactose and anhydrogalactose. It is a two-component, polyanionic colloid. The two components, designated as *kappa* and *lambda,* occur in varying proportions and degrees of polymerization and are associated with ammonium, calcium, potassium, or sodium ions or with combinations of these four. Physical properties of solutions of carrageenan may vary with variations in the proportions of the two components. It is obtained by extraction with water of members of the *Gigartinaceae* and *Solieriaceae* families of the class *Rhodophyceae* (red seaweed). Members of these families used in the production of carrageenan may include *Chondrus crispus, C. ocellatus, Eucheuma cottonii, E. spinosum, Gigartina acicularis, G. pistillata, G. radula,* and *G. stellata.*

Carrageenan occurs as a yellowish to colorless, coarse to fine powder which is practically odorless and has a mucilaginous taste. One gram dissolves in 100 ml. of water at a temperature of about 80°, forming a viscous, clear or slightly opalescent solution which flows readily. It disperses in water more readily if first moistened with alcohol, glycerin, or a saturated solution of sucrose in water.

Note: Carrageenan modified by increasing the concentration of one of the naturally occurring salts (ammonium, calcium, potassium, or sodium) to such a level that it is the dominant salt may also be used as a food additive.

IDENTIFICATION

A. Add 4 grams to 200 ml. of water and heat the mixture in a water bath at a temperature of about 80°, with constant stirring, until a viscous solution results. Replace any water lost by evaporation and allow it to cool to room temperature. It becomes more viscous and may form a gel.

B. To 50 ml. of the solution or gel obtained in *Identification test A* add 100 mg. of potassium chloride and 50 mg. of sodium chloride, mix well, reheat and cool. A short-textured gel forms. (Not positive for the pure *lambda* fraction of carrageenan.)

C. To 5 ml. of the solution or gel obtained in *Identification test A* add 1 drop of a 1 in 100 solution of methylene blue. A fibrous precipitate forms.

SPECIFICATIONS

Ash (Total). Not more than 35.0 percent.

Ash (Acid-insoluble). Not more than 1.0 percent.

Loss on drying. Not more than 12 percent.

Sulfate. Between 20 percent and 40 percent on the dry weight basis.

Limits of Impurities

Arsenic (as As). Not more than 3 parts per million (0.0003 percent).

Heavy metals (as Pb). Not more than 40 parts per million (0.004 percent).

Lead. Not more than 10 parts per million (0.001 percent).

TESTS

Ash (Total). Transfer about 2 grams, accurately weighed, into a previously ignited, tared, silica or platinum crucible. Heat the sample with a suitable infrared heat lamp, increasing the intensity gradually, until it is completely charred, and then continue for an additional 30 minutes. Transfer the crucible and charred sample into a muffle furnace and ignite at about 550° for 1 hour, then cool in a desiccator and weigh. Repeat the ignition in the muffle furnace until a constant weight is attained. If a carbon-free ash is not obtained after the first ignition, moisten the charred spots with a 1 in 10 solution of ammonium nitrate and dry under an infrared heat lamp before reigniting.

Ash (Acid-insoluble). Proceed as directed in the general method, page 869.

Loss on drying, page 931. Dry at 105° for 12 hours.

Sulfate. Transfer about 500 mg., previously dried at 105° for 12 hours and accurately weighed, into a 100-ml. Kjeldahl flask. Add 10 ml. of nitric acid and heat gently for 30 minutes, adding more of the acid, if necessary, to prevent evaporation to dryness, and to yield a volume of about 3 ml. at the end of the heating. Cool the mixture to room temperature and decompose the excess nitric acid by the addition of formaldehyde T.S., dropwise, heating, if necessary, until no brown fumes continue to be evolved. Continue the heating until the volume of the reaction mixture is reduced to about 5 ml., then cool, and transfer the residue quantitatively with the aid of water into a 400-ml. beaker, dilute it to about 100 ml., and filter, if necessary, to produce a clear solution. Dilute the solution to about 200 ml., and add 1 ml. of hydrochloric acid. Heat to boiling and add, dropwise, with constant stirring, an excess (about 6 ml.) of hot barium chloride T.S. Heat the mixture for 1 hour on a steam bath, collect the precipitate of barium sulfate on a filter, wash it until free from chloride, dry, ignite, and weigh. The weight of the barium sulfate so obtained, multiplied by 0.4116, gives the equivalent of sulfate (SO_4).

Arsenic. A *Sample Solution* prepared as directed for organic compounds meets the requirements of the *Arsenic Test*, page 865.

Heavy metals. Prepare and test a 500-mg. sample as directed in *Method II* under the *Heavy Metals Test*, page 920, using 20 mcg. of lead ion (Pb) in the control (*Solution A*).

Lead. A *Sample Solution* prepared as directed for organic compounds meets the requirements of the *Lead Limit Test*, page 929, using 10 mcg. of lead ion (Pb) in the control.

Packaging and storage. Store in well-closed containers.

Functional use in foods. Emulsifier; stabilizer; thickener; gelling agent.

CARROT SEED OIL

DESCRIPTION

The volatile oil obtained by steam distillation from the crushed seeds of *Daucus carota* L. (Fam. *Umbelliferae*). It is a light yellow to amber liquid, having a pleasant aromatic odor. It is soluble in most fixed oils, and is soluble, with opalescence, in mineral oil. It is practically insoluble in glycerin and in propylene glycol.

SPECIFICATIONS

Acid value. Not more than 5.0.

Angular rotation. Between $-4°$ and $-30°$.

Refractive index. Between 1.480 and 1.491 at 20°.

Saponification value. Between 9 and 58.

Solubility in alcohol. Passes test.

Specific gravity. Between 0.900 and 0.943.

TESTS

Acid value. Determine as directed in the general method, page 893.

Angular rotation. Determine in a 100-mm. tube as directed under *Optical Rotation*, page 939.

Refractive index, page 945. Determine with an Abbé or other refractometer of equal or greater accuracy.

Saponification value. Determine as directed in the general method, page 896, using about 5 grams, accurately weighed.

Solubility in alcohol. Proceed as directed in the general method, page 899. One ml. dissolves in 0.5 ml. of 90 percent alcohol. The solution may become opalescent upon further dilution up to 10 ml.

Specific gravity. Determine by any reliable method (see page 5).

Packaging and storage. Store in full, tight, preferably glass, aluminum, tin-lined or other suitably lined containers in a cool place protected from light.

Functional use in foods. Flavoring agent.

CARVACROL

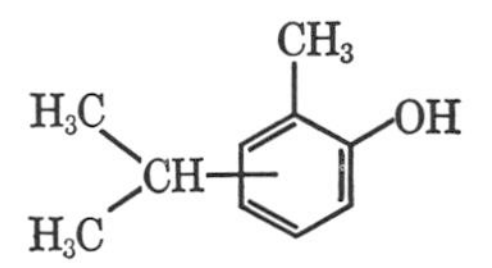

$C_{10}H_{14}O$ Mol. wt. 150.22

DESCRIPTION

A colorless to pale yellow liquid consisting mainly of a mixture of isomeric carvacrols (isopropyl *o*-cresols), and having a pungent, spicy odor resembling that of thymol. It is freely soluble in alcohol and in ether, but is insoluble in water.

SPECIFICATIONS

Assay. Not less than 98 percent, by volume, of phenols.

Refractive index. Between 1.521 and 1.526 at 20°.

Solubility in alcohol. Passes test.

Specific gravity. Between 0.974 and 0.979.

Limits of Impurities

Arsenic (as As). Not more than 3 parts per million (0.0003 percent).

Heavy metals (as Pb). Not more than 40 parts per million (0.004 percent).

Lead. Not more than 10 parts per million (0.001 percent).

TESTS

Assay. Proceed as directed under *Phenols,* page 898.

Refractive index, page 945. Determine with an Abbé or other refractometer of equal or greater accuracy.

Solubility in alcohol. Proceed as directed in the general method, page 899. One ml. dissolves in 4 ml. of 60 percent alcohol to form a clear solution.

Specific gravity. Determine by any reliable method (see page 5).

Arsenic. A *Sample Solution* prepared as directed for organic compounds meets the requirements of the *Arsenic Test,* page 865.

Heavy metals. Prepare and test a 500-mg. sample as directed in *Method II* under the *Heavy Metals Test,* page 920, using 20 mcg. of lead ion (Pb) in the control (*Solution A*).

Lead. A *Sample Solution* prepared as directed for organic compounds meets the requirements of the *Lead Limit Test,* page 929, using 10 mcg. of lead ion (Pb) in the control.

Packaging and storage. Store in full, tight, preferably glass, tin-lined or other suitably lined containers in a cool place protected from light.

Functional use in foods. Flavoring agent.

d-CARVONE

Dextro-Carvone; *d*-1-Methyl-4-isopropenyl-6-cyclohexen-2-one

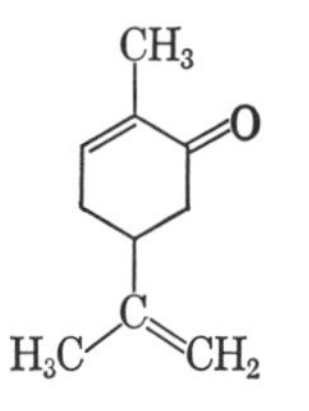

$C_{10}H_{14}O$ Mol. wt. 150.22

DESCRIPTION

d-Carvone is usually prepared by fractional distillation from caraway oil. It may be prepared in a similar manner from either dillseed oil or dillweed oil, but this type differs in odor and flavor from that derived from caraway oil. It is a colorless to light yellow liquid having an odor of caraway. It is soluble in propylene glycol, in fixed oils, and in mineral oil. It is miscible with alcohol, but relatively insoluble in glycerin.

SPECIFICATIONS

Assay. Not less than 95.0 percent of $C_{10}H_{14}O$.

Angular rotation. Between +56° and +60°.

Refractive index. Between 1.496 and 1.499 at 20°.

Solubility in alcohol. Passes test.

Specific gravity. Between 0.956 and 0.960.

Limits of Impurities

Arsenic (as As). Not more than 3 parts per million (0.0003 percent).

Heavy metals (as Pb). Not more than 40 parts per million (0.004 percent).

Lead. Not more than 10 parts per million (0.001 percent).

TESTS

Assay. Weigh accurately about 1 gram, and proceed as directed under *Aldehydes and Ketones-Hydroxylamine Method,* page 894, using 75.11 as the equivalence factor (*E*) in the calculation.

Angular rotation. Determine in a 100-mm. tube as directed under *Optical Rotation,* page 939.

Refractive index, page 945. Determine with an Abbé or other refractometer of equal or greater accuracy.

Solubility in alcohol. Proceed as directed in the general method, page 899. One ml. dissolves in 5 ml. of 60 percent alcohol.

Specific gravity. Determine by any reliable method (see page 5).

Arsenic. A *Sample Solution* prepared as directed for organic compounds meets the requirements of the *Arsenic Test,* page 865.

Heavy metals. Prepare and test a 500-mg. sample as directed in *Method II* under the *Heavy Metals Test,* page 920, using 20 mcg. of lead ion (Pb) in the control (*Solution A*).

Lead. A *Sample Solution* prepared as directed for organic compounds meets the requirements of the *Lead Limit Test,* page 929, using 10 mcg. of lead ion (Pb) in the control.

Packaging and storage. Store in full, tight, preferably glass, aluminum, tin-lined, or other suitably lined containers in a cool place protected from light.

Functional use in foods. Flavoring agent.

l-CARVONE

Levo-Carvone; *l*-1-Methyl-4-isopropenyl-6-cyclohexen-2-one

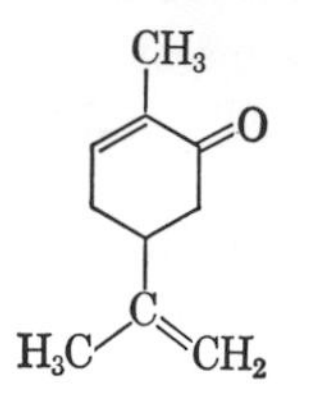

$C_{10}H_{14}O$ Mol. wt. 150.22

DESCRIPTION

l-Carvone occurs in several essential oils. It may be isolated from spearmint oil or synthesized commercially from *d*-limonene. It is a colorless to pale straw colored liquid having an odor of spearmint. It is soluble in propylene glycol, in fixed oils, and in mineral oil. It is miscible with alcohol, but insoluble in glycerin.

SPECIFICATIONS

Assay. Not less than 97.0 percent $C_{10}H_{14}O$.

Angular rotation. Between −57° and −62°.

Refractive index. Between 1.495 and 1.499 at 20°.

Solubility in alcohol. Passes test.

Specific gravity. Between 0.956 and 0.960.

TESTS

Assay. Weigh accurately about 1 gram, and proceed as directed under *Aldehydes and Ketones-Hydroxylamine Method,* page 894, using 75.11 as the equivalence factor (E) in the calculation.

Angular rotation. Determine in a 100-mm. tube as directed under *Optical Rotation,* page 939.

Refractive index, page 945. Determine with an Abbé or other refractometer of equal or greater accuracy.

Solubility in alcohol. Proceed as directed in the general method, page 899. One ml. dissolves in 2.0 ml. of 70 percent alcohol.

Specific gravity. Determine by any reliable method (see page 5).

Packaging and storage. Store in full, tight, preferably glass, aluminum, tin-lined, or other suitably lined containers in a cool place protected from light.

Functional use in foods. Flavoring agent.

β-CARYOPHYLLENE

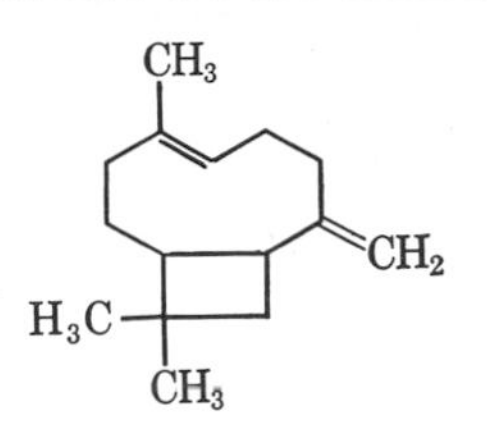

$C_{15}H_{24}$ Mol. wt. 204.36

DESCRIPTION

A mixture of sesquiterpenes differing slightly in structure and occurring in many essential oils, especially in clove oil. It is a colorless to slightly yellow oily liquid having a light clove-like odor. It is soluble in alcohol and in ether, but insoluble in water.

SPECIFICATIONS

Angular rotation. Between −5° and −10°.

Refractive index. Between 1.498 and 1.504 at 20°.

Solubility in alcohol. Passes test.

Specific gravity. Between 0.897 and 0.910.

Limits of Impurities

Arsenic (as As). Not more than 3 parts per million (0.0003 percent).

Heavy metals (as Pb). Not more than 40 parts per million (0.004 percent).

Lead. Not more than 10 parts per million (0.001 percent).

Free Phenols. Not more than 3 percent.

TESTS

Angular rotation. Determine in a 100-mm. tube as directed under *Optical Rotation*, page 939.

Refractive index, page 945. Determine with an Abbé or other refractometer of equal or greater accuracy.

Solubility in alcohol. Proceed as directed in the general method, page 899. One ml. dissolves in 4 ml. of 95 percent alcohol to form a clear solution.

Specific gravity. Determine by any reliable method (see page 5).

Arsenic. A *Sample Solution* prepared as directed for organic compounds meets the requirements of the *Arsenic Test*, page 865.

Heavy metals. Prepare and test a 500-mg. sample as directed in *Method II* under the *Heavy Metals Test*, page 920, using 20 mcg. of lead ion (Pb) in the control (*Solution A*).

Lead. A *Sample Solution* prepared as directed for organic compounds meets the requirements of the *Lead Limit Test,* page 929, using 10 mcg. of lead ion (Pb) in the control.

Free Phenols. Determine as directed in the general method, page 899.

Packaging and storage. Store in full, tight, preferably glass, tin-lined or other suitably lined containers in a cool place protected from light.

Functional use in foods. Flavoring agent.

CASCARILLA OIL

Sweetwood Bark Oil

DESCRIPTION

The volatile oil obtained by steam distillation of the dried bark of *Croton cascarilla* Benn. and of *Croton eluteria* Benn. (Fam. *Euphorbiaceae*). It is a light yellow to brown amber liquid, having a pleasant spicy odor. It is soluble in most fixed oils and in mineral oil, but it is practically insoluble in glycerin and in propylene glycol.

SPECIFICATIONS

Acid value. Between 3 and 10.

Angular rotation. Between $-1°$ and $+8°$.

Ester value after acetylation. Between 62 and 88.

Refractive index. Between 1.488 and 1.494 at 20°.

Saponification value. Between 8 and 20.

Solubility in alcohol. Passes test.

Specific gravity. Between 0.892 and 0.914.

TESTS

Acid value. Determine as directed in the general method, page 893.

Angular rotation. Determine in a 100-mm. tube as directed under *Optical Rotation,* page 939.

Ester value after acetylation. Proceed as directed under *Total Alcohols,* page 893, using about 2 grams of the dried acetylized oil, accurately weighed. Calculate the ester value after acetylation by the formula $A \times 28.05/B$, in which A is the number of ml. of 0.5 N alcoholic potassium hydroxide consumed in the saponification and B is the weight of the sample of acetylized oil in grams.

Refractive index, page 945. Determine with an Abbé or other refractometer of equal or greater accuracy.

Saponification value. Determine as directed in the general method, page 896, using 5 grams, accurately weighed.

Solubility in alcohol. Proceed as directed in the general method, page 899. One ml. dissolves in 0.5 ml. of 90 percent alcohol and remains in solution on dilution to 10 ml.

Specific gravity. Determine by any reliable method (see page 5).

Packaging and storage. Store in full, tight, preferably glass, aluminum, or tin-lined containers in a cool place protected from light.

Functional use in foods. Flavoring agent.

CASSIA OIL

Cinnamon Oil

DESCRIPTION

The volatile oil obtained by steam distillation from the leaves and twigs of *Cinnamomum cassia* Blume (Fam. *Lauraceae*), rectified by distillation. It is a yellowish or brownish liquid having the characteristic odor and taste of cassia cinnamon. Upon aging or exposure to air it darkens and thickens. It is soluble in glacial acetic acid and in alcohol.

SPECIFICATIONS

Assay. Not less than 80 percent, by volume, of total aldehydes.

Angular rotation. Between $-1°$ and $+1°$.

Refractive index. Between 1.602 and 1.614 at 20°.

Solubility in alcohol. Passes test.

Specific gravity. Between 1.045 and 1.063.

Limits of Impurities

Arsenic (as As). Not more than 3 parts per million (0.0003 percent).

Chlorinated compounds. Passes test.

Heavy metals (as Pb). Not more than 40 parts per million (0.004 percent).

Lead. Not more than 10 parts per million (0.001 percent).

Rosin or rosin oils. Passes test.

TESTS

Assay. Proceed as directed under *Aldehydes and Ketones-Neutral Sulfite Method*, page 895.

Angular rotation. Determine in a 100-mm. tube as directed under *Optical Rotation*, page 939.

Refractive index, page 945. Determine with an Abbé or other refractometer of equal or greater accuracy.

Solubility in alcohol. Proceed as directed in the general method, page 899. One ml. dissolves in 2 ml. of 70 percent alcohol.

Specific gravity. Determine by any reliable method (see page 5).

Arsenic. A *Sample Solution* prepared as directed for organic compounds meets the requirements of the *Arsenic Test,* page 865.

Chlorinated compounds. Proceed as directed in the general method, page 895.

Heavy metals. Prepare and test a 500-mg. sample as directed in *Method II* under the *Heavy Metals Test,* page 920, using 20 mcg. of lead ion (Pb) in the control (*Solution A*).

Lead. A *Sample Solution* prepared as directed for organic compounds meets the requirements of the *Lead Limit Test,* page 929, using 10 mcg. of lead ion (Pb) in the control.

Rosin or rosin oils. Shake a 2-ml. sample in a test tube with 5 to 10 ml. of solvent hexane, allow the liquids to separate, decant the hexane layer, which is but slightly colored, into another test tube, and shake it with an equal volume of cupric acetate solution (1 in 1000). The mixture does not assume a green color.

Packaging and storage. Store in full, tight, light-resistant containers. Avoid exposure to excessive heat.

Functional use in foods. Flavoring agent.

CASTOR OIL

DESCRIPTION

The fixed oil obtained from the seed of *Ricinus communis* L. (Fam. *Euphorbiaceae*). It is a pale yellowish or almost colorless, transparent, viscid liquid and has a faint, mild odor, and a bland, characteristic taste. It is soluble in alcohol, and is miscible with absolute alcohol, with glacial acetic acid, with chloroform, and with ether.

IDENTIFICATION

It is only partly soluble in solvent hexane (distinction from *most other fixed oils*), but it yields a clear liquid with an equal volume of alcohol (*foreign fixed oils*).

SPECIFICATIONS

Hydroxyl value. Between 160 and 168.

Iodine value. Between 83 and 88.

Saponification value. Between 176 and 182.

Specific gravity. Between 0.945 and 0.965.

Limits of Impurities

Arsenic (as As). Not more than 3 parts per million (0.0003 percent).

Free fatty acids. Passes test.

Heavy metals (as Pb). Not more than 10 parts per million (0.001 percent).

TESTS

Hydroxyl value. Determine as directed under *Method II* in the general procedure, page 904.

Iodine value. Determine by the *Wijs Method,* page 906, using about 300 mg., accurately weighed.

Saponification value. Determine as directed in the general method, page 914, using about 3 grams, accurately weighed.

Specific gravity. Determine as directed in the general method page 915.

Arsenic. A *Sample Solution* prepared as directed for organic compounds meets the requirements of the *Arsenic Test,* page 865.

Free fatty acids. Dissolve about 10 grams, accurately weighed, in 50 ml. of a mixture of equal volumes of alcohol and ether (which has been neutralized to phenolphthalein with 0.1 *N* sodium hydroxide) contained in a flask. Add 1 ml. of phenolphthalein T.S., and titrate with 0.1 *N* sodium hydroxide until the solution remains pink after shaking for 30 seconds. Not more than 7 ml. of 0.1 *N* sodium hydroxide is required for a 10.0-gram sample.

Heavy metals. Prepare and test a 2-gram sample as directed in *Method II* under the *Heavy Metals Test,* page 920, using 20 mcg. of lead ion (Pb) in the control (*Solution A*).

Packaging and storage. Store in tight containers, and avoid exposure to excessive heat.

Functional use in foods. Antisticking agent; release agent; component of protective coatings.

CEDAR LEAF OIL

Thuja Oil; White Cedar Leaf Oil

DESCRIPTION

The volatile oil obtained by steam distillation from the fresh leaves and branch ends of the eastern arborvitae, *Thuja occidentalis* L. (Fam. *Cupressaceae*). It is a colorless to yellow liquid having a strong camphoraceous and sage-like odor. It is soluble in most fixed oils, in mineral oil, and in propylene glycol. It is practically insoluble in glycerin.

SPECIFICATIONS

Assay. Not less than 60.0 percent of ketones, calculated as thujone ($C_{10}H_{16}O$).

Angular rotation. Between −10° and −14°.

Refractive index. Between 1.456 and 1.459 at 20°.

Solubility in alcohol. Passes test.

Specific gravity. Between 0.910 and 0.920.

TESTS

Assay. Weigh accurately about 1 gram, and proceed as directed under *Aldehydes and Ketones-Hydroxylamine Method*, page 894, using 76.10 as the equivalence factor (*E*) in the calculation.

Angular rotation. Determine in a 100-mm. tube as directed under *Optical Rotation*, page 939.

Refractive index, page 945. Determine with an Abbé or other refractometer of equal or greater accuracy.

Solubility in alcohol. Proceed as directed in the general method, page 899. One ml. dissolves in 3 ml. of 70 percent alcohol, occasionally becoming cloudy on dilution to 10 ml.

Specific gravity. Determine by any reliable method (see page 5).

Packaging and storage. Store in full, tight, preferably glass, or tin-lined containers, in a cool place protected from light.

Functional use in foods. Flavoring agent.

CELERY SEED OIL

DESCRIPTION

The volatile oil obtained by steam distillation of the fruit or seed of *Apium graveolens* L. It is a yellow to greenish brown liquid, having a pleasant aromatic odor. It is soluble in most fixed oils with the formation of a flocculent precipitate, and in mineral oil with turbidity. It is partly soluble in propylene glycol, and it is insoluble in glycerin.

SPECIFICATIONS

Acid value. Not more than 3.5.

Angular rotation. Between +48° and +78°.

Refractive index. Between 1.480 and 1.490 at 20°.

Saponification value. Between 35 and 75.

Solubility in alcohol. Passes test.

Specific gravity. Between 0.872 and 0.910.

TESTS

Acid value. Determine as directed in the general method, page 893.

Angular rotation. Determine in a 100-mm. tube as directed under *Optical Rotation*, page 939.

Refractive index, page 945. Determine with an Abbé or other refractometer of equal or greater accuracy.

Saponification value. Determine as directed in the general method, page 896, using 5 grams, accurately weighed.

Solubility in alcohol. Proceed as directed in the general method, page 899. One ml. dissolves in 8 ml. of 90 percent alcohol, usually with turbidity.

Specific gravity. Determine by any reliable method (see page 5).

Packaging and storage. Store in full, tight, glass, tin-lined, or aluminum containers in a cool place protected from light.

Functional use in foods. Flavoring agent.

CELLULOSE, MICROCRYSTALLINE

Cellulose Gel

DESCRIPTION

Microcrystalline cellulose is purified, partially depolymerized cellulose prepared by treating alpha cellulose, obtained as a pulp from fibrous plant material, with mineral acids. It occurs as a fine, white, odorless, crystalline powder. It consists of free-flowing, non-fibrous particles which may be compressed into self-binding tablets which disintegrate rapidly in water. It is insoluble in water, in dilute acids, and in most organic solvents. It is slightly soluble in sodium hydroxide T.S.

IDENTIFICATION

Mix 30 grams of the sample with 270 ml. of water in a high-speed (18,000 rpm) power blender for 5 minutes.

A. Transfer 100 ml. of the mixture to a 100-ml. graduate, and allow to stand for 3 hours: a white, opaque, bubble-free dispersion, which does not form a supernatant liquid at the surface, is obtained.

B. To 20 ml. of the mixture add a few drops of iodine T.S. and mix: no purplish to blue or blue color is produced.

SPECIFICATIONS

Loss on drying. Not more than 5 percent.

pH. Between 5.5 and 7.0.

Limits of Impurities

Arsenic (as As). Not more than 3 parts per million (0.0003 percent).

Heavy metals (as Pb). Not more than 10 parts per million (0.001 percent).

Residue on ignition. Not more than 0.05 percent.

Water-soluble substances. Not more than 0.16 percent.

TESTS

Loss on drying, page 931. Dry to constant weight at 105°.

pH. Shake about 5 grams with 40 ml. of water for 20 minutes, centrifuge, and determine the pH of the supernatant liquid by the *Potentiometric Method,* page 941.

Arsenic. A *Sample Solution* prepared as directed for organic compounds meets the requirements for the *Arsenic Test,* page 865.

Heavy metals. Prepare and test a 2-gram sample as directed in *Method II* under the *Heavy Metals Test,* page 920, using 20 mcg. of lead ion (Pb) in the control (*Solution A*).

Residue on ignition, page 945. Ignite 2 grams as directed in the general method.

Water-soluble substances. Shake 5 grams with 80 ml. of water for 10 minutes. Filter the mixture through Whatman No. 42 or equivalent filter paper into a tared beaker, evaporate the filtrate to dryness on a steam bath, dry at 105° for 1 hour, cool, and weigh.

Packaging and storage. Store in well-closed containers.

Functional use in foods. Anticaking agent; binding agent; disintegrating agent; dispersing agent; tableting aid.

CHAMOMILE OIL, ENGLISH

DESCRIPTION

The oil obtained by steam distillation of the dried flowers of the so-called English or Roman Chamomile, *Anthemis nobilis* L. It is a light blue, or light greenish blue, liquid with a strong aromatic odor, characteristic of the flowers. The color may change with age to greenish yellow, or yellow-brown. It is soluble in most fixed oils and it is almost completely soluble in mineral oil. It is soluble, with slight haziness, in propylene glycol, but it is insoluble in glycerin.

SPECIFICATIONS

Acid value. Not more than 15.0.

Ester value. Between 250 and 310.

Refractive index. Between 1.440 and 1.450 at 20°.

Solubility in alcohol. Passes test.

Specific gravity. Between 0.892 and 0.910.

TESTS

Acid value. Determine as directed in the general method, page 893.

Ester value. Determine as directed in the general method, page 897, using about 1 gram, accurately weighed.

Refractive index, page 945. Determine with an Abbé or other refractometer of equal or greater accuracy.

Solubility in alcohol. Proceed as directed in the general method, page 899. One ml. dissolves in 2 ml. of 80 percent alcohol, sometimes with a slight precipitate.

Specific gravity. Determine by any reliable method (see page 5).

Packaging and storage. Store in full, tight, glass or aluminum containers in a cool place protected from light.

Functional use in foods. Flavoring agent.

CHAMOMILE OIL, GERMAN

Hungarian Chamomile Oil

DESCRIPTION

The oil obtained by steam distillation of the flowers and stalks of *Matricaria chamomilla* L. It is a deep blue or bluish green liquid with a strong and characteristic odor and a bitter aromatic taste. When exposed to light or air, the blue color changes to green and finally to brown. Upon cooling, the oil may become viscous. It is soluble in most fixed oils and in propylene glycol. It is insoluble in glycerin and in mineral oil.

SPECIFICATIONS

Acid value. Between 5 and 50.

Ester value. Not more than 40.

Ester value after acetylation. Between 65 and 155.

Solubility in alcohol. Passes test.

Specific gravity. Between 0.910 and 0.950.

TESTS

Acid value. Determine as directed in the general method, page 893.

Ester value. Determine as directed in the general method, page 897, using about 5 grams, accurately weighed.

Ester value after acetylation. Acetylate a 10-ml. sample as directed under *Total Alcohols*, page 893. Weigh accurately about 1.5 grams of the dried, acetylated oil, and proceed as directed under *Ester Value*, page 897.

Solubility in alcohol. Proceed as directed in the general method, page 899. The oil does not usually dissolve clearly in 95 percent alcohol.

Specific gravity. Determine by any reliable method (see page 5).

Packaging and storage. Store in full, tight, glass, or aluminum containers in a cool place protected from light.

Functional use in foods. Flavoring agent.

CHOLIC ACID

Cholalic Acid; 3,7,12-Trihydroxycholanic Acid

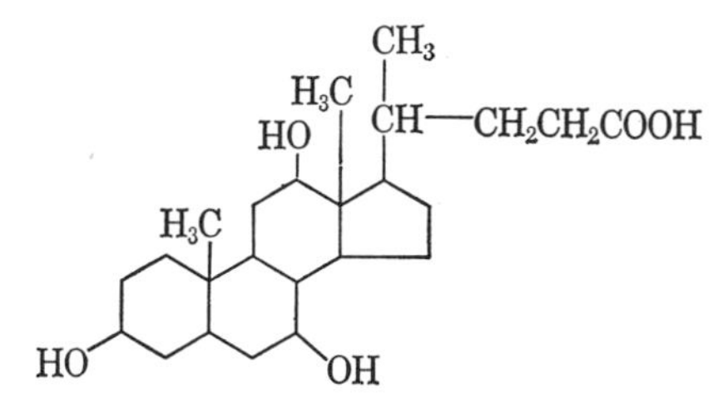

$C_{24}H_{40}O_5$ Mol. wt. 408.58

DESCRIPTION

Colorless plates or a white, crystalline powder having a bitter taste with a sweetish after-taste. One gram dissolves in about 30 ml. of alcohol or acetone, and in about 7 ml. of glacial acetic acid. It is very slightly soluble in water.

IDENTIFICATION

To 1 ml. of a 1 in 5000 solution in 50 percent acetic acid add 1 ml. of a solution of furfural (1 in 100). Cool in an ice bath for 5 minutes, add 15 ml. of dilute sulfuric acid (1 in 2), mix, and warm in a water bath at 70° for 10 minutes. Immediately cool in an ice bath and stir for 2 minutes. A blue color develops.

SPECIFICATIONS

Assay. Not less than 98.0 percent of $C_{24}H_{40}O_5$.

Melting range. Between 197° and 202°.

Specific rotation, $[\alpha]_D^{25°}$. Not less than +37°.

Limits of Impurities

Arsenic (as As). Not more than 3 parts per million (0.0003 percent).

Heavy metals (as Pb). Not more than 40 parts per million (0.004 percent).

Lead. Not more than 10 parts per million (0.001 percent).

Loss on drying. Not more than 0.5 percent.

Residue on ignition. Not more than 0.1 percent.

TESTS

Assay. Transfer about 400 mg., accurately weighed, into a 250-ml. Erlenmeyer flask, add 20 ml. of water and 40 ml. of alcohol, cover with a watch glass, heat gently on a steam bath until dissolved, and cool. Add 5 drops of phenolphthalein T.S., and titrate with 0.1 *N* sodium hydroxide, using a 10-ml. microburet, to the first pink color that persists for 15 seconds. Perform a blank determination (see page 2) and make any necessary correction. Each ml. of 0.1 *N* sodium hydroxide is equivalent to 40.86 mg. of $C_{24}H_{40}O_5$.

Melting range. Determine as directed in the general procedure, page 931.

Specific rotation, page 939. Determine in a solution in alcohol containing 200 mg. in each 10 ml.

Arsenic. A *Sample Solution* prepared as directed for organic compounds meets the requirements of the *Arsenic Test,* page 865.

Heavy metals. Prepare and test a 500-mg. sample as directed in *Method II* under the *Heavy Metals Test,* page 920, using 20 mcg. of lead ion (Pb) in the control (*Solution A*).

Lead. A *Sample Solution* prepared as directed for organic compounds meets the requirements of the *Lead Limit Test,* page 929, using 10 mcg. of lead ion (Pb) in the control.

Loss on drying, page 931. Dry at 140° under a vacuum of not more than 5 mm. of mercury for 4 hours.

Residue on ignition, page 945. Ignite 2 grams as directed in the general method.

Packaging and storage. Store in tight containers.

Functional use in foods. Emulsifying agent.

CHOLINE BITARTRATE

(2-Hydroxyethyl)trimethylammonium Bitartrate

$$[HOCH_2CH_2\overset{+}{N}(CH_3)_3]HC_4H_4O_6^-$$

$C_9H_{19}NO_7$ Mol. wt. 253.25

DESCRIPTION

A white, hygroscopic, crystalline powder having an acidic taste. It is odorless or may have a faint trimethylamine-like odor. It is freely soluble in water, slightly soluble in alcohol, and insoluble in ether, chloroform, and benzene.

IDENTIFICATION

A. Dissolve 500 mg. in 2 ml. of water, add 3 ml. of sodium hydroxide T.S., and heat to boiling. The odor of trimethylamine is detectable.

B. Dissolve 500 mg. in 2 ml. of iodine T.S. A reddish brown precipitate is immediately formed. Add 5 ml. of sodium hydroxide T.S. The precipitate dissolves and the solution becomes clear yellow. Heat the solution. A pale yellow precipitate forms and the odor of iodoform may be detected.

C. To 2 ml. of cobaltous chloride T.S. add 1 ml. of a 1 in 100 solution of the sample and 2 ml. of potassium ferrocyanide solution (1 in 50). An emerald-green color develops immediately.

SPECIFICATIONS

Assay. Not less than 98.0 percent of $C_9H_{19}NO_7$, calculated on the anhydrous basis.

Limits of Impurities

Arsenic (as As). Not more than 3 parts per million (0.0003 percent).

Heavy metals (as Pb). Not more than 20 parts per million (0.002 percent).

Lead. Not more than 10 parts per million (0.001 percent).

Residue on ignition. Not more than 0.1 percent.

Water. Not more than 0.5 percent.

TESTS

Assay. Transfer about 500 mg., accurately weighed, into a 250-ml. Erlenmeyer flask, add 50 ml. of glacial acetic acid, and warm on a steam bath until solution is complete. Cool, add 2 drops of crystal violet T.S., and titrate with 0.1 *N* perchloric acid in glacial acetic acid to a green end-point. Perform a blank determination (see page 2), and make any necessary correction. Each ml. of 0.1 *N* perchloric acid is equivalent to 25.36 mg. of $C_9H_{19}NO_7$.

Arsenic. A *Sample Solution* prepared as directed for organic compounds meets the requirements of the *Arsenic Test,* page 865.

Heavy metals. A solution of 1 gram in 25 ml. of water meets the requirements of the *Heavy Metals Test,* page 920, using 20 mcg. of lead ion (Pb) in the control (*Solution A*).

Lead. A *Sample Solution* prepared as directed for organic compounds meets the requirements of the *Lead Limit Test,* page 929, using 10 mcg. of lead ion (Pb) in the control.

Residue on ignition. Ignite 2 grams as directed in the general method, page 945.

Water. Determine by drying in a vacuum desiccator over phosphorus pentoxide for 4 hours or by the *Karl Fischer Titrimetric Method,* page 977, using a 2-gram sample dissolved in 50 ml. of methanol.

Packaging and storage. Store in tight containers.

Functional use in foods. Nutrient; dietary supplement.

CHOLINE CHLORIDE

(2-Hydroxyethyl)trimethylammonium Chloride

$$[HOCH_2CH_2\overset{+}{N}(CH_3)_3]Cl^-$$

$C_5H_{14}ClNO$ Mol. wt. 139.63

DESCRIPTION

Colorless or white crystals or crystalline powder, usually having a slight odor of trimethylamine. It is hygroscopic, and is very soluble in water and in alcohol.

IDENTIFICATION

A. It responds to *Identification tests A, B,* and *C* under *Choline Bitartrate,* page 190.

B. A 1 in 20 solution gives positive tests for *Chloride,* page 926.

SPECIFICATIONS

Assay. Not less than 98.0 percent of $C_5H_{14}ClNO$, calculated on the anhydrous basis.

Limits of Impurities

Arsenic (as As). Not more than 3 parts per million (0.0003 percent).

Heavy metals (as Pb). Not more than 20 parts per million (0.002 percent).

Lead. Not more than 10 parts per million (0.001 percent).

Residue on ignition. Not more than 0.05 percent.
Water. Not more than 0.5 percent.

TESTS

Assay. Transfer about 300 mg., accurately weighed, into a 250-ml. Erlenmeyer flask, add 50 ml. of glacial acetic acid, and warm on a steam bath until solution is complete. Cool, add 10 ml. of mercuric acetate T.S. and 2 drops of crystal violet T.S., and titrate with 0.1 *N* perchloric acid in glacial acetic acid to a green end-point. Perform a blank determination (see page 2), and make any necessary correction. Each ml. of 0.1 *N* perchloric acid is equivalent to 13.96 mg. of C_5H_{14}-ClNO.

Arsenic. A *Sample Solution* prepared as directed for organic compounds meets the requirements of the *Arsenic Test,* page 865.

Heavy metals. A solution of 1 gram in 25 ml. of water meets the requirements of the *Heavy Metals Test,* page 920, using 20 mcg. of lead ion (Pb) in the control (*Solution A*).

Lead. A *Sample Solution* prepared as directed for organic compounds meets the requirements of the *Lead Limit Test,* page 929, using 10 mcg. of lead ion (Pb) in the control.

Residue on ignition. Ignite 4 grams as directed in the general method, page 945.

Water. Determine by drying in a vacuum desiccator for 4 hours over phosphorus pentoxide or by the *Karl Fischer Titrimetric Method,* page 977.

Packaging and storage. Store in tight containers.
Functional use in foods. Nutrient; dietary supplement.

CINNAMALDEHYDE

Cinnamic Aldehyde; Cinnamal

C_6H_5—CH=CHCHO

C_9H_8O Mol. wt. 132.16

DESCRIPTION

Cinnamaldehyde is the main constituent of oils of cassia, cinnamon barks and roots. It is usually prepared synthetically. It is a yellow, strongly refractive liquid, having an odor resembling that of cinnamon oil, and a burning aromatic taste. It is affected by light. One gram dissolves in about 700 ml. of water. It is miscible with alcohol, with chloroform, with ether, and with fixed or volatile oils.

SPECIFICATIONS

Assay. Not less than 98.0 percent of C_9H_8O.

Acid value. Not more than 5.0.

Refractive index. Between 1.619 and 1.623 at 20°.

Solubility in alcohol. Passes test.

Specific gravity. Between 1.046 and 1.050.

Limits of Impurities

Arsenic (as As). Not more than 3 parts per million (0.0003 percent).

Chlorinated compounds. Passes test.

Heavy metals (as Pb). Not more than 40 parts per million (0.004 percent).

Hydrocarbons. Passes test.

Lead. Not more than 10 parts per million (0.001 percent).

TESTS

Assay. Weigh accurately about 1.5 grams, and proceed as directed under *Aldehydes*, page 894, using 66.08 as the equivalence factor (E) in the calculation.

Acid value. Determine as directed in the general procedure, page 893.

Refractive index, page 945. Determine with an Abbé or other refractometer of equal or greater accuracy.

Solubility in alcohol. Proceed as directed in the general method, page 899. One ml. dissolves in 5 ml. of 60 percent alcohol.

Specific gravity. Determine by any reliable method (see page 5).

Arsenic. A *Sample Solution* prepared as directed for organic compounds meets the requirements of the *Arsenic Test*, page 865.

Chlorinated compounds. Proceed as directed in the general method, page 895.

Heavy metals. Prepare and test a 500-mg. sample as directed in *Method II* under the *Heavy Metals Test*, page 920, using 20 mcg. of lead ion (Pb) in the control (*Solution A*).

Hydrocarbons. Measure 10 ml. from a pipet into a 100-ml. cassia flask and add 75 ml. of a freshly prepared solution of sodium bisulfite (12 in 100) previously heated to a temperature of 85°. Shake the flask vigorously until solution is complete, then add sufficient sodium bisulfite solution to raise the meniscus within the graduated portion of the neck. No oil separates.

Lead. A *Sample Solution* prepared as directed for organic compounds meets the requirements of the *Lead Limit Test*, page 929, using 10 mcg. of lead ion (Pb) in the control.

Packaging and storage. Store in full, tight, preferably glass, aluminum, tin-lined, or other suitably lined containers in a cool place protected from light.

Functional use in foods. Flavoring agent.

CINNAMIC ACID

3-Phenylpropenoic Acid

C_6H_5—CH═CHCOOH

$C_9H_8O_2$ Mol. wt. 148.16

DESCRIPTION

White, crystalline scales having a characteristic, honey-floral odor. One gram dissolves in about 2,000 ml. of water, in about 6 ml. of alcohol, and in about 12 ml. of chloroform. It is freely soluble in acetic acid, in acetone, in benzene, and in most oils. Its alkali salts are soluble in water.

SPECIFICATIONS

Assay. Not less than 99.0 percent of $C_9H_8O_2$.

Solidification point. Not lower than 133°.

Limits of Impurities

Arsenic (as As). Not more than 3 parts per million (0.0003 percent).

Chlorinated compounds (as Cl). Not more than 50 parts per million (0.005 percent).

Heavy metals (as Pb). Not more than 10 parts per million (0.001 percent).

Residue on ignition. Not more than 0.05 percent.

TESTS

Assay. Dissolve about 500 mg., previously dried in a desiccator for 3 hours over silica gel and accurately weighed, in 25 ml. of 50 percent alcohol previously neutralized with 0.1 *N* sodium hydroxide, add phenolphthalein T.S., and titrate with 0.1 *N* sodium hydroxide. Each ml. of 0.1 *N* sodium hydroxide is equivalent to 14.82 mg. of $C_9H_8O_2$.

Solidification point. Determine as directed in the general method, page 954.

Arsenic. A *Sample Solution* prepared as directed for organic compounds meets the requirements of the *Arsenic Test,* page 865.

Chlorinated compounds. Mix 1 gram with 500 mg. of sodium carbonate, add 15 ml. of water, evaporate the solution on a steam bath, and then ignite the residue at the lowest possible temperature until it is thoroughly charred. Extract the charred mass with a mixture of 20 ml. of water and 5.5 ml. of nitric acid, filter, and wash the residue with sufficient water to make 50 ml. Dilute 10 ml. of the filtrate to 25 ml. with water and add 1 ml. of silver nitrate T.S. Any turbidity does not exceed that shown in a control of equal volume containing 100 mg. of sodium carbonate, 1.1 ml. of nitric acid, and 10 mcg. of chloride ion (Cl).

Heavy metals. Volatilize 2 grams over a low flame. To the residue add 2 ml. of nitric acid and about 10 mg. of sodium carbonate, and evaporate to dryness on a water bath. Dissolve the residue in a mixture of 1 ml. of diluted acetic acid T.S. and 24 ml. of water. This solution meets the requirements of the *Heavy Metals Test*, page 920, using 20 mcg. of lead ion (Pb) in the control (*Solution A*).

Residue on ignition. Ignite 2 grams as directed in the general method, page 945.

Packaging and storage. Store in well-closed containers.

Functional use in foods. Flavoring agent.

CINNAMON BARK OIL, CEYLON

DESCRIPTION

The volatile oil obtained by steam distillation from the dried inner bark of the clipped cinnamon shrub, *Cinnamomum zeylanicum* Nees (Fam. *Lauraceae*). It is a yellow liquid with an odor of cinnamon and a spicy burning taste. It is soluble in most fixed oils and in propylene glycol. It is insoluble in glycerin and in mineral oil.

SPECIFICATIONS

Assay. Not less than 55.0 percent and not more than 78.0 percent of aldehydes, calculated as cinnamic aldehyde (C_9H_8O).

Angular rotation. Between −2° and 0°.

Refractive index. Between 1.573 and 1.591 at 20°.

Solubility in alcohol. Passes test.

Specific gravity. Between 1.010 and 1.030.

TESTS

Assay. Weigh accurately about 2.5 grams, and proceed as directed under *Aldehydes*, page 894, using 66.10 as the equivalence factor (*E*) in the calculation.

Angular rotation. Determine in a 100-mm. tube as directed under *Optical Rotation*, page 939.

Refractive index, page 945. Determine with an Abbé or other refractometer of equal or greater accuracy.

Solubility in alcohol. Proceed as directed in the general method, page 899. One ml. dissolves in 3 ml. of 70 percent alcohol.

Specific gravity. Determine by any reliable method (see page 5).

Packaging and storage. Store in full, tight, light-resistant glass, aluminum, or tin-lined containers in a cool place protected from light.

Functional use in foods. Flavoring agent.

CINNAMON LEAF OIL

DESCRIPTION

The volatile oil obtained by steam distillation from the leaves and twigs of the true cinnamon shrub *Cinnamomum zeylanicum* Nees. The commercial oils, according to the geographical origin, are designated as either cinnamon leaf oil, Ceylon, or cinnamon leaf oil, Seychelles, and the two types differ in physical and chemical properties. The oil is a light to dark brown liquid having a spicy cinnamon, clove-like odor and taste. It is soluble in most fixed oils and in propylene glycol. It is soluble, with cloudiness in mineral oil, but it is insoluble in glycerin.

SPECIFICATIONS

Assay. *Ceylon type:* Not less than 80 percent and not more than 88 percent, by volume, of phenols as eugenol; *Seychelles type:* Not less than 87 percent and not more than 96 percent, by volume, of phenols as eugenol.

Angular rotation. *Ceylon Type:* Between $-2°$ and $+1°$; *Seychelles Type:* Between $-2°$ and $0°$.

Refractive index. *Ceylon Type:* Between 1.529 and 1.537; *Seychelles Type:* Between 1.533 and 1.540 at 20°.

Solubility in alcohol. Passes test.

Specific gravity. *Ceylon Type:* Between 1.030 and 1.050; *Seychelles Type:* Between 1.040 and 1.060.

TESTS

Assay. Shake a suitable quantity of the oil with about 2 percent of powdered tartaric acid, and filter. Proceed with a sample of the filtered oil as directed under *Phenols*, page 898.

Angular rotation. Determine in a 100-mm. tube as directed under *Optical Rotation*, page 939.

Refractive index, page 945. Determine with an Abbé or other refractometer of equal or greater accuracy.

Solubility in alcohol. Proceed as directed in the general method, page 899. One ml. of the Ceylon Type oil dissolves in 1.5 ml. of 70 percent alcohol. One ml. of the Seychelles Type oil dissolves in 1 ml. of 70 percent alcohol. The solutions may cloud upon further dilution.

Specific gravity. Determine by any reliable method (see page 5).

Packaging and storage. Store in, full, tight, light-resistant, glass, aluminum, or tin-lined containers in a cool place protected from light.

Labeling. Label cinnamon leaf oil to indicate whether it is the Ceylon or Seychelles Type.

Functional use in foods. Flavoring agent.

CINNAMYL ACETATE

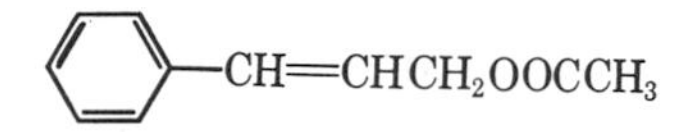

$C_{11}H_{12}O_2$ Mol. wt. 176.22

DESCRIPTION

A colorless to slightly yellow liquid having a sweet-balsamic-floral odor. It is practically insoluble in water and in glycerin; miscible with alcohol, with chloroform, and with ether.

SPECIFICATIONS

Assay. Not less than 98.0 percent of $C_{11}H_{12}O_2$.

Refractive index. Between 1.539 and 1.543 at 20°.

Solubility in alcohol. Passes test.

Specific gravity. Between 1.047 and 1.051.

Limits of Impurities

Acid value. Not more than 3.0.

Heavy metals (as Pb). Not more than 40 parts per million (0.004 percent).

TESTS

Assay. Weigh accurately about 1.2 grams, and proceed as directed under *Ester Determination*, page 896, using 88.11 as the equivalence factor (*e*) in the calculation.

Refractive index, page 945. Determine with an Abbé or other refractometer of equal or greater accuracy.

Solubility in alcohol. Proceed as directed in the general method, page 899. One ml. dissolves in 5 ml. of 70 percent alcohol.

Specific gravity. Determine by any reliable method (see page 5).

Acid value. Determine as directed in the general method, page 893.

Heavy metals. Prepare and test a 500-mg. sample as directed in *Method II* under the *Heavy Metals Test*, page 920, using 20 mcg. of lead ion (Pb) in the control (*Solution A*).

Packaging and storage. Store in tight, light-resistant containers.
Functional use in foods. Flavoring agent.

CINNAMYL ALCOHOL

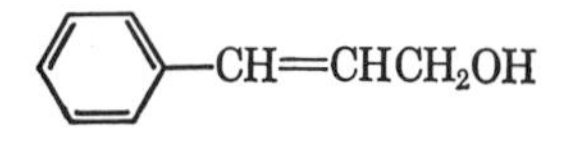

$C_9H_{10}O$ Mol. wt. 134.18

DESCRIPTION

A white, to slightly yellow solid having a characteristic balsamic odor. It is soluble in most fixed oils and in propylene glycol, but it is insoluble in glycerin and in mineral oil.

Note: The specifications in this monograph apply only to synthetic cinnamyl alcohol.

SPECIFICATIONS

Assay. Not less than 98.0 percent of $C_9H_{10}O$.

Solidification point. Not less than 31°.

Solubility in alcohol. Passes test.

Limits of Impurities

Aldehydes. Not more than 1.5 percent, calculated as cinnamic aldehyde (C_9H_8O).

Chlorinated compounds. Passes test.

TESTS

Assay. Proceed as directed under *Total Alcohols*, page 893. Weigh accurately about 1 gram of the acetylated alcohol for the saponification, and use 67.09 as the equivalence factor (*e*) in the calculation.

Solidification point. Determine as directed in the general method, page 954.

Solubility in alcohol. Proceed as directed in the general method, page 899. One gram dissolves in 1 ml. of 70 percent alcohol and remains in solution on dilution to 10 ml.

Aldehydes. Weigh accurately about 5 grams, and proceed as directed under *Aldehydes and Ketones-Hydroxylamine Method*, page 894, using 66.08 as the equivalence factor (*E*) in the calculation for cinnamic aldehyde (C_9H_8O).

Chlorinated compounds. Proceed as directed in the general method, page 895.

Packaging and storage. Store in full, tight, preferably glass, or suitably lined containers in a cool place protected from light. Aluminum containers should not be used.

Functional use in foods. Flavoring agent.

CINNAMYL ANTHRANILATE

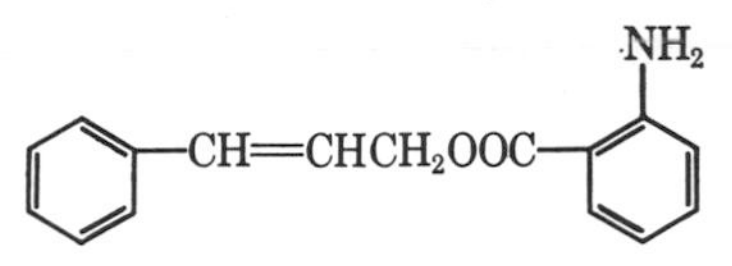

$C_{16}H_{15}NO_2$ Mol. wt. 253.30

DESCRIPTION

A reddish yellow powder having a balsamic, fruity note characteristic of anthranilates. It is soluble in alcohol, in ether, and in chloroform, but insoluble in water.

SPECIFICATIONS

Assay. Not less than 96.0 percent of $C_{16}H_{15}NO_2$.

Solidification point. Not less than 60°.

Solubility in alcohol. Passes test.

TESTS

Assay. Weigh accurately about 1.5 grams, and proceed as directed under *Ester Determination*, page 896, using 126.7 as the equivalence factor (*e*) in the calculation.

Solidification point. Determine as directed in the general method, page 954.

Solubility in alcohol. Proceed as directed in the general method, page 899. One gram dissolves in 20 ml. of 95 percent alcohol to form a clear solution.

Packaging and storage. Store in full, tight, preferably glass, tin-lined, or other suitably lined containers.

Functional use in foods. Flavoring agent.

CINNAMYL FORMATE

C_6H_5—CH=CHCH$_2$OOCH

$C_{10}H_{10}O_2$ Mol. wt. 162.19

DESCRIPTION

A colorless to slightly yellow liquid having a balsamic odor with a cinnamon-like background. It is miscible with alcohol, with chloroform, and with ether, but is practically insoluble in water.

SPECIFICATIONS

Assay. Not less than 92.0 percent of $C_{10}H_{10}O_2$.

Refractive index. Between 1.550 and 1.556 at 20°.

Solubility in alcohol. Passes test.

Specific gravity. Between 1.074 and 1.079.

Limits of Impurities

Acid value. Not more than 3.0.

TESTS

Assay. Weigh accurately about 1.1 grams, and proceed as directed under *Ester Determination*, page 896, using 81.10 as the equivalence factor (*e*) in the calculation.

Refractive index, page 945. Determine with an Abbé or other refractometer of equal or greater accuracy.

Solubility in alcohol. Proceed as directed in the general method, page 899. One ml. dissolves in 10 ml. of 70 percent and in 2 ml. of 80 percent alcohol to form clear solutions.

Specific gravity. Determine by any reliable method (see page 5).

Acid value. Determine as directed in the general method, page 893.

Packaging and storage. Store in full, tight, preferably glass, tin-lined, or other suitably lined containers in a cool place protected from light.

Functional use in foods. Flavoring agent.

CINNAMYL ISOVALERATE

C_6H_5—$CH{=}CHCH_2OOCCH_2CH(CH_3)CH_3$

$C_{14}H_{18}O_2$ Mol. wt. 218.30

DESCRIPTION

A colorless to pale yellow liquid having a combined spicy-fruity-floral odor. It is practically insoluble in water, in glycerin, and in propylene glycol; miscible with alcohol, with chloroform, and with ether.

SPECIFICATIONS

Assay. Not less than 95.0 percent of $C_{14}H_{18}O_2$.

Refractive index. Between 1.518 and 1.524 at 20°.

Solubility in alcohol. Passes test.

Specific gravity. Between 0.991 and 0.996.

Limits of Impurities

Acid value. Not more than 3.0.

TESTS

Assay. Weigh accurately about 1.5 grams, and proceed as directed under *Ester Determination*, page 896, using 109.15 as the equivalence factor (*e*) in the calculation.

Refractive index, page 945. Determine with an Abbé or other refractometer of equal or greater accuracy.

Solubility in alcohol. Proceed as directed in the general method, page 899. One ml. dissolves in 1 ml. of 90 percent alcohol.

Specific gravity. Determine by any reliable method (see page 5).

Acid value. Determine as directed in the general method, page 893.

Packaging and storage. Store in tight, light-resistant containers.

Functional use in foods. Flavoring agent.

CINNAMYL PROPIONATE

C_6H_5—CH=CHCH$_2$OOCCH$_2$CH$_3$

$C_{12}H_{14}O_2$ Mol. wt. 190.24

DESCRIPTION

A colorless to pale yellow liquid having a fruity-floral odor. It is practically insoluble in water, in glycerin, and in propylene glycol; miscible with alcohol, with chloroform, and with ether.

SPECIFICATIONS

Assay. Not less than 98.0 percent of $C_{12}H_{14}O_2$.

Refractive index. Between 1.532 and 1.537 at 20°.

Specific gravity. Between 1.029 and 1.033.

Limit of Impurities

Acid value. Not more than 3.0.

TESTS

Assay. Weigh accurately about 1.2 grams, and proceed as directed under *Ester Determination*, page 896, using 95.12 as the equivalence factor (*e*) in the calculation.

Refractive index, page 945. Determine with an Abbé or other refractometer of equal or greater accuracy.

Specific gravity. Determine by any reliable method (see page 5).

Acid value. Determine as directed in the general method, page 893.

Packaging and storage. Store in tight, light-resistant containers.

Functional use in foods. Flavoring agent.

CITRAL

Geranial; *trans*-3,7-Dimethyl-2,6-octadien-1-al

Neral; *cis*-3,7-Dimethyl-2,6-octadien-1-al

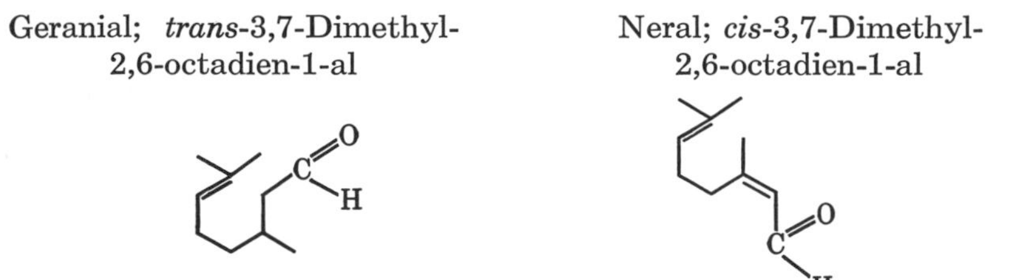

$C_{10}H_{16}O$ Mol. wt. 152.24

DESCRIPTION

Citral is the principal constituent of lemongrass oil and *Backhousia citriodora* oil. It is usually isolated from citral-containing oils by chemical means. It may also be prepared synthetically. Both synthetic and natural citral are composed of the two configurational isomers shown above. Citral is a pale yellow liquid having a strong lemon-like odor. It is soluble in fixed oils, in mineral oil, and in propylene glycol. It is insoluble in glycerin.

Note: The specifications in this monograph apply only to pure citral. Citral produced by direct fractional distillation of essential oils usually contains constituents other than citral.

SPECIFICATIONS

Assay. Not less than 96.0 percent of $C_{10}H_{16}O$.

Refractive index. Between 1.486 and 1.490 at 20°.

Solubility in alcohol. Passes test.

Specific gravity. Between 0.885 and 0.891.

Limits of Impurities

Sodium bisulfite solubility. Not less than 98 percent.

TESTS

Assay. Weigh accurately about 1 gram, and proceed as directed under *Aldehydes and Ketones-Hydroxylamine Method*, page 894, using 76.12 as the equivalence factor (E) in the calculation.

Refractive index, page 945. Determine with an Abbé or other refractometer of equal or greater accuracy.

Solubility in alcohol. Proceed as directed in the general method, page 899. One ml. dissolves in 7 ml. of 70 percent alcohol.

Specific gravity. Determine by any reliable method (see page 5).

Sodium bisulfite solubility. Pipet 10.0 ml. into a 150-ml. cassia flask, add 10 ml. of a freshly prepared solution containing 30 percent, by weight, of anhydrous sodium bisulfite ($NaHSO_3$). Shake the flask in a boiling water bath for 2 minutes. If the mixture has not solidified, continue shaking the flask in the boiling water bath for an additional 2 minutes. Add an additional 50 ml. of the sodium bisulfite solution, return the flask to the water bath and shake until the material is apparently clear. Add enough hot sodium bisulfite solution to the flask to raise the meniscus to the top of the graduated neck of the flask. Immerse the flask in the boiling water bath for 10 minutes. Remove the flask. Not more than 0.2 ml. of oil separates when the flask and contents are cooled to 25°.

Packaging and storage. Store in full, tight, preferably glass containers in a cool place protected from light. Citral may turn yellow and viscous even in closed containers.

Functional use in foods. Flavoring agent.

CITRIC ACID

$$
\begin{array}{c}
CH_2.COOH \\
| \\
HO.C.COOH \\
| \\
CH_2.COOH
\end{array}
$$

$C_6H_8O_7$ Mol. wt. 192.13

DESCRIPTION

Citric acid is anhydrous or contains one molecule of water of hydration. It occurs as colorless, translucent crystals or as a white, granular to fine crystalline powder. It is odorless, has a strongly acid taste, and the hydrous form is efflorescent in dry air. One gram is soluble in about 0.5 ml. of water, in about 2 ml. of alcohol, and in about 30 ml. of ether. A 1 in 10 solution gives positive tests for *Citrate*, page 926.

SPECIFICATIONS

Assay. Not less than 99.5 percent of $C_6H_8O_7$, calculated on the anhydrous basis.

Water. Anhydrous form, not more than 0.5 percent; hydrous form, not more than 8.8 percent.

Limits of Impurities

Arsenic (as As). Not more than 3 parts per million (0.0003 percent).
Heavy metals (as Pb). Not more than 10 parts per million (0.001 percent).
Oxalate. Passes test.
Readily carbonizable substances. Passes test.
Residue on ignition. Not more than 0.05 percent.

TESTS

Assay. Dissolve about 3 grams, accurately weighed, in 40 ml. of water, add phenolphthalein T.S., and titrate with 1 *N* sodium hydroxide. Each ml. of 1 *N* sodium hydroxide is equivalent to 64.04 mg. of $C_6H_8O_7$.

Water. Determine by the *Karl Fischer Titrimetric Method*, page 977.

Arsenic. A *Sample Solution* prepared as directed for organic compounds meets the requirements of the *Arsenic Test*, page 865.

Heavy metals. A solution of 2 grams in 25 ml. of water meets the requirements of the *Heavy Metals Test*, page 920, using 20 mcg. of lead ion (Pb) in the control (*Solution A*).

Oxalate. Neutralize 10 ml. of a 1 in 10 solution with ammonia T.S., add 5 drops of diluted hydrochloric acid T.S., cool, and add 2 ml. of calcium chloride T.S. No turbidity is produced.

Readily carbonizable substances, page 943. Transfer 1.0 gram, finely powdered, to a 22 × 175-mm. test tube, previously rinsed with

10 ml. of sulfuric acid T.S. and allowed to drain for 10 minutes. Add 10 ml. of sulfuric acid T.S., agitate the tube until solution is complete, and immerse the tube in a water bath at 90 ± 1° for 60 ± 0.5 minutes, keeping the level of the acid below the level of the water during the heating period. Cool the tube in a stream of water, and transfer the acid solution to a color-comparison tube. The color of the acid solution is not darker than that of the same volume of *Matching Fluid K* in a similar matching tube, viewing the tubes vertically against a white background.

Residue on ignition, page 945. Ignite 4 grams as directed in the general method.

Packaging and storage. Store in tight containers.

Labeling. Label to indicate whether it is anhydrous or hydrous.

Functional use in foods. Sequestrant; dispersing agent; acidifier; flavoring agent.

CITRONELLAL

3,7-Dimethyl-6-octen-1-al

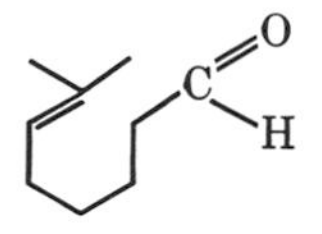

$C_{10}H_{18}O$ Mol. wt. 154.25

DESCRIPTION

An aldehyde obtained from natural oils such as citronella oil or prepared synthetically. It is a colorless to slightly yellow liquid having an intense lemon-citronella-rose type odor. It is soluble in alcohol and in most fixed oils, slightly soluble in mineral oil and in propylene glycol, and insoluble in glycerin and in water.

SPECIFICATIONS

Assay. Not less than 85.0 percent of aldehydes, calculated as citronellal ($C_{10}H_{18}O$).

Angular rotation. Between −1° and +11°.

Refractive index. Between 1.446 and 1.456 at 20°.

Solubility in alcohol. Passes test.

Specific gravity. Between 0.850 and 0.860.

Limits of Impurities

Acid value. Not more than 3.0.

TESTS

Assay. Weigh accurately about 1.1 grams, and proceed as directed under *Aldehydes and Ketones-Hydroxylamine Method*, page 894, using 77.13 as the equivalence factor (*E*) in the calculation. Allow to stand for 1 hour before titrating.

Angular rotation. Determine in a 100-mm. tube as directed under *Optical Rotation*, page 939.

Refractive index, page 945. Determine with an Abbé or other refractometer of equal or greater accuracy.

Solubility in alcohol. Proceed as directed in the general method, page 899. One ml. dissolves in 5 ml. of 70 percent alcohol and remains in solution upon further dilution.

Specific gravity. Determine by any reliable method (see page 5).

Acid value. Determine as directed in the general method, page 893.

Packaging and storage. Store in full, tight, preferably glass, tin-lined or other suitably lined containers in a cool place protected from light.

Functional use in foods. Flavoring agent.

CITRONELLOL

3,7-Dimethyl-6-octen-1-ol

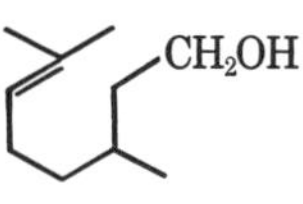

$C_{10}H_{20}O$ Mol. wt. 156.27

DESCRIPTION

Citronellol may be obtained by reduction of citronellal or geraniol, or by fractional distillation of geranium oil or citronella oil. It is a colorless oily liquid having a rose-like odor. It is soluble in most fixed oils, in mineral oil, and in propylene glycol. It is insoluble in glycerin. Some commercial grades of citronellol may be composed almost entirely of the specific chemical whose structural formula is shown above, while other commercial grades that conform to the limits of these specifications and of satisfactory purity may contain other isomeric and closely related terpenic alcohols.

SPECIFICATIONS

Assay. Not less than 90.0 percent of $C_{10}H_{20}O$.

Angular rotation. Between $-1°$ and $+5°$.

Refractive index. Between 1.454 and 1.462 at 20°.

Solubility in alcohol. Passes test.
Specific gravity. Between 0.850 and 0.860.
Limits of Impurities

Aldehydes. Not more than 1 percent, calculated as citronellal ($C_{10}H_{18}O$).

Arsenic (as As). Not more than 3 parts per million (0.0003 percent).

Esters. Not more than 1 percent, calculated as citronellyl acetate ($C_{12}H_{22}O_2$).

Heavy metals (as Pb). Not more than 40 parts per million (0.004 percent).

Lead. Not more than 10 parts per million (0.001 percent).

TESTS

Assay. Proceed as directed under *Total Alcohols*, page 893. Weigh accurately about 1.2 grams of the acetylated alcohol for the saponification, and use 78.13 as the equivalence factor (*e*) in the calculation.

Angular rotation. Determine in a 100-mm. tube as directed under *Optical Rotation*, page 939.

Refractive index, page 945. Determine with an Abbé or other refractometer of equal or greater accuracy.

Solubility in alcohol. Proceed as directed in the general method, page 899. One ml. dissolves in 2 ml. of 70 percent alcohol and remains in solution on dilution to 10 ml.

Specific gravity. Determine by any reliable method (see page 5).

Aldehydes. Weigh accurately about 5 grams, and proceed as directed under *Aldehydes and Ketones-Hydroxylamine Method*, page 894, using 77.13 as the equivalence factor (*E*) in the calculation for citronellal ($C_{10}H_{18}O$).

Arsenic. A *Sample Solution* prepared as directed for organic compounds meets the requirements of the *Arsenic Test*, page 865.

Esters. Weigh accurately about 5 grams, and proceed as directed under *Ester Determination*, page 896, using 99.15 as the equivalence factor (*e*) in the calculation for citronellyl acetate ($C_{12}H_{22}O_2$).

Heavy metals. Prepare and test a 500-mg. sample as directed in *Method II* under the *Heavy Metals Test*, page 920, using 20 mcg. of lead ion (Pb) in the control (*Solution A*).

Lead. A *Sample Solution* prepared as directed for organic compounds meets the requirements of the *Lead Limit Test*, page 929, using 10 mcg. of lead ion (Pb) in the control.

Packaging and storage. Store in full, tight, preferably glass, or tin-lined containers in a cool place protected from light.
Functional use in foods. Flavoring agent.

CITRONELLYL ACETATE

3,7-Dimethyl-6-octen-1-yl Acetate

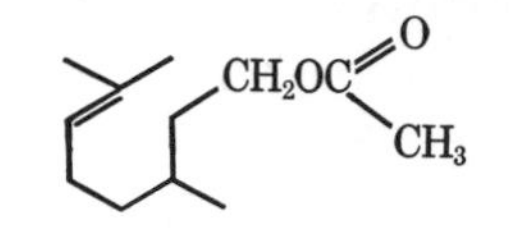

$C_{12}H_{22}O_2$ Mol. wt. 198.31

DESCRIPTION

A colorless liquid with a fruity odor. It is soluble in most fixed oils and in mineral oil. It is insoluble in glycerin and in propylene glycol. Some commercial grades of citronellyl acetate may be composed almost entirely of the specific chemical whose structural formula is shown above, while other commercial grades that conform to the limits of these specifications and of satisfactory purity may contain other isomeric and closely related terpenic esters.

SPECIFICATIONS

Assay. Not less than 92.0 percent of total esters, calculated as citronellyl acetate ($C_{12}H_{22}O_2$).

Angular rotation. Between $-1°$ and $+4°$.

Refractive index. Between 1.440 and 1.450 at 20°.

Solubility in alcohol. Passes test.

Specific gravity. Between 0.883 and 0.893.

Limits of Impurities

Acid value. Not more than 1.0.

TESTS

Assay. Weigh accurately about 1.4 grams, and proceed as directed under *Ester Determination*, page 896, using 99.15 as the equivalence factor (e) in the calculation.

Angular rotation. Determine in a 100-mm. tube as directed under *Optical Rotation*, page 939.

Refractive index, page 945. Determine with an Abbé or other refractometer of equal or greater accuracy.

Solubility in alcohol. Proceed as directed in the general method, page 899. One ml. dissolves in 9 ml. of 70 percent alcohol.

Specific gravity. Determine by any reliable method (see page 5).

Acid value. Determine as directed in the general method, page 893.

Packaging and storage. Store in full, tight, preferably glass, aluminum, or tin-lined containers in a cool place protected from light.

Functional use in foods. Flavoring agent.

CITRONELLYL BUTYRATE

3,7-Dimethyl-6-octen-1-yl Butyrate

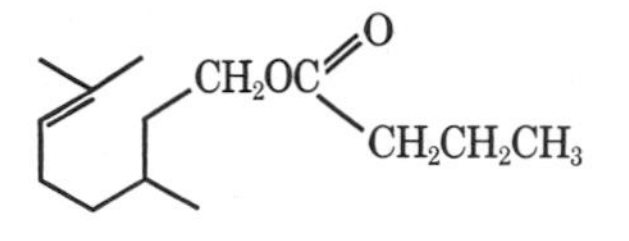

$C_{14}H_{26}O_2$ Mol. wt. 226.36

DESCRIPTION

A colorless liquid having a strong fruity-rosy odor. It is miscible with alcohol, with chloroform, and with ether, but practically insoluble in water. Some commercial grades of citronellyl butyrate may be composed almost entirely of the specific chemical whose structural formula is shown above, while other commercial grades that conform to the limits of these specifications and of satisfactory purity may contain other isomeric and closely related terpenic esters.

SPECIFICATIONS

Assay. Not less than 90.0 percent of $C_{14}H_{26}O_2$.

Angular rotation. Between +1°30′ and −1°30′.

Refractive index. Between 1.444 and 1.448 at 20°.

Solubility in alcohol. Passes test.

Specific gravity. Between 0.873 and 0.883.

Limits of Impurities

Acid value. Not more than 1.0.

Arsenic (as As). Not more than 3 parts per million (0.0003 percent).

Heavy metals (as Pb). Not more than 40 parts per million (0.004 percent).

Lead. Not more than 10 parts per million (0.001 percent).

TESTS

Assay. Weigh accurately about 1.5 grams, and proceed as directed under *Ester Determination*, page 896, using 113.2 as the equivalence factor (*e*) in the calculation.

Angular rotation. Determine in a 100-mm. tube as directed under *Optical Rotation*, page 939.

Refractive index, page 945. Determine with an Abbé or other refractometer of equal or greater accuracy.

Solubility in alcohol. Proceed as directed in the general method, page 899. One ml. dissolves in 6 ml. of 80 percent alcohol to form a clear solution.

Specific gravity. Determine by any reliable method (see page 5).

Acid value. Determine as directed in the general method, page 893.

Arsenic. A *Sample Solution* prepared as directed for organic compounds meets the requirements of the *Arsenic Test,* page 865.

Heavy metals. Prepare and test a 500-mg. sample as directed in *Method II* under the *Heavy Metals Test,* page 920, using 20 mcg. of lead ion (Pb) in the control (*Solution A*).

Lead. A *Sample Solution* prepared as directed for organic compounds meets the requirements of the *Lead Limit Test,* page 929, using 10 mcg. of lead ion (Pb) in the control.

Packaging and storage. Store in full, tight, preferably glass, tin-lined, or other suitably lined containers in a cool place protected from light.

Functional use in foods. Flavoring agent.

CITRONELLYL FORMATE

3,7-Dimethyl-6-octen-1-yl Formate

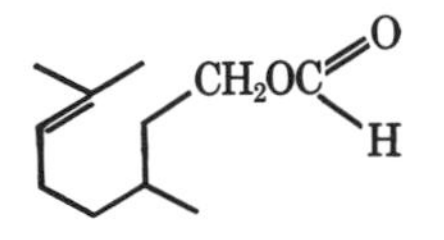

$C_{11}H_{20}O_2$ Mol. wt. 184.28

DESCRIPTION

A colorless liquid with a strong fruity, somewhat floral odor. It is soluble in most fixed oils and in mineral oil. It is sparingly soluble in propylene glycol, but it is insoluble in glycerin. Some commercial grades of citronellyl formate may be composed almost entirely of the specific chemical whose structural formula is shown above, while other commercial grades that conform to the limits of these specificaiions and of satisfactory purity may contain other isomeric and closely related terpenic esters.

SPECIFICATIONS

Assay. Not less than 86.0 percent of total esters, calculated as citronellyl formate ($C_{11}H_{20}O_2$).

Refractive index. Between 1.443 and 1.449 at 20°.

Solubility in alcohol. Passes test.

Specific gravity. Between 0.890 and 0.903.

Limits of Impurities

Acid value. Not more than 3.0.

TESTS

Assay. Weigh accurately about 1 gram, and proceed as directed under *Ester Determination*, page 896, using 92.14 as the equivalence factor (*e*) in the calculation.

Refractive index, page 945. Determine with an Abbé or other refractometer of equal or greater accuracy.

Solubility in alcohol. Proceed as directed in the general method, page 899. One ml. dissolves in 3 ml. of 80 percent alcohol, and remains in solution on dilution to 10 ml.

Specific gravity. Determine by any reliable method (see page 5).

Acid value. Determine as directed in the general method, page 893.

Packaging and storage. Store in full, tight, preferably glass, tin or other suitably lined containers in a cool place protected from light.

Functional use in foods. Flavoring agent.

CITRONELLYL ISOBUTYRATE

3,7-Dimethyl-6-octen-1-yl Isobutyrate

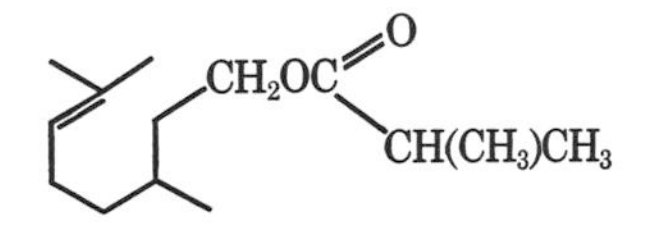

$C_{14}H_{26}O_2$ Mol. wt. 226.36

DESCRIPTION

A colorless liquid having a rosy-fruity odor. It is miscible with alcohol, with chloroform, and with ether, but is practically insoluble in water. Some commercial grades of citronellyl isobutyrate may be composed almost entirely of the specific chemical whose structural formula is shown above, while other commercial grades that conform to the limits of these specifications and of satisfactory purity may contain other isomeric and closely related terpenic esters.

SPECIFICATIONS

Assay. Not less than 92.0 percent of $C_{14}H_{26}O_2$.

Angular rotation. Between +1°30′ and −1°30′.

Refractive index. Between 1.440 and 1.448 at 20°.

Solubility in alcohol. Passes test.

Specific gravity. Between 0.870 and 0.880.

Limits of Impurities

Acid value. Not more than 1.0.

TESTS

Assay. Weigh accurately about 1.5 grams, and proceed as directed under *Ester Determination*, page 896, using 113.2 as the equivalence factor (e) in the calculation.

Angular rotation. Determine in a 100-mm. tube as directed under *Optical Rotation*, page 939.

Refractive index, page 945. Determine with an Abbé or other refractometer of equal or greater accuracy.

Solubility in alcohol. Proceed as directed in the general method, page 899. One ml. dissolves in 6 ml. of 80 percent alcohol to form a clear solution.

Specific gravity. Determine by any reliable method (see page 5).

Acid value. Determine as directed in the general method, page 893.

Packaging and storage. Store in full, tight, preferably glass, tin-lined, or other suitably lined containers in a cool place protected from light.

Functional use in foods. Flavoring agent.

CITRONELLYL PROPIONATE

3,7-Dimethyl-6-octen-1-yl Propionate

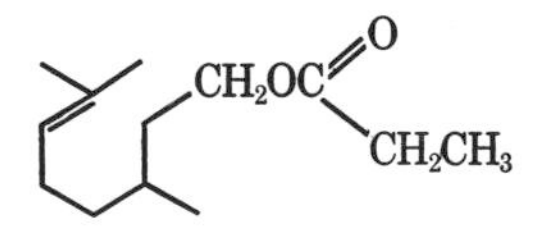

$C_{13}H_{24}O_2$ Mol. wt. 212.34

DESCRIPTION

A colorless liquid having a characteristic rose or fruit-like odor. It is miscible with alcohol, but practically insoluble in water. Some commercial grades of citronellyl propionate may be composed almost entirely of the specific chemical whose structural formula is shown above, while other commercial grades that conform to the limits of these specifications and of satisfactory purity may contain other isomeric and closely related terpenic esters.

SPECIFICATIONS

Assay. Not less than 90 percent of $C_{13}H_{24}O_2$.

Angular rotation. Between $+1°30'$ and $-1°30'$.

Refractive index. Between 1.443 and 1.449 at 20°.

Solubility in alcohol. Passes test.

Specific gravity. Between 0.877 and 0.886.

Limits of Impurities

Acid value. Not more than 1.0.

TESTS

Assay. Weigh accurately about 1.4 grams, and proceed as directed under *Ester Determination*, page 896, using 106.2 as the equivalence factor (*e*) in the calculation.

Angular rotation. Determine in a 100-mm. tube as directed under *Optical Rotation*, page 939.

Refractive index, page 945. Determine with an Abbé or other refractometer of equal or greater accuracy.

Solubility in alcohol. Proceed as directed in the general method, page 899. One ml. dissolves in 4 ml. of 80 percent alcohol to form a clear solution.

Specific gravity. Determine by any reliable method (see page 5).

Acid value. Determine as directed in the general method, page 893.

Packaging and storage. Store in full, tight, preferably glass, tin-lined or other suitably lined containers in a cool place protected from light.

Functional use in foods. Flavoring agent.

CLARY OIL

Clary Sage Oil; Oil of Muscatel

DESCRIPTION

The oil obtained by steam distillation from the flowering tops and leaves of the clary sage plant, *Salvia sclarea* L. (Fam. *Labiatae*). Commercial grades are produced in Russia and in France. It is a pale yellow to yellow liquid, having a herbaceous odor and a winey bouquet. It is soluble in most fixed oils, and in mineral oil up to 3 volumes, but becomes opalescent on further dilution. It is insoluble in glycerin and in propylene glycol.

SPECIFICATIONS

Assay. Not less than 48.0 percent and not more than 75.0 percent of esters, calculated as linalyl acetate ($C_{12}H_{20}O_2$).

Angular rotation. Between $-6°$ and $-20°$.

Refractive index. Between 1.458 and 1.473 at 20°.

Solubility in alcohol. Passes test.

Specific gravity. Between 0.886 and 0.929.

Limits of Impurities

Acid value. Not more than 2.5.

Arsenic (as As). Not more than 3 parts per million (0.0003 percent).

Heavy metals (as Pb). Not more than 40 parts per million (0.004 percent).

Lead. Not more than 10 parts per million (0.001 percent).

TESTS

Assay. Weigh accurately about 2 grams, and proceed as directed under *Ester Determination*, page 896, using 98.15 as the equivalence factor (*e*) in the calculation.

Angular rotation. Determine in a 100-mm. tube as directed under *Optical Rotation*, page 939.

Refractive index, page 945. Determine with an Abbé or other refractometer of equal or greater accuracy.

Solubility in alcohol. Proceed as directed in the general method, page 899. One ml. dissolves in 3 ml. of 90 percent alcohol, becoming opalescent upon further dilution.

Specific gravity. Determine by any reliable method (see page 5).

Acid value. Determine as directed in the general method, page 893.

Arsenic. A *Sample Solution* prepared as directed for organic compounds meets the requirements of the *Arsenic Test*, page 865.

Heavy metals. Prepare and test a 500-mg. sample as directed in *Method II* under the *Heavy Metals Test*, page 920, using 20 mcg. of lead ion (Pb) in the control (*Solution A*).

Lead. A *Sample Solution* prepared as directed for organic compounds meets the requirements of the *Lead Limit Test*, page 929, using 10 mcg. of lead ion (Pb) in the control.

Packaging and storage. Store in full, tight, preferably glass, aluminum, tin-lined, or galvanized iron containers in a cool place protected from light.

Functional use in foods. Flavoring agent.

CLOVE LEAF OIL

DESCRIPTION

The volatile oil obtained by steam distillation of the leaves of *Eugenia caryophyllata* Thunberg (*Eugenia aromatica* L. Baill.). (Fam. *Myrta-*

ceae). It is a pale yellow liquid. It is soluble in propylene glycol, in most fixed oils, with slight opalescence, and relatively insoluble in glycerin and in mineral oil.

SPECIFICATIONS

Assay. Not less than 84 percent and not more than 88 percent, by volume, of phenols as eugenol.

Angular rotation. Between −2° and 0°.

Refractive index. Between 1.531 and 1.535 at 20°.

Solubility in alcohol. Passes test.

Specific gravity. Between 1.036 and 1.046.

TESTS

Assay. Shake a suitable quantity of the oil with 2 percent of powdered tartaric acid for about 2 minutes and filter. Then, using a sample of the filtered oil, proceed as directed under *Phenols*, page 898, modified by heating the flask in a boiling water bath for 10 minutes, after shaking the oil with potassium hydroxide T.S. Remove from the boiling water bath, cool, and proceed as directed.

Angular rotation. Determine in a 100-mm. tube as directed under *Optical Rotation*, page 939.

Refractive index, page 945. Determine with an Abbé or other refractometer of equal or greater accuracy.

Solubility in alcohol. Proceed as directed in the general method, page 899. One ml. dissolves in 2 ml. of 70 percent alcohol. A slight opalescence may occur when additional solvent is added.

Specific gravity. Determine by any reliable method (see page 5).

Packaging and storage. Store in full, tight, light-resistant, glass, tin-lined, stainless or aluminum containers in a cool place protected from light.

Functional use in foods. Flavoring agent.

CLOVE OIL

DESCRIPTION

The volatile oil obtained by steam distillation from the dried flowerbuds of *Eugenia caryophyllata* Thunberg (*Eugenia aromatica* L. Baill.). (Fam. *Myrtaceae*). It is a colorless or pale yellow liquid having the characteristic clove odor and taste. It darkens and thickens upon aging or exposure to air.

SPECIFICATIONS

Assay. Not less than 85 percent, by volume, of phenols, as eugenol.

Angular rotation. Between −1° 30′ and 0°.

Refractive index. Between 1.527 and 1.535 at 20°.

Solubility in alcohol. Passes test.

Specific gravity. Between 1.038 and 1.060.

Limits of Impurities

Arsenic (as As). Not more than 3 parts per million (0.0003 percent).

Heavy metals (as Pb). Not more than 40 parts per million (0.004 percent).

Lead. Not more than 10 parts per million (0.001 percent).

Phenol. Passes test.

TESTS

Assay. Proceed as directed under *Phenols*, page 898.

Angular rotation. Determine in a 100-mm. tube as directed under *Optical Rotation*, page 939.

Refractive index, page 945. Determine with an Abbé or other refractometer of equal or greater accuracy.

Solubility in alcohol. Proceed as directed in the general method, page 899. One ml. dissolves in 2 ml. of 70 percent alcohol.

Specific gravity. Determine by any reliable method (see page 5).

Arsenic. A *Sample Solution* prepared as directed for organic compounds meets the requirements of the *Arsenic Test*, page 865.

Heavy metals. Prepare and test a 500-mg. sample as directed in *Method II* under the *Heavy Metals Test*, page 920, using 20 mcg. of lead ion (Pb) in the control (*Solution A*).

Lead. A *Sample Solution* prepared as directed for organic compounds meets the requirements of the *Lead Limit Test*, page 929, using 10 mcg. of lead ion (Pb) in the control.

Phenol. Shake 1 ml. of sample with 20 ml. of hot water. The water shows no more than a scarcely perceptible acid reaction with blue litmus paper. Cool the mixture, pass the water layer through a wetted filter, and treat the clear filtrate with 1 drop of ferric chloride T.S. The mixture has only a transient grayish green color, but not a blue or violet color.

Packaging and storage. Store in full, tight, light-resistant containers and avoid exposure to excessive heat.

Functional use in foods. Flavoring agent.

CLOVE STEM OIL

DESCRIPTION

The volatile oil obtained by steam distillation from the dried stems of the buds of *Eugenia caryophyllata* Thunberg (*Eugenia aromatica* L. Baill.). (Fam. *Myrtaceae*). It is a yellow to light brown liquid with a characteristic odor and taste. It is soluble in fixed oils, and in propylene glycol, but it is relatively insoluble in glycerin and in mineral oil.

SPECIFICATIONS

Assay. Not less than 89 percent and not more than 95 percent, by volume, of phenols as eugenol.

Angular rotation. Between −1.5° and 0°.

Refractive index. Between 1.534 and 1.538 at 20°.

Solubility in alcohol. Passes test.

Specific gravity. Between 1.048 and 1.056.

TESTS

Assay. Shake a suitable quantity of the oil with about 2 percent of powdered tartaric acid for about 2 minutes and filter. Then, using a sample of the filtered oil, proceed as directed under *Phenols*, page 898, modified by heating the flask in a boiling water bath for 10 minutes, after shaking the oil with potassium hydroxide T.S. Remove from the boiling water bath, cool, and proceed as directed.

Angular rotation. Determine in a 100-mm. tube as directed under *Optical Rotation*, page 939.

Refractive index, page 945. Determine with an Abbé or other refractometer of equal or greater accuracy.

Solubility in alcohol. Proceed as directed in the general method, page 899. One ml. dissolves in 2 ml. of 70 percent alcohol.

Specific gravity. Determine by any reliable method (see page 5).

Packaging and storage. Store preferably in full, tight, light-resistant glass, aluminum, or tin-lined containers in a cool place protected from light.

Functional use in foods. Flavoring agent.

COGNAC OIL, GREEN

Wine Yeast Oil

DESCRIPTION

The volatile oil obtained by steam distillation from wine lees. It is a green to bluish green liquid with the characteristic aroma of cognac. It is soluble in most fixed oils and in mineral oil. It is very slightly soluble in propylene glycol, and it is insoluble in glycerin.

SPECIFICATIONS

Acid value. Between 32 and 70.

Angular rotation. Between $-1°$ and $+2°$.

Ester value. Between 200 and 245.

Refractive index. Between 1.427 and 1.430 at 20°.

Solubility in alcohol. Passes test.

Specific gravity. Between 0.864 and 0.870.

TESTS

Acid value. Determine as directed in the general method, page 893.

Angular rotation. Determine in a 100-mm. tube as directed under *Optical Rotation*, page 939.

Ester value. Proceed as directed in the general method, page 897, using about 1 gram, accurately weighed.

Refractive index, page 945. Determine with an Abbé or other refractometer of equal or greater accuracy.

Solubility in alcohol. Proceed as directed in the general method, page 899. One ml. dissolves in 2 ml. of 80 percent alcohol.

Specific gravity. Determine by any reliable method (see page 5).

Packaging and storage. Store in full, tight containers in a cool place protected from light.

Functional use in foods. Flavoring agent.

COPAIBA OIL

DESCRIPTION

The volatile oil obtained by steam distillation of copaiba balsam, an exudate from the trunk of various South American species of *Copaifera* L. (Fam. *Leguminosae*). It is a colorless to slightly yellow liquid having the characteristic odor of copaiba balsam, and an aromatic, slightly bitter and pungent taste. It is soluble in alcohol, in most fixed oils, and in mineral oil. It is insoluble in glycerin and practically insoluble in propylene glycol.

SPECIFICATIONS

Angular rotation. Between $-7°$ and $-33°$.

Refractive index. Between 1.493 and 1.500 at 20°.

Specific gravity. Between 0.880 and 0.907.

Limits of Impurities

Gurjun oil. Passes test.

TESTS

Angular rotation. Determine in a 100-mm. tube as directed under *Optical Rotation*, page 939.

Refractive index, page 945. Determine with an Abbé or other refractometer of equal or greater accuracy.

Specific gravity. Determine by any reliable method (see page 5).

Gurjun oil. Add 5 to 6 drops of the sample to 10 ml. of glacial acetic acid containing 5 drops of nitric acid. No purple color develops in 2 minutes, indicating the absence of gurjun oil.

Packaging and storage. Store in full, tight, preferably glass, tin or other suitably lined, or aluminum containers in a cool place protected from light.

Functional use in foods. Flavoring agent.

COPPER GLUCONATE

$[CH_2OH(CHOH)_4COO]_2Cu$

$C_{12}H_{22}CuO_{14}$ Mol. wt. 453.84

DESCRIPTION

A fine, light blue powder. It is very soluble in water and is very slightly soluble in alcohol.

IDENTIFICATION

A. A 1 in 20 solution gives positive tests for *Copper*, page 926.

B. To 5 ml. of a warm solution (1 in 10) add 0.7 ml. of glacial acetic acid and 1 ml. of freshly distilled phenylhydrazine, heat on a steam bath for 30 minutes, and allow to cool. Induce crystallization by scratching the inner surface of the container with a glass stirring rod. Crystals of gluconic acid phenylhydrazide form.

SPECIFICATIONS

Assay. Not less than 98.0 percent and not more than the equivalent of 102.0 percent of $C_{12}H_{22}CuO_{14}$.

Limits of Impurities

Arsenic (as As). Not more than 3 parts per million (0.0003 percent).

Lead. Not more than 10 parts per million (0.001 percent).

Reducing substances. Not more than 1 percent.

TESTS

Assay. Dissolve about 1.5 grams, accurately weighed, in 100 ml.

of water in a 250-ml. Erlenmeyer flask, add 2 ml. of glacial acetic acid and 5 grams of potassium iodide, mix well, and titrate with 0.1 *N* sodium thiosulfate to a light yellow color. Add 2 grams of ammonium thiocyanate, mix, then add 3 ml. of starch T.S. and continue titrating to a milk-white end-point. Each ml. of 0.1 *N* sodium thiosulfate is equivalent to 45.38 mg. of $C_{12}H_{22}CuO_{14}$.

Arsenic. A solution of 1 gram in 35 ml. of water meets the requirements of the *Arsenic Test*, page 865.

Lead. A solution of 1 gram in 25 ml. of water meets the requirements of the *Lead Limit Test*, page 929.

Reducing substances. Transfer about 1 gram of the sample, accurately weighed, into a 250-ml. Erlenmeyer flask, dissolve in 10 ml. of water, add 25 ml. of alkaline cupric citrate T.S., and cover the flask with a small beaker. Boil gently for exactly 5 minutes and cool rapidly to room temperature. Add 25 ml. of a 1 in 10 solution of acetic acid, 10.0 ml. of 0.1 *N* iodine, 10 ml. of diluted hydrochloric acid T. S., and 3 ml. of starch T.S., and titrate with 0.1 *N* sodium thiosulfate to the disappearance of the blue color. Calculate the weight, in mg., of reducing substances (as D-glucose) by the formula $(V_1N_1 - V_2N_2)27$, in which V_1 and N_1 are the volume and normality, respectively, of the iodine solution, V_2 and N_2 are the volume and normality, respectively, of the sodium thiosulfate solution, and 27 is an empirically determined equivalence factor for D-glucose.

Packaging and storage. Store in well-closed containers.

Functional use in foods. Nutrient; dietary supplement.

CORIANDER OIL

DESCRIPTION

The volatile oil obtained by steam distillation from the dried ripe fruit of *Coriandrum sativum* L. (Fam. *Umbelliferae*). It is a colorless or pale yellow liquid, having the characteristic odor and taste of coriander.

SPECIFICATIONS

Angular rotation. Between +8° and +15°.

Refractive index. Between 1.462 and 1.472 at 20°.

Solubility in alcohol. Passes test.

Specific gravity. Between 0.863 and 0.875.

TESTS

Angular rotation. Determine in a 100-mm. tube as directed under *Optical Rotation*, page 939.

Refractive index, page 945. Determine with an Abbé or other refractometer of equal or greater accuracy.

Solubility in alcohol. Proceed as directed in the general method, page 899. One ml. dissolves in 3 ml. of 70 percent alcohol.

Specific gravity. Determine by any reliable method (see page 5).

Packaging and storage. Store in full, tight containers, protected from light. Avoid exposure to excessive heat.

Functional use in foods. Flavoring agent.

COSTUS ROOT OIL

DESCRIPTION

The volatile oil obtained by steam distillation from the dried, triturated roots of the herbaceous perennial plant *Saussurea lappa* Clarke (Fam. *Compositae*), or by a solvent extraction procedure followed by vacuum distillation of the resinoid extract. It is a light yellow to brown, viscous liquid, having a peculiar, persistent odor reminiscent of violet, orris, and vetivert. It is soluble in most fixed oils and in mineral oil. It is insoluble in glycerin and in propylene glycol.

SPECIFICATIONS

Acid value. Not more than 42.

Angular rotation. Between +10° and +36°.

Ester value. Between 90 and 150.

Refractive index. Between 1.512 and 1.523 at 20°.

Solubility in alcohol. Passes test.

Specific gravity. Between 0.995 and 1.039.

TESTS

Acid value. Determine as directed in the general method, page 893.

Angular rotation. Determine in a 100-mm. tube as directed under *Optical Rotation*, page 939.

Ester value. Determine as directed in the general method, page 897, using about 1 gram, accurately weighed.

Refractive index, page 945. Determine with an Abbé or other refractometer of equal or greater accuracy.

Solubility in alcohol. Proceed as directed in the general method, page 899. One ml. dissolves in 0.5 ml. of 90 percent alcohol, but the solution becomes cloudy upon further dilution and occasionally paraffin crystals may separate.

Specific gravity. Determine by any reliable method (see page 5).

Packaging and storage. Store in full, tight, preferably glass or aluminum containers in a cool place protected from light.

Functional use in foods. Flavoring agent.

CRESYL ACETATE

p-Tolyl Acetate

$CH_3-C_6H_4-OOCCH_3$

$C_9H_{10}O_2$ Mol. wt. 150.18

DESCRIPTION

A clear, colorless liquid having a strong floral odor. It is soluble in most fixed oils and in propylene glycol. It is insoluble in glycerin but is moderately soluble in mineral oil.

SPECIFICATIONS

Assay. Not less than 98.0 percent of $C_9H_{10}O_2$.

Refractive index. Between 1.499 and 1.502 at 20°.

Solubility in alcohol. Passes test.

Specific gravity. Between 1.044 and 1.050.

Limits of Impurities

Acid value. Not more than 1.0.

Arsenic (as As). Not more than 3 parts per million (0.0003 percent).

Free cresol. Not more than 1.0 percent.

Heavy metals (as Pb). Not more than 40 parts per million (0.004 percent).

Lead. Not more than 10 parts per million (0.001 percent).

TESTS

Assay. Weigh accurately about 1.2 grams, and proceed as directed under *Ester Determination* (High Boiling Solvent), page 896, using 75.09 as the equivalence factor (*e*) in the calculation.

Refractive index, page 945. Determine with an Abbé or other refractometer of equal or greater accuracy.

Solubility in alcohol. Proceed as directed in the general method, page 899. One ml. dissolves in 2 ml. of 70 percent alcohol.

Specific gravity. Determine by any reliable method (see page 5).

Acid value. Determine as directed in the general method, page 893, using phenol red T.S. as the indicator.

Arsenic. A *Sample Solution* prepared as directed for organic compounds meets the requirements of the *Arsenic Test*, page 865.

Free cresol

Ferric chloride solution. Add 1.5 grams of anhydrous ferric chloride to 850 ml. of chloroform in a 2-liter beaker. Add 100 ml. of ethylene glycol monobutyl ether. When the ferric chloride has dissolved, add 50 ml. of pyridine, mix, and filter through a Buchner funnel.

Procedure. Transfer a 5-ml. sample to a 15-mm. test tube, and add 10 ml. of the *Ferric chloride solution.* The color of the solution is no darker green than a solution of 5 ml. of a 1 percent solution of cresol in cresol-free methyl *p*-cresol with 10 ml. of the *Ferric chloride solution.*

Heavy metals. Prepare and test a 500-mg. sample as directed in *Method II* under the *Heavy Metals Test*, page 920, using 20 mcg. of lead ion (Pb) in the control (*Solution A*).

Lead. A *Sample Solution* prepared as directed for organic compounds meets the requirements of the *Lead Limit Test*, page 929, using 10 mcg. of lead ion (Pb) in the control.

Packaging and storage. Store in full, tight, preferably glass, aluminum, tin, or other suitably lined containers in a cool place protected from light.

Functional use in foods. Flavoring agent.

CUBEB OIL

DESCRIPTION

The volatile oil obtained by steam distillation from the mature, unripe, sun-dried fruit of the perennial vine *Piper cubeba* L. (Fam. *Piperaceae*). It is a colorless or light green to bluish-green liquid, having a spicy odor and a slightly acrid taste. It is soluble in most fixed oils and in mineral oil, but it is insoluble in glycerin and propylene glycol.

SPECIFICATIONS

Acid value. Not more than 2.0.

Angular rotation. Between $-12°$ and $-43°$.

Refractive index. Between 1.492 and 1.502 at 20°.

Saponification value. Not more than 8.

Solubility in alcohol. Passes test.

Specific gravity. Between 0.898 and 0.928.

TESTS

Acid value. Determine as directed in the general method, page 893.

Angular rotation. Determine in a 100-mm. tube as directed under *Optical Rotation*, page 939.

Refractive index, page 945. Determine with an Abbé or other refractometer of equal or greater accuracy.

Saponification value. Determine as directed in the general method, page 896, using about 5 grams, accurately weighed.

Solubility in alcohol. Proceed as directed in the general method, page 899. One ml. dissolves in 10 ml. of 90 percent alcohol.

Specific gravity. Determine by any reliable method (see page 5).

Packaging and storage. Store in full, tight, preferably glass, aluminum, or tin-lined containers in a cool place protected from light.

Functional use in foods. Flavoring agent.

CUMIN OIL

DESCRIPTION

The volatile oil obtained by steam distillation from the plant, *Cuminum cyminum* L. It is a light yellow to brown liquid, having a strong and somewhat disagreeable odor. It is relatively soluble in most fixed oils and in mineral oil. It is very soluble in glycerin and in propylene glycol.

SPECIFICATIONS

Assay. Not less than 45.0 percent and not more than 52.0 percent of aldehydes, calculated as cuminaldehyde ($C_{10}H_{12}O$).

Angular rotation. Between +3° and +8°.

Refractive index. Between 1.501 and 1.506 at 20°.

Solubility in alcohol. Passes test.

Specific gravity. Between 0.905 and 0.925.

TESTS

Assay. Weigh accurately about 1 gram, and proceed as directed under *Aldehydes*, page 894, using 74.10 as the equivalence factor (E) in the calculation. Allow the mixture to stand for 30 minutes at room temperature before titrating.

Angular rotation. Determine in a 100-mm. tube as directed under *Optical Rotation*, page 939.

Refractive index, page 945. Determine with an Abbé or other refractometer of equal or greater accuracy.

Solubility in alcohol. Proceed as directed in the general method, page 899. One ml. dissolves in 8 ml. of 80 percent alcohol. The solution may become hazy upon the addition of more alcohol.

Specific gravity. Determine by any reliable method (see page 5).

Packaging and storage. Store in full, tight, preferably glass, tin, or suitably lined containers in a cool place protected from light.

Functional use in foods. Flavoring agent.

CYCLAMEN ALDEHYDE

2-Methyl-3-(*p*-isopropylphenyl)-propionaldehyde

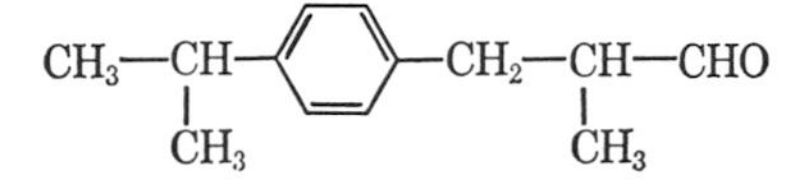

$C_{13}H_{18}O$ Mol. wt. 190.29

DESCRIPTION

A colorless to pale yellow liquid, having a strong floral odor. It is soluble in most fixed oils and in mineral oil. It is practically insoluble in propylene glycol, and it is insoluble in glycerin.

SPECIFICATIONS

Assay. Not less than 90.0 percent of $C_{13}H_{18}O$.

Refractive index. Between 1.503 and 1.508 at 20°.

Solubility in alcohol. Passes test.

Specific gravity. Between 0.946 and 0.952.

Limits of Impurities

Acid value. Not more than 5.0.

TESTS

Assay. Weigh accurately about 1.5 grams, and proceed as directed under *Aldehydes*, page 894, using 95.15 as the equivalence factor (E) in the calculation.

Refractive index, page 945. Determine with an Abbé or other refractometer of equal or greater accuracy.

Solubility in alcohol. Proceed as directed in the general method, page 899. One ml. dissolves in 1 ml. of 80 percent alcohol.

Specific gravity. Determine by any reliable method (see page 5).

Acid value. Determine as directed in the general method, page 893.

Packaging and storage. Store in full, tight, preferably glass, aluminum, tin-lined, or other suitably lined containers in a cool place protected from light.

Functional use in foods. Flavoring agent.

L-CYSTEINE MONOHYDROCHLORIDE

L-2-Amino-3-mercaptopropanoic Acid Monohydrochloride

$$HSCH_2\underset{|}{C}HCOOH \cdot H_2O$$
$$NH_2 \cdot HCl$$

$C_3H_7NO_2S.HCl.H_2O$ Mol. wt. 175.64

DESCRIPTION

A white, odorless, crystalline powder, having a characteristic acidic taste. It is freely soluble in water and in alcohol. The anhydrous form melts with decomposition at about 175°.

IDENTIFICATION

A. Dissolve 100 mg. in 5 ml. of water and add 10 ml. of cupric nitrate T.S. A bluish gray precipitate is formed.

B. A 1 in 20 solution gives positive tests for *Chloride*, page 926.

SPECIFICATIONS

Assay. Not less than 98.0 percent and not more than 102.0 percent of $C_3H_7NO_2S.HCl$ after drying.

Specific rotation, $[\alpha]_D^{20°}$. Between +5.0° and +8.0°, on the dried basis.

Limits of Impurities

Arsenic (as As). Not more than 3 parts per million (0.0003 percent).

Heavy metals (as Pb). Not more than 20 parts per million (0.002 percent).

Lead. Not more than 10 parts per million (0.001 percent).

Loss on drying. Not less than 8 percent and not more than 12 percent.

Residue on ignition. Not more than 0.1 percent.

TESTS

Assay. Transfer about 300 mg., previously dried as directed under *Loss on drying* and accurately weighed, into a 250-ml. glass-stoppered flask. Add 20 ml. of water, 4 grams of potassium iodide, 5 ml. of diluted hydrochloric acid T.S., and 25.0 ml. of 0.1 *N* iodine. Stopper the flask, allow the mixture to stand for 30 minutes in a dark place, and titrate the excess iodine with 0.1 *N* sodium thiosulfate. Perform a blank determination (see page 2). Each ml. of 0.1 *N* iodine is equivalent to 15.76 mg. of $C_3H_7NO_2S.HCl$.

Specific rotation, page 939. Determine in a solution containing 8 grams of a previously dried sample in sufficient 1 *N* hydrochloric acid to make 100 ml.

Arsenic. A *Sample Solution* prepared as directed for organic compounds meets the requirements of the *Arsenic Test*, page 865.

Heavy metals. Prepare and test a 1-gram sample as directed in *Method II* under the *Heavy Metals Test*, page 920, using 20 mcg. of lead ion (Pb) in the control (*Solution A*).

Lead. A *Sample Solution* prepared as directed for organic compounds meets the requirements of the *Lead Limit Test*, page 929, using 10 mcg. of lead ion (Pb) in the control.

Loss on drying, page 931. Dry at room temperature for 24 hours in a vacuum desiccator using a suitable desiccant and maintaining a pressure of not more than 5 mm. of Hg.

Residue on ignition, page 945. Ignite 1 gram as directed in the general method.

Packaging and storage. Store in well-closed, light-resistant containers.

Functional use in foods. Nutrient; dietary supplement.

L-CYSTINE

3,3′-Dithiobis(2-aminopropanoic acid)

$HOOCCH(NH_2)CH_2SSCH_2CH(NH_2)COOH$

$C_6H_{12}N_2O_4S_2$ Mol. wt. 240.30

DESCRIPTION

Colorless, practically odorless, white crystals. It is soluble in diluted mineral acids and in alkaline solutions. It is very slightly soluble in water and in alcohol.

SPECIFICATIONS

Assay. Not less than 98.0 percent and not more than the equivalent of 102.0 percent of $C_6H_{12}N_2O_4S_2$.

Nitrogen (Total). Between 11.5 percent and 11.9 percent.

Specific rotation, $[\alpha]_D^{20°}$. Between −215° and −225°.

Limits of Impurities

Arsenic (as As). Not more than 3 parts per million (0.0003 percent).

Heavy metals (as Pb). Not more than 40 parts per million (0.004 percent).

Iron. Not more than 50 parts per million (0.005 percent).

Lead. Not more than 10 parts per million (0.001 percent).

Loss on drying. Not more than 0.2 percent.

Residue on ignition. Not more than 0.1 percent.

TESTS

Assay

Sodium Cyanide Solution. On the day of use, dissolve 2.5 grams of sodium cyanide in 25 ml. of sodium hydroxide T.S., and dilute to 50 ml. with water.

Sodium Hydrosulfite Solution. Within 1 hour of use, dissolve 1 gram of sodium hydrosulfite, $Na_2S_2O_4$, in 25 ml. of sodium hydroxide T.S., and dilute to 50 ml. with water.

Sodium Naphthoquinone-4-sulfonate Solution. Within 1 hour of use, dissolve 150 mg. of sodium β-naphthoquinone-4-sulfonate, $C_{10}H_7NaO_4S$, in sufficient water to make 50 ml.

Sodium Sulfite Solution. On the day of use, dissolve 5 grams of sodium sulfite in 25 ml. of sodium hydroxide T.S., and dilute to 50 ml. with water.

Standard Preparation. Transfer about 100 mg. of U.S.P. L-Cystine Reference Standard, previously dried for 3 hours over phosphorus pentoxide and accurately weighed, into a 250-ml. volumetric flask, dissolve in 100 ml. of 0.1 *N* hydrochloric acid, dilute to volume with water, and mix. Transfer 20.0 ml. of this solution to a 100-ml. volumetric flask, dilute to volume with water, and mix.

Assay Preparation. Transfer about 100 mg. of the sample, previously dried for 3 hours over phosphorus pentoxide and accurately weighed, into a 250-ml. volumetric flask, dissolve in 100 ml. of 0.1 *N* hydrochloric acid, dilute to volume with water, and mix. Transfer 20.0 ml. of this solution to a 100-ml. volumetric flask, dilute to volume with water, and mix.

Procedure. Pipet 5 ml. each of the *Standard Preparation* and of the *Assay Preparation* into separate 25-ml. volumetric flasks. To each flask add 2 ml. of the *Sodium Cyanide Solution*, allow to stand for 10 minutes, then add 1 ml. of the *Sodium Naphthoquinone-4-sulfonate Solution*, followed in 10 seconds by 5 ml. of the *Sodium Sulfite Solution*. Allow the color to develop for 25 minutes, then add to each flask 2 ml. of 5 *N* sodium hydroxide and 1 ml. of the *Sodium Hydrosulfite Solution*. Dilute each flask to volume with water, mix, and concomitantly determine the absorbance of each solution at 500 mμ in 1-cm. cells, with a suitable spectrophotometer, using water as the blank. Calculate the quantity, in mg., of $C_6H_{12}N_2O_4S_2$ in the sample taken by the formula $1.25C\ (A_U/A_S)$, in which C is the concentration, in mcg. per ml., of the *Standard Preparation*, and A_U and A_S are the absorbances of the solutions from the *Assay Preparation* and the *Standard Preparation*, respectively.

Nitrogen (Total). Determine as directed under *Nitrogen Determination*, page 937, using 300 mg. of a sample previously dried and accurately weighed.

Specific rotation, page 939. Determine in a solution containing

2 grams of a previously dried sample in sufficient 1 *N* hydrochloric acid to make 100 ml.

Arsenic. A *Sample Solution* prepared as directed for organic compounds meets the requirements of the *Arsenic Test*, page 865.

Heavy metals. Prepare and test a 500-mg. sample as directed in *Method II* under the *Heavy Metals Test*, page 920, using 20 mcg. of lead ion (Pb) in the control (*Solution A*).

Iron. To the ash obtained in the test for *Residue on Ignition* add 2 ml. of dilute hydrochloric acid (1 in 2) and evaporate to dryness on a steam bath. Dissolve the residue in 1 ml. of hydrochloric acid and dilute with water to 50 ml. Dilute 10 ml. of this solution to 40 ml. with water and add 40 mg. of ammonium persulfate crystals and 10 ml. of ammonium thiocyanate T.S. Any red or pink color does not exceed that produced by 1.0 ml. of *Iron Standard Solution* (10 mcg. Fe) in an equal volume of a solution containing the quantities of the reagents used in the test.

Lead. A *Sample Solution* prepared as directed for organic compounds meets the requirements of the *Lead Limit Test*, page 929, using 10 mcg. of lead ion (Pb) in the control.

Loss on drying, page 931. Dry over silica gel for 4 hours.

Residue on ignition. Ignite 2 grams as directed in the general method, page 945.

Packaging and storage. Store in well-closed containers.

Functional use in foods. Nutrient; dietary supplement.

Δ-DECALACTONE

$$CH_3(CH_2)_4CHCH_2CH_2CH_2$$

```
CH3(CH2)4CHCH2CH2CH2
         |        |
         O--------C=O
```

$C_{10}H_{18}O_2$ Mol. wt. 170.25

DESCRIPTION

A colorless liquid having a coconut-fruity odor which becomes butter-like in lower concentrations. It is practically insoluble in water, but is very soluble in alcohol, in propylene glycol, and in vegetable oil.

SPECIFICATIONS

Assay. Not less than 98.0 percent of $C_{10}H_{18}O_2$.

Refractive index. Between 1.456 and 1.459 at 20°.

Saponification value. Between 323 and 333.

Limits of Impurities

Acid value. Not more than 5.0.

Unsaponifiable matter. Not more than 0.15 percent.

TESTS

Assay. The purity of Δ-decalactone is determined by gas-liquid chromatography (see page 886) using an instrument containing a thermal conductivity detector and helium as the carrier gas. The operating conditions of the apparatus may vary, depending upon the particular instrument used, but a suitable chromatogram is obtained with a glass, stainless steel, aluminum, or copper column, 2.74-meter × 4.8-mm. outside diameter, packed with 25 percent polyester (polydiethylene glycol glutarate, stabilized with 2 percent phosphoric acid, is satisfactory) on 60- to 80-mesh Chromosorb P or W, and operated at a constant temperature between 190° and 210°. If the detector is separately thermostated, it should be maintained at the column temperature or up to 25° hotter. The recorder should be equipped with an attenuator switch and should be operated in the 0–1 mv. range, with 1 second full scale deflection at a chart speed of ½ inch per second. A constant gas flow should be established and maintained throughout the determination. The inlet gas pressure, which will vary between columns and instruments, should not exceed 40 psi.

Procedure. With helium gas flowing through the apparatus, adjust the column and detector to the operating temperature, and record a base line. Inject a sample of 0.4 to 4 microliters into the apparatus, adjusting the sample size, if necessary, so that the major peak is not attenuated more than 8 times, preferably less, and obtain the chromatogram. Under the conditions described, and using a helium flow rate of about 15 ml. per minute, the Δ-decalactone is eluted in about 15 minutes, and an area of about 600 $cm.^2$ is generated at attenuation ×8. The area of the Δ-decalactone peak is not less than 98.0 percent of the total areas of all peaks.

Refractive index, page 945. Determine with an Abbé or other refractometer of equal or greater accuracy.

Saponification value. Weigh accurately about 1 gram, and determine as directed in the general method, page 896.

Acid value. Determine as directed in *Method II* under the general procedure, page 902.

Unsaponifiable matter. Determine as directed in the general method, page 915.

Packaging and storage. Store in tight containers.

Functional use in foods. Flavoring agent.

DECANAL

Aldehyde C-10; Capraldehyde

$CH_3(CH_2)_8CHO$

$C_{10}H_{20}O$ Mol. wt. 156.27

DESCRIPTION

Decanal is found in sweet orange, mandarin, grapefruit, rose, and other oils. It may be produced by oxidation of 1-decanol, or by reduction of decanoic acid. It is a colorless to light yellow liquid having a pronounced fat-like odor which develops a floral character on dilution. It is miscible with alcohol, with fixed oils, with mineral oil, and with propylene glycol, occasionally with turbidity. It is insoluble in glycerin.

SPECIFICATIONS

Assay. Not less than 92.0 percent of $C_{10}H_{20}O$.

Refractive index. Between 1.426 and 1.430 at 20°.

Specific gravity. Between 0.823 and 0.832.

Limits of Impurities

Acid value. Not more than 10.0.

TESTS

Assay. Weigh accurately about 1.5 grams, and proceed as directed under *Aldehydes*, page 894, using 78.14 as the equivalence factor (*E*) in the calculation.

Refractive index, page 945. Determine with an Abbé or other refractometer of equal or greater accuracy.

Specific gravity. Determine by any reliable method (see page 5).

Acid value. Determine as directed in the general method, page 893.

Packaging and storage. Store in full, tight, preferably glass, or aluminum containers in a cool place protected from light.

Functional use in foods. Flavoring agent.

DECANOIC ACID

Capric Acid

$CH_3(CH_2)_8COOH$

$C_{10}H_{20}O_2$ Mol. wt. 172.27

DESCRIPTION

White crystals having a characteristic, unpleasant, rancid odor. It is soluble in most organic solvents and practically insoluble in water.

SPECIFICATIONS

Acid value. Between 320 and 329.

Saponification value. Between 320 and 331.

Titer (Solidification Point). Between 27° and 32°.

Limits of Impurities

Arsenic (as As). Not more than 3 parts per million (0.0003 percent).

Heavy metals (as Pb). Not more than 10 parts per million (0.001 percent).

Iodine value. Not more than 0.6.

Residue on ignition. Not more than 0.1 percent.

Unsaponifiable matter. Not more than 0.2 percent.

Water. Not more than 0.2 percent.

TESTS

Acid value. Determine as directed under *Method I* in the general procedure, page 902.

Saponification value. Determine as directed in the general method, page 914, using about 2 grams, accurately weighed.

Titer (Solidification point). Determine as directed under *Solidification Point*, page 954.

Arsenic. A *Sample Solution* prepared as directed for organic compounds meets the requirements of the *Arsenic Test*, page 865.

Heavy metals. Prepare and test a 2-gram sample as directed in *Method II* under the *Heavy Metals Test*, page 920, using 20 mcg. of lead ion (Pb) in the control (*Solution A*).

Iodine value. Determine by the *Wijs Method*, page 906.

Residue on ignition, page 945. Ignite 10 grams as directed in the general method.

Unsaponifiable matter. Determine as directed in the general method, page 915.

Water. Determine by the *Karl Fischer Titrimetric Method*, page 977.

Packaging and storage. Store in well-closed containers.

Functional use in foods. Component in the manufacture of other food grade additives; defoaming agent.

1-DECANOL (NATURAL)

Decyl Alcohol; Alcohol C-10; Nonylcarbinol

$CH_3(CH_2)_8CH_2OH$

$C_{10}H_{22}O$ Mol. wt. 158.29

DESCRIPTION

A colorless liquid having a floral odor resembling that of orange flowers. It is soluble in alcohol, in ether, in mineral oil, in propylene glycol, and in most fixed oils. It is insoluble in water and in glycerin. (*Note:* The following *Specifications* apply only to 1-decanol derived from natural precursors.)

SPECIFICATIONS

Assay. Not less than 98.0 percent of $C_{10}H_{22}O$.

Refractive index. Between 1.435 and 0.439 at 20°.

Solidification point. Not less than 5°.

Solubility in alcohol. Passes test.

Specific gravity. Between 0.826 and 0.831.

Limits of Impurities

Acid value. Not more than 1.0.

TESTS

Assay. Proceed as directed under *Total Alcohols,* page 893. Weigh accurately about 1.2 grams of the acetylated oil for the saponification, and use 79.15 as the equivalence factor (*e*) in the calculation.

Refractive index, page 945. Determine with an Abbé or other refractometer of equal or greater accuracy.

Solidification point. Determine as directed in the general method, page 954.

Solubility in alcohol. Proceed as directed in the general method, page 899. One ml. dissolves in 3 ml. 60 percent alcohol to form a clear solution.

Specific gravity. Determine by any reliable method (see page 5).

Acid value. Determine as directed in the general method, page 893.

Packaging and storage. Store in full, tight, preferably glass, tin-lined or other suitably lined containers in a cool place protected from light. *Do not store in aluminum containers.*

Functional use in foods. Flavoring agent.

DEHYDROACETIC ACID

3-Acetyl-6-methyl-1,2-pyran-2,4(3*H*)-dione; Methylacetopyronone; DHA

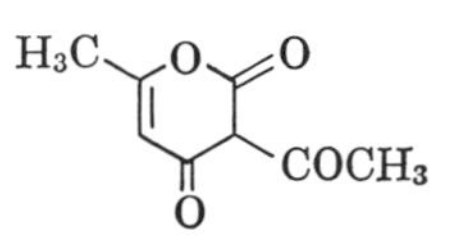

$C_8H_8O_4$ Mol. wt. 168.15

DESCRIPTION

A white or nearly white crystalline powder. It is odorless or almost odorless, and has a faint, acid taste. It is soluble in aqueous solutions of fixed alkalies, but is very slightly soluble in water. One gram dissolves in about 35 ml. of alcohol, 5 ml. of acetone, and 6 ml. of benzene.

SPECIFICATIONS

Assay. Not less than 98.0 percent of $C_8H_8O_4$, calculated on the anhydrous basis.

Melting range. Between 109° and 111°.

Limits of Impurities

Arsenic (as As). Not more than 3 parts per million (0.0003 percent).

Heavy metals (as Pb). Not more than 10 parts per million (0.001 percent).

Loss on drying. Not more than 1 percent.

Residue on ignition. Not more than 0.1 percent.

TESTS

Assay. Transfer about 500 mg., accurately weighed, to a 250-ml. Erlenmeyer flask, dissolve it in 75 ml. of neutral alcohol, add phenolphthalein T.S., and titrate with 0.1 *N* sodium hydroxide to a pink endpoint which persists for at least 30 seconds. Each ml. of 0.1 *N* sodium hydroxide is equivalent to 16.82 mg. of $C_8H_8O_4$.

Melting range. Determine as directed in the general procedure, page 931.

Arsenic. A *Sample Solution* prepared as directed for organic compounds meets the requirements of the *Arsenic Test*, page 865.

Heavy metals. Prepare and test a 2-gram sample as directed in *Method II* under the *Heavy Metals Test*, page 920, using 20 mcg. of lead ion (Pb) in the control (*Solution A*).

Loss on drying, page 931. Dry at 80° for 4 hours.

Residue on ignition. Ignite 2 grams as directed in the general method, page 945.

Packaging and storage. Store in well-closed containers.

Functional use in foods. Preservative.

DESOXYCHOLIC ACID

Deoxycholic Acid; 13α,12α-Dihydroxycholanic Acid

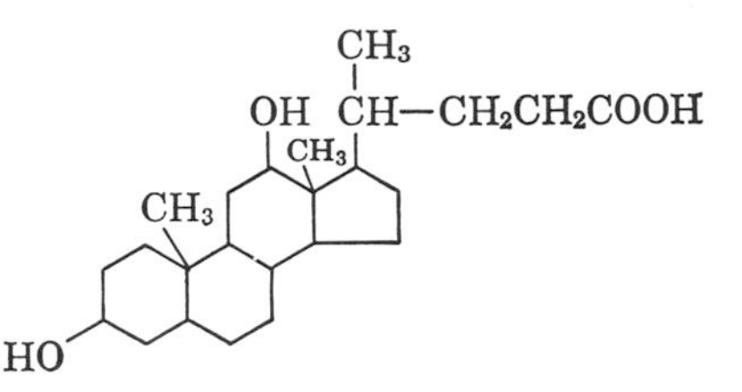

$C_{24}H_{40}O_4$ Mol. wt. 392.58

DESCRIPTION

A white crystalline powder. It is practically insoluble in water, slightly soluble in chloroform and in ether, soluble in acetone and solutions of alkali hydroxides and carbonates, and freely soluble in alcohol.

SPECIFICATIONS

Assay. Not less than 98.0 percent and not more than the equivalent of 102.0 percent of $C_{24}H_{40}O_4$ calculated on the anhydrous basis.

Melting range. Between 172° and 175°.

Limits of Impurities

Arsenic (as As). Not more than 3 parts per million (0.0003 percent).

Heavy metals (as Pb). Not more than 40 parts per million (0.004 percent).

Lead. Not more than 10 parts per million (0.001 percent).

Loss on drying. Not more than 1 percent.

Residue on ignition. Not more than 0.2 percent.

TESTS

Assay. Transfer about 500 mg., accurately weighed, into a 250-ml. Erlenmeyer flask and add 20 ml. of water and 40 ml. of alcohol. Cover the flask with a watch glass, heat the mixture gently on a steam bath until the sample is dissolved, and allow to cool to room temperature. To the solution add a few drops of phenolphthalein T.S. and titrate with 0.1 *N* sodium hydroxide to a pink end-point that persists for 15 seconds. Each ml. of 0.1 *N* sodium hydroxide is equivalent to 39.26 mg. of $C_{24}H_{40}O_4$.

Melting range. Determine as directed in the general procedure, page 931.

Arsenic. A *Sample Solution* prepared as directed for organic compounds meets the requirements of the *Arsenic Test,* page 865.

Heavy metals. Prepare and test a 500-mg. sample as directed in *Method II* under the *Heavy Metals Test,* page 920, using 20 mcg. of lead ion (Pb) in the control (*Solution A*).

Lead. A *Sample Solution* prepared as directed for organic compounds meets the requirements of the *Lead Limit Test,* page 929, using 10 mcg. of lead ion (Pb) in the control.

Loss on drying, page 931. Dry at 140° under a vacuum of not more than 5 mm. of mercury for 4 hours.

Residue on ignition, page 945. Ignite 1 gram as directed in the general method.

Packaging and storage. Store in tight containers.

Functional use in foods. Emulsifying agent.

DEXPANTHENOL

D(+)-Pantothenyl Alcohol; Panthenol

$HOCH_2C(CH_3)_2CH(OH)CONH(CH_2)_2CH_2OH$

$C_9H_{19}NO_4$ Mol. wt. 205.25

DESCRIPTION

The dextrorotatory isomer of the alcohol analogue of pantothenic acid. It occurs as a clear, viscous, somewhat hygroscopic liquid having a slight characteristic odor. Some crystallization may occur on standing. Its solutions are alkaline to litmus. It is freely soluble in water, in alcohol, in methanol, and in propylene glycol. It is soluble in chloroform and in ether, and slightly soluble in glycerin. [*Note:* The physiological activity of 1.0 gram of dexpanthenol is equivalent to 1.16 grams of dextro-calcium pantothenate.]

SPECIFICATIONS

Assay. Not less than 98.0 percent and not more than 102.0 percent of $C_9H_{19}NO_4$ (dexpanthenol), calculated on the anhydrous basis.

Specific rotation, $[\alpha]^{25°}_{D}$. Between +29.0° and +31.5° on the anhydrous basis.

Refractive index. Between 1.495 and 1.502 at 20°.

Limits of Impurities

Aminopropanol. Not more than 1 percent.

Arsenic (as As). Not more than 3 parts per million (0.0003 percent).

Heavy metals (as Pb). Not more than 10 parts per million (0.001 percent).

Residue on ignition. Not more than 0.1 percent.

Water. Not more than 1 percent.

TESTS

Assay. Transfer about 400 mg., accurately weighed, into a 300-ml. reflux flask fitted with a standard-taper glass joint, add 50.0 ml. of 0.1 *N* perchloric acid in glacial acetic acid, and reflux for 5 hours. Cool, covering the condenser with foil to prevent contamination by moisture, and rinse the condenser with glacial acetic acid. Add 5 drops of crystal violet T.S., and titrate with 0.1 *N* potassium acid phthalate in glacial acetic acid to a blue-green end-point. Perform a blank determination, and make any necessary correction (see page 5). Each ml. of 0.1 *N* perchloric acid is equivalent to 20.53 mg. of $C_9H_{19}NO_4$.

Specific rotation, page 939. Determine in a solution containing 500 mg., calculated on the anhydrous basis, in each 10 ml. of water.

Refractive index, page 945. Determine with an Abbé or other refractometer of equal or greater accuracy.

Aminopropanol. Transfer about 5 grams of the sample, accurately weighed into a 50-ml. flask, and dissolve in 10 ml. of water. Add bromothymol blue T.S., and titrate with 0.1 *N* sulfuric acid from a micro-buret to a yellow end-point. Each ml. of 0.1 *N* sulfuric acid is equivalent to 7.5 mg. of aminopropanol.

Arsenic. A *Sample Solution* prepared as directed for organic compounds meets the requirements of the *Arsenic Test,* page 865.

Heavy metals. Prepare and test a 2-gram sample as directed in *Method II* under the *Heavy Metals Test,* page 920, using 20 mcg. of lead ion (Pb) in the control (*Solution A*).

Residue on ignition. Ignite a 1-gram sample as directed in the general method, page 945.

Water. Determine by the *Karl Fischer Titrimetric Method,* page 977.

Packaging and storage. Store in tight containers.

Functional use in foods. Nutrient; dietary supplement.

DIACETYL

2,3-Butanedione; Dimethyldiketone; Dimethylglyoxal

$$CH_3-\overset{\overset{O}{\parallel}}{C}-\overset{\overset{O}{\parallel}}{C}-CH_3$$

$C_4H_6O_2$ Mol. wt. 86.09

DESCRIPTION

Diacetyl is usually prepared by special fermentation of glucose, or synthesized from methyl ethyl ketone. It is a clear yellow to yellowish green liquid with a strong, pungent odor. In very dilute solution it has a typical buttery odor and flavor. It is miscible with alcohol, with most fixed oils, and with propylene glycol. It is soluble in glycerin and in water, but it is insoluble in mineral oil.

SPECIFICATIONS

Assay. Not less than 97.0 percent of $C_4H_6O_2$.

Refractive index. Between 1.393 and 1.397 at 20°.

Solidification point. Between −2.0° and −4.0°.

Specific gravity. Between 0.979 and 0.985.

Limits of Impurities

Arsenic (as As). Not more than 3 parts per million (0.0003 percent).

Heavy metals (as Pb). Not more than 40 parts per million (0.004 percent).

Lead. Not more than 10 parts per million (0.001 percent).

TESTS

Assay. Weigh accurately about 500 mg., and proceed as directed under *Aldehydes and Ketones-Hydroxylamine Method*, page 894, using 21.52 as the equivalence factor (E) in the calculation.

Refractive index, page 945. Determine with an Abbé or other refractometer of equal or greater accuracy.

Solidification point. Determine as directed in the general method, page 954.

Specific gravity. Determine by any reliable method (see page 5).

Arsenic. A *Sample Solution* prepared as directed for organic compounds meets the requirements of the *Arsenic Test*, page 865.

Heavy metals. A solution of 500 mg. in 25 ml. of water meets the requirements of the *Heavy Metals Test*, page 920, using 20 mcg. of lead ion (Pb) in the control (*Solution A*).

Lead. A *Sample Solution* prepared as directed for organic compounds meets the requirements of the *Lead Limit Test*, page 929, using 10 mcg. of lead ion (Pb) in the control.

Packaging and storage. Store in full, tight, preferably glass, aluminum, tin-lined, or other suitably lined, light-resistant containers.

Functional use in foods. Flavoring agent.

DIACETYL TARTARIC ACID ESTERS OF MONO- AND DIGLYCERIDES

DESCRIPTION

The reaction product of partial glycerides of edible oils, fats, or fat-forming fatty acids with diacetyl tartaric anhydride. The esters range in appearance from sticky, viscous liquids through a fat-like consistency to a waxy solid, depending upon the iodine value of the oils or fats used in their manufacture. The diacetyl tartroyl esters have a faint acid odor and are miscible in all proportions with oils and fats. They are soluble in most common fat solvents, in methanol, in acetone, and in ethyl acetate, but insoluble in other alcohols, in acetic acid, and in water. They are dispersible in water and resistant to hydrolysis for moderate periods of time. The pH of a 3 percent dispersion in water is between 2 and 3.

IDENTIFICATION

To a solution of 500 mg. in 10 ml. of methanol add, dropwise, lead acetate T.S. A white, flocculent, practically insoluble precipitate forms.

SPECIFICATIONS

Assay for tartaric acid. Between 17.0 and 20.0 grams of tartaric acid ($C_4H_6O_6$) per 100 grams after saponification.

Acetic acid. Between 14.0 and 17.0 grams of CH_3COOH per 100 grams after saponification.

Acid value. Between 62 and 76.

Fatty acids, Total. Not less than 56.0 grams of total fatty acids per 100 grams after saponification.

Glycerin. Not less than 12.0 grams of $C_3H_8O_3$ per 100 grams after sapodification.

Saponification value. Between 380 and 425.

Limits of Impurities

Arsenic (as As). Not more than 3 parts per million (0.0003 percent).

Heavy metals (as Pb). Not more than 10 parts per million (0.001 percent).

Residue on ignition. Not more than 0.01 percent.

TESTS

Assay for tartaric acid

Standard Reference Curve. Transfer 100 mg. of reagent grade tartaric acid, accurately weighed, into a 100-ml. volumetric flask, dissolve it in about 90 ml. of water, add water to volume, and mix well. Transfer 3.0, 4.0, 5.0, and 6.0-ml. portions into separate 19 × 150-mm. matched cuvettes, and add sufficient water to make 10.0 ml. To each cuvette add 4.0 ml. of a freshly prepared 1 in 20 solution of sodium metavanadate and 1.0 ml. of acetic acid. *Use these solutions within 10 minutes after color development.* Prepare a blank in the same manner using 10 ml. of water in place of the tartaric acid solutions. Set the instrument at zero with the blank, and then determine the absorbance of the four solutions of tartaric acid at 520 mμ with a suitable spectrophotometer or a photoelectric colorimeter equipped with a 520 mμ filter. From the data thus obtained, prepare a reference curve by plotting the absorbances on the ordinate against the corresponding quantities, in mg., of the tartaric acid on the abscissa.

Assay preparation. Transfer about 4 grams of the sample, accurately weighed, into a 250-ml. Erlenmeyer flask, and add 80 ml. of approximately 0.5 *N* potassium hydroxide and 0.5 ml. of phenolphthalein T.S. Connect an air condenser at least 65 cm. in length to the flask, and heat the mixture on a hot plate for about 2.5 hours. Add to the hot mixture approximately 10 percent phosphoric acid until it is definitely acid to congo red test paper. Reconnect the air condenser, and heat until the fatty acids are liquified and clear. Cool and then transfer the mixture into a 250-ml. separator with the aid of small portions of water and chloroform. Extract the liberated fatty acids with three successive 25-ml. portions of chloroform, and collect the extracts in a second separator. Wash the combined chloroform extracts with two 25-ml. portions of water, and add the washings to the separator containing the water layer. Retain the combined chloroform extracts for the determination of *Total fatty acids.* Transfer the contents of the first separator to a 250-ml. beaker, heat on a steam bath to remove traces of chloroform, filter through acid-washed, fine texture filter paper into a 500-ml. volumetric flask, and finally dilute to volume with water (*Solution I*). Pipet 25.0 ml. of this solution into a 100-ml. volumetric flask, and dilute to volume with water (*Solution II*). Retain the rest of *Solution I* for the determination of *Glycerin.*

Procedure. Transfer 10.0 ml. of *Solution II* prepared under *Assay preparation* into a 19 × 150-mm. cuvette, and continue as directed under *Standard Reference Curve*, beginning with ". . . add 4.0 ml. of a freshly prepared 1 in 20 solution of sodium metavanadate . . ." From the reference curve determine the weight, in mg., of tartaric acid in the final dilution, multiply this by 20, and divide the result by the weight of the original sample to obtain the percent of tartaric acid.

Acetic acid. Determine as directed under *Volatile Acidity*, page 916, using a 4-gram sample, accurately weighed, and 30.03 as the equivalence factor, *E*.

Acid value. Transfer about 1 gram, accurately weighed, into a 125-ml. Erlenmeyer flask. Prepare a solvent by mixing 1 volume of benzene with 4 volumes of methanol, adding phenol red T.S., and neutralizing, if necessary. Dissolve the sample in about 25 ml. of this solvent by warming gently, if necessary. Titrate the solution with 0.1 *N* methanolic potassium hydroxide to a light red end-point. Perform a blank determination on a 25-ml. portion of the solvent, and make any necessary correction (see page 2). Calculate the acid value by the formula $56.1\ V \times N/W$, in which *V* is the volume, in ml., and *N* is the normality, respectively, of the methanolic potassium hydroxide, and *W* is the weight, in grams, of the sample taken.

Fatty acids, Total. Dry the combined chloroform extracts of fatty acids obtained in the *Assay for tartaric acid* by shaking with a few grams of anhydrous sodium sulfate. Filter the solution into a tared 250-ml. beaker, evaporate the chloroform on a steam bath, cool, and weigh.

Glycerin. Transfer 5.0 ml. of *Solution I* prepared in the *Assay for tartaric acid* into a 250-ml. glass-stoppered Erlenmeyer or iodine flask. Add to the flask 15 ml. of glacial acetic acid and 25.0 ml. of periodic acid solution, prepared by dissolving 2.7 grams of periodic acid (H_5IO_6) in 50 ml. of water, adding 950 ml. of glacial acetic acid, and mixing thoroughly; protect this solution from light. Shake the mixture for 1 or 2 minutes, allow it to stand for 15 minutes, add 15 ml. of potassium iodide solution (15 in 100) and 15 ml. of water, swirl, let stand 1 minute, and then titrate the liberated iodine with 0.1 *N* sodium thiosulfate using starch T.S. as the indicator. Perform a *residual blank titration* (see page 2) using water in place of the sample. The corrected volume is the number of ml. of 0.1 *N* sodium thiosulfate required for the glycerin and the tartaric acid in the sample represented by the 5 ml. of *Solution I*. From the percent determined in the *Assay for tartaric acid* calculate the volume of 0.1 *N* sodium thiosulfate required for the tartaric acid in the titration. The difference between the corrected volume and the calculated volume required for the tartaric acid is the number of ml. of 0.1 *N* sodium thiosulfate consumed due to the glycerin in the sample. One ml. of 0.1 *N* sodium thiosulfate is equivalent to 2.303 mg. of glycerin and to 7.505 mg. of tartaric acid.

Saponification value. Determine as directed in the general method, page 914, using about 2 grams, accurately weighed. Add 5 to 10 ml. of water to samples and blanks before saponification; otherwise sufficient salts precipitate during saponification to cause serious bumping and spattering.

Arsenic. A *Sample Solution* prepared as directed for organic compounds meets the requirements of the *Arsenic Test*, page 865.

Heavy metals. Prepare and test a 2-gram sample as directed in *Method II* under the *Heavy Metals Test*, page 920, using 20 mcg. of lead ion (Pb) in the control (*Solution A*).

Residue on ignition. Ignite 10 grams as directed in the general method, page 945.

Packaging and storage. Store in well-closed containers.

Functional use in foods. Emulsifier.

DIATOMACEOUS SILICA

DESCRIPTION

A white to gray or buff-colored powder consisting of processed siliceous skeletons of diatoms. It is insoluble in water, in acids (except hydrofluoric), and in dilute alkalies. The *natural* powder (gray to off-white) is air dried and classified by particle size; the *calcined* powder (pink to buff-colored) is air dried, classified, calcined, and classified; the *flux-calcined* powder (white) is air dried, classified, calcined with an alkaline sodium salt, and classified. When examined with a 100- to 200-power microscope, the typical diatom shapes are observed.

SPECIFICATIONS

Loss on drying. *Natural* powders: not more than 10 percent; *calcined* and *flux-calcined* powders: not more than 3 percent.

Loss on ignition. *Natural* powders: not more than 7 percent, on the dried basis; *calcined* and *flux-calcined* powders: not more than 2 percent, on the dried basis.

Nonsiliceous substances. Not more than 25 percent, on the dried basis.

pH. Passes test.

Limits of Impurities

Arsenic (as As). Not more than 10 parts per million (0.001 percent).

Lead. Not more than 10 parts per million (0.001 percent).

TESTS

Loss on drying, page 931. Dry at 105° for 2 hours.

Loss on ignition. Weigh accurately about 1 gram, and ignite to constant weight in a suitable tared crucible.

Nonsiliceous substances. Transfer about 200 mg., accurately weighed, into a tared platinum crucible, add 5 ml. of hydrofluoric acid and 2 drops of sulfuric acid (1 in 2), and evaporate gently to dryness. Cool, add 5 ml. of hydrofluoric acid, evaporate again to dryness, and then ignite to constant weight.

pH, page 941. Boil 10 grams with 100 ml. of water for 30 minutes, make up to 100 ml. with water, and filter through a fine-porosity sintered-glass funnel. The pH of the filtrate prepared with *natural* or *calcined* powders is between 5.0 and 10.0, and of that prepared with *flux-calcined* powders is between 8.0 and 11.0.

Arsenic. Transfer 10.0 grams of the sample into a 250-ml. beaker, add 50 ml. of 0.5 *N* hydrochloric acid, cover with a watch glass, and heat at 70° for 15 minutes. Cool, and decant through a Whatman No. 3 filter paper into a 100-ml. volumetric flask. Wash the slurry with three 10-ml. portions of hot water and the filter paper with 15 ml. of hot water, dilute to volume with water, and mix. A 3.0-ml. portion of this solution meets the requirements of the *Arsenic Test,* page 865.

Lead. A 10.0-ml. portion of the solution prepared in the *Arsenic Test* meets the requirements of the *Lead Limit Test,* page 929, using 10 mcg. of lead ion (Pb) in the control.

Packaging and storage. Store in well-closed containers.

Functional use in foods. Filter aid in food processing.

DIETHYL MALONATE

Ethyl Malonate; Malonic Ester

$$CH_2(COOCH_2CH_3)_2$$

$C_7H_{12}O_4$ Mol. wt. 160.17

DESCRIPTION

A colorless liquid with a slight, fruitlike odor. It is soluble in most fixed oils and in propylene glycol. It is insoluble in glycerin and in mineral oil.

SPECIFICATIONS

Assay. Not less than 98.0 percent of $C_7H_{12}O_4$.

Specific gravity. Between 1.053 and 1.056.

Solubility in alcohol. Passes test.

Refractive index. Between 1.413 and 1.416 at 20°.

Limits of Impurities

Acid value. Not more than 1.0.

TESTS

Assay. Weigh accurately about 1 gram, and proceed as directed under *Ester Determination*, page 896, using 40.04 as the equivalence factor (*e*) in the calculation.

Specific gravity. Determine by any reliable method (see page 5).

Solubility in alcohol. Proceed as directed in the general method, page 899. One ml. dissolves in 1.5 ml. of 60 percent alcohol.

Refractive index, page 945. Determine with an Abbé or other refractometer of equal or greater accuracy.

Acid value. Determine as directed in the general method, page 893.

Packaging and storage. Store in full, tight, preferably glass, tin, or suitably lined containers in a cool place, protected from light.

Functional use in foods. Flavoring agent.

DIETHYL SEBACATE

Ethyl Sebacate

$C_2H_5OOC(CH_2)_8COOC_2H_5$

$C_{14}H_{26}O_4$ Mol. wt. 258.36

DESCRIPTION

A colorless to light yellow liquid with a faint odor; insoluble in water; miscible with alcohol, ether, and other organic solvents.

SPECIFICATIONS

Assay. Not less than 98.0 percent of $C_{14}H_{26}O_4$.

Refractive index. Between 1.435 and 1.438 at 20°.

Specific gravity. Between 0.960 and 0.965.

Limits of Impurities

Acid value. Not more than 1.0.

TESTS

Assay. Weigh accurately about 1 gram, and proceed as directed under *Ester Determination*, page 896, using 64.59 as the equivalence factor (*e*) in the calculation.

Refractive index, page 945. Determine with an Abbé or other refractometer of equal or greater accuracy.

Specific gravity. Determine by any reliable method (see page 5).

Acid value. Determine as directed in the general method, page 893.

Packaging and storage. Store in amber glass, aluminum, or suitably lined containers in a cool place protected from light.

Functional use in foods. Flavoring agent.

DIETHYL SUCCINATE

Ethyl Succinate

$C_2H_5OOCCH_2CH_2COOC_2H_5$

$C_8H_{14}O_4$ Mol. wt. 174.20

DESCRIPTION

A clear, colorless, mobile liquid having a faint, pleasant odor. It is miscible in all proportions with alcohol and with ether. One ml. dissolves in about 50 ml. of water.

SPECIFICATIONS

Assay. Not less than 99.0 percent of $C_8H_{14}O_4$.

Limits of Impurities

Acidity (as succinic acid). Not more than 0.02 percent.

Diethyl maleate. Not more than 0.03 percent.

Heavy metals (as Pb). Not more than 40 parts per million (0.004 percent).

Water. Not more than 0.05 percent.

TESTS

Assay and limit of diethyl maleate. Determine the purity of diethyl succinate by gas-liquid chromatography (see page 886), using an instrument containing a thermal conductivity detector and helium as the carrier gas. The operating conditions of the apparatus may vary, depending upon the particular instrument used, but a suitable chromatogram is obtained with an aluminum column, approximately 2-meters × 6-mm., packed with 30 percent, by weight, Carbowax 20M on 60/80-mesh Chromosorb W, and operated at a constant temperature of about 200°. The injection port temperature should be about 250°, and the detector should be thermostated at about 260°. The recorder should be operated in the 0–1 mv. range, with the detector current set at 200 ma. The gas flow rate should be adjusted (about 55 ml. per minute) so that, after the injection of a 20-μl. sample, the retention time for diethyl succinate will be about 8.5 minutes, resulting in a retention time for diethyl maleate of about 10.8 minutes under the conditions described. After the chromatogram is obtained, measure the area under all peaks, and calculate the concentrations of diethyl succinate and diethyl maleate, in weight percent, by applying an appropriate correction factor. (*Note:* In the calculation of diethyl succinate content, the area and weight percent may be assumed to be identical.) Determine the calibration factor for diethyl maleate by analyzing a sample of known composition to which measured amounts of diethyl maleate have been added, following the procedure described above.

Acidity. Transfer 57 ml. (59 grams) into a 250-ml. Erlenmeyer flask, add phenolphthalein T.S., and titrate with 0.1 *N* alcoholic potassium hydroxide to a pink end-point that persists for at least 15 seconds. Each ml. of 0.1 *N* alcoholic potassium hydroxide is equivalent to 5.904 mg. of succinic acid ($C_4H_6O_4$).

Diethyl maleate. Determine as directed under *Assay and limit of diethyl maleate.*

Heavy metals. Prepare and test a 500-mg. sample as directed in *Method II* under the *Heavy Metals Test,* page 920, using 20 mcg. of lead ion (Pb) in the control (*Solution A*).

Water. Determine as directed in the *Karl Fischer Titrimetric Method,* page 977.

Packaging and storage. Store in tight containers.

Functional use in foods. Flavoring agent.

DILAURYL THIODIPROPIONATE

$(C_{12}H_{25}OOCCH_2CH_2)_2S$

$C_{30}H_{58}O_4S$ Mol. wt. 514.85

DESCRIPTION

White crystalline flakes having a characteristic sweetish, ester-like odor. It is insoluble in water, but soluble in most organic solvents.

SPECIFICATIONS

Assay. Not less than 99.0 percent of $C_{30}H_{58}O_4S$.

Solidification point. Not below 40°.

Limits of Impurities

Acidity (as thiodipropionic acid). Not more than 0.2 percent of $C_6H_{10}O_4S$.

Arsenic (as As). Not more than 3 parts per million (0.0003 percent).

Heavy metals (as Pb). Not more than 20 parts per million (0.002 percent).

Lead. Not more than 10 parts per million (0.001 percent).

TESTS

Assay. Transfer about 700 mg., accurately weighed, into a 250-ml. Erlenmeyer flask, and add 100 ml. of acetic acid and 50 ml. of alcohol.

Heat the mixture at a temperature of about 40° until the sample is completely dissolved, then add 3 ml. of hydrochloric acid and 4 drops of *p*-ethoxychrysoidin T.S., and immediately titrate the solution with 0.1 *N* bromine. When the end-point is approached (pink color), add 4 more drops of the indicator solution and continue the titration, dropwise, to a color change from red to pale yellow. Perform a blank determination (see page 2) and make any necessary correction. Each ml. of 0.1 *N* bromine is equivalent to 25.74 mg. of $C_{30}H_{58}O_4S$. Multiply the percent of thiodipropionic acid, determined in the *Acidity* test, by 2.89, and subtract this value from the percent of dilauryl thiodipropionate calculated from the titration. The difference is the percent purity of $C_{30}H_{58}O_4S$.

Solidification point. Determine as directed in the general procedure, page 954.

Acidity (as thiodipropionic acid). Transfer about 2 grams, accurately weighed, into a 250-ml. Erlenmeyer flask. Dissolve the sample in 50 ml. of a mixture composed of 1 part of methyl alcohol and 3 parts of benzene, add 5 drops of phenolphthalein T.S., and titrate with 0.1 *N* alcoholic potassium hydroxide. Each ml. of 0.1 *N* alcoholic potassium hydroxide is equivalent to 8.91 mg. of $C_6H_{10}O_4S$.

Arsenic. A *Sample Solution* prepared as directed for organic compounds meets the requirements of the *Arsenic Test*, page 865.

Heavy metals. Prepare and test a 1-gram sample as directed in *Method II* under the *Heavy Metals Test*, page 920, using 20 mcg. of lead on (Pb) in the control (*Solution A*).

Lead. A *Sample Solution* prepared as directed for organic compounds meets the requirements of the *Lead Limit Test*, page 929, using 10 mcg. of lead ion (Pb) in the control.

Packaging and storage. Store in well-closed containers.

Functional use in foods. Antioxidant.

DILL SEED OIL, EUROPEAN

DESCRIPTION

The volatile oil obtained by steam distillation from the crushed, dried fruit (or seeds) of *Anethum graveolens*, L. (Fam. *Umbelliferae*), grown in various European countries. It is a slightly yellowish to light yellow liquid with a caraway-like odor and flavor. It is soluble in most fixed oils and in mineral oil. It is soluble, with slight opalescence, in propylene glycol, but it is practically insoluble in glycerin.

SPECIFICATIONS

Assay. Not less than 42 percent and not more than 60 percent, by volume, of ketones as carvone.

Angular rotation. Between +70° and +82°.

Refractive index. Between 1.483 and 1.490 at 20°.

Solubility in alcohol. Passes test.

Specific gravity. Between 0.890 and 0.915.

Limits of Impurities

Arsenic (as As). Not more than 3 parts per million (0.0003 percent).

Heavy metals (as Pb). Not more than 40 parts per million (0.004 percent).

Lead. Not more than 10 parts per million (0.001 percent).

TESTS

Assay. Proceed as directed under *Aldehydes and Ketones—Neutral Sulfite Method*, page 895.

Angular rotation. Determine in a 100-mm. tube as directed under *Optical Rotation*, page 939.

Refractive index, page 945. Determine with an Abbé or other refractometer of equal or greater accuracy.

Solubility in alcohol. Proceed as directed in the general method, page 899. One ml. dissolves in 2 ml. of 80 percent alcohol, with slight opalescence which may not disappear on dilution to as much as 10 ml.

Specific gravity. Determine by any reliable method (see page 5).

Arsenic. A *Sample Solution* prepared as directed for organic compounds meets the requirements of the *Arsenic Test*, page 865.

Heavy metals. Prepare and test a 500-mg. sample as directed in *Method II* under the *Heavy Metals Test*, page 920, using 20 mcg. of lead ion (Pb) in the control (*Solution A*).

Lead. A *Sample Solution* prepared as directed for organic compounds meets the requirements of the *Lead Limit Test*, page 929, using 10 mcg. of lead ion (Pb) in the control.

Packaging and storage. Store in full, tight, preferably glass, aluminum, or other suitably lined containers in a cool place protected from light.

Functional use in foods. Flavoring agent.

DILL SEED OIL, INDIAN

Dill Seed Oil, East Indian; Dill Oil, Indian

DESCRIPTION

The volatile oil obtained by steam distillation from the crushed mature fruit of Indian dill, *Anethum sowa* D.C. (Fam. *Umbelliferae*). It is a light yellow to light brown liquid with a rather harsh caraway-like odor and flavor. It is soluble in most fixed oils and in mineral oil, occasionally with slight opalescence. It is sparingly soluble in propylene glycol and practically insoluble in glycerin.

SPECIFICATIONS

Assay. Not less than 20 percent and not more than 30 percent, by volume, of ketones as carvone.

Angular rotation. Between +40° and +58°.

Refractive index. Between 1.486 and 1.495 at 20°.

Solubility in alcohol. Passes test.

Specific gravity. Between 0.925 and 0.980.

Limits of Impurities

Arsenic (as As). Not more than 3 parts per million (0.0003 percent).

Heavy metals (as Pb). Not more than 40 parts per million (0.004 percent).

Lead. Not more than 10 parts per million (0.001 percent).

TESTS

Assay. Proceed as directed under *Aldehydes and Ketones—Neutral Sulfite Method,* page 895.

Angular rotation. Determine in a 100-mm. tube as directed under *Optical Rotation,* page 939.

Refractive index, page 945. Determine with an Abbé or other refractometer of equal or greater accuracy.

Solubility in alcohol. Proceed as directed in the general method, page 899. One ml. dissolves in 0.5 ml. of 90 percent alcohol and remains clear on dilution.

Specific gravity. Determine by any reliable method (see page 5).

Arsenic. A *Sample Solution* prepared as directed for organic compounds meets the requirements of the *Arsenic Test,* page 865.

Heavy metals. Prepare and test a 500-mg. sample as directed in *Method II* under the *Heavy Metals Test,* page 920, using 20 mcg. of lead ion (Pb) in the control (*Solution A*).

Lead. A *Sample Solution* prepared as directed for organic com pounds meets the requirements of the *Lead Limit Test,* page 929, using 10 mcg. of lead ion (Pb) in the control.

Packaging and storage. Store in full, tight, preferably glass, aluminum, or other suitably lined containers protected from light.

Functional use in foods. Flavoring agent.

DILLWEED OIL, AMERICAN

Dill Oil; Dill Herb Oil

DESCRIPTION

The volatile oil obtained by steam distillation from the freshly cut stalks, leaves and seeds of the plant, *Anethum graveolens*, L. It is a light yellow to yellow liquid. It is soluble in most fixed oils and in mineral oil. It is soluble, usually with opalescence or turbidity, in propylene glycol, but it is practically insoluble in glycerin.

SPECIFICATIONS

Assay. Usually not less than 28 percent and not more than 45 percent, by volume, of ketones as carvone.

Note: Oil obtained from early season distillation may show a carvone content as low as 25.0 percent and a correspondingly lower specific gravity, lower refractive index, and a higher angular rotation.

Angular rotation. Between +84° and +95°.

Refractive index. Between 1.480 and 1.485 at 20°.

Solubility in alcohol. Passes test.

Specific gravity. Between 0.884 and 0.900.

Limits of Impurities

Arsenic (as As). Not more than 3 parts per million (0.0003 percent).

Heavy metals (as Pb). Not more than 40 parts per million (0.004 percent).

Lead. Not more than 10 parts per million (0.001 percent).

TESTS

Assay. Proceed as directed under *Aldehydes and Ketones—Neutral Sulfite Method*, page 895.

Angular rotation. Determine in a 100-mm. tube as directed under *Optical Rotation*, page 939.

Refractive index, page 945. Determine with an Abbé or other refractometer of equal or greater accuracy.

Solubility in alcohol. Proceed as directed in the general method, page 899. One ml. dissolves in 1 ml. of 90 percent alcohol, frequently with opalescence which may not disappear on dilution to as much as 10 ml.

Specific gravity. Determine by any reliable method (see page 5).

Arsenic. A *Sample Solution* prepared as directed for organic compounds meets the requirements of the *Arsenic Test*, page 865.

Heavy metals. Prepare and test a 500-mg. sample as directed in *Method II* under the *Heavy Metals Test*, page 920, using 20 mcg. of lead ion (Pb) in the control (*Solution A*).

Lead. A *Sample Solution* prepared as directed for organic compounds meets the requirements of the *Lead Limit Test*, page 929, using 10 mcg. of lead ion (Pb) in the control.

Packaging and storage. Store in full, tight, preferably glass, aluminum, or tin-lined containers in a cool place protected from light.

Functional use in foods. Flavoring agent.

DIMETHYL ANTHRANILATE

Methyl *N*-Methyl Anthranilate

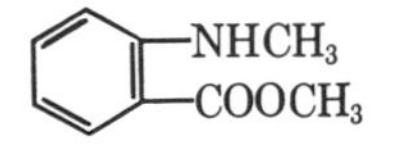

$C_9H_{11}NO_2$ Mol. wt. 165.19

DESCRIPTION

A pale yellow liquid having a bluish fluorescence and a grape-like odor. It may contain small amounts of methyl anthranilate which is limited by the upper assay tolerance and by the solidification point. It is practically insoluble in water and in glycerin, partially soluble in propylene glycol, and soluble in most fixed oils.

SPECIFICATIONS

Assay. Not less than 98.0 percent and not more than the equivalent of 101.3 percent of $C_9H_{11}NO_2$.

Refractive index. Between 1.578 and 1.581 at 20°.

Solidification point. Not less than 14°.

Solubility in alcohol. Passes test.

Specific gravity. Between 1.126 and 1.132.

TESTS

Assay. Weigh accurately about 1.1 grams, and proceed as directed under *Ester Determination*, page 896, using 82.60 as the equivalence factor (*e*) in the calculation.

Refractive index, page 945. Determine with an Abbé or other refractometer of equal or greater accuracy.

Solidification point. Determine as directed in the general method, page 954.

Solubility in alcohol. Proceed as directed in the general method, page 899. One ml. dissolves in 3 ml. of 80 percent alcohol and remains in solution upon dilution to 10 ml.

Specific gravity. Determine by any reliable method (see page 5).

Packaging and storage. Store in full, tight, preferably glass, aluminum, tin-lined, or other suitably lined containers in a cool place.
Functional use in foods. Flavoring agent.

DIMETHYL BENZYL CARBINOL

α,α-Dimethylphenethyl Alcohol

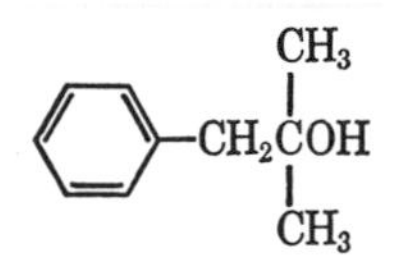

$C_{10}H_{14}O$ Mol. wt. 150.22

DESCRIPTION

A white crystalline solid which melts readily, and may exist in supercooled form as a colorless to pale yellow liquid having a floral odor. It is soluble in most fixed oils, in mineral oil, and in propylene glycol. It is insoluble in glycerin.

SPECIFICATIONS

Assay. Not less than 97.0 percent of $C_{10}H_{14}O$.

Refractive index. Between 1.514 and 1.517 at 20°, in supercooled liquid form.

Solidification point. Not less than 22°.

Solubility in alcohol. Passes test.

Specific gravity. Between 0.972 and 0.977.

Limits of Impurities

Acid value. Not more than 1.0.

TESTS

Assay. Acetylate a 10-ml. sample as directed under *Linalool Determination*, page 897. Weigh accurately about 1.5 grams of the acetylated alcohol, and proceed as directed under *Ester Determination*, page 896. Calculate the percent of dimethyl benzyl carbinol ($C_{10}H_{14}O$) in the original sample by the formula given for linalool, page 897, substituting 7.511 for 7.707 as the equivalence factor.

Refractive index, page 945. Determine with an Abbé or other refractometer of equal or greater accuracy.

Solidification point. Determine as directed in the general method, page 954.

Solubility in alcohol. Proceed as directed in the general method, page 899. One ml. dissolves in 3 ml. of 50 percent alcohol, and remains in solution on dilution to 10 ml.

Specific gravity. Determine by any reliable method (see page 5).

Acid value. Determine as directed in the general method, page 893.

Chlorinated compounds. Proceed as directed in the general method, page 895.

Packaging and storage. Store in full, tight, preferably glass, aluminum, tin-lined, or other suitably lined containers in a cool place protected from light.

Functional use in foods. Flavoring agent.

DIMETHYL BENZYL CARBINYL ACETATE

α,α-Dimethylphenethyl Acetate

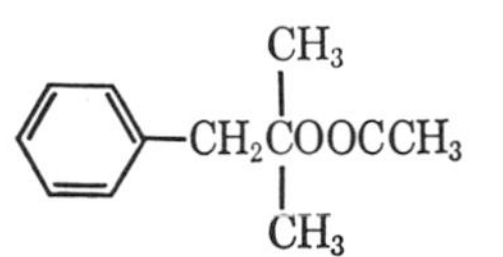

$C_{12}H_{16}O_2$ Mol. wt. 192.26

DESCRIPTION

A colorless liquid having a floral, fruity odor. It solidifies at room temperature. It is soluble in most fixed oils and in mineral oil. It is sparingly soluble in propylene glycol, but it is insoluble in glycerin.

SPECIFICATIONS

Assay. Not less than 98.0 percent of $C_{12}H_{16}O_2$.

Refractive index. Between 1.491 and 1.495 at 20°, in supercooled liquid form.

Solidification point. Not lower than 28°.

Solubility in alcohol. Passes test.

Specific gravity. Between 0.995 and 1.002, in supercooled liquid form.

Limits of Impurities

Acid value. Not more than 1.0.

Chlorinated compounds. Passes test.

TESTS

Assay. Weigh accurately about 1.3 grams, and proceed as directed under *Ester Determination*, page 896, using 96.13 as the equivalence factor (*e*) in the calculation.

Refractive index, page 945. Determine with an Abbé or other refractometer of equal or greater accuracy.

Solidification point. Determine as directed in the general method, page 954.

Solubility in alcohol. Proceed as directed in the general method, page 899. One ml. dissolves in 4 ml. of 70 percent alcohol.

Specific gravity. Determine by any reliable method (see page 5).

Acid value. Determine as directed in the general method, page 893.

Chlorinated compounds. Proceed as directed in the general method, page 895.

Packaging and storage. Store in full, tight, preferably glass, aluminum, tin-lined, or other suitably lined containers in a cool place protected from light.

Functional use in foods. Flavoring agent.

3,7-DIMETHYL-1-OCTANOL

Dimethyl Octanol; Tetrahydrogeraniol

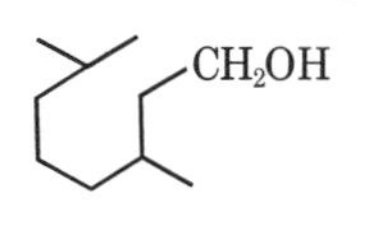

$C_{10}H_{22}O$ Mol. wt. 158.29

DESCRIPTION

A colorless liquid having a sweet rose-like odor. It is soluble in most fixed oils, in mineral oil, and in propylene glycol. It is insoluble in glycerin. Some commercial grades of 3,7-dimethyl-1-octanol may be composed almost entirely of the specific chemical whose structural formula is shown above, while other commercial grades that conform to the limits of these specifications and of satisfactory purity may contain other isomeric and closely related terpenic alcohols.

SPECIFICATIONS

Assay. Not less than 90.0 percent of total alcohols calculated as dimethyl octanol ($C_{10}H_{22}O$).

Refractive index. Between 1.435 and 1.445 at 20°.

Solubility in alcohol. Passes test.

Specific gravity. Between 0.826 and 0.842.

Limits of Impurities

Acid value. Not more than 1.0.

TESTS

Assay. Proceed as directed under *Total Alcohols*, page 893. Weigh accurately about 1.2 grams of the acetylated alcohol for the saponification, and use 79.15 as the equivalence factor (*e*) in the calculation.

Refractive index, page 945. Determine with an Abbé or other refractometer of equal or greater accuracy.

Solubility in alcohol. Proceed as directed in the general method, page 899. One ml. dissolves in 3 ml. of 70 percent alcohol.

Specific gravity. Determine by any reliable method (see page 5).

Acid value. Determine as directed in the general method, page 893.

Packaging and storage. Store in full, tight, preferably glass, iron, or suitably lined containers in a cool place protected from light. Aluminum containers should not be used.

Functional use in foods. Flavoring agent.

DIMETHYLPOLYSILOXANE

Dimethyl Silicone

DESCRIPTION

Dimethylpolysiloxane is a mixture of fully methylated linear siloxane polymers containing repeating units of the formula $[(CH_3)_2\text{-}SiO]$ and stabilized with trimethylsiloxy end-blocking units of the formula $[(CH_3)_3SiO—]$. It occurs as a clear, colorless, viscous liquid that is soluble in most aliphatic and aromatic hydrocarbon solvents but insoluble in water. [*Note:* Dimethylpolysiloxane is frequently used in commerce as such, or as a liquid containing silica (usually 4–5 percent), which must be removed by high-speed centrifugation (about 20,000 rpm) before testing the dimethylpolysiloxane for *Identification, Refractive index, Specific gravity,* and *Viscosity.* This monograph does not apply to aqueous emulsions containing emulsifying agents and preservatives, in addition to silica.]

IDENTIFICATION

Moisten about 100 mg. of the sample with a few drops of sulfuric acid and nitric acid in a platinum crucible, ignite at a red heat over a burner for about 10 minutes or until ashing is complete, and cool. Transfer the residue to a nickel crucible, fuse completely with 1 gram of sodium hydroxide, and cool. Dissolve the residue in 50 ml. of water, and filter. Place 1 drop of the filtrate on a sheet of filter paper, followed by 1 drop of ammonium molybdate T.S. and 1 drop of benzidine T.S., and place the paper over ammonium hydroxide. A greenish blue spot is produced.

SPECIFICATIONS

Loss on heating. Not more than 18 percent.

Refractive index. Between 1.400 and 1.404.

Specific gravity. Between 0.964 and 0.973.

Viscosity. Between 300 and 600 centistokes.

Limits of Impurities

Arsenic (as As). Not more than 3 parts per million (0.0003 percent).

Heavy metals (as Pb). Not more than 10 parts per million (0.001 percent).

TESTS

Loss on heating. Heat 15 grams of the sample in an open tared aluminum cup, having an internal surface of about 30 $cm.^2$, for 4 hours at 200° in a circulating air oven, cool, and weigh.

Refractive index, page 945. Determine with an Abbé or other refractometer of equal or greater accuracy.

Specific gravity. Determine by any reliable method (see page 5).

Viscosity. Determine as directed in the general method, page 970.

Arsenic. A *Sample Solution* prepared as directed for organic compounds meets the requirements of the *Arsenic Test*, page 865.

Heavy metals. Prepare and test a 2-gram sample as directed in *Method II* under the *Heavy Metals Test*, page 920, using 20 mcg. of lead ion (Pb) in the control (*Solution A*). (*Note:* If silica is present, it must be removed by filtration before the pH is adjusted.)

Packaging and storage. Store in tight containers.

Functional use in foods. Defoaming agent.

DIOCTYL SODIUM SULFOSUCCINATE

DSS

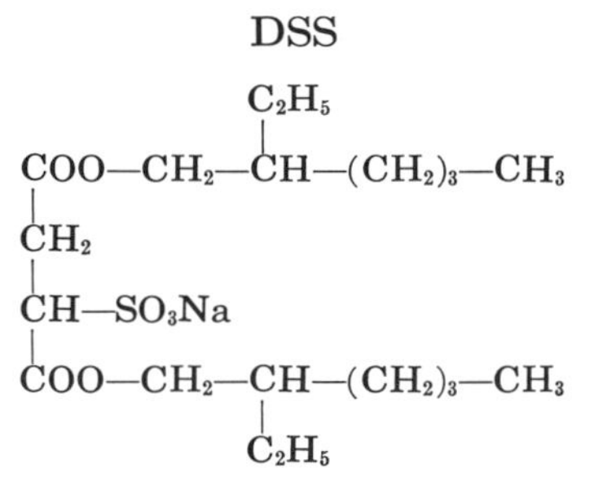

$C_{20}H_{37}NaO_7S$ Mol. wt. 444.56

DESCRIPTION

A white, wax-like, plastic solid, having a characteristic odor suggestive of octyl alcohol. It is free from the odor of other solvents. One gram dissolves slowly in about 70 ml. of water. It is freely soluble in alcohol and in glycerin, and is very soluble in solvent hexane.

SPECIFICATIONS

Assay. Not less than 99.0 percent of $C_{20}H_{37}NaO_7S$ after drying.

Clarity of solution. Passes test.

Residue on ignition. Between 15.5 percent and 16.2 percent.

Limits of Impurities

Arsenic (as As). Not more than 3 parts per million (0.0003 percent).

Heavy metals (as Pb). Not more than 10 parts per million (0.001 percent).

Loss on drying. Not more than 2 percent.

TESTS

Assay

Sample solution. Transfer 3.0 grams, previously dried at 105° for 2 hours, to a 400-ml. beaker, add 50 ml. of hot water, and stir until a paste is formed. Add 2 additional 50-ml. portions of hot water, and stir until the sample is completely dissolved. (*Note: Dioctyl sodium sulfosuccinate dissolves slowly in water and solution must be complete at this point.*) Cool the solution to room temperature, transfer it quantitatively, with the aid of water, to a 1000-ml. volumetric flask, dilute to volume with water, and mix.

Tetra-n-butylammonium iodide solution. Transfer 1.250 grams of tetra-*n*-butylammonium iodide to a 500-ml. volumetric flask, dilute to volume with water, and mix.

Salt solution. Dissolve 100 grams of anhydrous sodium sulfate and 10 grams of sodium carbonate in sufficient water to make 1000.0 ml.

Procedure. Pipet 25 ml. of *Tetra-n-butylammonium iodide solution* into a 250-ml. volumetric flask, and add 50.0 ml. of the *Salt solution*, 25 ml. of chloroform, and 0.4 ml. of bromophenol blue T.S. to the flask. Stopper and shake vigorously for 20 to 30 seconds or until all of the blue color goes into the chloroform layer. Titrate with the *Sample solution*, using a 50-ml. buret, until about 1 ml. from the end-point, and shake the stoppered flask vigorously for about 2 minutes. Continue the titration in 2-drop increments, shaking the stoppered flask vigorously for about 10 seconds after each addition, and then allow the flask to stand about 10 seconds or until the chloroform layer has separated. Continue the titration until the blue color is completely absent in the chloroform layer. Calculate the percent of $C_{20}H_{37}NaO_7S$ by the formula $2507/V$, in which V is the number of ml. of the *Sample solution* required to reach the end-point, and 2507 is a factor derived by the equation $[1.250 \times (25/500) \times (444.6/369.4) \times 100]/[3.000/1000]$.

Clarity of solution. Dissolve 25 grams in 94 ml. of alcohol. The solution does not develop a haze within 24 hours.

Residue on ignition. Ignite 1 gram as directed in the general procedure, page 945.

Arsenic. A *Sample Solution* prepared as directed for organic compounds meets the requirements of the *Arsenic Test*, page 865.

Heavy metals. Ignite 2 grams in a platinum crucible until free from carbon, cool, moisten the residue with 1 ml. of hydrochloric acid, and evaporate to dryness on a steam bath. Add 2 ml. of diluted acetic acid T.S., digest on a steam bath for 5 minutes, filter into a 50-ml. Nessler tube, and wash the residue with sufficient water to make 25 ml. This solution meets the requirements of the *Heavy Metals Test*, page 920, using 20 mcg. of lead ion (Pb) in the control (*Solution A*).

Loss on drying, page 931. Dry at 105° for 2 hours.

Packaging and storage. Store in well-closed containers.

Functional use in foods. Emulsifying and wetting agent.

DISODIUM EDTA

Disodium Ethylenediaminetetraacetate; Disodium(Ethylenedinitrilo)tetraacetate; Disodium Edetate

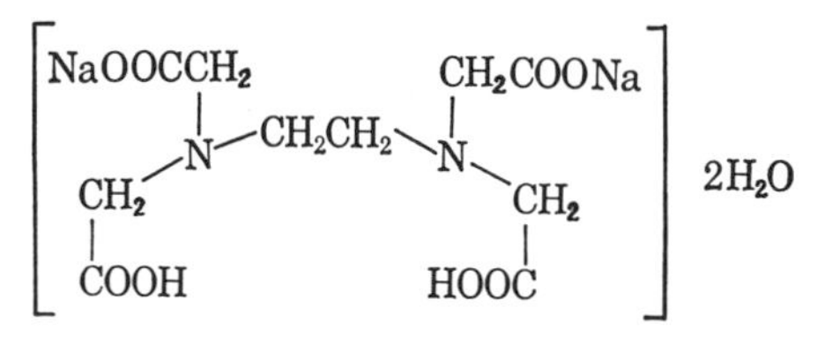

$C_{10}H_{14}N_2Na_2O_8 . 2H_2O$ Mol. wt. 372.24

DESCRIPTION

A white, crystalline powder. It is soluble in water.

IDENTIFICATION

A. A 1 in 20 solution responds to the flame test for *Sodium*, page 928.

B. To 5 ml. of water in a test tube add 2 drops of ammonium thiocyanate T.S. and 2 drops of ferric chloride T.S. To the deep red solution so obtained add about 50 mg. of the sample, and mix. The deep red color disappears.

SPECIFICATIONS

Assay. Not less than 99.0 percent of $C_{10}H_{14}N_2Na_2O_8 . 2H_2O$.

pH of a 1 in 100 solution. Between 4.3 and 4.7.

Limits of Impurities

Arsenic (as As). Not more than 3 parts per million (0.0003 percent).

Heavy metals (as Pb). Not more than 20 parts per million (0.002 percent).

Lead. Not more than 10 parts per million (0.001 percent).
Nitrilotriacetic acid. Passes test.

TESTS

Assay

Assay Preparation. Transfer about 5 grams of the sample, accurately weighed, into a 250-ml. volumetric flask, dissolve in water, dilute to volume, and mix.

Procedure. Place about 200 mg. of chelometric standard calcium carbonate, accurately weighed, in a 400-ml. beaker, add 10 ml. of water, and swirl to form a slurry. Cover the beaker with a watch glass, and introduce 2 ml. of diluted hydrochloric acid T.S. from a pipet inserted between the lip of the beaker and the edge of the watch glass. Swirl the contents of the beaker to dissolve the calcium carbonate. Wash down the sides of the beaker, the outer surface of the pipet, and the watch glass, and dilute to about 100 ml. with water. While stirring, preferably with a magnetic stirrer, add about 30 ml. of the *Assay Preparation* from a 50-ml. buret, then add 15 ml. of sodium hydroxide T.S. and 300 mg. of hydroxy naphthol blue indicator, and continue the titration to a blue end-point. Calculate the weight, in mg., of $C_{10}H_{14}N_2Na_2O_8.2H_2O$ in the sample taken by the formula 929.8 (W/V), in which W is the weight, in mg., of calcium carbonate, and V is the volume, in ml., of the *Assay Preparation* consumed in the titration.

pH of a 1 in 100 solution. Determine by the *Potentiometric Method*, page 941.

Arsenic. A *Sample Solution* prepared as directed for organic compounds meets the requirements of the *Arsenic Test*, page 865.

Heavy metals. Prepare and test a 1-gram sample as directed in *Method II* under the *Heavy Metals Test*, page 920, using 20 mcg. of lead ion (Pb) in the control (*Solution A*).

Lead. A *Sample Solution* prepared as directed for organic compounds meets the requirements of the *Lead Limit Test*, page 929, using 10 mcg. of lead ion (Pb) in the control.

Nitrilotriacetic acid

Stock test solution. Transfer 10.0 grams of the sample into a 100-ml. volumetric flask, dissolve in 40 ml. of potassium hydroxide solution (1 in 10), dilute to volume with water, and mix.

Diluted stock test solution. Pipet 10.0 ml. of the *Stock test solution* into a 100-ml. volumetric flask, dilute to volume with water, and mix.

Test preparation. Pipet 20.0 ml. of the *Diluted stock test solution* into a 150-ml. beaker, add 1 ml. of potassium hydroxide solution (1 in 10), 2 ml. of ammonium nitrate solution (1 in 10), and about 50 mg. of eriochrome black T indicator, and titrate with cadmium nitrate solution (3 in 100) to a red end-point. Record the volume, in ml., of the titrant required as V, and discard the solution.

Pipet 20.0 ml. of the *Diluted stock test solution* into a 100-ml. volumetric flask, and add the volume, *V*, of cadmium nitrate solution (3 in 100) required in the initial titration, plus 0.05 ml. in excess. Add 1.5 ml. of potassium hydroxide solution (1 in 10), 10 ml. of ammonium nitrate solution (1 in 10), and 0.5 ml. of methyl red T.S., then dilute to volume with water, and mix.

Stock standard solution. Transfer 1.0 gram of nitrilotriacetic acid into a 100-ml. volumetric flask, dissolve in 10 ml. of potassium hydroxide solution (1 in 10), dilute to volume with water, and mix.

Diluted stock standard solution. Pipet 1.0 ml. of the *Stock standard solution* and 10.0 ml. of the *Stock test solution* into a 100-ml. volumetric flask, dilute to volume with water, and mix.

Standard preparation. Proceed as directed under *Test preparation,* using *Diluted stock standard solution* where *Diluted stock test solution* is specified.

Procedure. Rinse a polarographic cell with a portion of the *Standard preparation,* then add a suitable volume to the cell, immerse it in a constant-temperature bath maintained at 25° ± 0.5°, and de-aerate by bubbling oxygen-free nitrogen through the solution for 10 minutes. Insert the dropping mercury electrode of a suitable polarograph, and record the polarogram from −0.6 to −1.2 volts at a sensitivity of 0.006 microampere per mm., using a saturated calomel electrode as the reference electrode. In the same manner, polarograph a portion of the *Test preparation.* The diffusion current observed with the *Test preparation* is not greater than 10 percent of the difference between the diffusion currents observed with the *Standard preparation* and the *Test preparation,* respectively. (*Note:* An extra polarographic wave appearing ahead of the nitrilotriacetic acid-cadmium complex wave is probably due to uncomplexed cadmium. This wave should be ignored in measuring the diffusion current.)

Packaging and storage. Store in well-closed containers.

Functional use in foods. Preservative; sequestrant; stabilizer.

DISODIUM GUANYLATE

Sodium 5′-Guanylate; Disodium Guanosine-5′-monophosphate

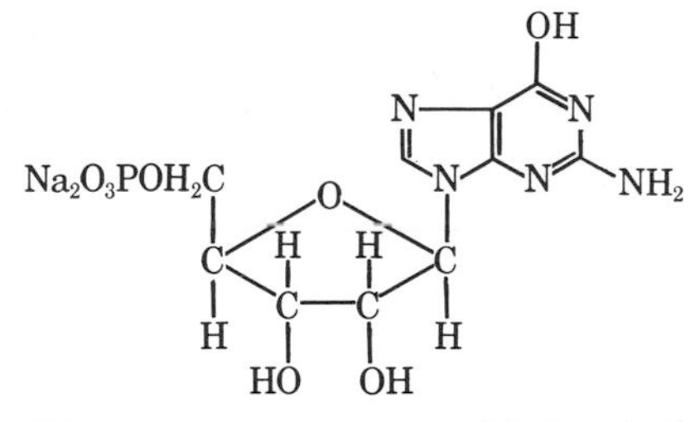

$C_{10}H_{12}N_5Na_2O_8P . xH_2O$ Mol. wt. (anhydrous) 407.20

DESCRIPTION

Disodium guanylate contains approximately 7 molecules of water of crystallization. It occurs as colorless or white crystals or as a white, crystalline powder, having a characteristic taste. It is soluble in water, sparingly soluble in alcohol, and practically insoluble in ether.

IDENTIFICATION

A 1 in 50,000 solution of the sample in 0.01 *N* hydrochloric acid exhibits an absorbance maximum at 256 ± 2 mμ, page 957.

SPECIFICATIONS

Assay. Not less than 97.0 percent and not more than the equivalent of 102.0 percent of $C_{10}H_{12}N_5Na_2O_8P$, calculated on the dried basis.

pH of a 1 in 20 solution. Between 7.0 and 8.5.

Clarity and color of solution. Passes test.

Loss on drying. Not more than 25 percent.

Limits of Impurities

Amino acids. Passes test.

Ammonium salts. Passes test.

Arsenic (as As). Not more than 3 parts per million (0.0003 percent).

Heavy metals (as Pb). Not more than 20 parts per million (0.002 percent).

Lead. Not more than 10 parts per million (0.001 percent).

Other nucleotides. Passes test.

TESTS

Assay. Transfer about 500 mg. of the sample, accurately weighed, into a 1000-ml. volumetric flask, dissolve in 0.01 *N* hydrochloric acid, dilute to volume with 0.01 *N* hydrochloric acid, and mix. Transfer 10.0 ml. of this solution into a 250-ml. volumetric flask, dilute to volume with 0.01 *N* hydrochloric acid, and mix. Determine the absorbance of this solution and of a similar solution of F.C.C. Disodium Guanylate Reference Standard, at a concentration of 20 mcg. per ml., in 1-cm. cells, at the maximum at about 260 mμ, with a suitable spectrophotometer, using 0.01 *N* hydrochloric acid as the blank. Calculate the quantity, in mg., of $C_{10}H_{12}N_5Na_2O_8P$ in the sample taken by the formula $25C \times A_U/A_S$, in which C is the exact concentration of the Reference Standard solution, in mcg. per ml., A_U is the absorbance of the sample solution, and A_S is the absorbance of the Reference Standard solution.

pH of a 1 in 20 solution. Determine by the *Potentiometric Method*, page 941.

Clarity and color of solution. A 100-mg. portion of the sample dissolved in 10 ml. of water is colorless and shows no more than a trace of turbidity.

Loss on drying, page 931. Dry at 120° for 4 hours.

Amino acids. To 5 ml. of a 1 in 1000 solution of the sample add 1 ml. of ninhydrin T.S., and heat for 3 minutes. No color is produced.

Ammonium salts. Transfer about 100 mg. of the sample into a small test tube, and add 50 mg. of magnesium oxide and 1 ml. of water. Moisten a piece of red litmus paper with water, suspend it in the tube, cover the mouth of the tube, and heat in a water bath for 5 minutes. The litmus paper does not change to blue.

Arsenic. A *Sample Solution* prepared as directed for organic compounds meets the requirements of the *Arsenic Test*, page 865.

Heavy metals. Prepare and test a 1-gram sample as directed in *Method II* under the *Heavy Metals Test*, page 920, using 20 mcg. of lead ion (Pb) in the control (*Solution A*).

Lead. A *Sample Solution* prepared as directed for organic compounds meets the requirements of the *Lead Limit Test*, page 929, using 10 mcg. of lead ion (Pb) in the control.

Other nucleotides. Prepare a strip of Whatman No. 2 or equivalent filter paper about 20 × 40 cm., and draw a line across the narrow dimension about 5 cm. from one end. Using a micropipet, apply on the center of the line 10 microliters of a 1 in 100 solution of the sample in water, and dry the paper in air. Fill the trough of an apparatus suitable for descending chromatography (see page 882) with a 160:3:40 mixture of saturated ammonium sulfate solution, *tert*-butyl alcohol, and 0.025 *N* ammonia, respectively, and suspend the strip in the chamber, placing the end of the strip in the trough at a distance about 1 cm. from the pencil line. Seal the chamber, and allow the chromatogram to develop until the solvent front descends to a distance about 30 cm. from the starting line. Remove the strip from the chamber, dry in air, and observe under shortwave (254 mμ) ultraviolet light in the dark. Only one spot is visible.

Packaging and storage. Store in well-closed containers.

Functional use in foods. Flavor enhancer.

DISODIUM INOSINATE

Sodium 5′-Inosinate; Disodium Inosine-5′-monophosphate

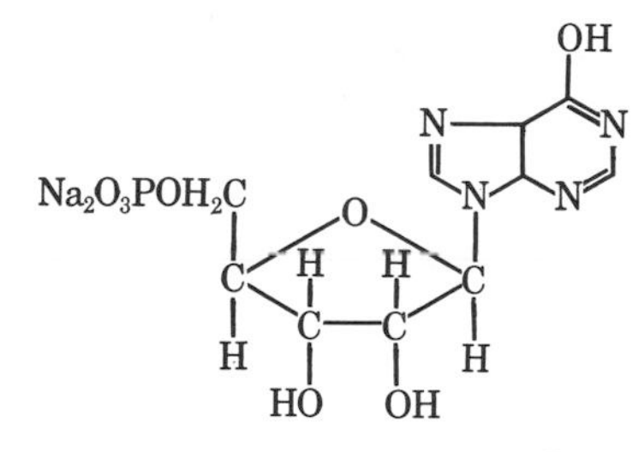

$C_{10}H_{11}N_4Na_2O_8P.xH_2O$ Mol. wt. (anhydrous) 392.19

DESCRIPTION

Disodium inosinate contains approximately 7.5 molecules of water of crystallization. It occurs as colorless or white crystals or as a white, crystalline powder, having a characteristic taste. It is soluble in water, sparingly soluble in alcohol, and practically insoluble in ether.

IDENTIFICATION

A 1 in 50,000 solution of the sample in 0.01 *N* hydrochloric acid exhibits an absorbance maximum at 250 ± 2 mμ, page 957. The ratio A_{250}/A_{260} is between 1.55 and 1.65, and the ratio A_{280}/A_{260} is between 0.20 and 0.30.

SPECIFICATIONS

Assay. Not less than 97.0 percent and not more than the equivalent of 102.0 percent of $C_{10}H_{11}N_4Na_2O_8P$, calculated on the anhydrous basis.

pH of a 1 in 20 solution. Between 7.0 and 8.5.

Clarity and color of solution. Passes test.

Water. Not more than 28.5 percent.

Limits of Impurities

Amino acids. Passes test.

Ammonium salts. Passes test.

Arsenic (as As). Not more than 3 parts per million (0.0003 percent).

Barium. Not more than 150 parts per million (0.015 percent).

Heavy metals (as Pb). Not more than 20 parts per million (0.002 percent).

Lead. Not more than 10 parts per million (0.001 percent).

Other nucleotides. Passes test.

TESTS

Assay. Transfer about 500 mg. of the sample, accurately weighed, into a 1000-ml. volumetric flask, dissolve in 0.01 *N* hydrochloric acid, dilute to volume with 0.01 *N* hydrochloric acid, and mix. Transfer

10.0 ml. of this solution into a 250-ml. volumetric flask, dilute to volume with 0.01 *N* hydrochloric acid, and mix. Determine the absorbance of this solution and of a similar solution of F.C.C. Disodium Inosinate Reference Standard, at a concentration of 20 mcg. per ml., in 1-cm. cells, at the maximum at about 250 mμ, with a suitable spectrophotometer, using 0.01 *N* hydrochloric acid as the blank. Calculate the quantity, in mg., of $C_{10}H_{11}N_4Na_2O_8P$ in the sample taken by the formula $25C \times A_U/A_S$, in which C is the exact concentration of the Reference Standard solution, in mcg. per ml., A_U is the absorbance of the sample solution, and A_S is the absorbance of the Reference Standard solution.

pH of a 1 in 20 solution. Determine by the *Potentiometric Method*, page 941.

Clarity and color of solution. A 500-mg. portion of the sample dissolved in 10 ml. of water is colorless and shows no more than a trace of turbidity.

Water. Determine by the *Karl Fischer Titrimetric Method*, page 977.

Amino acids. To 5 ml. of a 1 in 1000 solution of the sample add 1 ml. of ninhydrin T.S. No color is produced.

Ammonium salts. Transfer about 100 mg. of the sample into a small test tube, and add 50 mg. of magnesium oxide and 1 ml. of water. Moisten a piece of red litmus paper with water, suspend it in the tube, cover the mouth of the tube, and heat in a water bath for 5 minutes. The litmus paper does not change to blue.

Arsenic. A *Sample Solution* prepared as directed for organic compounds meets the requirements of the *Arsenic Test*, page 865.

Barium. Dissolve 1 gram of the sample in 100 ml. of water, filter, and add 5 ml. of diluted sulfuric acid T.S. to the filtrate. Any turbidity is not greater than that produced in a similar solution containing 1.5 ml. of *Barium Standard Solution* (150 mcg. Ba).

Heavy metals. Prepare and test a 1-gram sample as directed in *Method II* under the *Heavy Metals Test*, page 920, using 20 mcg. of lead ion (Pb) in the control (*Solution A*).

Lead. A *Sample Solution* prepared as directed for organic compounds meets the requirements of the *Lead Limit Test*, page 929, using 10 mcg. of lead ion (Pb) in the control.

Other nucleotides. Prepare a strip of Whatman No. 2 or equivalent filter paper about 20 $\times$ 40 cm., and draw a line across the narrow dimension about 5 cm. from one end. Using a micropipet, apply on the center of the line 10 microliters of a 1 in 100 solution of the sample in water, and dry the paper in air. Fill the trough of an apparatus suitable for descending chromatography (see page 882) with a 160:3:40 mixture of saturated ammonium sulfate solution, *tert*-butyl alcohol, and 0.025 *N* ammonia, respectively, and suspend the strip in the chamber, placing the end of the strip in the trough at a distance about 1 cm. from the pencil line. Seal the chamber, and allow the chromatogram to develop until the solvent front descends to a distance about 30 cm. from the starting line. Remove the strip from the chamber, dry in air, and observe under shortwave (254 mμ) ultraviolet light in the dark. Only one spot is visible.

Packaging and storage. Store in well-closed containers.

Functional use in foods. Flavor enhancer.

Δ-DODECALACTONE

```
CH3(CH2)6CHCH2CH2CH2
         |        |
         O--------C=O
```

$C_{12}H_{22}O_2$ Mol. wt. 198.31

DESCRIPTION

A colorless liquid having a coconut-fruity odor which becomes butter-like in lower concentrations. It is practically insoluble in water, but is very soluble in alcohol, in propylene glycol, and in vegetable oil.

SPECIFICATIONS

Assay. Not less than 99.0 percent of $C_{12}H_{22}O_2$.

Refractive index. Between 1.458 and 1.461 at 20°.

Saponification value. Between 278 and 286.

Limits of Impurities

Acid value. Not more than 8.0.

Unsaponifiable matter. Not more than 0.11 percent.

TESTS

Assay. Determine as directed for *Δ-Decalactone*, page 229. The Δ-dodecalactone is eluted in about 25 minutes, and an area of about 600 $cm.^2$ is generated at attenuation ×8. The area of the Δ-dodecalactone peak is not less than 99.0 percent of the total areas of all peaks.

Refractive index, page 945. Determine with an Abbé or other refractometer of equal or greater accuracy.

Saponification value. Weigh accurately about 1 gram, and determine as directed in the general method, page 896.

Acid value. Determine as directed in *Method II* under the general procedure, page 902.

Unsaponifiable matter. Determine as directed in the general method, page 915.

Packaging and storage. Store in tight containers.

Functional use in foods. Flavoring agent.

ERYTHORBIC ACID

D-Araboascorbic Acid

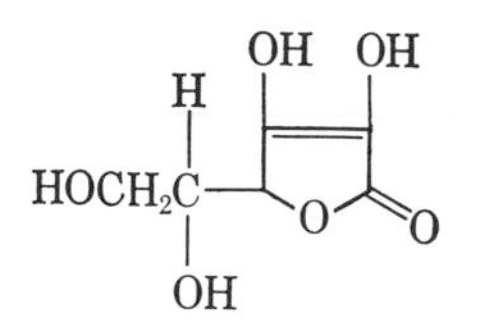

$C_6H_8O_6$ Mol. wt. 176.13

DESCRIPTION

White or slightly yellow crystals or powder. On exposure to light it gradually darkens. In the dry state it is reasonably stable in air, but in solution it rapidly deteriorates in the presence of air. It melts between 164° and 171° with decomposition. One gram is soluble in about 2.5 ml. of water and in about 20 ml. of alcohol. It is slightly soluble in glycerin.

IDENTIFICATION

A. A 1 in 50 solution slowly reduces alkaline cupric tartrate T.S. at 25°, but more readily upon heating.

B. To 2 ml. of a 1 in 50 solution add a few drops of sodium nitroferricyanide T.S., followed by 1 ml. of approximately 0.1 *N* sodium hydroxide. A transient blue color is produced immediately.

C. Dissolve about 15 mg. in 15 ml. of a trichloroacetic acid solution (1 in 20), add about 200 mg. of activated charcoal, and shake the mixture vigorously for 1 minute. Filter through a small fluted filter, refiltering if necessary to obtain a clear filtrate. To 5 ml. of the clear filtrate add 1 drop of pyrrole, agitate the mixture until the pyrrole is dissolved, then heat in a water bath at 50°. A blue color develops.

SPECIFICATIONS

Assay. Not less than 99.0 percent of $C_6H_8O_6$.

Specific rotation, $[\alpha]_D^{25°}$. Between −16.5° and −18.0°.

Limits of Impurities

Arsenic (as As). Not more than 3 parts per million (0.0003 percent).

Heavy metals (as Pb). Not more than 20 parts per million (0.002 percent).

Lead. Not more than 10 parts per million (0.001 percent).

Residue on ignition. Not more than 0.3 percent.

TESTS

Assay. Dissolve about 400 mg., accurately weighed, in a mixture of 100 ml. of water, recently boiled and cooled, and 25 ml. of diluted sulfuric acid T.S. Titrate the solution immediately with 0.1 *N* iodine, adding starch T.S. near the end-point. Each ml. of 0.1 *N* iodine is equivalent to 8.806 mg. of $C_6H_8O_6$.

Specific rotation. Transfer about 2.5 grams, accurately weighed, into a 25-ml. volumetric flask, dissolve it in about 20 ml. of water, and dilute to volume. Determine the specific rotation as directed under *Optical Rotation*, page 939.

Arsenic. A *Sample Solution* prepared as directed for organic compounds meets the requirements of the *Arsenic Test*, page 865.

Heavy metals. A solution of 1 gram in 25 ml. of water meets the requirements of the *Heavy Metals Test*, page 920, using 20 mcg. of lead ion (Pb) in the control (*Solution A*).

Lead. A *Sample Solution* prepared as directed for organic compounds meets the requirements of the *Lead Limit Test*, page 929, using 10 mcg. of lead ion (Pb) in the control.

Residue on ignition, page 945. Ignite 1 gram as directed in the general method.

Packaging and storage. Store in tight, light-resistant containers.

Functional use in foods. Preservative; antioxidant.

ESTRAGOLE

p-Allylanisole

$CH_3O-C_6H_4-CH_2CH{=}CH_2$

$C_{10}H_{12}O$ Mol. wt. 148.21

DESCRIPTION

A colorless to light yellow liquid having an anise-like odor. It is soluble in alcohol, but practically insoluble in water.

SPECIFICATIONS

Refractive index. Between 1.517 and 1.522 at 20°.

Solubility in alcohol. Passes test.

Specific gravity. Between 0.957 and 0.965.

TESTS

Refractive index, page 945. Determine with an Abbé or other refractometer of equal or greater accuracy.

Solubility in alcohol. Proceed as directed in the general method, page 899. One ml. dissolves in 6 ml. of 80 percent alcohol to form clear solutions.

Specific gravity. Determine by any reliable method (see page 5).

Packaging and storage. Store in full, tight, preferably glass, tin-lined or other suitably lined containers in a cool place protected from light.

Functional use in foods. Flavoring agent.

ETHOXYLATED MONO- and DIGLYCERIDES

Polyoxyethylene (20) Mono- and Diglycerides of Fatty Acids; Polyglycerate 60

DESCRIPTION

A mixture of stearate, palmitate, and lesser amounts of myristate partial esters of glycerin condensed with approximately 20 moles of ethylene oxide per mole of alpha-monoglyceride reaction mixture having an average molecular weight of 535 ($\pm$10 percent). It occurs as a pale slightly yellow colored, oily liquid or semi-gel having a faint, characteristic odor and a mildly bitter taste. It is soluble in water, in alcohol, and in xylene. It is partially soluble in mineral oil and in vegetable oils. It conforms to the regulations of the Federal Food and Drug Administration pertaining to specifications for fats or fatty acids derived from edible sources. [*Note:* If the product is manufactured by direct esterification of glycerin with a mixture of primary stearic, palmitic, and myristic acids, the intermediate product (before reaction with ethylene oxide) has an acid value of not greater than 0.3 and a water content not greater than 0.2 percent.]

IDENTIFICATION

A. To 5 ml. of a 1 in 20 solution in water, add 5 ml. of sodium hydroxide T.S., boil for a few minutes, cool, and acidify with hydrochloric acid T.S. The solution is strongly opalescent.

B. A mixture of 46 volumes of the sample with 54 volumes of water at 40° or below yields a gelatinous mass.

SPECIFICATIONS

Oxyethylene content (apparent). Not less than 60.5 percent and not more than 65.0 percent, calculated as ethylene oxide (C_2H_4O), on the anhydrous basis.

Hydroxyl value. Between 65 and 80.

Saponification value. Between 65 and 75.

Stearic, palmitic, and myristic acids. Between 31 and 33 grams per 100 grams of sample.

Limits of Impurities

Acid value. Not more than 2.

Arsenic (as As). Not more than 3 parts per million (0.0003 percent).

Heavy metals (as Pb). Not more than 10 parts per million (0.001 percent).

Water. Not more than 1 percent.

TESTS

Oxyethylene content (apparent). Weigh accurately about 70 mg., and proceed as directed under *Oxyethylene Determination*, page 910.

Hydroxyl value. Determine as directed under *Method II* in the general procedure, page 904.

Saponification value. Determine as directed in the general method, page 914, using about 6 grams, accurately weighed.

Stearic, palmitic, and myristic acids. Isolate the fatty acids as directed in the test for *Lauric acid* under *Polysorbate 20*, page 632, and determine the weight of the acids. The product so obtained has an *Acid Value* between 199 and 211 (*Method I*, page 902) and a *Solidification Point*, page 954, not below 50°.

Acid value. Determine as directed under *Method II* in the general procedure, page 902.

Arsenic. A *Sample Solution* prepared as directed for organic compounds meets the requirements of the *Arsenic Test*, page 865.

Heavy metals. Prepare and test a 2-gram sample as directed in *Method II* under the *Heavy Metals Test*, page 920, using 20 mcg. of lead ion (Pb) in the control (*Solution A*).

Water. Determine by the *Karl Fischer Titrimetric Method*, page 977.

Packaging and storage. Store in well-closed containers.

Functional use in foods. Dough conditioner.

ETHOXYQUIN

6-Ethoxy-1,2-dihydro-2,2,4-trimethylquinoline

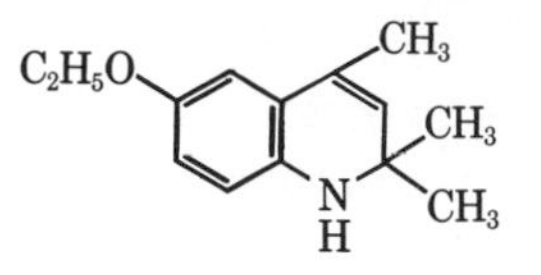

$C_{14}H_{19}NO$ Mol. wt. (monomer) 217.31

DESCRIPTION

Ethoxyquin is a mixture consisting predominantly of the monomer ($C_{14}H_{19}NO$). It occurs as a clear liquid that may darken with age without affecting its antioxidant activity. Its specific gravity is about 1.02, and its refractive index is about 1.57.

SPECIFICATIONS

Assay. Not less than 90.0 percent of $C_{14}H_{19}NO$.

Limits of Impurities

Arsenic (as As). Not more than 3 parts per million (0.0003 percent).

Heavy metals (as Pb). Not more than 10 parts per million (0.001 percent).

TESTS

Assay. Transfer about 200 mg. of the sample, accurately weighed, into a 150-ml. beaker containing 50 ml. of glacial acetic acid, and immediately titrate with 0.1 *N* perchloric acid in glacial acetic acid, determining the end-point potentiometrically. Perform a blank determination, and make any necessary correction (see page 2). Each ml. of 0.1 *N* perchloric acid is equivalent to 21.73 mg. of $C_{14}H_{19}NO$ (monomer).

Arsenic. A *Sample Solution* prepared as directed for organic compounds meets the requirements of the *Arsenic Test*, page 865.

Heavy metals. Prepare and test a 2-gram sample as directed under *Method II* in the *Heavy Metals Test*, page 920, using 20 mcg. of lead ion (Pb) in the control (*Solution A*).

Packaging and storage. Store in tightly closed carbon steel or black iron (not rubber, neoprene, or nylon) containers in a cool, dark place. Prolonged exposure to sunlight causes polymerization.

Functional use in foods. Antioxidant for apples and pears.

ETHYL ACETATE

$CH_3COOC_2H_5$

$C_4H_8O_2$ Mol. wt. 88.11

DESCRIPTION

Ethyl acetate is a transparent, colorless liquid, with a fragrant, refreshing, slightly acetous odor, and an acetous, burning taste. One ml. dissolves in about 10 ml. of water. It is miscible with alcohol, ether, fixed oils, or volatile oils. It is readily volatilized even at low temperature and is flammable; when burned, a yellow flame and an acetous odor are produced.

SPECIFICATIONS

Assay. Not less than 99.0 percent of $C_4H_8O_2$, the remainder consisting chiefly of alcohol and water.

Distillation range. Between 76° and 77.5°.

Specific gravity. Between 0.894 and 0.898.

Limits of Impurities

Acidity. Passes test.

Arsenic (as As). Not more than 3 parts per million (0.0003 percent).

Butylic and amylic derivatives. Passes test.

Heavy metals (as Pb). Not more than 10 parts per million (0.001 percent).

Methyl compounds. Passes test.

Nonvolatile residue. Not more than 0.02 percent.

Readily carbonizable substances. Passes test.

TESTS

Assay. Transfer about 1.5 grams, accurately weighed in a tared, stoppered weighing bottle, to a suitable flask, add 50.0 ml. of 0.5 *N* sodium hydroxide, and heat on a steam bath under a reflux condenser for 1 hour. Allow it to cool, add phenolphthalein T.S., and titrate the excess sodium hydroxide with 0.5 *N* hydrochloric acid. Perform a blank determination (see page 2), and make any necessary correction. Each ml. of 0.5 *N* sodium hydroxide is equivalent to 44.06 mg. of $C_4H_8O_2$.

Distillation range. Proceed as directed in the general method, page 890.

Specific gravity. Determine by any reliable method (see page 5).

Acidity. A solution of 2 ml. in 10 ml. of neutralized alcohol requires not more than 0.1 ml. of 0.1 *N* sodium hydroxide for neutralization using 2 drops of phenolphthalein T.S. as the indicator.

Arsenic. A *Sample Solution* prepared as directed for organic compounds meets the requirements of the *Arsenic Test*, page 865.

Butylic and amylic derivatives. Allow 10 ml. to evaporate spontaneously from clean, odorless blotting paper. The final odor does not resemble that of pineapple or banana.

Heavy metals. Evaporate 2.5 ml. (2-gram sample) to dryness with 10 mg. of sodium carbonate, heat gently to volatilize any organic matter, and dissolve the residue in 25 ml. of water. This solution meets the requirements of the *Heavy Metals Test*, page 920, using 20 mcg. of lead ion (Pb) in the control (*Solution A*).

Methyl compounds. Place 20 ml. in a 500-ml. separator, add a solution of 20 grams of sodium hydroxide in 50 ml. of water, stopper the separator, and wrap it securely in a towel for protection against the heat of the reaction. Shake the mixture vigorously for about 5 minutes, cautiously opening the stopcock from time to time to permit the escape of air. Continue the shaking vigorously until a homogeneous liquid results, then distil and collect about 25 ml. of the distillate. To 1 drop of the distillate, add 1 drop of dilute phosphoric acid (1 in 20) and 1 drop of potassium permanganate solution (1 in 20). Mix, allow to stand for 1 minute, and add sodium bisulfite solution (1 in 20) dropwise, until the color is discharged. If a brown color remains, add 1 drop of the dilute phosphoric acid. To the colorless solution add 5 ml. of a freshly prepared solution of chromotropic acid (1 in 2000) in 75 percent sulfuric acid, and heat on a steam bath for 10 minutes at 60°. No violet color appears.

Nonvolatile residue. Evaporate 10 grams, or more, in a tared glass or porcelain dish on a steam bath, dry at 105° for 1 hour, and weigh.

Readily carbonizable substances. Pour 2 ml. carefully upon 10 ml. of sulfuric acid T.S. so as to form separate layers. No discoloration is developed within 15 minutes.

Packaging and storage. Store in tight containers and avoid excessive heat.

Functional use in foods. Flavoring agent.

ETHYL ACETOACETATE

Acetoacetic Ester; Ethyl 3-Oxobutanoate

$$CH_3COCH_2CO_2C_2H_5 \rightleftarrows CH_3C(OH){=}CHCOOC_2H_5$$

$C_6H_{10}O_3$ Mol. wt. 130.14

DESCRIPTION

An equilibrium mixture of the keto- and enol-forms of ethyl acetoacetate. It is a clear, colorless to very light yellow, mobile liquid having a characteristic, agreeable odor. It is miscible in all proportions with alcohol, with ether, and with ethyl acetate. One ml. dissolves in about 12 ml. of water.

SPECIFICATIONS

Assay. Not less than 97.5 percent of $C_6H_{10}O_3$.

Refractive index. Between 1.418 and 1.421 at 20°.

Specific gravity. Between 1.022 and 1.027.

Limits of Impurities

Acidity (as acetic acid). Not more than 0.2 percent.

Arsenic (as As). Not more than 3 parts per million (0.0003 percent).

Heavy metals (as Pb). Not more than 10 parts per million (0.001 percent).

Ignition residue. Not more than 0.01 percent.

TESTS

Assay. Introduce 50 ml. of freshly distilled pyridine into a 250-ml. Erlenmeyer flask, and add about 450 mg. of the sample, accurately weighed. Stopper the flask, and swirl the mixture to effect complete solution. Add a few drops of thymolphthalein T.S., and titrate with 0.1 *N* sodium methoxide in pyridine to the first appearance of a blue end-point. Perform a blank determination (see page 2). During the titration direct a gentle stream of nitrogen into the flask through a short piece of 6-mm. glass tubing attached near the tip of the buret. Each ml. of 0.1 *N* sodium methoxide in pyridine is equivalent to 13.01 mg. of $C_6H_{10}O_3$.

Refractive index, page 945. Determine with an Abbé or other refractometer of equal or greater accuracy.

Specific gravity. Determine by any reliable method (see page 5)

Acidity. Transfer 10 ml. of water, recently boiled and then cooled to about 5°, into a 250-ml. Erlenmeyer flask containing 50 ml. of alcohol, add about 0.5 ml. of bromocresol purple T.S., and neutralize the mixture with 0.1 *N* sodium hydroxide to the appearance of a blue end-point. Introduce 25.0 ml. of the sample, previously cooled to about 5°, into the flask, and titrate with 0.1 *N* sodium hydroxide to a blue end-point that persists for at least 30 seconds. Not more than 8.7 ml. is required.

Arsenic. A *Sample Solution* prepared as directed for organic compounds meets the requirements of the *Arsenic Test*, page 865.

Heavy metals. Prepare and test a 2-gram sample as directed in *Method II* under the *Heavy Metals Test*, page 920, using 20 mcg. of lead ion (Pb) in the control (*Solution A*).

Ignition residue. Transfer a 50-gram sample into a tared 125-ml. platinum dish, heat until the vapors are ignited, withdraw the flame, protect the combustion from drafts, and allow the vapors to continue to burn spontaneously. Transfer the dish into a muffle furnace maintained at about 900° until all carbonaceous material has been removed, then cool in a desiccator, and weigh.

Packaging and storage. Store in tight, light-resistant containers.

Functional use in foods. Flavoring agent.

ETHYL ACRYLATE

$CH_2{=}CHCOOC_2H_5$

$C_5H_8O_2$ Mol. wt. 100.12

DESCRIPTION

A clear, colorless, mobile liquid having an intense, harsh, fruity odor. *It acts as a lachrymator.* It is miscible in all proportions with alcohol and with ether. One ml. is soluble in about 50 ml. of water.

SPECIFICATIONS

Assay. Not less than 99.5 percent of $C_5H_8O_2$.

Specific gravity. Between 0.916 and 0.919.

Limits of Impurities

Acidity (as acrylic acid). Not more than 0.005 percent.

Antioxidants. Not more than 220 parts per million (0.022 percent of hydroquinone and/or hydroquinone monomethylether).

Water. Not more than 0.05 percent.

TESTS

Assay. Determine the percent of ethyl acrylate by gas-liquid chromatography (see page 886) using an instrument containing a thermal conductivity detector. Prepare a 2.5-meter × 6-mm. column packed with 25 percent, by weight, of Carbowax 20M on 42/60 GC-22. Observe the following operating conditions during the determination: *Sample,* 10 μl.; *Injector temperature,* about 300°; *Column temperature,* 50° to 225°, programmed at a rate of 5.6° per minute; *Detector temperature,* about 300°; *Detector current,* 200 ma.; and *Helium flow,* about 100 ml. per minute. Under the conditions described the retention time for ethyl acrylate is 9 minutes. The area of the ethyl acrylate peak is not less than 99.5 percent of the total area of all peaks.

Specific gravity. Determine by any reliable method (see page 5).

Acidity. Transfer 50 ml. (46 grams) of the sample into a 250-ml. Erlenmeyer flask, and add 50 ml. of methanol previously neutralized to bromothymol blue T.S. with 0.1 *N* alcoholic potassium hydroxide.

Add an additional 5 or 6 drops of the bromothymol blue T.S., and titrate with 0.1 *N* alcoholic potassium hydroxide to a bluish green endpoint. Each ml. of 0.1 *N* alcoholic potassium hydroxide is equivalent to 7.21 mg. of $C_3H_4O_2$.

Antioxidants

Preliminary examination of the sample. Wash a 25-ml. portion of the sample with 25 ml. of sodium hydroxide solution (1 in 10). Any yellow or brown coloration in the extract indicates the presence of hydroquinone, in which case both of the procedures below (*A* and *B*) must be followed to determine the antioxidant content. If the sodium hydroxide extract remains colorless, the first procedure (*A*) need not be run, and the antioxidant content is determined by the second procedure (*B*) alone.

A. Determination of hydroquinone.

Carbonyl-free Methanol. To 500 ml. of anhydrous methanol add 5 grams of 2,4-dinitrophenylhydrazine, heat the mixture under a reflux condenser for 2 hours, and then recover the methanol by distillation. Store the carbonyl-free methanol in tight containers.

2,4-Dinitrophenylhydrazine Solution. Dissolve 100 mg. of 2,4-dinitrophenylhydrazine in 50 ml. of *Carbonyl-free Methanol,* add 4 ml. of hydrochloric acid, and dilute to 100 ml. with water.

Sodium Carbonate Solution. Dissolve 530 mg. of sodium carbonate in sufficient water to make 100 ml.

Pyridine-Diethanolamine Solution. Mix 5 ml. of diethanolamine with 500 ml. of freshly distilled pyridine.

Calibration Curve. Transfer 25 mg. of hydroquinone, accurately weighed, into a 100-ml. volumetric flask, add sufficient butyl acetate to volume, and mix thoroughly (250 mcg. per ml.). Prepare a series of standards by transferring 1.0-, 2.0-, 3.0-, 4.0-, and 6.0-ml. portions of this solution into separate 50-ml. volumetric flasks, and dilute each aliquot to 50.0 ml. with butyl acetate. One ml. of each of these standards contains 5, 10, 15, 20, and 30 mcg., respectively, of hydroquinone. Transfer 1.0 ml. of each solution into separate 25-ml. glass-stoppered graduates, and continue as directed in the *Procedure,* beginning with "add 2.0 ml. of water. . . ." Plot a calibration curve of absorbance *versus* mcg. of hydroquinone. Fifteen mcg. of hydroquinone should be equivalent to approximately 0.30 units of absorbance and the curve should intersect the origin.

Procedure. Using a hypodermic syringe, transfer 0.2 ml. of the sample, accurately weighed, into a 25-ml. glass-stoppered graduate, add 2.0 ml. of water, stopper the graduate, and mix the contents well without allowing contact between the liquid and the stopper. Add to the mixture 0.5 ml. of the *Sodium Carbonate Solution,* and immediately shake gently for 5 seconds avoiding contact between the solution and the stopper. Immediately add 1.0 ml. of a 15 percent, volume in volume, solution of sulfuric acid, shake as previously directed, and add

1-ml. of the *Dinitrophenylhydrazine Solution.* Stopper the graduate and place it in a water bath, maintained at a temperature between 70° and 72°, for 1 hour. Shake samples 3 times during the heating period. Cool the graduate to room temperature, dilute the contents to 15 ml. with water, add 5.8 ml. of benzene, stopper, shake vigorously, and then allow the phases to separate. Transfer 2.0 ml. of the benzene layer, using a suitable pipet, into a test tube, add 10.0 ml. of *Pyridine-Diethanolamine Solution,* and mix. Transfer a portion of this solution into a 2-cm. cell, and determine the absorbance at 620 mμ with a suitable spectrophotometer using as a blank 1.0 ml. of butyl acetate treated in the same manner as the sample except that 5.0 ml. of benzene is used for the extraction instead of 5.8 ml. From the previously prepared *Calibration Curve* read the mcg. of hydroquinone and/or benzoquinone corresponding to the absorbance of the solution from the sample, and record this value as w. Calculate the parts per million of hydroquinone (*p.p.m. HQ*) in the sample by the formula $1000\ w/W$, in which W is the weight of the sample taken, in mg.

B. Determination of hydroquinone monomethyl ether.

Antioxidant-free Ethyl Acrylate. Wash a suitable volume of the sample with three separate similar sized volumes of sodium hydroxide solution (1 in 10). After the last washing add a small amount of sodium chloride, if necessary, to remove any turbidity that may be present.

Calibration Curve. Transfer 25.0 mg. of hydroquinone monomethyl ether, accurately weighed, into a 100-ml. volumetric flask, add *Antioxidant-free Ethyl Acrylate* to volume, and shake to effect complete solution (250 mcg. per ml.). Prepare a series of standards by transferring 1.0-, 5.0-, 10.0-, and 20.0-ml. portions of this solution into separate 25-ml. volumetric flasks, dilute to volume with *Antioxidant-free Ethyl Acrylate,* and mix. One ml. of each of the standards contains 10, 50, 100, and 200 mcg., respectively, of hydroquinone monomethyl ether. Transfer 5.0 ml. of each solution into separate 50-ml. volumetric flasks, dilute each to volume with isooctane, and mix. Determine the absorbance of each solution in a 1-cm. silica cell at 292 mμ with a suitable spectrophotometer using a 1 in 10 dilution of *Antioxidant-free Ethyl Acrylate* as the blank. Plot a calibration curve of absorbance *versus* mcg. of hydroquinone monomethyl ether. The curve should be linear and should intersect the origin.

Procedure. Transfer 5.0 ml. of the sample, accurately weighed, into a 50-ml. volumetric flask, dilute to volume with isooctane, and mix. Determine the absorbance of this solution in a 1-cm. silica cell at 292 mμ with a suitable spectrophotometer, using a 1 in 10 dilution of *Antioxidant-free Ethyl Acrylate* in isooctane as the blank. From the previously prepared *Calibration Curve* read the mcg. of hydroquinone monomethyl ether corresponding to the absorbance of the sample solution and record this value as w. Calculate the parts per million of hydroquinone monomethyl ether (*p.p.m. HMME*) in the sample by the formula *w*/*W* in which *W* is the weight of the sample taken, in grams. [*Note:* If the first sodium hydroxide extract obtained under *Preliminary examination of the sample* (or under *Antioxidant-free Ethyl Acrylate*) showed a yellow coloration, the true *p.p.m. HMME* is obtained by subtracting the *p.p.m. HQ*, obtained under section *A*, from the apparent *p.p.m. HMME.*]

Water. Determine by the *Karl Fisher Titrimetric Method*, page 977.

Packaging and storage. Store in tight containers in a cool place.

Functional use in foods. Flavoring agent.

ETHYL ALCOHOL

Alcohol; Ethanol

C_2H_6O Mol. wt. 46.07

DESCRIPTION

A clear, colorless, mobile liquid having a slight, characteristic odor and a burning taste. It is miscible with water, with ether, and with chloroform. It boils at about 78° and is flammable. Its refractive index at 20° is about 1.364.

(*Note:* This monograph applies only to undenatured alcohol.)

SPECIFICATIONS

Assay. Not less than 94.9 percent by volume (92.3 percent by weight) of C_2H_6O.

Solubility in water. Passes test.

Limits of Impurities

Acidity (as acetic acid). Not more than 30 parts per million (0.003 percent).

Alkalinity (as NH_3). Not more than 30 parts per million (0.003 percent).

Fusel oil. Passes test.

Heavy metals (as Pb). Not more than 1 part per million (0.0001 percent).

Ketones, isopropyl alcohol. Passes test.

Methanol. Passes test.

Nonvolatile residue. Not more than 30 parts per million (0.003 percent).

Substances darkened by sulfuric acid. Passes test.

Substances reducing permanganate. Passes test.

TESTS

Assay. Its specific gravity, determined by any reliable method (see page 5), is not greater than 0.8096 at 25°/25° (equivalent to 0.8161 at 15.56°/15.56°).

Solubility in water. Transfer 50 ml. of the sample to a 100-ml. glass-stoppered graduate, dilute to 100 ml. with water, and mix. Place the graduate in a water bath maintained at 10°, and allow to stand for 30 minutes. No haze or turbidity develops.

Acidity. Transfer 10 ml. of the sample to a glass-stoppered flask containing 25 ml. of water, add 0.5 ml. of phenolphthalein T.S., and add 0.02 *N* sodium hydroxide to the first appearance of a pink color that persists after shaking for 30 seconds. Add 25 ml. (about 20 grams) of the sample, mix, and titrate with 0.02 *N* sodium hydroxide until the pink color is restored. Not more than 0.5 ml. is required.

Alkalinity. Add 2 drops of methyl red T.S. to 25 ml. of water, add 0.02 *N* sulfuric acid until a red color just appears, then add 25 ml. (about 20 grams) of the sample, and mix. Not more than 0.2 ml. of 0.02 *N* sulfuric acid is required to restore the red color.

Fusel oil. Mix 10 ml. of the sample with 1 ml. of glycerin and 1 ml. of water, and allow to evaporate from a piece of clean, odorless absorbent paper. No foreign odor is perceptible when the last traces of alcohol leave the paper.

Heavy metals. Evaporate 25 ml. (about 20 grams) of the sample to dryness on a steam bath in a glass evaporating dish. Cool, add 2 ml. of hydrochloric acid, and slowly evaporate to dryness again on the steam bath. Moisten the residue with 1 drop of hydrochloric acid, add 10 ml. of hot water, and digest for 2 minutes. Cool, and dilute to 25 ml. with water. This solution meets the requirements of the *Heavy Metals Test*, page 920, using 20 mcg. of lead ion (Pb) in the control (*Solution A*).

Ketones, isopropyl alcohol. Transfer 1 ml. of the sample, 3 ml. of water, and 10 ml. of mercuric sulfate T.S. to a test tube, mix, and heat in a boiling water bath. No precipitate forms within 3 minutes.

Methanol. Transfer 1 drop of the sample to a test tube, add 1 drop of dilute phosphoric acid (1 in 20) and 1 drop of potassium permanganate solution (1 in 20), mix, and allow to stand for 1 minute. Add sodium bisulfite solution (1 in 10), dropwise, until the permanganate color is discharged. If a brown color remains, add 1 drop of the phosphoric acid solution. To the colorless solution add 5 ml. of freshly prepared chromotropic acid T.S., and heat on a water bath at 60° for 10 minutes. No violet color develops.

Nonvolatile residue. Evaporate 125 ml. (about 100 grams) of the sample to dryness in a tared dish on a steam bath, dry the residue at 105° for 30 minutes, cool, and weigh.

Substances darkened by sulfuric acid. Transfer 10 ml. of sulfuric acid into a small Erlenmeyer flask, cool to 10°, and add 10 ml. of the sample, dropwise, with constant agitation. The mixture is colorless or has no more color than either the acid or sample before mixing.

Substances reducing permanganate. Transfer 20 ml. of the sample, previously cooled to 15°, to a glass-stoppered cylinder, add 0.1 ml. of 0.1 *N* potassium permanganate, mix, and allow to stand for 5 minutes. The pink color is not entirely discharged.

Packaging and storage. Store in tight containers, remote from fire.

Functional use in foods. Extraction solvent; vehicle.

ETHYL *p*-ANISATE

Ethyl *p*-Methoxybenzoate

$CH_3O-C_6H_4-COOC_2H_5$

$C_{10}H_{12}O_3$ Mol. wt. 180.20

DESCRIPTION

A colorless to slightly yellow liquid having a light fruity-anise odor. It is soluble in alcohol, in chloroform, and in ether, but insoluble in water.

SPECIFICATIONS

Assay. Not less than 97.0 percent of $C_{10}H_{12}O_3$.

Refractive index. Between 1.522 and 1.526 at 20°

Solubility in alcohol. Passes test.

Specific gravity. Between 1.101 and 1.104.

Limits of Impurities

Acid value. Not more than 1.0.

TESTS

Assay. Weigh accurately about 1.2 grams, and proceed as directed under *Ester Determination,* page 896, using 90.10 as the equivalence factor (*e*) in the calculation.

Refractive index, page 945. Determine with an Abbé or other refractometer of equal or greater accuracy.

Solubility in alcohol. Proceed as directed in the general method, page 899. One ml. dissolves in 7 ml. of 60 percent alcohol to form clear solutions.

Specific gravity. Determine by any reliable method (see page 5).

Acid value. Determine as directed in the general method, page 893.

Packaging and storage. Store in full, tight, preferably glass, tin-lined or other suitably lined, containers in a cool place protected from light.

Functional use in foods. Flavoring agent.

ETHYL ANTHRANILATE

Ethyl *o*-Aminobenzoate

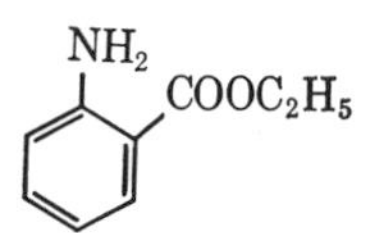

$C_9H_{11}NO_2$ Mol. wt. 165.19

DESCRIPTION

A clear, colorless to amber-colored liquid having a floral, orange blossom-like odor. It is soluble in alcohol, in propylene glycol, and in most fixed oils. One part is soluble in 2 parts of 70 percent alcohol.

SPECIFICATIONS

Assay. Not less than 96.0 percent of $C_9H_{11}NO_2$.

Refractive index. Between 1.563 and 1.566 at 20°.

Solidification point. Not below 13°.

Specific gravity. Between 1.115 and 1.120.

Limits of Impurities

Acid value. Not more than 1.0.

TESTS

Assay. Weigh accurately about 1.5 grams, and proceed as directed under *Ester Determination*, page 896, using 82.60 as the equivalence factor (*e*) in the calculation.

Refractive index, page 945. Determine with an Abbé or other refractometer of equal or greater accuracy.

Solidification point. Determine as directed in the general method, page 954.

Specific gravity. Determine by any reliable method (see page 5).

Acid value. Determine as directed in the general method, page 893.

Packaging and storage. Store in full, tight containers in a cool place protected from light.

Functional use in foods. Flavoring agent.

ETHYL BENZOATE

$C_6H_5-COOC_2H_5$

$C_9H_{10}O_2$ Mol. wt. 150.18

DESCRIPTION

A colorless liquid having a fruity odor. It is soluble in most fixed oils, in mineral oil, and in propylene glycol, but it is insoluble in glycerin.

SPECIFICATIONS

Assay. Not less than 98.0 percent of $C_9H_{10}O_2$.

Refractive index. Between 1.503 and 1.506 at 20°.

Solubility in alcohol. Passes test.

Specific gravity. Between 1.043 and 1.046.

Limits of Impurities

Acid value. Not more than 1.0.

Chlorinated compounds. Passes test.

TESTS

Assay. Weigh accurately about 900 mg., and proceed as directed under *Ester Determination*, page 896, using 75.09 as the equivalence factor (*e*) in the calculation.

Refractive index, page 945. Determine with an Abbé or other refractometer of equal or greater accuracy.

Solubility in alcohol. Proceed as directed in the general method, page 899. One ml. dissolves in 6 ml. of 60 percent alcohol.

Specific gravity. Determine by any reliable method (see page 5).

Acid value. Determine as directed in the general method, page 893.

Chlorinated compounds. Proceed as directed in the general method, page 895.

Packaging and storage. Store in full, tight, preferably glass, tin, or other suitably lined containers in a cool place protected from light.

Functional use in foods. Flavoring agent.

2-ETHYLBUTYRALDEHYDE

$$\begin{array}{c} C_2H_5 \\ | \\ CH_3CH_2CHCHO \end{array}$$

$C_6H_{12}O$ Mol. wt. 100.16

DESCRIPTION

A clear, colorless, mobile liquid having a characteristic pungent odor. It is miscible in all proportions with alcohol and with ether. One ml. dissolves in about 50 ml. of water.

SPECIFICATIONS

Assay. Not less than 95.0 percent of $C_6H_{12}O$.

Distillation range. Not less than 95 percent distils between 100° and 120°.

Specific gravity. Between 0.808 and 0.814.

Limits of Impurities

Acidity (as 2-ethylbutyric acid). Not more than 2.0 percent.

TESTS

Assay. Transfer 65 ml. of 0.5 *N* hydroxylamine hydrochloride and 50.0 ml. of 0.5 *N* triethanolamine into a suitable heat-resistant pressure bottle provided with a tight closure that can be securely fastened. Replace the air in the bottle by passing a gentle stream of nitrogen for two minutes through a glass tube positioned so that the end is just above the surface of the liquid. To the mixture in the pressure bottle add about 1 gram of the sample, accurately weighed, by means of a suitable weighing pipet. Wrap the bottle securely in a canvas bag and place it in a hot water bath maintained at a temperature of $98 \pm 2°$ for one hour. Remove the bottle from the bath, allow it to cool to room temperature, remove from the canvas bag, and uncap cautiously to prevent any loss of the contents. Titrate with 0.5 *N* sulfuric acid to a greenish blue end-point. Perform a residual blank titration (see *Blank Tests,* page 2). Each ml. of 0.5 *N* sulfuric acid is equivalent to 50.08 mg. of $C_6H_{12}O$.

Distillation range. Distil 100 ml. as directed in the general method, page 890.

Specific gravity. Determine by any reliable method (see page 5).

Acidity. Transfer 36 ml. (29 grams) into a 250-ml. Erlenmeyer flask, add phenolphthalein T.S., and titrate with 0.5 *N* alcoholic potassium hydroxide to a pink end-point that persists for at least 15 seconds. Each ml. of 0.5 *N* alcoholic potassium hydroxide is equivalent to 58.08 mg. of $C_6H_{12}O_2$.

Packaging and storage. Store in tight containers.

Functional use in foods. Flavoring agent.

ETHYL BUTYRATE

$CH_3CH_2CH_2COOC_2H_5$

$C_6H_{12}O_2$ Mol. wt. 116.16

DESCRIPTION

Ethyl butyrate may be prepared by the esterification of normal butyric acid with alcohol. It is a colorless liquid with a fruity odor. It is soluble in fixed oils, in mineral oil and in propylene glycol. It is relatively insoluble in glycerin.

SPECIFICATIONS

Assay. Not less than 98.0 percent of $C_6H_{12}O_2$.

Refractive index. Between 1.391 and 1.394 at 20°.

Solubility in alcohol. Passes test.

Specific gravity. Between 0.870 and 0.877.

Limits of Impurities

Acid value. Not more than 1.0.

TESTS

Assay. Weigh accurately about 1 gram, and proceed as directed under *Ester Determination*, page 896, using 58.08 as the equivalence factor (*e*) in the calculation.

Refractive index, page 945. Determine with an Abbé or other refractometer of equal or greater accuracy.

Solubility in alcohol. Proceed as directed in the general method, page 899. One ml. dissolves in 3 ml. of 60 percent alcohol.

Specific gravity. Determine by any reliable method (see page 5).

Acid value. Determine as directed in the general method, page 893.

Packaging and storage. Store in full, tight, glass, or well-tinned containers in a cool place protected from light.

Functional use in foods. Flavoring agent.

2-ETHYLBUTYRIC ACID

$$\begin{array}{c} \quad\quad\quad C_2H_5 \\ \quad\quad\quad | \\ CH_3CH_2CHCOOH \end{array}$$

$C_6H_{12}O_2$ Mol. wt. 116.16

DESCRIPTION

A clear, colorless, limpid liquid having a mildly rancid odor. It is miscible in all proportions with alcohol and with ether. One ml. dissolves in about 65 ml. of water.

SPECIFICATIONS

Assay. Not less than 98.0 percent of $C_6H_{12}O_2$.

Distillation range. Between 190° and 200°.

Specific gravity. Between 0.917 and 0.922.

Limits of Impurities

Heavy metals (as Pb). Not more than 40 parts per million (0.004 percent).

Water. Not more than 0.2 percent.

TESTS

Assay. Transfer 50 ml. of isopropanol and 50 ml. of water into a 250-ml. Erlenmeyer flask, add a few drops of phenolphthalein T.S., and neutralize the mixture with 0.1 *N* sodium hydroxide. Add to the flask about 2.2 grams of the sample, accurately weighed, using a suitable weighing pipet. Titrate with 0.5 *N* sodium hydroxide to a pink endpoint that persists for at least 15 seconds. Each ml. of 0.5 *N* sodium hydroxide is equivalent to 58.08 mg. of $C_6H_{12}O_2$.

Distillation range. Distil 100 ml. as directed in the general method, page 890.

Specific gravity. Determine by any reliable method (see page 5).

Heavy metals. Prepare and test a 500-mg. sample as directed in *Method II* under the *Heavy Metals Test*, page 920, using 20 mcg. of lead ion (Pb) in the control (*Solution A*).

Water. Determine by the *Karl Fischer Titrimetric Method*, page 977.

Packaging and storage. Store in tight containers.

Functional use in foods. Flavoring agent.

ETHYLCELLULOSE

DESCRIPTION

Ethylcellulose is the ethyl ether of cellulose in the form of a free-flowing, white to light tan powder. It is heat-labile, and exposure to high temperatures (240°) causes color degradation and loss of properties. It is practically insoluble in water, in glycerin, and in propylene glycol, but is soluble in varying proportions in certain organic solvents, depending upon the ethoxyl content. Ethylcellulose containing less than 46–48 percent of ethoxyl groups is freely soluble in tetrahydrofuran, in methyl acetate, in chloroform, and in aromatic hydrocarbon-alcohol mixtures. Ethylcellulose containing 46–48 percent or more of ethoxyl groups is freely soluble in alcohol, in methanol, in toluene, in chloroform, and in ethyl acetate. A 1 in 20 aqueous suspension of the sample is neutral to litmus.

IDENTIFICATION

Dissolve 5 grams of the sample in 95 grams of an 80:20 (w/w) mixture of toluene-alcohol. A clear, stable, slightly yellow solution is formed. Pour a few ml. of the solution onto a glass plate, and allow the solvent to evaporate. A thick, tough, continuous, clear film remains. The film is flammable.

SPECIFICATIONS

Assay. Not less than 44.0 percent and not more than 50.0 percent of ethoxyl groups (—OC_2H_5), after drying (equivalent to not more than 2.6 ethoxyl groups per anhydroglucose unit).

Viscosity. Not less than 90 percent and not more than 110 percent of that stated on the label, in centipoises, except that the 4-centipoise viscosity grade may vary between 3 and 5.5 centipoises, and the 7-centipoise viscosity grade may vary between 6 and 8 centipoises.

Limits of Impurities

Arsenic (as As). Not more than 3 parts per million (0.0003 percent).

Heavy metals (as Pb). Not more than 40 parts per million (0.004 percent).

Lead. Not more than 10 parts per million (0.001 percent).

Loss on drying. Not more than 3 percent.

Residue on ignition. Not more than 0.4 percent.

TESTS

Assay. Place about 50 mg. of the sample, previously dried at 105° for 2 hours, in a tared gelatin capsule, weigh accurately, transfer the capsule and its contents into the boiling flask of a methoxyl determination apparatus, and proceed as directed under *Methoxyl Determination*, page 935. Each ml. of 0.1 *N* sodium thiosulfate is equivalent to 751 mcg. of ethoxyl groups (—OC_2H_5).

Viscosity

Solvent systems. For ethylcellulose containing less than 46–48 percent of ethoxyl groups, prepare a solvent consisting of a 60:40 (w/w) mixture of toluene-alcohol; for ethylcellulose containing 46–48 percent or more of ethoxyl groups, prepare a solvent consisting of an 80:20 (w/w) mixture of toluene-alcohol.

Procedure. Transfer 5.0 grams of the sample, previously dried at 105° for 2 hours and accurately weighed, into a bottle containing 95 ± 0.5 grams of the appropriate solvent system. Shake or tumble the bottle until the sample is completely dissolved, and then adjust the temperature of the solution to 25° ± 0.1°. Determine the viscosity as directed under *Viscosity of Methylcellulose*, page 971, but make all determinations at 25° instead of 20° as directed therein.

Arsenic. A *Sample Solution* prepared as directed for organic compounds meets the requirements of the *Arsenic Test,* page 865.

Heavy metals. Prepare and test a 500-mg. sample as directed in *Method II* under the *Heavy Metals Test,* page 920, using 20 mcg. of lead ion (Pb) in the control (*Solution A*).

Lead. A *Sample Solution* prepared as directed for organic compounds meets the requirements of the *Lead Limit Test,* page 929, using 10 mcg. of lead ion (Pb) in the control.

Loss on drying, page 931. Dry at 105° for 2 hours.

Residue on ignition. Ignite 1 gram as directed in the general method, page 945.

Packaging and storage. Store in well-closed containers.

Functional use in foods. Protective coating component for vitamin and mineral tablets; binder and filler in dry vitamin preparations.

ETHYL CINNAMATE

Ethyl 3-Phenylpropenoate

C_6H_5—CH=CHCOOC_2H_5

$C_{11}H_{12}O_2$ Mol. wt. 176.22

DESCRIPTION

A colorless, oily liquid having a faint cinnamon-like odor sometimes described as a sweet, balsamic honey note. It is practically insoluble in water and in glycerin; miscible with alcohol and with ether.

SPECIFICATIONS

Assay. Not less than 99.0 percent of $C_{11}H_{12}O_2$.

Refractive index. Between 1.558 and 1.560 at 20°.

Solubility in alcohol. Passes test.

Specific gravity. Between 1.045 and 1.048.

Limits of Impurities

Acid value. Not more than 1.0.

TESTS

Assay. Weigh accurately about 1.2 grams, and proceed as directed under *Ester Determination*, page 896, using 88.11 as the equivalence factor (e) in the calculation.

Refractive index, page 945. Determine with an Abbé or other refractometer of equal or greater accuracy.

Solubility in alcohol. Proceed as directed in the general method, page 899. One ml. dissolves in 5 ml. of 70 percent alcohol.

Specific gravity. Determine by any reliable method (see page 5).

Acid value. Determine as directed in the general method, page 893.

Packaging and storage. Store in tight, light-resistant containers.

Functional use in foods. Flavoring agent.

ETHYLENE DICHLORIDE

Dichloroethane

$C_2H_4Cl_2$ Mol. wt. 98.96

DESCRIPTION

A clear, colorless, flammable, oily liquid having a chloroform-like odor and a sweet taste. It is slightly soluble in water and is soluble in alcohol, in ether, in acetone, and in carbon tetrachloride. Its refractive index at 20° is about 1.445.

SPECIFICATIONS

Distillation range. Between 82° and 85°.

Specific gravity. Between 1.245 and 1.255.

Limits of Impurities

Acidity (as HCl). Not more than 10 parts per million (0.001 percent).

Free halogens. Passes test.

Heavy metals (as Pb). Not more than 1 part per million (0.0001 percent).

Nonvolatile residue. Not more than 20 parts per million (0.002 percent).

Water. Not more than 0.03 percent.

TESTS

Distillation range. Determine as directed in the general method, page 890.

Specific gravity. Determine by any reliable method (see page 5).

Acidity. Transfer 25 ml. of alcohol to a 100-ml. glass-stoppered flask, add 2 drops of phenolphthalein T.S., and titrate with 0.01 *N* sodium hydroxide to the first appearance of a slight pink color. Add 25 ml. (about 31 grams) of the sample, mix, and titrate with 0.01 *N* sodium hydroxide until the faint pink color is restored. Not more than 0.85 ml. is required.

Free halogens. Shake 10 ml. of the sample vigorously for 2 minutes with 10 ml. of 10 percent potassium iodide solution and 1 ml. of starch T.S. A blue color does not appear in the water layer.

Heavy metals. Evaporate 16 ml. (about 20 grams) of the sample to dryness (*Caution:* use hood) on a steam bath in a glass evaporating dish. Cool, add 2 ml. of hydrochloric acid, and slowly evaporate to dryness again on the steam bath. Moisten the residue with 1 drop of hydrochloric acid, add 10 ml. of hot water, and digest for 2 minutes. Filter if necessary through a small filter, wash the evaporating dish and the filter with about 10 ml. of water, and dilute to 25 ml. with water. This solution meets the requirements of the *Heavy Metals Test,* page 920, using 20 mcg. of lead ion (Pb) in the control (*Solution A*).

Nonvolatile residue. Evaporate 80 ml. (about 100 grams) of the sample to dryness (*Caution:* use hood) in a tared dish on a steam bath, dry the residue at 105° for 30 minutes, cool, and weigh.

Water. Determine by the *Karl Fischer Titrimetric Method,* page 977.

Packaging and storage. Store in tight containers.

Functional use in foods. Extraction solvent.

ETHYL FORMATE

$HCOOC_2H_5$

$C_3H_6O_2$ Mol. wt. 74.08

DESCRIPTION

A colorless, flammable liquid, having a characteristic odor. It is soluble in most fixed oils and in propylene glycol. It is sparingly soluble in mineral oil, but it is insoluble in glycerin. It is soluble with gradual decomposition in water.

SPECIFICATIONS

Assay. Not less than 95.0 percent of $C_3H_6O_2$.

Refractive index. Between 1.359 and 1.363 at 20°

Solubility in alcohol. Passes test.

Specific gravity. Between 0.916 and 0.921.

Limits of Impurities

Free acid (as formic acid). Not more than 0.1 percent.

TESTS

Assay. Weigh accurately about 500 mg., and proceed as directed under *Ester Determination,* page 896, using 37.04 as the equivalence factor (*e*) in the calculation. Modify the procedure by allowing the mixture to stand at room temperature for 15 minutes, instead of heating on a steam bath for 1 hour.

Refractive index, page 945. Determine with an Abbé or other refractometer of equal or greater accuracy.

Solubility in alcohol. Proceed as directed in the general method, page 899. One ml. dissolves in 0.5 ml. of 50 percent alcohol.

Specific gravity. Determine by any reliable method (see page 5).

Free acid. Transfer about 5 grams, accurately weighed, into a glass-stoppered flask containing a solution of 500 mg. of potassium iodate and 2 grams of potassium iodide in 50 ml. of water. Titrate the liberated iodine with 0.1 *N* sodium thiosulfate, using starch T.S. as the indicator. Each ml. of 0.1 *N* sodium thiosulfate is equivalent to 4.603 mg. of CH_2O_2.

Packaging and storage. Store in full, tight, preferably glass, tin-lined, or other suitably lined containers in a cool place protected from light.

Functional use in foods. Flavoring agent.

ETHYL HEPTANOATE

Ethyl Heptoate

$CH_3(CH_2)_5COOC_2H_5$

$C_9H_{18}O_2$ Mol. wt. 158.24

DESCRIPTION

A colorless liquid having a fruity, wine-like odor and taste with a burning after-taste. It is miscible with alcohol, with chloroform, with most mixed oils, and forms an azeotropic mixture with water containing 72 percent by weight of ethyl heptanoate boiling at 98.5°. It is slightly soluble in propylene glycol and insoluble in glycerin.

SPECIFICATIONS

Assay. Not less than 98.0 percent of $C_9H_{18}O_2$.

Refractive index. Between 1.411 and 1.415 at 20°.

Solubility in alcohol. Passes test.

Specific gravity. Between 0.867 and 0.872.

Limits of Impurities

Acid value. Not more than 1.0.

Arsenic (as As). Not more than 3 parts per million (0.0003 percent).

Heavy metals (as Pb). Not more than 40 parts per million (0.004 percent).

Lead. Not more than 10 parts per million (0.001 percent).

TESTS

Assay. Weigh accurately about 1 gram, and proceed as directed under *Ester Determination*, page 896, using 79.12 as the equivalence factor (*e*) in the calculation.

Refractive index, page 945. Determine with an Abbé or other refractometer of equal or greater accuracy.

Solubility in alcohol. Proceed as directed in the general method, page 899. One ml. dissolves in 3 ml. of 70 percent alcohol to form a clear solution.

Specific gravity. Determine by any reliable method (see page 5).

Acid value. Determine as directed in the general method, page 893.

Arsenic. A *Sample Solution* prepared as directed for organic compounds meets the requirements of the *Arsenic Test*, page 865.

Heavy metals. Prepare and test a 500-mg. sample as directed in *Method II* under the *Heavy Metals Test*, page 920, using 20 mcg. of lead ion (Pb) in the control (*Solution A*).

Lead. A *Sample Solution* prepared as directed for organic compounds meets the requirements of the *Lead Limit Test*, page 929, using 10 mcg. of lead ion (Pb) in the control.

Packaging and storage. Store in full, tight, preferably glass, tin-lined or other suitably lined containers in a cool place protected from light.

Functional use in foods. Flavoring agent.

ETHYL HEXANOATE

Ethyl Caproate; Ethyl Capronate

$CH_3(CH_2)_4COOC_2H_5$

$C_8H_{16}O_2$ Mol. wt. 144.21

DESCRIPTION

A colorless liquid having a strong fruity odor. It is soluble in most fixed oils and in mineral oil. It is sparingly soluble in propylene glycol, and it is insoluble in glycerin.

SPECIFICATIONS

Assay. Not less than 98.0 percent of $C_8H_{16}O_2$.

Refractive index. Between 1.406 and 1.409 at 20°.

Solubility in alcohol. Passes test.

Specific gravity. Between 0.867 and 0.871.

Limits of Impurities

Acid value. Not more than 1.0.

TESTS

Assay. Weigh accurately about 1 gram, and proceed as directed under *Ester Determination*, page 896, using 72.10 as the equivalence factor (e) in the calculation.

Refractive index, page 945. Determine with an Abbé or other refractometer of equal or greater accuracy.

Solubility in alcohol. Proceed as directed in the general method, page 899. One ml. dissolves in 2 ml. of 70 percent alcohol.

Specific gravity. Determine by any reliable method (see page 5).

Acid value. Determine as directed in the general method, page 893.

Packaging and storage. Store in glass, or suitably lined containers, in a cool place protected from light.

Functional use in foods. Flavoring agent.

ETHYL ISOVALERATE

Ethyl 2-Methylbutyrate

$(CH_3)_2CHCH_2COOC_2H_5$

$C_7H_{14}O_2$ Mol. wt. 130.19

DESCRIPTION

A clear, colorless liquid having a strong fruity odor which upon dilution is suggestive of apple. It is miscible with alcohol and with most fixed oils, and it is soluble in propylene glycol and in mineral oil. It is soluble in about 350 parts of water.

SPECIFICATIONS

Assay. Not less than 98.0 percent of $C_7H_{14}O_2$.

Refractive index. Between 1.395 and 1.399 at 20°.

Specific gravity. Between 0.862 and 0.866.

Limits of Impurities

Acid value. Not more than 2.0.

TESTS

Assay. Weigh accurately about 1.5 grams, and proceed as directed under *Ester Determination*, page 896, using 65.10 as the equivalence factor (*e*) in the calculation.

Refractive index, page 945. Determine with an Abbé or other refractometer of equal or greater accuracy.

Specific gravity. Determine by any reliable method (see page 5).

Acid value. Determine as directed in the general method, page 893.

Packaging and storage. Store in full, tight containers in a cool place protected from light.

Functional use in foods. Flavoring agent.

ETHYL LACTATE

Ethyl 2-Hydroxypropionate

$CH_3CHOHCOOC_2H_5$

$C_5H_{10}O_3$ Mol. wt. 118.13

DESCRIPTION

A colorless liquid having a characteristic odor. It is very soluble in water, in alcohol, in ether, and in chloroform.

SPECIFICATIONS

Assay. Not less than 98.0 percent of $C_5H_{10}O_3$.

Refractive index. Between 1.410 and 1.420 at 20°.

Specific gravity. Between 1.029 and 1.032.

Limits of Impurities

Acid value. Not more than 1.0.

Heavy metals (as Pb). Not more than 40 parts per million (0.004 percent).

TESTS

Assay. Weigh accurately about 700 mg., and proceed as directed under *Ester Determination*, page 896, using 59.07 as the equivalence factor (*e*) in the calculation.

Refractive index, page 945. Determine with an Abbé or other refractometer of equal or greater accuracy.

Specific gravity. Determine by any reliable method (see page 5).

Acid value. Determine as directed in the general method, page 893.

Heavy metals. Prepare and test a 500-mg. sample as directed in *Method II* under the *Heavy Metals Test*, page 920, using 20 mcg. of lead ion (Pb) in the control (*Solution A*).

Packaging and storage. Store in tight, light-resistant containers.
Functional use in foods. Flavoring agent.

ETHYL LAURATE

Ethyl Dodecanoate

$CH_3(CH_2)_{10}COOC_2H_5$

$C_{14}H_{28}O_2$ Mol. wt. 228.38

DESCRIPTION

A colorless, oily liquid having a light fruity-floral odor. It is miscible with alcohol, with chloroform, and with ether, but insoluble in water.

SPECIFICATIONS

Assay. Not less than 98.0 percent of $C_{14}H_{28}O_2$.
Refractive index. Between 1.430 and 1.434 at 20°.
Solidification point. Not less than −10°.
Solubility in alcohol. Passes test.
Specific gravity. Between 0.858 and 0.862.
Limits of Impurities
Acid value. Not more than 1.0.

TESTS

Assay. Weigh accurately about 1.5 grams, and proceed as directed under *Ester Determination*, page 896, using 114.2 as the equivalence factor (*e*) in the calculation.

Refractive index, page 945. Determine with an Abbé or other refractometer of equal or greater accuracy.

Solidification point. Determine as directed in the general method, page 954.

Solubility in alcohol. Proceed as directed in the general method, page 899. One ml. dissolves in 9 ml. of 80 percent alcohol to form clear solutions.

Specific gravity. Determine by any reliable method (see page 5).

Acid value. Determine as directed in the general method, page 893.

Packaging and storage. Store in tight, preferably glass, tin-lined or other suitably lined containers in a cool place protected from light.
Functional use in foods. Flavoring agent.

ETHYL MALTOL

3-Hydroxy-2-ethyl-4-pyrone

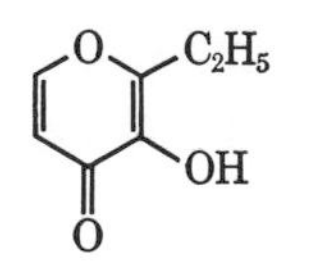

$C_7H_8O_3$ Mol. wt. 140.14

DESCRIPTION

A white, crystalline powder having a characteristic odor and a sweet, fruit-like flavor in dilute solution. One gram dissolves in about 55 ml. of water, 10 ml. of alcohol, 17 ml. of propylene glycol, and 5 ml. of chloroform. It melts at about 90°.

IDENTIFICATION

The infrared absorption spectrum of a 1 in 50 solution of Ethyl Maltol in chloroform, determined in a 0.1-mm. cell, exhibits maxima only at the same wavelengths as that of F.C.C. Ethyl Maltol Reference Standard, similarly measured.

SPECIFICATIONS

Assay. Not less than 99.0 percent of $C_7H_8O_3$.

Limits of Impurities

Arsenic (as As). Not more than 3 parts per million (0.0003 percent).

Heavy metals (as Pb). Not more than 20 parts per million (0.002 percent).

Lead. Not more than 10 parts per million (0.001 percent).

Residue on ignition. Not more than 0.2 percent.

Water. Not more than 0.5 percent.

TESTS

Assay

Standard Solution. Weigh accurately about 50 mg. of F.C.C. Ethyl Maltol Reference Standard, dissolve it in sufficient 0.1 *N* hydrochloric acid to make 250.0 ml., and mix. Transfer 5.0 ml. of this solution into a 100-ml. volumetric flask, dilute to volume with 0.1 *N* hydrochloric acid, and mix.

Assay Solution. Weigh accurately about 50 mg. of the sample, dissolve it in sufficient 0.1 *N* hydrochloric acid to make 250.0 ml., and mix. Transfer 5.0 ml. of this solution to a 100-ml. volumetric flask, dilute to volume with 0.1 *N* hydrochloric acid, and mix.

Procedure. Determine the absorbance of each solution in a 1-cm.

cell at the wavelength of maximum absorption at about 276 mμ, with a suitable spectrophotometer, using 0.1 *N* hydrochloric acid as the blank. Calculate the quantity, in mg., of $C_7H_8O_3$ in the sample taken by the formula $5C(A_U/A_S)$, in which *C* is the concentration, in mcg. per ml., of F.C.C. Ethyl Maltol Reference Standard in the *Standard Solution*, A_U is the absorbance of the *Assay Solution*, and A_S is the absorbance of the *Standard Solution*.

Arsenic. A *Sample Solution* prepared as directed for organic compounds meets the requirements of the *Arsenic Test*, page 865.

Heavy metals. Prepare and test a 1-gram sample as directed in *Method II* under the *Heavy Metals Test*, page 920, using 20 mcg. of lead ion (Pb) in the control (*Solution A*).

Lead. A *Sample Solution* prepared as directed for organic compounds meets the requirements of the *Lead Limit Test*, page 929, using 10 mcg. of lead ion (Pb) in the control.

Residue on ignition. Ignite a 1-gram sample as directed in the general method, page 945.

Water. Determine by the *Karl Fischer Titrimetric Method*, page 977.

Packaging and storage. Store in tight containers.

Functional use in foods. Flavoring agent.

ETHYL METHYL PHENYLGLYCIDATE

Aldehyde C-16; Strawberry Aldehyde

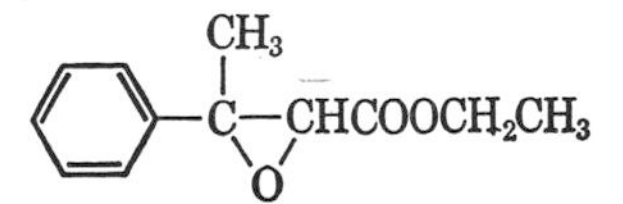

$C_{12}H_{14}O_3$ Mol. wt. 206.24

DESCRIPTION

Ethyl methyl phenylglycidate, a glycidic acid ester, not an aldehyde, is usually prepared by the reaction of acetophenone and the ethyl ester of monochloroacetic acid in the presence of an alkaline condensing agent. It is a colorless to pale yellow liquid with a strong fruity odor, suggestive of strawberries. It is soluble in fixed oils and in propylene glycol, but it is relatively insoluble in glycerin and in mineral oil.

SPECIFICATIONS

Assay. Not less than 98.0 percent of $C_{12}H_{14}O_3$.

Refractive index. Between 1.504 and 1.513 at 20°.

Solubility in alcohol. Passes test.

Specific gravity. Between 1.086 and 1.112.

Limits of Impurities

Acid value. Not more than 2.0.

Arsenic (as As). Not more than 3 parts per million (0.0003 percent).

Heavy metals (as Pb). Not more than 40 parts per million (0.004 percent).

Lead. Not more than 10 parts per million (0.001 percent).

TESTS

Assay. Weigh accurately about 1 gram, and proceed as directed under *Ester Determination*, page 896, using 103.1 as the equivalence factor (e) in the calculation.

Refractive index, page 945. Determine with an Abbé or other refractometer of equal or greater accuracy.

Solubility in alcohol. Proceed as directed in the general method, page 899. One ml. dissolves in 3 ml. of 70 percent alcohol.

Specific gravity. Determine by any reliable method (see page 5).

Acid value. Determine as directed in the general method, page 893.

Arsenic. A *Sample Solution* prepared as directed for organic compounds meets the requirements of the *Arsenic Test*, page 865.

Heavy metals. Prepare and test a 500-mg. sample as directed in *Method II* under the *Heavy Metals Test*, page 920, using 20 mcg. of lead ion (Pb) in the control (*Solution A*).

Lead. A *Sample Solution* prepared as directed for organic compounds meets the requirements of the *Lead Limit Test*, page 929, using 10 mcg. of lead ion (Pb) in the control.

Packaging and storage. Store in full, tight, preferably glass, tin-lined, or aluminum containers in a cool place protected from light.

Functional use in foods. Flavoring agent

ETHYL NONANOATE

Ethyl Pelargonate

$CH_3(CH_2)_7COOC_2H_5$

$C_{11}H_{22}O_2$ Mol. wt. 186.30

DESCRIPTION

A clear, colorless liquid having a fruity, fat-like odor suggestive of cognac. It is miscible with alcohol and with propylene glycol. It is practically insoluble in water. One part is soluble in 10 parts of 70 percent alcohol.

SPECIFICATIONS

Assay. Not less than 98.0 percent of $C_{11}H_{22}O_2$.

Refractive index. Between 1.420 and 1.424 at 20°.

Specific gravity. Between 0.863 and 0.867.

Limits of Impurities

Acid value. Not more than 3.0.

TESTS

Assay. Weigh accurately about 1.5 grams, and proceed as directed under *Ester Determination*, page 896, using 93.15 as the equivalence factor (*e*) in the calculation.

Refractive index, page 945. Determine with an Abbé or other refractometer of equal or greater accuracy.

Specific gravity. Determine by any reliable method (see page 5).

Acid value. Determine as directed in the general method, page 893.

Packaging and storage. Store in full, tight containers in a cool place protected from light.

Functional use in foods. Flavoring agent.

ETHYL OCTANOATE

Ethyl Caprylate; Ethyl Octoate

$CH_3(CH_2)_6COOC_2H_5$

$C_{10}H_{20}O_2$ Mol. wt. 172.27

DESCRIPTION

A colorless liquid having a fruity-floral odor. It is soluble in most fixed oils and in mineral oil. It is slightly soluble in propylene glycol, but it is insoluble in glycerin.

SPECIFICATIONS

Assay. Not less than 98.0 percent of $C_{10}H_{20}O_2$.

Refractive index. Between 1.417 and 1.419 at 20°.

Solubility in alcohol. Passes test.

Specific gravity. Between 0.865 and 0.869.

Limits of Impurities

Acid value. Not more than 1.0.

TESTS

Assay. Weigh accurately about 1.1 grams, and proceed as directed under *Ester Determination*, page 896, using 86.13 as the equivalence factor (*e*) in the calculation.

Refractive index, page 945. Determine with an Abbé or other refractometer of equal accuracy.

Solubility in alcohol. Proceed as directed in the general method, page 899. One ml. dissolves in 4 ml. of 70 percent alcohol.

Specific gravity. Determine by any reliable method (see page 5).

Acid value. Determine as directed in the general method, page 893.

Packaging and storage. Store in glass, or suitably lined containers, in a cool place protected from light.

Functional use in foods. Flavoring agent.

ETHYL OXYHYDRATE (So-Called)

Rum Ether (So-called)

DESCRIPTION

A distillate obtained from the reaction products of alcohol with aqueous acetic acid and other acids. It occurs as a clear, colorless liquid having a sharp, rum-like odor. It is miscible with alcohol, with propylene glycol, and with glycerin.

SPECIFICATIONS

Alcohol content. Not less than 14.0 percent by volume at 15.56°.

Ester value. Not less than 25.

Limits of Impurities

Arsenic (as As). Not more than 3 parts per million (0.0003 percent).

Methanol-formaldehyde. Not more than 1.5 percent.

Heavy metals (as Pb). Not more than 10 parts per million (0.001 percent).

TESTS

Alcohol content. Mix 25.0 ml. of the sample with an equal volume of water in a separator, saturate with sodium chloride, and extract with three 25-ml. portions of solvent hexane. Extract the combined solvent hexane extracts with three 10-ml. portions of a saturated solution of sodium chloride, and then discard the solvent hexane solutions. Combine the saline solutions in a suitable distillation flask, and distil, collecting 25 ml. of distillate. The specific gravity of the distillate is not greater than 0.9814, indicating an alcohol content of not less than 14.0 percent by volume.

Ester value. Weigh accurately about 3 grams, and proceed as directed under *Ester Value*, page 897. If the expected Ester Value exceeds 100, a sample of about 1 gram, accurately weighed, should be used.

Arsenic. A *Sample Solution* prepared as directed for organic compounds meets the requirements of the *Arsenic Test*, page 865.

Methanol-formaldehyde

Chromotropic acid solution. Dissolve 5 grams of chromotropic acid sodium salt in sufficient water to make 100 ml., and filter, if necessary, to obtain a clear solution.

Potassium permanganate solution. Dissolve 3.0 grams of potassium permanganate and 15.0 ml. of phosphoric acid in sufficient water to make 100.0 ml. Discard after 30 days.

Standard preparation. Prepare a solution containing 0.025 percent, by volume, of methanol in 5.5 percent ethanol.

Test preparation. Dilute an accurately measured volume of the sample to provide a total alcohol concentration of between 5 and 6 percent. Dilute an accurately measured volume of this solution with 5.5 percent ethanol to provide a methanol concentration of about 0.05 percent.

Procedure. Transfer 1.0 ml. each of *Standard preparation*, of the *Test preparation*, and of 5.5 percent ethanol into separate 50-ml. volumetric flasks containing 2.0 ml. of *Potassium permanganate solution* previously chilled in an ice bath. Allow to stand for 30 minutes in the ice bath, then add sufficient dry sodium bisulfite to decolorize the solution, add 1.0 ml. of *Chromotropic acid solution*, and mix. Slowly add, with swirling, 15 ml. of sulfuric acid, and heat in a water bath at 60° to 75° for 15 minutes. Cool, dilute to volume with water, and mix. Determine the absorbance of each solution in a 1-cm. cell at 575 mμ, with a suitable spectrophotometer, using the blank to set the instrument. Calculate the percent, by volume, of methanol-formaldehyde in the sample by the formula $0.025F(A_U/A_S)$, in which F is the dilution factor for the sample, A_U is the absorbance of the solution from the *Test preparation*, and A_S is the absorbance of the solution from the *Standard preparation*.

Heavy metals. Prepare and test a 2-gram sample as directed in *Method II* under the *Heavy Metals Test*, page 920, using 20 mcg. of lead ion (Pb) in the control (*Solution A*).

Packaging and storage. Store in tight containers.

Functional use in foods. Flavoring agent.

ETHYL PHENYLACETATE

$C_6H_5-CH_2COOC_2H_5$

$C_{10}H_{12}O_2$ Mol. wt. 164.20

DESCRIPTION

A colorless, or nearly colorless liquid having a pleasant, sweet odor suggestive of honey. It is soluble in alcohol and in most fixed oils. It is insoluble in glycerin, in mineral oil, in propylene glycol, and in water.

SPECIFICATIONS

Assay. Not less than 98.0 percent of $C_{10}H_{12}O_2$.

Refractive index. Between 1.496 and 1.500 at 20°.

Solubility in alcohol. Passes test.

Specific gravity. Between 1.027 and 1.032.

Limits of Impurities

Acid value. Not more than 1.0.

Chlorinated compounds. Passes test.

TESTS

Assay. Weigh accurately about 1 gram, and proceed as directed under *Ester Determination*, page 896, using 82.10 as the equivalence factor (*e*) in the calculation.

Refractive index, page 945. Determine with an Abbé or other refractometer of equal or greater accuracy.

Solubility in alcohol. Proceed as directed in the general method, page 899. One ml. dissolves in 3 ml. of 70 percent alcohol.

Specific gravity. Determine by any reliable method (see page 5).

Acid value. Determine as directed in the general method, page 893.

Chlorinated compounds. Proceed as directed in the general method, page 895.

Packaging and storage. Store in tight, preferably glass, aluminum tin-lined, or steel containers in a cool dry place protected from light.

Functional use in foods. Flavoring agent.

ETHYL PHENYLGLYCIDATE

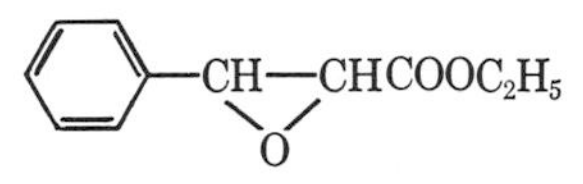

$C_{11}H_{12}O_3$ Mol. wt. 192.22

DESCRIPTION

A colorless to slightly yellow liquid having a strong strawberry-like odor and taste. It is soluble in alcohol, in chloroform, and in ether, but practically insoluble in water.

SPECIFICATIONS

Assay. Not less than 98.0 percent of $C_{11}H_{12}O_3$.

Refractive index. Between 1.516 and 1.521 at 20°.

Solubility in alcohol. Passes test.

Specific gravity. Between 1.120 and 1.125.

TESTS

Assay. Weigh accurately about 1.2 grams, and proceed as directed under *Ester Determination*, page 896, using 96.11 as the equivalence factor (*e*) in the calculation.

Refractive index, page 945. Determine with an Abbé or other refractometer of equal or greater accuracy.

Solubility in alcohol. Proceed as directed in the general method, page 899. One ml. dissolves in 6 ml. of 70 percent and in 1 ml. of 80 percent alcohol to form clear solutions.

Specific gravity. Determine by any reliable method (see page 5).

Packaging and storage. Store in tight, preferably glass, tin-lined, or other suitably lined containers in a cool place protected from light.

Functional use in foods. Flavoring agent.

ETHYL PROPIONATE

$CH_3CH_2COOC_2H_5$

$C_5H_{10}O_2$ Mol. wt. 102.13

DESCRIPTION

A colorless, transparent liquid having a fruity odor suggestive of rum. It is miscible with alcohol and with propylene glycol, and it is soluble in most fixed oils and in mineral oil. It is soluble in 5 parts of 70 percent alcohol and in 42 parts of water.

SPECIFICATIONS

Assay. Not less than 97.0 percent of $C_5H_{10}O_2$.

Refractive index. Between 1.383 and 1.385 at 20°.

Specific gravity. Between 0.886 and 0.889.

Limits of Impurities

Acid value. Not more than 2.0.

Heavy metals (as Pb). Not more than 10 parts per million (0.001 percent).

TESTS

Assay. Weigh accurately about 1 gram, and proceed as directed under *Ester Determination*, page 896, using 51.07 as the equivalence factor (*e*) in the calculation.

Refractive index, page 945. Determine with an Abbé or other refractometer of equal or greater accuracy.

Specific gravity. Determine by any reliable method (see page 5).

Acid value. Proceed as directed in the general method, page 893.

Heavy metals. Prepare and test a 2-gram sample as directed in *Method II* under the *Heavy Metals Test*, page 920, using 20 mcg. of lead ion (Pb) in the control (*Solution A*).

Packaging and storage. Store in full, tight containers in a cool place protected from light.

Functional use in foods. Flavoring agent.

ETHYL SALICYLATE

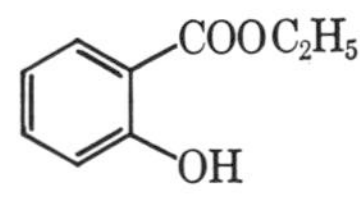

$C_9H_{10}O_3$ Mol. wt. 166.18

DESCRIPTION

A colorless liquid having a characteristic, wintergreen-like odor. It is soluble in alcohol, in acetic acid, and in most fixed oils, but very slightly soluble in water and in glycerin.

SPECIFICATIONS

Assay. Not less than 99.0 percent of $C_9H_{10}O_3$.

Refractive index. Between 1.520 and 1.523 at 20°.

Solubility in alcohol. Passes test.

Specific gravity. Between 1.126 and 1.130.

Limits of Impurities

Acid value. Not more than 1.0.

TESTS

Assay. Weigh accurately about 1 gram, and proceed as directed under *Ester Determination*, page 896, using 83.09 as the equivalence factor (e) in the calculation, but reflux for 2 hours and use phenol red T.S. as the indicator.

Refractive index, page 945. Determine with an Abbé or other refractometer of equal or greater accuracy.

Solubility in alcohol. Proceed as directed in the general method, page 899. One ml. dissolves in 4 ml. of 80 percent alcohol to form a clear solution.

Specific gravity. Determine by any reliable method (see page 5).

Acid value. Determine as directed in the general method, page 893, using phenol red T.S. as the indicator.

Packaging and storage. Store in full, tight, preferably glass, tin-lined, or other suitably lined containers in a cool place protected from light.

Functional use in foods. Flavoring agent.

ETHYL VANILLIN

3-Ethoxy-4-hydroxybenzaldehyde

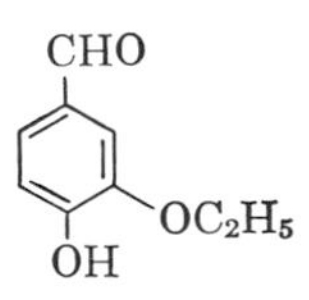

$C_9H_{10}O_3$ Mol. wt. 166.18

DESCRIPTION

Fine, white or slightly yellowish crystals, having a strong vanilla-like odor and taste. It is affected by light. Its solutions are acid to litmus. One gram dissolves in about 100 ml. of water at 50°. It is freely soluble in alcohol, in chloroform, in ether, in solutions of alkali hydroxides, and in propylene glycol.

IDENTIFICATION

A. Warm about 100 mg. with 1 ml. of 25 percent hydrochloric acid until complete solution is effected. Cool, add about 1 ml. of hydrogen peroxide T.S., and allow the mixture to stand, with frequent shaking, until precipitation is complete (10 to 20 minutes). Add an equal volume of benzene and shake thoroughly. The benzene layer is violet in color.

B. It is extracted completely from its solution in ether by shaking with a saturated solution of sodium bisulfite, from which it is precipitated by acids.

C. Add lead subacetate T.S. to a cold solution of the sample. A white precipitate, which is sparingly soluble in hot water, but soluble in acetic acid, is produced.

SPECIFICATIONS

Assay. Not less than 98.0 percent and not more than 101.0 percent of $C_9H_{10}O_3$ after drying.

Melting range. Between 76° and 78°.

Limits of Impurities

Arsenic (as As). Not more than 3 parts per million (0.0003 percent).

Heavy metals (as Pb). Not more than 10 parts per million (0.001 percent).

Loss on drying. Not more than 0.5 percent.

Residue on ignition. Not more than 0.05 percent.

TESTS

Assay. Transfer about 300 mg. of the sample, previously dried over phosphorus pentoxide for 4 hours and accurately weighed, into a 125-ml. Erlenmeyer flask, and dissolve in 50 ml. of dimethylformamide. Add 3 drops of thymol blue T.S., and titrate with 0.1 *N* sodium methoxide, using a magnetic stirrer and taking precautions against absorption of atmospheric carbon dioxide. Perform a blank determination (see page 2). Each ml. of 0.1 *N* sodium methoxide is equivalent to 16.62 mg. of $C_9H_{10}O_3$.

Melting range. After drying over phosphorus pentoxide for 4 hours, determine as directed in the general procedure, page 931.

Arsenic. A *Sample Solution* prepared as directed for organic compounds meets the requirements of the *Arsenic Test*, page 865.

Heavy metals. Prepare and test a 2-gram sample as directed in *Method II* under the *Heavy Metals Test*, page 920, using 20 mcg. of lead ion (Pb) in the control (*Solution A*).

Loss on drying, page 931. Dry over phosphorus pentoxide for 4 hours.

Residue on ignition, page 945. Ignite 2 grams as directed in the general procedure.

Packaging and storage. Store in tight, light-resistant containers.

Functional use in foods. Flavoring agent.

EUCALYPTUS OIL

DESCRIPTION

The volatile oil obtained by steam distillation from the fresh leaves of *Eucalyptus globulus* Labillardiere, and other species of *Eucalyptus* L'Heritier (Fam. *Myrtaceae*). It is a colorless, or pale yellow liquid, having a characteristic aromatic, somewhat camphoraceous odor and a pungent, spicy, cooling taste.

SPECIFICATIONS

Assay. Not less than 70.0 percent of cineole ($C_{10}H_{18}O$).

Refractive index. Between 1.458 and 1.470 at 20°.

Solubility in alcohol. Passes test.

Specific gravity. Between 0.905 and 0.925.

Limits of Impurities

Arsenic (as As). Not more than 3 parts per million (0.0003 percent).

Heavy metals (as Pb). Not more than 40 parts per million (0.004 percent).

Lead. Not more than 10 parts per million (0.001 percent).

Phellandrene. Passes test.

TESTS

Assay. Transfer about 3 grams, previously dried with anhydrous sodium sulfate and accurately weighed, into a 25 × 150-mm. test tube. Add to the sample 2.100 grams of melted *o*-cresol that is pure and dry, having a solidification point of 30° or higher (*Note: Moisture in the o-cresol may cause low results*). Stir the mixture with the thermometer to induce crystallization and note the highest temperature reading obtained. Warm the tube gently until the contents are completely melted, then insert the test tube into the apparatus assembled as directed under *Solidification Point*, page 954. Allow the mixture to cool slowly until crystallization starts, or until the temperature has fallen to the point previously noted. Stir the mixture vigorously with the thermometer, rubbing the sides of the test tube with an up and down motion to induce crystallization. Continue the stirring and rubbing so long as the temperature rises. Take the highest temperature obtained as the solidification point. Repeat the procedure until two results agreeing within 0.1° are obtained. Calculate the percent of cineole from the *Percentage of Cineole* table, page 898.

Refractive index, page 945. Determine with an Abbé or other refractometer of equal or greater accuracy.

Solubility in alcohol. Proceed as directed in the general method, page 899. One ml. dissolves in 5 ml. of 70 percent alcohol.

Specific gravity. Determine by any reliable method (see page 5).

Arsenic. A *Sample Solution* prepared as directed for organic compounds meets the requirements of the *Arsenic Test*, page 865.

Heavy metals. Prepare and test a 500-mg. sample as directed in *Method II* under the *Heavy Metals Test*, page 920, using 20 mcg. of lead ion (Pb) in the control (*Solution A*).

Lead. A *Sample Solution* prepared as directed for organic compounds meets the requirements of the *Lead Limit Test*, page 929, using 10 mcg. of lead ion (Pb) in the control.

Phellandrene. Mix 2.5 ml. of sample with 5 ml. of solvent hexane, add 5 ml. of a solution of sodium nitrite made by dissolving 5 grams of sodium nitrite in 8 ml. of water, then gradually add 5 ml. of glacial acetic acid. No crystals form in the mixture within 10 minutes.

Packaging and storage. Store in well-filled, tight containers in a cool place protected from light.

Functional use in foods. Flavoring agent.

EUGENOL

4-Allyl-2-methoxyphenol; Eugenic Acid; 4-Allylguaiacol

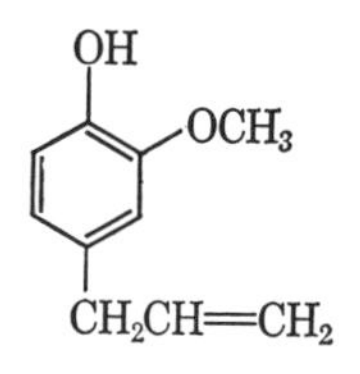

$C_{10}H_{12}O_2$ Mol. wt. 164.20

DESCRIPTION

Eugenol is the main constituent of carnation, cinnamon leaf, and clove oils. It is obtained from clove oil and other sources. It is a colorless to pale yellow liquid, having a strongly aromatic odor of clove, and a pungent, spicy taste. It darkens and thickens upon exposure to air. It is slightly soluble in water, and is miscible with alcohol, with chloroform, with ether, and with fixed oils.

SPECIFICATIONS

Assay. Not less than 100 percent, by volume, of phenols as eugenol.

Distillation range. Not less than 95 percent distils between 250° and 255°.

Refractive index. Between 1.540 and 1.542 at 20°.

Solubility in alcohol. Passes test.

Specific gravity. Between 1.064 and 1.070.

Limits of Impurities

Arsenic (as As). Not more than 3 parts per million (0.0003 percent).

Heavy metals (as Pb). Not more than 40 parts per million (0.004 percent).

Hydrocarbons. Passes test.

Lead. Not more than 10 parts per million (0.001 percent).

Phenol. Passes test.

TESTS

Assay. Proceed as directed under *Phenols*, page 898.

Distillation range. Proceed as directed in the general method, page 890.

Refractive index, page 945. Determine with an Abbé or other refractometer of equal or greater accuracy.

Solubility in alcohol. Proceed as directed in the general method, page 899. One ml. dissolves in 2 ml. of 70 percent alcohol.

Specific gravity. Determine by any reliable method (see page 5).

Arsenic. A *Sample Solution* prepared as directed for organic compounds meets the requirements of the *Arsenic Test*, page 865.

Heavy metals. Prepare and test a 500-mg. sample as directed in *Method II* under the *Heavy Metals Test*, page 920, using 20 mcg. of lead ion (Pb) in the control (*Solution A*).

Hydrocarbons. Dissolve 1 ml. in 20 ml. of 0.5 *N* sodium hydroxide in a stoppered 50-ml. tube, add 18 ml. of water, and mix. A clear mixture results immediately, but it may become turbid when exposed to air.

Lead. A *Sample Solution* prepared as directed for organic compounds meets the requirements of the *Lead Limit Test*, page 929, using 10 mcg. of lead ion (Pb) in the control.

Phenol. Shake 1 ml. with 20 ml. of water, filter, and add 1 drop of ferric chloride T.S. to 5 ml. of the clear filtrate. The mixture exhibits a transient grayish green color, but not a blue or violet color.

Packaging and storage. Store in full, tight, light-resistant containers in a cool place protected from light.

Functional use in foods. Flavoring agent.

FENNEL OIL

DESCRIPTION

The volatile oil obtained by steam distillation from the dried ripe fruit of *Foeniculum vulgare* Miller (Fam. *Umbelliferae*). It is a colorless or pale yellow liquid, having the characteristic odor and taste of fennel.

Note: If solid material has separated, carefully warm the sample until it is completely liquefied, and mix it before using.

SPECIFICATIONS

Angular rotation. Between +12° and +24°.

Refractive index. Between 1.528 and 1.538 at 20°.

Solidification point. Not lower than 3°.

Solubility in alcohol. Passes test.

Specific gravity. Between 0.953 and 0.973.

Limits of Impurities

Arsenic (as As). Not more than 3 parts per million (0.0003 percent).

Heavy metals (as Pb). Not more than 40 parts per million (0.004 percent).

Lead. Not more than 10 parts per million (0.001 percent).

TESTS

Angular rotation. Determine in a 100-mm. tube as directed under *Optical Rotation*, page 939.

Refractive index, page 945. Determine with an Abbé or other refractometer of equal or greater accuracy.

Solidification point. Determine as directed in the general method, page 954.

Solubility in alcohol. Proceed as directed in the general method, page 899. One ml. dissolves in 1 ml. of 90 percent alcohol.

Specific gravity. Determine by any reliable method (see page 5).

Arsenic. A *Sample Solution* prepared as directed for organic compounds meets the requirements of the *Arsenic Test*, page 865.

Heavy metals. Prepare and test a 500-mg. sample as directed in *Method II* under the *Heavy Metals Test*, page 920, using 20 mcg. of lead ion (Pb) in the control (*Solution A*).

Lead. A *Sample Solution* prepared as directed for organic compounds meets the requirements of the *Lead Limit Test*, page 929, using 10 mcg. of lead ion (Pb) in the control.

Packaging and storage. Store in full, tight containers in a cool place protected from light.

Functional use in foods. Flavoring agent.

FERRIC PHOSPHATE

Iron Phosphate; Ferric Orthophosphate

$FePO_4 . xH_2O$ Mol. wt. (anhydrous) 150.82

DESCRIPTION

Ferric phosphate contains from two to four molecules of water of hydration. It occurs as an odorless, yellowish white to buff colored powder. It is insoluble in water and in acetic acid, but is soluble in mineral acids.

IDENTIFICATION

Dissolve 1 gram in 5 ml. of dilute hydrochloric acid (1 in 2), and add an excess of sodium hydroxide T.S. A reddish brown precipitate forms. Boil the mixture, filter to remove the iron, and strongly acidify a portion of the filtrate with hydrochloric acid. Cool, mix with an equal volume of magnesia mixture T.S., and treat with a slight excess of ammonia T.S. An abundant white precipitate forms. This precipitate, after being washed, turns greenish yellow when treated with a few drops of silver nitrate T.S.

SPECIFICATIONS

Assay. Not less than 26.0 percent and not more than 30.0 percent of Fe.

Loss on ignition. Not more than 32.5 percent.

Limits of Impurities

Arsenic (as As). Not more than 3 parts per million (0.0003 percent).
Fluoride. Not more than 50 parts per million (0.005 percent).
Lead. Not more than 10 parts per million (0.001 percent).
Mercury. Not more than 3 parts per million (0.0003 percent).

TESTS

Assay. Dissolve about 3.5 grams of the sample, accurately weighed, in 75 ml. of dilute hydrochloric acid (1 in 2), heat to boiling, and boil for about 5 minutes. Cool, transfer into a 100-ml. volumetric flask, dilute to volume with the dilute hydrochloric acid, and mix. To 25.0 ml. of this solution add 100 ml. of the dilute hydrochloric acid, boil again for 5 minutes, and to the boiling solution add stannous chloride T.S., dropwise, with stirring, until the iron is just reduced as indicated by the disappearance of the yellow color. Add 2 drops in excess (but no more) of the stannous chloride T.S., dilute with about 50 ml. of water, and cool to room temperature. While stirring vigorously, add 15 ml. of a saturated solution of mercuric chloride, and then allow to stand for 5 minutes. Add 15 ml. of a sulfuric acid-phosphoric acid mixture, prepared by slowly adding 75 ml. of sulfuric acid to 300 ml. of water, cooling, adding 75 ml. of phosphoric acid, and then diluting to 500 ml. with water, and mix. Add 0.5 ml. of barium diphenylamine sulfonate T.S., and titrate with 0.1 *N* potassium dichromate to a reddish violet end-point. Each ml. of 0.1 *N* potassium dichromate is equivalent to 5.585 mg. of Fe.

Loss on ignition. Ignite at 800° for 1 hour

Arsenic. Assemble the special distillation apparatus as shown in Fig. 4 on page 866 of the general test. Transfer 2 grams of the sample, 50 ml. of hydrochloric acid, and 5 grams of cuprous chloride into the distilling flask (*B*). Reassemble the distillation apparatus and apply gentle suction to flask *F* to produce a continuous stream of bubbles. Heat the solution in flask *B* to boiling and distil until between 30 and 35 ml. of distillate has been collected in flask *D*. Quantitatively transfer the distillate to a 100-ml. volumetric flask with the aid of water, dilute to volume with water, and mix (*Sample Solution*). Prepare *Standard* and *Blank Solutions* in the same manner, using 6.0 ml. of *Standard Arsenic Solution* (page 865) in place of the sample in the *Standard Solution*, and 6.0 ml. of water in the *Blank Solution*. Transfer 50.0 ml. of the *Sample Solution* into the generator flask (Fig. 2, page 865), add 2 ml. of potassium iodide solution (15 in 100), and continue as directed in the *Procedure* under *Arsenic Test*, page 865, beginning with "(add) 0.5 ml. of *Stannous Chloride Solution*, and mix." Modify the *Procedure* by using 5.0 grams of Devarda's metal in place of the 3.0 grams of 20-mesh granular zinc, and maintain the temperature of the reaction mixture in the generator flask between 25° and 27°. Treat 50.0 ml. each of the *Standard Solution* and of the *Blank Solution* in the same manner and under the same conditions. Determine the absorbance at 525 mμ produced by each solution as directed under *Procedure*. Calculate the arsenic content (in parts per million) of the sample by the formula $3 \times (A_U - A_B)/(A_S - A_B)$, in which A_U is the absorbance produced by the *Sample Solution*, A_S is the absorbance produced by the *Standard Solution*, and A_B is the absorbance produced by the *Blank Solution*. [*Note:* If A_B exceeds 0.300, different samples of reagent grade cuprous chloride and Devarda's metal should be tested for arsenic content by the procedure described herein, and lots of these reagents should be selected which will give blank readings that do not exceed 0.300.]

Fluoride. Weigh accurately 1.0 gram, and proceed as directed in the *Fluoride Limit Test*, page 917.

Lead

Citrate-Cyanide Wash Solution. To 50 ml. of water add 50 ml. of *Ammonium Citrate Solution* (page 929) and 4 ml. of *Potassium Cyanide Solution* (page 929), mix, and adjust the pH, if necessary, with stronger ammonia T.S. to 9.0.

pH 2.5 Buffer Solution. To 25.0 ml. of 0.2 *M* potassium biphthalate add 37.0 ml. of 0.1 *N* hydrochloric acid, and dilute to 100.0 ml.

Dithizone-Carbon Tetrachloride Solution. Dissolve 10 mg. of dithizone in 1000 ml. of carbon tetrachloride. Prepare this solution fresh for each determination.

pH 2.5 Wash Solution. To 500 ml. of dilute nitric acid (1 in 100) add ammonia T.S. until the pH of the mixture is 2.5, then add 10 ml. of *pH 2.5 Buffer Solution,* and mix.

Ammonia-Cyanide Wash Solution. To 35 ml. of *pH 2.5 Wash Solution* add 4 ml. of *Ammonia-Cyanide Solution,* and mix.

> *Note:* Other solutions required are described under the *Lead Limit Test,* page 929.

Procedure. Dissolve 200 mg. of the sample, accurately weighed, in 10 ml. of dilute hydrochloric acid (1 in 2), prepare a control containing 2.0 ml. of *Diluted Standard Lead Solution* and 10 ml. of the dilute hydrochloric acid, and carry both solutions through the following procedure: Add 25 ml. of *Ammonium Citrate Solution,* heat on a steam bath for a few minutes, add 7 ml. of stronger ammonia T.S., and cool. Adjust the pH, if necessary, to 9.0, using the appropriate volumes of either stronger ammonia T.S. or hydrochloric acid, and transfer to a separator. Extract with 5-ml. portions of *Dithizone Extraction Solution* until the extraction solution retains its original color, and combine the extracts in a second separator. Wash the combined extracts by shaking for 30 seconds with 10 ml. of *Citrate-Cyanide Wash Solution,* then wash the wash solution with 3 ml. of *Dithizone Extraction Solution.* Combine the chloroform layers, add 20 ml. of dilute nitric acid (1 in 100), and shake for 30 seconds. Separate the layers, and shake the chloroform layer with an additional 5 ml. of the dilute nitric acid. Combine the acid washes in a small beaker, and adjust the pH with ammonia T.S. to 2.5 ± 0.2 (by means of a glass electrode). Transfer the solution to a separator, add 2 ml. of *pH 2.5 Buffer Solution,* and shake the solution for 30 seconds with 30 ml. of *Dithizone-Carbon Tetrachloride Solution.* Wash the combined carbon tetrachloride layers with 10 ml. of *pH 2.5 Wash Solution,* and combine the aqueous layers. Dislodge any drops of carbon tetrachloride remaining on the surface of the aqueous layer, and draw off and discard the carbon tetrachloride layer. (*Note:* Avoid any delay in completing the test after beginning the following extraction, since the color fades. Have the spectrophotometer ready to use, and carry the samples through the remainder of the procedure one at a time.) Add 4 ml. of *Ammonia-Cyanide Solution,* mix, and extract at once with 5-ml. portions of *Dithizone-Carbon Tetrachloride Solution* until the carbon tetrachloride shows no further pink color. Wash the combined extracts with 4 ml. of *Ammonia-Cyanide Wash Solution,* dry the stem of the separator, and drain the carbon tetrachloride through a plug of cotton to remove the last trace of water. Determine the absorbances of both solutions in 1-cm. cells at 520 mμ with a suitable spectrophotometer, using carbon tetrachloride as the blank. The absorbance of the sample solution does not exceed that of the control solution.

Mercury

Dithizone Extraction Solution. Dissolve 10 mg. of dithizone in 1000 ml. of chloroform.

Mercury Stock Solution. Transfer 1.3575 grams of mercuric chloride into a 1000-ml. volumetric flask, dissolve in 1 *N* sulfuric acid, dilute to volume with 1 *N* sulfuric acid, and mix. Each ml. contains 1 mg. of Hg.

Diluted Standard Mercury Solution. Transfer 3.0 ml. of the *Mercury Stock Solution* into a 1000-ml. volumetric flask, dilute to volume with 1 *N* sulfuric acid, and mix.

Procedure. Dissolve 1 gram of the sample in 30 ml. of 1 *N* perchloric acid by heating. Cool, adjust the volume to 30 ml., if necessary, and filter through No. 00 filter paper. Prepare a control containing 1.0 ml. of *Diluted Standard Mercury Solution* (3 mcg. Hg) and 30 ml. of 1 *N* perchloric acid, and treat the sample solution and the control as follows: Transfer into a 250-ml. separator, add 2 ml. of 6 *N* acetic acid and 3 ml. of chloroform, and shake for 1 minute. Allow the layers to separate, carefully draw off the chloroform layer, and discard. To the aqueous solution add 5.0 ml. of *Dithizone Extraction Solution,* and shake for 1 minute. Any color developed in the solution from the sample does not exceed that in the control.

Packaging and storage. Store in well-closed containers.

Functional use in foods. Nutrient; dietary supplement.

FERRIC PYROPHOSPHATE

Iron Pyrophosphate

$Fe_4(P_2O_7)_3 \cdot xH_2O$ Mol. wt. (anhydrous) 745.22

DESCRIPTION

A tan or yellowish white, odorless powder. It is insoluble in water but is soluble in mineral acids.

IDENTIFICATION

Dissolve 500 mg. in 5 ml. of dilute hydrochloric acid (1 in 2), and add an excess of sodium hydroxide T.S. A reddish brown precipitate forms. Allow the solution to stand for several minutes, and then filter, discarding the first few ml. To 5 ml. of the clear filtrate add 1 drop of bromophenol blue T.S., and titrate with 1 *N* hydrochloric acid to a green color. Add 10 ml. of a 1 in 8 solution of zinc sulfate, and readjust the pH to 3.8 (green color). A white precipitate forms (distinction from *orthophosphates*).

SPECIFICATIONS

Assay. Not less than 24.0 percent and not more than 26.0 percent of Fe.

Loss on ignition. Not more than 20 percent.

Limits of Impurities

Arsenic (as As). Not more than 3 parts per million (0.0003 percent).

Lead. Not more than 10 parts per million (0.001 percent).

Mercury. Not more than 3 parts per million (0.0003 percent).

TESTS

Assay. Proceed as directed in the *Assay* under *Ferric Phosphate*, page 309.

Loss on ignition. Ignite at 800° for 1 hour.

Arsenic. Prepare and test a 2-gram sample as directed in the test for *Arsenic* under *Ferric Phosphate*, page 865.

Lead. Proceed as directed in the test for *Lead* under *Ferric Phosphate*, page 309.

Mercury. Proceed as directed in the test for *Mercury* under *Ferric Phosphate*, page 309.

Packaging and storage. Store in well-closed containers.

Functional use in foods. Nutrient; dietary supplement.

FERROUS FUMARATE

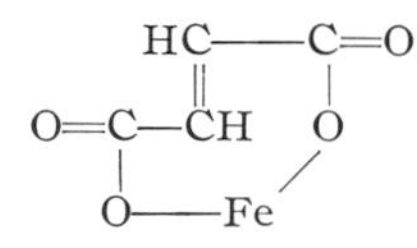

$C_4H_2FeO_4$ Mol. wt. 169.90

DESCRIPTION

An odorless, reddish orange to red-brown powder. It may contain soft lumps that produce a yellow streak when crushed.

IDENTIFICATION

A. To about 1.5 grams add 25 ml. of dilute hydrochloric acid (1 in 2), and dilute to 50 ml. with water. Heat to effect complete solution, then cool, filter on a fine-porosity sintered-glass crucible, wash the precipitate with dilute hydrochloric acid (3 in 100), saving the filtrate for *Identification test B*, and dry the precipitate at 105°. To 400 mg. of the dried precipitate add 3 ml. of water and 7 ml. of 1 *N* sodium hydroxide, and stir until solution is complete. Add diluted hydrochloric acid T.S., dropwise, until the solution is just acid to litmus, add 1 gram of *p*-nitrobenzyl bromide and 10 ml. of alcohol, and reflux the mixture for 2 hours. Cool, filter, and wash the precipitate with two small portions of a mixture of 2 parts of alcohol and 1 part of water, followed by two small portions of water. The precipitate, recrystallized from hot alcohol and dried at 105°, melts at about 152° (see page 931).

B. A portion of the filtrate obtained in the preceding test gives positive tests for *Iron*, page 927.

SPECIFICATIONS

Assay. Not less than 96.5 percent and not more than the equivalent of 101.0 percent of $C_4H_2FeO_4$.

Limits of Impurities

Arsenic (as As). Not more than 3 parts per million (0.0003 percent).

Ferric iron. Not more than 2.0 percent.

Lead. Not more than 10 parts per million (0.001 percent).

Loss on drying. Not more than 1 percent.

Mercury. Not more than 3 parts per million (0.0003 percent).

Sulfate. Not more than 0.2 percent.

TESTS

Assay. Transfer about 500 mg., accurately weighed, into a 500-ml. Erlenmeyer flask, add 25 ml. of dilute hydrochloric acid (2 in 5), and heat to boiling. Add, dropwise, a solution of 5.6 grams of stannous chloride in 50 ml. of dilute hydrochloric acid (3 in 10) until the yellow color disappears, and then add 2 drops in excess. Cool the solution in an ice bath to room temperature, add 8 ml. of mercuric chloride T.S., and allow to stand for 5 minutes. Add 200 ml. of water, 25 ml. of dilute sulfuric acid (1 in 2), and 4 ml. of phosphoric acid, then add orthophenanthroline T.S., and titrate with 0.1 *N* ceric sulfate. Each ml. of 0.1 *N* ceric sulfate is equivalent to 16.99 mg. of $C_4H_2FeO_4$.

Arsenic. Transfer 2 grams of the sample into a beaker and add 10 ml. of water and 10 ml. of sulfuric acid. Warm to precipitate the fumaric acid completely, cool, dilute with 30 ml. of water, and filter into a 100-ml. volumetric flask. Wash the precipitate with water, adding the washings to the flask, cool, dilute to volume with water, and mix. Transfer 50.0 ml. of this solution into an arsine generator, and proceed as directed under *Procedure* in the *Arsenic Test*, page 865, omitting the addition of dilute sulfuric acid (1 in 5).

Ferric iron. Transfer 2 grams of the sample into a 250-ml. glass-stoppered Erlenmeyer flask, add 25 ml. of water and 4 ml. of hydrochloric acid, and heat on a hot plate until solution is complete. Stopper the flask, and cool to room temperature. Add 3 grams of potassium iodide, stopper, swirl to mix, and allow to stand in the dark for 5 minutes. Remove the stopper, add 75 ml. of water, and titrate with 0.1 *N* sodium thiosulfate, adding starch T.S. near the end-point. Not more than 7.16 ml. of 0.1 *N* sodium thiosulfate is consumed.

Lead

[See page 929 for the preparation of other special solutions not described below.]

Citrate-Cyanide Wash Solution. To 50 ml. of water add 50 ml. of *Ammonium Citrate Solution* and 4 ml. of *Potassum Cyanide Solution,* mix, and adjust the pH, if necessary, with stronger ammonia T.S. to 9.0.

pH 2.5 Buffer Solution. To 25.0 ml. of 0.2 *M* potassium biphthalate add 37.0 ml. of 0.1 *N* hydrochloric acid, and dilute to 100.0 ml. with water.

Dithizone-Carbon Tetrachloride Solution. Dissolve 10 mg. of dithizone in 1000 ml. of carbon tetrachloride. Prepare this solution fresh for each determination.

pH 2.5 Wash Solution. To 500 ml. of dilute nitric acid (1 in 100) add ammonia T.S. until the pH of the mixture is 2.5, then add 10 ml. of *pH 2.5 Buffer Solution,* and mix.

Ammonia-Cyanide Wash Solution. To 35 ml. of *pH 2.5 Wash Solution* add 4 ml. of *Ammonia-Cyanide Solution,* and mix.

Procedure. Transfer 500 mg. of the sample, accurately weighed, into a 150-ml. beaker, and add 3 ml. of nitric acid and 5 ml. of perchloric acid. Prepare a control containing 5.0 ml. of *Diluted Standard Lead Solution,* 3 ml. of nitric acid, and 5 ml. of perchloric acid, and carry both sample and control solutions through the following procedure: Evaporate to dryness, cool, add 10 ml. of dilute hydrochloric acid (1 in 2), and heat on a steam bath until the residue is completely dissolved. Add 25 ml. of *Ammonium Citrate Solution,* heat on the steam bath for an additional few minutes, then add 7 ml. of stronger ammonia T.S., and cool. Adjust the pH, if necessary, to 9.0 (by means of a glass electrode or pH indicator paper), using either stronger ammonia T.S. or hydrochloric acid as necessary, and transfer into a separator. Extract with 5-ml. portions of *Dithizone Extraction Solution* until the extraction solution retains its original color, and combine the extracts in a second separator. Wash the combined extracts by shaking for 30 seconds with 10 ml. of *Citrate-Cyanide Wash Solution,* and then wash the wash solution with 3 ml. of *Dithizone Extraction Solution.* Combine the chloroform layers, add 20 ml. of dilute nitric acid (1 in 100), and shake for 30 seconds. Separate the layers, and shake the chloroform layer with an additional 5 ml. of the dilute nitric acid. Combine the acid washes in a small beaker, and adjust the pH with diluted ammonia T.S. to 2.5 ± 0.2 (by means of a glass electrode).

Transfer the solution into a separator, add 2 ml. of *pH 2.5 Buffer Solution,* and shake the solution for 30 seconds with 30 ml. of *Dithizone-Carbon Tetrachloride Solution.* Wash the combined carbon tetrachloride layers with 10 ml. of *pH 2.5 Wash Solution,* and combine the aqueous layers. Dislodge any drops of carbon tetrachloride remaining on the surface of the aqueous layer, and draw off and discard the carbon tetrachloride layer. (*Note:* Avoid any delay in completing the test after beginning the following extraction, since the color fades. Have the spectrophotometer ready to use, and carry the samples through the remainder of the procedure one at a time.) Add 4 ml. of the *Ammonia-Cyanide Solution,* mix, and then extract at once with 5-ml. portions of *Dithizone-Carbon Tetrachloride Solution* until the carbon tetrachloride shows no further pink color. Wash the combined extracts with 4 ml. of *Ammonia-Cyanide Wash Solution,* dry the stem of the separator, and drain the carbon tetrachloride through a plug of cotton to remove the last trace of water. Determine the absorbances of both solutions in 1-cm. cells at 520 mμ with a suitable spectrophotometer, using carbon tetrachloride as the blank. The absorbance of the sample solution does not exceed that of the control solution.

Loss on drying, page 931. Dry at 105° for 16 hours.

Mercury

Dithizone Extraction Solution. Dissolve 30 mg. of dithizone in 1000 ml. of chloroform, add 5 ml. of alcohol, and mix. Store in a refrigerator. Before use, shake a suitable volume with about half its volume of dilute nitric acid (1 in 100), discarding the nitric acid. Do not use if more than a month old.

Diluted Dithizone Extraction Solution. Just prior to use, dilute 5 ml. of *Dithizone Extraction Solution* with 25 ml. of chloroform.

Hydroxylamine Hydrochloride Solution. Dissolve 20 grams of hydroxylamine hydrochloride in sufficient water to make about 65 ml., transfer the solution into a separator, add a few drops of thymol blue T.S., and then add stronger ammonia T.S. until the solution assumes a yellow color. Add 10 ml. of sodium diethyldithiocarbamate solution (1 in 25), mix, and allow to stand for 5 minutes. Extract the solution with successive 10- to 15-ml. portions of chloroform until a 5-ml. test portion of the chloroform extract does not assume a yellow color when shaken with a dilute cupric sulfate solution. Add diluted hydrochloric acid T.S. until the extracted solution is pink, adding 1 or 2 drops more of thymol blue T.S., if necessary, then dilute to 100 ml. with water, and mix.

Mercury Stock Solution. Transfer 135.4 mg., accurately weighed, of mercuric chloride into a 100-ml. volumetric flask, dissolve in 1 *N* sulfuric acid, dilute to volume with the acid, and mix. Dilute 5.0 ml. of this solution to 500.0 ml. with 1 *N* sulfuric acid. Each ml. contains the equivalent of 10 mcg. of Hg.

Diluted Standard Mercury Solution. On the day of use, transfer 10.0 ml. of *Mercury Stock Solution* into a 100-ml. volumetric flask, dilute to volume with 1 *N* sulfuric acid, and mix. Each ml. contains the equivalent of 1 mcg. of Hg.

Sodium Citrate Solution. Dissolve 250 grams of sodium citrate dihydrate in 1000 ml. of water.

Sample Solution. Dissolve 1 gram of the sample in 30 ml. of diluted nitric acid T.S. by heating on a steam bath. Cool to room temperature in an ice bath, stir, and filter through S and S No. 589, or equivalent, filter paper which has been previously washed with diluted nitric acid T.S., followed by water. To the filtrate add 20 ml. of *Sodium Citrate Solution* and 1 ml. of *Hydroxylamine Hydrochloride Solution.*

Procedure. [Because mercuric dithizonate is light-sensitive, this procedure should be performed in subdued light.] Prepare a control containing 3.0 ml. of *Diluted Standard Mercury Solution* (3 mcg. Hg), 30 ml. of diluted nitric acid T.S., 5 ml. of *Sodium Citrate Solution,* and 1 ml. of *Hydroxylamine Hydrochloride Solution.* Treat the control and the *Sample Solution* as follows: Adjust the pH of each solution to 1.8 with stronger ammonia T.S., using a pH meter, and transfer the solutions into different separators. Extract with two 5-ml. portions of *Dithizone Extraction Solution,* and then extract again with 5 ml. of chloroform, discarding the aqueous solutions. Transfer the combined extracts from each separator into different separators, add 10 ml. of dilute hydrochloric acid (1 in 2) to each, shake well, and discard the chloroform layers. Extract the acid solutions with about 3 ml. of chloroform, shake well, and discard the chloroform layers. Add to each separator 0.1 ml. of 0.05 *M* disodium ethylenediaminetetraacetate, 2 ml. of 6 *N* acetic acid, mix, and then slowly add 5 ml. of stronger ammonia T.S. Stopper the separators, and cool under the stream of cold water. Dry the outside of the separators, pour the solutions (through the top of the separators) carefully, to avoid loss, into separate beakers, and adjust the pH of both solutions to 1.8 with ammonia T.S., using a pH meter. Return the sample and control solutions to their original separators, add 5.0 ml. of *Diluted Dithizone Extraction Solution,* and shake vigorously. Any color developed in the *Sample Solution* does not exceed that in the control.

Sulfate. Mix 1 gram of the sample with 100 ml. of water in a 250-ml. beaker, and heat on a steam bath, adding hydrochloric acid, dropwise, until complete solution is effected (about 2 ml. of the acid will be required). Filter the solution, if necessary, and dilute the clear solution or filtrate to 100 ml. with water. Heat to boiling, add 10 ml. of barium chloride T.S., warm on a steam bath for 2 hours, cover, and allow to stand overnight. If crystals of ferrous fumarate form, warm on a steam bath to dissolve them, then filter through paper, wash the residue with hot water, and transfer the paper containing the residue to a tared crucible. Char the paper, without burning, and ignite the crucible and its contents at 600° to constant weight. Each mg. of the residue is equivalent to 0.412 mg. (412 mcg.) of SO_4.

Packaging and storage. Store in well-closed containers.

Functional use in foods. Nutrient; dietary supplement.

FERROUS GLUCONATE

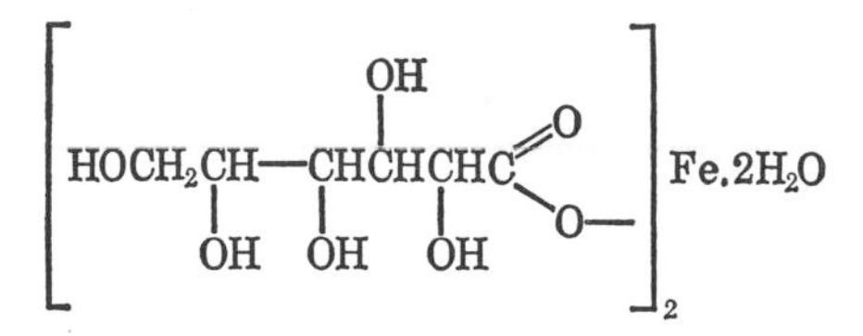

$C_{12}H_{22}FeO_{14}.2H_2O$ Mol. wt. 482.18

DESCRIPTION

Fine yellowish gray or pale greenish yellow powder or granules having a slight odor resembling that of burned sugar. One gram dissolves in about 10 ml. of water with slight heating. It is practically insoluble in alcohol. A 1 in 20 solution is acid to litmus.

IDENTIFICATION

A. To 5 ml. of a warm 1 in 10 solution of the sample add 0.65 ml. of glacial acetic acid and 1 ml. of freshly distilled phenylhydrazine, and heat the mixture on a steam bath for 30 minutes. Cool, and scratch the inner surface of the container with a glass stirring rod. Crystals of gluconic acid phenylhydrazide form.

B. A 1 in 20 solution gives positive tests for *Ferrous salts*, page 927.

SPECIFICATIONS

Assay. Not less than 95.0 percent of $C_{12}H_{22}FeO_{14}$, calculated on the dried basis.

Loss on drying. Between 6.5 and 10 percent.

Limits of Impurities

Arsenic (as As). Not more than 3 parts per million (0.0003 percent).

Chloride. Not more than 700 parts per million (0.07 percent).

Ferric iron. Not more than 2 percent.

Lead. Not more than 10 parts per million (0.001 percent).

Mercury. Not more than 3 parts per million (0.0003 percent).

Oxalic acid. Passes test.

Reducing sugars. Passes test.

Sulfate. Not more than 0.1 percent.

TESTS

Assay. Dissolve about 1.5 grams, accurately weighed, in a mixture of 75 ml. of water and 15 ml. of diluted sulfuric acid T.S. in a 300-ml. Erlenmeyer flask, and add 250 mg. of zinc dust. Close the flask with a stopper containing a Bunsen valve, allow to stand at room temperature for 20 minutes, then filter through a Gooch crucible containing an asbestos mat coated with a thin layer of zinc dust, and wash the crucible and contents with 10 ml. of diluted sulfuric acid T.S., followed by 10 ml. of water. Add orthophenanthroline T.S., and titrate the filtrate in the suction flask immediately with 0.1 *N* ceric sulfate. Perform a blank determination (see page 2), and make any necessary correction. Each ml. of 0.1 *N* ceric sulfate is equivalent to 44.62 mg. of $C_{12}H_{22}FeO_{14}$.

Loss on drying, page 931. Dry at 105° for 4 hours.

Arsenic. Place 2 grams of the sample in a 100-ml. round bottom flask fitted with a 24/40 standard taper joint. Add 40 ml. of dilute sulfuric acid (1 in 4) and 2 ml. of 30 percent potassium bromide solution, and connect immediately to a modified Bethge distillation apparatus (see Fig. 3, page 866), or other suitable apparatus having a reservoir with a water jacket which is cooled with ice water. Heat the flask over an Argand burner until the sample dissolves, and collect 25 ml. of distillate. Transfer the distillate to an arsine generator flask, and wash the condenser and reservoir several times with small portions of water. Add bromine T.S. until the distillate is slightly yellow, dilute to 35 ml. with water, and continue as directed in the *Procedure* under *Arsenic Test*, page 865, using 6.0 ml. of *Standard Arsenic Solution* in the preparation of the standard.

Chloride, page 879. Dissolve 1 gram in 100 ml. of water. Any turbidity produced by a 10-ml. portion of this solution does not exceed that shown in a control containing 70 mcg. of chloride ion (Cl).

Ferric iron. Dissolve about 5 grams, accurately weighed, in a mixture of 100 ml. of water and 10 ml. of hydrochloric acid in a 250-ml. glass-stoppered flask, add 3 grams of potassium iodide, shake well, and allow to stand in the dark for 5 minutes. Titrate any liberated iodine with 0.1 *N* sodium thiosulfate, using starch T.S. as the indicator. Each ml. of 0.1 *N* sodium thiosulfate is equivalent to 5.585 mg. of ferric iron.

Lead. Determine as directed in the test for *Lead* under *Ferrous Fumarate*, page 313.

Mercury. Determine as directed in the test for *Mercury* under *Ferrous Fumarate*, page 313.

Oxalic acid. Dissolve 1 gram in 10 ml. of water, add 2 ml. of hydrochloric acid, transfer to a separator, and extract successively with 50 and 20 ml. of ether. Combine the ether extracts, add 10 ml. of water, and evaporate the ether on a steam bath. Add 1 drop of acetic acid (36 percent) and 1 ml. of calcium acetate solution (1 in 20). No turbidity is produced within 5 minutes.

Reducing sugars. Dissolve 500 mg. in 10 ml. of water, warm, and make the solution alkaline with 1 ml. of ammonia T.S. Pass hydrogen sulfide gas into the solution to precipitate the iron, and allow the mixture to stand for 30 minutes to coagulate the precipitate. Filter, and wash the precipitate with two successive 5-ml. portions of water. Acidify the combined filtrate and washings with hydrochloric acid, and add 2 ml. of diluted hydrochloric acid T.S. in excess. Boil the solution until the vapors no longer darken lead acetate paper, and continue to boil, if necessary, until it has been concentrated to about 10 ml. Cool, add 5 ml. of sodium carbonate T.S. and 20 ml. of water, filter, and adjust the volume of the filtrate to 100 ml. To 5 ml. of the filtrate add 2 ml. of alkaline cupric tartrate T.S., and boil for 1 minute. No red precipitate is formed within 1 minute.

Sulfate, page 879. Any turbidity produced by a 200-mg. sample does not exceed that shown in a control containing 200 mcg. of sulfate (SO_4).

Packaging and storage. Store in tight containers.

Functional use in foods. Nutrient; dietary supplement; coloring adjunct.

FERROUS SULFATE

$FeSO_4 . 7H_2O$ Mol. wt. 278.01

DESCRIPTION

Pale, bluish green crystals or granules. It is odorless, has a saline, styptic taste, and is efflorescent in dry air. In moist air it oxidizes readily to form brownish yellow basic ferric sulfate. Its 1 in 10 solution is acid to litmus, having a pH of about 3.7, and gives positive tests for *Ferrous salts*, page 927, and for *Sulfate*, page 928. One gram dissolves in 1.5 ml. of water at 25° and in 0.5 ml. of boiling water. It is insoluble in alcohol.

SPECIFICATIONS

Assay. Not less than 99.5 percent and not more than the equivalent of 104.5 percent of $FeSO_4 . 7H_2O$.

Limits of Impurities

Arsenic (as As). Not more than 3 parts per million (0.0003 percent).

Lead. Not more than 10 parts per million (0.001 percent).

Mercury. Not more than 3 parts per million (0.0003 percent).

TESTS

Assay. Dissolve about 1 gram, accurately weighed, in a mixture of

25 ml. of diluted sulfuric acid T.S. and 25 ml. of recently boiled and cooled water, and titrate with 0.1 *N* potassium permanganate until a permanent pink color is produced. Each ml. of 0.1 *N* potassium permanganate is equivalent to 27.80 mg. of $FeSO_4.7H_2O$.

Arsenic. Place 2 grams of the sample in a 100-ml. round bottom flask fitted with a 24/40 standard taper joint. Add 40 ml. of dilute sulfuric acid (1 in 4) and 2 ml. of 30 percent potassium bromide solution, and connect immediately to a modified Bethge distillation apparatus (see Fig. 3, page 866), or other suitable apparatus having a reservoir with a water jacket which is cooled with ice water. Heat the flask over an Argand burner until the sample dissolves, and collect 25 ml. of distillate. Transfer the distillate to an arsine generator flask, and wash the condenser and reservoir several times with small portions of water. Add bromine T.S. until the distillate is slightly yellow, dilute to 35 ml. with water, and continue as directed in the *Procedure* under *Arsenic Test*, page 865, using 6.0 ml. of *Standard Arsenic Solution* in the preparation of the standard.

Lead. Determine as directed in the test for *Lead* under *Ferrous Fumarate*, page 313.

Mercury. Determine as directed in the test for *Mercury* under *Ferrous Fumarate*, page 313.

Packaging and storage. Store in tight containers.

Functional use in foods. Nutrient; dietary supplement.

FERROUS SULFATE, DRIED

$FeSO_4.xH_2O$ Mol. wt. (anhydrous) 151.90

DESCRIPTION

A grayish white to buff-colored powder consisting primarily of $FeSO_4.H_2O$, with varying amounts of $FeSO_4.4H_2O$. It dissolves slowly in water but is insoluble in alcohol. It gives positive tests for *Ferrous salts*, page 927, and for *Sulfate*, page 928.

SPECIFICATIONS

Assay. Not less than 86.0 percent and not more than 89.0 percent of $FeSO_4$.

Limits of Impurities

Arsenic (as As). Not more than 3 parts per million (0.0003 percent).

Insoluble substances. Not more than 0.05 percent.

Lead. Not more than 10 parts per million (0.001 percent).

Mercury. Not more than 3 parts per million (0.0003 percent).

TESTS

Assay. Proceed as directed in the *Assay* under *Ferrous Sulfate*, page 320. Each ml. of 0.1 *N* potassium permanganate is equivalent to 15.19 mg. of $FeSO_4$.

Insoluble substances. Dissolve 2 grams in 20 ml. of freshly boiled dilute sulfuric acid (1 in 100), heat to boiling, and then digest in a covered beaker on a steam bath for 1 hour. Filter through a tared filtering crucible, wash thoroughly, and dry at 105°. The weight of the insoluble residue does not exceed 1 mg.

Arsenic. Place 2 grams of the sample in a 100-ml. round bottom flask fitted with a 24/40 standard taper joint. Add 40 ml. of dilute sulfuric acid (1 in 4) and 2 ml. of 30 percent potassium bromide solution, and connect immediately to a modified Bethge distillation apparatus (see Fig. 3, page 866), or other suitable apparatus having a reservoir with a water jacket which is cooled with ice water. Heat the flask over an Argand burner until the sample dissolves, and collect 25 ml. of distillate. Transfer the distillate to an arsine generator flask, and wash the condenser and reservoir several times with small portions of water. Add bromine T.S. until the distillate is slightly yellow, dilute to 35 ml. with water, and continue as directed in the *Procedure* under *Arsenic Test*, page 865, using 6.0 ml. of *Standard Arsenic Solution* in the preparation of the standard.

Lead. Determine as directed in the test for *Lead* under *Ferrous Fumarate*, page 313.

Mercury. Determine as directed in the test for *Mercury* under *Ferrous Fumarate*, page 313.

Packaging and storage. Store in tight containers.

Functional use in foods. Nutrient; dietary supplement.

FIR NEEDLE OIL, CANADIAN

Balsam Fir Oil

DESCRIPTION

The volatile oil obtained by steam distillation from needles and twigs of *Abies balsamea* L., Mill (Fam. *Pinaceae*). It is a colorless to faintly yellow liquid, having a pleasant balsam-like odor. It is soluble in most fixed oils and in mineral oil. It is slightly soluble in propylene glycol, but it is insoluble in glycerin.

SPECIFICATIONS

Assay. Not less than 8.0 percent and not more than 16.0 percent of esters, calculated as bornyl acetate ($C_{12}H_{20}O_2$).

Angular rotation. Between $-19°$ and $-24°$.

Refractive index. Between 1.473 and 1.476 at 20°.

Solubility in alcohol. Passes test.

Specific gravity. Between 0.872 and 0.878.

TESTS

Assay. Weigh accurately about 5 grams, and proceed as directed under *Ester Determination*, page 896, using 98.15 as the equivalence factor (e) in the calculation.

Angular rotation. Determine in a 100-mm. tube as directed under *Optical Rotation*, page 939.

Refractive index, page 945. Determine with an Abbé or other refractometer of equal or greater accuracy.

Solubility in alcohol. Proceed as directed in the general method, page 899. One ml. dissolves in 4 ml. of 90 percent alcohol, occasionally with haziness.

Specific gravity. Determine by any reliable method (see page 5).

Packaging and storage. Store in full, tight, preferably glass, aluminum, tin-lined, or other suitable containers in a cool place protected from light.

Functional use in foods. Flavoring agent.

FIR NEEDLE OIL, SIBERIAN

Pine Needle Oil

DESCRIPTION

The volatile oil obtained by steam distillation from the needles and twigs of *Abies sibirica* Lebed. (Fam. *Pinaceae*). It is an almost colorless or faintly yellow liquid. It is soluble in most fixed oils and in mineral oil. It is insoluble in glycerin and in propylene glycol.

SPECIFICATIONS

Assay. Not less than 32.0 percent and not more than 44.0 percent of esters, calculated as bornyl acetate ($C_{12}H_{20}O_2$).

Angular rotation. Between $-33°$ and $-45°$.

Refractive index. Between 1.468 and 1.473 at 20°.

Solubility in alcohol. Passes test.

Specific gravity. Between 0.898 and 0.912.

TESTS

Assay. Weigh accurately about 2 grams, and proceed as directed under *Ester Determination*, page 896, using 98.15 as the equivalence factor (*e*) in the calculation.

Angular rotation. Determine in a 100-mm. tube as directed under *Optical Rotation*, page 939.

Refractive index, page 945. Determine with an Abbé or other refractometer of equal or greater accuracy.

Solubility in alcohol. Proceed as directed in the general method, page 899. One ml. dissolves in 1 ml. of 90 percent alcohol. Occasionally the solution may become hazy upon further dilution.

Specific gravity. Determine by any reliable method (see page 5).

Packaging and storage. Store in full, tight, preferably glass, aluminum, or tin-lined containers in a cool place protected from light.

Functional use in foods. Flavoring agent.

FOOD STARCH, MODIFIED

Modified Food Starch

DESCRIPTION

Modified food starches are products of the treatment of any of several grain- or root-based native starches (e.g., corn, sorghum, wheat, potato, tapioca, sago, etc.) with small amounts of certain chemical agents, which modify the physical characteristics of the native starches to produce desirable properties.

Starch molecules are polymers of anhydroglucose and occur in both linear and branched form. The degree of polymerization and, accordingly, the molecular weight of the naturally occurring starch molecules vary radically. Furthermore, they vary in the ratio of branched chain polymers (amylopectin) to linear chain polymers (amylose), both within a given type of starch and from one type to another. These factors, in addition to any type of chemical modification used, affect the viscosity, texture, and stability of the starch sols significantly.

Starch is chemically modified by mild degradation reactions or by reactions between the hydroxyl groups of the native starch and the reactant selected. One or more of the following processes are used: mild oxidation (bleaching), moderate oxidation, acid depolymerization, monofunctional esterification, polyfunctional esterification (cross-linking), monofunctional etherification, polyfunctional etherification (cross-linking), alkaline gelatinization, and certain combinations of these treatments. These methods of preparation can be used as a basis for classifying the starches thus produced (see *Additional Specifications* below). Generally, however, the products are called modified food starches, or food starch-modified.

Modified food starches are usually produced as white or nearly white, tasteless, odorless powders, as intact granules, and, if pregelatinized (i.e., subjected to heat treatment in the presence of water), as flakes, amorphous powders, or coarse particles. Modified food starches are insoluble in alcohol, in ether, and in chloroform. If not pregelatinized, they are practically insoluble in cold water. Upon heating in water, the granules usually begin to swell at temperatures between 45° and 80°, depending upon the botanical origin and the degree of modification. They gelatinize completely at higher temperatures. Pregelatinized starches hydrate in cold water.

IDENTIFICATION

A. Suspend about 1 gram of the sample in 20 ml. of water, and add a few drops of iodine T.S. A dark blue to red color is produced.

B. Place about 2.5 grams of the sample in a boiling flask, add 10 ml. of dilute hydrochloric acid (3 percent) and 70 ml. of water, mix, reflux for about 3 hours, and cool. Add 0.5 ml. of the resulting solution to 5 ml. of hot alkaline cupric tartrate T.S. A copious red precipitate is produced.

C. Examine a portion of the sample with a polarizing microscope in polarized light under crossed Nicol prisms. The typical polarization cross is usually observed, except in the case of pregelatinized starches.

GENERAL SPECIFICATIONS

Crude fat. Not more than 0.15 percent.

Loss on drying. *Cereal starch:* not more than 15 percent; *potato starch:* not more than 21 percent; *sago* and *tapioca starch:* not more than 18 percent.

pH of dispersions. Between 3.0 and 9.0.

Protein. Not more than 0.5 percent, except not more than 1 percent in modified high amylose starches.

Limits of Impurities

Arsenic (as As). Not more than 3 parts per million (0.0003 percent).

Heavy metals (as Pb). Not more than 40 parts per million (0.004 percent).

Lead. Not more than 5 parts per million (0.0005 percent).

Sulfur dioxide. Not more than 80 parts per million (0.008 percent).

ADDITIONAL SPECIFICATIONS. The modified food starches listed below according to method of preparation must meet all of the above *General Specifications* in addition to the specified methods of *Treatment* (the reagent for which, if not specifically limited, should not exceed the amount reasonably required to accomplish the intended modification) and any requirements for *Residuals Limitation.*

* * * * *

Mild oxidation (Bleached starch). The starches resulting from mild oxidation are not altered chemically; in all cases, extraneous color bodies are oxidized, solubilized, and removed by washing and filtration. These treatments may be used in combination with the other forms of treatment listed herein.

Treatment to produce bleached starch:	*Residuals limitation*
Active oxygen obtained from hydrogen peroxide, and/or peracetic acid, not to exceed 0.45% of active oxygen	—
Ammonium persulfate, not to exceed 0.075%, and sulfur dioxide, not to exceed 0.05%	—
Chlorine, as sodium hypochlorite, not to exceed 0.0082 pound (3.72 grams) of chlorine per pound (454 grams) of dry starch	—
Potassium permanganate, not to exceed 0.2%	Not more than 50 ppm (0.005%) of residual manganese (as Mn)
Sodium chlorite, not to exceed 0.5%	—

* * * * *

Moderate oxidation (Oxidized starch). The maximum specified treatment introduces about 1 carboxyl group per 28 anhydroglucose units. The starch is whitened, and its molecular weight and viscosity are reduced.

Treatment to produce oxidized starch:	*Residuals limitation*
Chlorine, as sodium hypochlorite, not to exceed 0.055 pound (25 grams) of chlorine per pound (454 grams) of dry starch	—

* * * * *

Acid depolymerization (Thin-boiling, or acid-modified starch). This treatment results in partial depolymerization, causing a reduction in viscosity. The treatment may be used in combination with the other treatments that follow.

Treatment to produce thin-boiling starch:	*Residuals limitation*
Hydrochloric acid and/or sulfuric acid	—

* * * * *

Monofunctional and/or polyfunctional esterification (starch esters). The starch esters are named individually, depending upon the method of preparation.

Treatment to produce starch acetate:	*Residuals limitation*
Acetic anhydride or vinyl acetate	Not more than 2.5% of acetyl groups introduced into finished product

Treatment to produce acetylated distarch adipate:	
Adipic anhydride, not to exceed 0.12%, and acetic anhydride	Not more than 2.5% of acetyl groups introduced into finished product
Treatment to produce starch phosphate:	
Monosodium orthophosphate	Not more than 0.4% of residual phosphate (calculated as P)
Treatment to produce starch sodium octenyl succinate:	
1-Octenyl succinic anhydride, not to exceed 3%	—
Treatment to produce starch aluminum octenyl succinate:	
1-Octenyl succinic anhydride, not to exceed 2%, and aluminum sulfate, not to exceed 2%	—
Treatment to produce distarch phosphate:	
Phosphorus oxychloride, not to exceed 0.1%	—
Sodium trimetaphosphate	Not more than 0.04% of residual phosphate (calculated as P)
Treatment to produce phosphated distarch phosphate:	
Sodium tripolyphosphate and sodium trimetaphosphate	Not more than 0.4% of residual phosphate (calculated as P)
Treatment to produce starch sodium succinate:	
Succinic anhydride, not to exceed 4%	—

* * * * *

Monofunctional and/or polyfunctional etherification (Starch ethers-hemiacetals, or ethers). The starch ethers are named individually, depending upon the method of preparation.

Treatment to produce distarchoxy propanol:	*Residuals limitation*
Acrolein, not to exceed 0.6%	—
Treatment to produce distarch glycerol:	
Epichlorohydrin, not to exceed 0.3%	—
Treatment to produce hydroxypropyl distarch glycerol:	
Epichlorohydrin, not to exceed 0.1%, combined with propylene oxide, not to exceed 10%	Not more than 5 ppm (0.0005%) of residual propylene chlorohydrin

Treatment to produce hydroxypropyl starch:	
Propylene oxide, not to exceed 25%	Not more than 5 ppm (0.0005%) of residual propylene chlorohydrin

* * * * *

Etherification and esterification (Starch ether-esters). The starch ether-esters are named individually, depending upon their method of preparation.

Treatment to produce acetylated distarch glycerol:	*Residuals limitation*
Acrolein, not to exceed 0.6%, and vinyl acetate, not to exceed 7.5%	Not more than 2.5% of acetyl groups introduced into finished product
Epichlorohydrin, not to exceed 0.3%, and acetic anhydride	Not more than 2.5% of acetyl groups introduced into finished product
Treatment to produce succinyl distarch glycerol:	
Epichlorohydrin, not to exceed 0.3%, and succinic anhydride, not to exceed 4%	—
Treatment to produce hydroxypropyl distarch phosphate:	
Phosphorus oxychloride, not to exceed 0.1%, and propylene oxide, not to exceed 10%	Not more than 5 ppm (0.0005%) of residual propylene chlorohydrin

* * * * *

Etherification with oxidation (Oxidized starch ethers)

Treatment to produce oxidized hydroxypropyl starch:	*Residuals limitation*
Chlorine, as sodium hypochlorite, not to exceed 0.055 pound (25 grams) of chlorine per pound (454 grams) of dry starch; active oxygen obtained from hydrogen peroxide, not to exceed 0.45%; and propylene oxide, not to exceed 25%	Not more than 5 ppm (0.0005%) of residual propylene chlorohydrin

* * * * *

Alkaline gelatinization (Gelatinized starch)

Treatment to produce gelatinized starch:	*Residuals limitation*
Sodium hydroxide, not to exceed 1%	—

* * * * *

TESTS (General Specifications)

Crude fat. Determine as directed in the general method, page 960.

Loss on drying, page 931. Dry a 5-gram sample in a vacuum oven, not exceeding 100 mm. of mercury, at 120° for 4 hours.

pH of dispersions. Mix 20 grams of the sample with 80 ml. of water, and agitate continuously at a moderate rate for 5 minutes. (In the case of pregelatinized starches, 3 grams should be suspended in 97 ml. of water.) Determine the pH of the resulting suspension by the *Potentiometric Method,* page 941. (*Note:* The distilled water used for sample dispersion should require not more than 0.05 ml. of 0.1 *N* acid or alkali per 200 ml. to obtain the methyl red or phenolphthalein end-point, respectively.)

Protein. Transfer about 10 grams of the sample, accurately weighed, into an 800-ml. Kjeldahl flask, and add 10 grams of anhydrous potassium or sodium sulfate, 300 mg. of cupric sulfate (or other suitable catalyst), and 60 ml. of sulfuric acid. Gently heat the mixture, keeping the flask inclined at about a 45° angle, and after frothing has ceased, boil briskly until the solution has remained clear for about 1 hour. Cool, add 300 ml. of water, mix, and cool again. Cautiously pour about 75 ml. (or enough to make the mixture strongly alkaline) of sodium hydroxide solution (2 in 5) down the inside of the flask so that it forms a layer under the acid solution, and then add a few pieces of granular zinc. Immediately connect the flask to a distillation apparatus consisting of a Kjeldahl connecting bulb and a condenser, the delivery tube of which extends well beneath the surface of an accurately measured excess of 0.1 *N* sulfuric acid (or 50 ml. of 0.4 percent boric acid solution) contained in a 500-ml. flask. Gently rotate the contents of the Kjeldahl flask to mix, and distil until all ammonia has passed into the absorbing acid solution (about 250 ml. of distillate). To the receiving flask add 0.25 ml. of methyl red-methylene blue T.S., and titrate the excess acid with 0.1 *N* sodium hydroxide. (*Note:* Titrate with 0.1 *N* sulfuric acid if boric acid was used to absorb the ammonia.) Perform a blank determination, substituting pure sucrose or dextrose for the sample, and make any necessary correction (see page 2). Each ml. of 0.1 *N* sulfuric acid consumed is equivalent to 1.401 mg. of nitrogen (N). Calculate the percent N in the sample, and then calculate the percent protein by multiplying the percent N by 6.25, in the case of starches obtained from corn, or by 5.7, in the case of starches obtained from wheat. Other factors may be applied as necessary for starches obtained from other sources.

Arsenic. A *Sample Solution* prepared as directed for organic compounds meets the requirements of the *Arsenic Test,* page 865.

Heavy metals. Prepare and test a 500-mg. sample as directed in *Method II* under the *Heavy Metals Test,* page 920, using 20 mcg. of lead ion (Pb) in the control (*Solution A*).

Lead. A *Sample Solution* prepared as directed for organic compounds meets the requirements of the *Lead Limit Test,* page 929, using 5 mcg. of lead ion (Pb) in the control.

Sulfur dioxide. Determine as directed in the general method, page 964.

TESTS (**Additional Specifications**)

Manganese. Determine the residual manganese in bleached starch prepared with potassium permanganate as directed in the general method, page 961.

Acetyl groups. Determine the content of acetyl groups in starch acetate, acetylated distarch adipate, and acetylated distarch glycerol as directed in the general method, page 959.

Phosphate. Determine the residual phosphate (calculated as P) in starch phosphate, distarch phosphate, and phosphated distarch phosphate as directed in the general method, page 961.

Propylene chlorohydrin. Determine the residual propylene chlorohydrin in hydroxypropyl distarch glycerol, hydroxypropyl starch, hydroxypropyl starch phosphate, and oxidized hydroxypropyl starch as directed in the general method, page 962.

Packaging and storage. Store in well-closed containers.

Functional use in foods. Thickener; colloidal stabilizer; binder.

FORMIC ACID

HCOOH

CH_2O_2 Mol. wt. 46.03

DESCRIPTION

A colorless, *highly corrosive* liquid having a characteristic pungent odor. It is miscible with water, with alcohol, with glycerin, and with ether. Its specific gravity is about 1.20.

IDENTIFICATION

To 5 ml. add 2 ml. of mercuric chloride T.S. and warm the mixture. A white precipitate of mercurous chloride forms.

SPECIFICATIONS

Assay. Not less than 85.0 percent of CH_2O_2.

Dilution test. Passes test.

Limits of Impurities

Acetic acid. Not more than 0.4 percent.

Arsenic (as As). Not more than 3 parts per million (0.0003 percent).

Heavy metals (as Pb). Not more than 10 parts per million (0.001 percent).

Sulfate. Not more than 40 parts per million (0.004 percent).

TESTS

Assay. Tare a small glass-stoppered Erlenmeyer flask containing about 15 ml. of water. Transfer about 1.5 ml. of the sample into the flask and weigh. Dilute the solution of the sample to 50 ml., add phenolphthalein T.S., and titrate with 1 *N* sodium hydroxide. Each ml. of 1 *N* sodium hydroxide is equivalent to 46.03 mg. of CH_2O_2.

Dilution test. Dilute 1 volume of the sample with 3 volumes of water. No turbidity is observed within 1 hour.

Acetic acid. Dilute 1 ml. (1.2 grams) to 100 ml. with water, transfer 50 ml. of this solution into a 250-ml. boiling flask, and add 5 grams of yellow mercuric oxide. Boil the mixture under a reflux condenser with continuous stirring for 2 hours, cool, filter, and wash the residue with about 25 ml. of water. To the combined filtrate and washings add phenolphthalein T.S. and titrate with 0.02 *N* sodium hydroxide. Not more than 2.0 ml. of 0.02 *N* sodium hydroxide is required to produce a pink color.

Arsenic. A solution of 1 gram in 10 ml. of water meets the requirements of the *Arsenic Test*, page 865.

Heavy metals. To 1.7 ml. (2 grams) in a beaker add about 10 mg. of sodium carbonate, evaporate to dryness on a steam bath, and dissolve the residue in 25 ml. of water. This solution meets the requirements of the *Heavy Metals Test*, page 920, using 20 mcg. of lead ion (Pb) in the control (*Solution A*).

Sulfate, page 879. To 2.1 ml. (2.5 grams) in a beaker add about 10 mg. of sodium carbonate and evaporate to dryness on a steam bath. Any turbidity produced by the residue does not exceed that shown in a control containing 100 mcg. of sulfate (SO_4).

Packaging and storage. Store in tight containers.

Functional use in foods. Flavoring adjunct.

FUMARIC ACID

trans-Butenedioic Acid; *trans*-1,2-Ethylenedicarboxylic Acid

HOOCCH
‖
HCCOOH

$C_4H_4O_4$ Mol. wt. 116.07

DESCRIPTION

White, odorless granules or crystalline powder. It is soluble in alcohol, slightly soluble in water and in ether, and very slightly soluble in chloroform.

SPECIFICATIONS

Assay. Not less than 99.5 percent of $C_4H_4O_4$, calculated on the anhydrous basis.

Limits of Impurities

Arsenic (as As). Not more than 3 parts per million (0.0003 percent).

Heavy metals (as Pb). Not more than 10 parts per million (0.001 percent).

Maleic Acid. Not more than 0.1 percent.

Residue on ignition. Not more than 0.1 percent.

Water. Not more than 0.5 percent.

TESTS

Assay. Transfer about 1 gram, accurately weighed, into a 250-ml. Erlenmeyer flask, add 50 ml. of methanol, and dissolve the sample by warming gently on a steam bath. Cool, add phenolphthalein T.S., and titrate with 0.5 *N* sodium hydroxide to the first appearance of a pink color that persists for at least 30 seconds. Perform a blank determination (see page 2) and make any necessary correction. Each ml. of 0.5 *N* sodium hydroxide is equivalent to 29.02 mg. of $C_4H_4O_4$.

Arsenic. A *Sample Solution* prepared as directed for organic compounds meets the requirements of the *Arsenic Test*, page 865.

Heavy metals. Dissolve 2 grams in a mixture of 10 ml. of water and 15 ml. of ammonia T.S. This solution meets the requirements of the *Heavy Metals Test*, page 920, using 20 mcg. of lead ion (Pb) in the control (*Solution A*).

Maleic Acid

Buffer Solution. Dissolve 53.5 grams of ammonium chloride in about 900 ml. of water, adjust the pH to 8.2 with approximately 0.3 *N* ammonium hydroxide, and dilute with water to 1000 ml.

Standard Solution. Transfer to a 100-ml. volumetric flask about 100 mg., accurately weighed, of maleic acid of the highest purity available, dissolve in about 10 ml. of water, then dilute to volume with water and mix.

Sample Solution. Transfer about 50 grams of the sample, accurately weighed, into a 250-ml. beaker, add 80 ml. of water, and stir for 10 minutes with a mechanical stirrer. Filter, using suction, and wash with about 40 ml. of water. Transfer the combined filtrate and washings to a 250-ml. beaker, add an additional 50-gram sample, accurately weighed, to the beaker, and repeat the stirring, filtration, and washing procedure. Transfer the combined filtrate and washings to a 250-ml. volumetric flask, add 2 drops of phenolphthalein T.S., then add sodium hydroxide T.S., with stirring, until a light pink color persists for at least 30 seconds, and dilute to volume with water.

Procedure. Transfer 10.0 ml. of the *Sample Solution* into a 100-ml. volumetric flask, add 20 ml. of *Buffer solution*, dilute to volume with water, and mix (*Solution A*). Rinse a polarographic cell with a portion of the solution, then add a suitable volume of the solution to the cell, immerse it in a water bath regulated at 24.5° to 25.5°, and deaerate by bubbling purified nitrogen through the solution for at least 6 minutes. Insert the dropping mercury electrode of a suitable polarograph, and record the polarogram from −1 to −2 volts, using a saturated calomel electrode as the reference electrode. Determine the height of the wave occurring at the half-wave potential near −1.36 volts. In the same manner polarograph a solution prepared by adding 10.0 ml. of the *Sample Solution*, 20 ml. of the *Buffer Solution*, and 2.0 ml. of the *Standard Solution* to a 100-ml. volumetric flask and diluting to volume with water (*Solution B*). Calculate the weight, in mg., of maleic acid in the total weight of sample taken by the formula $2500C \times A/(B - A)$, in which A is the wave height of *Solution A*, B is the wave height of *Solution B*, and C is the concentration, in mg. per ml., of added maleic acid in *Solution B*.

Residue on ignition. Ignite 2 grams as directed in the general method, page 945.

Water. Determine by the *Karl Fischer Titrimetric Method*, page 977.

Packaging and storage. Store in well-closed containers.

Functional use in foods. Acidifier; flavoring agent.

FURFURAL

2-Furaldehyde; Pyromucic Aldehyde

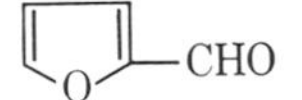

$C_5H_4O_2$ Mol. wt. 96.09

DESCRIPTION

Colorless to yellow oily liquid, which turns to reddish brown on prolonged storage. Odor typical of cyclic aldehydes. It is soluble in water and miscible with alcohol.

SPECIFICATIONS

Assay. Not less than 96.0 percent of $C_5H_4O_2$.

Refractive index. Between 1.522 and 1.528 at 20°.

Specific gravity. Between 1.154 and 1.158.

Limits of Impurities

Acid value. Not more than 1.0.

Heavy metals (as Pb). Not more than 40 parts per million (0.004 percent).

TESTS

Assay. Weigh accurately about 1.5 grams, and proceed as directed under *Aldehydes*, page 894, using 48.05 as the equivalence factor (E) in the calculation.

Refractive index, page 945. Determine with an Abbé or other refractometer of equal or greater accuracy.

Specific gravity. Determine by any reliable method (see page 5).

Acid value. Determine as directed in the general method, page 893.

Heavy metals. Prepare and test a 500-mg. sample as directed in *Method II* under the *Heavy Metals Test*, page 920, using 20 mcg. of lead ion (Pb) in the control (*Solution A*).

Packaging and storage. Store in tight containers, preferably glass, protected from light. Any material which has been stored for a prolonged period should be redistilled.

Functional use in foods. Flavoring agent.

GARLIC OIL

DESCRIPTION

The volatile oil obtained by steam distillation from the crushed bulbs or cloves of the common garlic plant, *Allium sativum* L. (Fam. *Liliaceae*). It is a clear reddish orange liquid, having a strong pungent odor, and flavor characteristic of garlic. It is soluble in most fixed oils and in mineral oil. It may be incompletely soluble in alcohol. It is insoluble in glycerin and in propylene glycol.

SPECIFICATIONS

Refractive index. Between 1.559 and 1.579 at 20°.

Specific gravity. Between 1.040 and 1.090.

TESTS

Refractive index, page 945. Determine with an Abbé or other refractometer of equal or greater accuracy.

Specific gravity. Determine by any reliable method (see page 5).

Packaging and storage. Store in full, tight, preferably glass, or aluminum containers in a cool place protected from light.

Functional use in foods. Flavoring agent.

GERANIOL

trans-3,7-Dimethyl-2,6-octadien-1-ol

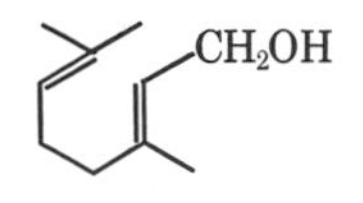

$C_{10}H_{18}O$ Mol. wt. 154.25

DESCRIPTION

Geraniol is the main constituent of many essential oils. It may be obtained by fractional distillation from Java type citronella and other essential oils, or by chemical synthesis. It is a colorless liquid having a rose-like odor. It is soluble in most fixed oils, in mineral oil, and in propylene glycol, but it is insoluble in glycerin. Some commercial grades of geraniol may be composed almost entirely of the specific chemical whose structural formula is shown above, while other commercial grades that conform to the limits of these specifications and of satisfactory purity may contain other isomeric and closely related terpenic alcohols.

SPECIFICATIONS

Assay. Not less than 88.0 percent of total alcohols, calculated as geraniol ($C_{10}H_{18}O$).

Angular rotation. Between −3° and +2°.

Refractive index. Between 1.469 and 1.478 at 20°.

Solubility in alcohol. Passes test.

Specific gravity. Between 0.870 and 0.885.

Limits of Impurities

Aldehydes. Not more than 1 percent, calculated as citronellal ($C_{10}H_{18}O$).

Esters. Not more than 1 percent, calculated as geranyl acetate ($C_{12}H_{20}O_2$).

TESTS

Assay. Proceed as directed under *Total Alcohols,* page 893. Weigh accurately about 1.2 grams of the acetylated alcohol for the saponification, and use 77.13 as the equivalence factor (*e*) in the calculation.

Angular rotation. Determine in a 100-mm. tube as directed under *Optical Rotation,* page 939.

Refractive index, page 945. Determine with an Abbé or other refractometer of equal or greater accuracy.

Solubility in alcohol. Proceed as directed in the general method, page 899. One ml. dissolves in 3 ml. of 70 percent alcohol. It remains soluble on dilution up to 10 ml.

Specific gravity. Determine by any reliable method (see page 5).

Aldehydes. Weigh accurately about 5 grams, and proceed as directed under *Aldehydes and Ketones-Hydroxylamine Method*, page 894, using 77.13 as the equivalence factor (E) in the calculation.

Esters. Weigh accurately about 5 grams, and proceed as directed under *Ester Determination*, page 896, using 98.15 as the equivalence factor (e) in the calculation.

Packaging and storage. Store in full, tight, preferably glass or tin-lined containers in a cool place protected from light.

Functional use in foods. Flavoring agent.

GERANIUM OIL, ALGERIAN

Rose Geranium Oil, Algerian

DESCRIPTION

The oil obtained by steam distillation from the leaves of *Pelargonium graveolens* L'Her. (Fam. *Geraniaceae*). It is a light yellow to deep yellow liquid, having a characteristic odor resembling rose and geraniol. It is soluble in most fixed oils and soluble, usually with opalescence, in mineral oil, and in propylene glycol. It is practically insoluble in glycerin.

SPECIFICATIONS

Assay. Not less than 13.0 percent and not more than 29.5 percent of esters, calculated as geranyl tiglate ($C_{15}H_{24}O_2$).

Acid value. Between 1.5 and 9.5.

Angular rotation. Between $-7°$ and $-13°$.

Ester value after acetylation. Between 203 and 234.

Refractive index. Between 1.464 and 1.472 at 20°.

Solubility in alcohol. Passes test.

Specific gravity. Between 0.886 and 0.898.

TESTS

Assay. Proceed as directed under *Ester Value*, page 897, using about 6 grams, accurately weighed. The ester value multiplied by 0.422 equals the percent of geranyl tiglate ($C_{15}H_{24}O_2$).

Acid value. Determine as directed in the general method, page 893, using about 5 grams, accurately weighed, and modifying the procedure by using 15 ml. of water, instead of alcohol, as diluent, and by agitating the mixture thoroughly during the titration to keep the oil in suspension.

Angular rotation. Determine in a 100-mm. tube as directed under *Optical Rotation*, page 939.

Ester value after acetylation. Proceed as directed under *Total Alcohols*, page 893, using about 1.9 grams of the acetylated oil, accurately weighed, for saponification. Calculate the ester value after acetylation by the formula $A \times 28.05/B$, in which A is the number of ml. of 0.5 N alcoholic potassium hydroxide consumed in the saponification, and B is the weight of acetylated oil in grams.

Refractive index, page 945. Determine with an Abbé or other refractometer of equal or greater accuracy.

Solubility in alcohol. Proceed as directed in the general method, page 899. One ml. dissolves in 3 ml. of 70 percent alcohol, but on further dilution with the alcohol opalescence may occur, sometimes followed by separation of paraffin particles.

Specific gravity. Determine by any reliable method (see page 5).

Packaging and storage. Store in full, tight, preferably glass, aluminum, or tin-lined containers in a cool place protected from light.

Functional use in foods. Flavoring agent.

GERANYL ACETATE

3,7-Dimethyl-2,6-octadien-1-yl Acetate

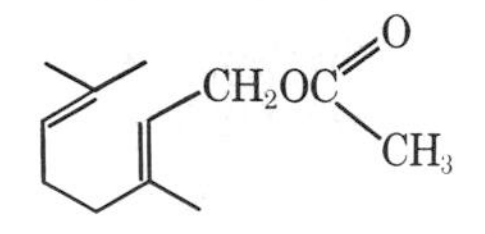

$C_{12}H_{20}O_2$ Mol. wt. 196.29

DESCRIPTION

Geranyl acetate is found in the oils of *Daucus carota* L., *Eucalyptus macarthurii* Deane, and other oils. It is obtained from geraniol by acetylation. It is a colorless liquid, having a pleasant flowery odor. It is soluble in most fixed oils and in mineral oil. It is moderately soluble in propylene glycol, but it is insoluble in glycerin. Some commercial grades of geranyl acetate may be composed almost entirely of the specific chemical whose structural formula is shown above, while other commercial grades that conform to the limits of these specifications and of satisfactory purity may contain other isomeric and closely related terpenic esters.

SPECIFICATIONS

Assay. Not less than 90.0 percent of total esters, calculated as geranyl acetate ($C_{12}H_{20}O_2$).

Angular rotation. Between −2.0° and +3.0°.

Refractive index. Between 1.458 and 1.464 at 20°.

Solubility in alcohol. Passes test.

Specific gravity. Between 0.900 and 0.914.

TESTS

Assay. Weigh accurately about 1 gram, and proceed as directed under *Ester Determination*, page 896, using 98.15 as the equivalence factor (*e*) in the calculation.

Angular rotation. Determine in a 100-mm. tube at 20° as directed under *Optical Rotation*, page 939.

Refractive index, page 945. Determine with an Abbé or other refractometer of equal or greater accuracy.

Solubility in alcohol. Proceed as directed in the general method, page 899. One ml. dissolves in 8 ml. of 70 percent alcohol.

Specific gravity. Determine by any reliable method (see page 5).

Packaging and storage. Store in full, tight, preferably glass, aluminum, or tin-lined containers in a cool place protected from light.

Functional use in foods. Flavoring agent.

GERANYL BENZOATE

3,7-Dimethyl-2,6-octadien-1-yl Benzoate

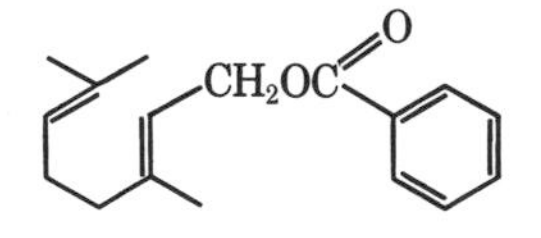

$C_{17}H_{22}O_2$ Mol. wt. 258.36

DESCRIPTION

A slightly yellowish liquid having a characteristic floral odor resembling that of ylang ylang oil. It is miscible with alcohol and with chloroform, but is practically insoluble in water. Some commercial grades of geranyl benzoate may be composed almost entirely of the specific chemical whose structural formula is shown above, while other commercial grades that conform to the limits of these specifications and of satisfactory purity may contain other isomeric and closely related terpenic esters.

SPECIFICATIONS

Assay. Not less than 95.0 percent of $C_{17}H_{22}O_2$.

Angular rotation. Between +1° and −1°.

Refractive index. Between 1.513 and 1.518 at 20°.

Solubility in alcohol. Passes test.

Specific gravity. Between 0.978 and 0.984.

Limits of Impurities

Acid value. Not more than 1.0.

TESTS

Assay. Weigh accurately about 1.5 grams, and proceed as directed under *Ester Determination*, page 896, using 129.2 as the equivalence factor (e) in the calculation.

Angular rotation. Determine in a 100-mm. tube as directed under *Optical Rotation*, page 939.

Refractive index, page 945. Determine with an Abbé or other refractometer of equal or greater accuracy.

Solubility in alcohol. Proceed as directed in the general method, page 899. One ml. dissolves in 4 ml. of 90 percent alcohol to form a clear solution.

Specific gravity. Determine by any reliable method (see page 5).

Acid value. Determine as directed in the general method, page 893.

Packaging and storage. Store in full, tight, preferably glass, tin-lined or other suitably lined containers in a cool place protected from light.

Functional use in foods. Flavoring agent.

GERANYL BUTYRATE

3,7-Dimethyl-2,6-octadien-1-yl Butyrate

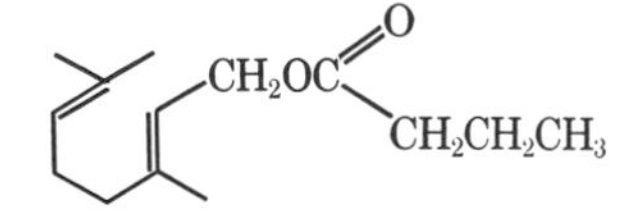

$C_{14}H_{24}O_2$ Mol. wt. 224.34

DESCRIPTION

A colorless to pale yellow liquid having a rose or fruit-like odor. It is soluble in most fixed oils and in mineral oil. It is insoluble in glycerin and in propylene glycol. Some commercial grades of geranyl butyrate may be composed almost entirely of the specific chemical whose structural formula is shown above, while other commercial grades that conform to the limits of these specifications and of satisfactory purity may contain other isomeric and closely related terpenic esters.

SPECIFICATIONS

Assay. Not less than 92.0 percent of $C_{14}H_{24}O_2$.

Refractive index. Between 1.455 and 1.462 at 20°.

Solubility in alcohol. Passes test.

Specific gravity. Between 0.889 and 0.904.

Limits of Impurities

Acid value. Not more than 1.0.

TESTS

Assay. Weigh accurately about 1 gram, and proceed as directed under *Ester Determination*, page 896, using 112.2 as the equivalence factor (*e*) in the calculation.

Refractive index, page 945. Determine with an Abbé or other refractometer of equal or greater accuracy.

Solubility in alcohol. Proceed as directed in the general method, page 899. One ml. dissolves in 4 ml. of 80 percent alcohol.

Specific gravity. Determine by any reliable method (see page 5).

Acid value. Determine as directed in the general method, page 893.

Packaging and storage. Store in full, tight, preferably glass, tin, or other suitably lined containers in a cool place protected from light.

Functional use in foods. Flavoring agent.

GERANYL FORMATE

3,7-Dimethyl-2,6-octadlen-1-yl Formate

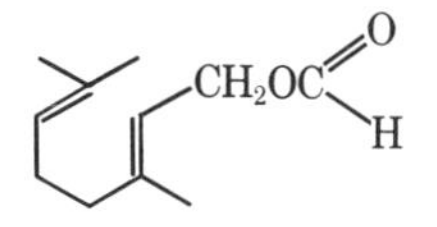

$C_{11}H_{18}O_2$ Mol. wt. 182.26

DESCRIPTION

A colorless to pale yellow liquid having a fresh, leafy rose odor. It is soluble in most fixed oils and in mineral oil. It is insoluble in glycerin and in propylene glycol. Some commercial grades of geranyl formate may be composed almost entirely of the specific chemical whose structural formula is shown above, while other commercial grades that conform to the limits of these specifications and of satisfactory purity may contain other isomeric and closely related terpenic esters.

SPECIFICATIONS

Assay. Not less than 85.0 percent of $C_{11}H_{18}O_2$.

Refractive index. Between 1.457 and 1.466 at 20°.

Solubility in alcohol. Passes test.

Specific gravity. Between 0.906 and 0.914.

Limits of Impurities

Acid value. Not more than 3.0.

TESTS

Assay. Weigh accurately about 1 gram, and proceed as directed under *Ester Determination*, page 896, using 91.13 as the equivalence factor (e) in the calculation.

Refractive index, page 945. Determine with an Abbé or other refractometer of equal or greater accuracy.

Solubility in alcohol. Proceed as directed in the general method, page 899. One ml. dissolves in 3 ml. of 80 percent alcohol.

Specific gravity. Determine by any reliable method (see page 5).

Acid value. Determine as directed in the general method, page 893.

Packaging and storage. Store in full, tight, preferably glass, tin, or other suitably lined containers in a cool place protected from light.

Functional use in foods. Flavoring agent.

GERANYL PHENYLACETATE

3,7-Dimethyl-2,6-octadien-1-yl Phenylacetate

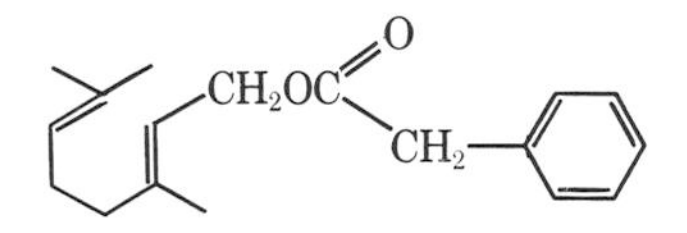

$C_{18}H_{24}O_2$ Mol. wt. 272.39

DESCRIPTION

A yellow liquid having a honey-rose odor. It is miscible with alcohol, with chloroform, and with ether, but is practically insoluble in water. Some commercial grades of geranyl phenylacetate may be composed almost entirely of the specific chemical whose structural formula is shown above, while other commercial grades that conform to the limits of these specifications and of satisfactory purity may contain other isomeric and closely related terpenic esters.

SPECIFICATIONS

Assay. Not less than 97.0 percent of $C_{18}H_{24}O_2$.

Angular rotation. Between +1° and −1°.

Refractive index. Between 1.507 and 1.511 at 20°.

Solubility in alcohol. Passes test.

Specific gravity. Between 0.971 and 0.978.

Limits of Impurities

Acid value. Not more than 2.0.

TESTS

Assay. Weigh accurately about 1.6 grams, and proceed as directed under *Ester Determination*, page 896, using 136.2 as the equivalence factor (e) in the calculation.

Angular rotation. Determine in a 100-mm. tube as directed under *Optical Rotation*, page 939.

Refractive index, page 945. Determine with an Abbé or other refractometer of equal or greater accuracy.

Solubility in alcohol. Proceed as directed in the general method, page 899. One ml. dissolves in 4 ml. of 90 percent alcohol to form a clear solution.

Specific gravity. Determine by any reliable method (see page 5).

Acid value. Determine as directed in the general method, page 893.

Packaging and storage. Store in full, tight, preferably glass, tin-lined or other suitably lined containers in a cool place protected from light.

Functional use in foods. Flavoring agent.

GERANYL PROPIONATE

3,7-Dimethyl-2,6-octadien-1-yl Propionate

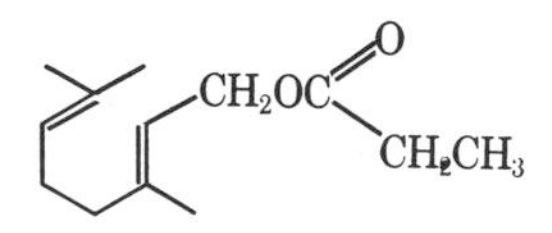

$C_{13}H_{22}O_2$ Mol. wt. 210.32

DESCRIPTION

An almost colorless liquid having a fruity, somewhat floral odor. It is soluble in most fixed oils and in mineral oil. It is insoluble in glycerin and in propylene glycol. Some commercial grades of geranyl propionate may be composed almost entirely of the specific chemical whose structural formula is shown above, while other commercial grades that conform to the limits of these specifications and of satisfactory purity may contain other isomeric and closely related terpenic esters.

SPECIFICATIONS

Assay. Not less than 92.0 percent of $C_{13}H_{22}O_2$.

Refractive index. Between 1.456 and 1.464 at 20°.

Solubility in alcohol. Passes test.

Specific gravity. Between 0.896 and 0.913.

Limits of Impurities

Acid value. Not more than 1.0.

TESTS

Assay. Weigh accurately about 1.2 grams, and proceed as directed under *Ester Determination*, page 896, using 105.2 as the equivalence factor (*e*) in the calculation.

Refractive index, page 945. Determine with an Abbé or other refractometer of equal or greater accuracy.

Solubility in alcohol. Proceed as directed in the general method, page 899. One ml. dissolves in 4 ml. of 80 percent alcohol.

Specific gravity. Determine by any reliable method (see page 5).

Acid value. Determine as directed in the general method, page 893.

Packaging and storage. Store in full, tight, preferably glass, aluminum, tin, or other suitably lined containers in a cool place protected from light.

Functional use in foods. Flavoring agent.

GIBBERELLIC ACID

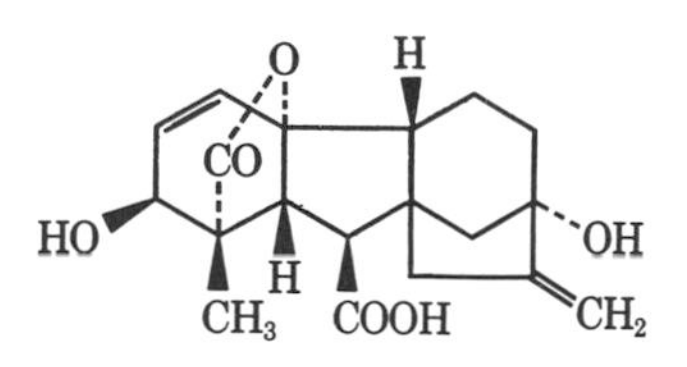

$C_{19}H_{22}O_6$ Mol. wt. 346.37

DESCRIPTION

A white to pale yellow, odorless or practically odorless, crystalline powder. It is slightly soluble in water and is soluble in alcohol and in acetone. It melts at about 234°.

IDENTIFICATION

Dissolve a few mg. of the sample in 2 ml. of sulfuric acid. A reddish solution having a green fluorescence is formed.

SPECIFICATIONS

Assay. Not less than 90.0 percent of $C_{19}H_{22}O_6$.

Specific rotation, $[\alpha]_D^{25°}$. Between +75.0° and +85.0°.

Limits of Impurities

Arsenic (as As). Not more than 3 parts per million (0.0003 percent).

Heavy metals (as Pb). Not more than 40 parts per million (0.004 percent).

Lead. Not more than 10 parts per million (0.001 percent).
Loss on drying. Not more than 1 percent.

TESTS

Assay

Standard preparation. Transfer an accurately weighed quantity of F.C.C. Gibberellic Acid Reference Standard, equivalent to about 25 mg. of pure gibberellic acid (corrected for phase purity and volatiles content), into a 50-ml. volumetric flask, dissolve in methanol, dilute to volume with methanol, and mix. Transfer 10.0 ml. of this solution into a second 50-ml. volumetric flask, dilute to volume with methanol, and mix.

Assay preparation. Transfer about 40 mg. of the sample, accurately weighed, into a 50-ml. volumetric flask, dissolve in methanol, dilute to volume with methanol, and mix. Transfer 10.0 ml. of this solution into a 100-ml. volumetric flask, dilute to volume with methanol, and mix.

Procedure. Transfer 5.0 ml. of the *Assay preparation* into a 25 × 200-mm. glass-stoppered tube, and transfer 4.0-ml. and 5.0-ml. portions of the *Standard preparation* into separate similar tubes. Place the tubes in a boiling water bath, evaporate to dryness, and then dry in an oven at 90° for 10 minutes. Remove the tubes from the oven, stopper, and allow to cool to room temperature. Dissolve the residue in each tube in 10.0 ml. of dilute sulfuric acid (8 in 10), heat in a boiling water bath for 10 minutes, and then cool in a 10° water bath for 5 minutes. Determine the absorbance of the solutions in 1-cm. cells at 535 mμ with a suitable spectrophotometer, using the dilute sulfuric acid as the blank. Record the absorbance of the solution from the *Assay preparation* as A_U. Note the absorbance of the two solutions prepared from the 4.0-ml. and 5.0-ml. aliquots of the *Standard preparation*, and record the absorbance of the final solution giving the value nearest to that of the sample as A_S; record as V the volume of the aliquot used in preparing this solution. Calculate the quantity, in mg., of $C_{19}H_{22}O_6$ in the sample taken by the formula $500C \times (V/5) \times (A_U/A_S)$, in which C is the exact concentration, in mg. per ml., of the *Standard preparation*.

Specific rotation, page 939. Determine in a solution in alcohol containing 100 mg. in each ml.

Arsenic. A *Sample Solution* prepared as directed for organic compounds meets the requirements of the *Arsenic Test*, page 865.

Heavy metals. Prepare and test a 500-mg. sample as directed in *Method II* under the *Heavy Metals Test*, page 920, using 20 mcg. of lead ion (Pb) in the control (*Solution A*).

Lead. A *Sample Solution* prepared as directed for organic compounds meets the requirements of the *Lead Limit Test*, page 929, using 10 mcg. of lead ion (Pb) in the control.

Loss on drying, page 931. Dry at 100° in vacuum for 7 hours.

Packaging and storage. Store in well-closed containers.

Functional use in foods. Enzyme activator.

GINGER OIL

DESCRIPTION

The volatile oil obtained by steam distillation of the dried ground rhizome of *Zingiber officianale*, Roscoe (Fam. *Zingiberaceae*). It is a light yellow to yellow liquid, having the aromatic, characteristic odor of ginger. It is soluble in most fixed oils and in mineral oil. It is soluble, usually with turbidity, in alcohol, but it is insoluble in glycerin and in propylene glycol.

SPECIFICATIONS

Angular rotation. Between −28° and −45°.

Refractive index. Between 1.488 and 1.494 at 20°.

Saponification value. Not more than 20.

Specific gravity. Between 0.871 and 0.882.

TESTS

Angular rotation. Determine in a 100-mm. tube as directed under *Optical Rotation*, page 939.

Refractive index, page 945. Determine with an Abbé or other refractometer of equal or greater accuracy.

Saponification value. Determine as directed in the general method, page 896, using about 5 grams, accurately weighed.

Specific gravity. Determine by any reliable method (see page 5).

Packaging and storage. Store in full, tight, glass, aluminum, or tin-lined containers in a cool place protected from light.

Functional use in foods. Flavoring agent.

GLUCONO DELTA-LACTONE

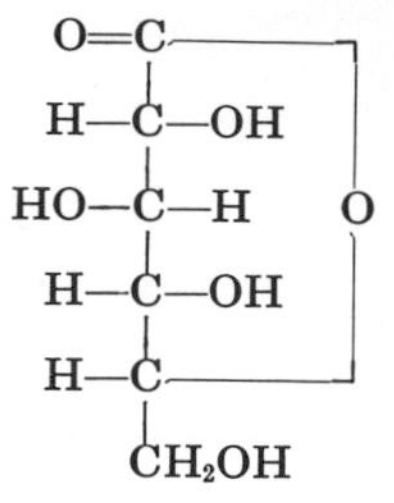

$C_6H_{10}O_6$ Mol. wt. 178.14

DESCRIPTION

A fine, white, practically odorless, crystalline powder. It is freely soluble in water and is sparingly soluble in alcohol. It decomposes at about 153°.

IDENTIFICATION

Dissolve about 500 mg. in 5 ml. of warm water in a test tube, add 1 ml. of freshly distilled phenylhydrazine, heat on a steam bath for 30 minutes, and allow to cool. Induce crystallization, if necessary, by scratching the inner surface of the tube with a glass rod. Crystals of gluconic acid phenylhydrazide form.

SPECIFICATIONS

Assay. Not less than 99.0 percent of $C_6H_{10}O_6$.

Limits of Impurities

Arsenic (as As). Not more than 3 parts per million (0.0003 percent).

Heavy metals (as Pb). Not more than 20 parts per million (0.002 percent).

Lead. Not more than 10 parts per million (0.001 percent).

Reducing substances (as D-glucose). Not more than 0.5 percent.

TESTS

Assay. Dissolve about 6 grams, accurately weighed, in 100 ml. of water in a 300-ml. Erlenmeyer flask, add 50.0 ml. of 1 *N* sodium hydroxide, and allow to stand for 15 minutes. Add phenolphthalein T.S., and titrate the excess alkali with 1 *N* hydrochloric acid. Perform a blank determination (see page 2). Each ml. of 1 *N* hydrochloric acid is equivalent to 178.1 mg. of $C_6H_{10}O_6$.

Arsenic. A *Sample Solution* prepared as directed for organic compounds meets the requirements of the *Arsenic Test*, page 865.

Heavy metals. A solution of 1 gram in 25 ml. of water meets the requirements of the *Heavy Metals Test*, page 920, using 20 mcg. of lead ion (Pb) in the control (*Solution A*).

Lead. A *Sample Solution* prepared as directed for organic compounds meets the requirements of the *Lead Limit Test*, page 929, using 10 mcg. of lead ion (Pb) in the control.

Reducing substances. Weigh accurately 10.0 grams into a 400-ml. beaker, dissolve the sample in 40 ml. of water, add phenolphthalein T.S., and neutralize with sodium hydroxide solution (1 in 2). Dilute to 50.0 ml. with water, and add 50 ml. of alkaline cupric tartrate T.S. Heat the mixture on an asbestos gauze over a Bunsen burner, regulating the flame so that boiling begins in 4 minutes, and continue the boiling for exactly 2 minutes. Filter through a Gooch crucible, wash the filter with 3 or more small portions of water, and place the crucible in an upright position in the original beaker. Add 5 ml. of water and 3 ml. of nitric acid to the crucible, mix with a stirring rod to ensure complete solution of the cuprous oxide, and wash the solution into a beaker with several ml. of water. To the beaker add sufficient bromine T.S. (5 to 10 ml.) until the color becomes yellow, and dilute with water to about 75 ml. Add a few glass beads, boil over a Bunsen burner until the bromine is completely removed, and cool. Slowly add ammonium hydroxide until a deep blue color appears, then adjust the pH to approximately 4 with glacial acetic acid, and dilute to about 100 ml. with water. Add 4 grams of potassium iodide, and titrate with 0.1 *N* sodium thiosulfate, adding starch T.S. just before the end-point is reached. Not more than 16.1 ml. is required.

Packaging and storage. Store in well-closed containers.

Functional use in foods. Acid; leavening agent; sequestrant.

L-GLUTAMIC ACID

Glutamic acid; L-2-Aminopentanedioic Acid

$HOOCCH_2CH_2CH(NH_2)COOH$

$C_5H_9NO_4$ Mol. wt. 147.13

DESCRIPTION

A white, practically odorless, free-flowing, crystalline powder. It is slightly soluble in water, forming acidic solutions. The pH of a saturated solution is about 3.2.

IDENTIFICATION

A. Dissolve about 150 mg. in a mixture of 4 ml. of water and 1 ml. of sodium hydroxide T.S., add 1 ml. of ninhydrin T.S. and 100 mg. of sodium acetate, and heat in a boiling water bath for 10 minutes. An intense, violet-blue color is formed.

B. The glutamic acid dissolves completely on stirring when either 5.6 ml. of 1 *N* hydrochloric acid or 6.8 ml. of 1 *N* sodium hydroxide is added to a suspension of 1 gram of the sample in 9 ml. of water.

SPECIFICATIONS

Assay. Not less than 99.0 percent of $C_5H_9NO_4$.

Specific rotation. $[\alpha]^{25°}_{546.1\ m\mu}$: Between +37.7° and +38.5°; $[\alpha]^{20°}_{D}$: between +31.5° and +32.2°.

Limits of Impurities

Arsenic (as As). Not more than 3 parts per million (0.0003 percent).

Chloride. Not more than 0.2 percent.

Heavy metals (as Pb). Not more than 20 parts per million (0.002 percent).

Lead. Not more than 10 parts per million (0.001 percent).

Loss on drying. Not more than 0.1 percent.

TESTS

Assay. Dissolve about 500 mg., accurately weighed, in 250 ml. of water, add bromothymol blue T.S., and titrate with 0.1 *N* sodium hydroxide to a blue end-point. Each ml. of 0.1 *N* sodium hydroxide is equivalent to 14.71 mg. of $C_5H_9NO_4$.

Specific rotation, page 939. $[\alpha]^{25°}_{546.1\ m\mu}$: Determine in a solution containing 11.8 grams in sufficient 1.5 *N* hydrochloric acid to make 100 ml.; $[\alpha]^{20°}_{D}$: determine in a solution containing 10 grams in sufficient 2 *N* hydrochloric acid to make 100 ml.

Arsenic. A *Sample Solution* prepared as directed for organic compounds meets the requirements of the *Arsenic Test,* page 865.

Chloride, page 879. Any turbidity produced by a 10-mg. sample does not exceed that shown in a control containing 20 mcg. of chloride ion (Cl).

Heavy metals. Prepare and test a 1-gram sample as directed in *Method II* under the *Heavy Metals Test,* page 920, using 20 mcg. of lead ion (Pb) in the control (*Solution A*).

Lead. A *Sample Solution* prepared as directed for organic compounds meets the requirements of the *Lead Limit Test,* page 929, using 10 mcg. of lead ion (Pb) in the control.

Loss on drying, page 931. Dry at 85° for 3 hours.

Packaging and storage. Store in well-closed containers.

Functional use in foods. Salt substitute; nutrient; dietary supplement.

GLUTAMIC ACID HYDROCHLORIDE

2-Aminopentanedioic Acid Hydrochloride

$$\left[\begin{array}{l}HOOC—CH_2—CH_2—\underset{\substack{| \\ {}^{+}NH_3}}{CH}—COOH\end{array}\right]Cl^-$$

$C_5H_9NO_4.HCl$ Mol. wt. 183.59

DESCRIPTION

A white, crystalline powder. One gram dissolves in about 3 ml. of water. It is almost insoluble in alcohol and in ether. Its solutions are acid to litmus.

IDENTIFICATION

A. To 1 ml. of a 1 in 3 solution add 1 ml. of barium hydroxide solution (1 in 50), filter, and add 10 ml. of alcohol. A white, crystalline precipitate of barium glutamate forms on standing.

B. To 1 ml. of a 1 in 30 solution add 1 ml. of ninhydrin T.S. and 100 mg. of sodium acetate, and boil for 10 minutes. An intense violet-blue color is produced.

SPECIFICATIONS

Assay. Not less than 99.0 percent and not more than the equivalent of 101.0 percent of $C_5H_9NO_4.HCl$ after drying.

Specific rotation. $[\alpha]^{25°}_{546.1\ m\mu}$: Between +30.2° and +30.9°; $[\alpha]^{20°}_{D}$: between +25.2° and +25.8°.

Limits of Impurities

Arsenic (as As). Not more than 3 parts per million (0.0003 percent).

Heavy metals (as Pb). Not more than 20 parts per million (0.002 percent).

Lead. Not more than 10 parts per million (0.001 percent).

Loss on drying. Not more than 0.5 percent.

Readily carbonizable substances. Passes test.

Residue on ignition. Not more than 0.25 percent.

TESTS

Assay. Dissolve about 300 mg., previously dried at 80° for 4 hours and accurately weighed, in 50 ml. of water, add bromothymol blue T.S., and titrate with 0.1 *N* sodium hydroxide. Each ml. of 0.1 *N* sodium hydroxide is equivalent to 9.180 mg. of $C_5H_9NO_4.HCl$.

Specific rotation, page 939. $[\alpha]^{25°}_{546.1\ m\mu}$: Determine in a solution containing 14.7 grams in sufficient 0.7 *N* hydrochloric acid to make 100 ml.; $[\alpha]^{20°}_{D}$: determine in a solution containing 10 grams in sufficient 2 *N* hydrochloric acid to make 100 ml.

Arsenic. A *Sample Solution* prepared as directed for organic compounds meets the requirements of the *Arsenic Test,* page 865.

Heavy metals. Prepare and test a 1-gram sample as directed in *Method II* under the *Heavy Metals Test*, page 920, using 20 mcg. of lead ion (Pb) in the control (*Solution A*).

Lead. A *Sample Solution* prepared as directed for organic compounds meets the requirements of the *Lead Limit Test*, page 929, using 10 mcg. of lead ion (Pb) in the control.

Loss on drying, page 931. Dry at 80° for 4 hours.

Readily carbonizable substances. Dissolve 500 mg. of the sample in 5 ml. of sulfuric acid T.S. The solution is colorless.

Residue on ignition. Ignite 1 gram as directed in the general method, page 945.

Packaging and storage. Store in well-closed, light-resistant containers.

Functional use in foods. Salt substitute; flavoring agent.

GLYCERIN

Glycerol

$CH_2OH.CHOH.CH_2OH$

$C_3H_8O_3$ Mol. wt. 92.10

DESCRIPTION

A clear, colorless, syrupy liquid, having a sweet taste. It has not more than a slight characteristic odor, which is neither harsh nor disagreeable. It is hygroscopic and its solutions are neutral. Glycerin is miscible with water and with alcohol. It is insoluble in chloroform, in ether, and in fixed and volatile oils.

IDENTIFICATION

Heat a few drops of glycerin in a test tube with about 500 mg. of potassium bisulfate. The characteristic, pungent vapors of acrolein are evolved.

SPECIFICATIONS

Assay. Not less than 95.0 percent of $C_3H_8O_3$.

Specific gravity. Not less than 1.249.

Limits of Impurities

Acrolein, glucose, and ammonium compounds. Passes test.

Arsenic (as As). Not more than 3 parts per million (0.0003 percent).

Butanetriols. Not more than 0.2 percent.

Chlorinated compounds (as Cl). Not more than 30 parts per million (0.003 percent).

Color. Passes test.

Fatty acids and esters. Passes test (limit about 0.1 percent, calculated as butyric acid).

Heavy metals (as Pb). Not more than 5 parts per million (0.0005 percent).

Readily carbonizable substances. Passes test.

Residue on ignition. Not more than 0.01 percent.

Assay

Sodium Periodate Solution. Dissolve 60 grams of sodium metaperiodate ($NaIO_4$) in sufficient water containing 120 ml. of 0.1 *N* sulfuric acid to make 1000 ml. Do not heat to dissolve the periodate. If the solution is not clear, filter through a sintered-glass filter. Store the solution in a glass-stoppered, light-resistant container. Test the suitability of this solution as follows: Pipet 10 ml. into a 250-ml. volumetric flask, dilute to volume with water, and mix. To about 550 mg. of glycerin dissolved in 50 ml. of water add 50 ml. of the diluted periodate solution with a pipet. For a blank, pipet 50 ml. of the solution into a flask containing 50 ml. of water. Allow the solutions to stand for 30 minutes, then add to each 5 ml. of hydrochloric acid and 10 ml. of potassium iodide T.S., and rotate to mix. Allow to stand for 5 minutes, add 100 ml. of water, and titrate with 0.1 *N* sodium thiosulfate, shaking continuously and adding starch T.S. near the end-point. The ratio of the volume of 0.1 *N* sodium thiosulfate required for the glycerin-periodate mixture to that required for the blank should be between 0.750 and 0.765.

Procedure. Transfer about 400 mg. of the sample, accurately weighed, into a 600-ml. beaker, dilute with 50 ml. of water, add bromothymol blue T.S., and acidify with 0.2 *N* H_2SO_4 to a definite green or greenish yellow color. Neutralize with 0.05 *N* sodium hydroxide to a definite blue end-point, free of green color. Prepare a blank containing 50 ml. of water, and neutralize in the same manner. Pipet 50 ml. of the *Sodium Periodate Solution* into each beaker, mix by swirling gently, cover with a watch glass, and allow to stand for 30 minutes at room temperature (not above 35°) in the dark or in subdued light. Add 10 ml. of a mixture consisting of equal volumes of ethylene glycol and water, and allow to stand for 20 minutes. Dilute each solution to about 300 ml. with water, and titrate with 0.1 *N* sodium hydroxide to a pH of 8.1 ± 0.1 for the sample and 6.5 ± 0.1 for the blank, using a pH meter previously calibrated with pH 4.0 Acid Phthalate Standard Buffer Solution (see page 984). Each ml. of 0.1 *N* sodium hydroxide, after correction for the blank, is equivalent to 9.210 mg. of glycerin ($C_3H_8O_3$).

Specific gravity. Determine by any reliable method (see page 5).

Acrolein, glucose, and ammonium compounds. Heat a mixture of 5 ml. of glycerin and 5 ml. of potassium hydroxide solution (1 in 10) at 60° for 5 minutes. It neither becomes yellow nor emits an odor of ammonia.

Arsenic. A *Sample Solution* prepared as directed for organic compounds meets the requirements of the *Arsenic Test,* page 865.

Butanetriols

Reagents. Reagents other than those described may be employed if resolution equivalent to that described in the *Procedure* is obtained.

Chromatographic Siliceous Earth. Use 80- to 100-mesh Chromosorb W or other comparable grade of purified chromatographic siliceous earth.

1,4-Butanediol. Purify the commercial product by vacuum distillation with a suitably packed column, collecting the portion distilling between 120° and 121° at 8 mm. of mercury.

1,2,4-Butanetriol. Purify the commercial product by vacuum distillation with a suitably packed column, collecting the portion distilling between 151° and 153° at 2 mm. of mercury.

Preparation of Column Material. Place about 500 grams of chromatographic siliceous earth in a large beaker, add sufficient 6 *N* hydrochloric acid to cover the material, and allow it to stand overnight. Decant the acid, wash the siliceous earth on a Buchner funnel with water until the wash water is neutral to pH indicator paper then wash with acetone until free from water, and spread out to dry in the air. Transfer the washed and dried material to a sintered glass funnel, cover with chloroform, stir, and remove the chloroform by aspiration. Repeat the washing with chloroform, and again dry in the air at room temperature.

Weigh 87.5 grams of the dried siliceous earth into a dish, and add sufficient acetone to form a slurry. Transfer 12.5 grams of polyoxyethylene (8) ethylenediamine into a beaker, and dissolve in acetone. Place the dish on a steam bath, and heat gently, with stirring, while adding the solution of polyoxyethylene (8) ethylenediamine. Continue heating until enough acetone has evaporated to cause the mixture to become free-flowing, and spread out to dry at room temperature. [*Note:* The column prepared with polyoxyethylene (8) ethylenediamine does not have long-term stability, particularly when used with a flame-ionization detector; it is more stable, however, when used with a thermal-ionization detector. To prolong stability in either case, the column should be kept sealed against exposure to air when not in use.]

Procedure. Weigh accurately about 10 grams of the sample, add 1 drop of *1,4-Butanediol,* accurately weighed, as the internal standard, dilute with 5 ml. of methanol, and mix. Inject a 10-μl. portion of this solution into a gas chromatographic apparatus equipped with a linear temperature programming device. The operating conditions of the apparatus may vary, depending upon the particular instrument used, but a suitable chromatogram is obtained with a copper column, 1.5 m. in length and 6.3 mm. in outside diameter, packed with the column material previously described. In addition, the carrier is helium, flowing at the rate of 100 ml. per minute; the injector block temperature is 320°, the detector block temperature is 250°, and the column temperature is programmed to rise from 150° to 180° at a rate of 5.6° per minute. The detector bridge current should be maintained at 250 ma. when the operating conditions described are employed.

The resolution factor, *R*, should be not less than 1.9 between the *threo*- and the *erythro*-butanetriols peaks, not less than 2.5 between the *erythro*-1,2,3-butanetriol and the glycerin peaks, and not less than 4.5 between the glycerin and the 1,2,4-butanetriol peaks. (These values for *R* are obtained when mixtures of equal quantities of glycerin and the butanetriols are determined in an apparatus programmed as described above.)

Prepare a 1 in 1000 solution in glycerin of *1,2,4-Butanetriol*, accurately weighed, and calculate the percent (*P*) of 1,2,4-butanetriol in the standard mixture. Weigh accurately about 10 grams of the standard mixture, add 1 drop of *1,4-Butanediol*, accurately weighed, as the internal standard, and dilute with 5 ml. of methanol. Inject about 10 μl. of this solution, and obtain a standard chromatogram under the same operating conditions as employed for the sample, applying attenuation of the detector signal as necessary. Under the conditions described, the 1,4-butanediol is eluted in about 8 minutes, and an area of about 10 $cm.^2$ is generated as compared to an area of 1.0–1.5 $cm.^2$ for the butanetriols when present in a concentration of about 0.1 percent. In addition, the following retention times have been obtained: 1.00 for 1,4-butanediol, 2.14 for *threo*-1,2,3-butanetriol, 2.52 for *erythro*-1,2,3-butanetriol, and 5.26 for 1,2,4-butanetriol. Retention times will vary if programming different from that described is used.

Measure the areas of the peaks produced by the 1,4-butanediol (a) and by the 1,2,4-butanetriol (A), and calculate the response factor (f) by the formula $WPa/100wA$, in which W is the exact weight of the standard mixture used for dilution with the methanol, and w is the exact weight of the drop of 1,4-butanediol internal standard added to the standard mixture.

Calculate the percent of each butanetriol in the sample by the formula $100fW'Ax/A'W$, in which f is the response factor previously determined, W' is the exact weight of 1,4-butanediol internal standard added to the sample solution, Ax is the area of the peak produced by each butanetriol, A' is the area of the 1,4-butanediol peak, and W is the weight of the sample. The sum of the percents found does not exceed 0.2.

Chlorinated compounds. Transfer 5.0 grams of glycerin into a dry 100-ml. round bottom, ground joint flask and add to it 15 ml. of morpholine. Connect the flask with a ground joint reflux condenser, and reflux the mixture gently for 3 hours. Rinse the condenser with 10 ml. of water, receiving the washing into the flask, and cautiously acidify with nitric acid. Transfer the solution to a suitable comparison tube, add 0.5 ml. of silver nitrate T.S., dilute with water to 50.0 ml., and mix thoroughly. Any turbidity does not exceed that produced by 150 mcg. of chloride (Cl) in an equal volume of solution containing the quantities of reagents used in the test, omitting the refluxing.

Color. The color of glycerin, when viewed downward against a white surface in a 50-ml. Nessler tube, is not darker than the color of a standard made by diluting 0.40 ml. of ferric chloride C.S. with water to 50 ml. and similarly viewed in a Nessler tube of approximately the same diameter and color as that containing the sample.

Fatty acids and esters. Mix a 40.0-ml. (50-gram) sample with 50 ml. of recently boiled water and 5.0 ml. of 0.5 *N* sodium hydroxide. Boil the mixture for 5 minutes, cool, add phenolphthalein T.S., and titrate the excess alkali with 0.5 *N* hydrochloric acid. Not more than 1 ml. of 0.5 *N* sodium hydroxide is consumed.

Heavy metals. Mix a 4.0-ml. (5-gram) sample with 2 ml. of 0.1 *N* hydrochloric acid, add water to make 25 ml., and proceed as directed under the *Heavy Metals Test*, page 920. Any color does not exceed that produced in a control (*Solution A*) containing 25 mcg. of lead ion (Pb).

Readily carbonizable substances, page 943. Rinse a glass-stoppered, 25-ml. cylinder with sulfuric acid T.S., and allow it to drain for 10 minutes. Add 5 ml. of glycerin and 5 ml. of sulfuric acid T.S., shake vigorously for 1 minute, and allow to stand for 1 hour. The mixture is no darker than *Matching Fluid H*.

Residue on ignition. Heat 50 grams in a tared, open dish, and ignite the vapors, allowing them to burn until the sample has been completely consumed. After cooling, moisten the residue with 0.5 ml. of sulfuric acid, and complete the ignition by heating for 15-minute periods at 800° ± 25° to constant weight.

Packaging and storage. Store in tight containers.

Functional use in foods. Humectant; solvent; bodying agent; plasticizer.

GLYCEROL ESTER OF PARTIALLY DIMERIZED ROSIN

DESCRIPTION

A hard, pale, amber-colored resin (color M or paler as determined by A.S.T.M. Designation D 509) produced by the esterification of partially dimerized rosin with food-grade glycerin and purified by steam stripping. It is soluble in acetone and in benzene, but is insoluble in water.

SPECIFICATIONS

Acid value. Between 3 and 8.

Drop softening point. Between 109° and 119°.

Limits of Impurities

Arsenic (as As). Not more than 3 parts per million (0.0003 percent).

Heavy metals (as Pb). Not more than 40 parts per million (0.004 percent).

Lead. Not more than 3 parts per million (0.0003 percent).

TESTS

Acid value. Determine as directed in the general procedure, page 945.

Drop softening point. Determine as directed in the general procedure, page 946, using a bath temperature of 125°.

Arsenic. Prepare a *Sample Solution* as directed in the general method under *Chewing Gum Base*, page 877. This solution meets the requirements of the *Arsenic Test*, page 865.

Heavy metals. Prepare and test a 500-mg. sample as directed in *Method II* under the *Heavy Metals Test*, page 920, using 20 mcg. of lead ion (Pb) in the control (*Solution A*).

Lead. Prepare a *Sample Solution* as directed in the general method under *Chewing Gum Base*, page 878. This solution meets the requirements of the *Lead Limit Test*, page 929, using 10 mcg. of lead ion (Pb) in the control.

Packaging and storage. Store in well-closed containers.

Functional use in foods. Masticatory substance in chewing gum base.

GLYCEROL ESTER OF PARTIALLY HYDROGENATED WOOD ROSIN

DESCRIPTION

A medium-hard, pale amber-colored resin (color N or paler as determined by A.S.T.M. Designation D 509) produced by the esterification of partially hydrogenated wood rosin with food-grade glycerin and purified by steam stripping. It is soluble in acetone and in benzene, but is insoluble in water and in alcohol.

SPECIFICATIONS

Acid value. Between 3 and 10.

Drop softening point. Between 79° and 88°.

Limits of Impurities

Arsenic (as As). Not more than 3 parts per million (0.0003 percent).

Heavy metals (as Pb). Not more than 40 parts per million (0.004 percent).

Lead. Not more than 3 parts per million (0.0003 percent).

TESTS

Acid value. Determine as directed in the general procedure, page 945.

Drop softening point. Determine as directed in the general procedure, page 946, using a bath temperature of 100°.

Arsenic. Prepare a *Sample Solution* as directed in the general method under *Chewing Gum Base*, page 877. This solution meets the requirements of the *Arsenic Test*, page 865.

Heavy metals. Prepare and test a 500-mg. sample as directed in *Method II* under the *Heavy Metals Test*, page 920, using 20 mcg. of lead ion (Pb) in the control (*Solution A*).

Lead. Prepare a *Sample Solution* as directed in the general method under *Chewing Gum Base*, page 878. This solution meets the requirements of the *Lead Limit Test*, page 929, using 10 mcg. of lead ion (Pb) in the control.

Packaging and storage. Store in well-closed containers.

Functional use in foods. Masticatory substance in chewing gum base.

GLYCEROL ESTER OF POLYMERIZED ROSIN

DESCRIPTION

A hard, pale amber-colored resin (color M or paler as determined by A.S.T.M. Designation D 509) produced by the esterification of polymerized rosin with food-grade glycerin and purified by steam stripping. It is soluble in acetone and in benzene, but is insoluble in water and in alcohol.

SPECIFICATIONS

Acid value. Between 3 and 12.

Ring-and-ball softening point. Between 80° and 126°.

Limits of Impurities

Arsenic (as As). Not more than 3 parts per million (0.0003 percent).

Heavy metals (as Pb). Not more than 40 parts per million (0.004 percent).

Lead. Not more than 3 parts per million (0.0003 percent).

TESTS

Acid value. Determine as directed in the general procedure, page 945.

Ring-and-ball softening point. Determine as directed in the general procedure, page 948.

Arsenic. Prepare a *Sample Solution* as directed in the general method under *Chewing Gum Base,* page 877. This solution meets the requirements of the *Arsenic Test,* page 865.

Heavy metals. Prepare and test a 500-mg. sample as directed in *Method II* under the *Heavy Metals Test,* page 920, using 20 mcg. of lead ion (Pb) in the control (*Solution A*).

Lead. Prepare a *Sample Solution* as directed in the general method under *Chewing Gum Base,* page 878. This solution meets the requirements of the *Lead Limit Test,* page 929, using 10 mcg. of lead ion (Pb) in the control.

Packaging and storage. Store in well-closed containers.

Functional use in foods. Masticatory substance in chewing gum base.

GLYCEROL ESTER OF TALL OIL ROSIN

DESCRIPTION

A pale amber-colored resin (color N or paler as determined by A.S.T.M. Designation D 509) produced by the esterification of tall oil rosin with food-grade glycerin and purified by steam stripping. It is soluble in acetone and in benzene, but is insoluble in water.

SPECIFICATIONS

Acid value. Between 5 and 12.

Ring-and-ball softening point. Between 80° and 88°.

Limits of Impurities

Arsenic (as As). Not more than 3 parts per million (0.0003 percent).

Heavy metals (as Pb). Not more than 40 parts per million (0.004 percent).

Lead. Not more than 3 parts per million (0.0003 percent).

TESTS

Acid value. Determine as directed in the general procedure, page 945.

Ring-and-ball softening point. Determine as directed in the general procedure, page 948.

Arsenic. Prepare a *Sample Solution* as directed in the general method under *Chewing Gum Base*, page 877. This solution meets the requirements of the *Arsenic Test*, page 865.

Heavy metals. Prepare and test a 500-mg. sample as directed in *Method II* under the *Heavy Metals Test*, page 920, using 20 mcg. of lead ion (Pb) in the control (*Solution A*).

Lead. Prepare a *Sample Solution* as directed in the general method under *Chewing Gum Base*, page 878. This solution meets the requirements of the *Lead Limit Test*, page 929, using 10 mcg. of lead ion (Pb) in the control.

Packaging and storage. Store in well-closed containers.

Functional use in foods. Masticatory substance in chewing gum base.

GLYCEROL ESTER OF WOOD ROSIN

DESCRIPTION

A hard, pale amber-colored resin (color N or paler as determined by A.S.T.M. Designation D 509) produced by the esterification of pale wood rosin with food-grade glycerin. When intended for use in chewing gum base, the product is usually purified by steam stripping, but when intended for use in adjusting the density of citrus oils for beverages, it is purified by countercurrent steam distillation. It is soluble in acetone and in benzene, but is insoluble in water.

SPECIFICATIONS

Acid value. Between 3 and 9.

Drop softening point. Between 88° and 96°.

Limits of Impurities

Arsenic (as As). Not more than 3 parts per million (0.0003 percent).

Heavy metals (as Pb). Not more than 40 parts per million (0.004 percent).

Lead. Not more than 3 parts per million (0.0003 percent).

TESTS

Acid value. Determine as directed in the general procedure, page 945.

Drop softening point. Determine as directed in the general procedure, page 946, using a bath temperature of 105°.

Arsenic. Prepare a *Sample Solution* as directed in the general method under *Chewing Gum Base*, page 877. This solution meets the requirements of the *Arsenic Test*, page 865.

Heavy metals. Prepare and test a 500-mg. sample as directed in *Method II* under the *Heavy Metals Test,* page 920, using 20 mcg. of lead ion (Pb) in the control (*Solution A*).

Lead. Prepare a *Sample Solution* as directed in the general method under *Chewing Gum Base,* page 878. This solution meets the requirements of the *Lead Limit Test,* page 920, using 10 mcg. of lead ion (Pb) in the control.

Packaging and storage. Store in well-closed containers.

Functional use in foods. Masticatory substance in chewing gum base; beverage stabilizer.

GLYCINE

Aminoacetic acid; Glycocoll

H_2NCH_2COOH

$C_2H_5NO_2$ Mol. wt. 75.07

DESCRIPTION

A white, odorless, crystalline powder, having a sweetish taste. Its solution is acid to litmus. One gram dissolves in about 4 ml. of water. It is very slightly soluble in alcohol and in ether.

IDENTIFICATION

A. To 5 ml. of a 1 in 10 solution add 5 drops of diluted hydrochloric acid T.S. and 5 drops of a solution of sodium nitrite (1 in 2). A vigorous evolution of a colorless gas is produced.

B. Add 1 ml. of ferric chloride T.S. to 2 ml. of a 1 in 10 solution. A red color is produced which disappears upon the addition of an excess of diluted hydrochloric acid T.S., and reappears upon the addition of an excess of stronger ammonia T.S.

C. To 2 ml. of a 1 in 10 solution add 1 drop of liquefied phenol and 5 ml. of sodium hypochlorite T.S. A blue color is produced.

SPECIFICATIONS

Assay. Not less than 98.5 percent and not more than the equivalent of 101.5 percent of $C_2H_5NO_2$ after drying.

Limits of Impurities

Arsenic (as As). Not more than 3 parts per million (0.0003 percent).

Heavy metals (as Pb). Not more than 20 parts per million (0.002 percent).

Lead. Not more than 5 parts per million (0.0005 percent).

Loss on drying. Not more than 0.2 percent.

Readily carbonizable substances. Passes test.
Residue on ignition. Not more than 0.1 percent.

TESTS

Assay. Transfer about 175 mg., previously dried at 105° for 2 hours and accurately weighed, to a 250-ml. flask. Dissolve the sample in 50 ml. of glacial acetic acid, add 2 drops of crystal violet T.S., and titrate with 0.1 *N* perchloric acid to a bluish green end-point. Perform a blank determination (see page 2) and make any necessary correction. Each ml. of 0.1 *N* perchloric acid is equivalent to 7.507 mg. of $C_2H_5NO_2$.

Arsenic. A *Sample Solution* prepared as directed for organic compounds meets the requirements of the *Arsenic Test*, page 865.

Heavy metals. A solution of 1 gram in 25 ml. of water meets the requirements of the *Heavy Metals Test*, page 920, using 20 mcg. of lead ion (Pb) in the control (*Solution A*).

Lead. A *Sample Solution* prepared as directed for organic compounds meets the requirements of the *Lead Limit Test*, page 929, using 5 mcg. of lead ion (Pb) in the control.

Loss on drying, page 931. Dry at 105° for 2 hours.

Readily carbonizable substances, page 943. Dissolve 500 mg. in 5 ml. of sulfuric acid T.S. The solution is no darker than *Matching Fluid A*.

Residue on ignition, page 945. Ignite 2 grams as directed in the general method.

Packaging and storage. Store in well-closed containers.
Functional use in foods. Nutrient and dietary supplement.

GRAPEFRUIT OIL, EXPRESSED

Grapefruit Oil, Coldpressed; Oil of Shaddock

DESCRIPTION

The oil obtained by expression from the fresh peel of the grapefruit, *Citrus paradisi* Macfayden (*Citrus decumana* L.) It is a yellow, sometimes reddish liquid, often showing a flocculent separation of waxy material. It is soluble in most fixed oils and in mineral oil, often with opalescence or cloudiness. It is slightly soluble in propylene glycol and insoluble in glycerin.

SPECIFICATIONS

Angular rotation. Between +91° and +96°.
Refractive index. Between 1.475 and 1.478 at 20°.

Residue on evaporation. Between 5 percent and 10 percent.
Specific gravity. Between 0.848 and 0.856.

Limits of Impurities

Arsenic (as As). Not more than 3 parts per million (0.0003 percent).
Heavy metals (as Pb). Not more than 40 parts per million (0.004 percent).
Lead. Not more than 10 parts per million (0.001 percent).

TESTS

Angular rotation. Determine in a 100-mm. tube as directed under *Optical Rotation*, page 939.

Refractive index, page 945. Determine with an Abbé or other refractometer of equal or greater accuracy.

Residue on evaporation. Proceed as directed in the general method, page 899, heating for 5 hours.

Specific gravity. Determine by any reliable method (see page 5).

Arsenic. A *Sample Solution* prepared as directed for organic compounds meets the requirements of the *Arsenic Test,* page 865.

Heavy metals. Prepare and test a 500-mg. sample as directed in *Method II* under the *Heavy Metals Test,* page 920, using 20 mcg. of lead ion (Pb) in the control (*Solution A*).

Lead. A *Sample Solution* prepared as directed for organic compounds meets the requirements of the *Lead Limit Test,* page 929, using 10 mcg. of lead ion (Pb) in the control.

Packaging and storage. Store in full, tight, preferably glass, or tin-lined containers in a cool place protected from light.
Functional use in foods. Flavoring agent.

GUAR GUM

DESCRIPTION

A gum obtained from the ground endosperms of *Cyamopsis tetragonolobus* (L.) Taub. (Fam. *Leguminosae*). It consists chiefly of a high molecular weight hydrocolloidal polysaccharide, composed of galactan and mannan units combined through glycosidic linkages, which may be described chemically as a galactomannan. It is a white to yellowish white, nearly odorless, powder. It is dispersible in either hot or cold water forming a sol, having a pH between 5.4 and 6.4, which may be converted to a gel by the addition of small amounts of sodium borate.

IDENTIFICATION

A. Transfer a 2-gram sample into a 400-ml. beaker, moisten it thoroughly with about 4 ml. of isopropyl alcohol, add with vigorous stirring 200 ml. of cold water, and continue the stirring until the gum is completely and uniformly dispersed. An opalescent, viscous solution is formed.

B. Transfer 100 ml. of the solution prepared in *Identification Test A* into another 400-ml. beaker, heat the mixture in a boiling water bath for about 10 minutes, and then cool to room temperature. No appreciable increase in viscosity is produced (*distinction from locust bean gum*).

SPECIFICATIONS

Galactomannans. Not less than 66.0 percent.

Limits of Impurities

Acid-insoluble matter. Not more than 7 percent.

Arsenic (as As). Not more than 3 parts per million (0.0003 percent).

Ash (**Total**). Not more than 1.5 percent.

Heavy metals (as Pb). Not more than 20 parts per million (0.002 percent).

Lead. Not more than 10 parts per million (0.001 percent).

Loss on drying. Not more than 15 percent.

Protein. Not more than 10 percent.

Starch. Passes test.

TESTS

Galactomannans. The difference between the sum of the percentages of *Acid insoluble matter*, *Total ash*, *Loss on drying*, and *Protein* and 100 represents the percent of *Galactomannans*.

Acid-insoluble matter. Transfer 1.5 grams, accurately weighed, into a 250-ml. beaker containing 150 ml. of water and 15 ml. of 1 percent sulfuric acid. Cover the beaker with a watch glass and heat the mixture on a steam bath for 6 hours rubbing down the wall of the beaker frequently with a rubber-tipped stirring rod and replacing any water lost by evaporation. At the end of the 6-hour heating period add about 500 mg. of a suitable filter aid, accurately weighed, and filter through a tared Gooch crucible provided with an asbestos pad. Wash the residue several times with hot water, dry the crucible and its contents at 105° for 3 hours, cool in a desiccator, and weigh. The difference between the weight of the filter aid and that of the residue is the weight of *Acid-insoluble matter*.

Arsenic. A *Sample Solution* prepared as directed for organic compounds meets the requirements of the *Arsenic Test*, page 865.

Ash (**Total**). Determine as directed in the general method, page 868.

Heavy metals. Prepare and test a 1-gram sample as directed in *Method II* under the *Heavy Metals Test*, page 920, using 20 mcg. of lead ion (Pb) in the control (*Solution A*).

Lead. A *Sample Solution* prepared as directed for organic compounds meets the requirements of the *Lead Limit Test*, page 929, using 10 mcg. of lead ion (Pb) in the control.

Loss on drying, page 931. Dry at 105° for 5 hours.

Protein. Transfer about 3.5 grams, accurately weighed, into a 500-ml. Kjeldahl flask and proceed as directed under *Nitrogen Determination*, page 937. The percent of nitrogen determined multiplied by 5.7 gives the percent of protein in the sample.

Starch. To a 1 in 10 solution of the gum add a few drops of iodine T.S. No blue color is produced.

Packaging and storage. Store in well-closed containers.

Functional use in foods. Stabilizer; thickener; emulsifier.

GUM GUAIAC

Guaiac Resin

DESCRIPTION

The resin of the wood of *Guajacum officinale* L. or of *Guajacum sanctum* L. (Fam. *Zygophyllaceae*). It occurs as irregular masses enclosing fragments of vegetable tissues, or in large, nearly homogeneous masses, and occasionally in more or less rounded or ovoid tears; externally brownish black to dusky brown, acquiring a greenish color on long exposure, the fractured surface having a glassy luster, the thin pieces being transparent and varying in color from brown to yellowish orange. The powder is moderate yellow brown, becoming olive-brown on exposure to the air. It has a balsamic odor and a slightly acrid taste. Gum guaiac dissolves incompletely but readily in alcohol, in ether, in chloroform, and in solutions of alkalies. It is slightly soluble in carbon disulfide and in benzene.

IDENTIFICATION

A. Add 1 drop of ferric chloride T.S. to 5 ml. of an alcoholic solution of the sample (1 in 100). A blue color is produced which gradually changes to green, finally becoming greenish yellow.

B. A mixture of 5 ml. of an alcoholic solution of the sample (1 in 100) and 5 ml. of water becomes blue upon shaking with 20 mg. of lead peroxide. Filter the solution, and boil a portion of the filtrate. The color disappears but may be restored by the addition of lead peroxide and shaking. Add a few drops of diluted hydrochloric acid T.S. to a second portion of the filtrate. The color is immediately discharged.

SPECIFICATIONS

Alcohol-insoluble residue. Not more than 15 percent.

Melting range. Between 85° and 90°.

Limits of Impurities

Arsenic (as As). Not more than 3 parts per million (0.0003 percent).

Ash (Total). Not more than 5 percent.

Ash (Acid-insoluble). Not more than 2 percent.

Heavy metals (as Pb). Not more than 40 parts per million (0.004 percent).

Lead. Not more than 10 parts per million (0.001 percent).

Rosin. Passes test.

TESTS

Alcohol-insoluble residue. Place 2 grams of the sample, finely powdered and accurately weighed, in a dry, tared extraction thimble, and extract it with alcohol in a suitable continuous extraction apparatus for 3 hours or until completely extracted. Dry the insoluble residue in the thimble for 4 hours at 105°, and weigh.

Melting range. Determine as directed in the general method, page 931.

Arsenic. A *Sample Solution* prepared as directed for organic compounds meets the requirements of the *Arsenic Test,* page 865.

Ash (Total). Determine as directed in the general method, page 868.

Ash (Acid-insoluble). Determine as directed in the general method, page 869.

Heavy metals. Prepare and test a 500-mg. sample as directed in *Method II* under the *Heavy Metals Test,* page 920, using 20 mcg. of lead ion (Pb) in the control (*Solution A*).

Lead. A *Sample Solution* prepared as directed for organic compounds meets the requirements of the *Lead Limit Test,* page 929, using 10 mcg. of lead ion (Pb) in the control.

Rosin. A 1 in 10 solution of the sample in petroleum ether is colorless, and when shaken with an equal quantity of a fresh solution of cupric acetate (1 in 200) is not more green than a similar solution of cupric acetate in petroleum ether.

Packaging and storage. Store in well-closed containers.

Functional use in foods. Preservative; antioxidant.

HEPTANAL

Aldehyde C-7; Heptaldehyde

$C_7H_{14}O$ $CH_3(CH_2)_5CHO$ Mol. wt. 114.19

DESCRIPTION

A colorless to slightly yellow liquid having a penetrating odor. It is miscible with alcohol and with ether, but only slightly soluble in water.

SPECIFICATIONS

Assay. Not less than 90.0 percent of $C_7H_{14}O$.

Bisulfite solubility. Not less than 95 percent.

Refractive index. Between 1.412 and 1.420 at 20°.

Solubility in alcohol. Passes test.

Specific gravity. Between 0.814 and 0.819.

Limits of Impurities

Acid value. Not more than 10.0.

TESTS

Assay. Weigh accurately about 700 mg., and proceed as directed under *Aldehydes*, page 894, using 57.10 as the equivalence factor (E) in the calculation.

Bisulfite solubility. Pipet 5 ml. of the sample into a 100-ml. cassia flask, and add 70 ml. of a 1 in 10, weight in weight, solution of sodium metabisulfite. Warm the mixture on a water bath to 50°–60°, and shake the flask vigorously for 15 minutes. When the liquids have separated completely, add sufficient sodium metabisulfite solution to raise the lower level of the oily layer within the graduated portion of the neck of the flask. Not more than 0.25 ml. of oil separates.

Refractive index, page 945. Determine with an Abbé or other refractometer of equal or greater accuracy.

Solubility in alcohol. Proceed as directed in the general method, page 899. One ml. dissolves in 2 ml. of 70 percent alcohol to form a clear solution.

Specific gravity. Determine by any reliable method (see page 5).

Acid value. Determine as directed in the general method, page 893.

Packaging and storage. Store in full, tight, preferably glass, tin-lined or other suitably lined containers in a cool place protected from light.

Functional use in foods. Flavoring agent.

2-HEPTANONE

Methyl Amyl Ketone

$CH_3CO(CH_2)_4CH_3$

$C_7H_{14}O$ Mol. wt. 114.19

DESCRIPTION

A clear, colorless, mobile liquid having a characteristic ketonic odor. It is miscible in all proportions with alcohol and ether. One ml. dissolves in about 250 ml. of water.

SPECIFICATIONS

Assay. Not less than 95.0 percent of $C_7H_{14}O$.

Distillation range. Between 147° and 154°.

Specific gravity. Between 0.813 and 0.818.

Limits of Impurities

Acidity (as acetic acid). Not more than 0.05 percent.

Residue on evaporation. Not more than 5 mg. per 100 ml.

Water. Not more than 0.3 percent.

TESTS

Assay. Transfer 65 ml. of 0.5 *N* hydroxylamine hydrochloride and 50.0 ml. of 0.5 *N* triethanolamine into a suitable heat-resistant pressure bottle provided with a tight closure that can be securely fastened. Replace the air in the bottle by passing a gentle stream of nitrogen for 2 minutes through a glass tube positioned so that the end is just above the surface of the liquid. To the mixture in the pressure bottle add about 1.2 grams of the sample, accurately weighed, using a suitable weighing pipet. Cap the bottle and allow it to stand at room temperature for one hour, shaking occasionally. Cool, if necessary, and uncap the bottle cautiously to prevent any loss of the contents. Titrate with 0.5 *N* sulfuric acid to a greenish blue end-point. Perform a residual blank titration (see page 2). Each ml. of 0.5 *N* sulfuric acid is equivalent to 57.10 mg. of $C_7H_{14}O$.

Distillation range. Distil 100 ml. as directed in the general method, page 890.

Specific gravity. Determine by any reliable method (see page 5).

Acidity. Transfer 74 ml. (60 grams) into a 250-ml. Erlenmeyer flask, add a few drops of phenolphthalein T.S., and titrate with 0.1 *N* alcoholic potassium hydroxide to a pink end-point that persists for at least 15 seconds. Each ml. of 0.1 *N* alcoholic potassium hydroxide is equivalent to 6.01 mg. of $C_2H_4O_2$.

Residue on evaporation, page 745. Evaporate 100 ml. in a tared platinum dish, on a steam bath, and dry at 105° for 30 minutes. Cool in a desiccator and weigh. The weight of the residue does not exceed 5 mg.

Water. Determine by the *Karl Fischer Titrimetric Method*, page 977, using freshly distilled pyridine instead of methanol as the solvent.

Packaging and storage. Store in tight containers.

Functional use in foods. Flavoring agent.

3-HEPTANONE

Ethyl Butyl Ketone

$CH_3(CH_2)_3COCH_2CH_3$

$C_7H_{14}O$ Mol. wt. 114.19

DESCRIPTION

A clear, colorless, mobile liquid having a characteristic ketonic odor. It is miscible in all proportions with alcohol and with ether. One ml. dissolves in about 70 ml. of water.

SPECIFICATIONS

Assay. Not less than 97.0 percent of $C_7H_{14}O$.

Distillation range. Between 143° and 151°.

Specific gravity. Between 0.813 and 0.818.

Limits of Impurities

Acidity (as acetic acid). Not more than 0.02 percent.

Water. Not more than 0.3 percent.

TESTS

Assay. Transfer 65 ml. of 0.5 *N* hydroxylamine hydrochloride and 50.0 ml. of 0.5 *N* triethanolamine into a suitable heat-resistant pressure bottle provided with a tight closure that can be securely fastened. Replace the air in the bottle by passing a gentle stream of nitrogen through a glass tube positioned so that the end is just above the surface of the liquid. To the mixture in the pressure bottle add about 1 gram of the sample, accurately weighed, using a suitable weighing pipet. Allow the bottle to stand at room temperature for one hour, shaking occasionally. Cool, if necessary, and uncap the bottle cautiously to prevent any loss of the contents. Titrate with 0.5 *N* sulfuric acid to a greenish blue end-point. Perform a residual blank titration (see *Blank Tests*, page 2). Each ml. of 0.5 *N* sulfuric acid is equivalent to 57.05 mg. of $C_7H_{14}O$.

Distillation range. Distil 100 ml. as directed in the general method, page 890.

Specific gravity. Determine by any reliable method (see page 5).

Acidity. Transfer 74 ml. (60 grams) into a 250-ml. Erlenmeyer flask, add a few drops of phenolphthalein T.S., and titrate with 0.1 *N* alcoholic potassium hydroxide to a pink end-point that persists for at least 15 seconds. Each ml. of 0.1 *N* alcoholic potassium hydroxide is equivalent to 6.01 mg. of $C_2H_4O_2$.

Water. Determine by the *Karl Fischer Titrimetric Method*, page 977, using freshly distilled pyridine instead of methanol as the solvent.

Packaging and storage. Store in tight containers.

Functional use in foods. Flavoring agent.

HEPTYL ALCOHOL

Enanthic Alcohol

$CH_3(CH_2)_5CH_2OH$

$C_7H_{16}O$ Mol. wt. 116.20

DESCRIPTION

A colorless liquid having a fatty-citrus odor. It is miscible with alcohol and with ether, and is very slightly soluble in water.

SPECIFICATIONS

Assay. Not less than 97.0 percent of $C_7H_{16}O$.

Refractive index. Between 1.423 and 1.427 at 20°.

Solubility in alcohol. Passes test.

Specific gravity. Between 0.820 and 0.824.

Limits of Impurities

Acid value. Not more than 1.0.

Aldehydes (as heptanal). Not more than 1.0 percent.

TESTS

Assay. Proceed as directed under *Total Alcohols*, page 893. Weigh accurately about 1 gram of the acetylated alcohol for the saponification, and use 58.10 as the equivalence factor (*f*) in the calculation.

Refractive index, page 945. Determine with an Abbé or other refractometer of equal or greater accuracy.

Solubility in alcohol. Proceed as directed in the general method, page 899. One ml. dissolves in 2 ml. of 60 percent alcohol to form clear solutions.

Specific gravity. Determine by any reliable method (see page 5).

Acid value. Determine as directed in the general method, page 893.

Aldehydes. Weigh accurately about 10 grams and proceed as directed under *Aldehydes*, page 894, using 57.10 as the equivalence factor (E) in the calculation.

Packaging and storage. Store in full, tight, preferably glass, tin-lined or other suitably lined containers in a cool place protected from light.

Functional use in foods. Flavoring agent.

HEPTYLPARABEN

n-Heptyl-*p*-hydroxybenzoate

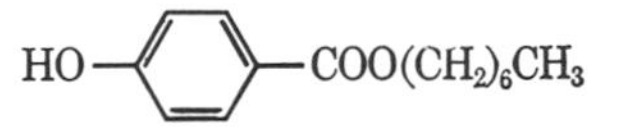

$C_{14}H_{20}O_3$ Mol. wt. 236.31

DESCRIPTION

Small, colorless crystals or a white crystalline powder. It is odorless or has a faint, characteristic odor and a slight burning taste. It is very slightly soluble in water but is freely soluble in alcohol and in ether.

IDENTIFICATION

Dissolve 500 mg. in 10 ml. of sodium hydroxide T.S., boil for 30 minutes, allow the solution to evaporate to a volume of about 5 ml., and cool. Acidify the solution with diluted sulfuric acid T.S., collect the crystals on a filter, wash several times with small portions of water, and dry in a desiccator over silica gel. The *p*-hydroxybenzoic acid so obtained melts between 212° and 217° (see page 931).

SPECIFICATIONS

Assay. Not less than 99.0 percent of $C_{14}H_{20}O_3$, calculated on the dried basis.

Melting range. Between 48° and 51°.

Limits of Impurities

Acidity. Passes test.

Arsenic (as As). Not more than 3 parts per million (0.0003 percent).

Heavy metals (as Pb). Not more than 10 parts per million (0.001 percent).

Loss on drying. Not more than 0.5 percent.

Residue on ignition. Not more than 0.05 percent.

TESTS

Assay. Transfer into a flask about 3.5 grams, accurately weighed, add 40.0 ml. of 1 *N* sodium hydroxide, and rinse the sides of the flask with water. Cover with a watch glass, boil gently for 1 hour, and cool. Add 5 drops of bromothymol blue T.S., and titrate the excess sodium hydroxide with 1 *N* sulfuric acid to pH 6.5 by matching the color of pH 6.5 phosphate buffer (see page 984) containing the same proportion of indicator. Perform a blank determination (see page 2). Each ml. of 1 *N* sodium hydroxide is equivalent to 236.3 mg. of $C_{14}H_{20}O_3$.

Melting range. Determine as directed in the general procedure, page 931.

Acidity. Heat 750 mg. with 15 ml. of water at 80° for 1 minute, cool, and filter. The filtrate is acid or neutral to litmus. To 10 ml. of the filtrate add 0.2 ml. of 0.1 *N* sodium hydroxide and 2 drops of methyl red T.S. The solution is yellow, without even a light cast of pink.

Arsenic. A *Sample Solution* prepared as directed for organic compounds meets the requirements of the *Arsenic Test*, page 865.

Heavy metals, page 920. Dissolve 2 grams in 23 ml. of acetone, and add 2 ml. of diluted acetic acid T.S., 2 ml. of water, and 10 ml. of hydrogen sulfide T.S. Any color does not exceed that produced in a control (*Solution A*) made with 23 ml. of acetone, 2 ml. of diluted acetic acid T.S., 2 ml. of *Standard Lead Solution* (20 mcg. Pb ion), and 10 ml. of hydrogen sulfide T.S.

Loss on drying, page 931. Dry in a desiccator over silica gel for 5 hours.

Residue on ignition. Ignite 2 grams as directed in the general method, page 945.

Packaging and storage. Store in tight containers.

Functional use in foods. Preservative; antimicrobial agent.

HEXANOIC ACID

Caproic Acid

$CH_3(CH_2)_4COOH$

$C_6H_{12}O_2$ Mol. wt. 116.16

DESCRIPTION

An acid usually prepared from coconut oil or obtained as a by-product of lauric acid production. It occurs as a colorless to very pale yellow, oily liquid having a characteristic cheese- or sweat-like odor.

It is miscible with alcohol, with most fixed oils, and with ether. One gram is soluble in about 250 ml. of water.

SPECIFICATIONS

Assay. Not less than 98.0 percent of $C_6H_{12}O_2$.

Refractive index. Between 1.415 and 1.418 at 20°.

Specific gravity. Between 0.923 and 0.928.

Titer (**Solidification point**). Not lower than −4.5°.

Limits of Impurities

Arsenic (as As). Not more than 3 parts per million (0.0003 percent).

Heavy metals (as Pb). Not more than 40 parts per million (0.004 percent).

Lead. Not more than 10 parts per million (0.001 percent).

TESTS

Assay. Transfer about 2 grams, accurately weighed, into a 250-ml. Erlenmeyer flask. Add about 75 ml. of water and phenolphthalein T.S., and titrate the mixture with 0.5 *N* sodium hydroxide to a pink end-point that persists for at least 15 seconds. Each ml. of 0.5 *N* sodium hydroxide is equivalent to 58.08 mg. of $C_6H_{12}O_2$.

Refractive index, page 945. Determine with an Abbé or other refractometer of equal or greater accuracy.

Specific gravity. Determine by any reliable method (see page 5).

Titer (**Solidification point**). Determine as directed in the general method, page 954.

Arsenic. A *Sample Solution* prepared as directed for organic compounds meets the requirements of the *Arsenic Test,* page 865.

Heavy metals. Prepare and test a 500-mg. sample as directed in *Method II* under the *Heavy Metals Test,* page 920, using 20 mcg. of lead ion (Pb) in the control (*Solution A*).

Lead. A *Sample Solution* prepared as directed for organic compounds meets the requirements of the *Lead Limit Test,* page 929, using 10 mcg. of lead ion (Pb) in the control.

Packaging and storage. Store in tight containers.

Functional use in foods. Flavoring agent.

HEXYL ALCOHOL (NATURAL)

1-Hexanol; Alcohol C-6

$CH_3(CH_2)_4CH_2OH$

$C_6H_{14}O$ Mol. wt. 102.18

DESCRIPTION

A clear, colorless, mobile liquid having a mild, slightly sweet, penetrating odor. It is miscible in all proportions with alcohol and with ether. One ml. dissolves in about 175 ml. of water.

(*Note:* The following *Specifications* apply only to hexyl alcohol derived from natural precursors.)

SPECIFICATIONS

Assay. Not less than 96.5 percent of $C_6H_{14}O$.

Distillation range. Between 153° and 160°.

Hydroxyl value. Not less than 530.

Iodine value. Not more than 1.2.

Specific gravity. Between 0.816 and 0.821.

Limits of Impurities

Acidity (as acetic acid). Not more than 0.01 percent.

TESTS

Assay

Phthalic Anhydride-Pyridine Solution. Add about 115 grams of phthalic anhydride to 700 ml. of freshly distilled pyridine contained in a 1-liter glass-stoppered, amber-colored bottle and shake vigorously until complete solution is effected.

Procedure. Transfer 25.0 ml. of *Phthalic Anhydride-Pyridine Solution* into a suitable heat-resistant pressure bottle provided with a tight closure that can be securely fastened. Introduce about 1 gram of the sample, accurately weighed, into the pressure bottle using a suitable weighing pipet. Cap the bottle, enclose it securely in a canvas bag, and heat for three hours in a water bath maintained at a temperature between 98° and 100°, keeping the water level in the bath at about the same height as the liquid in the bottle. Remove the bottle from the bath and allow it to cool to room temperature, open it cautiously to prevent loss of the contents, and transfer 50.0 ml. of 0.5 *N* sodium hydroxide into the bottle. (*This 50.0 ml. of 0.5* N *sodium hydroxide is not to be considered in the final calculation.*) Add 5 drops of 1 in 100 solution of phenolphthalein in pyridine to the mixture and titrate with 0.5 *N* sodium hydroxide to a pink end-point that persists for at least 15 seconds.

Perform a residual blank titration (see *Blank Tests*, page 2). Each ml. of 0.5 *N* sodium hydroxide is equivalent to 51.09 mg. of $C_6H_{14}O$.

Distillation range. Distil 100 ml. as directed in the general method, page 890.

Hydroxyl value. Calculate the hydroxyl value by multiplying the number of ml. of 0.5 *N* sodium hydroxide required in the *Assay* by 28.06 and dividing the result by the weight of the sample in grams.

Iodine value. Determine by the *Hanus Method*, page 905.

Specific gravity. Determine by any reliable method (see page 5).

Acidity. Transfer 73 ml. (60 grams) into a 250-ml. Erlenmeyer flask, add phenolphthalein T.S., and titrate with 0.1 *N* alcoholic potassium hydroxide to a pink end-point that persists for at least 15 seconds. Each ml. of 0.1 *N* alcoholic potassium hydroxide is equivalent to 6.01 mg. of $C_2H_4O_2$.

Packaging and storage. Store in tight containers.

Functional use in foods. Flavoring agent.

α-HEXYLCINNAMALDEHYDE

C_6H_5—CH=C(CH_2)$_5$$CH_3$ (with CHO on the C)

$C_{15}H_{20}O$ Mol. wt. 216.32

DESCRIPTION

A pale yellow liquid having a jasmin-like odor, particularly on dilution. It is soluble in most fixed oils and in mineral oil. It is insoluble in glycerin and in propylene glycol.

SPECIFICATIONS

Assay. Not less than 95.0 percent of $C_{15}H_{20}O$.

Refractive index. Between 1.548 and 1.552 at 20°.

Solubility in alcohol. Passes test.

Specific gravity. Between 0.953 and 0.959.

Limits of Impurities

Acid value. Not more than 5.0.

TESTS

Assay. Weigh accurately about 1.5 grams, and proceed as directed under *Aldehydes and Ketones-Hydroxylamine Method*, page 894, using 108.2 as the equivalence factor (*E*) in the calculation. Modify the procedure by allowing the sample and the blank to stand 30 minutes at room temperature, instead of for 1 hour.

Refractive index, page 945. Determine with an Abbé or other refractometer of equal or greater accuracy.

Solubility in alcohol. Proceed as directed in the general method, page 899. One ml. dissolves in 1 ml. of 90 percent alcohol.

Specific gravity. Determine by any reliable method (see page 5).

Acid value. Determine as directed in the general method, page 893.

Chlorinated compounds. Proceed as directed in the general method, page 895.

Packaging and storage. Store in full, tight, preferably glass, aluminum, tin-lined, or other suitably lined containers in a cool place protected from light. It is affected by air and can not be stored unless protected by a suitable antioxidant.

Functional use in foods. Flavoring agent.

HOPS OIL

DESCRIPTION

The volatile oil obtained by steam distillation of the freshly dried membranous cones of the female plants of *Humulus lupulus* L. or *Humulus americanus* Nutt. (Fam. *Moraceae*). It is a light yellow to greenish yellow liquid, having a characteristic aromatic odor. Age darkens the color, and the oil tends to become viscous. It is soluble in most fixed oils and, sometimes with opalescence, in mineral oil. It is practically insoluble in glycerin and in propylene glycol.

SPECIFICATIONS

Acid value. Not more than 11.0.

Angular rotation. Between $-2°$ and $+2°5'$.

Refractive index. Between 1.470 and 1.494 at 20°.

Saponification value. Between 14 and 69.

Solubility in alcohol. Passes test.

Specific gravity. Between 0.825 and 0.926.

TESTS

Acid value. Determine as directed in the general method, page 893, using about 5 grams, accurately weighed.

Angular rotation. Determine in a 100-mm. tube as directed under *Optical Rotation*, page 939.

Refractive index, page 945. Determine with an Abbé or other refractometer of equal or greater accuracy.

Saponification value. Determine as directed in the general method, page 896, using about 5 grams, accurately weighed.

Solubility in alcohol. Proceed as directed in the general method, page 899. One ml. usually is not soluble in 95 percent alcohol. Old oils are less soluble than fresh oils.

Specific gravity. Determine by any reliable method (see page 5).

Packaging and storage. Store in full, tight, preferably glass, aluminum, tin-lined, or other suitably lined containers in a cool place protected from light.

Functional use in foods. Flavoring agent.

HYDROCHLORIC ACID

HCl Mol. wt. 36.46

DESCRIPTION

A water solution of hydrogen chloride of varying concentrations. It is a clear, colorless or slightly yellowish, corrosive liquid having a pungent odor. It is miscible with water and with alcohol. Concentrations of hydrochloric acid commercially available are usually expressed in Baumé degrees (Be°) from which percentages of HCl and specific gravities can readily be derived. (See *Hydrochloric Acid Table*, page 922.) The usually available concentrations are 18°, 20°, 22°, and 23° Be. Concentrations above 8.5° Be (12.5 percent) fume in moist air, lose hydrogen chloride, and create a corrosive atmosphere. Because of these characteristics, suitable precautions must be observed during sampling and analysis to prevent losses. It gives positive tests for *Chloride*, page 926.

Note: Hydrochloric acid produced during the manufacture of chlorinated hydrocarbon insecticides is not considered to be of food grade quality and is not suitable for use in food processing.

SPECIFICATIONS

Assay. Not less than the minimum or within the range of Baumé degrees claimed or implied by the vendor.

Color. Passes test.

Concentration of HCl. Not less than the minimum or within the range specified or implied by the vendor.

Specific gravity. Not less than the minimum or within the range specified or implied by the vendor.

Limits of Impurities

Arsenic (as As). Not more than 1 part per million (0.0001 percent).

Heavy metals (as Pb). Not more than 5 parts per million (0.0005 percent).

Iron. Not more than 5 parts per million (0.0005 percent).

Nonvolatile residue. Not more than 0.5 percent.

Oxidizing substances (as Cl_2). Not more than 30 parts per million (0.003 percent).

Reducing substances (as SO_3). Not more than 70 parts per million (0.007 percent).

Sulfate. Not more than 0.5 percent.

TESTS

Assay. Transfer about 200 ml., previously cooled to a temperature below 15°, into a 250-ml. hydrometer cylinder. Insert a suitable Baumé hydrometer graduated at 0.1° intervals, adjust the temperature to 15.6° (60° F.), and note the reading at the bottom of the meniscus.

Color. It shows no more color than *Matching Fluid* A, page 944.

Concentration of HCl. Tare accurately a 125-ml. glass-stoppered Erlenmeyer flask containing 35.0 ml. of 1 *N* sodium hydroxide. Partially fill, without the use of vacuum, a 10-ml. serological pipet from near the bottom of a representative sample, remove any acid adhering to the outside and discard the first ml. flowing from the pipet. Hold the tip of the pipet just above the surface of the sodium hydroxide solution, and transfer between 2.5 and 3 ml. of the sample into the flask, leaving at least 1 ml. in the pipet. Stopper the flask, mix the contents, and weigh accurately to obtain the weight of the sample. Add methyl orange T.S. and titrate the excess sodium hydroxide with 1 *N* hydrochloric acid. Each ml. of 1 *N* sodium hydroxide is equivalent to 36.46 mg. of HCl. Alternatively the concentration of the HCl in the sample may be obtained from the specific gravity by reference to the *Hydrochloric Acid Table*, page 922.

Specific gravity. Determine at 15.6° (60° F.) with a hydrometer, calculate it from the Baumé degrees observed in the *Assay*, or obtain it by reference to the *Hydrochloric Acid Table*, page 922.

Reducing substances. Transfer 1 ml. of A.C.S. reagent grade hydrochloric acid into a 30-ml. test tube, dilute to 20 ml. with recently boiled and cooled water, and add 1 ml. of potassium iodide T.S., 1 ml. of starch T.S., and 2.0 ml. of 0.001 *N* iodine. Stopper the test tube and mix thoroughly. The blue color produced is not discharged by 1 ml. of the sample.

Arsenic. A dilution of 3 grams (2.6 ml.) in 10 ml. of water meets the requirements of the *Arsenic Test*, page 865, using as a control a mixture of 3 ml. of the *Standard Arsenic Solution* and 2.6 ml. of A.C.S. reagent grade hydrochloric acid.

Heavy metals. Evaporate 4 grams (3.5 ml.) to dryness on a steam bath, dissolve the residue in 2 ml. of diluted acetic acid T.S., and dilute to 25 ml. with water. This solution meets the requirements of the *Heavy Metals Test,* page 920, using 20 mcg. of lead ion (Pb) in the control (*Solution A*).

Iron. Dilute 4.3 ml. (5 grams) to a volume of 40 ml., and add about 40 mg. of ammonium persulfate and 10 ml. of ammonium thiocyanate T.S. Any red color does not exceed that produced by 2.5 ml. of *Iron Standard Solution* (25 mcg. Fe) in an equal volume of solution containing the same quantities of A.C.S. reagent grade hydrochloric acid and the reagents used in the test.

Nonvolatile residue. Transfer 1 gram into a tared glass dish, evaporate to dryness on a steam bath and then dry at 110° for 1 hour, cool in a desiccator, and weigh. The weight of the residue does not exceed 5 mg.

Oxidizing substances. Transfer 1 ml. into a 30-ml. test tube, dilute to 20 ml. with freshly boiled and cooled water, and add 1 ml. of potassium iodide T.S. and 1 ml. of starch T.S. Stopper the test tube and mix thoroughly. Any blue color does not exceed that produced in a control consisting of 1.0 ml. of 0.001 *N* iodine in an equal volume of water containing the same quantities of the same reagents and 1 ml. of A.C.S. reagent grade hydrochloric acid.

Sulfate, page 879. Dilute a 1-gram sample to 100.0 ml. with water, transfer 5.0 ml. of this dilution into a 50-ml. tall form Nessler tube, and dilute to 20 ml. with water. Add a drop of phenolphthalein T.S., neutralize the solution with ammonia T.S., and then add 1 ml. of hydrochloric acid T.S. To the clear solution, previously filtered, if necessary, add 3 ml. of barium chloride T.S., dilute to 50 ml. with water, and mix. Prepare a control consisting of 1 ml. of A.C.S. reagent grade hydrochloric acid and 250 mcg. of sulfate (SO_4) and the same quantities of the reagents used for the sample. Any turbidity shown in the sample does not exceed that of the control.

Packaging and storage. Store in tight containers.

Functional use in foods. Acid.

HYDROGEN PEROXIDE

H_2O_2 Mol. wt. 34.01

DESCRIPTION

A clear, colorless liquid having an odor resembling ozone. It is miscible with water. The grades of hydrogen peroxide suitable for food use usually have a concentration between 30 percent and 50 percent.

Note: Although hydrogen peroxide undergoes exothermic decomposition in the presence of dirt and other foreign materials, it is safe and stable under recommended conditions of handling and storage. Information on safe handling and use may be obtained from the supplier, or by consulting Chemical Safety Data Sheet SD53 published by the Manufacturing Chemists' Association.

SPECIFICATIONS

Assay. Not less than the labeled concentration or within the range stated on the label.

Limits of Impurities

Acidity (as H_2SO_4). Not more than 300 parts per million (0.03 percent).

Arsenic (as As). Not more than 3 parts per million (0.0003 percent).

Heavy metals (as Pb). Not more than 10 parts per million (0.001 percent).

Iron. Not more than 0.5 part per million (0.00005 percent).

Phosphate. Not more than 50 parts per million (0.005 percent).

Residue on evaporation. Not more than 60 parts per million (0.006 percent).

Tin. Not more than 10 parts per million (0.001 percent).

TESTS

Assay. Accurately weigh a volume of the sample equivalent to about 300 mg. of H_2O_2 into a 100-ml. volumetric flask, dilute to volume with water, and mix thoroughly. To a 20.0-ml. portion of this solution add 25 ml. of diluted sulfuric acid T.S., and titrate with 0.1 *N* potassium permanganate. Each ml. of 0.1 *N* potassium permanganate is equivalent to 1.701 mg. of H_2O_2.

Acidity. Dilute 9 ml. (10 grams) with 90 ml. of carbon dioxide-free water, add methyl red T.S., and titrate with 0.02 *N* sodium hydroxide. The volume of sodium hydroxide solution should not be more than 3 ml. greater than the volume required for a blank test on 90 ml. of the water used for dilution.

Arsenic. Add 1 ml. of ammonia T.S. to 1 ml. of the sample, evaporate to dryness on a steam bath, and dissolve the residue in 35 ml. of water. This solution meets the requirements of the *Arsenic Test*, page 865.

Heavy metals. Evaporate 1.8 ml. (2 grams) to dryness on a steam bath with 10 mg. of sodium chloride, and dissolve the residue in 25 ml. of water. The solution so obtained meets the requirements of the *Heavy Metals Test,* page 920, using 20 mcg. of lead ion (Pb) in the control (*Solution A*).

Iron. Evaporate 18 ml. (20 grams) to dryness on a steam bath with 10 mg. of sodium chloride, dissolve the residue in 2 ml. of hydrochloric acid, and dilute to 50 ml. with water. Add about 40 mg. of ammonium persulfate crystals and 10 ml. of ammonium thiocyanate T.S., and mix. Any red or pink color does not exceed that produced by 1.0 ml. of *Iron Standard Solution* (10 mcg. Fe) in an equal volume of solution containing the quantities of the reagents used in the test.

Phosphate. Evaporate 400 mg. to dryness on a steam bath. Dissolve the residue in 25 ml. of approximately 0.5 *N* sulfuric acid, add 1 ml. of ammonium molybdate solution [500 mg. of $(NH_4)_6Mo_7O_{24}\cdot 4H_2O$ in each 10 ml. of water] and 1 ml. of *p*-methylaminophenol sulfate T.S., and allow to stand for 2 hours. Any blue color does not exceed that produced by 2.0 ml. of *Phosphate Standard Solution* (20 mcg. PO_4) in an equal volume of solution containing the quantities of the reagents used in the test.

Residue on evaporation. Evaporate 25 grams to dryness in a tared porcelain or silica dish on a steam bath, and dry to constant weight at 105°. The weight of the residue does not exceed 1.5 mg.

Tin

Aluminum Chloride Solution. Dissolve 8.93 grams of aluminum chloride, $AlCl_3\cdot 6H_2O$, in sufficient water to make 1000 ml.

Gelatin Solution. On the day of use, dissolve 100 mg. of gelatin in 50 ml. of boiled water which has been cooled to between 50° and 60°.

Tin Stock Solution. Dissolve 250.0 mg. of lead-free tin foil in 10 to 15 ml. of hydrochloric acid, and dilute to 250.0 ml. with dilute hydrochloric acid (1 in 2).

Standard Solution. On the day of use, transfer 5.0 ml. of *Tin Stock Solution* into a 100-ml. volumetric flask, dilute to volume with water, and mix. Transfer 2.0 ml. of this solution (100 mcg. Sn) into a 250-ml. Erlenmeyer flask, and add 15 ml. of water, 5 ml. of nitric acid, and 2 ml. of sulfuric acid. Place a small stemless funnel in the mouth of the flask, and heat until strong fumes of sulfuric acid are evolved. Cool, add 5 ml. of water, evaporate again to strong fumes, and cool. Repeat the addition of water and heating to strong fumes, then add 15 ml. of water, heat to boiling, and cool. Dilute to about 35 ml. with water, add 1 drop of methyl red T.S. and 2.0 ml. of the *Aluminum Chloride Solution,* and mix. Make the solution just alkaline by the dropwise addition of stronger ammonia T.S., stirring gently, and then add 0.1 ml. in excess. (*Caution:* To avoid dissolving the aluminum hydroxide precipitate, do not add more than 0.1 ml. in excess of the ammonia solution.) Centrifuge for about 15 minutes at 4000 rpm, and then decant the supernatant liquid as completely as possible without disturbing the precipitate. Dissolve the precipitate in 5 ml. of dilute hydrochloric acid (1 in 2), add 1.0 ml. of the *Gelatin Solution*, and dilute to 20.0 ml. with a saturated solution of aluminum chloride.

Sample Solution. Transfer 9 ml. (10 grams) of the sample into a 250-ml. Erlenmeyer flask, and add 15 ml. of water, 5 ml. of nitric acid, and 2 ml. of sulfuric acid. Mix, and heat gently on a hot plate to initiate and maintain a vigorous decomposition. When decomposition is complete, place a small stemless funnel in the mouth of the flask, and continue as directed for the *Standard Solution*, beginning with ". . . and heat until strong fumes of sulfuric acid are evolved."

Procedure. Rinse a polarographic cell or other vessel with a portion of the *Standard Solution*, then add a suitable volume to the cell, immerse it in a constant temperature bath maintained at 35° ± 0.2°, and de-aerate by bubbling oxygen-free nitrogen or hydrogen through the solution for at least 10 minutes. Insert the dropping mercury electrode of a suitable polarograph, and record the polarogram from −0.2 to −0.7 v. and at a sensitivity of 0.0003 μamp. per millimeter, using a saturated calomel reference electrode. In the same manner, polarograph a portion of the *Sample Solution* at the same current sensitivity. The height of the wave produced by the *Sample Solution* is not greater than that produced by the *Standard Solution* at the same half-wave potential.

Packaging and storage. Store in a cool place in containers with a vent in the stopper.

Functional use in foods. Bleaching, oxidizing agent; starch modifier; preservative.

HYDROXYCITRONELLAL

7-Hydroxy-3,7-dimethyl Octanal

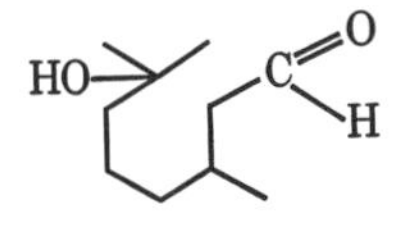

$C_{10}H_{20}O_2$ Mol. wt. 172.27

DESCRIPTION

A colorless liquid with a sweet, floral, lily-like odor. It is soluble in most fixed oils and in propylene glycol. It is relatively insoluble in glycerin and in mineral oil.

SPECIFICATIONS

Assay. Not less than 95.0 percent of $C_{10}H_{20}O_2$.

Refractive index. Between 1.447 and 1.450 at 20°.

Solubility in alcohol. Passes test.
Solubility in bisulfite. Passes test.
Specific gravity. Between 0.918 and 0.923.
Limits of Impurities
Acid value. Not more than 5.0.

TESTS

Assay. Weigh accurately about 1.3 grams, and proceed as directed under *Aldehydes*, page 894, using 86.13 as the equivalence factor (E) in the calculation.

Refractive index, page 945. Determine with an Abbé or other refractometer of equal or greater accuracy.

Solubility in alcohol. Proceed as directed in the general method, page 899. One ml. dissolves in 1 ml. of 50 percent alcohol.

Solubility in bisulfite. Transfer 5 ml. to a 100-ml. cassia flask. Add 70 ml. of a 10 percent, by weight, solution of sodium metabisulfite. Shake the mixture vigorously for 15 minutes. Add the sodium metabisulfite solution to raise the liquid into the graduated neck of the flask. No oil separates, and the solution is no more than slightly cloudy.

Specific gravity. Determine by any reliable method (see page 5).

Acid value. Determine as directed in the general method, page 893.

Packaging and storage. Store in full, tight, glass, aluminum, tin-lined, or iron containers in a cool place protected from light. Galvanized iron containers should not be used.

Functional use in foods. Flavoring agent.

HYDROXYCITRONELLAL DIMETHYL ACETAL

7-Hydroxy-3,7-dimethyl Octanal: Acetal

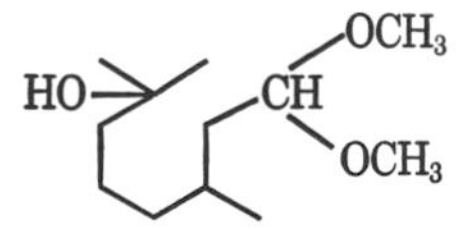

$C_{12}H_{26}O_3$ Mol. wt. 218.34

DESCRIPTION

A colorless liquid with a light floral odor. It is soluble in most fixed oils, in mineral oil, and in propylene glycol. It is insoluble in glycerin.

SPECIFICATIONS

Assay. Not less than 95.0 percent of $C_{12}H_{26}O_3$.

Refractive index. Between 1.441 and 1.444 at 20°.

Solubility in alcohol. Passes test.

Specific gravity. Between 0.925 and 0.930.

Limits of Impurities

Acid value. Not more than 1.0.

Free Aldehydes. Not more than 3.0 percent, calculated as hydroxycitronellal ($C_{10}H_{20}O_2$).

TESTS

Assay. Weigh accurately about 1.5 grams, and proceed as directed under *Acetals*, page 892, but reflux between 5 and 10 minutes. Record the number of ml. of 0.5 N alcoholic potassium hydroxide consumed per gram of sample as A. Weigh accurately a separate 5-gram portion of the sample, and proceed as directed under *Aldehydes and Ketones-Hydroxylamine Method*, page 894, recording the number of ml. of 0.5 N hydrochloric acid liberated per gram of sample as B. Calculate the percent of $C_{12}H_{26}O_3$ by the formula $10.92(A - B)$.

Refractive index, page 945. Determine with an Abbé or other refractometer of equal or greater accuracy.

Solubility in alcohol. Proceed as directed in the general method, page 899. One ml. dissolves in 2 ml. of 50 percent alcohol.

Specific gravity. Determine by any reliable method (see page 5).

Acid value. Determine as directed in the general method, page 893.

Free aldehydes. Weigh accurately about 5 grams, and proceed as directed under *Aldehydes and Ketones-Hydroxylamine Method*, page 894, using 86.13 as the equivalence factor (E) for hydroxycitronellal ($C_{10}H_{20}O_2$).

Packaging and storage. Store in full, tight, preferably glass, aluminum, or suitably lined containers in a cool place protected from light.

Functional use in foods. Flavoring agent.

HYDROXYLATED LECITHIN

DESCRIPTION

Hydroxylated lecithin is derived from a complex mixture of acetone-insoluble phosphatides from soybean and other plant lecithins, consisting chiefly of phosphatidyl choline, phosphatidyl ethanolamine, and phosphatidyl inositol, as well as other minor phospholipids and glycolipids mixed with varying amounts of triglycerides, fatty acids, sterols, and carbohydrates. The mixture is treated with hydrogen peroxide, benzoyl peroxide, and lactic acid to produce a hydroxylated product having an iodine value approximately 10 percent lower than that of the starting material. Hydroxylated lecithin may vary in consistency from fluid to plastic, depending upon the content of free fatty acid and soybean oil and whether or not it contains diluents. It is light yellow in color and has a characteristic "bleached" odor. It is partially soluble in water but hydrates readily to form emulsions; it is more dispersible and hydrates more readily than crude lecithin.

SPECIFICATIONS

Acetone-insoluble matter (phosphatides). Not less than 50 percent.

Acid value. Not more than 70.

Benzene-insoluble matter. Not more than 0.3 percent.

Iodine value. Between 85 and 95.

Peroxide value. Not more than 100.

Water. Not more than 1.5 percent.

Limits of Impurities

Arsenic (as As). Not more than 3 parts per million (0.0003 percent).

Heavy metals (as Pb). Not more than 40 parts per million (0.004 percent).

Lead. Not more than 10 parts per million (0.001 percent).

TESTS

Acetone-insoluble matter (phosphatides)

Purification of Phosphatides. Dissolve 5 grams of phosphatides from previous *Acetone-insoluble matter* determinations in 10 ml. of petroleum ether, and add 25 ml. of acetone to the solution. Transfer approximately equal portions of the precipitate to each of two 40-ml. centrifuge tubes using additional portions of acetone to facilitate the transfer. Stir thoroughly, dilute to 40 ml. with acetone, stir again, chill for 15 minutes in an ice bath, stir again, and then centrifuge for 5 minutes. Decant the acetone, crush the solids with a stirring rod, refill the tube with acetone, stir, chill, centrifuge, and decant as before. The solids after the second centrifugation require no further purification and may be used for preparing the *Phosphatide-Acetone Solution.* Five grams of the purified phosphatides are required to saturate about 16 liters of acetone.

Phosphatide-Acetone Solution. Add a quantity of purified phosphatides to sufficient acetone, previously cooled to a temperature of about 5°, to form a saturated solution, and maintain the mixture at this temperature for 2 hours, shaking it vigorously at 15-minute intervals. Decant the solution through a rapid filter paper avoiding the transfer of any undissolved solids to the paper and conducting the filtration under refrigerated conditions (not above 5°).

Procedure. If it is plastic or semisolid, soften a portion of the sample by warming it in a water bath at a temperature not exceeding 60° and then mixing it thoroughly. Transfer 2 grams of a well-mixed sample, accurately weighed, into a 40-ml. centrifuge tube, previously tared with a glass stirring rod, and add 15 ml. of *Phosphatide-Acetone Solution* from a buret. Warm the mixture in a water bath until the sample melts, but avoid evaporation of the acetone. Stir until the sample is completely disintegrated and dispersed, and then transfer the tube into an ice bath, chill for 5 minutes, remove from the ice bath, and add about one-half of the required volume of *Phosphatide-Acetone Solution,* previously chilled for 5 minutes in an ice bath. Stir the mixture to complete dispersion of the sample, dilute to 40 ml. with chilled *Phosphatide-Acetone Solution* (5°), again stir and return the tube and contents to the ice bath for 15 minutes. At the end of the 15-minute chilling period stir again while still in the ice bath, remove the stirring rod, temporarily supporting it in a vertical upside-down position, and centrifuge the mixture immediately at about 2000 rpm for 5 minutes. Decant the supernatant liquid from the centrifuge tube, crush the centrifuged solids with the same stirring rod previously used, and refill the tube to the 40-ml. mark with chilled (5°) *Phosphatide-Acetone Solution,* and repeat the chilling, stirring, centrifugation, and decantation procedure previously followed. After the second centrifugation and decantation of the supernatant acetone, again crush the solids with the assigned stirring rod, and place the tube and its contents in a horizontal position at room temperature until the excess acetone has evaporated. Mix the residue again, dry the centrifuge tube and its contents at 105° for 45 minutes in a forced draft oven, cool, and weigh. Calculate the percent of acetone-insoluble matter by the formula $(100R/S) - B$, in which R is the weight of residue, S is the weight of the sample, and B is the percent of *Benzene-insoluble matter* determined as directed in this monograph.

Acid value. If it is plastic or semisolid, soften a portion of the sample by warming it in a water bath at a temperature not exceeding 60° and then mix it thoroughly. Transfer about 2 grams of a well-mixed sample into a 250-ml. Erlenmeyer flask and dissolve it in 50 ml. of petroleum ether. To this solution add 50 ml. of alcohol, previously neutralized to phenolphthalein with 0.1 N sodium hydroxide, and mix well. Add phenolphthalein T.S. and titrate with 0.1 N sodium hydroxide to a pink end-point which persists for 5 seconds. Calculate the number of mg. of potassium hydroxide required to neutralize the acids in 1 gram of the sample by multiplying the number of ml. of 0.1 N sodium hydroxide consumed in the titration by 5.6 and dividing the result by the weight of the sample.

Benzene-insoluble matter. If plastic or semisolid, soften a portion of the sample by warming it at a temperature not exceeding 60° and then mix it thoroughly. Weigh 10 grams of a previously well-mixed sample into a 250-ml. wide-mouth Erlenmeyer flask, add 100 ml. of benzene, and shake until the sample is dissolved. Filter the solution through a 30-ml. Corning "C" porosity or equivalent filtering funnel which previously has been dried at 105° for 1 hour, cooled in a desiccator, and weighed. Wash the flask with two successive 25-ml. portions of benzene and pass the washings through the filter. Dry the funnel at 105° for 1 hour, cool to room temperature in a desiccator, and weigh. From the gain in weight of the funnel calculate the percent of the benzene-insoluble matter in the sample.

Iodine value. Determine by the *Wijs Method*, page 906.

Peroxide value. Weigh accurately about 10 grams of the sample, add 30 ml. of a 3:2 mixture of glacial acetic acid-chloroform, and mix. Add 1 ml. of a saturated solution of potassium iodide, mix, and allow to stand for 10 minutes. Add 100 ml. of water, begin titrating with 0.05 N sodium thiosulfate, adding starch T.S. as the end-point is approached, and continue the titration until the blue starch color has just disappeared. Perform a blank determination (see page 2), and make any necessary correction. Calculate the peroxide value, as mcg. of peroxide per kg. of sample, by the formula $S \times N \times 1000/W$, in which S is the net volume, in ml., of sodium thiosulfate solution required for the sample, N is the exact normality of the sodium thiosulfate solution, and W is the weight, in grams, of the sample taken.

Water. Determine by the *Toluene Distillation Method*, page 979, using a 100-gram sample, accurately weighed.

Arsenic. A *Sample Solution* prepared as directed for organic compounds meets the requirements of the *Arsenic Test*, page 865.

Heavy metals. Prepare and test a 500-mg. sample as directed in *Method II* under the *Heavy Metals Test*, page 920, using 20 mcg. of lead ion (Pb) in the control (*Solution A*).

Lead. A *Sample Solution* prepared as directed for organic compounds meets the requirements of the *Lead Limit Test*, page 929, using 10 mcg. of lead ion (Pb) in the control.

Packaging and storage. Store in well-closed containers.
Functional use in foods. Emulsifier; clouding agent.

HYDROXYPROPYL CELLULOSE

DESCRIPTION

Hydroxypropyl cellulose is a cellulose ether containing hydroxypropyl substitution. It occurs as a white powder. It is soluble in water and in certain organic solvents.

SPECIFICATIONS

Assay. Not more than 80.5 percent of hydroxypropoxyl groups (—$OCH_2CHOHCH_3$) after drying, equivalent to not more than 4.6 hydroxypropyl groups per anhydroglucose unit.

pH of a 2 percent solution. Between 5.0 and 8.0.

Viscosity of a 10 percent solution. Not less than 145 centipoises.

Limits of Impurities

Arsenic (as As). Not more than 3 parts per million (0.0003 percent).

Heavy metals (as Pb). Not more than 40 parts per million (0.004 percent).

Lead. Not more than 10 parts per million (0.001 percent).

Loss on drying. Not more than 5 percent.

Residue on ignition. Not more than 0.5 percent.

TESTS

Assay. Weigh accurately about 85 mg. of the sample, previously dried at 105° for 3 hours, and determine the hydroxypropoxyl content as directed under *Hydroxypropoxyl Determination*, page 923.

pH of a 2 percent solution. Determine by the *Potentiometric Method*, page 941.

Viscosity of a 10 percent solution. Transfer an accurately weighed sample, equivalent to 20 grams of hydroxypropyl cellulose on the dried basis, into a tared sample container, and proceed as directed under *Viscosity of Sodium Carboxymethylcellulose*, page 973.

Arsenic. A *Sample Solution* prepared as directed for organic compounds meets the requirements of the *Arsenic Test*, page 865.

Heavy metals. Prepare and test a 500-mg. sample as directed in *Method II* under the *Heavy Metals Test*, page 920, adding 1 ml. of hydroxylamine hydrochloride solution (1 in 5) to the solution of the residue. Any color does not exceed that produced in a control (*Solution A*) containing 20 mcg. of lead ion (Pb).

Lead. A *Sample Solution* prepared as directed for organic compounds meets the requirements of the *Lead Limit Test*, page 929, using 10 mcg. of lead ion (Pb) in the control.

Loss on drying, page 931. Dry at 105° for 3 hours.

Residue on ignition. Ignite 1 gram as directed in the general method, page 945.

Packaging and storage. Store in well-closed containers.

Functional use in foods. Emulsifier; film former; protective colloid; stabilizer; suspending agent; thickener.

HYDROXYPROPYL METHYLCELLULOSE

Propylene Glycol Ether of Methylcellulose

DESCRIPTION

Hydroxypropyl methylcellulose is the propylene glycol ether of methylcellulose in which both the hydroxypropyl and methyl groups are attached to the anhydroglucose rings of cellulose by ether linkages. Several product types are available which are defined by varying combinations of methoxyl and hydroxypropoxyl content. It occurs as a white, fibrous powder or as granules. It is soluble in water and in certain organic solvent systems. Aqueous solutions are surface active, form films upon drying, and undergo a reversible transformation from sol to gel upon heating and cooling, respectively.

IDENTIFICATION

A. Add 1 gram to 100 ml. of water. It swells and disperses to form a clear to opalescent mucilaginous solution, depending upon the intrinsic viscosity, which is stable in the presence of most electrolytes.

B. Add 1 gram to 100 ml. of boiling water and stir the mixture. A slurry is formed which, when cooled to 20°, dissolves to form a clear or opalescent mucilaginous solution.

C. Pour a few ml. of the solution prepared for *Identification Test B* onto a glass plate, and allow the water to evaporate. A thin, self-sustaining film results.

SPECIFICATIONS

Assay for hydroxypropoxyl groups. Within the range claimed by the vendor for any product type between a minimum of 3.0 percent and a maximum of 12.0 percent of hydroxypropoxyl groups ($—OCH_2$-$CHOHCH_2$).

Assay for methoxyl groups. Within the range claimed by the vendor for any product type between a minimum of 19.0 percent and a maximum of 30.0 percent of methoxyl groups ($—OCH_3$).

Viscosity. The viscosity of a solution containing 2 grams in each 100 grams of solution is not less than 80 percent and not more than 120 percent of that stated on the label for viscosity types of 100 centipoises or less, and not less than 75 percent and not more than 140 percent of that stated on the label for viscosity types higher than 100 centipoises.

Limits of Impurities

Arsenic (as As). Not more than 3 parts per million (0.0003 percent).

Heavy metals (as Pb). Not more than 10 parts per million (0.001 percent).

Loss on drying. Not more than 5 percent.

Residue on ignition. Not more than 1.5 percent for products with viscosities of 50 centipoises or above; not more than 3.0 percent for products with viscosities below 50 centipoises.

TESTS

Assay for hydroxypropoxyl groups. Proceed as directed for *Hydroxypropoxyl Determination*, page 923.

Assay for methoxyl groups. Place about 50 mg., previously dried at 105° for 3 hours, in an empty, tared gelatin capsule, weigh accurately, place the capsule and contents in the boiling flask of a methoxyl apparatus, and proceed as directed for *Methoxyl Determination*, page 935. Each ml. of 0.1 *N* sodium thiosulfate is equivalent to 0.5172 mg. of methoxyl groups ($—OCH_3$). Correct the percent of methoxyl groups ($—OCH_3$) thus determined by the formula $A - (B \times 0.93 \times 31/75)$ in which A is the total $—OCH_3$ groups determined in the *Assay*, B is the percent of $—OCH_2CHOHCH_3$ determined in the *Assay for hydroxypropoxyl groups*, and 0.93 is an average obtained by determining, on a large number of samples, the propylene produced from the reaction of hydriodic acid with hydroxypropoxyl groups during the assay for methoxyl groups. The result represents the actual percent of methoxyl groups ($—OCH_3$) determined in the sample taken for the assay.

Viscosity. Weigh accurately a sample, equivalent to 2 grams of solids on the dried basis, transfer to a wide-mouth, 250-ml. centrifuge bottle, and add 98 grams of water previously heated to between 80° and 90°. Stir with a mechanical stirrer for 10 minutes, then place the bottle in an ice bath until solution is complete, adjust the weight of the solution to 100 grams if necessary, and centrifuge it to expel any entrapped air. Adjust the temperature of the solution to 20° ± 0.1°, and determine the viscosity as directed under *Viscosity of Methylcellulose*, page 971.

Arsenic. A *Sample Solution* prepared as directed for organic compounds meets the requirements of the *Arsenic Test*, page 865.

Heavy metals. Prepare and test a 2-gram sample as directed in *Method II* under the *Heavy Metals Test*, page 920, adding 1 ml. of hydroxylamine hydrochloride solution (1 in 5) to the solution of the residue. Any color does not exceed that produced in a control (*Solution A*) containing 20 mcg. of lead ion (Pb).

Loss on drying, page 931. Dry a 3-gram sample at 105° for 2 hours.

Residue on ignition. Ignite 1 gram as directed in the general method, page 945.

Packaging and storage. Store in well-closed containers.

Functional use in foods. Thickening agent; stabilizer; emulsifier.

INDOLE

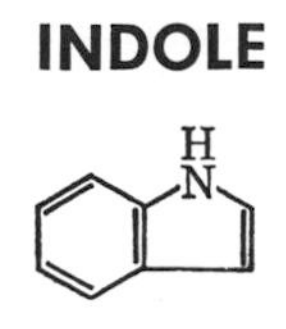

C_8H_7N Mol. wt. 117.15

DESCRIPTION

A white, lustrous, flaky, crystalline product having an unpleasant odor in high concentration, but free from fecal quality. The odor becomes floral in higher dilutions. It is soluble in most fixed oils and in propylene glycol. It is insoluble in glycerin and in mineral oil.

SPECIFICATIONS

Solidification point. Not lower than 51°.

Solubility in alcohol. Passes test.

TESTS

Solidification point. Proceed as directed in the general method, page 954, using a sample which has been dried over sulfuric acid.

Solubility in alcohol. Proceed as directed in the general method, page 899. One gram dissolves in 3 ml. of 70 percent alcohol.

Packaging and storage. Store in full, tight, preferably amber glass bottles, tin-lined boxes, or tightly sealed paper-lined drums in a cool place protected from light. Discoloration which occurs on exposure to light or heat does not materially affect the odor.

Functional use in foods. Flavoring agent.

INOSITOL

1,2,3,5/4,6-Cyclohexanehexol; *i*-Inositol; *meso*-Inositol

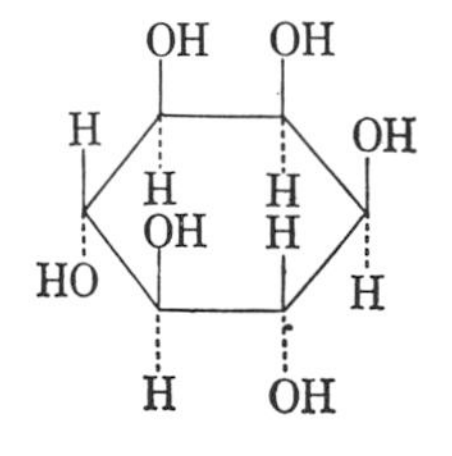

$C_6H_{12}O_6$ Mol. wt. 180.16

DESCRIPTION

It occurs as fine, white crystals or as a white crystalline powder. It is odorless, has a sweet taste, and is stable in air. Its solutions are neutral to litmus. It is optically inactive. One gram is soluble in 6 ml. of water. It is slightly soluble in alcohol, and is insoluble in ether and in chloroform.

IDENTIFICATION

A. To 1 ml. of a 1 in 50 solution in a porcelain evaporating dish, add 6 ml. of nitric acid, and evaporate to dryness on a water bath. Dissolve the residue in 1 ml. of water, add 0.5 ml. of a 1 in 10 solution of strontium acetate, and again evaporate to dryness on a steam bath. A violet color is produced.

B. The inositol hexaacetate obtained in the *Assay* melts between 212° and 216° (see page 931).

SPECIFICATIONS

Assay. Not less than 97.0 percent of $C_6H_{12}O_6$ after drying.

Melting range. Between 224° and 227°.

Limits of Impurities

Arsenic (as As). Not more than 3 parts per million (0.0003 percent).

Calcium. Passes test.

Chloride. Not more than 50 parts per million (0.005 percent).

Heavy metals (as Pb). Not more than 20 parts per million (0.002 percent).

Lead. Not more than 10 parts per million (0.001 percent).

Loss on drying. Not more than 0.5 percent.

Residue on ignition. Not more than 0.1 percent.

Sulfate. Not more than 60 parts per million (0.006 percent).

TESTS

Assay. Transfer about 200 mg., previously dried at 105° for 4 hours and accurately weighed, to a 250-ml. beaker, add 5 ml. of a mixture consisting of 1 part of diluted sulfuric acid T.S. in 50 parts of acetic anhydride, and cover the beaker with a watch glass. Heat on a steam bath for 20 minutes, then chill in an ice bath, and add 100 ml. of water. Boil for 20 minutes, allow to cool and transfer quantitatively, with the aid of a little water, to a 250-ml. separator. Extract the solution with 6 successive, 30- , 25- , 20- , 15- , 10- , and 10-ml. portions of chloroform, using the solvent to rinse the original flask. Collect the chloroform extracts in a second 250-ml. separator and wash the combined extracts with 10 ml. of water. Transfer the chloroform solution through a funnel containing a pledget of cotton into a 150-ml. tared Soxhlet flask. Wash the separator and funnel with 10 ml. of chloroform and add to the combined extracts. Evaporate to dryness on a steam bath, dry in an oven at 105° for 1 hour, cool in a desiccator, and weigh. The weight of the inositol hexaacetate obtained, multiplied by 0.4167, represents the equivalent of $C_6H_{12}O_6$.

Melting range. Determine as directed in the general procedure, page 931.

Arsenic. A *Sample Solution* prepared as directed for organic compounds meets the requirements of the *Arsenic Test,* page 865.

Calcium. To 10 ml. of a 1 in 10 solution add 1 ml. of ammonium oxalate T.S. The solution remains clear for at least 1 minute.

Chloride, page 879. Any turbidity produced by a 400-mg. sample does not exceed that shown in a control containing 20 mcg. of chloride ion (Cl).

Heavy metals. A solution of 1 gram in 25 ml. of water meets the requirements of the *Heavy Metals Test,* page 920, using 20 mcg. of lead ion (Pb) in the control (*Solution A*).

Lead. A *Sample Solution* prepared as directed for organic compounds meets the requirements of the *Lead Limit Test,* page 929, using 10 mcg. of lead ion (Pb) in the control.

Loss on drying, page 931. Dry at 105° for 4 hours.

Residue on ignition, page 945. Ignite 2 grams as directed in the general method.

Sulfate, page 879. Any turbidity produced by a 5-gram sample does not exceed that shown in a control containing 300 mcg. of sulfate (SO_4).

Packaging and storage. Store in well-closed containers.

Functional use in foods. Nutrient; dietary supplement.

α-IONONE

4(2,6,6-Trimethyl-2-cyclohexenyl)-3-butene-2-one

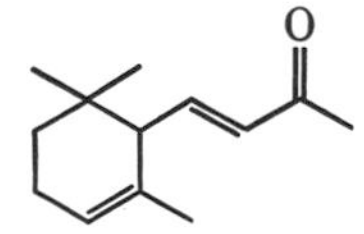

$C_{13}H_{20}O$ Mol. wt. 192.30

DESCRIPTION

A colorless to pale yellow liquid having a woody-violet odor. It is soluble in most fixed oils, in mineral oil, and in propylene glycol. It is insoluble in glycerin and in water.

SPECIFICATIONS

Assay. Not less than 99.0 percent of $C_{13}H_{20}O$.

Refractive Index. Between 1.497 and 1.502 at 20°.

Solubility in alcohol. Passes test.

Specific gravity. Between 0.927 and 0.933.

TESTS

Assay. Weigh accurately about 1.3 grams, and proceed as directed under *Aldehydes and Ketones-Hydroxylamine Method*, page 894, using 96.15 as the equivalence factor (E) in the calculation.

Refractive index, page 945. Determine with an Abbé or other refractometer of equal or greater accuracy.

Solubility in alcohol. Proceed as directed in the general method, page 899. One ml. dissolves in 10 ml. of 60 percent alcohol.

Specific gravity. Determine by any reliable method (see page 5).

Packaging and storage. Store in full, tight, preferably glass, tin, aluminum, or galvanized iron containers in a cool place protected from light.

Functional use in foods. Flavoring agent.

β-IONONE

4(2,6,6-Trimethyl-1-cyclohexenyl)-3-butene-2-one

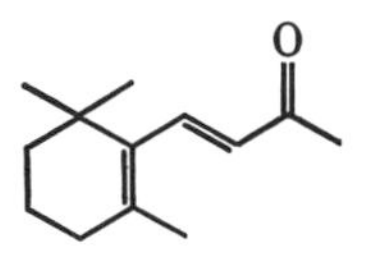

$C_{13}H_{20}O$ Mol. wt. 192.30

DESCRIPTION

A slightly yellow liquid having a more fruity and woody odor than α-ionone. It is soluble in most fixed oils, in mineral oil, and in propylene glycol. It is insoluble in glycerin and in water.

SPECIFICATIONS

Assay. Not less than 90.0 percent of $C_{13}H_{20}O$.

Refractive index. Between 1.517 and 1.522 at 20°.

Specific gravity. Between 0.940 and 0.947.

TESTS

Assay. Weigh accurately about 1.3 grams, and proceed as directed under *Aldehydes and Ketones-Hydroxylamine Method*, page 894, using 96.15 as the equivalence factor (E) in the calculation.

Refractive index, page 945. Determine with an Abbé or other refractometer of equal or greater accuracy.

Specific gravity. Determine by any reliable method (see page 5).

Packaging and storage. Store in full, tight, preferably glass, tin, aluminum, or galvanized iron containers in a cool place protected from light.

Functional use in foods. Flavoring agent.

IRON, ELECTROLYTIC

Fe At. wt. 55.85

DESCRIPTION

Electrolytic iron is elemental iron obtained by electrodeposition in the form of an amorphous, lusterless, grayish black powder. It is stable in dry air. It dissolves in dilute mineral acids with the evolution of hydrogen and the formation of solutions of the corresponding salts, which give positive tests for *Ferrous salts,* page 927.

SPECIFICATIONS

Assay. Not less than 97.0 percent of Fe.

Sieve analysis. Not less than 100 percent passes through a 100-mesh sieve; not less than 95 percent passes through a 325-mesh sieve.

Limits of Impurities

Acid-insoluble substances. Not more than 0.2 percent.
Arsenic (as As). Not more than 4 parts per million (0.004 percent).
Lead. Not more than 20 parts per million (0.0020 percent).
Mercury. Not more than 2 parts per million (0.0002 percent).

TESTS

Assay. Determine as directed under *Iron, Reduced,* page 394.

Sieve analysis. Determine as directed under *Sieve Analysis of Granular Metal Powders,* page 953.

Acid-insoluble substances. Dissolve 1 gram in 25 ml. of diluted sulfuric acid T.S., and heat on a steam bath until the evolution of hydrogen ceases. Filter through a tared filter crucible, wash with water until free from sulfate, dry at 105° for 1 hour, cool, and weigh.

Arsenic. Dissolve 1 gram in 25 ml. of diluted sulfuric acid T.S., heat on a steam bath until the evolution of hydrogen ceases, cool, and dilute to 35 ml. with water. This solution meets the requirements of the *Arsenic Test,* page 865, using 4.0 ml. of *Standard Arsenic Solution* (4 mcg. As) in the control.

Lead. Determine as directed under *Iron, Reduced,* page 394, but prepare the control, as directed in the *Procedure,* with 4.0 ml. of *Diluted Standard Lead Solution* (4 mcg. Pb).

Mercury. Determine as directed under *Iron, Reduced,* page 394, but use 2 grams of the sample and 40 ml. of *Sodium Citrate Solution* in preparing the *Sample Solution,* and prepare the *Diluted Standard Mercury Solution* as follows: Transfer 4.0 ml. of *Mercury Stock Solution* into a 250-ml. volumetric flask, dilute to volume with 1 *N* hydrochloric acid, and mix (1 ml. = 4 mcg. of Hg). Modify the first sentence of the *Procedure* to read: "Prepare a control by treating 1.0 ml. of *Diluted Standard Mercury Solution* (4 mcg. Hg) in the same manner. . ."

Packaging and storage. Store in well-closed containers.
Functional use in foods. Nutrient; dietary supplement.

IRON, REDUCED

Fe At. wt. 55.85

DESCRIPTION

Reduced iron is elemental iron obtained by chemical process in the form of a grayish black powder, all of which should pass through a 100-mesh sieve. It is lusterless or has not more than a slight luster. When viewed under a microscope having a magnifying power of 100 diameters, it appears as an amorphous powder, free from particles having a crystalline structure. It is stable in dry air. It dissolves in dilute mineral acids with the evolution of hydrogen and the formation of solutions of the corresponding salts which give positive tests for *Ferrous salts,* page 927.

SPECIFICATIONS

Assay. Not less than 96.0 percent of Fe.

Limits of Impurities

Acid-insoluble substances. Not more than 1.25 percent.
Arsenic (as As). Not more than 8 parts per million (0.0008 percent).
Lead. Not more than 25 parts per million (0.0025 percent).
Mercury. Not more than 5 parts per million (0.0005 percent).

TESTS

Assay. Transfer about 200 mg., accurately weighed, into a 300-ml. Erlenmeyer flask, add 50 ml. of diluted sulfuric acid T.S., and close the flask with a stopper containing a Bunsen valve, made by inserting a glass tube connected to a short piece of rubber tubing with a slit on the side and a glass rod inserted in the other end and arranged so that gases can escape, but air cannot enter. Heat on a steam bath until the iron is dissolved, cool the solution, dilute it with 50 ml. of recently boiled and cooled water, add 2 drops of orthophenanthroline T.S., and titrate with 0.1 *N* ceric sulfate until the red color changes to a weak blue. Each ml. of 0.1 *N* ceric sulfate is equivalent to 5.585 mg. of Fe.

Acid-insoluble substances. Dissolve 1 gram in 25 ml. of diluted sulfuric acid T.S., and heat on a steam bath until the evolution of hydrogen ceases. Filter through a tared filter crucible, collecting the filtrate in a 100-ml. volumetric flask, wash with water until free from sulfate, and dry at 105° for 1 hour. The weight of the residue does not exceed 12.5 mg. Dilute the filtrate to volume with water for use in the test for *Arsenic.*

Arsenic. Transfer 40 ml. of the filtrate (equivalent to 400 mg. of Fe) obtained in the test for *Acid-insoluble substances* into an arsine generator flask, and continue as directed under *Procedure* in the *Arsenic Test,* page 865, beginning with "Add 20 ml. of dilute sulfuric acid (1 in 5) . . . ."

Lead

Solutions. Prepare as directed in the *Lead* test under *Ferrous Fumarate,* page 313.

Procedure. Transfer 200 mg. of the sample into a 150-ml. beaker, and add 8 ml. of hydrochloric acid and 2 ml. of nitric acid. Prepare a control containing 5.0 ml. of *Diluted Standard Lead Solution* (5 mcg. Pb), 8 ml. of hydrochloric acid, and 2 ml. of nitric acid, and carry the sample and the control solutions through the following procedure: After the initial reaction subsides, evaporate to dryness on a steam bath, cool, and dissolve in 10 ml. of dilute hydrochloric acid (1 in 2), warming on the steam bath, if necessary, to effect solution. Add 25 ml. of *Ammonium Citrate Solution,* heat on the steam bath for an additional few minutes, then add 7 ml. of stronger ammonia T.S., and cool. Adjust the pH, if necessary, to 9.0 (by means of a glass electrode or pH indicator paper), using either stronger ammonia T.S. or hydrochloric acid, and transfer to a separator. Extract with 5-ml. portions of *Dithizone Extraction Solution* until the extraction solution retains its original color, and combine the extracts in a second separator. Wash the combined extracts by shaking for 30 seconds with 10 ml. of *Citrate-Cyanide Wash Solution,* and then wash the wash solution with 3 ml. of *Dithizone Extraction Solution.* Combine the chloroform layers, add 20 ml. of dilute nitric acid (1 in 100), and shake for 30 seconds. Separate the layers, and shake the chloroform layer with an additional 5 ml. of the dilute nitric acid. Combine the acid washes in a small beaker, and adjust the pH with diluted ammonia T.S. to 2.5 ± 0.2 (by means of a glass electrode). Transfer the solution into a separator, add 2 ml. of *pH 2.5 Buffer Solution,* and shake the solution for 30 seconds with 30 ml. of *Dithizone-Carbon Tetrachloride Solutions.* Wash the carbon tetrachloride layer with 10 ml. of *pH 2.5 Wash Solution,* discard the carbon tetrachloride, and combine the aqueous layers. Add 4 ml. of *Ammonia-Cyanide Solution,* mix, and extract at once with 5-ml. portions of *Dithizone-Carbon Tetrachloride Solution* until the carbon tetrachloride shows no further pink color. Wash the combined extracts with 4 ml. of *Ammonia-Cyanide Wash Solution,* dry the stem of the separator, and drain the carbon tetrachloride through a plug of cotton to remove the last trace of water. Determine the absorbances of both solutions in 1-cm. cells at 520 mμ with a suitable spectrophotometer, using carbon tetrachloride as the blank. The absorbance of the sample solution should not exceed that of the control.

Mercury

Dithizone Stock Solution. Dissolve 30 mg. of dithizone in 1000 ml. of chloroform, add 5 ml. of alcohol, and mix. Store in a refrigerator in a dark bottle. Prepare fresh each month.

Dithizone Extraction Solution. On the day of use, dilute 30 ml. of *Dithizone Stock Solution* to 100 ml. with chloroform.

Hydroxylamine Hydrochloride Solution. Prepare as directed in the test for *Mercury* under *Ferrous Fumarate*, page 313.

Mercury Stock Solution. Transfer 33.8 mg., accurately weighed, of mercuric chloride into a 100-ml. volumetric flask, dissolve in 1 *N* hydrochloric acid, dilute to volume with the acid, and mix. This solution contains the equivalent of 250 mcg. of Hg in each ml.

Diluted Standard Mercury Solution. Transfer 2.0 ml. of *Mercury Stock Solution* into a 100-ml. volumetric flask, dilute to volume with 1 *N* hydrochloric acid, and mix. Each ml. contains the equivalent of 5 mcg. of Hg.

Sodium Citrate Solution. Dissolve 250 grams of sodium citrate dihydrate in 1000 ml. of water.

Sample Solution. Transfer 1 gram of the sample into a 250-ml. beaker, add 20 ml. of dilute nitric acid (1 in 2), and digest on a steam bath for about 45 minutes. Add 5 ml. of dilute hydrochloric acid (1 in 3), and continue heating on the steam bath until the sample is dissolved. Cool to room temperature, and filter, if necessary, through a medium-porosity filter paper. Wash with a few ml. of water, add 20 ml. of *Sodium Citrate Solution* and 1 ml. of *Hydroxylamine Hydrochloride Solution* to the filtrate, and adjust the pH to 1.8 with stronger ammonia T.S.

Procedure. [Because mercuric dithizonate is light-sensitive, this procedure should be performed in subdued light.] Prepare a control by treating 1.0 ml. of *Diluted Standard Mercury Solution* (5 mcg. Hg) in the same manner and with the same reagents as directed for the preparation of the *Sample Solution.* Transfer the control and the *Sample Solution* into separate 250-ml. separators, and treat both solutions as follows: Extract with 5 ml. of *Dithizone Extraction Solution,* shaking the mixtures vigorously for 1 minute. Drain carefully, collecting the chloroform in another separator. If the chloroform does not show a pronounced green color due to excess reagent, add another 5 ml. of the extraction solution, shake again, and drain into the separator. Continue the extraction with 5-ml. portions, if necessary, collecting each successive extract in the second separator, until the final chloroform layer contains dithizone in marked excess. To the combined chloroform extracts add 15 ml. of dilute hydrochloric acid (1 in 3), shake the mixture vigorously for 1 minute, and discard the chloroform. Extract with 2 ml. of chloroform, drain carefully, and discard the chloroform. Add 1 ml. of 0.05 *M* disodium ethylenediaminetetraacetate and 2 ml. of 6 *N* acetic acid to the aqueous layer. Slowly add 5 ml. of ammonia T.S., and cool the separator. Transfer the solution into a 150-ml. beaker, adjust the pH to 1.8 with ammonia T.S. or dilute nitric acid (1 in 10), using a pH meter, and return the solution to the separator. Add 5.0 ml. of *Dithizone Extraction Solution*, and shake vigorously for 1 minute. Allow the layers to separate, insert a plug of cotton into the stem of the separator, and collect the dithizone extract in a test tube. Determine the absorbance of each solution in 1-cm. cells at 490 mμ with a suitable spectrophotometer, using chloroform as the blank. The absorbance of the *Sample Solution* does not exceed that of the control.

Packaging and storage. Store in well-closed containers.

Functional use in foods. Nutrient; dietary supplement.

ISOAMYL ACETATE

Amyl Acetate; β-Methyl Butyl Acetate

$CH_3COOC_5H_{11}$

$C_7H_{14}O_2$ Mol. wt. 130.19

DESCRIPTION

A colorless liquid with a fruity, pear-like odor composed of the acetates of mixed isomeric amyl alcohols, with the two isoamyl alcohols predominating. It is slightly soluble in water and is miscible with alcohol, ether, ethyl acetate, most fixed oils, and mineral oil. It is insoluble in glycerin and practically insoluble in propylene glycol.

SPECIFICATIONS

Assay. Not less than 95.0 percent of $C_7H_{14}O_2$.

Refractive index. Between 1.400 and 1.404 at 20°.

Solubility in alcohol. Passes test.

Specific gravity. Between 0.868 and 0.878.

Limits of Impurities

Acid value. Not more than 1.0.

Arsenic (as As). Not more than 3 parts per million (0.0003 percent).

Heavy metals (as Pb). Not more than 40 parts per million (0.004 percent).

Lead. Not more than 10 parts per million (0.001 percent).

TESTS

Assay. Weigh accurately about 800 mg., and proceed as directed under *Ester Determination*, page 896, using 65.10 as the equivalence factor (*e*) in the calculation.

Refractive index, page 945. Determine with an Abbé or other refractometer of equal or greater accuracy.

Solubility in alcohol. Proceed as directed in the general method, page 899. One ml. dissolves in 3 ml. of 60 percent alcohol to form a clear solution.

Specific gravity. Determine by any reliable method (see page 5).

Acid value. Determine as directed in the general method, page 893.

Arsenic. A *Sample Solution* prepared as directed for organic compounds meets the requirements of the *Arsenic Test*, page 865.

Heavy metals. Prepare and test a 500-mg. sample as directed in *Method II* under the *Heavy Metals Test*, page 920, using 20 mcg. of lead ion (Pb) in the control (*Solution A*).

Lead. A *Sample Solution* prepared as directed for organic compounds meets the requirements of the *Lead Limit Test*, page 929, using 10 mcg. of lead ion (Pb) in the control.

Packaging and storage. Store in full, tight, preferably glass, tin, or suitably lined containers in a cool place protected from light.

Functional use in foods. Flavoring agent.

ISOAMYL BUTYRATE

Amyl Butyrate

$CH_3(CH_2)_2COOC_5H_{11}$

$C_9H_{18}O_2$ Mol. wt. 158.24

DESCRIPTION

A colorless liquid having a strong, characteristic fruity odor. It is usually prepared by esterification of isoamyl alcohols with butyric acid. It is soluble in most fixed oils and in mineral oil. It is insoluble in glycerin and in propylene glycol.

SPECIFICATIONS

Assay. Not less than 98.0 percent of $C_9H_{18}O_2$.

Refractive index. Between 1.409 and 1.414 at 20°.

Solubility in alcohol. Passes test.

Specific gravity. Between 0.860 and 0.864.

Limits of Impurities

Acid value. Not more than 1.0.

TESTS

Assay. Weigh accurately about 1 gram, and proceed as directed under *Ester Determination*, page 896, using 79.12 as the equivalence factor (*e*) in the calculation.

Refractive index, page 945. Determine with an Abbé or other refractometer of equal or greater accuracy.

Solubility in alcohol. Proceed as directed in the general method, page 899. One ml. dissolves in 4 ml. of 70 percent alcohol.

Specific gravity. Determine by any reliable method (see page 5).

Acid value. Determine as directed in the general method, page 893.

Packaging and storage. Store in full, tight, preferably glass, tin, or suitably lined containers in a cool place protected from light.

Functional use in foods. Flavoring agent.

ISOAMYL FORMATE

Amyl Formate

$HCOOC_5H_{11}$

$C_6H_{12}O_2$ Mol. wt. 116.16

DESCRIPTION

Isoamyl formate is a mixture of isomeric amyl formates, predominantly isoamyl formate. It is a colorless liquid with a plum-like odor. It is soluble in most fixed oils, in mineral oil, and in propylene glycol. It is practically insoluble in glycerin.

SPECIFICATIONS

Assay. Not less than 92.0 percent of $C_6H_{12}O_2$.

Refractive index. Between 1.396 and 1.400 at 20°.

Solubility in alcohol. Passes test.

Specific gravity. Between 0.878 and 0.885.

Limits of Impurities

Acid value. Not more than 1.0.

TESTS

Assay. Weigh accurately about 1 gram, and proceed as directed under *Ester Determination*, page 896, using 58.08 as the equivalence factor (*e*) in the calculation.

Refractive index, page 945. Determine with an Abbé or other refractometer of equal or greater accuracy.

Solubility in alcohol. Proceed as directed in the general method, page 899. One ml. dissolves in 4 ml. of 60 percent alcohol and remains in solution on dilution to 10 ml.

Specific gravity. Determine by any reliable method (see page 5).

Acid value. Determine as directed in the general method, page 893.

Packaging and storage. Store in full, tight, preferably glass containers in a cool place protected from light.

Functional use in foods. Flavoring agent.

ISOAMYL HEXANOATE

Amyl Hexanoate; Isoamyl Caproate; Pentyl Hexanoate

$CH_3(CH_2)_4COOC_5H_{11}$

$C_{11}H_{22}O_2$ Mol. wt. 186.30

DESCRIPTION

A colorless liquid having a characteristic fruity odor, composed of the hexanoates of mixed isomeric amyl alcohols, with the two isoamyl alcohols predominating. It is soluble in alcohol, in fixed oils, and in mineral oil. It is insoluble in glycerin and in propylene glycol.

SPECIFICATIONS

Assay. Not less than 98.0 percent of $C_{11}H_{22}O_2$.

Refractive index. Between 1.418 and 1.422 at 20°.

Solubility in alcohol. Passes test.

Specific gravity. Between 0.858 and 0.863.

Limits of Impurities

Acid value. Not more than 1.0.

Arsenic (as As). Not more than 3 parts per million (0.0003 percent).

Heavy metals (as Pb). Not more than 40 parts per million (0.004 percent).

Lead. Not more than 10 parts per million (0.001 percent).

TESTS

Assay. Weigh accurately about 1.2 grams and proceed as directed under *Ester Determination,* page 896, using 93.15 as the equivalence factor (*e*) in the calculation.

Refractive index, page 945. Determine with an Abbé or other refractometer of equal or greater accuracy.

Solubility in alcohol. Proceed as directed in the general method, page 899. One ml. dissolves in 3 ml. of 80 percent alcohol to form a clear solution.

Specific gravity. Determine by any reliable method (see page 5).

Acid value. Determine as directed in the general method, page 893.

Arsenic. A *Sample Solution* prepared as directed for organic compounds meets the requirements of the *Arsenic Test,* page 865.

Heavy metals. Prepare and test a 500-mg. sample as directed in *Method II* under the *Heavy Metals Test,* page 920, using 20 mcg. of lead ion (Pb) in the control (*Solution A*).

Lead. A *Sample Solution* prepared as directed for organic compounds meets the requirements of the *Lead Limit Test,* page 929, using 10 mcg. of lead ion (Pb) in the control.

Packaging and storage. Store in full, tight, preferably glass, tin, galvanized, or suitably lined containers in a cool place protected from light.

Functional use in foods. Flavoring agent.

ISOAMYL ISOVALERATE

Amyl Valerate; Amyl Isovalerianate

$(CH_3)_2CHCH_2COOCH_2CH_2CH(CH_3)_2$

$C_{10}H_{20}O_2$ Mol. wt. 172.27

DESCRIPTION

A clear, colorless liquid having a fruity odor which upon dilution resembles apples. It is miscible with alcohol and with most fixed oils, and it is slightly soluble in propylene glycol. One part is soluble in 6 parts of 70 percent alcohol.

SPECIFICATIONS

Assay. Not less than 98.0 percent of $C_{10}H_{20}O_2$.

Refractive index. Between 1.411 and 1.414 at 20°.

Specific gravity. Between 0.854 and 0.857.

Limits of Impurities

Acid value. Not more than 2.0.

TESTS

Assay. Weigh accurately about 1.5 grams, and proceed as directed under *Ester Determination*, page 896, using 86.14 as the equivalence factor (*e*) in the calculation.

Refractive index, page 945. Determine with an Abbé or other refractometer of equal or greater accuracy.

Specific gravity. Determine by any reliable method (see page 5).

Acid value. Determine as directed in the general method, page 893.

Packaging and storage. Store in full, tight containers in a cool place protected from light.

Functional use in foods. Flavoring agent.

ISOAMYL SALICYLATE

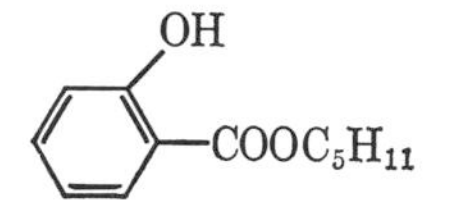

Amyl Salicylate

$C_{12}H_{16}O_3$ Mol. wt. 208.26

DESCRIPTION

A colorless liquid having a characteristic, pleasant odor. It is practically insoluble in glycerin, in propylene glycol, and in water, but is miscible with alcohol, with chloroform, with ether, and with most fixed oils.

SPECIFICATIONS

Assay. Not less than 98.0 percent of $C_{12}H_{16}O_3$.

Refractive index. Between 1.505 and 1.509 at 20°.

Solubility in alcohol. Passes test.

Specific gravity. Between 1.047 and 1.053.

Limits of Impurities

Acid value. Not more than 1.0.

TESTS

Assay. Weigh accurately about 1.3 grams, and proceed as directed under *Ester Determination*, page 896, using 104.13 as the equivalence factor (e) in the calculation, but reflux for 2 hours and use phenol red T.S. as the indicator.

Refractive index, page 945. Determine with an Abbé or other refractometer of equal of greater accuracy.

Solubility in alcohol. Proceed as directed in the general method, page 899. One ml. dissolves in 3 ml. of 90 percent alcohol and remains in solution upon further dilution.

Specific gravity. Determine by any reliable method (see page 5).

Acid value. Determine as directed in the general method, page 893, using phenol red T.S. as the indicator.

Packaging and storage. Store in full, tight, preferably glass, tin-lined, or other suitably lined containers in a cool place.

Functional use in foods. Flavoring agent.

ISOBORNYL ACETATE

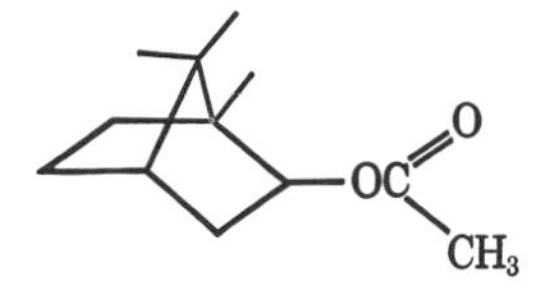

$C_{12}H_{20}O_2$ Mol. wt. 196.29

DESCRIPTION

A clear, colorless liquid when fresh, but it develops a very pale straw shade on storage. It has an agreeable camphoraceous odor somewhat like pine needles or hemlock. It is soluble in most fixed oils and in mineral oil. It is sparingly soluble in propylene glycol, and it is insoluble in glycerin and in water.

SPECIFICATIONS

Assay. Not less than 97.0 percent of $C_{12}H_{20}O_2$.

Angular rotation. Between $-1°$ and $+1°$.

Refractive index. Between 1.462 and 1.465 at 20°.

Solubility in alcohol. Passes test.

Specific gravity. Between 0.980 and 0.984.

Limits of Impurities

Acid value. Not more than 1.0.

TESTS

Assay. Weigh accurately about 1 gram, and proceed as directed under *Ester Determination*, page 896, using 98.15 as the equivalence factor (e) in the calculation.

Angular rotation. Determine in a 100-mm. tube as directed under *Optical Rotation*, page 939.

Refractive index, page 945. Determine with an Abbé or other refractometer of equal or greater accuracy.

Solubility in alcohol. Proceed as directed in the general method, page 899. One ml. dissolves in 3 ml. of 70 percent alcohol.

Specific gravity. Determine by any reliable method (see page 5).

Acid value. Determine as directed in the general method, page 893.

Packaging and storage. Store in full, tight, preferably glass, tin-lined, or aluminum containers in a cool place protected from light.

Functional use in foods. Flavoring agent.

ISOBUTYL ACETATE

$CH_3COOCH_2CH(CH_3)_2$

$C_6H_{12}O_2$ Mol. wt. 116.16

DESCRIPTION

A clear, colorless liquid having a fruity odor which upon dilution resembles bananas. It is soluble in alcohol, in propylene glycol, in most fixed oils, and in mineral oil. One part is soluble in about 180 parts of water.

SPECIFICATIONS

Assay. Not less than 90.0 percent of $C_6H_{12}O_2$.

Refractive index. Between 1.389 and 1.392 at 20°.

Specific gravity. Between 0.862 and 0.871.

Limits of Impurities

Acid value. Not more than 1.0.

TESTS

Assay. Weigh accurately about 1 gram, and proceed as directed under *Ester Determination*, page 896, using 58.08 as the equivalence factor (*e*) in the calculation.

Refractive index, page 945. Determine with an Abbé or other refractometer of equal or greater accuracy.

Specific gravity. Determine by any reliable method (see page 5).

Acid value. Determine as directed in the general method, page 893.

Packaging and storage. Store in full, tight containers in a cool place protected from light.

Functional use in foods. Flavoring agent.

ISOBUTYL ALCOHOL

Isobutanol

$(CH_3)_2CHCH_2OH$

$C_4H_{10}O$ Mol. wt. 74.12

DESCRIPTION

A clear, colorless, mobile liquid having a characteristic, penetrating vinous odor. It is miscible in all proportions with alcohol and ether. One ml. dissolves in about 140 ml. of water.

SPECIFICATIONS

Distillation range. Within a 1.5° range.

Specific gravity. Between 0.799 and 0.801.

Limits of Impurities

Acidity (as acetic acid). Not more than 0.003 percent.

Residue on evaporation. Not more than 1 mg. per 100 ml.

Water. Not more than 0.1 percent.

TESTS

Distillation range. Distil 100 ml. as directed in the general method, page 890. The difference between the temperature observed at the beginning and the end of the distillation does not exceed 1.5°. The boiling point of pure isobutyl alcohol at 760 mm. pressure is 107.9°.

Specific gravity. Determine by any reliable method (see page 5).

Acidity. Transfer 75 ml. (60 grams) into a 250-ml. Erlenmeyer flask, add phenolphthalein T.S., and titrate with 0.1 *N* alcoholic potassium hydroxide to a pink end-point that persists for at least 15 seconds. Each ml. of 0.1 *N* alcoholic potassium hydroxide is equivalent to 6.01 mg. of $C_2H_4O_2$.

Residue on evaporation, page 899. Evaporate 100 ml. to dryness in a tared platinum dish on a steam bath, and dry at 105° for 30 minutes. Cool in a desiccator and weigh. The weight of the residue does not exceed 1 mg.

Water. Determine by the *Karl Fischer Titrimetric Method,* page 977.

Packaging and storage. Store in tight containers.

Functional use in foods. Flavoring agent.

ISOBUTYL CINNAMATE

C_6H_5—CH═CHCOOCH$_2$CH(CH$_3$)CH$_3$

$C_{13}H_{16}O_2$ Mol. wt. 204.27

DESCRIPTION

A colorless liquid having a sweet, fruity, balsamic odor. It is miscible with alcohol, with chloroform, and with ether, but is practically insoluble in water.

SPECIFICATIONS

Assay. Not less than 98.0 percent of $C_{13}H_{16}O_2$.

Refractive index. Between 1.539 and 1.541 at 20°.

Solubility in alcohol. Passes test.

Specific gravity. Between 1.001 and 1.004.

Limits of Impurities

Acid value. Not more than 1.0.

TESTS

Assay. Weigh accurately about 1.4 grams, and proceed as directed under *Ester Determination*, page 896, using 102.14 as the equivalence factor (*e*) in the calculation.

Refractive index, page 945. Determine with an Abbé or other refractometer of equal or greater accuracy.

Solubility in alcohol. Proceed as directed in the general method, page 899. One ml. dissolves in 3 ml. of 80 percent alcohol to form a clear solution.

Specific gravity. Determine by any reliable method (see page 5).

Acid value. Determine as directed in the general method, page 893.

Packaging and storage. Store in full, tight, preferably glass, tin-lined or other suitably lined containers in a cool place protected from light.

Functional use in foods. Flavoring agent.

ISOBUTYLENE-ISOPRENE COPOLYMER

Butyl Rubber

DESCRIPTION

A synthetic copolymer containing from 0.5 to 2.0 molar percent of isoprene, the remainder, respectively, consisting of isobutylene. It is prepared by copolymerization of isobutylene and isoprene in methyl chloride solution, using aluminum chloride as catalyst. After completion of polymerization, the rubber particles are treated with hot water containing a suitable food-grade deagglomerating agent, such as stearic acid. Finally, the coagulum is dried to remove residual volatiles.

SPECIFICATIONS

Total unsaturation. Not less than 0.5 percent and not more than 2.0 percent, as isoprene.

Limits of Impurities

Arsenic (as As). Not more than 3 parts per million (0.0003 percent).

Heavy metals (as Pb). Not more than 40 parts per million (0.004 percent).

Lead. Not more than 3 parts per million (0.0003 percent).

TESTS

Total unsaturation. Determine as directed in the general method, page 878.

Arsenic. Prepare a *Sample Solution* as directed in the general method under *Chewing Gum Base*, page 877. This solution meets the requirements of the *Arsenic Test*, page 865.

Heavy metals. Prepare and test a 500-mg. sample as directed in *Method II* under the *Heavy Metals Test*, page 920, using 20 mcg. of lead ion (Pb) in the control (*Solution A*).

Lead. Prepare a *Sample Solution* as directed in the general method under *Chewing Gum Base*, page 878. This solution meets the requirements of the *Lead Limit Test*, page 929, using 10 mcg. of lead ion (Pb) in the control.

Packaging and storage. Store in well-closed containers.

Functional use in foods. Masticatory substance in chewing gum base.

ISOBUTYL PHENYLACETATE

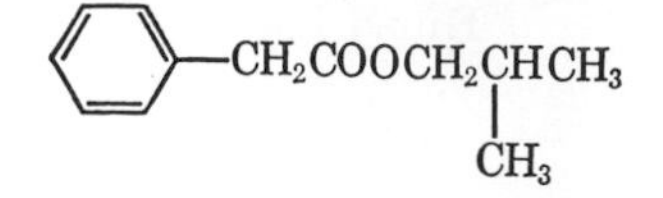

$C_{12}H_{16}O_2$ Mol. wt. 192.26

DESCRIPTION

A colorless liquid having a rose honey-like odor. It is soluble in most fixed oils, but it is insoluble in glycerin, in mineral oil, and in propylene glycol.

SPECIFICATIONS

Assay. Not less than 98.0 percent of $C_{12}H_{16}O_2$.

Refractive index. Between 1.486 and 1.488 at 20°.

Solubility in alcohol. Passes test.

Specific gravity. Between 0.984 and 0.988.

Limits of Impurities

Acid value. Not more than 1.0.

TESTS

Assay. Weigh accurately about 1.2 grams, and proceed as directed under *Ester Determination*, page 896, using 96.13 as the equivalence factor (*e*) in the calculation.

Refractive index, page 945. Determine with an Abbé or other refractometer of equal or greater accuracy.

Solubility in alcohol. Proceed as directed in the general method, page 899. One ml. dissolves in 2 ml. of 80 percent alcohol, and remains in solution on dilution to 10 ml.

Specific gravity. Determine by any reliable method (see page 5).

Acid value. Determine as directed in the general method, page 893.

Packaging and storage. Store in full, tight, preferably glass, aluminum, or tin-lined containers in a cool place protected from light.

Functional use in foods. Flavoring agent.

ISOBUTYL SALICYLATE

$$C_6H_4(OH)COOCH_2CH(CH_3)CH_3$$

$C_{11}H_{14}O_3$ Mol. wt. 194.23

DESCRIPTION

A colorless liquid having an orchid odor. It is soluble in most fixed oils and in mineral oil. It is practically insoluble in propylene glycol, and it is insoluble in glycerin.

SPECIFICATIONS

Assay. Not less than 98.0 percent of $C_{11}H_{14}O_3$.

Refractive index. Between 1.507 and 1.510 at 20°.

Solubility in alcohol. Passes test.

Specific gravity. Between 1.062 and 1.066.

Limits of Impurities

Acid value. Not more than 1.0.

TESTS

Assay. Weigh accurately about 1.3 grams, and proceed as directed under *Ester Determination*, page 896, using 97.12 as the equivalence factor (*e*) in the calculation. Modify the procedure by refluxing for 2 hours and by using phenol red T.S. as the indicator.

Refractive index, page 945. Determine with an Abbé or other refractometer of equal or greater accuracy.

Solubility in alcohol. Proceed as directed in the general method, page 899. One ml. dissolves in 9 ml. of 80 percent alcohol, and remains in solution on dilution to 10 ml.

Specific gravity. Determine by any reliable method (see page 5).

Acid value. Determine as directed in the general method, page 893, using phenol red T.S. as the indicator.

Packaging and storage. Store in full, tight, preferably glass, tin-lined, or other suitably lined containers in a cool place protected from light. Iron containers should not be used.

Functional use in foods. Flavoring agent.

ISOBUTYRALDEHYDE

$(CH_3)_2CHCHO$

C_4H_8O Mol. wt. 72.11

DESCRIPTION

A clear, colorless, mobile liquid having a characteristic, sharp pungent odor. It is miscible in all proportions with alcohol and with ether. One ml. dissolves in about 125 ml. of water.

SPECIFICATIONS

Assay. Not less than 98.0 percent of C_4H_8O.

Distillation range. Between 62.5° and 67° (first 97 percent).

Specific gravity. Between 0.783 and 0.788.

Limits of Impurities

Acidity (as butyric acid). Not more than 0.3 percent.

Water. Not more than 0.3 percent.

TESTS

Assay. Transfer 65 ml. of 0.5 *N* hydroxylamine hydrochloride and 50.0 ml. of 0.5 *N* triethanolamine into a suitable heat-resistant pressure bottle provided with a tight closure that can be securely fastened. Replace the air in the bottle by passing a gentle stream of nitrogen for two minutes through a glass tube positioned so that the end is just above the surface of the liquid. To the mixture in the pressure bottle add about 900 mg. of the sample contained in a sealed glass ampul and accurately weighed. Introduce several pieces of 8-mm. glass rod, cap the bottle, and shake vigorously to break the ampul. Allow the bottle to stand at room temperature, swirling occasionally, for 60 minutes. Cool, if necessary, and uncap the bottle cautiously to prevent any loss of the contents. Titrate with 0.5 *N* sulfuric acid to a greenish blue end-point. Perform a residual blank titration (see page 2). Each ml. of 0.5 *N* sulfuric acid is equivalent to 36.06 mg. of C_4H_8O.

Distillation range. Distil 100 ml. as directed in the general method, page 890.

Specific gravity. Determine by any reliable method (see page 5).

Acidity. Transfer 56 ml. (44 grams) into a 250-ml. Erlenmeyer flask through which a gentle stream of nitrogen has been passed for two minutes. Add phenolphthalein T.S. and titrate with 0.1 *N* alcoholic potassium hydroxide in an atmosphere of nitrogen to a pink end-point that persists for at least 15 seconds. Each ml. of 0.1 *N* alcoholic potassium hydroxide is equivalent to 8.81 mg. of $C_4H_8O_2$.

Water. Determine by the *Karl Fischer Titrimetric Method,* page 977, using freshly distilled pyridine instead of methanol as the solvent.

Packaging and storage. Store in tight containers.

Functional use in foods. Flavoring agent.

ISOBUTYRIC ACID

2-Methyl Propanoic Acid; Isopropylformic Acid

$$CH_3—\underset{\underset{CH_3}{|}}{C}HCOOH$$

$C_4H_8O_2$ Mol. wt. 88.11

DESCRIPTION

Isobutyric acid is a colorless liquid having a strong, penetrating odor suggestive of rancid butter. It is miscible with alcohol, propylene glycol, glycerin, mineral oil, and most fixed oils. It is soluble in water.

SPECIFICATIONS

Assay. Not less than 99.0 percent and not more than the equivalent of 101.0 percent of $C_4H_8O_2$.

Refractive index. Between 1.392 and 1.395 at 20°.

Specific gravity. Between 0.944 and 0.948.

Limits of Impurities

Heavy metals (as Pb). Not more than 40 parts per million (0.004 percent).

Reducing substances. Passes test.

TESTS

Assay. Transfer about 1.5 grams, accurately weighed, into a 250-ml. Erlenmeyer flask containing about 75 ml. of water, add phenolphthalein T.S., and titrate with 0.5 *N* sodium hydroxide. Each ml. of 0.5 *N* sodium hydroxide is equivalent to 44.06 mg. of $C_4H_8O_2$.

Refractive index, page 945. Determine with an Abbé or other refractometer of equal or greater accuracy.

Specific gravity. Determine by any reliable method (see page 5).

Heavy metals. Prepare and test a 500-mg. sample as directed in *Method II* under the *Heavy Metals Test*, page 920, using 20 mcg. of lead ion (Pb) in the control (*Solution A*).

Reducing substances. Dilute 2 ml. in a glass-stoppered flask with 50 ml. of water and 50 ml. of sulfuric acid, shaking the flask during the addition. While the solution is still warm, titrate with 0.1 *N* potassium permanganate. Not more than 1 ml. is required to produce a pink color that persists for 1 minute.

Packaging and storage. Store in tight, suitably lined containers, in a cool place, protected from light.

Functional use in foods. Flavoring agent.

ISOEUGENOL

2-Methoxy-4-propenylphenol

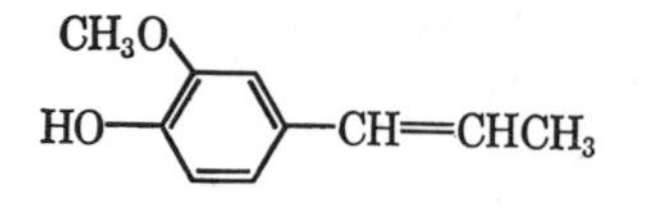

$C_{10}H_{12}O_2$ Mol. wt. 164.20

DESCRIPTION

A pale yellow, viscous liquid having a floral odor, reminiscent of carnation. It is soluble in most fixed oils and in propylene glycol. It is soluble, with turbidity, in mineral oil, but it is insoluble in glycerin.

SPECIFICATIONS

Assay. Not less than 99 percent of phenols, by volume.

Refractive index. Between 1.572 and 1.577 at 20°.

Solidification point. Not below 12°.

Solubility in alcohol. Passes test.

Specific gravity. Between 1.079 and 1.085.

TESTS

Assay. Proceed as directed under *Phenols*, page 898.

Refractive index, page 945. Determine with an Abbé or other refractometer of equal or greater accuracy.

Solidification point. Determine as directed in the general method, page 954.

Solubility in alcohol. Proceed as directed in the general method, page 899. One ml. dissolves in 5 ml. of 50 percent alcohol.

Specific gravity. Determine by any reliable method (see page 5).

Packaging and storage. Store in full, tight, preferably stainless steel, aluminum, suitably lined, or glass containers in a cool place protected from light.

Functional use in foods. Flavoring agent.

ISOEUGENYL ACETATE

2-Methoxy-4-propenyl Phenyl Acetate

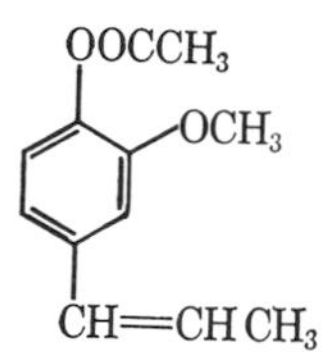

$C_{12}H_{14}O_3$ Mol. wt. 206.24

DESCRIPTION

White crystals having a spicy, clove-like odor. It is soluble in alcohol, in ether, and in chloroform, but practically insoluble in water.

SPECIFICATIONS

Assay. Not less than 98.0 percent of $C_{12}H_{14}O_3$.

Solidification point. Not less than 76°.

Solubility in alcohol. Passes test.

Limits of Impurities

Acid value. Not more than 2.0.

TESTS

Assay. Weigh accurately about 1.4 grams, and proceed as directed under *Ester Determination*, page 896, using 103.12 as the equivalence factor (*e*) in the calculation, but reflux for 4 hours and use phenol red T.S. as the indicator.

Solidification point. Determine as directed in the general method, page 954.

Solubility in alcohol. Proceed as directed in the general method, page 899. One gram dissolves in 27 ml. of 95 percent alcohol to form a clear solution.

Acid value. Determine as directed in the general method, page 893, using phenol red T.S. as the indicator.

Packaging and storage. Store in tight, light-resistant containers.

Functional use in foods. Flavoring agent.

DL-ISOLEUCINE

DL-2-Amino-3-methylvaleric Acid

$$\begin{array}{l} CH_3CH_2CHCHCOOH \\ \qquad\quad | \quad\; | \\ \qquad\;\; H_3C \;\; NH_2 \end{array}$$

$C_6H_{13}NO_2$ Mol. wt. 131.17

DESCRIPTION

A white, odorless, crystalline powder having a slightly bitter taste. It is soluble in water, but practically insoluble in alcohol and in ether. It melts with decomposition at about 292°. The pH of a 1 in 100 solution is between 5.5 and 7.0.

IDENTIFICATION

To 5 ml. of a 1 in 1000 solution add 1 ml. of triketohydrindene hydrate T.S. A bluish purple color is produced.

SPECIFICATIONS

Assay. Not less than 98.0 percent and not more than the equivalent of 102.0 percent of $C_6H_{13}NO_2$, calculated on the dried basis.

Limits of Impurities

Arsenic (as As). Not more than 3 parts per million (0.0003 percent).

Heavy metals (as Pb). Not more than 20 parts per million (0.002 percent).

Lead. Not more than 10 parts per million (0.001 percent).

Loss on drying. Not more than 0.3 percent.

Residue on ignition. Not more than 0.1 percent.

TESTS

Assay. Dissolve about 250 mg., accurately weighed, in 3 ml. of formic acid and 50 ml. of glacial acetic acid, add 2 drops of crystal violet T.S., and titrate with 0.1 *N* perchloric acid to the first appearance of a pure green color or until the blue color disappears completely. Each ml. of 0.1 *N* perchloric acid is equivalent to 13.12 mg. of $C_6H_{13}NO_2$.

Arsenic. A *Sample Solution* prepared as directed for organic compounds meets the requirements of the *Arsenic Test*, page 865.

Heavy metals. Prepare and test a 1-gram sample as directed in *Method II* under the *Heavy Metals Test*, page 920, using 20 mcg. of lead ion (Pb) in the control (*Solution A*).

Lead. A *Sample Solution* prepared as directed for organic compounds meets the requirements of the *Lead Limit Test*, page 929, using 10 mcg. of lead ion (Pb) in the control.

Loss on drying, page 931. Dry at 105° for 3 hours.

Residue on ignition, page 945. Ignite 1 gram as directed in the general method.

Packaging and storage. Store in well-closed containers.

Functional use in foods. Nutrient; dietary supplement.

L-ISOLEUCINE

L-2-Amino-3-methylvaleric Acid

$$\begin{array}{l} CH_3CH_2CHCHCOOH \\ \qquad\quad\;\; | \quad\; | \\ \qquad\;\; H_3C \;\; NH_2 \end{array}$$

$C_6H_{13}NO_2$ Mol. wt. 131.18

DESCRIPTION

Crystalline leaflets, or a white crystalline powder having a bitter taste. It is soluble in 25 parts of water, slightly soluble in hot alcohol, and soluble in diluted mineral acids and in alkaline solutions. It sublimes at between 168° and 170°, and melts with decomposition at about 284°. The pH of a 1 in 100 solution is between 5.5 and 7.0.

IDENTIFICATION

To 5 ml. of a 1 in 1000 solution add 1 ml. of triketohydrindene hydrate T.S. A reddish purple or bluish purple color is produced.

SPECIFICATIONS

Assay. Not less than 98.0 percent and not more than the equivalent of 102.0 percent of $C_6H_{13}NO_2$, calculated on the dried basis.

Specific rotation, $[\alpha]_D^{20°}$. Between +38.0° and +41.5°, on the dried basis.

Limits of Impurities

Arsenic (as As). Not more than 3 parts per million (0.0003 percent).

Heavy metals (as Pb). Not more than 20 parts per million (0.002 percent).

Lead. Not more than 10 parts per million (0.001 percent).

Loss on drying. Not more than 0.3 percent.

Residue on ignition. Not more than 0.2 percent.

TESTS

Assay. Dissolve about 250 mg., accurately weighed, in 3 ml. of formic acid and 50 ml. of glacial acetic acid, add 2 drops of crystal violet T.S., and titrate with 0.1 *N* perchloric acid to the first appearance of a pure green color or until the blue color disappears completely. Each ml. of 0.1 *N* perchloric acid is equivalent to 13.12 mg. of $C_6H_{13}NO_2$.

Specific rotation, page 939. Determine in a solution containing 4 grams of a previously dried sample in sufficient 6 *N* hydrochloric acid to make 100 ml.

Arsenic. A *Sample Solution* prepared as directed for organic compounds meets the requirements of the *Arsenic Test,* page 865.

Heavy metals. Prepare and test a 1-gram sample as directed in *Method II* under the *Heavy Metals Test,* page 920, using 20 mcg. of lead ion (Pb) in the control (*Solution A*).

Lead. A *Sample Solution* prepared as directed for organic compounds meets the requirements of the *Lead Limit Test,* page 929, using 10 mcg. of lead ion (Pb) in the control.

Loss on drying, page 931. Dry at 105° for 3 hours.

Residue on ignition, page 945. Ignite 1 gram as directed in the general method.

Packaging and storage. Store in well-closed containers.

Functional use in foods. Nutrient; dietary supplement.

ISOPROPYL ACETATE

$CH_3COOCH(CH_3)_2$

$C_5H_{10}O_2$ Mol. wt. 102.13

DESCRIPTION

A clear, colorless, mobile liquid having a characteristic odor. It is miscible in all proportions with alcohol and with ether. One ml. dissolves in about 35 ml. of water.

SPECIFICATIONS

Assay. Not less than 99.0 percent of $C_5H_{10}O_2$.

Distillation range. Between 86° and 90°.

Specific gravity. Between 0.866 and 0.869.

Limits of Impurities

Acidity (as acetic acid). Not more than 0.01 percent.

Heavy metals (as Pb). Not more than 40 parts per million (0.004 percent).

Residue on evaporation. Not more than 5 mg. per 100 ml.

Water. Not more than 0.1 percent.

TESTS

Assay. Transfer 25.0 ml. of 1 *N* potassium hydroxide into a suitable heat-resistant pressure bottle provided with a tight closure that can be securely fastened, and then add 10 ml. of isopropanol and a few pieces of glass rod. To the mixture in the pressure bottle add about 1.3 grams of the sample contained in a sealed glass ampul and accurately weighed. Cap the bottle, shake it vigorously to break the ampul, and allow it to stand at room temperature for 30 minutes. Uncap the bottle, add phenolphthalein T.S., and titrate with 0.5 *N* sulfuric acid to the disappearance of the pink color. Perform a residual blank titration (see *Blank Tests,* page 2). Each ml. of 0.5 *N* sulfuric acid is equivalent to 51.07 mg. of $C_5H_{10}O_2$.

Distillation range. Distil 100 ml. as directed in the general method, page 890.

Specific gravity. Determine by any reliable method (see page 5).

Acidity. Transfer 69 ml. (60 grams) into a 250-ml. Erlenmeyer flask, add phenolphthalein T.S., and titrate with 0.1 *N* alcoholic potassium hydroxide to a pink end-point that persists for at least 15 seconds. Each ml. of 0.1 *N* alcoholic potassium hydroxide is equivalent to 6 mg. of $C_2H_4O_2$.

Heavy metals. Prepare and test a 500-mg. sample as directed in *Method II* under the *Heavy Metals Test,* page 920, using 20 mcg. of lead ion (Pb) in the control (*Solution A*).

Residue on evaporation, page 945. Evaporate 100 ml. to dryness in a tared platinum dish on a steam bath, and dry at 105° for 30 minutes. Cool in a desiccator and weigh. The weight of the residue does not exceed 5 mg.

Water. Determine by the *Karl Fischer Titrimetric Method,* page 977.

Packaging and storage. Store in tight containers.

Functional use in foods. Flavoring agent.

ISOPROPYL ALCOHOL

2-Propanol; Isopropanol

$CH_3CHOHCH_3$

C_3H_8O Mol. wt. 60.10

DESCRIPTION

A clear, colorless, flammable liquid having a characteristic odor and a slightly bitter taste. It is miscible with water, with ethyl alcohol, ether, and with many other organic solvents. Its refractive index at 20° is about 1.377.

IDENTIFICATION

Add 3 ml. of water and 1 ml. of mercuric sulfate T.S. to 2 ml. of the sample contained in a test tube, and warm gently. A white or yellowish precipitate is formed.

SPECIFICATIONS

Assay. Not less than 99.7 percent of C_3H_8O, by weight.

Distillation range. Within a range of 1°, including 82.3°.

Solubility in water. Passes test.

Limits of Impurities

Acidity (as acetic acid). Not more than 10 parts per million (0.001 percent).

Heavy metals (as Pb). Not more than 1 part per million (0.0001 percent).

Nonvolatile residue. Not more than 10 parts per million (0.001 percent).

Substances reducing permanganate. Passes test.

Water. Not more than 0.2 percent.

TESTS

Assay. Its specific gravity, determined by any reliable method (see page 5), is not greater than 0.7840 at 25°/25° (equivalent to 0.7870 at 20°/20°).

Distillation range. Proceed as directed in the general method, page 890.

Solubility in water. Mix 10 ml. of the sample with 40 ml. of water. After 1 hour, the solution is as clear as an equal volume of water.

Acidity. Add 2 drops of phenolphthalein T.S. to 100 ml. of water, add 0.01 *N* sodium hydroxide to the first pink color that persists for at least 30 seconds, then add 50 ml. (about 39 grams) of the sample, and mix. Not more than 0.7 ml. of 0.1 *N* sodium hydroxide is required to restore the pink color.

Heavy metals. Evaporate 25 ml. (about 20 grams) of the sample to dryness on a steam bath in a glass evaporating dish. Cool, add 2 ml. of hydrochloric acid, and slowly evaporate to dryness again on the steam bath. Moisten the residue with 1 drop of hydrochloric acid, add 10 ml. of hot water, and digest for 2 minutes. Cool, and dilute to 25 ml. with water. This solution meets the requirements of the *Heavy Metals Test*, page 920, using 20 mcg. of lead ion (Pb) in the control (*Solution A*).

Nonvolatile residue. Evaporate 125 ml. (about 100 grams) of the sample to dryness in a tared dish on a steam bath, dry the residue at 105° for 30 minutes, cool, and weigh.

Substances reducing permanganate. Transfer 50 ml. of the sample into a 50-ml. glass-stoppered cylinder, add 0.25 ml. of 0.1 *N* potassium permanganate, mix, and allow to stand for 10 minutes. The pink color is not entirely discharged.

Water. Determine by the *Karl Fischer Titrimetric Method*, page 977.

Packaging and storage. Store in tight containers, remote from fire.

Functional use in foods. Extraction solvent.

ISOPULEGOL

p-Menth-8-en-3-ol

CH_3

OH

$CH_3—C═CH_2$

$C_{10}H_{18}O$ Mol. wt. 154.25

DESCRIPTION

A colorless liquid having a harsh, camphoraceous and mint-like odor with a rose leaf and geranium background.

SPECIFICATIONS

Assay. Not less than 95.0 percent of $C_{10}H_{18}O$.

Angular rotation. Between 0° and −7°.

Refractive index. Between 1.470 and 1.475 at 20°.

Solubility in alcohol. Passes test.

Specific gravity. Between 0.904 and 0.913.

Limits of Impurities

Acid value. Not more than 1.0.

Aldehydes. Not more than 1 percent, calculated as citronellal ($C_{10}H_{18}O$).

TESTS

Assay. Proceed as directed under *Total Alcohols*, page 893. Weigh accurately about 1.2 grams of the acetylated alcohol for the saponification, reflux it for 2 hours, and use 77.12 as the equivalence factor (*f*) in the calculation.

Angular rotation. Determine in a 100-mm. tube as directed under *Optical Rotation*, page 939.

Refractive index, page 945. Determine with an Abbé or other refractometer of equal or greater accuracy.

Solubility in alcohol. Proceed as directed in the general method, page 899. One ml. dissolves in 4 ml. of 60 percent alcohol to form a clear solution.

Specific gravity. Determine by any reliable method (see page 5).

Acid value. Determine as directed in the general method, page 893.

Aldehydes. Weigh accurately about 10 grams, and proceed as directed under *Aldehydes and Ketones-Hydroxylamine Method,* page 894, using 77.12 as the equivalence factor (E) in the calculation.

Packaging and storage. Store in full, tight containers, preferably glass, tin-lined or other suitably lined containers in a cool place protected from light.

Functional use in foods. Flavoring agent.

ISOVALERIC ACID

Isopropylacetic Acid

$(CH_3)_2CHCH_2COOH$

$C_5H_{10}O_2$ Mol. wt. 102.13

DESCRIPTION

A colorless liquid having an acidic taste and a disagreeable, rancid-cheese-like odor. It is soluble in water, in alcohol, in ether, and in chloroform.

SPECIFICATIONS

Assay. Not less than 99.0 percent of $C_5H_{10}O_2$.

Refractive index. Between 1.403 and 1.405 at 20°.

Specific gravity. Between 0.9285 and 0.9310.

TESTS

Assay. Mix about 1.5 grams, accurately weighed, with 100 ml. of recently boiled and cooled water in a 250-ml. Erlenmeyer flask, add phenolphthalein T.S., and titrate with 0.5 *N* sodium hydroxide to the appearance of a faint pink end-point which persists for at least 30 seconds. Each ml. of 0.5 *N* sodium hydroxide is equivalent to 51.07 mg. of $C_5H_{10}O_2$.

Refractive index, page 945. Determine with an Abbé or other refractometer of equal or greater accuracy.

Specific gravity. Determine by any reliable method (see page 5).

Packaging and storage. Store in tight, light-resistant containers.

Functional use in foods. Flavoring agent.

JUNIPER BERRIES OIL

DESCRIPTION

The volatile oil obtained by steam distillation from the dried ripe fruit of the plant *Juniperus communis* L. var. *erecta* Pursh (Fam. *Cupressaceae*). It is a colorless, or faintly greenish, or yellowish liquid with a characteristic odor and an aromatic bitter taste. It is soluble in most fixed oils and in mineral oil. It is insoluble in glycerin and in propylene glycol. The oil tends to polymerize on long storage.

SPECIFICATIONS

Angular rotation. Between $-15°$ and $0°$.

Refractive index. Between 1.474 and 1.484 at 20°.

Specific gravity. Between 0.854 and 0.879.

TESTS

Angular rotation. Determine in a 100-mm. tube as directed under *Optical Rotation*, page 939.

Refractive index, page 945. Determine with an Abbé or other refractometer of equal or greater accuracy.

Specific gravity. Determine by any reliable method (see page 5).

Packaging and storage. Store in full, tight, preferably glass, tin-lined, or galvanized containers in a cool place protected from light.

Functional use in foods. Flavoring agent.

KAOLIN

DESCRIPTION

A purified clay consisting mainly of alumina, silica, and water. It occurs as a fine, white to yellowish white or grayish powder having an earthy taste. It becomes darker and has a distinct clay-like odor when moistened. It is insoluble in water, in alcohol, in dilute acids, and in alkali solutions.

IDENTIFICATION

Mix 1 gram of the sample with 10 ml. of water and 5 ml. of sulfuric acid in a porcelain dish, and evaporate until the water is removed. Continue heating until dense white fumes of sulfur trioxide are evolved, then cool, and cautiously add 20 ml. of water. Boil for a few minutes, and filter. A gray residue of silica remains on the filter. To a portion of the filtrate add ammonia T.S. A gelatinous, white precipitate of aluminum hydroxide is produced that is insoluble in an excess of ammonia T.S.

SPECIFICATIONS

Loss on ignition. Not more than 15 percent.

Limits of Impurities

Acid-soluble substances. Not more than 2 percent.
Arsenic (as As). Not more than 3 parts per million (0.0003 percent).
Carbonate. Passes test.
Heavy metals (as Pb). Not more than 40 parts per million (0.004 percent).
Iron. Passes test.
Lead. Not more than 10 parts per million (0.001 percent).
Sulfide. Passes test.

TESTS

Loss on ignition. Ignite a 2-gram sample, accurately weighed, in a tared crucible at 575° ± 25° to constant weight, cool, and weigh.

Acid-soluble substances. Digest a 1-gram sample with 20 ml. of diluted hydrochloric acid T.S. for 15 minutes, and filter. Evaporate 10 ml. of the filtrate to dryness in a tared dish, ignite gently, cool, and weigh.

Sample Solution for the Determination of Arsenic, Heavy Metals, and Lead. Transfer 10.0 grams of the sample into a 250-ml. flask, and add 50 ml. of 05 *N* hydrochloric acid. Attach a reflux condenser to the flask, heat on a steam bath for 30 minutes, cool, and let the undissolved material settle. Decant the supernatant liquid through Whatman No. 3 filter paper, or equivalent, into a 100-ml. volumetric flask, retaining as much as possible of the insoluble material in the beaker. Wash the slurry and beaker with three 10-ml. portions of hot water, decanting each washing through the filter into the flask. Finally, wash the filter paper with 15 ml. of hot water, cool the filtrate to room temperature, dilute to volume with water, and mix.

Arsenic. A 10-ml. portion of the *Sample Solution* meets the requirements of the *Arsenic Test*, page 865.

Carbonate. Mix a 1-gram sample with 10 ml. of water, cool, and keep cool while adding 5 ml. of sulfuric acid. No effervescence occurs during the addition of the acid.

Heavy metals. A 5-ml. portion of the *Sample Solution* diluted to 25 ml. with water meets the requirements of the *Heavy Metals Test*, page 920, using 20 mcg. of lead ion (Pb) in the control (*Solution A*).

Iron. Mix a 2-gram sample with 10 ml. of water in a mortar, and add 500 mg. of sodium salicylate. No more than a light reddish tint is produced.

Lead. A 10-ml. portion of the *Sample Solution* meets the requirements of the *Lead Limit Test*, page 929, using 10 mcg. of lead ion (Pb) in the control.

Sulfide. Add a 1-gram sample to 25 ml. of water in a 250-ml. flask, then add 15 ml. of dilute hydrochloric acid T.S., and immediately cover the top of the flask with filter paper moistened with lead acetate T.S. Heat to boiling, and boil for several minutes. The paper does not show any brown coloration.

Packaging and storage. Store in well-closed containers.

Functional use in foods. Anticaking agent.

KARAYA GUM

Sterculia Gum

DESCRIPTION

A dried gummy exudation from *Sterculia urens* Roxburgh and other species of *Sterculia* (Fam. *Sterculiaceae*), or from *Cochlospermum gossypium* A. P. De Condolle, or other species of *Cochlospermum* Kunth (Fam. *Bixaceae*). It occurs in tears of variable size or in broken irregular pieces having a somewhat crystalline appearance. It is pale yellow to pinkish brown, translucent, and horny, and is sometimes admixed with a few darker fragments and occasional pieces of bark. The gum has a slightly acetous odor and a mucilaginous and slightly acetous taste. In the powdered form it is light gray to pinkish gray. Karaya gum is insoluble in alcohol, but it swells in water to form a gel.

IDENTIFICATION

A. Add 2 grams to 50 ml. of water. It swells to form a granular, stiff, slightly opalescent mucilage.

B. Add a few drops of Millon's Reagent to a 1 in 100 solution of the gum. A white curdy precipitate forms.

SPECIFICATIONS

Viscosity of a 1 percent solution. Not less than the minimum or within the range claimed by the vendor.

Limits of Impurities

Arsenic (as As). Not more than 3 parts per million (0.0003 percent).

Ash (**Acid-insoluble**). Not more than 1.0 percent.

Foreign gums. Passes test.

Heavy metals (as Pb). Not more than 40 parts per million (0.004 percent).

Insoluble matter. Not more than 3 percent.

Lead. Not more than 10 parts per million (0.001 percent).

Loss on drying. Not more than 20 percent.

Starch. Passes test.

TESTS

Viscosity. Transfer a 4-gram sample, finely powdered, into the container of a stirring apparatus equipped with blades capable of being adjusted to about 1,000 rpm. Add 10 ml. of alcohol to the sample, swirl to wet the gum uniformly, and then add 390 ml. of water, avoiding the formation of lumps. Stir the mixture for 7 minutes, pour the resulting dispersion into a 500-ml. bottle, insert a stopper, and allow to stand for about 12 hours in a water bath at 25°. Determine the apparent viscosity at this temperature with a model LVF Brookfield or equivalent type viscometer (see *Viscosity of Sodium Carboxymethylcellulose*, page 973) using a suitable spindle, speed, and factor.

Arsenic. A *Sample Solution* prepared as directed for organic compounds meets the requirements of the *Arsenic Test*, page 865.

Ash (Acid-insoluble). Determine as directed in the general method, page 869.

Foreign gums. The gum swells in 60 percent alcohol (*distinction from other gums*).

Heavy metals. Prepare and test a 500-mg. sample as directed in *Method II* under the *Heavy Metals Test*, page 920, using 20 mcg. of lead ion (Pb) in the control (*Solution A*).

Insoluble matter. Transfer about 5 grams, accurately weighed, into a 250-ml. Erlenmeyer flask, add a mixture of equal parts of diluted hydrochloric acid T.S. and water, cover the flask with a watch glass, and boil gently until the mixture loses its viscosity. Filter the solution through a tared filtering crucible, wash the residue with water until the washings are free from acid, dry at 105° for 1 hour, and weigh.

Lead. A *Sample Solution* prepared as directed for organic compounds meets the requirements of the *Lead Limit Test*, page 929, using 10 mcg. of lead ion (Pb) in the control.

Loss on drying, page 931. Powder an unground sample until it passes through a No. 40 sieve, mix well before weighing, and dry at 105° for 5 hours.

Starch. To a 1 in 10 solution of the gum add a few drops of iodine T.S. No blue color is produced.

Packaging and storage. Store in well-closed containers.

Functional use in foods. Stabilizer; thickener; emulsifier.

KELP

Pacific Kelp

DESCRIPTION

The dehydrated seaweed obtained from the giant kelp *Macrocystis pyrifera* and related species. It is dark green to olive brown in color and has a salty, characteristic taste.

SPECIFICATIONS

Ash (Total). Not more than 35 percent.

Iodine content. Between 0.15 percent and 0.22 percent.

Loss on drying. Not more than 13 percent.

Limits of Impurities

Arsenic (as As). Not more than 3 parts per million (0.0003 percent).

Heavy metals (as Pb). Not more than 40 parts per million (0.004 percent).

Lead. Not more than 10 parts per million (0.001 percent).

TESTS

Ash (Total). Determine as directed in the general method, page 868.

Iodine content. Transfer about 2 grams of the sample, accurately weighed, into a large porcelain crucible, and mix thoroughly with 10 grams of potassium carbonate. Ignite the sample in a muffle furnace, starting with low heat, and then ignite at 500–600° for 20 minutes or until combustion is complete. Dissolve the ash in about 200 ml. of boiling water, filter, and wash the filter paper with two 15-ml. portions of boiling water, adding the washings to the filtrate. Cool to room temperature, neutralize to methyl red T.S. with approximately 20 ml. of 85 percent phosphoric acid diluted with 20 ml. of water, and then add 1 ml. in excess. Cool the reaction mixture on an ice bath, and add bromine T.S. (about 5 ml.) until a permanent yellow color is obtained. Gently boil the solution to remove all free bromine, adding water if necessary to maintain a volume of 200 ml. or more. Boil for an additional 5 minutes after the bromine color has been completely dissipated. Add a few mg. of salicylic acid, stir, and cool to about 20°. Add 1 ml. of the diluted phosphoric acid solution and 5 ml. of potassium iodide T.S., and titrate immediately with 0.01 *N* sodium thiosulfate, using starch T.S. as the indicator. Each ml. of 0.01 *N* sodium thiosulfate is equivalent to 211.5 mcg. of I.

Loss on drying, page 931. Dry at 105° for 4 hours.

Arsenic. A *Sample Solution* prepared as directed for organic compounds meets the requirements of the *Arsenic Test,* page 865.

Heavy metals. Prepare and test a 500-mg. sample as directed in *Method II* under the *Heavy Metals Test,* page 920, using 20 mcg. of lead ion (Pb) in the control (*Solution A*).

Lead. A *Sample Solution* prepared as directed for organic compounds meets the requirements of the *Lead Limit Test,* page 929, using 10 mcg. of lead ion (Pb) in the control.

Packaging and storage. Store in well-closed containers.

Functional use in foods. Nutrient; dietary supplement.

LABDANUM OIL

DESCRIPTION

The volatile oil obtained by steam distillation from crude labdanum gum extracted from the perennial shrub *Cistus ladaniferus* L. (Fam. *Cistaceae*). It is a golden yellow, viscous liquid, having a powerful balsamic odor, which on dilution is reminiscent of ambergris. It turns dark brown on standing. It is soluble in most fixed oils and in mineral oil. It is insoluble in glycerin and in propylene glycol.

SPECIFICATIONS

Acid value. Between 18 and 86.

Angular rotation. Between $+0°15'$ and $+7°$.

Ester value. Between 31 and 86.

Refractive index. Between 1.492 and 1.507 at 20°.

Solubility in alcohol. Passes test.

Specific gravity. Between 0.905 and 0.993.

TESTS

Acid value. Determine as directed in the general method, page 894.

Angular rotation. Determine in a 100-mm. tube as directed under *Optical Rotation*, page 939.

Ester value. Determine as directed in the general method, page 897, using about 1 gram, accurately weighed.

Refractive index, page 945. Determine with an Abbé or other refractometer of equal or greater accuracy.

Solubility in alcohol. Proceed as directed in the general method, page 899. One ml. dissolves in 0.5 ml. of 90 percent alcohol but the solution usually becomes opalescent or turbid on further dilution.

Specific gravity. Determine by any reliable method (see page 5).

Packaging and storage. Store in full, tight, preferably glass, or aluminum containers in a cool place protected from light.

Functional use in foods. Flavoring agent.

LACTATED MONO-DIGLYCERIDES

DESCRIPTION

A mixture of partial lactic and fatty acid esters of glycerin. It varies in consistency from a soft to a hard, waxy solid. It is dispersible in hot water, and is moderately soluble in hot isopropanol, in xylene, and in cottonseed oil.

IDENTIFICATION

Transfer into a 25-ml. glass-stoppered test tube 1 ml. of the solution of the sample remaining after titrating with 0.1 *N* potassium hydroxide in the determination of *Total lactic acid*, add 0.1 ml. of cupric sulfate solution (1 gram of $CuSO_4.5H_2O$ in 25 ml. of water) and 6 ml. of sulfuric acid, and mix. Stopper loosely, heat in a boiling water bath for 5 minutes, then cool in an ice bath for 5 minutes. Remove from the ice bath, add 0.1 ml. of *p*-phenylphenol solution (75 mg. dissolved in 5 ml. of sodium hydroxide T.S.), and mix. Allow to stand at room temperature for 1 minute, then heat in a boiling water bath for 1 minute. A deep blue-violet color indicates the presence of lactic acid.

SPECIFICATIONS

The following specifications should conform to the representations of the vendor: **1-Monoglyceride content, Total lactic acid, Acid value, Free glycerin,** and **Water.**

Limits of Impurities

Arsenic (as As). Not more than 3 parts per million (0.0003 percent).

Heavy metals (as Pb). Not more than 10 parts per million (0.001 percent).

TESTS

1-Monoglyceride content. Determine as directed in the general method, page 907.

Total lactic acid. Transfer an accurately weighed portion of the melted sample, equivalent to between 140 and 170 mg. of lactic acid, into a 250-ml. Erlenmeyer flask. Pipet 20 ml. of 0.5 *N* alcoholic potassium hydroxide into the flask, connect an air condenser at least 65 cm. in length, and reflux for 30 minutes. Run a blank determination using the same volume of alkali. Add 20 ml. of water to each flask, then disconnect the condensers, evaporate to a volume of 20 ml., and cool to about 40°. Add methyl red T.S. to each flask, and titrate the blank with 0.5 *N* hydrochloric acid. Add exactly the same volume of 0.5 *N* hydrochloric acid to the sample flask, with swirling. To each flask add 50 ml. of hexane, swirl vigorously to dissolve the fatty acids in the sample flask, then transfer quantitatively the contents of each flask into separate 250-ml. separators and shake for 30 seconds. Collect the aqueous phases in 300-ml. Erlenmeyer flasks, wash the hexane solutions with 50 ml. of water, and combine the wash solutions with the original aqueous phases in the Erlenmeyer flasks, discarding the hexane solutions. Add phenolphthalein T.S. and titrate with 0.1 *N* potassium hydroxide to a pink color which persists for at least 30 seconds. Each ml. of 0.1 *N* potassium hydroxide is equivalent to 9.008 mg. of lactic acid ($C_3H_6O_3$).

Acid value. Determine as directed under *Method II* in the general procedure, page 902.

Free glycerin. Determine as directed in the general method, page 904.

Water. Determine by the *Karl Fischer Titrimetric Method,* page 977.

Arsenic. A *Sample Solution* prepared as directed for organic compounds meets the requirements of the *Arsenic Test,* page 865.

Heavy metals. Prepare and test a 2-gram sample as directed in *Method II* under the *Heavy Metals Test,* page 920, using 20 mcg. of lead ion (Pb) in the control (*Solution A*).

Packaging and storage. Store in well-closed containers.

Functional use in foods. Emulsifier; stabilizer.

LACTIC ACID

2-Hydroxypropionic Acid

DESCRIPTION

A colorless or yellowish, nearly odorless, syrupy liquid consisting of a mixture of lactic acid ($C_3H_6O_3$) and lactic acid lactate ($C_6H_{10}O_5$). It is obtained by the lactic fermentation of sugars or is prepared synthetically. It is usually available in solutions containing the equivalent of from 50 to 90 percent of lactic acid. It is hygroscopic, and when concentrated by boiling, the acid condenses to form lactic acid lactate, 2-(lactoyloxy)propanoic acid, which on dilution and heating hydrolyzes to lactic acid. It is miscible with water and with alcohol, and it gives positive tests for *Lactate,* page 927.

SPECIFICATIONS

Assay. Not less than 95.0 percent and not more than 105.0 percent of the labeled concentration of $C_3H_6O_3$.

Limits of Impurities

Arsenic (as As). Not more than 3 parts per million (0.0003 percent).

Chloride. Not more than 0.2 percent.

Citric, oxalic, phosphoric, or tartaric acid. Passes test.

Heavy metals (as Pb). Not more than 10 parts per million (0.001 percent).
Iron. Not more than 10 parts per million (0.001 percent).
Residue on ignition. Not more than 0.1 percent.
Sugars. Passes test.
Sulfate. Not more than 0.25 percent.

TESTS

Assay. Weigh accurately a portion of the sample equivalent to about 3 grams of lactic acid, transfer to a 250-ml. flask, add 50.0 ml. of 1 *N* sodium hydroxide, mix, and boil for 20 minutes. Add phenolphthalein T.S., titrate the excess alkali in the hot solution with 1 *N* sulfuric acid, and perform a blank determination (see page 2). Each ml. of 1 *N* sodium hydroxide is equivalent to 90.08 mg. of $C_3H_6O_3$.

Arsenic. A *Sample Solution* prepared as directed for organic compounds meets the requirements of the *Arsenic Test,* page 865.

Chloride. Dissolve about 5 grams, accurately weighed, in 50 ml. of water, and neutralize to litmus paper with sodium hydroxide solution (1 in 4). Add 2 ml. of potassium chromate T.S., and titrate with 0.1 *N* silver nitrate to the first appearance of a red tinge. Each ml. of 0.1 *N* silver nitrate is equivalent to 3.545 mg. of Cl.

Citric, oxalic, phosphoric, or tartaric acid. Dilute 1 gram to 10 ml. with water, add 40 ml. of calcium hydroxide T.S., and boil for 2 minutes. No turbidity is produced.

Heavy metals. A solution of 2 grams in 25 ml. of water meets the requirements of the *Heavy Metals Test,* page 920, using 20 mcg. of lead ion (Pb) in the control (*Solution A*).

Iron. To the ash obtained in the test for *Residue on ignition* add 2 ml. of dilute hydrochloric acid (1 in 2), and evaporate to dryness on a steam bath. Dissolve the residue in 1 ml. of hydrochloric acid, dilute to 40 ml. with water, and add about 40 mg. of ammonium persulfate crystals and 10 ml. of ammonium thiocyanate T.S. Any red or pink color does not exceed that produced by 20 ml. of *Iron Standard Solution* (20 mcg. Fe) in an equal volume of solution containing the quantities of reagents used in the test.

Residue on ignition, page 945. Ignite 2 grams as directed in the general method.

Sugars. Add 5 drops of the sample to 10 ml. of hot alkaline cupric tartrate T.S. No red precipitate is formed.

Sulfate. Transfer about 50 grams, accurately weighed, into a 600-ml. beaker, dissolve in 200 ml. of water, and neutralize to between pH 4.5 and 6.5 with sodium hydroxide solution (1 in 2), making the final adjustment with a more dilute alkali solution. Filter, if necessary, and heat the filtrate or clear solution to just below the boiling point.

Add 10 ml. of barium chloride T.S., stirring vigorously, boil the mixture gently for 5 minutes, and allow to stand for at least 2 hours, or preferably overnight. Collect the precipitate of barium sulfate on a tared Gooch crucible, wash until free from chloride, dry, and ignite at 600° to constant weight. The weight of barium sulfate so obtained, multiplied by 0.412, represents the weight of SO_4 in the sample taken.

Packaging and storage. Store in tight containers.

Functional use in foods. Acid.

LACTYLATED FATTY ACID ESTERS OF GLYCEROL AND PROPYLENE GLYCOL

Propylene Glycol Lactostearate

DESCRIPTION

A mixture of partial lactic and fatty acid esters of propylene glycol and glycerin produced by the lactylation of a product obtained by reacting edible fats or oils with propylene glycol. It varies in consistency from a soft solid to a hard, waxy solid. It is dispersible in hot water and is moderately soluble in hot isopropanol, in benzene, in chloroform, and in soybean oil.

IDENTIFICATION

Place about 150 mg. of melted sample in a 16 × 125-mm. tube equipped with a screw cap having a Teflon liner, and add 4 ml. of absolute methanol, 4 drops of a 25 percent sodium methoxide solution in absolute methanol, and a boiling chip. Cap the tube, reflux for 15 minutes, and cool to room temperature. Add 8 drops of a 15 percent potassium acid sulfate solution, 4 ml. of water, and 4 ml. of *n*-hexane, cap the tube, shake for 1 minute, and centrifuge for 30–60 seconds. Decant and discard the *n*-hexane layer, and repeat the extraction with three additional 4-ml. portions of *n*-hexane, discarding each extract. Transfer the aqueous alcoholic phase from the tube to a 50-ml. round bottom glass-stoppered flask, place the flask in a 50–55° water bath, and evaporate to near dryness (about 0.5 ml.) with a rotary film evaporator under full water aspirator vacuum. (*Caution:* Do not heat above 55°.) Remove the flask from the evaporator, add 1 ml. of a 1:1 methanol-0.5 *N* hydrochloric acid solution, swirl for several minutes, and decant the clear solution into a small flask. Inject a portion of this solution into a suitable gas chromatograph (see page 886), and obtain the chromatogram, observing the following operating conditions or equivalent conditions: *Detector,* flame ionization; *Carrier gas,* helium, flowing at a rate of 50 ml. per minute; *Column,* 1.8-m. × 3-mm. (i.d.) packed with 80/100-mesh Porapak Q (ethylvinylbenzene-divinylbenzene polymer porous beads); *Column temperature,* 175–210°, heated at a rate of 4° per minute and holding at 210° until the glycerin is eluted; *Inlet port temperature,* 310°; *Detector temperature,* 385°; *Recorder,* 0–1 mv. range, with 1 second full-scale deflection at a chart speed of 6.5 mm. per minute; *Sample size,* 2–3 μl. From the chromatogram so obtained, identify the peaks by their relative positions on the chart. The major peaks, representing propylene glycol, methyl lactate, lactic acid, and glycerin in the order listed, may be identified with suitable reference substances. Major peaks may also be identified by their relative retention times using a suitable internal standard.

SPECIFICATIONS

Acid value. Not more than 12.0.

Water-insoluble combined lactic acid. Between 14.0 percent and 18.0 percent.

The following specifications should conform to the representations of the vendor: **1-Monoglyceride content, Total lactic acid, Free lactic acid, Free glycerin,** and **Water.**

Limits of Impurities

Arsenic (as As). Not more than 3 parts per million (0.0003 percent).

Heavy metals (as Pb). Not more than 10 parts per million (0.001 percent).

TESTS

Acid value. Determine as directed under *Method II* in the general procedure, page 902.

Water-insoluble combined lactic acid. Transfer about 3 grams of the sample, accurately weighed, into a 250-ml. separator with the aid of 100 ml. of benzene, and wash with three 30-ml. portions of water, discarding the washings. Transfer the benzene layer to a 250-ml. glass-stoppered Erlenmeyer flask, wash the separator with a few ml. of benzene, and completely evaporate the combined benzene solution to dryness. Pipet 50.0 ml. of 0.7 *N* alcoholic potassium hydroxide into the flask, attach an air condenser, boil, gently on a steam bath for 30 minutes or until the sample is completely saponified, and remove the flask from the steam bath. Immediately remove the air condenser, and allow the solution to cool until it begins to jell. Add 75.0 ml. of 0.5 *N* hydrochloric acid, mix, and transfer the solution into a 500-ml. separator, washing the flask with two 15-ml. portions of water. Cool to 35° or lower, and extract with 100 ml. of diethyl ether. Transfer the water layer to a second 500-ml separator, and wash the diethyl ether with two 20-ml. portions of water, adding the wash water to the original aqueous phase in the second separator. Retain the ether solution. Extract the aqueous phase with a second 100-ml. portion of diethyl ether, and transfer the aqueous phase to a 500-ml.

Erlenmeyer flask. Combine and wash the ether extracts with five 20-ml. portions of water, and add the wash water to the flask. To the combined aqueous phases in the flask add 1 ml. of phenolphthalein T.S., and titrate with 0.5 N sodium hydroxide to the first appearance of a slight pink color. Perform a blank determination (see page 2), and calculate the percent of water-insoluble combined lactic acid in the sample by the formula $(S - B)(N)(9.008)/W$, in which $(S - B)$ represents the difference, in ml., between the volumes of 0.5 N sodium hydroxide required for the sample and blank, respectively, N is the exact normality of the sodium hydroxide solution, and W is the weight of the sample, in grams.

1-Monoglyceride content. Determine as directed in the general method, page 907.

Total lactic acid. Transfer about 3 grams of the sample, accurately weighed, into a 250-ml. glass-stoppered flask, pipet 50.0 ml. of 0.7 N alcoholic potassium hydroxide into the flask, attach an air condenser, and boil gently on a steam bath for 30 minutes or until the sample is completely saponified. Remove the flask from the steam bath, immediately remove the air condenser, and allow the solution to cool until it begins to jell. Add 75.0 ml. of 0.5 N hydrochloric acid, mix, and transfer the solution to a 500-ml. separator, washing the flask with two 15-ml. portions of water. Cool to 35° or lower, and extract with 100 ml. of diethyl ether. Transfer the aqueous layer to a second 500-ml. separator, and wash the ether layer with two 20-ml. portions of water, adding the wash water to the original aqueous phase in the second separator. Retain the ether solution. Extract the aqueous phase with a second 100-ml. portion of diethyl ether, and transfer the aqueous phase to a 500-ml. Erlenmeyer flask. Combine and wash the ether extracts with five 20-ml. portions of water, and add the wash water to the flask. To the combined aqueous phases in the Erlenmeyer flask, add 1 ml. of phenolphthalein T.S., and titrate with 0.5 N sodium hydroxide to the first appearance of a slight pink color. Perform a blank determination (see page 2), and calculate the percent of total lactic acid by the formula $(S - B)(N)(9.008)/W$, in which $(S - B)$ represents the difference, in ml., between the volumes of 0.5 N sodium hydroxide required for the sample and blank, respectively, N is the exact normality of the sodium hydroxide solution, and W is the weight of the sample, in grams.

Free lactic acid. Transfer about 15 grams of the sample, accurately weighed, into a beaker, dissolve in about 75 ml. of benzene, and transfer the solution into a 500-ml. glass-stoppered graduate. Wash the beaker with about 125 ml. of benzene in divided portions, adding the washings to the graduate. Add 200 ml. of water to the graduate, and shake vigorously for 1 minute. After 125 ml. or more of the aqueous phase has separated, pipet 100.0 ml. of the aqueous phase into an Erlenmeyer flask, add 1 ml. of phenolphthalein T.S., and titrate with 0.5 *N* sodium hydroxide to the first appearance of a slight pink color. Calculate the percent of free lactic acid in the sample by the formula $(V)(N)(9.008)/(0.5W)$, in which *V* is the volume, in ml., of 0.5 *N* sodium hydroxide required, *N* is the exact normality of the sodium hydroxide solution, and *W* is the weight of the sample, in grams.

Free glycerin. Determine as directed in the general method, page 904.

Water. Determine by the *Karl Fischer Titrimetric Method*, page 977.

Arsenic. A *Sample Solution* prepared as directed for organic compounds meets the requirements of the *Arsenic Test*, page 865.

Heavy metals. Prepare and test a 2-gram sample as directed in *Method II* under the *Heavy Metals Test*, page 920, using 20 mcg. of lead ion (Pb) in the control (*Solution A*).

Packaging and storage. Store in well-closed containers.

Functional use in foods. Emulsifier, stabilizer, whipping agent; plasticizer; surface-active agent.

LACTYLIC ESTERS OF FATTY ACIDS

DESCRIPTION

Lactylic esters of fatty acids are mixed fatty acid esters of lactic acid and its polymers, with minor quantities of free lactic acid, polylactic acid, and fatty acids. They vary in consistency from liquids to hard, waxy solids. They are dispersible in hot water and are soluble in organic solvents and in vegetable oils. They conform to the regulations of the federal Food and Drug Administration pertaining to specifications for fats or fatty acids derived from edible sources.

IDENTIFICATION

A. Transfer into a 25-ml. glass-stoppered test tube 1 ml. of the solution obtained in the test for *Total lactic acid* after titrating with 0.1 *N* potassium hydroxide. Add 0.1 ml. of cupric sulfate solution (1 gram of $CuSO_4 . 5H_2O$ in 25 ml. of water) and 6 ml. of sulfuric acid, and mix. Stopper loosely, heat in a boiling water bath for 5 minutes, then cool in an ice bath for 5 minutes, and remove from the bath. Add 0.1 ml. of *p*-phenylphenol T.S., mix, allow to stand at room temperature for 1 minute, and then heat in a boiling water bath for 1 minute. A deep, blue-violet color indicates the presence of lactic acid.

B. Assemble a suitable apparatus for ascending thin-layer chromatography. Prepare a slurry of chromatographic silica gel containing about 13 percent of calcium sulfate as the binder (use 1 gram of $CaSO_4$

to each 2 ml. of water), apply a uniformly thin layer to glass plates of convenient size, dry in the air for 10 minutes, and activate by drying at 100° for 1 hour. Store the cool plates in a clean, dry place until ready for use.

Transfer 1 gram of the sample into a 10-ml. volumetric flask, dissolve, and dilute to volume with chloroform. Transfer 250 mg. of stearic acid into another 10-ml. volumetric flask, dissolve, and dilute to volume with chloroform.

Spot 2 μl. of the sample solution and 1 μl. of the stearic acid solution approximately 1.5 cm. from the bottom of the plate, allow the spots to dry, and then place the plate in a suitable chromatographic chamber containing a mixture of 4 volumes of acetone, 4 volumes of acetic acid, and 92 volumes of hexane. Develop by ascending chromatography until the solvent front travels 15 cm. beyond the sample spot. Remove the plate from the chamber, dry thoroughly in air, and spray evenly with a saturated solution of chromium trioxide in sulfuric acid. Immediately place the sprayed plate on a hot plate maintained at about 200° in a hood, char until white fumes of sulfur trioxide cease, and cool on an asbestos mat at room temperature. The spots from the sample are located according to the following R_f values: stearic acid, 1.00; fatty acid, 1.00; acylated monolactic acid, 0.84; acylated dilactic acid, 0.76; acylated trilactic acid, 0.68; and tetralactic acid, 0.62.

SPECIFICATIONS

The following specifications should conform to the representations of the vendor: **Acylated monolactic acid, Acylated polylactic acid, Free fatty acid, Total lactic acid, Acid value, Saponification value,** and **Water.**

Limits of Impurities

Arsenic (as As). Not more than 3 parts per million (0.0003 percent).

Heavy metals (as Pb). Not more than 10 parts per million (0.001 percent).

TESTS

Assay for Acylated lactic acid, Acylated polylactic acid, and Free fatty acid. This assay is performed by gas-liquid chromatography (see page 886) using an instrument containing a thermal conductivity or flame ionization detector and helium as the carrier gas. The operating conditions of the apparatus may vary, depending upon the particular instrument used, but a suitable chromatogram is obtained with a 1.2-m. × 6.3-mm. column packed with 20 percent SE-30 or SE-52, or other comparable grades of silicone rubber gums, on Chromosorb P or W or Diatoport S, or other comparable grades of diatomaceous material. The column should be programmed between 150° and 310°, using a heating rate of 4° per minute, the inlet port temperature should be 335°, and the detector temperature should be 315°. The recorder should be equipped with an attenuator switch and should be operated in the 0 to 1 millivolt range, with 1 second full scale deflection at a chart speed of 12.7 mm. per minute. A constant gas flow rate of about 54 ml. per minute should be established and maintained throughout the determination.

Diazomethane Reagent. [*Caution: Diazomethane is both toxic and potentially explosive. Its preparation and use should be carried out in a hood.*] Place 1 ml. of potassium hydroxide solution (4 in 10), followed by 2.5 ml. of methanol, in a 25-ml. distilling flask fitted with a dropping funnel and an efficient spiral water-cooled condenser set downward for distillation. Connect the condenser to a 50-ml. receiving flask which is cooled in ice and vented to the hood. Heat the distilling flask in a water bath to 65°, add 2 ml. of ether through the dropping funnel, saturating the distillation apparatus with ether vapor, and close the stopcock. Place in the dropping funnel a solution containing 2.15 grams of *N*-methyl-*N*-nitroso-*p*-toluene sulfonamide in 13 ml. of ether, and adjust the stopcock so that the rate of distillation is about equal to the rate of addition from the funnel. When the funnel is empty, add another 2 ml. of ether, and continue the distillation until the distillate is colorless. The ether-alcohol solution of diazomethane so obtained should be used immediately or stored at −10° until used.

Procedure. To approximately 50 mg. of the sample add the *Diazomethane Reagent* until a yellow color persists. Carefully evaporate the ether at 50° under a stream of clean, dry nitrogen. Inject 0.5 to 2.0 μl. of the melted methyl esters so obtained into the gas chromatographic apparatus, using a 10 μl. capacity Hamilton fixed needle or equivalent. The sample size should be adjusted so that the major peak is not attenuated more than ×8. From the chromatogram so obtained, identify the peaks by their relative position on the chart. The esters, appearing in the order of increasing number of carbon atoms in the fatty acid and in order of increasing length of the polymer, are eluted as follows: myristate, palmitate, stearate, palmitoyl lactylate (2-palmitoyloxypropionate), stearoyl lactylate (2-stearoyloxypropionate), palmitoyl lactoyl lactylate, stearoyl lactoyl lactylate, palmitoyl dilactoyl lactylate, stearoyl dilactoyl lactylate, palmitoyl trilactoyl lactylate, stearoyl trilactoyl lactylate, and palmitoyl tetralactoyl lactylate. Other esters may be determined by interpolation of a conventional carbon number-retention plot.

Determine the composition of the sample, using the area normalization method, by the formula $\%_i = 100\ A_i/\Sigma(A_i + \ldots A_n)$, in which i represents the component of interest, A_i is the equalized area for the component of interest, and $\Sigma(A_i + \ldots A_n)$ is the sum of the equalized areas.

If free and polylactic acids are present, as determined below, the results should be corrected by multiplying $\%_i$ by [(100 − % free and polylactic acid)/100].

Total free and polylactic acids. Weigh accurately about 500 mg. of the sample, previously melted, transfer into a 50-ml. glass-stoppered separator with the aid of 15 ml. of benzene, and add 10 ml. of water. Invert the funnel 10 times, and allow to stand until the layers have separated. Filter the aqueous layer through a plug of glass wool into a 125-ml. flask, wash the benzene with two 10-ml. portions of water, and combine the aqueous layers. To the flask add 5.0 ml. of 0.1 *N* sodium hydroxide, and then heat on a steam bath for 15 minutes under a nitrogen atmosphere. Add phenolphthalein T.S., and titrate with 0.1 *N* hydrochloric acid to the disappearance of the pink color. Conduct a blank determination, using 30 ml. of water and 5.0 ml. of 0.1 *N* sodium hydroxide, and calculate the percent of free and polylactic acids in the sample by the formula $(B - S) \times 9.0/W$, in which $B - S$ represents the difference, in ml., between the volumes of 0.1 *N* hydrochloric acid required for the blank and the sample, respectively, 9.0 is an equivalence factor for the lactic acid, and W is the weight, in grams, of the sample.

Total lactic acid. Transfer an accurately weighed portion of the melted sample, equivalent to between 140 mg. and 170 mg. of lactic acid, into a 250-ml. Erlenmeyer flask, and add to the flask 20.0 ml. of 0.5 *N* alcoholic potassium hydroxide. Connect an air condenser, at least 65 cm. in length, to the flask, and reflux for 30 minutes. Add 20 ml. of water through the condenser, disconnect the condenser, and prepare a blank containing 20.0 ml. of the alkali and 20 ml. of water. Evaporate the contents of each flask to a volume of about 20 ml., cool to about 40°, add methyl red T.S. to the flask containing the blank, and titrate the blank with 0.5 *N* hydrochloric acid. Add exactly the same volume of the acid to the sample flask, with swirling, and then add to the sample flask 50 ml. of hexane, swirling vigorously to dissolve the fatty acids. Transfer quantitatively the contents of the sample flask into a 250-ml. separator, and shake for 30 seconds. Collect the aqueous phase in an Erlenmeyer flask, wash the hexane solution with 50 ml. of water, and combine the wash solution with the original aqueous phase in the flask, discarding the hexane solution. Add phenolphthalein T.S., and titrate with 0.1 *N* potassium hydroxide to a pink color which persists for at least 30 seconds. Each ml. of 0.1 *N* potassium hydroxide is equivalent to 9.008 mg. of lactic acid ($C_3H_6O_3$).

Acid value. Determine as directed under *Method II* in the general procedure, page 902.

Saponification value. Determine as directed in the general procedure, page 914.

Water. Determine by the *Karl Fischer Titrimetric Method*, page 977.

Arsenic. A *Sample Solution* prepared as directed for organic compounds meets the requirements of the *Arsenic Test*, page 865.

Heavy metals. Prepare and test a 2-gram sample as directed in *Method II* under the *Heavy Metals Test*, page 920, using 20 mcg. of lead ion (Pb) in the control (*Solution A*).

Packaging and storage. Store in tight, plastic-lined containers in a cool, dry place.

Functional use in foods. Emulsifier; surface-active agent.

LANOLIN, ANHYDROUS

Wool Fat

DESCRIPTION

A purified, yellowish white, semisolid, fat-like substance extracted from the wool of sheep. It is insoluble in water, but mixes with about twice its weight of water without separation. It is soluble in chloroform and in ether.

SPECIFICATIONS

Acid value. Not more than 1.12.

Iodine value. Between 18 and 36.

Melting range. Between 36° and 42°.

Limits of Impurities

Arsenic (as As). Not more than 3 parts per million (0.0003 percent).

Heavy metals (as Pb). Not more than 40 parts per million (0.004 percent).

Lead. Not more than 3 parts per million (0.0003 percent).

Loss on heating. Not more than 0.5 percent.

TESTS

Acid value. Determine as directed under *Method I* in the general procedure, page 902.

Iodine value. Determine by the *Wijs Method*, page 906.

Melting range. Determine as directed in the general procedure, page 931.

Arsenic. Prepare a *Sample Solution* as directed in the general method under *Chewing Gum Base*, page 877. This solution meets the requirements of the *Arsenic Test*, page 865.

Heavy metals. Prepare and test a 500-mg. sample as directed in *Method II* under the *Heavy Metals Test*, page 920, using 20 mcg. of lead ion (Pb) in the control (*Solution A*).

Lead. Prepare a *Sample Solution* as directed in the general method under *Chewing Gum Base*, page 878. This solution meets the requirements of the *Lead Limit Test*, page 929, using 10 mcg. of lead ion (Pb) in the control.

Loss on heating. Heat a 5-gram sample on a steam bath, with frequent stirring, to constant weight.

Packaging and storage. Store in well-closed containers, preferably at a temperature not exceeding 30°.

Functional use in foods. Masticatory substance in chewing gum base.

LAUREL LEAF OIL

Bay Leaf Oil

DESCRIPTION

The oil obtained by steam distillation from the leaves of *Laurus nobilis* L. (Fam. *Lauraceae*). It is a light yellow to yellow liquid, having an aromatic and spicy odor. It is soluble in most fixed oils, and is soluble with cloudiness in mineral oil and in propylene glycol. It is insoluble in glycerin.

Note: The oil from Laurus nobilis L. should not be confused with that of the West Indian bay tree, or California bay laurel.

SPECIFICATIONS

Acid value. Not more than 3.0.

Angular rotation. Between −10° and −19°.

Refractive index. Between 1.465 and 1.470 at 20°.

Saponification value. Between 15 and 45.

Saponification value after acetylation. Between 36 and 85.

Solubility in alcohol. Passes test.

Specific gravity. Between 0.905 and 0.929.

TESTS

Acid value. Determine as directed in the general procedure, page 893.

Angular rotation. Determine in a 100-mm. tube as directed under *Optical Rotation*, page 939.

Refractive index, page 945. Determine with an Abbé or other refractometer of equal or greater accuracy.

Saponification value. Determine as directed in the general method, page 896, using about 5 grams, accurately weighed.

Saponification value after acetylation. Proceed as directed under *Total Alcohols*, page 893, using about 2.5 grams of acetylated oil, accurately weighed. Calculate the saponification value by the formula $(28.05 \times A)/B$, in which A is the number of ml. of 0.5 N alcoholic potassium hydroxide consumed in the titration, and *B* is the weight of the acetylated oil in grams.

Solubility in alcohol. Proceed as directed in the general method, page 899. One ml. dissolves in 1 ml. of 80 percent alcohol, and remains in solution upon dilution to 10 ml.

Specific gravity. Determine by any reliable method (see page 5).

Packaging and storage. Store in full, tight, preferably glass, tin-lined, or aluminum containers in a cool place protected from light.

Functional use in foods. Flavoring agent.

LAURIC ACID

Dodecanoic Acid

$CH_3(CH_2)_{10}COOH$

$C_{12}H_{24}O_2$ Mol. wt. 200.32

DESCRIPTION

A solid organic acid obtained from coconut oil and other vegetable fats. It occurs as a white or faintly yellowish, somewhat glossy, crystalline solid or powder. It is practically insoluble in water, but is soluble in alcohol, in chloroform, and in ether.

SPECIFICATIONS

Acid value. Between 252 and 287.

Saponification value. Between 253 and 287.

Titer (Solidification Point). Between 26° and 44°.

Limits of Impurities

Arsenic (as As). Not more than 3 parts per million (0.0003 percent).

Heavy metals (as Pb). Not more than 10 parts per million (0.001 percent).

Iodine value. Not more than 3.0.

Residue on ignition. Not more than 0.1 percent.

Unsaponifiable matter. Not more than 0.3 percent.

Water. Not more than 0.2 percent.

TESTS

Acid value. Determine as directed under *Method I* in the general procedure, page 902.

Saponification value. Determine as directed in the general method, page 914, using about 3 grams, accurately weighed.

Titer (Solidification Point). Determine as directed under *Solidification Point*, page 954.

Arsenic. A *Sample Solution* prepared as directed for organic compounds meets the requirements of the *Arsenic Test*, page 865.

Heavy metals. Prepare and test a 2-gram sample as directed in *Method II* under the *Heavy Metals Test*, page 920, using 20 mcg. of lead ion (Pb) in the control (*Solution A*).

Iodine value. Determine by the *Wijs Method*, page 906.

Residue on ignition, page 945. Ignite 10 grams as directed in the general method.

Unsaponifiable matter. Determine as directed in the general method, page 915.

Water. Determine by the *Karl Fischer Titrimetric Method*, page 977.

Packaging and storage. Store in well-closed containers.

Functional use in foods. Component in the manufacture of other food grade additives; defoaming agent.

LAURYL ALCOHOL (NATURAL)

Alcohol C-12; 1-Dodecanol

$CH_3(CH_2)_{10}CH_2OH$

$C_{12}H_{26}O$ Mol. wt. 186.34

DESCRIPTION

A colorless liquid at temperatures above 21°, having a characteristic fatty odor. It is soluble in most fixed oils, in mineral oil, and in propylene glycol. It is insoluble in glycerin.

(*Note:* The following *Specifications* apply only to lauryl alcohol derived from natural precursors.)

SPECIFICATIONS

Assay. Not less than 97.0 percent of $C_{12}H_{26}O$.

Refractive index. Between 1.440 and 1.444 at 20°.

Solidification point. Not lower than 21°.

Solubility in alcohol. Passes test.

Specific gravity. Between 0.830 and 0.836.

Limits of Impurities

Acid value. Not more than 1.0.

TESTS

Assay. Proceed as directed under *Total Alcohols*, page 893. Weigh accurately about 1.5 grams of the acetylated alcohol for the saponification, and use 93.17 as the equivalence factor (*e*) in the calculation.

Refractive index, page 945. Determine with an Abbé or other refractometer of equal or greater accuracy.

Solidification point. Determine as directed in the general method, page 954.

Solubility in alcohol. Proceed as directed in the general method, page 899. One ml. dissolves in 3 ml. of 70 percent alcohol and remains in solution upon dilution to 10 ml.

Specific gravity. Determine by any reliable method (see page 5)

Acid value. Determine as directed in the general method, page 893.

Packaging and storage. Store in full, tight, preferably glass, or suitably lined containers in a cool place protected from light. Aluminum containers should not be used.

Functional use in foods. Flavoring agent.

LAURYL ALDEHYDE

Aldehyde C-12; Dodecanal

$CH_3(CH_2)_{10}CHO$

$C_{12}H_{24}O$ Mol. wt. 184.32

DESCRIPTION

A colorless to light yellow liquid, having a characteristic fatty odor, which solidifies at low temperatures. It is soluble in alcohol, in most fixed oils, in mineral oil, and, occasionally with slight turbidity, in propylene glycol. It is insoluble in glycerin and in water.

SPECIFICATIONS

Assay. Not less than 92.0 percent of $C_{12}H_{24}O$.

Refractive index. Between 1.433 and 1.439 at 20°.

Specific gravity. Between 0.826 and 0.836.

Limits of Impurities

Acid value. Not more than 1.0.

TESTS

Assay. Weigh accurately about 1.2 grams, and proceed as directed under *Aldehydes*, page 894, using 92.16 as the equivalence factor (E) in the calculation.

Refractive index, page 945. Determine with an Abbé or other refractometer of equal or greater accuracy.

Specific gravity. Determine by any reliable method (see page 5).

Acid value. Determine as directed in the general method, page 893.

Packaging and storage. Store in full, tight, preferably glass, or aluminum containers at a temperature between 70° and 100° F., protected from light. Cool storage may produce rapid polymerization. Long storage is inadvisable.

Functional use in foods. Flavoring agent.

LAVANDIN OIL, ABRIAL

DESCRIPTION

An oil obtained by steam distillation of the fresh flowering tops of a hybrid, *Lavandula abrialis* unofficial (Fam. *Labiatae*), of true lavender, *Lavandula officinalis*, or of spike lavender, *Lavandula latifolia*. It is a pale yellow to yellow liquid having a slight camphoraceous odor which is strongly suggestive of lavender. It is soluble in most fixed oils and in propylene glycol. It is soluble with opalescence in mineral oil, but it is relatively insoluble in glycerin.

SPECIFICATIONS

Assay. Not less than 28.0 percent and not more than 35.0 percent of esters, calculated as linalyl acetate ($C_{12}H_{20}O_2$).

Angular rotation. Between −2° and −5°.

Refractive index. Between 1.460 and 1.464 at 20°.

Solubility in alcohol. Passes test.

Specific gravity. Between 0.885 and 0.893.

TESTS

Assay. Weigh accurately about 3 grams, and proceed as directed under *Ester Determination*, page 896, using 98.15 as the equivalence factor (*e*) in the calculation.

Angular rotation. Determine in a 100-mm. tube as directed under *Optical Rotation*, page 939.

Refractive index, page 945. Determine with an Abbé or other refractometer of equal or greater accuracy.

Solubility in alcohol. Proceed as directed in the general method, page 899. One ml. dissolves in 2 ml. of 70 percent alcohol. A slight opalescence sometimes develops on further dilution.

Specific gravity. Determine by any reliable method (see page 5).

Packaging and storage. Store in full, tight, preferably glass, tin-lined, or good quality galvanized containers in a cool place protected from light.

Functional use in foods. Flavoring agent.

LAVENDER OIL

DESCRIPTION

The volatile oil obtained by steam distillation from the fresh flowering tops of *Lavandula officinalis* Chaix ex Villars (*Lavandula vera* De Candolle) (Fam. *Labiatae*). It is a colorless or yellow liquid, having the characteristic odor and taste of lavender flowers.

SPECIFICATIONS

Assay. Not less than 35.0 percent of esters, calculated as linalyl acetate ($C_{12}H_{20}O_2$).

Angular rotation. Between −3° and −10°.

Refractive index. Between 1.459 and 1.470 at 20°.

Solubility in alcohol. Passes test.

Specific gravity. Between 0.875 and 0.888.

Limits of Impurities

Alcohol. Passes test.

Foreign water-soluble esters. Passes test.

TESTS

Assay. Weigh accurately about 5 grams, and proceed as directed under *Ester Determination*, page 896, using 98.15 as the equivalence factor (*e*) in the calculation.

Angular rotation. Determine in a 100-mm. tube as directed under *Optical Rotation*, page 939.

Refractive index, page 945. Determine with an Abbé or other refractometer of equal or greater accuracy.

Solubility in alcohol. Proceed as directed in the general method, page 899. One ml. dissolves in 4 ml. of 70 percent alcohol.

Specific gravity. Determine by any reliable method (see page 5).

Alcohol. Transfer 5 ml. to a narrow, graduated, glass-stoppered, 10-ml. cylinder, add 5 ml. of water and shake. The volume of the oil does not diminish.

Foreign water-soluble esters. Mix 20 ml. of the sample with 40 ml. of 5 percent alcohol in a glass-stoppered, 100-ml. cylinder. When the mixture has cleared, pipet 30 ml. of the alcohol layer into a 125-ml. Erlenmeyer flask. Add phenolphthalein T.S., and neutralize the solution with 0.5 *N* sodium hydroxide. Add 5.0 ml. of 0.5 *N* sodium hydroxide and heat the mixture on a boiling water bath under a reflux condenser for 1 hour. Allow the mixture to cool, and titrate the excess alkali with 0.5 *N* hydrochloric acid. Not less than 4.7 ml. of the acid is required to neutralize the mixture.

Packaging and storage. Store in full, tight containers in a cool place protected from light.

Functional use in foods. Flavoring agent.

LECITHIN

DESCRIPTION

Food grade lecithin is obtained from soybeans. It is a complex mixture of acetone-insoluble phosphatides which consist chiefly of phosphatidyl choline, phosphatidyl ethanolamine, phosphatidyl serine, and phosphatidyl inositol, combined with various amounts of other substances such as triglycerides, fatty acids, and carbohydrates. Refined grades of lecithin may contain any of these components in varying proportions and combinations depending on the type of fractionation used. In its oil-free form, the preponderance of triglycerides and fatty acids are removed and the product contains 90 percent or more of soy phosphatides representing all or certain fractions of the total phosphatide complex. The consistency of both natural grades and refined grades of lecithin may vary from plastic to fluid, depending upon free fatty acid and soybean oil content and the presence or absence of other diluents. Its color varies from light yellow to brown depending upon whether it is bleached or unbleached. It is odorless or has a characteristic, slight nut-like odor and a bland taste. Edible diluents, such as cocoa butter and vegetable oils, often replace soybean oil to improve functional and flavor characteristics. Lecithin is only partially soluble in water, but it readily hydrates to form emulsions. The oil-free phosphatides are soluble in fatty acids, but are practically insoluble in fixed oils. When all soy phosphatide fractions are present, lecithin is partially soluble in alcohol, and practically insoluble in acetone.

SPECIFICATIONS

Acetone-insoluble matter (phosphatides). Not less than 50.0 percent.

Acid value. Not more than 36.

Benzene-insoluble matter. Not more than 0.3 percent.

Water. Not more than 1.5 percent.

Limits of Impurities

Arsenic (as As). Not more than 3 parts per million (0.0003 percent).

Heavy metals (as Pb). Not more than 40 parts per million (0.004 percent).

Lead. Not more than 10 parts per million (0.001 percent).

TESTS

Acetone-insoluble matter (Phosphatides)

Purification of Phosphatides. Dissolve 5 grams of phosphatides from previous *Acetone insoluble matter* determinations in 10 ml. of petroleum ether, and add 25 ml. of acetone to the solution. Transfer approximately equal portions of the precipitate to each of two 40-ml.

centrifuge tubes using additional portions of acetone to facilitate the transfer. Stir thoroughly, dilute to 40 ml. with acetone, stir again, chill for 15 minutes in an ice bath, stir again, and then centrifuge for 5 minutes. Decant the acetone, crush the solids with a stirring rod, refill the tube with acetone, stir, chill, centrifuge, and decant as before. The solids after the second centrifugation require no further purification and may be used for preparing the *Phosphatide-Acetone Solution.* Five grams of the purified phosphatides are required to saturate about 16 liters of acetone.

Phosphatide-Acetone Solution. Add a quantity of purified phosphatides to sufficient acetone, previously cooled to a temperature of about 5°, to form a saturated solution, and maintain the mixture at this temperature for 2 hours, shaking it vigorously at 15-minute intervals. Decant the solution through a rapid filter paper avoiding the transfer of any undissolved solids to the paper and conducting the filtration under refrigerated conditions (not above 5°).

Procedure. If it is plastic or semisolid, soften a portion of the lecithin by warming it in a water bath at a temperature not exceeding 60° and then mixing it thoroughly. Transfer 2 grams of a well-mixed sample, accurately weighed, into a 40-ml. centrifuge tube, previously tared with a glass stirring rod, and add 15 ml. of *Phosphatide-Acetone Solution* from a buret. Warm the mixture in a water bath until the lecithin melts, but avoid evaporation of the acetone. Stir until the sample is completely disintegrated and dispersed, and then transfer the tube into an ice bath, chill for 5 minutes, remove from the ice bath, and add about one-half of the required volume of *Phosphatide-Acetone Solution,* previously chilled for 5 minutes in an ice bath. Stir the mixture to complete dispersion of the sample, dilute to 40 ml. with chilled *Phosphatide-Acetone Solution* (5°), again stir and return the tube and contents to the ice bath for 15 minutes. At the end of the 15-minute chilling period stir again while still in the ice bath, remove the stirring rod, temporarily supporting it in a vertical upside-down position, and centrifuge the mixture immediately at about 2000 rpm for 5 minutes. Decant the supernatant liquid from the centrifuge tube, crush the centrifuged solids with the same stirring rod previously used, and refill the tube to the 40-ml. mark with chilled (5°) *Phosphatide-Acetone Solution,* and repeat the chilling, stirring, centrifugation, and decantation procedure previously followed. After the second centrifugation and decantation of the supernatant acetone, again crush the solids with the assigned stirring rod, and place the tube and its contents in a horizontal position at room temperature until the excess acetone has evaporated. Mix the residue again, dry the centrifuge tube and its contents at 105° for 45 minutes in a forced draft oven, cool, and weigh. Calculate the percent of acetone-insoluble substances by the formula $(100R/S) - B$, in which R is the weight of residue, S is the weight of the sample, and B is the percent of *Benzene-insoluble matter* determined as directed in this monograph.

Acid value. If it is plastic or semisolid, soften a portion of the lecithin by warming it in a water bath at a temperature not exceeding 60° and then mix it thoroughly. Transfer about 2 grams of a well-mixed sample into a 250-ml. Erlenmeyer flask and dissolve it in 50 ml. of petroleum ether. To this solution add 50 ml. of alcohol, previously neutralized to phenolphthalein with 0.1 *N* sodium hydroxide, and mix well. Add phenolphthalein T.S. and titrate with 0.1 *N* sodium hydroxide to a pink end-point which persists for 5 seconds. Calculate the number of mg. of potassium hydroxide required to neutralize the acids in 1 gram of the sample by multiplying the number of ml. of 0.1 *N* sodium hydroxide consumed in the titration by 5.6 and dividing the result by the weight of the sample.

Benzene-insoluble matter. If plastic or semisolid, soften a portion of the lecithin by warming it at a temperature not exceeding 60° and then mix it thoroughly. Weigh 10 grams of a previously well-mixed sample into a 250-ml. wide-mouth Erlenmeyer flask, add 100 ml. of benzene, and shake until the lecithin is dissolved. Filter the solution through a 30-ml. Corning "C" porosity or equivalent filtering funnel which previously has been dried at 105° for 1 hour, cooled in a desiccator, and weighed. Wash the flask with two successive 25-ml. portions of benzene and pass the washings through the filter. Dry the funnel at 105° for 1 hour, cool to room temperature in a desiccator, and weigh. From the gain in weight of the funnel calculate the percent of the benzene-insoluble matter in the sample.

Water. Determine by the *Toluene Distillation Method*, page 979, using a 100-gram sample, accurately weighed.

Arsenic. A *Sample Solution* prepared as directed for organic compounds meets the requirements of the *Arsenic Test*, page 865.

Heavy metals. Prepare and test a 500-mg. sample as directed in *Method II* under the *Heavy Metals Test*, page 920, using 20 mcg. of lead ion (Pb) in the control (*Solution A*).

Lead. A *Sample Solution* prepared as directed for organic compounds meets the requirements of the *Lead Limit Test*, page 929, using 10 mcg. of lead ion (Pb) in the control.

Packaging and storage. Store in well-closed containers.

Functional use in foods. Antioxidant; emulsifier.

LEMONGRASS OIL

DESCRIPTION

A volatile oil prepared by steam distillation of freshly cut and partially dried cymbopogon grasses indigenous to tropical and subtropical areas. Two types of lemongrass oil are commercially available. The *East Indian* type, also known as Cochin, Native, and British Indian Lemongrass Oil, is usually a dark yellow to light brownish red liquid having a pronounced heavy lemon-like odor. The *West Indian* type, also known as Madagascar, Guatemala, or other country of origin Lemongrass Oil, is light yellow to light brown in color and has a lemon-like odor of a lighter character than the East Indian type oil. Lemongrass oils are soluble in mineral oil, freely soluble in propylene glycol, but practically insoluble in water and in glycerin. The East Indian variety dissolves readily in alcohol, but the West Indian oil yields cloudy solutions.

SPECIFICATIONS

Assay. Not less than 75 percent, by volume, of aldehydes as citral.

Angular rotation. Between −3° and +1°.

Refractive index. Between 1.483 and 1.489.

Solubility in alcohol. Passes test.

Specific gravity. *East Indian* type: between 0.894 and 0.904; *West Indian* type: between 0.869 and 0.894.

Steam-volatile oil. Not less than 93 percent, by volume.

Limits of Impurities

Arsenic (as As). Not more than 3 parts per million (0.0003 percent).

Heavy metals (as Pb). Not more than 40 parts per million (0.004 percent).

Lead. Not more than 10 parts per million (0.001 percent).

TESTS

Assay. Mix 50.0 ml. of the sample with 500 mg. of tartaric acid, shake for 5 minutes, and filter. Dry the filtered oil over anhydrous sodium sulfate, and then pipet 10.0 ml. of the clear, treated oil into a 150-ml. cassia flask. (*Note:* Retain the remaining oil for the *Steam-volatile oil* test.) Add 75 ml. of a 30 percent solution of sodium bisulfite, stopper the flask, and shake until a semisolid to solid sodium bisulfite addition product has formed. Allow the mixture to stand at room temperature for 5 minutes, then loosen the stopper, and immerse the flask in a water bath heated to between 85° and 90°. Maintain the water bath at this temperature, shaking the flask occasionally, until the addition product dissolves, and then continue the heating and intermittent shaking for another 30 minutes. When the liquids have separated completely, add enough of the sodium bisulfite solution to raise the lower level of the oily layer within the graduated portion of the neck of the flask. Calculate the percent, by volume, of the citral by the formula $100 - (V \times 10)$, in which V is the number of ml. of separated oil in the graduated neck of the cassia flask.

Angular rotation. Determine in a 100-mm. tube as directed under *Optical Rotation*, page 939.

Refractive index, page 945. Determine with an Abbé or other refractometer of equal or greater accuracy.

Solubility in alcohol. Proceed as directed in the general method, page 899. *East Indian* type: one ml. dissolves in 3 ml. of 70 percent alcohol, usually with slight turbidity. *West Indian* type: yields a cloudy solution with 70, 80, 90, and 95 percent alcohol.

Specific gravity. Determine by any reliable method (see page 5).

Steam-volatile oil. Proceed as directed under *Volatile Oil Content*, page 901, using 25.0 ml. of the oil prepared as directed in the *Assay*.

Arsenic. A *Sample Solution* prepared as directed for organic compounds meets the requirements of the *Arsenic Test*, page 865.

Heavy metals. Prepare and test a 500-mg. sample as directed in *Method II* under the *Heavy Metals Test*, page 920, using 20 mcg. of lead ion (Pb) in the control (*Solution A*).

Lead. A *Sample Solution* prepared as directed for organic compounds meets the requirements of the *Lead Limit Test*, page 929, using 10 mcg. of lead ion (Pb) in the control.

Packaging and storage. Store in full, tight, preferably glass, aluminum, tin-lined, or other suitably lined containers in a cool place, protected from light.

Functional use in foods. Flavoring agent.

LEMON OIL

DESCRIPTION

A volatile oil obtained by expression, without the aid of heat, from the fresh peel of the fruit of *Citrus limon* L. Burmann filius (Fam. *Rutaceae*), with or without the previous separation of the pulp and the peel. It is a pale to deep yellow, or greenish yellow liquid, having the characteristic odor and taste of the outer part of fresh lemon peel. It is miscible with dehydrated alcohol and with glacial acetic acid.

Note: Do not use lemon oil that has a terebinthine odor.

SPECIFICATIONS

Assay. *California type:* not less than 2.2 percent and not more than 3.8 per cent of aldehydes, calculated as citral ($C_{10}H_{16}O$); *Italian type:* not less than 3.0 percent and not more than 5.5 percent of aldehydes, calculated as citral ($C_{10}H_{16}O$).

Angular rotation. Between +57° and +65.6°.

Refractive index. Between 1.473 and 1.476 at 20°.

Solubility in alcohol. Passes test.

Specific gravity. Between 0.849 and 0.855.

Ultraviolet absorbance. *California type:* Not less than 0.2; *Italian type:* Not less than 0.49.

Limits of Impurities

Arsenic (as As). Not more than 3 parts per million (0.0003 percent).

Foreign oils. Passes test.

Heavy metals (as Pb). Not more than 40 parts per million (0.004 percent).

Lead. Not more than 10 parts per million (0.001 percent).

TESTS

Assay. Weigh accurately a 5-ml. sample, and proceed as directed under *Aldehydes*, page 894, using 50 ml. of hydroxylamine hydrochloride solution, previously adjusted to a pH of about 3.4. Allow the mixture to stand for 15 minutes, with occasional shaking, before titrating, and use 76.12 as the equivalence factor (E) in the calculation.

Angular rotation. Determine in a 100-mm. tube as directed under *Optical Rotation*, page 939.

Refractive index, page 945. Determine with an Abbé or other refractometer of equal or greater accuracy.

Solubility in alcohol. Proceed as directed in the general method, page 899. One ml. dissolves in 3 ml. of alcohol.

Specific gravity. Determine by any reliable method (see page 5).

Ultraviolet absorbance. Proceed as directed under *Ultraviolet Absorbance of Citrus Oils*, page 900, using about 250 mg. of sample, accurately weighed. The maximum absorbance occurs at 315 ± 3 mμ.

Arsenic. A *Sample Solution* prepared as directed for organic compounds meets the requirements of the *Arsenic Test*, page 865.

Foreign oils. Transfer 50 ml. to a Ladenburg flask having 4 bulbs of the following approximate diameters: 6 cm., 3.5 cm., 3.0 cm., and 2.5 cm., respectively, in ascending order. The distance from the bottom of the flask to the side-arm is 20 cm. Distil the oil at the rate of 1 drop per second until the distillate measures 5 ml. The angular rotation of the distillate is not more than 6° less than that of the original oil, and the refractive index is not less than 0.001 and not more than 0.003 lower than that of the original oil at 20°.

Heavy metals. Prepare and test a 500-mg. sample as directed in *Method II* under the *Heavy Metals Test*, page 920, using 20 mcg. of lead ion (Pb) in the control (*Solution A*).

Lead. A *Sample Solution* prepared as directed for organic compounds meets the requirements of the *Lead Limit Test*, page 929, using 10 mcg. of lead ion (Pb) in the control.

Packaging and storage. Store in full, tight containers. Avoid exposure to excessive heat.

Functional use in foods. Flavoring agent.

DL-LEUCINE

DL-2-Amino-4-methylvaleric Acid

$$\begin{array}{l} CH_3CHCH_2CHCOOH \\ \quad\;\; | \qquad\quad | \\ \quad CH_3 \quad\; NH_2 \end{array}$$

$C_6H_{13}NO_2$ Mol. wt. 131.18

DESCRIPTION

Small white crystals or a crystalline powder. It is odorless and has a slightly bitter taste. It is freely soluble in water, slightly soluble in alcohol, and insoluble in ether. It melts with decomposition at about 290°. The pH of a 1 in 100 solution is between 5.5 and 7.0.

IDENTIFICATION

To 5 ml. of a 1 in 1000 solution add 1 ml. of triketohydrindene hydrate T.S. A reddish purple or bluish purple color is produced.

SPECIFICATIONS

Assay. Not less than 98.0 percent and not more than the equivalent of 102.0 percent of $C_6H_{13}NO_2$, calculated on the dried basis.

Limits of Impurities

Arsenic (as As). Not more than 3 parts per million (0.0003 percent).

Heavy metals (as Pb). Not more than 20 parts per million (0.002 percent).

Lead. Not more than 10 parts per million (0.001 percent).

Loss on drying. Not more than 0.3 percent.

Residue on ignition. Not more than 0.1 percent.

TESTS

Assay. Dissolve about 400 mg., accurately weighed, in 3 ml. of formic acid and 50 ml. of glacial acetic acid. Add 2 drops of crystal violet T.S., and titrate with 0.1 *N* perchloric acid to the first appearance of a pure green color or until the blue color disappears completely. Each ml. of 0.1 *N* perchloric acid is equivalent to 13.12 mg. of $C_6H_{13}NO_2$.

Arsenic. A *Sample Solution* prepared as directed for organic compounds meets the requirements of the *Arsenic Test*, page 865.

Heavy metals. Prepare and test a 1-gram sample as directed in

Method II under the *Heavy Metals Test*, page 920, using 20 mcg. of lead ion (Pb) in the control (*Solution A*).

Lead. A *Sample Solution* prepared as directed for organic compounds meets the requirements of the *Lead Limit Test*, page 929, using 10 mcg. of lead ion (Pb) in the control.

Loss on drying, page 931. Dry at 105° for 3 hours.

Residue on ignition, page 945. Ignite 1 gram as directed in the general method.

Packaging and storage. Store in well-closed containers.

Functional use in foods. Nutrient; dietary supplement.

L-LEUCINE

L-2-Amino-4-methylvaleric Acid

$$CH_3{-}\underset{\displaystyle CH_3}{\underset{|}{C}}HCH_2\underset{\displaystyle NH_2}{\underset{|}{C}}H{-}COOH$$

$C_6H_{13}NO_2$ Mol. wt. 131.18

DESCRIPTION

Small, white, lustrous plates, or a white crystalline powder. One gram dissolves in about 40 ml. of water and in about 100 ml. of acetic acid. It is sparingly soluble in alcohol, but soluble in dilute hydrochloric acid and in solutions of alkali hydroxides and carbonates. It sublimes at about 150°.

SPECIFICATIONS

Assay. Not less than 98.5 percent of $C_6H_{13}NO_2$ after drying.

Nitrogen (Total). Not less than 10.6 percent.

Specific rotation. $[\alpha]_D^{25°}$: Between +15.0° and +16.0°; $[\alpha]_D^{20°}$: between +14.9° and +16.5°.

Limits of Impurities

Arsenic (as As). Not more than 3 parts per million (0.0003 percent).

Heavy metals (as Pb). Not more than 30 parts per million (0.003 percent).

Iron. Not more than 50 parts per million (0.005 percent).

Lead. Not more than 10 parts per million (0.001 percent).

Loss on drying. Not more than 0.1 percent.

Methionine. Passes test.

Residue on ignition. Not more than 0.1 percent.

Tyrosine. Passes test.

TESTS

Assay. Transfer about 400 mg., previously dried at 105° for two hours and accurately weighed, into a 250-ml. flask. Dissolve the sample in about 50 ml. of acetic acid, add 2 drops of crystal violet T.S., and titrate with 0.1 *N* perchloric acid to a bluish green end-point. Perform a blank determination (see page 2) and make any necessary correction. Each ml. of 0.1 *N* perchloric acid is equivalent to 13.12 mg. of $C_6H_{13}NO_2$.

Nitrogen (Total). Proceed as directed under *Nitrogen Determination*, page 937, using about 300 mg. of the sample previously dried and accurately weighed.

Specific rotation, page 939. $[\alpha]_D^{25°}$: Determine in a solution containing 2 grams in sufficient 6 *N* hydrochloric acid to make 100 ml.; $[\alpha]_D^{20°}$: determine in a solution containing 4 grams in sufficient 6 *N* hydrochloric acid to make 100 ml.

Arsenic. A *Sample Solution* prepared as directed for organic compounds meets the requirements for the *Arsenic Test*, page 865.

Heavy metals. Prepare and test a 670-mg. sample as directed in *Method II* under the *Heavy Metals Test*, page 920, using 20 mcg. of lead ion (Pb) in the control (*Solution A*).

Iron. To the ash obtained in the test for *Residue on Ignition* add 2 ml. of dilute hydrochloric acid (1 in 2), and evaporate to dryness on a steam bath. Dissolve the residue in 1 ml. of hydrochloric acid and dilute with water to 50 ml. Dilute 10 ml. of this solution to 40 ml. with water and add 40 mg. of ammonium persulfate crystals and 10 ml. of ammonium thiocyanate T.S. Any red or pink color does not exceed that produced by 1.0 ml. of *Iron Standard Solution* (10 mcg. Fe) in an equal volume of a solution containing the quantities of the reagents used in the test.

Lead. A *Sample Solution* prepared as directed for organic compounds meets the requirements of the *Lead Limit Test*, page 929, using 10 mcg. of lead ion (Pb) in the control.

Loss on drying, page 931. Dry at 105° for 2 hours.

Methionine. Dissolve 100 mg. in 0.3 ml. of sulfuric acid saturated with anhydrous cupric sulfate. No yellow color is produced within 2 minutes.

Residue on ignition. Ignite 1 gram as directed in the general method, page 945.

Tyrosine. Dissolve 100 mg. in 3 ml. of diluted sulfuric acid T.S. and add to it 3 ml. of a 1 in 10 solution of mercuric sulfate in diluted sulfuric acid T.S. and 0.5 ml. of a 1 in 20 solution of sodium nitrite. No red or pink color appears within 15 minutes.

Packaging and storage. Store in well-closed containers.

Functional use in foods. Nutrient; dietary supplement.

LIME OIL, DISTILLED

DESCRIPTION

The volatile oil obtained by distillation from the juice, or the whole crushed fruit, of *Citrus aurantifolia* Swingle. It is a colorless to greenish yellow liquid. It is soluble in most fixed oils and in mineral oil. It is insoluble in glycerin and in propylene glycol.

SPECIFICATIONS

Aldehydes. Between 0.5 percent and 2.5 percent of aldehydes, calculated as citral ($C_{10}H_{16}O$).

Angular rotation. Between +34° and +47°.

Refractive index. Between 1.474 and 1.477 at 20°.

Solubility in alcohol. Passes test.

Specific gravity. Between 0.855 and 0.863.

Limits of Impurities

Arsenic (as As). Not more than 3 parts per million (0.0003 percent).

Heavy metals (as Pb). Not more than 40 parts per million (0.004 percent).

Lead. Not more than 10 parts per million (0.001 percent).

TESTS

Aldehydes. Weigh accurately about 5 grams, and proceed as directed under *Aldehydes*, page 894, using 76.12 as the equivalence factor (*E*) in the calculation. Allow the mixture to stand at room temperature for 15 minutes before titrating.

Angular rotation. Determine in a 100-mm. tube as directed under *Optical Rotation*, page 939.

Refractive index, page 945. Determine with an Abbé or other refractometer of equal or greater accuracy.

Solubility in alcohol. Proceed as directed in the general method, page 899. One ml. dissolves in 5 ml. of 90 percent alcohol.

Specific gravity. Determine by any reliable method (see page 5).

Arsenic. A *Sample Solution* prepared as directed for organic compounds meets the requirements of the *Arsenic Test*, page 865.

Heavy metals. Prepare and test a 500-mg. sample as directed in *Method II* under the *Heavy Metals Test*, page 920, using 20 mcg. of lead ion (Pb) in the control (*Solution A*).

Lead. A *Sample Solution* prepared as directed for organic compounds meets the requirements of the *Lead Limit Test*, page 929, using 10 mcg. of lead ion (Pb) in the control.

Packaging and storage. Store in full, tight, preferably glass, tin-lined, galvanized, or black iron containers in a cool place protected from light.

Functional use in foods. Flavoring agent.

LIMESTONE, GROUND

DESCRIPTION

Ground limestone consists essentially of calcium carbonate. It is obtained by crushing, grinding, and classifying of naturally occurring limestone benefited by flotation and/or air classification. It is produced as a fine, white to off-white, microcrystalline powder. It is odorless and tasteless and is stable in air. It is practically insoluble in water and in alcohol. The presence of any ammonium salt or carbon dioxide increases its solubility in water, but the presence of any alkali hydroxide reduces the solubility. It dissolves with effervescence in diluted acetic acid T.S., in diluted hydrochloric acid T.S., and in diluted nitric acid T.S., and the resulting solutions, after boiling, give positive tests for *Calcium*, page 926.

SPECIFICATIONS

Assay. Not less than 94.0 percent of $CaCO_3$ after drying.

Limits of Impurities

Acid-insoluble substances. Not more than 2.5 percent.
Arsenic (as As). Not more than 3 parts per million (0.0003 percent).
Fluoride. Not more than 50 parts per million (0.005 percent).
Heavy metals (as Pb). Not more than 40 parts per million (0.004 percent).
Lead. Not more than 3 parts per million (0.0003 percent).
Loss on drying. Not more than 2 percent.
Magnesium and alkali salts. Not more than 3.5 percent.
Mercury. Not more than 0.5 parts per million (0.00005 percent).

TESTS

Assay. Transfer about 200 mg., previously dried at 200° for 4 hours and accurately weighed, into a 400 ml. beaker, add 10 ml. of water, and swirl to form a slurry. Cover the beaker with a watch glass, and introduce 2 ml. of diluted hydrochloric acid T.S. from a pipet inserted between the lip of the beaker and the edge of the watch glass. Swirl the contents of the beaker to dissolve the sample. Wash down the sides of the beaker, the outer surface of the pipet, and the watch glass, and dilute to about 100 ml. with water. While stirring, preferably with a magnetic stirrer, add about 30 ml. of 0.05 *M* disodium ethylenediaminetetraacetate from a 50 ml. buret, then add 15 ml. of potassium hydroxide T.S. and 300 mg. of hydroxy naphthol blue indicator, and continue the titration to a blue end-point. Each ml. of 0.05 *M* disodium ethylenediaminetetraacetate is equivalent to 5.004 mg. of $CaCO_3$.

Acid-insoluble substances. Suspend 5 grams in 25 ml. of water, cautiously add with agitation 25 ml. of dilute hydrochloric acid (1 in 2), then add water to make a volume of about 200 ml. Heat the solution to boiling, cover, digest on a steam bath for 1 hour, cool, and filter. Wash the precipitate with water until the last washing shows no chloride with silver nitrate T.S., and then ignite it. The weight of the residue does not exceed 125 mg.

Arsenic. A solution of 1 gram in 10 ml. of diluted hydrochloric acid T.S. meets the requirements of the *Arsenic Test*, page 865.

Fluoride. Determine as directed under *Method II* in the general procedure, page 918.

Sample Solution for the Determination of Heavy Metals and Lead. Cautiously dissolve 5 grams in 25 ml. of dilute hydrochloric acid (1 in 2) and evaporate to dryness on a steam bath. Dissolve the residue in about 15 ml. of water and dilute to 25 ml. (1 ml. = 200 mg.).

Heavy metals. Neutralize 2.5 ml. (500 mg.) of the *Sample Solution* with sodium hydroxide T.S., using phenolphthalein as the indicator, and dilute to 25 ml. This solution meets the requirements of the *Heavy Metals Test*, page 920, using 20 mcg. of lead ion (Pb) in the control (*Solution A*).

Lead. A 5-ml. portion of the *Sample Solution* (1 gram) meets the requirements of the *Lead Limit Test*, page 929, using 3 mcg. of lead ion (Pb) in the control.

Loss on drying, page 931. Dry at 200° for 4 hours.

Magnesium and alkali salts. Mix 1 gram with 40 ml. of water, carefully add 5 ml. of hydrochloric acid, mix, and boil for 1 minute. Rapidly add 40 ml. of oxalic acid T.S., and stir vigorously until precipitation is well established. Immediately add 2 drops of methyl red T.S., then add ammonia T.S., dropwise, until the mixture is just alkaline, and cool. Transfer the mixture to a 100-ml. cylinder, dilute with water to 100 ml., let it stand for 4 hours or overnight, then decant the clear, supernatant liquid through a dry filter paper. To 50 ml. of the clear filtrate in a platinum dish add 0.5 ml. of sulfuric acid, and evaporate the mixture on a steam bath to a small volume. Carefully evaporate the remaining liquid to dryness over a free flame, and continue heating until the ammonium salts have been completely decomposed and volatilized. Finally, ignite the residue to constant weight. The weight of the residue does not exceed 17.5 mg.

Mercury. Determine as directed under *Mercury Limit Test*, page 934, using the following as the *Sample Preparation:* Transfer 4.0 grams of the sample into a 50-ml. beaker, cautiously dissolve in 10 ml. of dilute hydrochloric acid solution (1 in 2), add 2 drops of phenolphthalein T.S., and slowly neutralize, with constant stirring, with sodium hydroxide T.S. Add 1 ml. of dilute sulfuric acid solution (1 in 5) and 1 ml. of potassium permanganate solution (1 in 25), cover the beaker with a watch glass, boil for a few seconds, and cool.

Packaging and storage. Store in well-closed containers.

Functional use in foods. Texturizing and release agent and modifier for chewing gum base and chewing gum.

d-LIMONENE

d-p-Mentha-1,8-diene; Cinene

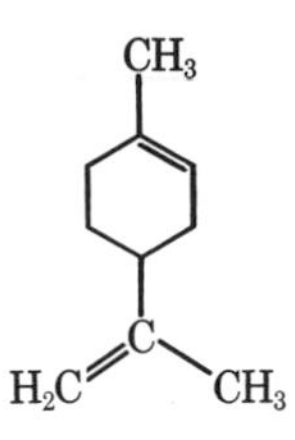

$C_{10}H_{15}$ Mol. wt. 136.25

DESCRIPTION

A colorless liquid having a pleasant, mildly citrus odor, free from camphoraceous and turpentine-like notes. It is obtained from citrus oils and should be substantially free from other terpenes. It is miscible with alcohol, with most fixed oils, and with mineral oil. It is very slightly soluble in glycerin and is insoluble in water and in propylene glycol.

SPECIFICATIONS

Angular rotation. Between +96° and +104°.

Peroxide value. Not more than 2.0.

Refractive index. Between 1.471 and 1.474 at 20°.

Specific gravity. Between 0.838 and 0.843.

Limits of Impurities

Arsenic (as As). Not more than 3 parts per million (0.0003 percent).
Heavy metals (as Pb). Not more than 40 parts per million (0.004 percent).
Lead. Not more than 10 parts per million (0.001 percent).

TESTS

Angular rotation. Determine in a 100-mm. tube as directed under *Optical Rotation*, page 939.

Peroxide value. To 50 ml. of a mixture of 3 volumes of glacial acetic acid and 2 volumes of chloroform add 10 ml. of the sample. To this solution add 1 ml. of a saturated solution of potassium iodide, allow to stand for exactly 1 minute with gentle shaking, and then introduce 100 ml. of water and a few drops of starch T.S. Titrate immediately with 0.1 *N* sodium thiosulfate. Each ml. of 0.1 *N* sodium thiosulfate, multiplied by 5, equals the peroxide value, expressed in millimoles of peroxide per liter of the sample.

Refractive index, page 945. Determine with an Abbé or other refractometer of equal or greater accuracy.

Specific gravity. Determine by any reliable method (see page 5).

Arsenic. A *Sample Solution* prepared as directed for organic compounds meets the requirements of the *Arsenic Test,* page 865.

Heavy metals. Prepare and test a 500-mg. sample as directed in *Method II* under the *Heavy Metals Test,* page 920, using 20 mcg. of lead ion (Pb) in the control (*Solution A*).

Lead. A *Sample Solution* prepared as directed for organic compounds meets the requirements of the *Lead Limit Test,* page 929, using 10 mcg. of lead ion (Pb) in the control.

Packaging and storage. Store in full, tight, preferably glass, aluminum, tin-lined, or other suitably lined containers in a cool place protected from light. *d*-Limonene readily oxidizes and develops undesirable odors in the presence of air. Therefore, the volume of air in the container must be kept at a minimum, and the use of a suitable preservative is recommended.

Functional use in foods. Flavoring agent.

LINALOE WOOD OIL

DESCRIPTION

The volatile oil obtained by steam distillation from the wood of *Bursera delpechiana* Poiss. (Fam. *Burseraceae*) and other *Bursera* species. It is a colorless to yellow liquid, having a pleasant flowery odor. It is soluble in most fixed oils and in propylene glycol. It is soluble in mineral oil, but becomes opalescent or turdid on dilution. It is insoluble in glycerin.

SPECIFICATIONS

Assay. Not less than 85.0 percent of alcohols, calculated as linalool ($C_{10}H_{18}O$).

Acid value. Not more than 3.0.

Angular rotation. Between −5° and −13°.

Ester value. Between 40 and 75.

Refractive index. Between 1.459 and 1.463 at 20°.

Solubility in alcohol. Passes test.

Specific gravity. Between 0.876 and 0.883.

TESTS

Assay. Proceed as directed under *Linalool Determination,* page 897, using about 1.5 grams of acetylated oil, accurately weighed, for the saponification.

Acid value. Determine as directed in the general method, page 893.

Angular rotation. Determine in a 100-mm. tube as directed under *Optical Rotation*, page 939.

Ester value. Determine as directed in the general method, page 897, using about 2.5 grams, accurately weighed.

Refractive index, page 945. Determine with an Abbé or other refractometer of equal or greater accuracy.

Solubility in alcohol. Proceed as directed in the general method, page 899. One ml. dissolves in 5 ml. of 60 percent alcohol.

Specific gravity. Determine by any reliable method (see page 5).

Packaging and storage. Store in full, tight, preferably glass, aluminum, or tin-lined containers in a cool place protected from light.

Functional use in foods. Flavoring agent.

LINALOOL

3,7-Dimenthyl-1-1,6-octadien-3-ol

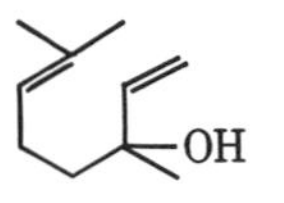

$C_{10}H_{18}O$ Mol. wt. 154.25

DESCRIPTION

Linalool is a naturally occurring terpene alcohol found in the volatile oils obtained from various flowers, fruits, grasses, leaves, roots, seeds, and woods. It has been obtained from a number of these oils. It may be prepared by fractionation of saponified Brazilian Bois de Rose oil. Commercial synthetic linalool has been prepared by isomerization of geraniol and by other methods. It is a colorless liquid having a pleasant floral odor. It is soluble in fixed oils, in mineral oil, and in propylene glycol. It is insoluble in glycerin. Some commercial grades of linalool may be composed almost entirely of the specific chemical whose structural formula is shown above, while other commercial grades that conform to the limits of these specifications and of satisfactory may contain other isomeric and closely related terpenic alcohols.

SPECIFICATIONS

Assay. Not less than 92.0 percent of $C_{10}H_{18}O$.

Angular rotation. Between $-2°$ and $+2°$.

Refractive index. Between 1.461 and 1.465 at 20°.

Solubility in alcohol. Passes test.

Specific gravity. Between 0.858 and 0.867.

Limits of Impurities

Esters. Not more than 0.5 percent, calculated as linalyl acetate ($C_{12}H_{20}O_2$).

TESTS

Assay. Proceed as directed under *Linalool Determination*, page 897, using about 1.2 grams of acetylized oil, accurately weighed.

Angular rotation. Determine in a 100-mm. tube as directed under *Optical Rotation*, page 939.

Refractive index, page 945. Determine with an Abbé or other refractometer of equal or greater accuracy.

Solubility in alcohol. Proceed as directed in the general method, page 899. One ml. dissolves in 4 ml. of 60 percent alcohol.

Specific gravity. Determine by any reliable method (see page 5).

Esters. Weigh accurately about 10 grams, and proceed as directed under *Ester Determination*, page 896, using 98.15 as the equivalence factor (*e*) in the calculation.

Packaging and storage. Store in full, tight, glass, tin-lined, galvanized, or other suitably lined containers in a cool place.

Functional use in foods. Flavoring agent.

LINALYL ACETATE

3,7-Dimethyl-1,6-octadien-3-yl Acetate

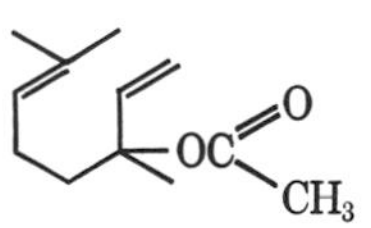

$C_{12}H_{20}O_2$ Mol. wt. 196.29

DESCRIPTION

Linalyl acetate occurs in bergamot, petitgrain, and other oils. It may be prepared by acetylation and fractionation of Brazilian Bois de Rose Oil, and it may also be prepared synthetically. It is a colorless to slightly yellow liquid having a pleasant floral odor. It is miscible with alcohol, and is soluble in fixed oils and in mineral oil. It is sparingly soluble in propylene glycol. It is insoluble in glycerin and in water. Some commercial grades of linalyl acetate may be composed almost entirely of the specific chemical whose structural formula is shown above, while other commercial grades that conform to the limits of these specifications and of satisfactory purity may contain other isomeric and closely related terpenic esters.

SPECIFICATIONS

Assay. Not less than 90.0 percent of esters, calculated as linalyl acetate ($C_{12}H_{20}O_2$).

Angular rotation. Between −1° and +1°.

Refractive index. Between 1.449 and 1.457 at 20°.

Solubility in alcohol. Passes test.

Specific gravity. Between 0.895 and 0.914.

Limits of impurities

Acid value. Not more than 1.0.

TESTS

Assay. Weigh accurately about 1 gram, and proceed as directed under *Ester Determination*, page 896, using 98.15 as the equivalence factor (*e*) in the calculation. The volume of 0.5 *N* alcoholic potassium hydroxide consumed by the sample should be corrected for the *Acid Value*.

Angular rotation. Determine in a 100-mm. tube at 20° as directed under *Optical Rotation*, page 939.

Refractive index, page 945. Determine with an Abbé or other refractometer of equal or greater accuracy.

Solubility in alcohol. Proceed as directed in the general method, page 899. One ml. dissolves in 5 ml. of 70 percent alcohol.

Specific gravity. Determine by any reliable method (see page 5).

Acid value. Determine as directed in the general method, page 893.

Packaging and storage. Store in full, tight, glass, tin-lined, galvanized, or other suitably lined containers in a cool place protected from light.

Functional use in foods. Flavoring agent.

LINALYL ACETATE, SYNTHETIC

3,7-Dimethyl-1,6-octadien-3-yl Acetate

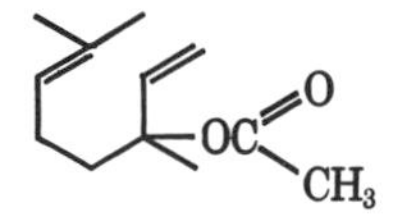

$C_{12}H_{20}O_2$ Mol. wt. 196.29

DESCRIPTION

Linalyl acetate, synthetic, prepared by chemical synthesis, is chemically identical with the pure natural product. It is a colorless liquid with a pleasant floral odor. It is soluble in most fixed oils, but is only slightly soluble in propylene glycol. It is insoluble in glycerin and in water. It is free from geranyl acetate and terpinyl acetate.

SPECIFICATIONS

Assay. Not less than 97.0 percent of $C_{12}H_{20}O_2$.

Refractive index. Between 1.449 and 1.452 at 20°.

Specific gravity. Between 0.895 and 0.908.

Limits of Impurities

Acid value. Not more than 1.0.

TESTS

Assay. Weigh accurately about 1 gram, and proceed as directed under *Ester Determination*, page 896, using 98.15 as the equivalence factor (*e*) in the calculation.

Refractive index, page 945. Determine with an Abbé or other refractometer of equal or greater accuracy.

Specific gravity. Determine by any reliable method (see page 5).

Acid value. Determine as directed in the general method, page 893.

Packaging and storage. Store in full, tight, preferably glass, aluminum, galvanized, tin-lined, or other suitably lined containers in a cool place protected from light.

Functional use in foods. Flavoring agent.

LINALYL BENZOATE

3,7-Dimethyl-1,6-octadien-3-yl Benzoate

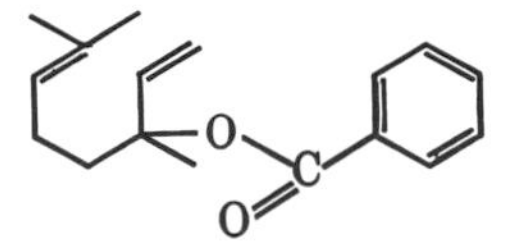

$C_{17}H_{22}O_2$ Mol. wt. 258.36

DESCRIPTION

A yellow to brownish yellow liquid having a characteristic tuberose-like odor. It is soluble in alcohol, in chloroform, and in ether, but practically insoluble in water. Some commercial grades of linalyl benzoate may be composed almost entirely of the specific chemical whose structural formula is shown above, while other commercial grades that conform to the limits of these specifications and of satisfactory purity may contain other isomeric and closely related terpenic esters.

SPECIFICATIONS

Assay. Not less than 75.0 percent of $C_{17}H_{22}O_2$.

Refractive index. Between 1.505 and 1.520 at 20°.

Solubility in alcohol. Passes test.

Specific gravity. Between 0.980 and 0.999.

Limits of Impurities

Acid value. Not more than 5.0.

TESTS

Assay. Weigh accurately about 1.5 grams, and proceed as directed under *Ester Determination*, page 896, using 129.18 as the equivalence factor (e) in the calculation. Reflux for 4 hours before titrating.

Refractive index, page 745. Determine with an Abbé or other refractometer of equal or greater accuracy.

Solubility in alcohol. Proceed as directed in the general method, page 899. One ml. dissolves in 1 ml. of 90 percent alcohol to form a clear solution.

Specific gravity. Determine by any reliable method (see page 5).

Acid value. Determine as directed in the general method, page 893.

Packaging and storage. Store in tight, full, preferably glass, tin-lined or other suitably lined containers in a cool place protected from light.

Functional use in foods. Flavoring agent.

LINALYL ISOBUTYRATE

3,7-Dimethyl-2,6-octadien-3-yl Isobutyrate

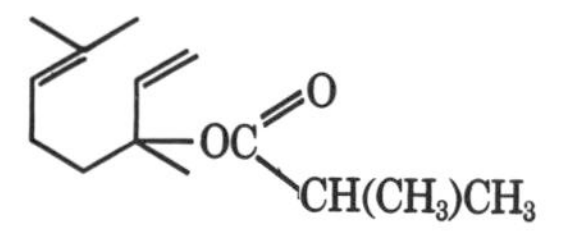

$C_{14}H_{24}O_2$ Mol. wt. 224.34

DESCRIPTION

A colorless to slightly yellow liquid having a fruity odor. It is miscible with alcohol, with ether, and with chloroform, but is practically insoluble in water. Some commercial grades of linalyl isobutyrate may be composed almost entirely of the specific chemical whose structural formula is shown above, while other commercial grades that conform to the limits of these specifications and of satisfactory purity may contain other isomeric and closely related terpenic esters.

SPECIFICATIONS

Assay. Not less than 95.0 percent of $C_{14}H_{24}O_2$.

Refractive index. Between 1.446 and 1.451 at 20°.

Solubility in alcohol. Passes test.

Specific gravity. Between 0.882 and 0.888.

Limits of Impurities

Acid value. Not more than 1.0.

TESTS

Assay. Weigh accurately about 1.5 grams, and proceed as directed under *Ester Determination* (*High Boiling Solvent*), page 896, using 112.17 as the equivalence factor (*e*) in the calculation.

Refractive index, page 945. Determine with an Abbé or other refractometer of equal or greater accuracy.

Solubility in alcohol. Proceed as directed in the general method, page 899. One ml. dissolves in 3 ml. of 80 percent alcohol to form a clear solution.

Specific gravity. Determine by any reliable method (see page 5).

Acid value. Determine as directed in the general method, page 893.

Packaging and storage. Store in full, tight, preferably glass, tin-lined or other suitably lined containers in a cool place protected from light.

Functional use in foods. Flavoring agent.

LINALYL PROPIONATE

3,7-Dimethyl-2,6-octadien-3-yl Propionate

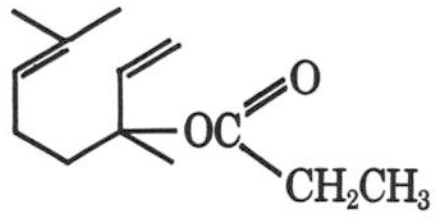

OC(=O)CH₂CH₃

$C_{13}H_{22}O_2$ Mol. wt. 210.32

DESCRIPTION

An almost colorless liquid, having a sweet floral odor similar to bergamot oil. It is soluble in most fixed oils and in mineral oil. It is slightly soluble in propylene glycol, but it is insoluble in glycerin. Some commercial grades of linalyl propionate may be composed almost entirely of the specific chemical whose structural formula is shown above, while other commercial grades that conform to the limits of these specifications and of satisfactory purity may contain other isomeric and closely related terpenic esters.

SPECIFICATIONS

Assay. Not less than 92.0 percent of $C_{13}H_{22}O_2$.

Angular rotation. Between $-1°$ and $+1°$.

Refractive index. Between 1.450 and 1.455 at 20°

Solubility in alcohol. Passes test.

Specific gravity. Between 0.895 and 0.902.

Limits of Impurities

Acid value. Not more than 1.0.

TESTS

Assay. Weigh accurately about 1 gram, and proceed as directed under *Ester Determination*, page 896, using 105.2 as the equivalence factor (*e*) in the calculation.

Angular rotation. Determine in a 100-mm. tube as directed under *Optical Rotation*, page 939.

Refractive index, page 945. Determine with an Abbé or other refractometer of equal or greater accuracy.

Solubility in alcohol. Proceed as directed in the general method, page 899. One ml. dissolves in 7 ml. of 70 percent alcohol.

Specific gravity. Determine by any reliable method (see page 5).

Acid value. Determine as directed in the general method, page 893.

Packaging and storage. Store in full, tight, preferably glass, aluminum, tin-lined, or other suitably lined containers in a cool place protected from light.

Functional use in foods. Flavoring agent.

LOCUST BEAN GUM

Carob Bean Gum

DESCRIPTION

A gum obtained from the ground endosperms of *Ceratonia siliqua* (L.) Taub. (Fam. *Leguminosae*). It consists chiefly of a high molecular weight hydrocolloidal polysaccharide, composed of galactan and mannan units combined through glycosidic linkages, which may be described chemically as a galactomannan. It is a white to yellowish white, nearly odorless, powder. It is dispersible in either hot or cold water forming a sol, having a pH between 5.4 and 7.0, which may be converted to a gel by the addition of small amounts of sodium borate.

IDENTIFICATION

A. Transfer a 2-gram sample into a 400-ml. beaker, moisten it with about 4 ml. of isopropyl alcohol, add with vigorous stirring 200 ml. of cold water, and continue the stirring until the gum is uniformly dispersed. An opalescent, slightly viscous solution is formed.

B. Transfer 100 ml. of the solution prepared in *Identification test A* into another 400-ml. beaker, heat the mixture in a boiling water bath for about 10 minutes, and then cool to room temperature. An appreciable increase in viscosity is produced (*distinction from guar gum*).

SPECIFICATIONS

Galactomannans. Not less than 73.0 percent.

Limits of Impurities

Acid-insoluble matter. Not more than 5 percent.

Arsenic (as As). Not more than 3 parts per million (0.0003 percent).

Ash (**Total**). Not more than 1.2 percent.

Heavy metals (as Pb). Not more than 20 parts per million (0.002 percent).

Lead. Not more than 10 parts per million (0.001 percent).

Loss on drying. Not more than 15 percent.

Protein. Not more than 8 percent.

Starch. Passes test.

TESTS

Galactomannans. The difference between the sum of the percentages of *Acid-insoluble matter*, *Total ash*, *Loss on drying*, and *Protein* and 100 represents the per cent of *Galactomannans*.

Acid-insoluble matter. Transfer 1.5 grams, accurately weighed, into a 250-ml. beaker containing 150 ml. of water and 15 ml. of 1 percent sulfuric acid. Cover the beaker with a watch glass and heat the mixture on a steam bath for 6 hours, rubbing down the wall of the beaker frequently with a rubber-tipped stirring rod and replacing any water lost by evaporation. Then add about 500 mg. of a suitable filter aid, accurately weighed, and filter through a tared Gooch crucible. Wash the residue several times with hot water, dry the crucible and its contents at 105° for 3 hours, cool in a desiccator, and weigh. The difference between the weight of the filter aid and that of the residue is the weight of the *Acid-insoluble matter*.

Arsenic. A *Sample Solution* prepared as directed for organic compounds meets the requirements of the *Arsenic Test*, page 865.

Ash (Total). Determine as directed in the general method, page 868.

Heavy metals. Prepare and test a 1-gram sample as directed in *Method II* under the *Heavy Metals Test,* page 920, using 20 mcg. of lead ion (Pb) in the control (*Solution A*).

Lead. A *Sample Solution* prepared as directed for organic compounds meets the requirements of the *Lead Limit Test,* page 929, using 10 mcg. of lead ion (Pb) in the control.

Loss on drying, page 931. Dry at 105° for 5 hours.

Protein. Transfer about 3.5 grams, accurately weighed, into a 500-ml. Kjeldahl flask and proceed as directed under *Nitrogen Determination,* page 937. The percent of nitrogen determined multiplied by 5.7 gives the percent of protein in the sample.

Starch. To a 1 in 10 solution of the gum add a few drops of iodine T.S. No blue color is produced.

Packaging and storage. Store in well-closed containers.

Functional use in foods. Stabilizer; thickener; emulsifier.

LOVAGE OIL

DESCRIPTION

The volatile oil obtained by steam distillation of the fresh root of the plant *Levisticum officinale* L. Koch syn. *Angelica levisticum,* Baillon (Fam. *Umbelliferae*). It is a yellow-greenish brown to deep brown liquid, having a strong characteristic aromatic odor and taste. It is soluble in most fixed oils, slightly soluble, with opalescence, in mineral oil, but it is relatively insoluble in glycerin and in propylene glycol.

Note: This oil becomes darker and more viscous under the influence of air and light.

SPECIFICATIONS

Acid value. Between 2.0 and 16.0.

Angular rotation. Between −1° and +5°.

Refractive index. Between 1.536 and 1.554 at 20°.

Saponification value. Between 238 and 258.

Solubility in alcohol. Passes test.

Specific gravity. Between 1.034 and 1.057.

TESTS

Acid value. Determine as directed in the general method, page 893.

Angular rotation. Determine in a 100-mm. tube as directed under *Optical Rotation,* page 939.

Refractive index, page 945. Determine with an Abbé or other refractometer of equal or greater accuracy.

Saponification value. Determine as directed in the general method, page 896, using 1.5 grams accurately weighed.

Solubility in alcohol. Proceed as directed in the general method, page 899. One ml. dissolves in 4 ml. of 80 percent alcohol, sometimes with slight turbidity. The age of the oil has an adverse effect upon solubility.

Specific gravity. Determine by any reliable method (see page 5).

Packaging and storage. Store in full, tight, glass, aluminum, tin-lined, or other suitably lined containers in a cool place protected from light.

Functional use in foods. Flavoring agent.

L-LYSINE MONOHYDROCHLORIDE

2,6-Diaminohexanoic Acid Hydrochloride

$NH_2(CH_2)_4CH(NH_2)COOH.HCl$

$C_6H_{14}N_2O_2.HCl$ Mol. wt. 182.65

DESCRIPTION

A white or nearly white, practically odorless, free-flowing, crystalline powder. It is freely soluble in water, but is almost insoluble in alcohol and in ether. It melts at about 260° with decomposition.

IDENTIFICATION

A. Heat 5 ml. of a 1 in 100 solution with 1 ml. of ninhydrin T.S. A violet color is produced.

B. A 1 in 20 solution gives positive tests for *Chloride*, page 926.

SPECIFICATIONS

Assay. Not less than 98.0 percent of $C_6H_{14}N_2O_2.HCl$, calculated on the dried basis.

Specific rotation. $[\alpha]_D^{25°}$: Between +19.2° and +20.8° (equivalent to +24.0° to +26.0° as free lysine); $[\alpha]_D^{20°}$: between +19.0° and +21.5° (as free lysine).

Limits of Impurities

Arsenic (as As). Not more than 3 parts per million (0.0003 percent).

Heavy metals (as Pb). Not more than 10 parts per million (0.001 percent).

Loss on drying. Not more than 1 percent.

Residue on ignition. Not more than 0.2 percent.

TESTS

Assay. Transfer about 275 mg., accurately weighed, into a 150-ml. beaker, and dissolve in 10 ml. of mercuric acetate T.S., heating on a steam bath to effect solution. Cool, add 75 ml. of glacial acetic acid, and titrate with 0.1 *N* perchloric acid, determining the end-point potentiometrically. Each ml. of 0.1 *N* perchloric acid is equivalent to 9.133 mg. of $C_6H_{14}N_2O_2.HCl$.

Specific rotation, page 939. $[\alpha]_D^{25°}$: Determine in a solution containing 2.5 grams in sufficient 6 *N* hydrochloric acid to make 100 ml.; $[\alpha]_D^{20°}$: determine in a solution containing 8 grams in sufficient 6 *N* hydrochloric acid to make 100 ml.

Arsenic. A *Sample Solution* prepared as directed for organic compounds meets the requirements of the *Arsenic Test,* page 865.

Heavy metals. Prepare and test a 2-gram sample as directed in *Method II* under the *Heavy Metals Test,* page 920, using 20 mcg. of lead ion (Pb) in the control (*Solution A*).

Loss on drying, page 931. Dry in vacuum at 60° for 4 hours.

Residue on ignition. Ignite 2 grams as directed in the general method, page 945.

Packaging and storage. Store in well-closed containers.

Functional use in foods. Nutrient; dietary supplement.

MACE OIL

DESCRIPTION

The volatile oil obtained by steam distillation from the ground, dried arillode of the ripe seed of *Myristica fragrans* Houtt. (Fam. *Myristicaceae*). Two types of oil, the East Indian and the West Indian, are commercially available. It is a colorless to pale yellow liquid, having the characteristic odor and taste of nutmeg. It is soluble in most fixed oils and in mineral oil, but it is insoluble in glycerin and in propylene glycol.

SPECIFICATIONS

Angular rotation. *East Indian* type: Between +2° and +30°; *West Indian* type: Between +20° and +45°.

Refractive index. *East Indian* type: Between 1.474 and 1.488; *West Indian* type: Between 1.469 and 1.480 at 20°.

Solubility in alcohol. Passes test.

Specific gravity. *East Indian* type: Between 0.880 and 0.930; *West Indian* type: Between 0.854 and 0.880.

TESTS

Angular rotation. Determine in a 100-mm. tube as directed under *Optical Rotation*, page 939.

Refractive index, page 945. Determine with an Abbé or other refractometer of equal or greater accuracy.

Solubility in alcohol. Proceed as directed in the general method, page 899. One ml. dissolves in 4 ml. of 90 percent alcohol.

Specific gravity. Determine by any reliable method (see page 5).

Packaging and storage. Store in full, tight, glass, tin-lined, or other suitably lined containers in a cool place protected from light.

Labeling. Label mace oil to indicate whether it is the East Indian or West Indian type.

Functional use in foods. Flavoring agent.

MAGNESIUM CARBONATE

DESCRIPTION

Magnesium carbonate is a basic hydrated magnesium carbonate or a normal hydrated magnesium carbonate. It occurs as light, white, friable masses, or as a bulky, white powder. It is odorless, and is stable in air. It is practically insoluble in water, to which, however, it imparts a slightly alkaline reaction. It is insoluble in alcohol, but is dissolved by dilute acids with effervescence. When treated with diluted hydrochloric acid T.S., it dissolves with effervescence and the resulting solution gives positive tests for *Magnesium*, page 927.

SPECIFICATIONS

Assay. The equivalent of not less than 40.0 percent and not more than 43.5 percent of MgO.

Limits of Impurities

Acid-insoluble substances. Not more than 500 parts per million (0.05 percent).

Arsenic (as As). Not more than 3 parts per million (0.0003 percent).

Calcium oxide. Not more than 0.6 percent.

Heavy metals (as Pb). Not more than 30 parts per million (0.003 percent).

Lead. Not more than 10 parts per million (0.001 percent).

Soluble salts. Not more than 1 percent.

TESTS

Assay. Dissolve about 1 gram, accurately weighed, in 30.0 ml. of 1 *N* sulfuric acid, add methyl orange T.S., and titrate the excess acid with

1 *N* sodium hydroxide. From the volume of 1 *N* sulfuric acid consumed, deduct the volume of 1 *N* sulfuric acid corresponding to the content of calcium oxide in the weight of the sample taken for the assay. The difference is the volume of 1 *N* sulfuric acid equivalent to the magnesium oxide present. Each ml. of 1 *N* sulfuric acid is equivalent to 20.16 mg. of MgO and to 28.04 mg. of CaO.

Acid-insoluble substances. Mix 5.0 grams with 75 ml. of water, add hydrochloric acid in small portions, with agitation, until no more of the sample dissolves, and boil for 5 minutes. If an insoluble residue remains, filter, wash well with water until the last washing is free from chloride, ignite, cool, and weigh.

Arsenic. A solution of 1 gram in 10 ml. of diluted hydrochloric acid T.S. meets the requirements of the *Arsenic Test*, page 865.

Calcium oxide. Dissolve about 1 gram, accurately weighed, in a mixture of 3 ml. of sulfuric acid and 22 ml. of water. Add 50 ml. of alcohol, and allow the mixture to stand overnight. If crystals of magnesium sulfate separate, warm the mixture to about 50° to dissolve them. Filter through a Gooch crucible containing an asbestos mat that previously has been washed with diluted sulfuric acid T.S., water, and alcohol, and ignited and weighed. Wash the crystals on the mat several times with a mixture of 2 volumes of alcohol and 1 volume of diluted sulfuric acid T.S. Ignite the crucible and contents at a dull red heat, cool, and weigh. The weight of calcium sulfate so obtained, multiplied by 0.4119, gives the equivalent of calcium oxide in the sample taken for the test.

Heavy metals. Dissolve 667 mg. in 10 ml. of diluted hydrochloric acid T.S., and evaporate the solution to dryness on a steam bath. Toward the end of the evaporation stir frequently to disintegrate the residue so that finally a dry powder is obtained. Dissolve the residue in 20 ml. of water, and evaporate to dryness in the same manner as before. Redissolve the residue in 25 ml. of water, and filter if necessary. This solution meets the requirements of the *Heavy Metals Test*, page 920, using 20 mcg. of lead ion (Pb) in the control (*Solution A*).

Lead. A solution of 1 gram in 10 ml. of diluted hydrochloric acid T.S. meets the requirements of the *Lead Limit Test*, page 929, using 10 mcg. of lead ion (Pb) in the control.

Soluble salts. Mix 2.0 grams with 100 ml. of a mixture of equal volumes of *n*-propyl alcohol and water. Heat the mixture to the boiling point with constant stirring, cool to room temperature, add water to make 100 ml., and filter. Evaporate 50 ml. of the filtrate on a steam bath to dryness, and dry at 105° for 1 hour. The weight of the residue does not exceed 10 mg.

Packaging and storage. Store in well-closed containers.

Functional use in foods. Alkali; drying agent; color-retention agent; anticaking agent; carrier.

MAGNESIUM CHLORIDE

$MgCl_2.6H_2O$ Mol. wt. 203.30

DESCRIPTION

Colorless, odorless flakes or crystals. It is very deliquescent. It is very soluble in water and freely soluble in alcohol. A 1 in 10 solution gives positive tests for *Magnesium*, page 927, and for *Chloride*, page 926.

SPECIFICATIONS

Assay. Not less than 99.0 percent and not more than the equivalent of 105.0 percent of $MgCl_2.6H_2O$.

Limits of Impurities

Ammonium. Not more than 50 parts per million (0.005 percent).

Arsenic (as As). Not more than 3 parts per million (0.0003 percent).

Heavy metals (as Pb). Not more than 10 parts per million (0.001 percent).

Sulfate. Not more than 200 parts per million (0.02 percent).

TESTS

Assay. Dissolve about 450 mg., accurately weighed, in 25 ml. of water, add 5 ml. of ammonia-ammonium chloride buffer T.S. and 0.1 ml. of eriochrome black T.S., and titrate with 0.05 *M* disodium ethylenediaminetetraacetate until the solution is blue in color. Each ml. of 0.05 *M* disodium ethylenediaminetetraacetate is equivalent to 10.16 mg. of $MgCl_2.6H_2O$.

Ammonium. Dissolve 1 gram in 90 ml. of water, and slowly add 10 ml. of a freshly boiled and cooled solution of sodium hydroxide (1 in 10). Allow to settle, then decant 20 ml. of the supernatant liquid into a color comparison tube, dilute to 50 ml. with water, and add 2 ml. of Nessler's reagent. Any color does not exceed that produced by 10 mcg. of ammonium (NH_4) ion in 48 ml. of water and 2 ml. of the sodium hydroxide solution.

Arsenic. A solution of 1 gram in 35 ml. of water meets the requirements of the *Arsenic Test*, page 865.

Heavy metals. A solution of 2 grams in 25 ml. of water meets the requirements of the *Heavy Metals Test*, page 920, using 20 mcg. of lead ion (Pb) in the control (*Solution A*).

Sulfate, page 879. Any turbidity produced by a 1-gram sample does not exceed that shown in a control containing 200 mcg. of sulfate (SO_4).

Packaging and storage. Store in tight containers.

Functional use in foods. Color-retention agent; firming agent.

MAGNESIUM HYDROXIDE

$Mg(OH)_2$ Mol. wt. 58.32

DESCRIPTION

A white, bulky powder. It dissolves in dilute acids, but is practically insoluble in water and in alcohol. A 1 in 20 solution in diluted hydrochloric acid T.S. gives positive tests for *Magnesium*, page 927.

SPECIFICATIONS

Assay. Not less than 95.0 percent of $Mg(OH)_2$ after drying.

Loss on ignition. Between 30.0 percent and 33.0 percent.

Limits of Impurities

Alkalies (Free) and soluble salts. Passes test.

Arsenic (as As). Not more than 3 parts per million (0.0003 percent).

Calcium oxide. Not more than 1 percent.

Heavy metals (as Pb). Not more than 40 parts per million (0.004 percent).

Lead. Not more than 10 parts per million (0.001 percent).

Loss on drying. Not more than 2 percent.

TESTS

Assay. Transfer about 400 mg., previously dried at 105° for 2 hours and accurately weighed, into an Erlenmeyer flask. Add 25.0 ml. of 1 *N* sulfuric acid, and, after solution is complete, add methyl red T.S. and titrate the excess acid with 1 *N* sodium hydroxide. From the volume of 1 *N* sulfuric acid consumed, deduct the volume of 1 *N* sulfuric acid corresponding to the content of calcium oxide in the sample taken for the assay. The difference is the volume of 1 *N* sulfuric acid equivalent to the $Mg(OH)_2$ in the sample of magnesium hydroxide taken. Each ml. of 1 *N* sulfuric acid is equivalent to 29.16 mg. of $Mg(OH)_2$ and to 28.04 mg. of CaO.

Loss on ignition. Transfer about 500 mg., accurately weighed, to a tared platinum crucible, and ignite, increasing the heat gradually, to constant weight.

Alkalies (Free) and soluble salts. Boil 2 grams with 100 ml. of water for 5 minutes in a covered beaker, then filter while hot. Titrate 50 ml. of the cooled filtrate with 0.1 *N* sulfuric acid, using methyl red T.S. as the indicator. Not more than 2 ml. of the acid is consumed. Evaporate 25 ml. of the filtrate to dryness, and dry at 105° for 3 hours. Not more than 10 mg. of residue remains.

Arsenic. A solution of 1 gram in 25 ml. of diluted hydrochloric acid T.S. meets the requirements of the *Arsenic Test*, page 865.

Calcium Oxide. Dissolve about 500 mg., accurately weighed, in a mixture of 3 ml. of sulfuric acid and 22 ml. of water. Add 50 ml. of alcohol, and allow the mixture to stand overnight. Warm the mixture to about 50°, if necessary, to dissolve any crystals of magnesium sulfate, and filter through a Gooch crucible containing an asbestos mat which has been previously washed with diluted sulfuric acid T.S., water, and alcohol, and ignited. Wash the crystals on the mat several times with a mixture of 3 volumes of alcohol and 1 volume of water. Ignite the crucible and contents at a dull red heat, cool, and weigh. The weight of calcium sulfate thus obtained, multiplied by 0.4119, gives the equivalent of calcium oxide (CaO).

Heavy metals. Dissolve 1 gram in 10 ml. of diluted hydrochloric acid T.S., and evaporate to dryness on a steam bath. Toward the end of the evaporation, stir the residue frequently, disintegrate it to obtain a dry powder, dissolve the powder in 20 ml. of water, and filter. A 10-ml. portion of the filtrate meets the requirements of the *Heavy Metals Test*, page 920, using 20 mcg. of lead ion (Pb) in the control (*Solution A*).

Lead. A solution of 1 gram in 20 ml. of diluted hydrochloric acid T.S. meets the requirements of the *Lead Limit Test*, page 929, using 10 mcg. of lead ion (Pb) in the control.

Loss on drying, page 931. Dry at 105° for 2 hours.

Packaging and storage. Store in tight containers.

Functional use in foods. Alkali; drying agent; color-retention agent.

MAGNESIUM OXIDE

MgO Mol. wt. 40.30

DESCRIPTION

A very bulky, white powder known as light magnesium oxide or a relatively dense, white powder known as heavy magnesium oxide. Five grams of light magnesium oxide occupy a volume of approximately 40 to 50 ml., while 5 grams of heavy magnesium oxide occupy a volume of approximately 10 to 20 ml. It is practically insoluble in water and is insoluble in alcohol. It is soluble in dilute acids. A solution of magnesium oxide in diluted hydrochloric acid T.S. gives positive tests for *Magnesium*, page 927.

SPECIFICATIONS

Assay. Not less than 96.0 percent of MgO after ignition.

Limits of Impurities

Acid-insoluble substances. Not more than 0.1 percent.
Alkalies (free) and soluble salts. Passes test.
Arsenic (as As). Not more than 3 parts per million (0.0003 percent).
Calcium oxide. Not more than 1.5 percent.

Heavy metals (as Pb). Not more than 40 parts per million (0.004 percent).

Lead. Not more than 10 parts per million (0.001 percent).

Loss on ignition. Not more than 10 percent.

TESTS

Assay. Ignite about 500 mg. to constant weight at 800° to 825° in a tared platinum crucible, weigh the residue accurately, dissolve it in 30.0 ml. of 1 *N* sulfuric acid, boil gently to remove any carbon dioxide, cool, add methyl orange T.S., and titrate the excess acid with 1 *N* sodium hydroxide. From the volume of 1 *N* sulfuric acid consumed deduct the volume of 1 *N* sulfuric acid corresponding to the content of calcium oxide in the magnesium oxide taken for the assay. The difference is the volume of 1 *N* sulfuric acid equivalent to the MgO in the portion of magnesium oxide taken. Each ml. of 1 *N* sulfuric acid is equivalent to 20.15 mg. of MgO and to 28.04 mg. of CaO.

Acid-insoluble substances. Mix 2 grams with 75 ml. of water, add hydrochloric acid in small portions, with agitation, until no more dissolves, and boil for 5 minutes. If an insoluble residue remains, filter, wash well with water until the last washing is free from chloride, ignite, cool, and weigh.

Alkalies (free) and soluble salts. Boil 2 grams with 100 ml. of water for 5 minutes in a covered beaker, and filter while hot. Add methyl red T.S., and titrate 50 ml. of the cooled filtrate with 0.1 *N* sulfuric acid. Not more than 2 ml. of the acid is consumed. Evaporate 25 ml. of the filtrate to dryness, and dry at 105° for 1 hour. Not more than 10 mg. of residue remains.

Arsenic. A solution of 1 gram in 20 ml. of diluted hydrochloric acid T.S. meets the requirements of the *Arsenic Test*, page 865.

Calcium Oxide. Dissolve about 400 mg., accurately weighed, in a mixture of 3 ml. of sulfuric acid and 22 ml. of water. Add 50 ml. of alcohol, and allow the mixture to stand overnight. If crystals of magnesium sulfate separate, warm the mixture to about 50° to dissolve them. Filter through a Gooch crucible containing an asbestos mat that previously has been washed with diluted sulfuric acid T.S., water, and alcohol, and ignited and weighed. Wash the crystals on the mat several times with a mixture of 2 volumes of alcohol and 1 volume of diluted sulfuric acid T.S. Ignite the crucible and contents at a dull red heat, cool, and weigh. The weight of calcium sulfate obtained, multiplied by 0.4119, gives the equivalent of calcium oxide in the sample taken for the test.

Heavy metals. Dissolve 500 mg. in 20 ml. of diluted hydrochloric acid T.S., and evaporate the solution to dryness on a steam bath. Toward the end of the evaporation stir frequently to disintegrate the residue so that finally a dry powder is obtained. Dissolve the residue in 20 ml. of water and evaporate to dryness in the same manner as before. Redissolve the residue in 20 ml. of water and filter if necessary. This solution meets the requirements of the *Heavy Metals Test*, page 920, using 20 mcg. of lead ion (Pb) in the control (*Solution A*).

Lead. A solution of 1 gram in 20 ml. of diluted hydrochloric acid T.S. meets the requirements of the *Lead Limit Test*, page 929, using 10 mcg. of lead ion (Pb) in the control.

Loss on ignition. Weigh accurately about 500 mg. in a tared covered platinum crucible. Ignite at between 800° and 825° for 15 minutes, cool and weigh.

Packaging and storage. Store in tight containers.

Labeling. Label magnesium oxide to indicate whether it is light magnesium oxide or heavy magnesium oxide.

Functional use in foods. Alkali; neutralizer.

MAGNESIUM PHOSPHATE, DIBASIC

Dimagnesium Phosphate

$MgHPO_4.3H_2O$ Mol. wt. 174.33

DESCRIPTION

A white, odorless crystalline powder. It is slightly soluble in water and insoluble in alcohol, but is soluble in dilute acids.

IDENTIFICATION

A. Dissolve about 200 mg. in 10 ml. of diluted nitric acid T.S. and add, dropwise, ammonium molybdate T.S. A greenish yellow precipitate of ammonium phosphomolybdate forms which is soluble in ammonia T.S.

B. Dissolve 100 mg. in 0.5 ml. of diluted acetic acid T.S. and 20 ml. of water. Add 1 ml. of ferric chloride T.S., let stand for 5 minutes, and filter. The filtrate gives a positive test for *Magnesium*, page 927.

SPECIFICATIONS

Assay. Not less than 96.0 percent of $Mg_2P_2O_7$, calculated on the ignited basis.

Loss on ignition. Between 29 and 36 percent.

Limits of Impurities

Arsenic (as As). Not more than 3 parts per million (0.0003 percent).

Fluoride. Not more than 10 parts per million (0.001 percent)

Heavy metals (as Pb). Not more than 30 parts per million (0.003 percent).

Lead. Not more than 5 parts per million (0.0005 percent).

TESTS

Assay. Weigh accurately about 500 mg. of the residue obtained in the test for *Loss on ignition,* dissolve it in a mixture of 50 ml. of water and 1 ml. of hydrochloric acid, dilute to 100.0 ml. with water, and mix. Transfer 50.0 ml. of this solution into a 250-ml. Erlenmeyer flask, add 10 ml. of ammonia-ammonium chloride buffer T.S. and 12 drops of eriochrome black T.S., and titrate with 0.1 *M* disodium ethylenediaminetetraacetate until the wine-red color changes to pure blue. Each ml. of 0.1 *M* disodium ethylenediaminetetraacetate is equivalent to 22.25 mg. of $Mg_2P_2O_7$.

Loss on ignition. Weigh accurately about 1 gram, and ignite, preferably in a muffle furnace, at 800° ± 25° to constant weight.

Arsenic. A solution of 1 gram in 5 ml. of diluted hydrochloric acid T.S. meets the requirements of the *Arsenic Test,* page 865.

Fluoride. Transfer 5.0 grams of the sample into a 200-ml. distilling flask connected with a condenser and carrying a thermometer and a dropping funnel equipped with a stopcock. Dissolve the sample in 25 ml. of dilute sulfuric acid (1 in 4), add 6 glass beads, and connect the apparatus for distillation, using a 600-ml. beaker to collect the distillate. Add 40 ml. of the dilute sulfuric acid to the flask through the dropping funnel, then fill the funnel with water, heat the solution to boiling, and continue heating until the temperature reaches 165°. Adjust the stopcock of the dropping funnel so that the temperature is maintained at 165° ± 5°, and continue the distillation until about 300 ml. has been collected. Rinse the condenser and condenser arm with water, collecting the rinsings in the beaker. Add sodium hydroxide T.S. to the distillate to make it alkaline to litmus paper, and then add 5 ml. in excess. Add 5 ml. of 30 percent hydrogen peroxide and 6 glass beads to the beaker, boil until a volume of about 30 ml. is reached, and cool. Transfer the condensed distillate, including the glass beads, into a 125-ml. distilling flask connected with a condenser and carrying a thermometer and a capillary tube, both of which must extend into the liquid. Add 30 ml. of perchloric acid, and continue as directed under the *Fluoride Limit Test, Method I,* page 917, beginning with "Connect a small dropping funnel or a steam generator to the capillary tube."

Heavy metals, page 920. Suspend 1.33 grams in 20 ml. of water, and add hydrochloric acid, dropwise, until the sample just dissolves. Adjust the pH to between 3 and 4, filter, and dilute the filtrate to 40 ml. with water. For the control (*Solution A*), add 20 mcg. of lead ion (Pb) to 10 ml. of the filtrate, and dilute to 40 ml. For the sample (*Solution B*), dilute the remaining 30 ml. of the filtrate to 40 ml. Add 10 ml. of hydrogen sulfide T.S. to each solution, and allow to stand for 5 minutes. *Solution B* is no darker than *Solution A*.

Lead. Dissolve 1 gram in 20 ml. of diluted hydrochloric acid T.S., evaporate the solution to a volume of about 10 ml. on a steam bath, dilute to about 20 ml. with water, and cool. This solution meets the requirements of the *Lead Limit Test*, page 929, using 5 mcg. of lead ion (Pb) in the control.

Packaging and storage. Store in well-closed containers.

Functional use in foods. Nutrient; dietary supplement.

MAGNESIUM PHOSPHATE, TRIBASIC

Trimagnesium Phosphate

$Mg_3(PO_4)_2 . xH_2O$ Mol. wt. (anhydrous) 262.86

DESCRIPTION

Tribasic magnesium phosphate may contain 4, 5, or 8 molecules of water of hydration. It occurs as a white, odorless, tasteless crystalline powder. It is readily soluble in dilute mineral acids but is almost insoluble in water.

IDENTIFICATION

A. Dissolve about 200 mg. in 10 ml. of diluted nitric acid T.S. and add, dropwise, ammonium molybdate T.S. A greenish yellow precipitate of ammonium phosphomolybdate forms which is soluble in ammonia T.S.

B. Dissolve 100 mg. in 0.7 ml. of diluted acetic acid T.S. and 20 ml. of water. Add 1 ml. of ferric chloride T.S., let stand for 5 minutes, and filter. The filtrate gives a positive test for *Magnesium*, page 927.

SPECIFICATIONS

Assay. Not less than 98.0 percent and not more than the equivalent of 101.5 percent of $Mg_3(PO_4)_2$, calculated on the ignited basis.

Titration value. Passes test.

Loss on heating. $Mg_3(PO_4)_2 . 4H_2O$, between 15 and 23 percent; $Mg_3(PO_4)_2 . 5H_2O$, between 20 and 27 percent; $Mg_3(PO_4)_2 . 8H_2O$, between 30 and 37 percent.

Limits of Impurities

Arsenic (as As). Not more than 3 parts per million (0.0003 percent).

Fluoride. Not more than 10 parts per million (0.001 percent).

Heavy metals (as Pb). Not more than 30 parts per million (0.003 percent).

Lead. Not more than 5 parts per million (0.0005 percent).

TESTS

Assay. Weigh accurately about 200 mg. of the sample, and dissolve it in a mixture of 25 ml. of water and 10 ml. of diluted nitric acid T.S. Filter, if necessary, wash any precipitate, then dissolve the precipitate by the addition of 1 ml. of diluted nitric acid T.S. Adjust the temperature to about 50°, add 75 ml. of ammonium molybdate T.S., and maintain the temperature at about 50° for 30 minutes, stirring occasionally. Allow to stand for 16 hours or overnight at room temperature. Wash the precipitate once or twice with water by decantation, using from 30 to 40 ml. each time, and pour these two washings through a filter. Transfer the precipitate to the same filter, and wash with potassium nitrate solution (1 in 100) until the last washing is not acid to litmus paper. Transfer the precipitate and filter to the precipitation vessel, add 50.0 ml. of 1 *N* sodium hydroxide, agitate until the precipitate is dissolved, add 3 drops of phenolphthalein T.S., and then titrate the excess alkali with 1 *N* sulfuric acid. Each ml. of 1 *N* sodium hydroxide is equivalent to 5.714 mg. of $Mg_3(PO_4)_2$.

Titration value. Ignite about 3 grams at about 425° to constant weight, and dissolve, by warming, 2.0 grams of the ignited salt in 50.0 ml. of 1 *N* hydrochloric acid. Cool, add methyl orange T.S., and slowly titrate the excess of 1 *N* hydrochloric acid with 1 *N* sodium hydroxide to a yellow color, shaking the mixture vigorously during the titration. Not less than 29.0 ml. and not more than 30.8 ml. of 1 *N* hydrochloric acid is consumed.

Loss on heating. Weigh accurately about 1 gram and heat at about 425° to constant weight.

Arsenic. A solution of 1 gram in 10 ml. of diluted hydrochloric acid T.S. meets the requirements of the *Arsenic Test*, page 865.

Fluoride. Determine as directed in the *Fluoride Limit Test* under *Magnesium Phosphate, Dibasic*, page 475.

Heavy metals, page 920. Suspend 1.33 grams in 20 ml. of water, and add hydrochloric acid, dropwise, until the sample just dissolves. Adjust the pH to between 3 and 4, filter, and dilute the filtrate to 40 ml. with water. For the control (*Solution A*), add 20 mcg. of lead ion (Pb) to 10 ml. of the filtrate, and dilute to 40 ml. For the sample (*Solution B*), dilute the remaining 30 ml. of the filtrate to 40 ml. Add 10 ml. of hydrogen sulfide T.S. to each solution, and allow to stand for 5 minutes. *Solution B* is no darker than *Solution A*.

Lead. Dissolve 1 gram in 20 ml. of diluted hydrochloric acid T.S., evaporate the solution to a volume of about 10 ml. on a steam bath, dilute to about 20 ml. with water, and cool. This solution meets the requirements of the *Lead Limit Test*, page 929, using 5 mcg. of lead ion (Pb) in the control.

Packaging and storage. Store in well-closed containers.

Functional use in foods. Nutrient; dietary supplement.

MAGNESIUM SILICATE

DESCRIPTION

A synthetic form of magnesium silicate in which the molar ratio of magnesium oxide to silicon dioxide is approximately 2:5. It occurs as a very fine, white, odorless, tasteless powder, free from grittiness. It is insoluble in water and in alcohol, but is readily decomposed by mineral acids. The pH of a 1 in 20 slurry is between 6.3 and 9.5.

IDENTIFICATION

A. Mix about 500 mg. with 10 ml. of diluted hydrochloric acid T.S., filter, and neutralize the filtrate to litmus paper with ammonia T.S. The neutralized filtrate responds to the tests for *Magnesium*, page 927.

B. Prepare a bead by fusing a few crystals of sodium ammonium phosphate on a platinum loop in the flame of a Bunsen burner. Place the hot, transparent bead in contact with a sample, and again fuse. Silica floats about in the bead, producing, upon cooling, an opaque bead with a web-like structure.

SPECIFICATIONS

Assay. Not less than 15.0 percent of MgO and not less than 67.0 percent of SiO_2, calculated on the anhydrous basis.

Limits of Impurities

Arsenic (as As). Not more than 3 parts per million (0.0003 percent).

Fluoride. Not more than 20 parts per million (0.002 percent).

Free alkali (as NaOH). Not more than 1 percent.

Heavy metals (as Pb). Not more than 40 parts per million (0.004 percent).

Lead. Not more than 10 parts per million (0.001 percent).

Soluble salts. Not more than 2 percent.

Water. Not more than 10 percent.

TESTS

Assay for magnesium oxide. Weigh accurately about 1.5 grams, and transfer it into a 250-ml. conical flask. Add 50.0 ml. of 1 *N* sulfuric acid, and digest on a steam bath for 1 hour. Cool to room temperature, add methyl orange T.S., and titrate the excess acid with 1 *N* sodium hydroxide. Each ml. of 1 *N* sulfuric acid is equivalent to 20.15 mg. of MgO.

Assay for silicon dioxide. Transfer about 700 mg., accurately weighed, into a 150-ml. beaker. Add 20 ml. of 1 *N* sulfuric acid, and heat on a steam bath for 1 hour and 30 minutes. Decant the supernatant liquid through an ashless filter paper, and wash the residue, by decantation, three times with hot water. Treat the residue with 25 ml. of water and digest on a steam bath for 15 minutes. Finally, transfer the residue to the filter and wash thoroughly with hot water. Transfer the filter paper and its contents to a platinum crucible. Heat to dryness, incinerate, then ignite strongly for 30 minutes, cool, and weigh. Moisten the residue with water, and add 6 ml. of hydrofluoric acid and 3 drops of sulfuric acid. Evaporate to dryness, ignite for 5 minutes, cool, and weigh. The loss in weight represents the weight of SiO_2.

Sample Solution for the Determination of Arsenic, Heavy Metals, and Lead. Transfer 10.0 grams of the sample into a 250-ml. flask, and add 50 ml. of 0.5 *N* hydrochloric acid. Attach a reflux condenser to the flask, heat on a steam bath for 30 minutes, cool, and let the undissolved material settle. Decant the supernatant liquid through Whatman No. 3 filter paper, or equivalent, into a 100-ml. volumetric flask, retaining as much as possible of the insoluble material in the beaker. Wash the slurry and beaker with three 10-ml. portions of hot water, decanting each washing through the filter into the flask. Finally, wash the filter paper with 15 ml. of hot water, cool the filtrate to room temperature, dilute to volume with water, and mix.

Arsenic. A 10-ml. portion of the *Sample Solution* meets the requirements of the *Arsenic Test*, page 865.

Fluoride. Weigh accurately 2.5 grams, and proceed as directed in the *Fluoride Limit Test*, page 917.

Free alkali. Add 2 drops of phenolphthalein T.S. to 20 ml. of diluted filtrate prepared in the test for *Soluble salts*, representing 1 gram of magnesium silicate. If a pink color is produced, not more than 2.5 ml. of 0.1 *N* hydrochloric acid is required to discharge it.

Heavy metals. A 5-ml. portion of the *Sample Solution* diluted to 25 ml. with water meets the requirements of the *Heavy Metals Test*, page 920, using 20 mcg. of lead ion (Pb) in the control (*Solution A*).

Lead. A 10-ml. portion of the *Sample Solution* meets the requirements of the *Lead Limit Test*, page 929, using 10 mcg. of lead ion (Pb) in the control.

Soluble salts. Boil 10 grams with 150 ml. of water for 15 minutes. Cool to room temperature, and add water to restore the original volume. Allow the mixture to stand for 15 minutes, and filter until clear. To 75 ml. of the clear filtrate add 25 ml. of water. Evaporate 50 ml. of this solution, representing 2.5 grams of magnesium silicate, in a tared platinum dish on a steam bath to dryness, and ignite gently to constant weight. The weight of the residue does not exceed 50 mg.

Water. Weigh accurately about 1 gram in a tared platinum crucible provided with a cover. Gradually apply heat to the crucible at first, then strongly ignite to constant weight.

Packaging and storage. Store in well-closed containers.

Functional use in foods. Anticaking agent.

MAGNESIUM STEARATE

DESCRIPTION

Magnesium stearate is a compound of magnesium with variable proportions of stearic and palmitic acids. It occurs as a fine, white, bulky, powder, having a faint, characteristic odor. It is unctuous, and is free from grittiness. It is insoluble in water, in alcohol, and in ether. It conforms to the regulations of the federal Food and Drug Administration pertaining to specifications for salts of fatty acids and fatty acids derived from edible fats sources.

IDENTIFICATION

A. Heat 1 gram with a mixture of 25 ml. of water and 5 ml. of hydrochloric acid. Fatty acids are liberated, floating as an oily layer on the surface of the liquid. The water layer gives positive tests for *Magnesium*, page 927.

B. Mix 25 grams of the sample with 200 ml. of hot water, then add 60 ml. of diluted sulfuric acid T.S., and heat the mixture, with frequent stirring, until the fatty acids separate cleanly as a transparent layer. Wash the fatty acids with boiling water until free from sulfate, collect them in a small beaker, and warm on a steam bath until the water has separated and the fatty acids are clear. Allow the acids to cool, pour off the water layer, then melt the acids, filter into a dry beaker, and dry at 105° for 20 minutes. The solidification point of the fatty acids so obtained is not below 54° (see page 931).

SPECIFICATIONS

Assay. Not less than the equivalent of 6.8 percent and not more than the equivalent of 8.0 percent of MgO.

Limits of Impurities

Arsenic (as As). Not more than 3 parts per million (0.0003 percent).

Heavy metals (as Pb). Not more than 40 parts per million (0.004 percent).

Lead. Not more than 10 parts per million (0.001 percent).

Loss on drying. Not more than 4 percent.

TESTS

Assay. Boil about 1 gram, accurately weighed, with 50.0 ml. of 0.1

N sulfuric acid for 10 minutes, or until the fatty acid layer is clear, adding water if necessary to maintain the original volume. Cool, filter, and wash the filter and flask thoroughly with water until the last washing is not acid to litmus. Add methyl orange T.S., and titrate the excess sulfuric acid with 0.1 *N* sodium hydroxide. Each ml. of 0.1 *N* sulfuric acid is equivalent to 2.015 mg. of MgO.

Arsenic. Mix 1 gram of the sample with 10 ml. of hydrochloric acid and 8 drops of bromine T.S., and heat on a steam bath until a transparent layer of melted fatty acid forms. Add 50 ml. of water, boil down to about 25 ml., and filter while hot. Cool, neutralize with a 1 in 2 solution of sodium hydroxide, and dilute to 35 ml. with water. This solution meets the requirements of the *Arsenic Test*, page 865.

Heavy metals, page 920. Place 750 mg. of the sample in a porcelain dish, place 250 mg. of the sample in a second dish for the control, and to each add 5 ml. of a 1 in 4 solution of magnesium nitrate in alcohol. Cover the dishes with 7.6 cm. short stem funnels so that the stems are straight up. Heat for 30 minutes on a hot plate at the low setting, then heat for 30 minutes at the medium setting, and cool. Remove the funnels, add 20 mcg. of lead ion (Pb) to the control, and heat each dish over an Argand burner until most of the carbon is burned off. Cool, add 10 ml. of nitric acid, and transfer the solutions into 250-ml. beakers. Add 5 ml. of 70 percent perchloric acid, evaporate to dryness, then add 2 ml. of hydrochloric acid to the residues, and wash down the inside of the beakers with water. Evaporate carefully to dryness again, swirling near the dry point to avoid spattering. Repeat the hydrochloric acid treatment, then cool, and dissolve the residues in about 10 ml. of water. To each solution add 1 drop of phenolphthalein T.S. and sufficient sodium hydroxide T.S. until the solutions just turn pink, and then add diluted hydrochloric acid T.S. until the solutions become colorless. Add 1 ml. of diluted acetic acid T.S. and a small amount of charcoal to each solution, and filter through Whatman No. 2, or equivalent, filter paper into 50-ml. Nessler tubes. Wash with water, dilute to 40 ml., and add 10 ml. of hydrogen sulfide T.S. to each tube. The color in the solution of the sample does not exceed that produced in the control.

Lead, page 929. Ignite 500 mg. in a silica crucible in a muffle furnace at 475° to 500° for 15 to 20 minutes. Cool, add 3 drops of nitric acid, evaporate over a low flame to dryness, and re-ignite at 475° to 500° for 30 minutes. Dissolve the residue in 1 ml. of a mixture of equal parts by volume of nitric acid and water, and wash into a separator with several successive portions of water. Add 3 ml. of *Ammonium Citrate Solution* and 0.5 ml. of *Hydroxylamine Hydrochloride Solution*, and make alkaline to phenol red T.S. with stronger ammonia T.S. Add 10 ml. of *Potassium Cyanide Solution*. Immediately extract the solution with successive 5-ml. portions of *Dithizone Extraction Solution*, draining off each extract into another separator, until the last portion of dithizone solution retains its green color. Shake the combined extracts for 30 seconds with 20 ml. of dilute nitric acid (1 in 100), and discard the chloroform layer. Add to the acid solution exactly 4 ml. of *Ammonia-Cyanide Solution* and 2 drops of *Hydroxylamine Hydrochloride Solution.* Add 10 ml. of *Standard Dithizone Solution,* and shake the mixture for 30 seconds. Filter the chloroform layer through an acid-washed filter paper into a Nessler tube, and compare the color with that of a standard prepared as follows: to 20 ml. of dilute nitric acid (1 in 100), add 5 mcg. of lead ion (Pb), 4 ml. of *Ammonia-Cyanide Solution* and 2 drops of *Hydroxylamine Hydrochloride Solution,* and shake for 30 seconds with 10 ml. of *Standard Dithizone Solution.* Filter through an acid-washed filter paper into a Nessler tube. The color of the sample solution does not exceed that in the control.

Loss on drying, page 931. Dry at 105° to constant weight, using 2-hour increments of heating.

Packaging and storage. Store in well-closed containers.

Functional use in foods. Anticaking agent; binder; emulsifier.

MAGNESIUM SULFATE

Epsom Salt

$MgSO_4.7H_2O$ Mol. wt. 246.47

DESCRIPTION

Small, colorless crystals, usually needle-like, with a cooling, saline, bitter taste. It is freely soluble in water, slowly soluble in glycerin, and sparingly soluble in alcohol. It effloresces in warm, dry air. Its solutions are neutral. A 1 in 20 solution gives positive tests for *Magnesium,* page 927, and for *Sulfate,* page 928.

SPECIFICATIONS

Assay. Not less than 99.5 percent of $MgSO_4$ after ignition.

Loss on ignition. Between 40 and 52 percent.

Limits of Impurities

Arsenic (as As). Not more than 3 parts per million (0.0003 percent).

Heavy metals (as Pb). Not more than 10 parts per million (0.001 percent).

Selenium. Not more than 30 parts per million (0.003 percent).

TESTS

Assay. Weigh accurately about 500 mg. of the residue obtained in the test for *Loss on ignition,* dissolve it in a mixture of 50 ml. of water and 1 ml. of hydrochloric acid, dilute to 100.0 ml. with water, and mix. Transfer 50.0 ml. of this solution into a 250-ml. Erlenmeyer flask, add 10 ml. of ammonia-ammonium chloride buffer T.S. and 12 drops of eriochrome black T.S., and titrate with 0.1 *M* disodium ethylenediaminetetraacetate until the wine-red color changes to pure blue. Each ml. of 0.1 *M* disodium ethylenediaminetetraacetate is equivalent to 12.04 mg. of $MgSO_4$.

Loss on ignition. Weigh accurately about 1 gram in a crucible, heat at 105° for 2 hours, then ignite in a muffle furnace at 450° ± 25° to constant weight.

Arsenic. A solution of 1 gram in 10 ml. of water meets the requirements of the *Arsenic Test*, page 865.

Heavy metals. A solution of 2 grams in 25 ml. of water meets the requirements of the *Heavy Metals Test*, page 920, using 20 mcg. of lead ion (Pb) in the control (*Solution A*).

Selenium. A solution of 2 grams in 40 ml. of dilute hydrochloric acid (1 in 2) meets the requirements of the *Selenium Limit Test*, page 953.

Packaging and storage. Store in well-closed containers.

Functional use in foods. Nutrient; dietary supplement.

MALIC ACID

DL-Malic Acid; Hydroxysuccinic Acid

HOCHCOOH
|
CH_2COOH

$C_4H_6O_5$ Mol. wt. 134.09

DESCRIPTION

White or nearly white, crystalline powder or granules having a strongly acid taste. One gram dissolves in 0.8 ml. of water and in 1.4 ml. of alcohol. Its solutions are optically inactive.

SPECIFICATIONS

Assay. Not less than 99.5 percent of $C_4H_6O_5$.

Melting range. Between 130° and 132°.

Limits of Impurities

Arsenic (as As). Not more than 3 parts per million (0.0003 percent).

Fumaric acid. Not more than 0.5 percent.

Heavy metals (as Pb). Not more than 20 parts per million (0.002 percent).

Lead. Not more than 10 parts per million (0.001 percent).

Maleic acid. Not more than 0.05 percent.
Residue on ignition. Not more than 0.1 percent.
Water-insoluble matter. Not more than 0.1 percent.

TESTS

Assay. Dissolve about 2 grams, accurately weighed, in 40 ml. of recently boiled and cooled water, add phenolphthalein T.S., and titrate with 1 *N* sodium hydroxide to the first appearance of a faint pink color which persists for at least 30 seconds. Each ml. of 1 *N* sodium hydroxide is equivalent to 67.04 mg. of $C_4H_6O_5$.

Melting range. Determine as directed in the general procedure, page 931.

Arsenic. A *Sample Solution* prepared as directed for organic compounds meets the requirements of the *Arsenic Test*, page 865.

Fumaric and maleic acids

Buffer Solution A. Dissolve 74.5 grams of potassium chloride in 500 ml. of water in a 1000-ml. volumetric flask, add 100 ml. of hydrochloric acid, and dilute to volume with water.

Buffer Solution B. Dissolve 171.0 grams of dibasic potassium phosphate, $K_2HPO_4.3H_2O$, in 1000 ml. of water, and add monobasic potassium phosphate, KH_2PO_4, until the pH is exactly 7.0.

Maxima Suppressor. Dissolve, with the aid of a magnetic stirrer, 1 gram of gelatin in 65 ml. of hot, boiled water, and, after cooling, add 35 ml. of anhydrous ethanol as a preservative.

Standard Solution. Weigh accurately about 20 grams of the sample, 100 mg. of fumaric acid of the highest purity available, and 10 mg. of maleic acid of the highest purity available, and transfer into a 500-ml. volumetric flask. Add 300 ml. of sodium hydroxide T.S. and a few drops of phenolphthalein T.S., and then continue the neutralization with sodium hydroxide T.S. to a faint pink color that persists for at least 30 seconds. Dilute to volume with water, and mix.

Sample Solution. Transfer about 4 grams of the sample, accurately weighed, into a 100-ml. volumetric flask, and dissolve in 25 ml. of water. Add phenolphthalein T.S., and neutralize with sodium hydroxide T.S. as directed for the *Standard Solution.* Dilute to volume with water, and mix.

Procedure. Transfer 25.0-ml. portions of the *Sample Solution* into separate 100-ml. volumetric flasks. Dilute one flask (*Sample A*) to volume with *Buffer Solution A.* To the other flask (*Sample B*) add 50 ml. of *Buffer Solution B,* and dilute to volume with water.

Rinse a polarograph cell with a portion of *Sample A,* add a suitable volume of the solution to the cell, immerse it in a water bath regulated at 24.5° to 25.5°, add 2 drops of the *Maxima Suppressor,* and then de-aerate by bubbling nitrogen through the solution for at least 5 minutes. Insert the dropping mercury electrode (negative polarity) of a suitable polarograph, adjust the current sensitivity as necessary, and record the polarogram from −0.1 to −0.8 volt at the rate of 0.2 volt per minute, using a saturated calomel electrode as the reference electrode. Transfer 25.0 ml. of the *Standard Solution* into a 100-ml. volumetric flask, and dilute to volume with *Buffer Solution A*. Obtain the polarogram of this standard (*Standard A*) in the same manner as directed for *Sample A*. In each polarogram, determine the height of the maleic acid plus fumaric acid wave occurring at the half-wave potential near −0.56 volt, recording that for the sample as i_u and that for the standard as i_s.

In the same manner, obtain polarograms from *Sample B* and *Standard B*, except record the polarogram from −1.05 to −1.7 volts at the rate of 0.1 volt per minute. In each polarogram, determine the height of the maleic acid wave occurring at the half-wave potential near −1.33 volts, recording that for the sample as i_u' and that for the standard as i_s'.

Calculation. Calculate the weight, in mg., of combined maleic acid and fumaric acid in the sample by the formula $500C \times i_u/(i_s - i_u)$, in which C is the concentration, in mg. per ml., of combined maleic acid and fumaric acid in the *Standard Solution*. Similarly, calculate the weight, in mg., of maleic acid in the sample by the formula $500C' \times i_u'/(i_s' - i_u')$, in which C' is the concentration, in mg. per ml., of maleic acid in the *Standard Solution*. Finally, calculate the weight of fumaric acid in the sample from the difference in these values.

Heavy metals. A solution of 1 gram in 25 ml. of water meets the requirements of the *Heavy metals Test*, page 920, using 20 mcg. of lead ion (Pb) in the control (*Solution A*).

Lead. A *Sample Solution* prepared as directed for organic compounds meets the requirements of the *Lead Limit Test*, page 929, using 10 mcg. of lead ion (Pb) in the control.

Residue on ignition. Ignite 2 grams as directed in the general method, page 945.

Water-insoluble matter. Dissolve 25 grams in 100 ml. of water, and filter through a tared Gooch crucible. Wash the filter with hot water, dry at 100° to constant weight, cool, and weigh.

Packaging and storage. Store in well-closed containers.

Functional use in foods. Acidifier; flavoring agent.

MALTOL

3-Hydroxy-2-methyl-4-pyrone

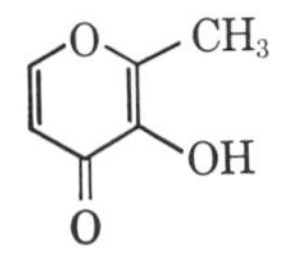

$C_6H_6O_3$ Mol. wt. 126.11

DESCRIPTION

A white, crystalline powder having a characteristic caramel-butterscotch odor, and suggestive of a fruity-strawberry aroma in dilute solution. One gram dissolves in about 82 ml. of water, 21 ml. of alcohol, 80 ml. of glycerin, and 28 ml. of propylene glycol. A 1 in 100,000 solution in 0.1 *N* hydrochloric acid exhibits an absorbance maximum at 274 ± 2 mμ.

SPECIFICATIONS

Assay. Not less than 99.0 percent of $C_6H_6O_3$.

Melting range. Between 160° and 164°.

Limits of Impurities

Arsenic (as As). Not more than 3 parts per million (0.0003 percent).

Heavy metals (as Pb). Not more than 20 parts per million (0.002 percent).

Lead. Not more than 10 parts per million (0.001 percent).

Residue on ignition. Not more than 0.2 percent.

Water. Not more than 0.5 percent.

TESTS

Assay

Standard Solution. Weigh accurately about 50 mg. of F.C.C. Maltol Reference Standard, dissolve it in sufficient 0.1 *N* hydrochloric acid to make 250.0 ml., and mix. Transfer 5.0 ml. of this solution into a 100-ml. volumetric flask, dilute to volume with 0.1 *N* hydrochloric acid, and mix.

Assay Solution. Weigh accurately about 50 mg. of the sample, dissolve it in sufficient 0.1 *N* hydrochloric acid to make 250.0 ml., and mix. Transfer 5.0 ml. of this solution into a 100-ml. volumetric flask, dilute to volume with 0.1 *N* hydrochloric acid, and mix.

Procedure. Determine the absorbance of each solution in a 1-cm. quartz cell at the wavelength of maximum absorption at about 274 mμ, with a suitable spectrophotometer, using 0.1 *N* hydrochloric acid as the blank. Calculate the quantity, in mg., of $C_6H_6O_3$ in the sample taken by the formula $5C(A_U/A_S)$, in which C is the concentration, in mcg. per ml., of F.C.C. Maltol Reference Standard in the *Standard Solution*, A_U is the absorbance of the *Assay Solution*, and A_S is the absorbance of the *Standard Solution*.

Melting range. Determine as directed in *Procedure for Class Ia*, page 931.

Arsenic. A *Sample Solution* prepared as directed for organic compounds meets the requirements of the *Arsenic Test*, page 865.

Heavy metals. Prepare and test a 1-gram sample as directed in *Method II* under the *Heavy Metals Test*, page 920, using 20 mcg. of lead ion (Pb) in the control (*Solution A*).

Lead. A *Sample Solution* prepared as directed for organic compounds meets the requirements of the *Lead Limit Test*, page 929, using 10 mcg. of lead ion (Pb) in the control.

Residue on ignition. Ignite 1 gram as directed in the general method, page 945.

Water. Determine by the *Karl Fischer Titrimetric Method*, page 977.

Packaging and storage. Store in tight containers.

Functional use in foods. Flavoring agent.

MANDARIN OIL, EXPRESSED

DESCRIPTION

The oil obtained by expression of the peels of the ripe fruit of the Mandarin Orange, *Citrus reticulata* Blanco var. *Mandarin*. It is a clear, dark orange to reddish yellow, or brownish orange liquid with a pleasant orange-like odor. It often shows a bluish fluorescence in diffused light. Oils produced from unripe fruit often show a green color. It is soluble in most fixed oils and in mineral oil. It is slightly soluble in propylene glycol, but it is insoluble in glycerin.

SPECIFICATIONS

Assay. Not less than 0.4 percent and not more than 1.8 percent of aldehydes, calculated as decylaldehyde ($C_{10}H_{20}O$).

Angular rotation. Between +63° and +78°.

Refractive index. Between 1.473 and 1.477 at 20°.

Residue on evaporation. Between 2 percent and 5 percent.

Specific gravity. Between 0.847 and 0.853.

Limits of Impurities

Arsenic (as As). Not more than 3 parts per million (0.0003 percent).

Heavy metals (as Pb). Not more than 40 parts per million (0.004 percent).

Lead. Not more than 10 parts per million (0.001 percent).

TESTS

Assay. Weigh accurately about 10 grams, and proceed as directed under *Aldehydes*, page 894, using 156.26 as the equivalence factor (*E*) in the calculation. Allow the mixture to stand for 30 minutes at room temperature before titrating.

Angular rotation. Determine in a 100-mm. tube as directed under *Optical Rotation*, page 939.

Refractive index, page 945. Determine with an Abbé or other refractometer of equal or greater accuracy.

Residue on evaporation. Proceed as directed in the general method, page 945, using about 5 grams, accurately weighed, and heating for 5 hours.

Specific gravity. Determine by any reliable method (see page 5).

Arsenic. A *Sample Solution* prepared as directed for organic compounds meets the requirements of the *Arsenic Test*, page 865.

Heavy metals. Prepare and test a 500-mg. sample as directed in *Method II* under the *Heavy Metals Test*, page 920, using 20 mcg. of lead ion (Pb) in the control (*Solution A*).

Lead. A *Sample Solution* prepared as directed for organic compounds meets the requirements of the *Lead Limit Test*, page 929, using 10 mcg. of lead ion (Pb) in the control.

Packaging and storage. Store in full, tight, preferably glass, tin-lined, galvanized, or other suitably lined containers in a cool place protected from light.

Functional use in foods. Flavoring agent.

MANGANESE CHLORIDE

$MnCl_2 . 4H_2O$ Mol. wt. 197.91

DESCRIPTION

Large, irregular, pink, translucent crystals. It is freely soluble in water at room temperature and very soluble in hot water.

IDENTIFICATION

A 1 in 20 solution gives positive tests for *Manganese*, page 927, and for *Chloride*, page 926.

SPECIFICATIONS

Assay. Not less than 98.0 percent and not more than the equivalent of 102.0 percent of $MnCl_2.4H_2O$.

pH of a 5 percent solution. Between 4.0 and 6.0.

Limits of Impurities

Arsenic (as As). Not more than 3 parts per million (0.0003 percent).

Heavy metals (as Pb). Not more than 10 parts per million (0.001 percent).

Insoluble matter. Not more than 50 parts per million (0.005 percent).

Iron. Not more than 5 parts per million (0.0005 percent).

Substances not precipitated by sulfide. Not more than 0.2 percent, after ignition.

Sulfate. Not more than 50 parts per million (0.005 percent).

TESTS

Assay. Transfer about 4 grams, accurately weighed, into a 250-ml. volumetric flask, dissolve in water, dilute to volume with water, and mix. Transfer 25.0 ml. of this solution into a 400-ml. beaker, and add 10 ml. of a 1 in 10 solution of hydroxylamine hydrochloride, 25 ml. of 0.05 *M* disodium ethylenediaminetetraacetate measured from a buret, 25 ml. of ammonia-ammonium chloride buffer T.S., and 5 drops of eriochrome black T.S. Heat the solution to between 55° and 65°, and titrate from the buret to a blue end-point. Each ml. of 0.05 *M* disodium ethylenediaminetetraacetate is equivalent to 9.896 mg. of $MnCl_2.4H_2O$.

pH of a 5 percent solution. Determine by the *Potentiometric Method,* page 941.

Arsenic. A solution of 1 gram in 35 ml. of water meets the requirements of the *Arsenic Test,* page 865.

Heavy metals. A solution of 2 grams in 25 ml. of water meets the requirements of the *Heavy Metals Test,* page 920, using 20 mcg. of lead ion (Pb) in the control (*Solution A*).

Insoluble matter. Dissolve about 20 grams, accurately weighed, in 200 ml. of water, and allow to stand on a steam bath for 1 hour. Filter through a tared sintered glass crucible, wash thoroughly with hot water, dry at 105° for 1 hour, cool, and weigh.

Iron. Dissolve 2.0 grams in 20 ml. of water, add 1 ml. of hydrochloric acid, and dilute to 50 ml. with water. Add about 40 mg. of ammonium persulfate crystals and 3 ml. of ammonium thiocyanate T.S. Any red or pink color does not exceed that produced by 1.0 ml. of *Iron Standard Solution* (10 mcg. Fe) in an equal volume of a solution containing the quantities of the reagents used in the test.

Substances not precipitated by sulfide. Dissolve 2.0 grams in about 90 ml. of water, add 4 ml. of ammonium hydroxide, heat to 80°, and pass hydrogen sulfide through the solution to completely precipitate the manganese. Dilute to 100 ml., mix, and allow the precipitate to settle. Decant the supernatant liquid through a filter, and evaporate 50 ml. of the filtrate to dryness in a tared dish. Add 0.5 ml. of sulfuric acid, ignite to constant weight, cool, and weigh.

Sulfate. Dissolve 10.0 grams in 100 ml. of water, add 1 ml. of diluted hydrochloric acid T.S., mix, and filter. Heat to boiling, then add 10 ml. of barium chloride T.S., and allow to stand overnight. Filter out any precipitate in a tared crucible, wash, ignite gently, cool, and weigh. The weight of the ignited precipitate should not be more than 1.2 mg. greater than the weight obtained in a complete blank test.

Packaging and storage. Store in well-closed containers.

Functional use in foods. Nutrient; dietary supplement.

MANGANESE GLUCONATE

$[CH_2OH(CHOH)_4COO]_2Mn.3H_2O$

$C_{12}H_{22}MnO_{14}.2H_2O$ Mol. wt. 481.27

DESCRIPTION

A slightly pink colored powder. It is very soluble in hot water and is very slightly soluble in alcohol.

IDENTIFICATION

A. A 1 in 20 solution gives positive tests for *Manganese*, page 927.

B. It meets the requirements of *Identification test B* under *Copper Gluconate*, page 219.

SPECIFICATIONS

Assay. Not less than 98.0 percent of $C_{12}H_{22}MnO_{14}.2H_2O$.

Limits of Impurities

Arsenic (as As). Not more than 3 parts per million (0.0003 percent).

Heavy metals (as Pb). Not more than 40 parts per million (0.004 percent).

Lead. Not more than 10 parts per million (0.001 percent).

Reducing substances. Not more than 0.5 percent.

TESTS

Assay. Dissolve about 600 mg., accurately weighed, in 50 ml. of water in a 250-ml. porcelain casserole, add 1 gram of hydroxylamine hydrochloride, 10 ml. of ammonia-ammonium chloride buffer T.S., and 5 drops of eriochrome black T.S., and titrate with 0.05 *M* disodium ethylenediaminetetraacetate to a deep blue color. Each ml. of 0.05 *M* disodium ethylenediaminetetraacetate is equivalent to 24.06 mg. of $C_{12}H_{22}MnO_{14}.2H_2O$.

Arsenic. A solution of 1 gram in 35 ml. of water meets the requirements of the *Arsenic Test*, page 865.

Heavy metals. A solution of 500 mg. in 25 ml. of water meets the requirements of the *Heavy Metals Test*, page 920, using 20 mcg. of lead ion (Pb) in the control (*Solution A*).

Lead. A solution of 1 gram in 25 ml. of water meets the requirements of the *Lead Limit Test*, page 929, using 10 mcg. of lead ion (Pb) in the control.

Reducing substances. Determine as directed in the test for *Reducing Substances* under *Copper Gluconate*, page 219.

Packaging and storage. Store in well-closed containers.

Functional use in foods. Nutrient; dietary supplement.

MANGANESE GLYCEROPHOSPHATE

$C_3H_7MnO_6P.xH_2O$ Mol. wt. (anhydrous) 225.00

DESCRIPTION

A white or pinkish white powder. It is odorless and is nearly tasteless. One gram dissolves in about 5 ml. of citric acid solution (1 in 4). It is slightly soluble in water, and is insoluble in alcohol.

IDENTIFICATION

A. A 1 in 20 solution in diluted hydrochloric acid T.S. gives positive tests for *Manganese*, page 927.

B. Heat a mixture of 100 mg. of the sample with 500 mg. of potassium bisulfate. Pungent vapors of acrolein are evolved.

SPECIFICATIONS

Assay. Not less than 98.0 percent of $C_3H_7MnO_6P$ after drying.

Loss on drying. Not more than 12 percent.

Limits of Impurities

Arsenic (as As). Not more than 3 parts per million (0.0003 percent).

Heavy metals (as Pb). Not more than 40 parts per million (0.004 percent).

Lead. Not more than 10 parts per million (0.001 percent).

TESTS

Assay. Dissolve about 1 gram, previously dried at 110° to constant weight and accurately weighed, in 1.5 ml. of nitric acid and 5 ml. of warm water. Dilute to 125 ml., add 2.0 grams of dibasic ammonium phosphate and a few drops of methyl red T.S., and heat to boiling. While the solution is boiling, slowly add stronger ammonia T.S., dropwise and with constant stirring, until alkaline, and then add 2.0 ml. in excess. Let stand 2 hours at room temperature. Filter through a tared Gooch crucible, and wash the precipitate with dilute ammonia T.S. (1 in 100). Dry at 105°, ignite at a bright red heat, cool in a desiccator, and weigh. Each gram of manganese pyrophosphate so obtained is equivalent to 1.585 grams of $C_3H_7MnO_6P$.

Loss on drying, page 931. Dry at 110° to constant weight.

Arsenic. A solution of 1 gram in 10 ml. of diluted hydrochloric acid T.S. meets the requirements of the *Arsenic Test,* page 865.

Heavy metals. Dissolve 500 mg. in 5 ml. of diluted hydrochloric acid T.S., and dilute to 25 ml. with water. This solution meets the requirements of the *Heavy Metals Test,* page 920, using 20 mcg. of lead ion (Pb) in the control (*Solution A*).

Lead. Mix 1 gram with 3 ml. of dilute nitric acid (1 in 2) and 10 ml. of water, and boil until brown fumes appear. Add 10 ml. of water, boil for 2 minutes, cool, and dilute to 100 ml. with water. A 25-ml. portion of this solution meets the requirements of the *Lead Limit Test,* page 929, using 25 ml. of *Ammonium Citrate Solution,* 1 ml. of *Potassium Cyanide Solution,* 0.5 ml. of *Hydroxylamine Hydrochloride Solution,* and 2.5 mcg. of lead ion (Pb).

Packaging and storage. Store in well-closed containers.

Functional use in foods. Nutrient; dietary supplement.

MANGANESE HYPOPHOSPHITE

$Mn(PH_2O_2)_2 \cdot xH_2O$ Mol. wt. (anhydrous) 184.91

DESCRIPTION

A pink, granular or crystalline powder which is stable in the air. It is odorless, and nearly tasteless. One gram dissolves in about 6.5 ml. of water at 25° or in about 6 ml. of boiling water. It is insoluble in alcohol. A 1 in 20 solution gives positive tests for *Manganese,* page 927, and for *Hypophosphite,* page 927.

Caution should be observed in mixing Manganese Hypophosphite with nitrates, chlorates, or other oxidizing agents, as an explosion may occur if it is triturated or heated.

SPECIFICATIONS

Assay. Not less than 97.0 percent of $Mn(PH_2O_2)_2$ after drying.

Loss on drying. Not more than 9 percent.

Limits of Impurities

Arsenic (as As). Not more than 3 parts per million (0.0003 percent).

Heavy metals (as Pb). Not more than 40 parts per million (0.004 percent).

Lead. Not more than 10 parts per million (0.001 percent).

TESTS

Assay. Transfer about 120 mg., previously dried at 105° for 1 hour and accurately weighed, into a 100-ml. volumetric flask, dissolve in water, and dilute with water to volume. Transfer 50.0 ml. of this solution into a 250-ml. glass-stoppered iodine flask, add 50.0 ml. of 0.1 *N* bromine and 20 ml. of diluted sulfuric acid T.S., and stopper the flask. Place a few ml. of a saturated solution of potassium iodide in the lip around the stopper, shake the flask well, and allow to stand for 3 hours. Place the flask in an ice bath for 5 minutes, then carefully remove the stopper and allow the potassium iodide solution to be drawn into the flask. Add 2 grams of potassium iodide dissolved in 10 ml. of recently boiled water, shake the flask, and titrate the liberated iodine with 0.1 *N* sodium thiosulfate, using starch T.S. as the indicator. Each ml. of 0.1 *N* bromine is equivalent to 2.311 mg. of $Mn(PH_2O_2)_2$.

Loss on drying, page 931. Dry at 105° for 1 hour.

Arsenic. Dissolve 1 gram in 15 ml. of water, add 3 ml. of nitric acid, evaporate to dryness on a steam bath, and dissolve the residue in 35 ml. of water. This solution meets the requirements of the *Arsenic Test,* page 865.

Heavy metals. Dissolve 500 mg. in 15 ml. of water, add 3 ml. of nitric acid, evaporate to dryness on a steam bath, and dissolve the residue in 25 ml. of water. This solution meets the requirements of the *Heavy Metals Test,* page 920, using 20 mcg. of lead ion (Pb) in the control (*Solution A*).

Lead. Dissolve 250 mg. in 10 ml. of water, add 2 ml. of dilute nitric acid (1 in 2), and boil until brown fumes appear. Add 10 ml. of water, boil for 2 minutes, then cool and dilute with water to about 25 ml. This solution meets the requirements of the *Lead Limit Test,* page 929, using 25 ml. of *Ammonium Citrate Solution,* 1 ml. of *Potassium Cyanide Solution,* 0.5 ml. of *Hydroxylamine Hydrochloride Solution,* and 2.5 mcg. of lead ion (Pb).

Packaging and storage. Store in well-closed containers.

Functional use in foods. Nutrient; dietary supplement.

MANGANESE SULFATE

$MnSO_4.H_2O$ Mol. wt. 169.01

DESCRIPTION

A pale pink, granular, odorless powder. It is freely soluble in water and is insoluble in alcohol.

IDENTIFICATION

A 1 in 10 solution gives positive tests for *Manganese*, page 927, and for *Sulfate*, page 928.

SPECIFICATIONS

Assay. Not less than 98.0 percent and not more than the equivalent of 102.0 percent of $MnSO_4.H_2O$.

Loss on heating. Between 10 and 13 percent.

Limits of Impurities

Arsenic (as As). Not more than 3 parts per million (0.0003 percent).

Heavy metals (as Pb). Not more than 40 parts per million (0.004 percent).

Lead. Not more than 10 parts per million (0.001 percent).

TESTS

Assay. Transfer about 4 grams, accurately weighed, into a 250-ml. volumetric flask, dissolve in water, dilute to volume with water, and mix. Transfer a 25.0-ml. portion of this solution into a 400-ml. beaker, and add 10 ml. of a 1 in 10 solution of hydroxylamine hydrochloride, 25 ml. of 0.05 *M* disodium ethylenediaminetetraacetate measured from a buret, 25 ml. of ammonia-ammonium chloride buffer T.S., and 5 drops of eriochrome black T.S. Heat the solution to between 55° and 65°, and titrate from the buret to a blue end-point. Each ml. of 0.05 *M* disodium ethylenediaminetetraacetate is equivalent to 8.450 mg. of $MnSO_4.H_2O$.

Loss on heating. Heat about 1 gram, accurately weighed, in a crucible tared in a stoppered weighing bottle, to constant weight at 400°–500°. Cool in a desiccator, transfer to the stoppered weighing bottle, and weigh.

Arsenic. A solution of 1 gram in 35 ml. of water meets the requirements of the *Arsenic Test*, page 865.

Heavy metals. A solution of 500 mg. in 25 ml. of water meets the requirements of the *Heavy Metals Test*, page 920, using 20 mcg. of lead ion (Pb) in the control (*Solution A*).

Lead. Dissolve 1 gram in 3 ml. of dilute nitric acid (1 in 2) and 10 ml. of water, and boil for 2 minutes. Cool, and dilute to 100 ml. with water. A 25-ml. portion of this solution meets the requirements of the *Lead Limit Test*, page 929, using 25 ml. of *Ammonium Citrate Solution*, 1 ml. of *Potassium Cyanide Solution*, 0.5 ml. of *Hydroxylamine Hydrochloride Solution*, and 2.5 mcg. of lead ion (Pb).

Packaging and storage. Store in well-closed containers.

Functional use in foods. Nutrient; dietary supplement.

MANNITOL

D-Mannitol; Mannite;

1,2,3,4,5,6-Hexanehexol

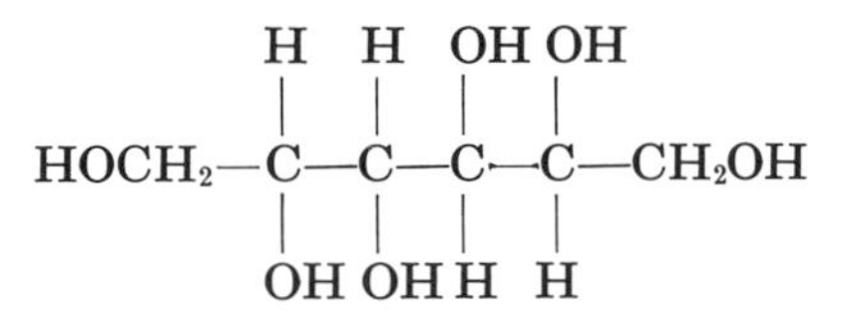

$C_6H_{14}O_6$ Mol. wt. 182.17

DESCRIPTION

A white, crystalline solid consisting of D-Mannitol and a small quantity of sorbitol. It is odorless and has a sweet taste. It is soluble in water, very slightly soluble in alcohol, and practically insoluble in most other common organic solvents.

IDENTIFICATION

A. Add 5 drops of a saturated solution to 1 ml. of ferric chloride T.S. Add 5 drops of water to a second tube containing 1 ml. of ferric chloride T.S. Add 5 drops of sodium hydroxide solution (1 in 5) to each tube. A brown precipitate of ferric hydroxide forms in the tube containing no mannitol, and a yellow precipitate forms in the tube containing mannitol. Shake the tubes vigorously. A clear solution results in the tube containing mannitol, but the precipitate persists in the other tube. Additional sodium hydroxide solution does not cause precipitation in the tube containing mannitol, but further precipitation occurs in the other.

B. Place a 500-mg. sample in a test tube, and add 3 ml. of acetic anhydride and 1 ml. of pyridine. Heat the mixture in a boiling water bath for 15 minutes, with frequent shaking, or until solution is complete, continue heating for 5 minutes, and cool. Add 20 ml. of water, mix well, allow to stand for 5 minutes, and collect the precipitate on a sintered-glass filter. The mannitol hexaacetate so obtained, after drying at 60° under vacuum for 1 hour or after recrystallization from ether, melts between 119° and 124° (see page 931).

SPECIFICATIONS

Assay. Not less than 96.0 percent and not more than the equivalent of 101.0 percent of $C_6H_{14}O_6$ (mannitol), calculated on the dried basis.

Melting range. Between 165° and 168°.

Specific rotation, $[\alpha]_D^{25°}$. Between +23.3° and +24.3°.

Limits of Impurities

Arsenic (as As). Not more than 3 parts per million (0.0003 percent).

Chloride. Not more than 70 parts per million (0.007 percent).

Heavy metals (as Pb). Not more than 10 parts per million (0.001 percent).

Loss on drying. Not more than 0.3 percent.

Reducing sugars. Passes test.

Sulfate. Not more than 100 parts per million (0.01 percent).

TESTS

Assay

Adsorbent. Mix intimately, as by mechanical rolling or tumbling for 12 hours, 9 parts, by weight, of very fine chromatographic fuller's earth and 2 parts, by weight, of chromatographic siliceous earth. Use only that portion which passes a 100-mesh sieve.

Chromatographic column. Insert a pledget of cotton on the removable porous plate of a slightly tapered, 38 × 230-mm. chromatographic tube, the constricted end of which is fitted to a 500-ml. filtering flask. Apply a vacuum of between 20 and 30 mm. of mercury, and maintain the vacuum within this range until the chromatogram is developed. Add the *Adsorbent* through a powder funnel in a steady stream until the tube is filled. Tap the tube from bottom to top, around its circumference, with a wooden rod to ensure uniform packing. Adjust the column height to within about 20 mm. of the top of the tube, and level the top of the column with a rubber stopper.

Developing solvent. Mix 15 parts, by volume, of water with 85 parts, by volume, of isopropyl alcohol.

Standard sorbitol solution. Weigh accurately 450.0 mg. of sorbitol of greater than 99.0 percent purity, and transfer into a 100-ml. volumetric flask. Dissolve in about 60 ml. of the *Developing solvent,* dilute to volume with the *Developing solvent,* and mix.

Assay preparation. Weigh accurately 450.0 mg. of the sample, and transfer into a 100-ml. volumetric flask. Dissolve in about 60 ml. of the *Developing solvent,* heating gently to effect solution. Add 5.0 ml. of the *Standard sorbitol solution,* cool to room temperature, dilute to volume with *Developing solvent,* and mix.

Procedure. Pour 20 ml. of the *Developing solvent* onto the prepared chromatographic column, leaving a 5-mm. layer of the solvent above the top of the column. Pipet 10.0 ml. of the *Assay preparation* into the tube, and, after it has practically disappeared into the column, add two 10-ml. portions of the *Developing solvent*, allowing each portion to just enter the column. (*Caution:* Do not allow the column to go dry.) Attach a separator containing 305 ml. of the *Developing solvent* to the tube by means of a rubber stopper, and adjust the flow of eluate to maintain a solvent layer about 5 to 10 mm. above the top of the column.

After all of the *Developing solvent* has passed through the column, maintain the vacuum until the column shrinks from the walls of the tube but does not become dry, and then disconnect the source of vacuum rapidly so as to cause a flow of air between the column and the walls of the tube. Place a pledget of cotton on the surface of the column, invert the tube, and tap the tube on a solid surface to loosen the column. Extrude the column onto a strip of aluminum foil, using a cork stopper attached to a wooden rod inserted through the constricted end of the tube. Streak the entire length of the column, from top to bottom, at three places equidistant around its circumference, using a solution consisting of 1 gram of potassium permanganate and 10 grams of sodium hydroxide dissolved in 100 ml. of water.

The position of the polyol zones on the chromatogram is indicated by a color change of the streaks from green to brown, with the leading edge of the mannitol zone located about 15 cm. from the top of the column. The leading edge of the sorbitol zone appears about 6 cm. from the top of the column. (Sorbitol, which is present in F.C.C. mannitol at too low a concentration to be detectable on the potassium permanganate streaks, is added in order to yield a readily detectable polyol zone.) The trailing end of the mannitol zone will discolor the potassium permanganate streaks very slowly and indistinctly; therefore, the distinct leading edge of the sorbitol zone must be used as a guide in cutting the zones from the column. Mark the leading edges of the sorbitol and mannitol zones, and, using a sharp blade, cut the column at the leading edge of the sorbitol zone and about 20 mm. below the leading edge of the mannitol zone. Discard the top and bottom portions of the column. Cut the mannitol zone into small pieces, and allow it to dry overnight, protected from drafts.

Powder the dried mannitol zone with a spatula, pack it into a clean, dry chromatographic tube as directed under *Chromatographic column*, and elute the mannitol by passing 150 ml. of warm water through the column. Transfer the eluate into a 200-ml. volumetric flask, dilute to volume with water, and mix. Transfer 50.0 ml. of this solution into a 250-ml. iodine flask, add 50.0 ml. of sodium periodate solution (prepared by dissolving 4.5 grams of sodium periodate in 500 ml. of water, add 2 ml. of sulfuric acid, and diluting to 1000 ml. with water), and heat for 20 minutes in a water bath maintained at 80° to 85°. Cool to room temperature in an ice bath, add 3 grams of sodium bicarbonate and 10 ml. of potassium iodide solution (1 in 5), and titrate immediately with 0.05 *N* sodium arsenite, adding starch T.S. as the end-point is approached. Perform a blank determination, using water in place of the mannitol eluate, and make any necessary corrections (see page 2). Each ml. of 0.05 *N* sodium arsenite is equivalent to 910.9 mcg. of mannitol ($C_6H_{14}O_6$).

Melting range. Determine as directed in the general procedure, page 931.

Specific rotation. Transfer 10.0 grams of the sample, accurately weighed, into a 100-ml. volumetric flask, and add sodium borate equivalent to 12.8 grams of $Na_2B_4O_7 . 10H_2O$, the purity of which is determined by assay as follows: dissolve about 3 grams of sodium borate decahydrate, accurately weighed, in 50 ml. of water, add methyl red T.S., and titrate with 0.5 *N* hydrochloric acid, each ml. of which is equivalent to 95.34 mg. of $Na_2B_4O_7 . 10H_2O$. To the flask add sufficient water to make about 90 ml, allow to stand for 1 hour with occasional shaking, then dilute to volume with water, and mix. Determine the specific rotation of this solution as directed in the general procedure, page 939.

Arsenic. A *Sample Solution* prepared as directed for organic compounds meets the requirements of the *Arsenic Test*, page 865.

Chloride, page 879. Any turbidity produced by a 200-mg. sample does not exceed that shown in a control containing 14 mcg. of chloride ion (Cl).

Heavy metals. A solution of 2 grams in 25 ml. of water meets the requirements of the *Heavy Metals Test*, page 920, using 20 mcg. of lead ion (Pb) in the control (*Solution A*).

Loss on drying, page 931. Dry at 105° for 4 hours.

Reducing sugars. Add 1 ml. of a saturated solution to 5 ml. of alkaline cupric citrate T.S., and heat for 5 minutes in a boiling water bath. No more than a very slight precipitation occurs.

Sulfate, page 879. Any turbidity produced by a 2-gram sample does not exceed that shown in a control containing 200 mcg. of sulfate (SO_4).

Packaging and storage. Store in well-closed containers.

Functional use in foods. Nutrient; dietary supplement; texturizing agent.

MARJORAM OIL, SPANISH

DESCRIPTION

A volatile oil obtained by steam distillation from the flowering plant *Thymus mastichina* L. (Fam. *Labiatae*). It is a slightly yellow liquid, having a camphoraceous note. It is soluble in most fixed oils, but it is insoluble in glycerin, in propylene glycol, and in mineral oil.

SPECIFICATIONS

Assay. Not less than 49 percent and not more than 65 percent of cineole.

Acid value. Not more than 2.0.

Angular rotation. Between $-5°$ and $+10°$.

Refractive index. Between 1.463 and 1.468 at 20°.

Saponification value. Between 5 and 20.

Solubility in alcohol. Passes test.

Specific gravity. Between 0.904 and 0.920.

TESTS

Assay. Dry a sample over anhydrous sodium sulfate, and transfer 3 grams of the dried oil, accurately weighed, into a test tube as directed for *Solidification Point*, page 954. Add 2.1 grams of melted *o*-cresol. The *o*-cresol must be pure and dry, and have a solidification point not below 30°. Insert the thermometer, stir, and warm the tube gently until the mixture is completely melted. Proceed as directed under *Solidification Point*, page 954. Repeat the procedure until two successive readings agree within 0.1°. Compute the percent of cineole from the *Percentage of Cineole* table, page 898.

Acid value. Determine as directed in the general method, page 893.

Angular rotation. Determine in a 100-mm. tube as directed under *Optical Rotation*, page 939.

Refractive index, page 945. Determine with an Abbé or other refractometer of equal or greater accuracy.

Saponification value. Determine as directed in the general method, page 896, using about 10 grams, accurately weighed.

Solubility in alcohol. Proceed as directed in the general method, page 899. One ml. dissolves in 1 ml. of 80 percent alcohol, and remains in solution on further addition of the alcohol to a total volume of 10 ml.

Specific gravity. Determine by any reliable method (see page 5).

Packaging and storage. Store in full, tight, preferably glass, aluminum, tin-lined, or other suitably lined containers in a cool place, protected from light.

Functional use in foods. Flavoring agent.

MASTICATORY SUBSTANCES, NATURAL

Coagulated or Concentrated Latices of Vegetable Origin

DESCRIPTION

The masticatory substances of vegetable origin are comprised of the gums from the trees of *Sapotaceae*, *Apocynaceae*, *Moraceae*, and *Euphorbiaceae* as listed below. The coagulated material varies in color from white to brown, depending on its moisture content and heat treatment during purification. The gums are purified by extensive treatment either alone or in combination with other gums or food-grade materials. They are heat treated and then clarified by centrifuging or by any other appropriate means of filtration.

Family	Genus and species
Sapotaceae	
Chicle	*Manilkara zapotilla* Gilly and *Manilkara chicle* Gilly
Chiquibul	*Manilkara zapotilla* Gilly
Crown gum	*Manilkara zapotilla* Gilly and *Manilkara chicle* Gilly
Gutta hang kang	*Palaquium leiocarpum* Boerl. and *Palaquium oblongifolium* Burck
Gutta Katiau	*Palaquium ganua moteleyana* Clarke (also known as *Sideroxylon glabrescens*)
Massaranduba balata (and the solvent-free resin extract of Massaranduba balata)	*Manilkara huberi* (Ducke) Chevalier
Massaranduba chocolate	*Manilkara solimoesensis* Gilly
Nispero	*Manilkara zapotilla* Gilly and *Manilkara chicle* Gilly
Rosidinha (rosadinha)	*Micropholis* (also known as *Sideroxylon*) spp.
Venezuelan chicle	*Manilkara williamsii* Standley and related spp.
Apocynaceae	
Jelutong	*Dyera costulata* Hook. F. and *Dyera lowii* Hook. F.
Leche caspi (sorva)	*Couma macrocarpa* Barb. Rodr.
Pendare	*Couma macrocarpa* Barb. Rodr. and *Couma utilis* (Mart.) Muell. Arg.
Perillo	*Couma macrocarpa* Barb. Rodr. and *Couma utilis* (Mart.) Muell. Arg.
Moraceae	
Leche de vaca	*Brosimum utile* (H.B.K.) Pittier and *Poulsenia* spp.; also *Lacmellea standleyi* (Woodson), *Monachino* (Apocynaceae)
Niger Gutta	*Ficus platyphylla* Del.
Tunu (tuno)	*Castilla fallax* Cook

Euphorbiaceae

Chilte . . . . *Cnidoscolus* (also known as *Jatropha*) *elasticus* Lundell and *Cnidoscolus tepiquensis* (Cost. and Gall.) McVaugh

Natural rubber (latex solids) . . . . . *Hevea brasiliensis*

SPECIFICATIONS

The following specifications, where applicable, should conform to the representations of the vendor: cleanliness, color, texture, odor, and loss on drying.

Limits of Impurities

Arsenic (as As). Not more than 3 parts per million (0.0003 percent).

Heavy metals (as Pb). Not more than 40 parts per million (0.004 percent).

Lead. Not more than 3 parts per million (0.0003 percent).

TESTS

Loss on drying, page 931. Dry a 10-gram sample at 145° for 1 hour.

Arsenic. Prepare a *Sample Solution* as directed in the general method under *Chewing Gum Base,* page 877. This solution meets the requirements of the *Arsenic Test,* page 865.

Heavy metals. Prepare and test a 500-mg. sample as directed in *Method II* under the *Heavy Metals Test,* page 920, using 20 mcg. of lead ion (Pb) in the control (*Solution A*).

Lead. Prepare a *Sample Solution* as directed in the general method under *Chewing Gum Base,* page 878. This solution meets the requirements of the *Lead Limit Test,* page 929, using 10 mcg. of lead ion (Pb) in the control.

Packaging and storage. Store in well-closed containers.

Functional use in foods. Masticatory substance in chewing gum base.

MENTHOL

3-*p*-Menthanol

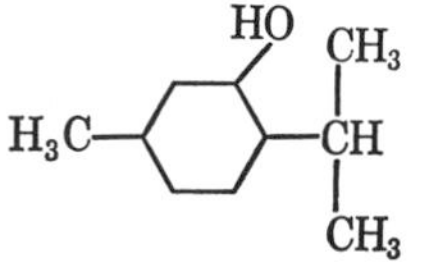

$C_{10}H_{20}O$ Mol. wt. 156.27

DESCRIPTION

Menthol is an alcohol obtained from various mint oils, or prepared synthetically. It may be either levorotatory (*l*-menthol) from natural or synthetic sources, or racemic (*dl*-menthol) produced synthetically. It occurs as colorless, hexagonal crystals, usually needle-like, or in fused masses, or as a crystalline powder. It has a pleasant, peppermint-like odor. It is very soluble in alcohol and is freely soluble in fixed and volatile oils. It is slightly soluble in water.

IDENTIFICATION

Triturate a sample with about an equal weight of camphor, of chloral hydrate, or of phenol. The mixture liquefies.

SPECIFICATIONS

Melting range of *l*-menthol. Between 41° and 43°.

Solidification point of *dl*-menthol. Between 27° and 28°.

Specific rotation, $[\alpha]_D^{25°}$. For *l-menthol:* Between −45° and −51°; for *dl-menthol:* between −2° and +2°.

Limits of Impurities

Arsenic (as As). Not more than 3 parts per million (0.0003 percent).
Heavy metals (as Pb). Not more than 40 parts per million (0.004 percent).
Lead. Not more than 10 parts per million (0.001 percent).
Nonvolatile residue. Not more than 500 parts per million (0.05 percent).

Readily oxidizable substances in *dl*-menthol. Passes test.

TESTS

Melting range of *l*-menthol. Determine as directed in the general procedure, page 931.

Solidification point of *dl*-menthol. Determine as directed in the general method, page 954, using a sample previously dried in a desiccator over silica gel for 24 hours and adjusting the temperature of the cooling bath to a temperature between 23° and 25°. *dl*-Menthol solidifies between 27° and 28°. Continue the stirring. After a few minutes the temperature quickly rises to 30.5° to 32°.

Specific rotation. Proceed as directed under *Optical Rotation,* page 939, using a solution containing 1 gram in each 10 ml. of alcohol.

Arsenic. A *Sample Solution* prepared as directed for organic compounds meets the requirements of the *Arsenic Test,* page 865.

Heavy metals. Prepare and test a 500-mg. sample as directed in *Method II* under the *Heavy Metals Test,* page 920, using 20 mcg. of lead ion (Pb) in the control (*Solution A*).

Lead. A *Sample Solution* prepared as directed for organic compounds meets the requirements of the *Lead Limit Test,* page 929, using 10 mcg. of lead ion (Pb) in the control.

Nonvolatile residue. Heat 2 grams, accurately weighed, in a tared porcelain dish on a steam bath until volatilized. Dry the residue at 105° for 1 hour. The residue weighs not more than 1 mg.

Readily oxidizable substances in *dl*-menthol. Transfer 500 mg. of *dl*-menthol into a clean, dry test tube, add 10 ml. of potassium permanganate solution prepared by diluting 3 ml. of 0.1 *N* potassium permanganate with water to 100 ml. Place the test tube in a beaker of water maintained between 45° and 50°. At 30-second intervals, quickly remove the test tube from the bath and shake. The color of potassium permanganate is still apparent after 5 minutes.

Packaging and storage. Store in full, tight containers in a cool place protected from light.

Labeling. Label menthol to indicate whether it is levorotatory or racemic.

Functional use in foods. Flavoring agent.

DL-METHIONINE

DL-2-Amino-4-(methylthio)butyric Acid

$$CH_3SCH_2CH_2\underset{\displaystyle NH_2}{\underset{|}{C}}HCOOH$$

$C_5H_{11}NO_2S$ Mol. wt. 149.21

DESCRIPTION

White, crystalline platelets or powder having a characteristic odor. One gram dissolves in about 30 ml. of water. It is soluble in dilute acids and in solutions of alkali hydroxides. It is very slightly soluble in alcohol, and practically insoluble in ether. It is optically inactive. The pH of a 1 in 100 solution is between 5.6 and 6.1.

IDENTIFICATION

A. Add 25 mg. of a previously dried sample to 1 ml. of a saturated solution of anhydrous cupric sulfate in sulfuric acid. A yellow color appears.

B. To 10 ml. of a 1 in 1000 solution add successively, shaking after each addition, 1 ml. of a 1 in 5 solution of sodium hydroxide, 1 ml. of a 1 in 100 glycine solution, and 0.3 ml. of a freshly prepared 1 in 10 solution of sodium nitroferricyanide. Keep the mixture at about 40° for 10 minutes, cool in an ice bath for 2 minutes, then add 2 ml. of 20 percent hydrochloric acid and shake the mixture. A reddish-purple color appears.

C. To 1 ml. of a 1 in 30 solution add 1 ml. of ninhydrin T.S. and 100 mg. of sodium acetate, and heat to boiling. An intense violet-blue color is formed (distinction from hydroxy analog).

SPECIFICATIONS

Assay. Not less than 99.0 percent of $C_5H_{11}NO_2S$, calculated on the dried basis.

Limits of Impurities

Arsenic (as As). Not more than 3 parts per million (0.0003 percent).
Heavy metals (as Pb). Not more than 20 parts per million (0.002 percent).
Lead. Not more than 10 parts per million (0.001 percent).
Loss on drying. Not more than 0.5 percent.
Residue on ignition. Not more than 0.1 percent.

TESTS

Assay. Transfer about 300 mg., accurately weighed, into a glass-stoppered flask. Add 100 ml. of water, 5 grams of dibasic potassium phosphate, 2 grams of monobasic potassium phosphate, and 2 grams of potassium iodide, and mix well to dissolve. Add exactly 50 ml. of 0.1 *N* iodine, stopper the flask, mix well, allow to stand for 30 minutes, and then titrate the excess iodine with 0.1 *N* sodium thiosulfate. Perform a blank determination (see page 2) and make any necessary correction. Each ml. of 0.1 *N* iodine is equivalent to 7.461 mg. of $C_5H_{11}NO_2S$.

Arsenic. A *Sample Solution* prepared as directed for organic compounds meets the requirements of the *Arsenic Test,* page 865.

Heavy metals. Prepare and test a 1-gram sample as directed in *Method II* under the *Heavy Metals Test,* page 920, using 20 mcg. of lead ion (Pb) in the control (*Solution A*).

Lead. A *Sample Solution* prepared as directed for organic compounds meets the requirements of the *Lead Limit Test,* page 929, using 10 mcg. of lead ion (Pb) in the control.

Loss on drying, page 931. Dry at 105° for 2 hours.

Residue on ignition, page 945. Ignite 1 gram as directed in the general method.

Packaging and storage. Store in well-closed, light-resistant containers.

Functional use in foods. Nutrient; dietary supplement.

L-METHIONINE

L-2-Amino-4-(methylthio)butyric Acid

$$CH_3SCH_2CH_2\underset{\displaystyle NH_2}{\underset{|}{C}}HCOOH$$

$C_5H_{11}NO_2S$ Mol. wt. 149.21

DESCRIPTION

Colorless or white lustrous plates, or a white crystalline powder. It has a slight, characteristic odor. It is soluble in water, in alkali solutions, and in dilute mineral acids. It is slightly soluble in alcohol and practically insoluble in ether.

IDENTIFICATION

L-Methionine responds to *Identification Tests A, B* and *C* under DL-*Methionine*, page 504.

SPECIFICATIONS

Assay. Not less than 99.0 percent of $C_5H_{11}NO_2S$, calculated on the dried basis.

Specific rotation. $[\alpha]_D^{25°}$: Between $-6.8°$ and $-8.2°$; $[\alpha]_D^{20°}$: between $+21.0°$ and $+25.0°$.

Limits of Impurities

Arsenic (as As). Not more than 3 parts per million (0.0003 percent).

Heavy metals (as Pb). Not more than 20 parts per million (0.002 percent).

Lead. Not more than 10 parts per million (0.001 percent).

Loss on drying. Not more than 0.5 percent.

Residue on ignition. Not more than 0.1 percent.

TESTS

Assay. Transfer about 300 mg., accurately weighed, into a glass-stoppered flask. Add 100 ml. of water, 5 grams of dibasic potassium phosphate, 2 grams of monobasic potassium phosphate, and 2 grams of potassium iodide, and mix well to dissolve. Add exactly 50 ml. of 0.1 *N* iodine, stopper the flask, mix well, allow to stand for 30 minutes, and then titrate the excess iodine with 0.1 *N* sodium thiosulfate. Perform a blank determination (see page 2) and make any necessary correction. Each ml. of 0.1 *N* iodine is equivalent to 7.461 mg. of $C_5H_{11}NO_2S$.

Specific rotation, page 939. $[\alpha]_D^{25°}$: Determine in a solution containing 4 grams in sufficient water to make 100 ml.; $[\alpha]_D^{20°}$: determine in a solution containing 2 grams in sufficient 6 *N* hydrochloric acid to make 100 ml.

Arsenic. A *Sample Solution* prepared as directed for organic compounds meets the requirements of the *Arsenic Test*, page 865.

Heavy metals. Prepare and test a 1-gram sample as directed in *Method II* under the *Heavy Metals Test*, page 920, using 20 mcg. of lead ion (Pb) in the control (*Solution A*).

Lead. A *Sample Solution* prepared as directed for organic compounds meets the requirements of the *Lead Limit Test*, page 929, using 10 mcg. of lead ion (Pb) in the control.

Loss on drying, page 931. Dry at 105° for 2 hours.

Residue on ignition, page 945. Ignite 1 gram as directed in the general method.

Packaging and storage. Store in well-closed, light-resistant containers.

Functional use in foods. Nutrient; dietary supplement.

p-METHOXYBENZALDEHYDE

Anisic Aldehyde; *p*-Anisaldehyde

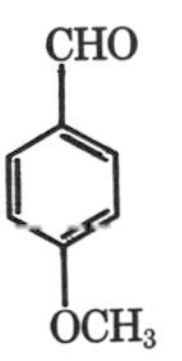

$C_8H_8O_2$ Mol. wt. 136.15

DESCRIPTION

A colorless to slightly yellow liquid having a characteristic hawthorn-like odor. It is miscible with alcohol, with ether, and with most fixed oils. It is soluble in propylene glycol, but is practically insoluble in glycerin, in water, and in mineral oil.

SPECIFICATIONS

Assay. Not less than 97.5 percent of $C_8H_8O_2$.

Refractive index. Between 1.571 and 1.574 at 20°.

Solubility in alcohol. Passes test.

Specific gravity. Between 1.119 and 1.123.

Limits of Impurities

Acid value. Not more than 6.0.

Arsenic (as As). Not more than 3 parts per million (0.0003 percent).

Chlorinated compounds. Passes test.

Heavy metals (as Pb). Not more than 40 parts per million (0.004 percent).

Lead. Not more than 10 parts per million (0.004 percent).

TESTS

Assay. Weigh accurately about 1.2 grams, and proceed as directed under *Aldehydes*, page 893, but allow the sample and the blank to stand at room temperature for 15 minutes. Use 68.08 as the equivalence factor (E) in the calculation.

Refractive index, page 945. Determine with an Abbé or other refractometer of equal or greater accuracy.

Solubility in alcohol. Proceed as directed in the general method, page 899. One ml. dissolves in 7 ml. of 50 percent alcohol to form a clear solution.

Specific gravity. Determine by any reliable method (see page 5).

Acid value. Determine as directed in the general method, page 893.

Arsenic. A *Sample Solution* prepared as directed for organic compounds meets the requirements of the *Arsenic Test*, page 865.

Chlorinated compounds. Determine as directed in the general method, page 895.

Heavy metals. Prepare and test a 500-mg. sample as directed in *Method II* under the *Heavy Metals Test*, page 920, using 20 mcg. of lead ion (Pb) in the control (*Solution A*).

Lead. A *Sample Solution* as directed for organic compounds meets the requirements of the *Lead Limit Test*, page 929, using 10 mcg. of lead ion (Pb) in the control.

Packaging and storage. Store in full, tight, preferably glass or tin-lined containers in a cool place protected from light.

Functional use in foods. Flavoring agent.

4′-METHYL ACETOPHENONE

Methyl *p*-Tolyl Ketone

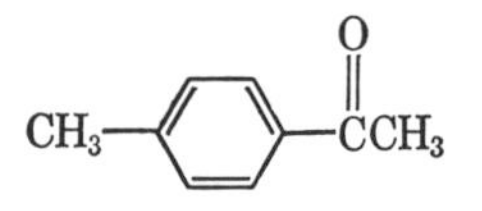

$C_9H_{10}O$ Mol. wt. 134.18

DESCRIPTION

A colorless, or nearly colorless liquid having a fruity-floral odor resembling acetophenone. It is soluble in most fixed oils and in propylene glycol. It is slightly soluble in mineral oil, but it is insoluble in glycerin.

SPECIFICATIONS

Assay. Not less than 98.0 percent of $C_9H_{10}O$.

Refractive index. Between 1.532 and 1.535 at 20°.

Solubility in alcohol. Passes test.

Specific gravity. Between 1.001 and 1.004.

Limits of Impurities

Chlorinated compounds. Passes test.

TESTS

Assay. Weigh accurately about 1 gram, and proceed as directed under *Aldehydes and Ketones-Hydroxylamine Method*, page 894, using 67.09 as the equivalence factor (E) in the calculation.

Refractive index, page 945. Determine with an Abbé or other refractometer of equal or greater accuracy.

Solubility in alcohol. Proceed as directed in the general method, page 899. One ml. dissolves in 10 ml. of 50 percent alcohol.

Specific gravity. Determine by any reliable method (see page 5).

Chlorinated compounds. Proceed as directed in the general method, page 895.

Packaging and storage. Store in full, tight, preferably glass aluminum, tin-lined, or iron containers.

Functional use in foods. Flavoring agent.

METHYL ALCOHOL

Methanol

CH_3OH Mol. wt. 32.04

DESCRIPTION

A clear, colorless, flammable liquid having a characteristic odor. It is miscible with water, with ethyl alcohol, and with ether. Its refractive index at 20° is about 1.329.

SPECIFICATIONS

Assay. Not less than 99.85 percent of CH_3OH, by weight.

Distillation range. Within a range of 1°, including 64.6° ± 0.1°.

Solubility in water. Passes test.

Limits of Impurities

Acetone and aldehydes. Not more than 30 parts per million (0.003 percent).

Acidity (as formic acid). Not more than 15 parts per million (0.0015 percent).

Alkalinity (as NH_3). Not more than 3 parts per million (0.0003 percent).

Heavy metals (as Pb). Not more than 1 part per million (0.0001 percent).

Nonvolatile residue. Not more than 10 parts per million (0.001 percent).

Substances darkened by sulfuric acid. Passes test.

Substances reducing permanganate. Passes test.

Water. Not more than 0.1 percent.

TESTS

Assay. Its specific gravity, determined by any reliable method (see page 5), is not greater than 0.7893 at 25°/25° (equivalent to 0.7928 at 20°/20°).

Distillation range. Proceed as directed in the general method, page 890.

Solubility in water. Mix 15 ml. of the sample with 45 ml. of water. After 1 hour, the solution is as clear as an equal volume of water.

Acetone and aldehydes. To 1.25 ml. (about 1 gram) of the sample add 3.75 ml. of water and 5.0 ml. of alkaline mercuric-potassium iodide T.S. Any turbidity does not exceed that produced in a standard containing 30 mcg. of acetone.

Acidity. To a mixture of 10 ml. of alcohol and 25 ml. of water add 0.5 ml. of phenolphthalein T.S., and titrate with 0.02 *N* sodium hydroxide to the first pink color that persists for at least 30 seconds. Add 19 ml. (about 15 grams) of the sample, mix, and titrate with 0.02 *N* sodium hydroxide until the pink color is restored. Not more than 0.25 ml. is required.

Alkalinity. Add 1 drop of methyl red T.S. to 25 ml. of water, add 0.02 *N* sulfuric acid until a red color just appears, then add 29 ml. (about 22.5 grams) of the sample, and mix. Not more than 0.2 ml. of 0.02 *N* sulfuric acid is required to restore the red color.

Heavy metals. Evaporate 25 ml. (about 20 grams) of the sample to dryness on a steam bath in a glass evaporating dish. Cool, add 2 ml. of hydrochloric acid, and slowly evaporate to dryness again on the steam bath. Moisten the residue with 1 drop of hydrochloric acid, add 10 ml. of hot water, and digest for 2 minutes. Cool, and dilute to 25 ml. with water. This solution meets the requirements of the *Heavy Metals Test*, page 920, using 20 mcg. of lead ion (Pb) in the control (*Solution A*).

Nonvolatile residue. Evaporate 125 ml. (about 100 grams) of the sample to dryness in a tared dish on a steam bath, dry the residue at 105° for 30 minutes, cool, and weigh.

Readily carbonizable substances, page 943. A mixture of 25 ml. of sulfuric acid T.S. (cooled to 10°) and 25 ml. of the sample has no more color than 3.5 ml. of platinum-cobalt C.S., diluted to 50 ml. with water (equivalent to not more than 35 A.P.H.A. color units).

Substances reducing permanganate. Transfer 20 ml. of the sample, previously cooled to 15°, to a glass-stoppered cylinder, add 0.1 ml. of 0.1 *N* potassium permanganate, mix, and allow to stand for 5 minutes. The pink color is not entirely discharged.

Water. Determine by the *Karl Fischer Titrimetric Method,* page 977.

Packaging and storage. Store in tight containers, remote from heat, sparks, and open flames.

Functional use in foods. Extraction solvent.

p-METHYL ANISOLE

p-Cresyl Methyl Ether; Methyl *p*-Cresol

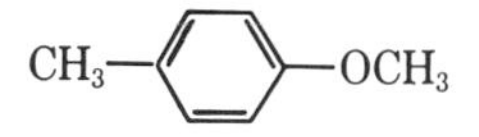

$C_8H_{10}O$ Mol. wt. 122.17

DESCRIPTION

A colorless liquid having a pungent odor suggestive of ylang ylang. It is soluble in most fixed oils and in mineral oil, but is insoluble in glycerin and in propylene glycol.

SPECIFICATIONS

Refractive index. Between 1.510 and 1.513 at 20°.

Solubility in alcohol. Passes test.

Specific gravity. Between 0.966 and 0.970.

Limits of Impurities

Cresol. Not more than 0.5 percent.

TESTS

Refractive index, page 945. Determine with an Abbé or other refractometer of equal or greater accuracy.

Solubility in alcohol. Proceed as directed in the general method, page 899. One ml. dissolves in 3 ml. of 80 percent alcohol, and remains in solution upon further dilution.

Specific gravity. Determine by any reliable method (see page 5).

Cresol. Proceed as directed under *Free Phenols,* page 899, using 54.06 as the equivalence factor (*f*) in the calculation.

Packaging and storage. Store in full, tight, preferably glass, aluminum, tin-lined, or other suitably lined containers in a cool place protected from light.

Functional use in foods. Flavoring agent.

METHYL ANTHRANILATE

Methyl 2-Aminobenzoate

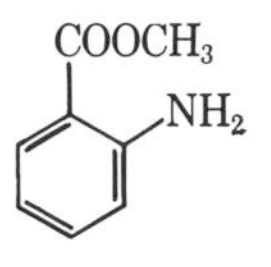

$C_8H_9NO_2$ Mol. wt. 151.17

DESCRIPTION

Methyl anthranilate is found in neroli oil and in citrus and other oils. It is prepared synthetically by esterification of anthranilic acid. It is a colorless to pale yellow liquid with a bluish fluorescence. It has a grape-like odor. It is soluble in most fixed oils and in propylene glycol, and is partly soluble in mineral oil. It is insoluble in glycerin.

SPECIFICATIONS

Assay. Not less than 98.0 percent of $C_8H_9NO_2$.

Refractive index. Between 1.582 and 1.584 at 20°, in supercooled liquid form.

Solidification point. Not less than 23.8°.

Solubility in alcohol. Passes test.

Specific gravity. Between 1.161 and 1.169.

TESTS

Assay. Weigh accurately about 1 gram, and proceed as directed under *Ester Determination*, page 896, using 75.59 as the equivalence factor (*e*) in the calculation.

Refractive index, page 945. Determine with an Abbé or other refractometer of equal or greater accuracy.

Solidification point. Determine as directed in the general method, page 954, supercooling the sample to 19° to 20°, but not below 18°, and seeding the melt to induce crystallization.

Solubility in alcohol. Proceed as directed in the general method, page 899. One ml. dissolves in 5 ml. of 60 percent alcohol, and remains in solution on dilution up to 10 ml. with the alcohol.

Specific gravity. Determine by any reliable method (see page 5).

Packaging and storage. Store in full, tight, glass, aluminum, or tin-lined containers in a cool place protected from light. Prolonged storage or exposure to light may cause discoloration.

Functional use in foods. Flavoring agent.

METHYL BENZOATE

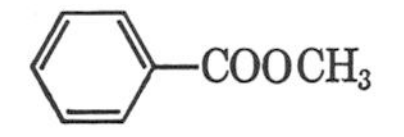

$C_8H_8O_2$ Mol. wt. 136.14

DESCRIPTION

A colorless liquid, having a fruity odor resembling cananga oil. It is soluble in most fixed oils, in mineral oil, and in propylene glycol. It is insoluble in glycerin.

SPECIFICATIONS

Assay. Not less than 98.0 percent of $C_8H_8O_2$.

Refractive index. Between 1.514 and 1.518 at 20°.

Solubility in alcohol. Passes test.

Specific gravity. Between 1.082 and 1.088.

Limits of Impurities

Acid value. Not more than 1.0.

Chlorinated compounds. Passes test.

TESTS

Assay. Weigh accurately about 900 mg., and proceed as directed under *Ester Determination*, page 896, using 68.07 as the equivalence factor (*e*) in the calculation.

Refractive index, page 945. Determine with an Abbé or other refractometer of equal or greater accuracy.

Solubility in alcohol. Proceed as directed in the general method, page 899. One ml. dissolves in 4 ml. of 60 percent alcohol.

Specific gravity. Determine by any reliable method (see page 5).

Acid value. Determine as directed in the general method, page 893.

Chlorinated compounds. Proceed as directed in the general method, page 895.

Packaging and storage. Store in full, tight, preferably glass, tin, or other suitably lined containers in a cool place protected from light.

Functional use in foods. Flavoring agent.

α-METHYLBENZYL ACETATE

Methyl Phenylcarbinyl Acetate; α-Phenyl Ethyl Acetate

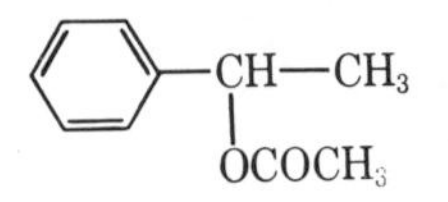

$C_{10}H_{12}O_2$ Mol. wt. 164.20

DESCRIPTION

A colorless liquid having a characteristic odor suggesting gardenia. It is soluble in most fixed oils and is freely soluble in glycerin and in mineral oil. It is insoluble in water.

SPECIFICATIONS

Assay. Not less than 97.0 percent of $C_{10}H_{12}O_2$.

Refractive index. Between 1.493 and 1.497 at 20°.

Solubility in alcohol. Passes test.

Specific gravity. Between 1.023 and 1.026.

Limits of Impurities

Acid value. Not more than 2.0.
Chlorinated compounds. Passes test.

TESTS

Assay. Weigh accurately about 1 gram, and proceed as directed under *Ester Determination*, page 896, using 82.10 as the equivalence factor (*e*) in the calculation.

Refractive index, page 945. Determine with an Abbé or other refractometer of equal or greater accuracy.

Solubility in alcohol. Proceed as directed in the general method, page 899. One ml. dissolves in 7 ml. of 60 percent alcohol.

Specific gravity. Determine by any reliable method (see page 5).

Acid value. Determine as directed in the general method, page 893.

Chlorinated compounds. Proceed as directed in the general method, page 895.

Packaging and storage. Store in full, tight, preferably glass, aluminum, tin-lined, or other suitably lined containers, or steel containers in a cool place protected from light.

Functional use in foods. Flavoring agent.

α-METHYLBENZYL ALCOHOL

Methyl Phenylcarbinol; α-Phenethyl Alcohol

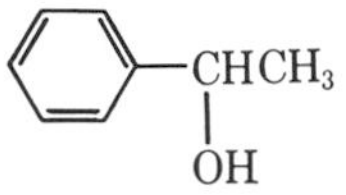

$C_8H_{10}O$ Mol. wt. 122.17

DESCRIPTION

A colorless liquid having a mild hyacinth-gardenia odor. It congeals below room temperature. It is soluble in most fixed oils and in propylene glycol, and is freely soluble in glycerin and in mineral oil.

SPECIFICATIONS

Assay. Not less than 99.0 percent of $C_8H_{10}O$.

Refractive index. Between 1.525 and 1.529 at 20°.

Solidification point. Not lower than 19°.

Solubility in alcohol. Passes test.

Specific gravity. Between 1.009 and 1.014.

Limits of Impurities

Ketones (as acetophenone). Not more than 1.0 percent.

TESTS

Assay. Proceed as directed under *Total Alcohols*, page 893. Weigh accurately about 1 gram of the acetylated alcohol for the saponification, and use 61.08 as the equivalence factor (e) in the calculation.

Refractive index, page 945. Determine with an Abbé or other refractometer of equal or greater accuracy.

Solidification point. Determine as directed in the general method, page 954.

Solubility in alcohol. Proceed as directed in the general method, page 899. One ml. dissolves in 3 ml. of 50 percent alcohol.

Specific gravity. Determine by any reliable method (see page 5).

Ketones. Weigh accurately about 10 grams, and proceed as directed under *Aldehydes and Ketones-Hydroxylamine Method*, page 894, using 60.07 as the equivalence factor (E) in the calculation.

Packaging and storage. Store in full, tight, preferably glass, galvanized, or iron containers in a cool place protected from light.

Functional use in foods. Flavoring agent.

METHYLCELLULOSE

DESCRIPTION

Methylcellulose is the methyl ether of cellulose in the form of a white, fibrous powder or granules. It is soluble in water and in a limited number of organic solvent systems. Aqueous solutions of methylcellulose are surface active, form films upon drying, and undergo a reversible transformation from sol to gel upon heating and cooling, respectively.

IDENTIFICATION

A. Add 1 gram to 100 ml. of water. It swells and disperses to form a clear to opalescent mucilaginous solution, depending upon the intrinsic viscosity, which is stable in the presence of most electrolytes and of alcohol in concentrations up to 40 percent.

B. Heat a few ml. of the solution prepared for *Identification test A*. The solution becomes cloudy, and a flaky precipitate appears which redissolves as the solution cools.

C. Pour a few ml. of the solution prepared for *Identification test A* onto a glass plate, and allow the water to evaporate. A thin, self-sustaining film results.

SPECIFICATIONS

Assay. Not less than 27.5 percent nor more than 31.5 percent of methoxyl groups ($—OCH_3$) after drying.

Viscosity. The apparent viscosity of a solution containing 2 grams in each 100 ml. is not less than 80 percent and not more than 120 percent of that stated on the label for viscosity types of 100 centipoises or less; and not less than 75 percent and not more than 140 percent of that stated on the label for viscosity types higher than 100 centipoises.

Limits of Impurities

Arsenic (as As). Not more than 3 parts per million (0.0003 percent).

Heavy metals (as Pb). Not more than 10 parts per million (0.001 percent).

Loss on drying. Not more than 5 percent.

Residue on ignition. Not more than 1.5 percent.

TESTS

Assay. Place about 50 mg. of the dried sample, obtained as directed under *Loss on drying*, in an empty, tared gelatin capsule, weigh accurately, place the capsule and contents in the boiling flask of a methoxyl apparatus, and proceed as directed for *Methoxyl Determination*, page 937.

Viscosity. Weigh accurately a sample, equivalent to 2 grams of solids on the dried basis, transfer to a wide-mouth, 250-ml. centrifuge bottle, and add 98 grams of water previously heated to between 80°

and 90°. Stir with a mechanical stirrer for 10 minutes, then place the bottle in an ice bath until solution is complete, adjust the weight of the solution to 100 grams, if necessary, and centrifuge it to expel any entrapped air. Adjust the temperature of the solution to 20° ± 0.1°, and determine the viscosity as directed under *Viscosity of Methylcellulose*, page 971.

Arsenic. A *Sample Solution* prepared as directed for organic compounds meets the requirements of the *Arsenic Test*, page 865.

Heavy metals. Prepare and test a 2-gram sample as directed in *Method II* under the *Heavy Metals Test*, page 920, adding 1 ml. of hydroxylamine hydrochloride solution (1 in 5) to the solution of the residue. Any color does not exceed that produced in a control (*Solution A*) containing 20 mcg. of lead ion (Pb) and 1 ml. of the hydroxylamine hydrochloride solution.

Loss on drying, page 931. Dry a 3-gram sample at 105° for 2 hours.

Residue on ignition. Ignite 1 gram as directed in the general method, page 945.

Packaging and storage. Store in well-closed containers.

Functional use in foods. Thickener; stabilizer; emulsifier; bodying agent; bulking agent; binder; film former.

α-METHYLCINNAMALDEHYDE

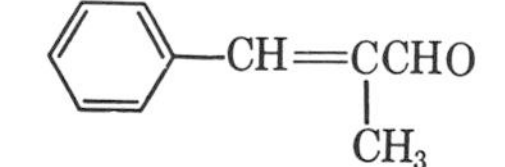

$C_{10}H_{10}O$ Mol. wt. 146.19

DESCRIPTION

A yellow liquid, having a characteristic cinnamon-like odor. It is soluble in most fixed oils and in propylene glycol. It is insoluble in glycerin and in mineral oil.

SPECIFICATIONS

Assay. Not less than 97.0 percent of $C_{10}H_{10}O$.

Refractive index. Between 1.602 and 1.607 at 20°.

Solubility in alcohol. Passes test.

Specific gravity. Between 1.035 and 1.039.

Limits of Impurities

Acid value. Not more than 5.0.

TESTS

Assay. Weigh accurately about 1 gram, and proceed as directed under *Aldehydes*, page 894, using 73.10 as the equivalence factor (*E*) in the calculation.

Refractive index, page 945. Determine with an Abbé or other refractometer of equal or greater accuracy.

Solubility in alcohol. Proceed as directed in the general method, page 899. One ml. dissolves in 2 ml. of 70 percent alcohol, and remains in solution upon dilution to 10 ml.

Specific gravity. Determine by any reliable method (see page 5).

Acid value. Determine as directed in the general method, page 893.

Packaging and storage. Store in full, tight, preferably glass, aluminum, tin-lined, or other suitably lined containers in a cool place protected from light.

Functional use in foods. Flavoring agent.

METHYL CINNAMATE

C_6H_5—CH=CHCOOCH_3

$C_{10}H_{10}O_2$ Mol. wt. 162.19

DESCRIPTION

A white to slightly yellow solid, having a fruity balsamic odor. It is soluble in glycerin, in most fixed oils, in mineral oil, and in propylene glycol.

SPECIFICATIONS

Assay. Not less than 98.0 percent of $C_{10}H_{10}O_2$.

Solidification point. Not lower than 33.8°.

Solubility in alcohol. Passes test.

Limits of Impurities

Acid value. Not more than 2.0.

Arsenic (as As). Not more than 3 parts per million (0.0003 percent).

Chlorinated compounds. Passes test.

Heavy metals (as Pb). Not more than 40 parts per million (0.004 percent).

Lead. Not more than 10 parts per million (0.001 percent).

TESTS

Assay. Weigh accurately about 1 gram, and proceed as directed under *Ester Determination*, page 896, using 81.10 as the equivalence factor (e) in the calculation.

Solidification point. Determine as directed in the general method, page 954.

Solubility in alcohol. Proceed as directed in the general method, page 899. One gram dissolves in 4 ml. of 80 percent alcohol.

Acid value. Determine as directed in the general procedure, page 893.

Arsenic. A *Sample Solution* prepared as directed for organic compounds meets the requirements of the *Arsenic Test*, page 865.

Chlorinated compounds. Proceed as directed in the general method, page 895.

Heavy metals. Prepare and test a 500-mg. sample as directed in *Method II* under the *Heavy Metals Test*, page 920, using 20 mcg. of lead ion (Pb) in the control (*Solution A*).

Lead. A *Sample Solution* prepared as directed for organic compounds meets the requirements of the *Lead Limit Test*, page 929, using 10 mcg. of lead ion (Pb) in the control.

Packaging and storage. Store in full, tight, preferably glass, aluminum, or tin-lined containers in a cool place protected from light. After prolonged storage or repeated melting, the product may have a lowered solidification point and become opalescent.

Functional use in foods. Flavoring agent.

METHYL CYCLOPENTENOLONE

3-Methylcyclopentane-1,2-dione

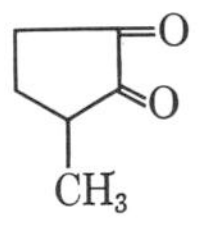

$C_6H_8O_2$ Mol. wt. 112.13

DESCRIPTION

A white crystalline powder with a characteristic nutty odor, and suggestive of a maple-licorice aroma in dilute solution. Soluble in alcohol and in propylene glycol; slightly soluble in most fixed oils. One part dissolves in about 72 parts water.

SPECIFICATIONS

Melting range. Between 104° and 108°.

Solubility in alcohol. Passes test.

Limits of Impurities

Heavy metals (as Pb). Not more than 40 parts per million (0.004 percent).

TESTS

Melting range. Determine as directed in the general procedure, page 931.

Solubility in alcohol. Proceed as directed in the general method, page 899. One gram dissolves in 5 ml. of 90 percent alcohol.

Heavy metals. Prepare and test a 500-mg. sample as directed in *Method II* under the *Heavy Metals Test*, page 920, using 20 mcg. of lead ion (Pb) in the control (*Solution A*).

Packaging and storage. Store in tight containers. Avoid contact with iron.

Functional use in foods. Flavoring agent.

METHYLENE CHLORIDE

Dichloromethane

CH_2Cl_2 Mol. wt. 84.93

DESCRIPTION

A clear, colorless, nonflammable liquid having an odor resembling that of chloroform. It is soluble in about 50 parts of water and is miscible with alcohol, with acetone, with chloroform, with ether, and with carbon tetrachloride. Its refractive index at 20° is about 1.424.

SPECIFICATIONS

Distillation range. Between 39.5° and 40.5°.

Specific gravity. Between 1.318 and 1.323.

Limits of Impurities

Acidity (as HCl). Not more than 10 parts per million (0.001 percent).

Free halogens. Passes test.

Foreign odor. Passes test.

Heavy metals (as Pb). Not more than 1 part per million (0.0001 percent).

Nonvolatile residue. Not more than 20 parts per million (0.002 percent).

Water. Not more than 0.02 percent.

TESTS

Distillation range. Proceed as directed in the general method, page 890.

Specific gravity. Determine by any reliable method (see page 5).

Acidity. Transfer 100 ml. (about 132 grams) of the sample into a separator, add 100 ml. of neutralized water, and shake vigorously for 2 minutes. Allow the layers to separate, transfer the aqueous phase into an Erlenmeyer flask, add 4 drops of bromothymol blue T.S., and titrate with 0.01 *N* sodium hydroxide. Not more than 3.6 ml. is required.

Free halogens. Transfer 10 ml. of the sample to a separator, add 25 ml. of water, and shake vigorously for 1 minute. Allow the layers to separate completely, and then remove and discard the lower sample layer. To the aqueous phase add 1 ml. of potassium iodide T.S. and a few drops of starch T.S., and allow to stand for 5 minutes. A blue color does not develop.

Foreign odor. Pour a few ml. of the sample onto a clean filter paper, observe the odor at once, and then allow to evaporate in air at room temperature. No foreign odor is perceptible after the last traces of the sample have evaporated from the paper.

Heavy metals. Evaporate 15 ml. (about 20 grams) of the sample to dryness in a glass evaporating dish (*Caution:* use hood) on a steam bath. Cool, add 2 ml. of hydrochloric acid, and slowly evaporate to dryness again on the steam bath. Moisten the residue with 1 drop of hydrochloric acid, add 10 ml. of hot water, and digest for 2 minutes. Filter if necessary through a small filter, wash the evaporating dish and the filter with about 10 ml. of water, and dilute to 25 ml. with water. This solution meets the requirements of the *Heavy Metals Test,* page 920, using 20 mcg. of lead ion (Pb) in the control (*Solution A*).

Nonvolatile residue. Evaporate 38 ml. (about 50 grams) of the sample to dryness (*Caution:* use hood) in a tared dish on a steam bath, dry the residue at 105° for 30 minutes, cool, and weigh.

Water. Determine by the *Karl Fischer Titrimetric Method,* page 977.

Packaging and storage. Store in tight containers.

Functional use in foods. Extraction solvent.

METHYL ESTER OF ROSIN, PARTIALLY HYDROGENATED

DESCRIPTION

A light amber-colored liquid resin produced by the esterification of rosin with methanol, followed by partial hydrogenation and purification by steam stripping. It is soluble in acetone and in benzene, but is insoluble in water.

SPECIFICATIONS

Acid value. Between 4 and 8.

Refractive index. Between 1.517 and 1.520 at 20°.

Viscosity. Between 23 and 76 poises.

Limits of Impurities

Arsenic (as As). Not more than 3 parts per million (0.0003 percent).

Heavy metals (as Pb). Not more than 40 parts per million (0.004 percent).

Lead. Not more than 3 parts per million (0.0003 percent).

TESTS

Acid value. Determine as directed in the general procedure, page 945.

Refractive index, page 945. Determine with an Abbé or other refractometer of equal or greater accuracy.

Viscosity. Determine as directed in the general procedure, page 952.

Arsenic. Prepare a *Sample Solution* as directed in the general method under *Chewing Gum Base*, page 877. This solution meets the requirements of the *Arsenic Test*, page 865.

Heavy metals. Prepare and test a 500-mg. sample as directed in *Method II* under the *Heavy Metals Test*, page 920, using 20 mcg. of lead ion (Pb) in the control (*Solution A*).

Lead. Prepare a *Sample Solution* as directed in the general method under *Chewing Gum Base*, page 878. This solution meets the requirements of the *Lead Limit Test*, page 929, using 10 mcg. of lead ion (Pb) in the control.

Packaging and storage. Store in well-closed containers.

Functional use in foods. Masticatory substance in chewing gum base.

METHYL ETHYLCELLULOSE

DESCRIPTION

Methyl ethylcellulose is the methyl ether of ethylcellulose in which both the methyl and ethyl groups are attached to the anhydroglucose units by ether linkages. It occurs as a white or pale cream-colored, fibrous solid or powder. It is practically odorless and disperses in cold water to form aqueous solutions which undergo a reversible transformation from sol to gel upon heating and cooling, respectively.

IDENTIFICATION

A. Add 1 gram to 100 ml. of water. It disperses to form an opalescent, fibrous sol.

B. Heat a few ml. of the sol prepared for *Identification Test A* to about 60°. The sol becomes cloudy and a gelatinous precipitate forms which redissolves upon cooling.

C. The remaining sol from *Identification Test A*, when whipped as for egg white in a kitchen type mixer, produces a stable air/liquid foam.

SPECIFICATIONS

Assay. Not less than 14.5 percent and not more than 19.0 percent of ethoxyl groups ($—OC_2H_5$) and not less than 3.5 percent and not more than 6.5 percent of methoxyl groups ($—OCH_3$).

Viscosity. The apparent viscosity of a solution containing the equivalent of 2.5 grams of dry sample in each 100 grams of solution is not less than 80 percent and not more than 120 percent of that stated on the label or otherwise represented by the vendor. The usual range of viscosity types is between 20 and 60 centipoises.

Limits of Impurities

Arsenic (as As). Not more than 3 parts per million (0.0003 percent).

Heavy metals (as Pb). Not more than 20 parts per million (0.002 percent).

Lead. Not more than 5 parts per million (0.0005 percent).

Loss on drying. *Fibrous form:* not more than 15 percent; *Powdered form:* not more than 10 percent.

Residue on ignition. Not more than 0.6 percent.

TESTS

Assay for ethoxyl groups. Proceed as directed for *Hydroxypropoxyl Determination*, page 923. Each ml. of 0.02 *N* sodium hydroxide is equivalent to 0.9 mg. of ethoxyl groups ($—OC_2H_5$).

Assay for methoxyl groups. Place about 50 mg., previously dried at 105°, in a tared gelatin capsule, and weigh accurately. Proceed as directed for *Methoxyl Determination*, page 935, but calculate the total alkoxyl content as ethoxyl groups ($—OC_2H_5$). Each ml. of 0.1 *N* sodium thiosulfate is equivalent to 0.7510 mg. of ethoxyl groups ($—OC_2H_5$). Now calculate the methoxyl groups ($—OCH_3$) by the formula $(A - B) \times 31/45$, in which A is the total alkoxyl content calculated as $—OC_2H_5$, B is the $—OC_2H_5$ determined in the *Assay for ethoxyl groups*, 31 is the molecular weight of $—OCH_3$, and 45 is the molecular weight of $—OC_2H_5$.

Viscosity. Transfer an accurately weighed sample, equivalent to 5.0 grams on the dried basis, into a 250-ml. beaker. Adjust the rotor of a variable-speed stirrer about an inch above the sample, add 195 ml. of recently boiled and cooled water, and stir at a speed that will avoid undue aeration. Continue the stirring for about 1.5 hours, then either set aside for 3 hours or overnight, or centrifuge to expel any entrapped air. Adjust the temperature to $20° \pm 0.1°$ and determine as directed under *Viscosity of Methylcellulose*, page 971, using a *Viscometer for High Viscosity.*

Arsenic. A *Sample Solution* prepared as directed for organic compounds meets the requirements of the *Arsenic Test*, page 865.

Heavy metals. Prepare and test a 1-gram sample as directed in *Method II* under the *Heavy Metals Test*, page 920, adding 1 ml. of a 1 in 5 hydroxylamine hydrochloride solution to the solution of the residue. Any color does not exceed that produced in a control (*Solution A*) containing 20 mcg. of lead ion (Pb) and 1 ml. of the hydroxylamine hydrochloride solution.

Lead. A *Sample Solution* prepared as directed for organic compounds meets the requirements of the *Lead Limit Test*, page 929, using 5 mcg. of lead ion (Pb) in the control.

Loss on drying, page 931. Dry a 3-gram sample at 105° for 4 hours.

Residue on ignition. Ignite 1 gram as directed in the general method, page 945.

Packaging and storage. Store in well-closed containers.

Functional use in foods. Emulsifier; stabilizer; foaming agent.

METHYL EUGENOL

Eugenyl Methyl Ether; 1,2-Dimethoxy-4-allylbenzene

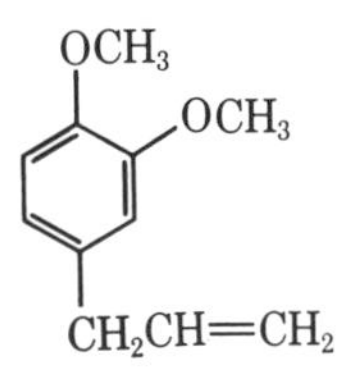

$C_{11}H_{14}O_2$ Mol. wt. 178.23

DESCRIPTION

A colorless to pale yellow liquid having a delicate clove-carnation odor. It is soluble in most fixed oils and in mineral oil, but is insoluble in glycerin and in propylene glycol.

SPECIFICATIONS

Refractive index. Between 1.532 and 1.536 at 20°.

Solubility in alcohol. Passes test.

Specific gravity. Between 1.032 and 1.036.

Limits of Impurities

Eugenol. Not more than 1.0 percent.

TESTS

Refractive index, page 945. Determine with an Abbé or other refractometer of equal or greater accuracy.

Solubility in alcohol. Proceed as directed in the general method, page 899. One ml. dissolves in 2 ml. of 70 percent alcohol, and remains in solution upon dilution to 10 ml.

Specific gravity. Determine by any reliable method (see page 5).

Eugenol. Proceed as directed under *Free Phenols,* page 899, using 82.11 as the equivalence factor (f) in the calculation.

Packaging and storage. Store in full, tight, preferably glass, aluminum, tin-lined, or other suitably lined containers in a cool place protected from light. Iron containers should not be used.

Functional use in foods. Flavoring agent.

METHYL FORMATE

$HCOOCH_3$

$C_2H_4O_2$ Mol. wt. 60.05

DESCRIPTION

A colorless, flammable liquid having a pleasant odor. It is miscible with water and with practically all organic solvents. It boils at about 32°, and its specific gravity is about 0.96. *Caution: Methyl formate is very flammable. Do not use where it may be ignited.*

SPECIFICATIONS

Assay. Not less than 97.0 percent of $C_2H_4O_2$.

Limits of Impurities

Acidity (as formic acid). Not more than 0.04 percent.

Arsenic (as As). Not more than 3 parts per million (0.0003 percent).

Heavy metals (as Pb). Not more than 10 parts per million (0.001 percent).

Nonvolatile residue. Not more than 20 parts per million (0.002 percent).

Water. Not more than 0.05 percent.

TESTS

Assay. Transfer 100.0 ml. of 0.5 *N* sodium hydroxide from a buret into a 100-ml. volumetric flask. For the blank, similarly transfer 100.0 ml. into a 500-ml. Erlenmeyer flask. Weigh the volumetric flask and its contents, pipet 2 ml. of the sample into the flask, mix, and reweigh to obtain the weight of the sample added. Quantitatively transfer the contents of the volumetric flask into a second 500-ml. Erlenmeyer flask with the aid of several small portions of water. Add sufficient water to the blank flask to match the volume of the sample flask, then add phenolphthalein T.S. to each flask, and titrate the excess alkali with 0.5 *N* hydrochloric acid (see page 2). Each ml. of 0.5 *N* hydrochloric acid is equivalent to 30.03 mg. of $C_2H_4O_2$.

Acidity. Transfer 50 ml. of the sample into a 300-ml. Erlenmeyer flask containing 1 ml. of bromothymol blue T.S. and 50 ml. of methanol which has been previously titrated with 0.1 *N* alcoholic potassium hydroxide to the first appearance of a blue color, and titrate the sample with the alkali to the same blue color. Not more than 4.2 ml. is required.

Sample Solution for the Determination of Arsenic and Heavy Metals. Evaporate a 4.2-ml. (4-gram) sample to dryness with 20 mg. of sodium carbonate, heat gently to volatilize any organic matter, and dissolve the residue in 40 ml. of water.

Arsenic. A 10-ml. portion of the *Sample Solution* meets the requirements of the *Arsenic Test*, page 865.

Heavy metals. A 20-ml. portion of the *Sample Solution*, diluted to 25 ml. with water, meets the requirements of the *Heavy Metals Test*, page 920, using 20 mcg. of lead ion (Pb) in the control (*Solution A*).

Nonvolatile residue. Dry a 125-ml. platinum dish at 105° for at least 15 minutes, cool in a desiccator, and weigh. Evaporate four 50-ml. portions of the sample in the dish on a steam bath, but do not allow the dish to go dry between additions of the sample. After all of the sample has evaporated, heat in an oven at 105° for 30 minutes, cool in a desiccator, and weigh. Not more than 3.9 mg. of residue remains.

Water. Determine by the *Karl Fischer Titrimetric Method*, page 977, using 50 ml. of methanol and 50 ml. of the sample. Not more than 24 mg. of water is found.

Packaging and storage. Store in tight containers, and observe the safety precautions stated on the label.

Functional use in foods. Fumigant.

6-METHYL-5-HEPTEN-2-ONE

Methyl Heptenone

$$\begin{array}{l} CH_3C{=}CHCH_2CH_2COCH_3 \\ \quad\;\; | \\ \quad\; CH_3 \end{array}$$

$C_8H_{14}O$ Mol. wt. 126.20

DESCRIPTION

A slightly yellow liquid having a sharp citrus-lemongrass odor. It is miscible with alcohol, with chloroform, and with ether, but it is practically insoluble in water.

SPECIFICATIONS

Assay. Not less than 95.0 percent of $C_8H_{14}O$.

Refractive index. Between 1.438 and 1.442 at 20°.

Solubility in alcohol. Passes test.

Specific gravity. Between 0.846 and 0.851.

TESTS

Assay. Weigh accurately about 700 mg., and proceed as directed under *Aldehyde and Ketones-Hydroxylamine Method*, page 894, using 63.10 as the equivalence factor (E) in the calculation.

Refractive index, page 945. Determine with an Abbé or other refractometer of equal or greater accuracy.

Solubility in alcohol. Proceed as directed in the general method, page 899. One ml. dissolves in 2 ml. of 70 percent alcohol to form a clear solution.

Specific gravity. Determine by any reliable method (see page 5).

Packaging and storage. Store in full, tight, preferably glass, tin-lined or other suitably lined containers in a cool place protected from light.

Functional use in foods. Flavoring agent.

METHYL ISOEUGENOL

4-Allyl-1,2-dimethoxy Benzene; Isoeugenyl Methyl Ether; 4-Propenyl Veratrole

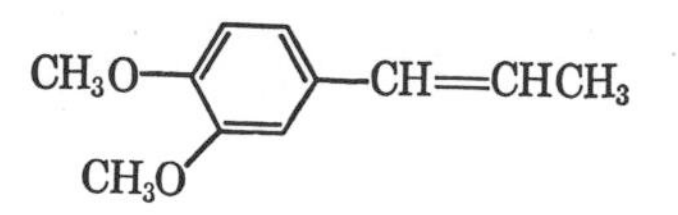

$C_{11}H_{14}O_2$ Mol. wt. 178.23

DESCRIPTION

A colorless to pale yellow liquid, having a delicate clove-carnation odor. It is soluble in most fixed oils. It is practically insoluble in mineral oil, and it is insoluble in glycerin and in propylene glycol.

SPECIFICATIONS

Refractive index. Between 1.566 and 1.569 at 20°.

Solubility in alcohol. Passes test.

Specific gravity. Between 1.047 and 1.053.

Limits of Impurities

Isoeugenol. Not more than 1.0 percent.

TESTS

Refractive index, page 945. Determine with an Abbé or other refractometer of equal or greater accuracy.

Solubility in alcohol. Proceed as directed in the general method, page 899. One ml. dissolves in 2 ml. of 70 percent alcohol, and remains in solution upon dilution to 10 ml.

Specific gravity. Determine by any reliable method (see page 5).

Isoeugenol. Proceed as directed under *Free Phenols,* page 899, using 82.10 as the equivalence factor (f) in the calculation for isoeugenol ($C_{10}H_{12}O_2$).

Packaging and storage. Store in full, tight, preferably glass, aluminum, tin-lined, or other suitably lined containers in a cool place protected from light. Iron containers should not be used.

Functional use in foods. Flavoring agent.

METHYL β-NAPHTHYL KETONE

2′-Acetonaphthone

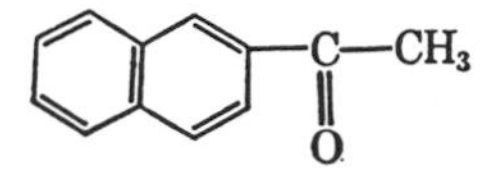

$C_{12}H_{10}O$ Mol. wt. 170.21

DESCRIPTION

A white or nearly white crystalline solid, having an odor suggestive of orange blossom. It is soluble in most fixed oils, and is slightly soluble in mineral oil and in propylene glycol. It is insoluble in glycerin.

SPECIFICATIONS

Assay. Not less than 99.0 percent of $C_{12}H_{10}O$.

Solidification point. Not lower than 53°.

Solubility in alcohol. Passes test.

TESTS

Assay. Weigh accurately about 1.5 grams, and proceed as directed under *Aldehydes and Ketones-Hydroxylamine Method*, page 894, using 85.10 as the equivalence factor (E) in the calculation.

Solidification point. Determine as directed in the general method, page 954.

Solubility in alcohol. Proceed as directed in the general method, page 899. One gram dissolves in 5 ml. of 95 percent alcohol.

Packaging and storage. Store in fiber or pressboard drums, wooden barrels, glass bottles, or suitably lined tin cans in a cool place protected from light.

Functional use in foods. Flavoring agent.

METHYL 2-OCTYNOATE

Methyl Heptine Carbonate

$CH_3(CH_2)_4C{\equiv}CCOOCH_3$

$C_9H_{14}O_2$ Mol. wt. 154.21

DESCRIPTION

A colorless to slightly yellow liquid, having a powerful and unpleasant odor. When diluted its odor is similar to violets. It is soluble in most fixed oils and in mineral oil. It is slightly soluble in propylene glycol, but it is insoluble in glycerin.

SPECIFICATIONS

Assay. Not less than 96.0 percent of $C_9H_{14}O_2$.

Refractive index. Between 1.446 and 1.449 at 20°.

Solubility in alcohol. Passes test.

Specific gravity. Between 0.919 and 0.924.

Limits of Impurities

Acid value. Not more than 1.0.

Chlorinated compounds. Passes test.

TESTS

Assay. Weigh accurately about 1 gram, and proceed as directed under *Ester Determination*, page 896, using 77.11 as the equivalence factor (*e*) in the calculation.

Refractive index, page 945. Determine with an Abbé or other refractometer of equal or greater accuracy.

Solubility in alcohol. Proceed as directed in the general method, page 899. One ml. dissolves in 5 ml. of 70 percent alcohol.

Specific gravity. Determine by any reliable method (see page 5).

Acid value. Determine as directed in the general method, page 893.

Chlorinated compounds. Proceed as directed in the general method, page 895.

Packaging and storage. Store in full, tight, preferably glass, aluminum, tin-lined, or other suitably lined containers in a cool place protected from light.

Functional use in foods. Flavoring agent.

METHYLPARABEN

Methyl *p*-Hydroxybenzoate

$HO-C_6H_4-COOCH_3$

$C_8H_8O_3$ Mol. wt. 152.15

DESCRIPTION

Small, colorless crystals or a white, crystalline powder. It is odorless or has a faint, characteristic odor and a slight, burning taste. One gram dissolves in about 400 ml. of water at 25°, in about 50 ml. of water at 80°, in 2.5 ml. of alcohol, in about 7 ml. of ether, and in about 4 ml. of propylene glycol. It is slightly soluble in glycerin, fixed oils, benzene, and carbon tetrachloride.

IDENTIFICATION

Dissolve 500 mg. in 10 ml. of sodium hydroxide T.S., boil for 30 minutes, allow the solution to evaporate to a volume of about 5 ml., and cool. Acidify the solution with diluted sulfuric acid T.S., collect the crystals on a filter, wash several times with small portions of water, and dry in a desiccator over silica gel. The *p*-hydroxybenzoic acid so obtained melts between 212° and 217° (see page 931).

SPECIFICATIONS

Assay. Not less than 99.0 percent of $C_8H_8O_3$, calculated on the dried basis.

Melting range. Between 125° and 128°.

Limits of Impurities

Acidity. Passes test.

Arsenic (as As). Not more than 3 parts per million (0.0003 percent).

Heavy metals (as Pb). Not more than 10 parts per million (0.001 percent).

Loss on drying. Not more than 0.5 percent.

Residue on ignition. Not more than 0.05 percent.

TESTS

Assay. Transfer about 2 grams, accurately weighed, into a flask, add 40.0 ml. of 1 *N* sodium hydroxide, and rinse the sides of the flask with water. Cover with a watch glass, boil gently for 1 hour, and cool. Add 5 drops of bromothymol blue T.S., and titrate the excess sodium hydroxide with 1 *N* sulfuric acid to pH 6.5 by matching the color of pH 6.5 phosphate buffer (see page 984) containing the same proportion of indicator. Perform a blank determination (see page 2). Each ml. of 1 *N* sodium hydroxide is equivalent to 152.2 mg. of $C_8H_8O_3$.

Melting range. Determine as directed in the general procedure, page 931.

Acidity. Heat 750 mg. with 15 ml. of water at 80° for 1 minute, cool, and filter. The filtrate is acid or neutral to litmus. To 10 ml. of the filtrate add 0.2 ml. of 0.1 *N* sodium hydroxide and 2 drops of methyl red T.S. The solution is yellow, without even a light cast of pink.

Arsenic. A *Sample Solution* prepared as directed for organic compounds meets the requirements of the *Arsenic Test*, page 865.

Heavy metals, page 920. Dissolve 2 grams in 23 ml. of acetone, and add 2 ml. of diluted acetic acid T.S., 2 ml. of water, and 10 ml. of hydrogen sulfide T.S. Any color does not exceed that produced in a control (*Solution A*) made with 23 ml. of acetone, 2 ml. of diluted acetic acid

T.S., 2 ml. of *Standard Lead Solution* (20 mcg. Pb ion), and 10 ml. of hydrogen sulfide T.S.

Loss on drying, page 931. Dry over silica gel for 5 hours.

Residue on ignition. Ignite 4 grams as directed in the general method, page 945.

Packaging and storage. Store in well-closed containers.

Functional use in foods. Preservative; antimicrobial agent.

4-METHYL-2-PENTANONE

Methyl Isobutyl Ketone

$CH_3COCH_2CH(CH_3)_2$

$C_6H_{12}O$ Mol. wt. 100.16

DESCRIPTION

A clear, colorless, mobile liquid having a characteristic ketonic odor. It is miscible in all proportions with alcohol and with ether. One ml. dissolves in about 50 ml. of water.

SPECIFICATIONS

Assay. Not less than 99.0 percent of $C_6H_{12}O$.

Distillation range. Between 114° and 117°.

Specific gravity. Between 0.796 and 0.799.

Limits of Impurities

Acidity (as acetic acid). Not more than 0.01 percent.

Heavy metals (as Pb). Not more than 40 parts per million (0.004 percent).

Water. Not more than 0.1 percent.

TESTS

Assay. Transfer 65 ml. of 0.5 *N* hydroxylamine hydrochloride and 50.0 ml. of 0.5 *N* triethanolamine into a suitable heat-resistant pressure bottle provided with a tight closure that can be securely fastened. Replace the air in the bottle by passing a gentle stream of nitrogen for 2 minutes through a glass tube positioned so that the end is just above the surface of the liquid. To the mixture in the pressure bottle add about 1.2 grams of the sample, contained in a sealed glass ampul and accurately weighed. Add several pieces of 8-mm. glass rod, stopper the bottle, and shake vigorously to break the ampul. Allow the mixture to stand at room temperature for 60 minutes, swirling occasionally. Cool, if necessary, and uncap the bottle cautiously to prevent any loss of the contents. Titrate with 0.5 *N* sulfuric acid to a greenish blue end-point. Perform a residual blank titration (see page 2). Each ml. of 0.5 *N* sulfuric acid is equivalent to 50.08 mg. of $C_6H_{12}O$.

Distillation range. Distil 100 ml. as directed in the general method, page 890.

Specific gravity. Determine by any reliable method (see page 5).

Acidity. Transfer 75 ml. (60 grams) into a 250-ml. Erlenmeyer flask, add a few drops of phenolphthalein T.S., and titrate with 0.1 *N* alcoholic potassium hydroxide to a pink end-point that persists for at least 15 seconds. Each ml. of 0.1 *N* alcoholic potassium hydroxide is equivalent to 6.01 mg. of $C_2H_4O_2$.

Heavy metals. Prepare and test a 500-mg. sample as directed in *Method II* under the *Heavy Metals Test*, page 920, using 20 mcg. of lead ion (Pb) in the control (*Solution A*).

Water. Determine by the *Karl Fischer Titrimetric Method*, page 977, using freshly distilled pyridine instead of methanol as the solvent.

Packaging and storage. Store in tight containers.

Functional use in foods. Flavoring agent.

METHYL PHENYLACETATE

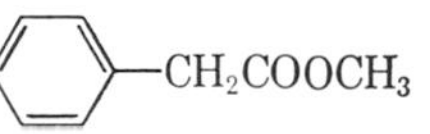

$C_9H_{10}O_2$ Mol. wt. 150.18

DESCRIPTION

A colorless or nearly colorless liquid, having a strong odor suggestive of honey. It is soluble in alcohol and in most fixed oils. It is insoluble in glycerin, in mineral oil, in propylene glycol, and in water.

SPECIFICATIONS

Assay. Not less than 98.0 percent of $C_9H_{10}O_2$.

Refractive index. Between 1.505 and 1.509 at 20°.

Solubility in alcohol. Passes test.

Specific gravity. Between 1.061 and 1.067.

Limits of Impurities

Acid value. Not more than 1.0.

Chlorinated compounds. Passes test.

TESTS

Assay. Weigh accurately about 1 gram, and proceed as directed under *Ester Determination*, page 896, using 75.09 as the equivalence factor (*e*) in the calculation.

Refractive index, page 945. Determine with an Abbé or other refractometer of equal or greater accuracy.

Solubility in alcohol. Proceed as directed in the general method, page 899. One ml. dissolves in 6 ml. of 60 percent alcohol.

Specific gravity. Determine by any reliable method (see page 5).

Acid value. Determine as directed in the general method, page 893.

Chlorinated compounds. Proceed as directed in the general method, page 895.

Packaging and storage. Store in full, tight, preferably glass, aluminum, tin-lined, or other suitably lined, or steel containers in a cool place protected from light.

Functional use in foods. Flavoring agent.

METHYL SALICYLATE

Wintergreen Oil; Gaultheria Oil

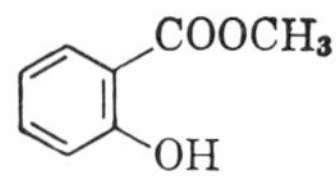

$C_8H_8O_3$ Mol. wt. 152.15

DESCRIPTION

Methyl salicylate is produced synthetically or is obtained by maceration and subsequent distillation with steam from the leaves of *Gualtheria procumbens* L. (Fam. *Ericaceae*) or from the bark of *Betula lenta* L. (Fam. *Betulaceae*). It is a colorless, yellowish, or reddish liquid having the characteristic odor and taste of wintergreen. It boils, with decomposition, between 219° and 224°. It is soluble in alcohol and in glacial acetic acid, and it is very slightly soluble in water.

IDENTIFICATION

Shake 1 drop with about 5 ml. of water, and add 1 drop of ferric chloride T.S. A deep violet color is produced.

SPECIFICATIONS

Assay. Not less than 98.0 percent of $C_8H_8O_3$.

Acid value. Not more than 0.5.

Angular rotation. Synthetic methyl salicylate is optically inactive. Wintergreen oil is slightly levorotatory, exhibiting a rotation of not more than −1.5°.

Refractive index. Between 1.535 and 1.538 at 20°.

Solubility in alcohol. Passes test.

Specific gravity. Between 1.180 and 1.185 for the synthetic product, and between 1.176 and 1.182 for the natural product.

Limits of Impurities

Arsenic (as As). Not more than 3 parts per million (0.0003 percent).

Heavy metals (as Pb). Not more than 40 parts per million (0.004 percent).

Lead. Not more than 10 parts per million (0.001 percent).

TESTS

Assay. Weigh accurately about 2 grams, and proceed as directed under *Ester Determination,* page 896, using 76.08 as the equivalence factor (*e*) in the calculation. Modify the procedure by using 50.0 ml. of 0.5 *N* alcoholic potassium hydroxide and by refluxing on the steam bath for 2 hours.

Acid value. Determine as directed in the general method, page 893, using bromocresol purple T.S. instead of phenolphthalein T.S. as the indicator.

Angular rotation. Determine in a 100-mm. tube as directed under *Optical Rotation,* page 939.

Refractive index, page 945. Determine with an Abbé or other refractometer of equal or greater accuracy.

Solubility in alcohol. Proceed as directed in the general method, page 899. One ml. dissolves in 7 ml. of 70 percent alcohol. The solution of the natural product may have not more than a slight cloudiness.

Specific gravity. Determine by any reliable method (see page 5).

Arsenic. A *Sample Solution* prepared as directed for organic compounds meets the requirements of the *Arsenic Test,* page 865.

Heavy metals. Prepare and test a 500-mg. sample as directed in *Method II* under the *Heavy Metals Test,* page 920, using 20 mcg. of lead ion (Pb) in the control (*Solution A*).

Lead. A *Sample Solution* prepared as directed for organic compounds meets the requirements of the *Lead Limit Test,* page 929, using 10 mcg. of lead ion (Pb) in the control.

Packaging and storage. Store in full, tight containers in a cool place protected from light.

Labeling. Label methyl salicylate to indicate whether it was made synthetically or distilled from either of the plants mentioned in the *Description.*

Functional use in foods. Flavoring agent.

2-METHYLUNDECANAL

Aldehyde C-12 MNA; Methyl *n*-Nonyl Acetaldehyde

$$CH_3(CH_2)_8\underset{\displaystyle CH_3}{\underset{|}{C}}HCHO$$

$C_{12}H_{24}O$ Mol. wt. 184.32

DESCRIPTION

A colorless to slightly yellow liquid having a characteristic fatty odor which, upon dilution, assumes a floral note. It is soluble in fixed oils, in mineral oil, in alcohol, and, sometimes with slight turbidity, in propylene glycol. It is insoluble in glycerin.

SPECIFICATIONS

Assay. Not less than 94.0 percent of $C_{12}H_{24}O$.

Refractive index. Between 1.431 and 1.436 at 20°.

Specific gravity. Between 0.822 and 0.830.

Limits of Impurities

Acid value. Not more than 10.0.

TESTS

Assay. Weigh accurately about 1.2 grams, and proceed as directed under *Aldehydes*, page 894, using 92.16 as the equivalence factor (E) in the calculation.

Refractive index, page 945. Determine with an Abbé or other refractometer of equal or greater accuracy.

Specific gravity. Determine by any reliable method (see page 5).

Acid value. Determine as directed in the general method, page 893.

Packaging and storage. Store in tight, full containers in a cool place protected from light.

Functional use in foods. Flavoring agent.

MINERAL OIL, WHITE

Liquid Petrolatum

DESCRIPTION

A mixture of refined liquid hydrocarbons, essentially paraffinic and naphthenic in nature, obtained from petroleum. It occurs as a colorless, transparent, oily liquid, free or nearly free from fluorescence. It is odorless and tasteless when cold, and develops not more than a faint odor of petroleum when heated. It is insoluble in water and in alcohol, soluble in volatile oils, and is miscible with most fixed oils, but not with castor oil. It may contain any antioxidant permitted in food by the federal Food and Drug Administration, in an amount not greater than that required to produce its intended effect.

SPECIFICATIONS

Specific gravity. Not less than that stated or within the range claimed by the vendor.

Viscosity. Not less than that stated or within the range claimed by the vendor.

Limits of Impurities

Readily carbonizable substances. Passes test.

Sulfur compounds. Passes test.

Ultraviolet absorption (polynuclear hydrocarbons). Passes test.

TESTS

Specific gravity. Determine by any reliable method (see page 5).

Viscosity. Determine by any reliable method.

Readily carbonizable substances. Place 5 ml. of the sample in a glass-stoppered test tube that previously has been rinsed with chromic acid cleaning mixture (200 grams of sodium dichromate dissolved in about 100 ml. of water to which 1500 ml. of sulfuric acid has been added slowly with stirring), then rinsed with water, and dried. Add 5 ml. of sulfuric acid containing between 94.5 and 94.9 percent of H_2SO_4, and heat in a boiling water bath for 10 minutes. After the test tube has been in the bath for 30 seconds, remove it quickly and, while holding the stopper in place, give three vigorous vertical shakes over an amplitude of about 5 inches. Repeat every 30 seconds. Do not keep the test tube out of the bath longer than 3 seconds for each shaking period. At the end of 10 minutes from the time when first placed in the water bath, remove the test tube. The sample remains unchanged in color and the acid does not become darker than standard color produced by mixing in a similar test tube 3 ml. of ferric chloride C.S., 1.5 ml. of cobaltous chloride C.S., and 0.5 ml. of cupric sulfate C.S., this mixture being overlaid with 5 ml. of mineral oil.

Sulfur compounds. Prepare a saturated solution of lead monoxide in a 1 in 5 solution of sodium hydroxide, and mix 2 drops of the clear solution with 4 ml. of the sample and 2 ml. of absolute alcohol. The mixture, after being heated at 70° for 10 minutes and then cooled, is no darker than a blank consisting of 4 ml. of mineral oil and 2 ml. of absolute alcohol.

Ultraviolet absorption. It meets the ultraviolet absorbance specifications required by the federal Food and Drug Administration for mineral oil.

Packaging and storage. Store in tight containers.

Functional use in foods. Binder; defoaming agent; lubricant; release agent; fermentation aid; protective coating.

MONO- AND DIGLYCERIDES

DESCRIPTION

Mono- and diglycerides consist of mixtures of glycerol mono- and diesters, with minor amounts of triesters, of edible fats or oils or edible fat-forming fatty acids. Those commercially available vary in consistency from yellow liquids, through ivory-colored plastics, to hard, ivory-colored solids having a bland odor and taste. They are insoluble in water, but are soluble in alcohol, in ethyl acetate, and in chloroform and other chlorinated hydrocarbons.

SPECIFICATIONS

The following specifications should conform to the representations of the vendor: **1-Monoglyceride content, Total monoglycerides, Hydroxyl value, Iodine value,** and **Saponification value.**

Limits of Impurities

Acid value. Not more than 6.

Arsenic (as As). Not more than 3 parts per million (0.0003 percent).

Free glycerin. Not more than 7 percent.

Heavy metals (as Pb). Not more than 10 parts per million (0.001 percent).

Residue on ignition. Not more than 0.5 percent.

TESTS

1-Monoglyceride content. Determine as directed in the general method, page 907.

Total monoglycerides. Determine as directed in the general method, page 908.

Hydroxyl value. Determine as directed under *Method II* in the general procedure, page 904.

Iodine value. Determine by the *Wijs Method*, page 906.

Saponification value. Determine as directed in the general method, page 914, using about 4 grams, accurately weighed.

Acid value. Determine as directed under *Method II* in the general procedure, page 902.

Arsenic. A *Sample Solution* prepared as directed for organic compounds meets the requirements of the *Arsenic Test*, page 865.

Free glycerin. Determine as directed in the general method, page 904.

Heavy metals. Prepare and test a 2-gram sample as directed in *Method II* under the *Heavy Metals Test,* page 920, using 20 mcg. of lead ion (Pb) in the control (*Solution A*).

Residue on ignition. Ignite 5 grams as directed in the general method, page 945.

Packaging and storage. Store in well-closed containers.

Functional use in foods. Emulsifier; stabilizer.

MONOAMMONIUM L-GLUTAMATE

Monoammonium Glutamate; Ammonium Glutamate

$C_5H_{12}N_2O_4 . H_2O$ Mol. wt. 182.18

DESCRIPTION

A white, practically odorless, free-flowing, crystalline powder. It is freely soluble in water but is practically insoluble in common organic solvents.

IDENTIFICATION

A. To 1 ml. of a 1 in 30 solution add 1 ml. of triketohydrindene hydrate T.S. and 100 mg. of sodium acetate, and heat in a boiling water bath for 10 minutes. An intense violet-blue color is formed.

B. To 10 ml. of a 1 in 10 solution add 5.6 ml. of 1 *N* hydrochloric acid. A white crystalline precipitate of glutamic acid forms on standing. When 6 ml. of 1 *N* hydrochloric acid is added to the turbid solution, the glutamic acid dissolves on stirring.

C. A 1 in 10 solution gives positive tests for *Ammonium,* page 926.

SPECIFICATIONS

Assay. Not less than 99.0 percent of $C_5H_{12}N_2O_4 . H_2O$.

Specific rotation, $[\alpha]^{25°}_{546.1\ m\mu}$. Between +30.1° and +31.6°.

pH of a 1 in 20 solution. Between 6.0 and 7.0.

Limits of Impurities

Arsenic (as As). Not more than 3 parts per million (0.0003 percent).

Heavy metals (as Pb). Not more than 20 parts per million (0.002 percent).

Lead. Not more than 10 parts per million (0.001 percent).

Loss on drying. Not more than 0.5 percent.

Residue on ignition. Not more than 0.1 percent.

TESTS

Assay. Dissolve about 250 mg., accurately weighed, in 100 ml. of glacial acetic acid. A few drops of water may be added prior to the addition of the acetic acid to effect faster dissolution of the sample. Titrate with 0.1 *N* perchloric acid in glacial acetic acid, determining the end-point potentiometrically. Each ml. of 0.1 *N* perchloric acid is equivalent to 9.109 mg. of $C_5H_{12}N_2O_4 \cdot H_2O$.

Specific rotation. Determine in a solution containing 14.6 grams in sufficient 2.3 *N* hydrochloric acid to make 100 ml.

pH of a 1 in 20 solution. Determine by the *Potentiometric Method,* page 941.

Arsenic. A *Sample Solution* prepared as directed for organic compounds meets the requirements of the *Arsenic Test,* page 865.

Heavy metals. Prepare and test a 1-gram sample as directed in *Method II* under the *Heavy Metals Test,* page 920, using 20 mcg. of lead ion (Pb) in the control (*Solution A*).

Lead. A *Sample Solution* prepared as directed for organic compounds meets the requirements of the *Lead Limit Test,* page 929, using 10 mcg. of lead ion (Pb) in the control.

Loss on drying, page 931. Dry at 50° for 4 hours.

Residue on ignition, page 945. Ignite 1 gram as directed in the general method.

Packaging and storage. Store in tight containers.

Functional use in foods. Flavor enhancer; salt substitute.

MONOGLYCERIDE CITRATE

DESCRIPTION

A mixture of glyceryl monooleate and its citric acid monoester, manufactured by the reaction of glyceryl monooleate with citric acid under controlled conditions. It occurs as a soft, white to ivory colored, waxy solid having a lard-like consistency and a bland odor and taste. It is soluble in most common fat solvents and in alcohol, but is insoluble in water.

SPECIFICATIONS

Acid value. Between 70 and 100.

Total citric acid. Between 14.0 percent and 17.0 percent.

Saponification value. Between 260 and 265.

Limits of Impurities

Arsenic (as As). Not more than 3 parts per million (0.0003 percent).

Heavy metals (as Pb). Not more than 10 parts per million (0.001 percent).

Residue on ignition. Not more than 0.3 percent.

Water. Not more than 0.2 percent.

TESTS

Acid value. Determine as directed under *Method II* in the general procedure, page 902.

Total citric acid

Standard Solution. Transfer about 35 mg. of sodium citrate dihydrate, accurately weighed, into a 100-ml. volumetric flask, dissolve and dilute to volume with water, and mix. Calculate the concentration (C), in mcg. per ml., of citric acid in the final solution by the formula $(1000 \times 0.6533W)/(100)$, in which W is the weight, in mg., of the sodium citrate dihydrate taken, and 0.6533 is a factor converting sodium citrate dihydrate to citric acid.

Sample Solution. Transfer about 150 mg. of the sample, accurately weighed, into a saponification flask, add 50 ml. of 4 percent alcoholic potassium hydroxide solution, and reflux for 1 hour. Acidify the reaction mixture with hydrochloric acid to a pH of 2.8–3.2, transfer to a 400-ml. beaker, and evaporate to dryness on a steam bath. Quantitatively transfer the contents of the beaker into a separator, using no more than 50 ml. of water, and then extract with three 50-ml. portions of petroleum ether (b.p. 30–60°), discarding the extracts. Transfer the water layer to a 100-ml. volumetric flask, dilute to volume with water, and mix.

Procedure. Pipet 2.0 ml. each of the *Standard Solution* and of the *Sample Solution* into separate 40-ml. graduated centrifuge tubes, and to each tube add 2 ml. of dilute sulfuric acid (1 in 2) and 11 ml. of water. Boil for 3 minutes, cool, and add 5 ml. of bromine T.S. to each tube. Dilute to the 20-ml. mark, allow to stand for 10 minutes, and centrifuge. Transfer 4.0 ml. of each solution into separate 19 × 110-mm. test tubes, add 1 ml. of water, 0.5 ml. of dilute sulfuric acid (1 in 2), and 0.3 ml. of 1 *M* potassium bromide, and shake. Add 0.3 ml. of 1.5 *N* potassium permanganate, shake, and allow to stand for 2 minutes. Add 1 ml. of a saturated solution of ferrous sulfate, shake, allow to stand for 2 minutes, and then dilute to 10 ml. with water. Add 10.0 ml. of *n*-hexane (previously washed with sulfuric acid, followed by a water wash, and then dried over anhydrous sodium sulfate), shake vigorously for 2 minutes, and then centrifuge at a low speed for 1 minute. Transfer 5.0 ml. of the hexane extract into a 20 × 145-mm. tube containing 10.0 ml. of sodium sulfide solution (4 grams of $Na_2S \cdot 9H_2O$ in each 100 ml. of water), and shake vigorously briefly (3 oscillations only). Centrifuge the mixture at low speed for 1 minute. Immediately determine the absorbance of each aqueous layer in a 1-cm. cell at 450 mμ with a suitable spectrophotometer, using a reagent blank in the reference cell. Calculate the quantity, in mg., of citric acid in the sample taken by the formula $0.1C \times A_U/A_S$, in which C is as defined under *Standard Solution*, A_U is the absorbance of the final solution from the *Sample Solution*, and A_S is the absorbance of the final solution from the *Standard Solution*.

Saponification value. Determine as directed in the general method, page 914.

Arsenic. A *Sample Solution* prepared as directed for organic compounds meets the requirements of the *Arsenic Test*, page 865.

Heavy metals. Prepare and test a 2-gram sample as directed in *Method II* under the *Heavy Metals Test*, page 920, using 20 mcg. of lead ion (Pb) in the control (*Solution A*).

Residue on ignition. Ignite 1 gram as directed in the general method, page 945.

Water. Determine by the *Karl Fischer Titrimetric Method*, page 977.

Packaging and storage. Store in well-closed containers.

Functional use in foods. Synergist and solubilizer for antioxidants.

MONOPOTASSIUM L-GLUTAMATE

Monopotassium Glutamate; Potassium Glutamate; MPG

$C_5H_8KNO_4 . H_2O$ Mol. wt. 203.24

DESCRIPTION

A white, practically odorless, free-flowing, crystalline powder. It is hygroscopic, is freely soluble in water, and is slightly soluble in alcohol.

IDENTIFICATION

A. To 1 ml. of a 1 in 30 solution add 1 ml. of triketohydrindene hydrate T.S. and 100 mg. of sodium acetate, and heat in a boiling water bath for 10 minutes. An intense, violet-blue color is formed.

B. To 10 ml. of a 1 in 10 solution add 5.6 ml. of 1 *N* hydrochloric acid. A white, crystalline precipitate of glutamic acid forms on standing. When 6 ml. of 1 *N* hydrochloric acid is added to the turbid solution, the glutamic acid dissolves on stirring.

C. A 1 in 10 solution gives positive tests for *Potassium*, page 928.

SPECIFICATIONS

Assay. Not less than 99.0 percent of $C_5H_8KNO_4.H_2O$.

Specific rotation. $[\alpha]^{25°}_{546.1\ m\mu}$: Between +27.7° and +28.3°; $[\alpha]^{20°}_{D}$: between +22.5° and +24.0°.

pH of a 1 in 50 solution. Between 6.7 and 7.3.

Limits of Impurities

Arsenic (as As). Not more than 3 parts per million (0.0003 percent).

Chloride. Not more than 0.1 percent.

Heavy metals (as Pb). Not more than 20 parts per million (0.002 percent).

Lead. Not more than 10 parts per million (0.001 percent).

Loss on drying. Not more than 0.1 percent.

TESTS

Assay. Dissolve about 250 mg., accurately weighed, in 100 ml. of glacial acetic acid. A few drops of water may be added prior to the addition of the acetic acid to effect faster dissolution of the sample. Titrate with 0.1 *N* perchloric acid in glacial acetic acid, determining the end-point potentiometrically. Each ml. of 0.1 *N* perchloric acid is equivalent to 10.16 mg. of $C_5H_8KNO_4.H_2O$.

Specific rotation, page 939. $[\alpha]^{25°}_{546.1\ m\mu}$: Determine in a solution containing 16.3 grams in sufficient 2.3 *N* hydrochloric acid to make 100 ml.; $[\alpha]^{20°}_{D}$: determine in a solution containing 10 grams in sufficient 2 *N* hydrochloric acid to make 100 ml.

pH of a 1 in 50 solution. Determine by the *Potentiometric Method*, page 941.

Arsenic. A *Sample Solution* prepared as directed for organic compounds meets the requirements of the *Arsenic Test*, page 865.

Chloride, page 879. Any turbidity produced by a 20-mg. sample does not exceed that shown in a control containing 20 mcg. of chloride ion (Cl).

Heavy metals. Prepare and test a 1-gram sample as directed in *Method II* under the *Heavy Metals Test*, page 920, using 20 mcg. of lead ion (Pb) in the control (*Solution A*).

Lead. A *Sample Solution* prepared as directed for organic compounds meets the requirements of the *Lead Limit Test*, page 929, using 10 mcg. of lead ion (Pb) in the control.

Loss on drying, page 931. Dry at 60° in a vacuum for 2 hours.

Packaging and storage. Store in tight containers.

Functional use in foods. Flavor enhancer; salt substitute.

MONOSODIUM L-GLUTAMATE

Monosodium Glutamate; Sodium Glutamate; MSG

$C_5H_8NNaO_4.H_2O$ Mol. wt. 187.13

DESCRIPTION

White, practically odorless, free-flowing crystals or crystalline powder. It is freely soluble in water, and is sparingly soluble in alcohol. It may have either a slightly sweet or a slightly salty taste.

IDENTIFICATION

A. To 1 ml. of a 1 in 30 solution add 1 ml. of triketohydrindene hydrate T.S. and 100 mg. of sodium acetate, and heat in a boiling water bath for 10 minutes. An intense violet-blue color is formed.

B. To 10 ml. of a 1 in 10 solution add 5.6 ml. of 1 *N* hydrochloric acid. A white crystalline precipitate of glutamic acid forms on standing. When 6 ml. of 1 *N* hydrochloric acid is added to the turbid solution, the glutamic acid dissolves on stirring.

C. A 1 in 10 solution gives positive tests for *Sodium*, page 928.

SPECIFICATIONS

Assay. Not less than 99.0 percent of $C_5H_8NNaO_4.H_2O$.

Clarity and color of solution. Passes test.

Specific rotation. $[\alpha]^{25°}_{546.1\ m\mu}$: Between +29.7° and +30.2°; $[\alpha]^{20°}_{D}$: between +24.8° and +25.3°.

pH of a 1 in 20 solution. Between 6.7 and 7.2.

Limits of Impurities

Arsenic (as As). Not more than 3 parts per million (0.0003 percent).

Chloride. Not more than 0.2 percent.

Heavy metals (as Pb). Not more than 20 parts per million (0.002 percent).

Lead. Not more than 10 parts per million (0.001 percent).

Loss on drying. Not more than 0.3 percent.

TESTS

Assay. Dissolve about 250 mg., accurately weighed, in 100 ml. of glacial acetic acid. A few drops of water may be added prior to the addition of the acetic acid to effect faster dissolution of the sample. Titrate with 0.1 *N* perchloric acid in glacial acetic acid, determining the end-point potentiometrically. Each ml. of 0.1 *N* perchloric acid is equivalent to 9.356 mg. of $C_5H_8NNaO_4.H_2O$.

Clarity and color of solution. A solution of 1 gram of the sample in 10 ml. of water is colorless and has no more turbidity (see page 959) than a standard mixture prepared as follows: Dilute 0.2 ml. of

Standard chloride solution (see page 879) to 20 ml. with water, add 1 ml. of dilute nitric acid (1 in 3), 0.2 ml. of a 1 in 50 solution of dextrin, and 1 ml. of a 1 in 50 solution of silver nitrate, mix, and allow to stand for 15 minutes.

Specific rotation, page 939. $[\alpha]^{25°}_{546.1\ m\mu}$: Determine in a solution containing 15 grams in sufficient 2.3 *N* hydrochloric acid to make 100 ml.; $[\alpha]^{20°}_{D}$: determine in a solution containing 10 grams in sufficient 2 *N* hydrochloric acid to make 100 ml.

pH of a 1 in 20 solution. Determine by the *Potentiometric Method*, page 941.

Arsenic. A *Sample Solution* prepared as directed for organic compounds meets the requirements of the *Arsenic Test*, page 865.

Chloride, page 879. Any turbidity produced by a 10-mg. sample does not exceed that shown in a control containing 20 mcg. of chloride ion (Cl).

Heavy metals. Prepare and test a 1-gram sample as directed in *Method II* under the *Heavy Metals Test*, page 920, using 20 mcg. of lead ion (Pb) in the control (*Solution A*).

Lead. A *Sample Solution* prepared as directed for organic compounds meets the requirements of the *Lead Limit Test*, page 929, using 10 mcg. of lead ion (Pb) in the control.

Loss on drying, page 931. Dry at 60° in a vacuum for 2 hours.

Packaging and storage. Store in tight containers.

Functional use in foods. Flavor enhancer.

MYRISTIC ACID

Tetradecanoic Acid

$CH_3(CH_2)_{12}COOH$

$C_{14}H_{28}O_2$ Mol. wt. 228.38

DESCRIPTION

A solid organic acid obtained from coconut oil and other fats. It occurs as a hard, white or faintly yellowish, somewhat glossy and crystalline solid, or as a white or yellowish white powder. Myristic acid is practically insoluble in water, but is soluble in alcohol, in chloroform, and in ether.

SPECIFICATIONS

Acid value. Between 242 and 249.

Saponification value. Between 242 and 251.

Titer (Solidification Point). Between 48° and 55.5°.

Limits of Impurities

Arsenic (as As). Not more than 3 parts per million (0.0003 percent).

Heavy metals (as Pb). Not more than 10 parts per million (0.001 percent).

Iodine value. Not more than 1.0.

Residue on ignition. Not more than 0.1 percent.

Unsaponifiable matter. Not more than 1 percent.

Water. Not more than 0.2 percent.

TESTS

Acid value. Determine as directed under *Method I* in the general procedure, page 902.

Saponification value. Determine as directed in the general method, page 914, using about 3 grams, accurately weighed.

Titer (Solidification Point). Determine as directed under *Solidification Point*, page 954.

Arsenic. A *Sample Solution* prepared as directed for organic compounds meets the requirements of the *Arsenic Test*, page 865.

Heavy metals. Prepare and test a 2-gram sample as directed in *Method II* under the *Heavy Metals Test*, page 920, using 20 mcg. of lead ion (Pb) in the control (*Solution A*).

Iodine value. Determine by the *Wijs Method*, page 906.

Residue on ignition, page 945. Ignite 2 grams as directed in the general method.

Unsaponifiable matter, page 915. Determine as directed in the general method.

Water. Determine by the *Karl Fischer Titrimetric Method*, page 977.

Packaging and storage. Store in well-closed containers.

Functional use in foods. Component in the manufacture of other food grade additives; defoaming agent.

MYRRH OIL

DESCRIPTION

The volatile oil obtained by steam distillation from myrrh gum obtained from several species of *Commiphora* (Fam. *Burseraceae*). It is a light brown or green liquid, having the characteristic odor of the gum. It is soluble in most fixed oils, but is only slightly soluble in mineral oil. It is insoluble in glycerin and in propylene glycol. It becomes darker in color and more viscous under the influence of air and light.

SPECIFICATIONS

Acid value. Between 2 and 13.

Angular rotation. Between −60° and −83°.

Refractive index. Between 1.519 and 1.528 at 20°.

Saponification value. Between 9 and 35.

Solubility in alcohol. Passes test.

Specific gravity. Between 0.985 and 1.014.

TESTS

Acid value. Determine as directed in the general method, page 893.

Angular rotation. Determine in a 100-mm. tube as directed under *Optical Rotation*, page 939.

Refractive index, page 945. Determine with an Abbé or other refractometer of equal or greater accuracy.

Saponification value. Determine as directed in the general method, page 896, using about 5 grams, accurately weighed.

Solubility in alcohol. Proceed as directed in the general method, page 899. One ml. dissolves in 10 ml. of 90 percent alcohol, occasionally with opalescence or turbidity.

Specific gravity. Determine by any reliable method (see page 5).

Packaging and storage. Store in full, tight, preferably glass containers in a cool place protected from light.

Functional use in foods. Flavoring agent.

NEROL

cis-3,7-Dimethyl-2,6-octadien-1-ol

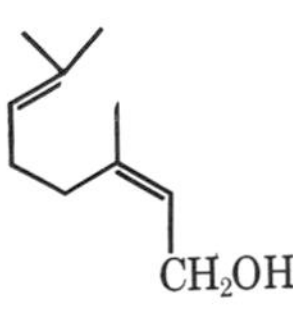

$C_{10}H_8O$ Mol. wt. 154.25

DESCRIPTION

A colorless liquid having a fresh, sweet rose-like character. It is miscible with alcohol, with chloroform, and with ether, but it is practically insoluble in water. Commercial grades of nerol that conform to these specifications may contain varying amounts of geraniols and sometimes other terpenic alcohols along with the specific chemical whose structural formula is shown above.

SPECIFICATIONS

Assay. Not less than 95.0 percent of total alcohols, calculated as nerol ($C_{10}H_{18}O$).

Refractive index. Between 1.473 and 1.478 at 20°.

Solubility in alcohol. Passes test.

Specific gravity. Between 0.875 and 0.880.

TESTS

Assay. Proceed as directed under *Total Alcohols,* page 893. Weigh accurately about 1.2 grams of the acetylated alcohol for the saponification, and use 77.13 as the equivalence factor (e) in the calculation.

Refractive index, page 945. Determine with an Abbé or other refractometer of equal or greater accuracy.

Solubility in alcohol. Proceed as directed in the general method, page 899. One ml. dissolves in 9 ml. of 50 percent alcohol to form a clear solution.

Specific gravity. Determine by any reliable method (see page 5).

Packaging and storage. Store in full, tight, preferably glass, tin-lined or other suitably lined containers in a cool place protected from light.

Functional use in foods. Flavoring agent.

NEROLIDOL

3,7,11-Trimethyl-1,6,10-dodecatrien-3-ol

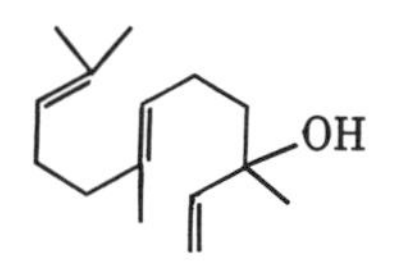

$C_{15}H_{26}O$ Mol. wt. 222.37

DESCRIPTION

A colorless to straw colored liquid, having a faint floral odor suggestive of rose and apple. It may be isolated from suitable essential oils, or prepared synthetically. It is soluble in most fixed oils, in mineral oil, and in propylene glycol. It is insoluble in glycerin.

SPECIFICATIONS

Assay. Not less than 90.0 percent of total alcohols, calculated as nerolidol ($C_{15}H_{26}O$).

Angular rotation. *Natural:* Between +11° and +14°; *Synthetic:* optically inactive.

Refractive index. Between 1.478 and 1.483 at 20°.

Solubility in alcohol. Passes test.

Specific gravity. Between 0.870 and 0.880.

Limits of Impurities

Esters. Not more than 0.5 percent, calculated as nerolidyl acetate ($C_{17}H_{29}O_2$).

TESTS

Assay. Proceed as directed under *Linalool Determination*, page 897, using 1.5 grams of the dried acetylated oil, and reflux with 0.5 *N* alcoholic potassium hydroxide for 6 hours. Calculate the total alcohols by the formula: $[100(b - S) \times 111.19]/[W - 0.042(b - S)]$, in which b is the number of ml. of 0.5 *N* hydrochloric acid consumed in the residual titration, S is the number of ml. of 0.5 *N* hydrochloric acid consumed in the titration of the sample, and W is the weight, in mg., of the acetylated oil sample.

Angular rotation. Determine in a 100-mm. tube as directed under *Optical Rotation*, page 939.

Refractive index, page 945. Determine with an Abbé or other refractometer of equal or greater accuracy.

Solubility in alcohol. Proceed as directed in the general method, page 899. One ml. dissolves in 4 ml. of 70 percent alcohol.

Specific gravity. Determine by any reliable method (see page 5).

Esters. Weigh accurately about 10 grams, and proceed as directed under *Ester Determination*, page 896, using 132.7 as the equivalence factor (e) in the calculation for nerolidyl acetate ($C_{17}H_{29}O_2$).

Packaging and storage. Store in full, tight, preferably glass, stainless steel, tin, or suitably lined containers in a cool place protected from light.

Functional use in foods. Flavoring agent.

NIACIN

Nicotinic Acid; 3-Pyridinecarboxylic Acid

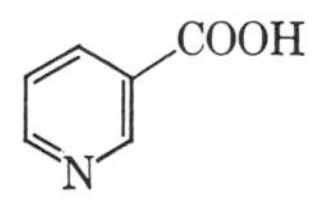

$C_6H_5NO_2$ Mol. wt. 123.11

DESCRIPTION

White or light yellow crystals or a crystalline powder. It is odorless or has a slight odor. One gram dissolves in 60 ml. of water. It is freely soluble in boiling water and in boiling alcohol, and also in solutions of alkali hydroxides and carbonates. It is almost insoluble in ether.

IDENTIFICATION

A. Triturate a sample with twice its weight of 2,4-dinitrochlorobenzene. Gently heat about 10 mg. of the mixture in a test tube until melted, and continue the heating for a few seconds longer. Cool, and add 3 ml. of alcoholic potassium hydroxide T.S. A deep red to deep wine-red color is produced.

B. Dissolve about 50 mg. in 20 ml. of water, neutralize to litmus paper with 0.1 *N* sodium hydroxide, and add 3 ml. of cupric sulfate T.S. A blue precipitate gradually forms.

C. The infrared absorption spectrum of a potassium bromide dispersion of the sample, previously dried at 105° for 1 hour, exhibits maxima only at the same wavelengths as that of a similar preparation of N.F. Niacin Reference Standard.

D. Determine the absorbance of a solution of the sample containing 20 mcg. in each ml. of water, in a 1-cm. cell at 237 mμ and 262 mμ, using water as the blank. The ratio A_{237}/A_{262} is between 0.35 and 0.39.

SPECIFICATIONS

Assay. Not less than 99.5 percent and not more than 101.0 percent of $C_6H_5NO_2$, calculated on the dried basis.

Melting range. Between 234° and 238°.

Limits of Impurities

Heavy metals (as Pb). Not more than 20 parts per million (0.002 percent).

Loss on drying. Not more than 1 percent.

Residue on ignition. Not more than 0.1 percent.

TESTS

Assay. Dissolve about 300 mg., accurately weighed, in about 50 ml. of water, add phenolphthalein T.S., and titrate with 0.1 *N* sodium hydroxide. Perform a blank determination (see page 2). Each ml. of 0.1 *N* sodium hydroxide is equivalent to 12.31 mg. of $C_6H_5NO_2$.

Melting range. Determine as directed in the general procedure, page 931.

Heavy metals. Mix 1 gram with 2 ml. of diluted acetic acid T.S., add water to make 25 ml., heat gently until solution is complete, and cool. This solution meets the requirements of the *Heavy Metals Test*, page 920, using 20 mcg. of lead ion (Pb) in the control (*Solution A*).

Loss on drying, page 931. Dry at 105° for 1 hour.

Residue on ignition, page 945. Ignite 1 gram as directed in the general method.

Packaging and storage. Store in well-closed containers.

Functional use in foods. Nutrient; dietary supplement.

NIACINAMIDE

Nicotinamide

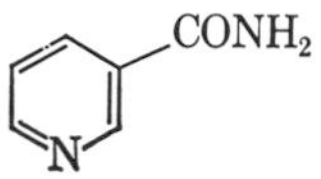

$C_6H_6N_2O$ Mol. wt. 122.13

DESCRIPTION

A white, crystalline powder. It is odorless or nearly so, and has a bitter taste. Its solutions are neutral to litmus. One gram dissolves in about 1 ml. of water, in about 1.5 ml. of alcohol, and in about 10 ml. of glycerin.

IDENTIFICATION

A. Transfer about 20 mg. to a 1000-ml. volumetric flask, dissolve and dilute with water to volume, mix, and determine the absorbance of the solution in 1-cm. cells at 245 mμ and at 262 mμ with a suitable spectrophotometer, using water as the blank. The ratio A_{245}/A_{262} is 0.65 ± 0.02.

B. To 20 mg. in a test tube add 1 pellet of sodium hydroxide T.S., and heat gently over an open flame. The odor of ammonia is perceptible. Upon heating more vigorously, the odor of pyridine is perceptible.

SPECIFICATIONS

Assay. Not less than 98.5 percent and not more than the equivalent of 101.0 percent of $C_6H_6N_2O$ after drying.

Melting range. Between 128° and 131°.

Limits of Impurities

Heavy metals (as Pb). Not more than 30 parts per million (0.003 percent).

Loss on drying. Not more than 0.5 percent.

Readily carbonizable substances. Passes test.

Residue on ignition. Not more than 0.1 percent.

TESTS

Assay. Dissolve about 300 mg., previously dried over silica gel for 4 hours and accurately weighed, in 20 ml. of glacial acetic acid, warming slightly if necessary to effect solution. Add 100 ml. of benzene and 2 drops of crystal violet T.S., and titrate with 0.1 *N* perchloric acid. Perform a blank determination (see page 2) and make any necessary correction. Each ml. of 0.1 *N* perchloric acid is equivalent to 12.21 mg. of $C_6H_6N_2O$.

Melting range. Determine as directed in the general procedure, page 931.

Heavy metals. Dissolve 667 mg. in 10 ml. of water, add 5 ml. of 1 *N* hydrochloric acid, and dilute to 25 ml. with water. This solution meets the requirements of the *Heavy Metals Test*, page 920, using 20 mcg. of lead ion (Pb) in the control (*Solution A*).

Loss on drying, page 931. Dry over silica gel for 4 hours.

Readily carbonizable substances, page 945. Dissolve 200 mg. in 5 ml. of sulfuric acid T.S. The solution has no more color than *Matching Fluid A*.

Residue on ignition, page 945. Ignite 1 gram as directed in the general method.

Packaging and storage. Store in tight containers.

Functional use in foods. Nutrient; dietary supplement.

NIACINAMIDE ASCORBATE

DESCRIPTION

A complex of ascorbic acid ($C_6H_8O_6$) and niacinamide ($C_6H_6N_2O$). It occurs as a lemon-yellow colored powder which is odorless or has a very slight odor. It may gradually darken upon exposure to air. One gram is soluble in 3.5 ml. of water and in about 20 ml. of alcohol. It is very slightly soluble in chloroform and in ether, sparingly soluble in glycerin, and practically insoluble in benzene.

IDENTIFICATION

A. A 1 in 50 solution responds to the *Identification tests* under *Ascorbic Acid*, page 66.

B. An 80-mg. sample responds to *Identification test B* under *Niacinamide*, page 551.

SPECIFICATIONS

Assay. Not less than 73.5 percent of ascorbic acid ($C_6H_8O_6$) and not less than 24.5 percent of niacinamide ($C_6H_6N_2O$), calculated on the anhydrous basis. The total of ascorbic acid and niacinamide is not less than 99.0 percent.

Melting range. Between 141° and 145°.

Limits of Impurities

Heavy metals (as Pb). Not more than 25 parts per million (0.0025 percent).

Loss on drying. Not more than 0.5 percent.

Residue on ignition. Not more than 0.1 percent.

TESTS

Assay for ascorbic acid. Proceed as directed in the *Assay* under *Ascorbic Acid*, page 66.

Assay for niacinamide. Proceed as directed in the *Assay* under *Niacinamide*, page 551, but use an undried sample and make the calculation on the anhydrous basis.

Melting range, page 931. Determine as directed in *Procedure for Class Ia*.

Heavy metals. Prepare and test an 800-mg. sample as directed in *Method II* under the *Heavy Metals Test*, page 920, using 20 mcg. of lead ion (Pb) in the control (*Solution A*).

Loss on drying, page 931. Dry to constant weight at 75°.

Residue on ignition, page 945. Ignite 2 grams as directed in the general method.

Packaging and storage. Store in tight, light-resistant containers.

Functional use in foods. Nutrient; dietary supplement.

γ-NONALACTONE

Aldehyde C-18 (So-called)

$$CH_3(CH_2)_4\underset{|}{C}HCH_2\underset{|}{C}H_2$$
$$O\text{———}C{=}O$$

$C_9H_{16}O_2$ Mol. wt. 156.23

DESCRIPTION

A nearly colorless to yellow liquid having a strong odor suggestive of coconut. It is soluble in most fixed oils, in mineral oil, and in propylene glycol. It is practically insoluble in glycerin.

SPECIFICATIONS

Assay. Not less than 98.0 percent of $C_9H_{16}O_2$.

Refractive index. Between 1.446 and 1.450 at 20°.

Solubility in alcohol. Passes test.

Specific gravity. Between 0.958 and 0.966.

Limits of Impurities

Acid value. Not more than 2.0.

TESTS

Assay. Weigh accurately about 1 gram, and proceed as directed under *Ester Determination*, page 896, using 78.12 as the equivalence factor (*e*) in the calculation. Correct the number of ml. of 0.5 *N* alcoholic potassium hydroxide consumed in the saponification for the acid value.

Refractive index, page 945. Determine with an Abbé or other refractometer of equal or greater accuracy.

Solubility in alcohol. Proceed as directed in the general method, page 899. One ml. dissolves in 5 ml. of 50 percent alcohol.

Specific gravity. Determine by any reliable method (see page 5).

Acid value. Determine as directed in the general method, page 893.

Packaging and storage. Store in full, tight, preferably glass, aluminum, or tin-lined containers in a cool place protected from light.

Functional use in foods. Flavoring agent.

NONANAL

Aldehyde C-9; Pelargonic Aldehyde

$CH_3(CH_2)_7CHO$

$C_9H_{18}O$ Mol. wt. 142.24

DESCRIPTION

A colorless to light yellow liquid having a strong fatty odor which develops an orange and rose character on being diluted. It is soluble in alcohol, in most fixed oils, in mineral oil, and in propylene glycol. It is practically insoluble in glycerin.

SPECIFICATIONS

Assay. Not less than 92.0 percent of $C_9H_{18}O$.

Refractive index. Between 1.422 and 1.429 at 20°.

Specific gravity. Between 0.820 and 0.830.

Limits of Impurities

Acid value. Not more than 10.0.

TESTS

Assay. Weigh accurately about 1.5 grams, and proceed as directed under *Aldehydes*, page 894, using 71.12 as the equivalence factor (E) in the calculation.

Refractive index, page 945. Determine with an Abbé or other refractometer of equal or greater accuracy.

Specific gravity. Determine by any reliable method (see page 5).

Acid value. Determine as directed in the general method, page 893.

Packaging and storage. Store in full, tight, preferably glass, or aluminum containers in a cool place protected from light. Do not store for long periods of time. Dilution with alcohol improves stability during storage.

Functional use in foods. Flavoring agent.

NONYL ACETATE

$CH_3COO(CH_2)_8CH_3$

$C_{11}H_{22}O_2$ Mol. wt. 186.30

DESCRIPTION

A colorless liquid having a floral-fruity odor. It is freely soluble in absolute alcohol and in ether, but insoluble in water.

SPECIFICATIONS

Assay. Not less than 97.0 percent of $C_{11}H_{22}O_2$.

Refractive index. Between 1.422 and 1.426 at 20°.

Solubility in alcohol. Passes test.

Specific gravity. Between 0.864 and 0.868.

Limits of Impurities

Acid value. Not more than 1.0.

TESTS

Assay. Weigh accurately about 1.2 grams, and proceed as directed under *Ester Determination*, page 896, using 93.15 as the equivalence factor (e) in the calculation.

Refractive index, page 945. Determine with an Abbé or other refractometer of equal or greater accuracy.

Solubility in alcohol. Proceed as directed in the general method, page 899. One ml. dissolves in 6 ml. of 70 percent alcohol to form a clear solution.

Specific gravity. Determine by any reliable method (see page 5).

Acid value. Determine as directed in the general method, page 893.

Packaging and storage. Store in full, tight, preferably glass, tin-lined or other suitably lined containers in a cool place protected from light.

Functional use in foods. Flavoring agent.

NONYL ALCOHOL

1-Nonanol; Alcohol C-9

$CH_3(CH_2)_7CH_2OH$

$C_9H_{20}O$ Mol. wt. 144.26

DESCRIPTION

A colorless liquid having a rose-citrus odor. It is miscible with alcohol, with chloroform, and with ether, but it is insoluble in water.

SPECIFICATIONS

Assay. Not less than 97.0 percent of $C_9H_{20}O$.

Refractive index. Between 1.431 and 1.435 at 20°.

Solubility in alcohol. Passes test.

Specific gravity. Between 0.824 and 0.830.

Limits of Impurities

Acid value. Not more than 1.0.

TESTS

Assay. Proceed as directed under *Total Alcohols*, page 893. Weigh accurately about 1.2 grams of the acetylated oil for the saponification, and use 72.13 as the equivalence factor (*e*) in the calculation.

Refractive index, page 945. Determine with an Abbé or other refractometer of equal or greater accuracy.

Solubility in alcohol. Proceed as directed in the general method, page 899. One ml. dissolves in 3 ml. of 60 percent alcohol to form a clear solution.

Specific gravity. Determine by any reliable method (see page 5).

Acid value. Determine as directed in the general method, page 893.

Packaging and storage. Store in full, tight, preferably glass, tin-lined or other suitably lined containers in a cool place protected from light.

Functional use in foods. Flavoring agent.

NUTMEG OIL

Myristica Oil

DESCRIPTION

The volatile oil obtained by steam distillation from the dried kernels of the ripe seed of *Myristica fragrans* Houttuyn (Fam. *Myristicaceae*). Two types of oil, the East Indian and the West Indian, are commercially available. It is a colorless or pale yellow liquid, having the characteristic odor and taste of nutmeg. It is soluble in alcohol.

SPECIFICATIONS

Angular rotation. *East Indian:* Between +8° and +30°; *West Indian:* Between +25° and +45°.

Refractive index. *East Indian:* Between 1.474 and 1.488; *West Indian:* Between 1.469 and 1.476 at 20°.

Residue on evaporation. *East Indian:* Not more than 60 mg. per 3 ml.; *West Indian:* Not more than 50 mg. per 3 ml.

Solubility in alcohol. Passes test.

Specific gravity. *East Indian:* Between 0.880 and 0.910; *West Indian:* Between 0.854 and 0.880.

Limits of Impurities

Arsenic (as As). Not more than 3 parts per million (0.0003 percent).

Heavy metals (as Pb). Not more than 40 parts per million (0.004 percent).

Lead. Not more than 10 parts per million (0.001 percent).

TESTS

Angular rotation. Determine in a 100-mm. tube as directed under *Optical Rotation*, page 939.

Refractive index, page 945. Determine with an Abbé or other refractometer of equal or greater accuracy.

Residue on evaporation. Proceed as directed in the general method, page 899, using 3 ml. of sample. Heat on a steam bath for 5 hours, and then heat at 105° for 1 hour.

Solubility in alcohol. Proceed as directed in the general method, page 899. One ml. of East Indian oil dissolves in 3 ml. of 90 percent alcohol. One ml. of West Indian oil dissolves in 4 ml. of 90 percent alcohol.

Specific gravity. Determine by any reliable method (see page 5).

Arsenic. A *Sample Solution* prepared as directed for organic compounds meets the requirements of the *Arsenic Test*, page 865.

Heavy metals. Prepare and test a 500-mg. sample as directed in *Method II* under the *Heavy Metals Test*, page 920, using 20 mcg. of lead ion (Pb) in the control (*Solution A*).

Lead. A *Sample Solution* prepared as directed for organic compounds meets the requirements of the *Lead Limit Test*, page 929, using 10 mcg. of lead ion (Pb) in the control.

Packaging and storage. Store in full, tight containers in a cool place protected from light.

Labeling. Label myristica oil to indicate whether it is the East Indian or West Indian type.

Functional use in foods. Flavoring agent.

OCTANAL

Aldehyde C-8; Caprylic Aldehyde

$CH_3(CH_2)_6CHO$

$C_8H_{16}O$ Mol. wt. 128.22

DESCRIPTION

A colorless to light yellow liquid having a sharp fatty and fruity odor. It is soluble in alcohol, in most fixed oils, in mineral oil, and in propylene glycol. It is insoluble in glycerin.

SPECIFICATIONS

Assay. Not less than 92.0 percent of $C_8H_{16}O$.

Refractive index. Between 1.417 and 1.425 at 20°.

Specific gravity. Between 0.818 and 0.830.

Limits of Impurities

Acid value. Not more than 10.0.

TESTS

Assay. Weigh accurately about 1.5 grams, and proceed as directed under *Aldehydes*, page 894, using 64.11 as the equivalence factor (E) in the calculation.

Refractive index, page 945. Determine with an Abbé or other refractometer of equal or greater accuracy.

Specific gravity. Determine by any reliable method (see page 5).

Acid value. Determine as directed in the general method, page 893.

Packaging and storage. Store in full, tight, preferably glass, aluminum, or suitably lined containers in a cool place protected from light. Do not store for long periods of time. Dilution with alcohol improves stability in storage.

Functional use in foods. Flavoring agent.

OCTANOIC ACID

Caprylic Acid

$CH_3(CH_2)_6COOH$

$C_8H_{16}O_2$ Mol. wt. 144.21

DESCRIPTION

A colorless oily liquid having a slight, unpleasant odor and a burning rancid taste. It is slightly soluble in water and soluble in most organic solvents. Its specific gravity is about 0.910.

SPECIFICATIONS

Acid value. Between 366 and 396.

Saponification value. Between 366 and 398.

Titer (Solidification point). Between 8° and 15°.

Limits of Impurities

Arsenic (as As). Not more than 3 parts per million (0.0003 percent).

Heavy metals (as Pb). Not more than 10 parts per million (0.001 percent).

Iodine value. Not more than 2.0.

Residue on ignition. Not more than 0.1 percent.

Unsaponifiable matter. Not more than 0.2 percent.

Water. Not more than 0.4 percent.

TESTS

Acid value. Determine as directed under *Method I* in the general procedure, page 902.

Saponification value. Determine as directed in the general method, page 914, using about 2 grams, accurately weighed.

Titer (Solidification point). Determine as directed under *Solidification Point*, page 954.

Arsenic. A *Sample Solution* prepared as directed for organic compounds meets the requirements of the *Arsenic Test*, page 865.

Heavy metals. Prepare and test a 2-gram sample as directed in *Method II* under the *Heavy Metals Test*, page 920, using 20 mcg. of lead ion (Pb) in the control (*Solution A*).

Iodine value. Determine by the *Wijs Method*, page 906.

Residue on ignition, page 945. Ignite 10 grams as directed in the general method.

Unsaponifiable matter. Determine as directed in the general method, page 915.

Water. Determine by the *Karl Fischer Titrimetric Method*, page 977.

Packaging and storage. Store in tight containers.

Functional use in foods. Component in the manufacture of other food grade additives; defoaming agent.

1-OCTANOL (NATURAL)

Alcohol C-8; Octyl Alcohol; Capryl Alcohol

$CH_3(CH_2)_6CH_2OH$

$C_8H_{18}O$ Mol. wt. 130.23

DESCRIPTION

A colorless liquid having a sharp, fatty odor. It is soluble in most fixed oils, in mineral oil, and in propylene glycol. It is insoluble in glycerin.

Note: The following *Specifications* apply only to 1-octanol derived from natural precursors.

SPECIFICATIONS

Assay. Not less than 98.0 percent of $C_8H_{18}O$.

Refractive index. Between 1.428 and 1.431 at 20°.

Solubility in alcohol. Passes test.

Specific gravity. Between 0.822 and 0.830.

Limits of Impurities

Acid value. Not more than 1.0.

TESTS

Assay. Proceed as directed under *Total Alcohols*, page 893. Weigh accurately about 1.4 grams of the acetylated alcohol for the saponification, and use 65.12 as the equivalence factor (*e*) in the calculation.

Refractive index, page 945. Determine with an Abbé or other refractometer of equal or greater accuracy.

Solubility in alcohol. Proceed as directed in the general method, page 899. One ml. dissolves in 5 ml. of 50 percent alcohol.

Specific gravity. Determine by any reliable method (see page 5).

Acid value. Determine as directed in the general method, page 893.

Packaging and storage. Store in full, tight, preferably glass, or suitably lined containers in a cool place protected from light. Aluminum containers should not be used.

Functional use in foods. Flavoring agent.

OCTYL ACETATE

$CH_3COO(CH_2)_7CH_3$

$C_{10}H_{20}O_2$ Mol. wt. 172.27

DESCRIPTION

A colorless liquid having a fruity odor somewhat suggestive of orange and jasmine. It is miscible with alcohol, oils, and other organic solvents, but is insoluble in water.

SPECIFICATIONS

Assay. Not less than 98.0 percent of $C_{10}H_{20}O_2$.

Refractive index. Between 1.418 and 1.421 at 20°.

Solubility in alcohol. Passes test.

Specific gravity. Between 0.865 and 0.868.

Limits of Impurities

Acid value. Not more than 1.0.

TESTS

Assay. Weigh accurately about 1.1 grams, and proceed as directed under *Ester Determination*, page 896, using 86.14 as the equivalence factor (*e*) in the calculation.

Refractive index, page 945. Determine with an Abbé or other refractometer of equal or greater accuracy.

Solubility in alcohol. Proceed as directed in the general method, page 899. One ml. dissolves in 4 ml. of 70 percent alcohol to form a clear solution.

Specific gravity. Determine by any reliable method (see page 5).

Acid value. Determine as directed in the general method, page 893.

Packaging and storage. Store in full, tight, preferably glass, tin-lined or other suitably lined containers in a cool place protected from light.

Functional use in foods. Flavoring agent.

OCTYL FORMATE

$C_8H_{17}OOCH$

$C_9H_{18}O_2$ Mol. wt. 158.24

DESCRIPTION

A colorless liquid having a fruity odor. It is soluble in most fixed oils, in mineral oil, and in propylene glycol. It is practically insoluble in glycerin.

SPECIFICATIONS

Assay. Not less than 96.0 percent of $C_9H_{18}O_2$.

Refractive index. Between 1.418 and 1.420 at 20°.

Solubility in alcohol. Passes test.

Specific gravity. Between 0.869 and 0.872.

Limits of Impurities

Acid value. Not more than 1.0.

TESTS

Assay. Weigh accurately about 1 gram, and proceed as directed under *Ester Determination*, page 896, using 79.12 as the equivalence factor (*e*) in the calculation.

Refractive index, page 945. Determine with an Abbé or other refractometer of equal or greater accuracy.

Solubility in alcohol. Proceed as directed in the general method, page 899. One ml. dissolves in 5 ml. of 70 percent alcohol, and remains in solution on dilution to 10 ml.

Specific gravity. Determine by any reliable method (see page 5).

Acid value. Determine as directed in the general method, page 893.

Packaging and storage. Store in full, tight, preferably glass containers in a cool place protected from light.

Functional use in foods. Flavoring agent.

OLEIC ACID

cis-9-Octadecenoic Acid

$CH_3(CH_2)_7CH=CH(CH_2)_7COOH$

$C_{18}H_{34}O_2$ Mol. wt. 282.47

DESCRIPTION

An unsaturated acid obtained from fats. Oleic acid is a colorless to pale yellow, oily liquid when freshly prepared, but upon exposure to air it gradually absorbs oxygen and darkens. It has a characteristic lard-like odor and taste. When strongly heated in air, it is decomposed with the production of acrid vapors. Its specific gravity is about 0.895. It is practically insoluble in water, but is miscible with alcohol, with ether, with benzene, and with fixed and volatile oils. It conforms to the regulations of the federal Food and Drug Administration pertaining to specifications for fats or fatty acids derived from edible sources.

SPECIFICATIONS

Acid value. Between 196 and 204.

Iodine value. Between 83 and 103.

Saponification value. Between 196 and 206.

Titer (**Solidification Point**). Not above 10°.

Limits of Impurities

Arsenic (as As). Not more than 3 parts per million (0.0003 percent).

Heavy metals (as Pb). Not more than 10 parts per million (0.001 percent).

Residue on ignition. Not more than 0.01 percent.

Unsaponifiable matter. Not more than 2 percent.

Water. Not more than 0.4 percent.

TESTS

Acid value. Determine as directed under *Method I* in the general procedure, page 902.

Saponification value. Determine as directed in the general method, page 914, using about 3 grams, accurately weighed.

Iodine value. Determine by the *Wijs Method*, page 906.

Titer (**Solidification Point**). Determine as directed under *Solidification Point*, page 954.

Arsenic. A *Sample Solution* prepared as directed for organic compounds meets the requirements of the *Arsenic Test*, page 865.

Heavy metals. Prepare and test a 2-gram sample as directed in *Method II* under the *Heavy Metals Test*, page 920, using 20 mcg. of lead ion (Pb) in the control (*Solution A*).

Residue on ignition, page 945. Ignite 10 grams as directed in the general method.

Unsaponifiable matter, page 915. Determine as directed in the general method.

Water. Determine by the *Karl Fischer Titrimetric Method*, page 977.

Packaging and storage. Store in tight containers.

Functional use in foods. Component in the manufacture of other food grade additives; defoaming agent; lubricant; binder.

OLIBANUM OIL

Oil of Frankincense

DESCRIPTION

The volatile oil distilled from a gum obtained from the trees, *Boswellia carterii* Birdw., and other *Boswellia* species (Fam. *Burseraceae*). It is a pale yellow liquid, having a balsamic odor with a faint lemon note. It is soluble in most fixed oils and, with a slight haze, in mineral oil. It is insoluble in glycerin and in propylene glycol.

SPECIFICATIONS

Acid value. Not more than 4.0.

Angular rotation. Between −15° and +35°.

Ester value. Between 4 and 40.

Refractive index. Between 1.465 and 1.482 at 20°.

Solubility in alcohol. Passes test.

Specific gravity. Between 0.862 and 0.889.

TESTS

Acid value. Determine as directed in the general method, page 893.

Angular rotation. Determine in a 100-mm. tube as directed under *Optical Rotation*, page 939.

Ester value. Determine as directed in the general method, page 896, using about 5 grams, accurately weighed.

Refractive index, page 945. Determine with an Abbé or other refractometer of equal or greater accuracy.

Solubility in alcohol. Proceed as directed in the general method, page 899. One ml. dissolves in 6 ml. of 90 percent alcohol, occasionally with opalescence.

Specific gravity. Determine by any reliable method (see page 5).

Packaging and storage. Store in full, tight, preferably glass, aluminum, or tin-lined containers in a cool place protected from light.

Functional use in foods. Flavoring agent.

ONION OIL

DESCRIPTION

A volatile oil obtained by steam distillation of the bulbs of *Allium cepa* L. (Fam. *Liliaceae*). It is a clear amber yellow to amber orange liquid having a strong pungent odor and taste characteristic of onion. It is soluble in most fixed oils, in mineral oil, and in alcohol. It is insoluble in glycerin and in propylene glycol. *Note:* Onion oil is purchased mainly on the basis of its odor and flavor which render definitive specifications of little value.

SPECIFICATIONS

Refractive index. Between 1.549 and 1.570 at 20°.

Specific gravity. Between 1.050 and 1.135.

TESTS

Refractive index, page 945. Determine with an Abbé or other refractometer of equal or greater accuracy.

Specific gravity. Determine by any reliable method (see page 5).

Packaging and storage. Store in full, tight, preferably glass or aluminum containers in a cool place protected from light.

Functional use in foods. Flavoring agent.

ORANGE OIL

Sweet Orange Oil

DESCRIPTION

The volatile oil obtained by expression from the fresh peel of the ripe fruit of *Citrus sinensis* L. Osbeck (Fam. *Rutaceae*). It is an intensely yellow, orange, or deep orange liquid, having the characteristic odor and taste of the outer part of fresh, sweet orange peel. It is miscible with dehydrated alcohol, and with carbon disulfide. It is soluble in glacial acetic acid.

Note: Do not use sweet orange oil that has a terebinthine odor.

SPECIFICATIONS

Assay. Not less than 1.2 percent and not more than 2.5 percent of aldehydes, calculated as decyl aldehyde ($C_{10}H_{20}O$).

Angular rotation. Between +94° and +99°.

Refractive index. Between 1.472 and 1.474 at 20°.

Specific gravity. Between 0.842 and 0.846.

Ultraviolet absorbance. *California type:* Not less than 0.130; *Florida type:* Not less than 0.240.

Limits of Impurities

Arsenic (as As). Not more than 3 parts per million (0.0003 percent).

Foreign oils. Passes test.

Heavy metals (as Pb). Not more than 40 parts per million (0.004 percent).

Lead. Not more than 10 parts per million (0.001 percent).

Washed citrus oils. Passes tests.

TESTS

Assay. Weigh accurately a 10-ml. sample, and proceed as directed under *Aldehydes*, page 894, using 50 ml. of hydroxylamine hydrochloride solution, previously adjusted to a pH of about 3.4. Allow the mixture to stand for 15 minutes, with occasional shaking, before titrating, and use 78.14 as the equivalence factor (E) in the calculation.

Angular rotation. Determine in a 100-mm. tube as directed under *Optical Rotation*, page 939.

Refractive index, page 945. Determine with an Abbé or other refractometer of equal or greater accuracy.

Specific gravity. Determine by any reliable method (see page 5).

Ultraviolet absorbance. Proceed as directed under *Ultraviolet Absorbance of Citrus Oils*, page 900, using about 250 mg. of sample, accurately weighed. The maximum absorbance occurs at 330 $\pm$ 3 mμ.

Arsenic. A *Sample Solution* prepared as directed for organic compounds meets the requirements of the *Arsenic Test*, page 920.

Foreign oils. Proceed as directed for *Foreign Oils* under *Lemon Oil*, page 448. The angular rotation of the distillate does not differ from that of the original oil by more than 2 degrees, and the refractive index of the distillate at 20° is not less than 0.0005 and not more than 0.0018 lower than that of the original oil at 20°.

Heavy metals. Prepare and test a 500-mg. sample as directed in *Method II* under the *Heavy Metals Test*, page 920, using 20 mcg. of lead ion (Pb) in the control (*Solution A*).

Lead. A *Sample Solution* prepared as directed for organic compounds meets the requirements of the *Lead Limit Test*, page 929, using 10 mcg. of lead ion (Pb) in the control.

Washed citrus oils

A. Evaporate 3 ml. in a tared glass crystallizing dish on a steam bath for 5 hours. Continue heating at 105° for 2 hours. Cool in a desiccator, and weigh. Not less than 43 mg. of residue remains.

B. A one-ml. sample does not form a clear solution with 2 ml. of 90 percent alcohol.

Packaging and storage. Store in full, tight containers. Avoid exposure to excessive heat.

Functional use in foods. Flavoring agent.

ORANGE OIL, BITTER

DESCRIPTION

The volatile oil obtained by expression, without the use of heat, from the fresh peel of the fruit of *Citrus aurantium* L. (Fam. *Rutaceae*).

It is a pale yellow or yellowish brown liquid, with the characteristic aromatic odor of the Seville orange, and an aromatic somewhat bitter taste. It is miscible with absolute alcohol, and with an equal volume of glacial acetic acid. It is soluble in fixed oils and in mineral oil. It is slightly soluble in propylene glycol, but it is relatively insoluble in glycerin. It is affected by light, and its alcohol solutions are neutral to litmus.

SPECIFICATIONS

Aldehydes. Not less than 0.5 percent and not more than 1.0 percent of aldehydes, calculated as *n*-decyl aldehyde ($C_{10}H_{20}O$).

Angular rotation. Between +88° and +98°.

Refractive index. Between 1.472 and 1.476 at 20°.

Residue on evaporation. Between 2 percent and 5 percent.

Specific gravity. Between 0.845 and 0.851.

Limits of Impurities

Arsenic (as As). Not more than 3 parts per million (0.0003 percent).

Heavy metals (as Pb). Not more than 40 parts per million (0.004 percent).

Lead. Not more than 10 parts per million (0.001 percent).

TESTS

Aldehydes. Weigh accurately about 10 grams, and proceed as directed under *Aldehydes*, page 894, using 78.14 as the equivalence factor (*E*) in the calculation. Allow the mixture to stand for 30 minutes at room temperature before titrating.

Angular rotation. Determine in a 100-mm. tube as directed under *Optical Rotation*, page 939.

Refractive index, page 945. Determine with an Abbé or other refractometer of equal or greater accuracy.

Residue on evaporation. Proceed as directed in the general method, page 899, using 5 grams of sample, and heat for 4.5 hours.

Specific gravity. Determine by any reliable method (see page 5).

Arsenic. A *Sample Solution* prepared as directed for organic compounds meets the requirements of the *Arsenic Test*, page 865.

Heavy metals. Prepare and test a 500-mg. sample as directed in *Method II* under the *Heavy Metals Test*, page 920, using 20 mcg. of lead ion (Pb) in the control (*Solution A*).

Lead. A *Sample Solution* prepared as directed for organic compounds meets the requirements of the *Lead Limit Test*, page 929, using 10 mcg. of lead ion (Pb) in the control.

Packaging and storage. Store in full, tight, preferably glass or tin-lined containers in a cool place protected from light.

Functional use in foods. Flavoring agent.

ORIGANUM OIL, SPANISH

DESCRIPTION

The volatile oil obtained by steam distillation from the flowering herb, *Thymus capitatus* Hoffm. et Link, and various species of *Origanum*. It is a yellowish red to dark brownish red liquid, having a pungent spicy odor suggestive of thyme oil. It is soluble in most fixed oils, and in propylene glycol. It is soluble, with turbidity, in mineral oil, but it is insoluble in glycerin.

SPECIFICATIONS

Assay. Not less than 60 percent and not more than 75 percent, by volume, of phenols.

Angular rotation. Between $-2°$ and $+3°$.

Refractive index. Between 1.502 and 1.508 at 20°.

Solubility in alcohol. Passes test.

Specific gravity. Between 0.935 and 0.960.

TESTS

Assay. Shake a suitable quantity of sample with about 2 percent of powdered tartaric acid, and filter. Proceed as directed under *Phenols*, page 898.

Angular rotation. Determine in a 100-mm. tube as directed under *Optical Rotation*, page 939. Occasionally the oil is too dark to read in a 100-mm. tube.

Refractive index, page 945. Determine with an Abbé or other refractometer of equal or greater accuracy.

Solubility in alcohol. Proceed as directed in the general method, page 899. One ml. is soluble in 2 ml. of 70 percent alcohol. The solution may become cloudy on dilution.

Specific gravity. Determine by any reliable method (see page 5).

Packaging and storage. Store in full, tight, preferably glass, aluminum, or tin-lined containers in a cool place protected from light. A precipitate may form in galvanized containers, and the oil darkens in iron drums.

Functional use in foods. Flavoring agent.

ORRIS ROOT OIL

DESCRIPTION

The volatile oil obtained by steam distillation from the peeled, dried, and aged rhizomes of *Iris pallida* Lam. (Fam. *Iridaceae*). At room temperature it is a light yellow to brown yellow mass, which melts between 38° and 50° to form a yellow to yellow brown liquid. It is soluble in most fixed oils, in mineral oil, and in propylene glycol. It is insoluble in glycerin.

SPECIFICATIONS

Assay. Not less than 9.0 percent and not more than 20.0 percent of ketones, calculated as irone ($C_{14}H_{22}O$).

Acid value. Between 175 and 235.

Ester value. Between 4 and 35.

Melting range. Between 38° and 50°.

TESTS

Assay. Weigh accurately about 1 gram, and proceed as directed under *Aldehydes*, page 894, using 103.2 as the equivalence factor (*E*) in the calculation. Allow the mixture to stand 1 hour at room temperature before titrating.

Acid value. Determine as directed in the general method, page 893.

Ester value. Determine as directed in the general method, page 897, using about 1 gram, accurately weighed.

Melting range. Determine as directed in the general procedure (*Class II*), page 931.

Packaging and storage. Store in full, tight, preferably glass, tin-lined or other suitably lined aluminum containers in a cool place protected from light.

Functional use in foods. Flavoring agent.

OXYSTEARIN

DESCRIPTION

Oxystearin is a mixture of the glycerides of partially oxidized stearic and other fatty acids. It occurs as a tan to light brown, fatty or wax-like substance having a bland taste. It is soluble in ether, in solvent hexane, and in chloroform.

SPECIFICATIONS

Acid value. Not more than 15.

Hydroxyl value. Between 30 and 45.

Iodine value. Not more than 15.

Refractive index (butyro). Between 59 and 61 at 48° (equivalent to 1.465–1.467 on the Abbé scale.)

Saponification value. Between 225 and 240.

Limits of Impurities

Arsenic (as As). Not more than 3 parts per million (0.0003 percent).

Heavy metals (as Pb). Not more than 10 parts per million (0.001 percent).

Unsaponifiable matter. Not more than 0.8 percent.

TESTS

Acid value. Determine as directed under *Method II* in the general procedure, page 902.

Hydroxyl value. Determine as directed under *Method II* in the general procedure, page 904, using about 5 grams, accurately weighed.

Iodine value. Determine by the *Wijs Method*, page 906.

Refractive index, page 945. Melt the sample, filter through filter paper, and determine the refractive index at 48° with an Abbé or butyro refractometer.

Saponification value. Determine as directed in the general method, page 914, using about 3 grams, accurately weighed.

Arsenic. A *Sample Solution* prepared as directed for organic compounds meets the requirements of the *Arsenic Test*, page 865.

Heavy metals. Prepare and test a 2-gram sample as directed in *Method II* under the *Heavy Metals Test*, page 920, using 20 mcg. of lead ion (Pb) in the control (*Solution A*).

Unsaponifiable matter. Determine as directed in the general method, page 915.

Packaging and storage. Store in well-closed containers.

Functional use in foods. Crystallization inhibitor in salad and cooking oils; sequestrant; defoaming agent.

PALMAROSA OIL

Geranium Oil, East Indian; Geranium Oil, Turkish

DESCRIPTION

The volatile oil obtained by steam distillation from the partially dried grass *Cymbopogon martini* Stapf. var. *motia*. It is a light yellow to yellow oil which is often hazy and brownish. It is soluble in most fixed oils and in propylene glycol. It is soluble, usually with opalescence or turbidity, in mineral oil. It is practically insoluble in glycerin.

SPECIFICATIONS

Assay for alcohols. Not less than 88.0 percent and not more than 94.0 percent of total alcohols, calculated as geraniol ($C_{10}H_{18}O$).

Assay for esters. Not less than 4.0 percent and not more than 18.0 percent of esters, calculated as geranyl acetate ($C_{12}H_{20}O_2$).

Angular rotation. Between −2° and +3°.

Refractive index. Between 1.473 and 1.478 at 20°.

Solubility in alcohol. Passes test.

Specific gravity. Between 0.879 and 0.892.

TESTS

Assay for alcohols. Proceed as directed under *Total Alcohols*, page 893. Weigh accurately about 1 gram of the acetylated oil for the saponification, and use 77.13 as the equivalence factor (*e*) in the calculation.

Assay for esters. Weigh accurately about 5 grams, and proceed as directed under *Ester Determination*, page 896, using 98.15 as the equivalence factor (*e*) in the calculation.

Angular rotation. Determine in a 100-mm. tube as directed under *Optical Rotation*, page 939.

Refractive index, page 945. Determine with an Abbé or other refractometer of equal or greater accuracy.

Solubility in alcohol. Proceed as directed in the general method, page 899. One ml. dissolves in 2 ml. of 70 percent alcohol.

Specific gravity. Determine by any reliable method (see page 5).

Packaging and storage. Store in full, tight, preferably glass, or tin-lined containers in a cool place protected from light.

Functional use in foods. Flavoring agent.

PALMITIC ACID

Hexadecanoic Acid

$C_{16}H_{32}O_2$ Mol. wt. 256.43

DESCRIPTION

A mixture of solid organic acids obtained from fats consisting chiefly of palmitic acid ($C_{16}H_{32}O_2$) with varying amounts of stearic acid ($C_{18}H_{36}O_2$). It occurs as a hard, white or faintly yellowish, somewhat glossy and crystalline solid, or as a white or yellowish white powder. It has a slight characteristic odor and taste. Palmitic acid is practically insoluble in water. It is soluble in alcohol, in ether, and in chloroform. It conforms to the regulations of the federal Food and Drug Administration pertaining to specifications for fats or fatty acids derived from edible sources.

SPECIFICATIONS

Acid value. Between 204 and 220.

Saponification value. Between 205 and 221.

Titer (Solidification point). Between 53.3° and 62°.

Limits of Impurities

Arsenic (as As). Not more than 3 parts per million (0.0003 percent).

Heavy metals (as Pb). Not more than 10 parts per million (0.001 percent).

Iodine value. Not more than 2.0.

Residue on ignition. Not more than 0.1 percent.

Unsaponifiable matter. Not more than 1.5 percent.

Water. Not more than 0.2 percent.

TESTS

Acid value. Determine as directed under *Method I* in the general procedure, page 902.

Saponification value. Determine as directed in the general method, page 914, using about 3 grams, accurately weighed.

Titer (Solidification Point). Determine as directed under *Solidification Point*, page 954.

Arsenic. A *Sample Solution* prepared as directed for organic compounds meets the requirements of the *Arsenic Test*, page 865.

Heavy metals. Prepare and test a 2-gram sample as directed in *Method II* under the *Heavy Metals Test*, page 920, using 20 mcg. of lead ion (Pb) in the control (*Solution A*).

Iodine value. Determine by the *Wijs Method*, page 906.

Residue on ignition, page 945. Ignite 2 grams as directed in the general method.

Unsaponifiable matter, page 915. Determine as directed in the general method.

Water. Determine by the *Karl Fischer Titrimetric Method*, page 977.

Packaging and storage. Store in well-closed containers.

Functional use in foods. Component in the manufacture of other food grade additives; defoaming agent.

DL-PANTHENOL

DL-Pantothenyl Alcohol; Racemic Pantothenyl Alcohol

$HOCH_2C(CH_3)_2CH(OH)CONH(CH_2)_2CH_2OH$

$C_9H_{19}NO_4$ Mol. wt. 205.25

DESCRIPTION

A racemic mixture of the dextrorotatory (active) and levorotatory (inactive) isomers of panthenol, the alcohol analogue of pantothenic acid. It occurs as a white to creamy white crystalline powder having a slight, characteristic odor. Its solutions are neutral or alkaline to litmus. It is freely soluble in water, in alcohol, and in propylene glycol. It is soluble in chloroform and in ether, and slightly soluble in glycerin. [*Note:* The physiological activity of DL-panthenol is one-half that of dexpanthenol (D-panthenol).]

SPECIFICATIONS

Assay. Not less than 99.0 percent and not more than 102.0 percent of $C_9H_{19}NO_4$ (DL-panthenol), calculated on the dried basis.

Melting range. Between 64.5° and 68.5°.

Limits of Impurities

Aminopropanol. Not more than 0.1 percent.

Arsenic (as As). Not more than 3 parts per million (0.0003 percent).

Heavy metals (as Pb). Not more than 10 parts per million (0.001 percent).

Loss on drying. Not more than 0.5 percent.

Residue on ignition. Not more than 0.1 percent.

TESTS

Assay. Transfer about 400 mg., accurately weighed, into a 300-ml. reflux flask fitted with a standard-taper glass joint, add 50.0 ml. of 0.1 *N* perchloric acid in glacial acetic acid, and reflux for 5 hours. Cool, covering the condenser with foil to prevent contamination by moisture, and rinse the condenser with glacial acetic acid. Add 5 drops of crystal violet T.S., and titrate with 0.1 *N* potassium acid phthalate in glacial acetic acid to a blue-green end-point. Perform a blank determination, and make any necessary correction (see page 2). Each ml. of 0.1 *N* perchloric acid is equivalent to 20.53 mg. of $C_9H_{19}NO_4$.

Melting range. Determine as directed in the general procedure, page 931.

Aminopropanol. Transfer about 10 grams of the sample, accurately weighed, into a 50-ml. flask, and dissolve in 25 ml. of water. Add bromothymol blue T.S., and titrate with 0.01 *N* sulfuric acid from a micro-buret to a yellow end-point. Each ml. of 0.01 *N* sulfuric acid is equivalent to 0.75 mg. (750 mcg.) of aminopropanol.

Arsenic. A *Sample Solution* prepared as directed for organic compounds meets the requirements of the *Arsenic Test,* page 865.

Heavy metals. Prepare and test a 2-gram sample as directed in *Method II* under the *Heavy Metals Test,* page 920, using 20 mcg. of lead ion (Pb) in the control (*Solution A*).

Loss on drying, page 931. Dry at 56° for 4 hours in vacuum over phosphorus pentoxide.

Residue on ignition. Ignite a 1-gram sample as directed in the general method, page 945.

Packaging and storage. Store in tight containers.

Functional use in foods. Nutrient; dietary supplement.

PAPAIN

DESCRIPTION

Papain is a purified proteolytic substance derived from *Carica papaya* L. (Fam. *Caricacea*). It occurs as a white to light tan amorphous powder. It is soluble in water, the solution being colorless to light yellow and somewhat opalescent. It is practically insoluble in alcohol, in chloroform, and in ether. It may be adjusted, if necessary, to conform to the Codex requirement for papain activity by admixture with papain containing a higher or lower enzyme activity, or with lactose or other suitable diluents.

SPECIFICATIONS

Assay. Not less than 6000 N.F. Units of papain activity per mg.

(*Note:* One N.F. Unit of papain activity is the activity which releases the equivalent of 1 mcg. of tyrosine from a specified casein substrate under the conditions of the *Assay,* using the enzyme concentration which liberates 40 mcg. of tyrosine per ml. of *Standard Solution.*)

Loss on drying. Not more than 7 percent.

pH of a 1 in 50 solution. Between 4.8 and 6.2.

Limits of Impurities

Arsenic (as As). Not more than 3 parts per million (0.0003 percent).

Heavy metals (as Pb). Not more than 10 parts per million (0.001 percent).

TESTS

Assay

Sodium Phosphate Solution (0.05 *M*). Dissolve 7.1 grams of anhydrous dibasic sodium phosphate, Na_2HPO_4, in water, dilute to 1000 ml. with water, and add 1 drop of toluene as preservative.

Citric Acid Solution (0.05 *M*). Dissolve 10.5 grams of citric acid monohydrate, $C_6H_8O_7.H_2O$, in water, dilute to 1000 ml. with water, and add 1 drop of toluene as preservative.

Casein Substrate. On the day of use, disperse 1 gram of Hammersten type casein in 50 ml. of the *Sodium Phosphate Solution*, and heat in a boiling water bath for 30 minutes, stirring occasionally. Cool to room temperature, and add the *Citric Acid Solution* to a pH of 6.0 ± 0.1, stirring rapidly and continuously during the addition to prevent precipitation of the casein. Finally, dilute to 100 ml. with water.

Phosphate-Cysteine-EDTA Buffer Solution. On the day of use, transfer 3.55 grams of anhydrous dibasic sodium phosphate, Na_2HPO_4, into a 500-ml. volumetric flask, dissolve in 400 ml. of water, add 7.0 grams of disodium ethylenediaminetetraacetate dihydrate, $C_{10}H_{14}N_2Na_2O_8.2H_2O$, and 3.05 grams of cysteine hydrochloride monohydrate, $C_3H_7NO_2S.HCl.H_2O$, and mix. Adjust the pH to 6.0 ± 0.1 with 1 *N* hydrochloric acid or 1 *N* sodium hydroxide, and dilute to volume with water.

Trichloroacetic Acid Solution. Dissolve 30 grams of trichloroacetic acid, $C_2HCl_3O_2$, in 100 ml. of water.

Standard Solution. Transfer 100.0 mg. of N.F. Papain Reference Standard, accurately weighed, into a 100-ml. volumetric flask, dissolve in the *Phosphate-Cysteine-EDTA Buffer Solution*, dilute to volume with the solution, and mix. Pipet 2 ml. of this solution into a 50-ml. volumetric flask, dilute to volume with the buffer solution, and mix. Use within 30 minutes after preparation.

Assay Solution. Dissolve an accurately weighed amount of the sample, equivalent to about 100 mg. of N.F. Papain Reference Standard, in sufficient *Phosphate-Cysteine-EDTA Buffer Solution* to make 100 ml. Dilute 2.0 ml. of this solution to 50.0 ml. with the buffer solution.

Procedure. Into each of twelve 18 × 150-mm. test tubes, pipet 5 ml. of the *Casein Substrate*, place them in a 40° water bath, and allow 10 minutes for the solutions to reach the bath temperature. Into each of duplicate tubes, both of which are labeled S_1, pipet 1 ml. of the *Standard Solution* and 1 ml. of the *Phosphate-Cysteine-EDTA Buffer Solution* (*P-C-EDTA Buffer*), mix by swirling, note the zero time, stopper, and replace in the bath. Into each of two other tubes, labeled S_2, pipet 1.5 ml. of the *Standard Solution* and 0.5 ml. of the *P-C-EDTA Buffer*, and proceed as with the S_1 tubes. Repeat this procedure for two tubes labeled S_3, to which 2.0 ml. of the *Standard Solution* is added, and for two tubes labeled U_2, to which 1.5 ml. of the *Assay Solution* and 0.5 ml. of the *P-C-EDTA Buffer* are added. After exactly 60 minutes, add to all twelve tubes 3.0 ml. of the *Trichloroacetic Acid Solution*, and shake vigorously. With the four tubes to which no standard or sample solutions were added, prepare blanks by pipetting, respectively, 1.0 ml. of the *Standard Solution* plus 1.0 ml. of the *P-C-EDTA Buffer*, 1.5 ml. of the *Standard Solution* plus 0.5 ml. of the *P-C-EDTA Buffer*, 2.0 ml. of the *Standard Solution*, and 1.5 ml. of the *Assay Solution* plus 0.5 ml. of the *P-C-EDTA Buffer*. Replace all tubes in the 40° bath for 30 to 40 minutes, allowing the precipitated protein to coagulate completely, and then filter through Whatman No. 40 or equivalent filter paper, discarding the first 3 ml. of filtrate. Determine the absorbance of each filtrate, which must be perfectly clear, at 280 mμ against its respective blank. Plot the readings for S_1, S_2, and S_3 against the enzyme concentration of each corresponding concentration. By interpolation from this standard curve, correcting for dilution factors, calculate the potency of the sample, in units per mg., by the formula $C \times (100/W) \times (50/2) \times (10/1.5) \times A$, in which C is the concentration, in mg. per ml., obtained from the standard curve, W is the weight of the sample, in mg., and A is the activity of the Reference Standard in units per mg.

Loss on drying, page 931. Dry in a vacuum oven at 60° for 4 hours.

pH of a 1 in 50 solution. Determine by the *Potentiometric Method*, page 941.

Arsenic. A *Sample Solution* prepared as directed for organic compounds meets the requirements of the *Arsenic Test*, page 865.

Heavy metals. Prepare and test a 2-gram sample as directed under *Method II* in the *Heavy Metals Test*, page 920, using 20 mcg. of lead ion (Pb) in the control (*Solution A*).

Packaging and storage. Store in tight, light-resistant containers in a cool, dry place.

Functional use in foods. Proteolytic enzyme.

PARSLEY SEED OIL

DESCRIPTION

The oil obtained by steam distillation of the ripe seed of *Petroselinum sativum* Hoffm. (Fam. *Umbelliferae*). It is a yellow to light brown liquid, having a rather harsh odor. It is soluble in most fixed oils, and in mineral oil. It is slightly soluble in propylene glycol, but it is insoluble in glycerin.

SPECIFICATIONS

Acid value. Not more than 4.0.

Angular rotation. Between −4° and −10°.

Refractive index. Between 1.513 and 1.522 at 20°.

Saponification value. Between 2 and 10.
Solubility in alcohol. Passes test.
Specific gravity. Between 1.040 and 1.080.

TESTS

Acid value. Determine as directed in the general method, page 893.

Angular rotation. Determine in a 100-mm. tube as directed under *Optical Rotation*, page 939.

Refractive index, page 945. Determine with an Abbé or other refractometer of equal or greater accuracy.

Saponification value. Determine as directed in the general method, page 896, using 5 grams, accurately weighed.

Solubility in alcohol. Proceed as directed in the general method, page 899. One ml. dissolves in 6 ml. of 80 percent alcohol, occasionally with slight haziness.

Specific gravity. Determine by any reliable method (see page 5).

Packaging and storage. Store in full, preferably glass, tin-lined, or other suitably lined containers in a cool place protected from light.

Functional use in foods. Flavoring agent.

PECTIN

DESCRIPTION

A hydrophyllic colloidal carbohydrate obtained from the dilute acid extract of the inner portion of citrus fruit rinds or from apple pomace. Food-grade pectin is usually classified according to its degree of esterification, or methoxy content. Except for the sugar that is added for standardization of gel properties, it consists chiefly of partially methylated polygalacturonic acid units, i.e., a linear galacturonoglycan of $\alpha(1 \rightarrow 4)$-linked D-galactopyranosyluronic acid.

In *low ester* pectin, the remaining carboxylic acid groups exist in the form of the free acid or as its ammonium, potassium, or sodium salts, and in some types the acid amide; its useful properties may vary with the proportion of methoxyl and amide substitutions and with the degree of polymerization.

In *high ester* pectin, portions of the carboxyl groups occur as methyl esters, and the remaining carboxylic acid groups exist in the form of the free acid or as its ammonium, potassium, or sodium salts; its useful properties may vary with the degree of methylation and with the degree of polymerization.

Both forms of pectin are usually standardized with dextrose or other sugars, to "100 gel power" in the case of *low ester* pectin, or to "150 jelly grade" in the case of *high ester* pectin. They may also contain added buffer salts such as sodium citrate or sodium bicarbonate. Both forms may be supplied without the added sugars as "unstandardized" pectin.

Pectin usually occurs as a practically odorless, yellowish white, coarse to fine powder having a mucilaginous taste. It dissolves in water, forming an opalescent colloidal solution. It is practically insoluble in alcohol.

IDENTIFICATION

A. To a 1 in 100 solution of the sample in water add an equal volume of alcohol. A translucent, gelatinous precipitate is formed (difference from most gums).

B. To 10 ml. of a 1 in 100 solution of the sample in water add 1 ml. of thorium nitrate solution (1 in 10), stir, and allow to stand for 2 minutes. A stable precipitate or gel forms (difference from most gums).

C. To 5 ml. of a 1 in 100 solution of the sample in water add 1 ml. of sodium hydroxide T.S., and allow to stand at room temperature for 15 minutes. A gel or semi-gel forms (difference from tragacanth and other gums).

D. Acidify the gel from the preceding test with 1 ml. of hydrochloric acid T.S., and shake well. A voluminous, colorless, gelatinous precipitate forms, which, upon boiling, becomes white and flocculent (pectic acid).

SPECIFICATIONS

Ash (Total). Not more than 10 percent.

Degree of amide substitution of low ester pectin. Not more than 40 percent.

Degree of esterification of low ester pectin. Not more than 50 percent.

Degree of esterification of high ester pectin. Not less than 50 percent.

Galacturonic acid. Not less than 35 percent.

Gel power of low ester pectin. Not less than 95 percent and not more than 105 percent of the labeled gel power.

Jelly grade of high ester pectin. Not less than 95 percent and not more than 105 percent of the labeled jelly grade.

Limits of Impurities

Acid-insoluble ash. Not more than 1 percent.

Arsenic (as As). Not more than 3 parts per million (0.0003 percent).

Heavy metals (as Pb). Not more than 40 parts per million (0.004 percent).
Lead. Not more than 10 parts per million (0.001 percent).
Loss on drying. Not more than 12 percent.
Sodium methyl sulfate. Not more than 0.1 percent.

TESTS

Ash (Total). Determine as directed in the general method, page 868.

Degree of amide substitution and **degree esterification of low ester pectin.** Transfer 5.0 grams of *low ester* pectin, accurately weighed, into a beaker, and stir for 10 minutes with a mixture of 5 ml. of hydrochloric acid and 100 ml. of 60 percent isopropyl alcohol. Filter the mixture through a dry, tared, coarse sintered-glass filter tube (30–60 ml. capacity), and wash with six 15-ml. portions of the acid-alcohol mixture, followed by 60 percent isopropyl alcohol, until the filtrate is free from chloride. Finally, wash with 20 ml. of 60 percent isopropyl alcohol, dry at 105° for 1 hour, cool, and weigh. Transfer exactly one-tenth of the total net weight of the dried sample (representing 500 mg. of the original unwashed sample) into a 250-ml. Erlenmeyer flask, and moisten the sample with 2 ml. of alcohol. Add 100 ml. of carbon dioxide-free water, stopper, and swirl occasionally until the sample is completely dissolved. Add 5 drops of phenolphthalein T.S., and titrate with 0.1 *N* sodium hydroxide, recording the volume required, in ml., as V_1 (*initial titer*). Add 20.0 ml. of 0.5 *N* sodium hydroxide, stopper, shake vigorously, and allow to stand for 15 minutes. Add 20.0 ml. of 0.5 *N* hydrochloric acid, shake until the pink color disappears, then add 3 drops of phenolphthalein T.S., and titrate with the 0.1 *N* sodium hydroxide to a faint pink color that persists after vigorous shaking. Record the volume of 0.1 *N* sodium hydroxide required, in ml., as V_2 (*saponification titer*). Quantitatively transfer the contents of the flask into a 500-ml. distillation flask fitted with a Kjeldahl trap and a water-cooled condenser, the delivery tube of which extends well beneath the surface of a mixture of 150 ml. of carbon dioxide-free water and 20.0 ml. of 0.1 *N* hydrochloric acid in a receiving flask. To the distillation flask add 20 ml. of sodium hydroxide solution (1 in 10), seal the connections, and then begin heating carefully to avoid excessive foaming. Continue heating until 80–120 ml. of distillate has been collected. Add a few drops of methyl red T.S., to the receiving flask, and titrate the excess acid with 0.1 *N* sodium hydroxide, recording the volume required, in ml., as *S*. Perform a blank determination on 20.0 ml. of 0.1 *N* hydrochloric acid, and record the volume required, in ml., as *B*. Record the *amide titer* $(B - S)$ as V_3. The sum of V_1, V_2, and V_3 is recorded as V_t (*total titer*).

Calculate the degree of amide substitution by the formula $100 \times V_3/V_t$.

Calculate the degree of esterification by the formula $100 \times V_2/V_t$.

Degree of esterification of high ester pectin. Transfer 5.0 grams of *high ester* pectin, accurately weighed, into a beaker, and stir for 10 minutes with a mixture of 5 ml. of hydrochloric acid and 100 ml. of 60 percent isopropyl alcohol. Filter the mixture through a dry, tared, coarse sintered-glass filter tube (30–60 ml. capacity), and wash with six 15-ml. portions of the acid-alcohol mixture, followed by 60 percent isopropyl alcohol, until the filtrate is free from chloride. Finally, wash with 20 ml. of 60 percent isopropyl alcohol, dry at 105° for 1 hour, cool, and weigh. Transfer exactly one-tenth of the total net weight of the dried sample (representing 500 mg. of the original unwashed sample) into a 250-ml. Erlenmeyer flask, and moisten the sample with 2 ml. of alcohol. Add 100 ml. of carbon dioxide-free water, stopper, and swirl occasionally until the sample is completely dissolved. Add 5 drops of phenolphthalein T.S., and titrate with 0.1 *N* sodium hydroxide, recording the volume required, in ml., as v_1 (*initial titer*). Add 20.0 ml. of 0.5 *N* sodium hydroxide, stopper, and shake vigorously for 15 minutes. Add 20.0 ml. of 0.5 *N* hydrochloric acid, shake until the pink color disappears, then add 3 drops of phenolphthalein T.S., and titrate with 0.1 *N* sodium hydroxide to a faint pink color that persists after vigorous shaking. Record the volume of 0.1 *N* sodium hydroxide required, in ml., as v_2 (*saponification titer*). Calculate the degree of esterification by the formula $100 \times v_2/(v_1 + v_2)$.

Galacturonic acid. Calculate the weight, in mg., of galacturonic acid ($C_5H_9O_5COOH$) in the original, 500-mg. unwashed sample of *low ester* pectin by the formula $V_t \times 19.41$, in which V_t is the *total titer* determined in the test for *Degree of amide substitution and degree of esterification of low ester pectin.*

Calculate the weight, in mg., of galacturonic acid ($C_5H_9O_5COOH$) in the original, 500-mg. unwashed sample of *high ester* pectin by the formula $(v_1 + v_2) \times 19.41$, in which v_1 and v_2 are the *initial titer* and *saponification titer*, respectively, determined in the test for *Degree of esterification of high ester pectin.*

Gel power of low ester pectin

Test gel. If the pectin is standardized, weigh accurately a 6.00-gram sample, or if the pectin is unstandardized, weigh accurately a sample, in grams, equal to 600 divided by the labeled gel power. Mix the sample with 40 grams of sucrose, add the mixture to 425 ml. of water, in a saucepan, previously mixed with 5.0 ml. of citric acid solution (543 grams of $C_6H_8O_7$ per 1000 ml. of water) and 10.0 ml. of sodium citrate solution (60 grams of $C_6H_5Na_3O_7.2H_2O$ per 1000 ml. of water), and continue stirring until the sample is dispersed. Heat while stirring, bring to a boil, then add 140 grams of sucrose, continue boiling, and stir until dissolved. Add 25.0 ml. of calcium chloride solution (22.05 grams of $CaCl_2.2H_2O$ per 1000 ml. of water) to the mixture while stirring, and continue to boil until a net weight of 600 grams is reached. Remove from the heat, and allow the foam to rise.

Quickly skim off the foam, and immediately pour into two Ridgelimeter glasses (see Kertesz, A. T., "The Pectic Substances", page 498, Interscience, New York, 1951), previously prepared by wrapping the glass rims with masking tape so that the tape forms an extension of the rims and extends about 12 mm. above the top of the glass. Add enough of the sample mixture so that each glass is filled to within about 2 mm. of the upper edge of the tape. Place caps over the tape, and allow the gels to stand for 18–24 hours in a cabinet maintained at 25°.

Procedure. Determine the percentage of sag of the *Test gel* as follows: Remove the tape from each glass, and cut the gel along the edge of the glass by means of a thin wire, such as a cheese cutter. Discard the cut-off section, and then carefully rotate and invert the glass to turn out the gel onto the glass plate of the Ridgelimeter. Exactly 2 minutes after the gel has been on the plate, bring the micrometer screw of the Ridgelimeter into contact with the gel surface, and record the scale reading, in percent sag, for both samples. Average the two readings to obtain the average % sag. Calculate the gel power by the formula $(600/W) \times [2.00 - (\% \text{ sag} + 4.5)/25.0]$, in which W is the weight of the sample taken, in grams. (*Note:* For valid gel power values, the gels should contain from 30 to 32 percent solids, the pH should be between 2.9 and 3.1, and the percent sag readings should agree within 0.6 of each other and fall within the range of 16.0 to 25.0.)

Jelly grade of high ester pectin

Test jelly. If the pectin is standardized, weigh accurately a 4.33-gram sample, or if the pectin is unstandardized, weigh accurately a sample, in grams, equal to 650 divided by the labeled grade. Mix the sample with 40 grams of sucrose, add the mixture, while stirring, to 405 ml. of water contained in a stainless steel saucepan, and continue stirring until the sample is dispersed (1 to 2 minutes). Heat while stirring, and bring to a full boil. Add 606 grams of sucrose, stir, and continue to boil until a net weight of 1015 grams is reached. Remove from the heat, and allow to cool to 95°. Prepare three Ridgelimeter glasses (see Kertesz, A. T., "The Pectic Substances", page 498, Interscience, New York, 1951), previously prepared by wrapping the glass rims with masking tape so that the tape forms an extension of the rims and extends about 12 mm. above the top of the glass. Add 2.0 ml. of 48.8 percent tartaric acid solution to each glass, and then add enough of the sample mixture so that each glass is filled to within about 2 mm. of the upper edge of the tape. Stir while filling the glasses in order to mix the sample with the tartaric acid solution. Place caps over the tape, and allow the jellies to stand for 18 to 24 hours in a cabinet maintained at 25°.

Procedure. Determine the percentage of sag of the *Test jelly* as follows: Remove the tape from each glass, and cut the jelly along the edge of the glass by means of a thin wire, such as a cheese cutter. Discard the cut-off section, and then carefully rotate and invert the glass to turn out the jelly onto the glass plate of the Ridgelimeter, taking precautions to minimize stresses and to prevent rupturing the jelly. Exactly 2 minutes after the jelly has been on the plate bring the micrometer screw of the Ridgelimeter into contact with the jelly surface, and record the scale reading, in percent sag, for each of the three samples. Average the three readings to obtain the average % sag. Calculate the jelly grade by the formula $(650/W) \times [2.00 - (\% \text{ sag}/23.5)]$, in which W is the weight of the sample taken, in grams. (*Note:* For valid grade values, the jellies should contain from 64.8 to 65.2 percent solids, the pH should be between 2.2 and 2.4, and the percent sag readings of the three jellies should agree within 0.6 percent of each other and fall within the range of 20.0 to 28.0 percent.)

Acid-insoluble ash. Determine as directed in the general method, page 869.

Arsenic. A *Sample Solution* prepared as directed for organic compounds meets the requirements of the *Arsenic Test,* page 865.

Heavy metals. Prepare and test a 500-mg. sample as directed in *Method II* under the *Heavy Metals Test,* page 920, using 20 mcg. of lead ion (Pb) in the control (*Solution A*).

Lead. A *Sample Solution* prepared as directed for organic compounds meets the requirements of the *Lead Limit Test,* page 929, using 10 mcg. of lead ion (Pb) in the control.

Loss on drying, page 931. Dry at 105° for 2 hours.

Sodium methyl sulfate

Barium molybdate. Dissolve 29.0 grams of sodium molybdate, $Na_2MoO_4.2H_2O$, and 24.0 grams of barium chloride, $BaCl_2.2H_2O$, in separate 1000-ml. volumes of water, heat both solutions to 70–80°, and then slowly add the barium chloride solution to the sodium molybdate solution while stirring. Allow the precipitate to settle, decant off the liquid, and wash the precipitate with three 100-ml. portions of warm (70°) water, decanting each washing. Dissolve the precipitate in 200 ml. of 2 *N* hydrochloric acid, dilute to about 1000 ml. with water, add 1 ml. of bromothymol blue T.S., and mix. Heat to 70–80°, add 150 ml. of 2 *N* ammonia, with mixing, and titrate with 2 *N* ammonia to a green end-point. Remove the liquid by decantation, and wash the precipitate with three 100-ml. portions of water, decating each washing through a Buchner funnel. Mix the precipitate with 100 ml. of water, and pour the mixture into the funnel. Wash the filter cake with several portions of water, and dry at 110° overnight.

Buffer solution. Dissolve 31.0 grams of boric acid and 8.55 grams of sodium chloride in sufficient water to make 1000 ml.

Sodium sulfate solution. Transfer 100.0 mg. of anhydrous sodium sulfate into a 100-ml. volumetric flask, dissolve in water, dilute to volume with water, and mix.

Standard preparations. Transfer 1.0, 2.0, 3.0, 4.0, and 5.0 ml. of the *Sodium sulfate solution* into separate 100-ml. volumetric flasks, add 1 ml. of 70 percent perchloric acid and 2–3 drops of paranitrophenol solution (1 in 1000 in methanol) to each flask, and then add ammonium hydroxide, dropwise, to the first appearance of a yellow color. Add just sufficient dilute hydrochloric acid (1 in 50) to discharge the yellow color, then add 10.0 ml. of the *Buffer solution* and 50.0 ml. of methanol, dilute to volume with water, and mix. Continue with each standard as directed under *Sample preparation,* beginning with "Pipet 20.0 ml. of this solution into a 125-ml. Erlenmeyer flask. . . ."

Sample preparation. Transfer 5.00 grams of the sample into a 100-ml. beaker, mix well with about 5 grams of powdered cellulose, and transfer the mixture into a 22 × 80-mm. extraction thimble. Plug the thimble with cotton, and extract overnight on a steam bath in a suitable continuous extraction apparatus, using 100 ml. of a 3:1 mixture of methanol-chloroform as the solvent. Add 500 mg. each of barium carbonate and sodium bicarbonate to the extract, and shake mechanically at medium agitation for 1 hour. Filter through Whatman No. 40 or equivalent paper into a 125-ml. long-neck boiling flask, and wash the flask and filter with two 10-ml. portions of chloroform, taking precautions to ensure that any solids do not pass into the filtrate. If it is turbid, filter the filtrate again. Add a glass bead to the boiling flask, evaporate the filtrate to dryness on a steam bath, and cool. Rinse the sides of the flask with 10 ml. of nitric acid and 1 ml. of 70 percent perchloric acid, and evaporate at high heat until the dense white perchloric acid fumes just disappear from the bowl of the flask. Cool, rinse the sides of the flask with 10 ml. of water, add 2–3 drops of paranitrophenol solution (1 in 1000 in methanol), and then add ammonium hydroxide, dropwise, to the first appearance of a yellow color. Add just sufficient dilute hydrochloric acid (1 in 50) to discharge the yellow color, and then transfer the solution into a 100-ml. volumetric flask with the aid of 10.0 ml. of the *Buffer solution* and 10 ml. of water. Add 50.0 ml. of methanol to the volumetric flask, dilute to volume with water, and mix.

Pipet 20.0 ml. of this solution into a 125-ml. Erlenmeyer flask containing 200 mg. of *Barium molybdate.* Stopper, shake for 1 hour, and filter through Whatman No. 40 or equivalent filter paper into a 50-ml. Erlenmeyer flask. Pipet 10.0 ml. of the filtrate into a 50-ml. volumetric flask, and add 10 ml. of water, 7 ml. of hydrochloric acid, 3 ml. of potassium thiocyanate solution (1 in 10), and 15 ml. of acetone. Mix well, then heat at 60–70° in a water bath for 30 minutes, cool, dilute to volume with water, and mix.

Run a blank determination with 5-grams of powdered cellulose in the same manner, using the same quantities of the same reagents as in the treatment of the sample.

Procedure. Determine the absorbance of the five *Standard preparations* and of the *Sample preparation,* against the blank, in 1-cm. cells at

460 mμ, using a suitable spectrophotometer. Plot the absorbances of the *Standard preparations* versus ml. of *Sodium sulfate solution* used in preparing each standard, and then convert the volumes to percent of sodium methyl sulfate, each 1.0 ml. of *Sodium sulfate solution* being equivalent to 0.0188 percent of sodium methyl sulfate. From the absorbance of the *Sample preparation,* determine the percent of sodium methyl sulfate in the 5-gram pectin sample taken for analysis by means of the standard curve.

Packaging and storage. Store in well-closed containers.

Functional use in foods. Gelling agent; thickener; stabilizer; emulsifier.

PENNYROYAL OIL

Pennyroyal Oil, Imported

DESCRIPTION

The volatile oil obtained by steam distillation from the fresh or partly dried plant *Mentha pulegium* L. (Fam. *Labiatae*). It is a light yellow to yellow aromatic liquid, having a mint-like odor. It is soluble in most fixed oils and in propylene glycol. It is soluble, with slight cloudiness, in mineral oil, but it is practically insoluble in glycerin.

SPECIFICATIONS

Assay. Not less than 88 percent and not more than 96 percent, by volume, of ketones.

Angular rotation. Between +18° and +25°.

Refractive index. Between 1.483 and 1.488 at 20°.

Solubility in alcohol. Passes test.

Specific gravity. Between 0.928 and 0.940.

TESTS

Assay. Proceed as directed under *Aldehydes and Ketones-Neutral Sulfite Method,* page 895.

Angular rotation. Determine in a 100-mm. tube as directed under *Optical Rotation,* page 939.

Refractive index, page 945. Determine with an Abbé or other refractometer of equal or greater accuracy.

Solubility in alcohol. Proceed as directed in the general method, page 899. One ml. dissolves in 2 ml. of 70 percent alcohol.

Specific gravity. Determine by any reliable method (see page 5).

Packaging and storage. Store in full, tight, preferably glass, tin-lined, or suitably galvanized containers in a cool place protected from light.

Functional use in foods. Flavoring agent.

PENTAERYTHRITOL ESTER OF PARTIALLY HYDROGENATED WOOD ROSIN

DESCRIPTION

A hard, amber-colored resin (color K or paler as determined by A.S.T.M. Designation D 509) produced by the esterification of partially hydrogenated wood rosin with pentaerythritol and purified by steam stripping. It is soluble in acetone and in benzene, but is insoluble in water.

SPECIFICATIONS

Acid value. Between 7 and 18.

Drop softening point. Between 102° and 110°.

Limits of Impurities

Arsenic (as As). Not more than 3 parts per million (0.0003 percent).

Heavy metals (as Pb). Not more than 40 parts per million (0.004 percent).

Lead. Not more than 3 parts per million (0.0003 percent).

TESTS

Acid value. Determine as directed in the general procedure, page 945.

Drop softening point. Determine as directed in the general procedure, page 946, using a bath temperature of 120°.

Arsenic. Prepare a *Sample Solution* as directed in the general method under *Chewing Gum Base*, page 887. This solution meets the requirements of the *Arsenic Test*, page 865.

Heavy metals. Prepare and test a 500-mg. sample as directed in *Method II* under the *Heavy Metals Test*, page 920, using 20 mcg. of lead ion (Pb) in the control (*Solution A*).

Lead. Prepare a *Sample Solution* as directed in the general method under *Chewing Gum Base*, page 878. This solution meets the requirements of the *Lead Limit Test*, page 929, using 10 mcg. of lead ion (Pb) in the control.

Packaging and storage. Store in well-closed containers.

Functional use in foods. Masticatory substance in chewing gum base.

PENTAERYTHRITOL ESTER OF WOOD ROSIN

DESCRIPTION

A hard, pale amber-colored resin (color M or paler as determined by A.S.T.M. Designation D 509) produced by the esterification of pale wood rosin with pentaerythritol and purified by steam stripping. It is soluble in acetone and in benzene, but is insoluble in water and in alcohol.

SPECIFICATIONS

Acid value. Between 6 and 16.

Drop softening point. Between 109° and 116°.

Limits of Impurities

Arsenic (as As). Not more than 3 parts per million (0.0003 percent).

Heavy metals (as Pb). Not more than 40 parts per million (0.004 percent).

Lead. Not more than 3 parts per million (0.0003 percent).

TESTS

Acid value. Determine as directed in the general procedure, page 945.

Drop softening point. Determine as directed in the general procedure, page 946, using a bath temperature of 125°.

Arsenic. Prepare a *Sample Solution* as directed in the general method under *Chewing Gum Base*, page 877. This solution meets the requirements of the *Arsenic Test*, page 865.

Heavy metals. Prepare and test a 500-mg. sample as directed in *Method II* under the *Heavy Metals Test*, page 920, using 20 mcg. of lead ion (Pb) in the control (*Solution A*).

Lead. Prepare a *Sample Solution* as directed in the general method under *Chewing Gum Base*, page 878. This solution meets the requirements of the *Lead Limit Test*, page 929, using 10 mcg. of lead ion (Pb) in the control.

Packaging and storage. Store in well-closed containers.

Functional use in foods. Masticatory substance in chewing gum base.

2,3-PENTANEDIONE

Acetyl Propionyl

$$CH_3{-}CH_2{-}\overset{\overset{\large O}{\|}}{C}{-}\overset{\overset{\large O}{\|}}{C}{-}CH_3$$

$C_5H_8O_2$ Mol. wt. 100.12

DESCRIPTION

A clear yellow to yellowish green liquid with a strong, pungent odor, which becomes typically buttery upon dilution. It is miscible with alcohol, with propylene glycol, and with fixed oils. It is soluble in mineral oil, but practically insoluble in glycerin and water.

SPECIFICATIONS

Assay. Not less than 93.0 percent of $C_5H_8O_2$.

Refractive index. Between 1.402 and 1.406 at 20°.

Solubility in alcohol. Passes test.

Specific gravity. Between 0.952 and 0.962.

Limits of Impurities

Heavy metals (as Pb). Not more than 40 parts per million (0.004 percent).

TESTS

Assay. Weigh accurately about 400 mg., and proceed as directed under *Aldehydes and Ketones-Hydroxylamine Method*, page 894, using 50.06 as the equivalence factor (*E*) in the calculation.

Refractive index, page 945. Determine with an Abbé or other refractometer of equal or greater accuracy.

Solubility in alcohol. Proceed as directed in the general method, page 899. One ml. dissolves in 3 ml. of 50 percent alcohol.

Specific gravity. Determine by any reliable method (see page 5).

Heavy metals. Prepare and test a 500-mg. sample as directed in *Method II* under the *Heavy Metals Test,* page 920, using 20 mcg. of lead ion (Pb) in the control (*Solution A*).

Packaging and storage. Store in full, tight containers, preferably glass, aluminum, tin, or tin-lined, in a cool place protected from light.

Functional use in foods. Flavoring agent.

2-PENTANONE

Methyl Propyl Ketone

$CH_3COCH_2CH_2CH_3$

$C_5H_{10}O$ Mol. wt. 86.13

DESCRIPTION

A clear, colorless, mobile liquid having a characteristic, sharp, floral, penetrating odor. It is miscible in all proportions with alcohol and with ether. One ml. dissolves in about 25 ml. of water.

SPECIFICATIONS

Assay. Not less than 95.0 percent of $C_5H_{10}O$.

Distillation range. Between 97° and 106°.

Specific gravity. Between 0.801 and 0.806.

Limits of Impurities

Acidity (as acetic acid). Not more than 0.02 percent.

Heavy metals (as Pb). Not more than 40 parts per million (0.004 percent).

Water. Not more than 0.2 percent.

TESTS

Assay. Transfer 65 ml. of 0.5 *N* hydroxylamine hydrochloride and 50.0 ml. of 0.5 *N* triethanolamine into a suitable heat-resistant pressure bottle provided with a tight closure that can be securely fastened. Replace the air in the bottle by passing a gentle stream of nitrogen for two minutes through a glass tube positioned so that the end is just above the surface of the liquid. Add about 1 gram of the sample, accurately weighed, using a suitable weighing pipet. Cap the bottle and allow the mixture to stand at room temperature for 15 minutes, swirling occasionally. Cool, if necessary, and uncap the bottle cautiously to prevent any loss of the contents. Titrate with 0.5 *N* sulfuric acid to a greenish blue end-point. Perform a residual blank titration (see *Blank Tests*, page 2). Each ml. of 0.5 *N* sulfuric acid is equivalent to 43.06 mg. of $C_5H_{10}O$.

Distillation range. Distil 100 ml. as directed in the general method, page 890.

Specific gravity. Determine by any reliable method (see page 5).

Acidity. Transfer 74 ml. (60 grams) into a 250-ml. Erlenmeyer flask, add a few drops of phenolphthalein T.S., and titrate with 0.1 *N* alcoholic potassium hydroxide to a pink end-point that persists for at least 15 seconds. Each ml. of 0.1 *N* alcoholic potassium hydroxide is equivalent to 6.01 mg. of $C_2H_4O_2$.

Heavy metals. Prepare and test a 500-mg. sample as directed in *Method II* under the *Heavy Metals Test,* page 920, using 20 mcg. of lead ion (Pb) in the control (*Solution A*).

Water. Determine by the *Karl Fischer Titrimetric Method,* page 977, using freshly distilled pyridine instead of methanol as the solvent.

Packaging and storage. Store in tight containers.

Functional use in foods. Flavoring agent.

PEPPERMINT OIL

DESCRIPTION

An essential oil obtained by steam distillation from the fresh overground parts of the flowering plant of *Mentha piperita* L. (Fam. *Labiatae*), rectified by distillation, and neither partially nor wholly demontholized. It is a colorless or pale yellow liquid, having a strong, penetrating odor of peppermint, and a pungent taste, followed by a sensation of coldness when air is drawn into the mouth.

IDENTIFICATION

Mix in a dry test tube 3 drops of the oil with 5 ml. of a solution of nitric acid in glacial acetic acid (1 in 300), and place the tube in a beaker of boiling water. A blue color develops within 5 minutes, which, on continued heating, deepens and shows a copper-colored fluorescence, and then fades, leaving a golden-yellow solution.

SPECIFICATIONS

Assay for total esters. Not less than 5.0 percent of esters, calculated as menthyl acetate ($C_{12}H_{22}O_2$).

Assay for total menthol. Not less than 50.0 percent of menthol ($C_{10}H_{20}O$).

Angular rotation. Between $-18°$ and $-32°$.

Refractive index. Between 1.459 and 1.465 at 20°.

Solubility in alcohol. Passes test.

Specific gravity. Between 0.896 and 0.908.

Limits of Impurities

Arsenic (as As). Not more than 3 parts per million (0.0003 percent).

Dimethyl sulfide. Passes test.

Heavy metals (as Pb). Not more than 40 parts per million (0.004 percent).

Lead. Not more than 10 parts per million (0.001 percent).

TESTS

Assay for total esters. Weigh accurately about 10 grams, and proceed as directed under *Ester Determination*, page 896, using 99.16 as the equivalence factor (e) in the calculation.

Assay for total menthol. Proceed as directed under *Total Alcohols*, page 893, using a 5-gram sample of the acetylized oil. Calculate the percentage of total menthol by the formula $7.814A(1 - 0.0021E)/(B - 0.021\,A)$, in which A is the difference between the number of ml. of 0.5 N hydrochloric acid required in the titration and the number of ml. of 0.5 N hydrochloric acid required in the residual blank titration, B is the weight of the sample of the acetylized oil, and E is the percentage of total esters determined and calculated as menthyl acetate ($C_{12}H_{22}O_2$).

Angular rotation. Determine in a 100-mm. tube as directed under *Optical Rotation*, page 939.

Refractive index, page 945. Determine with an Abbé or other refractometer of equal or greater accuracy.

Solubility in alcohol. Proceed as directed in the general method, page 899. One ml. dissolves in 3 ml. of 70 percent alcohol.

Specific gravity. Determine by any reliable method (see page 5).

Arsenic. A *Sample Solution* prepared as directed for organic compounds meets the requirements of the *Arsenic Test*, page 865.

Dimethyl sulfide. Distil 1 ml. from a sample of 25 ml., and carefully superimpose the distillate on 5 ml. of mercuric chloride T.S. in a test tube. A white film does not form at the zone of contact within 1 minute.

Heavy metals. Prepare and test a 500-mg. sample as directed in *Method II* under the *Heavy Metals Test*, page 920, using 20 mcg. of lead ion (Pb) in the control (*Solution A*).

Lead. A *Sample Solution* prepared as directed for organic compounds meets the requirements of the *Lead Limit Test*, page 929, using 10 mcg. of lead ion (Pb) in the control.

Packaging and storage. Store in full, tight containers in a cool place protected from light.

Functional use in foods. Flavoring agent.

PEPPER OIL, BLACK

DESCRIPTION

The volatile oil obtained by steam distillation from the dried, unripened fruit of the plant *Piper nigrum* L. (Fam. *Piperaceae*). It is an almost colorless to slightly greenish liquid, having the characteristic odor of pepper and a relatively mild taste. It is soluble in most fixed oils, in mineral oil, and in propylene glycol. It is sparingly soluble in glycerin.

SPECIFICATIONS

Angular rotation. Between −1° and −23°.

Refractive index. Between 1.479 and 1.488 at 20°.

Solubility in alcohol. Passes test.

Specific gravity. Between 0.864 and 0.884.

Limits of Impurities

Arsenic (as As). Not more than 3 parts per million (0.0003 percent).

Heavy metals (as Pb). Not more than 40 parts per million (0.004 percent).

Lead. Not more than 10 parts per million (0.001 percent).

TESTS

Angular rotation. Determine in a 100-mm. tube as directed under *Optical Rotation*, page 939.

Refractive index, page 945. Determine with an Abbé or other refractometer of equal or greater accuracy.

Solubility in alcohol. Proceed as directed in the general method, page 899. One ml. dissolves in 3 ml. of 95 percent alcohol.

Specific gravity. Determine by any reliable method (see page 5).

Arsenic. A *Sample Solution* prepared as directed for organic compounds meets the requirements of the *Arsenic Test*, page 865.

Heavy metals. Prepare and test a 500-mg. sample as directed in *Method II* under the *Heavy Metals Test*, page 920, using 20 mcg. of lead ion (Pb) in the control (*Solution A*).

Lead. A *Sample Solution* prepared as directed for organic compounds meets the requirements of the *Lead Limit Test*, page 929, using 10 mcg. of lead ion (Pb) in the control.

Packaging and storage. Store preferably in full, tight, glass containers in a cool place protected from light.

Functional use in foods. Flavoring agent.

PEPSIN

DESCRIPTION

Pepsin is a substance containing a proteolytic enzyme obtained from the glandular layer of the fresh stomach of the hog, *Sus scrofa* L. var. *domesticus* Gray (Fam. *Suidae*). It occurs as lustrous, transparent or translucent scales; as granular or spongy masses, ranging in color from weak yellow to light brown; or as a fine, white to weak yellow, amorphous powder, free from offensive odor, and having a slightly acid or salty taste. It is not more than slightly hygroscopic. Dry pepsin is not injured by heating to 100°. The activity of pepsin in solution is destroyed by alkalies or by temperatures exceeding 70°.

Pepsin is freely soluble in water, the solution being more or less opalescent. It is practically insoluble in alcohol, in chloroform, and in ether. A 1 in 50 solution of pepsin is acid to litmus.

IDENTIFICATION

A solution of pepsin yields precipitates with solutions of tannic acid or gallic acid and with solutions of the salts of many heavy metals. On heating a solution of pepsin in acidified water to 100°, it becomes milky or yields a light, flocculent precipitate and loses all proteolytic power.

SPECIFICATIONS

Assay. It digests not less than 3,000 times and not more than 3,500 times its weight of coagulated egg albumen.

Limits of Impurities

Arsenic (as As). Not more than 3 parts per million (0.0003 percent).

Heavy metals (as Pb). Not more than 10 parts per million (0.001 percent).

TESTS

Assay. Mix 35 ml. of 1.0 *N* hydrochloric acid with 385 ml. of water. Dissolve 100 mg. of pepsin, accurately weighed, in 150 ml. of this solution. Prepare a similar solution of N.F. Pepsin Reference Standard, accurately weighed. Boil one or more hen eggs for 15 minutes to provide coagulated albumen for the assay. Cool them rapidly to room temperature by immersion in cold water; remove the shell and pellicle and all of the yolk and at once rub the albumen through a clean, dry No. 40 sieve, rejecting the first portion that passes through the sieve. Place 10 grams of the succeeding well mixed portion in each of 3 widemouth bottles of about 100-ml. capacity. Immediately add 35 ml. of the dilute acid at one time or in portions and, by suitable means, thoroughly disintegrate the particles of albumen. Place the bottles in a water bath at 52°. After the contents of the bottles have reached that temperature, add 5.0 ml. of the acidified solution of pepsin to one bottle, 4.30 ml. of the same solution and 0.7 ml. of the dilute acid to another bottle, and 5.0 ml. of the Reference Standard solution to the third bottle. At once stopper the bottles securely, invert them 3 times, and maintain them at 52° for 2 hours and 30 minutes, agitating the contents equally every 10 minutes by inverting the bottles once. Remove the bottles from the bath, pour the contents into 100-ml. conically shaped measuring vessels, having diameters not exceeding 1 cm. at the bottom and complying in other respects with the water and sediment tube A.S.T.M. Standard Method D 96-68, graduated from 0 to 0.5 ml. in 0.05-ml. graduations; from 0.5 to 2 ml. in 0.1-ml. graduations; from 2 to 3 ml. in 0.2-ml. graduations; from 3 to 5 ml. in 0.5-ml. graduations; from 5 to 10 ml. in 1-ml. graduations; from 10 to 25 ml. in 5-ml. graduations; and with graduation marks at 50, 75, and 100 ml. points. Transfer the undigested egg albumen which adheres to the sides of the bottles to the respective measuring vessels with the aid of small portions of water until 50 ml. has been used for each. Mix the contents of each measuring vessel, and allow them to stand for 30 minutes. The volume of the undissolved albumen in the measuring vessel corresponding to 5.0 ml. of the solution of pepsin being assayed does not exceed the volume of the undissolved albumen in the measuring vessel corresponding to 5.0 ml. of the Reference Standard solution, and the volume of the undissolved albumen in the measuring vessel corresponding to 4.30 ml. of the solution of pepsin being assayed is not less than the volume of the undissolved albumen in the measuring vessel corresponding to 5.0 ml. of the Reference Standard solution. (*Note:* Other measuring vessels than the type described in this monograph may be used if they are of such design and graduation as to permit accurate measurement of the residue.)

Arsenic. A *Sample Solution* prepared as directed for organic compounds meets the requirements of the *Arsenic Test,* page 865.

Heavy metals. Prepare and test a 2-gram sample as directed in *Method II* under the *Heavy Metals Test,* page 920, using 20 mcg. of lead ion (Pb) in the control (*Solution A*).

Packaging and storage. Store in tight containers and avoid exposure to excessive heat.

Functional use in foods. Proteolytic enzyme.

PETITGRAIN OIL, PARAGUAY

DESCRIPTION

The volatile oil obtained by steam distillation from the leaves and small twigs of the bitter orange tree, *Citrus aurantium* L. subspecies *amara.* It is a yellow to brownish yellow liquid having a somewhat harsh, bitter-sweet, floral odor. It is soluble in most fixed oils and is soluble, with opalescence or turbidity, in mineral oil and in propylene glycol. It is relatively insoluble in glycerin.

SPECIFICATIONS

Assay. Not less than 45.0 percent and not more than 60 percent of esters, calculated as linalyl acetate ($C_{12}H_{20}O_2$).

Angular rotation. Between $-4°$ and $+1°$.

Refractive index. Between 1.455 and 1.462 at 20°.

Solubility in alcohol. Passes test.

Specific gravity. Between 0.878 and 0.889.

TESTS

Assay. Weigh accurately about 2 grams, and proceed as directed under *Ester Determination*, page 896, using 98.15 as the equivalence factor (*e*) in the calculation.

Angular rotation. Determine in a 100-mm. tube as directed under *Optical Rotation*, page 939.

Refractive index, page 945. Determine with an Abbé or other refractometer of equal or greater accuracy.

Solubility in alcohol. Proceed as directed in the general method, page 899. One ml. dissolves in 4 ml. of 70 percent alcohol. The solution usually develops opalescence or turbidity upon further dilution.

Specific gravity. Determine by any reliable method (see page 5).

Packaging and storage. Store in full, tight, preferably glass, tin-lined, or aluminum containers in a cool place protected from light.

Functional use in foods. Flavoring agent.

PETROLATUM

White Petrolatum; Yellow Petrolatum

DESCRIPTION

A purified mixture of semisolid hydrocarbons obtained from petroleum, occurring as an unctuous mass, and varying in color from white to yellowish or light amber. It is transparent in thin layers and has not more than a slight fluorescence, even after being melted. It is free or nearly free from odor and taste. It is insoluble in water and is almost insoluble in cold or hot alcohol and in cold absolute alcohol. It is soluble in ether, in solvent hexane, and in most fixed and volatile oils, and is freely soluble in benzene, in carbon disulfide, in chloroform, and in turpentine oil. It may contain any antioxidant permitted by the federal Food and Drug Administration, in an amount not greater than that required to produce its intended effect.

SPECIFICATIONS

Consistency. Passes test (between 100 and 275).

Melting range. Between 38° and 60°.

Specific gravity. Between 0.815 and 0.880 at 60°.

Limits of Impurities

Acidity or alkalinity. Passes test.

Color. Passes test.
Fixed oils, fats, and rosin. Passes test.
Organic acids. Passes test.
Residue on ignition. Passes test.
Ultraviolet absorption (polynuclear hydrocarbons). Passes test.

TESTS

Consistency

Apparatus. Determine the consistency of petrolatum by means of a penetrometer fitted with a polished cone-shaped metal plunger weighing 150 grams, having a detachable steel tip of the following dimensions: the tip of the cone has an angle of 30°, the point being truncated to a diameter of 0.38 ± 0.03 mm., the base of the tip is 8.38 ± 0.05 mm. in diameter, and the length of the tip is 15 ± 0.25 mm. The remaining portion of the cone has an angle of 90°, is 28.2 mm. in height, and has a maximum diameter at the base of 65.1 mm. The containers for the test are flat-bottomed metal or glass cylinders that are 102 ± 6 mm. in diameter and not less than 60 mm. in height.

Procedure. Melt a quantity of the sample at 82 ± 2.5°, and pour into one or more of the containers, filling to within 6 mm. of the rim. Cool at 25 ± 2.5° over a period of not less than 16 hours, protecting from drafts. Two hours before the test, place the containers in a water bath at 25 ± 0.5°. If the room temperature is below 23.5° or above 26.5°, adjust the temperature of the cone to 25 ± 0.5° by placing it in the water bath.

Without disturbing the surface of the sample, place the container on the penetrometer table, and lower the cone until the tip just touches the top surface of the sample at a spot 25 mm. to 38 mm. from the edge of the container. Adjust the zero setting, and quickly release the plunger, then hold it free for 5 seconds. Secure the plunger, and read the total penetration from the scale. Make 3 or more trials, each so spaced that there is no overlapping of the areas of penetration. When the penetration exceeds 20 mm., use a separate container of the sample for each trial. Read the penetration to the nearest 0.1 mm. Calculate the average of the three or more readings, and conduct further trials to a total of ten if the individual results differ from the average by more than ±3 percent. The final average of the trials is not less than 10.0 mm. and not more than 27.5 mm., indicating a consistency value between 100 and 275.

Melting range, page 931. Determine as directed in *Procedure for Class III.*

Specific gravity. Determine by any reliable method (see page 5).

Acidity or alkalinity. Introduce 35 grams of the sample into a 250-ml. separator, add 100 ml. of boiling water, and shake vigorously for 5 minutes. After the petrolatum and water have separated, draw off the water into a casserole, wash the sample in the separator with two 50-ml. portions of boiling water, and add the washings to the casserole. To the accumulated 200 ml. of water add 1 drop of phenolphthalein T.S., and boil. The solution does not acquire a pink color. If the addition of phenolphthalein produces no pink color, add 0.1 ml. of methyl orange T.S. No red or pink color is produced.

Color. Melt about 10 grams on a steam bath, and pour about 5 ml. of the liquid into a 150 × 50-mm. clear-glass bacteriological test tube, keeping the sample melted. The petrolatum is not darker than a solution made by mixing 3.8 ml. of ferric chloride C.S. and 1.2 ml. of cobaltous chloride C.S. in a similar tube, the comparison of the two being made in reflected light against a white background, holding the sample tube directly against the background at such an angle that there is no fluorescence.

Fixed oils, fats, and rosin. Digest 10 grams of the sample at 100° with 10 grams of sodium hydroxide and 50 ml. of water for 30 minutes. Separate the water layer, and add to it an excess of diluted sulfuric acid T.S. No oily or solid matter separates.

Organic acids. Weigh 20 grams of the sample, add 50 ml. of alcohol, previously neutralized to phenolphthalein T.S. with sodium hydroxide, and 50 ml. of water, agitate thoroughly, and heat to boiling. Add 1 ml. of phenolphthalein T.S., and titrate rapidly, with vigorous agitation, to the production of a sharp pink end-point, noting the change in the alcohol-water layer. Not more than 0.4 ml. of 0.1 *N* sodium hydroxide is required.

Residue on ignition. Heat 4 grams of the sample in an open porcelain or platinum dish over a Bunsen flame. It volatilizes without emitting any acrid odor, and on ignition yields not more than 0.05 percent of residue.

Ultraviolet absorption. It meets the ultraviolet absorbance specifications required by the federal Food and Drug Administration for petrolatum.

Packaging and storage. Store in tight containers.

Functional use in foods. Defoaming agent; lubricant; protective coating; release agent.

α-PHELLANDRENE

p-Mentha-1,5-diene

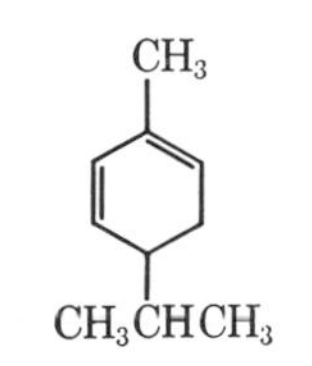

$C_{10}H_{16}$ Mol. wt. 136.24

DESCRIPTION

A colorless to slightly yellow liquid having a herbaceous odor with a mint-like background. It is freely soluble in alcohol, but insoluble in water.

SPECIFICATIONS

Angular rotation. Between $-60°$ and $-100°$.

Refractive index. Between 1.473 and 1.477 at 20°.

Solubility in alcohol. Passes test.

Specific gravity. Between 0.840 and 0.855.

Limits of Impurities

Arsenic (as As). Not more than 3 parts per million (0.0003 percent).

Heavy metals (as Pb). Not more than 40 parts per million (0.004 percent).

Lead. Not more than 10 parts per million (0.001 percent).

TESTS

Angular rotation. Determine in a 100-mm. tube as directed under *Optical Rotation*, page 939.

Refractive index, page 945. Determine with an Abbé or other refractometer of equal or greater accuracy.

Solubility in alcohol. Proceed as directed in the general method, page 899. One ml. dissolves in 1 ml. of 95 percent alcohol to form a clear solution.

Specific gravity. Determine by any reliable method (see page 5).

Arsenic. A *Sample Solution* prepared as directed for organic compounds meets the requirements of the *Arsenic Test*, page 865.

Heavy metals. Prepare and test a 500-mg. sample as directed in *Method II* under the *Heavy Metals Test*, page 920, using 20 mcg. of lead ion (Pb) in the control (*Solution A*).

Lead. A *Sample Solution* prepared as directed for organic compounds meets the requirements of the *Lead Limit Test*, page 929, using

10 mcg. of lead ion (Pb) in the control.

Packaging and storage. Store in full, tight, preferably glass, tin-lined or other suitably lined containers in a cool place protected from light.

Functional use in foods. Flavoring agent.

PHENETHYL ACETATE

2-Phenylethyl Acetate

C_6H_5—$CH_2CH_2OOCCH_3$

$C_{10}H_{12}O_2$ Mol. wt. 164.20

DESCRIPTION

A colorless liquid with a floral odor. It is soluble in most fixed oils, in mineral oil, and in propylene glycol. It is insoluble in glycerin and in water.

SPECIFICATIONS

Assay. Not less than 98.0 percent of $C_{10}H_{12}O_2$.

Refractive index. Between 1.497 and 1.501 at 20°.

Solubility in alcohol. Passes test.

Specific gravity. Between 1.030 and 1.034.

Limits of Impurities

Acid value. Not more than 1.0.

TESTS

Assay. Weigh accurately about 1 gram, and proceed as directed under *Ester Determination*, page 896, using 82.10 as the equivalence factor (e) in the calculation.

Refractive index, page 945. Determine with an Abbé or other refractometer of equal or greater accuracy.

Solubility in alcohol. Proceed as directed in the general method, page 899. One ml. dissolves in 2 ml. of 70 percent alcohol, and remains clear upon dilution to 10 ml.

Specific gravity. Determine by any reliable method (see page 5).

Acid value. Determine as directed in the general method, page 893.

Packaging and storage. Store in full, tight, preferably glass, aluminum, or tin-lined containers in a cool place protected from light.

Functional use in foods. Flavoring agent.

PHENETHYL ALCOHOL

2-Phenylethyl Alcohol

C_6H_5—CH_2CH_2OH

$C_8H_{10}O$ Mol. wt. 122.17

DESCRIPTION

A colorless liquid with a roselike odor. It is soluble in most fixed oils, in glycerin, and in propylene glycol. It is slightly soluble in mineral oil.

SPECIFICATIONS

Odor. Passes test.

Refractive index. Between 1.531 and 1.534 at 20°.

Solubility in alcohol. Passes test.

Solubility in water. Passes test.

Specific gravity. Between 1.017 and 1.020.

Limits of Impurities

Chlorinated compounds. Passes test.

TESTS

Odor. Mix thoroughly 2 ml. of the sample with 20 ml. of ice cold, odorless water. No off-odor should be discernible in the mixture.

Refractive index, page 945. Determine with an Abbé or other refractometer of equal or greater accuracy.

Solubility in alcohol. Proceed as directed in the general method, page 899. One ml. dissolves in 2 ml. of 50 percent alcohol and remains clear on dilution to 10 ml.

Solubility in water. Transfer 2.0 ml. into a glass-stoppered 100-ml. graduated cylinder, fill to the mark with water, and shake at least 15 seconds. After the air bubbles have risen, the solution should be free from oil droplets, and should be no more turbid than 100 ml. of water to which 1 drop of hydrochloric acid, 2 ml. of barium chloride T.S., and 0.2 ml. of 0.1 *N* sulfuric acid have been added with shaking.

Specific gravity. Determine by any reliable method (see page 5).

Chlorinated compounds. Proceed as directed in the general method, page 895.

Packaging and storage. Store in full, tight, preferably galvanized iron, glass, or suitably lined containers in a cool place protected from light.

Functional use in foods. Flavoring agent.

PHENETHYL ISOBUTYRATE

$$C_6H_5-CH_2CH_2OOCCH(CH_3)CH_3$$

$C_{12}H_{16}O_2$ Mol. wt. 192.26

DESCRIPTION

A colorless to slightly yellow liquid having a fruity and somewhat roselike odor. It is soluble in alcohol, but practically insoluble in water.

SPECIFICATIONS

Assay. Not less than 98.0 percent of $C_{12}H_{16}O_2$.

Refractive index. Between 1.486 and 1.490 at 20°.

Solubility in alcohol. Passes test.

Specific gravity. Between 0.987 and 0.990.

Limits of Impurities

Acid value. Not more than 1.0.

TESTS

Assay. Weigh accurately about 1.2 grams, and proceed as directed under *Ester Determination*, page 896, using 96.13 as the equivalence factor (*e*) in the calculation.

Refractive index, page 945. Determine with an Abbé or other refractometer of equal or greater accuracy.

Solubility in alcohol. Proceed as directed in the general method, page 899. One ml. dissolves in 3 ml. of 80 percent alcohol to form a clear solution.

Specific gravity. Determine by any reliable method (see page 5).

Acid value. Determine as directed in the general method, page 893.

Packaging and storage. Store in full, tight, preferably glass, tin-lined or other suitably lined containers in a cool place protected from light.

Functional use in foods. Flavoring agent.

PHENETHYL ISOVALERATE

$$C_6H_5-CH_2CH_2OOCCH_2CH(CH_3)CH_3$$

$C_{13}H_{18}O_2$ Mol. wt. 206.29

DESCRIPTION

A colorless to slightly yellow liquid having a fruity or roselike odor. It is miscible with alcohol, but insoluble in water.

SPECIFICATIONS

Assay. Not less than 98.0 percent of $C_{13}H_{18}O_2$.

Refractive index. Between 1.484 and 1.486 at 20°.

Solubility in alcohol. Passes test.

Specific gravity. Between 0.973 and 0.976.

Limits of Impurities

Acid value. Not more than 1.0.

TESTS

Assay. Weigh accurately about 1.5 grams, and proceed as directed under *Ester Determination*, page 896, using 103.15 as the equivalence factor (e) in the calculation.

Refractive index, page 945. Determine with an Abbé or other refractometer of equal or greater accuracy.

Solubility in alcohol. Proceed as directed in the general method, page 899. One ml. dissolves in 11 ml. of 70 percent alcohol to form a clear solution.

Specific gravity. Determine by any reliable method (see page 5).

Acid value. Determine as directed in the general method, page 893.

Packaging and storage. Store in full, tight, preferably glass, tin-lined or other suitably lined containers in a cool place protected from light.

Functional use in foods. Flavoring agent.

PHENETHYL PHENYLACETATE

$C_6H_5-CH_2CH_2OOCCH_2-C_6H_5$

$C_{16}H_{16}O_2$ Mol. wt. 240.30

DESCRIPTION

A colorless to slightly yellow liquid having a rose-hyacinth type odor. It may solidify at temperatures below 26°. It is soluble in alcohol, but insoluble in water.

SPECIFICATIONS

Assay. Not less than 98.0 percent of $C_{16}H_{16}O_2$.

Solidification point. Not less than 26°.

Solubility in alcohol. Passes test.

Specific gravity. Between 1.079 and 1.082.

Limits of Impurities

Acid value. Not more than 1.0.

TESTS

Assay. Weigh accurately about 1.5 grams, and proceed as directed under *Ester Determination*, page 896, using 120.15 as the equivalence factor (*e*) in the calculation.

Solidification point. Determine as directed in the general method, page 954.

Solubility in alcohol. Proceed as directed in the general method, page 899. One ml. dissolves in 4 ml. of 90 percent alcohol to form a clear solution.

Specific gravity. Determine by any reliable method (see page 5).

Acid value. Determine as directed in the general method, page 893.

Packaging and storage. Store in full, tight, preferably glass, tin-lined or other suitably lined containers in a cool place protected from light.

Functional use in foods. Flavoring agent.

PHENETHYL SALICYLATE

HO

C_6H_5—CH_2CH_2OOC—C_6H_4

$C_{15}H_{14}O_3$ Mol. wt. 242.28

DESCRIPTION

White crystals having a balsamic-rose odor. It is soluble in alcohol, but insoluble in water.

SPECIFICATIONS

Assay. Not less than 98.0 percent of $C_{15}H_{14}O_3$.

Solidification point. Not lower than 41°.

Solubility in alcohol. Passes test.

Limits of Impurities

Acid value. Not more than 1.0.

TESTS

Assay. Weigh accurately about 1.5 grams, and proceed as directed under *Ester Determination*, page 896, using phenol red T.S. as the indicator and 121.14 as the equivalence factor (e) in the calculation.

Solidification point. Determine as directed in the general method, page 954.

Solubility in alcohol. Proceed as directed in the general method, page 899. One gram dissolves in 20 ml. of 95 percent alcohol to form a clear solution.

Acid value. Determine as directed in the general method, page 893, using phenol red T.S. as the indicator.

Packaging and storage. Store in well-closed containers.

Functional use in foods. Flavoring agent.

PHENOXYETHYL ISOBUTYRATE

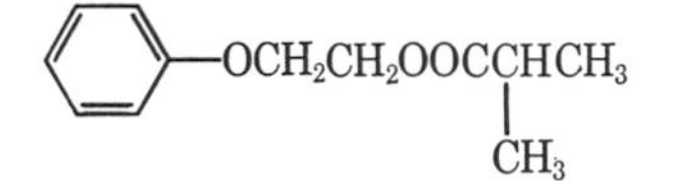

$C_{12}H_{16}O_3$ Mol. wt. 208.26

DESCRIPTION

A clear colorless liquid having a characteristic honey-rose type odor.

It is miscible with alcohol, with chloroform, and with ether, but it is practically insoluble in water.

SPECIFICATIONS

Assay. Not less than 97.0 percent of $C_{12}H_{16}O_3$.

Refractive index. Between 1.492 and 1.495 at 20°.

Solubility in alcohol. Passes test.

Specific gravity. Between 1.044 and 1.048.

Limits of Impurities

Acid value. Not more than 1.0.

TESTS

Assay. Weigh accurately about 1.4 grams, and proceed as directed under *Ester Determination*, page 896, using 104.13 as the equivalence factor (e) in the calculation.

Refractive index, page 945. Determine with an Abbé or other refractometer of equal or greater accuracy.

Solubility in alcohol. Proceed as directed in the general method, page 899. One ml. dissolves in 2 ml. of 70 percent alcohol to form a clear solution.

Specific gravity. Determine by any reliable method (see page 5).

Acid value. Determine as directed in the general method, page 893.

Packaging and storage. Store in full, tight, preferably glass, tin-lined or other suitably lined containers in a cool place protected from light.

Functional use in foods. Flavoring agent.

PHENYLACETALDEHYDE

α-Toluic Aldehyde

C_6H_5—CH_2CHO

C_8H_8O Mol. wt. 120.15

DESCRIPTION

A colorless to slightly yellow, oily liquid having a harsh odor, which on dilution becomes suggestive of hyacinth. It tends to become more viscous on aging. It is soluble in most fixed oils and in propylene glycol. It is insoluble in glycerin and in mineral oil.

SPECIFICATIONS

Assay. Not less than 90.0 percent of C_8H_8O.

Refractive index. Between 1.524 and 1.532 at 20°.

Solubility in alcohol. Passes test.

Specific gravity. Between 1.025 and 1.035.

Limits of Impurities

Acid value. Not more than 5.0.

TESTS

Assay. Weigh accurately about 1 gram, and proceed as directed under *Aldehydes*, page 894, using 60.08 as the equivalence factor (E) in the calculation.

Refractive index, page 945. Determine with an Abbé or other refractometer of equal or greater accuracy.

Solubility in alcohol. Proceed as directed in the general method, page 899. One ml. dissolves in 2 ml. of 80 percent alcohol.

Specific gravity. Determine by any reliable method (see page 2).

Acid value. Determine as directed in the general method, page 893.

Packaging and storage. Store in full, tight, preferably glass, or aluminum containers in a cool place protected from light. Phenyl acetaldehyde may polymerize and become viscous and precipitate crystals on long storage.

Functional use in foods. Flavoring agent.

PHENYLACETALDEHYDE DIMETHYL ACETAL

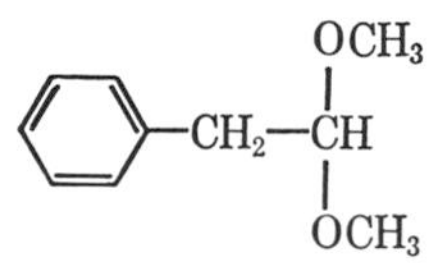

$C_{10}H_{14}O_2$ Mol. wt. 166.22

DESCRIPTION

A colorless liquid with a strong characteristic odor. It is soluble in most fixed oils and in propylene glycol. It is insoluble in glycerin and in mineral oil.

SPECIFICATIONS

Assay. Not less than 95.0 percent of $C_{10}H_{14}O_2$.

Refractive index. Between 1.493 and 1.496 at 20°.

Solubility in alcohol. Passes test.
Specific gravity. Between 1.000 and 1.006.
Limits of Impurities
Acid value. Not more than 1.0.
Chlorinated compounds. Passes test.
Free aldehydes. Not more than 1 percent, calculated as phenyl acetaldehyde (C_8H_8O).

TESTS

Assay. Weigh accurately about 1 gram, and proceed as directed under *Acetals*, page 892, using 83.11 as the equivalence factor (f) in the calculation.

Refractive index, page 945. Determine with an Abbé or other refractometer of equal or greater accuracy.

Solubility in alcohol. Proceed as directed in the general method, page 899. One ml. dissolves in 2 ml. of 70 percent alcohol and remains clear upon dilution to 10 ml.

Specific gravity. Determine by any reliable method (see page 5).

Acid value. Determine as directed in the general method, page 893.

Chlorinated compounds. Proceed as directed in the general method, page 895.

Free aldehydes. Weigh accurately about 5 grams, and proceed as directed under *Aldehydes*, page 894, using 60.07 as the equivalence factor (E) in the calculation for phenyl acetaldehyde (C_8H_8O).

Packaging and storage. Store in full, tight, preferably glass or aluminum containers in a cool place protected from light.
Functional use in foods. Flavoring agent.

PHENYLACETIC ACID

α-Toluic Acid

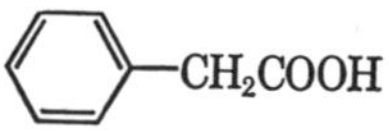

C_6H_5—CH_2COOH

$C_8H_8O_2$ Mol. wt. 136.15

DESCRIPTION

A glistening white, crystalline solid with a persistent disagreeable odor, which in diluted solutions becomes somewhat suggestive of geranium leaf and rose. It is soluble in most fixed oils and in glycerin. It is slightly soluble in water, but it is insoluble in mineral oil.

SPECIFICATIONS

Assay. Not less than 99.0 percent of $C_8H_8O_2$.

Melting range. Between 76° and 78°.

Limits of Impurities

Arsenic (as As). Not more than 3 parts per million (0.0003 percent).

Heavy metals (as Pb). Not more than 40 parts per million (0.004 percent).

Lead. Not more than 10 parts per million (0.001 percent).

TESTS

Assay. Dissolve about 500 mg., previously dried over sulfuric acid for 3 hours and accurately weighed, in 25 ml. of diluted alcohol (1 in 2) which has been neutralized with 0.1 *N* sodium hydroxide, using phenolphthalein T.S. as indicator. Titrate the solution with 0.1 *N* sodium hydroxide to a pink end-point. Each ml. of 0.1 *N* sodium hydroxide is equivalent to 13.62 mg. of $C_8H_8O_2$.

Melting range, page 931. Determine as directed for *Class Ia.*

Arsenic. A *Sample Solution* prepared as directed for organic compounds meets the requirements of the *Arsenic Test,* page 865.

Heavy metals. Prepare and test a 500-mg. sample as directed in *Method II* under the *Heavy Metals Test,* page 920, using 20 mcg. of lead ion (Pb) in the control (*Solution A*).

Lead. A *Sample Solution* prepared as directed for organic compounds meets the requirements of the *Lead Limit Test,* page 929, using 10 mcg. of lead ion (Pb) in the control.

Packaging and storage. Store in full, tight, preferably glass, paper-lined fiberboard, pressed cardboard, or wooden containers.

Functional use in foods. Flavoring agent.

DL-PHENYLALANINE

DL-α-Amino-β-phenylpropionic Acid

$C_6H_5CH_2CH(NH_2)COOH$

$C_9H_{11}NO_2$ Mol. wt. 165.19

DESCRIPTION

White, odorless, crystalline platelets. It is soluble in water, in dilute mineral acids, and in solutions of alkali hydroxides. It is very slightly soluble in alcohol. It is optically inactive.

IDENTIFICATION

A. Heat 5 ml. of a 1 in 1000 solution with 1 ml. of triketohydrindene hydrate T.S. A reddish purple color is produced.

B. Heat 5 ml. of a 1 in 100 solution with a few drops of potassium dichromate T.S. A characteristic odor is evolved.

C. To a 10 mg. sample add 500 mg. of potassium nitrate and 2 ml. of sulfuric acid and heat the mixture on a water bath for 20 minutes. Cool, add 2 ml. of hydroxylamine T.S., immerse in ice water for 10 minutes, and then add 10 ml. of sodium hydroxide T.S. A reddish violet color is produced.

SPECIFICATIONS

Assay. Not less than 98.0 percent and not more than 102.0 percent of $C_9H_{11}NO_2$ after drying.

Nitrogen (Total). Between 8.3 percent and 8.65 percent.

Limits of Impurities

Ammonium salts (as NH_3). Not more than 300 parts per million (0.03 percent).

Arsenic (as As). Not more than 3 parts per million (0.0003 percent).

Chloride. Not more than 200 parts per million (0.02 percent).

Heavy metals (as Pb). Not more than 20 parts per million (0.002 percent).

Iron. Not more than 50 parts per million (0.005 percent).

Lead. Not more than 10 parts per million (0.001 percent).

Loss on drying. Not more than 0.3 percent.

Residue on ignition. Not more than 0.3 percent.

Sulfate. Not more than 400 parts per million (0.04 percent).

TESTS

Assay. Transfer about 500 mg., previously dried at 105° for 2 hours and accurately weighed, into a 250-ml. flask. Dissolve the sample in 75 ml. of glacial acetic acid, add 2 drops of crystal violet T.S., and titrate with 0.1 *N* perchloric acid to a bluish green end-point. Perform a blank determination (see page 2) and make any necessary correction. Each ml. of 0.1 *N* perchloric acid is equivalent to 16.52 mg. of $C_9H_{11}NO_2$.

Nitrogen (Total). Determine as directed under *Nitrogen Determination*, page 937, using about 300 mg. of a sample, previously dried and accurately weighed.

Ammonium salts

Ammonium chloride standard solution. Dissolve 78.5 mg. of ammonium chloride in sufficient water to make 250.0 ml., and dilute 10.0 ml. of this solution to 100.0 ml. with water. The dilute solution contains the equivalent of 10 mcg. of NH_3 in each ml.

Procedure. Place 5 grams of sodium hydroxide pellets and 300 ml. of water in a 500 ml. distillation flask. Remove ammonia from the solution by distilling until 25 ml. of the distillate gives no color with 0.5 ml. of alkaline mercuric-potassium iodide T.S. Allow the solution in the distillation flask to cool and add 50 mg. of phenylalanine. Distil, collecting two 25-ml. fractions in 50-ml. Nessler tubes. Add 0.5 ml. of alkaline mercuric-potassium iodide T.S. to the distillates and to a series of standards, prepared by dilution of the *Ammonium chloride standard solution*, containing the equivalent of 0, 5, 10, 15, and 20 mcg. of ammonia in 25 ml. of solution. After allowing 10 minutes for color development, determine the amount of ammonia present in the distillates by comparing their color intensities with those of the standards. If the second 25 ml. of distillate from the phenylalanine sample contains appreciable ammonia, distil and collect additional 25-ml. fractions and determine their ammonia content.

Arsenic. A *Sample Solution* prepared as directed for organic compounds meets the requirements of the *Arsenic Test*, page 865.

Chloride, page 879. Any turbidity produced by a 100-mg. sample does not exceed that shown in a control containing 20 mcg. of chloride ion (Cl).

Heavy metals. Prepare and test a 1-gram sample as directed in *Method II* under the *Heavy Metals Test*, page 920, using 20 mcg. of lead ion (Pb) in the control (*Solution A*).

Iron. To the ash obtained in the test for *Residue on ignition* add 2 ml. of dilute hydrochloric acid (1 in 2), and evaporate to dryness on a steam bath. Dissolve the residue in 1 ml. of hydrochloric acid and dilute with water to 50 ml. Dilute 10 ml. of this solution to 40 ml. with water and add 40 mg. of ammonium persulfate crystals and 10 ml. of ammonium thiocyanate T.S. Any red or pink color does not exceed that produced by 1.0 ml. of *Iron Standard Solution* (10 mcg. Fe) in an equal volume of a solution containing the quantities of the reagents used in the test.

Lead. A *Sample Solution* prepared as directed for organic compounds meets the requirements of the *Lead Limit Test*, page 929, using 10 mcg. of lead ion (Pb) in the control.

Loss on drying, page 931. Dry at 105° for 2 hours.

Residue on ignition, page 945. Ignite 1 gram as directed in the general method.

Sulfate, page 879. Any turbidity produced by a 500-mg. sample does not exceed that shown in a control containing 200 mcg. of sulfate (SO_4).

Packaging and storage. Store in well-closed containers.

Functional use in foods. Nutrient; dietary supplement.

L-PHENYLALANINE

L-α-Amino-β-phenylpropionic Acid

$C_6H_5CH_2CH(NH_2)COOH$

$C_9H_{11}NO_2$ Mol. wt. 165.19

DESCRIPTION

Colorless or white plate-like crystals or a white crystalline powder having a slight characteristic odor and a slightly bitter taste. One gram is soluble in about 35 ml. of water. It is slightly soluble in alcohol, in dilute mineral acids, and in alkali hydroxide solutions. It melts with decomposition at about 283°. The pH of a 1 in 100 solution is between 5.4 and 6.0.

IDENTIFICATION

A. Heat 5 ml. of a 1 in 1000 solution with 1 ml. of triketohydrindene hydrate T.S. A reddish purple color is produced.

B. Heat 5 ml. of a 1 in 100 solution with a few drops of potassium dichromate T.S. A characteristic odor is evolved.

C. To a 10 mg. sample add 500 mg. of potassium nitrate and 2 ml. of sulfuric acid and heat the mixture on a water bath for 20 minutes. Cool, add 2 ml. of hydroxylamine T.S., immerse in ice water for 10 minutes, and then add 10 ml. of sodium hydroxide T.S. A reddish violet color is produced.

SPECIFICATIONS

Assay. Not less than 98.5 percent and not more than the equivalent of 102.0 percent of $C_9H_{11}NO_2$, calculated on the dried basis.

Specific rotation, $[\alpha]_D^{20°}$. Between −33.0° and −35.2°, on the dried basis.

Limits of Impurities

Arsenic (as As). Not more than 3 parts per million (0.0003 percent).

Heavy metals (as Pb). Not more than 20 parts per million (0.002 percent).

Lead. Not more than 10 parts per million (0.001 percent).

Loss on drying. Not more than 0.3 percent.

Residue on ignition. Not more than 0.1 percent.

TESTS

Assay. Dissolve about 300 mg., accurately weighed, in 3 ml. of formic acid and 50 ml. of glacial acetic acid, add 2 drops of crystal violet T.S., and titrate with 0.1 *N* perchloric acid to a bluish green end-point. Each ml. of 0.1 *N* perchloric acid is equivalent to 16.52 mg. of $C_9H_{11}NO_2$.

Specific rotation, page 939. Determine in a solution containing 2 grams of previously dried sample in sufficient water to make 100 ml.

Arsenic. A *Sample Solution* prepared as directed for organic compounds meets the requirements of the *Arsenic Test,* page 865.

Heavy metals. Prepare and test a 1-gram sample as directed in *Method II* under the *Heavy Metals Test,* page 920, using 20 mcg. of lead ion (Pb) in the control.

Lead. A *Sample Solution* prepared as directed for organic compounds meets the requirements of the *Lead Limit Test,* page 929, using 10 mcg. of lead ion (Pb) in the control.

Loss on drying, page 931. Dry at 105° for 3 hours.

Residue on ignition, page 945. Ignite 1 gram as directed in the general method.

Packaging and storage. Store in well-closed, light-resistant containers.

Functional use in foods. Nutrient; dietary supplement.

3-PHENYL-1-PROPANOL

Phenylpropyl Alcohol; Hydrocinnamyl Alcohol

$C_6H_5-CH_2CH_2CH_2OH$

$C_9H_{12}O$ Mol. wt. 136.20

DESCRIPTION

A colorless, slightly viscous liquid having a characteristic sweet hyacinth-mignonette odor. It is soluble in most fixed oils and in propylene glycol. It is insoluble in glycerin and in mineral oil.

SPECIFICATIONS

Assay. Not less than 98.0 percent of $C_9H_{12}O$.

Refractive index. Between 1.524 and 1.528 at 20°.

Solubility in alcohol. Passes test.

Specific gravity. Between 0.998 and 1.002.

Limits of Impurities

Aldehydes. Not more than 0.5 percent, calculated as phenyl propyl aldehyde ($C_9H_{10}O$).

TESTS

Assay. Proceed as directed under *Total Alcohols,* page 893. Weigh accurately about 1 gram of the acetylated alcohol for the saponification, and use 68.10 as the equivalence factor (*e*) in the calculation.

Refractive index, page 945. Determine with an Abbé or other refractometer of equal or greater accuracy.

Solubility in alcohol. Proceed as directed in the general method, page 899. One ml. dissolves in 1 ml. of 70 percent alcohol.

Specific gravity. Determine by any reliable method (see page 5).

Aldehydes. Weigh accurately about 5 grams, and proceed as directed under *Aldehydes*, page 894, using 67.09 as the equivalence factor (*E*) in the calculation for phenyl propyl aldehyde ($C_9H_{10}O$).

Packaging and storage. Store in full, tight, preferably glass, tin-lined, or good quality iron containers in a cool place protected from light. Aluminum should not be used.

Functional use in foods. Flavoring agent.

2-PHENYLPROPIONALDEHYDE

Hydratropic Aldehyde; α-Methyl Phenylacetaldehyde

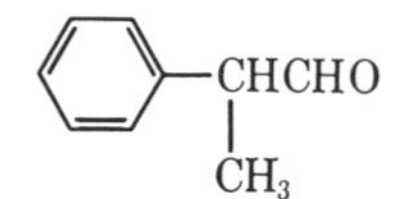

$C_9H_{10}O$ Mol. wt. 134.18

DESCRIPTION

A pale yellow to water-white liquid having a characteristic floral odor. It is soluble in most fixed oils and in mineral oil. It is insoluble in glycerin, but is sparingly soluble in propylene glycol.

SPECIFICATIONS

Assay. Not less than 95.0 percent of $C_9H_{10}O$.

Refractive index. Between 1.515 and 1.520 at 20°.

Specific gravity. Between 0.998 and 1.006.

Limits of Impurities

Acid value. Not more than 5.0.

TESTS

Assay. Weigh accurately about 1 gram, and proceed as directed under *Aldehydes*, page 894, using 67.09 as the equivalence factor (*E*) in the calculation.

Refractive index, page 945. Determine with an Abbé or other refractometer of equal or greater accuracy.

Specific gravity. Determine by any reliable method (see page 5).

Acid value. Determine as directed in the general method, page 893.

Packaging and storage. Store in full, tight containers in a cool place protected from light.

Functional use in foods. Flavoring agent.

3-PHENYLPROPIONALDEHYDE

Hydrocinnamaldehyde; Phenylpropyl Aldehyde

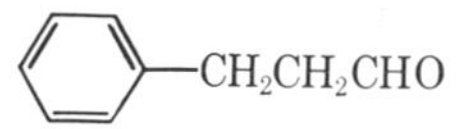

$C_9H_{10}O$ Mol. wt. 134.18

DESCRIPTION

A colorless to slightly yellow liquid having a strong, pungent, floral character of a hyacinth type. It is miscible with alcohol and with ether, but insoluble in water.

SPECIFICATIONS

Assay. Not less than 90 percent, by volume, of aldehydes.

Refractive index. Between 1.520 and 1.532 at 20°.

Solubility in alcohol. Passes test.

Specific gravity. Between 1.010 and 1.020.

Limits of Impurities

Acid value. Not more than 10.0.

Chlorinated compounds. Passes test.

TESTS

Assay. Pipet 5 ml. of the sample into a 100-ml. cassia flask, and add 70 ml. of a 1 in 10, weight in weight, solution of sodium metabisulfite. Warm the mixture on a water bath to 50°–60°, and shake the flask vigorously for 15 minutes. When the liquids have separated completely, add sufficient sodium metabisulfite solution to raise the lower level of the oily layer within the graduated portion of the neck of the flask. Not more than 0.5 ml. of oil separates.

Refractive index, page 945. Determine with an Abbé or other refractometer of equal or greater accuracy.

Solubility in alcohol. Proceed as directed in the general method, page 899. One ml. dissolves in 7 ml. of 60 percent alcohol to form a clear solution.

Specific gravity. Determine by any reliable method (see page 5).

Acid value. Determine as directed in the general method, page 893.

Chlorinated compounds. Proceed as directed in the general method, page 895.

Packaging and storage. Store in full, tight, preferably glass, tin-lined or other suitably lined containers in a cool place protected from light.

Functional use in foods. Flavoring agent.

2-PHENYLPROPIONALDEHYDE DIMETHYL ACETAL

Hydratropic Aldehyde Dimethyl Acetal

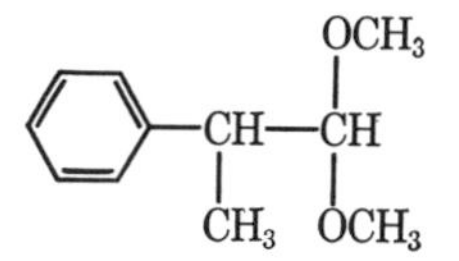

$C_{11}H_{16}O_2$ Mol. wt. 180.25

DESCRIPTION

A colorless to slightly yellow liquid having a mushroom-like odor. It is soluble in alcohol and in ether, but practically insoluble in water.

SPECIFICATIONS

Assay. Not less than 95.0 percent of $C_{11}H_{16}O_2$.

Refractive index. Between 1.492 and 1.497 at 20°.

Solubility in alcohol. Passes test.

Specific gravity. Between 0.989 and 0.994.

Limits of Impurities

Aldehydes. Not more than 3.0 percent, as 2-phenylpropionaldehyde.

TESTS

Assay. Weigh accurately about 2 grams, and proceed as directed under *Acetals*, page 892, using 90.13 as the equivalence factor (*f*) in the calculation.

Refractive index, page 945. Determine with an Abbé or other refractometer of equal or greater accuracy.

Solubility in alcohol. Proceed as directed in the general method, page 899. One ml. dissolves in 7 ml. of 60 percent and in 3 ml. of 70 percent alcohol to form clear solutions.

Specific gravity. Determine by any reliable method (see page 5).

Aldehydes. Weigh accurately about 10 grams, and proceed as directed for aldehydes under *Aldehydes and Ketones-Hydroxylamine Method,* page 894, using 67.09 as the equivalence factor (E) in the calculation.

Packaging and storage. Store in full, tight, preferably glass, tin-lined, or other suitably lined containers in a cool place protected from light.

Functional use in foods. Flavoring agent.

3-PHENYLPROPYL ACETATE

C_6H_5—$CH_2CH_2CH_2OOCCH_3$

$C_{11}H_{14}O_2$ Mol. wt. 178.23

DESCRIPTION

A colorless liquid having a characteristic combination spicy and floral odor. It is soluble in alcohol, but practically insoluble in water.

SPECIFICATIONS

Assay. Not less than 98.0 percent of $C_{11}H_{14}O_2$.

Refractive index. Between 1.494 and 1.497 at 20°.

Solubility in alcohol. Passes test.

Specific gravity. Between 1.012 and 1.015.

Limits of Impurities

Acid value. Not more than 1.0.

TESTS

Assay. Weigh accurately about 1.2 grams, and proceed as directed under *Ester Determination,* page 896, using 89.12 as the equivalence factor (e) in the calculation.

Refractive index, page 945. Determine with an Abbé or other refractometer of equal or greater accuracy.

Solubility in alcohol. Proceed as directed in the general method, page 899. One ml. dissolves in 3 ml. of 70 percent alcohol to form a clear solution.

Specific gravity. Determine by any reliable method (see page 5).

Acid value. Determine as directed in the general method, page 893.

Packaging and storage. Store in full, tight, preferably glass, tin-lined or other suitably lined containers in a cool place protected from light.

Functional use in foods. Flavoring agent.

PHOSPHORIC ACID

Orthophosphoric Acid

H_3PO_4 Mol. wt. 98.00

DESCRIPTION

A colorless, odorless solution of H_3PO_4, usually available in concentrations ranging from 75.0 percent to 85.0 percent. It is miscible with water and with alcohol. A 1 in 10 solution gives positive tests for *Phosphate*, page 928.

SPECIFICATIONS

Assay. Not less than the minimum or within the range of percent claimed by the vendor.

Limits of Impurities

Arsenic (as As). Not more than 3 parts per million (0.0003 percent).

Fluoride. Not more than 10 parts per million (0.001 percent).

Heavy metals (as Pb). Not more than 10 parts per million (0.001 percent).

TESTS

Assay. Weigh accurately about 1.5 grams in a tared, glass-stoppered flask, and dilute to 120 ml. with water. Add 0.5 ml. of thymolphthalein T.S., mix, and titrate with 1 *N* sodium hydroxide to the first appearance of a blue color. Each ml. of 1 *N* sodium hydroxide is equivalent to 49.00 mg. of H_3PO_4.

Arsenic. A solution of 1 gram in 35 ml. of water meets the requirements of the *Arsenic Test*, page 865.

Fluoride. Proceed as directed in the *Fluoride Limit Test*, page 917.

Heavy metals. A solution of 2 grams in 10 ml. of water meets the requirements of the *Heavy Metals Test*, page 920, using 20 mcg. of lead ion (Pb) in the control (*Solution A*).

Packaging and storage. Store in tight containers.

Functional use in foods. Acid; sequestrant.

PIMENTA LEAF OIL

Pimento Leaf Oil

DESCRIPTION

The volatile oil obtained by steam distillation from the leaves of the evergreen shrub *Pimenta officinalis* Lindl. (Fam. *Myrtaceae*). It is a pale yellow to light brownish yellow liquid when freshly distilled, becoming darker with age. In contact with iron, it acquires a blue shade turning to dark brown on extended contact. It has a spicy odor. It is soluble in propylene glycol, and it is soluble, with slight opalescence, in most fixed oils. It is relatively insoluble in glycerin and in mineral oil.

SPECIFICATIONS

Assay. Not less than 80 percent and not more than 91 percent, by volume, of phenols.

Angular rotation. Between −2° and +0.5°.

Refractive index. Between 1.531 and 1.536 at 20°.

Solubility in alcohol. Passes test.

Specific gravity. Between 1.037 and 1.050.

Limits of Impurities

Arsenic (as As). Not more than 3 parts per million (0.0003 percent).

Heavy metals (as Pb). Not more than 40 parts per million (0.004 percent).

Lead. Not more than 10 parts per million (0.001 percent).

TESTS

Assay. Shake a suitable quantity of the oil with about 2 percent of powdered tartaric acid for about 2 minutes, then filter. Using a sample of the filtered oil, proceed as directed under *Phenols*, page 898, modified by heating the flask on a boiling water bath for 10 minutes and cooling, after shaking the mixture of oil and 1 *N* potassium hydroxide.

Angular rotation. Determine in a 100-mm. tube as directed under *Optical Rotation*, page 939.

Refractive index, page 945. Determine with an Abbé or other refractometer of equal or greater accuracy.

Solubility in alcohol. Proceed as directed in the general method, page 899. One ml. dissolves in 2 ml. of 70 percent alcohol; a slight opalescence may occur when additional solvent is added.

Specific gravity. Determine by any reliable method (see page 5).

Arsenic. A *Sample Solution* prepared as directed for organic compounds meets the requirements of the *Arsenic Test*, page 865.

Heavy metals. Prepare and test a 500-mg. sample as directed in *Method II* under the *Heavy Metals Test*, page 920, using 20 mcg. of lead ion (Pb) in the control (*Solution A*).

Lead. A *Sample Solution* prepared as directed for organic compounds meets the requirements of the *Lead Limit Test*, page 929, using 10 mcg. of lead ion (Pb) in the control.

Packaging and storage. Store in full, tight, preferably glass, aluminum, stainless steel, or tin-lined containers in a cool place protected from light.

Functional use in foods. Flavoring agent.

PIMENTA OIL

Pimento Oil; Allspice Oil

DESCRIPTION

The volatile oil distilled from the fruit of *Pimenta officinalis*, Lindley (Fam. *Myrtaceae*). It is a colorless, yellow, or reddish yellow liquid, which becomes darker with age. It has the characteristic odor and taste of allspice. It is affected by light.

SPECIFICATIONS

Assay. Not less than 65 percent, by volume, of phenols.

Angular rotation. Between $-4°$ and $0°$.

Refractive index. Between 1.527 and 1.540 at 20°

Solubility in alcohol. Passes test.

Specific gravity. Between 1.018 and 1.048.

Limits of Impurities

Arsenic (as As). Not more than 3 parts per million (0.0003 percent).

Heavy metals (as Pb). Not more than 40 parts per million (0.004 percent).

Lead. Not more than 10 parts per million (0.001 percent).

TESTS

Assay. Proceed as directed under *Phenols*, page 898, modified by heating on a steam bath for 10 minutes, after shaking for 5 minutes. Then cool and let stand overnight, or until the liquids are clear.

Angular rotation. Determine in a 100-mm. tube as directed under *Optical Rotation*, page 939.

Refractive index, page 945. Determine with an Abbé or other refractometer of equal or greater accuracy.

Solubility in alcohol. Proceed as directed in the general method, page 899. One ml. dissolves in 2 ml. of 70 percent alcohol.

Specific gravity. Determine by any reliable method (see page 5).

Arsenic. A *Sample Solution* prepared as directed for organic compounds meets the requirements of the *Arsenic Test*, page 865.

Heavy metals. Prepare and test a 500-mg. sample as directed in *Method II* under the *Heavy Metals Test*, page 920, using 20 mcg. of lead ion (Pb) in the control (*Solution A*).

Lead. A *Sample Solution* prepared as directed for organic compounds meets the requirements of the *Lead Limit Test*, page 929, using 10 mcg. of lead ion (Pb) in the control.

Packaging and storage. Store in full, tight, preferably glass, aluminum, or tin-lined containers in a cool place protected from light.

Functional use in foods. Flavoring agent.

PINE NEEDLE OIL, DWARF

Pine Needle Oil

DESCRIPTION

The volatile oil obtained by steam distillation from the fresh leaves of *Pinus mugo* Turra var. *pumilio* (Haenke) Zenari (Fam. *Pinaceae*). It is a colorless or yellow liquid, having a pleasant aromatic odor and a bitter, pungent taste.

SPECIFICATIONS

Assay. Not less than 3.0 percent and not more than 10.0 percent of esters, calculated as bornyl acetate ($C_{12}H_{20}O_2$).

Angular rotation. Between −5° and −15°.

Distillation range. Not more than 10 percent distils below 165°.

Refractive index. Between 1.475 and 1.480 at 20°.

Solubility in alcohol. Passes test.

Specific gravity. Between 0.853 and 0.871.

TESTS

Assay. Weigh accurately about 10 grams, and proceed as directed under *Ester Determination*, page 896, using 98.15 as the equivalence factor (*e*) in the calculation.

Angular rotation. Determine in a 100-mm. tube as directed under *Optical Rotation*, page 939.

Distillation range. Proceed as directed in the general method, page 890.

Refractive index, page 945. Determine with an Abbé or other refractometer of equal or greater accuracy.

Solubility in alcohol. Proceed as directed in the general method, page 899. One ml. dissolves in 10 ml. of 90 percent alcohol, often with turbidity.

Specific gravity. Determine by any reliable method (see page 5).

Packaging and storage. Store in full, tight, preferably glass, aluminum, or tin-lined containers in a cool place protected from light.

Functional use in foods. Flavoring agent.

PINE NEEDLE OIL, SCOTCH

DESCRIPTION

A volatile oil obtained by steam distillation from the needles of *Pinus sylvestris* L. (Fam. *Pinaceae*). It is a colorless or yellowish liquid with an aromatic, turpentine-like odor. It is soluble in most fixed oils, soluble, with faint opalescence, in mineral oil, and slightly soluble in propylene glycol. It is practically insoluble in glycerin.

SPECIFICATIONS

Assay. Not less than 1.5 percent and not more than 5.0 percent of esters, calculated as bornyl acetate ($C_{12}H_{20}O_2$).

Angular rotation. Between −4° and +10°.

Refractive index. Between 1.473 and 1.479 at 20°.

Solubility in alcohol. Passes test.

Specific gravity. Between 0.857 and 0.885.

TESTS

Assay. Weigh accurately about 10 grams, and proceed as directed under *Ester Determination*, page 896, using 98.15 as the equivalence factor (*e*) in the calculation.

Angular rotation. Determine in a 100-mm. tube as directed under *Optical Rotation*, page 939.

Refractive index, page 945. Determine with an Abbé or other refractometer of equal of greater accuracy.

Solubility in alcohol. Proceed as directed in the general method, page 899. One ml. dissolves in 6 ml. of 90 percent alcohol, occasionally with slight opalescence.

Specific gravity. Determine by any reliable method (see page 5).

Packaging and storage. Store in full, tight, preferably glass, aluminum, or tin-lined containers in a cool place protected from light.

Functional use in foods. Flavoring agent.

PIPERONAL

3,4-Methylenedioxybenzaldehyde; Heliotropine; Piperonyl Aldehyde

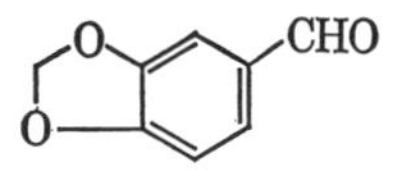

$C_8H_6O_3$ Mol. wt. 150.13

DESCRIPTION

Piperonal is found in oils of *Spirea ulmaria* L., *Doriphora sassafras* Endl., and other oils. It is prepared by oxidation of isosafrole. It is a white crystalline substance with a sweet floral odor resembling heliotrope and free from safrole by-odor. It is very soluble in alcohol, soluble in most fixed oils and in propylene glycol, slightly soluble in mineral oil and insoluble in glycerin and in water.

SPECIFICATIONS

Assay. Not less than 99.0 percent of $C_8H_6O_3$.

Solidification point. Not less than 35°.

Solubility in alcohol. Passes test.

Limits of Impurities

Arsenic (as As). Not more than 3 parts per million (0.0003 percent).

Heavy metals (as Pb). Not more than 40 parts per million (0.004 percent).

Lead. Not more than 10 parts per million (0.001 percent).

TESTS

Assay. Weigh accurately about 1.5 grams, and proceed as directed under *Aldehydes*, page 894, using 75.06 as the equivalence factor (E) in the calculation.

Solidification point. Proceed as directed in the general method, page 954.

Solubility in alcohol. Proceed as directed in the general method, page 899. One gram dissolves in 4 ml. of 70 percent alcohol.

Arsenic. A *Sample Solution* prepared as directed for organic compounds meets the requirements of the *Arsenic Test,* page 865.

Heavy metals. Prepare and test a 500-mg. sample as directed in *Method II* under the *Heavy Metals Test,* page 920, using 20 mcg. of lead ion (Pb) in the control (*Solution A*).

Lead. A *Sample Solution* prepared as directed for organic compounds meets the requirements of the *Lead Limit Test,* page 929, using 10 mcg. of lead ion (Pb) in the control.

Packaging and storage. Store in fiberboard, or pressboard drums, wooden barrels, glass containers, or suitably lined metal containers in a cool, dry place protected from light. To prevent discoloration avoid direct contact with metal.

Functional use in foods. Flavoring agent.

POLOXAMER 331

α-Hydro-*omega*-hydroxy-poly(oxyethylene)poly(oxypropylene) (51–57 moles)poly(oxyethylene) Block Copolymer, Avg. mol. wt. 3,800

DESCRIPTION

Poloxamer 331 is a block copolymer condensate of ethylene oxide and propylene oxide having an average molecular weight of 3,800. It occurs as a practically colorless liquid having a specific gravity of about 1.02 and a refractive index of about 1.452. It is very slightly soluble in water at 25°, but is freely soluble at 0°; it is freely soluble in alcohol, but is insoluble in propylene glycol and in ethylene glycol.

SPECIFICATIONS

Molecular weight. Between 3,500 and 4,125.

Cloud point of a 1 in 10 solution. Between 9° and 12°.

Hydroxyl value. Between 27.2 and 32.1.

pH of a 1 in 2 solution. Between 6.0 and 7.4.

Limits of Impurities

Arsenic (as As). Not more than 3 parts per million (0.0003 percent).

Heavy metals (as Pb). Not more than 5 parts per million (0.0005 percent).

TESTS

Molecular weight. Determine the *Hydroxyl value*, H, and then calculate the molecular weight by the formula $(56{,}100) \times (2/H)$.

Cloud point. Prepare a 10 percent solution of the sample in water at a temperature below the expected cloud point, and transfer about 100 ml. of this solution into a 50 × 120-mm. test tube. Immerse the tube in a water bath, previously cooled to at least 10° below the expected cloud point, so that the water level is a few mm. above that of the test solution. Place a suitable thermometer (see page 966) in the test solution, and position it so that the immersion line will be at the surface of the liquid. Stir the solution slowly with a mechanical stirrer (about 200 rpm), and heat gradually so that the test solution is heated at a rate of about 1° per minute. Do not allow the temperature of the water bath to rise more than 10° above that of the test solution at any time. Continue heating in this manner, and record the temperature (cloud point) at which the test solution becomes cloudy.

Hydroxyl value

Distilled pyridine. Distil pyridine over phthalic anhydride (about 60 grams for each 1000 ml.), discarding the first 25 ml. and the last 50 ml. of distillate from each 1000 ml. distilled.

Phenolphthalein indicator. Prepare a 1 percent solution of phenolphthalein in undistilled pyridine.

Phthalation reagent. Dissolve 14.4 grams of phthalic anhydride in sufficient *Distilled pyridine* to make 100 ml., mix vigorously, and store in a brown bottle. Allow to stand at least 2 hours before use. Determine the suitability of the reagent as follows: Mix 10.0 ml. of the reagent with 25 ml. of undistilled pyridine and 50 ml. of water, allow to stand for 15 minutes, then add a few drops of *Phenolphthalein indicator*, and titrate with 0.5 N sodium hydroxide. Multiply the volume, in ml., of the alkali solution required by its exact normality; if the result is not within the range 18.8–20.0, adjust the concentration of the reagent accordingly.

Procedure. Transfer about 15 grams of the sample, accurately weighed, into a 250-ml. hydroxyl flask, and add 25.0 ml. of *Phthalation reagent* to the flask, using a pipet previously rinsed with the reagent and touching the tip of the pipet against the protrusion of the flask approximately 15 seconds after the pipet has drained. In the same manner, transfer the same volume of the reagent into a second flask to serve as the blank. Add a few glass beads to each flask, swirl to dissolve the sample, and reflux for 1 hour. Cool the flasks to room temperature, and wash each condenser with two 10-ml. portions of undistilled pyridine. Disconnect the condensers, add 10 ml. of water to each flask, stopper, swirl, and allow to stand for 10 minutes. To each flask add 50.0 ml. of approximately 0.66 N sodium hydroxide, then add 0.5 ml. of *Phenolphthalein indicator,* and titrate with 0.5 N sodium hydroxide to the first pink color that persists for at least 15 seconds. Calculate the uncorrected hydroxyl value, h, by the formula $(B - S) \times (N \times 56.1/W)$, in which B is the volume, in ml., of 0.5 N sodium hydroxide required for the blank, S is the volume required for the sample, N is the exact normality of the sodium hydroxide solution, and W is the weight of the sample, in grams. Correct the results, if the sample contains significant acidity or alkalinity, as follows: Dissolve approximately the same amount of the sample, accurately weighed, as used above in 40 ml. of undistilled pyridine, and add 60 ml. of water and 0.5 ml. of *Phenolphthalein indicator.* If the solution is colorless, titrate with 0.1 N sodium hydroxide to a light pink end-point, recording the volume, in ml., required as v. If the solution is pink, titrate with 0.1 N hydrochloric acid to the disappearance of the pink color, recording the volume, in ml., required as v'. Calculate the acidity correction factor, A, by the formula $v \times n \times 56.1/w$, in which n is the exact normality of the sodium hydroxide solution, and w is the weight of the sample, in grams. Calculate the alkalinity correction factor, A', by the formula $v' \times n' \times 56.1/w$, in which n' is the exact normality of the hydrochloric acid solution. Finally, calculate the corrected hydroxyl value, H, by the formula $h + A$, or $h - A'$, whichever is appropriate.

pH. Dissolve a volume of the sample in an equal volume of methanol, and determine the pH as directed in the *Potentiometric Method*, page 941.

Arsenic. A *Sample Solution* prepared as directed for organic compounds meets the requirements of the *Arsenic Test,* page 865.

Heavy metals. Prepare and test a 4-gram sample as directed in *Method II* under the *Heavy Metals Test,* page 920, using 20 mcg. of lead ion (Pb) in the control (*Solution A*).

Packaging and storage. Store in tight containers.

Functional use in foods. Solubilizing and stabilizing agent in flavor concentrates.

POLOXAMER 407

α-Hydro-*omega*-hydroxy-poly(oxyethylene)poly(oxypropylene) (63–71 moles)poly(oxyethylene) Block Copolymer, Avg. mol. wt. 12,500

DESCRIPTION

Poloxamer 407 is a copolymer condensate of ethylene oxide and propylene oxide having an average molecular weight of 12,500. It occurs as a white solid having a melting range of about 52° to 56°. It is freely soluble in alcohol and in water but is insoluble in propylene glycol and in ethylene glycol.

SPECIFICATIONS

Molecular weight. Between 9,760 and 13,200.

Cloud point of a 1 in 10 solution. Above 100°.

Hydroxyl value. Between 8.5 and 11.5.

pH of a 1 in 40 solution. Between 6.0 and 7.4.

Limits of Impurities

Arsenic (as As). Not more than 3 parts per million (0.0003 percent).

Heavy metals (as Pb). Not more than 5 parts per million (0.0005 percent).

TESTS

Molecular weight. Determine the *Hydroxyl value, H,* and then calculate the molecular weight by the formula $(56{,}100) \times (2/H)$.

Cloud point. Dissolve about 5 grams of the sample in 50 ml. of water in a test tube, place the tube in a boiling water bath, and heat to 100°. The solution does not become cloudy.

Hydroxyl value

Distilled pyridine, Phenolphthalein indicator, and *Phthalation reagent.* Prepare as directed in the monograph for *Poloxamer 331,* page 621.

Procedure. Proceed as directed for *Procedure* under *Hydroxyl value* in the monograph for *Poloxamer 331,* page 621, using a sample of about 45 grams accurately weighed, and adding 25 ml. of *Distilled pyridine* to both the sample flask and the blank flask before refluxing.

pH. Determine by the *Potentiometric Method,* page 941.

Arsenic. A *Sample Solution* prepared as directed for organic compounds meets the requirements of the *Arsenic Test,* page 865.

Heavy metals. Prepare and test a 4-gram sample as directed in *Method II* under the *Heavy Metals Test*, page 920, using 20 mcg. of lead ion (Pb) in the control (*Solution A*).

Packaging and storage. Store in tight containers.

Functional use in foods. Solubilizing and stabilizing agent in flavor concentrates.

POLYETHYLENE

DESCRIPTION

A white, translucent, partially crystalline and partially amorphous resin produced by the direct polymerization of ethylene at high temperatures and high pressure. Various grades and types, differing from one another in molecular weight, molecular weight distribution, degree of chain branching, and extent of crystallinity, are available. It is soluble in hot benzene, but is insoluble in water.

SPECIFICATIONS

Molecular weight. Between 2,000 and 21,000.

Limits of Impurities

Arsenic (as As). Not more than 3 parts per million (0.0003 percent).

Heavy metals (as Pb). Not more than 40 parts per million (0.004 percent).

Lead. Not more than 3 parts per million (0.0003 percent).

Volatiles. Not more than 0.5 percent.

TESTS

Molecular weight. Determine as directed in the general method, page 874.

Arsenic. Prepare a *Sample Solution* as directed in the general method under *Chewing Gum Base*, page 877. This solution meets the requirements of the *Arsenic Test*, page 865.

Heavy metals. Prepare and test a 500-mg. sample as directed in *Method II* under the *Heavy Metals Test*, page 920, using 20 mcg. of lead ion (Pb) in the control (*Solution A*).

Lead. Prepare a *Sample Solution* as directed in the general method under *Chewing Gum Base*, page 878. This solution meets the requirements of the *Lead Limit Test*, page 929, using 10 mcg. of lead ion (Pb) in the control.

Volatiles. Dry a 4-gram sample for 45 minutes at 105° as directed under *Loss on drying*, page 931. (*Caution:* To reduce explosion hazard, pass carbon dioxide or nitrogen into the lower part of the drying oven at a rate of about 100 ml. per minute.)

Packaging and storage. Store in well-closed containers.

Functional use in foods. Masticatory substance in chewing gum base.

POLYETHYLENE GLYCOLS

PEG

DESCRIPTION

Polyethylene glycols are addition polymers of ethylene oxide and water, ranging in molecular weight from about 200 to about 9500 and having the general formula $HOCH_2(CH_2OCH_2)_nCH_2OH$, in which n represents the average number of oxyethylene groups. Commercially available polyethylene glycols are usually designated by a number that roughly corresponds to the average molecular weight. Polyethylene glycols having an average molecular weight of 600 or below occur as clear, colorless or practically colorless, viscous, slightly hygroscopic liquids that are miscible with water. Polyethylene glycols having an average molecular weight of 1000 or above are freely soluble in water and occur as creamy white, waxy solids or as flakes resembling paraffin. The polyethylene glycols are soluble in many organic solvents, including aliphatic ketones and alcohols, chloroform, glycol ethers, esters, and aromatic hydrocarbons; they are insoluble in ether and in most aliphatic hydrocarbons. As their molecular weight increases, water solubility, vapor pressure, hygroscopicity, and solubility in organic solvents decrease, while solidification point, specific gravity, flash point, and viscosity increase. The pH of a 1 in 20 solution of any of these polyethylene glycols is between 4.0 and 7.5.

SPECIFICATIONS

The following specifications should conform to the representations of the vendor: **Average molecular weight, Solidification point,** and **Viscosity.**

Limits of Impurities

Arsenic (as As). Not more than 3 parts per million (0.0003 percent).

Ethylene glycol and diethylene glycol. Not more than 0.2 percent (combined total).

Heavy metals (as Pb). Not more than 10 parts per million (0.001 percent).

Residue on ignition. Not more than 0.1 percent.

TESTS

Average molecular weight

Phthalic anhydride solution. Place 49.0 grams of phthalic anhydride in an amber bottle and dissolve it in 300 ml. of pyridine that has been freshly distilled over phthalic anhydride. Shake the bottle vigorously until solution is effected, and allow to stand overnight before using.

Sample preparation for liquid polyethylene glycols. Carefully introduce 25.0 ml. of the *Phthalic anhydride solution* into a clean, dry, heat-resistant pressure bottle. To the bottle add an accurately weighed amount of the sample equivalent to its expected average molecular weight divided by 160. (Thus, a sample of about 1.3 grams would be taken for PEG 200, or about 3.8 grams for PEG 600.) Stopper the bottle, and wrap it securely in a fabric bag.

Sample preparation for solid polyethylene glycols. Carefully introduce 25.0 ml. of the *Phthalic anhydride solution* into a clean, dry, heat-resistant pressure bottle. To the bottle add an accurately weighed amount of the sample, previously melted, equivalent to its expected molecular weight divided by 160; because of limited solubility, however, do not use more than 25 grams of any sample. Add 25 ml. of pyridine, freshly distilled over phthalic anhydride, swirl to effect solution, stopper the bottle, and wrap it securely in a fabric bag.

Procedure. Immerse the sample bottle in a water bath, maintained between 96° and 100°, to the same depth as that of the mixture in the bottle. Heat in the water bath for 30 to 60 minutes, using 60 minutes for PEG's having molecular weights of 3000 or higher, then remove the bottle from the bath and allow it to cool to room temperature. Uncap the bottle carefully to release any pressure, remove the bottle from the fabric bag, add 5 drops of a 1 in 100 solution of phenolphthalein in pyridine, and titrate with 0.5 N sodium hydroxide to the first pink color that persists for 15 seconds, recording the volume, in ml., of 0.5 N sodium hydroxide required as S. Perform a blank determination on 25.0 ml. of the *Phthalic anhydride solution* plus any additional pyridine added to the sample bottle, and record the volume, in ml., of 0.5 N sodium hydroxide required as B. Calculate the average molecular weight of the sample by the formula $[2000W]/[(B - S)(N)]$, in which W is the weight of the sample, in grams, $(B - S)$ is the difference between the volume of 0.5 N sodium hydroxide consumed by the blank and by the sample, and N is the exact normality of the sodium hydroxide solution.

Solidification point. Determine as directed in the general method, page 954.

Viscosity. Determine as directed under *Viscosity of Dimethylpolysiloxane*, page 970, maintaining the constant-temperature bath at the temperature specified by the vendor, and using a viscometer of such a capillary diameter as is appropriate for the viscosity of the sample being tested.

Arsenic. A *Sample Solution* prepared as directed for organic compounds meets the requirements of the *Arsenic Test*, page 865.

Ethylene glycol and diethylene glycol

POLYETHYLENE GLYCOLS HAVING AVERAGE MOLECULAR WEIGHTS BELOW 450

Apparatus. Use a suitable gas chromatograph (see page 886) equipped with a hydrogen flame ionization detector (Varian Aerograph 600D or equivalent), containing a 1.5-meter × 3-mm. (inside diameter) stainless steel column packed with sorbitol 12 percent, by weight, on 60/80-mesh nonacid-washed diatomaceous earth (Chromosorb W, or equivalent).

Operating conditions. The operating parameters may vary, depending upon the particular instrument used, but a suitable chromatogram may be obtained using the following conditions: *Column temperature,* 165°; *Inlet temperature,* 260°; *Carrier gas,* nitrogen (or other suitable inert gas), flowing at a rate of 70 ml. per minute; *Recorder,* −0.5 to +1.05 mv., full span, 1 second full response time; *Hydrogen and air flow to burner,* optimize to give maximum sensitivity.

Standard solutions. Prepare chromatographic standards by dissolving accurately weighed amounts of commercial ethylene glycol and diethylene glycol, previously purified by distillation if necessary, in water. Suitable concentrations range from 1 to 6 mg. of each glycol per ml.

Sample preparation. Transfer about 4 grams of the sample, accurately weighed, into a 10-ml. volumetric flask, dilute to volume with water, and mix.

Procedure. Inject a 2-μl. portion of each of the *Standard solutions* into the chromatograph, and obtain the chromatogram for each solution. Under the stated conditions, the elution time is approximately 2.0 minutes for ethylene glycol and 6.5 minutes for diethylene glycol. Measure the peak heights, and record the values as follows: A = height, in mm., of the ethylene glycol peak; B = weight, in mg., of ethylene glycol per ml. of the *Standard solution;* C = height, in mm., of the diethylene glycol peak; and D = weight, in mg., of diethylene glycol per ml. of the *Standard solution.*

Similarly, inject a 2-μl. portion of the *Sample preparation* into the chromatograph, and obtain the chromatogram, recording the height of the ethylene glycol peak as E and that of the diethylene glycol peak as F. Calculate the percent of ethylene glycol in the sample by the formula $(E \times B)/(A \times \text{sample weight, in grams})$; calculate the percent of diethylene glycol in the sample by the formula $(F \times D)/(C \times \text{sample weight, in grams})$. Not more than 0.2 percent of total ethylene and diethylene glycols is found.

POLYETHYLENE GLYCOLS HAVING AVERAGE MOLECULAR WEIGHTS OF 450 OR HIGHER

Sample preparation. Dissolve 50.0 grams of the sample in 75 ml. of diphenyl ether in a 250-ml. distillation flask. Slowly distil at a pressure of 1–2 mm. of mercury into a receiver graduated to 100 ml. in 1-

ml. subdivisions, until 25 ml. of distillate has been collected. Add 25.0 ml. of water to the distillate, shake vigorously, and allow the layers to separate. Cool the container in an ice bath to solidify the diphenyl ether and to facilitate its removal. Filter the water layer through filter paper into a 50-ml. glass-stoppered graduated cylinder, and to the filtrate add an equal volume of freshly distilled acetonitrile.

Standard preparation. Transfer 50.0 mg. of diethylene glycol to a 25-ml. volumetric flask, dilute to volume with a 1:1 mixture of freshly distilled acetonitrile and water, and mix.

Procedure. Transfer 10.0 ml. each of the *Sample preparation* and of the *Standard preparation* into separate 50-ml. flasks each containing 15 ml. of ceric ammonium nitrate T.S., and mix. Within 2 to 5 minutes, determine the absorbance of each solution in a 1-cm. cell at 450 mμ, with a suitable spectrophotometer, using a blank consisting of 15 ml. of ceric ammonium nitrate T.S. and 10 ml. of a 1:1 mixture of acetonitrile and water. The absorbance of the solution from the *Sample preparation* does not exceed that from the *Standard preparation.*

Heavy metals. Prepare and test a 2-gram sample as directed in *Method II* under the *Heavy Metals Test,* page 920, using 20 mcg. of lead ion (Pb) in the control (*Solution A*).

Residue on ignition. Ignite 10 grams as directed in the general method, page 945.

Packaging and storage. Store in tight containers.

Functional use in foods. Dispersing, coating, binding, plasticizing agent; lubricant; flavoring adjuvant.

POLYGLYCEROL ESTERS OF FATTY ACIDS

DESCRIPTION

Polyglycerol esters of fatty acids are mixed partial esters formed by reacting polymerized glycerols with edible fats, oils, or fatty acids. Minor amounts of mono-, di-, and triglycerides, free glycerol and polyglycerols, free fatty acids, and sodium salts of fatty acids may be present. The polyglycerols vary in degree of polymerization, which is specified by a number (such as tri-, penta-, deca-, etc.) that is related to the average number of glycerol residues per polyglycerol molecule. A specified polyglycerol consists of a distribution of molecular species characteristic of its nominal degree of polymerization. By varying the proportions as well as the nature of the fats or fatty acids to be reacted with the polyglycerols, a large and diverse class of products may be obtained. They include light yellow to amber colored, oily to very viscous liquids; light tan to medium brown plastic or soft solids; and light tan to brown, hard, waxy solids. The esters range from very hydrophilic to very lipophilic, but as a class tend to be dispersible in water and soluble in organic solvents and oils. They conform to the regulations of the federal Food and Drug Administration pertaining to specifications for fats or fatty acids derived from edible sources.

SPECIFICATIONS

The following specifications should conform to the representations of the vendor: **Acid value, Hydroxyl value, Iodine value, Residue on ignition, Saponification value,** and **Sodium salts of fatty acids.**

Limits of Impurities

Arsenic (as As). Not more than 3 parts per million (0.0003 percent).

Heavy metals (as Pb). Not more than 10 parts per million (0.001 percent).

TESTS

Acid value. Determine as directed under *Method II* in the general procedure, page 902.

Hydroxyl value. Transfer to a 300-ml. Erlenmeyer flask an accurately weighed sample approximately equivalent to 561 divided by the expected hydroxyl value, ±10 percent. Mix 9 volumes of pyridine with 1 volume of acetic anhydride, pipet 25.0 ml. of this solution into the sample flask, and pipet 25.0 ml. into a separate 300-ml. Erlenmeyer flask to serve as the blank. Add boiling stones to each flask, and fit the flasks with air condensers, lubricating the joints only with a few drops of pyridine. Reflux the sample solution gently by heating on a hot plate, confining the vapors in the lower portion of the condenser, and continue refluxing for 45 minutes. Do not heat the blank. Cool the sample flask to room temperature, and rinse the condenser, the condenser tip, and sides of the flask with 25 ml. of pyridine. Add about 50 ml. of 0.55 N sodium hydroxide to each flask, mix by swirling for about 45 seconds, then add 1 ml. of phenolphthalein T.S. and 75 ml. of isopropanol, and continue the titration with stirring to the first pink color that persists for at least 30 seconds. Calculate the hydroxyl value by the formula $AV + [(56.1)(B - S)(N/W)]$, in which AV is the *Acid value*, determined as directed above; $(B - S)$ is the difference between the volume, in ml., of 0.55 N sodium hydroxide required for the blank and for the sample, respectively; N is the exact normality of the sodium hydroxide solution; and W is the weight of the sample, in grams.

Iodine value. Determine by the *Wijs Method*, page 906.

Residue on ignition. Determine as directed in the general method, page 945.

Saponification value. Determine as directed in the general method, page 914, using an accurately weighed amount of the sample approximately equivalent to 700 divided by the expected saponification value.

Sodium salts of fatty acids. Dissolve about 5 grams of the sample, accurately weighed, in 75 ml. of glacial acetic acid, add 2 drops of crystal violet T.S., and titrate with 0.1 *N* perchloric acid in glacial acetic acid to an emerald green end-point. Perform a blank determination, and make any necessary correction, recording the net volume of perchloric acid consumed as *V*. Calculate the number of mg. of potassium hydroxide equivalent to the sodium soaps per gram of the sample by the formula $(56.1) \times (VN/W)$, in which *N* is the exact normality of the perchloric acid, and *W* is the weight of the sample, in grams.

Arsenic. A *Sample Solution* prepared as directed for organic compounds meets the requirements of the *Arsenic Test*, page 865.

Heavy metals. Prepare and test a 2-gram sample as directed in *Method II* under the *Heavy Metals Test*, page 920, using 20 mcg. of lead ion (Pb) in the control (*Solution A*).

Packaging and storage. Store in well-closed containers.
Functional use in foods. Emulsifier.

POLYISOBUTYLENE

DESCRIPTION

A synthetic polymer produced by the low-temperature polymerization of isobutylene in liquid ethylene, methyl chloride, or hexane, using an aluminum chloride catalyst. After completion of polymerization, volatile components are removed by raising the temperature of the reaction mixture. Low molecular weight grades are soft and gummy; high molecular weight grades are tough and elastic. All grades are light in color, odorless, and tasteless, and are soluble in benzene and in diisobutylene but insoluble in water.

SPECIFICATIONS

Molecular weight (Flory). Not less than 37,000.

Limits of Impurities

Arsenic (as As). Not more than 3 parts per million (0.0003 percent).
Heavy metals (as Pb). Not more than 40 parts per million (0.004 percent).
Lead. Not more than 3 parts per million (0.0003 percent).
Volatiles. Not more than 0.3 percent.

TESTS

Molecular weight. Determine as directed in the general method, page 875.

Arsenic. Prepare a *Sample Solution* as directed in the general method under *Chewing Gum Base,* page 877. This solution meets the requirements of the *Arsenic Test,* page 865.

Heavy metals. Prepare and test a 500-mg. sample as directed in *Method II* under the *Heavy Metals Test,* page 920, using 20 mcg. of lead ion (Pb) in the control (*Solution A*).

Lead. Prepare a *Sample Solution* as directed in the general method under *Chewing Gum Base,* page 878. This solution meets the requirements of the *Lead Limit Test,* page 929, using 10 mcg. of lead ion (Pb) in the control.

Volatiles. Dry a 5-gram sample for 2 hours at 105° as directed under *Loss on drying,* page 931. (*Caution:* To reduce explosion hazard, pass carbon dioxide or nitrogen into the lower part of the drying oven at a rate of about 100 ml. per minute.)

Packaging and storage. Store low molecular weight grades in metal drums; store higher molecular weight grades in talc-coated cartons, or wrap in polyethylene film.

Functional use in foods. Masticatory substance in chewing gum base.

POLYSORBATE 20

Polyoxyethylene (20) Sorbitan Monolaurate

DESCRIPTION

Polysorbate 20 is a mixture of laurate partial esters of sorbitol and sorbitol anhydrides condensed with approximately 20 moles of ethylene oxide (C_2H_4O) for each mole of sorbitol and its mono- and dianhydrides. It is a lemon to amber colored liquid having a faint, characteristic odor and a warm, somewhat bitter taste. It is soluble in water, in alcohol, in ethyl acetate, in methanol, and in dioxane, but is insoluble in mineral oil and in mineral spirits.

IDENTIFICATION

To 5 ml. of a 1 in 20 solution add 5 ml. of sodium hydroxide T.S. boil for a few minutes, cool, and acidify with diluted hydrochloric acid T.S. The solution is strongly opalescent.

SPECIFICATIONS

Assay for oxyethylene content. Not less than 70.0 percent and not more than 74.0 percent of oxyethylene groups ($—C_2H_4O—$), equivalent to between 97.3 percent and 103.0 percent of polysorbate 20, calculated on the anhydrous basis.

Hydroxyl value. Between 96 and 108.

Lauric acid. Between 15 and 17 grams per 100 grams of sample.

Saponification value. Between 40 and 50.

Limits of Impurities

Acid value. Not more than 2.

Arsenic (as As). Not more than 3 parts per million (0.0003 percent).

Heavy metals (as Pb). Not more than 10 parts per million (0.001 percent).

Residue on ignition. Not more than 0.15 percent.

Water. Not more than 3 percent.

TESTS

Assay for oxyethylene content. Weigh accurately a 65-mg. sample, and proceed as directed in the general method, page 910.

Hydroxyl value. Determine as directed under *Method II* in the general procedure, page 904.

Lauric acid. Transfer about 25 grams of the sample, accurately weighed, into a 500-ml. round-bottom boiling flask, add 250 ml. of alcohol and 7.5 grams of potassium hydroxide, and mix. Connect a suitable condenser to the flask, reflux the mixture for 1 to 2 hours, then transfer to an 800-ml. beaker, rinsing the flask with about 100 ml. of water and adding it to the beaker. Heat on a steam bath to evaporate the alcohol, adding water occasionally to replace the alcohol, and evaporate until the odor of alcohol can no longer be detected. Adjust the final volume to about 250 ml. with hot water. Neutralize the soap solution with dilute sulfuric acid (1 in 2), add 10 percent in excess, and heat, while stirring, until the fatty acid layer separates. Transfer the fatty acids into a 500-ml. separator, wash with three or four 20-ml. portions of hot water, and combine the washings with the original aqueous layer from the saponification. Extract the combined aqueous layer with three 50-ml. portions of petroleum ether, add the extracts to the fatty acid layer, evaporate to dryness in a tared dish, cool, and weigh. The lauric acid so obtained has an *Acid value* between 250 and 275 (*Method I*, page 931).

Saponification value. Determine as directed in the general method, page 914, using about 8 grams, accurately weighed.

Acid value. Determine as directed under *Method II* in the general procedure, page 902.

Arsenic. A *Sample Solution* prepared as directed for organic compounds meets the requirements of the *Arsenic Test*, page 865.

Heavy metals. Prepare and test a 2-gram sample as directed in *Method II* under the *Heavy Metals Test*, page 920, using 20 mcg. of lead ion (Pb) in the control (*Solution A*).

Residue on ignition. Ignite 5 grams as directed in the general method, page 945.

Water. Determine by the *Karl Fischer Titrimetric Method,* page 977.

Packaging and storage. Store in tight containers.

Functional use in foods. Emulsifier; stabilizer.

POLYSORBATE 60

Polyoxyethylene (20) Sorbitan Monostearate

DESCRIPTION

Polysorbate 60 is a mixture of stearate and palmitate partial esters of sorbitol and sorbitol anhydrides condensed with approximately 20 moles of ethylene oxide (C_2H_4O) for each mole of sorbitol and its mono- and dianhydrides. It is a lemon to orange colored, oily liquid or semigel having a faint characteristic odor and a warm, somewhat bitter taste. It is soluble in water, in aniline, in ethyl acetate, and in toluene, but is insoluble in mineral oil and in vegetable oils. It conforms to the regulations of the federal Food and Drug Administration pertaining to specifications for fats or fatty acids derived from edible sources.

IDENTIFICATION

A. To 5 ml. of a 1 in 20 solution add 5 ml. of sodium hydroxide T.S., boil for a few minutes, cool, and acidify with diluted hydrochloric acid T.S. The solution is strongly opalescent.

B. A mixture of 60 volumes of polysorbate 60 with 40 volumes of water at 25° or below yields a gelatinous mass.

SPECIFICATIONS

Assay for oxyethylene content. Not less than 65.0 percent and not more than 69.5 percent of oxyethylene groups (—C_2H_4O—), equivalent to between 97.0 percent and 103.0 percent of polysorbate 60, calculated on the anhydrous basis.

Hydroxyl value. Between 81 and 96.

Stearic and palmitic acids. Between 24 and 26 grams per 100 grams of sample.

Saponification value. Between 45 and 55.

Limits of Impurities

Acid value. Not more than 2.

Arsenic (as As). Not more than 3 parts per million (0.0003 percent).

Heavy metals (as Pb). Not more than 10 parts per million (0.001 percent).

Residue on ignition. Not more than 0.25 percent.
Water. Not more than 3 percent.

TESTS

Assay for oxyethylene content. Weigh accurately a 65-mg. sample, and proceed as directed in the general method, page 910.

Hydroxyl value. Determine as directed under *Method II* in the general procedure, page 904.

Stearic and palmitic acids. Isolate the fatty acids as directed in the test for *Lauric acid* under *Polysorbate 20,* page 632, and determine the weight of the acids. The product so obtained has an *Acid value* between 200 and 212 (*Method I,* page 902) and a *Solidification Point,* page 954, not below 52°.

Saponification value. Determine as directed in the general method, page 914, using about 8 grams, accurately weighed.

Acid value. Determine as directed under *Method II* in the general procedure, page 902.

Arsenic. A *Sample Solution* prepared as directed for organic compounds meets the requirements of the *Arsenic Test,* page 865.

Heavy metals. Prepare and test a 2-gram sample as directed in *Method II* under the *Heavy Metals Test,* page 920, using 20 mcg. of lead ion (Pb) in the control (*Solution A*).

Residue on ignition. Ignite 5 grams as directed in the general method, page 945.

Water. Determine by the *Karl Fischer Titrimetric Method,* page 977.

Packaging and storage. Store in tight containers.
Functional use in foods. Emulsifier; stabilizer.

POLYSORBATE 65

Polyoxyethylene (20) Sorbitan Tristearate

DESCRIPTION

Polysorbate 65 is a mixture of stearate and palmitate partial esters of sorbitol and its anhydrides condensed with approximately 20 moles of ethylene oxide (C_2H_4O) for each mole of sorbitol and its mono- and dianhydrides. It is a tan, waxy solid having a faint, characteristic odor and a waxy, somewhat bitter taste. It is soluble in mineral oil and in vegetable oils, mineral spirits, acetone, ether, dioxane, alcohol, and in methanol, and is dispersible in water and in carbon tetrachloride. It conforms to the regulations of the federal Food and Drug Administration pertaining to specifications for fats or fatty acids derived from edible sources.

IDENTIFICATION

To 5 ml. of a 1 in 20 solution add 5 ml. of sodium hydroxide T.S., boil for a few minutes, cool, and acidify with diluted hydrochloric acid T.S. The solution is strongly opalescent.

SPECIFICATIONS

Assay for oxyethylene content. Not less than 46.0 percent and not more than 50.0 percent of oxyethylene groups (—C_2H_4O—), equivalent to between 96.0 percent and 104.0 percent of polysorbate 65, calculated on the anhydrous basis.

Hydroxyl value. Between 44 and 60.

Stearic and palmitic acids. Between 42 and 44 grams per 100 grams of sample.

Saponification value. Between 88 and 98.

Limits of Impurities

Acid value. Not more than 2.

Arsenic (as As). Not more than 3 parts per million (0.0003 percent).

Heavy metals (as Pb). Not more than 10 parts per million (0.001 percent).

Residue on ignition. Not more than 0.25 percent.

Water. Not more than 3 percent.

TESTS

Assay for oxyethylene content. Weigh accurately a 90-mg. sample, and proceed as directed in the general method, page 910.

Hydroxyl value. Determine as directed under *Method II* in the general procedure, page 904.

Stearic and palmitic acids. Isolate the fatty acids as directed in the test for *Lauric acid* under *Polysorbate 20*, page 632, and determine the weight of the acids. The product so obtained has an *Acid value* between 200 and 212 (*Method I*, page 902) and a *Solidification Point*, page 954, not below 52°.

Saponification value. Determine as directed in the general method, page 914, using about 6 grams, accurately weighed.

Acid value. Determine as directed under *Method II* in the general procedure, page 902.

Arsenic. A *Sample Solution* prepared as directed for organic compounds meets the requirements of the *Arsenic Test*, page 865.

Heavy metals. Prepare and test a 2-gram sample as directed in *Method II* under the *Heavy Metals Test*, page 920, using 20 mcg. of lead ion (Pb) in the control (*Solution A*).

Residue on ignition. Ignite 5 grams as directed in the general method, page 945.

Water. Determine by the *Karl Fischer Titrimetric Method*, page 977.

Packaging and storage. Store in tight containers.
Functional use in foods. Emulsifier; stabilizer.

POLYSORBATE 80

Polyoxyethylene (20) Sorbitan Monooleate

DESCRIPTION

Polysorbate 80 is a mixture of oleate partial esters of sorbitol and sorbitol anhydrides condensed with approximately 20 moles of ethylene oxide (C_2H_4O) for each mole of sorbitol and its mono- and dianhydrides. It is a yellow to orange colored, oily liquid having a faint, characteristic odor and a warm, somewhat bitter taste. It is very soluble in water, producing an odorless, nearly colorless solution, and is soluble in alcohol, in fixed oils, in ethyl acetate, and in toluene. It is insoluble in mineral oil. It conforms to the regulations of the federal Food and Drug Administration pertaining to specifications for fats or fatty acids derived from edible sources.

IDENTIFICATION

A. To 5 ml. of a 1 in 20 solution add 5 ml. of sodium hydroxide T.S., boil for a few minutes, cool, and acidify with diluted hydrochloric acid T.S. The solution is strongly opalescent.

B. To a 1 in 20 solution add bromine T.S., dropwise. The bromine is decolorized.

C. A mixture of 60 volumes of polysorbate 80 with 40 volumes of water at 25° or below yields a gelatinous mass.

SPECIFICATIONS

Assay for oxyethylene content. Not less than 65.0 percent and not more than 69.5 percent of oxyethylene groups ($—C_2H_4O—$), equivalent to between 96.5 percent and 103.5 percent of polysorbate 80, calculated on the anhydrous basis.

Hydroxyl value. Between 65 and 80.

Oleic acid. Between 22 and 24 grams per 100 grams of sample.

Saponification value. Between 45 and 55.

Limits of Impurities

Acid value. Not more than 2.
Arsenic (as As). Not more than 3 parts per million (0.0003 percent).
Heavy metals (as Pb). Not more than 10 parts per million (0.001 percent).
Residue on ignition. Not more than 0.15 percent.
Water. Not more than 3 percent.

TESTS

Assay for oxyethylene content. Weigh accurately a 65-mg. sample, and proceed as directed in the general method, page 910.

Hydroxyl value. Determine as directed under *Method II* in the general procedure, page 904.

Oleic acid. Isolate the fatty acids as directed in the test for *Lauric acid* under *Polysorbate 20*, page 632, and determine the weight of the acid. The product so obtained has an *Acid value* between 196 and 206 (*Method I*, page 902) and an *Iodine value* between 80 and 92 (*Wijs Method*, page 906).

Saponification value. Determine as directed in the general method, page 914, using about 8 grams, accurately weighed.

Acid value. Determine as directed under *Method II* in the general procedure, page 902.

Arsenic. A *Sample Solution* prepared as directed for organic compounds meets the requirements of the *Arsenic Test*, page 865.

Heavy metals. Prepare and test a 2-gram sample as directed in *Method II* under the *Heavy Metals Test*, page 920, using 20 mcg. of lead ion (Pb) in the control (*Solution A*).

Residue on ignition. Ignite 5 grams as directed in the general method, page 945.

Water. Determine by the *Karl Fischer Titrimetric Method*, page 977.

Packaging and storage. Store in tight containers.

Functional use in foods. Emulsifier; stabilizer.

POLYVINYL ACETATE

DESCRIPTION

A clear, water-white to pale yellow solid resin prepared by the polymerization of vinyl acetate. After completion of polymerization, the resin is freed of traces of residual catalyst (usually a peroxide), monomer, and/or solvent by vacuum drying, steam sparging, washing, or any combination of these treatments. The resin is soluble in benzene and in acetone, but is insoluble in water.

SPECIFICATIONS

Molecular weight. Not less than 2000.

Limits of Impurities

Arsenic (as As). Not more than 3 parts per million (0.0003 percent).

Free acetic acid. Not more than 0.05 percent.

Heavy metals (as Pb). Not more than 40 parts per million (0.004 percent).

Lead. Not more than 3 parts per million (0.0003 percent).

Volatiles. Not more than 1 percent.

TESTS

Molecular weight. Determine as directed in the general method, page 875.

Arsenic. Prepare a *Sample Solution* as directed in the general method under *Chewing Gum Base*, page 877. This solution meets the requirements of the *Arsenic Test*, page 865.

Free acetic acid. Transfer 10.0 grams of the sample into a 250-ml. glass-stoppered Erlenmeyer flask, dissolve in 75 ml. of benzene, add 60 ml. of alcohol, and mix. Add phenolphthalein T.S., and titrate with 0.02 *N* methanolic potassium hydroxide to a faint pink end-point. Perform a blank determination (see page 2), and make any necessary correction. Each ml. of 0.02 *N* methanolic potassium hydroxide is equivalent to 1.201 mg. of $C_2H_4O_2$.

Heavy metals. Prepare and test a 500-mg. sample as directed in *Method II* under the *Heavy Metals Test*, page 920, using 20 mcg. of lead ion (Pb) in the control (*Solution A*).

Lead. Prepare a *Sample Solution* as directed in the general method under *Chewing Gum Base*, page 878. This solution meets the requirements of the *Lead Limit Test*, page 929, using 10 mcg. of lead ion (Pb) in the control.

Volatiles. Dry a sample of about 1.5 grams at 100° for 2 hours in vacuum as directed under *Loss on drying*, page 931.

Packaging and storage. Store in well-closed containers.

Functional use in foods. Masticatory substance in chewing gum base.

POTASSIUM ACID TARTRATE

Potassium Bitartrate; Cream of Tartar

KOOCCH(OH)CH(OH)COOH

$C_4H_5KO_6$ Mol. wt. 188.18

DESCRIPTION

Potassium acid tartrate is a salt of L(+)-tartaric acid. It occurs as colorless or slightly opaque crystals, or as a white, crystalline powder, having a pleasant, acid taste. A saturated solution is acid to litmus. One gram dissolves in 165 ml. of water at 25°, in 16 ml. of boiling water, and in 8820 ml. of alcohol.

IDENTIFICATION

A. When sufficiently heated, it chars and emits flammable vapors, having an odor resembling that of burning sugar. At a higher temperature and with free access to air, the carbon of the black residue is consumed, and there remains a white, fused mass of potassium carbonate which imparts a reddish purple color to a nonluminous flame.

B. A saturated solution yields a yellowish orange precipitate with sodium cobaltinitrite T.S.

C. Neutralize a saturated solution with sodium hydroxide T.S. in a test tube, add silver nitrate T.S., then just sufficient ammonia T.S. to dissolve the white precipitate, and boil the solution. Silver is deposited on the inner surface of the tube, forming a mirror.

SPECIFICATIONS

Assay. Not less than 99.0 percent and not more than the equivalent of 101.0 percent of $C_4H_5KO_6$ after drying.

Limits of Impurities

Ammonia. Passes test.

Arsenic (as As). Not more than 3 parts per million (0.0003 percent).

Heavy metals (as Pb). Not more than 20 parts per million (0.002 percent).

Insoluble matter. Passes test.

Lead. Not more than 10 parts per million (0.001 percent).

TESTS

Assay. Weigh accurately about 6 grams, previously dried at 105° for 3 hours, dissolve it in 100 ml. of boiling water, add phenolphthalein T.S., and titrate with 1 *N* sodium hydroxide. Each ml. of 1 *N* sodium hydroxide is equivalent to 188.2 mg. of $C_4H_5KO_6$.

Ammonia. Heat 500 mg. with 5 ml. of sodium hydroxide T.S. No odor of ammonia is detected.

Arsenic. A *Sample Solution* prepared as directed for organic compounds meets the requirements of the *Arsenic Test*, page 865.

Heavy metals. Prepare and test a 1-gram sample as directed in *Method II* under the *Heavy Metals Test*, page 920, using 20 mcg. of lead ion (Pb) in the control (*Solution A*).

Insoluble matter. Agitate 500 mg. with 3 ml. of ammonia T.S. No undissolved residue remains.

Lead. Dissolve 1 gram in 3 ml. of dilute nitric acid (1 in 2), boil for 1 minute, cool, and dilute to 20 ml. with water. This solution meets the requirements of the *Lead Limit Test*, page 929, using 10 mcg. of lead ion (Pb) in the control.

Packaging and storage. Store in tight containers.

Functional use in foods. Acid; buffer.

POTASSIUM ALGINATE

Algin

$(C_6H_7O_6K)_n$ Equiv. wt., *Calculated*, 214.22
Equiv. wt., *Actual* (Avg.), 238.00

DESCRIPTION

The potassium salt of alginic acid (see *Alginic Acid*, page 26) occurs as a white to yellowish, fibrous or granular powder. It is nearly odorless and tasteless. It dissolves in water to form a viscous, colloidal solution. It is insoluble in alcohol and in hydroalcoholic solutions in which the alcohol content is greater than about 30 percent by weight. It is insoluble in chloroform, in ether, and in acids having a pH lower than about 3.

IDENTIFICATION

A. To 5 ml. of a 1 in 100 solution add 1 ml. of calcium chloride T.S. A voluminous, gelatinous precipitate is formed.

B. To 10 ml. of a 1 in 100 solution add 1 ml. of diluted sulfuric acid T.S. A heavy gelatinous precipitate is formed.

C. Potassium alginate meets the requirements of *Identification Test C* under *Alginic Acid*, page 26.

D. Extract the *Ash* from potassium alginate with diluted hydrochloric acid T.S. and filter. The filtrate gives positive tests for *Potassium*, page 928.

SPECIFICATIONS

Assay. It yields, on the dried basis, not less than 16.5 percent and not more than 19.5 percent of carbon dioxide (CO_2) corresponding to between 89.25 and 105.50 percent of potassium alginate (Equiv. wt. 238.00).

Ash. Between 19 and 29 percent on the dried basis.

Limits of Impurities

Arsenic (as As). Not more than 3 parts per million (0.0003 percent).

Heavy metals (as Pb). Not more than 40 parts per million (0.004 percent).

Insoluble matter. Not more than 0.2 percent.

Lead. Not more than 10 parts per million (0.001 percent).

Loss on drying. Not more than 15 percent.

TESTS

Assay. Proceed as directed in the *Alginates Assay*, page 863. Each ml. of 0.25 *N* sodium hydroxide consumed in the assay is equivalent to 28.75 mg. of potassium alginate (Equiv. wt. 238.00).

Ash. Determine as directed under *Ash* in the monograph on *Alginic Acid*, page 26.

Arsenic. A *Sample Solution* prepared as directed for organic compounds meets the requirements of the *Arsenic Test*, page 865.

Heavy metals. Prepare and test a 500-mg. sample as directed in *Method II* under the *Heavy Metals Test*, page 920, but use a platinum crucible for the ignition. Any color does not exceed that produced in a control (*Solution A*) containing 20 mcg. of lead ion (Pb).

Insoluble matter. Determine as directed under *Insoluble matter* in the monograph on *Alginic Acid*, page 26.

Lead. A *Sample Solution* prepared as directed for organic compounds meets the requirements of the *Lead Limit Test*, page 929, using 10 mcg. of lead ion (Pb) in the control.

Loss on drying, page 931. Dry at 105° for 4 hours.

Packaging and storage. Store in well-closed containers.

Functional use in foods. Stabilizer; thickener; emulsifier.

POTASSIUM BICARBONATE

$KHCO_3$ Mol. wt. 100.12

DESCRIPTION

Colorless, transparent, monoclinic prisms or a white, granular powder. It is odorless and is stable in air. Its solutions are neutral or alkaline to phenolphthalein T.S. One gram dissolves in 2.8 ml. of water. It is almost insoluble in alcohol. A 1 in 10 solution gives positive tests for *Potassium*, page 928, and for *Bicarbonate*, page 926.

SPECIFICATIONS

Assay. Not less than 99.0 percent and not more than the equivalent of 101.0 percent of $KHCO_3$, calculated on the dried basis.

Limits of Impurities

Arsenic (as As). Not more than 3 parts per million (0.0003 percent).

Heavy metals (as Pb). Not more than 10 parts per million (0.001 percent).

Loss on drying. Not more than 0.25 percent.

Normal carbonate. Passes test.

TESTS

Assay. Dissolve about 4 grams, accurately weighed, in 25 ml. of water, add methyl orange T.S., and titrate with 1 *N* sulfuric acid. Each ml. of 1 *N* sulfuric acid is equivalent to 100.1 mg. of $KHCO_3$.

Arsenic. A solution of 1 gram in 4 ml. of diluted hydrochloric acid T.S. meets the requirements of the *Arsenic Test*, page 865.

Heavy metals. Dissolve 2 grams in 5 ml. of water and 8 ml. of diluted hydrochloric acid T.S., boil gently for 1 minute, and dilute to 25 ml. with water. This solution meets the requirements of the *Heavy Metals Test*, page 920, using 20 mcg. of lead ion (Pb) in the control (*Solution A*).

Loss on drying, page 931. Dry over silica gel for 4 hours.

Normal carbonate. Dissolve 1 gram of the sample without agitation in 20 ml. of water at a temperature not above 5°, and add 2 ml. of 0.1 *N* hydrochloric acid and 2 drops of phenolphthalein T.S. The solution does not assume more than a faint pink color immediately.

Packaging and storage. Store in well-closed containers.

Functional use in foods. Alkali; leavening agent.

POTASSIUM BROMATE

$KBrO_3$ Mol. wt. 167.00

DESCRIPTION

White crystals or a granular powder. It is soluble in water and slightly soluble in alcohol. The pH of a 1 in 20 solution is between 5 and 9.

IDENTIFICATION

A. A 1 in 20 solution imparts a violet color to a nonluminous flame.

B. To a 1 in 20 solution add sulfurous acid dropwise. A yellow color is produced which disappears upon the addition of an excess of sulfurous acid.

SPECIFICATIONS

Assay. Not less than 99.0 percent and not more than the equivalent of 101.0 percent of $KBrO_3$ after drying.

Limits of Impurities

Arsenic (as As). Not more than 3 parts per million (0.0003 percent).

Heavy metals (as Pb). Not more than 10 parts per million (0.001 percent).

Loss on drying. Not more than 0.5 percent.

TESTS

Assay. Dissolve about 100 mg., previously dried to constant weight over a suitable desiccant and accurately weighed, in 50 ml. of water contained in a 250-ml. glass-stoppered Erlenmeyer flask. Add 3 grams of potassium iodide, followed by 3 ml. of hydrochloric acid. Allow the mixture to stand for 5 minutes, add 100 ml. of cold water, and titrate the liberated iodine with 0.1 *N* sodium thiosulfate, adding starch T.S. as the end-point is approached. Perform a blank determination (see page 2). Each ml. of 0.1 *N* sodium thiosulfate consumed is equivalent to 2.783 mg. of $KBrO_3$.

Arsenic. Dissolve 1 gram in a mixture of 5 ml. of hydrochloric acid and 5 ml. of water, and evaporate the solution until crystals appear. Cool, dissolve the residue in water, and dilute to 35 ml. This solution meets the requirements of the *Arsenic Test*, page 865.

Heavy metals. Dissolve 2 grams in 10 ml. of water, add 10 ml. of hydrochloric acid, and evaporate to dryness on a steam bath. Dissolve the residue in 10 ml. of hydrochloric acid, again evaporate to dryness, and then dissolve the residue in 25 ml. of water. This solution meets the requirements of the *Heavy Metals Test*, page 920, using 20 mcg. of lead ion (Pb) in the control (*Solution A*).

Loss on drying, page 931. Dry over a suitable desiccant to constant weight.

Packaging and storage. Store in well-closed containers.

Functional use in foods. Maturing agent; dough conditioner.

POTASSIUM CARBONATE

K_2CO_3 Mol. wt. 138.21

DESCRIPTION

Potassium carbonate is anhydrous or contains 1.5 molecules of water of crystallization. The anhydrous form occurs as a white, granular powder and the hydrated form as small, white, translucent crystals or granules. It is odorless, has a strongly alkaline taste, is very deliquescent, and its solutions are alkaline. One gram dissolves in 1 ml. of water at 25°, and in about 0.7 ml. of boiling water. It is insoluble in alcohol. A 1 in 10 solution gives positive tests for *Potassium*, page 928, and for *Carbonate*, page 926.

SPECIFICATIONS

Assay. Not less than 99.0 percent of K_2CO_3 after drying.

Loss on drying. K_2CO_3 (anhydrous), not more than 1 percent; $K_2CO_3 . 1\frac{1}{2}H_2O$ (hydrated), between 10 and 16.5 percent.

Limits of Impurities

Arsenic (as As). Not more than 3 parts per million (0.0003 percent).
Heavy metals (as Pb). Not more than 20 parts per million (0.002 percent).
Insoluble substances. Passes test.
Lead. Not more than 10 parts per million (0.001 percent).

TESTS

Assay. Weigh accurately, in a stoppered weighing-bottle, about 1 gram of the dried sample obtained in the test for *Loss on drying*, dissolve it in 50.0 ml. of 1 *N* sulfuric acid, add methyl orange T.S., and titrate the excess acid with 1 *N* sodium hydroxide. Each ml. of 1 *N* sulfuric acid is equivalent to 69.11 mg. of K_2CO_3.

Loss on drying, page 931. Dry about 3 grams, accurately weighed, at 180° for 4 hours.

Arsenic. A solution of 1 gram cautiously dissolved in diluted hydrochloric acid T.S. (about 5 ml.) meets the requirements of the *Arsenic Test*, page 865.

Heavy metals. To 1 gram add 2 ml. of water and 6 ml. of diluted hydrochloric acid T.S., boil for 1 minute, and dilute to 25 ml. with water. This solution meets the requirements of the *Heavy Metals Test*, page 920, using 20 mcg. of lead ion (Pb) in the control (*Solution A*).

Insoluble substances. No residue is left on dissolving 1 gram in 20 ml. of water.

Lead. A solution of 1 gram cautiously dissolved in diluted hydrochloric acid T.S. (about 5 ml.) meets the requirements of the *Lead Limit Test*, page 929, using 10 mcg. of lead ion (Pb) in the control.

Packaging and storage. Store in tight containers.

Functional use in foods. Alkali.

POTASSIUM CARBONATE SOLUTION

DESCRIPTION

A clear or slightly turbid, colorless, alkaline solution which absorbs carbon dioxide when exposed to the air forming potassium bicarbonate. It gives positive tests for *Potassium*, page 928, and for *Carbonate*, page 926.

SPECIFICATIONS

Assay. Not less than 97.0 percent and not more than 103.0 percent, by weight, of the labeled amount of K_2CO_3.

Limits of Impurities

Arsenic (as As). Not more than 3 parts per million (0.0003 percent), calculated on the basis of K_2CO_3 determined in the *Assay*.

Heavy metals (as Pb). Not more than 20 parts per million (0.002 percent), calculated on the basis of K_2CO_3 determined in the *Assay*.

Lead. Not more than 10 parts per million (0.001 percent), calculated on the basis of the K_2CO_3 determined in the *Assay*.

TESTS

Assay. Based on the stated or labeled percent of K_2CO_3, weigh accurately a volume of the solution equivalent to about 1 gram of potassium carbonate, and add it to 50.0 ml. of 1 *N* sulfuric acid. Add methyl orange T.S., and titrate the excess acid with 1 *N* sodium hydroxide. Each ml. of 1 *N* sulfuric acid is equivalent to 69.11 mg. of K_2CO_3.

Arsenic. Dilute the equivalent of 1 gram of K_2CO_3, calculated on the basis of the *Assay*, to 10 ml. with water, and cautiously neutralize to litmus with diluted hydrochloric acid T.S. This solution meets the requirements of the *Arsenic Test*, page 865.

Heavy metals. Dilute the equivalent of 1 gram of K_2CO_3, calculated on the basis of the *Assay*, with a mixture of 5 ml. of water and 6 ml. of diluted hydrochloric acid T.S., and heat to boiling. Cool, and dilute to 25 ml. of water. This solution meets the requirements of the *Heavy Metals Test*, page 920, using 20 mcg. of lead ion (Pb) in the control (*Solution A*).

Lead. Dilute the equivalent of 1 gram of K_2CO_3, calculated on the basis of the *Assay*, with a mixture of 6 ml. of diluted hydrochloric acid T.S. and 5 ml. of water. This solution meets the requirements of the *Lead Limit Test*, page 929, using 10 mcg. of lead ion (Pb) in the control.

Packaging and storage. Store in tight containers.

Solutions usually available. A concentration, weight in weight, of about 50.0 percent.

Functional use in foods. Alkali.

POTASSIUM CHLORIDE

KCl Mol. wt. 74.56

DESCRIPTION

Colorless, elongated, prismatic, or cubical crystals, or a white, granular powder. It is odorless, has a saline taste, and is stable in air. Its solutions are neutral to litmus. One gram dissolves in 2.8 ml. of water at 25°, and in about 2 ml. of boiling water. It is insoluble in alcohol. A 1 in 20 solution gives positive tests for *Potassium*, page 928, and for *Chloride*, page 926.

SPECIFICATIONS

Assay. Not less than 99.0 percent of KCl after drying.

Limits of Impurities

Acidity or alkalinity. Passes test.

Arsenic (as As). Not more than 3 parts per million (0.0003 percent).

Heavy metals (as Pb). Not more than 10 parts per million (0.001 percent).
Iodide or bromide. Passes test.
Loss on drying. Not more than 1 percent.
Sodium. Passes test.

TESTS

Assay. Dry about 250 mg. at 105° for 2 hours, weigh accurately, and dissolve in 50 ml. of water in a glass-stoppered flask. Add, while agitating, 50.0 ml. of 0.1 *N* silver nitrate, 3 ml. of nitric acid, and 5 ml. of nitrobenzene, shake vigorously, add 2 ml. of ferric ammonium sulfate T.S., and titrate the excess silver nitrate with 0.1 *N* ammonium thiocyanate. Each ml. of 0.1 *N* silver nitrate is equivalent to 7.456 mg. of KCl.

Acidity or alkalinity. To a solution of 5 grams in 50 ml. of recently boiled and cooled water add 3 drops of phenolphthalein T.S. No pink color is produced. Then add 0.3 ml. of 0.02 *N* sodium hydroxide. A pink color is produced.

Arsenic. A solution of 1 gram in 35 ml. of water meets the requirements of the *Arsenic Test*, page 865.

Heavy metals. A solution of 2 grams in 25 ml. of water meets the requirements of the *Heavy Metals Test*, page 920, using 20 mcg. of lead ion (Pb) in the control (*Solution A*).

Iodide or bromide. Dissolve 2 grams of the sample in 6 ml. of water, add 1 ml. of chloroform, and then add, dropwise and with constant agitation, 5 ml. of a mixture of equal parts of chlorine T.S. and water. The chloroform is free from even a transient violet or permanent orange color.

Loss on drying, page 931. Dry at 105° for 2 hours.

Sodium. A 1 in 20 solution, tested on a platinum wire, does not impart a pronounced yellow color to a nonluminous flame.

Packaging and storage. Store in well-closed containers.

Functional use in foods. Nutrient; dietary supplement; gelling agent; salt substitute; yeast food.

POTASSIUM CITRATE

$C_6H_5K_3O_7 . H_2O$ Mol. wt. 324.42

DESCRIPTION

Transparent crystals, or a white, granular powder. It is odorless, has a cooling, saline taste, and is deliquescent when exposed to moist air. One gram dissolves in about 0.5 ml. of water. It is almost insoluble in alcohol. A 1 in 10 solution gives positive tests for *Potassium*, page 928, and for *Citrate*, page 926.

SPECIFICATIONS

Assay. Not less than 99.0 percent of $C_6H_5K_3O_7$ after drying.

Loss on drying. Between 3 percent and 6 percent.

Limits of Impurities

Alkalinity. Passes test.

Arsenic (as As). Not more than 3 parts per million (0.0003 percent).

Heavy metals (as Pb). Not more than 10 parts per million (0.001 percent).

TESTS

Assay. Dissolve about 250 mg., previously dried at 180° for 4 hours and accurately weighed, in 40 ml. of glacial acetic acid, warming slightly to effect solution. Cool the solution to room temperature, add 2 drops of crystal violet T.S., and titrate with 0.1 *N* perchloric acid. Perform a blank determination (see page 2) and make any necessary correction. Each ml. of 0.1 *N* perchloric acid is equivalent to 10.214 mg. of $C_6H_5K_3O_7$.

Loss on drying, page 931. Dry at 180° for 4 hours.

Alkalinity. A 1 in 20 solution is alkaline to litmus, but after the addition of 0.2 ml. of 0.1 *N* sulfuric acid to 10 ml. of this solution no pink color is produced by the addition of 1 drop of phenolphthalein T.S.

Arsenic. A *Sample Solution* prepared as directed for organic compounds meets the requirements of the *Arsenic Test*, page 865.

Heavy metals. A solution of 2 grams in 25 ml. of water meets the requirements of the *Heavy Metals Test*, page 920, using 20 mcg. of lead ion (Pb) in the control (*Solution A*).

Packaging and storage. Store in tight containers.

Functional use in foods. Miscellaneous and general purpose; buffer; sequestrant.

POTASSIUM GIBBERELLATE

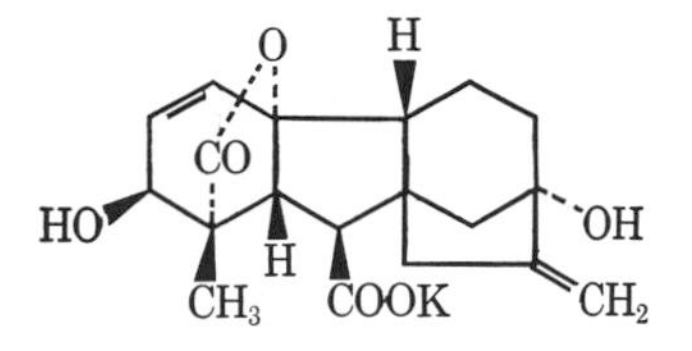

$C_{19}H_{21}KO_6$ Mol. wt. 384.45

DESCRIPTION

A white to slightly off-white, odorless or practically odorless, crystalline powder. It is soluble in water, in alcohol, and in acetone. The pH of a 1 in 20 solution is about 6. It is deliquescent.

IDENTIFICATION

A. Dissolve a few mg. of the sample in 2 ml. of sulfuric acid. A reddish solution having a green fluorescence is formed.

B. A 1 in 10 solution of the sample gives positive tests for *Potassium*, page 928.

SPECIFICATIONS

Assay. Not less than 80.0 percent and not more than 87.0 percent of $C_{19}H_{21}KO_6$, equivalent to between 72.1 percent and 78.4 percent of $C_{19}H_{22}O_6$ (gibberellic acid).

Loss on drying. Between 5 and 13 percent.

Residue on ignition. Between 19 and 23 percent.

Specific rotation, $[\alpha]_D^{25°}$. Between +43.0° and +60.0°.

Limits of Impurities

Arsenic (as As). Not more than 3 parts per million (0.0003 percent).

Heavy metals (as Pb). Not more than 40 parts per million (0.004 percent).

Lead. Not more than 10 parts per million (0.001 percent).

TESTS

Assay

Standard preparation. Transfer an accurately weighed quantity of F.C.C. Gibberellic Acid Reference Standard, equivalent to about 25 mg. of pure gibberellic acid (corrected for phase purity and volatiles content), into a 50-ml. volumetric flask, dissolve in methanol, dilute to volume with methanol, and mix. Transfer 10.0 ml. of this solution into a second 50-ml. volumetric flask, dilute to volume with methanol, and mix.

Assay preparation. Transfer about 65 mg. of the sample, accurately weighed, into a 50-ml. volumetric flask, dissolve in methanol, dilute to volume with methanol, and mix. Transfer 10.0 ml. of this solution into a 100-ml. volumetric flask, dilute to volume with methanol, and mix.

Procedure. Transfer 5.0 ml. of the *Assay preparation* into a 25 × 200-mm. glass-stoppered tube, and transfer 4.0-ml. and 5.0-ml. portions of the *Standard preparation* into separate similar tubes. Place the tubes in a boiling water bath, evaporate to dryness, and then dry in an oven at 90° for 10 minutes. Remove the tubes from the oven, stopper, and allow to cool to room temperature. Dissolve the residue in each tube in 10.0 ml. of dilute sulfuric acid (8 in 10), heat in a boiling water bath for 10 minutes, and then cool in a 10° water bath for 5 minutes. Determine the absorbance of the solutions in 1-cm. cells at 535 mμ with a suitable spectrophotometer, using the dilute sulfuric acid as the blank. Record the absorbance of the solution from the *Assay preparation* as A_U. Note the absorbance of the two solutions prepared from the 4.0-ml. and 5.0-ml. aliquots of the *Standard preparation*, and record the absorbance of the final solution giving the value nearest to that of the sample as A_S; record as V the volume of the aliquot used in preparing this solution. Calculate the quantity, in mg., of $C_{19}H_{21}KO_6$ in the sample taken by the formula $500 \times (C/0.8983) \times (V/5) \times (A_U/A_S)$, in which C is the exact concentration, in mg. per ml., of the *Standard preparation*, and 0.8983 is the ratio of the molecular weight of potassium gibberellate to that of gibberellic acid.

Loss on drying, page 931. Dry at 100° in vacuum for 7 hours.

Residue on ignition. Ignite a 1-gram sample as directed in the general method, page 945.

Specific rotation, page 939. Determine in a solution containing 50 mg. in each ml.

Arsenic. A *Sample Solution* prepared as directed for organic compounds meets the requirements of the *Arsenic Test*, page 865.

Heavy metals. Prepare and test a 500-mg. sample as directed in *Method II* under the *Heavy Metals Test*, page 920, using 20 mcg. of lead ion (Pb) in the control (*Solution A*).

Lead. A *Sample Solution* prepared as directed for organic compounds meets the requirements of the *Lead Limit Test*, page 929, using 10 mcg. of lead ion (Pb) in the control.

Packaging and storage. Store in tight containers protected from light.

Functional use in foods. Enzyme activator.

POTASSIUM GLYCEROPHOSPHATE

$C_3H_7K_2O_6P \cdot 3H_2O$ Mol. wt. 302.31

DESCRIPTION

Potassium glycerophosphate is a pale yellow, syrupy liquid containing 3 molecules of water of hydration, or it is prepared as a colorless to pale yellow, syrupy solution having a concentration of 50 percent to 75 percent. It is very soluble in water, and its solutions are alkaline to litmus paper.

IDENTIFICATION

A. A 1 in 10 solution gives positive tests for *Potassium*, page 928.

B. Heat a mixture of 100 mg. of the sample with 500 mg. of potassium bisulfate. Pungent vapors of acrolein are evolved.

SPECIFICATIONS

Assay. $C_3H_7K_2O_6P \cdot 3H_2O$, not less than 80.0 percent of $C_3H_7K_2O_6P$; *potassium glycerophosphate solutions*, not less than 95.0 percent and not more than 105.0 percent of the labeled concentration of $C_3H_7K_2O_6P$.

Limits of Impurities

Arsenic (as As). Not more than 3 parts per million (0.0003 percent).

Heavy metals (as Pb). Not more than 20 parts per million (0.002 percent).

Lead. Not more than 5 parts per million (0.0005 percent).

TESTS

Assay. Weigh accurately a portion of the sample equivalent to about 4 grams of $C_3H_7K_2O_6P$, dissolve it in 30 ml. of water, add methyl orange T.S., and titrate with 0.5 *N* hydrochloric acid. Each ml. of 0.5 *N* hydrochloric acid is equivalent to 124.13 mg. of $C_3H_7K_2O_6P$.

Arsenic. A *Sample Solution* prepared as directed for organic compounds meets the requirements of the *Arsenic Test*, page 865.

Heavy metals. A solution of 1 gram in 25 ml. of water meets the requirements of the *Heavy Metals Test*, page 920, using 20 mcg. of lead ion (Pb) in the control (*Solution A*).

Lead. A *Sample Solution* prepared as directed for organic compounds meets the requirements of the *Lead Limit Test*, page 929, using 5 mcg. of lead ion (Pb) in the control.

Packaging and storage. Store in tight containers.

Functional use in foods. Nutrient; dietary supplement.

POTASSIUM HYDROXIDE

Caustic Potash

KOH Mol. wt. 56.11

DESCRIPTION

White, or nearly white, pellets, flakes, sticks, fused masses, or other forms. Upon exposure to air, it readily absorbs carbon dioxide and moisture, and deliquesces. One gram dissolves in 1 ml. of water, in about 3 ml. of alcohol, and in about 2.5 ml. of glycerin. It is very soluble in boiling alcohol. A 1 in 25 solution gives positive tests for *Potassium*, page 928.

SPECIFICATIONS

Assay. Not less than 85.0 percent of total alkali, calculated as KOH.

Limits of Impurities

Arsenic (as As). Not more than 3 parts per million (0.0003 percent).

Carbonate (as K_2CO_3). Not more than 3.5 percent.

Heavy metals (as Pb). Not more than 30 parts per million (0.003 percent).

Insoluble substances. Passes test.

Lead. Not more than 10 parts per million (0.001 percent).

Mercury. Not more than 1 part per million (0.0001 percent).

TESTS

Assay. Dissolve about 1.5 grams, accurately weighed, in 40 ml. of recently boiled and cooled water, cool to 15°, add phenolphthalein T.S., and titrate with 1 *N* sulfuric acid. At the discharge of the pink color, record the volume of acid required, then add methyl orange T.S. and continue the titration until a persistent pink color is produced. Record the total volume of acid required for the titration. Each ml. of 1 *N* sulfuric acid is equivalent to 56.11 mg. of total alkali, calculated as KOH.

Arsenic. Dissolve 1 gram in about 10 ml. of water, cautiously neutralize to litmus paper with sulfuric acid, and cool. This solution meets the requirements of the *Arsenic Test*, page 865.

Carbonate. Each ml. of 1 *N* sulfuric acid required between the phenolphthalein and methyl orange end-points in the *Assay* is equivalent to 138.2 mg. of K_2CO_3.

Heavy metals. Dissolve 670 mg. in a mixture of 5 ml. of water and 5 ml. of diluted hydrochloric acid T.S. Heat to boiling, cool, dilute to 25 ml. with water, and filter. This solution meets the requirements of the *Heavy Metals Test*, page 920, using 20 mcg. of lead ion (Pb) in the control (*Solution A*).

Insoluble substances. A 1 in 20 solution is complete, clear, and colorless.

Lead. Dissolve 1 gram in a mixture of 5 ml. of water and 11 ml. of diluted hydrochloric acid T.S., and cool. This solution meets the requirements of the *Lead Limit Test*, page 929, using 10 mcg. of lead ion (Pb) in the control.

Mercury. Determine as directed under *Mercury Limit Test*, page 934, using the following as the *Sample Preparation:* Transfer 2.0 grams of the sample into a 50-ml. beaker, dissolve in 10 ml. of water, add 2 drops of phenolphthalein T.S., and slowly neutralize, with constant stirring, with dilute hydrochloric acid solution (1 in 2). Add 1 ml. of dilute sulfuric acid solution (1 in 5) and 1 ml. of potassium permanganate solution (1 in 25), cover the beaker with a watch glass, boil for a few seconds, and cool.

Packaging and storage. Store in tight containers.

Functional use in foods. Alkali.

POTASSIUM HYDROXIDE SOLUTION

DESCRIPTION

A clear or slightly turbid, colorless or slightly colored, strongly caustic, hygroscopic solution which absorbs carbon dioxide when exposed to the air forming potassium carbonate. It gives positive tests for *Potassium*, page 928.

SPECIFICATIONS

Assay. Not less than 97.0 percent and not more than 103.0 percent, by weight, of the labeled amount of KOH calculated as total alkali.

Limits of Impurities

Arsenic (as As). Not more than 3 parts per million (0.0003 percent) calculated on the basis of KOH determined in the *Assay*.

Carbonate (as K_2CO_3). Not more than 3.5 percent of the KOH determined in the *Assay*.

Heavy metals (as Pb). Not more than 30 parts per million (0.003 percent) calculated on the basis of KOH determined in the *Assay*.

Lead. Not more than 10 parts per million (0.001 percent) based on the KOH determined in the *Assay*.

Mercury. Not more than 1 part per million (0.0001 percent) calculated on the basis of the KOH determined in the *Assay*.

TESTS

Assay. Based on the stated or labeled percent of KOH, weigh accurately a volume of the solution equivalent to about 1.5 grams of potassium hydroxide, and dilute it to 40 ml. with recently boiled and cooled water. Continue as directed in the *Assay* under *Potassium Hydroxide*, page 652, beginning with "...cool to 15°...."

Arsenic. Dilute the equivalent of 1 gram of KOH, calculated on the basis of the *Assay*, to 10 ml. with water, cautiously neutralize to litmus paper with sulfuric acid, and cool. This solution meets the requirements of the *Arsenic Test*, page 865.

Carbonate. Each ml. of 1 *N* sulfuric acid required between the phenolphthalein and methyl orange end-points in the *Assay* is equivalent to 138.2 mg. of K_2CO_3.

Heavy metals. Dilute the equivalent of 670 mg. of KOH, calculated on the basis of the *Assay*, with a mixture of 5 ml. of water and 5 ml. of diluted hydrochloric acid T.S., and heat to boiling. Cool, dilute to 25 ml. with water, and filter. This solution meets the requirements of the *Heavy Metals Test*, page 920, using 20 mcg. of lead ion (Pb) in the control (*Solution A*).

Lead. Dilute the equivalent of 1 gram of KOH, calculated on the basis of the *Assay*, with a mixture of 5 ml. of water and 11 ml. of diluted hydrochloric acid T.S. This solution meets the requirements of the *Lead Limit Test*, page 929, using 10 mcg. of lead ion (Pb) in the control.

Mercury. Determine as directed under *Mercury Limit Test*, page 934, using the following as the *Sample Preparation:* Transfer an accurately weighed amount of the sample, equivalent to 2.0 grams of KOH, into a 50-ml. beaker, add 10 ml. of water and 2 drops of phenolphthalein T.S., and slowly neutralize, with constant stirring, with dilute hydrochloric acid solution (1 in 2). Add 1 ml. of dilute sulfuric acid solution (1 in 5) and 1 ml. of potassium permanganate solution (1 in 25), cover the beaker with a watch glass, boil for a few seconds, and cool.

Packaging and storage. Store in tight containers.

Solutions usually available. A nominal concentration, weight in weight, of 50 percent.

Functional use in foods. Alkali.

POTASSIUM IODATE

KIO_3 Mol. wt. 214.00

DESCRIPTION

A white, odorless, crystalline powder. One gram dissolves in about 15 ml. of water. It is insoluble in alcohol. The pH of a 1 in 20 solution is between 5 and 8.

IDENTIFICATION

To 1 ml. of a 1 in 10 solution of the sample add 1 drop of starch T.S. and a few drops of 20 percent hypophosphorous acid. A transient blue color appears.

SPECIFICATIONS

Assay. Not less than 99.0 percent and not more than the equivalent of 101.0 percent of KIO_3 after drying.

Limits of Impurities

Arsenic (as As). Not more than 3 parts per million (0.0003 percent).

Chlorate. Passes test (limit about 0.01 percent).

Heavy metals (as Pb). Not more than 10 parts per million (0.001 percent).

Iodide. Passes test (limit about 20 parts per million).

Loss on drying. Not more than 0.5 percent.

TESTS

Assay. Weigh accurately about 1.2 grams, previously dried at 105° for 3 hours, dissolve it in about 50 ml. of water in a 100-ml. volumetric flask, dilute to volume with water, and mix. Transfer 10.0 ml. into a 250-ml. glass-stoppered flask, add 40 ml. of water, 3 grams of potassium iodide, and 10 ml. of dilute hydrochloric acid (3 in 10), and stopper the flask. Allow to stand for 5 minutes, add 100 ml. of cold water, and titrate the liberated iodine with 0.1 *N* sodium thiosulfate, adding starch T.S. near the end-point. Perform a blank determination (see page 2) and make any necessary correction. Each ml. of 0.1 *N* sodium thiosulfate is equivalent to 3.567 mg. of KIO_3.

Arsenic. Dissolve 1 gram in a mixture of 2 ml. of hydrochloric acid and 1 ml. of sulfuric acid, and evaporate to fumes of sulfur trioxide. Add 1 ml. of hydrochloric acid, again evaporate to fumes, and dissolve the residue in 10 ml. of water. Heat on a steam bath, discharge any yellow color remaining with hydrazine sulfate, cool, and dilute with water to 35 ml. This solution meets the requirements of the *Arsenic Test*, page 865.

Chlorate. To 2 grams in a beaker add 2 ml. of sulfuric acid. The sample remains white, and no odor or gas is evolved.

Heavy metals, page 920. Mix 2 grams of the sample with 10 ml. of hydrochloric acid (*Solution B*). Prepare a standard (*Solution A*) by adding a few mg. of potassium chloride to 10 ml. of hydrochloric acid. Cautiously evaporate both solutions to dryness, then repeat the acid treatment on both the sample and standard residues twice, using 5-ml. portions of hydrochloric acid and evaporating to dryness each time. Dissolve the residues in 10 ml. of water, heat on a steam bath, and discharge any yellow color remaining in the sample solution with hydrazine sulfate. Cool each solution and neutralize to phenolphthalein with 0.1 *N* sodium hydroxide. Transfer the solutions to 50-ml. Nessler tubes, add 2.0 ml. of *Standard Lead Solution* (20 mcg. Pb) to the standard, and dilute both solutions with water to 25 ml. To each tube add 6 ml. of diluted acetic acid T.S. and 10 ml. of freshly prepared hydrogen sulfide T.S., allow to stand for 5 minutes, and view downward over a white surface. The color of *Solution B* is no darker than that of *Solution A*.

Iodide. Dissolve 1 gram in 10 ml. of water and add 1 ml. of diluted sulfuric acid T.S. and 1 drop of starch T.S. No blue color is formed.

Loss on drying, page 931. Dry at 105° for 3 hours.

Packaging and storage. Store in well-closed containers.

Functional use in foods. Maturing agent; dough conditioner.

POTASSIUM IODIDE

KI Mol. wt. 166.01

DESCRIPTION

Hexahedral crystals, either transparent and colorless or somewhat opaque and white, or a white, granular powder. It is stable in dry air, but slightly hygroscopic in moist air. One gram is soluble in 0.7 ml. of water at 25°, in 0.5 ml. of boiling water, in 2 ml. of glycerin, and in 22 ml. of alcohol. The pH of a 1 in 20 solution is between 6 and 10. A 1 in 10 solution responds to the tests for *Potassium*, page 928, and for *Iodide*, page 927.

SPECIFICATIONS

Assay. Not less than 99.0 percent and not more than the equivalent of 101.5 percent of KI after drying.

Limits of Impurities

Arsenic (as As). Not more than 3 parts per million (0.0003 percent).

Heavy metals (as Pb). Not more than 10 parts per million (0.001 percent).

Iodate. Not more than 4 parts per million (0.0004 percent).

Loss on drying. Not more than 1 percent.

Nitrate, nitrite, and ammonia. Passes test.

Thiosulfate and barium. Passes test.

TESTS

Assay. Dissolve about 500 mg., previously dried at 105° for 4 hours and accurately weighed, in about 10 ml. of water, and add 35 ml. of hydrochloric acid and 5 ml. of chloroform. Titrate with 0.05 *M* potassium iodate until the purple color of iodine disappears from the chloroform. Add the last portions of the iodate solution dropwise, agitating vigorously and continuously. After the chloroform has been decolorized, allow the mixture to stand for 5 minutes. If the chloroform develops a purple color, titrate further with the iodate solution. Each ml. of 0.05 *M* potassium iodate is equivalent to 16.60 mg. of KI.

Arsenic. A solution of 1 gram in 10 ml. of water meets the requirements of the *Arsenic Test*, page 865.

Heavy metals. A solution of 2 grams in 25 ml. of water meets the requirements of the *Heavy Metals Test*, page 920, using 20 mcg. of lead ion (Pb) in the control (*Solution A*).

Iodate. Dissolve 1.1 grams of the sample in sufficient ammonia- and carbon dioxide-free water to make 10 ml. of solution, and transfer to a color-comparison tube. Add 1 ml. of starch T.S. and 0.25 ml. of 1 *N* sulfuric acid, mix, and compare the color with that of a control containing, in each 10 ml., 100 mg. of potassium iodide, 1 ml. of standard iodate solution [prepared by diluting 1 ml. of a 1 in 2500 solution of potassium iodate to 100 ml. with water], 1 ml. of starch T.S., and 0.25 ml. of 1 *N* sulfuric acid. Any color produced in the solution of the sample does not exceed that in the control.

Loss on drying, page 931. Dry at 105° for 4 hours.

Nitrate, nitrite, and ammonia. Dissolve 1 gram of the sample in 5 ml. of water in a 40-ml. test tube, and add 5 ml. of sodium hydroxide T.S. and about 200 mg. of aluminum wire. Insert a plug of cotton in the upper portion of the tube, and place a piece of moistened red litmus paper over the mouth of the tube. Heat in a steam bath for about 15 minutes. No blue coloration of the paper is discernible.

Thiosulfate and barium. Dissolve 500 mg. of the sample in 10 ml. of ammonia- and carbon dioxide-free water, and add 2 drops of diluted sulfuric acid. No turbidity develops within 1 minute.

Packaging and storage. Store in well-closed containers.

Functional use in foods. Nutrient; dietary supplement.

POTASSIUM METABISULFITE

Potassium Pyrosulfite

$K_2S_2O_5$ Mol. wt. 222.33

DESCRIPTION

White or colorless free-flowing crystals, crystalline powder, or granules, usually having an odor of sulfur dioxide. It gradually oxidizes in air to the sulfate. It is soluble in water and is insoluble in alcohol. Its solutions are acid to litmus. A 1 in 10 solution gives positive tests for *Potassium*, page 928, and for *Sulfite*, page 928.

SPECIFICATIONS

Assay. Not less than 90.0 percent of $K_2S_2O_5$.

Limits of Impurities

Arsenic (as As). Not more than 3 parts per million (0.0003 percent).

Heavy metals (as Pb). Not more than 10 parts per million (0.001 percent).

Iron. Not more than 10 parts per million (0.001 percent).

Selenium. Not more than 30 parts per million (0.003 percent).

TESTS

Assay. Weigh accurately about 250 mg., add it to exactly 50 ml. of 0.1 *N* iodine contained in a glass-stoppered flask, and stopper the flask. Allow to stand for 5 minutes, add 1 ml. of hydrochloric acid, and titrate the excess iodine with 0.1 *N* sodium thiosulfate, using starch T.S. as the indicator. Each ml. of 0.1 *N* iodine is equivalent to 5.558 mg. of $K_2S_2O_5$.

Arsenic. Dissolve 1 gram in 5 ml. of water, add 0.5 ml. of sulfuric acid, evaporate to about 1 ml. on a steam bath, and dilute to 35 ml. with water. This solution meets the requirements of the *Arsenic Test*, page 865.

Heavy metals. Dissolve 2 grams in 20 ml. of water, add 5 ml. of hydrochloric acid, evaporate to about 1 ml. on a steam bath, and dissolve the residue in 25 ml. of water. This solution meets the requirements of the *Heavy Metals Test*, page 920, using 20 mcg. of lead ion (Pb) in the control (*Solution A*).

Iron. To 1 gram of the sample add 2 ml. of hydrochloric acid, and evaporate to dryness on a steam bath. Dissolve the residue in 2 ml. of hydrochloric acid and 20 ml. of water, add a few drops of bromine T.S., and boil the solution to remove the bromine. Cool, dilute with water to 25 ml., then add 50 mg. of ammonium persulfate and 5 ml. of ammonium thiocyanate T.S. Any red or pink color does not exceed that produced in a control containing 1.0 ml. of *Iron Standard Solution* (10 mcg. Fe).

Selenium. Prepare and test a 2-gram sample as directed in the *Selenium Limit Test*, page 953.

Packaging and storage. Store in well-filled, tight containers, and avoid exposure to excessive heat.

Functional use in foods. Preservative; antioxidant.

POTASSIUM NITRATE

KNO_3 Mol. wt. 101.11

DESCRIPTION

Colorless, transparent prisms, or white granules or crystalline powder. It is odorless, has a salty taste, and produces a cooling sensation in the mouth. It is slightly hygroscopic in moist air. Its solutions are neutral to litmus. One gram dissolves in 3 ml. of water at 25°, in 0.5 ml. of boiling water, and in about 620 ml. of alcohol. A 1 in 10 solution gives positive tests for *Potassium*, page 928, and for *Nitrate*, page 927.

SPECIFICATIONS

Assay. Not less than 99.0 percent of KNO_3 after drying.

Limits of Impurities

Arsenic (as As). Not more than 3 parts per million (0.0003 percent).
Chlorate. Passes test.
Heavy metals (as Pb). Not more than 20 parts per million (0.002 percent).
Lead. Not more than 10 parts per million (0.001 percent).
Loss on drying. Not more than 1 percent.

TESTS

Assay. Weigh accurately about 400 mg., previously dried at 105° for 4 hours, dissolve in 10 ml. of hydrochloric acid in a small beaker or porcelain dish, and evaporate to dryness on a steam bath. Dissolve the residue in 10 ml. of hydrochloric acid, and again evaporate to dryness, continuing the heat until the residue, when dissolved in water, is neutral to litmus. Transfer the residue with the aid of 25 ml. of water to a glass-stoppered flask, add exactly 50 ml. of 0.1 *N* silver nitrate, then add 3 ml. of nitric acid and 3 ml. of nitrobenzene, and shake vigorously. Add ferric ammonium sulfate T.S., and titrate the excess silver nitrate with 0.1 *N* ammonium thiocyanate. Each ml. of 0.1 *N* silver nitrate is equivalent to 10.11 mg. of KNO_3.

Arsenic. Dissolve 1 gram in 3 ml. of water, add 2 ml. of sulfuric acid, and evaporate to strong fumes of sulfur trioxide. Cool, wash down the sides of the container with water, and heat again to strong fuming. Repeat the washing and fuming three more times, then cool and dilute to 35 ml. with water. This solution meets the requirements of the *Arsenic Test*, page 865.

Chlorate. Sprinkle about 100 mg. of a dry sample upon 1 ml. of sulfuric acid. The mixture does not become yellow.

Heavy metals. A solution of 2 grams in 25 ml. of water meets the requirements of the *Heavy Metals Test*, page 920, using 20 mcg. of lead ion (Pb) and 1 gram of the sample in the control (*Solution A*).

Lead. A solution of 1 gram in 10 ml. of water meets the requirements of the *Lead Limit Test*, page 929, using 10 mcg. of lead ion (Pb) in the control.

Loss on drying, page 931. Dry at 105° for 4 hours.

Packaging and storage. Store in tight containers.

Functional use in foods. Preservative; color fixative in meat and meat products.

POTASSIUM NITRITE

KNO_2 Mol. wt. 85.11

DESCRIPTION

Small, white or yellowish, deliquescent granules or cylindrical sticks. It is very soluble in water, but is sparingly soluble in alcohol. A 1 in 10 solution is alkaline to litmus, and it gives positive tests for *Potassium*, page 928, and for *Nitrite*, page 928.

SPECIFICATIONS

Assay. Not less than 90.0 percent of KNO_2.

Limits of Impurities

Arsenic (as As). Not more than 3 parts per million (0.0003 percent).

Heavy metals (as Pb). Not more than 20 parts per million (0.002 percent).

Lead. Not more than 10 parts per million (0.001 percent).

TESTS

Assay. Transfer about 1.2 grams, accurately weighed, into a 100-ml. volumetric flask, dissolve in water, dilute to volume, and mix. Pipet 10 ml. of this solution into a mixture of 50.0 ml. of 0.1 *N* potassium permanganate, 100 ml. of water, and 5 ml. of sulfuric acid, keeping the tip of the pipet well below the surface of the liquid. Warm the solution to 40°, allow it to stand for 5 minutes, and add 25.0 ml. of 0.1 *N* oxalic acid. Heat the mixture to about 80°, and titrate with 0.1 *N* potassium permanganate. Each ml. of 0.1 *N* potassium permanganate is equivalent to 4.255 mg. of KNO_2.

Arsenic. Dissolve 1 gram in 10 ml. of diluted sulfuric acid T.S., boil gently for 1 minute, cool, and dilute to 35 ml. with water. This solution meets the requirements of the *Arsenic Test*, page 865.

Heavy metals. Dissolve 1 gram in 15 ml. of diluted hydrochloric acid T.S., and evaporate to dryness on a steam bath. To the residue add 2 ml. of hydrochloric acid, again evaporate to dryness, and dissolve the residue in 25 ml. of water. This solution meets the requirements of the

Heavy Metals Test, page 920, using 20 mcg. of lead ion (Pb) in the control (*Solution A*).

Lead. A solution of 1 gram in 10 ml. of water meets the requirements of the *Lead Limit Test,* page 929, using 10 mcg. of lead ion (Pb) in the control.

Packaging and storage. Store in tight containers.

Functional use in foods. Color fixative in meat and meat products.

POTASSIUM PHOSPHATE, DIBASIC

Dipotassium Monophosphate; Dipotassium Phosphate

K_2HPO_4 Mol. wt. 174.18

DESCRIPTION

A colorless or white, granular salt which is deliquescent when exposed to moist air. One gram is soluble in about 3 ml. of water. It is insoluble in alcohol. The pH of a 1 percent solution is about 9. A 1 in 20 solution gives positive tests for *Potassium,* page 928, and for *Phosphate,* page 928.

SPECIFICATIONS

Assay. Not less than 98.0 percent of K_2HPO_4 after drying.

Limits of Impurities

Arsenic (as As). Not more than 3 parts per million (0.0003 percent).

Fluoride. Not more than 10 parts per million (0.001 percent).

Heavy metals (as Pb). Not more than 20 parts per million (0.002 percent).

Insoluble substances. Not more than 0.2 percent.

Lead. Not more than 5 parts per million (0.0005 percent).

Loss on drying. Not more than 2 percent.

TESTS

Assay. Transfer about 6.5 grams of the sample, previously dried at 105° for 4 hours and accurately weighed, into a 250-ml. beaker, add 50.0 ml. of 1 *N* hydrochloric acid and 50 ml. of water, and stir until the sample is completely dissolved. Place the electrodes of a suitable pH meter in the solution, and titrate the excess acid with 1 *N* sodium hydroxide to the inflection point occurring at about pH 4. Record the buret reading, and calculate the volume (*A*) of 1 *N* hydrochloric acid consumed by the sample. Continue the titration with 1 *N* sodium hydroxide until the inflection point occurring at about pH 8.8 is reached, record the buret reading, and calculate the volume (B) of 1 N sodium hydroxide required in the titration between the two inflection points (pH 4 to pH 8.8). When A is equal to, or less than, B, each ml. of the volume A of 1 N hydrochloric acid is equivalent to 174.2 mg. of K_2HPO_4. When A is greater than B, each ml. of the volume $2B - A$ of 1 N sodium hydroxide is equivalent to 174.2 mg. of K_2HPO_4.

Arsenic. A solution of 1 gram in 10 ml. of water meets the requirements of the *Arsenic Test*, page 865.

Fluoride. Place 5 grams of the sample, 25 ml. of water, 50 ml. of perchloric acid, 5 drops of silver nitrate solution (1 in 2), and a few glass beads in a 250-ml. distilling flask connected with a condenser and carrying a thermometer and a capillary tube, both of which must extend into the liquid. Connect a small dropping funnel, filled with water, or a steam generator to the capillary tube. Support the flask on an asbestos mat with a hole which exposes about one-third of the flask to the flame. Distil into a 250-ml. flask until the temperature reaches 135°. Add water from the funnel or introduce steam through the capillary to maintain the temperature between 135° and 140°. Continue the distillation until 225–240 ml. has been collected, then dilute to 250 ml. with water, and mix.

Place a 50-ml. aliquot of this solution in a 100-ml. Nessler tube. In another similar Nessler tube place 50 ml. of water as a control. Add to each tube 0.1 ml. of a filtered solution of sodium alizarinsulfonate (1 in 1000) and 1 ml. of freshly prepared hydroxylamine hydrochloride solution (1 in 4000), and mix well. Add, dropwise, and with stirring, 0.05 N sodium hydroxide to the tube containing the distillate until its color just matches that of the control, which is faintly pink. Then add to each tube exactly 1 ml. of 0.1 N hydrochloric acid, and mix well. From a buret, graduated in 0.05-ml., add slowly to the tube containing the distillate enough thorium nitrate solution (1 in 4000) so that, after mixing, the color of the liquid just changes to a faint pink. Note the volume of the solution added, add exactly the same volume to the control, and mix. Now add to the control sodium fluoride T.S. (10 mcg. F per ml.) from a buret to make the colors of the two tubes match after dilution to the same volume. Mix well, and allow all air bubbles to escape before making the final color comparison. Check the endpoint by adding 1 or 2 drops of sodium fluoride T.S. to the control. A distinct change in color should take place. Note the volume of sodium fluoride added. The volume of sodium fluoride T.S. required for the control solution should not exceed 1.0 ml.

Heavy metals. A solution of 1 gram in 25 ml. of water meets the requirements of the *Heavy Metals Test*, page 920, using 20 mcg. of lead ion (Pb) in the control (*Solution A*), using glacial acetic acid to adjust the pH of the sample solution.

Insoluble substances. Dissolve 10 grams in 100 ml. of hot water, and filter through a tared filtering crucible. Wash the insoluble residue with hot water, dry at 105° for 2 hours, cool, and weigh.

Lead. A solution of 1 gram in 20 ml. of water meets the requirements of the *Lead Limit Test*, page 929, using 5 mcg. of lead ion (Pb) in the control.

Loss on drying, page 931. Dry at 105° for 4 hours.

Packaging and storage. Store in tight containers.

Functional use in foods. Buffer; sequestrant; yeast food.

POTASSIUM PHOSPHATE, MONOBASIC

Potassium Biphosphate; Potassium Dihydrogen Phosphate; Monopotassium Phosphate

KH_2PO_4 Mol. wt. 136.09

DESCRIPTION

Colorless crystals or a white granular or crystalline powder. It is odorless and is stable in air. It is freely soluble in water but is insoluble in alcohol. The pH of a 1 in 100 solution is between 4.2 and 4.7. A 1 in 20 solution gives positive tests for *Potassium*, page 928, and for *Phosphate*, page 928.

SPECIFICATIONS

Assay. Not less than 98.0 percent of KH_2PO_4 after drying.

Limits of Impurities

Arsenic (as As). Not more than 3 parts per million (0.0003 percent).

Fluoride. Not more than 10 parts per million (0.001 percent).

Heavy metals (as Pb). Not more than 20 parts per million (0,002 percent).

Insoluble substances. Not more than 0.2 percent.

Lead. Not more than 5 parts per million (0.0005 percent).

Loss on drying. Not more than 1 percent.

TESTS

Assay. Transfer about 5 grams of the sample, previously dried at 105° for 4 hours and accurately weighed, into a 250-ml. beaker, add 100 ml. of water and 5.0 ml. of 1 *N* hydrochloric acid, and stir until the sample is completely dissolved. Place the electrodes of a suitable pH meter in the solution, and slowly titrate the excess acid, stirring constantly, with 1 *N* sodium hydroxide to the inflection point occurring at about pH 4. Record the buret reading, and calculate the volume (*A*), if any, of 1 *N* hydrochloric acid consumed by the sample. Continue the titration with 1 *N* sodium hydroxide until the inflection point occurring at about pH 8.8 is reached, record the buret reading, and calculate the volume (B) of 1 N sodium hydroxide required in the titration between the two inflection points (pH 4 and pH 8.8). Each ml. of the volume $B - A$ of 1 N sodium hydroxide is equivalent to 136.1 mg. of KH_2PO_4.

Arsenic. A solution of 1 gram in 10 ml. of water meets the requirements of the *Arsenic Test*, page 865.

Fluoride. Proceed as directed in the test for *Fluoride* under *Potassium Phosphate, Dibasic*, page 661.

Heavy metals. A solution of 1 gram in 25 ml. of water meets the requirements of the *Heavy Metals Test*, page 920, using 20 mcg. of lead ion (Pb) in the control (*Solution A*).

Insoluble substances. Dissolve 10 grams in 100 ml. of hot water, and filter through a tared filtering crucible. Wash the residue with hot water, dry at 105° for 2 hours, cool, and weigh.

Lead. A solution of 1 gram in 20 ml. of water meets the requirements of the *Lead Limit Test*, page 929, using 5 mcg. of lead ion (Pb) in the control.

Loss on drying, page 931. Dry at 105° for 4 hours.

Packaging and storage. Store in tight containers.
Functional use in foods. Buffer; sequestrant; yeast food.

POTASSIUM PHOSPHATE, TRIBASIC

Tripotassium Phosphate

K_3PO_4 Mol. wt. 212.28

DESCRIPTION

Tribasic potassium phosphate is anhydrous or may contain 1 molecule of water of hydration. It occurs as white, odorless, hygroscopic crystals or granules. It is freely soluble in water, but is insoluble in alcohol. The pH of a 1 in 100 solution is about 11.5. A 1 in 20 solution gives positive tests for *Potassium*, page 928, and for *Phosphate*, page 928.

SPECIFICATIONS

Assay. Not less than 97.0 percent of K_3PO_4, calculated on the ignited basis.

Loss on ignition. K_3PO_4 (anhydrous), not more than 3 percent; $K_3PO_4 . H_2O$ (monohydrate), between 8 and 20 percent.

Limits of Impurities

Arsenic (as As). Not more than 3 parts per million (0.0003 percent).

Fluoride. Not more than 10 parts per million (0.001 percent).

Heavy metals (as Pb). Not more than 20 parts per million (0.002 percent).

Insoluble substances. Not more than 0.2 percent.

Lead. Not more than 5 parts per million (0.0005 percent).

TESTS

Assay. Dissolve an accurately weighed quantity of the sample, equivalent to about 8 grams of anhydrous K_3PO_4, in 40 ml. of water in a 400-ml. beaker, and add 100.0 ml. of 1 *N* hydrochloric acid. Pass a stream of carbon dioxide-free air, in fine bubbles, through the solution for 30 minutes to expel carbon dioxide, covering the beaker loosely to prevent any loss by spraying. Wash the cover and sides of the beaker with a few ml. of water, and place the electrodes of a suitable pH meter in the solution. Titrate the solution with 1 *N* sodium hydroxide to the inflection point occurring at about pH 4, then calculate the volume (A) of 1 *N* hydrochloric acid consumed. Protect the solution from absorbing carbon dioxide, and continue the titration with 1 *N* sodium hydroxide until the inflection point occurring at about pH 8.8 is reached. Calculate the volume (B) of 1 *N* sodium hydroxide consumed in this titration. When A is equal to, or greater than, $2B$, each ml. of the volume B of 1 *N* sodium hydroxide is equivalent to 212.3 mg. of K_3PO_4. When A is less than $2B$, each ml. of the volume $A - B$ of 1 *N* sodium hydroxide is equivalent to 212.3 mg. of K_3PO_4.

Loss on ignition. Ignite at about 800° for 30 minutes.

Arsenic. A solution of 1 gram in 10 ml. of water meets the requirements of the *Arsenic Test*, page 865.

Fluoride. Proceed as directed in the test for *Fluoride* under *Potassium Phosphate, Dibasic*, page 661.

Heavy metals. A solution of 1 gram in 25 ml. of water meets the requirements of the *Heavy Metals Test*, page 920, using 20 mcg. of lead ion (Pb) in the control (*Solution A*).

Insoluble substances. Dissolve 10 grams in 100 ml. of hot water, and filter through a tared filtering crucible. Wash the insoluble residue with hot water, dry at 105° for 2 hours, cool, and weigh.

Lead. A solution of 1 gram in 20 ml. of water meets the requirements of the *Lead Limit Test*, page 929, using 5 mcg. of lead ion (Pb) in the control.

Packaging and storage. Store in tight containers.

Functional use in foods. Emulsifier.

POTASSIUM POLYMETAPHOSPHATE

Potassium Metaphosphate; Potassium Kurrol's Salt

$(KPO_3)x$

DESCRIPTION

Potassium polymetaphosphate is a straight chain polyphosphate having a high degree of polymerization. It occurs as a white, odorless powder. It is insoluble in water, but is soluble in dilute solutions of sodium salts.

IDENTIFICATION

A. Finely powder about 1 gram of the sample, and add it slowly to 100 ml. of a 1 in 50 solution of sodium chloride while stirring vigorously. A gelatinous mass is formed.

B. Mix 500 mg. with 10 ml. of nitric acid and 50 ml. of water, boil for about 30 minutes, and cool. The resulting solution gives positive tests for *Potassium*, page 928, and for *Phosphate*, page 928.

SPECIFICATIONS

Assay. Not less than 59.0 percent and not more than 61.0 percent of P_2O_5.

Viscosity. Between 6.5 and 15 centipoises.

Limits of Impurities

Arsenic (as As). Not more than 3 parts per million (0.0003 percent).

Fluoride. Not more than 10 parts per million (0.001 percent).

Heavy metals (as Pb). Not more than 20 parts per million (0.002 percent).

Lead. Not more than 5 parts per million (0.0005 percent).

TESTS

Assay. Mix about 300 mg., accurately weighed, with 15 ml. of nitric acid and 30 ml. of water, boil for 30 minutes, cool, and dilute with water to about 100 ml. Heat at 60°, add an excess of ammonium molybdate T.S., and heat at 50° for 30 minutes. Filter, and wash the precipitate with dilute nitric acid (1 in 36), followed by potassium nitrate solution (1 in 100) until the filtrate is no longer acid to litmus. Dissolve the precipitate in 50.0 ml. of 1 *N* sodium hydroxide, add phenolphthalein T.S., and titrate the excess sodium hydroxide with 1 *N* sulfuric acid. Each ml. of 1 *N* sodium hydroxide is equivalent to 3.086 mg. of P_2O_5.

Viscosity. Dissolve 300 mg. of the sample in 200 ml. of a solution of sodium pyrophosphate (3.5 grams of $Na_4P_2O_7$ dissolved in 1000 ml. of water), using a magnetic stirrer. When solution is complete, or after 30 minutes, whichever occurs first, transfer 10 ml. into an Ostwald-

Fenske viscometer, and determine the time, T, in seconds, required for the liquid to flow from the upper to the lower mark in the capillary tube. Calculate the viscosity, in centipoises, by the formula Tv/dt, in which t is the time, in seconds, required for a glycerin-water mixture of known viscosity, v, and specific gravity, d, to flow from the upper to the lower mark of the capillary tube during calibration of the viscometer under similar conditions.

Arsenic. A solution of 1 gram in 15 ml. of diluted hydrochloric acid T.S. meets the requirements of the *Arsenic Test*, page 865.

Fluoride. Proceed as directed in the test for *Fluoride* under *Potassium Phosphate, Dibasic*, page 661.

Heavy metals. Warm 1 gram with 10 ml. of diluted hydrochloric acid T.S. until no more dissolves, dilute with water to 25 ml., and filter. This solution meets the requirements of the *Heavy Metals Test*, page 920, using 20 mcg. of lead ion (Pb) in the control (*Solution A*).

Lead. A solution of 1 gram in 10 ml. of diluted hydrochloric acid T.S. meets the requirements of the *Lead Limit Test*, page 929, using 5 mcg. of lead ion (Pb) in the control.

Packaging and storage. Store in well-closed containers.

Functional use in foods. Fat emulsifier; moisture-retaining agent.

POTASSIUM PYROPHOSPHATE

Tetrapotassium Pyrophosphate

$K_4P_2O_7$ Mol. wt. 330.34

DESCRIPTION

Colorless crystals, or a white, granular solid. It is hygroscopic. It is very soluble in water, but is insoluble in alcohol. The pH of a 1 in 100 solution is about 10.5.

IDENTIFICATION

A. A 1 in 20 solution gives positive tests for *Potassium*, page 928.

B. To 1 ml. of a 1 in 100 solution add a few drops of silver nitrate T.S. A white precipitate is formed which is soluble in diluted nitric acid T.S.

SPECIFICATIONS

Assay. Not less than 95.0 percent of $K_4P_2O_7$, calculated on the ignited basis.

Limits of Impurities

Arsenic (as As). Not more than 3 parts per million (0.0003 percent).

Fluoride. Not more than 10 parts per million (0.001 percent).
Heavy metals (as Pb). Not more than 20 parts per million (0.002 percent).
Insoluble substances. Not more than 0.1 percent.
Lead. Not more than 5 parts per million (0.0005 percent).
Loss on ignition. Not more than 0.5 percent.

TESTS

Assay. Dissolve about 600 mg., accurately weighed, in 100 ml. of water in a 400-ml. beaker, and adjust the pH of the solution to exactly 3.8 with hydrochloric acid, using a pH meter. Add 50 ml. of a 1 in 8 solution of zinc sulfate (125 grams of $ZnSO_4 . 7H_2O$ dissolved in water, diluted to 1000 ml., filtered, and adjusted to pH 3.8), and allow to stand for 2 minutes. Titrate the liberated acid with 0.1 *N* sodium hydroxide until a pH of 3.8 is again reached. After each addition of sodium hydroxide near the end-point, time should be allowed for any precipitated zinc hydroxide to redissolve. Each ml. of 0.1 *N* sodium hydroxide is equivalent to 16.52 mg. of $K_4P_2O_7$.

Arsenic. A solution of 1 gram in 35 ml. of water meets the requirements of the *Arsenic Test*, page 865.

Fluoride. Proceed as directed in the test for *Fluoride* under *Potassium Phosphate, Dibasic*, page 661.

Heavy metals. A solution of 1 gram in 25 ml. of water meets the requirements of the *Heavy Metals Test*, page 920, using 20 mcg. of lead ion (Pb) in the control (*Solution A*).

Insoluble substances. Dissolve 10 grams in 100 ml. of hot water, and filter through a tared filtering crucible. Wash the insoluble residue with hot water, dry at 105° for 2 hours, cool, and weigh.

Lead. A solution of 1 gram in 20 ml. of water meets the requirements of the *Lead Limit Test*, page 929, using 5 mcg. of lead ion (Pb) in the control.

Loss on ignition. Ignite at about 800° for 30 minutes.

Packaging and storage. Store in tight containers.
Functional use in foods. Emulsifier; texturizer.

POTASSIUM SORBATE

2,4-Hexadienoic Acid, Potassium Salt

$C_6H_7KO_2$ Mol. wt. 150.22

DESCRIPTION

White crystals, crystalline powder, or pellets. It melts at about

270° with decomposition. It is soluble in alcohol and freely soluble in water.

IDENTIFICATION

A. A 1 in 10 solution gives positive tests for *Potassium*, page 928.

B. To 2 ml. of a 1 in 10 solution add a few drops of bromine T.S. The color is discharged.

SPECIFICATIONS

Assay. Not less than 98.0 percent and not more than the equivalent of 101.0 percent of $C_6H_7KO_2$, calculated on the dried basis.

Limits of Impurities

Acidity (as sorbic acid). Passes test (about 1 percent).

Alkalinity (as K_2CO_3). Passes test (about 1 percent).

Arsenic (as As). Not more than 3 parts per million (0.0003 percent).

Heavy metals (as Pb). Not more than 10 parts per million (0.001 percent).

Loss on drying. Not more than 1 percent.

TESTS

Assay. Dissolve about 250 mg., accurately weighed, in 40 ml. of glacial acetic acid in a 250-ml. glass-stoppered Erlenmeyer flask, warming if necessary to effect solution. Cool to room temperature, add 2 drops of crystal violet T.S., and titrate with 0.1 *N* perchloric acid in glacial acetic acid to a blue-green end-point which persists for at least 30 seconds. Perform a blank determination (see page 2) and make any necessary correction. Each ml. of 0.1 *N* perchloric acid is equivalent to 15.02 mg. of $C_6H_7KO_2$.

Acidity or alkalinity. Dissolve 1.1 grams in 20 ml. of water and add 3 drops of phenolphthalein T.S. If the solution is colorless, titrate with 0.1 *N* sodium hydroxide to a pink color that persists for 15 seconds. Not more than 1.1 ml. is required. If the solution is pink in color, titrate with 0.1 *N* hydrochloric acid. Not more than 0.8 ml. is required to discharge the pink color.

Arsenic. A *Sample Solution* prepared as directed for organic compounds meets the requirements of the *Arsenic Test*, page 865.

Heavy metals. Prepare and test a 2-gram sample as directed in *Method II* under the *Heavy Metals Test*, page 920, using 20 mcg. of lead ion (Pb) in the control (*Solution A*).

Loss on drying, page 931. Dry at 105° for 3 hours.

Packaging and storage. Store in tight containers.

Functional use in foods. Preservative.

POTASSIUM SULFATE

K_2SO_4 Mol. wt. 174.26

DESCRIPTION

Colorless or white crystals or crystalline powder having a bitter, saline taste. One gram dissolves in about 8.5 ml. of water. It is insoluble in alcohol. The pH of a 1 in 20 solution is between 5.5 and 8.5. A 1 in 10 solution gives positive tests for *Potassium*, page 928, and for *Sulfate*, page 928.

SPECIFICATIONS

Assay. Not less than 99.0 percent of K_2SO_4.

Limits of Impurities

Arsenic (as As). Not more than 3 parts per million (0.0003 percent).

Heavy metals (as Pb). Not more than 10 parts per million (0.001 percent).

Selenium. Not more than 30 parts per million (0.003 percent).

TESTS

Assay. Dissolve about 500 mg., accurately weighed, in 200 ml. of water, add 1 ml. of hydrochloric acid, and heat to boiling. Gradually add, in small portions and while stirring constantly, an excess of hot barium chloride T.S. (about 8 or 9 ml.), and heat the mixture on a steam bath for 1 hour. Collect the precipitate on a filter, wash until free from chloride, dry, ignite, and weigh. The weight of the barium sulfate so obtained, multiplied by 0.7466, indicates its equivalent of K_2SO_4.

Arsenic. A solution of 1 gram in 35 ml. of water meets the requirements of the *Arsenic Test*, page 865.

Heavy metals. A solution of 3 grams in 25 ml. of water meets the requirements of the *Heavy Metals Test*, page 920, using 20 mcg. of lead ion (Pb) and 1 gram of the sample in the control (*Solution A*).

Selenium. A solution of 2 grams in 40 ml. of dilute hydrochloric acid (1 in 2) meets the requirements of the *Selenium Limit Test*, page 953.

Packaging and storage. Store in well-closed containers.

Functional use in foods. Water corrective; miscellaneous.

POTASSIUM SULFITE

K_2SO_3 Mol. wt. 158.26

DESCRIPTION

A white, odorless, granular powder. It undergoes oxidation in air.

One gram dissolves in about 3.5 ml. of water. It is slightly soluble in alcohol. A 1 in 20 solution gives positive tests for *Potassium*, page 928, and for *Sulfite*, page 928.

SPECIFICATIONS

Assay. Not less than 90.0 percent of K_2SO_3.

Alkalinity (as K_2CO_3). Between 0.25 percent and 0.45 percent.

Limits of Impurities

Arsenic (as As). Not more than 3 parts per million (0.0003 percent).

Heavy metals (as Pb). Not more than 10 parts per million (0.001 percent).

Selenium. Not more than 30 parts per million (0.003 percent).

TESTS

Assay. Dissolve about 750 mg., accurately weighed, in a mixture of 100.0 ml. of 0.1 *N* iodine and 5 ml. of diluted hydrochloric acid T.S., and titrate the excess iodine with 0.1 *N* sodium thiosulfate, adding starch T.S. as the indicator. Each ml. of 0.1 *N* iodine is equivalent to 7.913 mg. of K_2SO_3.

Alkalinity. Dissolve 1 gram in 20 ml. of water, add 25 ml. of 3 percent hydrogen peroxide, previously neutralized to methyl red T.S., mix thoroughly, cool to room temperature, and titrate with 0.02 *N* hydrochloric acid. Perform a blank determination (see page 2) using 25 ml. of the neutralized hydrogen peroxide solution. Each ml. of 0.02 *N* hydrochloric acid is equivalent to 1.38 mg. of K_2CO_3.

Arsenic. Dissolve 1 gram in 5 ml. of water, add 1 ml. of sulfuric acid, evaporate to copious fumes of sulfur trioxide, cool, and dilute to 35 ml. with water. This solution meets the requirements of the *Arsenic Test*, page 865.

Heavy metals. Dissolve 2 grams in 10 ml. of water, add 4 ml. of hydrochloric acid, and evaporate to dryness on a steam bath. To the residue add 5 ml. of hot water and 1 ml. of hydrochloric acid, and again evaporate to dryness. Dissolve the residue in water and dilute to 25 ml. This solution meets the requirements of the *Heavy Metals Test*, page 920, using 20 mcg. of lead ion (Pb) in the control (*Solution A*).

Selenium. Prepare and test a 2-gram sample as directed in the *Selenium Limit Test*, page 953.

Packaging and storage. Store in tight containers.

Functional use in foods. Preservative; antioxidant.

POTASSIUM TRIPOLYPHOSPHATE

Pentapotassium Triphosphate; Potassium Triphosphate

$K_5P_3O_{10}$ Mol. wt. 448.42

DESCRIPTION

White granules or a white powder. It is hygroscopic and is very soluble in water. The pH of a 1 in 100 solution is between 9.2 and 10.1.

IDENTIFICATION

A. A 1 in 20 solution gives positive tests for *Potassium*, page 928.

B. To 1 ml. of a 1 in 100 solution add a few drops of silver nitrate T.S. A white precipitate is formed which is soluble in diluted nitric acid T.S.

SPECIFICATIONS

Assay. Not less than 85.0 percent of $K_5P_3O_{10}$.

Loss on drying. Not more than 0.7 percent.

Limits of Impurities

Arsenic (as As). Not more than 3 parts per million (0.0003 percent).

Fluoride. Not more than 10 parts per million (0.001 percent).

Heavy metals (as Pb). Not more than 10 parts per million (0.001 percent).

Insoluble substances. Not more than 2 percent.

Lead. Not more than 5 parts per million (0.0005 percent).

TESTS

Assay. Proceed as directed in the *Assay* under *Sodium Tripolyphosphate*, page 780. Calculate the quantity, in mg., of $K_5P_3O_{10}$ in the sample taken by the formula $0.650 \times 25V$.

Loss on drying, page 731. Dry at about 105° for 1 hour.

Arsenic. A solution of 1 gram in 35 ml. of water meets the requirements of the *Arsenic Test*, page 865.

Fluoride. Proceed as directed in the *Fluoride Limit Test*, page 917.

Heavy metals. A solution of 2 grams in 25 ml. of water meets the requirements of the *Heavy Metals Test*, page 920, using 20 mcg. of lead ion (Pb) in the control (*Solution A*).

Insoluble substances. Dissolve 10 grams in 100 ml. of hot water, and filter through a tared filtering crucible. Wash the insoluble residue with hot water, dry at 105° for 2 hours, cool, and weigh.

Lead. A solution of 1 gram in 20 ml. of water meets the requirements of the *Lead Limit Test*, page 929, using 5 mcg. of lead ion (Pb) in the control.

Packaging and storage. Store in tight containers.

Functional use in foods. Texturizer.

L-PROLINE

L-2-Pyrrolidinecarboxylic Acid

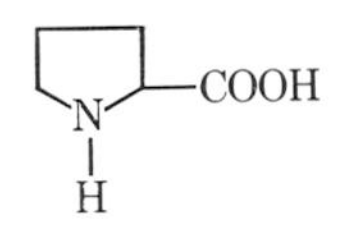

$C_5H_9NO_2$ Mol. wt. 115.13

DESCRIPTION

White crystals or a crystalline powder. It is odorless and has a slightly sweet taste. It is very soluble in water and in alcohol, but insoluble in ether.

IDENTIFICATION

To 5 ml. of a 1 in 1000 solution add 1 ml. of triketohydrindene hydrate T.S. A yellow color is produced.

SPECIFICATIONS

Nitrogen (Total). Not less than 12.0 percent and not more than 12.3 percent of N, is equivalent to not less than 98.5 percent and not more than 102.0 percent of $C_5H_9NO_2$, calculated on the dried basis.

Specific rotation, $[\alpha]_D^{20°}$. Between −84.0° and −86.0°, on the dried basis.

Limits of Impurities

Arsenic (as As). Not more than 3 parts per million (0.0003 percent).

Heavy metals (as Pb). Not more than 20 parts per million (0.002 percent).

Lead. Not more than 10 parts per million (0.001 percent).

Loss on drying. Not more than 0.3 percent.

Residue on ignition. Not more than 0.1 percent.

TESTS

Nitrogen (Total). Determine as directed under *Nitrogen Determination,* page 937, using about 200 mg. of the sample, previously dried and accurately weighed. Each ml. of 0.1 *N* sulfuric acid is equivalent to 1.401 mg. of N and to 11.51 mg. of $C_5H_9NO_2$.

Specific rotation, page 939. Determine in a solution containing 4 grams of a previously dried sample in sufficient water to make 100 ml.

Arsenic. A *Sample Solution* prepared as directed for organic compounds meets the requirements of the *Arsenic Test*, page 865.

Heavy metals. Prepare and test a 1-gram sample as directed in *Method II* under the *Heavy Metals Test*, page 920, using 20 mcg. of lead ion (Pb) in the control (*Solution A*).

Lead. A *Sample Solution* prepared as directed for organic compounds meets the requirements of the *Lead Limit Test*, page 929, using 10 mcg. of lead ion (Pb) in the control.

Loss on drying, page 931. Dry at 105° for 3 hours.

Residue on ignition, page 945. Ignite 1 gram as directed in the general method.

Packaging and storage. Store in well-closed, light-resistant containers.

Functional use in foods. Nutrient; dietary supplement.

PROPENYLGUAETHOL

1-Ethoxy-2-hydroxy-4-propenylbenzene

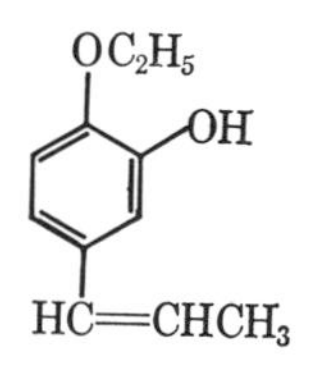

$C_{11}H_{14}O_2$ Mol. wt. 178.23

DESCRIPTION

A white crystalline powder having a vanillin-like odor. One gram is soluble in 20 ml. of 95 percent alcohol. It is soluble in most vegetable oils, but practically insoluble in water.

SPECIFICATIONS

Melting range. Between 85° and 88°.

Limits of Impurities

Heavy metals (as Pb). Not more than 10 parts per million (0.001 percent).

Residue on ignition. Not more than 0.1 percent.

TESTS

Melting range. Determine as directed in the general method, page 931.

Heavy metals. Prepare and test a 2-gram sample as directed in *Method II* under the *Heavy Metals Test*, page 920, using 20 mcg. of lead ion (Pb) in the control (*Solution A*).

Residue on ignition. Ignite 2 grams as directed in the general method, page 945.

Packaging and storage. Store in tight, light-resistant containers.

Functional use in foods. Flavoring agent.

PROPIONALDEHYDE

CH_3CH_2CHO

C_3H_6O Mol. wt. 58.08

DESCRIPTION

A clear, colorless, mobile liquid having a characteristic, sharp, pungent odor. It is miscible in all proportions with alcohol, with ether, and with water.

SPECIFICATIONS

Assay. Not less than 97.0 percent of C_3H_6O.

Distillation range. Between 46° and 50° (first 97 percent).

Specific gravity. Between 0.800 and 0.805.

Limits of Impurities

Acidity (as propionic acid). Not more than 0.1 percent.

Heavy metals (as Pb). Not more than 40 parts per million (0.004 percent).

Water. Not more than 2.5 percent.

TESTS

Assay. Transfer 65 ml. of 0.5 *N* hydroxylamine hydrochloride and 50.0 ml. of 0.5 *N* triethanolamine into a suitable heat-resistant pressure bottle provided with a tight closure that can be securely fastened. Replace the air in the bottle by passing a gentle stream of nitrogen for 2 minutes through a glass tube positioned so that the end is just above the surface of the liquid. To the mixture in the pressure bottle add about 750 mg. of sample contained in a sealed glass ampul and accurately weighed. Introduce several pieces of 8-mm. glass rod, cap the bottle, and shake vigorously to break the ampul. Allow the bottle to stand at room temperature, swirling occasionally, for 30 minutes. Cool, if necessary, and uncap the bottle cautiously to prevent loss of the contents. Titrate with 0.5 *N* sulfuric acid to a greenish blue endpoint. Perform a residual blank titration (see *Blank Tests*, page 2). Each ml. of 0.5 *N* sulfuric acid is equivalent to 29.04 mg. of C_3H_6O.

Distillation range. Distil as directed in the general method, page 890.

Specific gravity. Determine by any reliable method (see page 5).

Acidity. Transfer 50 ml. of methanol into a 250-ml. Erlenmeyer flask through which a gentle stream of nitrogen has previously been passed for 2 minutes. Add phenolphthalein T.S. to the methanol and neutralize it with 0.02 *N* alcoholic potassium hydroxide. Introduce 20 ml. (16 grams) of the sample using a suitable transfer pipet. Add 3 or 4 additional drops of phenolphthalein T.S., and titrate with 0.1 *N* sodium hydroxide to a pink end-point that persists for at least 15 seconds. During the titration direct a gentle stream of nitrogen into the flask through a short piece of 6-mm. glass tubing fastened near the tip of the buret. Each ml. of 0.1 *N* sodium hydroxide is equivalent to 7.41 mg. of $C_3H_6O_2$.

Heavy metals. Prepare and test a 500-mg. sample as directed in *Method II* under the *Heavy Metals Test*, page 920, using 20 mcg. of lead ion (Pb) in the control (*Solution A*).

Water. Determine by the *Karl Fischer Titrimetric Method*, page 977, using freshly distilled pyridine instead of methanol as the solvent.

Packaging and storage. Store in tight containers.

Functional use in foods. Flavoring agent.

PROPIONIC ACID

CH_3CH_2COOH

$C_3H_6O_2$ Mol. wt. 74.08

DESCRIPTION

An oily liquid having a slightly pungent, rancid odor. It is miscible with water and with alcohol and various other organic solvents.

SPECIFICATIONS

Assay. Not less than 99.5 percent of $C_3H_6O_2$.

Distillation range. Between 138.5° and 142.5°.

Specific gravity. Between 0.993 and 0.997 at 20°/20°.

Limits of Impurities

Aldehydes (as propionaldehyde). Passes test (limit about 0.05 percent).

Arsenic (as As). Not more than 3 parts per million (0.0003 percent).

Heavy metals (as Pb). Not more than 10 parts per million (0.001 percent).

Nonvolatile residue. Not more than 0.01 percent.

Readily oxidizable substances (as formic acid). Passes test (limit about 0.05 percent).

Water. Not more than 0.15 percent.

TESTS

Assay. Mix about 1.5 grams, accurately weighed, with 100 ml. of recently boiled and cooled water in a 250-ml. Erlenmeyer flask, add phenolphthalein T.S., and titrate with 0.5 *N* sodium hydroxide to the first appearance of a faint pink end-point which persists for at least 30 seconds. Each ml. of 0.5 *N* sodium hydroxide is equivalent to 37.04 mg. of $C_3H_6O_2$.

Distillation range. Determine as directed in the general procedure, page 890.

Specific gravity. Determine by any reliable method (see page 5).

Aldehydes. Transfer 10.0 ml. of the sample into a 250-ml. glass-stoppered Erlenmeyer flask containing 50 ml. of water and 10.0 ml. of a 1 in 8 solution of sodium bisulfite, stopper the flask, and shake vigorously. Allow the mixture to stand for 30 minutes, then titrate with 0.1 *N* iodine to the same brownish yellow end-point obtained with a blank treated with the same quantities of the same reagents (see page 2). The difference between the volume of 0.1 *N* iodine required for the blank and that required for the sample is not more than 1.75 ml.

Arsenic. A *Sample Solution* prepared as directed for organic compounds meets the requirements of the *Arsenic Test*, page 865.

Heavy metals. A solution of 2 grams in 25 ml. of water meets the requirements of the *Heavy Metals Test*, page 920, using 20 mcg. of lead ion (Pb) in the control (*Solution A*).

Nonvolatile residue. Transfer 100 ml. into a tared 125-ml. platinum evaporating dish, previously heated at 105° to constant weight, and evaporate the sample to dryness on a steam bath. Heat the dish at 105° for 30 minutes or to constant weight, cool in a desiccator, and weigh.

Readily oxidizable substances. Dissolve 15 grams of sodium hydroxide in 50 ml. of water, cool, add 6 ml. of bromine, stirring to effect complete solution, and dilute to 2000 ml. with water. Transfer 25.0 ml. of this solution into a 250-ml. glass-stoppered Erlenmeyer flask containing 100 ml. of water, and add 10 ml. of a 1 in 5 solution of sodium acetate and 10.0 ml. of the sample. Allow to stand for 15 minutes, add 5 ml. of a 1 in 4 solution of potassium iodide and 10 ml. of hydrochloric acid, and titrate with 0.1 *N* sodium thiosulfate just to the disappearance of the brown color. Perform a blank determination (see page 2). The difference between the volume of 0.1 *N* sodium thiosulfate required for the blank and that required for the sample is not more than 2.2 ml.

Water. Determine by the *Karl Fischer Titrimetric Method*, page 977.

Packaging and storage. Store in well-closed containers.

Functional use in foods. Preservative; mold and rope inhibitor.

p-PROPYL ANISOLE

Dihydroanethole

$CH_3O-C_6H_4-CH_2CH_2CH_3$

$C_{10}H_{14}O$ Mol. wt. 150.22

DESCRIPTION

A colorless to pale yellow liquid having a characteristic anise-type odor with a background of sassafras. It is soluble in most fixed oils and in mineral oil, but is insoluble in glycerin and in propylene glycol.

SPECIFICATIONS

Refractive index. Between 1.502 and 1.506 at 20°.

Solubility in alcohol. Passes test.

Specific gravity. Between 0.940 and 0.943.

TESTS

Refractive index, page 945. Determine with an Abbé or other refractometer of equal or greater accuracy.

Solubility in alcohol. Proceed as directed in the general method, page 899. One ml. dissolves in 5 ml. of 80 percent alcohol and remains in solution upon further dilution.

Specific gravity. Determine by any reliable method (see page 5).

Packaging and storage. Store in tight, full containers in a cool place protected from light.

Functional use in foods. Flavoring agent.

PROPYLENE GLYCOL

1,2-Propanediol; 1,2-Dihydroxypropane; Methyl Glycol

$CH_3.CH(OH).CH_2OH$

$C_3H_8O_2$ Mol. wt. 76.10

DESCRIPTION

A clear, colorless, viscous liquid having a slight, characteristic taste.

It is practically odorless. It absorbs moisture when exposed to moist air. It is miscible with water, acetone, and chloroform in all proportions. It is soluble in ether and will dissolve many essential oils, but is immiscible with fixed oils.

IDENTIFICATION

A. Mix 500 mg. with 3.6 grams of triphenylchloromethane and 5 ml. of pyridine, and heat under a reflux condenser on a steam bath for 1 hour. Cool, dissolve the mixture in 100 ml. of warm acetone, and stir well with 100 mg. of activated charcoal. Filter, evaporate the filtrate to about 50 ml., and allow to stand overnight in a refrigerator. Filter off the crystals, recrystallize until free from pyridine (three times) and dry in a current of air. The crystals so obtained melt at about 176° (see page 931).

B. Heat gently 1 ml. with 500 mg. of potassium bisulfate. A fruity odor is evolved, and when the solution is heated to dryness, no sharp, acrid odor of acrolein is perceptible.

SPECIFICATIONS

Assay. Not less than 97.5 percent, by weight, of $C_3H_8O_2$.

Distillation range. Betwen 185° and 189°.

Specific gravity. Between 1.035 and 1.037.

Limits of Impurities

Acidity. Passes test.

Arsenic (as As). Not more than 3 parts per million (0.0003 percent).

Heavy metals (as Pb). Not more than 10 parts per million (0.001 percent).

Residue on ignition. Not more than 0.07 percent.

Water. Not more than 0.2 percent.

TESTS

Assay. Transfer into a 100-ml. volumetric flask about 300 mg., accurately weighed by means of a weighing pipet, dilute to volume with water, and mix. Transfer 10.0 ml. of this solution into a 125-ml. Erlenmeyer flask, add 5.0 ml. of 0.1 *M* periodic acid, swirl, and let stand for 15 minutes. Add 10 ml. of a saturated solution of sodium bicarbonate, followed by 15.0 ml. of 0.1 *N* sodium arsenite and 1 ml. of a potassium iodide solution (1 in 20), and mix. Add enough sodium bicarbonate so that at the end-point several grams will remain undissolved, and titrate with 0.1 *N* iodine, using a 10-ml. microburet and continuing the titration to a faint yellow color. Perform a blank determination (see page 2) and make any necessary corrections. Each ml. of 0.1 *N* iodine is equivalent to 3.805 mg. of $C_3H_8O_2$.

Distillation range. Determine as directed in the general procedure, page 890.

Specific gravity. Determine by any reliable method (see page 5).

Acidity. Add 1 ml. of phenolphthalein T.S. to 50 ml. of water, then add 0.1 *N* sodium hydroxide until the solution remains pink for 30 seconds. To this solution add 10 ml. of the sample, accurately measured, and titrate with 0.1 *N* sodium hydroxide until the original pink color returns and remains for 30 seconds. Not more than 0.2 ml. of 0.1 *N* sodium hydroxide is required.

Arsenic. A *Sample Solution* prepared as directed for organic compounds meets the requirements of the *Arsenic Test*, page 865.

Heavy metals. A solution of 2 grams in 25 ml. of water meets the requirements of the *Heavy Metals Test*, page 920, using 20 mcg. of lead ion (Pb) in the control (*Solution A*).

Residue on ignition, page 945. Heat a 50-gram sample in a tared 100-ml. shallow dish until it ignites, and allow it to burn without further application of heat in a place free from drafts. Cool, moisten the residue with 5 ml. of sulfuric acid, and ignite at about 800° for 15 minutes.

Water. Determine by the *Karl Fischer Titrimetric Method*, page 977.

Packaging and storage. Store in tight containers.

Functional use in foods. Solvent; wetting agent; humectant.

PROPYLENE GLYCOL ALGINATE

Hydroxypropyl Alginate; Algin Derivative

$(C_9H_{14}O_7)_n$ Equiv. wt., *Calculated*, 234.21

DESCRIPTION

The propylene glycol ester of alginic acid (see *Alginic Acid*, page 26) varies in composition according to its degree of esterification and the percentages of free and neutralized carboxyl groups in the molecule. It occurs as a white to yellowish, fibrous or granular powder. It is practically odorless and tasteless. It dissolves in water, in solutions of dilute organic acids, and, depending upon the degree of esterification, in hydroalcoholic mixtures containing up to 60 percent by weight of alcohol to form stable, viscous colloidal solutions at a pH of 3.

IDENTIFICATION

A. It meets the requirements of *Identification test C* under *Alginic Acid*, page 26, but does not respond to *Identification tests A* and *B* under *Sodium Alginate*, page 721.

B. Transfer 20 ml. of the saponified solution obtained in the determination of *Esterified carboxyl groups* into a 250-ml. Erlenmeyer flask, add 50 ml. of 0.1 *M* periodic acid, swirl, and allow to stand for 30 minutes. Add 2 grams of potassium iodide, titrate with 0.1 *N* sodium thiosulfate to a faint yellow color, and then dilute the mixture to 200 ml. To 10 ml. of this dilution add 5 ml. of hydrochloric acid and 10 ml. of modified Schiff's reagent. A pink color develops in about 20 minutes (formaldehyde). To another 10 ml. of the solution add 1 ml. of a saturated solution of piperazine hydrate and 0.5 ml. of sodium nitroferricyanide T.S. A blue color develops (*acetaldehyde*). *Note:* Oxidation of propylene glycol alginate yields formaldehyde and acetaldehyde.

SPECIFICATIONS

Assay. It yields, on the dried basis, not less than 16 percent and not more than 20 percent of carbon dioxide (CO_2).

Ash. Not more than 10 percent.

Free carboxyl groups. Not more than 35 percent.

Esterified carboxyl groups. Between 40 and 85 percent.

Neutralized carboxyl groups. Between 10 and 45 percent.

Limits of Impurities

Arsenic (as As). Not more than 3 parts per million (0.0003 percent).
Heavy metals (as Pb). Not more than 40 parts per million (0.004 percent).
Insoluble matter. Not more than 0.2 percent.
Lead. Not more than 10 parts per million (0.001 percent).
Loss on drying. Not more than 20 percent.

TESTS

Assay. Proceed as directed in the *Alginates Assay*, page 863.

Free carboxyl groups. Transfer about 1 gram, previously dried at 105° for 4 hours and accurately weighed, into a 600-ml. beaker. Dissolve the sample in 200 ml. of water, and, while stirring mechanically with a glass stirrer, titrate with 0.1 *N* sodium hydroxide to a pH of 7.0 determining the end-point potentiometrically. Calculate the percent of the free carboxyl groups by the formula: [(No. ml. 0.1 *N* NaOH consumed) × 44]/%CO_2 × wt. of sample in grams.

Ash. Determine as directed under *Ash* in the monograph on *Alginic Acid*, page 26.

Esterified carboxyl groups. Transfer quantitatively the solution obtained in the determination of *Free carboxyl groups* into a 1-liter Erlenmeyer flask, add a few drops of phenolphthalein T.S. and 50.0 ml. of 0.1 *N* sodium hydroxide. Stopper the flask, swirl the solution, and then allow it to stand for 30 minutes at room temperature. Titrate the excess sodium hydroxide to a faint pink end-point with 0.1 *N* hydrochloric acid. Transfer the solution to a 600-ml. beaker and complete the titration to a pH of 7.0, determining the end-point potentiometrically. Calculate the percent of esterified carboxyl groups by the formula: [(No. ml. 0.1 *N* NaOH consumed) × 44]/% CO_2 × wt. of sample in grams.

Neutralized carboxyl groups. Calculate the percent of neutralized carboxyl groups by subtracting the sum of the percent of *Free carboxyl groups* and the percent of *Esterified carboxyl groups* from 100 percent.

Arsenic. A *Sample Solution* prepared as directed for organic compounds meets the requirements of the *Arsenic Test*, page 865.

Heavy metals. Prepare and test a 500-mg. sample as directed in *Method II* under the *Heavy Metals Test*, page 920, but use a platinum crucible for the ignition. Any color does not exceed that produced in a control (*Solution A*) containing 20 mcg. of lead ion (Pb).

Insoluble matter. Determine as directed under *Insoluble matter* in the monograph on *Alginic Acid*, page 26.

Lead. A *Sample Solution* prepared as directed for organic compounds meets the requirements of the *Lead Limit Test*, page 929, using 10 mcg. of lead ion (Pb) in the control.

Loss on drying, page 931. Dry at 105° for 4 hours.

Packaging and storage. Store in well-closed containers.

Functional use in foods. Stabilizer; thickener; emulsifier.

PROPYLENE GLYCOL MONOSTEARATE

DESCRIPTION

A mixture of propylene glycol mono- and diesters of stearic and palmitic acids. It occurs as white beads or flakes having a bland odor and taste. It is insoluble in water, but is soluble in alcohol, in ethyl acetate, and in chloroform and other chlorinated hydrocarbons.

SPECIFICATIONS

Total monoester content. Not less than the minimum percent claimed by the vendor.

Hydroxyl value, Iodine value, and **Saponification value.** Not greater than the values stated or within the range claimed by the vendor.

Limits of Impurities

Acid value. Not more than 20.

Arsenic (as As). Not more than 3 parts per million (0.0003 percent).

Free propylene glycol. Not more than 1.5 percent.

Heavy metals (as Pb). Not more than 10 parts per million (0.001 percent).

Soap (as potassium stearate). Not more than 7 percent.

TESTS

Total monoester content. Determine the percent of free

propylene glycol (*G*) in the sample as directed under *Free Glycerin or Propylene Glycol*, page 904, and determine the *Hydroxyl Value* (*H*) as directed in *Method II* of the general procedure, page 904. Calculate the hydroxyl equivalent of free propylene glycol (*F*) by the formula $561.1G/38$.

Separate the fatty acids as described in the test for *Lauric acid* under *Polysorbate 20*, page 632, and determine the *Acid Value* (*AV*) of the acids as directed in *Method I* of the general procedure, page 902.

Calculate the average molecular weight (*Mol. wt.*) of the monoester by the formula $(56109/AV) + 76.10 - 18.02$. Finally, calculate the percent of total monoester in the original sample by the formula $(H - F) \times Mol.\ wt./561$.

Hydroxyl value. Determine as directed under *Method II* in the general procedure, page 904, using about 2 grams, accurately weighed.

Iodine value. Determine by the *Wijs Method*, page 906.

Saponification value. Weigh accurately about 4 grams, and determine as directed in the general method, page 914.

Acid value. Determine as directed in *Method II* under *Acid Value*, page 902.

Arsenic. A *Sample Solution* prepared as directed for organic compounds meets the requirements of the *Arsenic Test*, page 865.

Free propylene glycol. Determine as directed under *Free Glycerin or Propylene Glycol*, page 904.

Heavy metals. Prepare and test a 2-gram sample as directed in *Method II* under the *Heavy Metals Test*, page 920, using 20 mcg. of lead ion (Pb) in the control (*Solution A*).

Soap. Weigh accurately about 5 grams, and proceed as directed in the general method, page 915, using 31.0 as the equivalence factor (*e*) in the calculation.

Packaging and storage. Store in well-closed containers.

Functional use in foods. Emulsifier; stabilizer.

PROPYL GALLATE

Gallic Acid, Propyl Ester

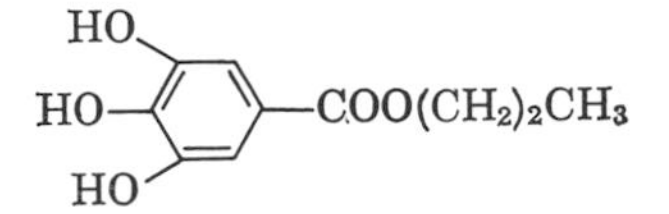

$C_{10}H_{12}O_5$ Mol. wt. 212.20

DESCRIPTION

A fine, white to nearly white, odorless powder having a slightly bitter taste. It is slightly soluble in water and freely soluble in alcohol and in ether.

IDENTIFICATION

Place about 5 grams of the sample and several boiling chips in a 500-ml. round-bottom flask, connect a water-cooled condenser to the flask, and introduce a steady stream of nitrogen into the flask, maintaining the flow of nitrogen at all times during the remainder of the procedure. Pour 100 ml. of 1 *N* sodium hydroxide through the top of the condenser, heat the solution to boiling, boil for 30 minutes, and cool. Place the reaction flask in an ice bath, and slowly, with occasional swirling, add dilute sulfuric acid (10 percent) until a pH of 2–3 is obtained (using pH paper). Filter the precipitate through a sintered-glass crucible, wash with a minimum amount of water, and dry at 110° for 2 hours. The gallic acid so obtained melts at about 240° with decomposition (see page 931).

SPECIFICATIONS

Assay. Not less than 98.0 percent and not more than the equivalent of 102.0 percent of $C_{10}H_{12}O_5$ after drying.

Melting range. Between 146° and 150°.

Limits of Impurities

Arsenic (as As). Not more than 3 parts per million (0.0003 percent).
Heavy metals (as Pb). Not more than 10 parts per million (0.001 percent).
Loss on drying. Not more than 0.5 percent.
Residue on ignition. Not more than 0.1 percent.

TESTS

Assay. Transfer about 200 mg., previously dried at 110° for 4 hours and accurately weighed, to a 400-ml. beaker, dissolve it in 150 ml. of water, and heat to boiling. With constant and vigorous stirring, add 50 ml. of bismuth nitrate T.S., continue stirring and heating until precipitation is complete, and cool. Filter the yellow precipitate on a tared sintered glass crucible, wash it with cold dilute nitric acid (1 in 300), and dry at 110° to constant weight. The weight of the precipitate so obtained, multiplied by 0.4866, represents its equivalent of $C_{10}H_{12}O_5$.

Melting range. Determine as directed in the general procedure, page 931, after drying at 110° for 4 hours.

Arsenic. A *Sample Solution* prepared as directed for organic compounds meets the requirements of the *Arsenic Test,* page 865.

Heavy metals. Prepare and test a 2-gram sample as directed in *Method II* under the *Heavy Metals Test,* page 920, using 20 mcg. of lead ion (Pb) in the control (*Solution A*).

Loss on drying, page 931. Dry at 110° for 4 hours.

Residue on ignition, page 945. Ignite 2 grams as directed in the general method.

Packaging and storage. Store in well-closed containers.

Functional use in foods. Antioxidant.

PROPYLPARABEN

Propyl *p*-Hydroxybenzoate

$HO-C_6H_4-COO(CH_2)_2CH_3$

$C_{10}H_{12}O_3$ Mol. wt. 180.21

DESCRIPTION

Small, colorless crystals or a white powder. One gram dissolves in about 2500 ml. of water at 25°, in about 400 ml. of boiling water, in about 1.5 ml. of alcohol, and in about 3 ml. of ether.

IDENTIFICATION

Dissolve about 500 mg. in 10 ml. of sodium hydroxide T.S., and boil for 30 minutes, allowing the solution to evaporate to a volume of about 5 ml. Cool the mixture, and carefully acidify with diluted sulfuric acid T.S. Collect the precipitate on a filter when cool, wash it several times with small portions of water, and dry in a desiccator over silica gel. The liberated *p*-hydroxybenzoic acid melts between 212° and 217° (see page 931).

SPECIFICATIONS

Assay. Not less than 99.0 percent of $C_{10}H_{12}O_3$, calculated on the dried basis.

Melting range. Between 95° and 98°.

Limits of Impurities

Acidity. Passes test.

Arsenic (as As). Not more than 3 parts per million (0.0003 percent).

Heavy metals (as Pb). Not more than 10 parts per million (0.001 percent).

Loss on drying. Not more than 0.5 percent.

Residue on ignition. Not more than 0.05 percent.

TESTS

Assay. Place in a flask about 2 grams, accurately weighed, add 40.0 ml. of 1 *N* sodium hydroxide, and rinse the sides of the flask with water. Cover with a watch glass, boil gently for 1 hour, and cool. Add 5 drops of bromothymol blue T.S., and titrate the excess sodium hydroxide with

1 *N* sulfuric acid to match the color of pH 6.5 phosphate buffer (see page 984) containing the same proportion of indicator. Perform a blank determination (see page 2). Each ml. of 1 *N* sodium hydroxide is equivalent to 180.2 mg. of $C_{10}H_{12}O_3$.

Melting range. Determine as directed in the general procedure, page 931.

Acidity. Heat 750 mg. with 15 ml. of water at 80° for 1 minute, cool, and filter. The filtrate is acid or neutral to litmus. To 10 ml. of the filtrate add 0.2 ml. of 0.1 *N* sodium hydroxide and 2 drops of methyl red T.S. The solution is yellow, without even a light cast of pink.

Arsenic. A *Sample Solution* prepared as directed for organic compounds meets the requirements of the *Arsenic Test*, page 865.

Heavy metals, page 920. Dissolve 2 grams in 23 ml. of acetone, and add 2 ml. of diluted acetic acid T.S., 2 ml. of water, and 10 ml. of hydrogen sulfide T.S. Any color does not exceed that produced in a control made with 23 ml. of acetone, 2 ml. of diluted acetic acid T.S., 2 ml. of *Standard Lead Solution* (20 mcg. Pb ion), and 10 ml. of hydrogen sulfide T.S.

Loss on drying, page 931. Dry over silica gel for 5 hours.

Residue on ignition. Ignite 4 grams as directed in the general method, page 945.

Packaging and storage. Store in well-closed containers.

Functional use in foods. Preservative; antimicrobial agent.

PVP

Polyvinylpyrrolidone; Povidone; Poly [1-(2-oxo-1-pyrrolidinyl)-ethylene]

DESCRIPTION

PVP is a polymer of purified 1-vinyl-2-pyrrolidone produced catalytically. It occurs as a white to tan powder, free from objectionable odor. It is soluble in water, in alcohol, and in chloroform and is insoluble in ether. The pH of a 1 in 20 solution is between 3 and 7.

IDENTIFICATION

A. To 10 ml. of a 1 in 50 solution of the sample add 20 ml. of 1 *N* hydrochloric acid and 5 ml. of potassium dichromate T.S. An orange-yellow precipitate is produced.

B. Add 5 ml. of a 1 in 50 solution of the sample to 75 mg. of cobalt nitrate and 300 mg. of ammonium thiocyanate dissolved in 2 ml. of water, mix, and then make the resulting solution acid with diluted hydrochloric acid T.S. A pale blue precipitate forms.

C. To 5 ml. of a 1 in 200 solution of the sample add a few drops of iodine T.S. A deep red color is produced.

SPECIFICATIONS

Nitrogen. Not less than 12.2 percent and not more than 13.0 percent, calculated on the anhydrous basis.

Relative viscosity of a 1 in 100 solution. Between 1.188 and 1.325, equivalent to an average molecular weight of 40,000. (*Note:* The relative viscosity of a 1 in 100 solution of PVP used as a clarifying agent in beer production is between 3.225 and 5.652, equivalent to an average molecular weight of 360,000.)

Limits of Impurities

Aldehydes (as acetaldehyde). Not more than 0.5 percent.

Arsenic (as As). Not more than 1 part per million (0.0001 percent).

Ash (**Total**). Not more than 200 parts per million (0.02 percent).

Heavy metals (as Pb). Not more than 10 parts per million (0.001 percent).

Unsaturation (as vinylpyrrolidone). Not more than 1.0 percent.

Water. Not more than 5 percent.

TESTS

Nitrogen. Determine as directed under *Nitrogen Determination*, page 937, using a sample of about 1 gram, accurately weighed.

Relative viscosity. Transfer an accurately weighed portion of the sample, equivalent to 1 gram of anhydrous PVP, into a 250-ml. Erlenmeyer flask, and calculate the amount of water to be added to make a 1.0 percent solution. Allow the mixture to stand at room temperature, with occasional swirling, until solution is complete, and then allow to stand for 1 hour longer. Filter through a dry sintered-glass filter funnel of coarse porosity, then pipet 10.0 ml. of the filtrate into a Cannon-Fenske viscometer, and place the viscometer in a water bath maintained at 25° ± 0.05°. After allowing the sample solution and pipet to warm in the water bath for 10 minutes, draw the solution by means of very gentle suction up through the capillary until the meniscus is from 3 to 4 mm. above the upper etched mark. Release the vacuum, and, when the meniscus reaches the upper etched mark, begin timing the flow through the capillary. Record the exact time when the meniscus reaches the lower etched mark, and calculate the flow time to the nearest 0.1 second. Repeat this operation until at least three readings are obtained. The readings must agree within 0.3 second; if not, repeat the determination with additional 10-ml. portions of the sample solution after recleaning the viscosity pipet with sulfuric acid-dichromate cleaning solution. Calculate the average flow time for the sample solution, and then obtain the flow time in a similar manner for 10 ml. of filtered water for the same viscosity pipet. Calculate the relative viscosity of the sample by dividing the average flow time for the sample solution by that of the water sample.

Aldehydes. Transfer about 10 grams of the sample, accurately weighed, into a 250-ml. round bottom flask containing 80 ml. of 25 percent sulfuric acid, reflux for about 45 minutes under a water-cooled condenser, and then distil about 100 ml. into a receiver containing 20.0 ml. of 1 *N* hydroxylamine hydrochloride previously adjusted to pH 3.1. Titrate the contents of the receiver with 0.1 *N* sodium hydroxide to a pH of 3.1, using a pH meter, and perform a blank determination (see page 2). Each ml. of 0.1 *N* sodium hydroxide is equivalent to 4.405 mg. of C_2H_4O.

Arsenic. A *Sample Solution* prepared with 3 grams of the sample as directed for organic compounds meets the requirements of the *Arsenic Test*, page 865.

Ash (Total). Weigh accurately about 10 grams, and proceed as directed in the general method, page 868.

Heavy metals. Prepare and test a 2-gram sample as directed in *Method II* under the *Heavy Metals Test*, page 920, using 20 mcg. of lead ion (Pb) in the control (*Solution A*).

Unsaturation. Dissolve about 4 grams of the sample, accurately weighed, in 30 ml. of water in a 125-ml. round-bottom flask, add 500 mg. of sodium acetate, mix, and begin titrating with 0.1 *N* iodine. When the iodine color no longer fades, add 3 additional ml. of the titrant, and allow the solution to stand for 5 to 10 minutes. Add starch T.S., and titrate the excess iodine with 0.1 *N* sodium thiosulfate. Perform a blank determination (see page 2), using the same volume of 0.1 *N* iodine, accurately measured, as was used for the sample. Each ml. of 0.1 *N* iodine is equivalent to 55.56 mg. of vinylpyrrolidone.

Water. Determine by the *Karl Fischer Titrimetric Method*, page 977.

Packaging and storage. Store in tight containers.

Functional use in foods. Clarifying agent; stabilizer; bodying agent; tableting adjuvant; dispersant.

PYRIDOXINE HYDROCHLORIDE

5-Hydroxy-6-methyl-3,4-pyridinedimethanol Hydrochloride;
Vitamin B_6 Hydrochloride

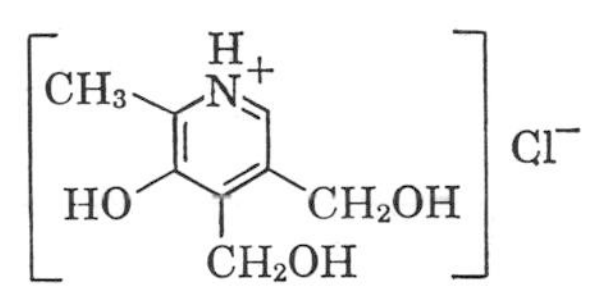

$C_8H_{11}NO_3.HCl$ Mol. wt. 205.64

DESCRIPTION

Colorless or white crystals or a white, crystalline powder. It is stable in air, but is slowly affected by sunlight. Its solutions are acid to litmus, having a pH of about 3. One gram dissolves in 5 ml. of water, and in about 100 ml. of alcohol. It is insoluble in ether. It melts at about 206° with some decomposition.

IDENTIFICATION

A. Place 1 ml. of a solution containing about 100 mcg. in each ml. into each of two test tubes marked *A* and *B*, respectively, and add to each tube 2 ml. of a 1 in 5 sodium acetate solution. To tube *A* add 1 ml. of water, and to tube *B* add 1 ml. of a 1 in 25 boric acid solution, and mix. Cool both tubes to about 20°, and rapidly add to each tube 1 ml. of a 1 in 200 solution of 2,6-dichloroquinonechlorimide in alcohol. A blue color is produced in *A*, which fades rapidly and becomes red in a few minutes, but no blue color is produced in *B*.

B. To 2 ml. of a 1 in 200 solution add 0.5 ml. of phosphotungstic acid T.S. A white precipitate is formed.

C. It gives positive tests for *Chloride*, page 926.

SPECIFICATIONS

Assay. Not less than 98.0 percent of $C_8H_{11}NO_3.HCl$, calculated on the dried basis.

Chloride content. Not less than 16.9 percent and not more than 17.6 percent of Cl, calculated on the dried basis.

Limits of Impurities

Heavy metals (as Pb). Not more than 30 parts per million (0.003 percent).

Loss on drying. Not more than 0.5 percent.

Residue on ignition. Not more than 0.1 percent.

TESTS

Assay. Dissolve about 170 mg., accurately weighed, in a mixture of 5 ml. of glacial acetic acid and 5 ml. of acetic anhydride, and boil gently to effect solution. Add 30 ml. of acetic anhydride, and boil gently for 3 minutes. Cool the solution to room temperature, add 30 ml. of benzene and 5 drops of neutral red T.S., and titrate with 0.1 *N* perchloric acid to the red-to-blue color transition. Perform a blank determination (see page 2), and make any necessary correction. Each ml. of 0.1 *N* perchloric acid is equivalent to 20.56 mg. of $C_8H_{11}NO_3$.-HCl.

Chloride content. Dissolve about 500 mg. of the sample, accurately weighed, in 50 ml. of methanol in a glass-stoppered flask. Add 5 ml. of glacial acetic acid and 2 to 3 drops of eosin Y T.S., and titrate with 0.1 *N* silver nitrate. Each ml. of 0.1 *N* silver nitrate is equivalent to 3.545 mg. of Cl.

Heavy metals. A solution of 670 mg. in 25 ml. of water meets the requirements of the *Heavy Metals Test*, page 920, using 20 mcg. of lead ion (Pb) in the control (*Solution A*).

Loss on drying, page 931. Dry in a vacuum over silica gel for 4 hours.

Residue on ignition. Ignite 2 grams as directed in the general method, page 945.

Packaging and storage. Store in tight, light-resistant containers, and avoid exposure to sunlight.

Functional use in foods. Nutrient; dietary supplement.

QUININE HYDROCHLORIDE

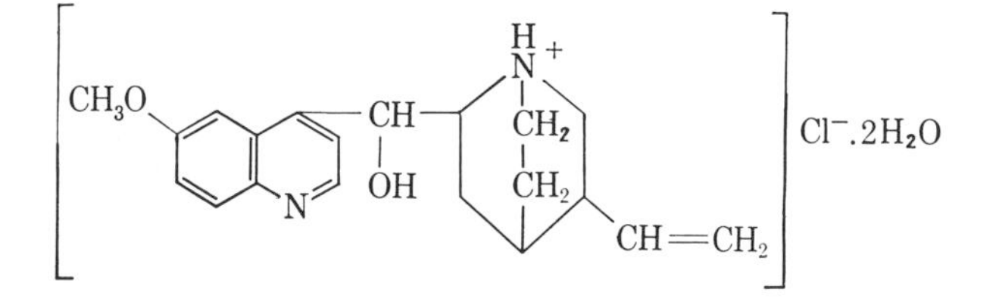

$C_{20}H_{24}N_2O_2.HCl.2H_2O$ Mol. wt. 396.92

DESCRIPTION

White, silky, glistening needles. It is odorless, has a very bitter taste, and effloresces when exposed to warm air. Its solutions are neutral or alkaline to litmus. One gram dissolves in 16 ml. of water, in 1 ml. of alcohol, in about 7 ml. of glycerin, and in about 1 ml. of chloroform. It is very slightly soluble in ether.

IDENTIFICATION

A. To 5 ml. of a 1 in 1000 solution of the sample add 1 or 2 drops of bromine T.S. followed by 1 ml. of ammonia T.S. The liquid acquires an emerald-green color due to the formation of thalleioquin.

B. A 1 in 100 solution is levorotatory (see page 939).

C. It gives positive tests for *Chloride*, page 926.

SPECIFICATIONS

Assay. Not less than 99.0 percent and not more than 101.5 percent of $C_{20}H_{24}N_2O_2.HCl$, calculated on the dried basis.

Specific rotation, $[\alpha]_D^{25°}$. Between −247° and −252°.

Limits of Impurities

Arsenic (as As). Not more than 3 parts per million (0.0003 percent).

Barium. Passes test.

Chloroform-alcohol insoluble substances. Passes test.

Heavy metals (as Pb). Not more than 10 parts per million (0.001 percent).

Loss on drying. Not more than 10 percent.

Other cinchona alkaloids. Passes test.

Readily carbonizable substances. Passes test.

Residue on ignition. Not more than 0.15 percent.

Sulfate. Not more than 500 parts per million (0.05 percent).

TESTS

Assay. Dissolve about 150 mg. of the sample, accurately weighed, in 20 ml. of acetic anhydride, add 2 drops of malachite green T.S. and 5.5 ml. of mercuric acetate T.S., and titrate with 0.1 *N* perchloric acid from a microburet to a yellow end-point. Perform a blank determination (see page 2). Each ml. of 0.1 *N* perchloric acid is equivalent to 17.99 mg. of $C_{20}H_{24}N_2O_2.HCl$.

Specific rotation, page 939. Determine in a solution containing 200 mg. in 10 ml. of 0.1 *N* hydrochloric acid.

Arsenic. A *Sample Solution* prepared as directed for organic compounds meets the requirements of the *Arsenic Test*, page 865.

Barium. To 10 ml. of a hot solution of the sample (1 in 20) add 1 ml. of diluted sulfuric acid T.S. No turbidity is produced.

Chloroform-alcohol insoluble substances. One gram dissolves completely in 7 ml. of a mixture of 2 volumes of chloroform and 1 volume of absolute alcohol.

Heavy metals. Prepare and test a 2-gram sample as directed in *Method II* under the *Heavy Metals Test*, page 920, using 20 mcg. of lead ion (Pb) in the control (*Solution A*).

Loss on drying, page 931. Dry at 120° for 3 hours.

Other cinchona alkaloids. Dissolve about 2.5 grams in 60 ml. of water in a separator, add 10 ml. of ammonia T.S., extract the mixture successively with 30 ml. and 20 ml. of chloroform, and evaporate the combined chloroform extracts to dryness on a steam bath. Dissolve 1.5 grams of the residue in 25 ml. of alcohol, dilute the solution with 50 ml. of hotwater, add 1 *N* sulfuric acid (about 5 ml.) until the solution is acid, using 2 drops of methyl red T.S. as the indicator, and neutralize the excess of acid with 1 *N* sodium hydroxide. Evaporate the solution to dryness on a steam bath, powder the residue, and agitate it in a test tube with 20 ml. of water at 65° for 30 minutes. Cool the mixture to 15°, macerate it at this temperature for 2 hours with occasional shaking, and then filter it through a filter paper (8 to 10 cm.). Transfer 5 ml. of the filtrate, at a temperature of 15°, to a test tube and mix it gently, with shaking, with 6 ml. of ammonia T.S. (which must contain between 10 and 10.2 percent of NH_3, have a temperature of 15°, and be added at once). A clear liquid is produced.

Readily carbonizable substances, page 943. Dissolve 100 mg. in 2 ml. of sulfuric acid T.S. The solution is no darker than *Matching Fluid M*.

Residue on ignition. Ignite 1 gram as directed in the general method, page 945.

Sulfate, page 879. Any turbidity produced by a 500-mg. sample does not exceed that shown in a control containing 250 mcg. of sulfate (SO_4).

Packaging and storage. Store in tight, light-resistant containers.

Functional use in foods. Flavoring agent.

QUININE SULFATE

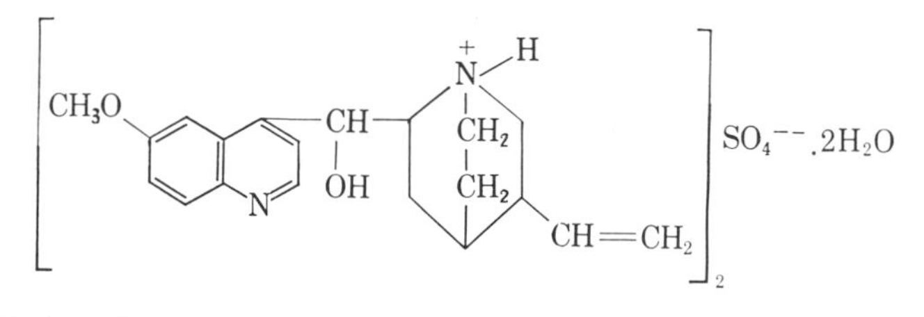

$(C_{20}H_{24}N_2O_2)_2 . H_2SO_4 . 2H_2O$ Mol. wt. 782.96

DESCRIPTION

Fine, white, needle-like crystals, usually lusterless, making a light and readily compressible mass. It is odorless and has a persistent, very bitter taste. It darkens on exposure to light. Its saturated solution is neutral or alkaline to litmus. One gram dissolves in about 500 ml. of water and in about 120 ml. of alcohol at 25°, in about 35 ml. of water at 100°, and in about 10 ml. of alcohol at 80°. It is slightly soluble in chloroform and in ether, but is freely soluble in a mixture of 2 volumes of chloroform and 1 volume of absolute alcohol.

IDENTIFICATION

A. Acidify a saturated solution of the sample with diluted sulfuric acid T.S. The resulting solution has a vivid blue fluorescence and is levorotatory (see page 939).

B. To 5 ml. of a 1 in 1000 solution add 1 or 2 drops of bromine T.S. followed by 1 ml. of ammonia T.S. The liquid acquires an emerald-green color due to the formation of thalleioquin.

C. A 1 in 50 solution made with the aid of a few drops of hydrochloric acid gives positive tests for *Sulfate*, page 928.

SPECIFICATIONS

Assay. Not less than 99.0 percent and not more than 101.0 percent of $(C_{20}H_{24}N_2O_2)_2 \cdot H_2SO_4$, calculated on the dried basis.

Specific rotation, $[\alpha]_D^{25°}$. Between $-240°$ and $-244°$.

Limits of Impurities

Arsenic (as As). Not more than 3 parts per million (0.0003 percent).

Chloroform-alcohol insoluble substances. Not more than 0.1 percent.

Heavy metals (as Pb). Not more than 10 parts per million (0.001 percent).

Loss on drying. Not more than 5 percent.

Other cinchona alkaloids. Passes test.

Readily carbonizable substances. Passes test.

Residue on ignition. Not more than 0.05 percent.

TESTS

Assay. Dissolve about 200 mg. of the sample, accurately weighed, in 20 ml. of acetic anhydride, add 2 drops of malachite green T.S., and titrate with 0.1 *N* perchloric acid from a microburet to a yellow end-point. Perform a blank determination (see page 2). Each ml. of 0.1 *N* perchloric acid is equivalent to 24.90 mg. of $(C_{20}H_{24}N_4O_2)_2 \cdot H_2SO_4$.

Specific rotation, page 939. Determine in a solution containing 200 mg. in 10 ml. of 0.1 *N* hydrochloric acid.

Arsenic. A *Sample Solution* prepared as directed for organic compounds meets the requirements of the *Arsenic Test*, page 865.

Chloroform-alcohol insoluble substances. Warm 2 grams of the sample with 15 ml. of a mixture of 2 volumes of chloroform and 1 volume of absolute alcohol at 50° for 10 minutes. Filter through a tared, sintered-glass filter, using gentle suction, and wash the filter with five 10-ml. portions of the chloroform-alcohol mixture. Dry at 105° for 1 hour, cool, and weigh.

Heavy metals. Prepare and test a 2-gram sample as directed in *Method II* under the *Heavy Metals Test*, page 920, using 20 mcg. of lead ion (Pb) in the control (*Solution A*).

Loss on drying, page 931. Dry at 120° for 3 hours.

Other cinchona alkaloids. Agitate 1.8 grams, previously dried at 50° for 2 hours, with 20 ml. of water at 65° for 30 minutes. Cool the mixture to 15°, macerate it at this temperature for 2 hours with occasional shaking, and then filter it through a filter paper (8 to 10 cm.). Transfer 5 ml. of the filtrate, at a temperature of 15°, to a test tube and mix it gently, without shaking, with 6 ml. of ammonia T.S. (which must contain between 10 and 10.2 percent of NH_3, have a temperature of 15°, and be added at once). A clear liquid is produced.

Readily carbonizable substances, page 943. Dissolve 200 mg. in 5 ml. of sulfuric acid T.S. The solution is no darker than *Matching Fluid M*.

Residue on ignition. Ignite 2 grams as directed in the general method, page 945.

Packaging and storage. Store in well-closed, light-resistant containers.

Functional use in foods. Flavoring agent.

RHODINOL

DESCRIPTION

An almost colorless liquid having a typical pronounced rose-like odor. Rhodinol is a mixture of terpenic alcohols in which 1-citronellol predominates. It is usually derived from Réunion geranium oil. It is soluble in most fixed oils, in mineral oil, and in propylene glycol. It is insoluble in glycerin.

SPECIFICATIONS

Assay. Not less than 82.0 percent of total alcohols, calculated as rhodinol ($C_{10}H_{20}O$).

Angular rotation. Between −4° and −9°.

Refractive index. Between 1.463 and 1.473 at 20°.

Solubility in alcohol. Passes test.

Specific gravity. Between 0.860 and 0.880.

Limits of Impurities

Esters. Not more than 1.0 percent, calculated as rhodinyl acetate ($C_{12}H_{22}O_2$).

TESTS

Assay. Proceed as directed under *Total Acohols*, page 893. Weigh accurately about 1.2 grams of the acetylated alcohol for the saponification, and use 78.14 as the equivalence factor (*e*) in the calculation.

Angular rotation. Determine in a 100-mm. tube as directed under *Optical Rotation*, page 939.

Refractive index, page 945. Determine with an Abbé or other refractometer of equal or greater accuracy.

Solubility in alcohol. Proceed as directed in the general method, page 899. One ml. dissolves in 1.2 ml. of 70 percent alcohol.

Specific gravity. Determine by any reliable method (see page 5).

Esters. Weigh accurately about 5 grams, and proceed as directed under *Ester Determination*, page 896, using 99.15 as the equivalence factor (*e*) in the calculation for rhodinyl acetate ($C_{12}H_{22}O_2$).

Packaging and storage. Store in full, tight, preferably glass, or tin-lined containers in a cool place protected from light.

Functional use in foods. Flavoring agent.

RHODINYL ACETATE

DESCRIPTION

A colorless to slightly yellow liquid having a light and fresh rose-like odor. It is a mixture of acetates of terpene alcohols derived from geranium oil in which *levo*-citronellyl acetate predominates. It is soluble in most fixed oils and in mineral oil. It is insoluble in glycerin and in propylene glycol.

SPECIFICATIONS

Assay. Not less than 87.0 percent of esters, calculated as rhodinyl acetate ($C_{12}H_{22}O_2$).

Angular rotation. Between $-2°$ and $-6°$.

Refractive index. Between 1.453 and 1.458 at 20°.

Solubility in alcohol. Passes test.

Specific gravity. Between 0.895 and 0.908.

Limits of Impurities

Acid value. Not more than 1.0.

TESTS

Assay. Weigh accurately about 1.3 grams, and proceed as directed under *Ester Determination*, page 896, using 99.15 as the equivalence factor (*e*) in the calculation.

Angular rotation. Determine in a 100-mm. tube as directed under *Optical Rotation*, page 939.

Refractive index, page 945. Determine with an Abbé or other refractometer of equal or greater accuracy.

Solubility in alcohol. Proceed as directed in the general method, page 899. One ml. dissolves in 2 ml. of 80 percent alcohol, and remains in solution upon dilution to 10 ml.

Specific gravity. Determine by any reliable method (see page 5).

Acid value. Determine as directed in the general method, page 893.

Packaging and storage. Store in full, tight, preferably glass, aluminum, tin-lined, or other suitably lined containers in a cool place protected from light.

Functional use in foods. Flavoring agent.

RHODINYL FORMATE

DESCRIPTION

A colorless to slightly yellow liquid having a characteristic combined leafy and rose-like odor. It is a mixture of formates of terpene alcohols derived from geranium oil in which *levo*-citronellyl formate predominates. It is soluble in most fixed oils and in mineral oil. It is insoluble in glycerin and in propylene glycol.

SPECIFICATIONS

Assay. Not less than 85.0 percent of esters, calculated as rhodinyl formate ($C_{11}H_{20}O_2$).

Refractive index. Between 1.453 and 1.458 at 20°.

Solubility in alcohol. Passes test.

Specific gravity. Between 0.901 and 0.908.

Limits of Impurities

Acid value. Not more than 2.0.

TESTS

Assay. Weigh accurately about 1.2 grams, and proceed as directed under *Ester Determination*, page 896, using 92.14 as the equivalence factor (*e*) in the calculation.

Refractive index, page 945. Determine with an Abbé or other refractometer of equal or greater accuracy.

Solubility in alcohol. Proceed as directed in the general method, page 899. One ml. dissolves in 2 ml. of 80 percent alcohol to form a clear solution.

Specific gravity. Determine by any reliable method (see page 2).

Acid value. Determine as directed in the general method, page 893.

Packaging and storage. Store in full, tight, preferably glass, tin-lined or other suitably lined containers in a cool place protected from light.

Functional use in foods. Flavoring agent.

RIBOFLAVIN

Vitamin B_2

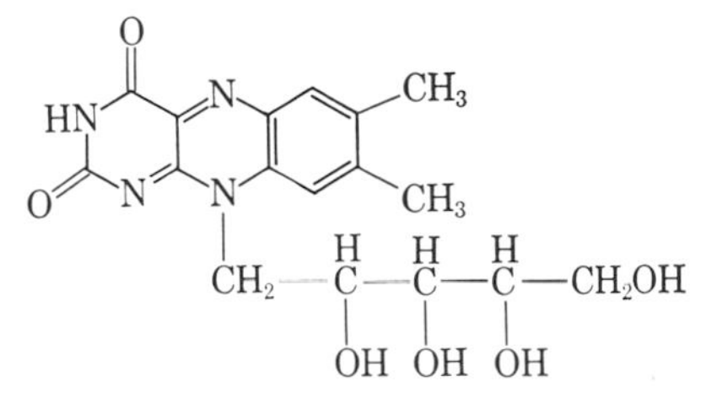

$C_{17}H_{20}N_4O_6$ Mol. wt. 376.37

DESCRIPTION

A yellow to orange-yellow, crystalline powder having a slight odor. It melts at about 280°, and its saturated solution is neutral to litmus. When dry, it is not affected by diffused light, but when in solution, light induces deterioration. One gram dissolves in from 3000 to about 20,000 ml. of water, the variations being due to difference in the internal crystalline structure. It is less soluble in alcohol than in water. It is insoluble in ether and in chloroform, but is very soluble in dilute solutions of alkalies.

IDENTIFICATION

A solution of 1 mg. in 100 ml. of water is pale greenish yellow by transmitted light, and has an intense yellowish green fluorescence which disappears upon the addition of mineral acids or alkalies.

SPECIFICATIONS

Assay. Not less than 98.0 percent and not more than the equivalent of 102.0 percent of $C_{17}H_{20}N_4O_6$, calculated on the dried basis.

Specific rotation, $[\alpha]_D^{25}$. Between −112° and −122°, calculated on the dried basis.

Limits of Impurities

Loss on drying. Not more than 1.5 percent.

Lumiflavin. Passes test.
Residue on ignition. Not more than 0.3 percent.

TESTS

Assay. (*Note:* Conduct this assay so that the solutions are protected from direct sunlight at all stages.) Place about 50 mg., accurately weighed, in a 1000-ml. volumetric flask containing about 50 ml. of water. Add 5 ml. of 6 *N* acetic acid and sufficient water to make about 800 ml. Heat on a steam bath, protected from light, with frequent agitation until dissolved. Cool to about 25°, add water to volume, and mix. Dilute this solution with water, quantitatively and stepwise, to bring it within the operating sensitivity of the fluorometer used.

In the same manner prepare a standard solution to contain, in each ml., a quantity, accurately weighed, of U.S.P. Riboflavin Reference Standard equivalent to that of the solution prepared as directed in the preceding paragraph, and measure the intensity of its fluorescence in a fluorometer at about 460 mμ. Immediately after the reading, add to the solution about 10 mg. of sodium hydrosulfite, stirring with a glass rod until dissolved, and at once measure the fluorescence again. The difference between the two readings represents the intensity of the fluorescence due to the Standard.

Similarly, measure the intensity of the fluorescence of the final solution of the riboflavin being assayed at about 460 mμ, before and after the addition of sodium hydrosulfite. Calculate the quantity of $C_{17}H_{20}N_4O_6$ in the final solution of riboflavin by the formula $C(I_U/I_S)$, in which C is the concentration, in mcg. per ml., of U.S.P. Riboflavin Reference Standard in the final solution of the Standard and I_U and I_S are the corrected fluorescence values observed for the solutions of the riboflavin and the Standard, respectively.

Specific rotation, page 939. Weigh accurately about 50 mg., and dissolve it in a mixture of 2 ml. of a 1 in 250 solution of sodium hydroxide in alcohol and sufficient cold, carbon dioxide-free water to make exactly 10 ml. of solution, and complete the determination of the rotation in a 100-mm. tube within 30 minutes after preparing the solution.

Loss on drying, page 931. Dry at 105° for 2 hours.

Lumiflavin. Prepare alcohol-free chloroform as follows: Shake 20 ml. of chloroform gently but thoroughly with 20 ml. of water for 3 minutes, draw off the chloroform layer, and wash twice more with 20-ml. portions of water. Finally filter the chloroform through a dry filter paper, shake it well for 5 minutes with 5 grams of powdered anhydrous sodium sulfate, allow the mixture to stand for 2 hours, and decant or filter the clear chloroform.

Shake 25 mg. of the sample with 10 ml. of the alcohol-free chloroform for 5 minutes, and filter. The filtrate has no more color than an equal volume of a solution made by diluting 3 ml. of 0.1 *N* potassium dichromate with water to make 1000 ml.

Residue on ignition, page 945. Ignite 1 gram as directed in the general method.

Packaging and storage. Store in tight, light-resistant containers.

Functional use in foods. Nutrient; dietary supplement.

RIBOFLAVIN 5′-PHOSPHATE SODIUM

Riboflavin 5′-Phosphate Ester Monosodium Salt

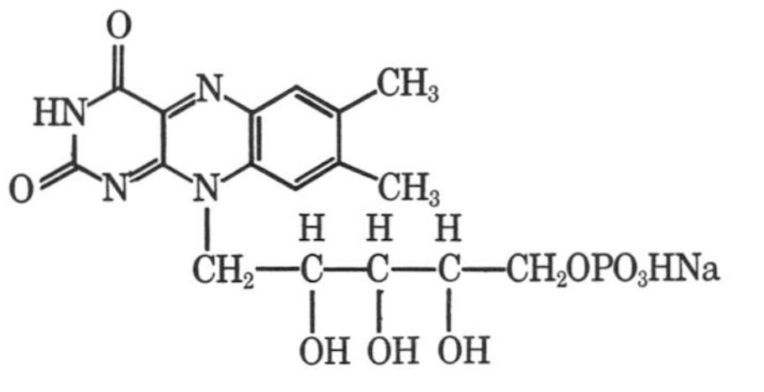

$C_{17}H_{20}N_4NaO_9P.2H_2O$ Mol. wt. 514.37

DESCRIPTION

A fine, orange-yellow, crystalline powder having a slight odor. One gram dissolves in about 30 ml. of water. When dry, it is not affected by diffused light, but when in solution light induces deterioration rapidly. It is hygroscopic.

IDENTIFICATION

A solution of 1.5 mg. in 100 ml. of water responds to the *Identification Test* under *Riboflavin,* page 697.

SPECIFICATIONS

Assay. Not less than the equivalent of 70.0 percent and not more than the equivalent of 75.0 percent of riboflavin ($C_{17}H_{20}N_4O_6$).

pH of a 1 in 100 solution. Between 5.0 and 6.5.

Residue on ignition. Not more than 25 percent.

Specific rotation, $[\alpha]_D^{25°}$. Between +37.0° and +42.0°, calculated on the dried basis.

Limits of Impurities

Free phosphate. Not more than 1 percent, calculated as PO_4.

Free riboflavin. Not more than 6 percent.

Loss on drying. Not more than 7 percent.

Riboflavin diphosphate. Not more than 6 percent, calculated as riboflavin.

TESTS

Assay. (*Note:* Use low-actinic glassware, and conduct this assay so that all solutions are protected from direct sunlight at all stages.)

Assay preparation. Transfer about 50 mg. of the sample, accurately weighed, into a 250-ml. Erlenmeyer flask, add 20 ml. of pyridine and 75 ml. of water, and dissolve the sample by frequent shaking. Transfer the solution to a 1000-ml. volumetric flask, dilute to volume with water, and mix. Transfer 10.0 ml. of this solution into a second 1000-ml. volumetric flask, add sufficient 0.1 *N* sulfuric acid (about 4 ml.) so that the final pH of the solution is between 5.9 and 6.1, dilute to volume with water, and mix.

Standard preparation. Transfer about 35 mg. of U.S.P. Riboflavin Reference Standard, accurately weighed, into a 250-ml. Erlenmeyer flask, add 20 ml. of pyridine and 75 ml. of water, and dissolve the riboflavin by frequent shaking. Transfer the solution to a 1000-ml. volumetric flask, dilute to volume with water, and mix. Transfer 10.0 ml. of this solution into a second 1000-ml. volumetric flask, add sufficient 0.1 *N* sulfuric acid (about 4 ml.) so that the final pH of the solution is between 5.9 and 6.1, dilute to volume with water, and mix.

Procedure. Using a suitable fluorometer, determine the intensity of the fluorescence of each solution at about 460 mμ, recording the fluorescence of the *Assay preparation* as I_U and that of the *Standard preparation* as I_S. Calculate the quantity, in mg. of $C_{17}H_{20}N_4O_6$ in the sample taken by the formula $100C \times I_U/I_S$, in which C is the exact concentration, in mcg. per ml., of the *Standard preparation,* corrected for loss on drying.

pH of a 1 in 100 solution. Determine by the *Potentiometric Method,* page 941.

Residue on ignition. Ignite a 1-gram sample as directed in the general method, page 945.

Specific rotation, page 939. Transfer about 750 mg., accurately weighed, into a 50-ml. volumetric flask, dissolve in and dilute to volume with 20 percent hydrochloric acid, and mix. Determine the rotation in a 1-decimeter tube within 15 minutes.

Free phosphate

Standard preparation. Transfer 220.0 mg. of monobasic potassium phosphate, KH_2PO_4, into a 1000-ml. volumetric flask, dissolve in and dilute to volume with water, and mix. Transfer 20.0 ml. of this solution into a 100-ml. volumetric flask, dilute to volume with water, and mix.

Test preparation. Transfer 300.0 mg. of the sample into a 100-ml. volumetric flask, dissolve in and dilute to volume with water, and mix.

Acid molybdate solution. Dilute 25 ml. of ammonium molybdate solution [7 grams of $(NH_4)_6Mo_7O_{24}.4H_2O$) in sufficient water to make 100 ml.] to 200 ml. with water, and then add slowly 25 ml. of 7.5 *N* sulfuric acid.

Ferrous sulfate solution. Just before use, prepare a 10 percent aqueous ferrous sulfate solution containing 2 ml. of 7.5 *N* sulfuric acid per 100 ml. of final solution.

Procedure. Transfer 10.0 ml. each of the *Standard preparation* and of the *Test preparation* into separate 50-ml. Erlenmeyer flasks, add 10.0 ml. of *Acid molybdate solution* and 5.0 ml. of *Ferrous sulfate solution* to each flask, and mix. Determine the absorbance of each solution in a 1-cm. cell at 700 mμ with a suitable spectrophotometer, using as the blank a mixture of 10.0 ml. of water, 10.0 ml. of *Acid molybdate solution,* and 5.0 ml. of *Ferrous sulfate solution.* The absorbance of the solution from the *Test preparation* is not greater than that of the *Standard preparation.*

Free riboflavin and riboflavin diphosphate. (*Note:* Use low-actinic glassware, and conduct this test so that all solutions are protected from direct sunlight at all stages.)

Standard preparation. Transfer 35.0 mg. of U.S.P. Riboflavin Reference Standard into a 250-ml. Erlenmeyer flask, add 20 ml. of pyridine and 75 ml. of water, and dissolve the riboflavin by frequent shaking. Transfer the solution into a 1000-ml. volumetric flask, dilute to volume with water, and mix. Transfer 20.0 ml. of this solution into a second 1000-ml. volumetric flask, adjust the pH to 6.0 by the addition of 8 ml. of 0.1 *N* sulfuric acid, dilute to volume with water, and mix. Finally, transfer 25.0 ml. of the last solution into a 100-ml. volumetric flask, dilute to volume with dioxane-water mixture (1:3), and mix. This solution contains 0.175 mcg. of riboflavin per ml.

pH 7 Buffer solution. Dissolve 15.6 grams of monobasic sodium phosphate ($NaH_2PO_4 . 2H_2O$) in about 100 ml. of water, add 59.3 ml. of 1 *N* sodium hydroxide, and dilute to 2000 ml. with water. Check the pH with a pH meter, and adjust to 7.0 if necessary.

Test preparation. Dissolve 100.0 mg. of the sample in 10.0 ml. of *pH 7 Buffer solution.* Prepare a strip of Whatman chromatography paper, Type 3-mm., medium flow rate, or other equivalent paper suitable for electrophoresis, and saturate the paper with *pH 7 Buffer solution.* Using a micropipet, apply 0.01 ml. of the sample solution along a narrow line on the cathode side of the paper strip contained in a suitable paper electrophoresis chamber. Apply a potential of approximately 250 v., allow electrophoresis to continue for 6 hours, and then remove the paper from the chamber. Detect any free riboflavin and/or riboflavin diphosphate by observing the strip in daylight or under ultraviolet light. Free riboflavin, if present, will appear as a band nearest to the starting line, and riboflavin diphosphate will appear farthest from the starting line. (*Caution:* The riboflavin will be destroyed if exposed to the ultraviolet light for more than a few seconds.) Cut off the respective bands, place them in separate 250-ml. Erlenmeyer flasks containing 35.0 ml. of dioxane-water mixture (1:3), and allow to stand until the spots are completely eluted from the strips.

Procedure. Using a suitable fluorometer, determine the intensity of the fluorescence of each sample solution and of the *Standard preparation* at about 460 mμ. The fluoresence of the sample solution containing the eluted riboflavin band and riboflavin diphosphate band, respectively, is not greater than that produced by the *Standard preparation.*

Loss on drying, page 931. Dry at 100° in vacuum over phosphorus pentoxide for 5 hours.

Packaging and storage. Store in tight, light-resistant containers.

Functional use in foods. Nutrient; dietary supplement.

RICE BRAN WAX

DESCRIPTION

A refined wax obtained from rice bran. It is hard, slightly crystalline, and ranges in color from tan to light brown. It is soluble in chloroform and in benzene, but is insoluble in water.

SPECIFICATIONS

Free fatty acids. Not more than 10 percent.

Iodine value. Not more than 20.

Melting range. Between 75° and 80°.

Saponification value. Between 75 and 120.

Limits of Impurities

Arsenic (as As). Not more than 3 parts per million (0.0003 percent).

Heavy metals (as Pb). Not more than 40 parts per million (0.004 percent).

Lead. Not more than 3 parts per million (0.0003 percent).

TESTS

Free fatty acids. Determine as directed in the general method procedure, page 903.

Iodine value. Determine by the *Wijs Method,* page 906.

Melting range. Determine as directed for *Class II* substances in the general procedure, page 931.

Saponification value. Determine as directed in the general procedure, page 914.

Arsenic. Prepare a *Sample Solution* as directed in the general method under *Chewing Gum Base,* page 877. This solution meets the requirements of the *Arsenic Test,* page 865.

Heavy metals. Prepare and test a 500-mg. sample as directed in

Method II under *Heavy Metals Test*, page 920, using 20 mcg. of lead ion (Pb) in the control (*Solution A*).

Lead. Prepare a *Sample Solution* as directed in the general method under *Chewing Gum Base*, page 878. This solution meets the requirements of the *Lead Limit Test*, page 929, using 10 mcg. of lead ion (Pb) in the control.

Packaging and storage. Store in well-closed containers.

Functional use in foods. Masticatory substance in chewing gum base; coating agent.

ROSE OIL

DESCRIPTION

The volatile oil obtained by steam distillation from the fresh flowers of *Rosa gallica* L., *Rosa damascena* Miller, *Rosa alba* L., *Rosa centifolia* L. and varieties of these species (Fam. *Rosaceae*). It is a colorless or yellow liquid, having the characteristic odor and taste of rose. At 25° it is a viscous liquid. Upon gradual cooling it changes to a translucent, crystalline mass, which may be liquefied by warming.

SPECIFICATIONS

Angular rotation. Between −1° and −4°.

Refractive index. Between 1.457 and 1.463 at 30°.

Solubility. Passes test.

Specific gravity. Between 0.848 and 0.863 at 30°/15°.

TESTS

Angular rotation. Determine in a 100-mm. tube as directed under *Optical Rotation*, page 939.

Refractive index, page 945. Determine with an Abbé or other refractometer of equal or greater accuracy.

Solubility. One ml. is miscible with 1 ml. of chloroform without turbidity. Add 20 ml. of 90 percent alcohol to this mixture. The resulting liquid is neutral or acid to moistened litmus paper, and upon standing at 20°, deposits crystals within 5 minutes.

Specific gravity. Determine by any reliable method (see page 5).

Packaging and storage. Store in full, tight containers in a cool place protected from light.

Functional use in foods. Flavoring agent.

ROSEMARY OIL

DESCRIPTION

The volatile oil obtained by steam distillation from the fresh flowering tops of *Rosemarinus officinalis* L. (Fam. *Labiatae*). It is a colorless or pale yellow liquid, having the characteristic odor of rosemary and a warm camphoraceous taste.

SPECIFICATIONS

Assay for esters. Not less than 1.5 percent of esters, calculated as bornyl acetate ($C_{12}H_{20}O_2$).

Assay for total borneol. Not less than 8.0 percent borneol ($C_{10}H_{18}O$).

Angular rotation. Between $-5°$ and $+10°$.

Refractive index. Between 1.464 and 1.476 at 20°.

Solubility in alcohol. Passes test.

Specific gravity. Between 0.894 and 0.912.

TESTS

Assay for esters. Weigh accurately about 10 ml., and proceed as directed under *Ester Determination*, page 896, using 98.15 as the equivalence factor (e) in the calculation.

Assay for total borneol. Proceed as directed under *Total Alcohols*, page 893, using 5 ml. of the dried, acetylized oil, accurately weighed, for the saponification. Calculate the percent of total borneol by the formula: $7.712A(1 - 0.0021E)/(B - 0.021A)$, in which A is the difference between the number of ml. of 0.5 N hydrochloric acid required for the sample and the number of ml. of 0.5 N hydrochloric acid required for the residual blank titration, B is the weight of the acetylized oil taken, and E is the percentage of esters calculated as bornyl acetate ($C_{12}H_{20}O_2$).

Angular rotation. Determine in a 100-mm. tube as directed under *Optical Rotation*, page 939.

Refractive index, page 945. Determine with an Abbé or other refractometer of equal or greater accuracy.

Solubility in alcohol. Proceed as directed in the general method, page 899. One ml. dissolves in 1 ml. of 90 percent alcohol. Upon further dilution, the solution may become turbid.

Specific gravity. Determine by any reliable method (see page 5).

Packaging and storage. Store in full, tight containers. Avoid exposure to excessive heat.

Functional use in foods. Flavoring agent.

RUE OIL

DESCRIPTION

The volatile oil obtained by steam distillation from the fresh blossoming plants, *Ruta graveolens* L., *Ruta montana* L., or *Ruta bracteosa* L. (Fam. *Rutaceae*). It is a yellow to yellow-amber liquid, having a characteristic fatty odor. It is soluble in most fixed oils and in mineral oil, but it is relatively insoluble in glycerin and in propylene glycol.

SPECIFICATIONS

Assay. Not less than 90.0 percent of ketones, calculated as methyl nonyl ketone ($C_{11}H_{22}O$).

Angular rotation. Between −1° and +3°.

Refractive index. Between 1.430 and 1.440 at 20°.

Solidification point. Between 7.5° and 10.5°.

Solubility in alcohol. Passes test.

Specific gravity. Between 0.826 and 0.838.

TESTS

Assay. Weigh accurately about 1 gram, and proceed as directed under *Aldehydes and Ketones-Hydroxylamine Method*, page 894, using 85.10 as the equivalence factor (*E*) in the calculation.

Angular rotation. Determine in a 100-mm. tube as directed under *Optical Rotation*, page 939.

Refractive index, page 945. Determine with an Abbé or other refractometer of equal or greater accuracy.

Solidification point. Determine as directed in the general method, page 954.

Solubility in alcohol. Proceed as directed in the general method, page 899. One ml. dissolves in 4 ml. of 70 percent alcohol, occasionally with opalescence or precipitation of solids.

Specific gravity. Determine by any reliable method (see page 5).

Packaging and storage. Store in full, tight, preferably glass, tin-lined, or aluminum containers in a cool place protected from light.

Functional use in foods. Flavoring agent.

SACCHARIN

o-Benzosulfimide; Gluside; 1,2-Benzisothiazolin-3-one 1,1-Dioxide

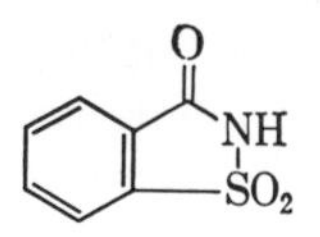

$C_7H_5NO_3S$ Mol. wt. 183.18

DESCRIPTION

White crystals or a white, crystalline powder. It is odorless or has a faint, aromatic odor. In dilute solutions, it is about 500 times as sweet as sucrose. Its solutions are acid to litmus. One gram is soluble in 290 ml. of water at 25°, in 25 ml. of boiling water, and in 30 ml. of alcohol. It is slightly soluble in chloroform and in ether, and is readily dissolved by dilute solutions of ammonia, solutions of alkali hydroxides, or solutions of alkali carbonates with the evolution of carbon dioxide.

IDENTIFICATION

A. Dissolve about 100 mg. in 5 ml. of sodium hydroxide solution (1 in 20), evaporate to dryness, and gently fuse the residue over a small flame until it no longer evolves ammonia. After the residue has cooled, dissolve it in 20 ml. of water, neutralize the solution with diluted hydrochloric acid T.S., and filter. The addition of a drop of ferric chloride T.S. to the filtrate produces a violet color.

B. Mix 20 mg. with 40 mg. of resorcinol, add 10 drops of sulfuric acid, and heat the mixture in a liquid bath at 200° for 3 minutes. After cooling, add 10 ml. of water and an excess of sodium hydroxide T.S. A fluorescent green liquid results.

SPECIFICATIONS

Assay. Not less than 98.0 percent and not more than the equivalent of 101.0 percent of $C_7H_5NO_3S$ after drying.

Melting range. Between 226° and 230°.

Limits of Impurities

Arsenic (as As). Not more than 3 parts per million (0.0003 percent).
Benzoic and salicylic acids. Passes test.
Heavy metals (as Pb). Not more than 10 parts per million (0.001 percent).
Loss on drying. Not more than 1 percent.
Readily carbonizable substances. Passes test.
Residue on ignition. Not more than 0.2 percent.
Selenium. Not more than 30 parts per million (0.003 percent).

TESTS

Assay. Dissolve about 500 mg., previously dried at 105° for 2 hours and accurately weighed, in 75 ml. of hot water, cool quickly, add phenolphthalein T.S., and titrate with 0.1 *N* sodium hydroxide. Each ml. of 0.1 *N* sodium hydroxide is equivalent to 18.32 mg. of $C_7H_5NO_3S$.

Melting range, page 931. Determine as directed in *Procedure for Class Ia.*

Arsenic. A *Sample Solution* prepared as directed for organic compounds meets the requirements of the *Arsenic Test,* page 865.

Benzoic and salicylic acids. To 10 ml. of a hot, saturated solution add ferric chloride T.S., dropwise. No precipitate or violet color appears in the liquid.

Heavy metals. Prepare and test a 2-gram sample as directed in *Method II* under the *Heavy Metals Test,* page 920, using 20 mcg. of lead ion (Pb) in the control (*Solution A*).

Loss on drying, page 931. Dry at 105° for 2 hours.

Readily carbonizable substances, page 943. Dissolve 200 mg. in 5 ml. of sulfuric acid T.S., and keep at a temperature of 48° to 50° for 10 minutes. The color is no darker than *Matching Fluid A.*

Residue on ignition, page 945. Ignite 1 gram as directed in the general method.

Selenium. Prepare and test a 2-gram sample as directed in the *Selenium Limit Test,* page 953.

Packaging and storage. Store in well-closed containers.

Functional use in foods. Nonnutritive sweetener.

SAGE OIL, DALMATIAN

DESCRIPTION

The oil obtained by steam distillation from the partially dried leaves of the plant, *Salvia officinalis* L. It is a yellowish or greenish yellow liquid, having a warm, camphoraceous and thujone-like odor and flavor. It is soluble in most fixed oils and in mineral oil. Frequently the solutions in mineral oil are opalescent. It is slightly soluble in propylene glycol, but it is practically insoluble in glycerin.

SPECIFICATIONS

Assay. Not less than 50.0 percent of ketones, calculated as thujone ($C_{10}H_{16}O$).

Angular rotation. Between +2° and +29°.

Ester value after acetylation. Between 25 and 60.

Refractive index. Between 1.457 and 1.469 at 20°.

Saponification value. Between 5 and 20.

Solubility in alcohol. Passes test.

Specific gravity. Between 0.903 and 0.925.

Limits of Impurities

Arsenic (as As). Not more than 3 parts per million (0.0003 percent).

Heavy metals (as Pb). Not more than 40 parts per million (0.004 percent).

Lead. Not more than 10 parts per million (0.001 percent).

TESTS

Assay. Weigh accurately about 1 gram, and proceed as directed under *Aldehydes and Ketones-Hydroxylamine Method*, page 894, using 76.12 as the equivalence factor (E) in the calculation.

Angular rotation. Determine in a 100-mm. tube as directed under *Optical Rotation*, page 939.

Ester value after acetylation. Proceed as directed under *Total Alcohols*, page 893, using about 2.5 grams of the acetylized oil. Calculate the ester value after acetylation by the formula $A \times 28.05/B$, in which A is the number of ml. of 0.5 N alcoholic potassium hydroxide consumed in the saponification of the acetylized oil, and B is the weight, in grams, of the acetylized oil taken as the sample.

Refractive index, page 945. Determine with an Abbé or other refractometer of equal or greater accuracy.

Saponification value. Determine as directed in the general method, page 896, using about 5 grams, accurately weighed.

Solubility in alcohol. Proceed as directed in the general method, page 899. One ml. dissolves in 1 ml. of 80 percent alcohol.

Specific gravity. Determine by any reliable method (see page 5).

Arsenic. A *Sample Solution* prepared as directed for organic compounds meets the requirements of the *Arsenic Test*, page 865.

Heavy metals. Prepare and test a 500-mg. sample as directed in *Method II* under the *Heavy Metals Test*, page 920, using 20 mcg. of lead ion (Pb) in the control (*Solution A*).

Lead. A *Sample Solution* prepared as directed for organic compounds meets the requirements of the *Lead Limit Test*, page 929, using 10 mcg. of lead ion (Pb) in the control.

Packaging and storage. Store in full, tight, preferably glass, tin-lined, or galvanized iron containers in a cool place protected from light.

Functional use in foods. Flavoring agent.

SAGE OIL, SPANISH

DESCRIPTION

The volatile oil obtained by distillation from the plants of *Salvia lavandulaefolia* Vahl. or *Salvia hispanorium* Lag., (Fam. *Labiatae*). It is a colorless to slightly yellow oil, having a camphoraceous odor with a cineole top note. It is soluble in most fixed oils and in glycerin. It is soluble, usually with opalescence, in mineral oil and in propylene glycol.

SPECIFICATIONS

Angular rotation. Between −3° and +24°.

Refractive index. Between 1.468 and 1.473 at 20°.

Saponification value. Between 14 and 57.

Saponification value after acetylation. Between 56 and 98.

Solubility in alcohol. Passes test.

Specific gravity. Between 0.909 and 0.932.

TESTS

Angular rotation. Determine in a 100-mm. tube as directed under *Optical Rotation*, page 939.

Refractive index, page 945. Determine with an Abbé or other refractometer of equal or greater accuracy.

Saponification value. Determine as directed in the general method, page 896, using about 5 grams, accurately weighed.

Saponification value after acetylation. Acetylate a 10-ml. sample as directed under *Total Alcohols*, page 893. Weigh accurately about 2.5 grams of the dried, acetylated oil, and proceed as directed under *Saponification Value*, page 896, using the weight, in grams, of the acetylated oil for W in the calculation formula.

Solubility in alcohol. Proceed as directed in the general method, page 899. One ml. dissolves in 2 ml. of 80 percent alcohol. The solution may become opalescent upon dilution.

Specific gravity. Determine by any reliable method (see page 5).

Packaging and storage. Store in full, tight, preferably glass, or tin-lined containers in a cool place protected from light.

Functional use in foods. Flavoring agent.

SANDALWOOD OIL, EAST INDIAN

DESCRIPTION

The volatile oil obtained by steam distillation from the dried, ground roots and wood of *Santalum album* L. (Fam. *Santalaceae*). It is a pale yellow to yellow, somewhat viscous oily liquid, having a strong, persistent characteristic odor. It is soluble in most fixed oils, in propylene glycol, and in mineral oil, sometimes with haziness. It is insoluble in glycerin.

SPECIFICATIONS

Assay. Not less than 90.0 percent of alcohol, calculated as santalol ($C_{15}H_{24}O$).

Angular rotation. Between −15° and −20°.

Refractive index. Between 1.500 and 1.510 at 20°.

Solubility in alcohol. Passes test.

Specific gravity. Between 0.965 and 0.980.

TESTS

Assay. Proceed as directed under *Total Alcohols*, page 893. Weigh accurately about 1.2 grams of the acetylated alcohol for the saponification, and use 110.2 as the equivalence factor (*e*) in the calculation.

Angular rotation. Determine in a 100-mm. tube as directed under *Optical Rotation*, page 939.

Refractive index, page 945. Determine with an Abbé or other refractometer of equal or greater accuracy.

Solubility in alcohol. Proceed as directed in the general method, page 899. One ml. dissolves in 5 ml. of 70 percent alcohol and remains in solution on dilution to 10 ml.

Specific gravity. Determine by any reliable method (see page 5).

Packaging and storage. Store in full, tight, preferably glass, aluminum, or suitably lined containers in a cool place protected from light.

Functional use in foods. Flavoring agent.

SANTALOL

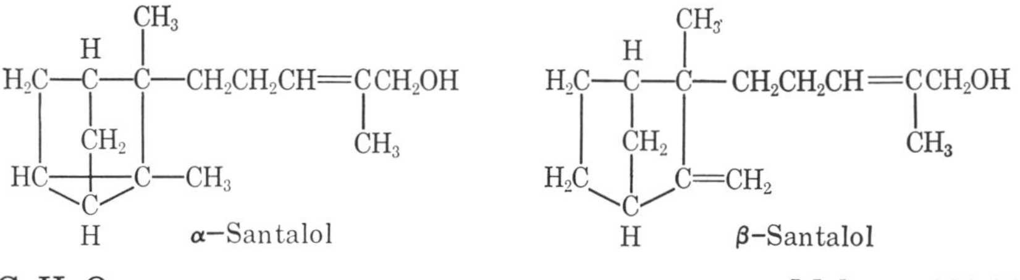

$C_{15}H_{24}O$ Mol. wt. 220.36

DESCRIPTION

A mixture of α- and β-isomers of santalol obtained from sandalwood oils. It is a colorless to slightly yellow viscous liquid having a sandalwood-like odor. It is very soluble in alcohol, in fixed oils, in mineral oil, and in propylene glycol, but insoluble in water and in glycerin.

SPECIFICATIONS

Assay. Not less than 95.0 percent of alcohols, calculated as santalol ($C_{15}H_{24}O$).

Angular rotation. Between −11° and −19°.

Refractive index. Between 1.505 and 1.509 at 20°.

Solubility in alcohol. Passes test.

Specific gravity. Between 0.965 and 0.975.

TESTS

Assay. Proceed as directed under *Total Alcohols*, page 893. Weigh accurately about 1.6 grams of the acetylated alcohol for the saponification, and use 110.2 as the equivalence factor (e) in the calculation.

Angular rotation. Determine in a 100-mm. tube as directed under *Optical Rotation*, page 939.

Refractive index, page 945. Determine with an Abbé or other refractometer of equal or greater accuracy.

Solubility in alcohol. Proceed as directed in the general method, page 899. One ml. dissolves in 4 ml. of 70 percent alcohol to form a clear solution.

Specific gravity. Determine by any reliable method (see page 5).

Packaging and storage. Store in full, tight, preferably glass, tin-lined or other suitably lined containers in a cool place protected from light.

Functional use in foods. Flavoring agent.

SANTALYL ACETATE

$C_{17}H_{26}O_2$ Mol. wt. 262.39

DESCRIPTION

A mixture of α- and β-isomers obtained by the acetylation of santalol. It is a colorless to slightly yellow liquid having an odor similar to sandalwood oil. It is soluble in alcohol, but practically insoluble in water.

SPECIFICATIONS

Assay. Not less than 95.0 percent of esters, calculated as santalyl acetate ($C_{17}H_{26}O_2$).

Refractive index. Between 1.488 and 1.491 at 20°.

Solubility in alcohol. Passes test.

Specific gravity. Between 0.980 and 0.986.

Limits of Impurities

Acid value. Not more than 1.0.

TESTS

Assay. Weigh accurately about 1.6 grams, and proceed as directed under *Ester Determination*, page 896, using 131.2 as the equivalence factor (*e*) in the calculation.

Refractive index, page 945. Determine with an Abbé or other refractometer of equal or greater accuracy.

Solubility in alcohol. Proceed as directed in the general method, page 899. One ml. dissolves in 9 ml. of 80 percent alcohol to form a clear solution.

Specific gravity. Determine by any reliable method (see page 5).

Acid value. Determine as directed in the general method, page 893.

Packaging and storage. Store in full, tight, preferably glass, tin-lined or other suitably lined containers in a cool place protected from light.

Functional use in foods. Flavoring agent.

SAVORY OIL (SUMMER VARIETY)

Summer Savory Oil

DESCRIPTION

The volatile oil obtained by steam distillation from the whole dried plant *Satureia hortensis* L. (Fam. *Labiatae*). It is a light yellow to dark brown liquid having a spicy aromatic note suggestive of thyme or origanum. It is soluble in most fixed oils and in mineral oil, but it is practically insoluble in glycerin and in propylene glycol.

SPECIFICATIONS

Assay. Not less than 20.0 percent and not more than 57.0 percent of phenols as carvacrol ($C_{10}H_{14}O$).

Angular rotation. Between $-5°$ and $+4°$.

Refractive index. Between 1.486 and 1.505 at 20°.

Saponification value. Not more than 6.

Solubility in alcohol. Passes test.

Specific gravity. Between 0.875 and 0.954.

TESTS

Assay. Proceed as directed under *Phenols*, page 898.

Angular rotation. Determine in a 100-mm. tube as directed under *Optical Rotation*, page 939.

Refractive index, page 945. Determine with an Abbé or other refractometer of equal or greater accuracy.

Saponification value. Determine as directed in the general method, page 896, using 5 grams, accurately weighed.

Solubility in alcohol. Proceed as directed in the general method, page 899. One ml. usually dissolves in 2 ml. of 80 percent alcohol. Some oils may be slightly hazy in 10 ml. of 90 percent alcohol.

Specific gravity. Determine by any reliable method (see page 5).

Packaging and storage. Store in full, tight, preferably glass, aluminum, tin-lined, or other suitably lined containers in a cool place protected from light.

Functional use in foods. Flavoring agent.

DL-SERINE

DL-2-Amino-3-hydroxypropanoic Acid

$$\begin{array}{ccccc} H_2C & — & CH & — & COOH \\ | & & | & & \\ HO & & NH_2 & & \end{array}$$

$C_3H_7NO_3$ Mol. wt. 105.10

DESCRIPTION

White crystals or a crystalline powder. It is soluble in water, but insoluble in alcohol and in ether. It melts with decomposition at about 246° using a closed capillary tube and a bath preheated to 225°. It is optically inactive.

IDENTIFICATION

To 5 ml. of a 1 in 1000 solution add 1 ml. of triketohydrindene hydrate T.S. A bluish purple or purple color is produced.

SPECIFICATIONS

Assay. Not less than 98.0 percent and not more than the equivalent of 102.0 percent of $C_3H_7NO_3$, calculated on the dried basis.

Limits of Impurities

Arsenic (as As). Not more than 3 parts per million (0.0003 percent)

Heavy metals (as Pb). Not more than 20 parts per million (0.002 percent).

Lead. Not more than 10 parts per million (0.001 percent).

Loss on drying. Not more than 0.3 percent.

Residue on ignition. Not more than 0.1 percent.

TESTS

Assay. Transfer about 300 mg., accurately weighed, into a 100-ml. beaker, add 25 ml. of water, and heat on a steam bath until dissolved. Add 5 ml. of formaldehyde T.S., previously neutralized to phenolphthalein T.S. by the addition of 0.1 *N* sodium hydroxide. Cool the mixture, add a few drops phenolphthalein T.S., and titrate with 0.1 *N* sodium hydroxide to a pink end-point that persists for at least 30 seconds. Each ml. of 0.1 *N* sodium hydroxide is equivalent to 10.51 mg. of $C_3H_7NO_3$.

Arsenic. A *Sample Solution* prepared as directed for organic compounds meets the requirements of the *Arsenic Test*, page 865.

Heavy metals. Prepare and test a 1-gram sample as directed in *Method II* under the *Heavy Metals Test*, page 920, using 20 mcg. of lead ion (Pb) in the control (*Solution A*).

Lead. A *Sample Solution* prepared as directed for organic compounds meets the requirements of the *Lead Limit Test*, page 929, using 10 mcg. of lead ion (Pb) in the control.

Loss on drying, page 931. Dry at 105° for 3 hours.

Residue on ignition, page 945. Ignite 1 gram as directed in the general method.

Packaging and storage. Store in well-closed containers.

Functional use in foods. Nutrient; dietary supplement.

L-SERINE

L-2-Amino-3-hydroxypropanoic Acid

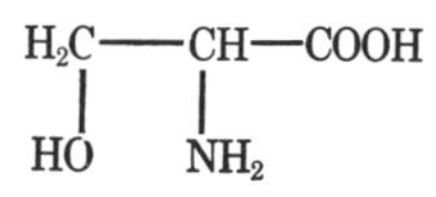

$C_3H_7NO_3$ Mol. wt. 105.10

DESCRIPTION

A white crystalline powder without odor and having a sweet taste. It is soluble in water, but insoluble in alcohol and in ether. It melts with decomposition at about 228°.

IDENTIFICATION

A. To 5 ml. of a 1 in 1000 solution add 1 ml. of triketohydrindene hydrate T.S. A reddish-purple or purple color is produced.

B. Dissolve about 500 mg. in 10 ml. of water, add 200 mg. of periodic acid, and heat. The odor of formaldehyde is produced.

SPECIFICATIONS

Assay. Not less than 98.0 percent and not more than the equivalent of 102.0 percent of $C_3H_7NO_3$, calculated on the dried basis.

Specific rotation, $[\alpha]_D^{20°}$. Between +13.5° and +16°, on the dried basis.

Limits of Impurities

Arsenic (as As). Not more than 3 parts per million (0.0003 percent).

Heavy metals (as Pb). Not more than 20 parts per million (0.002 percent).

Lead. Not more than 10 parts per million (0.001 percent).

Loss on drying. Not more than 0.3 percent.

Residue on ignition. Not more than 0.1 percent.

TESTS

Assay. Transfer about 300 mg., accurately weighed, into a 100-ml. beaker, add 25 ml. of water, and heat on a steam bath until dissolved. Add 5 ml. of formaldehyde T.S., previously neutralized to phenolphthalein T.S. by the addition of 0.1 *N* sodium hydroxide, and cool. Add a few drops of phenolphthalein T.S., and titrate with 0.1 *N* sodium hydroxide to a pink end-point that persists for at least 30 seconds. Each ml. of 0.1 *N* sodium hydroxide is equivalent to 10.51 mg. of $C_3H_7NO_3$.

Specific rotation, page 939. Determine in a solution containing 10 grams of a previously dried sample in sufficient 2 *N* hydrochloric acid to make 100 ml.

Arsenic. A *Sample Solution* prepared as directed for organic compounds meets the requirements of the *Arsenic Test,* page 865.

Heavy metals. Prepare and test a 1-gram sample as directed in *Method II* under the *Heavy Metals Test,* page 920, using 20 mcg. of lead ion (Pb) in the control (*Solution A*).

Lead. A *Sample Solution* prepared as directed for organic compounds meets the requirements of the *Lead Limit Test,* page 929, using 10 mcg. of lead ion (Pb) in the control.

Loss on drying, page 931. Dry at 105° for 3 hours.

Residue on ignition, page 945. Ignite 1 gram as directed in the general method.

Packaging and storage. Store in well-closed containers.

Functional use in foods. Nutrient; dietary supplement.

SILICON DIOXIDE

Silica Aerogel; Hydrated Silica

DESCRIPTION

Silica aerogel is a white, fluffy, powdered or granular microcellular silica. *Hydrated silica* is a precipitated, hydrated silicon dioxide occurring as a fine, white, amorphous powder, or as beads or granules. Both forms of silicon dioxide are insoluble in water and in alcohol and other organic solvents, but are soluble in hot phosphoric or hydrofluoric acids and in solutions of alkalies at 80° to 100°.

SPECIFICATIONS

Assay

Silica aerogel: Not less than 90.0 percent of SiO_2.

Hydrated silica: Not less than 89.0 percent of SiO_2, calculated on the dried basis.

Loss on drying. *Hydrated silica:* not more than 6 percent.

Loss on ignition. Not more than 6 percent.

Limits of Impurities

Arsenic (as As). Not more than 3 parts per million (0.0003 percent).

Heavy metals (as Pb). Not more than 30 parts per million (0.003 percent).

Lead. Not more than 10 parts per million (0.001 percent).

Soluble ionizable salts (as Na_2SO_4). Not more than 5 percent.

TESTS

Assay. Transfer about 2 grams, accurately weighed, into a tared platinum crucible, ignite at 600° for 2 hours, cool in a desiccator, and weigh. Moisten the residue with 7 or 8 drops of alcohol, add 3 drops of sulfuric acid, and add enough hydrofluoric acid to cover the wetted sample. Evaporate to dryness on a hot plate, using medium heat (95–105°), then add 5 ml. of hydrofluoric acid, swirl the dish carefully to wash down the sides, and again evaporate to dryness. Ignite the dried residue to a red heat over a Meker burner, cool in a desiccator, and weigh. The difference between the total weight loss and the weight loss after ignition at 600° represents the weight of SiO_2 in the sample taken.

Loss on drying, page 931. Dry hydrated silica at 105° for 2 hours.

Loss on ignition. Transfer about 5 grams, accurately weighed, into a suitable tared crucible, and ignite at 600° to constant weight.

Sample Solution for the Determination of Arsenic, Heavy Metals, and Lead. Transfer 10.0 grams of the sample into a 250-ml. beaker, add 50 ml. of 0.5 *N* hydrochloric acid, cover with a watch glass, and heat slowly to boiling. Boil gently for 15 minutes, cool, and let the undissolved material settle. Decant the supernatant liquid through a Whatman No. 3 filter paper, or equivalent, into a 100-ml. volumetric flask, retaining as much as possible of the insoluble material in the beaker. Wash the slurry and beaker with three 10-ml. portions of hot water, decanting each washing through the filter into the flask. Finally, wash the filter paper with 15 ml. of hot water, cool the filtrate to room temperature, dilute to volume with water, and mix.

Arsenic. A 10-ml. portion of the *Sample Solution* meets the requirements of the *Arsenic Test*, page 865.

Heavy metals. Dilute 20.0 ml. of the *Sample Solution* to 30.0 ml. with water. A 10-ml. portion of the dilution meets the requirements of the *Heavy Metals Test*, page 920, using 20 mcg. of lead ion (Pb) in the control (*Solution A*).

Lead. A 10-ml. portion of the *Sample Solution* meets the requirements of the *Lead Limit Test*, page 929, using 10 mcg. of lead ion (Pb) in the control.

Soluble ionizable salts. Weigh accurately 12.5 grams of the sample, and stir it with 240 ml. of water for at least 5 minutes with a high speed mixer. Transfer the mixture into a 250-ml. graduate, and wash the mixer container with water, adding the washings to the graduate to make 250 ml. Stopper the graduate, and invert it several times to mix the slurry. The conductivity of the slurry, determined with a suitable conductance bridge assembly, is not greater than that produced by a control solution containing 750 mg. of anhydrous sodium sulfate in each 250 ml.

Packaging and storage. Store in well-closed containers.

Functional use in foods. Anticaking agent; defoaming agent.

SODIUM ACETATE

$C_2H_3NaO_2 . 3H_2O$ Mol. wt. 136.08

DESCRIPTION

Colorless, transparent crystals or a granular crystalline powder. It is odorless or has a faint, acetous odor. It effloresces in warm, dry air. One gram dissolves in about 0.8 ml. of water and in about 19 ml. of alcohol. A 1 in 20 solution gives positive tests for *Sodium*, page 928, and for *Acetate*, page 925.

SPECIFICATIONS

Assay. Not less than 99.0 percent and not more than the equivalent of 101.0 percent of $C_2H_3NaO_2$ after drying.

Loss on drying. Between 36 percent and 41 percent.

Limits of Impurities

Alkalinity (as Na_2CO_3). Not more than 0.05 percent.

Arsenic (as As). Not more than 3 parts per million (0.0003 percent).

Heavy metals (as Pb). Not more than 10 parts per million (0.001 percent).

Potassium compounds. Passes test.

TESTS

Assay. Weigh accurately about 400 mg. of the sample obtained in the test for *Loss on drying*, dissolve it in 40 ml. of glacial acetic acid, add 2 drops of crystal violet T.S., and titrate with 0.1 *N* perchloric acid in glacial acetic acid. Perform a blank determination (see page 2), and make any necessary correction. Each ml. of 0.1 *N* perchloric acid is equivalent to 8.203 mg. of $C_2H_3NaO_2$.

Loss on drying, page 931. Dry at 80° overnight and follow by drying at 120° for 4 hours.

Alkalinity. Dissolve 2 grams in about 20 ml. of water, and add 3 drops of phenolphthalein T.S. If a pink color is produced, not more than 0.1 ml. of 0.1 *N* sulfuric acid is required to discharge it.

Arsenic. A solution of 1 gram in 10 ml. of water meets the requirements of the *Arsenic Test*, page 865.

Heavy metals. A solution of 2 grams in 25 ml. of water meets the requirements of the *Heavy Metals Test*, page 920, using glacial acetic acid to adjust the pH of *Solution B*, and using 20 mcg. of lead ion (Pb) in the control (*Solution A*).

Potassium compounds. Mix a few drops of sodium bitartrate T.S. with 5 ml. of a clear, saturated solution of the sample. No turbidity is produced within 5 minutes.

Packaging and storage. Store in tight containers.

Functional use in foods. Buffer.

SODIUM ACETATE, ANHYDROUS

$C_2H_3NaO_2$ Mol. wt. 82.03

DESCRIPTION

A white, odorless, granular powder. It is hygroscopic. One gram dissolves in about 2 ml. of water. A 1 in 20 solution gives positive tests for *Sodium*, page 928, and for *Acetate*, page 925.

SPECIFICATIONS

Assay. Not less than 99.0 percent and not more than the equivalent of 101.0 percent of $C_2H_3NaO_2$ after drying.

Limits of Impurities

Alkalinity (as NaOH). Not more than 0.2 percent.

Arsenic (as As). Not more than 3 parts per million (0.0003 percent).

Heavy metals (as Pb). Not more than 10 parts per million (0.001 percent).

Loss on drying. Not more than 2 percent.

Potassium compounds. Passes test.

TESTS

Assay. Weigh accurately about 400 mg. of the sample obtained in the test for *Loss on drying*, dissolve it in 40 ml. of glacial acetic acid, add 2 drops of crystal violet T.S., and titrate with 0.1 *N* perchloric acid in glacial acetic acid. Perform a blank determination (see page 2), and make any necessary correction. Each ml. of 0.1 *N* perchloric acid is equivalent to 8.203 mg. of $C_2H_3NaO_2$.

Alkalinity. Dissolve 2 grams in 20 ml. of water, and add phenolphthalein T.S. If a pink color is produced, not more than 1.0 ml. of 0.1 *N* sulfuric acid is required to discharge it.

Arsenic. A solution of 1 gram in 10 ml. of water meets the requirements of the *Arsenic Test*, page 865.

Heavy metals. A solution of 2 grams in 25 ml. of water meets the requirements of the *Heavy Metals Test*, page 920, using glacial acetic acid to adjust the pH of *Solution B*, and using 20 mcg. of lead ion (Pb) in the control (*Solution A*).

Loss on drying, page 931. Dry at 80° overnight and follow by drying at 120° for 4 hours.

Potassium compounds. Mix a few drops of sodium bitartrate T.S. with 5 ml. of a clear, saturated solution of the sample. No turbidity is produced within 5 minutes.

Packaging and storage. Store in tight containers.

Functional use in foods. Buffer.

SODIUM ACID PYROPHOSPHATE

Disodium Pyrophosphate; Disodium Dihydrogen Pyrophosphate

$Na_2H_2P_2O_7$ Mol. wt. 221.94

DESCRIPTION

White, fused masses or free-flowing powder. It is freely soluble in water. The pH of a 1 in 100 solution is about 4.

IDENTIFICATION

A. A 1 in 20 solution gives positive tests for *Sodium*, page 928.

B. To 1 ml. of a 1 in 100 solution add a few drops of silver nitrate T.S. A white precipitate is formed which is soluble in diluted nitric acid T.S.

SPECIFICATIONS

Assay. Not less than 95.0 percent of $Na_2H_2P_2O_7$.

Neutralizing value. Not less than 72.

Limits of Impurities

Arsenic (as As). Not more than 3 parts per million (0.0003 percent).

Fluoride. Not more than 10 parts per million (0.001 percent).

Heavy metals (as Pb). Not more than 20 parts per million (0.002 percent).

Insoluble substances. Not more than 0.6 percent.

Lead. Not more than 5 parts per million (0.0005 percent).

Loss on drying. Not more than 0.5 percent.

TESTS

Assay. Dissolve about 500 mg., accurately weighed, in 100 ml. of water in a 400-ml. beaker. Adjust the pH of the solution to 3.8 with hydrochloric acid, using a pH meter, then add 50 ml. of a 1 in 8 solution of zinc sulfate (125 grams of $ZnSO_4 . 7H_2O$ dissolved in water, diluted to 1000 ml., filtered, and adjusted to pH 3.8), and allow to stand for 2 minutes. Titrate the liberated acid with 0.1 *N* sodium hydroxide until a pH of 3.8 is again reached. After each addition of sodium hydroxide near the end-point, time should be allowed for any precipitated zinc hydroxide to redissolve. Each ml. of 0.1 *N* sodium hydroxide is equivalent to 11.10 mg. of $Na_2H_2P_2O_7$. (*Note:* The 0.1 *N* sodium hydroxide used in this titration must be standardized against sodium pyrophosphate, $Na_4P_2O_7$, that has been recrystallized three times from water and dried at 400° to constant weight.)

Neutralizing value. Dissolve 840.1 mg., accurately weighed, in about 50 ml. of water in a 250-ml. beaker, and titrate potentiometrically with 0.1 *N* sodium hydroxide to a pH of 9.9, using a glass electrode previously standardized to a pH of 9.15. Not less than 72 ml. of 0.1 *N* sodium hydroxide is required.

Arsenic. A solution of 1 gram in 10 ml. of water meets the requirements of the *Arsenic Test*, page 865.

Fluoride. Proceed as directed in the *Fluoride Limit Test*, page 917.

Heavy metals. A solution of 1 gram in 25 ml. of water meets the requirements of the *Heavy Metals Test*, page 920, using 20 mcg. of lead ion (Pb) in the control (*Solution A*).

Insoluble substances. Dissolve 10 grams in 100 ml. of hot water, and filter through a tared filtering crucible. Wash the insoluble residue with hot water, dry at 105° for 2 hours, cool, and weigh.

Lead. A solution of 1 gram in 20 ml. of water meets the requirements of the *Lead Limit Test*, page 929, using 5 mcg. of lead ion (Pb) in the control.

Loss on drying, page 931. Dry at 120° for 3 hours.

Packaging and storage. Store in tight containers.

Functional use in foods. Buffer; leavening agent; sequestrant.

SODIUM ALGINATE

Algin

$(C_6H_7O_6Na)_n$

Equiv. wt., *Calculated*, 198.11
Equiv. wt., *Actual* (Avg.), 222.00

DESCRIPTION

The sodium salt of alginic acid (see *Alginic Acid*, page 26), occurs as a white to yellowish, fibrous or granular powder. It is nearly odorless and tasteless. It dissolves in water to form a viscous, colloidal solution. It is insoluble in alcohol and in hydroalcoholic solutions in which the alcohol content is greater than about 30 percent by weight. It is insoluble in chloroform, in ether, and acids having a pH lower than about 3.

IDENTIFICATION

A. To 5 ml. of a 1 in 100 solution add 1 ml. of calcium chloride T.S. A voluminous, gelatinous precipitate is formed.

B. To 10 ml. of a 1 in 100 solution add 1 ml. of diluted sulfuric acid T.S. A heavy gelatinous precipitate is formed.

C. Sodium alginate meets the requirements of *Identification Test C* under *Alginic Acid*, page 26.

D. Extract the *Ash* from sodium alginate with diluted hydrochloric acid T.S. and filter. The filtrate gives positive tests for *Sodium*, page 928.

SPECIFICATIONS

Assay. It yields, on the dried basis, not less than 18.0 percent and not more than 21.0 percent of carbon dioxide (CO_2) corresponding to between 90.80 percent and 106.0 percent of sodium alginate (Equiv. wt. 222.00).

Ash. Between 18.0 percent and 27.0 percent on the dried basis.

Limits of Impurities

Arsenic (as As). Not more than 3 parts per million (0.0003 percent).

Heavy metals (as Pb). Not more than 40 parts per million (0.004 percent).

Insoluble matter. Not more than 0.2 percent.

Lead. Not more than 10 parts per million (0.001 percent).

Loss on drying. Not more than 15 percent.

TESTS

Assay. Proceed as directed in the *Alginates Assay*, page 26. Each ml. of 0.25 *N* sodium hydroxide consumed in the assay is equivalent to 27.75 mg. of sodium alginate (Equiv. wt. 222.00).

Ash. Determine as directed under *Ash* in the monograph on *Alginic Acid*, page 26.

Arsenic. A *Sample Solution* prepared as directed for organic compounds meets the requirements of the *Arsenic Test*, page 865.

Heavy metals. Prepare and test a 500-mg. sample as directed in *Method II* under the *Heavy Metals Test*, page 920, but use a platinum crucible for the ignition. Any color does not exceed that produced in a control (*Solution A*) containing 20 mcg. of lead ion (Pb).

Insoluble matter. Determine as directed under *Insoluble matter* in the monograph on *Alginic Acid*, page 26.

Lead. A *Sample Solution* prepared as directed for organic compounds meets the requirements of the *Lead Limit Test*, page 929, using 10 mcg. of lead ion (Pb) in the control.

Loss on drying, page 931. Dry at 105° for 4 hours.

Packaging and storage. Store in well-closed containers.

Functional use in foods. Stabilizer; thickener; emulsifier.

SODIUM ALUMINUM PHOSPHATE, ACIDIC

$NaAl_3H_{14}(PO_4)_8 . 4H_2O$ Mol. wt. 949.88

or or

$Na_3Al_2H_{15}(PO_4)_8$ Mol. wt. 897.82

DESCRIPTION

A white, odorless powder. It is insoluble in water, but is soluble in hydrochloric acid. A 1 in 10 solution in dilute hydrochloric acid (1 in 2) gives positive tests for *Aluminum*, page 925, and for *Phosphate*, page 928, and it responds to the flame test for *Sodium*, page 928.

SPECIFICATIONS

Assay. Not less than 95.0 percent of $NaAl_3H_{14}(PO_4)_8 . 4H_2O$, or not less than 95.0 percent of $Na_3Al_2H_{15}(PO_4)_8$.

Loss on ignition. $NaAl_3H_{14}(PO_4)_8 . 4H_2O$, between 19.5 and 21 percent; $Na_3Al_2H_{15}(PO_4)_8$, between 15 and 16 percent.

Neutralizing value. Not less than 100.

Limits of Impurities

Arsenic (as As). Not more than 3 parts per million (0.0003 percent).

Fluoride. Not more than 25 parts per million (0.0025 percent).

Heavy metals (as Pb). Not more than 40 parts per million (0.004 percent).

Lead. Not more than 10 parts per million (0.001 percent)

TESTS

Assay. Transfer about 2.5 grams, accurately weighed, into a 250-ml. volumetric flask, add 15 ml. of hydrochloric acid and one glass bead, and boil gently for about 5 minutes. Cool, dilute to volume with water, and mix. Transfer 10.0 ml. of this solution to a 250-ml. beaker, add phenolphthalein T.S., and neutralize with ammonia T.S. Add dilute hydrochloric acid (1 in 2) until the precipitate just dissolves, then dilute to 100 ml. with water and heat to 70°–80°. Add 10 ml. of 8-hydroxyquinoline T.S. and sufficient ammonium acetate T.S. until a yellow precipitate forms, then add 30 ml. in excess. Digest at 70° for 30 minutes, filter through a previously dried and weighed Gooch crucible, and wash thoroughly with hot water. Dry at 105° for 2 hours, cool, and weigh. Each mg. of the precipitate so obtained corresponds to 0.689 mg. of $NaAl_3H_{14}(PO_4)_8 . 4H_2O$, or to 0.977 mg. of $Na_3Al_2H_{15}(PO_4)_8$.

Loss on ignition. Ignite at 750° to 800° for 2 hours.

Neutralizing value. Transfer 840.1 mg., accurately weighed, into a 375-ml. casserole, and add 20 grams of sodium chloride, 5 ml. of sodium citrate solution (1 in 10), and 25 ml. of water. Stir vigorously at once for about 30 seconds, then add exactly 120 ml. of 0.1 *N* sodium hydroxide, bring the suspension to a boil in exactly 2 minutes, and boil for 5 minutes. While the solution is still boiling hot, add exactly 0.05 ml. of phenolphthalein T.S., and titrate the excess alkali with 0.2 *N* hydrochloric acid until the pink color has almost disappeared. Boil the solution for 1 minute, and titrate again with 0.2 *N* hydrochloric acid until the pink color just disappears. Calculate the neutralizing value, as parts of $NaHCO_3$ equivalent to 100 parts of sodium aluminum phosphate, by the formula $120 - 2V$, in which V is the total volume of 0.2 *N* hydrochloric acid consumed.

Arsenic. A solution of 1 gram in 10 ml. of dilute hydrochloric acid (1 in 2) meets the requirements of the *Arsenic Test*, page 865.

Fluoride. Weigh accurately 2.0 grams, and proceed as directed in the *Fluoride Limit Test*, page 917.

Heavy metals. Dissolve 500 mg. in 2.5 ml. of diluted hydrochloric acid T.S., and add water to make 25 ml. This solution meets the requirements of the *Heavy Metals Test*, page 920, using 20 mcg. of lead ion (Pb) in the control (*Solution A*).

Lead. A solution of 1 gram in 5 ml. of diluted hydrochloric acid T.S. meets the requirements of the *Lead Limit Test*, page 929, using 10 mcg. of lead ion (Pb) in the control.

Packaging and storage. Store in well-closed containers.

Functional use in foods. Leavening agent.

SODIUM ALUMINUM PHOSPHATE, BASIC

Kasal

DESCRIPTION

A white, odorless powder comprised of an autogenous mixture of an alkaline sodium aluminum phosphate [approximately $Na_8Al_2(OH)_2$-$(PO_4)_4$], with about 30 percent dibasic sodium phosphate. It is soluble in hydrochloric acid; the sodium phosphate moiety is soluble in water whereas the sodium aluminum phosphate moiety is only sparingly soluble in water. A 1 in 10 solution in dilute hydrochloric acid (1 in 2) gives a positive test for *Aluminum*, page 925, and for *Phosphate*, page 928, and it responds to the flame test for *Sodium*, page 928.

SPECIFICATIONS

Assay. Not less than 9.5 percent and not more than 12.5 percent of Al_2O_3, calculated on the ignited basis.

Loss on ignition. Not more than 9 percent.

Limits of Impurities

Arsenic (as As). Not more than 3 parts per million (0.0003 percent).

Fluoride. Not more than 25 parts per million (0.0025 percent).

Heavy metals (as Pb). Not more than 40 parts per million (0.004 percent).

Lead. Not more than 10 parts per million (0.001 percent).

TESTS

Assay. Transfer about 2.5 grams, accurately weighed, into a 250-ml. volumetric flask, add 15 ml. of hydrochloric acid and one glass bead, and boil gently for about 5 minutes. Cool, dilute to volume with water, and mix. Transfer 10.0 ml. of this solution to a 250-ml. beaker, add phenolphthalein T.S., and neutralize with ammonia T.S. Add dilute hydrochloric acid (1 in 2) until the precipitate just dissolves, then dilute to 100 ml. with water and heat to 70°–80°. Add 10 ml. of 8-hydroxyquinoline T.S. and sufficient ammonium acetate T.S. until a yellow precipitate forms, then add 30 ml. in excess. Digest at 70° for 30 minutes, filter through a previously dried and weighed Gooch crucible, and wash thoroughly with hot water. Dry at 105° for 2 hours, cool, and weigh. Each mg. of the precipitate so obtained corresponds to 0.111 mg. of Al_2O_3.

Loss on ignition. Ignite at 750° to 800° for 2 hours.

Arsenic. A solution of 1 gram in 10 ml. of dilute hydrochloric acid (1 in 2) meets the requirements of the *Arsenic Test*, page 865.

Fluoride. Weigh accurately 2.0 grams, and proceed as directed in the *Fluoride Limit Test*, page 917.

Heavy metals. Dissolve 500 mg. in 2.5 ml. of diluted hydrochloric acid T.S., and add water to make 25 ml. This solution meets the requirements of the *Heavy Metals Test,* page 920, using 20 mcg. of lead ion (Pb) in the control (*Solution A*).

Lead. A solution of 1 gram in 5 ml. of diluted hydrochloric acid T.S. meets the requirements of the *Lead Limit Test,* page 929, using 10 mcg. of lead ion (Pb) in the control.

Packaging and storage. Store in well-closed containers.

Functional use in foods. Emulsifier.

SODIUM ASCORBATE

Vitamin C Sodium; Sodium L-Ascorbate

$C_6H_7NaO_6$ Mol. wt. 198.11

DESCRIPTION

A white or almost white crystalline solid. One gram is soluble in 2 ml. of water. The pH of a 1 in 10 solution is about 7.5.

IDENTIFICATION

A. A 1 in 50 solution slowly reduces alkaline cupric tartrate T.S. at 25°, but more readily upon heating.

B. To 2 ml. of a 1 in 50 solution of the sample acidified with 0.5 ml. of 0.1 *N* hydrochloric acid add 4 drops of methylene blue T.S., and warm to 40°. The deep blue color is practically discharged within 3 minutes.

C. Dissolve 15 mg. of the sample in 15 ml. of a 1 in 20 solution of trichloroacetic acid, add about 200 mg. of activated charcoal, shake vigorously for 1 minute, and filter through a small fluted filter, returning the filtrate, if necessary, until clear. To 5 ml. of the filtrate add 1 drop of pyrrole, and agitate gently until dissolved, and then heat in a water bath at 50°. A blue color develops.

D. It gives positive tests for *Sodium,* page 928.

SPECIFICATIONS

Assay. Not less than 99.0 percent and not more than the equivalent of 101.0 percent of $C_6H_7NaO_6$ after drying.

Specific rotation, $[\alpha]_D^{25°}$. Between +103° and +108°.

Limits of Impurities

Arsenic (as As). Not more than 3 parts per million (0.0003 percent).

Heavy metals (as Pb). Not more than 20 parts per million (0.002 percent).

Lead. Not more than 10 parts per million (0.001 percent).

Loss on drying. Not more than 0.25 percent.

TESTS

Assay. Dissolve about 400 mg., previously dried over phosphorus pentoxide for 24 hours and accurately weighed, in a mixture of 100 ml. of water, recently boiled and cooled, and 25 ml. of diluted sulfuric acid T.S., and titrate with 0.1 *N* iodine, adding starch T.S. near the end-point. Each ml. of 0.1 *N* iodine is equivalent to 9.905 mg. of $C_6H_7NaO_6$.

Specific rotation, page 939. Determine in a solution containing 1 gram in each 10 ml.

Arsenic. A *Sample Solution* prepared as directed for organic compounds meets the requirements of the *Arsenic Test,* page 865.

Heavy metals. Prepare and test a 1-gram sample as directed in *Method II* under the *Heavy Metals Test,* page 920, using 20 mcg. of lead ion (Pb) in the control (*Solution A*).

Lead. A *Sample Solution* prepared as directed for organic compounds meets the requirements of the *Lead Limit Test,* page 929, using 10 mcg. of lead ion (Pb) in the control.

Loss on drying, page 931. Dry over phosphorus pentoxide for 24 hours.

Packaging and storage. Store in tight, light-resistant containers.

Functional use in foods. Antioxidant; nutrient; dietary supplement.

SODIUM BENZOATE

$C_7H_5NaO_2$ Mol. wt. 144.11

DECRIPTION

White, odorless or nearly odorless, granules, crystalline powder or flakes. One gram dissolves in 2 ml. of water, in 75 ml. of alcohol, and in 50 ml. of 90 percent alcohol. It gives positive tests for *Sodium,* page 928, and for *Benzoate,* page 926.

SPECIFICATIONS

Assay. Not less than 99.0 percent of $C_7H_5NaO_2$, calculated on the dried basis.

Limits of Impurities

Alkalinity (as NaOH). Not more than 0.04 percent.

Arsenic (as As). Not more than 3 parts per million (0.0003 percent).

Heavy metals (as Pb). Not more than 10 parts per million (0.001 percent).

Water. Not more than 1.5 percent.

TESTS

Assay. Transfer about 600 mg., accurately weighed, to a 250-ml. beaker, add 100 ml. of glacial acetic acid, and stir until the sample is completely dissolved. Add crystal violet T.S., and titrate with 0.1 *N* perchloric acid in glacial acetic acid. Each ml. of 0.1 *N* perchloric acid is equivalent to 14.41 mg. of $C_7H_5NaO_2$.

Alkalinity. Dissolve 2 grams in 20 ml. of hot water, and add 2 drops of phenolphthalein T.S. If a pink color is produced, not more than 0.2 ml. of 0.1 *N* sulfuric acid is required to discharge it.

Arsenic. A *Sample Solution* prepared as directed for organic compounds meets the requirements of the *Arsenic Test*, page 865.

Heavy metals. Dissolve 4 grams in 40 ml. of water, add dropwise, with vigorous stirring, 10 ml. of diluted hydrochloric acid T.S., and filter. A 25-ml. portion of the filtrate meets the requirements of the *Heavy Metals Test*, page 920, using 20 mcg. of lead ion (Pb) in the control (*Solution A*).

Water. Determine by the *Karl Fischer Titrimetric Method*, page 977.

Packaging and storage. Store in well-closed containers.

Functional use in foods. Preservative; antimicrobial agent.

SODIUM BICARBONATE

Baking Soda

$NaHCO_3$ Mol. wt. 84.01

DESCRIPTION

A white crystalline powder. It is stable in dry air, but slowly decomposes in moist air. Its solutions, when freshly prepared with cold water, without shaking, are alkaline to litmus. The alkalinity increases as the solutions stand, are agitated, or are heated. One gram dissolves in 10 ml. of water. It is insoluble in alcohol. A 1 in 10 solution gives positive tests for *Sodium*, page 928, and for *Bicarbonate*, page 926.

SPECIFICATIONS

Assay. Not less than 99.0 percent of $NaHCO_3$ after drying.

Limits of Impurities

Ammonia. Passes test.
Arsenic (as As). Not more than 3 parts per million (0.0003 percent).
Heavy metals (as Pb). Not more than 5 parts per million (0.0005 percent).
Insoluble substances. Passes test.
Loss on drying. Not more than 0.25 percent.

TESTS

Assay. Weigh accurately about 3 grams, previously dried over silica gel for 4 hours, dissolve it in 25 ml. of water, add methyl orange T.S., and titrate with 1 *N* sulfuric acid. Each ml. of 1 *N* sulfuric acid is equivalent to 84.01 mg. of $NaHCO_3$.

Ammonia. Heat 1 gram in a test tube. No odor of ammonia is detected.

Arsenic. A solution of 1 gram in 5 ml. of diluted hydrochloric acid T.S. meets the requirements of the *Arsenic Test,* page 865.

Heavy metals. Dissolve 4 grams in 10 ml. of diluted hydrochloric acid T.S., boil gently for 1 minute, and dilute to 25 ml. with water. This solution meets the requirements of the *Heavy Metals Test,* page 920, using 20 mcg. of lead ion (Pb) in the control (*Solution A*).

Insoluble substances. One gram dissolves completely in 20 ml. of water to give a clear solution.

Loss on drying, page 931. Dry over silica gel for 4 hours.

Packaging and storage. Store in well-closed containers.

Functional use in foods. Alkali; leavening agent.

SODIUM BISULFATE

Sodium Acid Sulfate; Nitre Cake

$NaHSO_4$ Mol. wt. 120.06

DESCRIPTION

White, odorless crystals or granules. It is soluble in water, and its solutions are strongly acid. It is decomposed by alcohol into sodium sulfate and free sulfuric acid. It gives positive tests for *Sodium,* page 928, and for *Sulfate,* page 928.

SPECIFICATIONS

Assay. Not less than 35.0 percent and not more than 39.0 percent of available H_2SO_4, equivalent to not less than 85.4 percent and not more than 95.2 percent of $NaHSO_4$.

Limits of Impurities

Arsenic (as As). Not more than 3 parts per million (0.0003 percent).

Heavy metals (as Pb). Not more than 30 parts per million (0.003 percent).

Lead. Not more than 10 parts per million (0.001 percent).

Loss on drying. Not more than 0.8 percent.

Selenium. Not more than 30 parts per million (0.003 percent).

Water-insoluble substances. Not more than 0.05 percent.

TESTS

Assay. Dissolve about 5 grams, accurately weighed, in about 125 ml. of water, add phenolphthalein T.S., and titrate with 1 *N* sodium hydroxide. Each ml. of 1 *N* sodium hydroxide is equivalent to 49.04 mg. of H_2SO_4, or to 120.06 mg. of $NaHSO_4$.

Arsenic. A solution of 1 gram in 35 ml. of water meets the requirements of the *Arsenic Test*, page 865.

Heavy metals. A solution of 670 mg. in 25 ml. of water meets the requirements of the *Heavy Metals Test*, page 920, using 20 mcg. of lead ion (Pb) in the control (*Solution A*).

Lead. A solution of 1 gram in 25 ml. of water meets the requirements of the *Lead Limit Test*, page 929, using 10 mcg. of lead ion (Pb) in the control.

Loss on drying, page 931. Dry in a desiccator over phosphorus pentoxide for 24 hours.

Selenium. A solution of 2 grams in 40 ml. of dilute hydrochloric acid (1 in 2) meets the requirements of the *Selenium Limit Test*, page 953.

Water-insoluble substances. Dissolve 50 grams in 300 ml. of hot water in a 600-ml. beaker, allow the insoluble matter to settle, and filter by decanting through a tared Gooch crucible, washing the insoluble matter into the crucible with additional hot water. Dry at 100° to 110° for 1 hour, cool in a desiccator, and weigh.

Packaging and storage. Store in tight containers.

Functional use in foods. Acid.

SODIUM BISULFITE

DESCRIPTION

Sodium bisulfite consists of sodium bisulfite ($NaHSO_3$) and sodium metabisulfite ($Na_2S_2O_5$) in varying proportions. It occurs as white or yellowish white crystals or granular powder having an odor of sulfur dioxide. It is unstable in air. One gram dissolves in 4 ml. of water. It is slightly soluble in alcohol. A 1 in 10 solution gives positive tests for *Sodium*, page 928, and for *Sulfite*, page 928.

SPECIFICATIONS

Assay. Not less than 58.5 percent and not more than 67.4 percent of SO_2.

Limits of Impurities

Arsenic (as As). Not more than 3 parts per million (0.0003 percent).

Heavy metals (as Pb). Not more than 10 parts per million (0.001 percent).

Iron. Not more than 50 parts per million (0.005 percent).
Selenium. Not more than 30 parts per million (0.003 percent).

TESTS

Assay. Weigh accurately about 200 mg., add it to exactly 50 ml. of 0.1 *N* iodine contained in a glass-stoppered flask, and stopper the flask. Allow to stand for 5 minutes, add 1 ml. of hydrochloric acid, and titrate the excess iodine with 0.1 *N* sodium thiosulfate, adding starch T.S. as the indicator. Each ml. of 0.1 *N* iodine is equivalent to 3.203 mg. of SO_2.

Arsenic. Dissolve 1 gram in 5 ml. of water, add 0.5 ml. of sulfuric acid, evaporate to about 1 ml. on a steam bath, and dilute to 35 ml. with water. This solution meets the requirements of the *Arsenic Test,* page 865.

Heavy metals. Dissolve 2 grams in 10 ml. of water, add 5 ml. of hydrochloric acid, evaporate to dryness on a steam bath, and dissolve the residue in 25 ml. of water. This solution meets the requirements of the *Heavy Metals Test,* page 920, using 20 mcg. of lead ion (Pb) in the control (*Solution A*).

Iron. To 500 mg. of the sample add 2 ml. of hydrochloric acid, and evaporate to dryness on a steam bath. Dissolve the residue in 2 ml. of hydrochloric acid and 20 ml. of water, add a few drops of bromine T.S., and boil the solution to remove the bromine. Cool, dilute with water to 25 ml., then add 50 mg. of ammonium persulfate and 5 ml. of ammonium thiocyanate T.S. Any red or pink color does not exceed that produced in a control containing 2.5 ml. of *Iron Standard Solution* (25 mcg. Fe).

Selenium. Prepare and test a 2-gram sample as directed in the *Selenium Limit Test,* page 953.

Packaging and storage. Store in well-filled, tight containers, and avoid exposure to excessive heat.
Functional use in foods. Preservative.

SODIUM CARBONATE

$Na_2CO_3 . xH_2O$ Mol. wt. (anhydrous) 105.99

DESCRIPTION

Sodium carbonate is anhydrous or may contain 1 or 10 molecules of water of hydration. It occurs as colorless crystals or as a white, granular or crystalline powder. It is freely soluble in water, and its solutions are alkaline to litmus. The anhydrous salt is hygroscopic, and the two hydrates are efflorescent. The decahydrate melts at about 32°. All forms give positive tests for *Sodium*, page 928, and for *Carbonate*, page 926.

SPECIFICATIONS

Assay. Not less than 99.5 percent of Na_2CO_3 after drying.

Loss on drying. Na_2CO_3 (anhydrous), not more than 1 percent; $Na_2CO_3 . H_2O$, between 12 and 15 percent; $Na_2CO_3 . 10H_2O$, between 55 and 65 percent.

Limits of Impurities

Arsenic (as As). Not more than 3 parts per million (0.0003 percent).

Heavy metals (as Pb). Not more than 10 parts per million (0.001 percent).

TESTS

Assay. Weigh accurately about 2 grams of the dried salt, obtained as directed under *Loss on drying*, dissolve in 50 ml. of water, add methyl orange T.S., and titrate with 1 *N* sulfuric acid. Each ml. of 1 *N* sulfuric acid is equivalent to 53.00 mg. of Na_2CO_3.

Loss on drying, page 931. Dry about 3 grams of the anhydrous salt or the monohydrate, accurately weighed, at 250° to 300° to constant weight. For the decahydrate, weigh accurately about 8 grams, heat it first at about 70°, then gradually raise the temperature and finally dry at 250° to 300° to constant weight.

Arsenic. A solution of 1 gram in 5 ml. of diluted hydrochloric acid T.S. meets the requirements of the *Arsenic Test*, page 865.

Heavy metals. Mix 2 grams with 5 ml. of water and 10 ml. of diluted hydrochloric acid T.S., boil for 1 minute, cool, and dilute to 25 ml. with water. This solution meets the requirements of the *Heavy Metals Test*, page 920, using 20 mcg. of lead ion (Pb) in the control (*Solution A*).

Packaging and storage. Store the anhydrous salt and the decahydrate in tight containers; the monohydrate may be stored in well-closed containers.

Functional use in foods. Alkali.

SODIUM CARBOXYMETHYLCELLULOSE

CMC; Cellulose Gum

DESCRIPTION

It occurs as a white to cream colored powder or as granules. The powder is hygroscopic. A 1 in 100 aqueous suspension has a pH between 6.5 and 8.5. It is readily dispersed in water to form colloidal solutions. It is insoluble in most solvents.

IDENTIFICATION

Add about 1 gram of powdered sample to 50 ml. of warm water, while stirring, to produce a uniform dispersion. Continue the stirring until a colloidal solution is produced, and then cool to room temperature.

A. To about 10 ml. of the solution add 10 ml. of cupric sulfate T.S. A fluffy, bluish white precipitate is formed.

B. The filtrate from *Identification test A* gives positive tests for *Sodium*, page 928.

SPECIFICATIONS

Assay. Not less than 99.5 percent of sodium carboxymethylcellulose, calculated on the dried basis.

Degree of substitution. Not more than 0.95 carboxymethyl groups ($—CH_2COOH$) per anhydroglucose unit after drying.

Sodium. Not more than 9.5 percent after drying.

Viscosity of a 2 percent, weight in weight, solution. Not less than 25 centipoises.

Limits of Impurities

Arsenic (as As). Not more than 3 parts per million (0.0003 percent).

Heavy metals (as Pb). Not more than 40 parts per million (0.004 percent).

Lead. Not more than 10 parts per million (0.001 percent).

Loss on drying. Not more than 10 percent.

TESTS

Assay. Calculate the percent of sodium carboxymethylcellulose by subtracting from 100 the percents of *Sodium Chloride* and *Sodium Glycolate* determined as follows:

Sodium Chloride. Weigh accurately about 5 grams of the sample into a 250-ml. beaker, add 50 ml. of water and 5 ml. of 30 percent hydrogen peroxide, and heat on a steam bath for 20 minutes, stirring occasionally to ensure complete dissolution. Cool, add 100 ml. of water and 10 ml. of nitric acid, and titrate with 0.05 N silver nitrate to a potentiometric end-point, using silver and mercurous sulfate-potassium sulfate electrodes and stirring constantly. Calculate the percent of sodium chloride in the sample by the formula $584.4VN/(100 - b)W$, in which V and N represent the volume, in ml., and the normality, respectively, of the silver nitrate, b is the percent of *Loss on drying*, determined separately, W is the weight of the sample, in grams, and 584.4 is an equivalence factor for sodium chloride.

Sodium Glycolate. Transfer about 500 mg., accurately weighed, of the sample into a 100-ml. beaker, moisten thoroughly with 5 ml. of acetic acid, followed by 5 ml. of water, and stir with a glass rod until solution is complete (usually about 15 minutes). Slowly add 50 ml. of acetone, with stirring, then add 1 gram of sodium chloride, and stir for several minutes to ensure complete precipitation of the carboxymethylcellulose. Filter through a soft, open textured paper, previously wetted with a small amount of acetone, and collect the filtrate in a 100-ml. volumetric flask. Use an additional 30 ml. of acetone to facilitate transfer of the solids and to wash the filter cake, then dilute to volume with acetone, and mix.

Prepare a series of standard solutions as follows: Transfer 100 mg. of glycolic acid, previously dried in a desiccator at room temperature overnight and accurately weighed, into a 100-ml. volumetric flask, dissolve in water, dilute to volume with water, and mix. Use this solution within 30 days. Transfer 1.0 ml., 2.0 ml., 3.0 ml., and 4.0 ml. of the solution into separate 100-ml. volumetric flasks, add sufficient water to each flask to make 5 ml., then add 5 ml. of glacial acetic acid, and dilute to volume with acetone.

Transfer 2.0 ml. of the sample solution and 2.0 ml. of each standard solution into separate 25-ml. volumetric flasks, and prepare a blank flask containing 2.0 ml. of a solution containing 5 percent each of glacial acetic acid and water in acetone. Place the uncovered flasks in a boiling water bath for exactly 20 minutes to remove the acetone, remove from the bath, and cool. Add to each flask 5.0 ml. of 2,7-dihydroxynaphthalene T.S., mix thoroughly, add an additional 15 ml., and again mix thoroughly. Cover the mouth of each flask with a small piece of aluminum foil. Place the flasks upright in a boiling water bath for 20 minutes, then remove from the bath, cool, dilute to volume with sulfuric acid, and mix.

Determine the absorbance of each solution at 540 mμ, with a suitable spectrophotometer, against the blank, and prepare a standard curve using the absorbances obtained from the standard solutions. From the standard curve and the absorbance of the sample, determine the weight (w), in mg., of glycolic acid in the sample, and calculate the percent of sodium glycolate in the sample by the formula $12.9w/(100 - b)W$, in which 12.9 is a factor converting glycolic acid to sodium glycolate, b is the percent of *Loss on drying*, determined separately, and W is the weight of the sample, in grams.

Degree of Substitution. Weigh accurately about 200 mg. of the sample, previously dried at 105° for 3 hours, and transfer it into a 250-ml. glass-stoppered Erlenmeyer flask. Add 75 ml. of glacial acetic acid, connect the flask with a water-cooled condenser, and reflux gently on a hot plate for 2 hours. Cool, transfer the solution to a 250-ml. beaker with the aid of 50 ml. of glacial acetic acid, and titrate with 0.1 N perchloric acid in dioxane while stirring with a magnetic stirrer. Determine the end-point potentiometrically with a pH meter equipped with a standard glass electrode and a calomel electrode modified as follows: Discard the aqueous potassium chloride solution contained in the electrode, rinse and fill with the supernatant liquid obtained by shaking thoroughly 2 grams each of potassium chloride and silver chloride (or silver oxide) with 100 ml. of methanol, then add a few crystals of potassium chloride and silver chloride (or silver oxide) to the electrode.

Record the ml. of 0.1 *N* perchloric acid versus millivolts (0–700 mv range), and continue the titration to a few ml. beyond the end-point. Plot the titration curve and read the volume (A), in ml., of 0.1 *N* perchloric acid at the inflection point.

Calculate the degree of substitution (*DS*) by the formula $[16.2A/G]/[1.000 - (8.0A/G)]$, in which A is the volume, in ml., of 0.1 *N* perchloric acid required, G is the weight, in mg., of the sample taken, 16.2 is one-tenth the molecular weight of one anhydroglucose unit, and 8.0 is one-tenth the molecular weight of one sodium carboxymethyl group.

Sodium. From the weight of the sample and the number of ml. of 0.1 *N* perchloric acid consumed in the determination of *Degree of Substitution,* calculate the percent of sodium. Each ml. of 0.1 *N* perchloric acid is equivalent to 2.299 mg. of Na.

Viscosity. Determine as directed under *Viscosity of Sodium Carboxymethylcellulose,* page 973.

Arsenic. A *Sample Solution* prepared as directed for organic compounds meets the requirements of the *Arsenic Test,* page 865.

Heavy metals. Prepare and test a 500-mg. sample as directed in *Method II* under the *Heavy Metals Test,* page 920, adding 1 ml. of hydroxylamine hydrochloride solution (1 in 5) to the solution of the residue. Any color does not exceed that produced in a control (*Solution A*) containing 20 mcg. of lead ion (Pb) and 1 ml. of the hydroxylamine hydrochloride solution.

Lead. A *Sample Solution* prepared as directed for organic compounds meets the requirements of the *Lead Limit Test,* page 929, using 10 mcg. of lead ion (Pb) in the control.

Loss on drying, page 931. Dry to constant weight at 105°.

Packaging and storage. Store in well-closed containers.

Functional use in foods. Thickening agent; stabilizer.

SODIUM CHLORIDE

NaCl Mol. wt. 58.44

DESCRIPTION

Sodium chloride is a transparent to opaque white crystalline solid of variable particle size. It is odorless and has a characteristic saline taste. It remains dry in air at a relative humidity below 75 percent, but becomes deliquescent at relative humidities above this value. It may contain not more than 2 percent of suitable anticaking or crystal-modifying agents approved for these purposes by the federal Food and Drug Administration. One gram is soluble in 2.8 ml. of water at 25°, in 2.7 ml. of boiling water, and in about 10 ml. of glycerin. A 1 in 20 solution of sodium chloride in water gives positive tests for *Sodium*, page 928, and for *Chloride*, page 926.

SPECIFICATIONS

Assay. Not less than 97.5 percent of NaCl after drying.

Sodium ferrocyanide. Not more than 13 parts per million (0.0013 percent) of $Na_4Fe(CN)_6$.

Limits of Impurities

Arsenic (as As). Not more than 3 parts per million (0.0003 percent).

Heavy metals (as Pb). Not more than 5 parts per million (0.0005 percent).

Loss on drying. Not more than 1 percent.

TESTS

Assay. Weigh accurately about 250 mg., previously dried at 105° for 2 hours, and dissolve it in 50 ml. of water in a glass-stoppered flask. Add, while agitating, 3 ml. of nitric acid, 5 ml. of nitrobenzene, 50.0 ml. of 0.1 *N* silver nitrate, and 2 ml. ferric ammonium sulfate T.S. Shake well, and titrate the excess silver nitrate with 0.1 *N* ammonium thiocyanate. Each ml. of 0.1 *N* silver nitrate is equivalent to 5.844 mg. of NaCl.

Sodium ferrocyanide. Dissolve 9.62 grams of the sample in 80 ml. of water in a 150-ml. glass-stoppered cylinder or flask. Prepare a standard solution containing 125 mcg. of $Na_4Fe(CN)_6$ in each ml. by dissolving 99.5 mg. of $Na_4Fe(CN)_6 . 10H_2O$ in 500.0 ml. of water, then transfer 1.0 ml. of this solution into a similar 150-ml. container for the control. To each container add 2 ml. of ferrous sulfate T.S. and 1 ml. of diluted sulfuric acid T.S., dilute to 100 ml. with water, and mix. Transfer 50-ml. portions of the respective solutions into matched color-comparison tubes. The sample solution shows no more blue color than the control.

Arsenic. A solution of 1 gram in 10 ml. of water meets the requirements of the *Arsenic Test*, page 865.

Heavy metals. A solution of 4 grams in 25 ml. of water meets the requirements of the *Heavy Metals Test*, page 920, using 20 mcg. of lead ion (Pb) in the control (*Solution A*).

Loss on drying, page 931. Dry at 105° for 2 hours.

Packaging and storage. Store in well-closed containers.

Functional use in foods. Nutrient; preservative; flavoring agent and intensifier.

SODIUM CITRATE

$C_6H_5Na_3O_7 . 2H_2O$ Mol. wt. 294.10

DESCRIPTION

Sodium citrate is anhydrous or contains two molecules of water of crystallization. It occurs as colorless crystals or as a white, crystalline powder. One gram of the dihydrate dissolves in 1.5 ml. of water at 25° and in 0.6 ml. of boiling water. It is insoluble in alcohol. A 1 in 20 solution gives positive tests for *Sodium*, page 928, and for *Citrate*, page 926.

SPECIFICATIONS

Assay. Not less than 99.0 percent of $C_6H_5Na_3O_7$, calculated on the anhydrous basis.

Water. *Anhydrous sodium citrate*, not more than 1 percent; *sodium citrate dihydrate*, between 10 and 13 percent.

Limits of Impurities

Alkalinity. Passes test.

Arsenic (as As). Not more than 3 parts per million (0.0003 percent).

Heavy metals (as Pb). Not more than 10 parts per million (0.001 percent).

TESTS

Assay. Transfer about 350 mg., accurately weighed, to a 250-ml. beaker. Add 100 ml. of glacial acetic acid, stir until completely dissolved, and titrate with 0.1 *N* perchloric acid in glacial acetic acid, determining the end-point potentiometrically. Each ml. of 0.1 *N* perchloric acid is equivalent to 8.602 mg. of $C_6H_5Na_3O_7$.

Water. Determine by drying at 180° for 18 hours (see page 931), or by the *Karl Fischer Titrimetric Method*, page 977.

Alkalinity. A solution of 1 gram in 20 ml. of water is alkaline to litmus paper, but after the addition of 0.2 ml. of 0.1 *N* sulfuric acid no pink color is produced by 1 drop of phenolphthalein T.S.

Arsenic. A *Sample Solution* prepared as directed for organic compounds meets the requirements of the *Arsenic Test*, page 865.

Heavy metals. A solution of 2 grams in 25 ml. of water meets the requirements of the *Heavy Metals Test*, page 920, using 20 mcg. of lead ion (Pb) in the control (*Solution A*).

Packaging and storage. Store in tight containers.

Functional use in foods. Buffer; sequestrant; nutrient for cultured buttermilk.

SODIUM DEHYDROACETATE

Sodium 3-(1-Hydroxyethylidene)-6-methyl-1,2-pyran-2,4(3*H*)-dione

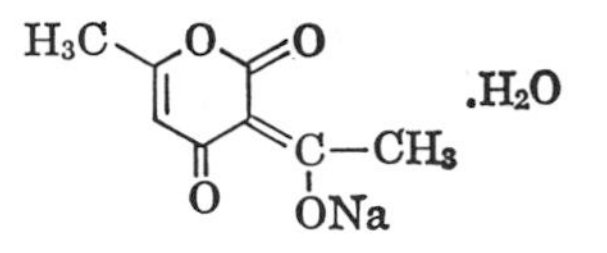

$C_8H_7NaO_4.H_2O$ Mol. wt. 208.15

DESCRIPTION

A white or nearly white, odorless powder having a slight characteristic taste. One gram dissolves in about 3 ml. of water, 2 ml. of propylene glycol, and 7 ml. of glycerin.

IDENTIFICATION

Dissolve about 1.5 grams in 10 ml. of water, add 5 ml. of diluted hydrochloric acid T.S., collect the crystals with suction, wash with 10 ml. of water, and dry between 75° and 80° for 4 hours. The crystals melt between 109° and 111° (see page 931).

SPECIFICATIONS

Assay. Not less than 98.0 percent of $C_8H_7NaO_4$, calculated on the anhydrous basis.

Water. Between 8.5 and 10 percent.

Limits of Impurities

Arsenic (as As). Not more than 3 parts per million (0.0003 percent).

Heavy metals (as Pb). Not more than 10 parts per million (0.001 percent).

TESTS

Assay. Transfer about 500 mg., accurately weighed, to a 125-ml. Erlenmeyer flask, dissolve it in 25 ml. of glacial acetic acid containing 1 drop of a 1 in 100 solution of *p*-naphtholbenzein in glacial acetic acid which has been previously neutralized to a blue color, and titrate with 0.1 *N* perchloric acid to the original blue color. Each ml. of 0.1 *N* perchloric acid is equivalent to 19.01 mg. of $C_8H_7NaO_4$.

Water. Determine by the *Karl Fischer Titrimetric Method*, page 977.

Arsenic. A *Sample Solution* prepared as directed for organic compounds meets the requirements of the *Arsenic Test*, page 865.

Heavy metals. Prepare and test a 2-gram sample as directed in *Method II* under the *Heavy Metals Test*, page 920, using 20 mcg. of lead ion (Pb) in the control (*Solution A*).

Packaging and storage. Store in well-closed containers.

Functional use in foods. Preservative.

SODIUM DIACETATE

Sodium Hydrogen Diacetate

$CH_3COONa.CH_3COOH.xH_2O$

$C_4H_7NaO_4.xH_2O$ Mol. wt. (anhydrous) 142.09

DESCRIPTION

Sodium diacetate is a molecular compound of sodium acetate and acetic acid. It is a white, hygroscopic, crystalline solid having an odor of acetic acid. One gram is soluble in about 1 ml. of water. The pH of a 1 in 10 solution is between 4.5 and 5.0. A 1 in 10 solution gives positive tests for *Acetate*, page 925, and for *Sodium*, page 928.

SPECIFICATIONS

Assay. Not less than 39.0 percent and not more than 41.0 percent of free acetic acid (CH_3COOH), and not less than 58.0 percent and not more than 60.0 percent of sodium acetate (CH_3COONa).

Water. Not more than 2 percent.

Limits of Impurities

Arsenic (as As). Not more than 3 parts per million (0.0003 percent).

Heavy metals (as Pb). Not more than 10 parts per million (0.001 percent).

Readily oxidizable substances (as formic acid). Not more than 0.2 percent.

TESTS

Assay

Free Acetic Acid. Weigh accurately about 4 grams, dissolve it in 50 ml. of water, add phenolphthalein T.S., and titrate with 1*N* sodium hydroxide. Each ml. of 1 *N* sodium hydroxide is equivalent to 60.05 mg. of CH_3COOH.

Sodium Acetate Content. Weigh accurately about 500 mg., dissolve it in 50 ml. of glacial acetic acid, and titrate with 0.1 *N* perchloric acid, determining the end-point potentiometrically. Each ml. of 0.1 *N* perchloric acid is equivalent to 8.203 mg. of CH_3COONa.

Water. Determine by the *Karl Fischer Titrimetric Method*, page 977.

Arsenic. A solution of 1 gram in 35 ml. of water meets the requirements of the *Arsenic Test*, page 865.

Heavy metals. A solution of 2 grams in 25 ml. of water meets the requirements of the *Heavy Metals Test*, page 920, using 20 mcg. of lead ion (Pb) in the control (*Solution A*).

Readily oxidizable substances. Dissolve 1.0 gram in about 50 ml. of water, add 10 ml. of diluted sulfuric acid T.S., and heat the solution to between 80° and 90°. Titrate the hot solution with 0.1 *N* potassium permanganate to a faint pink color that persists for at least 15 seconds. Each ml. of 0.1 *N* potassium permanganate is equivalent to 2.301 mg. of CH_2O_2.

Packaging and storage. Store in tight containers.

Functional use in foods. Sequestrant; preservative; mold and rope inhibitor.

SODIUM ERYTHORBATE

$C_6H_7NaO_6 . H_2O$ Mol. wt. 216.12

DESCRIPTION

White, odorless, crystalline powder or granules. In the dry state it is reasonably stable in air, but in solution it deteriorates in the presence of air, trace metals, heat, and light. One gram dissolves in about 7 ml. of water. The pH of a 1 in 20 solution is between 5.5 and 8.0.

IDENTIFICATION

A. A 1 in 50 solution slowly reduces alkaline cupric tartrate T.S. at 25°, but more readily upon heating.

B. To 2 ml. of a 1 in 50 solution acidified with 0.5 ml. of 0.1 *N* hydrochloric acid add a few drops of sodium nitroferricyanide T.S., followed by 1 ml. of 0.1 *N* sodium hydroxide. A transient blue color is produced immediately.

C. It gives positive tests for *Sodium*, page 928.

SPECIFICATIONS

Assay. Not less than 98.0 percent of $C_6H_7NaO_6 . H_2O$.

Specific rotation, $[\alpha]_D^{25°}$. Between +95.5° and +98.0°.

Limits of Impurities

Arsenic (as As). Not more than 3 parts per million (0.0003 percent).

Heavy metals (as Pb). Not more than 30 parts per million (0.003 percent).

Lead. Not more than 10 parts per million (0.001 percent).

Oxalate. Passes test.

TESTS

Assay. Dissolve about 400 mg., accurately weighed, in a mixture of 100 ml. of water, recently boiled and cooled, and 25 ml. of diluted sulfuric acid T.S., and immediately titrate with 0.1 *N* iodine, adding starch T.S. near the end-point. Each ml. of 0.1 *N* iodine is equivalent to 10.81 mg. of $C_6H_7NaO_6 . H_2O$.

Specific rotation, page 939. Determine in a solution containing 1 gram in each 10 ml.

Arsenic. A *Sample Solution* prepared as directed for organic compounds meets the requirements of the *Arsenic Test,* page 865.

Heavy metals. Prepare and test a 670-mg. sample as directed in *Method II* under the *Heavy Metals Test,* page 920, using 20 mcg. of lead ion (Pb) in the control (*Solution A*).

Lead. A *Sample Solution* prepared as directed for organic compounds meets the requirements of the *Lead Limit Test,* page 929, using 10 mcg. of lead ion (Pb) in the control.

Oxalate. To a solution of 1 gram in 10 ml. of water add 2 drops of glacial acetic acid and 5 ml. of a 1 in 10 solution of calcium acetate. The solution remains clear.

Packaging and storage. Store in tight, light-resistant containers.

Functional use in foods. Preservative; antioxidant.

SODIUM FERRIC PYROPHOSPHATE

Sodium Iron Pyrophosphate

$Na_8Fe_4(P_2O_7)_5 . xH_2O$ Mol. wt. (anhydrous) 1277.00

DESCRIPTION

A white to tan odorless powder. It is insoluble in water but is soluble in hydrochloric acid.

IDENTIFICATION

Dissolve 500 mg. in 5 ml. of dilute hydrochloric acid (1 in 2), and add an excess of sodium hydroxide T.S. A reddish brown precipitate forms. Age the solution for several minutes, and then filter, discarding the first few ml. To 5 ml. of the clear filtrate add 1 drop of bromophenol blue T.S., and titrate with 1 *N* hydrochloric acid to a green color. Add 10 ml. of a 1 in 8 solution of zinc sulfate, and readjust the pH to 3.8 (green color). A white precipitate forms (distinction from *orthophosphates*).

SPECIFICATIONS

Assay. Not less than 14.5 percent and not more than 16.0 percent of Fe.

Loss on ignition. Not more than 8 percent.

Limits of Impurities

Fluoride. Not more than 50 parts per million (0.005 percent).

Lead. Not more than 10 parts per million (0.001 percent).

Mercury. Not more than 3 parts per million (0.0003 percent).

TESTS

Assay. Proceed as directed in the *Assay* under *Ferric Phosphate*, page 309.

Loss on ignition. Ignite at 800° for 1 hour.

Fluoride. Weigh accurately 1.0 gram, and proceed as directed in the *Fluoride Limit Test*, page 917.

Lead. Proceed as directed in the test for *Lead* under *Ferric Phosphate*, page 309.

Mercury. Proceed as directed in the test for *Mercury* under *Ferric Phosphate*, page 309.

Packaging and storage. Store in well-closed containers.

Functional use in foods. Nutrient; dietary supplement.

SODIUM FERROCYANIDE

Yellow Prussiate of Soda

$Na_4Fe(CN)_6.10H_2O$ Mol. wt. 484.06

DESCRIPTION

Yellow crystals or crystalline powder. It is soluble in water, but is practically insoluble in most organic solvents.

SPECIFICATIONS

Assay. Not less than 99.0 percent of $Na_4Fe(CN)_6.10H_2O$.

Limits of Impurities

Chloride. Not more than 0.2 percent.

Free moisture. Not more than 1 percent.

Insoluble matter. Not more than 300 parts per million (0.03 percent).

Sulfate. Not more than 700 parts per million (0.07 percent).

TESTS

Assay. Transfer about 3 grams, accurately weighed, into a 400-ml. beaker, dissolve in 225 ml. of water, and add cautiously about 25 ml. of sulfuric acid T.S. Add, with stirring, 1 drop of orthophenanthroline T.S., and titrate with 0.1 *N* ceric sulfate until the color changes sharply from orange to pure yellow. Each ml. of 0.1 *N* ceric sulfate is equivalent to 96.81 mg. of $Na_4Fe(CN)_6.10H_2O$.

Chloride, page 879. Dissolve 100 mg. of the sample in 100 ml. of water. Any turbidity produced by a 10-ml. portion of this solution does not exceed that shown in a control containing 20 mcg. of chloride ion (Cl).

Free moisture. Heat 20 grams of the sample at 105° for 6 hours, cool in a desiccator, and weigh. Grind the dried sample rapidly, then heat 3 grams of the powder to constant weight at 105°, and calculate the total water content (W). Calculate the percent of free moisture in the sample by the formula $W - 0.3721A$, in which A is the percent of $Na_4Fe(CN)_6.10H_2O$ found in the *Assay*.

Insoluble matter. Dissolve 50 grams of the sample in 300 ml. of hot water, and filter off the insoluble matter on a tared Gooch crucible. Wash the residue thoroughly with hot water, dry the crucible in an oven at 105°, cool in a desiccator, and weigh.

Sulfate, page 879. Any turbidity produced by a 500-mg. sample does not exceed that shown in a control containing 350 mcg. of sulfate (SO_4).

Packaging and storage. Store in tight containers.

Functional use in foods. Anticaking agent for sodium chloride.

SODIUM GLUCONATE

$CH_2OH(CHOH)_4COONa$

$C_6H_{11}NaO_7$ Mol. wt. 218.14

DESCRIPTION

A white to tan, granular to fine, crystalline powder. It is very soluble in water and is sparingly soluble in alcohol. It is insoluble in ether.

IDENTIFICATION

A. A 1 in 20 solution gives positive tests for *Sodium*, page 928.

B. To 5 ml. of a warm solution (1 in 10) add 0.7 ml. of glacial acetic acid and 1 ml. of freshly distilled phenylhydrazine, heat on a steam bath for 30 minutes, and cool. Induce crystallization by scratching the inner surface of the container with a glass stirring rod. Crystals of gluconic acid phenylhydrazide form.

SPECIFICATIONS

Assay. Not less than 98.0 percent of $C_6H_{11}NaO_7$.

Limits of Impurities

Arsenic (as As). Not more than 3 parts per million (0.0003 percent).

Heavy metals (as Pb). Not more than 20 parts per million (0.002 percent).

Lead. Not more than 10 parts per million (0.001 percent).

Reducing substances. Not more than 0.5 percent.

TESTS

Assay. Transfer about 150 mg., accurately weighed, into a clean, dry 200-ml. Erlenmeyer flask, add 75 ml. of glacial acetic acid and dissolve by heating on a hot plate. Cool, add quinaldine red T.S., and titrate with 0.1 *N* perchloric acid in glacial acetic acid, using a 10-ml. microburet, to a colorless end-point. Each ml. of 0.1 *N* perchloric acid is equivalent to 21.81 mg. of $C_6H_{11}NaO_7$.

Arsenic. A solution of 1 gram in 35 ml. of water meets the requirements of the *Arsenic Test*, page 865.

Heavy metals. A solution of 1 gram in 25 ml. of water meets the requirements of the *Heavy Metals Test*, page 920, using 20 mcg. of lead ion (Pb) in the control (*Solution A*).

Lead. A solution of 1 gram in 25 ml. of water meets the requirements of the *Lead Limit Test*, page 929, using 10 mcg. of lead ion (Pb) in the control.

Reducing substances. Determine as directed in the test for *Reducing substances* under *Copper Gluconate*, page 219.

Packaging and storage. Store in well-closed containers.

Functional use in foods. Nutrient; dietary supplement; sequestrant.

SODIUM HYDROXIDE

Caustic Soda

NaOH — Mol. wt. 40.00

DESCRIPTION

White, or nearly white, pellets, flakes, sticks, fused masses, or other forms. Upon exposure to air, it readily absorbs carbon dioxide and moisture. One gram dissolves in 1 ml. of water. It is freely soluble in alcohol. A 1 in 25 solution gives positive tests for *Sodium*, page 928.

SPECIFICATIONS

Assay. Not less than 95.0 percent of total alkali, calculated as NaOH.

Limits of Impurities

Arsenic (as As). Not more than 3 parts per million (0.0003 percent).
Carbonate (as Na_2CO_3). Not more than 3 percent.
Heavy metals (as Pb). Not more than 30 parts per million (0.003 percent).
Insoluble substances and organic matter. Passes test.
Lead. Not more than 10 parts per million (0.001 percent).
Mercury. Not more than 1 part per million (0.0001 percent).

TESTS

Assay. Dissolve about 1.5 grams, accurately weighed, in 40 ml. of recently boiled and cooled water, cool to 15°, add phenolphthalein T.S., and titrate with 1 *N* sulfuric acid. At the discharge of the pink color, record the volume of acid required, then add methyl orange T.S. and continue the titration until a persistent pink color is produced. Record the total volume of acid required for the titration. Each ml. of 1 *N* sulfuric acid is equivalent to 40.00 mg. of total alkali, calculated as NaOH.

Arsenic. Dissolve 1 gram in about 10 ml. of water, cautiously neutralize to litmus paper with sulfuric acid, and cool. This solution meets the requirements of the *Arsenic Test*, page 865.

Carbonate. Each ml. of 1 *N* sulfuric acid required between the phenolphthalein and methyl orange end-points in the *Assay* is equivalent to 106.0 mg. of Na_2CO_3.

Heavy metals. Dissolve 670 mg. in a mixture of 5 ml. of water and 7 ml. of diluted hydrochloric acid T.S. Heat to boiling, cool, dilute to 25 ml. with water, and filter. This solution meets the requirements of the *Heavy Metals Test*, page 920, using 20 mcg. of lead ion (Pb) in the control (*Solution A*).

Insoluble substances and organic matter. A 1 in 20 solution is complete, clear, and colorless.

Lead. Dissolve 1 gram in a mixture of 5 ml. of water and 11 ml. of diluted hydrochloric acid T.S., and cool. This solution meets the requirements of the *Lead Limit Test*, page 929, using 10 mcg. of lead ion (Pb) in the control.

Mercury. Determine as directed under *Mercury Limit Test*, page 934, using the following as the *Sample Preparation:* Transfer 2.0 grams of the sample into a 50-ml. beaker, dissolve in 10 ml. of water, add 2 drops of phenolphthalein T.S., and slowly neutralize, with constant stirring, with dilute hydrochloric acid solution (1 in 2). Add 1 ml. of dilute sulfuric acid solution (1 in 5) and 1 ml. of potassium permanganate solution (1 in 25), cover the beaker with a watch glass, boil for a few seconds, and cool.

Packaging and storage. Store in tight containers.

Functional use in foods. Alkali.

SODIUM HYDROXIDE SOLUTION

DESCRIPTION

Sodium hydroxide solutions are usually available in nominal concentrations of 50 percent and 73 percent of NaOH, weight in weight, having freezing points of about 15° and 63°, respectively. These solutions are clear or slightly turbid, colorless or slightly colored, strongly caustic and hygroscopic, and when exposed to the air they absorb carbon dioxide forming sodium carbonate. Solutions of sodium hydroxide give positive tests for *Sodium*, page 928.

SPECIFICATIONS

Assay. Not less than 97.0 percent and not more than 103.0 percent, by weight, of the labeled amount of NaOH calculated as total alkalinity.

Limits of Impurities

Arsenic (as As). Not more than 3 parts per million (0.0003 percent) calculated on the basis of NaOH determined in the *Assay*.

Carbonate (as Na_2CO_3). Not more than 3 percent of the NaOH determined in the *Assay*.

Heavy metals (as Pb). Not more than 30 parts per million (0.003 percent) calculated on the basis of NaOH determined in the *Assay*.

Lead. Not more than 10 parts per million (0.001 percent) based on the NaOH determined in the *Assay*.

Mercury. Not more than 1 part per million (0.0001 percent) calculated on the basis of the NaOH determined in the *Assay*.

TESTS

Assay. Based on the stated or labeled percent of NaOH, weigh accurately a volume of the solution equivalent to about 1.5 grams of sodium hydroxide, and dilute it to 40 ml. with recently boiled and cooled water. Continue as directed in the *Assay* under *Sodium Hydroxide*, page 743, beginning with "...cool to 15°...."

Arsenic. Dilute the equivalent of 1 gram of NaOH, calculated on the basis of the *Assay*, to 10 ml. with water, cautiously neutralize to litmus paper with sulfuric acid, and cool. This solution meets the requirements of the *Arsenic Test*, page 865.

Carbonate. Each ml. of 1 *N* sulfuric acid required between the phenolphthalein and methyl orange end-points in the *Assay* is equivalent to 106.0 mg. of Na_2CO_3.

Heavy metals. Dilute the equivalent of 670 mg. of NaOH, calculated on the basis of the *Assay*, with a mixture of 5 ml. of water and 7 ml. of diluted hydrochloric acid T.S., and heat to boiling. Cool, dilute to 25 ml. with water, and filter. This solution meets the requirements of the *Heavy Metals Test*, page 920, using 20 mcg. of lead ion (Pb) in the control (*Solution A*).

Lead. Dilute the equivalent of 1 gram of NaOH, calculated on the basis of the *Assay*, with a mixture of 5 ml. of water and 11 ml. of diluted hydrochloric acid T.S. This solution meets the requirements of the *Lead Limit Test*, page 929, using 10 mcg. of lead ion (Pb) in the control.

Mercury. Determine as directed under *Mercury Limit Test*, page 934, using the following as the *Sample Preparation:* Transfer an accurately weighed amount of the sample, equivalent to 2.0 grams of NaOH, into a 50-ml. beaker, add 10 ml. of water and 2 drops of phenolphthalein T.S., and slowly neutralize, with constant stirring, with dilute hydrochloric acid solution (1 in 2). Add 1 ml. of dilute sulfuric acid solution (1 in 5) and 1 ml. of potassium permanganate solution (1 in 25), cover the beaker with a watch glass, boil for a few seconds, and cool.

Packaging and storage. Store in tight containers.

Functional use in foods. Alkali.

SODIUM LAURYL SULFATE

DESCRIPTION

Sodium lauryl sulfate is a mixture of sodium alkylsulfates consisting chiefly of sodium lauryl sulfate [$CH_3(CH_2)_{10}CH_2OSO_3Na$]. It occurs as small, white or light yellow crystals having a slight, characteristic odor. One gram dissolves in 10 ml. of water, forming an opalescent solution. A 1 in 10 solution gives positive tests for *Sodium*, page 928, and, after acidification with hydrochloric acid and boiling gently for 20 minutes, responds to the tests for *Sulfate*, page 928.

SPECIFICATIONS

Assay. Not less than 59.0 percent of total alcohols.

Limits of Impurities

Alkalinity (as NaOH). Passes test (about 0.25 percent).

Arsenic (as As). Not more than 3 parts per million (0.0003 percent).

Combined sodium chloride and sodium sulfate. Not more than 8 percent.

Heavy metals (as Pb). Not more than 20 parts per million (0.002 percent).

Lead. Not more than 5 parts per million (0.0005 percent).

Unsulfated alcohols. Not more than 4 percent.

TESTS

Assay. Transfer about 5 grams, accurately weighed, to an 800-ml. Kjeldahl flask, and add 150 ml. of water, 50 ml. of hydrochloric acid and a few boiling chips. Attach a reflux condenser to the flask, heat carefully to avoid excessive frothing, and then boil for about 4 hours. Cool the flask, rinse the condenser with ether, collecting the ether in the flask, and transfer the contents to a 500-ml. separator, rinsing the flask twice with ether and adding the washings to the separator. Extract the solution with two 75-ml. portions of ether, evaporate the combined ether extracts in a tared beaker on a steam bath, dry the residue at 105° for 30 minutes, cool, and weigh. The residue represents the total alcohols.

Alkalinity. Dissolve 1 gram in 100 ml. of water, add phenol red T.S., and titrate with 0.1 *N* hydrochloric acid. Not more than 0.6 ml. is required for neutralization.

Arsenic. A *Sample Solution* prepared as directed for organic compounds meets the requirements of the *Arsenic Test,* page 865.

Combined sodium chloride and sodium sulfate.

Sodium chloride. Dissolve about 5 grams, accurately weighed, in about 50 ml. of water. Neutralize the solution with dilute nitric acid (1 in 20), using litmus paper as the indicator, add 2 ml. of potassium chromate T.S., and titrate with 0.1 *N* silver nitrate. Each ml. of 0.1 *N* silver nitrate is equivalent to 5.844 mg. of sodium chloride.

Sodium sulfate. Transfer about 1 gram, accurately weighed, to a 400-ml. beaker, add 10 ml. of water, heat the mixture, and stir until completely dissolved. To the hot solution add 100 ml. of alcohol, cover, and digest at a temperature just below the boiling point for 2 hours. Filter, while hot, through a Gooch crucible, and wash the precipitate with 100 ml. of hot alcohol. Dissolve the precipitate in the crucible by washing with about 150 ml. of water, collecting the washings in a beaker. Acidify with 10 ml. of hydrochloric acid, heat to boiling, add 25 ml. of barium chloride T.S., and allow to stand overnight. Collect the precipitate of barium sulfate on a tared Gooch crucible, wash until free from chloride, dry, ignite, and weigh. The weight of barium sulfate so obtained, multiplied by 0.6086, represents the weight of Na_2SO_4.

Heavy metals. A solution of 1 gram in 25 ml. of water meets the requirements of the *Heavy Metals Test,* page 920, using 20 mcg. of lead ion (Pb) in the control (*Solution A*).

Lead. A *Sample Solution* prepared as directed for organic compounds meets the requirements of the *Lead Limit Test,* page 929, using 5 mcg. of lead ion (Pb) in the control.

Unsulfated alcohols. Dissolve about 10 grams, accurately weighed, in 100 ml. of water, and add 100 ml. of alcohol. Transfer the solution to a separator, and extract with three 50-ml. portions of solvent hexane. If an emulsion forms, sodium chloride may be added to promote separation of the two layers. Wash the combined solvent hexane extracts with three 50-ml. portions of water, and dry with anhydrous sodium sulfate. Filter the solvent hexane extract into a tared beaker, evaporate on a steam bath until the odor of solvent hexane no longer is perceptible, dry the residue at 105° for 30 minutes, cool, and weigh. The residue represents the unsulfated alcohols.

Packaging and storage. Store in well-closed containers.

Functional use in foods. Surfactant.

SODIUM METABISULFITE

Sodium Pyrosulfite

$Na_2S_2O_5$ Mol. wt. 190.10

DESCRIPTION

Colorless crystals or a white to yellowish crystalline powder having an odor of sulfur dioxide. It is freely soluble in water and slightly soluble in alcohol. Its solutions are acid to litmus. A 1 in 10 solution gives positive tests for *Sodium*, page 928, and for *Sulfite*, page 928.

SPECIFICATIONS

Assay. Not less than 90.0 percent of $Na_2S_2O_5$.

Limits of Impurities

Arsenic (as As). Not more than 3 parts per million (0.0003 percent).

Heavy metals (as Pb). Not more than 20 parts per million (0.002 percent).

Iron. Not more than 20 parts per million (0.002 percent).

Lead. Not more than 10 parts per million (0.001 percent).

Selenium. Not more than 30 parts per million (0.003 percent).

TESTS

Assay. Weigh accurately about 200 mg., add it to exactly 50 ml. of 0.1 *N* iodine contained in a glass-stoppered flask, and stopper the flask. Allow to stand for 5 minutes, add 1 ml. of hydrochloric acid, and titrate the excess iodine with 0.1 *N* sodium thiosulfate, adding starch T.S. as the indicator. Each ml. of 0.1 *N* iodine is equivalent to 4.752 mg. of $Na_2S_2O_5$.

Arsenic. Dissolve 1 gram in 5 ml. of water, add 0.5 ml. of sulfuric acid, evaporate to about 1 ml. on a steam bath, and dilute to 35 ml. with water. This solution meets the requirements of the *Arsenic Test*, page 865.

Heavy metals. Dissolve 1 gram in 10 ml. of water, add 5 ml. of hydrochloric acid, evaporate to dryness on a steam bath, and dissolve the residue in 25 ml. of water. This solution meets the requirements of the *Heavy Metals Test*, page 920, using 20 mcg. of lead ion (Pb) in the control (*Solution A*).

Iron. To 500 mg. of the sample add 2 ml. of hydrochloric acid, and evaporate to dryness on a steam bath. Dissolve the residue in 2 ml. of hydrochloric acid and 20 ml. of water, add a few drops of bromine T.S., and boil the solution to remove the bromine. Cool, dilute with water to 25 ml., then add 50 mg. of ammonium persulfate and 5 ml. of ammonium thiocyanate T.S. Any red or pink color does not exceed that produced in a control containing 1.0 ml. of *Iron Standard Solution* (10 mcg. Fe).

Lead. Dissolve 1 gram in 10 ml. of water, add 5 ml. of hydrochloric acid, evaporate to dryness on a steam bath, and dissolve the residue in about 20 ml. of water. This solution meets the requirements of the *Lead Limit Test*, page 929, using 10 mcg. of lead ion (Pb) in the control.

Selenium. Prepare and test a 2-gram sample as directed in the *Selenium Limit Test*, page 953.

Packaging and storage. Store in well-filled, tight containers, and avoid exposure to excessive heat.

Functional use in foods. Preservative; antioxidant.

SODIUM METAPHOSPHATE

Sodium Polyphosphate; Glassy Sodium Phosphate; Sodium Hexametaphosphate; Sodium Tetraphosphate; Graham's Salt; Kurrol's Salt; Sodium Trimetaphosphate; Sodium Tetrametaphosphate; Insoluble Sodium Metaphosphate

DESCRIPTION

This monograph covers a group of commercial phosphates, either crystalline or amorphous, ranging in composition from $(NaPO_3)_x$ through $Na_xH_2P_xO_{3x+1}$ to $Na_{x+2}P_xO_{3x+1}$. These compounds are usually identified by the P_2O_5 content or by the Na_2O/P_2O_5 ratio. They occur as colorless, glassy, transparent platelets or granules, or as a powder. Except for insoluble sodium metaphosphate, they are normally hygroscopic and are soluble in water.

The following three classes of sodium metaphosphate are available commercially:

Class A. Amorphous sodium polyphosphate, often referred to as "sodium hexametaphosphate," having an Na_2O/P_2O_5 mole ratio of about 1.1. The pH of a 1 in 100 solution is about 7.

Class B. Amorphous sodium polyphosphate, often referred to as "sodium tetraphosphate," having an Na_2O/P_2O_5 mole ratio of about 1.3. The pH of a 1 in 100 solution is about 7.8.

Class C. The amorphous and crystalline metaphosphates, having an Na_2O/P_2O_5 mole ratio of 1. The pH of a 1 in 100 solution is about 6.0.

IDENTIFICATION

A. A 1 in 20 solution gives positive tests for *Sodium*, page 928.

B. Dissolve 100 mg. in 5 ml. of hot diluted nitric acid T.S., warm on a steam bath for 10 minutes, and cool. Neutralize to litmus paper with sodium hydroxide T.S., and add silver nitrate T.S. A yellow precipitate is formed which is soluble in diluted nitric acid T.S.

SPECIFICATIONS

Assay. *Class A*, not less than 66.5 percent and not more than 68.0 percent of P_2O_5; *Class B*, not less than 62.8 percent and not more than 64.5 percent of P_2O_5; *Class C*, not less than 68.7 percent and not more than 70.0 percent of P_2O_5.

Limits of Impurities

Arsenic (as As). Not more than 3 parts per million (0.0003 percent).

Fluoride. Not more than 10 parts per million (0.001 percent).

Heavy metals (as Pb). Not more than 10 parts per million (0.001 percent).

Insoluble substances. Not more than 0.1 percent for soluble forms of sodium metaphosphate.

Loss on ignition. Not more than 1 percent.

TESTS

Assay. Dissolve about 250 mg., accurately weighed, in 25 ml. of water, add 15 ml. of nitric acid, heat to boiling, and boil for at least 30 minutes. Dilute the solution to about 100 ml., neutralize with ammonium hydroxide to methyl orange, then heat to 60° and add an excess of ammonium molybdate T.S. with vigorous stirring. Heat to 50°, allow to stand for 30 minutes with occasional stirring, and filter. Wash the precipitate with dilute nitric acid (1 in 36), followed by potassium nitrate solution (1 in 100) until the filtrate is no longer acid to litmus. Dissolve the precipitate in 50.0 ml. of 1 *N* sodium hydroxide, add phenolphthalein T.S., and titrate the excess sodium hydroxide with 1 *N* sulfuric acid. Each ml. of 1 *N* sodium hydroxide is equivalent to 3.086 mg. of P_2O_5.

Arsenic. A solution of 1 gram in 35 ml. of water meets the requirements of the *Arsenic Test*, page 865.

Fluoride. Proceed as directed in the *Fluoride Limit Test*, page 917.

Heavy metals. A solution of 2 grams in 25 ml. of water meets the requirements of the *Heavy Metals Test*, page 920, using 20 mcg. of lead ion (Pb) in the control (*Solution A*).

Insoluble substances. Dissolve 10 grams in 100 ml. of hot water, and filter through a tared filtering crucible. Wash the insoluble residue with hot water, dry at 105° for 2 hours, cool, and weigh.

Loss on ignition. Ignite at a dull red heat for 30 minutes.

Packaging and storage. Store in tight containers.

Functional use in foods. Emulsifier; sequestrant; texturizer.

SODIUM METHYLATE

Sodium Methoxide

CH_3ONa Mol. wt. 54.03

DESCRIPTION

A white, amorphous, hygroscopic, free-flowing powder. It reacts with oxygen and carbon dioxide and is decomposed by water. It is soluble in fats, in esters, and in alcohols. It decomposes without melting above 127°. *Caution: Sodium methylate and its solutions are caustic and flammable. Avoid contact with the eyes, skin, and clothing, and do not inhale vapors from sodium methylate solutions.*

SPECIFICATIONS

Assay. Not less than 97.0 percent of CH_3NaO.

Limits of Impurities

Arsenic (as As). Not more than 3 parts per million (0.0003 percent).

Heavy metals (as Pb). Not more than 25 parts per million (0.0025 percent).

Lead. Not more than 10 parts per million (0.001 percent).

Mercury. Not more than 1 part per million (0.0001 percent).

Sodium carbonate. Not more than 0.4 percent.

Sodium hydroxide. Not more than 1.1 percent.

TESTS

Note: The tests in the following section must be conducted with a minimum exposure of the sample to air. Preferably the tests should be conducted in a nitrogen hood.

Assay, Sodium carbonate, and Sodium hydroxide

Sample preparation. Select two tared weighing bottles, each approximately 30 mm. in diameter and 80 mm. high, nearly fill each with the sample, which should weigh between 12 and 15 grams, securely fit the covers, and weigh.

Determination of alkalinity as CH_3ONa. Remove the top from one of the sample bottles, and quickly drop the bottle into a 500-ml. Erlenmeyer flask containing 200 ml. of ice-cold, carbon dioxide-free water, sliding the sample bottle gently down the side of the flask to prevent splashing. Immediately stopper the flask with a rubber stopper, and swirl until the sample dissolves. Wash the sample solution into a 250-ml. volumetric flask, and nearly dilute to volume with carbon dioxide-free water. Allow the solution to reach room temperature, then dilute to volume with water, and mix. Transfer 50.0 ml. of this solution into a 500-ml. glass-stoppered Erlenmeyer flask, add 150 ml. of carbon dioxide-free water and 5 ml. of barium chloride

T.S., stopper the flask, mix, and allow to stand for 5 minutes. Add 3 drops of phenolphthalein T.S., and titrate with 1 *N* hydrochloric acid to the disappearance of the pink color. Retain the titrated solution for the *Determination of sodium carbonate.* Calculate the % alkalinity as CH_3ONa (% *A*) by the formula $(V_1 \times N \times 5.403)/(W \times 0.2)$, in which V_1 is the volume, in ml., and *N* the exact normality, of the hydrochloric acid used, and *W* is the weight of the sample, in grams.

Determination of sodium carbonate. Add 2 drops of methyl orange T.S. to the solution retained above, and continue the titration with 1 *N* hydrochloric acid to a permanent pink color. Calculate the % Na_2CO_3 by the formula $(V_2 \times N \times 5.30)/(W \times 0.2)$, in which V_2 is the volume, in ml., of 1 *N* hydrochloric acid consumed in the second titration, and *N* and *W* are as defined above.

Determination of sodium hydroxide. The *Karl Fischer Titrimetric Method*, page 977, may be adapted for this determination at the discretion of the analyst, or the following procedure may be used:

Solution A. Add 400 ml. of colorless pyridine, containing no more than 0.05% of water, to a 500-ml. Florence flask filled with a 2-hole rubber stopper and a 7-mm. glass tube extending nearly to the bottom of the flask. Place the flask in a cooling bath of running water, and pass dry sulfur dioxide from an upright cylinder until 80 ± 0.5 grams has been added. Disconnect the hose from the delivery tube before closing the gas valve. Transfer the solution into a dry glass-stoppered bottle, add 400 ml. of absolute methanol, and mix. Store in a dark place.

Solution B. Add 75 grams of iodine to 900 ml. of absolute methanol contained in a dry glass-stoppered bottle, and shake until the iodine dissolves. Transfer to a dry automatic buret protected by drying tubes.

Standardization of Solution B. Add 15 ml. of *Solution A*, measured with a dry graduate, to a dry 125 ml. iodine flask, and titrate with *Solution B* to a brownish yellow color. Stopper the flask immediately to prevent moisture contamination and end-point fading. Disregard the volume of *Solution B* added. To the flask add 50.0 ml. of methanol-water standard solution, containing 1.0 mg. of H_2O per ml., and immediately titrate with *Solution B* to the same brownish yellow end-point. Calculate the equivalence factor, *F*, in mg. of water per ml. of *Solution B*. Restandardize on each day of use.

Procedure. Select two 120-ml. wide-mouth glass jars with plastic screw caps, wash with hydrochloric acid, rinse with water followed by isopropanol, and dry with a current of air. Bore a hole through an extra screw cap to accommodate the tip from the automatic buret. Place a magnetic stirring bar in each jar, and flush the jars with dry nitrogen to remove carbon dioxide. To each jar add 30 ml. of *Solution A* and 15 ml. of absolute methanol, and screw on the caps. Replace the screw cap of one of the jars with the extra cap, insert the buret tip, and begin stirring. Titrate with *Solution B* to a distinct brownish yellow color that persists for at least 5 minutes. Replace the original cap, and titrate the other solution in the same manner. Remove the caps from both jars, and to one of the jars add about 2 grams of the sample, accurately weighed, from the remaining weighing bottle prepared as directed under *Sample preparation.* Replace the cap on the second jar, and titrate the sample solution with *Solution B* to the brownish yellow color. Titrate the blank to the same color. Calculate the % H_2O in the sample by the formula $(F \times V \times 100)/(W \times 1000)$, in which F is the equivalence factor, in mg. per ml., of *Solution B*, V is the net volume of *Solution B* required in the titration of the sample, in ml., and W is the weight of the sample taken, in grams.

Calculations. Calculate the % NaOH by the formula 2.222 × [% H_2O − (% Na_2CO_3 × 0.170)]. Finally, calculate the % CH_3ONa by the formula [% A − (% NaOH × 1.350)].

Arsenic. Cautiously dissolve 1 gram of the sample in 10 ml. of water, neutralize to litmus paper with diluted sulfuric acid T.S., and dilute to 35 ml. with water. This solution meets the requirements of the *Arsenic Test*, page 865.

Heavy metals. Cautiously dissolve 800 mg. of the sample in 10 ml. of water, add 10 ml. of diluted hydrochloric acid T.S., and heat to boiling. Cool, and dilute to 25 ml. with water. This solution meets the requirements of the *Heavy Metals Test*, page 920, using 20 mcg. of lead ion (Pb) in the control (*Solution A*).

Lead. Cautiously dissolve 1 gram of the sample in 10 ml. of water, add 10 ml. of diluted hydrochloric acid, and heat to boiling. Cool, and dilute to 25 ml. with water. This solution meets the requirements of the *Lead Limit Test*, page 929, using 10 mcg. of lead ion (Pb) in the control.

Mercury. Determine as directed under *Mercury Limit Test*, page 934, using the following as the *Sample Preparation:* Cautiously dissolve 2 grams of the sample in 10 ml. of water in a small beaker, add 2 drops of phenolphthalein T.S., and slowly neutralize, with constant stirring, with dilute sulfuric acid solution (1 in 5). Add 1 ml. of dilute sulfuric acid solution (1 in 5) and 1 ml. of potassium permanganate solution (1 in 25), and mix.

Packaging and storage. Store in air-tight containers, and take all necessary precautions to prevent combustion during handling.

Functional use in foods. Catalyst for the transesterification of fats.

SODIUM NITRATE

$NaNO_3$ Mol. wt. 85.00

DESCRIPTION

Colorless, odorless crystals or crystalline granules. It is moderately deliquescent in moist air. It is freely soluble in water, and is sparingly soluble in alcohol. A 1 in 5 solution is neutral to litmus and gives positive tests for *Sodium*, page 928, and for *Nitrate*, page 927.

SPECIFICATIONS

Assay. Not less than 99.0 percent of $NaNO_3$ after drying.

Limits of Impurities

Arsenic (as As). Not more than 3 parts per million (0.0003 percent).

Heavy metals (as Pb). Not more than 10 parts per million (0.001 percent).

Total chlorine. Passes test (approx. 0.2 percent).

TESTS

Assay. Weigh accurately about 350 mg., previously dried at 105° for 4 hours, dissolve in 10 ml. of hydrochloric acid in a small beaker or porcelain dish, and evaporate to dryness on a steam bath. Dissolve the residue in 10 ml. of hydrochloric acid, and again evaporate to dryness, continuing the heating until the residue, when dissolved in water, is neutral to litmus. Transfer the residue with the aid of 25 ml. of water to a glass-stoppered flask, add exactly 50 ml. of 0.1 *N* silver nitrate, then add 3 ml. of nitric acid and 3 ml. of nitrobenzene, and shake vigorously. Add ferric ammonium sulfate T.S., and titrate the excess silver nitrate with 0.1 *N* ammonium thiocyanate. Each ml. of 0.1 *N* silver nitrate is equivalent to 8.50 mg. of $NaNO_3$.

Arsenic. Dissolve 1 gram in 1 ml. of water, add 2 ml. of sulfuric acid, and evaporate to strong fumes of sulfur trioxide. Cool, wash down the sides of the container with water, and heat again to strong fuming. Repeat the washing and fuming three more times, then cool and dilute to 35 ml. with water. This solution meets the requirements of the *Arsenic Test*, page 865.

Heavy metals. A solution of 3 grams in 25 ml. of water meets the requirements of the *Heavy Metals Test*, page 920, using 20 mcg. of lead ion (Pb) and 1 gram of the sample in the control (*Solution A*).

Total chlorine. Dissolve 1 gram in 100 ml. of water, add enough 6 percent sulfurous acid to give the solution a distinct odor of sulfur dioxide, boil gently until the odor of the sulfur dioxide is no longer apparent, and adjust the volume to 100 ml. by the addition of water. Add 1.0 ml. of 0.1 *N* silver nitrate followed by 3 ml. of nitric acid and 3 ml. of nitrobenzene, and shake vigorously. Add ferric ammonium sulfate T.S., and titrate the excess silver nitrate with 0.1 *N* ammonium thiocyanate. No more than 0.6 ml. of the 0.1 *N* silver nitrate is consumed.

Packaging and storage. Store in tight containers.

Functional use in foods. Color fixative in meat and meat products.

SODIUM NITRITE

$NaNO_2$ Mol. wt. 69.00

DESCRIPTION

It occurs as a white to slightly yellow, granular powder, or as white or nearly white, opaque, fused masses or sticks. It has a mild, saline taste and is deliquescent in air. Its solution are alkaline to litmus. One gram of sodium nitrite dissolves in 1.5 ml. of water, but it is sparingly soluble in alcohol. Its solutions give positive tests for *Sodium*, page 928, and for *Nitrite*, page 928.

SPECIFICATIONS

Assay. Not less than 97.0 percent of $NaNO_2$ after drying.

Limits of Impurities

Arsenic (as As). Not more than 3 parts per million (0.0003 percent).

Heavy metals (as Pb). Not more than 20 parts per million (0.002 percent).

Lead. Not more than 10 parts per million (0.001 percent).

Loss on drying. Not more than 0.25 percent.

TESTS

Assay. Dissolve about 3 grams, previously dried over silica gel for 4 hours and accurately weighed, in water to make 100.0 ml. Pipet 10 ml. of this solution into a mixture of 100.0 ml. of 0.1 *N* potassium permanganate, 50 ml. of water, and 5 ml. of sulfuric acid, keeping the tip of the pipet well below the surface of the liquid. Warm the solution to 40°, allow it to stand for 5 minutes, and add 25.0 ml. of 0.1 *N* oxalic acid. Heat the mixture to about 80°, and titrate with 0.1 *N* potassium permanganate. Each ml. of 0.1 *N* potassium permanganate is equivalent to 3.450 mg. of $NaNO_2$.

Arsenic. Dissolve 1 gram in 10 ml. of diluted sulfuric acid T.S., boil gently for 1 minute, cool, and dilute to 35 ml. with water. This solution meets the requirements of the *Arsenic Test*, page 865.

Heavy metals. Dissolve 1 gram in a mixture of 10 ml. of water and 2 ml. of hydrochloric acid, and evaporate to dryness on a steam bath. Add another 2-ml. portion of hydrochloric acid, again evaporate to dryness, and dissolve the residue in 25 ml. of water. This solution meets the requirements of the *Heavy Metals Test,* page 920, using 20 mcg. of the lead ion (Pb) in the control (*Solution A*).

Lead. A solution of 1 gram in 10 ml. of water meets the requirements of the *Lead Limit Test,* page 929, using 10 mcg. of lead ion (Pb) in the control.

Loss on drying, page 931. Dry over silica gel for 4 hours.

Packaging and storage. Store in tight containers.

Functional use in foods. Color fixative in meat and meat products.

SODIUM PHOSPHATE, DIBASIC

Disodium Monohydrogen Phosphate; Disodium Phosphate

Na_2HPO_4 Mol. wt. 141.96

DESCRIPTION

Dibasic sodium phosphate is anhydrous or contains 2 molecules of water of hydration. It occurs as a white powder or crystalline solid. The anhydrous form is hygroscopic. Both forms are freely soluble in water and insoluble in alcohol. A 1 in 20 solution gives positive tests for *Phosphate,* page 928, and for *Sodium,* page 928.

SPECIFICATIONS

Assay. Not less than 98.0 percent of Na_2HPO_4 after drying.

Loss on drying. Na_2HPO_4 (anhydrous), not more than 5 percent; $Na_2HPO_4 . 2H_2O$ (dihydrate), between 18 percent and 22 percent.

Limits of Impurities

Arsenic (as As). Not more than 3 parts per million (0.0003 percent).

Fluoride. Not more than 50 parts per million (0.005 percent).

Heavy metals (as Pb). Not more than 10 parts per million (0.001 percent).

Insoluble substances. Not more than 0.2 percent.

TESTS

Assay. Transfer about 6.5 grams of the sample, previously dried at 105° for 4 hours and accurately weighed, into a 250-ml. beaker, add 50.0 ml. of 1 *N* hydrochloric acid and 50 ml. of water, and stir until the sample is completely dissolved. Place the electrodes of a suitable pH meter in the solution, and titrate the excess acid with 1 *N* sodium hydroxide to the inflection point occurring at about pH 4. Record the buret reading, and calculate the volume (*A*) of 1 *N* hydrochloric acid consumed by the sample. Continue the titration with 1 *N* sodium hydroxide until the inflection point occurring at about pH 8.8 is reached, record the buret reading, and calculate the volume (B) of 1 N sodium hydroxide required in the titration between the two inflection points (pH 4 to pH 8.8). When A is equal to, or less than, B, each ml. of the volume A of 1 N hydrochloric acid is equivalent to 142.0 mg. of Na_2HPO_4. When A is greater than B, each ml. of the volume $2B - A$ of 1 N sodium hydroxide is equivalent to 142.0 mg. of Na_2HPO_4.

Loss on drying, page 931. Dry at 120° for 4 hours.

Arsenic. A solution of 1 gram in 35 ml. of water meets the requirements of the *Arsenic Test,* page 865.

Fluoride. Weigh accurately 1.0 gram, and proceed as directed in the *Fluoride Limit Test,* page 917.

Heavy metals. A solution of 2 grams in 25 ml. of water meets the requirements of the *Heavy Metals Test,* page 920, using 20 mcg. of lead ion (Pb) in the control (*Solution A*).

Insoluble substances. Dissolve 10 grams in 100 ml. of hot water, and filter through a tared filtering crucible (not glass). Wash the insoluble residue with hot water, dry at 105° for 2 hours, cool, and weigh.

Packaging and storage. Store in tight containers.

Functional use in foods. Emulsifier; texturizer; buffer; nutrient; dietary supplement.

SODIUM PHOSPHATE, MONOBASIC

Monosodium Phosphate; Sodium Biphosphate; Monosodium Dihydrogen Phosphate

NaH_2PO_4 Mol. wt. 119.98

DESCRIPTION

Monobasic sodium phosphate is anhydrous or contains 1 or 2 molecules of water of hydration. It is odorless and is slightly hygroscopic. The anhydrous form is a white, crystalline powder. The hydrated forms occur as white or transparent crystals or granules. All forms are freely soluble in water, but are insoluble in alcohol. The pH of a 1 in 100 solution is between 4.1 and 4.7. A 1 in 20 solution gives positive tests for *Phosphate,* page 928, and for *Sodium,* page 928.

SPECIFICATIONS

Assay. Not less than 98.0 percent and not more than the equivalent of 103.0 percent of NaH_2PO_4 after drying.

Loss on drying. NaH_2PO_4 (anhydrous), not more than 2 percent; $NaH_2PO_4 \cdot H_2O$ (monohydrate), between 10 and 15 percent; $NaH_2PO_4 \cdot 2H_2O$ (dihydrate), between 20 and 25 percent.

Limits of Impurities

Arsenic (as As). Not more than 3 parts per million (0.0003 percent).

Fluoride. Not more than 50 parts per million (0.005 percent).

Heavy metals (as Pb). Not more than 10 parts per million (0.001 percent).

Insoluble substances. Not more than 0.2 percent.

TESTS

Assay. Transfer about 5 grams of the sample, previously dried at 105° for 4 hours and accurately weighed, into a 250-ml. beaker, add 100 ml. of water and 50.0 ml. of 1 *N* hydrochloric acid, and stir until the sample is completely dissolved. Place the electrodes of a suitable pH meter in the solution, and slowly titrate the excess acid, stirring constantly, with 1 *N* sodium hydroxide to the inflection point occurring at about pH 4. Record the buret reading, and calculate the volume (*A*), if any, of 1 *N* hydrochloric acid consumed by the sample. Continue the titration with 1 *N* sodium hydroxide until the inflection point occurring at about pH 8.8 is reached, record the buret reading, and calculate the volume (*B*) of 1 *N* sodium hydroxide required in the titration between the two inflection points (pH 4 and pH 8.8). Each ml. of the volume $B - A$ of 1 *N* sodium hydroxide is equivalent to 120.0 mg. of NaH_2PO_4.

Loss on drying, page 931. Dry first at 60° for 1 hour, then at 105° for 4 hours.

Arsenic. A solution of 1 gram in 35 ml. of water meets the requirements of the *Arsenic Test*, page 865.

Fluoride. Weigh accurately 1.0 gram, and proceed as directed in the *Fluoride Limit Test*, page 917.

Heavy metals. A solution of 2 grams in 25 ml. of water meets the requirements of the *Heavy Metals Test*, page 920, using 20 mcg. of lead ion (Pb) in the control (*Solution A*).

Insoluble substances. Dissolve 10 grams in 100 ml. of hot water, and filter through a tared filtering crucible (not glass). Wash the insoluble residue with hot water, dry at 105° for 2 hours, cool, and weigh.

Packaging and storage. Store in tight containers.

Functional use in foods. Buffer; emulsifier; nutrient; dietary supplement.

SODIUM PHOSPHATE, TRIBASIC

Trisodium Phosphate

Na_3PO_4 Mol. wt. (anhydrous) 163.94

DESCRIPTION

Tribasic sodium phosphate is anhydrous or contains 1 or 12 molecules of water of hydration. The formula for the dodecahydrate is approximately $4(Na_3PO_4.12H_2O)NaOH$. It occurs as white, odorless crystals or granules or as a crystalline powder. It is freely soluble in water but is insoluble in alcohol. The pH of a 1 in 100 solution is between 11.5 and 12.0. A 1 in 20 solution gives positive tests for *Sodium*, page 928, and for *Phosphate*, page 928.

SPECIFICATIONS

Assay. Na_3PO_4 (anhydrous) and $Na_3PO_4.H_2O$ (monohydrate), not less than 97.0 percent of Na_3PO_4, calculated on the ignited basis; $4(Na_3PO_4.12H_2O)NaOH$ (dodecahydrate), not less than 92.0 percent of Na_3PO_4, calculated on the ignited basis.

Loss on ignition. Na_3PO_4 (anhydrous), not more than 2 percent; $Na_3PO_4.H_2O$ (monohydrate), between 8 and 11 percent; $Na_3PO_4.12H_2O$ (dodecahydrate), between 45 and 57 percent.

Limits of Impurities

Arsenic (as As). Not more than 3 parts per million (0.0003 percent).

Fluoride. Not more than 50 parts per million (0.005 percent).

Heavy Metals (as Pb). Not more than 10 parts per million (0.001 percent).

Insoluble substances. Not more than 0.2 percent.

TESTS

Assay. Dissolve an accurately weighed quantity of the sample, equivalent to between 5.5 and 6 grams of anhydrous Na_3PO_4, in 40 ml. of water in a 400-ml. beaker, and add 100.0 ml. of 1 *N* hydrochloric acid. Pass a stream of carbon dioxide-free air, in fine bubbles, through the solution for 30 minutes to expel carbon dioxide, covering the beaker loosely to prevent any loss by spraying. Wash the cover and sides of the beaker with a few ml. of water, and place the electrodes of a standard pH meter in the solution. Titrate the solution with 1 *N* sodium hydroxide to the inflection point occurring at about pH 4, then calculate the volume (A) of 1 *N* hydrochloric acid consumed. Protect the solution from absorbing carbon dioxide from the air, and continue the titration with 1 *N* sodium hydroxide until the inflection point occurring at about pH 8.8 is reached. Calculate the volume (B) of 1 *N* sodium hydroxide consumed in the titration. When A is equal to, or greater than, $2B$, each ml. of the volume B of 1 *N* sodium hydroxide is equivalent to 163.9 mg. of Na_3PO_4. When A is less than $2B$, each ml. of the volume $A - B$ of 1 *N* sodium hydroxide is equivalent to 163.9 mg. of Na_3PO_4.

Loss on ignition. Ignite at about 800° for 30 minutes after drying at 110° for 5 hours.

Arsenic. A solution of 1 gram in 35 ml. of water meets the requirements of the *Arsenic Test*, page 865.

Fluoride. Weigh accurately 1.0 gram, and proceed as directed in the *Fluoride Limit Test*, page 917.

Heavy metals. A solution of 2 grams in 25 ml. of water meets the requirements of the *Heavy Metals Test*, page 920, using 20 mcg. of lead ion (Pb) in the control (*Solution A*).

Insoluble substances. Dissolve 10 grams in 100 ml. of hot water, and filter through a tared filtering crucible (not glass). Wash the insoluble residue with hot water, dry at 105° for 2 hours, cool, and weigh.

Packaging and storage. Store in well-closed containers.

Functional use in foods. Buffer; emulsifier; nutrient; dietary supplement.

SODIUM POTASSIUM TARTRATE

Rochelle Salt

KOOCCH(OH).CH(OH).COONa.$4H_2O$

$C_4H_4KNaO_6.4H_2O$ Mol. wt. 282.23

DESCRIPTION

Sodium potassium tartrate is a salt of L(+)-tartaric acid. It occurs as colorless crystals, or a white, crystalline powder, having a cooling, saline taste. As it effloresces slightly in warm, dry air, the crystals are often coated with a white powder. One gram dissolves in 1 ml. of water. It is practically insoluble in alcohol.

IDENTIFICATION

A. Upon ignition, it emits the odor of burning sugar and leaves a residue which is alkaline to litmus and which effervesces with acids.

B. To 10 ml. of a 1 in 20 solution add 10 ml. of acetic acid. A white crystalline precipitate is formed within 15 minutes.

C. A 1 in 10 solution gives positive tests for *Tartrate*, page 928.

SPECIFICATIONS

Assay. Not less than 99.0 percent and not more than the equivalent of 102.0 percent of $C_4H_4KNaO_6.4H_2O$.

Water. Between 21 percent and 26 percent.

Limits of Impurities

Alkalinity. Passes test.

Arsenic (as As). Not more than 3 parts per million (0.0003 percent).

Heavy metals (as Pb). Not more than 10 parts per million (0.001 percent).

TESTS

Assay. Weigh accurately about 2 grams, and proceed as directed under *Alkali Salts of Organic Acids Assay*, page 864. Each ml. of 0.5 *N* sulfuric acid is equivalent to 70.55 mg. of $C_4H_4KNaO_6 . 4H_2O$.

Water. Determine by the *Karl Fischer Titrimetric Method*, page 977, using a 200-mg. sample and 35 ml. of methanol in the *Procedure*.

Alkalinity. A 1 in 20 solution is alkaline to litmus, but after the addition of 0.2 ml. of 0.1 *N* sulfuric acid to 10 ml. of this solution no pink color is produced by the addition of 1 drop of phenolphthalein T.S.

Arsenic. A *Sample Solution* prepared as directed for organic compounds meets the requirements of the *Arsenic Test*, page 865.

Heavy metals. Prepare and test a 2-gram sample as directed in *Method II* under the *Heavy Metals Test*, page 920, using 20 mcg. of lead ion (Pb) in the control (*Solution A*).

Packaging and storage. Store in tight containers.
Functional use in foods. Buffer; sequestrant.

SODIUM PROPIONATE

CH_3CH_2COONa

$C_3H_5NaO_2$ Mol. wt. 96.06

DESCRIPTION

Colorless, transparent crystals or a granular crystalline powder. It is odorless, or has a faint acetic-butyric odor. It is deliquescent in moist air. One gram is soluble in about 1 ml. of water at 25°, in about 0.65 ml. of boiling water, and in about 24 ml. of alcohol. The pH of a 1 in 10 solution is between 8.0 and 10.5.

IDENTIFICATION

A. A 1 in 20 solution gives positive tests for *Sodium*, page 928.

B. Upon ignition, it yields an alkaline residue which effervesces with acids.

C. Warm a small sample with sulfuric acid. Propionic acid, recognized by its odor, is evolved.

SPECIFICATIONS

Assay. Not less than 99.0 percent of $C_3H_5NaO_2$ after drying.

Limits of Impurities

Alkalinity (as Na_2CO_3). Passes test (about 0.15 percent).
Arsenic (as As). Not more than 3 parts per million (0.0003 percent).
Heavy metals (as Pb). Not more than 10 parts per million (0.001 percent).
Iron. Not more than 30 parts per million (0.003 percent).
Water. Not more than 1 percent.

TESTS

Assay. Weigh accurately about 250 mg., previously dried at 105° for 1 hour, and dissolve it in 40 ml. of glacial acetic acid, warming if necessary to effect solution. Cool to room temperature, add 2 drops of crystal violet T.S., and titrate with 0.1 *N* perchloric acid. Perform a blank determination (see page 2) and make any necessary correction. Each ml. of 0.1 *N* perchloric acid is equivalent to 9.606 mg. of C_3H_5-NaO_2.

Alkalinity. Dissolve 4 grams in 20 ml. of water, and add 3 drops of phenolphthalein T.S. If a pink color is produced, not more than 0.6 ml. of 0.1 *N* sulfuric acid is required to discharge it.

Arsenic. A *Sample Solution* prepared as directed for organic compounds meets the requirements of the *Arsenic Test,* page 865.

Heavy metals. A solution of 2 grams in 25 ml. of water meets the requirements of the *Heavy Metals Test,* page 920, using 20 mcg. of lead ion (Pb) in the control (*Solution A*).

Iron. Dissolve 300 mg. in 40 ml. of water, and add 2 ml. of hydrochloric acid, about 40 mg. of ammonium persulfate, and 10 ml. of ammonium thiocyanate T.S. Any red or pink color does not exceed that produced by 0.9 ml. of *Iron Standard Solution* (9 mcg. Fe) in an equal volume of solution containing the quantities of reagents used in the test.

Water. Determine by the *Karl Fischer Titrimetric Method,* page 977.

Packaging and storage. Store in tight containers.

Functional use in foods. Preservative; mold and rope inhibitor.

SODIUM PYROPHOSPHATE

Tetrasodium Diphosphate; Tetrasodium Pyrophosphate

$Na_4P_2O_7$ Mol. wt. 265.90

DESCRIPTION

Sodium pyrophosphate is anhydrous or contains 10 molecules of water of hydration. It occurs as a white, crystalline or granular powder. The decahydrate effloresces slightly in dry air. It is soluble in water, but is insoluble in alcohol. The pH of a 1 in 100 solution is about 10.

IDENTIFICATION

A. A 1 in 20 solution gives positive tests for *Sodium,* page 928.

B. To 1 ml. of a 1 in 100 solution add a few drops of silver nitrate T.S. A white or slightly yellow precipitate is formed which is soluble in diluted nitric acid T.S.

SPECIFICATIONS

Assay. Not less than 95.0 percent of $Na_4P_2O_7$, calculated on the ignited basis.

Loss on ignition. $Na_4P_2O_7$ (anhydrous), not more than 0.5 percent; $Na_4P_2O_7 . 10H_2O$ (decahydrate), between 38 and 42 percent.

Limits of Impurities

Arsenic (as As). Not more than 3 parts per million (0.0003 percent).

Fluoride. Not more than 50 parts per million (0.005 percent).

Heavy metals (as Pb). Not more than 10 parts per million (0.001 percent).

Insoluble substances. Not more than 0.2 percent.

TESTS

Assay. Dissolve an accurately weighed quantity of the sample, equivalent to 500 mg. of anhydrous $Na_4P_2O_7$, in 100 ml. of water in a 400-ml. beaker. Adjust the pH of the solution to 3.8 with hydrochloric acid, using a pH meter, then add 50 ml. of a 1 in 8 solution of zinc sulfate (125 grams of $ZnSO_4 . 7H_2O$ dissolved in water, diluted to 1000 ml., filtered, and adjusted to pH 3.8), and allow to stand for 2 minutes. Titrate the liberated acid with 0.1 *N* sodium hydroxide until a pH of 3.8 is again reached. After each addition of sodium hydroxide near the end-point, time should be allowed for any precipitated zinc hydroxide to redissolve. Each ml. of 0.1 *N* sodium hydroxide is equivalent to 13.30 mg. of $Na_4P_2O_7$.

Loss on ignition. Dry at 110° for 4 hours, and then ignite at about 800° for 30 minutes.

Arsenic. A solution of 1 gram in 35 ml. of water meets the requirements of the *Arsenic Test*, page 865.

Fluoride. Weigh accurately 1.0 gram, and proceed as directed in the *Fluoride Limit Test*, page 917.

Heavy metals. A solution of 2 grams in 25 ml. of water meets the requirements of the *Heavy Metals Test*, page 920, using 20 mcg. of lead ion (Pb) in the control (*Solution A*).

Insoluble substances. Dissolve 10 grams in 100 ml. of hot water, and filter through a tared filtering crucible. Wash the insoluble residue with hot water, dry at 105° for 2 hours, cool, and weigh.

Packaging and storage. Store in tight containers.

Functional use in foods. Emulsifier; buffer; nutrient; dietary supplement.

SODIUM SACCHARIN

1,2-Benzisothiazolin-3-one 1,1-Dioxide Sodium Salt; Sodium *o*-Benzosulfimide; Soluble Saccharin

$C_7H_4NNaO_3S.2H_2O$ Mol. wt. 241.20

DESCRIPTION

White crystals or a white, crystalline powder. It is odorless or has a faint, aromatic odor. It has an intensely sweet taste, even in dilute solutions. It is about 500 times as sweet as sucrose in dilute solutions. In powdered form, it effloresces to the extent that it usually contains only about one-third the amount of water indicated in its molecular formula. One gram is soluble in 1.5 ml. of water and in about 50 ml. of alcohol.

IDENTIFICATION

A. Dissolve about 100 mg. in 5 ml. of sodium hydroxide solution (1 in 20), evaporate to dryness, and gently fuse the residue over a small flame until it no longer evolves ammonia. After the residue has cooled, dissolve it in 20 ml. of water, neutralize the solution with diluted hydrochloric acid T.S., and filter. The addition of a drop of ferric chloride T.S. to the filtrate produces a violet color.

B. Mix 20 mg. with 40 mg. of resorcinol, add 10 drops of sulfuric acid, and heat the mixture in a liquid bath at 200° for 3 minutes. After cooling, add 10 ml. of water and an excess of sodium hydroxide T.S. A fluorescent green liquid results.

C. The residue obtained by igniting a 2-gram sample gives positive tests for *Sodium*, page 928.

D. Add 1 ml. of hydrochloric acid to 10 ml. of a 1 in 10 solution of the sample, wash the crystalline precipitate well with cold water, and dry at 105° for 2 hours. The saccharin thus obtained melts between 226° and 230° (see page 931).

SPECIFICATIONS

Assay. Not less than 98.0 percent and not more than the equivalent of 101.0 percent of $C_7H_4NNaO_3S$, calculated on the anhydrous basis.

Water. Not less than 3 percent and not more than 15 percent.

Limits of Impurities

Alkalinity. Passes test.

Arsenic (as As). Not more than 3 parts per million (0.0003 percent).

Benzoate and salicylate. Passes test.

Heavy metals (as Pb). Not more than 10 parts per million (0.001 percent).

Readily carbonizable substances. Passes test.

Selenium. Not more than 30 parts per million (0.003 percent).

TESTS

Assay. Transfer about 500 mg. of the sample, accurately weighed, into a separator with the aid of 10 ml. of water, add 2 ml. of diluted hydrochloric acid T.S., and extract the precipitated saccharin first with 30 ml., then with five 20-ml. portions, of a solvent composed of 9 volumes of chloroform and 1 volume of alcohol. Filter each extract through a small filter paper moistened with the solvent mixture, and evaporate the combined filtrates on a steam bath to dryness with the aid of a current of air. Dissolve the residue in 40 ml. of alcohol, add 40 ml. of water, mix, add 3 drops of phenolphthalein T.S., and titrate with 0.1 *N* sodium hydroxide. Perform a blank determination on a mixture of 40 ml. of alcohol and 40 ml. of water (see page 2). Each ml. of 0.1 *N* sodium hydroxide is equivalent to 20.52 mg. of C_7H_4N-NaO_3S.

Water. Determine by the *Karl Fischer Titrimetric Method,* page 977.

Alkalinity. A 1 in 10 solution is neutral or alkaline to litmus, but produces no red color with phenolphthalein T.S.

Arsenic. A *Sample Solution* prepared as directed for organic compounds meets the requirements of the *Arsenic Test,* page 865.

Benzoate and salicylate. To 10 ml. of a 1 in 20 solution, previously acidified with 5 drops of acetic acid, add 3 drops of ferric chloride T.S. No precipitate or violet color appears.

Heavy metals. Prepare and test a 2-gram sample as directed in *Method II* under the *Heavy Metals Test,* page 920, using 20 mcg. of lead ion (Pb) in the control (*Solution A*).

Readily carbonizable substances, page 943. Dissolve 200 mg. in 5 ml. of sulfuric acid T.S., and keep at a temperature of 48° to 50° for 10 minutes. The color is no darker than *Matching Fluid A.*

Selenium. Prepare and test a 2-gram sample as directed in the *Selenium Limit Test,* page 953.

Packaging and storage. Store in well-closed containers.

Functional use in foods. Nonnutritive sweetener.

SODIUM SESQUICARBONATE

$Na_2CO_3.NaHCO_3.2H_2O$ Mol. wt. 226.03

DESCRIPTION

White crystals, flakes, or a crystalline powder. It is soluble in water, and its solutions are alkaline to litmus. A 1 in 10 solution gives positive tests for *Sodium,* page 928 and for *Carbonate,* page 926.

SPECIFICATIONS

Assay. *Sodium bicarbonate:* not less than 35.5 percent and not more than 37.2 percent of $NaHCO_3$; *Sodium carbonate:* not less than 47.0 percent and not more than 48.5 percent of Na_2CO_3.

Water. Between 15.2 percent and 16.2 percent.

Limits of Impurities

Arsenic (as As). Not more than 3 parts per million (0.0003 percent).

Heavy metals (as Pb). Not more than 10 parts per million (0.001 percent).

Iron. Not more than 20 parts per million (0.002 percent).

Sodium chloride. Not more than 0.5 percent.

TESTS

Assay for sodium bicarbonate. Dissolve about 8.5 grams, accurately weighed, in 50 ml. of water in a 250-ml. beaker, and titrate with 1 *N* sodium hydroxide until a drop of the solution, when added to a drop of a 1 in 10 solution of silver nitrate on a spot plate, instantly produces a dark brown color. Each ml. of 1 *N* sodium hydroxide is equivalent to 84.01 mg. of $NaHCO_3$. Calculate the percent of sodium bicarbonate in the sample taken.

Assay for sodium carbonate. Determine the total alkalinity (as Na_2O) of the sample as follows: Dissolve about 4.2 grams, accurately weighed, in 100 ml. of water in a 250-ml. beaker, add methyl orange T.S., and titrate with 1 *N* sulfuric acid, stirring vigorously near the end-point to expel carbon dioxide. Each ml. of 1 *N* sulfuric acid is equivalent to 30.99 mg. of Na_2O. Calculate the percent of sodium oxide (% Na_2O) in the sample taken.

Calculate the percent of sodium carbonate in the sample by the formula $[\%Na_2O - (\%NaHCO_3 \times 0.3689)] \times 1.7099$, in which $\%NaHCO_3$ is the percent of sodium bicarbonate determined in the *Assay for sodium bicarbonate*, 0.3689 is a factor converting $NaHCO_3$ to Na_2O, and 1.7099 is a factor converting Na_2O to Na_2CO_3.

Water. Calculate the percent of water by subtracting from 100 the sum of the percents of *sodium bicarbonate, sodium carbonate,* and *sodium chloride* found in the sample.

Arsenic. A solution of 1 gram in 5 ml. of diluted hydrochloric acid T.S. meets the requirements of the *Arsenic Test,* page 865.

Heavy metals. Mix 2 grams with 5 ml. of water and 10 ml. of diluted hydrochloric acid T.S., boil for 1 minute, cool, and dilute to 25 ml. with water. This solution meets the requirements of the *Heavy Metals Test,* page 920, using 20 mcg. of lead ion (Pb) in the control (*Solution A*).

Iron. Dissolve 500 mg. in 10 ml. of diluted hydrochloric acid T.S., and dilute to 50 ml. with water. Add about 40 mg. of ammonium persulfate crystals and 10 ml. of ammonium thiocyanate T.S. Any red or pink color does not exceed that produced by 1.0 ml. of *Iron*

Standard Solution (10 mcg. Fe) in an equal volume of solution containing 2 ml. of hydrochloric acid and the quantities of ammonium persulfate and ammonium thiocyanate used in the test.

Sodium chloride. Dissolve about 10 grams, accurately weighed, in 50 ml. of water in a 250-ml. beaker, add sufficient nitric acid to make the solution slightly acid, then add 1 ml. of ferric ammonium sulfate T.S. and 1.00 ml. of 0.05 *N* ammonium thiocyanate, and titrate with 0.05 *N* silver nitrate, stirring constantly, until the red color is completely discharged. Finally, back titrate with 0.05 *N* ammonium thiocyanate until a faint reddish color is obtained. Subtract the total volume of 0.05 *N* ammonium thiocyanate added from the volume of 0.05 *N* silver nitrate required. Each ml. of 0.05 *N* silver nitrate is equivalent to 2.922 mg. of NaCl. Calculate the percent of sodium chloride in the sample taken.

Packaging and storage. Store in well-closed containers.

Functional use in foods. Alkalizer; neutralizer in dairy products.

SODIUM SILICOALUMINATE

Sodium Aluminosilicate

DESCRIPTION

A series of hydrated sodium aluminum silicates having an $Na_2O/Al_2O_3/SiO_2$ mole ratio of approximately 1/1/13.2, respectively. It occurs as a fine, white, amorphous powder, or as beads. It is odorless and tasteless. It is insoluble in water and in alcohol and other organic solvents, but at 80° to 100° is partially soluble in strong acids and solutions of alkali hydroxides. The pH of a 20 percent slurry, prepared with carbon dioxide-free water, is between 6.5 and 10.5.

SPECIFICATIONS

Assay

Silicon dioxide: Not less than 66.0 percent and not more than 71.0 percent of SiO_2 after drying.

Aluminum oxide: Not less than 9.0 percent and not more than 13.0 percent of Al_2O_3 after drying.

Sodium oxide: Not less than 5.0 percent and not more than 6.0 percent of Na_2O after drying.

Loss on drying. Not more than 8 percent.

Loss on ignition. Between 8 and 11 percent.

Limits of Impurities

Arsenic (as As). Not more than 3 parts per million (0.0003 percent).

Heavy metals (as Pb). Not more than 10 parts per million (0.001 percent).

TESTS

Assay

Silicon dioxide. Transfer about 500 mg., previously dried at 105° for 2 hours and accurately weighed, into a 250-ml. beaker, wash the sides of the beaker with a few ml. of water, and then add 30 ml. of 72 percent perchloric acid and 15 ml. of hydrochloric acid. Heat on a hot plate in a hood until dense white fumes are evolved, cool, add 15 ml. of hydrochloric acid, and heat again to dense white fumes. Cool, add 70 ml. of water, and filter through Whatman No. 40 or equivalent filter paper. Wash the filter paper and precipitate with hot water until free from perchloric acid, and then transfer the filter paper and precipitate into a tared platinum crucible. Char, and ignite at 900° to constant weight. Moisten the residue with a few drops of water, then add 15 ml. of hydrofluoric acid and 8 drops of sulfuric acid, and heat on a hot plate until white fumes of sulfur trioxide are evolved. Cool, add 5 ml. of water, 10 ml. of hydrofluoric acid, and 3 drops of sulfuric acid, and evaporate to dryness on the hot plate. Heat cautiously over an open flame until sulfur trioxide fumes have ceased, and ignite at 900° to constant weight. The weight loss after the addition of hydrofluoric acid represents the weight of SiO_2 in the sample taken. Retain the residue for the determination of *Aluminum oxide.*

Aluminum oxide. Fuse the residue obtained in the *Silicon dioxide* determination with 2 grams of potassium pyrosulfate for 5 minutes, cool, dissolve the fusion in water, and dilute to 250 ml. in a volumetric flask. Transfer 100.0 ml. of the solution into a 600-ml. beaker, add 100 ml. of water and 5 drops of bromothymol blue T.S., and heat to a slow boil. Add ammonium hydroxide, dropwise, until a blue color appears, then boil the solution for 5 minutes to expel the excess ammonia. Filter through Whatman No. 41, or equivalent, filter paper, and wash the precipitate with six portions of hot ammonium chloride solution (1 in 50). Transfer the filter and precipitate into a tared platinum crucible, char the paper, and ignite over a Meker burner to constant weight. The weight of the residue, corrected for the ash content of the filter paper and multiplied by 2.5, represents the weight of Al_2O_3 in the original sample.

Sodium oxide. Transfer about 500 mg. of the sample, previously dried at 105° for 2 hours and accurately weighed, into a tared platinum dish, and moisten with 8 to 10 drops of water. Add 25 ml. of 70 percent perchloric acid and 10 ml. of hydrofluoric acid, and heat on a hot plate in a hood until dense white fumes of perchloric acid appear. Add 10 ml. of hydrofluoric acid, heat again to dense white fumes, and dissolve the residue in sufficient water to make 250.0 ml.

Set a suitable flame photometer to a wavelength of 589 $m\mu$. Adjust the instrument to zero transmittance against water, then adjust it to

100.0 percent transmittance with a standard solution containing 200 mcg. of sodium, in the form of the chloride, per ml. Read the percent transmittance of three other standard solutions containing 50, 100, and 150 mcg. each of sodium per ml., and plot the standard curve as percent transmittance vs. concentration of sodium.

Place a portion of the sample solution in the photometer, read the percent transmittance in the same manner, and by reference to the standard curve determine the concentration (C) of sodium, in mcg. per ml., in the sample solution. Calculate the quantity, in mg., of Na_2O in the sample taken by the formula $(250C \times 1.348/1000) - F$, in which F, as determined below, is the quantity of sodium oxide equivalent to any sodium sulfate present in the sample.

Correction for sodium sulfate content. Weigh accurately 12.5 grams of the sample, previously dried at 105° for 2 hours, and stir it with 240 ml. of water for at least 5 minutes with a high speed mixer. Transfer the mixture into a 250-ml. graduate, and wash the mixer container with water, adding the washings to the graduate to make 250 ml. Stopper the graduate, invert it several times to mix the sample, and determine the conductivity of the slurry using a suitable conductance bridge assembly. By means of a standard curve, obtained from solutions containing 50, 100, 200, and 500 mg. of sodium sulfate in each 100 ml., determine the concentration (C'), in mg. per 100 ml., of sodium sulfate in the sample slurry, and calculate the correction factor (F) by the formula $0.437(2.5C' \times w/W)$, in which w is the weight of the sample taken for the *Sodium oxide* determination, and W is the weight of the sample taken for the preparation of the slurry.

Loss on drying, page 931. Dry at 105° for 2 hours.

Loss on ignition. Transfer about 5 grams, previously dried at 105° for 2 hours and accurately weighed, into a suitable tared crucible. and ignite at 900° to constant weight.

Sample Solution for the Determination of Arsenic and Heavy Metals. Transfer 10.0 grams of the sample into a 250-ml. beaker, add 50 ml. of 0.5 N hydrochloric acid, cover with a watch glass, and heat slowly to boiling. Boil gently for 15 minutes, cool, and let the undissolved material settle. Decant the supernatant liquid through Whatman No. 4, or equivalent, filter paper into a 100-ml. volumetric flask, retaining as much as possible of the insoluble material in the beaker. Wash the slurry and beaker with three 10-ml. portions of hot water, decanting each washing through the filter into the flask. Finally, wash the filter paper with 15 ml. of hot water, cool the filtrate to room temperature, dilute to volume with water, and mix.

Arsenic. A 10-ml. portion of the *Sample Solution* meets the requirements of the *Arsenic Test,* page 865.

Heavy metals. A 20-ml. portion of the *Sample Solution* meets the requirements of the *Heavy Metals Test*, page 920, using 20 mcg. of lead ion (Pb) in the control (*Solution A*).

Packaging and storage. Store in well-closed containers.

Functional use in foods. Anticaking agent.

SODIUM STEAROYL-2-LACTYLATE

DESCRIPTION

A mixture of sodium salts of stearoyl lactylic acids and minor proportions of other sodium salts of related acids, manufactured by the reaction of stearic acid and lactic acid, neutralized to the sodium salts. It is a slightly hygroscopic, cream colored powder having a mild, caramel-like odor. It is soluble in hot oil or fat and is dispersible in warm water. It conforms to the regulations of the federal Food and Drug Administration pertaining to specifications for fats or fatty acids derived from edible sources.

IDENTIFICATION

A. Heat 1 gram with a mixture of 25 ml. of water and 5 ml. of hydrochloric acid. Fatty acids are liberated, floating as an oily layer on the surface of the liquid. The water layer gives positive tests for *Sodium*, page 928.

B. Mix 25 grams of the sample with 50 grams of a 15 percent alcoholic potassium hydroxide solution in an Erlenmeyer flask, and reflux for 1 hour or until saponification is complete. Cool, add 150 ml. water, and mix. After complete solution of the soap, add 60 ml. of diluted sulfuric acid T.S., and heat the mixture, with frequent stirring, until the fatty acids separate cleanly as a transparent layer. Wash the fatty acids with boiling water until free from sulfate, collect them in a small beaker, and warm on a steam bath until the water has separated and the fatty acids are clear. Allow the acids to cool, pour off the water layer, then melt the acids, filter into a dry beaker and dry at 105° for 20 minutes. The solidification point of the fatty acids so obtained is not below 54° (see page 931).

SPECIFICATIONS

Acid value. Between 60 and 80.

Ester value. Between 150 and 190.

Total lactic acid. Between 31.0 percent and 34.0 percent.

Sodium content. Between 3.5 percent and 5.0 percent.

Limits of Impurities

Arsenic (as As). Not more than 3 parts per million (0.0003 percent).

Heavy metals (as Pb). Not more than 10 parts per million (0.001 percent).

TESTS

Acid value. Transfer about 1 gram, accurately weighed, to a 125-ml. Erlenmeyer flask, add 25 ml. of alcohol, previously neutralized to phenolphthalein T.S., and heat on a hot plate until the sample is dissolved. Cool, add 5 drops of phenolphthalein T.S., and titrate rapidly with 0.1 N sodium hydroxide to the first pink color that persists for at least 30 seconds. Calculate the acid value by the formula $56.1V \times N/W$, in which V is the volume, in ml., and N is the normality, respectively, of the sodium hydroxide solution, and W is the weight, in grams, of the sample taken. Retain the neutralized solution for the determination of *Ester value.*

Ester value. To the neutralized solution retained in the test for *Acid value* add 10.0 ml. of alcoholic potassium hydroxide solution prepared by dissolving 11.2 grams of potassium hydroxide in 250 ml. of alcohol and diluting with 25 ml. of water. Add 5 drops of phenolphthalein T.S., connect a suitable condenser, and reflux for 2 hours. Cool, add 5 additional drops of phenolphthalein T.S., and titrate the excess alkali with 0.1 N sulfuric acid. Perform a blank determination using 10.0 ml. of the alcoholic potassium hydroxide solution. Calculate the ester value by the formula $56.1(B - S)N/W$, in which $B - S$ represents the difference between the volumes of 0.1 N sulfuric acid required for the blank and sample, respectively, N is the normality of the sulfuric acid, and W is the weight, in grams, of the sample taken.

Total Lactic Acid

Standard curve. Dissolve 1.067 grams of lithium lactate in sufficient water to make 1000.0 ml. Transfer 10.0 ml. of this solution into a 100-ml. volumetric flask, dilute to volume with water, and mix. Transfer 1.0, 2.0. 4.0, 6.0, and 8.0, ml. of the diluted standard solution into separate 100-ml. volumetric flasks, dilute each flask to volume with water, and mix. These standards represent 1, 2, 4, 6, and 8 mcg. of lactic acid per ml., respectively. Transfer 1.0 ml. of each solution into separate test tubes, and continue as directed in the *Procedure*, beginning with "Add 1 drop of cupric sulfate T.S. . ." After color development and reading the absorbance values, construct a *Standard curve* by plotting absorbance versus mcg. of lactic acid.

Test preparation. Transfer about 200 mg. of the sample, accurately weighed, into a 125-ml. Erlenmeyer flask, add 10 ml. of 0.5 N alcoholic potassium hydroxide and 10 ml. of water, attach an air condenser, and reflux gently for 45 minutes. Wash the sides of the flask and the condenser with about 40 ml. of water, and heat on a steam bath until no odor of alcohol remains. Add 6 ml. of dilute sulfuric acid (1 in 2), heat until the fatty acids are melted, then cool to about 60°, and add 25 ml. of petroleum ether. Swirl the mixture gently, and transfer quantitatively to a separator. Collect the water layer in a 100-ml. volumetric flask, and wash the petroleum ether layer with two 20-ml. portions of water, adding the washings to the volumetric flask. Dilute to volume with water, and mix. Transfer 1.0 ml. of this solution into a second 100-ml. volumetric flask, dilute to volume with water, and mix.

Procedure. Transfer 1.0 ml. of the *Test preparation* into a test tube, and transfer 1.0 ml. of water to a second test tube to serve as the blank. Treat each tube as follows: Add 1 drop of cupric sulfate T.S., swirl gently, and then add rapidly from a buret 9.0 ml. of sulfuric acid. Loosely stopper the tube, and heat in a water bath at 90° for exactly 5 minutes. Cool immediately to below 20° in an ice bath for 5 minutes, add 3 drops of *p*-phenylphenol T.S., shake immediately, and heat in a water bath at 30° for 30 minutes, shaking the tube twice during this time to disperse the reagent. Heat the tube in a water bath at 90° for exactly 90 seconds, and then cool immediately to room temperature in an ice water bath. Determine the absorbance of the solution in a 1-cm. cell, at 570 mμ, with a suitable spectrophotometer, using the blank to set the instrument. Obtain the weight, in mcg., of lactic acid in the portion of the *Test preparation* taken for the *Procedure* by means of the *Standard curve.*

Sodium content

[*Note:* Ordinary glassware should not be used in this test because of possible contamination by sodium; use suitable plastic (e.g. polyethylene) vessels where necessary.]

Stock lanthanum solution. Transfer 5.86 grams of lanthanum oxide, La_2O_3, to a 100-ml. volumetric flask, wet with a few ml. of water, slowly add 25 ml. of hydrochloric acid, and swirl until the material is completely dissolved. Dilute to volume with water, and mix.

Stock sodium solution. Use a solution containing 1 mg. of Na in each ml. (1000 ppm Na). The solution may be obtained commercially or prepared as follows: transfer 1.271 grams of sodium chloride, previously dried at 105° for 2 hours and accurately weighed, to a 500-ml. volumetric flask, dilute to volume with water, and mix.

Standard preparations. Transfer 10.0 ml. of the *Stock lanthanum solution* to each of three 100-ml. volumetric flasks. Using a microliter syringe, transfer 0.20 ml. of the *Stock sodium solution* to the first flask, 0.40 ml. to the second flask, and 0.50 ml. to the third flask. Dilute each flask to volume with water, and mix. The flasks contain 2.0, 4.0, and 5.0 mcg. of Na per ml., respectively. Prepare these solutions fresh daily.

Sample preparation. Transfer about 250 mg. of the sample, accurately weighed, to a 30-ml. beaker, dissolve with heating in 10 ml. of alcohol, and quantitatively transfer the solution into a 25-ml. volumetric flask. Wash the beaker with two 5-ml. portions of alcohol, adding the washings to the flask, dilute to volume with alcohol, and mix. Transfer 2.5 ml. of the *Stock Lanthanum solution* to a second 25-ml. volumetric flask. Using a microliter syringe, transfer 0.25 ml. of the alcoholic solution of the sample to the second flask, dilute to volume with water, and mix.

Procedure. Concomitantly determine the absorbance of each *Standard preparation* and of the *Sample preparation* at 5890 Å, with a suitable atomic absorption spectrophotometer, following the operating parameters as recommended by the manufacturer of the instrument. Plot the absorbances of the *Standard preparations* vs. concentration of Na, in mcg. per ml., and from the curve so obtained determine the concentration, *C*, in mcg. per ml., of Na in the *Sample preparation.* Calculate the quantity, in mg., of Na in the sample taken by the formula 2.5*C*.

Arsenic. A *Sample Solution* prepared as directed for organic compounds meets the requirements of the *Arsenic Test*, page 865.

Heavy metals. Prepare and test a 2-gram sample as directed in *Method II* under the *Heavy Metals Test*, page 920, using 20 mcg. of lead ion (Pb) in the control (*Solution A*).

Packaging and storage. Store in tight containers in a cool, dry place.

Functional use in foods. Emulsifier; dough conditioner; stabilizer; whipping agent.

SODIUM STEARYL FUMARATE

NaOOCCH
‖
HCCOOC$_{18}$H$_{37}$

$C_{22}H_{39}NaO_4$ Mol. wt. 390.54

DESCRIPTION

A fine, white powder. It is slightly soluble in methanol but is practically insoluble in water.

SPECIFICATIONS

Assay. Not less than 99.0 percent and not more than the equivalent of 101.5 percent of $C_{22}H_{39}NaO_4$, calculated on the anhydrous basis.

Saponification value. Between 142.2 and 146.

Limits of Impurities

Arsenic (as As). Not more than 3 parts per million (0.0003 percent).

Heavy metals (as Pb). Not more than 20 parts per million (0.002 percent).

Lead. Not more than 10 parts per million (0.001 percent).

Sodium stearyl maleate. Not more than 0.25 percent.

Stearyl alcohol. Not more than 0.5 percent.
Water. Not more than 5 percent.

TESTS

Assay. Transfer about 250 mg., accurately weighed, into a 50-ml. Erlenmeyer flask, mix it with 1 ml. of chloroform, and add 20 ml. of glacial acetic acid to dissolve the sample. Add quinaldine red T.S., and titrate with 0.1 *N* perchloric acid in glacial acetic acid. Each ml. of 0.1 *N* perchloric acid is equivalent to 39.05 mg. of $C_{22}H_{39}NaO_4$.

Saponification value. Transfer about 450 mg. of sodium stearyl fumarate, accurately weighed, into a 300-ml. Erlenmeyer flask, and add 50.0 ml. of ethanolic potassium hydroxide solution, rinsing down the inside of the flask during the addition. [Prepare the ethanolic potassium hydroxide solutions as follows: dissolve about 5.5 grams of potassium hydroxide in absolute ethanol, heating if necessary to effect solution, and dilute to 1000 ml. with absolute ethanol. Prepare fresh daily, and filter if necessary to remove carbonate.] Reflux the mixture gently on a steam bath for at least 2 hours, swirling gently occasionally but avoid splashing the mixture up into the condenser. Rinse the condenser with 10 ml. of 70 percent alcohol, followed by three 10-ml. portions of water, collecting the rinsings in the flask. Cool, rinse the sides of the flask with two 10-ml. portions of 70 percent alcohol, add phenolphthalein T.S., and titrate with 0.1 *N* hydrochloric acid to the disappearance of any pink color. Perform a blank determination using the same amount of the ethanolic potassium hydroxide solution (see page 2). Calculate the saponification value by the formula $56.1 (B - S) \times N/W$, in which $B - S$ represents the difference between the volumes of 0.1 *N* hydrochloric acid required for the blank and the sample, respectively, *N* is the exact normality of the hydrochloric acid, and *W* is the weight, in grams, of the sample taken.

Arsenic. A *Sample Solution* prepared as directed for organic compounds meets the requirements of the *Arsenic Test*, page 865.

Heavy metals. Prepare and test a 1-gram sample as directed in *Method II* under the *Heavy Metals Test*, page 920, using 20 mcg. of lead ion (Pb) in the control (*Solution A*).

Lead. A *Sample Solution* prepared as directed for organic compounds meets the requirements of the *Lead Limit Test*, page 929, using 10 mcg. of lead ion (Pb) in the control.

Sodium stearyl maleate and stearyl alcohol

Apparatus. Assemble a suitable apparatus for ascending thin-layer chromatography (see page 886). Prepare a slurry of 24 grams of chromatographic silica gel G in 75 ml. of water, apply a uniformly thin layer to 23-cm. square glass plates, or other convenient size, and dry in the air at room temperature for 2 hours.

Sample Solution. Weigh accurately 200 mg. of the sample into a glass-stoppered 10-ml. volumetric flask, dilute to volume with a solution of 10 percent acetic acid in chloroform, and mix. The mixture may be heated carefully, if necessary, to dissolve the sample, and then cooled before diluting to volume with the solvent mixture.

Standard Solution A. Weigh accurately 10 mg. of sodium stearyl maleate into a 100-ml. volumetric flask, dilute to volume with 10 percent acetic acid in chloroform, and shake well.

Standard Solution B. Weigh accurately 20 mg. of stearyl alcohol into a 100-ml. volumetric flask, dilute to volume with 10 percent acetic acid in chloroform, and shake well.

Standard Solution C. Mix 25.0 ml. of *Standard Solution A* with 25.0 ml. of *Standard Solution B*, and shake well. This mixture represents 0.25 percent of sodium stearyl maleate and 0.5 percent of stearyl alcohol, based upon the weight (200 mg.) of the sample taken.

Procedure. Spot 10 μl. each of the *Sample Solution* and of *Standard Solution C* at the bottom of the plate. Allow the spots to dry, then place the plate in a suitable chromatographic chamber containing a mixture of 10 volumes of benzene, 10 volumes of hexane, and 1 volume of acetic acid, previously equilibrated, and develop by ascending chromatography for 30 minutes to effect one pass. Remove the plate from the tank, dry in the air for 10 minutes, and then heat in an oven at 90° for 2 minutes. After cooling to room temperature, replace the plate in the chamber for a second pass of 30 minutes. After the second pass, remove the plate from the chamber and dry in the air for 15 to 20 minutes. Spray evenly with a mixture consisting of 0.5 percent of potassium permanganate and 0.3 percent of sodium carbonate in water. Maleate and fumarate appear as yellow spots against a pink background. Spray with sulfuric acid and heat in an oven at 150° for the detection of stearyl alcohol.

Visually compare any spots from the sample against the R_f of the spots from the standards. The spots from the sample do not appear to be stronger than the respective spots from the standards.

Water. Determine by the *Karl Fischer Titrimetric Method*, page 977.

Packaging and storage. Store in well-closed containers.

Functional use in foods. Dough conditioner.

SODIUM SULFATE

Na_2SO_4 Mol. wt. 142.04

DESCRIPTION

Sodium sulfate is anhydrous or contains 10 molecules of water of crystallization. It occurs as colorless crystals or as a fine, white, crystalline powder. The decahydrate is efflorescent. Sodium sulfate is freely soluble in water and practically insoluble in alcohol. The pH of a 1 in 20 solution is between 8.0 and 10.0. A 1 in 20 solution gives positive tests for *Sodium*, page 928, and for *Sulfate*, page 928.

SPECIFICATIONS

Assay. Not less than 99.0 percent of Na_2SO_4 after drying.

Loss on drying. *Anhydrous form:* not more than 1 percent; *decahydrate:* between 51 percent and 57 percent.

Limits of Impurities

Arsenic (as As). Not more than 3 parts per million (0.0003 percent).

Heavy metals (as Pb). Not more than 10 parts per million (0.001 percent).

Selenium. Not more than 30 parts per million (0.003 percent).

TESTS

Assay. Dissolve about 500 mg., previously dried at 105° for 3 hours and accurately weighed, in 200 ml. of water, add 1 ml. of hydrochloric acid, and heat to boiling. Gradually add, in small portions at at time and while stirring constantly, an excess of hot barium chloride T.S. (about 10 ml.), and heat the mixture on a steam bath for 1 hour. Collect the precipitate on a filter, wash until free from chloride, dry, ignite, and weigh. The weight of the barium sulfate so obtained, multiplied by 0.6086, indicates its equivalent of Na_2SO_4.

Loss on drying, page 931. Dry at 105° for 3 hours.

Arsenic. A solution of 1 gram in 35 ml. of water meets the requirements of the *Arsenic Test*, page 865.

Heavy metals. A solution of 3 grams in 25 ml. of water meets the requirements of the *Heavy Metals Test*, page 920, using 20 mcg. of lead ion (Pb) and 1 gram of the sample in the control (*Solution A*).

Selenium. A solution of 2 grams in 40 ml. of dilute hydrochloric acid (1 in 2) meets the requirements of the *Selenium Limit Test*, page 953.

Packaging and storage. Store in well-closed container.

Functional use in foods. Agent in caramel production.

SODIUM SULFITE

Exsiccated Sodium Sulfite

Na_2SO_3 Mol. wt. 126.04

DESCRIPTION

A white, or tan to slightly pink, odorless or nearly odorless powder having a cooling, saline sulfurous taste. It undergoes oxidation in air. Its solutions are alkaline to litmus and to phenolphthalein. One gram dissolves in about 4 ml. of water. It is sparingly soluble in alcohol. A 1 in 20 solution gives positive tests for *Sodium*, page 928, and for *Sulfite*, page 928.

SPECIFICATIONS

Assay. Not less than 95.0 percent of Na_2SO_3.

Limits of Impurities

Arsenic (as As). Not more than 3 parts per million (0.0003 percent).

Heavy metals (as Pb). Not more than 10 parts per million (0.001 percent).

Selenium. Not more than 30 parts per million (0.003 percent).

TESTS

Assay. Weigh accurately about 250 mg., add it to exactly 50 ml. of 0.1 *N* iodine contained in a glass-stoppered flask, and stopper the flask. Allow to stand for 5 minutes, add 1 ml. of hydrochloric acid, and titrate the excess iodine with 0.1 *N* sodium thiosulfate, adding starch T.S. as the indicator. Each ml. of 0.1 *N* iodine is equivalent to 6.302 mg. of Na_2SO_3.

Arsenic. Dissolve 1 gram in 5 ml. of water, add 0.5 ml. of sulfuric acid, evaporate to about 1 ml. on a steam bath, and dilute to 35 ml. with water. This solution meets the requirements of the *Arsenic Test*, page 865.

Heavy metals. Dissolve 2 grams in 10 ml. of water, add 4 ml. of hydrochloric acid, and evaporate to dryness on a steam bath. To the residue add 5 ml. of hot water and 1 ml. of hydrochloric acid, and again evaporate to dryness. Dissolve the residue in water and dilute to 25 ml. This solution meets the requirements of the *Heavy Metals Test*, page 920, using 20 mcg. of lead ion (Pb) in the control (*Solution A*).

Selenium. Prepare and test a 2-gram sample as directed under the *Selenium Limit Test*, page 953.

Packaging and storage. Store in tight containers.

Functional use in foods. Preservative; antioxidant.

SODIUM TARTRATE

Disodium Tartrate; Disodium *d*-Tartrate

$C_4H_4Na_2O_6 . 2H_2O$ Mol. wt. 230.08

DESCRIPTION

Sodium tartrate is the disodium salt of L(+)-tartaric acid. It occurs as transparent, colorless, odorless crystals. One gram dissolves in 3 ml. of water. It is insoluble in alcohol. The pH of a 1 in 20 solution is between 7 and 9. Upon ignition, it emits the odor of burning sugar and leaves a residue which is alkaline to litmus and which effervesces with acids. It gives positive tests for *Sodium*, page 928, and for *Tartrate*, page 928.

SPECIFICATIONS

Assay. Not less than 99.0 percent of $C_4H_4Na_2O_6$ after drying.

Loss on drying. Between 14 percent and 17 percent.

Limits of Impurities

Arsenic (as As). Not more than 3 parts per million (0.0003 percent).

Heavy metals (as Pb). Not more than 10 parts per million (0.001 percent).

Oxalate. Passes test (limit about 0.1 percent).

TESTS

Assay. Weigh accurately about 450 mg., previously dried at 150° for 3 hours, and transfer it to a 250-ml. beaker. Add 100 ml. of glacial acetic acid, and stir, preferably with a magnetic stirrer, until the sample is dissolved. Titrate with 0.1 *N* perchloric acid in glacial acetic acid, determining the end-point potentiometrically. Each ml. of 0.1 *N* perchloric acid is equivalent to 9.703 mg. of $C_4H_4Na_2O_6$.

Loss on drying, page 931. Dry at 150° for 3 hours.

Arsenic. A *Sample Solution* prepared as directed for organic compounds meets the requirements of the *Arsenic Test,* page 865.

Heavy metals. A solution of 2 grams in 25 ml. of water meets the requirements of the *Heavy Metals Test,* page 920, using 20 mcg. of lead ion (Pb) in the control (*Solution A*).

Oxalate. Dissolve 1 gram in 10 ml. of water, and add 5 drops of diluted acetic acid T.S. and 2 ml. of calcium chloride T.S. No turbidity develops within one hour.

Packaging and storage. Store in tight containers.

Functional use in foods. Sequestrant.

SODIUM THIOSULFATE

Sodium Hyposulfite

$Na_2S_2O_3.5H_2O$ Mol. wt. 248.17

DESCRIPTION

Large, colorless crystals or a coarse, crystalline powder. It is deliquescent in moist air and effloresces in dry air at a temperature above

33°. Its solutions are neutral or faintly alkaline to litmus. One gram dissolves in 0.5 ml. of water. It is insoluble in alcohol.

IDENTIFICATION

A. To a 1 in 10 solution add a few drops of iodine T.S. The color is discharged.

B. A 1 in 20 solution gives positive tests for *Sodium*, page 928, and for *Thiosulfate*, page 928.

SPECIFICATIONS

Assay. Not less than 99.0 percent of $Na_2S_2O_3$ after drying.

Water. Between 32 and 37 percent.

Limits of Impurities

Arsenic (as As). Not more than 3 parts per million (0.0003 percent).

Heavy metals (as Pb). Not more than 20 parts per million (0.002 percent).

Lead. Not more than 10 parts per million (0.001 percent).

Selenium. Not more than 30 parts per million (0.003 percent).

TESTS

Assay. Weigh accurately about 500 mg. of the dried sample obtained in the test for *Water*, dissolve it in 30 ml. of water, and titrate with 0.1 *N* iodine, using starch T.S. as the indicator. Each ml. of 0.1 *N* iodine is equivalent to 15.81 mg. of $Na_2S_2O_3$.

Water. Dry about 1 gram, accurately weighed, in a vacuum at 40° to 45° for 16 hours, cool, and weigh.

Arsenic. Dissolve 1.0 gram in about 5 ml. of water, add 9 ml. of nitric acid, and cautiously evaporate to dryness on a steam bath. Take up the residue in a few ml. of water, filter, wash thoroughly, and evaporate the combined filtrate and washings to dryness. Cool, and dissolve the residue in water to make 35 ml. This solution meets the requirements of the *Arsenic Test*, page 865.

Sample Solution for the Determination of Heavy Metals and Lead. Dissolve 5.0 grams in 40 ml. of water, slowly add 25 ml. of diluted hydrochloric acid T.S., and evaporate to dryness on a steam bath. Add 30 ml. of water to the residue, boil gently for 2 minutes, and filter. Heat the filtrate to boiling, add sufficient bromine T.S. to produce a clear solution, then add a slight excess of bromine. Boil to expel the excess bromine, cool, and dilute to 50.0 ml. with water.

Heavy metals. Dilute 10.0 ml. of the *Sample Solution* (1-gram sample) to 25 ml. with water, and proceed as directed under the *Heavy Metals Test*, page 920, using 20 mcg. of lead ion (Pb) in the control (*Solution A*).

Lead. A 10.0-ml. portion of the *Sample Solution* (1-gram sample) meets the requirements of the *Lead Limit Test*, page 929, using 10 mcg. of lead ion (Pb) in the control.

Selenium. Prepare and test a 2-gram sample as directed under the *Selenium Limit Test*, page 953.

Packaging and storage. Store in tight containers.

Functional use in foods. Sequestrant; antioxidant.

SODIUM TRIPOLYPHOSPHATE

Pentasodium Triphosphate; Triphosphate; Sodium Triphosphate

$Na_5P_3O_{10}$ Mol. wt. 367.86

DESCRIPTION

Sodium tripolyphosphate is anhydrous or contains 6 molecules of water of hydration. It occurs as white, slightly hygroscopic granules, or as a powder. It is freely soluble in water. The pH of a 1 in 100 solution is about 9.5.

IDENTIFICATION

A. A 1 in 20 solution gives positive tests for *Sodium*, page 928.

B. To 1 ml. of a 1 in 100 solution add a few drops of silver nitrate T.S. A white precipitate is formed which is soluble in diluted nitric acid T.S.

SPECIFICATIONS

Assay. Not less than 85.0 percent of $Na_5P_3O_{10}$.

Loss on drying. Not more than 0.7 percent.

Limits of Impurities

Arsenic (as As). Not more than 3 parts per million (0.0003 percent).

Fluoride. Not more than 50 parts per million (0.005 percent).

Heavy metals (as Pb). Not more than 10 parts per million (0.001 percent).

Insoluble substances. Not more than 0.1 percent.

Lead. Not more than 5 parts per million (0.0005 percent).

TESTS

Assay

Reagents and Solutions

Potassium Acetate Buffer (*pH 5.0*). Dissolve 78.5 grams of potassium acetate in 1000 ml. of water, and adjust the pH of the solution to 5.0

with acetic acid. Add a few mg. of mercuric iodide to inhibit mold growth.

0.3 M Potassium Chloride Solution. Dissolve 22.35 grams of potassium chloride in water, add 5 ml. of *Potassium Acetate Buffer*, dilute with water to 1000 ml., and mix. Add a few mg. of mercuric iodide.

0.6 M Potassium Chloride Solution. Dissolve 44.7 grams of potassium chloride in water, add 5 ml. of *Potassium Acetate Buffer*, dilute with water to 1000 ml., and mix. Add a few mg. of mercuric iodide.

1 M Potassium Chloride Solution. Dissolve 74.5 grams of potassium chloride in water, add 5 ml. of *Potassium Acetate Buffer*, dilute to 1000 ml. with water, and mix. Add a few mg. of mercuric iodide.

Chromatographic Column. Use a standard chromatographic column, 20- to 40-cm. in length, 20- to 28-mm. in inside diameter, with a sealed-in, coarse porosity fritted disk. If a stopcock is not provided, attach a stopcock having a 3- to 4-mm. diameter bore to the outlet of the column with a short length of flexible vinyl tubing.

Procedure. Close the column stopcock, fill the space between the fritted disk and the stopcock with water, and connect a vacuum line to the stopcock. Prepare a 1:1 water slurry of Dowex 1 $\times$ 8, chloride form, 100–200 or 200–400 mesh, or a comparable grade of styrene-divinylbenzene ion exchange resin, and decant off any very fine particles and any foam. Do this two or three times or until no more finely suspended material or foaming is observed. Fill the column with the slurry, and open the stopcock to allow the vacuum to pack the resin bed until the water level is slightly above the top of the resin, then immediately close the stopcock. Do not allow the liquid level to fall below the resin level at any time. Repeat this procedure until the packed resin column is 15 cm. (about 6 inches) above the fritted disk. Place one circle of tightly fitting glass fiber filter paper on top of the resin bed, then place a perforated polyethylene disk on top of the paper. Alternatively, a loosely packed plug of glass wool may be placed on top of the bed. Close the top of the column with a rubber stopper in which a 7.6 cm. length of capillary tubing (1.5-mm. i.d., 7 mm. o.d.) has been iserted through the center, so that about 12 mm. of the tubing extends through the bottom of the stopper. Connect the top of the capillary tubing to the stem of a 500-ml. separator with flexible vinyl tubing, and clamp the separator to a ring stand above the column. Wash the column by adding 100 ml. of water to the separator with all stopcocks closed. First open the separator stopcock, then open the column stopcock. The rate of flow should be about 5 ml. per minute. When the separator is empty, close the stopcock on the column, then close the separator stopcock.

Transfer about 500 mg. of the sample, accurately weighed, into a 250-ml. volumetric flask, dissolve and dilute to volume with water, and mix. Transfer 10.0 ml. of this solution into the separator, open both stopcocks, and allow the solution to drain into the column, rinsing the separator with 20 ml. of water. Discard the eluate.

Add 370 ml. of *0.3 M Potassium Chloride Solution* to the separator, and allow this solution to pass through the column, discarding the eluate. Add 250 ml. of *0.6 M Potassium Chloride Solution* to the column, allow the solution to pass through the column, and receive the eluate in a 400-ml. beaker. (To ensure a clean column for the next run, pass 100 ml. of *1 M Potassium Chloride Solution* through the column, and then follow with 100 ml. of water. Discard all washings.) To the beaker add 15 ml. of nitric acid, mix, and boil for 15 to 20 minutes. Add methyl orange T.S., and neutralize the solution with stronger ammonia T.S. Add 1 gram of ammonium nitrate crystals, stir to dissolve, and cool. Add 15 ml. of ammonium molybdate T.S., with stirring, and stir vigorously for 3 minutes, or allow to stand with occasional stirring for 10 to 15 minutes. Filter the contents of the beaker with suction through a 6–7 mm. paper pulp filter pad supported in a 25 mm. porcelain disk. The filter pad should be covered with a suspension of infusorial earth. After the contents of the beaker have been transferred to the filter, wash the beaker with five 10-ml. portions of a 1 in 100 solution of sodium or potassium nitrate, passing the washings through the filter, then wash the filter with five 5-ml. portions of the wash solution. Return the filter pad and the precipitate to the beaker, wash the funnel thoroughly with water into the beaker, and dilute to about 150 ml. Add 0.1 *N* sodium hydroxide from a buret until the yellow precipitate is dissolved, then add 5 to 8 ml. in excess. Add phenolphthalein T.S., and titrate the excess alkali with 0.1 *N* nitric acid. Finally, titrate with 0.1 *N* sodium hydroxide to the first appearance of the pink color. The difference between the total volume of 0.1 *N* sodium hydroxide added and the volume of nitric acid required represents the volume, *V*, in ml., of 0.1 *N* sodium hydroxide consumed by the phosphomolybdate complex. Calculate the quantity, in mg., of $Na_5P_3O_{10}$ in the sample taken by the formula $0.533 \times 25V$.

Loss on drying, page 931. Dry at 105° for 1 hour.

Arsenic. A solution of 1 gram in 35 ml. of water meets the requirements of the *Arsenic Test,* page 865.

Fluoride. Weigh accurately 1.0 gram, and proceed as directed in the *Fluoride Limit Test,* page 917.

Heavy metals. A solution of 2 grams in 25 ml. of water meets the requirements of the *Heavy Metals Test,* page 920, using 20 mcg. of lead ion (Pb) in the control (*Solution A*).

Insoluble substances. Dissolve 10 grams in 100 ml. of hot water, and filter through a tared filtering crucible. Wash the insoluble residue with hot water, dry at 105° for 2 hours, cool, and weigh.

Lead. A solution of 1 gram in 20 ml. of water meets the requirements of the *Lead Limit Test,* page 929, using 5 mcg. of lead ion (Pb) in the control.

Packaging and storage. Store in tight containers.
Functional use in foods. Texturizer.

SORBIC ACID

2,4-Hexadienoic Acid

$CH_3CH{=}CHCH{=}CHCOOH$

$C_6H_8O_2$ Mol. wt. 112.13

DESCRIPTION

A white, free-flowing powder with a characteristic odor. It is slightly soluble in water. One gram dissolves in about 10 ml. of alcohol and in about 20 ml. of ether.

IDENTIFICATION

A. To 2 ml. of a 1 in 10 solution of the sample in alcohol add a few drops of bromine T.S. The color is discharged.

B. A 1 in 400,000 solution in isopropanol exhibits an absorbance maximum at 254 ± 2 mμ.

SPECIFICATIONS

Assay. Not less than 99.0 percent and not more than the equivalent of 101.0 percent of $C_6H_8O_2$, calculated on the anhydrous basis.

Melting range. Between 132° and 135°.

Limits of Impurities

Arsenic (as As). Not more than 3 parts per million (0.0003 percent).
Heavy metals (as Pb). Not more than 10 parts per million (0.001 percent).
Residue on ignition. Not more than 0.2 percent.
Water. Not more than 0.5 percent.

TESTS

Assay. Dissolve about 250 mg., accurately weighed, in 50 ml. of anhydrous methanol which previously has been neutralized with 0.1 *N* sodium hydroxide, add phenolphthalein T.S., and titrate with 0.1 *N* sodium hydroxide to the first pink color which persists for at least 30 seconds. Each ml. of 0.1 *N* sodium hydroxide is equivalent to 11.21 mg. of $C_6H_8O_2$.

Melting range, page 931. Determine as directed for *Class Ia,* but heat at a rate of rise of 1° per minute until the melting is complete.

Arsenic. A *Sample Solution* prepared as directed for organic compounds meets the requirements of the *Arsenic Test,* page 865.

Heavy metals. Prepare and test a 2-gram sample as directed in

Method II under the *Heavy Metals Test*, page 920, using 20 mcg. of lead ion (Pb) in the control (*Solution A*).

Residue on ignition, page 945. Ignite 2 grams as directed in the general method.

Water. Determine by the *Karl Fischer Titrimetric Method*, page 977.

Packaging and storage. Store in tight containers protected from light, preferably at a temperature not exceeding 38°.

Functional use in foods. Preservative.

SORBITAN MONOSTEARATE

DESCRIPTION

A mixture of partial stearic and palmitic acid esters of sorbitol and its mono- and dianhydrides. It is manufactured by reacting edible commercial stearic acid (usually containing associated fatty acids, chiefly palmitic) with sorbitol. It is a light cream to tan colored, hard, waxy solid with a bland odor and taste. It is soluble at temperatures above its melting point in toluene, dioxane, carbon tetrachloride, ether, ethanol, methanol, and aniline. It is insoluble in cold water, and in mineral spirits and acetone, but is dispersible in warm water and soluble, with haze, above 50° in mineral oil and in ethyl acetate. It conforms to the regulations of the federal Food and Drug Administration pertaining to specifications for fats or fatty acids derived from edible sources.

IDENTIFICATION

A. The fatty acid residue obtained in the *Assay* has an *Acid Value* between 200 and 212 (*Method I*, page 902) and a *Solidification Point* not below 53°.

B. *Apparatus.* Assemble a suitable apparatus for ascending thin-layer chromatography. Prepare a slurry of chromatographic silica gel and water (1 gram to 2 ml.), apply a uniformly thin layer to glass plates of convenient size, dry in the air for 10 minutes, and activate by drying at 100° for 1 hour. Store the cooled plates in a desiccator over anhydrous calcium sulfate until ready for use.

Sample Solution. Transfer 500 mg. of the polyols obtained in the *Assay* into a 2-ml. volumetric flask, dissolve, dilute to volume with water, and mix.

Standard Solution. Transfer 50 mg. each of sorbitol, 1,4-sorbitan, and isosorbide into a 1-ml. volumetric flask, dissolve, and dilute to volume with water, and mix.

Procedure. By means of the template provided with the apparatus, mark the sample spot area 1.5 cm. from the bottom of the plate, then mark the solvent front line 15 cm. above the sample spot area, using a sharp needle to cut through the silica gel. In the same manner, mark a spot for the standard separated from the sample spot by about 1.5 cm. Using micropipets, spot 2 μl. of the *Sample Solution* and 1 μl. of the *Standard Solution* at the appropriate marks. Allow the spots to dry, then place the plate in a suitable chromatographic chamber containing a mixture of 100 volumes of acetone and 2 volumes of acetic acid as the developing solvent, and develop by ascending chromatography until the solvent front reaches the 15-cm. line (about 45 minutes). Remove the plate from the chamber, dry thoroughly in air, and spray evenly with sulfuric acid solution (1 in 2) until the surface is uniformly wet. (*Caution:* Do not overspray.) Immediately place the sprayed plate on a hot plate maintained at 200° in a hood. Char until white fumes of sulfur trioxide cease, and cool on an asbestos mat to room temperature. The spots from the sample are located at the same R_f values as are those of the polyols from the standard. The approximate R_f values are: sorbitol, 0.07; 1,4-sorbitan, 0.40; and isosorbide, 0.77.

SPECIFICATIONS

Assay. Not less than 29.5 grams and not more than 33.5 grams of polyols (as sorbitol and its mono- and dianhydrides) per 100 grams of sample, and not less than 71 grams and not more than 75 grams of fatty acids per 100 grams of sample.

Acid value. Between 5 and 10.

Hydroxyl value. Between 235 and 260.

Saponification value. Between 147 and 157.

Limits of Impurities

Arsenic (as As). Not more than 3 parts per million (0.0003 percent).

Heavy metals (as Pb). Not more than 10 parts per million (0.001 percent).

Water. Not more than 1.5 percent.

TESTS

Assay. Transfer about 25 grams of the sample, accurately weighed, into a 500-ml. round-bottom boiling flask, add 250 ml. of alcohol and 7.5 grams of potassium hydroxide, and mix. Connect a suitable condenser to the flask, reflux the mixture for 1 to 2 hours, then transfer to an 800-ml. beaker, rinsing the flask with about 100 ml. of water and adding it to the beaker. Heat on a steam bath to evaporate the alcohol, adding water occasionally to replace the alcohol, and evaporate until the odor of alcohol can no longer be detected. Adjust the final volume to about 250 ml. with hot water. Neutralize the soap solution with dilute sulfuric acid (1 in 2), add 10 percent in excess, and heat, while stirring, until the fatty acid layer separates. Transfer the fatty acids to a 500-ml. separator, wash with three or four 20-ml. portions of hot water to remove polyols, and combine the washings with the original aqueous polyol layer from the saponification. Extract the combined aqueous layer with three 20-ml. portions of petroleum ether, add the extracts to the fatty acid layer, evaporate to dryness in a tared dish, cool and weigh.

Neutralize the polyol solution with a 1 in 10 solution of potassium hydroxide to pH 7 using a suitable pH meter. Evaporate this solution to a moist residue, and separate the polyols from the salts by several extractions with hot alcohol. Evaporate the alcohol extracts on a steam bath to dryness in a tared dish, cool, and weigh. Avoid excessive drying and heating.

Acid value. Determine as directed under *Method II* in the general procedure, page 902.

Hydroxyl value. Determine as directed under *Method II* in the general procedure, page 904.

Saponification value. Determine as directed in the general procedure, page 914, using about 4 grams, accurately weighed.

Arsenic. A *Sample Solution* prepared as directed for organic compounds meets the requirements of the *Arsenic Test*, page 865.

Heavy metals. Prepare and test a 2-gram sample as directed in *Method II* under the *Heavy Metals Test*, page 920, using 20 mcg. of lead ion (Pb) in the control (*Solution A*).

Water. Determine by the *Karl Fischer Titrimetric Method*, page 977.

Packaging and storage. Store in well-closed containers.

Functional use in foods. Emulsifier; stabilizer; defoaming agent.

SORBITOL

D-Sorbitol; D-Glucitol; D-Sorbite; D-Sorbol; 1,2,3,4,5,6-Hexanehexol

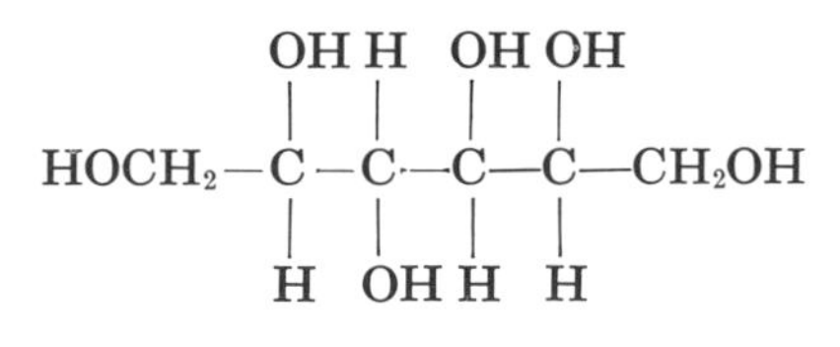

$C_6H_{14}O_6$ Mol. wt. 182.17

DESCRIPTION

White, hygroscopic powder, flakes, or granules having a sweet taste.

Its density is about 1.49. One gram dissolves in about 0.45 ml. of water. It is slightly soluble in alcohol, in methanol, and in acetic acid. Sorbitol can exist in any of several crystalline forms with melting points ranging from 89° to 101°.

IDENTIFICATION

Dissolve about 5 grams in 6 ml. of water, add 7 ml. of methanol, 1 ml. of benzaldehyde, and 1 ml. of hydrochloric acid, and shake in a mechanical shaker until crystals appear. Filter with the aid of suction, dissolve the crystals in 20 ml. of boiling water containing 1 gram of sodium bicarbonate, filter while hot, cool the filtrate, filter with suction, wash with 5 ml. of methanol-water mixture (1 in 2), and dry in air. The sorbitol monobenzylidene derivative so obtained melts between 174° and 179° (see page 931).

SPECIFICATIONS

Assay. Not less than 91.0 percent of sorbitol ($C_6H_{14}O_6$). Small amounts of mannitol and other polyhydric alcohols may be present.

Limits of Impurities

Arsenic (as As). Not more than 3 parts per million (0.0003 percent).
Chloride. Not more than 50 parts per million (0.005 percent).
Heavy metals (as Pb). Not more than 10 parts per million (0.001 percent).
Loss on drying. Not more than 1 percent.
Reducing sugars. Not more than 0.3 percent.
Residue on ignition. Not more than 0.1 percent.
Sulfate. Not more than 100 parts per million (0.01 percent).
Total sugars. Not more than 1 percent.

TESTS

Assay. Proceed as directed in the *Assay* under *Sorbitol Solution*, page 788.

Arsenic. A *Sample Solution* prepared as directed for organic compounds meets the requirements of the *Arsenic Test*, page 865.

Chloride, page 879. Any turbidity produced by a 400-mg. sample does not exceed that shown in a control containing 20 mcg. of chloride ion (Cl).

Heavy metals. A solution of 2 grams in 25 ml. of water meets the requirements of the *Heavy Metals Test*, page 920, using 20 mcg. of lead ion (Pb) in the control (*Solution A*).

Loss on drying, page 931. Dry at 80° and at a pressure not exceeding 5 mm. of mercury for 6 hours.

Reducing sugars. Dissolve 7.0 grams in 35 ml. of water in a 400-ml. beaker, and mix. Add 50 ml. of alkaline cupric tartrate T.S., cover the beaker with glass, heat the mixture at such a rate that it comes to a boil in about 4 minutes, and boil for exactly 2 minutes.

Collect the precipitated cuprous oxide in a tared Gooch crucible previously washed with hot water, alcohol, and ether, and dried at 100° for 30 minutes. Thoroughly wash the collected cuprous oxide on the filter with hot water, then with 10 ml. of alcohol, and finally with 10 ml. of ether, and dry at 100° for 30 minutes. The weight of the cuprous oxide does not exceed 50 mg.

Residue on ignition, page 945. Ignite 2 grams as directed in the general method.

Sulfate, page 879. Any turbidity produced by a 2-gram sample does not exceed that shown in a control containing 200 mcg. of sulfate (SO_4).

Total sugars. Transfer 2.1 grams into a 250-ml. flask fitted with a ground-glass joint, add 40 ml. of approximately 0.1 *N* hydrochloric acid, attach a reflux condenser, and reflux for 4 hours. Transfer the solution to a 400-ml. beaker, rinsing the flask with about 10 ml. of water, neutralize with 6 *N* sodium hydroxide, and continue as directed under *Reducing sugars*, beginning with "Add 50 ml. of alkaline cupric tartrate T.S. . . ." The weight of the cuprous oxide does not exceed 50 mg.

Packaging and storage. Store in tight containers.

Functional use in foods. Humectant; texturizing agent; sequestrant.

SORBITOL SOLUTION

DESCRIPTION

A water solution of sorbitol ($C_6H_{14}O_6$) containing a small amount of mannitol and other isomeric polyhydric alcohols. It is a clear, colorless, syrupy liquid, having a sweet taste. It is neutral to litmus. It is miscible with water, with glycerin, and with propylene glycol. It is soluble in alcohol and is practically insoluble in other common organic solvents.

IDENTIFICATION

To 6 ml. add 7 ml. of methanol, 1 ml. of benzaldehyde, and 1 ml. of hydrochloric acid, mix, and shake in a mechanical shaker until crystals appear. Filter with the aid of suction, dissolve the crystals in 20 ml. of boiling water containing 1 gram of sodium bicarbonate, filter while hot, cool the filtrate, filter with suction, wash with 5 ml. of methanol-water mixture (1 in 2), and dry in air. The sorbitol monobenzylidene derivative so obtained melts between 174° and 179° (see page 931).

SPECIFICATIONS

Assay. Not less than 64.0 percent of sorbitol ($C_6H_{14}O_6$).

Specific gravity. Not less than 1.285.
Water. Between 29 and 31 percent.

Limits of Impurities

Arsenic (as As). Not more than 3 parts per million (0.0003 percent)
Chloride. Not more than 50 parts per million (0.005 percent).
Heavy metals (as Pb). Not more than 10 parts per million (0.001 percent).
Reducing sugars. Not more than 0.21 percent.
Residue on ignition. Not more than 0.1 percent.
Sulfate. Not more than 100 parts per million (0.01 percent).
Total sugars. Not more than 0.7 percent.

TESTS

Assay

Adsorbent. Mix intimately, as by mechanical rolling or tumbling for 12 hours, 9 parts, by weight, of 200-mesh Florex AARVM and 2 parts, by weight, of Celite 545. Use only that portion which passes a 100-mesh sieve.

Developing solution. Mix 15 parts, by volume, of water with 85 parts, by volume, of reagent grade isopropyl alcohol.

Chromatographic column. Insert a pledget of cotton on the removable porous plate of a slightly tapered, 38 $\times$ 230-mm. chromatographic tube, the constricted end of which is fitted to a 500-ml. filtering flask. Apply a vacuum of between 25 and 50 mm. of mercury and maintain the pressure within this range until the chromatogram is developed. Add the *Adsorbent* in a steady stream through a powder funnel to the chromatographic tube until it is completely filled. Tap the tube around its circumference with a wooden rod, from bottom to top, to insure uniform packing. Adjust the height of the column, if necessary, to within about 15 mm. from the top of the tube, and level the top of the column with a rubber stopper.

> *Note:* Failure to pack the adsorbent uniformly in the chromatographic tube will yield unsatisfactory chromatograms. It is recommended that before attempting to assay a test sample the analyst determine the adequacy of his adsorbent-packing and procedure techniques by repeated assays of a fabricated sample consisting of sorbitol solution to which about 7 percent of mannitol has been added.

Sample Solution. Transfer an accurately weighed sample, equivalent to about 900 mg. of anhydrous sorbitol, into a 100-ml. volumetric flask, and dilute to volume with the *Developing Solution.*

Potassium Permanganate Solution. Dissolve 1 gram of potassium permanganate and 10 grams of sodium hydroxide in sufficient water to make 100 ml.

Sodium Periodate Solution. Dissolve 4.5 grams of sodium periodate

($NaIO_4$) in 500 ml. of water, add 2 ml. of sulfuric acid, and dilute to 1000 ml.

Procedure. Add sufficient *Developing Solution* to the chromatographic tube to wet the entire adsorbent column and to leave a 5-mm. layer above it. Transfer 10.0 ml. of the *Sample Solution* to the column and, when it has practically disappeared below the surface, add two successive 10-ml. portions of the *Developing Solution*, allowing each portion to just enter the column. Attach a separator, containing 320 ml. of the *Developing Solution*, to the tube with a rubber stopper, and adjust the flow to maintain a solvent layer about 5 mm. above the column.

After all of the *Developing Solution* has passed through the column, maintain the vacuum (25–50 mm. of mercury) until the column shrinks from the wall of the tube, but does not become dry. Disconnect the filtering flask from the aspirator to permit a flow of air to rise between the column and the tube wall.

Extrude the chromatogram on a strip of aluminum foil by placing a wad of cotton on top of the column, inverting the tube, and then tapping it on a solid surface several times to loosen the column. Complete the extrusion with the aid of a cork stopper attached to the end of a rod which can be inserted through the constricted end of the tube. Streak the entire length of the chromatogram with the *Potassium Permanganate Solution* applied with a small glass-fiber brush or fine-nozzle wash bottle at three places equidistant around its circumference.

The position of the polyol zones on the chromatogram is indicated by a color change in the potassium permanganate streaks from green to brown, with the leading edge of the sorbitol zone located between 12 and 15 cm. from the top of the column. Mannitol, if present, should appear just below the leading edge of the sorbitol zone and should show a distinct 1.5- to 2-cm. separation. Nonreducing sugars, if present in sufficient concentration, will appear above the sorbitol zone. Make a mark on each streak with a spatula or knife about 6 mm. below the leading edge of the sorbitol zone, then cut the chromatogram using the marks as a guide, and discard the lower portion.

If a zone is visible above the one containing the sorbitol, cut the chromatogram just below the leading edge of the slower moving zone and discard this portion. In the absence of a third zone, cut and discard the upper 3-cm. portion of the chromatogram.

Cut the sorbitol zone into small pieces and allow it to dry overnight at room temperature protected from drafts. Powder the dried sorbitol zone with a spatula, pack it in the chromatographic tube as directed for the preparation of the chromatographic column, and elute the sorbitol with 150 ml. of warm water. Transfer the eluate to a 250-ml. volumetric flask and dilute to volume with water.

Transfer 50.0 ml. of the diluted eluate into a 500-ml. iodine flask, add 50.0 ml. of the *Sodium Periodate Solution*, heat the mixture for 20 minutes in a water bath maintained at a temperature between 80°

and 85°, and then cool to room temperature in an ice bath. Add to the solution 2.5 grams of sodium bicarbonate and 10 ml. of a 1 in 5 solution of potassium iodide. Titrate immediately with 0.05 *N* sodium arsenite, using starch T.S. as the indicator. Perform a blank determination using water instead of the sorbitol eluate. Each ml. of 0.05 *N* sodium arsenite is equivalent to 0.9109 mg. of sorbitol ($C_6H_{14}O_6$).

Specific gravity. Determine by any reliable method (see page 5).

Water. Determine by the *Karl Fischer Titrimetric Method*, page 977.

Arsenic. A *Sample Solution* prepared as directed for organic compounds meets the requirements of the *Arsenic Test*, page 865.

Chloride, page 879. Any turbidity produced by a 400-mg. sample does not exceed that shown in a control containing 20 mcg. of chloride ion (Cl).

Heavy metals. A mixture of 2 grams with 25 ml. of water meets the requirements of the *Heavy Metals Test*, page 920, using 20 mcg. of lead ion (Pb) in the control (*Solution A*).

Reducing sugars. Transfer 10.0 grams, accurately weighed, to a 400-ml. beaker, with the aid of 35 ml. of water, and mix. Add 50 ml. of alkaline cupric tartrate T.S., cover the beaker with glass, heat the mixture at such a rate that it comes to a boil in approximately 4 minutes, and boil for exactly 2 minutes. Collect the precipitated cuprous oxide in a tared Gooch crucible previously washed successively with hot water, alcohol, and ether, and dried at 100° for 30 minutes. Thoroughly wash the collected cuprous oxide on the filter with hot water, then with 10 ml. of alcohol, and finally with 10 ml. of ether, and dry at 100° for 1 hour. The weight of the cuprous oxide does not exceed 50 mg.

Residue on ignition, page 945. Ignite 2 grams as directed under *Method II* (for liquids).

Sulfate, page 879. Any turbidity produced by a 2-gram sample does not exceed that shown in a control containing 200 mcg. of sulfate (SO_4).

Total sugars. Transfer 3.0 grams into a 250-ml. flask fitted with a ground-glass joint, add 40 ml. of approximately 0.1 *N* hydrochloric acid, attach a reflux condenser, and reflux for 4 hours. Transfer the solution to a 400-ml. beaker, rinsing the flask with about 10 ml. of water, neutralize with 6 *N* sodium hydroxide, and continue as directed under *Reducing sugars*, beginning with "Add 50 ml. of alkaline cupric tartrate T.S. . . ." The weight of the cuprous oxide does not exceed 50 mg.

Packaging and storage. Store in well-closed containers.

Functional use in foods. Humectant; texturizing agent; sequestrant.

SPEARMINT OIL

DESCRIPTION

The volatile oil obtained by steam distillation from the fresh overground parts of the flowering plant *Mentha spicata* L. (Common Spearmint), or of *Mentha cardiaca* Gerard ex Baker (Scotch Spearmint) (Fam. *Labiatae*). It is a colorless, yellow or greenish yellow liquid, having the characteristic odor and taste of spearmint.

SPECIFICATIONS

Assay. Not less than 55 percent, by volume, of ketones.

Angular rotation. Between −48° and −59°.

Reaction. Passes test.

Refractive index. Between 1.484 and 1.491 at 20°.

Solubility in alcohol. Passes test.

Specific gravity. Between 0.917 and 0.934.

Limits of Impurities

Arsenic (as As). Not more than 3 parts per million (0.0003 percent).

Heavy metals (as Pb). Not more than 40 parts per million (0.004 percent).

Lead. Not more than 10 parts per million (0.001 percent).

TESTS

Assay. Proceed as directed under *Aldehydes and Ketones-Neutral Sulfite Method,* page 895.

Angular rotation. Determine in a 100-mm. tube as directed under *Optical Rotation,* page 939.

Reaction. A recently prepared solution of the sample in 80 percent alcohol is neutral or only slightly acid to moistened litmus paper.

Refractive index, page 945. Determine with an Abbé or other refractometer of equal or greater accuracy.

Solubility in alcohol. Proceed as directed in the general method, page 899. One ml. dissolves in 1 ml. of 80 percent alcohol. On further dilution the solution may become turbid.

Specific gravity. Determine by any reliable method (see page 5).

Arsenic. A *Sample Solution* prepared as directed for organic compounds meets the requirements of the *Arsenic Test,* page 865.

Heavy metals. Prepare and test a 500-mg. sample as directed in *Method II* under the *Heavy Metals Test,* page 920, using 20 mcg. of lead ion (Pb) in the control (*Solution A*).

Lead. A *Sample Solution* prepared as directed for organic compounds meets the requirements of the *Lead Limit Test,* page 929, using 10 mcg. of lead ion (Pb) in the control.

Packaging and storage. Store in full, tight containers in a cool place protected from light.

Functional use in foods. Flavoring agent.

SPIKE LAVENDER OIL

DESCRIPTION

The volatile oil obtained by steam distillation from the flowers of *Lavandula latifolia*, Vill. (*Lavandula spica*, D.C.) (Fam. *Labiatae*). It is a pale yellow to yellow liquid, having a camphoraceous, lavender-like odor. It is soluble in most fixed oils and in propylene glycol. It is slightly soluble in glycerin and in mineral oil.

SPECIFICATIONS

Assay. Not less than 40.0 percent and not more than 50.0 percent of total alcohols, calculated as linalool ($C_{10}H_{18}O$).

Esters. Not less than 1.5 percent and not more than 3.0 percent of esters, calculated as linalyl acetate ($C_{12}H_{20}O_2$).

Angular rotation. Between $-5°$ and $+5°$.

Refractive index. Between 1.463 and 1.468 at 20°.

Solubility in alcohol. Passes test.

Specific gravity. Between 0.893 and 0.909.

TESTS

Assay. Proceed as directed under *Linalool Determination*, page 897, using about 1.5 grams of the acetylated oil, accurately weighed, for the saponification.

Esters. Weigh accurately about 10 grams, and proceed as directed under *Ester Determination*, page 896, using 98.15 as the equivalence factor (e) in the calculation.

Angular rotation. Determine in a 100-mm. tube as directed under *Optical Rotation*, page 939.

Refractive index, page 945. Determine with an Abbé or other refractometer of equal or greater accuracy.

Solubility in alcohol. Proceed as directed in the general method, page 899. One ml. dissolves in 3 ml. of 70 percent alcohol. The solution frequently becomes hazy upon further dilution.

Specific gravity. Determine by any reliable method (see page 5).

Packaging and storage. Store in full, tight, preferably glass, aluminum, tin-lined, or other suitably lined containers in a cool place protected from light.

Functional use in foods. Flavoring agent.

STANNOUS CHLORIDE

$SnCl_2 . 2H_2O$ Mol. wt. 225.63

DESCRIPTION

White or colorless crystals having no odor or a slight odor of hydrochloric acid. It is very soluble in water and is soluble in alcohol and in glacial acetic acid.

IDENTIFICATION

A. To a 1 in 20 solution of the sample in diluted hydrochloric acid T.S. add mercuric chloride T.S. dropwise. A white or grayish white precipitate is produced.

B. A 1 in 20 solution of the sample gives positive tests for *Chloride*, page 926.

SPECIFICATIONS

Assay. Not less than 98.0 percent and not more than the equivalent of 102.2 percent of $SnCl_2.2H_2O$.

Solubility in hydrochloric acid. Passes test.

Limits of Impurities

Arsenic (as As). Not more than 3 parts per million (0.0003 percent).

Other heavy metals (as Pb). Not more than 100 parts per million (0.01 percent).

Iron. Not more than 30 parts per million (0.003 percent).

Substances not precipitated by sulfide. Not more than 0.05 percent.

Sulfate. Passes test (about 0.003 percent).

TESTS

Assay. Transfer about 2 grams of the sample, accurately weighed, into a 250-ml. volumetric flask, dissolve in 15 ml. of hydrochloric acid, dilute to volume with water, and mix. Transfer 50.0 ml. of this solution into a 500-ml. flask, add 5 grams of sodium potassium tartrate, and mix. Make the solution alkaline to litmus with a cold saturated solution of sodium bicarbonate, and titrate at once with 0.1 *N* iodine, using starch T.S. as the indicator. Each ml. of 0.1 *N* iodine is equivalent to 11.28 mg. of $SnCl_2 . 2H_2O$.

Solubility in hydrochloric acid. A 5-gram portion of the sample dissolves completely in a mixture of 5 ml. of hydrochloric acid and 5 ml. of water, heating to 40°, if necessary, to effect solution.

Arsenic. Determine as directed under *Arsenic Test*, page 865, modifying the *Procedure* as follows: Mix 1.0 gram of the sample in the generator flask with 35 ml. of water and 4 grams of 20-mesh zinc. Add 20 ml. of dilute sulfuric acid (1 in 5), but omit the potassium iodide and stannous chloride reagent solutions. *Immediately* connect the generator flask to the lead acetate cotton scrubber unit and the absorber tube containing 3.0 ml. of *Silver Diethyldithiocarbamate Solution*, and allow the reaction to continue for 45 minutes as directed in the *Procedure*.

Other heavy metals. Dissolve 1 gram of the sample in a mixture of 2 ml. of hydrochloric acid and 3 ml. of nitric acid, and boil until solution is complete and brown fumes are no longer evolved. Cool, and dilute to 50 ml. with water. To 10 ml. of this solution add 8 ml. of sodium hydroxide solution (1 in 10), then cool, and dilute to 40 ml. with water. Prepare a control containing 2.0 ml. of *Standard Lead Solution* (20 mcg. Pb), 8 ml. of the sodium hydroxide solution, and 30 ml. of water. Add 10 ml. of hydrogen sulfide T.S. to each solution. Any color produced in the solution of the sample does not exceed that in the control.

Iron. Add 3 ml. of dilute hydrochloric acid (1 in 2) to the residue obtained in the test for *Substances not precipitated by sulfide*, cover with a watch glass, and digest on a steam bath for 15 minutes. Remove the cover, and evaporate to dryness on the steam bath. Dissolve the residue in a few ml. of water and 8 ml. of hydrochloric acid, dilute to 100 ml. with water, and mix. To 25 ml. of this solution add 25 ml. of water, 40 mg. of ammonium persulfate crystals, and 3 ml. of ammonium thiocyanate T.S. Any red or pink color does not exceed that produced by 1.5 ml. of *Iron Standard Solution* (15 mcg. of Fe) in an equal volume of solution containing the quantities of the reagents used in the test.

Substances not precipitated by sulfide. Dissolve 4 grams of the sample in 5 ml. of hydrochloric acid, dilute to 200 ml. with water, and pass hydrogen sulfide through the solution to completely precipitate the tin. Filter without washing, and evaporate 100 ml. of the filtrate to a few ml. in a tared dish. To the residue add 0.1–0.2 ml. of sulfuric acid, and evaporate to dryness. Ignite at $800 \pm 25°$ for 13 minutes, cool, and weigh. Retain the residue for the *Iron* test.

Sulfate. Dissolve 5 grams of the sample in 5 ml. of hydrochloric acid, dilute to 50 ml. with water, filter if not clear and heat the filtrate or clear solution to boiling. Add 5 ml. of barium chloride T.S., digest in a covered beaker on a steam bath for 2 hours, and allow to stand overnight. No precipitate forms.

Packaging and storage. Store in well-closed containers.

Functional use in foods. Reducing agent; antioxidant.

STEARIC ACID

Octadecanoic Acid

$C_{18}H_{36}O_2$ Mol. wt. 284.49

DESCRIPTION

A mixture of solid organic acids obtained from fats consisting chiefly of stearic acid ($C_{18}H_{36}O_2$) and palmitic acid ($C_{16}H_{32}O_2$). It occurs as a hard, white or faintly yellowish, somewhat glossy and crystalline solid, or as a white or yellowish white powder. It has a slight characteristic odor and taste resembling tallow. Stearic acid is practically insoluble in water. One gram dissolves in about 20 ml. of alcohol, in 2 ml. of chloroform, and in about 3 ml. of ether. It conforms to the regulations of the federal Food and Drug Administration pertaining to specifications for fats or fatty acids derived from edible sources.

SPECIFICATIONS

Acid value. Between 196 and 211.

Saponification value. Between 197 and 212.

Titer (**Solidification Point**). Between 54.5° and 69°.

Limits of Impurities

Arsenic (as As). Not more than 3 parts per million (0.0003 percent).

Heavy metals (as Pb). Not more than 10 parts per million (0.001 percent).

Iodine value. Not more than 7.

Residue on ignition. Not more than 0.1 percent.

Unsaponifiable matter. Not more than 1.5 percent.

Water. Not more than 0.2 percent.

TESTS

Acid value. Determine as directed under *Method I* in the general procedure, page 902.

Saponification value. Determine as directed in the general method, page 914, using about 3 grams, accurately weighed.

Titer (**Solidification Point**). Determine as directed under *Solidification Point*, page 954.

Arsenic. A *Sample Solution* prepared as directed for organic compounds meets the requirements of the *Arsenic Test*, page 865.

Heavy metals. Prepare and test a 2-gram sample as directed in *Method II* under the *Heavy Metals Test*, page 920, using 20 mcg. of lead ion (Pb) in the control (*Solution A*).

Iodine value. Determine by the *Wijs Method*, page 906.

Residue on ignition, page 945. Ignite 2 grams as directed in the general method.

Unsaponifiable matter, page 915. Determine as directed in the general method.

Water. Determine by the *Karl Fischer Titrimetric Method,* page 977.

Packaging and storage. Store in well-closed containers.

Functional use in foods. Component in the manufacture of other food grade additives; lubricant; defoaming agent.

STEARYL MONOGLYCERIDYL CITRATE

DESCRIPTION

A soft, practically tasteless, off-white to tan, waxy solid having a lard-like consistency. It is insoluble in water but is soluble in chloroform and in ethylene glycol. It is prepared by a controlled chemical reaction from citric acid, monoglycerides of fatty acids (obtained by the glycerolysis of edible fats and oils or derived from fatty acids), and stearyl alcohol. It conforms to the regulations of the federal Food and Drug Administration pertaining to specifications for fats or fatty acids derived from edible sources.

SPECIFICATIONS

Acid value. Between 40 and 52.

Total citric acid. Between 15.0 and 18.0 percent.

Saponification value. Between 215 and 255.

Limits of Impurities

Arsenic (as As). Not more than 3 parts per million (0.0003 percent).

Heavy metals (as Pb). Not more than 10 parts per million (0.001 percent).

Residue on ignition. Not more than 0.1 percent.

Water. Not more than 0.25 percent.

TESTS

Acid value. Determine as directed under *Method II* in the general procedure, page 902.

Total citric acid

Brominating Solution. Dissolve 19.84 grams of potassium bromide, 5.44 grams of potassium bromate, and 12 grams of sodium metavanadate, $NaVO_3$, in water by warming, and dilute to 1000 ml. with water. Filter if necessary.

Ferrous Sulfate Solution. Dissolve 44 grams of ferrous sulfate,

$FeSO_4.7H_2O$, in 1 *N* sulfuric acid, dilute to 100 ml. with 1 *N* sulfuric acid, and mix. Use within 5 days of preparation.

Sulfide Solution. On the day of use, dissolve 4 grams of thiourea in 100 ml. of a 1 in 50 solution of sodium borate, $Na_2B_4O_7.10H_2O$, and add 2 ml. of sodium sulfide T.S. Wait 30 minutes after the addition of the sodium sulfide T.S. before using.

Standard Solution. Transfer about 50 mg. of sodium citrate dihydrate, accurately weighed, into a 500-ml. volumetric flask, dissolve and dilute to volume with water, and mix. Transfer 15.0 ml. of this solution into a 100-ml. volumetric flask, dilute to volume with water, and mix. Calculate the concentration (C), in mcg. per ml., of citric acid in the final solution by the formula $(15 \times 1000 \times 0.6533W)/(100 \times 500)$, in which W is the weight, in mg., of the sodium citrate taken, and 0.6533 is the factor converting sodium citrate dihydrate to citric acid.

Sample Solution. Transfer about 250 mg. of the sample, accurately weighed, to a 250-ml. extraction flask, and add 15 ml. of 0.5 *N* sodium hydroxide, 5 ml. of alcohol, and a few glass beads. Connect the flask with a water-cooled condenser, and reflux for 3 hours. Immediately cool and neutralize to phenolphthalein T.S. with 0.5 *N* hydrochloric acid, then place the flask in an ice-bath and add 5 ml. of sulfuric acid T.S. Transfer the solution to a 125-ml. separator, extract with three 40-ml. portions of chloroform, and then extract the combined chloroform extracts in a 250-ml. separator with three 10-ml. portions of 0.5 *N* sulfuric acid, adding the acid extracts to a second 250-ml. separator. Wash the combined acid extracts with two 60-ml. portions of chloroform, and discard the chloroform washes. Filter the acid solution into a 500-ml. volumetric flask, neutralize slowly with 6 *N* sodium carbonate, and dilute to volume with water. Transfer 10.0 ml. of this solution into a 100-ml. volumetric flask, dilute to volume with water, and mix. Each ml. of the final solution contains approximately 10 mcg. of citric acid.

Procedure. Pipet 2 ml. each of the *Standard Solution* and of the *Sample Solution* into separate 40- or 45-ml. glass-stoppered centrifuge tubes, and add 3 ml. of water to each tube. Place 5 ml. of water in a third tube for the reagent blank. Place the tubes in an ice-bath, add 5 ml. of sulfuric acid T.S., mix thoroughly, and allow to stand for exactly 5 minutes. Remove the tubes from the ice-bath, and allow them to come to room temperature during the next 5 minutes. To each tube add 5 ml. of the *Brominating Solution*, then insert the stoppers, invert the tubes once or twice, and heat in a water bath at 30° for 20 minutes. Remove the tubes, add 1.5 ml. of *Ferrous Sulfate Solution*, invert again, and allow to stand for 5 minutes, shaking occasionally to ensure complete reduction of the excess free bromine in the tubes. Add 6.5 ml. of petroleum ether, shake for 2 or 3 minutes, and remove the water layer with a syringe. Wash the ether solutions with 15 ml. of water, then remove the water and filter the ether extracts into the original centrifuge tubes which have been previously rinsed with the

Sulfide Solution. Filter each ether extract through a tight plug of glass wool onto which has been placed a sufficient amount of anhydrous sodium sulfate to remove the last traces of water from the ether. Place 5.0 ml. of the filtrate in a clean, dry centrifuge tube, add 3 ml. of *Sulfide Solution,* shake vigorously for 1.5 minutes, and centrifuge. Decant about 0.5 ml. of the supernatant ether layer from each tube, then carefully transfer the ether solutions into 1-cm. cells and determine the absorbance of the extracts obtained from the *Standard Solution* and the *Sample Solution* at 500 mμ with a suitable spectrophotometer, using the reagent blank in the reference cell. Calculate the quantity, in mg., of citric acid in the sample taken by the formula $5C \times A_U/A_S$, in which C is the exact concentration, in mcg. per ml., of citric acid in the *Standard Solution,* A_U is the absorbance of the solution from the *Sample Solution,* and A_S is the absorbance of the solution from the *Standard Solution.*

Saponification value. Transfer about 1 gram, accurately weighed, into a 250-ml. Erlenmeyer flask, and add 25 ml. of ethylene glycol, 35.0 ml. of 0.5 *N* alcoholic potassium hydroxide, and a few glass beads. Reflux for one hour, using a water condenser, then rinse the condenser with water and cool. Add 1 ml. of phenolphthalein T.S., and titrate with 0.5 *N* hydrochloric acid. Perform a blank determination (see page 2) but do not reflux. The difference between the volumes, in ml., of 0.5 *N* hydrochloric acid consumed in the actual test and in the blank titration, multiplied by 28.05 and divided by the weight, in grams, of the sample taken, is the saponification value.

Arsenic. A *Sample Solution* prepared as directed for organic compounds meets the requirements of the *Arsenic Test,* page 865.

Heavy metals. Prepare and test a 2-gram sample as directed in *Method II* under the *Heavy Metals Test,* page 920, using 20 mcg. of lead ion (Pb) in the control.

Residue on ignition. Ignite 2 grams as directed in the general procedure, page 945.

Water. Determine by the *Karl Fischer Titrimetric Method,* page 977.

Packaging and storage. Store in well-closed containers.

Functional use in foods. Emulsion stabilizer.

SUCCINIC ACID

Butanedioic Acid

$HOOCCH_2CH_2COOH$

$C_4H_6O_4$ Mol. wt. 118.09

DESCRIPTION

Colorless or white, odorless crystals having an acid taste. One gram dissolves in 13 ml. of water at 25°, in 1 ml. of boiling water, in 18.5 ml. of alcohol, and in 20 ml. of glycerin.

SPECIFICATIONS

Assay. Not less than 99.0 percent of $C_4H_6O_4$.

Melting range. Between 185° and 190°.

Limits of Impurities

Arsenic (as As). Not more than 3 parts per million (0.0003 percent).

Heavy metals (as Pb). Not more than 10 parts per million (0.001 percent).

Residue on ignition. Not more than 0.025 percent.

TESTS

Assay. Dissolve about 250 mg., accurately weighed, in 25 ml. of recently boiled and cooled water, add phenolphthalein T.S., and titrate with 0.1 *N* sodium hydroxide to the first appearance of a faint pink color which persists for at least 30 seconds. Each ml. of 0.1 *N* sodium hydroxide is equivalent to 5.904 mg. of $C_4H_6O_4$.

Melting range. Determine as directed in the general procedure, page 931.

Arsenic. A *Sample Solution* prepared as directed for organic compounds meets the requirements of the *Arsenic Test*, page 865.

Heavy metals. A solution of 1.5 grams in 25 ml. of water meets the requirements of the *Heavy Metals Test*, page 920, using 15 mcg. of lead ion (Pb) in the control (*Solution* A).

Residue on ignition. Ignite 8 grams as directed in the general method, page 945.

Packaging and storage. Store in well-closed containers.

Functional use in foods. Buffer; neutralizing agent; miscellaneous.

SUCCINYLATED MONOGLYCERIDES

DESCRIPTION

A mixture of succinic acid esters of mono- and diglycerides produced by the succinylation of a product obtained by the glycerolysis of edible fats and oils, or by the direct esterification of glycerol with edible fat-forming fatty acids. It occurs as a waxy solid having an off-white color and a bland taste, melting at about 60°. It is soluble in warm methanol, in ethanol, and in *n*-propanol.

SPECIFICATIONS

Acid value. Between 70 and 120.

Hydroxyl value. Between 138 and 152.

Iodine value. Not more than 3.

Bound succinic acid. Not less than 14.8 percent.

Free succinic acid. Not more than 3 percent.

Total succinic acid. Between 14.8 percent and 25.6 percent.

Limits of Impurities

Arsenic (as As). Not more than 3 parts per million (0.0003 percent).

Heavy metals (as Pb). Not more than 10 parts per million (0.001 percent).

TESTS

Acid value. Determine as directed under *Method I* in the general procedure, page 902.

Hydroxyl value. Determine as directed under *Method II* in the general procedure, page 904.

Iodine value. Determine by the *Wijs Method*, page 906.

Free and bound succinic acid

0.02 N Sodium hydroxide in methanol. Dissolve 4.0 grams of sodium hydroxide in 1000 ml. of anhydrous methanol. Transfer 200.0 ml. of this solution to a 1000-ml. volumetric flask, dilute to volume with anhydrous methanol, and mix. Standardize the solution against dried succinic acid, using phenolphthalein T.S. as the indicator.

Procedure. Transfer about 125 mg. of the sample, accurately weighed, into a 250-ml. separator containing 100 ml. of benzene, and dissolve the sample by heating the separator with warm water. Treat the sample and a blank, consisting of 100 ml. of benzene in another separator, in the same manner as follows: cool the contents of the separator, add 50 ml. of water, and mix by inverting the separator about 20 times. Allow to stand for about 15 minutes, and then transfer the aqueous layer into a 125-ml. Erlenmeyer flask. Add 10 ml. of water to the separator, wash the benzene layer by inverting the separator five times, and add the washings to the 125-ml. flask. To the flask add five drops of phenolphthalein T.S., and titrate with 0.02 *N* sodium hydroxide in methanol. Perform a blank determination, and record the net volume of alkali, in ml., as V_1.

Transfer the benzene layer into a 500-ml. round-bottom flask, and rinse the separator with 10 ml. of benzene. Add a few boiling chips to the flask, and evaporate the benzene, preferably on a thin-film evaporator, under partial vacuum at about 60°. Dissolve the residue in the flask in 10 ml. of methanol, add 10 ml. of water and five drops of phenolphthalein T.S., and titrate with 0.02 N sodium hydroxide in methanol. Perform a blank determination, and record the net volume of alkali, in ml., as V_2.

Calculate the weight, in mg., of free succinic acid in the sample by the formula $(V_1/2)(N)(118.1)$, and calculate the weight, in mg., of bound succinic acid in the sample by the formula $(V_2)(N)(118.1)$, in which N is the exact normality of the sodium hydroxide solution.

Total succinic acid. The sum of the *Free succinic acid* and the *Bound succinic acid* represents the *Total succinic acid.*

Arsenic. A *Sample Solution* prepared as directed for organic compounds meets the requirements of the *Arsenic Test,* page 865.

Heavy metals. Prepare and test a 2-gram sample as directed in *Method II* under the *Heavy Metals Test,* page 920, using 20 mcg. of lead ion (Pb) in the control (*Solution A*).

Packaging and storage. Store in well-closed containers.

Functional use in food. Emulsifier; dough conditioners.

SULFURIC ACID

H_2SO_4 Mol. wt. 98.07

DESCRIPTION

A clear, colorless, oily liquid. It is very caustic and corrosive. It is miscible with water and with alcohol with the generation of much heat and contraction in volume. When mixed with other liquids, sulfuric acid should be added cautiously to the diluent. Some concentrations of sulfuric acid commercially available are expressed in Baumé degrees (Be°) and others (above 93.0 percent) as percent of H_2SO_4. The usually available concentrations are 60° and 66° Be, equivalent to 77.67 and 93.19 percent of H_2SO_4, respectively, and 98.0 percent H_2SO_4. It responds to the tests for *Sulfate,* page 928. Its specific gravity varies with the concentration of H_2SO_4 (see *Sulfuric Acid Table,* page 965).

SPECIFICATIONS

Assay. Not less than the minimum or within the range of Be°, or the percent of H_2SO_4, claimed or implied by the vendor.

Limits of Impurities

Arsenic (as As). Not more than 3 parts per million (0.0003 percent).

Chloride. Not more than 50 parts per million (0.005 percent).

Heavy metals (as Pb). Not more than 20 parts per million (0.002 percent).

Iron. Not more than 200 parts per million (0.02 percent).
Lead. Not more than 5 parts per million (0.0005 percent).
Nitrate. Not more than 10 parts per million (0.001 percent).
Reducing substances (as SO_2). Passes test [about 40 parts per million (0.004 percent)].
Selenium. Not more than 20 parts per million (0.002 percent).

TESTS

Assay. Transfer a 1-ml. sample into a small, tared, glass-stoppered Erlenmeyer flask, insert the stopper, weigh accurately, and cautiously add about 30 ml. of water. Cool the mixture, add methyl orange T.S., and titrate with 1 *N* sodium hydroxide. Each ml. of 1 *N* sodium hydroxide is equivalent to 49.04 mg. of H_2SO_4.

For concentrations of sulfuric acid below 93.0 percent, expressed in Baumé degrees, transfer about 200 ml., previously cooled to a temperature below 15°, into a 250-ml. hydrometer cylinder. Insert a suitable Baumé hydrometer graduated at 0.1° intervals, adjust the temperature to exactly 15.6° (60° F.), and note the reading at the bottom of the meniscus, estimating it to the nearest 0.05°. The percent of H_2SO_4 may be obtained by reference to the *Sulfuric Acid Table*, page 965.

Arsenic. A solution of 1 gram in 35 ml. of water meets the requirements of the *Arsenic Test*, page 865, using as a control a mixture of 3 ml. of the *Standard Arsenic Solution* (3 mcg. As) and 1 gram of A.C.S. reagent sulfuric acid.

Chloride, page 879. Transfer a volume equivalent to 5 grams of the acid into about 25 ml. of water contained in a 50-ml. volumetric flask, cool, and dilute to volume. Retain the unused portion for the *Heavy Metals*, *Iron*, and *Lead* tests. Any turbidity produced by 4 ml. of this solution (400 mg. sample) does not exceed that shown in a control containing 20 mcg. of chloride ion (Cl).

Heavy metals. Dilute 10 ml. of the solution (1-gram sample) prepared for the *Chloride* test to 25 ml. with water. This solution meets the requirements of the *Heavy Metals Test*, page 920, using 20 mcg. of lead ion (Pb) in the control (*Solution A*).

Iron. Dilute 1 ml. of the solution (100-mg. sample) prepared for the *Chloride* test to 40 ml. Add about 30 mg. of ammonium persulfate crystals and 10 ml. of ammonium thiocyanate T.S. Any red color does not exceed that produced by 2.0 ml. of *Iron Standard Solution* (20 mcg. Fe) in an equal volume of solution containing the same quantities of the reagents used in the test.

Lead. Dilute 10 ml. of the solution (1-gram sample) prepared for the *Chloride* test to 40 ml. This solution meets the requirements of the *Lead Limit Test*, page 929, using 5 mcg. of lead ion (Pb) in the control.

Nitrate

Standard Nitrate Solution. Transfer 8.022 grams of potassium nitrate, KNO_3, previously dried at 105° for 1 hour, into a 500-ml. volumetric flask, dissolve it in water, dilute to volume, and mix well. Slowly add from a buret 5.0 ml. of this solution to 400 ml. of A.C.S. reagent grade sulfuric acid, previously cooled to 5°, keeping the tip of the buret below the surface of the acid. After the solution has reached room temperature, transfer it into a 500-ml. volumetric flask, and dilute to volume with reagent grade sulfuric acid. Each ml. contains 100 mcg. of HNO_3.

Procedure. Into each of two 100-ml. Nessler tubes transfer 50 ml. of A.C.S. reagent grade sulfuric acid, add slowly 5 ml. of a freshly prepared 1 in 10 solution of ferrous sulfate, $FeSO_4 . 7H_2O$, mix with a glass rod, and cool in an ice bath to between 10° and 15°. To one tube of the cooled mixture add a 10-ml. sample, previously cooled to between 10° and 15°, and dilute to the 100-ml. mark with A.C.S. reagent grade sulfuric acid chilled to about the same temperature. Add the *Standard Nitrate Solution*, dropwise, from a micro-buret to the other tube, with frequent mixing, until the color of the control nearly matches that of the sample solution. Dilute the control solution to 100 ml. and continue adding the *Standard Nitrate Solution* to as exact a match in color intensity as possible when compared with the sample solution by looking down through the solutions against a white background illuminated by diffused light. Compute the weight of H_2SO_4 in the weight of the sample from the specific gravity and the volume taken (see *Sulfuric Acid Table*, page 965). Not more than 0.1 ml. of the *Standard Nitrate Solution* is required for each gram of H_2SO_4.

Reducing substances. Carefully dilute 8 grams with about 50 ml. of ice-cold water, keeping the solution cool during the addition. To the dilution add 0.1 ml. of 0.1 *N* potassium permanganate. The solution remains pink for not less than 5 minutes.

Selenium. Add 3 ml. of nitric acid to 3 grams of the sample, and heat until copious white fumes are evolved. Cool, carefully add about 5 ml. of water, and heat again to strong fuming. Cool, transfer the solution into a 200 × 25-mm. test tube with the aid of sufficient water to make 20 ml., and add 20 ml. of hydrochloric acid. This solution meets the requirements of the *Selenium Limit Test*, page 953.

Packaging and storage. Store in tight containers.

Functional use in foods. Acid.

TANGERINE OIL, EXPRESSED

DESCRIPTION

The oil obtained by expression from the peels of the ripe fruit of the Dancy tangerine, and from some other closely related varieties. It is a reddish orange to brownish orange liquid, with a pleasant orange-like odor. Oils produced from unripe fruit often show a green color. It is soluble in most fixed oils and in mineral oil, slightly soluble in propylene glycol, and relatively insoluble in glycerin.

SPECIFICATIONS

Aldehydes. Between 0.8 and 1.9 percent of aldehydes, calculated as *n*-decyl aldehyde ($C_{10}H_{20}O$).

Angular rotation. Between +88° and +96°.

Refractive index. Between 1.473 and 1.476 at 20°.

Residue on evaporation. Between 2.3 and 5.8 percent.

Specific gravity. Between 0.844 and 0.854.

Limits of Impurities

Arsenic (as As). Not more than 3 parts per million (0.0003 percent).

Heavy metals (as Pb). Not more than 40 parts per million (0.004 percent).

Lead. Not more than 10 parts per million (0.001 percent).

TESTS

Aldehydes. Weigh accurately about 10 grams, and proceed as directed under *Aldehydes*, page 894, using 78.13 as the equivalence factor (*E*) in the calculation. Allow the sample and the blank to stand at room temperature for 30 minutes, after adding the hydroxylamine hydrochloride solution.

Angular rotation. Determine in a 100-mm. tube as directed under *Optical Rotation*, page 939.

Refractive index, page 945. Determine with an Abbé or other refractometer of equal or greater accuracy.

Residue on evaporation. Proceed as directed in the general method, page 899, using 5 grams, and heat for 5 hours.

Specific gravity. Determine by any reliable method (see page 5).

Arsenic. A *Sample Solution* prepared as directed for organic compounds meets the requirements of the *Arsenic Test*, page 865.

Heavy metals. Prepare and test a 500-mg. sample as directed in *Method II* under the *Heavy Metals Test*, page 920, using 20 mcg. of lead ion (Pb) in the control (*Solution A*).

Lead. A *Sample Solution* prepared as directed for organic compounds meets the requirements of the *Lead Limit Test*, page 929, using 10 mcg. of lead ion (Pb) in the control.

Packaging and storage. Store in full, tight, preferably glass, tin-lined, galvanized, or other suitably lined containers in a cool place protected from light.

Functional use in foods. Flavoring agent.

TANNIC ACID

DESCRIPTION

A tannin usually obtained from nutgalls, the excrescences which form on the young twigs of *Quercus infectoria* Olivier, and allied species of *Quercus* L. (Fam. *Fagaceae*), or from the seed pods of Tara (*Caesalpinia spinosa*). It occurs as an amorphous powder, as glistening scales, or as spongy masses, varying in color from yellowish white to light brown. It is odorless or has a faint, characteristic odor and an astringent taste. Tannic acid is very soluble in water, in acetone, and in alcohol, but only slightly soluble in absolute alcohol. It is practically insoluble in benzene, in chloroform, in ether, and in solvent hexane. One gram dissolves in about 1 ml. of warm glycerin.

IDENTIFICATION

A. To a 1 in 10 solution add a small quantity of ferric chloride T.S. A bluish black color or precipitate forms.

B. A solution of tannic acid when added to a solution of either an alkaloidal salt, albumin, or gelatin produces a precipitate.

SPECIFICATIONS

Loss on drying. Not more than 12 percent.

Limits of Impurities

Arsenic. Not more than 3 parts per million (0.0003 percent).
Gums or dextrin. Passes test.
Heavy metals. Not more than 40 parts per million (0.004 percent).
Lead. Not more than 10 parts per million (0.001 percent).
Residue on ignition. Not more than 1 percent.
Resinous substances. Passes test.

TESTS

Loss on drying, page 931. Dry at 105° for 2 hours.

Arsenic. A *Sample Solution* prepared as directed for organic compounds meets the requirements of the *Arsenic Test,* page 865.

Gums or dextrin. Dissolve 1 gram in 5 ml. of water, filter, and to the filtrate add 10 ml. of alcohol. No turbidity is produced within 15 minutes.

Heavy metals. Transfer a 500-mg. sample into a 150-ml. beaker, and cautiously add 15 ml. of nitric acid and 5 ml. of 70 percent perchloric acid. Evaporate the mixture to dryness on a hot plate under a suitable hood, cool slightly, add 2 ml. of hydrochloric acid, and wash down the sides of the beaker with water. Carefully evaporate the solution to dryness on a hot plate, rotating the beaker to avoid spattering. Repeat the addition of 2 ml. of hydrochloric acid, washing down the sides of the beaker with water, and evaporation to dryness. Cool the residue, dissolve it in 10 ml. of water, add 1 drop of phenolphthalein T.S. and sufficient 1 *N* sodium hydroxide, dropwise, to produce a pink color. To this solution add 1 *N* hydrochloric acid, dropwise, until the pink color just disappears, then add 2 ml. of diluted acetic acid T.S., and transfer the solution into a 50-ml. Nessler tube. Dilute to 25 ml. with water and add 10 ml. of hydrogen sulfide T.S. After 10 minutes the color of the solution of the sample is no darker than that produced in a control of equal volume containing 20 mcg. of lead ion (Pb) and carried through the same procedure as the sample.

Lead. A *Sample Solution* prepared as directed for organic compounds meets the requirements of the *Lead Limit Test*, page 929, using 10 mcg. of lead ion (Pb) in the control.

Residue on ignition. Ignite 1 gram as directed in the general method, page 945.

Resinous substances. Dissolve 1 gram in 5 ml. of water, filter, and dilute the filtrate to 15 ml. No turbidity is produced.

Packaging and storage. Store in tight, light-resistant containers.

Functional use in foods. Clarifying agent.

TARRAGON OIL

Estragon Oil

DESCRIPTION

The volatile oil obtained by steam distillation from the leaves, stems and flowers of the plant, *Artemesia dracunculus* L. It is a pale yellow to amber liquid, having a delicate spicy odor similar to licorice and sweet basil, but characteristic of tarragon oil. It is soluble in most fixed oils, and in an equal volume of mineral oil, occasionally becoming hazy on further dilution. It is relatively insoluble in propylene glycol, and is insoluble in glycerin.

SPECIFICATIONS

Acid value. Not more than 2.0.

Angular rotation. Between +1.5° and +6.5°.

Refractive index. Between 1.504 and 1.520 at 20°.

Saponification value. Not more than 18.

Solubility in alcohol. Passes test.

Specific gravity. Between 0.914 and 0.956.

Limits of Impurities

Arsenic (as As). Not more than 3 parts per million (0.0003 percent).

Heavy metals (as Pb). Not more than 40 parts per million (0.004 percent).

Lead. Not more than 10 parts per million (0.001 percent).

TESTS

Acid value. Proceed as directed in the general method, page 893.

Angular rotation. Determine in a 100-mm. tube as directed under *Optical Rotation*, page 939.

Refractive index, page 945. Determine with an Abbé or other refractometer of equal or greater accuracy.

Saponification value. Proceed as directed in the general method, page 896, using about 5 grams, accurately weighed.

Solubility in alcohol. Proceed as directed in the general method, page 899. One ml. is soluble in 1 ml. of 90 percent alcohol.

Specific gravity. Determine by any reliable method (see page 5).

Arsenic. A *Sample Solution* prepared as directed for organic compounds meets the requirements of the *Arsenic Test*, page 865.

Heavy metals. Prepare and test a 500-mg. sample as directed in *Method II* under the *Heavy Metals Test*, page 920, using 20 mcg. of lead ion (Pb) in the control (*Solution A*).

Lead. A *Sample Solution* prepared as directed for organic compounds meets the requirements of the *Lead Limit Test*, page 929, using 10 mcg. of lead ion (Pb) in the control.

Packaging and storage. Store in full, tight, preferably glass containers in a cool place protected from light.

Functional use in foods. Flavoring agent.

TARTARIC ACID

L(+)-Tartaric Acid

```
      COOH
       |
    H—C—OH
       |
   HO—C—H
       |
      COOH
```

$C_4H_6O_6$ Mol. wt. 150.09

DESCRIPTION

Colorless or translucent crystals, or a white, fine to granular, crystalline powder. It is odorless, has an acid taste, and is stable in air. One gram dissolves in 0.8 ml. of water at 25°, in about 0.5 ml. of boiling water, and in about 3 ml. of alcohol. Its solutions are dextrorotatory, and they give positive tests for *Tartrate*, page 928.

SPECIFICATIONS

Assay. Not less than 99.7 percent of $C_4H_6O_6$ after drying.

Limits of Impurities

Arsenic (as As). Not more than 3 parts per million (0.0003 percent).
Heavy metals (as Pb). Not more than 10 parts per million (0.001 percent).
Loss on drying. Not more than 0.5 percent.
Oxalate. Passes test.
Residue on ignition. Not more than 0.05 percent.
Sulfate. Passes test.

TESTS

Assay. Weigh accurately about 2 grams, previously dried over phosphorus pentoxide for 3 hours, dissolve it in 40 ml. of water, add phenolphthalein T.S., and titrate with 1 *N* sodium hydroxide. Each ml. of 1 *N* sodium hydroxide is equivalent to 75.04 mg. of $C_4H_6O_6$.

Arsenic. A *Sample Solution* prepared as directed for organic compounds meets the requirements of the *Arsenic Test*, page 865.

Heavy metals. Prepare and test a 2-gram sample as directed in *Method II* under the *Heavy Metals Test*, page 920, using 20 mcg. of lead ion (Pb) in the control (*Solution A*).

Loss on drying, page 931. Dry over phosphorus pentoxide for 3 hours.

Oxalate. Nearly neutralize 10 ml. of a 1 in 10 solution with ammonia T.S., and add 10 ml. of calcium sulfate T.S. No turbidity is produced.

Residue on ignition. Ignite 4 grams as directed in the general method, page 945.

Sulfate. To 10 ml. of a 1 in 100 solution add 3 drops of hydrochloric acid and 1 ml. of barium chloride T.S. No turbidity is produced.

Packaging and storage. Store in well-closed containers.

Functional use in foods. Acid; sequestrant.

TERPENE RESIN, NATURAL

DESCRIPTION

A natural terpene occurring in some coal seams. The resin is separated from coal in froth flotation cells. The crude resin is leached with hexane, and the solution produced is freed of suspended matter by pressure filtration. The resin is concentrated in a pre-evaporator, and most of the solvent is removed in a melter-evaporator. The remaining solvent is removed in a spray dryer.

SPECIFICATIONS

Acid value. Less than 8.

Melting point. Not less than 155°.

Limits of Impurities

Arsenic (as As). Not more than 3 parts per million (0.0003 percent).

Heavy metals (as Pb). Not more than 40 parts per million (0.004 percent).

Lead. Not more than 3 parts per million (0.0003 percent).

TESTS

Acid value. Dissolve about 3 grams of the sample, accurately weighed, in 100 ml. of a mixture of 75 ml. of benzene and 36 ml. of alcohol previously neutralized to phenolphthalein T.S. with sodium hydroxide. Add 25 ml. of a saturated solution of sodium chloride, then add 10 grams in addition of sodium chloride and a few drops of phenolphthalein T.S., and titrate with 0.1 N alcoholic potassium hydroxide to the first pink color that persists for at least 30 seconds. Calculate the acid value by the formula $56.1V \times N/W$, in which V is the volume, in ml., and N is the normality, respectively, of the alcoholic potassium hydroxide solution, and W is the weight of the sample, in grams.

Melting point. Determine as directed for *Class Ib* substances in the general procedure, page 931.

Arsenic. Prepare a *Sample Solution* as directed in the general method under *Chewing Gum Base*, page 877. This solution meets the requirements of the *Arsenic Test*, page 865.

Heavy metals. Prepare and test a 500-mg. sample as directed in *Method II* under the *Heavy Metals Test*, page 920, using 20 mcg. of lead ion (Pb) in the control (*Solution A*).

Lead. Prepare a *Sample Solution* as directed in the general method under *Chewing Gum Base*, page 878. This solution meets the requirements of the *Lead Limit Test*, page 929, using 10 mcg. of lead ion (Pb) in the control.

Packaging and storage. Store in well-closed containers.

Functional use in foods. Masticatory substance in chewing gum base.

TERPENE RESIN, SYNTHETIC

DESCRIPTION

A synthetic resin composed essentially of polymers of alpha-pinene, beta-pinene (monomer isolated from sulfate-turpentine), and dipentene. The polymer is prepared by a batch or continuous process and is purified by steam and water washings. It is soluble in benzene, but is insoluble in water. Its color is less than 4 on the Gardner scale (measured in 50 percent mineral spirit solution).

SPECIFICATIONS

Acid value. Less than 5.

Saponification value. Less than 5.

Limits of Impurities

Arsenic (as As). Not more than 3 parts per million (0.0003 percent).

Heavy metals (as Pb). Not more than 40 parts per million (0.004 percent).

Lead. Not more than 3 parts per million (0.0003 percent).

TESTS

Acid value. Determine as directed in the general procedure, page 945.

Saponification value. Determine as directed in the general procedure, page 914.

Arsenic. Prepare a *Sample Solution* as directed in the general method under *Chewing Gum Base,* page 877. This solution meets the requirements of the *Arsenic Test,* page 865.

Heavy metals. Prepare and test a 500-mg. sample as directed in *Method II* under the *Heavy Metals Test,* page 920, using 20 mcg. of lead ion (Pb) in the control (*Solution A*).

Lead. Prepare a *Sample Solution* as directed in the general method under *Chewing Gum Base,* page 878. This solution meets the requirements of the *Lead Limit Test,* page 929, using 10 mcg. of lead ion (Pb) in the control.

Packaging and storage. Store in well-closed containers.

Functional use in foods. Masticatory substance in chewing gum base.

TERPINEOL

Menthen-1-ol-8

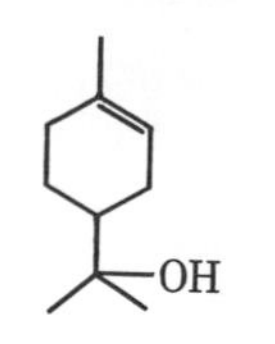

$C_{10}H_{18}O$ Mol. wt. 154.25

DESCRIPTION

A colorless, viscous liquid, having a lilac-like odor. It may be a mixture of *alpha*- (shown above), *beta*-, and *gamma*-isomers. It is soluble in mineral oil and slightly soluble in glycerin and in water.

SPECIFICATIONS

Angular rotation. Between −0° 10′ and +0° 10′.

Distillation range. Not less than 90 percent distils within a 5°-range between 214° and 224°.

Refractive index. Between 1.482 and 1.485 at 20°.

Solidification point. Not lower than +2°.

Solubility in alcohol. Passes test.

Specific gravity. Between 0.930 and 0.936.

TESTS

Angular rotation. Determine in a 100-mm. tube as directed under *Optical Rotation*, page 939.

Distillation range, page 890. Proceed as directed in the general method.

Refractive index, page 945. Determine with an Abbé or other refractometer of equal or greater accuracy.

Solidification point. Determine as directed in the general method, page 954.

Solubility in alcohol. Proceed as directed in the general method, page 899. One ml. dissolves in 2 ml. of 70 percent alcohol, in 4 ml. of 60 percent alcohol, and in 8 ml. of 50 percent alcohol.

Specific gravity. Determine by any reliable method (see page 5).

Packaging and storage. Store in full, tight, preferably glass, tin, aluminum, galvanized iron, or black iron containers in a cool place protected from light.

Functional use in foods. Flavoring agent.

TERPINYL ACETATE

Menthen-l-yl-8 Acetate

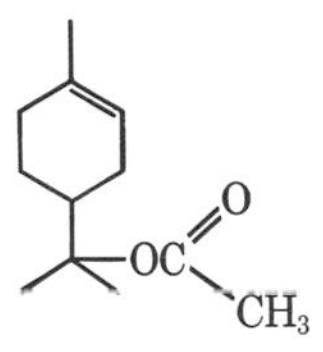

$C_{12}H_{20}O_2$ Mol. wt. 196.29

DESCRIPTION

A clear, colorless liquid having a lavender-like odor. It may be a mixture of *alpha-* (shown above), *beta-*, and *gamma*-isomers. It is soluble in most fixed oils, in mineral oil, and in propylene glycol. It is slightly soluble in glycerin.

SPECIFICATIONS

Assay. Not less than 97.0 percent of $C_{12}H_{20}O_2$.

Angular rotation. Between $-9°$ and $+9°$.

Refractive index. Between 1.464 and 1.467 at 20°.

Solubility in alcohol. Passes test.

Specific gravity. Between 0.953 and 0.962.

TESTS

Assay. Weigh accurately about 1 gram, and proceed as directed under *Ester Determination*, page 896, using 98.15 as the equivalence factor (e) in the calculation. Reflux for 4 hours.

Angular rotation. Determine in a 100-mm. tube as directed under *Optical Rotation*, page 939.

Refractive index, page 945. Determine with an Abbé or other refractometer of equal or greater accuracy.

Solubility in alcohol. Proceed as directed in the general method, page 899. One ml. dissolves in 5 ml. of 70 percent alcohol and remains in solution upon dilution to 10 ml.

Specific gravity. Determine by any reliable method (see page 5).

Packaging and storage. Store in full, tight, preferably glass, tin, galvanized iron, or other suitably lined containers in a cool place protected from light.

Functional use in foods. Flavoring agent.

TERPINYL PROPIONATE

Menthen-l-yl-8 Propionate

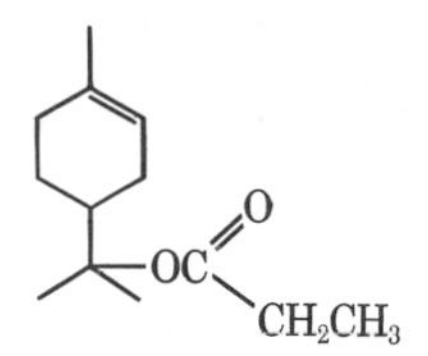

$C_{13}H_{22}O_2$ Mol. wt. 210.32

DESCRIPTION

A colorless to slightly yellow liquid having a somewhat floral, lavender-like odor. It may be a mixture of *alpha-* (shown above), *beta-*, or *gamma*-isomers. It is miscible with alcohol, with chloroform, with ether, with mineral oil, and with most fixed oils. It is soluble in glycerin and slightly soluble in propylene glycol.

SPECIFICATIONS

Assay. Not less than 95.0 percent of $C_{13}H_{22}O_2$.

Refractive index. Between 1.461 and 1.466 at 20°.

Solubility in alcohol. Passes test.

Specific gravity. Between 0.944 and 0.949.

Limits of Impurities

Acid value. Not more than 1.0.

TESTS

Assay. Weigh accurately about 1.5 grams, and proceed as directed under *Ester Determination* (*High Boiling Solvent*), page 896, using 105.16 as the equivalence factor (e) in the calculation.

Refractive index, page 945. Determine with an Abbé or other refractometer of equal or greater accuracy.

Solubility in alcohol. Proceed as directed in the general method, page 899. One ml. dissolves in 2 ml. of 80 percent alcohol to form a clear solution.

Specific gravity. Determine by any reliable method (see page 5).

Acid value. Determine as directed in the general method, page 893.

Packaging and storage. Store in full, tight, preferably glass, tin-lined, or other suitably lined containers in a cool place protected from light.

Functional use in foods. Flavoring agent.

THIAMINE HYDROCHLORIDE

Aneurine Hydrochloride; Thiamine Chloride; Vitamin B_1; Vitamin B_1 Hydrochloride

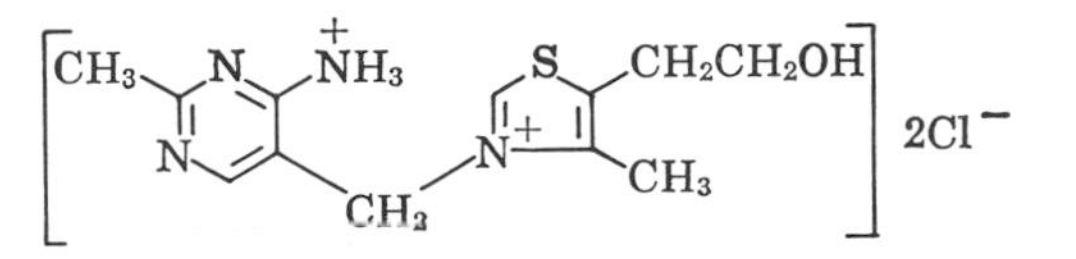

$C_{12}H_{17}ClN_4OS.HCl$ Mol. wt. 337.27

DESCRIPTION

Small, white crystals, or crystalline powder, usually having a slight, characteristic odor. When exposed to air, the anhydrous product rapidly absorbs about 4 percent of water. It melts at about 248° with some decomposition. One gram dissolves in about 1 ml. of water, and in about 100 ml. of alcohol. It is soluble in glycerin, and is insoluble in ether and in benzene.

IDENTIFICATION

A. The infrared absorption spectrum of a potassium bromide dispersion of the sample, previously dried at 105° for 2 hours, exhibits maxima only at the same wavelengths as that of a similar preparation of U.S.P. Thiamine Hydrochloride Reference Standard.

B. A 1 in 50 solution gives positive tests for *Chloride*, page 926.

SPECIFICATIONS

Assay. Not less than 98.0 percent and not more than the equivalent of 102.0 percent of $C_{12}H_{17}ClN_4OS.HCl$, calculated on the dried basis.

pH of a 1 in 100 solution. Between 2.7 and 3.4.

Limits of Impurities

Color of solution. Passes test.

Loss on drying. Not more than 5 percent.

Nitrate. Passes test.

Residue on ignition. Not more than 0.2 percent.

TESTS

Assay

Standard Preparation. Prepare as directed for *Standard Preparation* under *Thiamine Assay*, page 968.

Assay Preparation. Dissolve about 25 mg., accurately weighed, in sufficient *Acid Potassium Chloride Solution* (see page 968) to make 500 ml., and mix. Dilute 5 ml. of this solution, quantitatively and stepwise, using *Acid Potassium Chloride Solution*, to an estimated concentration of 0.2 mcg. per ml. Using this solution as the *Assay Preparation*, proceed as directed for *Procedure* under *Thiamine Assay*, page 968. Calculate the quantity, in mg., of $C_{12}H_{17}ClN_4OS.HCl$ in the sample taken by the formula $25(A - b)/(S - d)$, in which A, b, S, and d are as defined under *Calculation* in the *Thiamine Assay*.

pH. Determine by the *Potentiometric Method*, page 941.

Color of solution. Dissolve 1.0 gram in water to make 10 ml. This solution exhibits no more color than a dilution of 1.5 ml. of 0.1 *N* potassium dichromate in water to make 1000 ml.

Loss on drying, page 931. Dry at 105° for 2 hours.

Nitrate. To 2 ml. of a 1 in 50 solution add 2 ml. of sulfuric acid, cool, and superimpose 2 ml. of ferrous sulfate T.S. No brown ring is produced at the junction of the two layers.

Residue on ignition, page 945. Ignite 1 gram as directed in the general method.

Packaging and storage. Store in tight, light-resistant containers.

Functional use in foods. Nutrient; dietary supplement.

THIAMINE MONONITRATE

Thiamine Nitrate; Vitamin B_1 Mononitrate

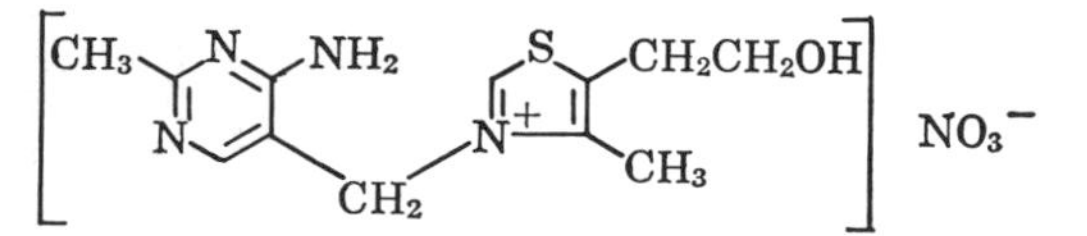

$C_{12}H_{17}N_5O_4S$ Mol. wt. 327.36

DESCRIPTION

White crystals or crystalline powder, usually having a slight, characteristic odor. One gram dissolves in about 35 ml. of water. It is slightly soluble in alcohol and in chloroform.

IDENTIFICATION

A. To 2 ml. of a 1 in 50 solution add 2 ml. of sulfuric acid, cool, and superimpose 2 ml. of ferrous sulfate T.S. A brown ring is produced at the junction of the two liquids.

B. Dissolve about 5 mg. of the sample in a mixture of 1 ml. of lead acetate T.S. and 1 ml. of sodium hydroxide solution (1 in 10). A yellow color is produced. Heat the mixture for several minutes on a steam bath. The color changes to brown and, on standing, a precipitate of lead sulfide separates.

C. A 1 in 10 solution of the sample yields a white precipitate with mercuric chloride T.S., and a red-brown precipitate with iodine T.S.

It is precipitated also by mercuric-potassium iodide T.S. and by trinitrophenol T.S.

D. Dissolve about 5 mg. of the sample in 5 ml. of 0.5 *N* sodium hydroxide, add 0.5 ml. of potassium ferricyanide T.S. and 5 ml. of isobutyl alcohol, shake vigorously for 2 minutes, and allow the layers to separate. When illuminated from above by a vertical beam of ultraviolet light and viewed at a right angle to this beam, the air-liquid meniscus shows a vivid blue fluorescence, which disappears when the mixture is slightly acidified, but reappears when it is again made alkaline.

SPECIFICATIONS

Assay. Not less than 98.0 percent and not more than the equivalent of 102.0 percent of $C_{12}H_{17}N_5O_4S$, calculated on the dried basis.

pH of a 1 in 50 solution. Between 6.0 and 7.5.

Limits of Impurities

Chloride. Not more than 600 parts per million (0.06 percent).

Loss on drying. Not more than 1 percent.

Residue on ignition. Not more than 0.2 percent.

TESTS

Assay

Standard Preparation. Prepare as directed for *Standard Preparation* under *Thiamine Assay*, page 968.

Assay Preparation. Using Thiamine Mononitrate instead of Thiamine Hydrochloride, prepare the *Assay Preparation* as directed in the *Assay* under *Thiamine Hydrochloride*, page 815, and proceed as directed for *Procedure* under *Thiamine Assay*, page 968. Calculate the quantity, in mg., of $C_{12}H_{17}N_5O_4S$ in the sample taken by the formula $25\ [0.9706\ (A - b)/(S - d)]$, in which 0.9706 is the ratio of the molecular weight of thiamine mononitrate to that of thiamine hydrochloride, and *A*, *b*, *S*, and *d* are as defined under *Calculation* in the *Thiamine Assay*.

pH. Determine by the *Potentiometric Method*, page 941.

Chloride, page 879. Any turbidity produced by a 25-mg. sample does not exceed that shown in a control containing 15 mcg. of chloride ion (Cl).

Loss on drying, page 931. Dry about 500 mg., accurately weighed, at 105° for 2 hours.

Residue on ignition, page 945. Ignite 1 gram as directed in the general method.

Packaging and storage. Store in tight, light-resistant containers.

Functional use in foods. Nutrient; dietary supplement.

L-THREONINE

L-2-Amino-3-hydroxybutyric Acid

```
CH3CHCHCOOH
    |  |
   HO  NH2
```

$C_4H_9NO_3$ Mol. wt. 119.12

DESCRIPTION

A white, odorless, crystalline powder having a slightly sweet taste. It is freely soluble in water, but insoluble in alcohol, ether, and in chloroform. It melts with decomposition at about 256°.

IDENTIFICATION

A. To 5 ml. of a 1 in 1000 solution add 1 ml. of triketohydrindene hydrate T.S. A reddish purple or purple color is produced.

B. To 5 ml. of a 1 in 10 solution add 5 ml. of a saturated solution of potassium periodate, and heat. Ammonia is evolved.

SPECIFICATIONS

Assay. Not less than 98.0 percent and not more than the equivalent of 102.0 percent of $C_4H_9NO_3$, calculated on the dried basis.

Specific rotation, $[\alpha]^{20°}_{D}$. Between −26.0° and −29.0°, on the dried basis.

Limits of Impurities

Arsenic (as As). Not more than 3 parts per million (0.0003 percent).

Heavy metals (as Pb). Not more than 20 parts per million (0.002 percent).

Lead. Not more than 10 parts per million (0.001 percent).

Loss on drying. Not more than 0.2 percent.

Residue on ignition. Not more than 0.1 percent.

TESTS

Assay. Dissolve about 200 mg., accurately weighed, in 3 ml. of formic acid and 50 ml. of glacial acetic acid, add 2 drops of crystal violet T.S., titrate with 0.1 *N* perchloric acid to a green end-point or until the blue color disappears completely. Each ml. of 0.1 *N* perchloric acid is equivalent to 11.91 mg. of $C_4H_9NO_3$.

Specific rotation, page 939. Determine in a solution containing 6 grams of a previously dried sample in sufficient water to make 100 ml.

Arsenic. A *Sample Solution* prepared as directed for organic compounds meets the requirements of the *Arsenic Test,* page 865.

Heavy metals. Prepare and test a 1-gram sample as directed in

Method II under the *Heavy Metals Test,* page 920, using 20 mcg. of lead ion (Pb) in the control (*Solution A*).

Lead. A *Sample Solution* prepared as directed for organic compounds meets the requirements of the *Lead Limit Test,* page 929, using 10 mcg. of lead ion (Pb) in the control.

Loss on drying, page 931. Dry at 105° for 3 hours.

Residue on ignition, page 945. Ignite 1 gram as directed in the general method.

Packaging and storage. Store in well-closed containers.

Functional use in foods. Nutrient; dietary supplement.

THYME OIL

DESCRIPTION

The volatile oil obtained by distillation from the flowering plant *Thymus vulgaris* L., or *Thymus zygis* L., and its var. *gracilis* Boissier (Fam. *Labiatae*). It is a colorless, yellow, or red liquid, with a characteristic pleasant odor, and a pungent, persistent taste. It is affected by light.

SPECIFICATIONS

Assay. Not less than 40 percent, by volume, of phenols.

Angular rotation. Levorotatory, but not more than −3°.

Refractive index. Between 1.495 and 1.505 at 20°.

Solubility in alcohol. Passes test.

Specific gravity. Between 0.915 and 0.935.

Limits of Impurities

Water-soluble phenols. Passes test.

TESTS

Assay. Proceed as directed under *Phenols,* page 898, allowing the mixture to stand overnight, then adding sufficient potassium hydroxide T.S. to raise the lower limit of the oily layer into the graduated portion of the neck of the flask. After the solution has become clear, adjust the temperature and read the volume of the residual liquid.

Angular rotation. Determine in a 100-mm. tube as directed under *Optical Rotation,* page 939.

Refractive index, page 945. Determine with an Abbé or other refractometer of equal or greater accuracy.

Solubility in alcohol. Proceed as directed in the general method, page 899. One ml. dissolves in 2 ml. of 80 percent alcohol.

Specific gravity. Determine by any reliable method (see page 5).

Water-soluble phenols. Shake a 1-ml. sample with 10 ml. of hot water, and after cooling, pass the water layer through a moistened filter. Not even a transient blue or violet color is produced in the filtrate upon the addition of 1 drop of ferric chloride T.S.

Packaging and storage. Store in full, tight, light-resistant containers in a cool place.

Functional use in foods. Flavoring agent.

dl-ALPHA TOCOPHEROL

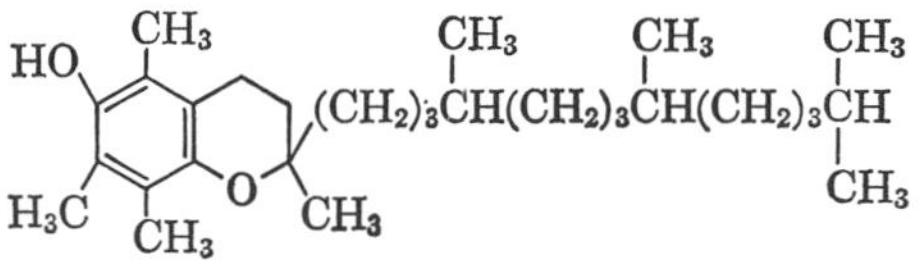

HO, CH_3, H_3C, CH_3, O, CH_3, $(CH_2)_3CH(CH_3)(CH_2)_3CH(CH_3)(CH_2)_3CH(CH_3)CH_3$

$C_{29}H_{50}O_2$ Mol. wt. 430.72

DESCRIPTION

A form of vitamin E. It occurs as a yellow to amber, nearly odorless, clear, viscous oil which oxidizes and darkens in air and, on exposure to light. It is insoluble in water, freely soluble in alcohol, and is miscible with ether, with chloroform, with acetone, and with vegetable oils.

IDENTIFICATION

A. Dissolve about 10 mg. in 10 ml. of absolute alcohol, add with swirling 2 ml. of nitric acid, and heat at about 75° for 15 minutes. A bright red to orange color develops.

B. Its absorptivity (see page 957), determined in alcohol at 292 $m\mu$, is between 7.1 and 7.6.

C. A 1 in 10 solution in chloroform has no significant angular rotation when determined in a 200-mm. tube, page 939.

SPECIFICATIONS

Assay. Not less than 97.0 percent of $C_{29}H_{50}O_2$.

Refractive index. Between 1.503 and 1.507 at 20°.

Limits of Impurities

Arsenic (as As). Not more than 3 parts per million (0.0003 percent).

Heavy metals (as Pb). Not more than 40 parts per million (0.004 percent).

Lead. Not more than 10 parts per million (0.001 percent).

TESTS

Assay. Dissolve about 50 mg., accurately weighed, in 100 ml. of 0.5 *N* alcoholic sulfuric acid. Add 20 ml. of water and 2 drops of diphenylamine T.S., and titrate with 0.01 *N* ceric sulfate at the rate of approximately 25 drops per 10 seconds, swirling or stirring constantly, until a blue end-point is reached which persists for 10 seconds. Perform a blank determination (see page 2) and make any necessary correction. Each ml. of 0.01 *N* ceric sulfate is equivalent to 2.154 mg. of $C_{29}H_{50}O_2$.

Refractive index, page 945. Determine with an Abbé or other refractometer of equal or greater accuracy.

Arsenic. A *Sample Solution* prepared as directed for organic compounds meets the requirements of the *Arsenic Test,* page 865.

Heavy metals. Place a 500-mg. sample in a silica crucible and proceed as directed in *Method II* under the *Heavy Metals Test,* page 920, using 20 mcg. of lead ion (Pb) in the control (*Solution A*).

Lead. A *Sample Solution* prepared as directed for organic compounds meets the requirements of the *Lead Limit Test,* page 929, using 10 mcg. of lead ion (Pb) in the control.

Packaging and storage. Store in tight, light-resistant containers.

Functional use in foods. Antioxidant; dietary supplement; nutrient.

TOCOPHEROLS CONCENTRATE, MIXED

DESCRIPTION

A source of vitamin E, obtained by the vacuum distillation of edible vegetable oils or their by-products. It may contain an edible vegetable oil added to adjust the required percent of total tocopherols, and the content of *d*-alpha tocopherol may be adjusted by suitable physical or chemical means. It occurs as a brownish red to red, nearly odorless, clear, viscous oil. It oxidizes and darkens slowly in air and on exposure to light. It is insoluble in water, soluble in alcohol, and is miscible with ether, with chloroform, with acetone, and with vegetable oils.

IDENTIFICATION

Dissolve about 50 mg. in 10 ml. of alcohol, add with swirling 2 ml. of nitric acid, and heat at about 75° for 15 minutes. A bright red to orange color develops.

SPECIFICATIONS

Assay. Not less than 34.0 percent of total tocopherols, of which not less than 50.0 percent consists of *d*-alpha tocopherol ($C_{29}H_{50}O_2$).

Limits of Impurities

Arsenic (as As). Not more than 3 parts per million (0.0003 percent).
Free fatty acids. Passes test (limit about 2.8 percent as oleic acid).
Heavy metals (as Pb). Not more than 40 parts per million (0.004 percent).
Lead. Not more than 10 parts per million (0.001 percent).

TESTS

Assay for total tocopherols. Dissolve an accurately weighed quantity, equivalent to about 15 mg. of total tocopherols, in sufficient absolute alcohol to make exactly 100 ml., and pipet 2.0 ml. of this solution into an opaque flask provided with a glass stopper. Add 1 ml. of alcoholic ferric chloride T.S., and begin timing the reaction, preferably with a stop watch. Add immediately, 2.5 ml. of α,α'-dipyridyl T.S., mix thoroughly with swirling, add exactly 19.5 ml. of absolute alcohol, and shake vigorously to insure complete mixing. After about 9½ minutes have elapsed from the beginning of the reaction, transfer part of the mixture to one of a pair of matched 1-cm. spectrophotometric cells. Exactly 10 minutes after the addition of the ferric chloride solution, determine the absorbance at 520 mμ with a suitable spectrophotometer, using absolute alcohol* as a blank. Perform a blank determination with the same quantities of the same reagents and in the same manner, but using 2 ml. of absolute alcohol in place of the 2 ml. of the sample solution. Subtract the absorbance determined for the blank from that determined for the sample, and record the difference as A_D. Calculate the total tocopherol content by the formula: $(A_D \times 28.2)/(L \times C_D) =$ mg. total tocopherols per gram, where A_D is the absorbance determined at 520 mμ corrected for the blank, L is the length of the cell in cm., C_D is the concentration of the sample expressed as grams of mixed tocopherols concentrate in each 100 ml. of the absolute alcohol solution, of which 2 ml. was taken for the assay, and 28.2 is the factor derived from most recent data on *alpha-* , *beta-* , *gamma-* , and *delta*-tocopherols present in mixed tocopherols concentrate.

Assay for non-alpha tocopherols. Dissolve an accurately weighed quantity of the sample, equivalent to about 30 mg. of non-alpha tocopherols, in sufficient absolute alcohol to make exactly 100 ml. Adjust all glassware and reagents to 25° ± 1° by immersing in a water bath adjusted to this temperature. Pipet 5.0 ml. of this solution into a 50-ml. glass-stoppered cylinder, add 0.2 ml. of glacial acetic acid, and mix thoroughly by swirling. Add 3.0 ml. of a freshly prepared solution of sodium nitrite (1 in 50) from a rapid delivery pipet, swirling vigorously for 5 seconds, and allow to stand for exactly 1 minute in the water bath. Add 2 ml. of a solution of potassium hydroxide (1 in 5) and mix thor-

* *Note:* The absorbance of the blank can sometimes be reduced, and the precision of the determination thereby improved, by purification of the absolute alcohol. Purification may be accomplished by adding 0.01 to 0.02 percent of potassium permanganate and of potassium hydroxide to the absolute alcohol and redistilling.

oughly, then add 10 ml. of water, about 200 mg. of anhydrous sodium sulfate, and 12.0 ml. of solvent hexane. Stopper the cylinder, shake vigorously for 30 seconds, and allow the layers to separate. Transfer a portion of the upper layer to one of a pair of matched 1-cm. spectrophotometric cells, and determine the absorbance at 410 mμ, placing in the other cell a blank prepared in exactly the same manner, but in which the 5.0 ml. of sample solution is replaced by 5 ml. of absolute alcohol. Calculate the non-alpha tocopherol content by the following formula: $(A_N \times 40.8)/(L \times C_N)$ = mg. of non-alpha tocopherols per gram, where A_N is the absorbance determined at 410 mμ, corrected for the blank, L is the length of the cell in cm., C_N is the concentration of the sample expressed as grams of mixed tocopherols concentrate in each 100 ml. of the absolute alcohol solution, of which 5 ml. was taken for the assay, and 40.8 is a factor derived from most recent data on *beta*- , *gamma*- , and *delta*-tocopherols in mixed tocopherols concentrate. Calculate the *d*-alpha tocopherol content by subtracting the number of mg. of non-alpha tocopherols per gram from the number of mg. total tocopherols per gram determined in the *Assay for total tocopherols.*

Arsenic. A *Sample Solution* prepared as directed for organic compounds meets the requirements of the *Arsenic Test,* page 865.

Free fatty acids. Dissolve 10 grams in 50 ml. of a mixture of equal volumes of alcohol and ether (which has been neutralized to phenolphthalein with 0.1 *N* sodium hydroxide) contained in a flask. Add 1 ml. of phenolphthalein T.S., and titrate with 0.1 *N* sodium hydroxide until the solution remains faintly pink after shaking for 30 seconds Not more than 10 ml. is required.

Heavy metals. Place 500 mg. in a silica crucible and proceed as directed in *Method II* under the *Heavy Metals Test,* page 920, using 20 mcg. of lead ion (Pb) in the control (*Solution A*).

Lead. A *Sample Solution* prepared as directed for organic compounds meets the requirements of the *Lead Limit Test,* page 929, using 10 mcg. of lead ion (Pb) in the control.

Packaging and storage. Store in tight, light-resistant containers.

Functional use in foods. Antioxidant; nutrient; dietary supplement.

d-ALPHA TOCOPHERYL ACETATE

d-Alpha Tocopherol Acetate

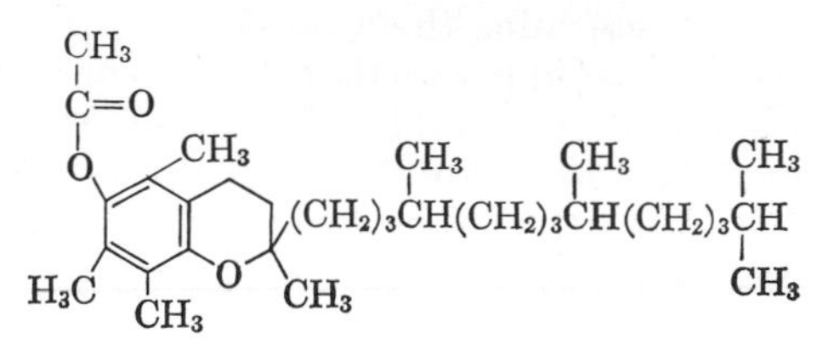

$C_{31}H_{52}O_3$ Mol. wt. 472.76

DESCRIPTION

A form of vitamin E prepared by crystallizing *d*-alpha tocopheryl acetate concentrate from a suitable solvent. It occurs as a yellow, nearly odorless, clear, viscous oil. It may solidify on standing in the cold, and melts at approximately 25°. It is unstable in the presence of alkalies, and is affected by light. It is insoluble in water, freely soluble in alcohol, and is miscible with ether, with chloroform, with acetone, and with vegetable oils.

IDENTIFICATION

A. To 10 ml. of the absolute alcohol solution of the hydrolyzed sample obtained in the *Assay*, add, with swirling, 2 ml. of nitric acid and heat at about 75° for 15 minutes. A bright red to orange color develops.

B. Its absorptivity (see page 959), determined in alcohol at 284 mμ, is between 4.0 and 4.4.

C. The angular rotation (see page 939) of a 1 in 10 solution in chloroform is approximately +0.25° when determined in a 200-mm. tube.

SPECIFICATIONS

Assay. Not less than 97.0 percent of $C_{31}H_{52}O_3$.

Refractive index. Between 1.494 and 1.498 at 20°.

Limits of Impurities

Arsenic (as As). Not more than 3 parts per million (0.0003 percent).

Heavy metals (as Pb). Not more than 40 parts per million (0.004 percent).

Lead. Not more than 10 parts per million (0.001 percent).

TESTS

Assay. Transfer about 250 mg., accurately weighed, into a 150-ml. light-resistant, round bottom, standard taper-neck flask, dissolve the sample in 25 ml. of absolute alcohol, and add 20 ml. of 5 *N* alcoholic sulfuric acid. Connect the flask with a standard taper reflux condenser, shield the solution from sunlight, and reflux in the all-glass apparatus for

3 hours. Cool the mixture to room temperature, transfer it quantitatively to a 200-ml. light-resistant volumetric flask, and dilute to volume with absolute alcohol. Transfer 50.0 ml. of this solution into a 250-ml. Erlenmeyer flask, and add 50 ml. of 0.5 *N* alcoholic sulfuric acid and 20 ml. of water. Add 2 drops of diphenylamine T.S., and titrate with 0.01 *N* ceric sulfate at a rate of approximately 25 drops per 10 seconds, swirling or stirring constantly, until a blue end-point is reached which persists for 10 seconds. Perform a blank determination using 100 ml. of 0.5 *N* alcoholic sulfuric acid, 20 ml. of water, and 2 drops of diphenylamine T.S., and make any necessary correction. Each ml. of 0.01 *N* ceric sulfate is equivalent to 2.364 mg. of $C_{31}H_{52}O_3$.

Refractive index, page 945. Determine with an Abbé or other refractometer of equal or greater accuracy.

Arsenic. A *Sample Solution* prepared as directed for organic compounds meets the requirements of the *Arsenic Test,* page 865.

Heavy metals. Place 500 mg. in a silica crucible and proceed as directed in *Method II* under the *Heavy Metals Test,* page 920, using 20 mcg. of lead ion (Pb) in the control (*Solution A*).

Lead. A *Sample Solution* prepared as directed for organic compounds meets the requirements of the *Lead Limit Test,* page 929, using 10 mcg. of lead ion (Pb) in the control.

Packaging and storage. Store in tight, light-resistant containers.

Functional use in foods. Nutrient; dietary supplement.

dl-ALPHA TOCOPHERYL ACETATE

dl-Alpha Tocopherol Acetate

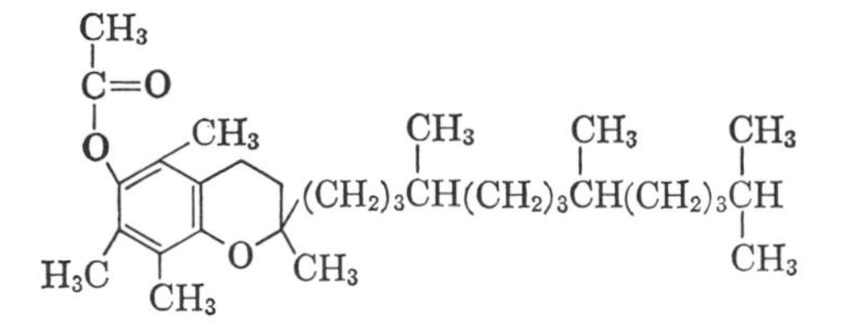

$C_{31}H_{52}O_3$ Mol. wt. 472.76

DESCRIPTION

A form of vitamin E. A yellow, nearly odorless, clear, viscous oil. It is unstable in the presence of alkalies, and is affected by light. It is insoluble in water, freely soluble in alcohol, and is miscible with ether, with chloroform, with acetone, and with vegetable oils.

IDENTIFICATION

A. To 10 ml. of the absolute alcohol solution of the hydrolyzed sample obtained in the *Assay,* add with swirling 2 ml. of nitric acid, and heat at about 75° for 15 minutes. A bright red to orange color develops.

B. Its absorptivity (see page 957), determined in alcohol at 284 mμ, is between 4.0 and 4.4.

C. The angular rotation (see page 939) of a 1 in 10 solution in chloroform is not significant when determined in a 200-mm. tube.

SPECIFICATIONS

Assay. Not less than 97.0 percent of $C_{31}H_{52}O_3$.

Refractive index. Between 1.494 and 1.498 at 20°.

Limits of Impurities

Arsenic (as As). Not more than 3 parts per million (0.0003 percent).

Heavy metals (as Pb). Not more than 40 parts per million (0.004 percent).

Lead. Not more than 10 parts per million (0.001 percent).

TESTS

Assay. Proceed as directed in the *Assay* under *d-Alpha Tocopheryl Acetate,* page 925. Each ml. of 0.01 *N* ceric sulfate is equivalent to 2.364 mg. of $C_{31}H_{52}O_3$.

Refractive index, page 945. Determine with an Abbé or other refractometer of equal or greater accuracy.

Arsenic. A *Sample Solution* prepared as directed for organic compounds meets the requirements of the *Arsenic Test,* page 865.

Heavy metals. Place 500 mg. in a silica crucible and proceed as directed in *Method II* under the *Heavy Metals Test,* page 920, using 20 mcg. of lead ion (Pb) in the control (*Solution A*).

Lead. A *Sample Solution* prepared as directed for organic compounds meets the requirements of the *Lead Limit Test,* page 929, using 10 mcg. of lead ion (Pb) in the control.

Packaging and storage. Store in tight, light-resistant containers.

Functional use in foods. Nutrient; dietary supplement.

d-ALPHA TOCOPHERYL ACETATE CONCENTRATE

d-Alpha Tocopherol Acetate Concentrate

DESCRIPTION

A source of vitamin E obtained by vacuum distillation and acetylation of edible vegetable oils or of the by-products of vegetable oil refining. The content of *d*-alpha tocopheryl acetate in *d*-Alpha Tocopheryl Acetate Concentrate may be adjusted by physical or chemical means. It occurs as a light brownish yellow, nearly odorless, clear, viscous oil. It is unstable in the presence of alkalies and of air. It is affected by light. It is insoluble in water, soluble in alcohol, and is miscible with ether, with chloroform, with acetone, and with vegetable oils.

IDENTIFICATION

To 10 ml. of the absolute alcohol solution of the unsaponifiable fraction obtained in the *Assay for total tocopheryl acetates*, add with swirling 2 ml. of nitric acid, and heat at about 75° for 15 minutes. A bright red to orange color develops.

SPECIFICATIONS

Assay. It contains not less than 25.0 percent of *d*-alpha tocopheryl acetate and may be so adjusted by the addition of an edible vegetable oil. Of the total tocopheryl acetates present not less than 64.0 percent consists of *d*-alpha tocopheryl acetate.

Limits of Impurities

Arsenic (as As). Not more than 3 parts per million (0.0003 percent).

Free fatty acids. Passes test (limit about 2.8 percent as oleic acid).

Heavy metals (as Pb). Not more than 40 parts per million (0.004 percent).

Lead. Not more than 10 parts per million (0.001 percent).

TESTS

Assay for total tocopheryl acetates. Transfer an accurately weighed quantity, equivalent to about 200 mg. of total tocopheryl acetates, into a 250-ml., light-resistant, round bottom standard taper-neck flask. Dissolve the sample in 50 ml. of absolute alcohol, connect the flask with a standard taper condenser, and reflux for at least 1 minute. While the solution is boiling, add through the condenser 1 gram of potassium hydroxide pellets, 1 at a time to avoid overheating. *Caution: Wear safety goggles.* Continue refluxing for 20 minutes, and without cooling add, dropwise, 2 ml. of hydrochloric acid through the condenser. Cool to room temperature, and transfer the contents of the flask to a 500-ml., light-resistant separator, rinsing the flask with 100 ml. of water and 100 ml. of ether. Shake vigorously, allow the layers to separate, and collect each of the two layers in separate flasks. Extract the water layer twice with 50-ml. portions of ether, and add these two ether extracts to the first ether extract. Wash the combined ether extracts with four 100-ml. portions of water, and then evaporate the ether solution on a water bath under vacuum or in an atmosphere of nitrogen until about 7 or 8 ml. of the liquid remain. Remove the flask from the water bath, and evaporate the last traces of ether without application of heat. Immediately dissolve the residue in absolute alcohol, transfer the solution to a 250-ml. volumetric flask, and dilute to exactly 250 ml. with absolute alcohol. *Express the concentration in terms of the number of grams of the original sample before saponification represented in each 100 ml. of the absolute alcohol solution.* Transfer an accurately measured volume of the solution, equivalent to about 15 mg. of total tocopheryl acetates, into a 100-ml., opaque volumetric flask, and add sufficient absolute alcohol to make exactly 100 ml. Pipet 2 ml. of this solution into a light-resistant flask provided with a glass stopper and proceed as directed in the *Assay for total tocopherols* under *Mixed Tocopherols Concentrate*, page 821, beginning with "Add 1 ml. of alcoholic ferric chloride T.S. . . ." Calculate the total tocopheryl acetate content by the following formula: $(A_D \times 32.0)/(L \times C_D)$ = mg. total tocopheryl acetates per gram where A_D is the absorbance determined at 520 mμ corrected for the blank, L is the length of the cell in cm., C_D is the concentration of the sample expressed as grams in each 100 ml. of the absolute alcohol solution, of which 2 ml. was taken for the assay, and 32.0 is the factor derived from most recent data on *alpha-* , *beta-* , *gamma-* , and *delta*-tocopherols in *d*-alpha tocopheryl acetate concentrate.

Assay for non-alpha tocopheryl acetates. Adjust all glassware and reagents to be used in the assay to 25° ± 1° by immersing in a water bath adjusted to this temperature. Prepare a sample solution from the unsaponifiable fraction obtained in the *Assay for total tocopheryl acetates*, using the alcohol solution contained in the 250-ml. volumetric flask, as follows: Transfer to a 100-ml. volumetric flask an accurately measured volume of the alcohol solution equivalent to about 30 mg. of non-alpha tocopheryl acetates, dilute to volume with absolute alcohol, and mix. Pipet 5 ml. of this solution into a 50-ml. glass-stoppered cylinder, and proceed as directed in the *Assay for non-alpha tocopherols* under *Mixed Tocopherols Concentrate*, page 821, beginning with " . . . add 0.2 ml. of glacial acetic acid . . . " Calculate the content of non-alpha tocopheryl acetates by the following formula: $(A_N \times 51.9)/(L \times C_N)$ = mg. non-alpha tocopheryl acetates per gram, where A_N is the absorbance determined at 410 mμ, L is the length of the cell in cm., C_N is the concentration of the sample expressed as grams in 100 ml. of the absolute alcohol solution, of which 5 ml. was taken for the assay, and 51.9 is a factor derived from most recent data on *beta-* , *gamma-* , and *delta*-tocopherols in *d*-alpha tocopheryl acetate concentrate. Calculate the *d*-alpha tocopheryl acetate content by subtracting the number of mg. of non-*alpha* tocopheryl acetates per gram from the number of mg. per gram obtained in the *Assay for total tocopheryl acetates.*

Arsenic. A *Sample Solution* prepared as directed for organic compounds meets the requirements of the *Arsenic Test*, page 865.

Free fatty acids. Dissolve 10 grams in 50 ml. of a mixture of equal volumes of alcohol and ether (which has been neutralized to phenolphthalein with 0.1 N sodium hydroxide) contained in a flask. Add 1 ml. of phenolphthalein T.S., and titrate with 0.1 N sodium hydroxide until the solution remains faintly pink after shaking for 30 seconds. Not more than 10 ml. is required.

Heavy metals. Place 500 mg. in a silica crucible and proceed as directed in *Method II* under the *Heavy Metals Test*, page 920, using 20 mcg. of lead ion (Pb) in the control (*Solution A*).

Lead. A *Sample Solution* prepared as directed for organic compounds meets the requirements of the *Lead Limit Test*, page 929, using 10 mcg. of lead ion (Pb) in the control.

Packaging and storage. Store in tight, light-resistant containers.

Functional use in foods. Nutrient; dietary supplement.

d-ALPHA TOCOPHERYL ACID SUCCINATE

d-Alpha Tocopherol Acid Succinate

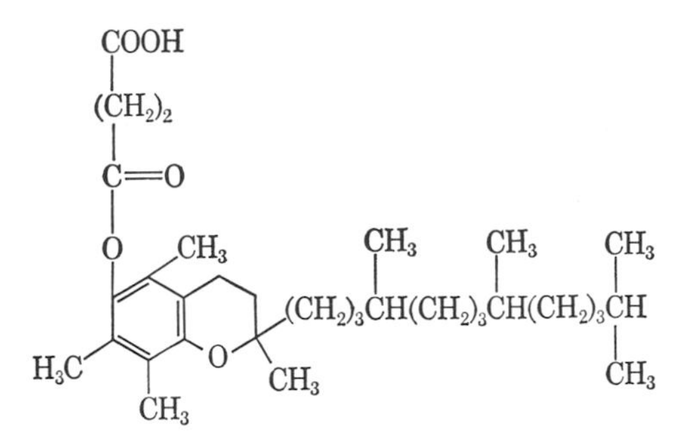

$C_{33}H_{54}O_5$ Mol. wt. 530.80

DESCRIPTION

A crystalline form of vitamin E prepared by crystallizing from a suitable solvent a concentrate of the acid succinate ester of *d*-alpha tocopherol prepared by physical and chemical treatment of edible vegetable oils or of the by-products of vegetable oil refining. It occurs as a white crystalline powder having little or no taste or odor. It is stable in air, but is unstable to alkali and to heat. It is insoluble in water, slightly soluble in aqueous alkali, soluble in alcohol, ether, acetone, and vegetable oils, and very soluble in chloroform.

IDENTIFICATION

A. To 10 ml. of the absolute alcohol solution of the unsaponifiable fraction obtained in the *Assay*, add, without swirling, 2 ml. of nitric acid, and heat at about 75° for 15 minutes. A bright red to orange color develops.

B. Its absorptivity (see page 957), determined in alcohol at 284 mμ, is between 3.5 and 4.0.

C. One gram requires for neutralization not less than 18.0 ml. and not more than 19.3 ml. of 0.1 *N* sodium hydroxide.

SPECIFICATIONS

Assay. Not less than 97.0 percent of $C_{33}H_{54}O_5$.

Melting range. Between 73° and 78°.

Limits of Impurities

Arsenic (as As). Not more than 3 parts per million (0.0003 percent).

Heavy metals (as Pb). Not more than 40 parts per million (0.004 percent).

Lead. Not more than 10 parts per million (0.001 percent).

TESTS

Assay. Transfer about 250 mg., accurately weighed, into a 250-ml. light-resistant, round bottom, standard taper-neck flask. Dissolve the sample in 50 ml. of absolute alcohol, connect the flask with a standard taper condenser, and reflux for at least 1 minute. While the solution is boiling, add through the condenser 1 gram of potassium hydroxide pellets, one at a time to avoid overheating. *Caution: Wear safety goggles.* Continue refluxing for 20 minutes, and without cooling add, dropwise, 2 ml. of hydrochloric acid through the condenser. Cool to room temperature, and transfer the contents of the flask to a 500-ml., light-resistant separator, rinsing the flask with 100 ml. of water and 100 ml. of ether. Shake vigorously, allow the layers to separate, and collect each of the two layers in separate flasks. Extract the water layer twice with 50-ml. portions of ether, and add these two ether extracts to the first ether extract. Wash the combined ether extracts with four 100-ml. portions of water, and then evaporate the ether solution on a water bath under vacuum or in an atmosphere of nitrogen until about 7 or 8 ml. of the liquid remains. Remove the flask from the water bath and evaporate the last traces of ether without application of heat. Immediately dissolve the residue in 0.5 *N* alcoholic sulfuric acid. Transfer the solution to a 200-ml. volumetric flask and dilute to volume with 0.5 *N* alcoholic sulfuric acid. Transfer 50.0 ml. of this solution into a 250-ml. Erlenmeyer flask, add 50 ml. of 0.5 *N* alcoholic sulfuric acid, and 20 ml. of water. Add 2 drops of diphenylamine T.S. and titrate with 0.01 *N* ceric sulfate at a rate of approximately 25 drops per 10 seconds, swirling or stirring constantly, until a blue end-point is reached which persists for 10 seconds. Perform a blank determination using 100 ml. of 0.5 *N* alcoholic sulfuric acid, 20 ml. of water, and 2 drops of diphenylamine T.S., and make any necessary correction. Each ml. of 0.01 *N* ceric sulfate is equivalent to 2.654 mg. of $C_{33}H_{54}O_5$.

Melting range. Determine as directed in the general procedure, page 931.

Arsenic. A *Sample Solution* prepared as directed for organic compounds meets the requirements of the *Arsenic Test,* page 865.

Heavy metals. Place 500 mg. in a silica crucible and proceed as directed in *Method II* under the *Heavy Metals Test*, page 920, using 20 mcg. of lead ion (Pb) in the control (*Solution A*).

Lead. A *Sample Solution* prepared as directed for organic compounds meets the requirements of the *Lead Limit Test*, page 929, using 10 mcg. of lead ion (Pb) in the control.

Packaging and storage. Store in tight, light-resistant containers.

Functional use in foods. Nutrient; dietary supplement.

p-TOLYL ISOBUTYRATE

p-Cresyl Isobutyrate

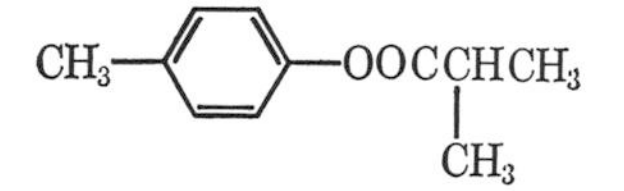

$C_{11}H_{14}O_2$ Mol. wt. 178.23

DESCRIPTION

A colorless liquid having a characteristic odor. It is soluble in alcohol, but practically insoluble in water.

SPECIFICATIONS

Assay. Not less than 95.0 percent of $C_{11}H_{14}O_2$.

Refractive index. Between 1.485 and 1.489 at 20°.

Solubility in alcohol. Passes test.

Specific gravity. Between 0.990 and 0.996.

Limits of Impurities

Acid value. Not more than 1.0.

TESTS

Assay. Weigh accurately about 1.2 grams, and proceed as directed under *Ester Determination* (*High Boiling Solvent*), page 896, using 89.12 as the equivalence factor (*e*) in the calculation.

Refractive index, page 945. Determine with an Abbé or other refractometer of equal or greater accuracy.

Solubility in alcohol. Proceed as directed in the general method, page 899. One ml. dissolves in 7 ml. of 70 percent alcohol to form clear solutions.

Specific gravity. Determine by any reliable method (see page 5).

Acid value. Determine as directed in the general method, page 893.

Packaging and storage. Store in full, tight, preferably glass, tin-lined or other suitably lined containers in a cool place protected from light.

Functional use in foods. Flavoring agent.

TRAGACANTH

DESCRIPTION

A dried gummy exudation obtained from *Astragalus gummifer* Labillardiere, or other Asiatic species of *Astragalus* (Fam. *Leguminosae*). *Unground Tragacanth* occurs as flattened, lamellated, frequently curved fragments or straight or spirally twisted linear pieces from 0.5 to 2.5 mm. in thickness. It is white to weak yellow in color, translucent, horny in texture, and having a short fracture. It is odorless and has an insipid, mucilaginous taste. It is rendered more easily pulverizable if heated to a temperature of 50°. *Powdered Tragacanth* is white to yellowish white in color. When examined microscopically in water mounts, it shows numerous angular fragments with circular or irregular lamellae, and starch grains up to 25 μ in diameter. It should show very few or no fragments of lignified vegetable tissue. One gram in 50 ml. of water swells to form a smooth, stiff, opalescent mucilage free from cellular fragments.

SPECIFICATIONS

Viscosity of a 1 percent solution. Not less than 250 centipoises.

Limits of Impurities

Arsenic (as As). Not more than 3 parts per million (0.0003 percent).

Ash (Total). Not more than 3.0 percent.

Ash (Acid-insoluble). Not more than 0.5 percent.

Heavy metals (as Pb). Not more than 40 parts per million (0.004 percent).

Karaya gum. Passes test.

Lead. Not more than 10 parts per million (0.001 percent).

TESTS

Viscosity. Transfer a 4.0-gram sample, finely powdered, into the container of a stirring apparatus equipped with blades capable of revolving at 10,000 rpm. Add 10 ml. of alcohol to the sample, swirl to wet the gum uniformly, and then add 390 ml. of water avoiding the formation of lumps. Immediately stir the mixture for 7 minutes, pour the resulting dispersion into a 500-ml. bottle, insert a stopper, and allow to stand for about 24 hours in a water bath at 25°. Determine the apparent viscosity at this temperature with a Model LVF

Brookfield or equivalent type viscometer (see *Viscosity of Sodium Carboxymethylcellulose*, page 973) using Spindle No. 2 at 30 rpm and a factor of 10.

Arsenic. A *Sample Solution* prepared as directed for organic compounds meets the requirements of the *Arsenic Test*, page 865.

Ash (Total). Determine as directed in the general method, page 868.

Ash (Acid-insoluble). Determine as directed in the general method, page 869.

Heavy metals. Prepare and test a 500-mg. sample as directed in *Method II* under the *Heavy Metals Test*, page 920, using 20 mcg. of lead ion (Pb) in the control (*Solution A*).

Karaya gum. Boil 1 gram with 20 ml. of water until a mucilage is formed, add 5 ml. of hydrochloric acid and again boil the mixture for 5 minutes. No permanent pink or red color develops.

Lead. A *Sample Solution* prepared as directed for organic compounds meets the requirements of the *Lead Limit Test*, page 929, using 10 mcg. of lead ion (Pb) in the control.

Packaging and storage. Store in well-closed containers.

Functional use in foods. Stabilizer; thickener; emulsifier.

TRIACETIN

Glyceryl Triacetate

```
 H
 |
HCOOCCH3
 |
HCOOCCH3
 |
HCOOCCH3
 |
 H
```

$C_9H_{14}O_6$ Mol. wt. 218.21

DESCRIPTION

A colorless, somewhat oily liquid with a slight, fatty odor, and a bitter taste. It is soluble in water, and is miscible with alcohol, with ether, and with chloroform. It distils between 258° and 270°.

IDENTIFICATION

A. Heat a few drops in a test tube with about 500 mg. of potassium bisulfate. Pungent vapors of acrolein are evolved.

B. The solution resulting from the *Assay* gives positive tests for *Acetate*, page 925.

SPECIFICATIONS

Assay. Not less than 98.5 percent of $C_9H_{14}O_6$.

Refractive index. Between 1.429 and 1.431 at 25°

Specific gravity. Between 1.154 and 1.158.

Limits of Impurities

Acidity. Passes test.

Arsenic (as As). Not more than 3 parts per million (0.0003 percent).

Heavy metals (as Pb). Not more than 10 parts per million (0.001 percent).

Unsaturated compounds. Passes test.

Water. Not more than 0.2 percent.

TESTS

Assay. Transfer about 1 gram of the sample, accurately weighed, into a suitable pressure bottle, add 25.0 ml. of 1 *N* potassium hydroxide and 15 ml. of isopropanol, stopper the bottle, and wrap securely in a canvas bag. Place in a water bath maintained at 98° ± 2°, and heat for 1 hour, allowing the water in the bath to just cover the liquid in the bottle. Remove the bottle from the bath, cool in air to room temperature, then loosen the wrapper, uncap the bottle to release any pressure, and remove the wrapper. Add 6 to 8 drops of phenolphthalein T.S., and titrate the excess alkali with 0.5 *N* sulfuric acid just to the disappearance of the pink color Perform a blank determination (see page 2). Each ml. of 0.5 *N* sulfuric acid is equivalent to 36.37 mg. of $C_9H_{14}O_6$.

Refractive index, page 945. Determine at 25° with an Abbé or other refractometer of equal or greater accuracy.

Specific gravity. Determine by any reliable method (see page 5).

Acidity. Dilute 25 grams, accurately weighed, with 50 ml. of neutralized alcohol, add 5 drops of phenolphthalein T.S., and titrate with 0.02 *N* sodium hydroxide. Not more than 1 ml. is required.

Arsenic. A *Sample Solution* prepared as directed for organic compounds meets the requirements of the *Arsenic Test*, page 865.

Heavy metals. Prepare and test a 2-gram sample as directed in *Method II* under the *Heavy Metals Test*, page 920, using 20 mcg. of lead ion (Pb) in the control (*Solution A*).

Unsaturated compounds. To 10 ml. of the sample in a glass-stoppered tube add, dropwise, a solution of bromine in carbon tetrachloride (1 ml. in 100 ml.) until a permanent yellow color is produced, and allow to stand in a dark place for 18 hours. No turbidity or precipitate appears.

Water. Determine by the *Karl Fischer Titrimetric Method*, page 977.

Packaging and storage. Store in tight containers, and do not permit contact with metal.

Functional use in foods. Humectant; solvent.

TRIBUTYRIN

Glyceryl Tributyrate; Butyrin

$$
\begin{array}{c}
\text{H} \\
| \\
\text{H—C—OCOC}_3\text{H}_7 \\
| \\
\text{H—C—OCOC}_3\text{H}_7 \\
| \\
\text{H—C—OCOC}_3\text{H}_7 \\
| \\
\text{H}
\end{array}
$$

$C_{15}H_{26}O_6$ Mol. wt. 302.37

DESCRIPTION

A colorless, somewhat oily liquid having a characteristic odor and a bitter taste. It is soluble in alcohol, in chloroform, and in ether, but insoluble in water. It distils between 305° and 310°.

SPECIFICATIONS

Assay. Not less than 99.0 percent of $C_{15}H_{26}O_6$.

Specific gravity. Between 1.034 and 1.037 at 20°.

Limits of Impurities

Acidity (as acetic acid). Not more than 0.04 percent.

Arsenic (as As). Not more than 3 parts per million (0.0003 percent).

Heavy metals (as Pb). Not more than 10 parts per million (0.001 percent).

TESTS

Assay. Weigh accurately about 1.5 grams, and proceed as directed under *Ester Determination*, page 896, using 50.40 as the equivalence factor (*e*) in the calculation.

Specific gravity. Determine by any reliable method (see page 5).

Acidity. To 75 ml. of methanol, contained in a 250-ml. wide-mouth Erlenmeyer flask, add 1 ml. of bromothymol blue T.S., and titrate with 0.05 *N* sodium hydroxide to a blue end-point. Transfer a 20.0-ml. sample into the flask, calculate its weight from its specific gravity, and titrate back to the same blue end-point. Each ml. of 0.05 *N* sodium hydroxide is equivalent to 3 mg. of acetic acid ($C_2H_4O_2$).

Arsenic. A *Sample Solution* prepared as directed for organic compounds meets the requirements of the *Arsenic Test*, page 865.

Heavy metals. Prepare and test a 2-gram sample as directed in *Method II* under the *Heavy Metals Test*, page 920, using 20 mcg. of lead ion (Pb) in the control (*Solution A*).

Packaging and storage. Store in tight containers.

Functional use in foods. Flavoring agent.

TRICHLOROETHYLENE

Ethylene Trichloride

$$\begin{array}{ccccc} & & H & & Cl \\ & & | & & | \\ Cl & — & C & = & C — Cl \end{array}$$

C_2HCl_3 Mol. wt. 131.39

DESCRIPTION

A clear, colorless, mobile liquid having a sweet, chloroform-like odor. It is immiscible with water but is miscible with alcohol, ether, acetone, and carbon tetrachloride. Its refractive index at 20° is about 1.477. It may contain a suitable stabilizer.

SPECIFICATIONS

Distillation range. Between 86° and 88°.

Specific gravity. Between 1.454 and 1.458.

Limits of Impurities

Acidity (as HCl). Not more than 10 parts per million (0.001 percent).

Alkalinity (as NaOH). Not more than 10 parts per million (0.001 percent).

Free halogens. Passes test.

Heavy metals (as Pb). Not more than 1 part per million (0.0001 percent).

Nonvolatile residue. Not more than 10 parts per million (0.001 percent).

Water. Not more than 0.05 percent.

TESTS

Distillation range. Determine as directed in the general method, page 890.

Specific gravity. Determine by any reliable method (see page 5).

Acidity or alkalinity. Transfer 25 ml. of water and 2 drops of phenolphthalein T.S. to a 250-ml. glass-stoppered flask, and add 0.01 *N* sodium hydroxide to the first appearance of a slight pink color. Add 25 ml. (about 36 grams) of the sample, and shake for 30 seconds. If the pink color persists, titrate with 0.01 *N* hydrochloric acid, shaking repeatedly, until the pink color just disappears; not more than 0.9 ml. is required. If the pink color is discharged when the sample is added, titrate with 0.01 *N* sodium hydroxide until the faint pink color is restored; not more than 1.0 ml. is required.

Free halogens. Shake 10 ml. of the sample vigorously for 2 minutes with 10 ml. of potassium iodide solution (1 in 10) and 1 ml. of starch T.S. A blue color does not appear in the water layer.

Heavy metals. Evaporate 14 ml. (about 20 grams) of the sample to dryness (*Caution:* use hood) on a steam bath in a glass evaporating dish. Cool, add 2 ml. of hydrochloric acid, and slowly evaporate to dryness again on the steam bath. Moisten the residue with 1 drop of hydrochloric acid, add 10 ml. of hot water, and digest for 2 minutes. Filter if necessary through a small filter, wash the evaporating dish and the filter with about 10 ml. of water, and dilute to 25 ml. with water. This solution meets the requirements of the *Heavy Metals Test*, page 920, using 20 mcg. of lead ion (Pb) in the control (*Solution A*).

Nonvolatile residue. Evaporate 69 ml. (about 100 grams) of the sample to dryness (*Caution:* use hood) in a tared dish on a steam bath, dry the residue at 105° for 30 minutes, cool, and weigh.

Water. Determine by the *Karl Fischer Titrimetric Method*, page 977.

Packaging and storage. Store in tight containers.

Functional use in foods. Extraction solvent.

TRIETHYL CITRATE

Ethyl Citrate

$$HO-\underset{CH_2COOC_2H_5}{\overset{CH_2COOC_2H_5}{C}}-COOC_2H_5$$

$C_{12}H_{20}O_7$ Mol. wt. 276.29

DESCRIPTION

An odorless, practically colorless, oily liquid. It is slightly soluble in water, but is miscible with alcohol and with ether. Its specific gravity is approximately 1.137 at 20°.

SPECIFICATIONS

Assay. Not less than 99.0 percent of $C_{12}H_{20}O_7$.

Refractive index. Between 1.439 and 1.441.

Limits of Impurities

Acidity (as citric acid). Not more than 0.02 percent.

Arsenic (as As). Not more than 3 parts per million (0.0003 percent).

Heavy metals (as Pb). Not more than 10 parts per million (0.001 percent).

Water. Not more than 0.25 percent.

TESTS

Assay. Weigh accurately about 1.5 grams of the sample into a 500-ml. flask equipped with a standard taper ground joint, and add 25 ml. of isopropyl alcohol and 25 ml. of water. Pipet 50 ml. of 0.5 *N* sodium hydroxide into the mixture, add a few boiling chips, and attach a suitable water-cooled condenser. Reflux for 1.5 hours, then cool, wash down the condenser with about 20 ml. of water, add 5 drops of phenolphthalein T.S., and titrate the excess alkali with 0.5 *N* sulfuric acid. Perform a blank determination (see page 2). Each ml. of 0.5 *N* sulfuric acid is equivalent to 46.05 mg. of $C_{12}H_{20}O_7$.

Refractive index, page 945. Determine with an Abbé or other refractometer of equal or greater accuracy.

Acidity. Dissolve 3.2 grams, accurately weighed, in 20 ml. of neutralized alcohol, add phenolphthalein T.S., and titrate with 0.1 *N* sodium hydroxide. Not more than 10.0 ml. is required.

Arsenic. A *Sample Solution* prepared as directed for organic compounds meets the requirements of the *Arsenic Test,* page 865.

Heavy metals. Prepare and test a 2-gram sample as directed in *Method II* under the *Heavy Metals Test,* page 920, using 20 mcg. of lead ion (Pb) in the control (*Solution A*).

Water. Determine by the *Karl Fischer Titrimetric Method,* page 977.

Packaging and storage. Store in well-closed containers.

Functional use in foods. Sequestrant.

DL-TRYPTOPHAN

DL-α-Amino-3-indolepropionic Acid

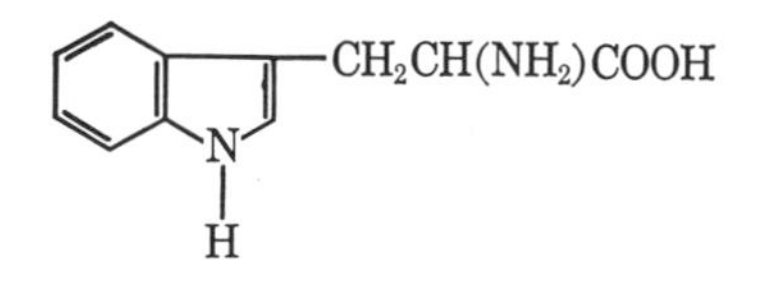

$C_{11}H_{12}N_2O_2$ Mol. wt. 204.23

DESCRIPTION

White, odorless, crystals or a crystalline powder. It is soluble in water and in dilute acids and alkalies. It is sparingly soluble in alcohol. It is optically inactive.

IDENTIFICATION

To a solution of DL-tryptophane in water add bromine T.S. A red color appears which may be extracted with amyl alcohol.

SPECIFICATIONS

Assay. Not less than 98.0 percent and not more than the equivalent of 102.0 percent of $C_{11}H_{12}N_2O_2$ after drying.

Nitrogen (Total). Between 13.4 percent and 13.9 percent.

Limits of Impurities

Ammonium salts (as NH_3). Not more than 300 parts per million (0.03 percent).

Arsenic (as As). Not more than 3 parts per million (0.0003 percent).

Chloride. Not more than 200 parts per million (0.02 percent).

Heavy metals (as Pb). Not more than 40 parts per million (0.004 percent).

Indole. Passes test.

Iron. Not more than 50 parts per million (0.005 percent).

Lead. Not more than 10 parts per million (0.001 percent).

Loss on drying. Not more than 0.3 percent.

Residue on ignition. Not more than 0.1 percent.

Sulfate. Not more than 400 parts per million (0.004 percent).

TESTS

Assay

Standard Solution. Transfer about 100 mg. of U.S.P. L-Tryptophane Reference Standard, previously dried at 105° for 3 hours and accurately weighed, into a 100-ml. volumetric flask, dissolve it in 50 ml. of hot water, cool, dilute to volume with water, and mix. Transfer 2.0 ml. of this solution into a second 100-ml. volumetric flask, dilute to volume with water, and mix.

Assay Solution. Transfer about 100 mg. of the sample, previously dried at 105° for 3 hours and accurately weighed, into a 100-ml. volumetric flask, dissolve it in 50 ml. of hot water, cool, dilute to volume with water, and mix. Transfer 2.0 ml. of this solution into a second 100-ml. volumetric flask, dilute to volume with water, and mix.

Procedure. Determine the absorbance of each solution in a 1-cm. quartz cell at the wavelength of maximum absorption at about 279 mμ, with a suitable spectrophotometer, using water as the blank. Calculate the quantity, in mg., of $C_{11}H_{12}N_2O_2$ in the sample taken by the formula $5C(A_U/A_S)$, in which C is the concentration, in mcg. per ml., of U.S.P. L-Tryptophane Reference Standard in the *Standard Solution*, A_U is the absorbance of the *Assay Solution*, and A_S is the absorbance of the *Standard Solution*.

Nitrogen (Total). Determine as directed under *Nitrogen Determination*, page 937, using about 300 mg. of the sample, previously dried and accurately weighed.

Ammonium salts

Ammonium chloride standard solution. Dissolve 78.5 mg. of ammonium chloride in sufficient water to make exactly 250 ml., and dilute

10.0 ml. of this solution to 100.0 ml. The dilute solution contains the equivalent of 10 mcg. of NH_3 in each ml.

Procedure. Place 5 grams of sodium hydroxide pellets and 300 ml. of water in a 500-ml. distillation flask. Remove ammonia from the solution by distilling until 25 ml. of the distillate gives no color with 0.5 ml. of alkaline mercuric-potassium iodide T.S. Allow the solution in the distillation flask to cool and add 50 mg. of tryptophane. Distil, collecting two 25-ml. fractions in 50-ml. Nessler tubes. Add 0.5 ml. of alkaline mercuric-potassium iodide T.S. to the distillates and to a series of standards, prepared by dilution of the *Ammonium chloride standard solution*, containing the equivalent of 0, 5, 10, 15, and 20 mcg. of ammonia in 25 ml. of solution. Stopper the tubes and invert them several times to mix their contents. After allowing 10 minutes for color development, determine the amount of ammonia present in the distillates by comparing their color intensities with those of the standards. If the second 25 ml. of distillate from the tryptophan sample contains appreciable ammonia, distil and collect additional 25-ml. fractions and determine their ammonia content.

Arsenic. A *Sample Solution* prepared as directed for organic compounds meets the requirements for the *Arsenic Test*, page 865.

Chloride, page 879. Any turbidity produced by a 100-mg. sample does not exceed that shown in a control containing 20 mcg. of chloride ion (Cl).

Heavy metals. A solution of 500 mg. in 20 ml. of water meets the requirements of the *Heavy Metals Test*, page 920, using 20 mcg. of lead ion (Pb) in the control (*Solution A*).

Indole. Dissolve 500 mg. in 1 ml. of diluted hydrochloric acid T.S. and dilute to 50 ml. Transfer 10 ml. of this solution to a test tube, add 2 ml. of *n*-amyl or isoamyl alcohol, and shake vigorously. The amyl alcohol layer has no more than a light yellow color (an orange or pink color indicates the presence of indole or other color-producing intermediates). Allow the remaining 40 ml. of the solution to stand overnight. A slightly pink color develops, but no precipitate forms.

Iron. To the ash obtained in the test for *Residue on ignition* add 2 ml. of dilute hydrochloric acid (1 in 12), and evaporate to dryness on a steam bath. Dissolve the residue in 1 ml. of hydrochloric acid and dilute with water to 50 ml. Dilute 10 ml. of this solution to 40 ml. with water and add 30 mg. of ammonium persulfate crystals and 10 ml. of ammonium thiocyanate T.S. Any red or pink color does not exceed that produced by 1.0 ml. of *Iron Standard Solution* (10 mcg. of Fe) in an equal volume of a solution containing the quantities of reagents used in the test.

Lead. A *Sample Solution* prepared as directed for organic compounds meets the requirements of the *Lead Limit Test*, page 929, using 10 mcg. of lead ion (Pb) in the control.

Loss on drying, page 931. Dry at 105° for 3 hours.

Residue on ignition. Ignite 2 grams as directed in the general method, page 945.

Sulfate, page 879. Any turbidity produced by a 500-mg. sample does not exceed that shown in a control containing 200 mcg. of sulfate (SO_4).

Packaging and storage. Store in well-closed containers.

Functional use in foods. Nutrient; dietary supplement.

L-TRYPTOPHAN

L-α-Amino-3-indolpropionic Acid

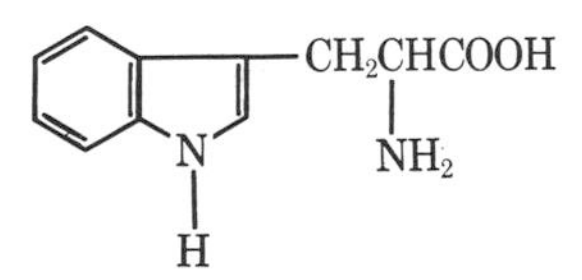

$C_{11}H_{12}N_2O_2$ Mol. wt. 204.23

DESCRIPTION

White to yellowish white crystals or a crystalline powder. It is odorless and has a slightly bitter taste. One gram dissolves in about 100 ml. of water. It is soluble in hot alcohol, in dilute hydrochloric acid, and in alkali hydroxide solutions.

IDENTIFICATION

Dissolve about 1 gram in 100 ml. of hydrochloric acid solution (1 in 5). To 1 ml. of this solution add 1 ml. of sodium sulfite solution (1 in 20). A yellow color is produced.

SPECIFICATIONS

Assay. Not less than 98.0 percent and not more than the equivalent of 102.0 percent of $C_{11}H_{12}N_2O_2$, calculated on the dried basis.

Specific rotation, $[\alpha]^{20°}_{D}$. Between −30.0° and −33.0°, on the dried basis.

Limits of Impurities

Arsenic (as As). Not more than 3 parts per million (0.0003 percent).

Heavy metals (as Pb). Not more than 20 parts per million (0.002 percent).

Lead. Not more than 10 parts per million (0.001 percent).

Loss on drying. Not more than 0.3 percent.

Residue on ignition. Not more than 0.1 percent.

TESTS

Assay. Dissolve about 300 mg., accurately weighed, in 3 ml. of formic acid and 50 ml. of glacial acetic acid, add 2 drops of crystal violet T.S., and titrate with 0.1 *N* perchloric acid to a green end-point or until the blue color disappears completely. Each ml. of 0.1 *N* perchloric acid is equivalent to 20.42 mg. of $C_{11}H_{12}N_2O_2$.

Specific rotation, page 939. Determine in a solution containing 1 gram of a previously dried sample in sufficient water to make 100 ml.

Arsenic. A *Sample Solution* prepared as directed for organic compounds meets the requirements of *Arsenic Test,* page 865.

Heavy metals. Prepare and test a 1-gram sample as directed in *Method II* under the *Heavy Metals Test,* page 920, using 20 mcg. of lead ion (Pb) in the control (*Solution A*).

Lead. A *Sample Solution* prepared as directed for organic compounds meets the requirements of the *Lead Limit Test,* page 929, using 10 mcg. of lead ion (Pb) in the control.

Loss on drying, page 931. Dry at 105° for 3 hours.

Residue on ignition, page 945. Ignite 1 gram as directed in the general method.

Packaging and storage. Store in well-closed, light-resistant containers.

Functional use in foods. Nutrient; dietary supplement.

L-TYROSINE

L-β-(*p*-Hydroxyphenyl)alanine

HO—C_6H_4—$CH_2CH(NH_2)COOH$

$C_9H_{11}NO_3$ Mol. wt. 181.19

DESCRIPTION

Colorless, silky needles, or a white crystalline powder. One gram is soluble in about 230 ml. of water. It is soluble in dilute mineral acids and in alkaline solutions. It is very slightly soluble in alcohol.

IDENTIFICATION

Heat 5 ml. of a 1 in 1000 solution with 1 ml. of triketohydrindene hydrate T.S. A reddish purple color is produced.

SPECIFICATIONS

Assay. Not less than 98.5 percent of $C_9H_{11}NO_3$ after drying.

Nitrogen. Between 7.5 percent and 7.8 percent.

Specific rotation. $[\alpha]^{25°}_D$: Between −9.8° and −11.2°; $[\alpha]^{20°}_D$: between −11.3° and −12.3°.

Limits of Impurities

Arsenic (as As). Not more than 3 parts per million (0.0003 percent).

Heavy metals (as Pb). Not more than 30 parts per million (0.003 percent).

Iron. Not more than 50 parts per million (0.005 percent).

Lead. Not more than 10 parts per million (0.001 percent).

Loss on drying. Not more than 0.3 percent.

Residue on ignition. Not more than 0.1 percent.

TESTS

Assay. Transfer about 400 mg., previously dried at 105° for 2 hours and accurately weighed, into a 250-ml. flask. Dissolve the sample in about 50 ml. of acetic acid, add 2 drops of crystal violet T.S., and titrate with 0.1 *N* perchloric acid to a bluish green end-point. Perform a blank determination (see page 2) and make any necessary correction. Each ml. of 0.1 *N* perchloric acid is equivalent to 18.12 mg. of $C_9H_{11}NO_3$.

Nitrogen (Total). Determine as directed under *Nitrogen Determination*, page 937, using about 300 mg. of the sample, previously dried and accurately weighed.

Specific rotation, page 939. $[\alpha]^{25°}_D$: Determine in a solution containing 4 grams in sufficient 1 *N* hydrochloric acid to make 100 ml.; $[\alpha]^{20°}_D$: determine in a solution containing 5 grams in sufficient 1 *N* hydrochloric acid to make 100 ml.

Arsenic. A *Sample Solution* prepared as directed for organic compounds meets the requirements of the *Arsenic Test*, page 865.

Heavy metals. A solution of 670 mg. in 25 ml. of water meets the requirements of the *Heavy Metals Test*, page 920, using 20 mcg. of lead ion (Pb) in the control (*Solution A*).

Iron. To the ash obtained in the test for *Residue on Ignition* add 2 ml. of dilute hydrochloric acid (1 in 2), and evaporate to dryness on a steam bath. Dissolve the residue in 1 ml. of hydrochloric acid and dilute with water to 50 ml. Dilute 10 ml. of this solution to 50 ml. with water and add 40 mg. of ammonium persulfate crystals and 10 ml. of ammonium thiocyanate T.S. Any red or pink color does not exceed that produced by 1.0 ml. of *Iron Standard Solution* (10 mcg. Fe) in an equal volume of a solution containing the quantities of the reagents used in the test.

Lead. A *Sample Solution* prepared as directed for organic compounds meets the requirements of the *Lead Limit Test*, page 929, using 10 mcg. of lead ion (Pb) in the control.

Loss on drying, page 931. Dry at 105° for 3 hours.

Residue on ignition. Ignite 2 grams as directed in the general method, page 945.

Packaging and storage. Store in well-closed containers.

Functional use in foods. Nutrient; dietary supplement.

γ-UNDECALACTONE

Aldehyde C-14 Pure (so-called); Peach Aldehyde

$$CH_3(CH_2)_6\underset{|}{C}HCH_2\underset{|}{C}H_2$$
$$O\text{———}C{=}O$$

$C_{11}H_{20}O_2$ Mol. wt. 184.28

DESCRIPTION

A colorless to yellow liquid having a strong fruity odor suggestive of peach, particularly upon dilution. It is soluble in most fixed oils, in mineral oil, and in propylene glycol. It is practically insoluble in glycerin.

SPECIFICATIONS

Assay. Not less than 98.0 percent of $C_{11}H_{20}O_2$.

Refractive index. Between 1.450 and 1.454 at 20°.

Solubility in alcohol. Passes test.

Solubility in alkali. Passes test.

Specific gravity. Between 0.942 and 0.945.

Limits of Impurities

Acid value. Not more than 5.0.

TESTS

Assay. Weigh accurately about 1 gram, and proceed as directed under *Ester Determination*, page 896, using 92.14 as the equivalence factor (e) in the calculation.

Refractive index, page 945. Determine with an Abbé or other refractometer of equal or greater accuracy.

Solubility in alcohol. Proceed as directed in the general method, page 899. One ml. dissolves in 5 ml. of 60 percent alcohol.

Solubility in alkali. Transfer 5.0 ml. into a 100-ml. cassia flask, add 70 ml. of 1 *N* potassium hydroxide, warm the mixture in a water bath at 50° to 60°, and shake for 15 minutes. Add sufficient 1 *N* potassium hydroxide to raise the level of the liquid into the graduated portion of the neck of the flask, and cool to room temperature. No oil layer forms and the solution is not more than slightly cloudy.

Specific gravity. Determine by any reliable method (see page 5).

Acid value. Determine as directed in the general method, page 893.

Packaging and storage. Store in full, tight, preferably glass, tin-lined, or aluminum containers in a cool place protected from light.

Functional use in foods. Flavoring agent.

UNDECANAL

Aldehyde C-11 Undecyclic; *n*-Undecyl Aldehyde

$CH_3(CH_2)_9CHO$

$C_{11}H_{22}O$ Mol. wt. 170.30

DESCRIPTION

A colorless to slightly yellow liquid having a sweet fatty odor. It is soluble in most fixed oils, in mineral oil, and in propylene glycol. It is practically insoluble in glycerin.

SPECIFICATIONS

Assay. Not less than 92.0 percent of $C_{11}H_{22}O$.

Refractive index. Between 1.430 and 1.435 at 20°.

Specific gravity. Between 0.825 and 0.832.

Limits of Impurities

Acid value. Not more than 10.0.

TESTS

Assay. Weigh accurately about 1 gram, and proceed as directed under *Aldehydes*, page 894, using 85.15 as the equivalence factor (E) in the calculation.

Refractive index, page 945. Determine with an Abbé or other refractometer of equal or greater accuracy.

Specific gravity. Determine by any reliable method (see page 5).

Acid value. Determine as directed in the general method, page 893.

Packaging and storage. Store in full, tight, preferably glass, or aluminum containers in a cool place protected from light.

Functional use in foods. Flavoring agent.

10-UNDECENAL

Aldehyde C-11(Undecylenic); Undecene-10-al

$CH_2{=}CH(CH_2)_8CHO$

$C_{11}H_{20}O$ Mol. wt. 168.28

DESCRIPTION

A colorless to light yellow liquid having a characteristic fatty, rose-like odor on dilution. It is soluble in most fixed oils, in mineral oil, and in propylene glycol. It is practically insoluble in glycerin and in water.

SPECIFICATIONS

Assay. Not less than 90.0 percent of $C_{11}H_{20}O$.

Refractive index. Between 1.441 and 1.447.

Specific gravity. Between 0.840 and 0.850.

Limits of Impurities

Acid value. Not more than 6.0.

TESTS

Assay. Weigh accurately about 1 gram and proceed as directed under *Aldehydes*, page 894, using 84.14 as the equivalence factor (E) in the calculation.

Refractive index, page 945. Determine with an Abbé or other refractometer of equal or greater accuracy.

Specific gravity. Determine by any reliable method (see page 5).

Acid value. Determine as directed in the general method, page 893.

Packaging and storage. Store in full, tight, preferably glass or aluminum containers in a cool place (above 10°), protected from light. Prolonged storage, particularly when undiluted, is not recommended.

Functional use in foods. Flavoring agent.

VALERIC ACID

Pentanoic Acid

$CH_3(CH_2)_3COOH$

$C_5H_{10}O_2$ Mol. wt. 102.13

DESCRIPTION

A clear, colorless, mobile liquid having an unpleasant, penetrating, rancid odor. It is miscible in all proportions with alcohol and with ether. One ml. dissolves in about 40 ml. of water.

SPECIFICATIONS

Assay. Not less than 99.0 percent of $C_5H_{10}O_2$.

Distillation range. Between 180° and 190°.

Specific gravity. Between 0.934 and 0.939.

Limits of Impurities

Aldehydes (as valeraldehyde). Not more than 0.3 percent of $C_5H_{10}O$.

Heavy metals (as Pb). Not more than 40 parts per million (0.004 percent).

Unsaturated compounds. Not more than 0.05 milliequivalents per gram.

Water. Not more than 0.1 percent.

TESTS

Assay. Transfer about 1.5 grams, accurately weighed, into a 250-ml. Erlenmeyer flask and add 100 ml. of a mixture of equal parts, by volume, of isopropanol and water previously neutralized with 0.1 *N* sodium hydroxide using phenolphthalein T.S. as the indicator. Titrate the mixture with 0.5 *N* sodium hydroxide. Each ml. of 0.5 *N* sodium hydroxide is equivalent to 51.07 mg. of $C_5H_{10}O_2$.

Distillation range. Distil 100 ml. as directed in the general method, page 890.

Specific gravity. Determine by any reliable method (see page 5).

Aldehydes

Mercurial Solution. To 915 ml. of water contained in a 2-liter bottle add 75 grams of potassium chloride, 120 grams of mercuric chloride, 321 grams of potassium iodide, and 500 ml. of a 40 in 100 solution of potassium hydroxide. Shake the bottle after each addition to ensure complete solution. The final solution may contain a slight amount of yellow precipitate which does not interfere with its effectiveness.

Agar Solution. Add 1 gram of agar to 100 ml. of boiling water and continue heating, with occasional swirling, until an essentially clear solution is obtained. Cool, dilute to 1000 ml. with water, add 35 mg. of mercuric iodide, and shake vigorously for a few seconds.

Procedure. Transfer 10.0 ml. of the sample into a 500-ml., glass-stoppered Erlenmeyer flask containing 50 ml. of the *Mercurial Solution* and 20 ml. of the *Agar Solution.* Stopper the flask and swirl the solution to effect complete solution, then allow to stand at room temperature for one hour. To the mixture add, with constant swirling, 25 ml. of glacial acetic acid and 50.0 ml. of 0.1 *N* iodine. Stopper the flask and shake it vigorously until the gray mercury precipitate is dissolved. Remove the stopper, rinse any adhering liquid into the flask and wash down the inside walls of the flask with water. Titrate with 0.1 *N* sodium thiosulfate adding starch T.S. when the brown color begins to fade, and continue the titration to the disappearance of the blue color. Perform a residual blank titration (see *Blank Tests*, page 2). Calculate the weight of the sample from the previously determined specific gravity. Each ml. of 0.1 *N* sodium thiosulfate is equivalent to 4.31 mg. of valeraldehyde ($C_5H_{10}O$).

Heavy metals. Prepare and test a 500-mg. sample as directed in *Method II* under the *Heavy Metals Test*, page 920, using 20 mcg. of lead ion (Pb) in the control (*Solution A*).

Unsaturated compounds

Bromine-Sodium Bromide Solution. Transfer 5.5 ml. of bromine into a 1-liter, glass-stoppered, reagent bottle containing 500 ml. of methanol and 100 grams of sodium bromide, mix thoroughly, and dilute to 1 liter with methanol. Fit the bottle with a 2-hole rubber stopper and through one of the holes insert a 25-ml. pipet with its tip extending below the surface of the liquid; through the other hole insert a short piece of glass tubing to which is attached an aspirator bulb.

Procedure. Introduce 10 ml. of a saturated solution of sodium bromide into a 500-ml. iodine flask provided with a vented ground-glass stopper. Add 1 gram of sodium bromide, mix thoroughly, and then add from the pipet, filled by pressure from the aspirator, 25 ml. of the *Bromine-Sodium Bromide Solution.* Cool the contents of the flask to about $-10°$ in an ice-brine bath. Transfer 10.0 ml. of the sample into the flask and mix thoroughly by swirling. Remove the flask from the bath and add to it 75 ml. of methanol, previously cooled to about $-10°$. Immerse the flask in a solid carbon dioxide-acetone bath, cool the contents to about $-65°$, and add 10 ml. of a 15 percent solution of potassium iodide through the vent in the ground-glass stopper. Titrate immediately with 0.1 *N* sodium thiosulfate to the disappearance of the yellow color. Perform a residual blank titration (see *Blank Tests*, page 2). Calculate the milliequivalents per gram by the formula: $(B - A) \times N/W$, in which A is the number of ml. of sodium thiosulfate required for the sample, B is the number of ml. required for the blank, N is the exact normality of the sodium thiosulfate used for the determination, and W is the weight of the sample calculated from its specific gravity previously determined.

Water. Determine by the *Karl Fischer Titrimetric Method*, page 977.

Packaging and storage. Store in tight containers.

Functional use in foods. Flavoring agent.

L-VALINE

L-2-Amino-3-methylbutyric Acid

```
CH3CHCHCOOH
   |  |
 H3C  NH2
```

$C_5H_{11}NO_2$ Mol. wt. 117.15

DESCRIPTION

A white, odorless, crystalline powder having a characteristic taste. It is freely soluble in water, but practically insoluble in alcohol and in ether. The pH of a 1 in 20 solution is between 5.5 and 7.0. In a closed capillary tube it melts at about 315°.

IDENTIFICATION

To 5 ml. of a 1 in 1000 solution add 1 ml. of triketohydrindene hydrate T.S. A reddish purple or bluish color is produced.

SPECIFICATIONS

Assay. Not less than 98.0 percent and not more than the equivalent of 102.0 percent of $C_5H_{11}NO_2$, calculated on the dried basis.

Specific rotation, $[\alpha]^{20°}_{D}$. Between +26.5° and +29.0°, on the dried basis.

Limits of Impurities

Arsenic (as As). Not more than 3 parts per million (0.0003 percent).

Heavy metals (as Pb). Not more than 20 parts per million (0.002 percent).

Lead. Not more than 10 parts per million (0.001 percent).

Loss on drying. Not more than 0.3 percent.

Residue on ignition. Not more than 0.1 percent.

TESTS

Assay. Dissolve about 200 mg., accurately weighed, in 3 ml. of formic acid and 50 ml. of glacial acetic acid, add 2 drops of crystal violet T.S., and titrate with 0.1 *N* perchloric acid to a green end-point or until the blue color disappears completely. Each ml. of 0.1 *N* perchloric acid is equivalent to 11.72 mg. of $C_5H_{11}NO_2$.

Specific rotation, page 939. Determine in a solution containing 8 grams of a previously dried sample in sufficient 6 *N* hydrochloric acid to make 100 ml.

Arsenic (as As). A *Sample Solution* prepared as directed for organic compounds meets the requirements of the *Arsenic Test*, page 865.

Heavy metals (as Pb). Prepare and test a 1-gram sample as directed in *Method II* under the *Heavy Metals Test*, page 920, using 20 mcg. of lead ion (Pb) in the control (*Solution A*).

Lead. A *Sample Solution* prepared as directed for organic compounds meets the requirements of the *Lead Limit Test,* page 929, using 10 mcg. of lead ion (Pb) in the control.

Loss on drying, page 931. Dry at 105° for 3 hours.

Residue on ignition, page 945. Ignite 1 gram as directed in the general method.

Packaging and storage. Store in well-closed containers.

Functional use in foods. Nutrient; dietary supplement.

VANILLIN

CHO

OCH_3

OH

4-Hydroxy-3-methoxybenzaldehyde

$C_8H_8O_3$ Mol. wt. 152.15

DESCRIPTION

Fine, white to slightly yellow crystals, usually needle-like, having an odor and taste suggestive of vanilla. It is affected by light. Its solutions are acid to litmus. One gram dissolves in about 100 ml. of water, in about 20 ml. of glycerin, and in 20 ml. of water at 80°. It is freely soluble in alcohol, in chloroform, in ether, and in solutions of fixed alkali hydroxides.

IDENTIFICATION

A. To 10 ml. of a cold, saturated solution add about 5 drops of ferric chloride T.S. A blue color is produced which changes to brown when the mixture is heated at about 80° for a few minutes.

B. Vanillin is extracted completely from its solution in ether by shaking with a saturated solution of sodium bisulfite, from which it is precipitated by acids.

SPECIFICATIONS

Assay. Not less than 97.0 percent and not more than the equivalent of 103.0 percent of $C_8H_8O_3$, calculated on the dried basis.

Melting range. Between 81° and 83°.

Limits of Impurities

Arsenic (as As). Not more than 3 parts per million (0.0003 percent).

Heavy metals (as Pb). Not more than 10 parts per million (0.001 percent).

Loss on drying. Not more than 0.5 percent.
Residue on ignition. Not more than 0.05 percent.

TESTS

Assay

Standard Solution. Transfer about 100 mg., accurately weighed, of U.S.P. Vanillin Reference Standard into a 250-ml. volumetric flask, add methanol to volume, and mix. Transfer 2.0 ml. of this solution into a 100-ml. volumetric flask, dilute to volume with methanol, and mix.

Assay Solution. Transfer about 100 mg., accurately weighed, of the sample into a 250-ml. volumetric flask, add methanol to volume, and mix. Transfer 2.0 ml. of this solution into a 100-ml. volumetric flask, dilute to volume with methanol, and mix.

Procedure. Determine the absorbance of each solution in a 1-cm. quartz cell at the wavelength of maximum absorption at about 308 mμ, with a suitable spectrophotometer, using methanol as the blank. Calculate the quantity, in mg., of $C_8H_8O_3$ in the sample taken by the formula $12.5C(A_U/A_S)$, in which C is the concentration, in mcg. per ml., of U.S.P. Vanillin Reference Standard in the *Standard Solution,* A_U is the absorbance of the *Assay Solution,* and A_S is the absorbance of the *Standard Solution.*

Melting range. Determine as directed in the general procedure, page 931.

Arsenic. A *Sample Solution* prepared as directed for organic compounds meets the requirements of the *Arsenic Test,* page 865.

Heavy metals. Prepare and test a 2-gram sample as directed in *Method II* under the *Heavy Metals Test,* page 920, using 20 mcg. of lead ion (Pb) in the control (*Solution A*).

Loss on drying, page 931. Dry over silica gel for 4 hours.

Residue on ignition. Ignite 2 grams as directed in the general method, page 945.

Packaging and storage. Store in tight, light-resistant containers

Functional use in foods. Flavoring agent.

VITAMIN A

DESCRIPTION

A suitable form or derivative of retinol ($C_{20}H_{30}O$; vitamin A alcohol). It usually consists of retinol or esters of retinol formed from edible fatty acids, principally acetic and palmitic acids, or mixtures of these. It may be diluted with edible oils, or it may be incorporated in solid edible carriers, extenders, or excipients. It may contain suitable preservatives, dispersants, and antioxidants, providing it is not to be used in foods in which such substances are prohibited. In liquid form it is a light yellow to red oil which may solidify on refrigeration. In solid form it may have the appearance of the diluent that has been added to it. It may be nearly odorless, or have a mild fishy odor, but no rancid odor or taste. In liquid form it is very soluble in chloroform and in ether, soluble in absolute alcohol and in vegetable oils, but it is insoluble in glycerin and in water. In solid form it may be dispersible in water. It is unstable to air and light.

IDENTIFICATION

A. Dissolve an amount equivalent to about 6 mcg. of retinol in 1 ml. of chloroform, and add 10 ml. of antimony trichloride T.S. A transient blue color appears at once.

B. Assemble an apparatus for *Thin-layer Chromatography* (see page 884), using chromatographic silica gel as the adsorbent and a mixture of 4 parts of cyclohexane and 1 part of ether as the solvent system. Prepare a *Standard Solution* by dissolving the contents of 1 capsule of U.S.P. Vitamin A Reference Solution in sufficient chloroform to make 25.0 ml.

If the vitamin A is in liquid form, dissolve a volume representing approximately 15,000 U.S.P. Units in sufficient chloroform to make 10 ml. If the vitamin A is in solid form, weigh a quantity representing approximately 15,000 U.S.P. Units, place in a separator, add 75 ml. of water, heat, if necessary, to dissolve the carrier, and cool. Shake vigorously for 1 minute, extract with 10 ml. of chloroform by shaking for 1 minute, and centrifuge to clarify the chloroform extract.

Apply at the starting point of the chromatogram 0.015 ml. of the *Standard Solution* and 0.01 ml. of the chloroform solution of the vitamin A sample. Develop the chromatogram in the chromatographic chamber, lined with filter paper dipping into the solvent mixture. When the solvent has ascended for a distance of 10 cm., remove the plate, allow it to dry in air, and spray it with antimony trichloride T.S. The blue spot formed is indicative of the presence of retinol. The approximate R_f values of the predominant spots, corresponding to the different forms of retinol are 0.1 for the alcohol, 0.45 for the acetate, and 0.7 for the palmitate.

SPECIFICATIONS

Assay. Not less than 95.0 percent of the vitamin A activity declared on the label.

Absorbance ratio (corrected/observed at 325 mμ). Not less than 0.85.

TESTS

Assay. Proceed as directed under *Vitamin A Assay*, page 974.

Absorbance ratio. Determine by the formula $[A_{325}]/A_{325}$, the terms of which are defined under *Calculation* in the *Vitamin A Assay*, page 974.

Packaging and storage. Store in tight containers, preferably under an atmosphere of an inert gas, protected from light.

Labeling. Label to indicate the form of the vitamin A, to declare the presence of any preservative, dispersant, antioxidant, or other added substance, and to declare the vitamin A activity in terms of the equivalent amount of retinol in mg. per gram and in U.S.P. Vitamin A Units. *Note:* One U.S.P. Vitamin A Unit is the specific biologic activity of 0.3 mcg. of the all-*trans* isomer of retinol.

Functional use in foods. Nutrient; dietary supplement.

VITAMIN D₂

Calciferol; Ergocalciferol

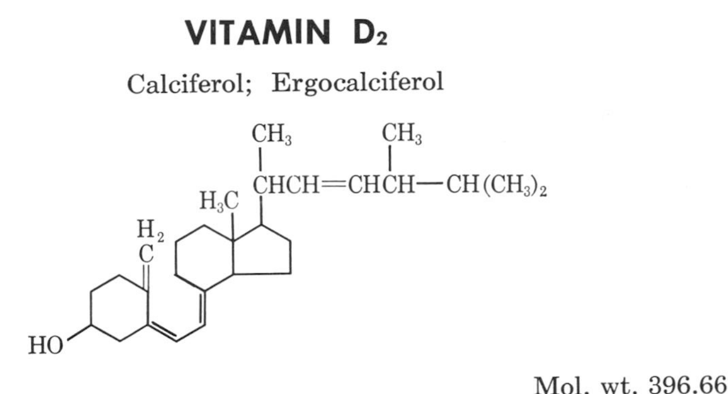

$C_{28}H_{44}O$ Mol. wt. 396.66

DESCRIPTION

White, odorless crystals. It is affected by air and by light. It is insoluble in water, but is soluble in alcohol, in chloroform, in ether, and in fatty oils.

IDENTIFICATION

A. To a solution of about 0.5 mg. in 5 ml. of chloroform add 0.3 ml. of acetic anhydride and 0.1 ml. of sulfuric acid, and shake vigorously. A bright red color is produced which rapidly changes through violet and blue to green.

B. Dissolve 50 mg. of the sample and 50 mg. of 3,5-dinitrobenzoyl chloride in separate 1-ml. portions of pyridine. Mix the solutions, warm on a steam bath for 10 minutes, add 5 ml. of water, filter, and wash the precipitate thoroughly with small portions of cold water. Recrystallize the precipitate twice from acetone, and dry in a vacuum desiccator for 2 hours. The dinitrobenzoyl derivative so obtained melts between 147° and 149° (see page 931).

C. The infrared absorption spectrum of a potassium bromide dispersion of the sample, in the range of 2 to 12 μ, exhibits maxima only at the same wavelengths as that of a similar preparation of U.S.P. Ergocalciferol Reference Standard.

D. The ultraviolet absorption spectrum of the sample in alcohol solution exhibits inflections at the same wavelengths as that of U.S.P. Ergocalciferol Reference Standard, similarly measured, and the respective absorptivities at 265 mμ do not differ by more than 3.0 percent.

SPECIFICATIONS

Melting range. Between 115° and 118°.

Specific rotation, $[\alpha]_D^{25°}$. Between +103° and +106°.

Limits of Impurities

Ergosterol. Passes test.

Reducing substances. Passes test.

TESTS

Melting range, page 931. Proceed as directed for *Class Ib.*

Specific rotation, page 939. Determine in a solution in alcohol containing 150 mg. in each 10 ml. Prepare the solution without delay, using a sample from a container opened not longer than 30 minutes, and determine the rotation within 30 minutes after the solution has been prepared.

Ergosterol. Dissolve 10 mg. in 2 ml. of 90 percent alcohol, add a solution of 20 mg. of digitonin in 2 ml. of 90 percent alcohol, mix, and allow to stand for 18 hours. No precipitate is formed.

Reducing substances. To 10 ml. of a 1 in 100 solution of the sample in absolute alcohol add 0.5 ml. of a 1 in 200 solution of blue tetrazolium in absolute alcohol. Add 0.5 ml. of a solution prepared by diluting 1 volume of 10 percent tetramethylammonium hydroxide with 9 volumes of absolute alcohol. Allow the mixture to stand for 5 minutes, then add 1 ml. of glacial acetic acid, and mix. Prepare a blank by treating 10 ml. of absolute alcohol in the same manner. The absorbance of the sample solution, determine at 525 mμ with a suitable spectrophotometer against the blank, is not greater than that obtained with a solution containing 0.2 mcg. per ml. of hydroquinone in absolute alcohol, similarly treated.

Packaging and storage. Store in hermetically sealed containers under nitrogen, in a cool place and protected from light.

Functional use in foods. Nutrient; dietary supplement.

VITAMIN D_3

Activated 7-Dehydrocholesterol; Cholecalciferol

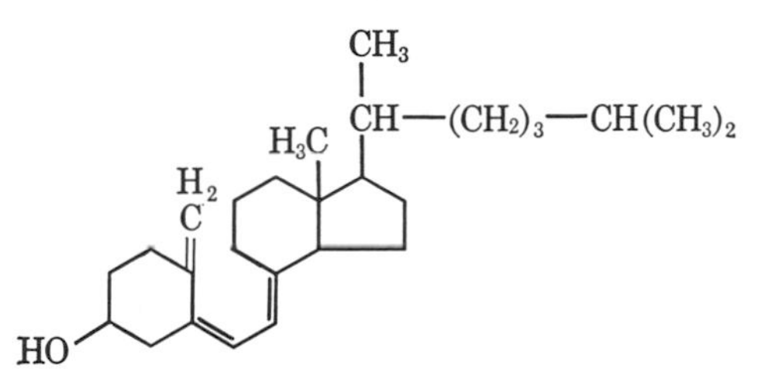

$C_{27}H_{44}O$ Mol. wt. 384.65

DESCRIPTION

White, odorless crystals. It is affected by air and by light. It is insoluble in water. It is soluble in alcohol, in chloroform, and in fatty oils.

IDENTIFICATION

A. To a solution of about 0.5 mg. in 5 ml. of chloroform add 0.3 ml. of acetic anhydride and 0.1 ml. of sulfuric acid, and shake vigorously. A bright red color, which rapidly changes through violet and blue to green is produced.

B. Dissolve 50 mg. of the sample and 50 mg. of 3,5-dinitrobenzoyl chloride in separate 1-ml. portions of anhydrous pyridine. Mix the solutions, warm on a steam bath for 10 minutes, add 5 ml. of water, filter, and wash the precipitate thoroughly with small portions of cold water. Recrystallize the precipitate twice from acetone, and dry in a vacuum desiccator for 2 hours. The dinitrobenzoyl derivative so obtained melts between 133° and 135° (see page 931).

C. The infrared absorption spectrum of a potassium bromide dispersion of the sample, in the range of 2 to 12 μ, exhibits maxima only at the same wavelengths as that of a similar preparation of U.S.P. Cholecalciferol Reference Standard.

D. The ultraviolet absorption spectrum of the sample in a solution in alcohol exhibits inflections at the same wavelengths as that of U.S.P. Cholecalciferol Reference Standard, similarly measured, and the respective absorptivities at the point of maximum absorbance occurring at about 265 mμ do not differ by more than 3.0 percent.

SPECIFICATIONS

Melting range. Between 84° and 88°.

Specific rotation, $[\alpha]_D^{25°}$. Between +105° and +112°

TESTS

Melting range, page 931. Proceed as directed for *Class Ib.*

Specific rotation, page 939. Determine in a solution in alcohol containing 50 mg. in each 10 ml. Prepare the solution without delay, using a sample from a container opened not longer than 30 minutes, and determine the rotation within 30 minutes after the solution has been prepared.

Packaging and storage. Store in hermetically sealed containers under nitrogen, in a cool place and protected from light.

Functional use in foods. Nutrient; dietary supplement.

XANTHAN GUM

DESCRIPTION

A high molecular weight polysaccharide gum produced by a pure-culture fermentation of a carbohydrate with *Xanthomonas campestris*, purified by recovery with isopropyl alcohol, dried, and milled. It contains D-glucose and D-mannose as the dominant hexose units, along with D-glucuronic acid, and is prepared as the sodium, potassium, or calcium salt. It occurs as a cream colored powder that is readily soluble in hot or cold water. Its solutions are neutral.

IDENTIFICATION

To 300 ml. of water, previously heated to 80° and stirred rapidly with a mechanical stirrer in a 400-ml. beaker, add, at the point of maximum agitation, a dry blend of 1.5 grams of the sample and 1.5 grams of locust bean gum. Stir until the mixture goes into solution, and then continue stirring for 30 minutes longer. Do not allow the water temperature to drop below 60° during stirring. Discontinue stirring, and allow the mixture to cool at room temperature for at least 2 hours. A firm, rubbery gel forms after the temperature drops below 40°, but no such gel forms in a 1 percent control solution of the sample prepared in the same manner but omitting the locust bean gum.

SPECIFICATIONS

Assay. It yields, on the dry basis, not less than 4.2 percent and not more than 5.0 percent of carbon dioxide (CO_2), corresponding to between 91.0 percent and 108.0 percent of xanthan gum.

Ash. Between 6.5 percent and 16 percent.

Loss on drying. Not more than 15 percent.

Pyruvic acid. Not less than 1.5 percent.

Viscosity. Passes test.

Limits of Impurities

Arsenic (as As). Not more than 3 parts per million (0.0003 percent).

Heavy metals (as Pb). Not more than 30 parts per million (0.003 percent).

Isopropyl alcohol. Not more than 750 parts per million (0.075 percent).

Lead. Not more than 5 parts per million (0.0005 percent).

TESTS

Assay. Proceed as directed under *Alginates Assay*, page 863, but use an undried sample of about 1.2 grams, accurately weighed.

Ash. Weigh accurately about 3 grams, previously dried at 105° for 4 hours, in a tared crucible, and incinerate at a low temperature, not exceeding a dull red heat, until free from carbon. Cool the crucible and its contents in a desiccator, weigh, and determine the weight of the ash.

Loss on drying, page 931. Dry at 105° for 2.5 hours.

Pyruvic acid

Sample preparation. Dissolve 600.0 mg. of the sample, accurately weighed, in sufficient water to make 100.0 ml., and transfer 10.0 ml. of the solution into a 50-ml. glass-stoppered flask. Pipet 20 ml. of 1 *N* hydrochloric acid into the flask, weigh the flask, and reflux for 3 hours, taking precautions to prevent loss of vapors. Cool to room temperature, and add water to make up for any weight loss during refluxing. Pipet 1.0 ml. of a 1 in 200 solution of 2,4-dinitrophenylhydrazine in 2 *N* hydrochloric acid into a 30-ml. separator, then add 2.0 ml. of the sample solution, mix, and allow to stand at room temperature for 5 minutes. Extract the mixture with 5 ml. of ethyl acetate, and discard the aqueous layer. Extract the hydrazone from the ethyl acetate with three 5-ml. portions of sodium carbonate T.S., collecting the extracts in a 50-ml. volumetric flask. Dilute to volume with sodium carbonate T.S., and mix.

Standard preparation. Transfer 45.0 mg. of pyruvic acid, accurately weighed, into a 500-ml. volumetric flask, dilute to volume with water, and mix. Transfer 10.0 ml. of this solution into a 50-ml. glass-stoppered flask, and continue as directed under *Sample preparation*, beginning with "Pipet 20 ml. of 1 *N* hydrochloric acid into the flask . . . ."

Procedure. Determine the absorbance of each solution with a suitable spectrophotometer in 1-cm. cells at the maximum at about 375 mμ, using sodium carbonate T.S. as the blank. The absorbance of the *Sample preparation* is equal to or greater than that of the *Standard preparation.*

Viscosity. Prepare two identical solutions, each containing 1 percent of the sample and 1 percent of potassium chloride in water, and stir for 2 hours. Determine the viscosity of one solution as directed under *Viscosity of Sodium Carboxymethylcellulose*, page 973, but maintain the temperature at 23.9° (75° F.). The viscosity (V_1) thus determined is not less than 600 centipoises. Determine the viscosity (V_2) of the other solution in the same manner, but maintain the temperature at 65.6° (150° F.). The ratio of the viscosities, V_1/V_2, is between 1.02 and 1.45.

Arsenic. A *Sample Solution* prepared as directed for organic compounds meets the requirements of the *Arsenic Test*, page 865.

Heavy metals. Prepare and test a 500-mg. sample as directed in *Method II* under the *Heavy Metals Test*, page 920, using a platinum crucible for the ignition and 15 mcg. of lead ion (Pb) in the control (*Solution A*).

Isopropyl alcohol. Transfer about 5 grams of the sample, accurately weighed, into a 1000-ml. round-bottom flask, add 200 ml. of water, connect the flask to a fractionating column, and distill 100 ml. directly into a 250-ml. round-bottom flask. Wash the condenser with two 10-ml. portions of water, adding the rinsings to the distillate. Pipet 25.0 ml. of a 1 in 10 solution of potassium dichromate in 20 percent sulfuric acid into the receiver flask, and connect the flask with a Liebig condenser. Heat in a boiling water bath for 5 minutes, then add through the top of the condenser 25 ml. of water and enough of a 48 percent solution of sodium hydroxide to change the orange dichromate color to the green chromate color. Wash the condenser with small portions of water to remove any adhering caustic. Connect the flask with the fractionating column, heat the solution to boiling, and connect the delivery tip of the fractionating column with a 250-ml. Erlenmeyer flask containing a hypoiodite solution prepared by mixing 50 ml. of 0.01 *N* iodine and 60 ml. of 1 *N* sodium hydroxide. Collect 75 ml. of distillate, then wash the condenser with several small portions of water, and remove the flask from the fractionating column. To the distillate add enough 6 *N* hydrochloric acid to make the solution acidic, and then titrate the excess iodine with 0.01 *N* sodium thiosulfate, using starch T.S. as the indicator and recording the volume of sodium thiosulfate consumed as *S*. Perform a blank titration by preparing the hypoiodide solution, acidifying, and titrating the iodine with 0.01 *N* sodium thiosulfate, recording the volume of sodium thiosulfate consumed as *B*. Calculate the isopropyl alcohol content in the sample taken, in parts per million, by the formula $(B - S) \times (N \times 100.1/W)$, in which *N* is the exact normality of the sodium thiosulfate solution, and *W* is the weight of the sample taken, in grams.

Lead. A *Sample Solution* prepared as directed for organic compounds meets the requirements of the *Lead Limit Test*, page 929, using 5 mcg. of lead ion (Pb) in the control.

Packaging and storage. Store in well-closed containers.

Functional use in foods. Stabilizer; thickener; emulsifier; suspending agent; bodying agent; foam enhancer.

ZINC SULFATE

$ZnSO_4 . 7H_2O$ Mol. wt. 287.54

DESCRIPTION

Colorless, transparent prisms or small needles, or a granular, crystalline powder. It is odorless and is efflorescent in dry air. Its solutions are acid to litmus. One gram dissolves in about 0.6 ml. of water and in about 2.5 ml. of glycerin. It is insoluble in alcohol. The pH of a 1 in 20 solution is between 4.4 and 6.0.

IDENTIFICATION

A 1 in 20 solution gives positive tests for *Zinc*, page 928, and for *Sulfate*, page 928.

SPECIFICATIONS

Assay. Not less than 99.0 percent and not more than 108.7 percent of $ZnSO_4 . 7H_2O$.

Acidity. Passes test.

Limits of Impurities

Alkalies and alkaline earths. Not more than 0.5 percent.

Arsenic (as As). Not more than 3 parts per million (0.0003 percent).

Heavy metals (as Pb). Not more than 10 parts per million (0.001 percent).

Selenium. Not more than 30 parts per million (0.003 percent).

TESTS

Assay. Dissolve about 300 mg., accurately weighed, in 100 ml. of water, add 5 ml. of ammonia-ammonium chloride buffer T.S. and 0.1 ml. of eriochrome black T.S., and titrate with 0.05 *M* disodium ethylenediaminetetraacetate until the solution is deep blue in color. Each ml. of 0.05 *M* disodium ethylenediaminetetraacetate is equivalent to 14.38 mg. of $ZnSO_4 . 7H_2O$.

Acidity. A 1 in 20 solution is not colored pink by methyl orange T.S.

Alkalies and alkaline earths. Transfer a 2-gram sample to a 200-ml. volumetric flask, dissolve in about 150 ml. of water, and precipitate the zinc completely with ammonium sulfide T.S. Dilute to volume with water, and mix. Filter through a dry filter, rejecting the first portion of the filtrate, and add a few drops of sulfuric acid to 100 ml. of the subsequent filtrate. Evaporate to dryness in a tared dish, ignite to constant weight, cool, and weigh. The weight of the residue does not exceed 5 mg.

Arsenic. A solution of 1 gram in 35 ml. of water meets the requirements of the *Arsenic Test*, page 865.

Heavy metals. Dissolve 500 mg. in 5 ml. of water, transfer to a color-comparison tube (*A*), add 10 ml. of potassium cyanide solution (1 in 10), mix, and allow the mixture to become clear. In a similar matched color-comparison tube (*B*) place 5 ml. of water, and add 0.50 ml. of *Standard Lead Solution* (5 mcg. of Pb) and 10 ml. of potassium cyanide solution (1 in 10). To each solution add 0.1 ml. of sodium sulfide T.S., mix, and allow to stand for 5 minutes. When viewed downward over a white surface, the solution in tube *A* is no darker than that in tube *B*.

Selenium. A solution of 2 grams in 40 ml. of dilute hydrochloric acid (1 in 2) meets the requirements of the *Selenium Limit Test*, page 953.

Packaging and storage. Store in tight containers.

Functional use in foods. Nutrient; dietary supplement.

General Tests and Apparatus

Alginates Assay 863
Alkali Salts of Organic Acids
Assay 864
Arsenic Test 865
Ash (Total) 868
Ash (Acid-Insoluble) 869
Calcium Pantothenate Assay. 869
Chewing Gum Base 873
Bound Styrene 873
Molecular Weight 874
Polyethylene 874
Polyisobutylene 875
Polyvinyl Acetate 875
Quinones 876
Residual Styrene 877
Sample Solution for Arsenic
Test 877
Sample Solution for Lead
Limit Test 878
Total Unsaturation 878
Chloride and Sulfate Limit
Tests 879
Chromatography 880
Column 880
Paper 881
Thin-Layer 884
Gas 886
Distillation Range 890
Essential Oils and Related
Substances 892
Acetals 892
Acid Value 893
Alcohols, Total 893
Aldehydes 894
Aldehydes and Ketones-
Hydroxylamine Method . 894
Aldehydes and Ketones-
Neutral Sulfite Method . . 895
Chlorinated Compounds ... 895
Esters 896
Ester Determination 896
Ester Determination
(High Boiling Solvent). 896
Saponification Value 896
Ester Value 897
Linalool Determination 897
Percentage of Cineole 898
Phenols 898
Phenols, Free 899
Residue on Evaporation 899
Solubility in Alcohol 899
Ultraviolet Absorbance of
Citrus Oils 900
Volatile Oil Content 901
Fats and Related Substances. 902
Acetyl Value 902
Acid Value 902
Fatty Acids, Free 903
Glycerin and Propylene
Glycol, Free 904
Hydroxyl Value 904
Iodine Value 905
Hanus Method 905
Wijs Method 906
1-Monoglycerides 907
Monoglycerides, Total 908
Oxyethylene Determination 910
Reichert-Meissl Value 912
Saponification Value 914
Soap 915
Specific Gravity 915
Unsaponifiable Matter 915
Volatile Acidity 916
Fluoride Limit Test 917
Heavy Metals Test 920
Hydrochloric Acid Table 922
Hydroxypropoxyl
Determination 923
Identification-General Tests. . 925
Lead Limit Test 929
Loss on Drying 931
Melting Range or Tempera-
ture 931
Mercury Limit Test 934
Methoxyl Determination 935
Nitrogen Determination 937
Optical Rotation 939
pH Determination 941
Readily Carbonizable
Substances 942
Refractive Index 945
Residue on Ignition 945
Rosins, Rosin Esters, and
Related Substances 945
Acid Value 945
Softening Point 946
Drop Method 946
Ring-and-Ball Method ... 948
Viscosity 952
Selenium Limit Test 953
Sieve Analysis of Granular
Metal Powders 953

Solidification Point.......... 954
Spectrophotometry, Colorimetry, and Turbidimetry..... 957
Starches and Related Substances................. 959
Acetyl Groups............ 959
Crude Fat................ 960
Manganese............... 961
Phosphorus............... 961
Propylene Chlorohydrin .. 962
Sulfur Dioxide............ 964
Sulfuric Acid Table.......... 965
Thermometers.............. 966
Thiamine Assay............. 968
Tocopherols................ 969
Viscosity of Dimethylpolysiloxane.................. 970
Viscosity of Methylcellulose.. 971
Viscosity of Sodium Carboxymethylcellulose.... 973
Vitamin A Assay............ 974
Volumetric Apparatus....... 976
Water Determination........ 977
Karl Fischer Titrimetric Method................ 977
Toluene Distillation Method................ 979
Weights and Balances....... 980

Solutions and Indicators

Colorimetric Solutions....... 983
Standard Buffer Solutions.... 983
Standard Solutions for the Preparation of Controls and Standards................ 984
Test Solutions (T.S.)........ 985
Volumetric Solutions........ 996
Indicators.................. 1004
Indicator Papers and Test Papers................... 1007

ALGINATES ASSAY

Apparatus. The apparatus required is shown in the accompanying diagram. It consists essentially of a soda lime column, *A*, a mercury valve, *B*, connected through a side arm, *C*, to a reaction flask, *D*, by means of a rubber connection. Flask *D* is a 100-ml. round-bottomed, long-neck boiling flask, resting in a suitable heating mantle, *E*.

The reaction flask is provided with a reflux condenser, *F*, to which is fitted a delivery tube, *G*, of 40-ml. capacity, having a stopcock, *H*. The reflux condenser terminates in a trap, *I*, containing 25 grams of 20-mesh zinc or tin, which can be connected with an absorption tower, *J*.

The absorption tower consists of a 45-cm. tube fitted with a medium-porosity fritted glass disk sealed to the inner part above the side arm and having a delivery tube sealed to it extending down to the end of the tube. A trap, consisting of a bulb of approximately 100-ml. capacity, is blown above the fritted disk and the outer portion of a ground spherical joint is sealed on above the bulb. A 250-ml. Erlenmeyer flask, *K*, is connected to the bottom of the absorption tower. The top of the tower is connected to a soda lime tower, *L*, which is connected to a suitable pump to provide vacuum and air supply, the choice of which is made by a 3-way stopcock, *M*. The volume of air or vacuum is controlled by a capillary-tube regulator or needle valve, *N*.

All joints are size 35/25, ground spherical type.

Procedure. Transfer a sample of about 250 mg., previously dried in vacuum for 4 hours at 60° and accurately weighed, into the reaction flask, *D*, add 25 ml. of 0.1 *N* hydrochloric acid, insert several boiling chips, and connect the flask to the reflux condenser, *F*, using syrupy phosphoric acid as a lubricant. (*Note:* Stopcock grease may be used for the other connections.) Check the system for air leaks by forcing mercury up into the inner tube of the mercury valve, *B*, to a height of about 5 cm. Turn off the pressure using the stopcock, *M*. If the mercury level does not fall appreciably after 1 to 2 minutes, the apparatus may be considered to be free from leaks. Draw carbon dioxide-free air through the apparatus at a rate of 3000 to 6000 ml. per hour. Raise the heating mantle, *E*, to the flask, heat the sample to boiling, and boil gently for 2 minutes. Turn off and lower the mantle, and allow the sample to cool for 15 minutes. Charge the delivery tube, *G*, with 23 ml. of hydrochloric acid. Disconnect the absorption tower, *L*, rapidly transfer 25.0 ml. of 0.25 *N* sodium hydroxide into the tower, add 5 drops of *n*-butanol, and again connect the absorption tower. Draw carbon dioxide-free air through the apparatus at the rate of about 2000 ml. per hour, add the hydrochloric acid to the reaction flask through the delivery tube, raise the heating mantle, and heat the reaction mixture to boiling. After 2 hours, discon-

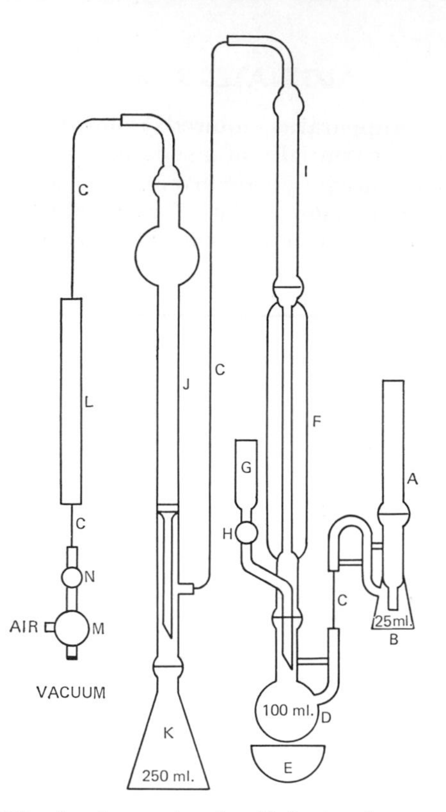

Fig. 1—Apparatus for Alginates Assay

tinue the current of air and heating. Force the sodium hydroxide solution down into the flask, *K*, using gentle air pressure, and then rinse down the absorption tower with three 15-ml. portions of water, forcing each washing into the flask with air pressure. Remove the flask, and add to it 10 ml. of a 10 percent solution of barium chloride ($BaCl_2 . 2H_2O$). Stopper the flask, shake gently for about 2 minutes, add phenolphthalein T.S., and titrate with 0.1 *N* hydrochloric acid. Perform a blank determination (see page 2). Each ml. of 0.25 *N* sodium hydroxide consumed is equivalent to 5.5 mg. of carbon dioxide (CO_2).

ALKALI SALTS OF ORGANIC ACIDS ASSAY

This assay is not applicable to organic alkali salts containing sulfur or halogens.

Unless otherwise directed, transfer about 2 grams of the sample, accurately weighed, to a porcelain crucible and carefully ignite until the material is completely charred. (*Caution: The ignited salt should be protected from contact with the free flame at all times.*) After cooling, place the crucible in a beaker, add 50 ml. of water and 50.0 ml. of 0.5 *N* sulfuric acid, and disperse the carbonized mass with a glass rod. Cover the beaker and boil the mixture for 30 minutes. Filter, wash the residue with hot water until the washings are neutral to litmus, and cool. To the combined filtrate and washings add phenolphthalein T.S. and titrate the excess acid with 0.5 *N* sodium hydroxide. The weight of the alkali salt is obtained by multiplying the volume of the acid consumed by the equivalence factor of the salt being analyzed.

A 400-mg. sample and 0.1 *N* acid and sodium hydroxide may be employed satisfactorily in this assay.

ARSENIC TEST

Silver Diethyldithiocarbamate Colorimetric Method

All reagents used in this test should be very low in arsenic content.

Apparatus. The general apparatus, shown in Fig. 2, is to be used unless otherwise specified in an individual monograph. It consists of a 125-ml. arsine generator flask (*a*) fitted with a scrubber unit (*c*) and an absorber tube (*e*), with a 24/40 standard-taper joint (*b*) and a ball-and-socket joint (*d*), secured with a No. 12 clamp, connecting the units. Alternatively, an apparatus embodying the principle of the general assembly described and illustrated may be used.

Note: The special assemblies shown in Fig. 3 and in Fig. 4 are to be used only when specified in certain monographs.

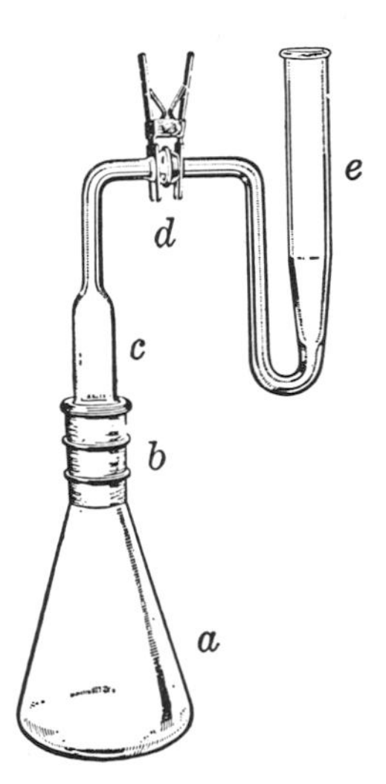

Fig. 2—General Apparatus for Arsenic Test (Courtesy of the Fisher Scientific Co., Pittsburgh, Pa.)

Standard Arsenic Solution. Weigh accurately 132.0 mg. of arsenic trioxide that has been finely pulverized and dried for 24 hours over a suitable desiccant, and dissolve it in 5 ml. of sodium hydroxide solution (1 in 5). Neutralize the solution with diluted sulfuric acid T.S., add 10 ml. in excess, and dilute to 1000.0 ml. with recently boiled water. Transfer 10.0 ml. of this solution into a 1000-ml. volumetric flask, add 10 ml. of diluted sulfuric acid T.S., dilute to volume with recently boiled water, and mix. Use this final solution, which contains 1 mcg. of arsenic (As) in each ml., within 3 days.

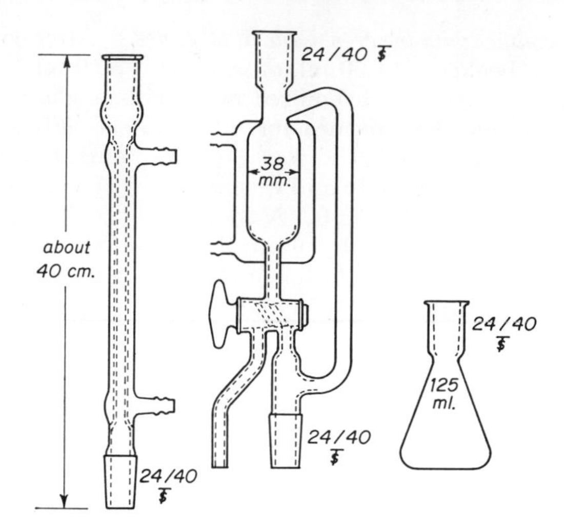

Fig. 3—Modified Bethge Apparatus for the Distillation of Arsenic Tribromide

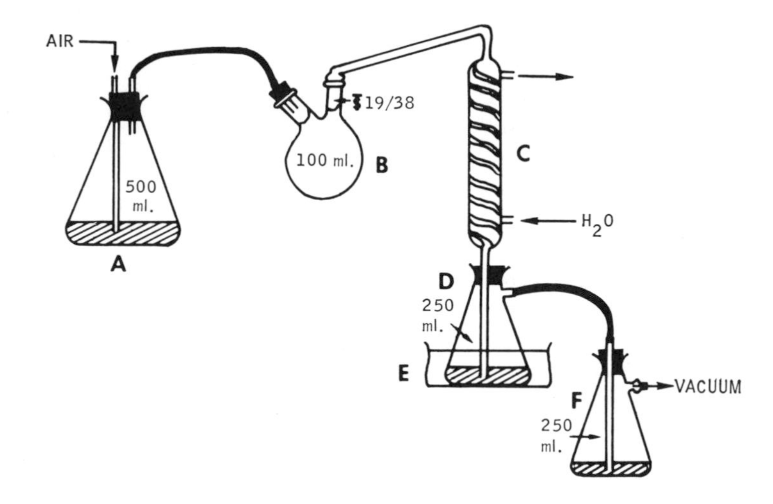

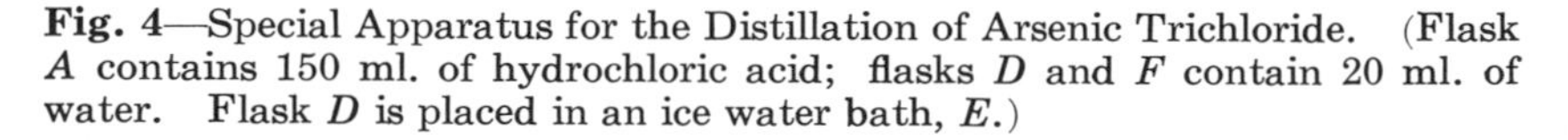

Fig. 4—Special Apparatus for the Distillation of Arsenic Trichloride. (Flask *A* contains 150 ml. of hydrochloric acid; flasks *D* and *F* contain 20 ml. of water. Flask *D* is placed in an ice water bath, *E*.)

Silver Diethyldithiocarbamate Solution. Dissolve 1 gram of recrystallized silver diethyldithiocarbamate, $(C_2H_5)_2NCSSAg$, in 200 ml. of reagent grade pyridine. Store this solution in a light-resistant container and use within 1 month.

Silver diethyldithiocarbamate is available commercially or may be prepared as follows: Dissolve 1.7 grams of reagent grade silver nitrate in 100 ml. of water. In a separate container, dissolve 2.3 grams of sodium diethyldithiocarbamate, $(C_2H_5)_2NCSSNa.3H_2O$, in 100 ml. of water, and filter. Cool both solutions to about 15°, mix the two solutions, while stirring, collect the yellow precipitate in a medium-porosity sintered-glass crucible or funnel, and wash with about 200 ml. of cold water.

Recrystallize the reagent, whether prepared as directed above or obtained commercially, as follows: Dissolve in freshly distilled pyridine, using about 100 ml. of solvent for each gram of reagent, and filter. Add an equal volume of cold water to the pyridine solution, while stirring. Filter off the precipitate, using suction, wash with cold water, and dry in vacuum at room temperature for 2 to 3 hours. The dry salt is pure yellow in color and should show no change in character after 1 month when stored in a light-resistant container. Discard any material that changes in color or develops a strong odor.

Stannous Chloride Solution. Dissolve 40 grams of reagent grade stannous chloride dihydrate, $SnCl_2.2H_2O$, in 100 ml. of hydrochloric acid. Store the solution in glass containers and use within 3 months.

Lead Acetate-Impregnated Cotton. Soak cotton in a saturated solution of reagent grade lead acetate, squeeze out the excess solution, and dry in a vacuum at room temperature.

Sample Solution. The solution obtained by treating the sample as directed in an individual monograph is used directly as the *Sample Solution* in the *Procedure.* Sample solutions of organic compounds are prepared in the generator flask (*a*), unless otherwise directed, according to the following general procedure:

> *Caution: Some substances may react unexpectedly with explosive violence when digested with hydrogen peroxide. Appropriate safety precautions must be employed at all times.*
> *Note: If halogen-containing compounds are present, use a lower temperature while heating the sample with sulfuric acid, do not boil the mixture, and add the peroxide, with caution, before charring begins, to prevent loss of trivalent arsenic.*

Transfer 1.0 gram of the sample into the generator flask, add 5 ml. of sulfuric acid and a few glass beads, and digest on a hot plate in a fume hood until charring begins. (Additional sulfuric acid may be necessary to completely wet some samples, but the total volume added should not exceed about 10 ml.) After the sample has been initially decomposed by the acid, add with caution, dropwise, 30 percent hydrogen peroxide, allowing the reaction to subside and reheating between drops. The first few drops must be added very slowly with sufficient mixing to prevent a rapid reaction, and heating should be discontinued if foaming becomes excessive. Swirl the solution in the flask to prevent unreacted substance from caking on the walls or bottom of the flask during digestion. *Maintain oxidizing conditions at all times during the digestion by adding small quantities of the peroxide whenever the mixture turns brown or darkens.* Continue the digestion until the organic matter is destroyed, fumes of sulfuric acid are copiously evolved, and the solution becomes colorless. Cool, add cautiously 10 ml. of water, again evaporate to strong fuming, and cool. Add cautiously 10 ml. of water, mix, wash the sides of the flask with a few ml. of water, and dilute to 35 ml.

Procedure. If the *Sample Solution* was not prepared in the generator flask, transfer to the flask a volume of the solution, prepared as directed, equivalent to 1.0 gram of the substance being tested, and add water to make 35 ml. Add 20 ml. of dilute sulfuric acid (1 in 5), 2 ml. of potassium iodide T.S., and 0.5 ml. of *Stannous Chloride Solution,* and mix. Allow the mixture to stand for 30 minutes at room temperature. Pack the scrubber tube (*c*) with two plugs of *Lead Acetate-Impregnated Cotton,* leaving a small air space between the two plugs, lubricate joints *b* and *d* with stopcock grease, if necessary, and connect the scrubber unit with the absorber tube (*e*). Transfer 3.0 ml. of *Silver Diethyldithiocarbamate Solution* to the absorber tube, add 3.0 grams of granular zinc (20-mesh) to the mixture in the flask, and immediately insert the standard-taper joint in the flask. Allow the evolution of hydrogen and color development to proceed at room temperature (25° ± 3°) for 45 minutes, swirling the flask gently at 10-minute intervals. (The addition of a small amount of isopropanol to the generator flask may improve the uniformity of the rate of gas evolution). Disconnect the absorber tube from the generator and scrubber units, and transfer the *Silver Diethyldithiocarbamate Solution* to a 1-cm. absorption cell. Determine the absorbance at 525 mμ with a suitable spectrophotometer or colorimeter, using *Silver Diethyldithiocarbamate Solution* as the blank. The absorbance due to any red color from the solution of the sample does not exceed that produced by 3.0 ml. of *Standard Arsenic Solution* (3 mcg. As) when treated in the same manner and under the same conditions as the sample. The room temperature during the generation of arsine from the standard should be held to within ±2° of that observed during the determination of the sample.

Interferences. Metals or salts of metals such as chromium, cobalt, copper, mercury, molybdenum, nickel, palladium, and silver are said to interfere with the evolution of arsine. Antimony, which forms stibine, is the only metal likely to produce a positive interference in the color development with the silver diethyldithiocarbamate. Stibine forms a red color which has a maximum absorbance at 510 mμ, but at 525 mμ the absorbance of the antimony complex is so diminished that the results of the determination would not be altered significantly.

ASH (TOTAL)

Unless otherwise directed, weigh accurately about 3 grams of the sample in a tared crucible, ignite at a low temperature (about 550°), not to exceed very dull redness, until free from carbon, cool in a desiccator, and weigh. If a carbon-free ash is not obtained, wet the charred mass with hot water, collect the insoluble residue on an ashless filter paper, and ignite the residue and filter paper until the ash is white or nearly so. Finally, add the filtrate, evaporate it to dryness, and heat the whole to a dull redness. If a carbon-free ash is still not obtained, cool the crucible, add 15 ml. of alcohol, break up the ash with a glass rod, then burn off the alcohol, again heat the whole to a dull redness, cool and weigh.

ASH (ACID-INSOLUBLE)

Boil the ash obtained as directed under *Total Ash,* above, with 25 ml. of diluted hydrochloric acid T.S. for 5 minutes, collect the insoluble matter on a tared Gooch crucible or ashless filter, wash with hot water, ignite, and weigh. Calculate the percentage of acid-insoluble ash from the weight of sample taken.

CALCIUM PANTOTHENATE ASSAY

Standard Stock Solution of Calcium Pantothenate

Dissolve 50 mg. of U.S.P. Calcium Pantothenate Reference Standard, previously dried and stored in the dark over phosphorus pentoxide and accurately weighed while protected from absorption of moisture during the weighing, in about 500 ml. of water in a 1000-ml. volumetric flask. Add 10 ml. of 0.2 *N* acetic acid and 100 ml. of sodium acetate solution (1 in 60), then add water to volume. Each ml. represents 50 mcg. of U.S.P. Calcium Pantothenate Reference Standard. Store under toluene in a refrigerator.

Standard Preparation

On the day of the assay, dilute a measured volume of *Standard Stock Solution of Calcium Pantothenate* with sufficient water to contain, in each ml., between 0.01 and 0.04 mcg. of calcium pantothenate, the exact concentration being such that the responses obtained as directed under *Procedure,* using 2.0 and 4.0 ml. of the *Standard Preparation,* are within the linear portion of the log-concentration response curve.

Assay Preparation

Prepare a solution containing approximately the equivalent of the calcium pantothenate concentration in the *Standard Preparation.*

Basal Medium Stock Solution

Acid-hydrolyzed Casein Solution	25	ml.
Cystine-Tryptophane Solution	25	ml.

Polysorbate 80 Solution	0.25	ml.
Dextrose, Anhydrous	10	grams
Sodium Acetate, Anhydrous	5	grams
Adenine-Guanine-Uracil Solution	5	ml.
Riboflavin-Thiamine Hydrochloride-Biotin Solution	5	ml.
Para-aminobenzoic Acid-Niacin-Pyridoxine Hydrochloride Solution	5	ml.
Salt Solution A	5	ml.
Salt Solution B	5	ml.

Dissolve the anhydrous dextrose and sodium acetate in the solutions previously mixed, and adjust to a pH of 6.8 with sodium hydroxide T.S. Finally, add water to make 250 ml.

Acid-Hydrolyzed Casein Solution. Mix 100 grams of vitamin-free casein with 500 ml. of dilute hydrochloric acid (1 in 2), and reflux the mixture for 8 to 12 hours. Remove the hydrochloric acid from the mixture by distillation under reduced pressure until a thick paste remains. Redissolve the resulting paste in water, adjust the solution to a pH of 3.5 ± 0.1 with sodium hydroxide solution, and add water to make 1000 ml. Add 20 grams of activated charcoal, stir for 1 hour, and filter. Repeat the treatment with activated charcoal. Store under toluene in a refrigerator at a temperature not below 10°. Filter the solution if a precipitate forms under storage.

Cystine-Tryptophan Solution. Suspend 4.0 grams of L-cystine and 1.0 gram of L-tryptophan (or 2.0 grams of DL-tryptophan) in 700 to 800 ml. of water, heat to 70° to 80°, and add dilute hydrochloric acid (1 in 2) dropwise, with stirring, until the solids are dissolved. Cool, and add water to make 1000 ml. Store under toluene in a refrigerator at a temperature not below 10°.

Adenine-Guanine-Uracil Solution. Dissolve 200 mg. each of adenine sulfate, guanine hydrochloride, and uracil, with the aid of heat, in 10 ml. of dilute hydrochloric acid (1 in 3), cool, and add water to make 200 ml. Store under toluene in a refrigerator.

Salt Solution A. Dissolve 25 grams of monobasic potassium phosphate and 25 grams of dibasic potassium phosphate in water to make 500 ml. Add 5 drops of hydrochloric acid, and store under toluene.

Salt Solution B. Dissolve 10 grams of magnesium sulfate, 500 mg. of sodium chloride, 500 mg. of ferrous sulfate, and 500 mg. of manganese sulfate in water to make 500 ml. Add 5 drops of hydrochloric acid, and store under toluene.

Polysorbate 80 Solution. Dissolve 25 grams of polysorbate 80 in sufficient alcohol to make 250 ml.

Riboflavin-Thiamine Hydrochloride-Biotin Solution. Prepare a solution containing in each ml., 20 mcg. of riboflavin, 10 mcg. of thiamine hydrochloride, and 0.04 mcg. of biotin, by dissolving riboflavin, crystalline thiamine hydrochloride, and biotin in 0.02 *N* acetic acid. Store, protected from light, under toluene in a refrigerator.

Para-aminobenzoic Acid-Niacin-Pyridoxine Hydrochloride Solution. Prepare a solution in neutral 25 percent alcohol to contain 10 mcg. of para-aminobenzoic acid, 50 mcg. of niacin, and 40 mcg. of pyridoxine hydrochloride in each ml. Store in a refrigerator.

Stock Culture of Lactobacillus Plantarum

Dissolve 2.0 grams of water-soluble yeast extract in 100 ml. of water, add 500 mg. of anhydrous dextrose, 500 mg. of anhydrous sodium acetate, and 1.5 grams of agar, and heat the mixture, with stirring, on a steam bath until the agar dissolves. Add approximately 10-ml. portions of the hot solution to test tubes, suitably close or cover the tubes, sterilize at 121°, and allow the tubes to cool in an upright position. Prepare stab cultures in 3 or more of the tubes, using a pure culture of *Lactobacillus plantarum,** incubating for 16 to 24 hours at any selected temperature between 30° and 37° but held constant to within ±0.5°, and finally store in a refrigerator. Prepare a fresh stab of the stock culture every week, and do not use for inoculum if the culture is more than 1 week old.

Culture Medium

To each of a series of test tubes containing 5.0 ml. of the *Basal Medium Stock Solution* add 5.0 ml. of water containing 0.2 mcg. of calcium pantothenate. Plug the tubes with cotton, sterilize in an autoclave at 121°, and cool.

Inoculum

Make a transfer of cells from the *Stock Culture of Lactobacillus plantarum* to a sterile tube containing 10 ml. of *Culture Medium.* Incubate this culture for 16 to 24 hours at any selected temperature between 30° and 37° but held constant to within ±0.5°. The cell suspension so obtained is the inoculum.

Procedure

To similar test tubes add, in duplicate, 1.0 and/or 1.5, 2.0, 3.0, 4.0, and 5.0 ml., respectively, of the *Standard Preparation.* To each tube and to 4 similar tubes containing no *Standard Preparation* add 5.0 ml. of *Basal Medium Stock Solution* and sufficient water to make 10 ml.

To similar test tubes add, in duplicate, volumes of the *Assay Preparation* corresponding to 3 or more of the levels listed above for the *Standard Preparation,* including the levels of 2.0, 3.0, and 4.0 ml. To each tube add 5.0 ml. of the *Basal Medium Stock Solution* and sufficient water to make 10 ml. Place one complete set of Standard and Assay tubes together in one tube rack and the duplicate set in a second rack or section of a rack.

* American Type Culture Collection No. 8014 is suitable. This strain formerly was known as *Lactobacillus arabinosus* 17-5.

Cover the tubes of both series suitably to prevent contamination, and sterilize in an autoclave at 121° for 5 minutes. Cool, and add 1 drop of *Inoculum* to each tube, except 2 of the 4 tubes containing no *Standard Preparation* (to serve as the uninoculated blanks). Incubate the tubes at a temperature between 30° and 37°, held constant to within ±0.5° until, following 16 to 24 hours of incubation, there has been no substantial increase in turbidity in the tubes containing the highest level of standard during a 2-hour period.

Determine the transmittance of the tubes in the following manner. Mix the contents of each tube, and transfer to an optical container if necessary. Place the container in a spectrophotometer that has been set at a specific wave length between 540 and 660 mμ, and read the transmittance when a steady state is reached. This steady state is observed a few seconds after agitation when the galvanometer reading remains constant for 30 seconds or more. Allow approximately the same time interval for the reading on each tube.

With the transmittance set at 1.00 for the uninoculated blank, read the transmittance of the inoculated blank. With the transmittance set at 1.00 for the inoculated blank, read the transmittance for each of the remaining tubes. If there is evidence of contamination with a foreign microorganism, disregard the results of the assay.

Calculation

Prepare a standard concentration-response curve as follows: For each level of the standard, calculate the response from the sum of the duplicate values of the transmittance as the difference, $y = 2.00 - \Sigma$ (of transmittance). Plot this response on the ordinate of cross-section paper against the logarithm of the ml. of *Standard Preparation* per tube on the abscissa, using for the ordinate either an arithmetic or a logarithmic scale, whichever gives the better approximation to a straight line. Draw the straight line or smooth curve that best fits the plotted points.

Calculate the response, y, adding together the two transmittances for each level of the *Assay Preparation*. Read from the standard curve the logarithm of the volume of the *Standard Preparation* corresponding to each of those values of y that fall within the range of the lowest and highest points plotted for the standard. Subtract from each logarithm so obtained the logarithm of the volume, in ml., of the *Assay Preparation* to obtain the difference, x, for each level. Average the values of x for each of three or more levels to obtain $\bar{x} = M'$, the log-relative potency of the *Assay Preparation*. Determine the quantity, in mg., of U.S.P. Calcium Pantothenate Reference Standard corresponding to the calcium pantothenate taken for assay as antilog M = antilog $(M' + \log R)$ where R is the number of mg. of calcium pantothenate taken for assay.

CHEWING GUM BASE

BOUND STYRENE

Abbé-type refractometer. Use an instrument, having fourth decimal place accuracy, that can be placed in a nearly horizontal position for measurement of the refractive index of solids. An Amici-type compensating prism for achromatization is necessary unless a sodium vapor lamp is used as a light source.

Ethanol-toluene azeotrope. Mix 70 volumes of ethanol or of formula 2B ethanol with 30 volumes of toluene, reflux for 4 hours over calcium oxide, and then distil, discarding the first and last portions and collecting only that portion distilling within a range of 1°. (*Note:* Refluxing and distilling are not necessary if anhydrous 2B ethanol or absolute grain alcohol is used.)

Sample preparation. Sheet out a sample of the polymer to a thickness of 0.5 mm., and cut the sheeted sample into strips approximately 13 mm. wide and 25 mm. long. Fasten one strip to each leg of a "spider", consisting of a 13-mm. square of sheet aluminum or stainless steel having a Nichrome wire leg about 38 mm. long attached to each corner. Place the spider and strips in a 400-ml. flask containing 60 ml. of *Ethanol-toluene azeotrope*, positioning the spider so that each sample strip is in contact on all sides with the solvent. Extract for 1 hour at a temperature at which the solvent boils gently, then replace the solvent with another 60-ml. portion of *Ethanol-toluene azeotrope*, and extract for an additional hour. Remove the spider and sample strips from the flask, and dry them at 100° to constant weight in a vacuum oven at a pressure of about 10 mm. of mercury. (*Caution:* The samples must be extracted and dried thoroughly, but overheating, which would cause plastication, must be avoided.)

Remove the extracted and dried strips from the spider, and press the strips between aluminum foil (0.025–0.08 mm. thick, having good tear strength) at 100° for 3 to 10 minutes (preferably not more than 5 minutes), using any suitable apparatus to produce a uniform thickness not exceeding 0.5 mm. If the pressing is done between flat platens, without a cavity, use a force between about 500 and 1500 pounds, increasing the applied force proportionately if several strips are pressed at one time; if cavity pressing plates are used, close the platens without applying pressure and preheat for 1 minute, then apply a force of about 11 tons for 3 minutes, remove the specimens from the press, and allow them to cool.

Procedure. Cut the pressed sample in half with sharp scissors, and peel off one piece of the foil. Cut off a strip about 6 mm. wide and 12 mm. long with the scissors so that one of the narrower ends is freshly cut.

Check the adjustment of the refractometer by means of a glass test plate pressed firmly against the prism, using a drop of α-bromonaphthalene as the contact liquid. The small light source should be collimated. The best readings are obtained with the glass test piece if the light is diffused through crumpled tissue paper. After this adjustment, clean the prism well with lens paper moistened with alcohol. The refractive index of the glass test piece and of the test specimen must be measured at a known constant temperature, preferably 25°.

Place the test sample on the prism with the cut edge toward the light source approximately where the edge of the glass test piece was positioned. Remove the tissue paper from the light source, press the specimen firmly on the prism, and wait at least 1 minute for the sample to attain temperature equilibrium. The upper prism may be closed lightly on the specimen if adequate light can still be focused on the end of the specimen. Unless the specimen is very thin, however, this operation can damage the prism or its mounting. Adjust the compensating prism until a sharp dividing line between light and dark fields with minimum color is obtained. Test the contact between the specimen and the prism by pressing the test specimen against the prism: there should be no change in the position of the boundary line during this test. Move the hand control, from the light into the dark field, until the boundary line just reaches the crosshairs, and make at least three readings. If the readings differ by more than 0.0001 refractive index unit, further readings should be made. Repeat the process of obtaining readings with another portion of the sample having a freshly cut edge, and average the mean values of the two sets of readings thus obtained. If the two mean values do not differ by more than 0.0002 refractive index unit, report the average as the results of the calculations. If necessary, correct the refractive index measurements to 25° using the formula $N_{25} = N_t + 0.00037(t - 25)$, in which N_{25} is the refractive index at 25°, and N_t is the refractive index at the temperature, t, of measurement.

Calculate the percent of bound styrene in the test sample by the formula $23.50 + 1164(N_{25} - 1.53456) - 3497(N_{25} - 1.53456)^2$. Alternatively, the percent of bound styrene may be determined by reference to suitable tables.

MOLECULAR WEIGHT

Polyethylene

Sample solutions. Dissolve 1 gram of the sample, accurately weighed, in 95 ml. of tetrahydronaphthalene, filter into a 100-ml. volumetric flask, dilute to volume with the solvent, and mix (*Solution 1*). Transfer 50.0 ml. of *Solution 1* into a tared dish, evaporate on a steam bath for about 1 hour, and then complete the evaporation to dryness by heating in a vacuum oven at 70° for 2 hours or to constant weight. Calculate the concentration, C_1, in grams per 100 ml., of

Solution 1. Prepare *Solutions* 2 and *3*, respectively, by diluting 5.0-ml. and 10.0-ml. portions of *Solution 1* to 50.0 ml. with the solvent, and then calculate the concentration of each (C_2 and C_3, respectively).

Procedure. Determine the flow time, in seconds, of the solvent (t_o) and of the three *Sample solutions* (t_1, t_2, and t_3, respectively) in a Cannon-Fenske viscometer immersed in a constant-temperature bath maintained at 130°. Calculate the specific viscosity, η_{sp}, of each *Sample solution* by the formula $(t/t_o)-1$, and then calculate the reduced viscosity of each by the formula η_{sp}/C. Plot the reduced viscosity of each solution against concentration, and extrapolate to zero concentration to obtain the intrinsic viscosity, $[\eta]$. Finally, calculate the molecular weight of the polyethylene by the formula $([\eta]/K)^{1/a}$, in which K is 5.1×10^{-4}, and a is 0.725.

Polyisobutylene (Flory Method)

Sample solutions. Dissolve 1 gram of the sample, accurately weighed, in 95 ml. of diisobutylene, filter into a 100-ml. volumetric flask, dilute to volume with the solvent, and mix (*Solution 1*). Transfer 50.0 ml. of *Solution 1* into a tared dish, evaporate on a steam bath for about 1 hour, and then complete the evaporation to dryness by heating in a vacuum oven at 70° for 2 hours or to constant weight. Calculate the concentration, C_1, in grams per 100 ml., of *Solution 1*. Prepare *Solutions 2* and *3*, respectively, by diluting 5.0-ml. and 10.0-ml. portions of *Solution 1* to 50.0 ml. with the solvent, and then calculate the concentration of each (C_2 and C_3, respectively).

Procedure. Determine the flow time, in seconds, of the solvent (t_o) and of the three *Sample solutions* (t_1, t_2, and t_3, respectively) in a Cannon-Fenske viscometer immersed in a constant-temperature bath maintained at 20°. Calculate the specific viscosity, η_{sp}, of each *Sample solution* by the formula $(t/t_o)-1$, and then calculate the reduced viscosity of each by the formula η_{sp}/C. Plot the reduced viscosity of each solution against concentration, and extrapolate to zero concentration to obtain the intrinsic viscosity, $[\eta]$. Finally, calculate the molecular weight of the polyisobutylene by the formula $([\eta]/K)^{1/a}$, in which K is 3.60×10^{-4}, and a is 0.64.

Polyvinyl Acetate

Sample solutions. Dissolve 1 gram of the sample, accurately weighed, in 95 ml. of acetone, filter into a 100-ml. volumetric flask, dilute to volume with the solvent, and mix (*Solution 1*). Transfer 50.0 ml. of *Solution 1* into a tared dish, evaporate on a steam bath for about 1 hour, and then complete the evaporation to dryness by heating in a vacuum oven at 70° for 2 hours or to constant weight. Calculate the concentration, C_1, in grams per 100 ml., of *Solution 1*. Prepare *Solutions 2* and *3*, respectively, by diluting 5.0-ml. and 10.0-ml. portions of *Solution 1* to 50.0 ml. with the solvent, and then calculate the concentration of each (C_2 and C_3, respectively).

Procedure. Determine the flow time, in seconds, of the solvent (t_o) and of the three *Sample solutions* (t_1, t_2, and t_3, respectively) in a Cannon-Fenske viscometer immersed in a constant-temperature bath maintained at 25°. Calculate the specific viscosity, η_{sp}, of each *Sample solution* by the formula $(t/t_o) - 1$, and then calculate the reduced viscosity of each by the formula η_{sp}/C. Plot the reduced viscosity of each solution against concentration, and extrapolate to zero concentration to obtain the intrinsic viscosity, $[\eta]$. Finally, calculate the molecular weight of the polyvinyl acetate by the formula $([\eta]/K)^{1/a}$, in which K is 1.88×10^{-4}, and a is 0.69.

QUINONES

Standard preparations. Transfer 25.0 mg. of hydroquinone into a 100-ml. volumetric flask, dissolve in water, dilute to volume with water, and mix. Transfer 1.0-, 2.0-, 3.0-, 4.0-, and 6.0-ml. aliquots of this solution into a series of 100-ml. volumetric flasks, dilute each to volume with water, and mix. Transfer 2.0 ml. of each of these solutions and 3.0 ml. of water into a series of 25-ml. graduates, add 0.5 ml. of 0.1 *N* sodium carbonate to each, and continue as directed under *Sample preparations*, beginning with "... shake immediately, then add 1.0 ml. of 15 percent sulfuric acid..."

Sample preparations. Place 30 grams of freshly coagulated and washed sample in a two-necked 250-ml. flask, add 100 ml. of water, and heat at 66° for 2 hours. (*Caution:* Do not boil.) Cool to room temperature, and transfer 5.0 ml. of the extract into a 25-ml. glass-stoppered graduate. Transfer 5.0 ml. of water into a second graduate to serve as the blank. To each graduate add 1.0 ml. of 15 percent sulfuric acid. To the graduate containing the sample extract add 0.5 ml. of 0.1 *N* sodium carbonate, shake immediately, and then add 1.0 ml. of 15 percent sulfuric acid. (*Note:* The elapsed time for this operation should not exceed 15 seconds.) To each graduate add 1.0 ml. of 2,4-dinitrophenylhydrazine solution (dissolve 100 mg. of 2,4-dinitrophenylhydrazine in 50 ml. of carbonyl-free methanol, add 4 ml. of hydrochloric acid, and dilute to 100 ml. with water), stopper, and heat at 70° in a water bath for 1 hour. Cool to room temperature, then add to each graduate 13 ml. of water and 5.0 ml. of benzene, stopper, and shake vigorously. Allow the phases to separate, and pipet 2.0 ml. of the benzene layer from each graduate into corresponding test tubes containing 10 ml. of a 1 in 100 solution of diethanolamine in pyridine. Shake each tube, and allow the color to develop for 10 minutes.

Procedure. Determine the absorbance of the *Sample preparation* in a 1-cm. cell at 620 mμ, with a suitable spectrophotometer, against the reagent blank. Determine the absorbance of each *Standard preparation* in the same manner. Prepare a *Standard curve* by plotting absorbance of each *Standard preparation* against mcg. of quinone.

From the *Standard curve*, read the quantity, in mcg., of quinones (as benzoquinone) in the *Sample preparation*, and record the value thus obtained as Q. Calculate the quantity of quinones (as benzoquinone), in parts per million, in the sample by the formula $20Q/W$, in which W is the weight of the sample taken, in grams.

RESIDUAL STYRENE

Transfer about 10 grams of freshly coagulated and washed rubber, accurately weighed, into a 250-ml. iodine flask, add 50 ml. of water, and then add 25 ml. of synthetic methanol containing 100 parts per million of *p-tert*-butyl catechol. Connect the flask to a suitable distillation apparatus, and distil the mixture, collecting the first 25 ml. of distillate in a 250-ml. iodine flask. Rinse the condenser with 20 ml. of the methanol, and add the rinsings to the recovery flask. Add 20.0 ml. of 0.1 *N* bromine from a buret, mix, cool to 30°, and then rapidly add 15 ml. of 18 percent sulfuric acid. Stopper the flask, shake, and add water to the funnel lip of the flask to serve as a vapor seal. Allow the flask to stand for 1 minute. If no yellow color remains after standing, add successive 10-ml. portions of 0.1 *N* bromine from the buret until a slight yellow color persists for 1 minute. Add the solution around the stopper and lift gently so that the solution enters the flask around the stopper. Wash the funnel lip with water in the same manner, and seal again with water. One minute after the last addition of 0.1 *N* bromine, add 10 ml. of potassium iodide T.S. to the funnel lip, and lift the stopper to allow the solution to enter the flask around the stopper. Shake the contents of the flask, and titrate the liberated iodine with 0.1 *N* sodium thiosulfate to a faint yellow color. Add 1 ml. of starch T.S., and continue the titration to the first appearance of a clear solution. Calculate the percentage of residual styrene in the sample taken by the formula $[(D \times E) - (F \times G)] \times 0.208$, in which D is the volume, in ml., and E the exact normality of the 0.1 *N* bromine used, and F is the volume, in ml., and G the exact normality of the 0.1 *N* sodium thiosulfate used; the value 0.208 is an equivalence factor related to styrene.

SAMPLE SOLUTION FOR ARSENIC TEST

Transfer a 1-gram sample, accurately weighed, into a Kjeldahl flask, rest the open end of the flask in a Kjeldahl fume bulb attached to a water aspirator, add 5 ml. of sulfuric acid and 4 ml. of 30 percent hydrogen peroxide, and digest over a small flame. (See *Caution* statement under *Arsenic Test*, page 865.) Continue adding the peroxide in 2-ml. portions, allowing the reaction to subside between additions, until all organic matter is destroyed, fumes of sulfuric acid are copiously evolved, and the solution becomes colorless. *Maintain oxidizing conditions at all times during the digestion by peroxide whenever the mixture turns brown or darkens.* (The amount of peroxide required to completely digest the samples will vary, but as much as 200 ml.

may be required in some cases, depending upon the nature of the material.) Cool, cautiously add 10 ml. of water, again evaporate to strong fuming, and cool. Transfer the solution into an arsine generator flask, wash the Kjeldahl flask and bulb with water, adding the washings to the generator flask, and dilute to 35 ml. with water.

SAMPLE SOLUTION FOR LEAD LIMIT TEST

Transfer a 3.3-gram sample, accurately weighed, into a porcelain dish or casserole, heat on a hot plate until completely charred, then heat in a muffle furnace at 480° for 8 hours or overnight, and cool. Cautiously add 5 ml. of nitric acid, evaporate to dryness on a hot plate, then heat again in the muffle furnace at 480° for exactly 15 minutes, and cool. Extract the ash with two 10-ml. portions of water, filtering each extract into a separator. Leach any insoluble material on the filter with 6 ml. of *Ammonium Citrate Solution,* 2 ml. of *Hydroxylamine Hydrochloride Solution,* and 5 ml. of water (see *Lead Limit Test,* page 929, for preparation of these solutions), adding the filtered washings to the separator. Continue as directed under *Procedure* in the *Lead Limit Test,* page 929, beginning with "To the separator add 2 drops of phenol red T.S. . . ."

TOTAL UNSATURATION

Transfer about 500 mg. of the sample, accurately weighed, into a 500-ml. iodine flask containing 100 ml. of filtered carbon tetrachloride. Stopper the flask securely, cover to protect the mixture from light, and shake in a mechanical shaker until the sample is completely dissolved (about 1.5 hours). Remove the flask from the shaker and the cover from the flask, and then add 5 ml. of a 1 in 5 solution of trichloroacetic acid in carbon tetrachloride, 25.0 ml. of 0.1 N iodine in carbon tetrachloride, and 25 ml. of a 3 in 10 solution of mercuric acetate in glacial acetic acid. Stopper the flask, mix the contents thoroughly by shaking, and allow to stand in a dark place for exactly 30 minutes. Add 50 ml. of potassium iodide T.S., immediately titrate with 0.1 N sodium thiosulfate, using starch T.S. as indicator, and record the volume, in ml., required as S. Perform a blank determination (see page 2), and record the volume of 0.1 N sodium thiosulfate required as B. Calculate the percent total unsaturation by the formula $(B - S) \times (N/W) \times 1.87$, in which N is the exact normality of the 0.1 N sodium thiosulfate solution, W is the weight of the sample taken, in grams, and 1.87 is an equivalence factor for isobutylene, which is the chief contributor of olefin linkages in the copolymer. (*Note:* The percentage thus obtained will be high if the copolymer contains an antioxidant that reacts as an unsaturated compound in this test procedure; in this case, the weight percent of the antioxidant must be determined by appropriate means and subtracted from the percent total unsaturation as determined herein.)

CHLORIDE AND SULFATE LIMIT TESTS

Where limits for chloride and sulfate are specified in the monograph, the sample solution and control should be compared in appropriate glass cylinders of the same dimensions and matched as closely as practicable with respect to their optical characteristics.

If the solution is not perfectly clear after acidification, filter through filter paper that has been washed free of chloride and sulfate. Add identical quantities of the precipitant (silver nitrate T.S. or barium chloride T.S.) in rapid succession to both the sample solution and the control solution.

Experience has shown that visual turbidimetric comparisons are best made between solutions containing from 10 to 20 mcg. of chloride ion (Cl) or from 200 to 400 mcg. of sulfate ion (SO_4) in 50 ml. Weights of samples are specified on this basis in the monographs in which these limits are included.

Chloride Test

Standard Chloride Solution. Dissolve 165 mg. of sodium chloride in water and dilute to 100.0 ml. Transfer 10.0 ml. of this solution to a 1000-ml. volumetric flask, dilute to volume with water, and mix. Each ml. of the final solution contains 10 mcg. of chloride ion (Cl).

Procedure. Unless otherwise directed, dissolve the specified amount of the test substance in 30 to 40 ml. of water, neutralize to litmus external indicator with nitric acid, if necessary, then add 1 ml. in excess. To the clear solution or filtrate add 1 ml. of silver nitrate T.S., dilute to 50 ml. with water, mix, and allow to stand for 5 minutes protected from direct sunlight. Compare the turbidity, if any, with that produced similarly in a control solution containing the required volume of *Standard Chloride Solution* and the quantities of the reagents used for the sample.

Sulfate Test

Standard Sulfate Solution. Dissolve 148 mg. of anhydrous sodium sulfate in water and dilute to 100.0 ml. Transfer 10.0 ml. of this solution to a 1000-ml. volumetric flask, dilute to volume with water, and mix. Each ml. of the final solution contains 10 mcg. of sulfate (SO_4).

Procedure. Unless otherwise directed, dissolve the specified amount of the test substance in 30 to 40 ml. of water, neutralize to litmus external indicator with hydrochloric acid, if necessary, then add 1 ml. of diluted hydrochloric acid T.S. To the clear solution or filtrate add 3 ml. of barium chloride T.S., dilute to 50 ml. with water, and mix. After 10 minutes compare the turbidity, if any, with that produced in a solution containing the required volume of *Standard Sulfate Solution* and the quantities of the reagents used for the sample.

CHROMATOGRAPHY

For the purposes of the Food Chemicals Codex, chromatography is defined as a procedure by which chemicals are separated by the passage of a mixture of them through a fixed bed possessing a varying but reversible affinity for the individual components. The general chromatographic technique requires that a solute undergo distribution between two phases, one of them fixed (stationary phase), the other moving (mobile phase). It is the mobile phase that transfers the solute through the medium until it eventually emerges separated from other solutes that are eluted earlier or later. Generally, the solute is transported through the separation medium by means of a flowing stream of liquid or a gas known as the "eluant." The stationary phase may act through adsorption, as in the case of adsorbents such as silica gel, cellulose fibers, and ion-exchange resins (cationic or anionic), or it may act by dissolving the solute, thus partitioning the latter between the stationary and mobile phases.

The types of chromatography employed in the F.C.C. are column, paper, thin-layer, and gas.

Column Chromatography

Adsorption Chromatography. The adsorbent (such as chromatographic fuller's earth or siliceous earth (kieselguhr), silica gel, and ion-exchange resins), as a dry solid or as a slurry, is packed into a glass or quartz tube of suitable dimensions having a restricted outflow orifice. A solution of the test substance in a small amount of solvent is added to the top of the column and allowed to flow into the adsorbent. The chemical components of the test solution are quantitatively removed from the solution and are adsorbed in a narrow transverse band at the top of the column. As additional solvent is allowed to flow through the column, either by gravity or by application of air pressure, each substance progresses down the column at a characteristic rate resulting in a spatial separation to give what is known as the *chromatogram*. The rate of movement for a given substance is affected by several variables, including the adsorptive power of the adsorbent, the nature of the solvent, and the temperature of the chromatographic system.

If the separated compounds are colored or if they fluoresce under ultraviolet light, the adsorbent column may be extruded and, by transverse cuts, the appropriate segments may then be isolated. The desired compounds are then extracted from each segment with a suitable solvent. If the compounds are colorless, they may be located by means of painting or spraying the extruded column with color-forming reagents.

In the elution chromatogram, solvents are allowed to flow through the column until a component of the mixture appears in the effluent solution, known as the "eluate." If additional substances are to be determined, they are eluted by continuing the first solvent or by passing other solvents through the column.

A modified procedure for adding the test mixture to the column is sometimes employed. The sample, in a solid form, is mixed with some of the adsorbent and added to the top of a column. The subsequent flow of solvent moves the components down the column in the usual manner.

Partition Chromatography. In partition chromatography the substances to be separated are partitioned between two immiscible liquids one of which, the stationary phase, is usually adsorbed on a solid adsorbent, thereby presenting a very large surface area to the flowing solvent or mobile phase. The exceedingly high number of liquid-liquid contacts allows an efficiency of separation not achieved in ordinary liquid-liquid extraction.

In practice, partition chromatography is carried out in much the same manner as adsorption chromatography; i.e., the mixture, dissolved in a small amount of the solvent, is added to the top of the column, and development and elution are accomplished with flowing solvents. The mobile solvent usually is saturated with the stationary solvent before use.

In some cases it is helpful to add the sample dissolved in the stationary solvent to a small amount of adsorbent, transferring this mixture to the top of the column.

Ion-exchange Resin Chromatography. Ion-exchange resins are solid polymers (usually derived from organic substances) that have been chemically reacted in such a manner as to provide ion active groups. The chemical behavior of ion-exchange resins is divided into two major classes based upon the character of these ion active groups: (1) *cation resins*, i.e., those capable of exchanging cations, and (2) *anion resins*, i.e., those capable of exchanging anions. In practice, ion-exchange chromatography is carried out in a similar manner as for adsorption chromatography. The sample is added at the top of the column, and inert ions elute from the column. By proper choice of the ion-exchange resin, impurities may be removed from a sample so that further analysis can be made on the eluate, or desired ions may be concentrated and then removed by washing the resin with suitable solvents and analyzed by appropriate methods.

Paper Chromatography

In paper chromatography the adsorbent is a sheet of paper of suitable texture and thickness. Chromatographic separation may proceed through the action of a single liquid phase in a process analogous to adsorption chromatography in columns, or two immiscible solvents may be employed for partition chromatography on paper. In the latter process, the mobile phase moves slowly over the stationary phase which covers the fibers of the paper or forms a complex with them. The ratio of the distance traveled on the paper sheets by a given compound to the distance traveled by the front of the mobile phase, from the point of application of the test substance, is designated as the R_f value of the compound. The ratio between the distances traveled by a given compound and a reference substance constitutes the R_r value.

Apparatus. The essential equipment for paper chromatography consists of the following:

Vapor-tight chamber. The chamber is constructed preferably of glass, stainless steel, or porcelain. It is provided with inlets for the addition of solvent or for releasing internal pressure, and it is designed to permit observation of the progress of the chromatographic run without being opened. Tall glass cylinders are convenient if they are made vapor-tight with suitable covers and a sealing compound.

Supporting rack. The rack serves as a support for the solvent troughs and the anti-siphoning rods. It is constructed of a corrosion-resistant material about 5 cm. shorter than the inside height of the chamber.

Solvent troughs. The troughs, made of glass, are designed to be longer than the width of the chromatographic sheets and to contain a volume of solvent greater than that required for one chromatographic run.

Anti-siphoning rods. Constructed of heavy glass, the rods are placed on the rack and arranged to run outside of, parallel to, and slightly above the edge of the glass trough.

Chromatographic sheets. Special chromatographic filter paper is cut to length approximately equal to the height of the chamber. The sheet is at least 2.5 cm. wide but not wider than the length of the trough. A fine pencil line is drawn horizontally across the filter paper at a distance from one end such that, when the sheet is suspended from the anti-siphoning rods with the upper end of the paper resting in the trough and the lower portion hanging free into the chamber, the line is located a few cm. below the rods. Care is necessary to avoid contaminating the paper by excessive handling or by contact with dirty surfaces.

Procedure for Descending Chromatography. Separation of substances by descending chromatography is accomplished by allowing the mobile phase to flow downward on the chromatographic sheet.

The substance or substances to be analyzed are dissolved in a suitable solvent. Convenient volumes of the resulting solution, normally containing 1 to 20 mcg. of the compound, are placed in 6- to 10-mm. spots along the pencil line not less than 3 cm. apart. If the total volume to be applied would produce spots of a diameter greater than 6 to 10 mm., it is applied in separate portions to the same spot, each portion being allowed to dry before the next is added.

The spotted chromatographic sheet is suspended in the chamber by use of the anti-siphoning rod and an additional heavy glass rod which holds the upper end of the sheet in the solvent trough. The bottom of the chamber is covered with a mixture containing both phases of the prescribed solvent system. It is important to ensure that the portion of the sheet hanging below the rods is freely suspended in the chamber without touching the rack or the chamber walls. The chamber is sealed to allow equilibration (saturation) of the chamber and the paper with solvent vapor. Any excess pressure is released as necessary. For large chambers, equilibration overnight may be necessary.

A volume of the mobile phase in excess of the volume required for complete development of the chromatogram is saturated with the immobile phase. After equilibration of the chamber, the prepared mobile solvent is introduced into the trough through the inlet. The inlet is closed and the mobile solvent phase is allowed to travel down the paper the desired distance. Precautions must be taken against allowing the solvent to run down the sheet when opening the chamber and removing the chromatogram. The location of the solvent front is quickly marked, and the sheets are dried.

The chromatogram is observed and measured directly or after suitable development to reveal the location of the spots of the isolated components of the mixture.

Procedure for Ascending Chromatography. In ascending chromatography the lower edge of the sheet (or strip) is dipped into the mobile phase, to permit the mobile phase to rise on the chromatographic sheet.

The test materials are applied to the chromatographic sheets as directed under *Procedure for Descending Chromatography*. Enough of both phases of the solvent mixture to cover the bottom of the chamber is added. Empty solvent troughs are placed on the bottom of the chamber, and the chromatographic sheets are suspended so that the end near which the spots have been added hangs free inside the empty trough.

The chamber is sealed, and equilibration is allowed to proceed as described under *Procedure for Descending Chromatography*. Then the solvent is added through the inlet to the trough in excess of the solvent required for complete moistening of the chromatographic sheet. The chamber is resealed. When the solvent front has reached the desired height, the chamber is opened and the sheet is removed, the location of the solvent front is quickly marked, and the sheets are dried.

Small cylinders may be used without troughs so that only the mobile phase is placed on the bottom. The chromatographic sheet is suspended during equilibration with the lower end just above the solvent and chromatography is started by lowering the sheet so that it touches the solvent.

Use of Reference Substances in Identity Tests. The ratio of the distance traveled on the medium by a given compound to the distance traveled by the front of the mobile phase, from the point of application of the test substance, is designated as the R_f value of the compound. The ratio between the distances traveled by a given compound and a reference substance constitutes the R_r value. Absolute R_f values are difficult to establish, since the observed R_f varies with the experimental conditions. They are, however, very useful in tentatively identifying chemical compounds. A more definite identification is accomplished by using as a reference substance an authentic specimen of the compound in question.

For this purpose three chromatograms are usually prepared on the same adsorbent medium: one from the substance to be identified, one from the authentic specimen, and one from a mixture of nearly equal amounts of the substance to be identified and the authentic specimen. Each chromatogram contains approximately the same quantity by weight of material to be chromatographed. If the substance to be identified and the authentic specimen are identical, all three chromatograms will have the same R_f value, and the mixed chromatogram yields a single spot; i.e., R_r is 1.0.

Location of the Spots. The spots produced by the chromatographed materials are usually located by direct inspection if the compounds are visible under white or ultraviolet light, or by inspection in white or ultraviolet light after treatment with reagents that will make the spots visible (reagents are most conveniently applied with an atomizer).

Thin-Layer Chromatography

Thin-layer chromatography is used for the rapid separation of compounds by means of a uniform layer of finely powdered material applied to a glass plate. The coated plate can be considered as an "open chromatographic column" and the separations achieved may be based upon adsorption, partition, or a combination of both effects, or upon exclusion, depending on the particular type of support, its preparation, and its use with different solvents. Ion-exchange films can be used for the fractionation of polar compounds.

The R_f values obtained by thin-layer chromatography are less reproducible than those obtained by paper chromatography; for this reason, it is usually necessary to prepare chromatograms of authentic specimens or reference standards, preferably in varied quantities, alongside the chromatograms of the sample. Presumptive identification can be effected by observation of 2 spots of identical R_f value and about equal magnitude. For semiquantitative estimation, a visual comparison of the size of the spots may be made, but more accurate quantitative measurements must be made by densitometry or careful removal of the spots from the plate, followed by elution with a suitable solvent and spectrophotometric measurement.

For two-dimensional thin-layer chromatography, the chromatographed plate is turned at a right angle and again chromatographed, usually in another chamber equilibrated with a different solvent system.

Apparatus. Acceptable apparatus and materials for thin-layer chromatography consist of the following:

Glass plates. Flat glass plates of uniform thickness throughout their areas.

Aligning tray. An aligning tray or other suitable flat surface is used to align and hold plates during application of the adsorbent.

Adsorbent. The adsorbent may consist of finely divided adsorbent materials for chromatography. It can be applied directly to the glass plate, or it can be bonded to the plate by means of plaster of Paris or with starch paste. The former will not yield as hard a surface as will the starch, but it is not affected by strongly oxidizing spray reagents. Pretreated chromatographic plates are available commercially.

Spreader. A suitable spreading device which, when moved over the glass plate, applies a uniform layer of adsorbent of desired thickness over the entire surface of the plate.

Storage rack. A rack of convenient size, to hold the prepared plates during drying and transportation.

Developing chamber. The chamber can accommodate one or more plates and can be properly closed and sealed as described under *Paper Chromatography.* It is fitted with a plate-support rack which can support the plates, back to back, when the lid of the chamber is in place.

An ultraviolet light source suitable for observations with short (254 mμ) and long (360 mμ) ultraviolet wavelengths may be required.

Procedure. Clean the plates scrupulously, as by immersion in a chromic acid cleansing mixture, rinsing them with copious quantities of water until the water runs off the plates without leaving any visible water or oily spots, and then dry.

Arrange the plate or plates on the aligning tray, and secure them so that they will not slip during the application of the adsorbent. Mix appropriate quantities of adsorbent and liquid, usually water, which when shaken for 30 seconds give a smooth slurry that will spread evenly with the aid of a spreader. Place the plates in the storage rack for a suitable period of time, and then dry at 105° for 30 minutes or as directed in the monograph.

Equilibrate the atmosphere in the chamber as described under *Procedure for Descending Chromatography* in the section on *Paper Chromatography.*

Apply the *Sample Solution* and the *Standard Solution* at points about 1.5 cm. apart and about 2 cm. from the lower edge of the plate (the lower edge is the first part over which the spreader moves in the application of the adsorbent layer), and allow to dry. A template will aid in determining the spot points and the 10 to 15-cm. distance through which the solvent front should move.

Unless otherwise directed in the individual monograph, place a mark 10 cm. above the spot point. Arrange the plate on the supporting rack (sample spots on the bottom), and introduce the rack into the developing chamber. The solvent in the chamber must be deep enough to reach the lower edge of the adsorbent, but must not touch the spot points. Seal the cover in place, and maintain the system until the solvent ascends to a point 10 to 15 cm. above the initial spots, this commonly requiring from 15 minutes to 1 hour. Remove the plates and dry them in air. Measure and record the distance of each spot from the point of origin. If further directed, spray the spots with the reagent specified, observe, and compare the sample with the standard chromatogram. Calculate R_f values as described under *Use of Reference Substances in Identity Tests* in the section on *Paper Chromatography*.

Gas Chromatography

Gas chromatography denotes the technique in which the moving phase is a *gas*. In the other forms of chromatography discussed in this section, the moving phase is a liquid. The stationary phase in all important chromatographic techniques is either a solid or a liquid, or a combination of both.

In gas-liquid chromatography, the substances to be separated are partitioned between a stationary liquid phase (substrate or solvent) and a mobile gas phase (carrier). The substrate, a nonvolatile liquid at the temperature of use, is immobilized as a thin film on a finely divided, inert solid support, such as chromatographic siliceous earth, crushed firebrick, glass beads, or even the inner wall of a small-diameter tube. The support normally does not retard the passage of the chemical or any of its minor components through the tube. If the latter is filled with the liquid-covered, finely divided solid, it is called a *packed* column (usually 3 to 6 mm. in internal diameter). If the inner wall of a small-diameter tube (usually 0.25 to 0.5 mm. in internal diameter) is coated with the liquid, it is called an open tubular or *capillary* column. In some techniques, the inner wall of such capillary tubing is coated with a finely divided solid support on which the liquid phase is placed, so that the difference between the two main types of gas chromatography columns is often not distinct.

In gas-solid chromatography, the identical situation holds except that the liquid phase is absent and the solid is an active adsorbent, such as alumina, silica gel, carbon or porous polymers.

In principle, the mobile gas phase continuously moves over the stationary liquid (*gas-liquid chromatography*) or solid (*gas-solid chromatography*) phase. When a vaporized substance is introduced into the gas stream at the head of the column, it is swept into the column and undergoes distribution between the gas and solid or liquid phases in a more or less stepwise manner. As the vapors pass through the column, some components dissolves in the liquid phase or are adsorbed upon the solid phase more readily than others. Each component moves through the column at a characteristic rate, depending on its partition coefficient with the stationary phase at the temperature employed, the concentration of the solid or liquid phase, and the flow rate of the carrier gas.

The behavior of a solute in such a partition process is most conveniently defined by a dimensionless factor called the *partition ratio, k.* The partition ratio is the ratio of the amount of solute dissolved in the stationary phase to the amount of solute dissolved in the mobile phase; i.e.,

$$k = \frac{\text{amount of solute in liquid phase}}{\text{amount of solute in gas phase}} \quad (1)$$

and, for Codex purposes, k is assumed constant throughout the column. The partition ratio is related also to the *time* that a solute spends in each of the phases; so that

$$k = \frac{\text{time in liquid phase}}{\text{time in gas phase}}. \quad (2)$$

Also, the partition ratio, k, can be related to the t_R factor (retention time). Obviously, the greater the total amount of liquid phase in the column, the more solute will dissolve in it and the greater will be the partition ratio by Equation (1). Similarly, an increase in the amount of liquid phase results in increased residence time in that phase, and the partition ratio increases (Equation 2). As has been noted, the gas phase simply serves to move the solute through the column between excursions into the liquid phase. Consequently, it is important to note that *all* solutes spend the *same time in the gas phase* in any particular column.

It is apparent that the value of the partition ratio, and, therefore, the time in the column depends upon the following solubility considerations: (a) the specific solute; (b) the specific liquid phase (solvent); (c) the amount of liquid phase; (d) the temperature; and (e) the gas flow rate. Therefore, a partition ratio exists for each column, solute, and temperature, and in order to reproduce the behavior of a particular solute, almost every experimental factor must be carefully reproduced, as is true of all other chromatographic techniques discussed in this section.

Apparatus and Procedure. The basic apparatus consists of a carrier gas supply, a sample injection port, a column, and a detector. The *carrier gas,* usually available in compressed form in a cylinder fitted with a suitable pressure-reducing valve, is conducted from the cylinder to the *sample injection port.* Since solutes to be chromatographed must be in the vapor phase, the injection port is heated to a temperature high enough to ensure rapid vaporization but not to cause thermal degradation of the chemical. Liquids, solutions, and gas samples are almost always injected by syringe through a silicone rubber septum in the injection port. The sample vapor mixes nearly instantaneously with the flowing carrier gas and is swept into the *column* (which is the main part of the chromatograph); or, in some instruments, the sample is injected directly onto the column packing. It is in the column that the different components of the vaporized sample are separated by virtue of their different interactions with the stationary column packing. The tubing that contains the packing is usually made of either metal or glass. Glass tubing is more inert and may be specified in certain monographs. The column also must be maintained at a selected temperature, which determines the time for the passage of the sample components, and, to a degree, the resolution and efficiency obtained with the particular column. For test substances containing many components of widely varying volatility, the *programmed temperature* technique may be used during all, or part, of the time required to elute all of the components. The temperature is increased continuously or intermittently at a specified rate by a suitable controller.

As the components emerge individually from the end of the column, they enter a *detector,* which indicates the amount of each component leaving the column. The detector also serves to indicate the time or gas volume required to reach the peak maximum that is characteristic for the particular experimental conditions being employed. The appearance time (retention time) of a compound is constant on a column, if operating conditions are constant. Control of the detector temperature to prevent condensation is as essential as temperature control of the injection port and column. Usually, the carrier gas is passed into a flow meter where the flow rate is determined. This step is necessary to reproduce a particular flow condition that has been found the most satisfactory for the resolution of a particular mixture. The detector is coupled to a suitable *automatic recording device,* such as a recording potentiometer. The resulting record, called the *chromatogram,* is a signal-time plot which may then be used to determine the identities and concentrations of the components.

The usual detector emits a signal proportional to the concentration of the solute in the carrier gas as it leaves the column. Thus, the chromatogram for each component appears as a bell-shaped (Gaussian) peak on a time axis. A detector arrangement that provides such a signal is called a *differential-type* detector, and the resulting curves accurately represent the distribution process as it has occurred during the residence time of the solutes in the column. The instrument should be operated so that the areas under these bell-shaped curves correspond to the amount of each solute present in the original mixture as it was introduced into the sample injection port.

Since gas chromatography is a separation method only, it cannot, without previous calibration, be used to identify compounds. To achieve qualitative analysis, the retention time or volume to the peak maximum of a known pure chemical must be determined. Then when a peak appears at that same time or volume under the same experimental conditions, the certainty of identification is reasonably high. Alternatively, the individual components may be collected in a cold trap as they emerge from the column and be subjected to independent analysis by instrumental or chemical methods.

The elution time or volume for air is an important quantity, since it corresponds closely to the gas space in the column. This value is frequently used to obtain absolute and relative retention values for characterization of compounds.

Individual components of a mixture may be identified best by means of their *relative retention*, α, i.e.,

$$\alpha = \frac{x_2 - x_a}{x_1 - x_a}, \tag{3}$$

where x_2 is the retention from zero time or injection point (as time, volume, or distance on the chromatogram) of the desired chemical, x_1 is the same for a standard material determined with the same column and temperature, and x_a is the retention for an inert component, such as air, which is not retarded in its passage through the column. Using these retention symbols, the partition ratio, k_2, for the compounds having a retention of x_2 is:

$$k_2 = \frac{x_2}{x_a} - 1. \tag{4}$$

Tables of α may be prepared, stating the standard material used to obtain x_1, which can then be referred to for identification purposes. All of the solubility factors mentioned above must be cited in such compilations.

As a measure of efficiency of the separation of two components in a mixture, the dimensionless *resolution factor*, R, is determined as

$$R = \frac{2(t_{R'} - t_{R''})}{(w_1 + w_2)}, \tag{5}$$

in which $t_{R'}$ and $t_{R''}$ are the respective retention times of the two components of the mixture, and w_1 and w_2 are the band widths determined by the intersection of the tangents of the inflection points of the Gaussian peaks with the baseline.

Quantitative results can be obtained from the areas under the peaks. The areas are determined by means of an automatic integrator, a planimeter, or by triangulation. For systems where equivalent accuracy can be obtained, peak heights may be substituted for areas. Peak area measurement is less accurate for small peaks and for those having short retentions.

Area percentages, % A_i, or a/a, of individual species within a chromatogram are used in purity analysis and are equal to 100 times the ratio of the peak area of the species, A_i, to the sum, ΣA_i, of all the areas of the peaks in the chromatogram. This procedure is termed *area normalization*. If the individual components and their relative responses to the detector are known, more accurate analysis may be made by multiplying the area for each peak by its response factor, and dividing the product by the sum of products for all peaks. Where response factors of the various components are nearly equal, the product of retention time and peak height may be substituted for peak area measurements to minimize graphical error.

When assays require quantitative comparison of one chromatogram with another, variations in the size of the sample injected may be a major source of error. This error may be eliminated by use of the *internal standard* method. This method is also used when nonvolatile compounds in the test substance do not elute and, hence, the sum of the peak areas or peak heights is not representative of the entire sample. If the internal standard is chemically similar to the substance being analyzed, minor variations in column and detector parameters can also be controlled. A known amount of a miscible standard substance with a retention time different from any of the sample components is added to the sample to be chromatographed. The area under the component peak is divided by the area of the standard peak, and the weight percent is calculated by comparing the result to those from known mixtures. In some procedures, the internal standard may be carried through the analysis prior to chromatography of the sample in order to control other quantitative aspects of the procedure.

DISTILLATION RANGE

Scope

This method is to be used for determining the distillation range of pure or nearly pure compounds or mixtures having a relatively narrow distillation range of about 40° C. or less. The result so determined is an indication of purity, not necessarily of identity. Products having a distillation range of greater than 40° may be determined by this method if a wide range thermometer, such as A.S.T.M.-E-1-2C or 3C, is specified in the individual monograph.

Definitions

Distillation range. The difference between the temperature observed at the start of a distillation and that observed at which a specified volume has distilled, or at which the dry point is reached.

Initial boiling point. The temperature indicated by the distillation thermometer at the instant the first drop of condensate leaves the end of the condenser tube.

Dry point. The temperature indicated at the instant the last drop of liquid evaporates from the lowest point in the distillation flask, disregarding any liquid on the side of the flask.

Apparatus

Distillation flask. A 200-ml. round-bottomed distilling flask of heat-resistant glass is preferred when sufficient sample (in excess of 100 ml.)

is available for the test. If a sample of less than 100 ml. must be used, a smaller flask having a capacity of at least double the volume of the liquid taken may be employed. The 200-ml. flask has a total length of 17.9 cm., and the inside diameter of the neck is 2.1 cm. Attached about midway on the neck, approximately 12 cm. from the bottom of the flask, is a side arm 12.7 cm. long and 5 mm. in internal diameter, which is inclined downward at an angle of 75° from the vertical.

Condenser. Use a straight glass condenser of heat-resistant tubing, 56 to 60 cm. long and equipped with a water jacket so that about 40 cm. of the tubing is in contact with the cooling medium. The lower end of the condenser may be bent to provide a delivery tube or it may be connected to a bent adapter which serves as the delivery tube.

Note: All-glass apparatus with standard-taper ground joints may be used alternatively if the assembly employed provides results equal to those obtained with the flask and condenser described above.

Receiver. The receiver is a 100-ml. cylinder which is graduated in 1-ml. subdivisions and calibrated "to contain." It is used for measuring the sample as well as for receiving the distillate.

Thermometer. An accurately standardized partial-immersion thermometer having the smallest practical subdivisions (not greater than 0.2° C.) is recommended in order to avoid the necessity for an emergent stem correction. Suitable thermometers are available as the A.S.T.M. Series 37C through 41C, and 102C through 107C, or as the MCA types R-1 through R-4 (see *Thermometers*, page 966).

Source of heat. A Bunsen burner is the preferred source of heat. An electric heater may be used, however, if it is shown to give results comparable to those obtained with the gas burner.

Shield. The entire burner and flask assembly should be protected from external air currents. Any efficient shield may be employed for this purpose.

Flask support. An asbestos board, 6.5 mm. in thickness and having a 10-cm. circular hole, is placed on a suitable ring or platform support and fitted loosely inside the shield to insure that hot gases from the source of heat do not come in contact with the sides or neck of the flask. A second 6.5-mm. asbestos board, at least 15 cm. square and provided with a 30-mm. circular hole, is placed on top of the first board. This board is used to hold the 200-ml. distillation flask which should be fitted firmly on the board so that direct heat is applied to the flask only through the opening in the board.

Procedure

Note: This procedure is to be used for liquids which distil above 50°, wherein the sample can be measured and received, and the condenser water used, at room temperature (20 to 30°). For materials boiling below 50°, cool the liquid to below 10° before sampling, receive the distillate in a water bath cooled to below 10°, and use water cooled to below 10° in the condenser.

Measure 100 ± 0.5 ml. of the liquid in the 100-ml. graduate, and transfer the sample together with an efficient antibumping device to the distilling flask. Do not use a funnel in the transfer nor allow any of the sample to enter the side arm of the flask. Place the flask on the asbestos boards, which are supported on a ring or platform, and place in position the shield for the flask and burner. Connect the flask and condenser, place the graduate under the outlet of the condenser tube, and insert the thermometer. The thermometer should be located in the center of the neck end so that the top of the bulb (when present, auxiliary bulb) is just below the bottom of the outlet to the side arm. Regulate the heating so that the first drop of liquid is collected within 5 to 10 minutes. Read the thermometer at the instant the first drop of distillate falls from the end of the condenser tube and record as the initial boiling point. Continue the distillation at the rate of 4 or 5 ml. of distillate per minute, noting the temperature as soon as the last drop of liquid evaporates from the bottom of the flask (dry point) or when the specified percentage has distilled over. Correct the observed temperature readings for any variation in the barometric pressure from the normal (760 mm.) by allowing 0.1° for each 2.7 mm. of variation, adding the correction if the pressure is lower, or subtracting if higher than 760 mm.

When a total-immersion thermometer is used, correct for the temperature of the emergent stem by the formula $0.00015 \times N (T - t)$, in which N represents the number of degrees of emergent stem from the bottom of the stopper, T, the observed temperatures of distillation, and t, the temperature registered by an auxiliary thermometer the bulb of which is placed midway of the emergent stem, adding the correction to the observed readings of the main thermometer.

ESSENTIAL OILS AND RELATED SUBSTANCES

ACETALS

Hydroxylamine Hydrochloride Solution. Prepare as directed under *Aldehydes*, page 894.

Procedure. Weigh accurately the quantity of the sample specified in the monograph, and transfer it into a 125-ml. Erlenmeyer flask. Add 30 ml. of *Hydroxylamine Hydrochloride Solution*, and reflux on a steam bath for exactly 60 minutes. Allow the condenser to drain into the flask for 5 minutes after removing the flask from the steam bath. Detach and rapidly cool the flask to room temperature. Add bromophenol blue T.S. as indicator, and titrate with 0.5 N alcoholic potassium hydroxide to pH 3.4, or to the same light color as produced in the original hydroxylamine hydrochloride solution on adding the indicator. Calculate the ml. of 0.5 N alcoholic potassium hydroxide consumed per gram of sample (A).

Using a separate portion of the sample, proceed as directed under *Aldehydes*, page 894. Calculate the ml. of 0.5 N alcoholic potassium hydroxide consumed per gram of sample (B).

Calculate the percent of acetals by the formula $(A - B) \times f$, in which f is the equivalence factor given in the monograph.

ACID VALUE

Dissolve about 10 grams of the sample, accurately weighed, in 50 ml. of alcohol, previously neutralized to phenolphthalein with 0.1 N sodium hydroxide. Add 1 ml. of phenolphthalein T.S. and titrate with 0.1 N sodium hydroxide until the solution remains faintly pink after shaking for 10 seconds, unless otherwise directed in the individual monograph. Calculate the acid value (AV) by the formula:

$$AV = \frac{5.61 \times S}{W}$$

in which S = the number of ml. of 0.1 N sodium hydroxide consumed in the titration of the sample and W = the weight of the sample in grams.

TOTAL ALCOHOLS

Unless otherwise stated in the monograph, transfer 10 grams of a solid sample, or 10 ml. of a liquid sample, accurately weighed, into a 100-ml. flask having a standard taper neck. Add 10 ml. of acetic anhydride and 1 gram of anhydrous sodium acetate, mix these materials, attach a reflux condenser to the flask, and reflux the mixture for 1 hour. Cool and add 50 ml. of water at a temperature between 50° and 60° through the condenser. Shake intermittently during a period of 15 minutes, cool to room temperature, transfer the mixture completely to a separator, allow the layers to separate, and then remove and reject the lower, aqueous layer. Wash the oil layer successively with 50 ml. of a saturated sodium chloride solution, 50 ml. of a 10 percent sodium carbonate solution, and 50 ml. of saturated sodium chloride solution. If the oil is still acid to moistened litmus paper, wash it with additional portions of sodium chloride solution until it is free from acid. Drain off the oil, dry it with anhydrous sodium sulfate, then filter it.

Weigh the quantity of acetylized oil specified in the monograph into a tared 125-ml. Erlenmeyer flask, add 10 ml. of neutral alcohol, 10 drops of phenolphthalein T.S., and 0.1 N alcoholic potassium hydroxide, dropwise, until a pink end-point is obtained. If more than 0.20 ml. is needed, reject the sample, and wash and test the remaining acetylized oil until its acid content is below this level. Prepare a blank for residual titration (see page 2), using the same volume of alcohol and indicator, and add 1 drop of 0.1 N alkali to produce a pink end-point. Measure 25.0 ml. of 0.5 N alcoholic potassium hydroxide into each of the flasks, reflux them simultaneously for one hour, cool, and titrate the contents of each flask with 0.5 N hydrochloric acid to the disappearance of the pink color. Calculate the percent of *Total Alcohols* by the formula:

$$A = \frac{(b - S)(100e)}{W - 21(b - S)}$$

in which b = the number of ml. of 0.5 N hydrochloric acid consumed in the residual blank titration, S = the number of ml. of 0.5 N hydrochloric acid consumed in the titration of the sample, e = the equivalence factor given in the monograph, and W = the weight of the sample of the acetylized oil in mg.

ALDEHYDES

Hydroxylamine Hydrochloride Solution. Dissolve 50 grams of hydroxylamine hydrochloride, preferably reagent grade or freshly recrystallized before using, in 90 ml. of water and dilute to 1000 ml. with aldehyde-free alcohol. Adjust the solution to a pH of 3.4 with 0.5 N alcoholic potassium hydroxide.

Procedure. Weigh accurately the quantity of sample specified in the monograph, and transfer it into a 125-ml. Erlenmeyer flask. Add 30 ml. of *Hydroxylamine Hydrochloride Solution*, mix thoroughly, and allow to stand at room temperature for 10 minutes, unless otherwise specified in the monograph. Titrate with 0.5 N alcoholic potassium hydroxide to a greenish yellow end-point that matches the color of 30 ml. of *Hydroxylamine Hydrochloride Solution* in a 125-ml. flask when the same volume of bromophenol blue T.S. has been added to each flask, or preferably titrate to a pH of 3.4 using a suitable pH meter. Calculate the percent of aldehyde (A) by the formula:

$$A = \frac{(S - b)(100E)}{W}$$

in which S = the number of ml. of 0.5 N alcoholic potassium hydroxide consumed in the titration of the sample, b = the number of ml. of 0.5 N alcoholic potassium hydroxide consumed in the titration of the blank, E = the equivalence factor given in the monograph, and W = the weight of the sample in mg.

ALDEHYDES AND KETONES—HYDROXYLAMINE METHOD

Hydroxylamine Solution. Dissolve 20 grams of hydroxylamine hydrochloride (reagent grade or preferably freshly recrystallized) in 40 ml. of water and dilute to 400 ml. with alcohol. Add, with stirring, 300 ml. of 0.5 N alcoholic potassium hydroxide, and filter. Use this solution within two days.

Procedure. Weigh accurately the quantity of the sample specified in the individual monograph, and transfer it into a 250-ml. glass-stoppered flask. Add 75.0 ml. of *Hydroxylamine Solution* to this flask and to a similar flask for a residual blank titration (see page 2). If the component to be determined is an *aldehyde*, stopper the flasks and allow them to stand at room temperature for 1 hour unless otherwise stated in the monograph. If the component to be determined is a *ketone*, attach the flask to a suitable condenser, and reflux the mixture for 1 hour unless otherwise stated in the monograph, and then cool to room temperature. Titrate both flasks to the same greenish yellow end-point using bromophenol blue T.S. as indicator, or preferably to a pH of 3.4 using a pH meter. (If the indicator is used, the end-point color must be the same as that produced when the blank is titrated to a pH of 3.4.) Calculate the percent of aldehyde or ketone by the formula:

$$AK = \frac{(b - S)(100E)}{W}$$

in which AK = percent of aldehyde or ketone, b = the number of ml. of 0.5 N hydrochloric acid consumed in the residual blank titration, S = the number of ml. of 0.5 N hydrochloric acid consumed in the titration of the sample, E = the equivalence factor given in the monograph, and W = the weight of the sample in mg.

ALDEHYDES AND KETONES—NEUTRAL SULFITE METHOD

Pipet a 10-ml. sample into a 100-ml. cassia flask fitted with a stopper, and add 50 ml. of a freshly prepared 30 in 100 solution of sodium sulfite. Add 2 drops of phenolphthalein T.S. and neutralize with a 50 percent (by volume) acetic acid solution. Heat the mixture in a boiling water bath, and shake the flask repeatedly, neutralizing the mixture from time to time by the addition of a few drops of the 50 percent acetic acid solution, stoppering the flask to prevent loss of volatile material. After no coloration appears upon the addition of a few more drops of phenolphthalein T.S. and heating for 15 minutes, cool to room temperature. When the liquids have separated completely, add sufficient sodium sulfite solution to raise the lower level of the oily layer within the graduated portion of the neck of the flask. Calculate the percent, by volume, of the aldehyde or ketone by the formula:

$$AK = 100 - (V \times 10)$$

in which AK = percent, by volume, of the aldehyde or ketone in the sample, and V = the number of ml. of separated oil in the graduated neck of the flask.

CHLORINATED COMPOUNDS

Wind a 1.5 × 5-cm. strip of 20-mesh copper gauze around the end of a copper wire. Heat the gauze in a nonluminous flame of a Bunsen burner until it glows without coloring the flame green. Permit the gauze to cool and re-ignite it several times until a good coat of oxide has formed. With a medicine dropper, apply 2 drops of the sample to the cooled gauze, ignite, and permit it to burn freely in the air. Again cool the gauze, add 2 more drops, and burn as before. Continue this process until a total of 6 drops has been added and ignited. Then hold the gauze in the outer edge of a Bunsen flame, adjusted to a height of 4 cm. Not even a transient green color is imparted to the flame. If at any of the additions the sample appears to be instantly vaporized, the test must be repeated from the beginning.

ESTERS

Ester Determination. Weigh accurately the quantity of the sample specified in the monograph, and transfer it into a 125-ml. Erlenmeyer flask containing a few boiling stones. Add to this flask, and, simultaneously, to a similar flask for a residual blank titration (see page 2), 25.0 ml. of 0.5 *N* alcoholic potassium hydroxide. Connect each flask to a reflux condenser, and reflux the mixtures on a steam bath for exactly 1 hour, unless otherwise directed in the monograph. Allow the mixtures to cool, add 10 drops of phenolphthalein T.S. to each flask, and titrate the excess alkali in each flask with 0.5 *N* hydrochloric acid. Calculate the percent of esters (E) in the sample by the formula:

$$E = \frac{(b - S)(100e)}{W}$$

in which b = the number of ml. of 0.5 *N* hydrochloric acid consumed in the residual blank titration, S = the number of ml. of 0.5 *N* hydrochloric acid consumed in the titration of the sample, e = the equivalence factor given in the monograph, and W = the weight of the sample in mg.

Ester Determination (High Boiling Solvent).

0.5 N Potassium Hydroxide Solution. Dissolve about 35 grams of potassium hydroxide in 75 ml. of water, add 1000 ml. of a suitable grade of monoethyl ether of diethylene glycol, and mix.

Procedure. Weigh accurately the quantity of the sample specified in the monograph, and transfer it into a 200-ml. Erlenmeyer flask having a standard-taper joint. To this flask and to a similar flask for a residual blank titration (see page 2), add 2 glass beads and 25.0 ml. of 0.5 *N Potassium Hydroxide Solution,* allowing exactly 1 minute for drainage from the buret or pipet. Attach an air condenser to each flask, reflux gently for 1 hour, and cool. Rinse down the condensers with about 50 ml. of water, then add phenolphthalein T.S. to each flask, and titrate the excess alkali with 0.5 *N* sulfuric acid to the disappearance of the pink color. Calculate the percent of esters (E) in the sample by the formula:

$$E = \frac{(b - S)(100e)}{W}$$

in which b is the number of ml. of 0.5 *N* sulfuric acid consumed in the blank determination, S is the number of ml. of 0.5 *N* sulfuric acid required in the titration of the sample, e is the equivalence factor given in the monograph, and W is the weight of the sample in mg.

Saponification Value. Proceed as directed for *Ester Determination* or *Ester Determination (High Boiling Solvent)*, as specified in the monograph. Calculate the saponification value (SV) by the formula:

$$SV = \frac{(b - S)(28.05)}{W}$$

in which b and S are as defined under *Ester Determination*, and W = weight of the sample in grams.

Ester Value. If the sample contains no free acids, the saponification value and the ester value are identical. If a determination of the *Acid Value* (AV) is specified in the monograph, calculate the ester value (EV) by the formula:

$$EV = SV - AV.$$

LINALOOL DETERMINATION

Transfer a 10-ml. sample, previously dried with sodium sulfate, into a 125-ml. glass-stoppered Erlenmeyer flask previously cooled in an ice bath. Add to the cooled oil 20 ml. of dimethyl aniline (monomethyl-free) and mix thoroughly. To the mixture add 8 ml. of acetyl chloride and 5 ml. of acetic anhydride, cool for several minutes, permit it to stand at room temperature for another 30 minutes, then immerse the flask in a water bath maintained at 40° ± 1° for 16 hours. Wash the acetylated oil with three 75-ml. portions of ice water, followed by successive washes with 25-ml. portions of 5 percent sulfuric acid, until the separated acid layer no longer becomes cloudy or emits an odor of dimethyl aniline when made alkaline. After removal of the dimethyl aniline, wash the acetylated oil first with 10 ml. of sodium carbonate T.S. and then with successive portions of water until the washings are neutral to litmus. Finally dry the acetylated oil with anhydrous sodium sulfate and proceed as directed for *Ester Determination* under *Esters*, page 896. Calculate the percent of linalool ($C_{10}H_{18}O$) by the formula:

$$L = \frac{7.707(b - S)}{W - 0.021(b - S)}$$

in which L = percent of linalool, b = the number of ml. of 0.5 N hydrochloric acid consumed in the residual blank titration, S = the number of ml. of 0.5 N hydrochloric acid consumed in the titration of the sample, and W = the weight of the sample in grams.

Note: When this method is applied to essential oils containing appreciable amounts of esters, perform an *Ester Determination*, page 896, on a sample of the *original oil* and calculate the percent of total linalool by the formula:

$$L = \frac{7.707(b - S)(1 - 0.0021E)}{W - 0.021(b - S)}$$

in which L = percent of linalool, E = the percent of esters, calculated as linalyl acetate ($C_{12}H_{20}O_2$) in the sample of the original oil, and b, S, and W are as defined in the preceding paragraph.

This entire procedure is applicable only to linalool and linalool-containing oils. It is not intended for the determination of other tertiary alcohols.

PERCENTAGE OF CINEOLE

Temp.	0.0	0.1	0.2	0.3	0.4	0.5	0.6	0.7	0.8	0.9
24	45.6	45.7	45.9	46.0	46.1	46.3	46.4	46.5	46.6	46.8
25	46.9	47.0	47.2	47.3	47.4	47.6	47.7	47.8	47.9	48.1
26	48.2	48.3	48.5	48.6	48.7	48.9	49.0	49.1	49.2	49.4
27	49.5	49.6	49.8	49.9	50.0	50.2	50.3	50.4	50.5	50.7
28	50.8	50.9	51.1	51.2	51.3	51.5	51.6	51.7	51.8	52.0
29	52.1	52.2	52.4	52.5	52.6	52.8	52.9	53.0	53.1	53.3
30	53.4	53.5	53.7	53.8	53.9	54.1	54.2	54.3	54.4	54.6
31	54.7	54.8	55.0	55.1	55.2	55.4	55.5	55.6	55.7	55.9
32	56.0	56.1	56.3	56.4	56.5	56.7	56.8	56.9	57.0	57.2
33	57.3	57.4	57.6	57.7	57.8	58.0	58.1	58.2	58.3	58.5
34	58.6	58.7	58.9	59.0	59.1	59.3	59.4	59.5	59.6	59.8
35	59.9	60.0	60.2	60.3	60.4	60.6	60.7	60.8	60.9	61.1
36	61.2	61.3	61.5	61.6	61.7	61.9	62.0	62.1	62.2	62.4
37	62.5	62.6	62.8	62.9	63.0	63.2	63.3	63.4	63.5	63.7
38	63.8	63.9	64.1	64.2	64.4	64.5	64.6	64.8	64.9	65.1
39	65.2	65.4	65.5	65.7	65.8	66.0	66.2	66.3	66.5	66.6
40	66.8	67.0	67.2	67.3	67.5	67.7	67.9	68.1	68.2	68.4
41	68.6	68.8	69.0	69.2	69.4	69.6	69.7	69.9	70.1	70.3
42	70.5	70.7	70.9	71.0	71.2	71.4	71.6	71.8	71.9	72.1
43	72.3	72.5	72.7	72.9	73.1	73.3	73.4	73.6	73.8	74.0
44	74.2	74.4	74.6	74.8	75.0	75.2	75.3	75.5	75.7	75.9
45	76.1	76.3	76.5	76.7	76.9	77.1	77.2	77.4	77.6	77.8
46	78.0	78.2	78.4	78.6	78.8	79.0	79.2	79.4	79.6	79.8
47	80.0	80.2	80.4	80.6	80.8	81.1	81.3	81.5	81.7	81.9
48	82.1	82.3	82.5	82.7	82.9	83.2	83.4	83.6	83.8	84.0
49	84.2	84.4	84.6	84.8	85.0	85.3	85.5	85.7	85.9	86.1
50	86.3	86.6	86.8	87.1	87.3	87.6	87.8	88.1	88.3	88.6
51	88.8	89.1	89.3	89.6	89.8	90.1	90.3	90.6	90.8	91.1
52	91.3	91.6	91.8	92.1	92.3	92.6	92.8	93.1	93.3	93.6
53	93.8	94.1	94.3	94.6	94.8	95.1	95.3	95.6	95.8	96.1
54	96.3	96.6	96.9	97.2	97.5	97.8	98.1	98.4	98.7	99.0
55	99.3	99.7	100.0							

PHENOLS

Pipet 10 ml. of the oil, which has been subjected to any treatment specified in the monograph, into a 100-ml. cassia flask, add 75 ml. of potassium hydroxide T.S., and shake vigorously for 5 minutes to ensure complete extraction of the phenol by the alkali solution. Allow the mixture to stand for about 30 minutes, then add sufficient potassium hydroxide T.S. to raise the oily layer into the graduated portion of the flask, stopper the flask, and allow it to stand overnight. Read the volume of insoluble oil to 0.05 ml. Calculate the percent, by volume, of phenols by the formula:

$$P = (10 - V) \times 10$$

in which P = percent of phenols, by volume, and V = observed volume of insoluble oil, in ml.

FREE PHENOLS

Transfer about 5 grams of the sample, accurately weighed, into a 150-ml. flask having a standard-taper neck. Pipet exactly 10 ml. of a 1 in 10 solution of acetic anhydride in anhydrous pyridine into the flask, and pipet exactly 10 ml. of this solution, preferably measured with the same pipet, into a second 150-ml. flask for the residual blank titration (see page 2). Connect the flasks to condensers, reflux for 1 hour, and cool to a temperature below 100°. Add 25 ml. of water to each flask through the condensers, and reflux again for 10 minutes. Cool the flasks, add phenolphthalein T.S., and titrate with 0.5 N potassium hydroxide. Calculate the percent of free phenols by the formula:

$$\text{Percent of Free Phenols} = \frac{(b - s) \times 100f}{W}$$

in which b is the number of ml. of 0.5 N potassium hydroxide consumed in the residual blank titration, s is the number of ml. of 0.5 N potassium hydroxide consumed in the titration of the sample, f is the equivalence factor given in the monograph, and W is the weight of the sample, in mg.

RESIDUE ON EVAPORATION

Weigh accurately the quantity of sample specified in the monograph, and transfer it into a suitable evaporating dish that has previously been heated on a steam bath, cooled to room temperature in a desiccator, and accurately weighed. Weigh the sample in the dish. Heat the evaporating dish containing the sample on the steam bath for the period of time specified in the monograph. Cool the dish and its contents to room temperature in a desiccator and weigh accurately. Calculate the residue as percent of the sample used.

SOLUBILITY IN ALCOHOL

Transfer a 1.0-ml. sample into a calibrated 10-ml. glass-stoppered cylinder graduated in 0.1 ml. subdivisions, and add slowly, in small portions, alcohol of the concentration specified in the monograph. Maintain the temperature at 20° and shake the cylinder thoroughly after each addition of alcohol. When a clear solution is first obtained, record the number of ml. of alcohol required. Continue the addition of the alcohol until a total of 10 ml. has been added. If opalescence or cloudiness occurs during these subsequent additions of alcohol, record the number of ml. of alcohol at which the phenomenon occurs.

ULTRAVIOLET ABSORBANCE OF CITRUS OILS

Transfer the quantity of the sample specified in the monograph into a 100-ml. volumetric flask, add alcohol to volume, and mix. Determine the ultraviolet absorption spectrum of the solution in the range of 260 mμ to 400 mμ in a 1-cm. cell with a suitable recording or manual spectrophotometer, using alcohol as the blank. If a manual instrument is used,

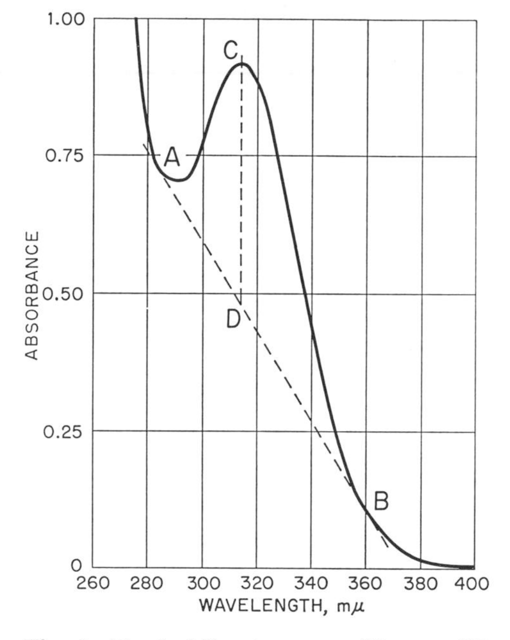

Fig. 5—Typical Spectrogram of Lemon Oil

read absorbances at 5-mμ intervals from 260 mμ to a point about 12 mμ from the expected maximum absorbance, then at 3-mμ intervals for 3 readings, and at 1-mμ intervals to a point about 5 mμ beyond the maximum, and then at 10-mμ intervals to 400 mμ. From these data, plot the absorbances as ordinates against wavelength on the abscissa, and draw the spectrogram. Draw a base-line tangent to the areas of minimum absorbance, as shown in the accompanying figure (which is typical of Lemon Oil), joining point *A* in the region of 280–300 mμ and a second point, *B*, in the region of 355–380 mμ. Locate the point of maximum absorbance, *C*, and from it drop a vertical line, perpendicular to the abscissa, that intersects line *AB* at *D*. Read from the ordinate the absorbances corresponding to points *C* and *D*, subtract the latter from the former, and correct the difference for the actual weight of oil taken, calculating to the basis of the sample weight specified in the monograph.

VOLATILE OIL CONTENT

This procedure is used, when specified in the individual monograph, for determining the volatile oil content of gums, resins, and essential oils.

Apparatus. The apparatus* is shown in the accompanying illustration. It consists of a 1000-ml. boiling flask, *A*, attached through trap *D* to a Liebig condenser, *C*, which is connected to a 25-ml. collector tube, *B*, graduated in 0.10-ml. units.

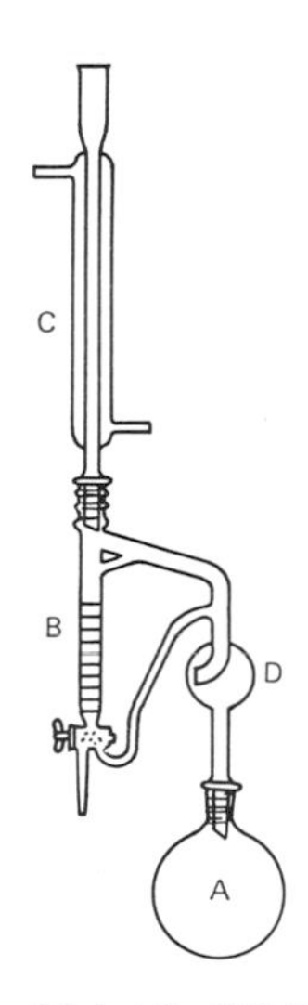

Fig. 6—Apparatus for Volatile Oil Content Determination

Procedure. Place 750 ml. of water in the boiling flask, boil for 10 minutes, and cool to 50°. Transfer the specified volume of the sample, prepared as directed in the monograph, into the flask, then immediately attach the remainder of the apparatus to the flask, and boil until the volume of distilled oil collected in the graduated collector tube remains constant. Avoid splashing the contents of the flask in order to prevent contamination of the distillate with nonvolatile material, and do not continue distillation for an extended time after the volume of distillate becomes constant. If the distilled oil is heavier than water, set the stopcock in the closed position to prevent return of the heavy distillate to the flask.

When distillation is complete, allow the contents of the collection tube to settle until the oil and water layers are separated completely. Allow the distillate to cool to room temperature, read its volume, and calculate therefrom the percent of volatile oil. (*Note:* When the volatile oil thus collected is to be used in additional tests, as may be

* Available as Catalog No. JD1135 from Scientific Glass Apparatus Company, 735 Broad St., Bloomfield, N.J. 07003.

specified in the monograph, the oil should be drained off, dried, and filtered before use.)

FATS AND RELATED SUBSTANCES

ACETYL VALUE

(Based on A.O.C.S. Method Cd 4-40)

The acetyl value is defined as the number of mg. of potassium hydroxide required to neutralize the acetic acid obtained by saponifying 1 gram of the acetylated sample.

Acetylation. Boil 50 ml. of the oil or melted fat with 50 ml. of freshly distilled acetic anhydride for 2 hours under a reflux condenser. Pour the mixture into a beaker containing 500 ml. of water, and boil for 15 minutes, bubbling a stream of nitrogen or carbon dioxide through the mixture to prevent bumping. Cool slightly, remove the water, add another 500 ml. of water, and boil again. Repeat for a third time with another 500-ml. portion of water, and remove the wash water, which should be neutral to litmus. Transfer the acetylated fat to a separator, and wash with two 200-ml. portions of warm water, separating as much as possible of the wash water each time. Transfer the washed sample to a beaker, add 5 grams of anhydrous sodium sulfate, and let stand for 1 hour, agitating occasionally to assist drying. Filter the oil through a dry filter paper, preferably in an oven at 100° to 110°, and keep the filtered oil in the oven until it is completely dry. The acetylated product should be a clear, brilliant oil.

Saponification. Weigh accurately from 2 to 2.5 grams each of the acetylated oil and of the original, untreated sample into separate 250-ml. Erlenmeyer flasks. Add to each flask 25.0 ml. of 0.5 *N* alcoholic potassium hydroxide, and continue as directed in the *Procedure* under *Saponification Value*, page 914, beginning with "Connect an air condenser. . . ." Record the saponification value of the untreated sample as S, and that of the acetylized oil as S', then calculate the acetyl value of the sample by the formula $(S' - S)/(1.000 - 0.00075S)$.

ACID VALUE

(Based on A.O.C.S. Methods Te 1a-64T and Cd 3a-63)

The acid value is defined as the number of mg. of potassium hydroxide required to neutralize the fatty acids in 1 gram of the test substance.

Method I. Unless otherwise directed, weigh accurately about 5 grams of the sample into a 500-ml. Erlenmeyer flask, and dissolve it in from 75 to 100 ml. of hot alcohol, previously boiled and neutralized to phenolphthalein T.S. with sodium hydroxide. Agitation and further heating may be necessary to effect complete solution of the sample. Add 0.5 ml. of phenolphthalein T.S., and titrate immediately, while shaking, with 0.5 N sodium hydroxide to the first pink color which persists for at least 30 seconds. Calculate the acid value by the formula $56.1V \times N/W$, in which V is the volume, in ml., and N is the normality, respectively, of the sodium hydroxide solution, and W is the weight, in grams, of the sample taken.

Method II. Prepare a solvent mixture consisting of equal parts, by volume, of isopropyl alcohol and toluene. Add 2 ml. of a 1 in 100 solution of phenolphthalein in isopropanol to 75 ml. of the mixture, and neutralize with alkali to a faint but permanent pink color. Weigh accurately the appropriate amount of sample indicated in the table below, dissolve it in the neutralized solvent mixture, warming if necessary, and shake vigorously while titrating with 0.1 N potassium hydroxide to the first faint pink color which persists for at least 30 seconds. Calculate the acid value by the formula $56.1V \times N/W$, in which V is the volume, in ml., and N is the normality, respectively, of the potassium hydroxide solution, and W is the weight, in grams, of the sample taken.

Acid Value	*Sample Wt., Grams*
0 to 1	20
1 to 4	10
4 to 15	2.5
15 to 50	0.5

FREE FATTY ACIDS

(Based on A.O.C.S. Method Ca 5a-40)

Unless otherwise directed, weigh accurately the appropriate amount of the sample, indicated in the table below, into a 250-ml. Erlenmeyer flask or other suitable container. Add 2 ml. of phenolphthalein T.S. to the specified amount of hot alcohol, neutralize with alkali to the first faint pink color that persists for 30 seconds, and then add the hot, neutralized alcohol to the sample container. Titrate with the appropriate normality of sodium hydroxide, shaking vigorously, to the first permanent pink color of the same intensity as that of the neutralized alcohol. Calculate the percent of free fatty acids in the sample by the formula VNe/W, in which V is the volume and N is the normality, respectively, of the sodium hydroxide used, W is the weight of the sample, in grams, and e is the equivalence factor given in the monograph.

F.F.A. Range, %	*Grams of Sample*	*Ml. of Alcohol*	*Strength of NaOH*
0.00 to 0.2	56.4 ± 0.2	50	0.1 *N*
0.2 to 1.0	28.2 ± 0.2	50	0.1 *N*
1.0 to 30.0	7.05 ± 0.05	75	0.25 *N*
30.0 to 50.0	7.05 ± 0.05	100	0.25–1.0 *N*
50.0 to 100	3.525 ± 0.001	100	1.0 *N*

FREE GLYCERIN OR PROPYLENE GLYCOL

(Based on A.O.C.S. Method Ca 14-56)

Reagents and Solutions

Use the *Periodic Acid Solution*, *Potassium Iodide Solution*, and the *Chloroform* as described under *1-Monoglycerides*, page 907.

Procedure

To the combined aqueous extracts obtained as directed under *1-Monoglycerides* add 50.0 ml. of *Periodic Acid Solution*. Run two blanks by adding 50.0 ml. of this reagent solution to two 500-ml. glass-stoppered Erlenmeyer flasks, each containing 75 ml. of water. Continue as directed in the *Procedure* under *1-Monoglycerides*, beginning with ". . . and allow to stand for at least 30 minutes but no longer than 90 minutes."

Calculation

Calculate the percent of free glycerin in the original sample by the formula $(b - S) \times N \times 2.30/W$, or calculate the percent of free propylene glycol by the formula $(b - S) \times N \times 3.81/W$, in which b is the number of ml. of sodium thiosulfate consumed in the blank determination, S is the number of ml. required in the titration of the aqueous extracts from the sample, N is the exact normality of the sodium thiosulfate, W is the weight, in grams, of the original sample taken, 2.30 is the molecular weight of glycerin divided by 40, and 3.81 is the molecular weight of propylene glycol divided by 20.

Note: If the aqueous extract contains more than 20 mg. of glycerin or more than 30 mg. of propylene glycol, dilute the extract in a volumetric flask and transfer a suitable aliquot into a 500-ml. glass-stoppered Erlenmeyer flask before proceeding with the test. The weight of the sample should be corrected in the calculation.

HYDROXYL VALUE

(Based on A.O.C.S. Methods Cd 4-40 and Cd 13-60)

The hydroxyl value is defined as the number of mg. of potassium hydroxide equivalent to the hydroxyl content of one gram of the unacetylated sample.

Method I. Proceed as directed under *Acetyl Value*, page 902, but calculate the hydroxyl value by the formula $(S' - S)/(1.000 - 0.00075S')$.

Method II. Unless otherwise directed, weigh accurately the appropriate amount of the sample indicated in the table below, transfer it into a 250-ml. glass-stoppered Erlenmeyer flask, and add 5.0 ml. of pyridine-acetic anhydride reagent (mix 3 volumes of freshly distilled pyridine with 1 volume of freshly distilled acetic anhydride).

Hydroxyl Value	*Sample Wt., Grams*
0 to 20	10
20 to 50	5
50 to 100	3
100 to 150	2
150 to 200	1.5
200 to 250	1.25
250 to 300	1.0
300 to 350	0.75

Pipet 5 ml. of the pyridine-acetic anhydride reagent into a second 250-ml. flask for the reagent blank. Heat the flasks for 1 hour on a steam bath under reflux condensers, then add 10 ml. of water through each condenser, heat for 10 minutes longer, and allow the flasks to cool to room temperature. Add 15 ml. of *n*-butyl alcohol, previously neutralized to phenolphthalein T.S. with 0.5 *N* alcoholic potassium hydroxide, through the condenser, then remove the condensers and wash the sides of the flasks with 10 ml. of *n*-butyl alcohol. To each flask add 1 ml. of phenolphthalein T.S., and titrate to a faint pink end-point with 0.5 *N* alcoholic potassium hydroxide, recording the ml. required for the sample as *S* and that for the blank as *B*. To correct for free acid, mix about 10 grams of the sample, accurately weighed, with 10 ml. of freshly distilled pyridine, previously neutralized to phenolphthalein, add 1 ml. of phenolphthalein T.S., and titrate to a faint end-point with 0.5 *N* alcoholic potassium hydroxide, recording the ml. required as *A*. Calculate the *Hydroxyl Value* by the formula $[B + (WA/C) - S] \times 56.1N/W$, in which *W* and *C* are the weights, in grams, of the samples taken for acetylation and for the free acid determination, respectively, and *N* is the exact normality of the alcoholic potassium hydroxide.

IODINE VALUE

The iodine value is a measure of unsaturation and is expressed as the number of grams of iodine absorbed, under the prescribed conditions, by 100 grams of the test substance.

Hanus Method

Iodobromide T.S. Dissolve 13.615 grams of iodine, with the aid of heat, in 825 ml. of glacial acetic acid that shows no reduction with dichromate and sulfuric acid. Cool, and titrate 25 ml. of the solution with 0.1 N sodium thiosulfate, recording the volume required as B. Prepare another solution containing 3 ml. (about 9 grams) of bromine in 200 ml. of glacial acetic acid. To 5 ml. of this solution add 10 ml. of potassium iodide T.S., and titrate with 0.1 N sodium thiosulfate, recording the volume required as C. Calculate the quantity, A, of the bromine solution required to double the halogen content of the remaining 800 ml. of iodine solution by the formula $800B/5C$. Mix the calculated volume, A, of the bromine solution with the iodine solution, and store in glass containers, protected from light.

Procedure. Weigh accurately the quantity of the sample specified in the monograph, transfer it into a 250-ml. iodine flask, and dissolve it in 10 ml. of chloroform. Add 25.0 ml. of iodobromide T.S., stopper the flask securely, and allow it to stand for exactly 30 minutes protected from light. Add 30 ml. of potassium iodide T.S. followed by 100 ml. of water, and titrate the liberated iodine with 0.1 N sodium thiosulfate, shaking thoroughly after each addition of the titrant. When the iodine color becomes quite pale, add 1 ml. of starch T.S. and continue the titration until the blue color is discharged. Perform a blank determination (see page 2), and calculate the iodine value by the formula $(B - S) \times 12.69N/W$, in which $B - S$ represents the difference between the volumes of 0.1 N sodium thiosulfate required for the blank and the sample, respectively, N is the exact normality of the sodium thiosulfate, and W is the weight, in grams, of the sample taken.

Wijs Method

Wijs Solution

Dissolve 13 grams of resublimed iodine in 1000 ml. of glacial acetic acid. Pipet 10.0 ml. of this solution into a 250-ml. flask, add 20 ml. of potassium iodide T.S. and 100 ml. of water, and titrate with 0.1 N sodium thiosulfate, adding starch T.S. near the end-point. Record the volume required as A. Set aside about 100 ml. of the iodine-acetic acid solution for future use. Pass chlorine gas, washed and dried with sulfuric acid, through the remainder of the solution until a 10.0 ml. portion requires not quite twice the volume of 0.1 N sodium thiosulfate consumed in the titration of the original iodine solution. A characteristic color change occurs when the desired amount of chlorine has been added. Alternatively, Wijs solution may be prepared by dissolving 16.5 grams of iodine monochloride, ICl, in 1000 ml. of glacial acetic acid. Store the solution in amber bottles sealed with paraffin until ready for use, and use within 30 days.

Total Halogen Content. Pipet 10.0 ml. of *Wijs Solution* into a 500-ml. Erlenmeyer flask containing 150 ml. of recently boiled and cooled water and 15 ml. of potassium iodide T.S. Titrate immediately with 0.1 N sodium thiosulfate, recording the volume required as B.

Halogen ratio. Calculate the I/Cl ratio by the formula $A/(B - A)$. The halogen ratio must be between 1.0 and 1.2. If the ratio is not within this range, the halogen content can be adjusted by the addition of the original solution or by passing more chlorine through the solution.

Procedure

The appropriate weight of the sample, in grams, is calculated by dividing the number 25 by the expected iodine value. Melt the sample, if necessary, and filter it through a dry filter paper. Transfer the accurately weighed quantity of the sample into a clean, dry, 500-ml. glass-stoppered bottle or flask containing 20 ml. of carbon tetrachloride, and pipet 25.0 ml. of *Wijs Solution* into the flask. The excess of iodine should be between 50 and 60 percent of the quantity added, that is, between 100 and 150 percent of the quantity absorbed. Swirl, and let stand in the dark for 30 minutes. Add 20 ml. of potassium iodide T.S. and 100 ml. of recently boiled and cooled water, and titrate the excess iodine with 0.1 N sodium thiosulfate, adding the titrant gradually and shaking constantly until the yellow color of the solution almost disappears. Add starch T.S., and continue the titration until the blue color disappears entirely. Toward the end of the titration, stopper the container and shake it violently so that any iodine remaining in solution in the carbon tetrachloride may be taken up by the potassium iodide solution. Concomitantly, conduct two determinations on blanks in the same manner and at the same temperature (see page 2). Calculate the iodine value by the formula $(B - S) \times 12.69N/W$, in which $B - S$ represents the difference between the volumes of sodium thiosulfate required for the blank and for the sample, respectively, N is the normality of the sodium thiosulfate, and W is the weight, in grams, of the sample taken.

1-MONOGLYCERIDES

(Based on A.O.C.S. Method Cd 11-57)

Reagents and Solutions

Periodic Acid Solution. Dissolve 5.4 grams of periodic acid, H_5IO_6, in 100 ml. of water, add 1900 ml. of glacial acetic acid, and mix. Store in a light-resistant, glass-stoppered bottle, or in a clear, glass-stoppered bottle protected from light.

Chloroform. Use chloroform meeting the following test: To each of three 500-ml. flasks add 50.0 ml. of *Periodic Acid Solution,* then add 50 ml. of chloroform and 10 ml. of water to two of the flasks and 50 ml. of water to the third. To each flask add 20 ml. of potassium iodide T.S., mix gently, and continue as directed in the *Procedure,* beginning with ". . . allow to stand at least 1 minute. . . ." The difference between the volume of 0.1 N sodium thiosulfate required in the titrations with and without the chloroform is not greater than 0.1 ml.

Procedure

Melt the sample, if not liquid, at a temperature not higher than 10° above its melting point, and mix thoroughly. Transfer an accurately weighed portion of the sample, equivalent to about 150 mg. of 1-monoglycerides, into a 100-ml. beaker (or weigh a sample equivalent to 20 mg. of glycerin or 30 mg. of propylene glycol if only *Free Glycerin* or *Propylene Glycol* is to be determined), and dissolve in 25 ml. of chloroform. Transfer the solution, with the aid of an additional 25 ml. of chloroform, into a separator, wash the beaker with 25 ml. of water, and add the washing to the separator. Stopper the separator tightly, shake vigorously for 30 to 60 seconds, and allow the layers to separate. (Add 1 to 2 ml. of glacial acetic acid to break emulsions formed due to the presence of soap.) Collect the aqueous layer in a 500-ml. glass-stoppered Erlenmeyer flask, and extract the chloroform solution again using two 25-ml. portions of water. Retain the combined aqueous extracts for the determination of *Free Glycerin* or *Propylene Glycol*, page 904. Transfer the chloroform to a 500-ml. glass-stoppered Erlenmeyer flask, and add 50.0 ml. of *Periodic Acid Solution* to this flask and to each of two blank flasks containing 50 ml. of chloroform and 10 ml. of water. Swirl the flasks during the addition of the reagent, and allow to stand for at least 30 minutes but no longer than 90 minutes. To each flask add 20 ml. of potassium iodide T.S., and allow to stand at least 1 minute but no longer than 5 minutes before titrating. Add 100 ml. of water, and titrate with 0.1 N sodium thiosulfate, using a magnetic stirrer to keep the solutions thoroughly mixed, to the disappearance of the brown iodine color, then add 2 ml. of starch T.S. and continue the titration to the disappearance of the blue color. Calculate the percent of 1-monoglycerides* in the sample by the formula $(B - S) \times N \times 17.927/W$, in which B is the number of ml. of sodium thiosulfate consumed in the blank determination, S is the number of ml. required in the titration of the sample, N is the exact normality of the sodium thiosulfate, W is the weight, in grams, of the sample taken, and 17.927 is the molecular weight of glyceryl monostearate divided by 20.

TOTAL MONOGLYCERIDES

Preparation of Silica Gel. Place about 10 grams of 100- to 200-mesh silica gel of a grade suitable for chromatographic work in a tared weighing bottle, cap immediately, and weigh accurately. Remove the cap, dry at 200° for 2 hours, cap immediately, and cool for 30 minutes. Raise the cap momentarily to equalize the pressure, then weigh again, reheat for 5 minutes at 200°, cool, and reweigh. Repeat this 5-minute drying cycle until two consecutive weights agree within 10 mg. Calcu-

* The monoglyceride may be calculated to some monoester other than glyceryl monostearate by dividing the molecular weight of the monoglyceride by 20 and substituting the value so obtained for 17.927 in the formula, using 17.80, for example, in calculating to the monooleate.

late the percent of water in the original silica gel (A) by the formula (loss in wt./sample wt.) × 100, then calculate the amount of water required to adjust the water content to 5 percent by the formula $W \times (5 - A)/95$, in which W is the weight, in grams, of the undried sample to be used.

Weigh accurately the appropriate amount of the undried silica gel to be used in the determination, transfer to a suitable blender or mixer, and add the calculated amount of water to give a final water content of 5 ± 0.1 percent. Blend for 1 hour to ensure complete water distribution, and store in a sealed container. Determine the water content of the adjusted silica gel as directed above, and readjust if necessary. (*Note:* Each new lot of silica gel should be checked for suitability by the analysis of a monoglyceride of known composition.)

Sample Preparation. Caution: To avoid rearrangement of partial glycerides, use extreme caution in applying heat to samples, and do not heat above 50°.

Samples melting below 50°. Melt the sample, if necessary, by warming for short periods below 50°, not exceeding a total of 30 minutes.

Samples melting above 50°. Grind about 10 grams in a mortar and pestle, chilling solid samples, if necessary, in solid carbon dioxide.

Weigh accurately about 1 gram of the prepared sample into a 100 ml. beaker, add 15 ml. of chloroform, and warm, if necessary, to effect solution. Use only minimum heat, and do not heat above 40°.

Preparation of Chromatographic Column. Connect a 19 × 290-mm. chromatographic tube, equipped with an outer 19/22 standard-taper joint at the top and a coarse fritted glass disk and inner 19/22 standard taper joint at the bottom, with an adapter consisting of an outer 19/22 joint connected to a Teflon stopcock. Do not grease the joints. Weigh 30 grams of the prepared silica gel into a 150-ml. beaker, add 50 to 60 ml. of petroleum ether, and stir slowly with a glass rod until all air bubbles are expelled. Transfer the slurry to the column through a powder funnel, and open the stopcock, allowing the liquid level to drop to about 2 cm. above the silica gel. Transfer any silica gel slurry remaining in the beaker into the column with a minimum amount of petroleum ether, then rinse the funnel and sides of the column. Drain the solvent through the stopcock until the level drops to 2 cm. above the silica gel, and remove the powder funnel.

Procedure. Carefully add the *Sample Preparation* to the prepared column. Open the stopcock, and adjust the flow rate to about 2 ml. per minute, discarding the eluate. Rinse the sample beaker with 5 ml. of chloroform, and add the rinsing to the column when the level drops to 2 cm. above the silica gel. Never allow the column to become dry on top, and maintain a flow rate of 2 ml. per minute throughout the elution. Avoid interruptions during elution which may cause pressure buildup and result in leakage through the stopcock or cracks in the silica gel packing.

Attach a 250-ml. reservoir separator, provided with a Teflon stopcock and a 19/22 standard taper drip tip inner joint, to the column. Add 200 ml. of benzene, elute, and discard the eluate, which contains the triglycerides fraction. When the level of benzene drops to 2 cm. above the silica gel, add 200 ml. of a 1 in 10 mixture of ether in benzene, elute, and discard the eluate, which contains the diglycerides and the free fatty acid fraction. When all of the ether-benzene solvent has been added from the separator and the level in the column drops to 2 cm. above the silica gel, add from 250 to 300 ml. of ether, and collect the monoglyceride fraction in a tared flask. Rinse the tip of the column into the flask with a few ml. of ether, and evaporate to dryness on a steam bath under a stream of nitrogen or dry air. Cool for at least 15 minutes, weigh, then reheat on the steam bath for 5 minutes in the same manner. Cool, reweigh, and repeat the 5-minute evaporation, cooling, and reweighing procedure until two consecutive weights agree within 2 mg. The weight of the residue represents the total monoglycerides in the sample taken.

OXYETHYLENE DETERMINATION

Apparatus

The apparatus for oxyethylene group determination is shown in the accompanying diagram. It consists of a boiling flask, *A*, fitted with a capillary side tube to provide an inlet for carbon dioxide and connected by a condenser with trap *B*, which contains an aqueous suspension of red phosphorus. The first absorption tube, *C*, contains a silver nitrate solution to absorb ethyl iodide. Absorption tube *D* is fitted with a 1.75-mm. spiral rod (23 turns, 8.5-mm. rise per turn), which is required to provide a longer contact of the evolved ethylene with the bromine solution. A standard-taper adapter and stopcock are connected to tube *D* to permit the transfer of the bromine solution into a titration flask without loss. A final trap, *E*, containing a potassium iodide solution, collects any bromine swept out by the flow of carbon dioxide.

Dimensions of the apparatus not readily determined from the diagram are as follows: carbon dioxide inlet capillary, 1-mm. inside diameter; flask *A*, 28-mm. diameter, 12/18 standard-taper joint; condenser, 9-mm. inside diameter; inlet to trap *B*, 2-mm. inside diameter; inlet to trap *C*, 7/15 standard-taper joint, 2-mm. inside diameter; trap *C*, 14-mm. inside diameter; trap *D*, inner tube, 8-mm. outside diameter, 2-mm. opening at bottom of spiral; outer tube, approximately 12.5-mm. inside diameter; side arm 7 cm. from top of inserted spiral, 3.5-mm. inside diameter, 2-mm. opening at bottom.

Reagents

Hydriodic Acid. Use special grade hydriodic acid suitable for alkoxyl determinations, or purify reagent grade as follows: Distil over red phosphorus in an all-glass apparatus, passing a slow stream of carbon dioxide through the apparatus until the distillation is terminated and the receiving flask has completely cooled. (*Caution: Use a safety shield and conduct the distillation in a hood.*)

Silver Nitrate Solution. Dissolve 15 grams of silver nitrate in 50 ml. of water, mix with 400 ml. of alcohol, and add a few drops of nitric acid.

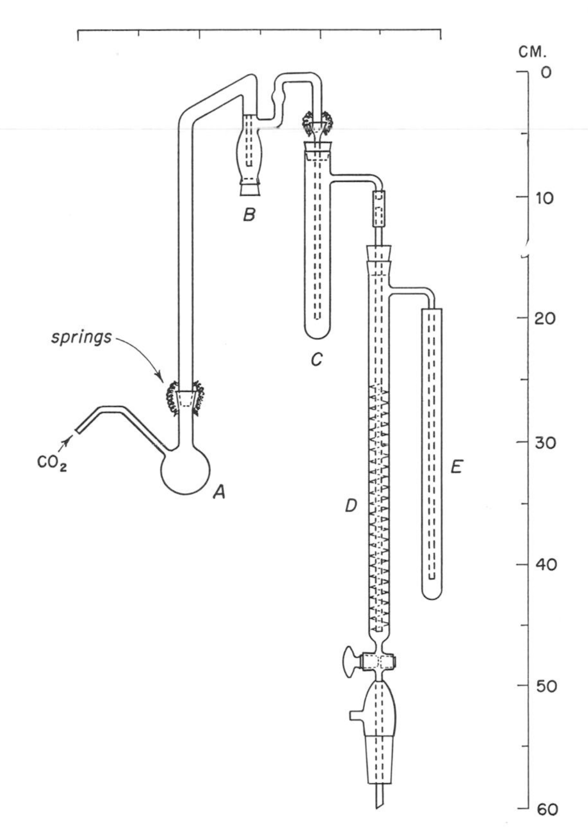

Fig. 7—Apparatus for Oxyethylene Determination

Bromine-Bromide Solution. Add 1 ml. of bromine to 300 ml. of glacial acetic acid saturated with dry potassium iodide (about 5 grams). Fifteen ml. of this solution requires about 40 ml. of 0.05 *N* sodium thiosulfate. Store in a brown bottle in a dark place, and standardize at least once a day during use.

Procedure

Fill trap *B* with enough of a suspension of 60 mg. of red phosphorus in 100 ml. of water to cover the inlet tube. Pipet 10 ml. of the *Silver Nitrate Solution* into tube *C* and 15 ml. of the *Bromine-Bromide Solution* into tube *D*, and place 10 ml. of a 1 in 10 solution of potassium iodide in trap *E*. Transfer an accurately weighed quantity of the sample specified in the monograph into the reaction flask, *A*, and add 10 ml. of *Hydriodic Acid* along with a few glass beads or boiling stones. Connect the flask to the condenser, and begin passing carbon dioxide through the apparatus at the rate of about 1 bubble per second. Heat the flask in an oil bath at 140° to 145°, and continue the reaction at this temperature for at least 40 minutes. Heating should be continued until the cloudy reflux in the condenser becomes clear and until the supernatant liquid in the silver nitrate tube, *C*, is almost completely clarified. Five minutes before the reaction is terminated, heat the *Silver Nitrate Solution* in tube *C* in a hot water bath at 50° to 60° to expel any dissolved olefin. At the completion of the decomposition, disconnect cautiously tubes *D* and *C* in the order named, then disconnect the carbon dioxide source and remove the oil bath. Connect tube *D* to a 500-ml. iodine flask containing 150 ml. of water and 10 ml. of a 1 in 10 solution of potassium iodide, run the *Bromine-Bromide Solution* into the flask, and rinse the tube and spiral with water. Add the potassium iodide solution from trap *E* to the flask, rinsing the side arm and tube with a few ml. of water, stopper the flask, and allow to stand for 5 minutes. Add 5 ml. of diluted sulfuric acid T.S., and titrate immediately with 0.05 *N* sodium thiosulfate, using 2 ml. of starch T.S. for the end-point. Transfer the silver nitrate solution from tube *C* into a flask, rinsing the tube with water, dilute to 150 ml. with water, and heat to boiling. Cool, and titrate with 0.05 *N* ammonium thiocyanate, using 3 ml. of ferric ammonium sulfate T.S. as the indicator. Perform a blank determination (see page 2). Calculate the percent of oxyethylene groups ($—CH_2CH_2O—$), as ethylene, by the formula $(B - S) \times N \times 2.203/W$, in which $B - S$ represents the difference between the volumes of sodium thiosulfate required for the blank and the sample solution, respectively, N is the normality of the sodium thiosulfate, W is the weight, in grams, of the sample taken, and 2.203 is an equivalence factor for oxyethylene. Calculate the percent of oxyethylene groups, as ethyl iodide, by the formula $(B' - S') \times N' \times 4.405/W$, in which $B' - S'$ represents the difference between the the volumes of ammonium thiocyanate required for the blank and the sample solution, respectively, N' is the normality of the ammonium thiocyanate, and 4.405 is an equivalence factor for oxyethylene. The sum of the values so obtained represents the percent of oxyethylene groups in the sample taken.

REICHERT-MEISSL VALUE

(Based on A.O.C.S. Method Cd 5-40)

The Reichert-Meissl value is a measure of soluble volatile fatty acids (chiefly butyric and caproic). It is expressed in terms of the number of ml. of 0.1 *N* sodium hydroxide required to neutralize the fatty acids obtained from a 5-gram sample under the specified conditions of the method.

Apparatus. Use a glass distillation apparatus of the same dimensions and construction as that shown in the accompanying illustration.

Reagents

Sodium Hydroxide Solution. Prepare a solution containing 50.0 percent by weight, of NaOH, and protect from contact with carbon dioxide. Allow the solution to settle and use only the clear liquid.

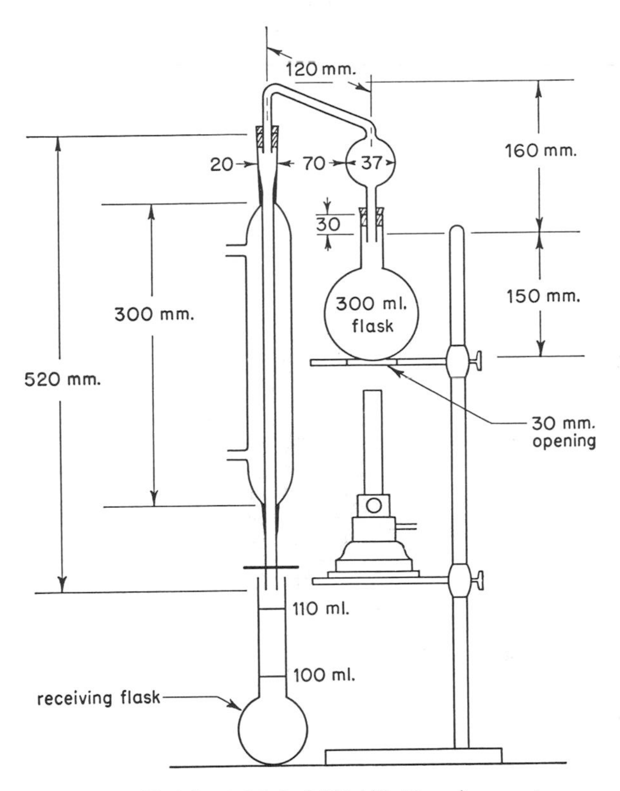

Fig. 8—Reichert-Meissl Distillation Apparatus

Glycerin-Sodium Hydroxide Mixture. Add 20 ml. of the *Sodium Hydroxide Solution* to 180 ml. of glycerin.

Procedure. Unless otherwise directed, weigh accurately about 5 grams of the sample, previously melted if necessary, into the 300-ml. distillation flask. Add 20.0 ml. of the *Glycerin-Sodium Hydroxide Mixture* and heat until the sample is completely saponified, as indicated by the mixture becoming perfectly clear. Shake the flask gently if any foaming occurs. Add 135 ml. of recently boiled and cooled water, dropwise at first to prevent foaming, then add 6 ml. of dilute sulfuric acid (1 in 5) and a few pieces of pumice stone or silicon carbide. Rest the flask on a piece of asbestos board having a center hole

5-cm. in diameter, and begin the distillation, regulating the flame so as to collect 110 ml. of distillate in 30 $\pm$ 2 minutes (measure time from the passage of the first drop of distillate from the condenser to the receiving flask), letting the distillate drip into the flask at a temperature not higher than 20°.

When 110 ml. has distilled, disconnect the receiving flask, and remove the flame. Mix the contents of the flask with gentle shaking and immerse almost completely for 15 minutes in water cooled to 15°. Filter the distillate through dry 9-cm. moderately retentive paper (S & S No. 589 White Ribbon or equivalent), add phenolphthalein T.S., and titrate 100 ml. of the filtrate with 0.1 N sodium hydroxide to the first pink color that remains unchanged for 2 to 3 minutes. Perform a blank determination (see page 2) using the same quantities of the same reagents, and calculate the Reichert-Meissl value by the formula $1.1 \times (S - B)$, in which S is the volume of 0.1 N sodium hydroxide required for the sample, and B is the volume required for the blank.

SAPONIFICATION VALUE

(Based on A.O.C.S. Methods Tl 1a-64 and Cd 3-25)

The saponification value is defined as the number of mg. of potassium hydroxide required to neutralize the free acids and saponify the esters in 1 gram of the test substance.

Procedure. Melt the sample, if necessary, and filter it through a dry filter paper to remove any traces of moisture. Unless otherwise directed, weigh accurately into a 250-ml. flask a sample of such size that the titration of the sample solution after saponification will require between 45 and 55 percent of the volume of 0.5 N hydrochloric acid required for the blank, and add to the flask 50.0 ml. of 0.5 N alcoholic potassium hydroxide. Connect an air condenser, at least 65 cm. in length, to the flask, and reflux gently until the sample is completely saponified (usually 30 minutes to 1 hour). Cool slightly, wash the condenser with a few ml. of water, add 1 ml. of phenolphthalein T.S., and titrate the excess potassium hydroxide with 0.5 N hydrochloric acid. Heat the contents of the flask to boiling, again titrate to the disappearance of any pink color that may have developed, and record the total volume of acid required. Perform a blank determination using the same amount of 0.5 N alcoholic potassium hydroxide (see page 2). Calculate the saponification value by the formula $56.1(B - S) \times N/W$, in which $B - S$ represents the difference between the volumes of 0.5 N hydrochloric acid required for the blank and the sample, respectively, N is the normality of the hydrochloric acid, and W is the weight, in grams, of the sample taken.

SOAP

Prepare a solvent mixture consisting of equal parts, by volume, of benzene and methanol, add bromophenol blue T.S., and neutralize with 0.5 *N* hydrochloric acid, or use neutralized acetone as the solvent. Weigh accurately the amount of sample specified in the individual monograph, dissolve it in 100 ml. of the neutralized solvent mixture and titrate with 0.5 *N* hydrochloric acid to a definite yellow end-point. Calculate the percent of soap in the sample by the formula VNe/W, in which V and N are the volume and normality, respectively, of the hydrochloric acid, W is the weight of the sample, in grams, and e is the equivalence factor given in the monograph.

SPECIFIC GRAVITY

The specific gravity of a fat or oil is determined at 25°, except when the substance is a solid at that temperature, in which case the specific gravity is determined at the temperature specified in the monograph, and is referred to water at 25°.

Clean a suitable pycnometer by filling it with a saturated solution of chromic acid (CrO_3) in sulfuric acid and allowing it to stand for at least 4 hours. Empty the pycnometer, rinse it thoroughly, then fill it with recently boiled water, previously cooled to about 20°, and place in a constant temperature bath at 25°. After 30 minutes, adjust the level of water to the proper point on the pycnometer, and stopper. Remove the pycnometer from the bath, wipe dry with a clean cloth free from lint, and weigh. Empty the pycnometer, rinse several times with alcohol and then with ether, allow to dry completely, remove any ether vapor, and weigh. Determine the weight of the contained water at 25° by subtracting the weight of the pycnometer from its weight when full.

Filter the oil or melted sample through filter paper to remove any impurities and the last traces of moisture, and cool to a few degrees below the temperature at which the determination is to be made. Fill the clean, dry pycnometer with the sample, and place it in the constant temperature bath at the specified temperature. After 30 minutes, adjust the level of the oil to the mark on the pycnometer, insert the stopper, wipe dry, and weigh. Subtract the weight of the empty pycnometer from its weight when filled with the sample, and divide the difference by the weight of the water contained at 25°. The quotient is the specific gravity at the temperature of observation, referred to water at 25°.

UNSAPONIFIABLE MATTER

(Based on A.O.C.S. Method Ca 6a-40)

This procedure determines those substances frequently found dissolved in fatty materials which cannot be saponified by alkali hydroxides but are soluble in the ordinary fat solvents.

Procedure. Weigh accurately 5.0 grams of the sample into a 250-ml. flask, add a solution of 2 grams of potassium hydroxide in 40 ml. of alcohol, and boil gently under a reflux condenser for 1 hour or until saponification is complete. Transfer the contents of the flask to a glass-stoppered extraction cylinder (approximately 30 cm. in length, 3.5 cm. in diameter, and graduated at 40, 80, and 130 ml.). Wash the flask with sufficient alcohol to make a volume of 40 ml. in the cylinder, and complete the transfer with warm and then cold water until the total volume is 80 ml. Finally, wash the flask with a few ml. of petroleum ether, add the washings to the cylinder, cool the contents of the cylinder to room temperature, and add 50 ml. of petroleum ether.

Insert the stopper, shake the cylinder vigorously for at least 1 minute, and allow both layers to become clear. Siphon the upper layer as completely as possible without removing any of the lower layer, collecting the ether fraction in a 500-ml. separator. Repeat the extraction and siphoning at least 6 times with 50-ml. portions of petroleum ether, shaking vigorously each time. Wash the combined extracts, with vigorous shaking, with 25-ml. portions of 10 percent alcohol until the wash water is neutral to phenolphthalein, and discard the washings. Transfer the ether extract to a tared beaker, and rinse the separator with 10 ml. of ether, adding the rinsings to the beaker. Evaporate the ether on a steam bath just to dryness, and dry the residue to constant weight, preferably at 75° to 80° under a vacuum of not more than 200 mm. of mercury, or at 100° for 30 minutes. Cool in a desiccator, and weigh to obtain the uncorrected weight of unsaponifiable matter.

Determine the quantity of fatty acids in the residue as follows: Dissolve the residue in 50 ml. of warm alcohol (containing phenolphthalein T.S. and previously neutralized with sodium hydroxide to a faint pink color), and titrate with 0.02 *N* sodium hydroxide to the same color. Each ml. of 0.02 *N* sodium hydroxide is equivalent to 5.659 mg. of fatty acids, calculated as oleic acid.

Subtract the calculated weight of fatty acids from the weight of the residue to obtain the corrected weight of unsaponifiable matter in the sample.

VOLATILE ACIDITY

Modified Hortvet-Sellier Method

Apparatus. Assemble a modified Hortvet-Sellier distillation apparatus as shown in Fig. 9, using a sufficiently large (approximately 38 × 203 mm.) inner Sellier tube and large distillation trap.

Procedure. Transfer the amount of sample, accurately weighed, specified in the monograph, into the inner tube of the assembly, and insert the tube in the outer flask containing about 300 ml. of recently

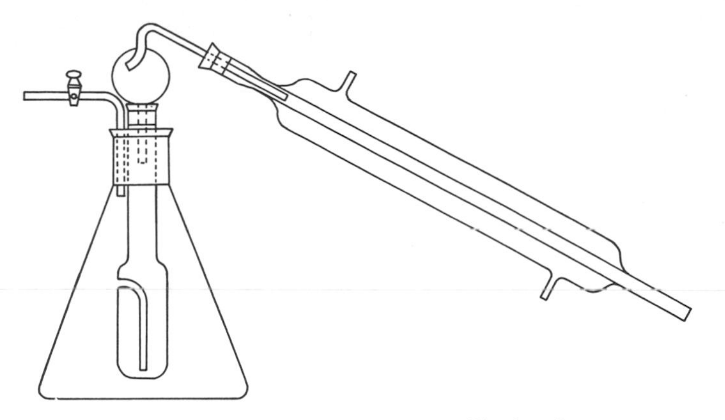

Fig. 9—Modified Hortvet-Sellier Distillation Apparatus

boiled hot water. To the sample add 10 ml. of approximately 4 *N* perchloric acid [35 ml. (60 grams) of 70 percent perchloric acid in 100 ml. of water], and connect the inner tube to a water-cooled condenser through the distillation trap. Distil by heating the outer flask so that 100 ml. of distillate is collected within 20 to 25 minutes. Collect the distillate in 100-ml. portions, add phenolphthalein T.S. to each portion, and titrate with 0.5 *N* sodium hydroxide. Continue the distillation until a 100-ml. portion of the distillate requires no more than 0.5 ml. of 0.5 *N* sodium hydroxide for neutralization. (*Caution:* Do not distil to dryness.) Calculate the weight, in mg., of volatile acids in the sample taken by the formula $V \times E$, in which V is the total volume, in ml., of 0.5 *N* sodium hydroxide consumed in the series of titrations, and E is the equivalence factor given in the monograph.

FLUORIDE LIMIT TEST

Method I (Thorium Nitrate Colorimetric Method)

This method should be used unless otherwise directed in the individual monograph.

Caution: When applying this test to organic compounds, the temperature at which the distillation is conducted must be rigidly controlled at all times to the recommended range of 135° to 140° to avoid the possibility of explosion. *Note:* To minimize the distillation blank resulting from fluoride leached from the glassware, the distillation apparatus should be treated as follows: treat the glassware with hot 10 percent sodium hydroxide solution, followed by flushing with tap water and rinsing with distilled water. At least once daily, treat in addition by boiling down 15–20 ml. of dilute sulfuric acid (1 in 2) until the still is filled with fumes; cool, pour off the acid, treat again with 10 percent sodium hydroxide solution, and rinse thoroughly. For further details, see sections 25.030 and 25.034 in *Official Methods of Analysis of the A.O.A.C.*, 11th Edition, 1970.

Unless otherwise directed, place a 5.0-gram sample and 30 ml. of water in a 125-ml. distillation flask connected with a condenser and carrying a thermometer and a capillary tube, both of which must extend unto the liquid. Slowly add, with continuous stirring, 10 ml. of perchloric acid, and then add 2 or 3 drops of silver nitrate solution (1 in 2) and a few glass beads. Connect a small dropping funnel or a steam generator to the capillary tube. Support the flask on an asbestos mat with a hole which exposes about one-third of the flask to the flame. Distil until the temperature reaches 135°. Add water from the funnel or introduce steam through the capillary, maintaining the temperature between 135° and 140° at all times. Continue the distillation until 100 ml. of distillate has been collected. After the 100-ml. portion (*Distillate A*) is collected, collect an additional 50-ml. portion of distillate (*Distillate B*) to ensure that all of the fluorine has been volatilized.

Place 50 ml. of *Distillate A* in a 50-ml. Nessler tube. In another similar Nessler tube place 50 ml. of water distilled through the apparatus as a control. Add to each tube 0.1 ml. of a filtered solution of sodium alizarinsulfaonte (1 in 1000) and 1 ml. of freshly prepared hydroxylamine hydrochloride solution (1 in 4000), and mix well. Add, dropwise, and with stirring, either 1 *N* or 0.05 *N* sodium hydroxide, depending upon the expected volume of volatile acid distilling over, to the tube containing the distillate until its color just matches that of the control, which is faintly pink. Then add to each tube 1.0 ml. of 0.1 *N* hydrochloric acid, and mix well. From a buret, graduated in 0.05 ml., add slowly to the tube containing the distillate enough thorium nitrate solution (1 in 4000) so that, after mixing, the color of the liquid just changes to a faint pink. Note the volume of the solution added, then add exactly the same volume to the control, and mix. Now add to the control solution sodium fluoride T.S. (10 mcg. F per ml.) from a buret to make the colors of the two tubes match after dilution to the same volume. Mix well, and allow all air bubbles to escape before making the final color comparison. Check the endpoint by adding 1 or 2 drops of sodium fluoride T.S. to the control. A distinct change in color should take place. Note the volume of sodium fluoride T.S. added.

Dilute *Distillate B* to 100 ml., and mix well. Place 50 ml. of this solution in a 50-ml. Nessler tube, and follow the procedure used for *Distillate A*. The total volume of sodium fluoride T.S. required for the solutions from both *Distillate A* and *Distillate B* should not exceed 2.5 ml.

Method II (**Ion Electrode Method A**)

Buffer Solution. Dissolve 36 grams of cyclohexylene dinitrilotetraacetic acid (CDTA) in sufficient 1 *M* sodium hydroxide to make 200 ml. Transfer 20 ml. of this solution (equivalent to 4 grams of disodium CDTA) into a 1000-ml. beaker containing 500 ml. of water, 57 ml. of glacial acetic acid, and 58 grams of sodium chloride, and stir to dissolve. Adjust the pH of the solution to between 5.0 and 5.5 by the addition of 5 *M* sodium hydroxide, then cool to room temperature, dilute to 1000 ml. with water, and mix.

Procedure. Unless otherwise directed in the individual monograph, place an 8.0-gram sample and 20 ml. of water in a 250-ml. distilling flask, cautiously add 20 ml. of perchloric acid, and then add 2 or 3 drops of silver nitrate solution (1 in 2) and a few glass beads. Following the directions, and observing the *Caution* and *Note*, as given under *Method I*, distil the solution until 200 ml. of distillate has been collected.

Transfer a 25.0-ml. aliquot of the distillate into a 250-ml. plastic beaker, and dilute to 100 ml. with the *Buffer Solution.* Place the fluoride ion and reference electrodes (or a combination fluoride electrode) of a suitable specific ion electrode apparatus (such as the Orion Model 407) in the solution, and adjust the calibration control until the indicator needle points to the center on the logarithmic concentration scale, allowing sufficient time for equilibration (about 20 minutes) and stirring constantly during the equilibration period and throughout the remainder of the procedure. Pipet 1.0 ml. of a solution containing 100 mcg. of fluoride ion (F) per ml. (prepared by dissolving 22.2 mg. of sodium fluoride, previously dried at 200° for 4 hours, in sufficient water to make 100.0 ml.) into the beaker, allow the electrode to come to equilibrium, and record the final reading on the logarithmic concentration scale. (*Note:* Follow the instrument manufacturer's instructions regarding precautions and interferences, electrode filling and check, temperature compensation, and calibration.)

Calculations. Calculate the fluoride content, in parts per million, of the sample taken by the formula $[(IA)/(R - I)] \times 100 \times [(200)/(25W)]$, in which I is the initial scale reading before the addition of the sodium fluoride solution; A is the concentration, in mcg. per ml., of fluoride in the sodium fluoride solution added to the sample solution; R is the final scale reading, after addition of the sodium fluoride solution; and W is the original weight of the sample, in grams.

Method III (Ion Electrode Method B)

Fluoride Standard Solution (5 mcg. F per ml.). Transfer 2.210 grams of sodium fluoride, previously dried at 110° for 2 hours and accurately weighed, into a 400-ml. plastic beaker, add 200 ml. of water, and stir until dissolved. Quantitatively transfer this solution into a 1000-ml. volumetric flask with the aid of water, dilute to volume with water, and mix. Store this stock solution in a plastic bottle. On the day of use, transfer 5.0 ml. of the stock solution into a 1000-ml. volumetric flask, dilute to volume with water, and mix.

Calibration Curve. Transfer into separate 250-ml. plastic beakers 1.0, 2.0, 3.0, 5.0, 10.0, and 15.0 ml. of the *Fluoride Standard Solution,* add 50 ml. of water, 5 ml. of 1 *N* hydrochloric acid, 10 ml. of 1 *M* sodium citrate, and 10 ml. of 0.2 *M* disodium ethylenediaminetetraacetate to each beaker, and mix. Transfer each solution into a 100-ml. volumetric flask, dilute to volume with water, and mix. Transfer a 50-ml. portion of each solution into a 125-ml. plastic beaker, and measure the potential of each solution with a suitable fluoride ion activity electrode (such as the Orion Model No. 94-09, with solid-state membrane), using a suitable reference electrode (such as the Orion Model No. 90-01, with single junction). Plot the calibration curve on 2-cycle semi-logarithmic paper (such as K & E No. 465130), with mcg. F per 100 ml. solution on the logarithmic scale.

Procedure. Transfer 1.00 gram of the sample into a 150-ml. glass beaker, add 10 ml. of water and, while stirring continuously, add 20 ml. of 1 *N* hydrochloric acid slowly to dissolve the sample. Boil rapidly for 1 minute, then transfer into a 250-ml. plastic beaker, and cool rapidly in ice water. Add 15 ml. of 1 *M* sodium citrate and 10 ml. of 0.2 *M* disodium ethylenediaminetetraacetate, and mix. Adjust the pH to 5.5 $\pm$ 0.1 with 1 *N* hydrochloric acid or 1 *M* sodium citrate, if necessary, then transfer into a 100-ml. volumetric flask, dilute to volume with water, and mix. Transfer a 50-ml. portion of this solution into a 125-ml. plastic beaker, and measure the potential of the solution with the apparatus described under *Calibration Curve.* Determine the fluoride content, in mcg., of the sample from the *Calibration Curve.*

HEAVY METALS TEST

This text is designed to limit the content of common metallic impurities that are colored by hydrogen sulfide (Ag, As, Bi, Cd, Cu, Hg, Pb, Sb, Sn) under the conditions specified. It demonstrates that the test substance is not grossly contaminated by such heavy metals and, within the precision of the test, that it does not exceed the *Heavy metals* limit given in the individual monograph in terms of the parts, by weight, of lead (Pb) per million parts of the test substance. It has been found that 20 mcg. of lead ion (Pb) in 50 ml. of solution is the optimum concentration for matching purposes by this method.

Method I is for the simpler and colorless substances and should be used in the *Procedure* unless otherwise directed in the individual monograph. *Method II* is specified for colored chemicals and those which, by virtue of their complex nature, interfere with the precipitation of heavy metals by sulfide.

Special Reagents

Ammonia T.S. Dilute 400 ml. of A.C.S. reagent grade ammonium hydroxide to 1000 ml. with water.

Hydrochloric Acid. Use reagent grade hydrochloric acid in preparing all solutions of hydrochloric acid employed in this test.

Lead Nitrate Stock Solution. Dissolve 159.8 mg. of lead nitrate, $Pb(NO_3)_2$, in 100 ml. of water containing 1 ml. of nitric acid, then dilute with water to 1000.0 ml. and mix. This solution should be prepared and stored in glass containers which are free from lead salts.

Standard Lead Solution. On the day of use, dilute 10 ml. of *Lead Nitrate Stock Solution,* accurately measured, with water to 100.0 ml. Each ml. of the solution so prepared contains the equivalent of 10 mcg. of lead ion (Pb).

Procedure

[*Note:* In the following procedures for Methods I and II, failure to adjust accurately the pH of the solutions within the specified limits may result in a significant loss of test sensitivity.]

Method I—Solution A. Pipet into a 50-ml. Nessler tube a volume of *Standard Lead Solution* containing the quantity of lead ion (Pb) equivalent to the heavy metals limit specified for the substance to be tested, and add water to make 25 ml. Adjust the pH to between 3.0 and 4.0 (short-range pH indicator paper) by the addition of diluted acetic acid T.S. or ammonia T.S., dilute with water to 40 ml., and mix.

Solution B. Place in a 50-ml. Nessler tube that matches the one used for *Solution A* 25 ml. of the solution prepared as directed in the individual monograph, adjust the pH to between 3.0 and 4.0 (short-range pH indicator paper) by the addition of diluted acetic acid T.S. or ammonia T.S., dilute to 40 ml. with water, and mix.

To each tube add 10 ml. of freshly prepared hydrogen sulfide T.S., mix, allow to stand for 5 minutes, and view downward over a white surface. The color of *Solution B* is no darker than that of *Solution A.*

Method II. Proceed as directed under *Method I,* but use the following in place of *Solution B:* Place the specified quantity of the sample, accurately weighed, in a suitable crucible, add sufficient sulfuric acid to wet the sample, and carefully ignite at a low temperature until thoroughly charred, covering the crucible loosely with a suitable lid during the ignition. After the substance is thoroughly carbonized, add 2 ml. of nitric acid and 5 drops of sulfuric acid, and cautiously heat until white fumes are evolved, then ignite, preferably in a muffle furnace, at 500° to 600° until the carbon is all burned off. Cool, add 4 ml. of dilute hydrochloric acid (1 in 2), cover, and digest on a steam bath for 10 to 15 minutes. Uncover, and slowly evaporate on a steam bath to dryness. Moisten the residue with 1 drop of hydrochloric acid, add 10 ml. of hot water, and digest for 2 minutes. Add dropwise ammonia T.S. until the solution is just alkaline to litmus paper, dilute with water to 25 ml., and adjust the pH to between 3.0 and 4.0 (short-range pH indicator paper) by the addition of diluted acetic acid T.S. Filter if necessary, wash the crucible and the filter with 10 ml. of water, dilute the combined filtrate and washing with water to 40 ml., and mix.

HYDROCHLORIC ACID TABLE[1]

Be.°	Sp. Gr.	% HCl	Be.°	Sp. Gr.	% HCl
1.00	1.0069	1.40	16.6	1.1292	25.56
2.00	1.0140	2.82	16.7	1.1301	25.72
3.00	1.0211	4.25	16.8	1.1310	25.89
4.00	1.0284	5.69	16.9	1.1319	26.05
5.00	1.0357	7.15	17.0	1.1328	26.22
5.25	1.0375	7.52	17.1	1.1336	26.39
5.50	1.0394	7.89	17.2	1.1345	26.56
5.75	1.0413	8.26	17.3	1.1354	26.73
6.00	1.0432	8.64	17.4	1.1363	26.90
6.25	1.0450	9.02	17.5	1.1372	27.07
6.50	1.0469	9.40	17.6	1.1381	27.24
6.75	1.0488	9.78	17.7	1.1390	27.41
7.00	1.0507	10.17	17.8	1.1399	27.58
7.25	1.0526	10.55	17.9	1.1408	27.75
7.50	1.0545	10.94	18.0	1.1417	27.92
7.75	1.0564	11.32	18.1	1.1426	28.09
8.00	1.0584	11.71	18.2	1.1435	28.26
8.25	1.0603	12.09	18.3	1.1444	28.44
8.50	1.0623	12.48	18.4	1.1453	28.61
8.75	1.0642	12.87	18.5	1.1462	28.78
9.00	1.0662	13.26	18.6	1.1471	28.95
9.25	1.0681	13.65	18.7	1.1480	29.13
9.50	1.0701	14.04	18.8	1.1489	29.30
9.75	1.0721	14.43	18.9	1.1498	29.48
10.00	1.0741	14.83	19.0	1.1508	29.65
10.25	1.0761	15.22	19.1	1.1517	29.83
10.50	1.0781	15.62	19.2	1.1526	30.00
10.75	1.0801	16.01	19.3	1.1535	30.18
11.00	1.0821	16.41	19.4	1.1544	30.35
11.25	1.0841	16.81	19.5	1.1554	30.53
11.50	1.0861	17.21	19.6	1.1563	30.71
11.75	1.0881	17.61	19.7	1.1572	30.90
12.00	1.0902	18.01	19.8	1.1581	31.08
12.25	1.0922	18.41	19.9	1.1590	31.27
12.50	1.0943	18.82	20.0	1.1600	31.45
12.75	1.0964	19.22	20.1	1.1609	31.64
13.00	1.0985	19.63	20.2	1.1619	31.82
13.25	1.1006	20.04	20.3	1.1628	32.01
13.50	1.1027	20.45	20.4	1.1637	32.19
13.75	1.1048	20.86	20.5	1.1647	32.38
14.00	1.1069	21.27	20.6	1.1656	32.56
14.25	1.1090	21.68	20.7	1.1666	32.75
14.50	1.1111	22.09	20.8	1.1675	32.93
14.75	1.1132	22.50	20.9	1.1684	33.12
15.00	1.1154	22.92	21.0	1.1694	33.31
15.25	1.1176	23.33	21.1	1.1703	33.50
15.50	1.1197	23.75	21.2	1.1713	33.69
15.75	1.1219	24.16	21.3	1.1722	33.88
16.0	1.1240	24.57	21.4	1.1732	34.07
16.1	1.1248	24.73	21.5	1.1741	34.26
16.2	1.1256	24.90	21.6	1.1751	34.45
16.3	1.1265	25.06	21.7	1.1760	34.64
16.4	1.1274	25.23	21.8	1.1770	34.83
16.5	1.1283	25.39	21.9	1.1779	35.02

Hydrochloric Acid Table—*continued*

Be.°	Sp. Gr.	% HCl	Be.°	Sp. Gr.	% HCl
22.0	1.1789	35.21	23.8	1.1963	38.95
22.1	1.1798	35.40	23.9	1.1973	39.18
22.2	1.1808	35.59	24.0	1.1983	39.41
22.3	1.1817	35.78	24.1	1.1993	39.64
22.4	1.1827	35.97	24.2	1.2003	39.86
22.5	1.1836	36.16	24.3	1.2013	40.09
22.6	1.1846	36.35	24.4	1.2023	40.32
22.7	1.1856	36.54	24.5	1.2033	40.55
22.8	1.1866	36.73	24.6	1.2043	40.78
22.9	1.1875	36.93	24.7	1.2053	41.01
23.0	1.1885	37.14	24.8	1.2063	41.24
23.1	1.1895	37.36	24.9	1.2073	41.48
23.2	1.1904	37.58	25.0	1.2083	41.72
23.3	1.1914	37.80	25.1	1.2093	41.99
23.4	1.1924	38.03	25.2	1.2103	42.30
23.5	1.1934	38.26	25.3	1.2114	42.64
23.6	1.1944	38.49	25.4	1.2124	43.01
23.7	1.1953	38.72	25.5	1.2134	43.40

[1] Courtesy of the Manufacturing Chemists' Association.

Specific Gravity determinations were made at 60° F., compared with water at 60° F.

From the Specific Gravities, the corresponding degrees Baumé were calculated by the following formula:

$$\text{Baumé} = 145 - \frac{145}{\text{Sp. Gr.}}$$

Baumé Hydrometers for use with this table must be graduated by the above formula, which should always be printed on the scale.

ALLOWANCE FOR TEMPERATURE

10–15° Be.—1/40° Be. or 0.0002 Sp. Gr. for 1° F.
15–22° Be.—1/30° Be. or 0.0003 Sp. Gr. for 1° F.
22–25° Be.—1/28° Be. or 0.00035 Sp. Gr. for 1° F

HYDROXYPROPOXYL DETERMINATION

Apparatus. The apparatus for hydroxypropoxyl group determination is shown in Fig. 10. The boiling flask, *D*, is fitted with an aluminum foil-covered Vigreaux column, *E*, on the sidearm and with a bleeder tube through the neck and to the bottom of the flask for the introduction of steam and nitrogen. A steam generator, *B*, is attached to the bleeder tube through tube *C*, and a condenser, *F*, is attached to the Vigreaux column. The boiling flask and steam generator are immersed in an oil bath, *A*, equipped with a thermoregulator such that a temperature of 155° and the desired heating rate may be maintained. The distillate is collected in a 150-ml. beaker, *G*, or other suitable container.

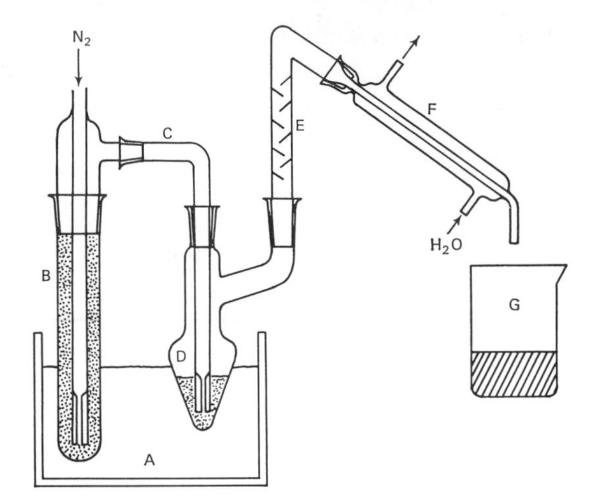

Fig. 10—Apparatus for Hydroxypropoxyl Determination

Procedure. Unless otherwise directed, transfer about 100 mg. of the sample, previously dried at 105° for 2 hours and accurately weighed, into the boiling flask, and add 10 ml. of chromium trioxide solution (60 grams in 140 ml. of water). Immerse the steam generator and the boiling flask in the oil bath (at room temperature) to the level of the top of the chromium trioxide solution. Start cooling water through the condenser, and pass nitrogen gas through the boiling flask at the rate of one bubble per second. Starting at room temperature, raise the temperature of the oil bath to 155° over a period of not less than 30 minutes, and maintain this temperature until the end of the determination. Distill until 50 ml. of distillate is collected. Detach the condenser from the Vigreaux column, and wash it with water, collecting the washings in the distillate container. Titrate the combined washings and distillate with 0.02 *N* sodium hydroxide to a pH of 7.0, using a pH meter set at the expanded scale. (*Note:* Phenolphthalein T.S. may be used for this titration if it is also used for all standards and blanks.) Record the volume, V_a, of the 0.02 *N* sodium hydroxide used. Add 500 mg. of sodium bicarbonate and 10 ml. of dilute sulfuric acid T.S., and then after evolution of carbon dioxide has ceased, add 1 gram of potassium iodide. Stopper the flask, shake the mixture, and allow it to stand in the dark for 5 minutes. Titrate the liberated iodine with 0.02 *N* sodium thiosulfate to the sharp disappearance of the yellow color, confirming the end-point by the addition of a few drops of starch T.S. Record the volume of 0.02 *N* sodium thiosulfate required as Y_a.

Make several reagent blank determinations, using only the chromium trioxide solution in the above procedure. The ratio of the sodium hydroxide titration (V_b) to the sodium thiosulfate titration (Y_b), corrected for variation in normalities, will give the acidity-to-oxidizing ratio, $V_b/Y_b = K$, for the chromium trioxide carried over in the distillation. The factor K should be constant for all determinations.

Make a series of blank determinations using 100 mg. of methylcellulose (containing no foreign material) in place of the sample, recording the average volume of 0.02 *N* sodium hydroxide required as V_m and the average volume of 0.02 *N* sodium thiosulfate required as Y_m.

Calculate the hydroxypropoxyl content of the sample, in mg., by the formula $75.0 \times [N_1(V_a - V_m) - kN_2(Y_a - Y_m)]$, in which N_1 is the exact normality of the 0.02 *N* sodium hydroxide solution, N_2 is the exact normality of the 0.02 *N* sodium thiosulfate solution, and $k = V_bN_1/Y_bN_2$.

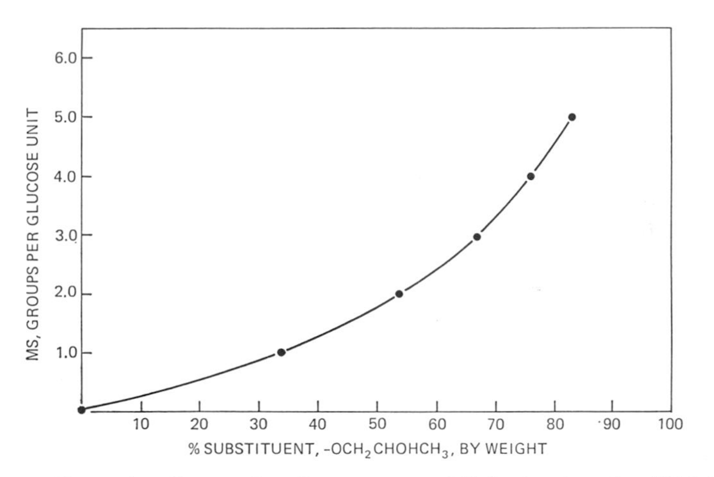

Fig. 11—Chart for Converting Percentage of Substitution, by Weight, of Hydroxypropoxyl Groups to Molecular Substitution per Glucose Unit

IDENTIFICATION—GENERAL TESTS

The tests under this heading are frequently referred to in the Codex for the presumptive identification of F.C.C. chemicals taken from labeled containers. These tests are not intended to be applied to mixtures unless so specified (see *Identification*, page 3).

Acetate. Acetic acid or acetates, when warmed with sulfuric acid and alcohol, form ethyl acetate, recognizable by its characteristic odor. With neutral solutions of acetates, ferric chloride T.S. produces a deep red color which is destroyed by the addition of a mineral acid.

Aluminum. Solutions of aluminum salts yield with ammonia T.S. a white, gelatinous precipitate which is insoluble in an excess of ammonia T.S. The same precipitate is produced by sodium hydroxide T.S. or sodium sulfide T.S., but it dissolves in an excess of either reagent.

Ammonium. Sodium hydroxide T.S. decomposes ammonium salts with the evolution of ammonia, recognizable by its odor and its alkaline effect upon moistened red litmus paper. The decomposition is accelerated by warming.

Benzoate. Neutral solutions of benzoates yield a salmon-colored precipitate with ferric chloride T.S. From moderately concentrated solutions of benzoate, diluted sulfuric acid T.S. precipitates free benzoic acid, which is readily soluble in ether.

Bicarbonate. See *Carbonate.*

Bisulfite. See *Sulfite.*

Bromide. Free bromine is liberated from solutions of bromides upon the addition of chlorine T.S., dropwise. When shaken with chloroform, the bromine dissolves, coloring the chloroform red to reddish brown. A yellowish white precipitate, which is insoluble in nitric acid and slightly soluble in ammonia T.S., is produced when solutions of bromides are treated with silver nitrate T.S.

Calcium. Insoluble oxalate salts are formed when solutions of calcium salts are treated in the following manner: Using 2 drops of methyl red T.S. as indicator, neutralize a solution of a calcium salt (1 in 20) with ammonia T.S., then add diluted hydrochloric acid T.S., dropwise, until the solution is acid. A white precipitate of calcium oxalate forms upon the addition of ammonium oxalate T.S. This precipitate is insoluble in acetic acid but dissolves in hydrochloric acid.

Calcium salts moistened with hydrochloric acid impart a transient yellowish red color to a nonluminous flame.

Carbonate. Carbonates and bicarbonates effervesce with acids, yielding a colorless gas which produces a white precipitate immediately when passed into calcium hydroxide T.S. Cold solutions of soluble carbonates are colored red by phenolphthalein T.S., whereas solutions of bicarbonates remain unchanged or are slightly changed.

Chloride. Solutions of chlorides yield with silver nitrate T.S. a white, curdy precipitate which is insoluble in nitric acid but soluble in a slight excess of ammonia T.S. Chlorine, recognizable by its distinctive odor, is evolved when solutions of chloride are warmed with potassium permanganate and diluted sulfuric acid T.S.

Citrate. When a few mg. of a citrate are added to a mixture of 15 ml. of pyridine and 5 ml. of acetic anhydride, a carmine red color is produced.

Cobalt. Solutions of cobaltous compounds yield a blue precipitate with sodium hydroxide T.S. This precipitate rapidly changes color, becoming olive-green, or rose-red if boiled soon after its formation. Solutions of cobalt salts yield a yellow precipitate when saturated with potassium chloride and treated with potassium nitrite and acetic acid.

Copper. When solutions of cupric compounds are acidified with hydrochloric acid, a red film of metallic copper is deposited upon a bright untarnished surface of metallic iron. An excess of ammonia

T.S., added to a solution of a cupric salt, produces first a bluish precipitate and then a deep blue-colored solution. Solutions of cupric salts yield with potassium ferrocyanide T.S. a reddish brown precipitate, insoluble in diluted acids.

Hypophosphite. Hypophosphites evolve spontaneously flammable phosphine when strongly heated. Solutions of hypophosphites yield a white precipitate with mercuric chloride T.S. This precipitate becomes gray when an excess of hypophosphite is present. Hypophosphite solutions, acidified with sulfuric acid and warmed with copper sulfate T.S., yield a red precipitate.

Iodide. Solutions of iodides, upon the addition of chlorine T.S., dropwise, liberate iodine which colors the solution yellow to red. Chloroform is colored violet when shaken with this solution. The iodine thus liberated gives a blue color with starch T.S. Silver nitrate T.S. produces in solutions of iodides a yellow, curdy precipitate which is insoluble in nitric acid and in ammonia T.S.

Iron. Solutions of ferrous and ferric compounds yield a black precipitate with ammonium sulfide T.S. This precipitate is dissolved by cold diluted hydrochloric acid T.S. with the evolution of hydrogen sulfide.

Ferric salts. Potassium ferrocyanide T.S. produces a dark blue precipitate in acid solutions of ferric salts. With an excess of sodium hydroxide T.S., a reddish brown precipitate is formed. Solutions of ferric salts produce with ammonium thiocyanate T.S. a deep red color which is not destroyed by diluted mineral acids.

Ferrous salts. Potassium ferricyanide T.S. produces a dark blue precipitate in solutions of ferrous salts. This precipitate, which is insoluble in dilute hydrochloric acid, is decomposed by sodium hydroxide T.S. Solutions of ferrous salts yield with sodium hydroxide T.S. a greenish white precipitate, the color rapidly changing to green and then to brown when shaken.

Lactate. When solutions of lactates are acidified with sulfuric acid, and potassium permanganate T.S. is added and the mixture heated, acetaldehyde, recognizable by its distinctive odor, is evolved.

Magnesium. Solutions of magnesium salts in the presence of ammonium chloride yield no precipitate with ammonium carbonate T.S., but a white, crystalline precipitate, which is insoluble in ammonia T.S., is formed upon the subsequent addition of sodium phosphate T.S.

Manganese. Solutions of manganous salts yield with ammonium sulfide T.S. a salmon-colored precipitate which dissolves in acetic acid.

Nitrate. When a solution of a nitrate is mixed with an equal volume of sulfuric acid, the mixture cooled, and a solution of ferrous sulfate superimposed, a brown color is produced at the junction of the two liquids. Brownish red fumes are evolved when a nitrate is heated with sulfuric acid and metallic copper. Nitrates do not decolorize acidified potassium permanganate T.S. (*distinction from nitrites*).

Nitrite. Nitrites yield brownish red fumes when treated with diluted mineral acids or acetic acid. A few drops of potassium iodide T.S. and a few drops of diluted sulfuric acid T.S. added to a solution of a nitrite liberate iodine which colors starch T.S. blue.

Peroxide. Solutions of peroxides slightly acidified with sulfuric acid yield a deep blue color upon the addition of potassium dichromate T.S. On shaking the mixture with an equal volume of ether and allowing the liquids to separate, the blue color is transferred to the ether layer.

Phosphate. Neutral solutions of orthophosphates yield with silver nitrate T.S. a yellow precipitate, which is soluble in diluted nitric acid T.S. or in ammonia T.S. With ammonium molybdate T.S., a yellow precipitate, which is soluble in ammonia T.S., is formed.

Potassium. Potassium compounds impart a violet color to a nonluminous flame if not masked by the presence of small quantities of sodium. In neutral, concentrated or moderately concentrated solutions of potassium salts, sodium bitartrate T.S. slowly produces a white, crystalline precipitate which is soluble in ammonia T.S. and in solutions of alkali hydroxides or carbonates. The precipitation may be accelerated by stirring or rubbing the inside of the test tube with a glass rod or by the addition of a small amount of glacial acetic acid or alcohol.

Sodium. Sodium compounds, after conversion to chloride or nitrate, yield with cobalt-uranyl acetate T.S. a golden-yellow precipitate, which forms after several minutes agitation. Sodium compounds impart an intense yellow color to a nonluminous flame.

Sulfate. Solutions of sulfates yield with barium chloride T.S. a white precipitate which is insoluble in hydrochloric and nitric acids. Sulfates yield with lead acetate T.S. a white precipitate which is soluble in ammonium acetate solution. Hydrochloric acid produces no precipitate when added to solutions of sulfates (*distinction from thiosulfates*).

Sulfite. When treated with diluted hydrochloric acid T.S. sulfites and bisulfites yield sulfur dioxide, recognizable by its characteristic odor. This gas blackens filter paper moistened with mercurous nitrate T.S.

Tartrate. When a few mg. of a tartrate are added to a mixture of 15 ml. of pyridine and 5 ml. of acetic anhydride, an emerald green color is produced.

Thiosulfate. Solutions of thiosulfates yield with hydrochloric acid a white precipitate which soon turns yellow, liberating sulfur dioxide, recognizable by its odor. The addition of ferric chloride T.S. to solutions of thiosulfates produces a dark violet color which quickly disappears.

Zinc. Zinc salts, in the presence of sodium acetate, yield a white precipitate with hydrogen sulfide. This precipitate, which is insoluble in acetic acid, is dissolved by diluted hydrochloric acid T.S. A similar precipitate is produced by ammonium sulfide T.S. in neutral or alkaline solutions. Solutions of zinc salts yield with potassium ferrocyanide T.S. a white precipitate which is insoluble in diluted hydrochloric acid T.S.

LEAD LIMIT TEST

Special Reagents

Select reagents having as low a lead content as practicable, and store all solutions in containers of borosilicate glass. Rinse all glassware thoroughly with warm dilute nitric acid (1 in 2) followed by water.

Ammonia-Cyanide Solution. Dissolve 2 grams of potassium cyanide in 15 ml. of stronger ammonia T.S. and dilute with water to 100 ml.

Ammonium Citrate Solution. Dissolve 40 grams of citric acid in 90 ml. of water, add 2 or 3 drops of phenol red T.S., then cautiously add stronger ammonia T.S. until the solution acquires a reddish color. Extract it with 20-ml. portions of *Dithizone Extraction Solution* until the dithizone solution retains its green color or remains unchanged.

Diluted Standard Lead Solution. (*1 mcg. Pb in 1 ml.*). Immediately before use, transfer 10.0 ml. of *Standard Lead Solution*, page 921, containing 10 mcg. of lead per ml., to a 100-ml. volumetric flask, dilute to volume with dilute nitric acid (1 in 100), and mix.

Dithizone Extraction Solution. Dissolve 30 mg. of dithizone in 1000 ml. of chloroform, add 5 ml. of alcohol, and mix. Store in a refrigerator. Before use, shake a suitable volume of the solution with about half its volume of dilute nitric acid (1 in 100), discarding the nitric acid. Do not use if more than one month old.

Hydroxylamine Hydrochloride Solution. Dissolve 20 grams of hydroxylamine hydrochloride in sufficient water to make about 65 ml., transfer the solution to a separator, add a few drops of thymol blue T.S., then add stronger ammonia T.S. until the solution assumes a yellow color. Add 10 ml. of sodium diethyldithiocarbamate solution (1 in 25), mix, and allow to stand for 5 minutes. Extract the solution with successive 10- to 15-ml. portions of chloroform until a 5-ml. test portion of the chloroform extract does not assume a yellow color when shaken with a dilute cupric sulfate solution. Add diluted hydrochloric acid T.S. until the extracted solution is pink, adding 1 or 2 drops more of thymol blue T.S. if necessary, then dilute with water to 100 ml. and mix.

Potassium Cyanide Solution. Dissolve 50 grams of potassium cyanide in sufficient water to make 100 ml. Remove the lead from the solution by extraction with successive portions of *Dithizone Extraction Solution* as described under *Ammonium Citrate Solution*, then extract any dithizone remaining in the cyanide solution by shaking with chloroform. Finally, dilute the cyanide solution with sufficient water so that each 100 ml. contains 10 grams of potassium cyanide.

Standard Dithizone Solution. Dissolve 10 mg. of dithizone in 1000 ml. of chloroform, keeping the solution in a glass-stoppered lead-free bottle suitably wrapped to protect it from light, and stored in a refrigerator.

Sample Solution

The solution obtained by treating the sample as directed in an individual monograph is used directly as the *Sample Solution* in the *Procedure.* Sample solutions of organic compounds are prepared, unless otherwise directed, according to the following general method:

> *Caution: Some substances may react unexpectedly with explosive violence when digested with hydrogen peroxide. Appropriate safety precautions must be employed at all times.*

Transfer 1.0 gram of the sample into a suitable flask, add 5 ml. of sulfuric acid and a few glass beads, and digest on a hot plate in a fume hood until charring begins. (Additional sulfuric acid may be necessary to completely wet some samples, but the total volume added should not exceed about 10 ml.) After the sample has been initially decomposed by the acid, add with caution, dropwise, 30 percent hydrogen peroxide, allowing the reaction to subside and reheating between drops. The first few drops must be added very slowly with sufficient mixing to prevent a rapid reaction, and heating should be discontinued if foaming becomes excessive. Swirl the solution in the flask to prevent unreacted substance from caking on the walls or bottom of the flask during the digestion. *Add small quantities of the peroxide when the solution begins to darken,* and continue the digestion until the organic matter is destroyed, fumes of sulfuric acid are copiously evolved, and the solution becomes colorless. Cool, add cautiously 10 ml. of water, again evaporate to strong fuming, and cool. Quantitatively transfer the solution into a separator with the aid of small quantities of water.

Procedure

Transfer the *Sample Solution,* prepared as directed in the individual monograph, into a separator and, unless otherwise directed, add 6 ml. of *Ammonium Citrate Solution* and 2 ml. of *Hydroxylamine Hydrochloride Solution.* (Use 10 ml. of the citrate solution when determining lead in iron salts.) To the separator add 2 drops of phenol red T.S., and make the solution just alkaline (red in color) by the addition of stronger ammonia T.S. Cool the solution, if necessary, under a stream of tap water, then add 2 ml. of *Potassium Cyanide Solution.* Immediately extract the solution with 5-ml. portions of *Dithizone Extraction Solution,* draining each extract into another separator, until the dithizone solution retains its green color. Shake the combined dithizone solutions for 30 seconds with 20 ml. of dilute nitric acid (1 in 100), discard the chloroform layer, add to the acid solution 5.0 ml. of *Standard Dithizone Solution* and 4 ml. of *Ammonia-Cyanide Solution,* and shake for 30 seconds. The purplish hue in the chloroform solution of the sample due to any lead dithizonate present does not exceed that in a control, containing the volume of *Diluted Standard Lead Solution* equivalent to the amount of lead specified in the monograph, when treated in the same manner as the sample.

LOSS ON DRYING

This procedure is used to determine the amount of volatile matter expelled under the conditions specified in the monograph. Since the volatile matter may include material other than adsorbed moisture, this test is designed for compounds in which the loss on drying may not definitely be attributable to water alone. For substances appearing to contain water as the only volatile constituent, the *Karl Fischer Titrimetric Method* provided under *Water*, page 977, is usually appropriate.

Procedure

Unless otherwise directed in the monograph, conduct the determination on 1 to 2 grams of the substance, previously mixed and accurately weighed. If the sample is in the form of large crystals, reduce the particle size to about 2 mm., quickly crushing to avoid absorption or loss of moisture. Tare a glass-stoppered shallow weighing bottle that has been dried for 30 minutes under the same conditions to be observed in the determination. Transfer the sample to the bottle, replace the cover, and weigh the bottle and its contents. By gentle sidewise shaking distribute the sample as evenly as possible to a depth of about 5 mm. for most substances and not over 10 mm. in the case of bulky materials. Place the loaded bottle in the drying chamber, removing the stopper and leaving it also in the chamber, and dry at the temperature and for the length of time specified. Upon opening the chamber, close the bottle promptly and allow it to come to room temperature, preferably in a desiccator, before weighing.

If the test substance melts at a temperature lower than that specified for the determination, preheat the bottle and its contents for 1 to 2 hours at a temperature 5° to 10° below the melting range, then continue drying at the specified temperature for the determination. When drying in a desiccator, exercise particular care to ensure that the desiccant is kept fully effective by frequent replacement.

MELTING RANGE OR TEMPERATURE

For purposes of this Codex, the melting range or temperature of a solid is defined as those points of temperature within which or the point at which the solid coalesces and is completely melted, when determined as directed below. Any apparatus or method capable of equal accuracy may be used. The accuracy should be checked frequently by the use of one or more of the six U.S.P. Melting Point Reference Standards, preferably the one that melts nearest the melting temperature of the compound to be tested.

Five procedures for the determination of melting range or temperature are given herein, varying in accordance with the nature of the substance. When no class is designated in the monograph, use the procedure for *Class I*.

The procedure known as the mixed melting point determination, whereby the melting range of a solid under test is compared with that of an intimate mixture of equal parts of the solid and an authentic specimen of it, may be used as a confirmatory identification test. Agreement of the observations on the original and the mixture usually constitutes reliable evidence of chemical identity.

Apparatus. The melting range apparatus consists of a glass container for a bath of colorless fluid, a suitable stirring device, an accurate thermometer (see page 966), and a controlled source of heat. The bath fluid is selected with a view to the temperature required, but light paraffin is used generally and certain liquid silicones are well adapted to the higher temperature ranges. The fluid is deep enough to permit immersion of the thermometer to its specified immersion depth so that the bulb is still about 2 cm. above the bottom of the bath. The heat may be supplied by an open flame or electrically. The capillary tube is about 10 cm. long and 0.8 to 1.2 mm. in internal diameter with walls 0.2 to 0.3 mm. in thickness.

The thermometer is preferably one that conforms to the specifications for A.S.T.M. E-1-1C or E-1-2C thermometers (see page 966), selected for the desired range or temperature.

Procedure for Class I. Reduce the sample to a very fine powder, and, unless otherwise directed, render it anhydrous when it contains water of hydration by drying it at the temperature specified in the monograph, or, when the substance contains no water of hydration, dry it over a suitable desiccant for 16–24 hours.

Charge a capillary glass tube, one end of which is sealed, with sufficient of the dry powder to form a column in the bottom of the tube 2.5 to 3.5 mm. high when packed down as closely as possible by moderate tapping on a solid surface.

Heat the bath until a temperature approximately 30° below the expected melting point is reached, attach the capillary tube to the thermometer, and adjust its height so that the material in the capillary is level with the thermometer bulb. Return the thermometer to the bath, continue the heating, with constant stirring, at a rate of rise of approximately 3° per minute until a temperature 3° below the expected melting point is attained, then carefully regulate the rate to about 1° to 2° per minute until melting is complete.

The temperature at which the column of the sample is observed to collapse definitely against the side of the tube at any point is defined as the beginning of melting, and the temperature at which the sample becomes liquid throughout is defined as the end of melting. The two temperatures fall within the limits of the melting range.

Procedure for Class Ia. Prepare the sample and charge the capillary glass tube as directed for *Class I*. Heat the bath until a temperature 10° $\pm$ 1° below the expected melting range is reached, then introduce the charged tube, and heat at a rate of rise of 3° ± 0.5° per minute until melting is complete. Record the melting range as for *Class I*.

Procedure for Class Ib. Place the sample in a closed container and cool to 10°, or lower, for at least 2 hours. Without previous powdering, charge the cooled material into the capillary tube as directed for *Class I*, then immediately place the charged tube in a vacuum desiccator and dry at a pressure not exceeding 20 mm. of mercury for 3 hours. Immediately upon removal from the desiccator, fire-seal the open end of the tube, and as soon as practicable proceed with the determination of the melting range as directed under *Class Ia*, beginning with "Heat the bath."

If the particle size of the material is too large for the capillary, precool the sample as above directed, then with as little pressure as possible gently crush the particles to fit the capillary, and immediately charge the tube.

Procedure for Class II. Carefully melt the material to be tested at as low a temperature as possible, and draw it into a capillary tube which is left open at both ends, to a depth of about 10 mm. Cool the charged tube at 10°, or lower, for 24 hours, or in contact with ice for at least 2 hours. Then attach the tube to the thermometer by means of a rubber band, adjust it in a water bath so that the upper edge of the material is 10 mm. below the water level, and heat as directed for *Class I* except, within 5° of the expected melting temperature, regulate the rate of rise of temperature to 0.5° to 1.0° per minute. The temperature at which the material is observed to rise in the capillary tube is the melting temperature.

Procedure for Class III. Melt a quantity of the substance slowly, while stirring, until it reaches a temperature of 90° to 92°. Remove the source of heat and allow the molten substance to cool to a temperature of 8° to 10° above the expected melting point. Chill the bulb of an A.S.T.M. 14C thermometer (see page 966) to 5°, wipe it dry, and while it is still cold dip it into the molten substance so that approximately the lower half of the bulb is submerged. Withdraw it immediately, and hold it vertically away from the heat until the wax surface dulls, then dip it for 5 minutes into a water bath having a temperature not higher than 16°.

Fix the thermometer securely in a test tube so that the lower point is 15 mm. above the bottom of the test tube. Suspend the test tube in a water bath adjusted to about 16°, and raise the temperature of the bath at the rate of 2° per minute to 30°, then change to a rate of 1° per minute, and note the temperature at which the first drop of melted substance leaves the thermometer. Repeat the determination twice on a freshly melted portion of the sample. If the variation of three determinations is less than 1°, take the average of the three as the melting point. If the variation of three determinations is greater than 1°, make two additional determinations and take the average of the five.

MERCURY LIMIT TEST

Mercury Detection Instrument. Use any suitable atomic absorption spectrophotometer (such as the Perkin-Elmer 303, Techtron AA 130, the Jarrell-Ash Maximum Versatility, or equivalent) equipped with a fast-response recorder and capable of measuring the radiation absorbed by mercury vapors at the mercury resonance line of 2536 Å. A simple mercury vapor meter or detector (such as the Beckman Model K-23 or equivalent) equipped with a variable span recorder is also satisfactory.

Aeration Apparatus. The apparatus (see accompanying diagram) consists of a flowmeter (*a*), capable of measuring at a flow rate of 1 cu. ft. per hour, connected via a 3-way stopcock (*b*), with Teflon plug, to 125-ml. gas washing bottles (*c* and *d*), followed by a drying tube packed with glass wool (*e*), and finally a suitable quartz liquid absorption cell (*f*), terminating with a vent (*g*). (*Note:* The absorption cell will vary in optical path length depending upon the type of mercury detection instrument used.) Bottle *c* is fitted with an extra-coarse fritted bubbler (Corning 31770 125 EC or equivalent), and the bottle is marked with a 60-ml. calibration line. The drying tube *e* is lightly packed with glass wool or magnesium perchlorate. Bottle *c* is used for the test solution, and bottle *d*, which remains empty throughout the procedure, is used to collect water droplets. Alternatively, an apparatus embodying the principle of the assembly described and illustrated may be used. The aerating medium may be either compressed air or compressed nitrogen.

Standard Preparation. Transfer 1.71 grams of mercuric nitrate, $Hg(NO_3)_2.H_2O$, to a 1000-ml. volumetric flask, dissolve in a mixture of 100 ml. of water and 2 ml. of nitric acid, dilute to volume with water, and mix. Discard after 1 month. Transfer 10.0 ml. of this solution to a second 1000-ml. volumetric flask, acidify with 5 ml. of dilute sulfuric acid solution (1 in 5), dilute to volume with water, and mix. Discard after 1 week. On the day of use, transfer 10.0 ml. of the second solution to a 100-ml. volumetric flask, acidify with 5 ml. of dilute sulfuric acid (1 in 5), dilute to volume with water, and mix. Each ml. of this solution contains 1 mcg. of Hg. Transfer 2.0 ml. of this solution (2 mcg. of Hg) to a 50-ml. beaker, and add 20 ml. of water, 1 ml. of dilute sulfuric acid solution (1 in 5), and 1 ml. of potassium permanganate solution (1 in 25). Cover the beaker with a watch glass, boil for a few seconds, and cool.

Sample Preparation. Prepare as directed in the individual monograph.

Procedure. Assemble the aerating apparatus as shown in the accompanying diagram, with bottles *c* and *d* empty and stopcock *b* in the bypass position. Connect the apparatus to the absorption cell (*f*) in the instrument, and adjust the air or nitrogen flow rate so that, in the following procedure, maximum absorption and reproducibility are obtained without excessive foaming in the test solution. Obtain a baseline reading at 2536 Å, following the manufacturer's instructions for operating the instrument. [Using the Perkin-Elmer 303 and a Beckman No. 75172 1-cm. absorption cell, for example, the following conditions are suitable: *slit width*, 4 (6.5 Å); *lamp current*, 10 mA; *recorder noise suppression*, 2 (approximately 90% of response in 1 second); and *scale expansion*, ×1. When the Techtron AA 120 is used, the following conditions are suitable: *slit width*, 100 μ; *lamp current*, 3 mA; *damping*, B; *scale expansion*, ×1; and *mode*, absorbance.] Treat the *Standard preparation* as follows: Destroy the excess permanganate by adding a 1 in 10 solution of hydroxylamine hydrochloride, dropwise, until the solution is colorless. Immediately wash the solution into bottle *c* with water, and dilute to the 60-ml. mark with water. Add 2 ml. of 10 percent stannous chloride solution (prepared fresh each week by dissolving 20 grams of $SnCl_2.2H_2O$ in 40 ml. of warm hydrochloric acid and diluting with 160 ml. of water), and immediately reconnect bottle *c* to the aerating apparatus. Turn stopcock *b* from the bypass to the aerating position, and obtain the reading on the recorder. Disconnect bottle *c* from the aerating apparatus, discard the *Standard preparation* mixture, wash bottle *c* with water, and repeat the foregoing procedure using the *Sample preparation*: any absorbance produced by the *Sample preparation* does not exceed that produced by the *Standard preparation*.

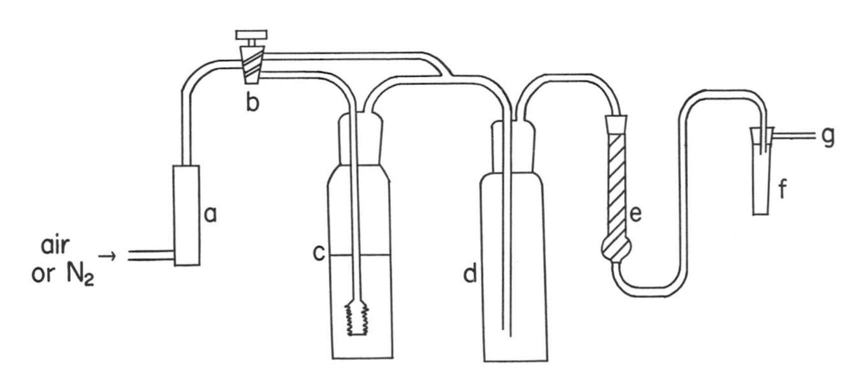

Fig. 12—Aeration Apparatus for Mercury Limit Test

METHOXYL DETERMINATION

Apparatus. The apparatus for methoxyl determination, as shown in Fig. 13, consists of a boiling flask, *A*, fitted with a capillary side arm to provide an inlet for carbon dioxide and connected to a column, *B*, which separates aqueous hydriodic acid from the more volatile methyl iodide. After the methyl iodide passes through a suspension of aqueous red phosphorus in the scrubber trap, *C*, it is absorbed in the bromine-acetic

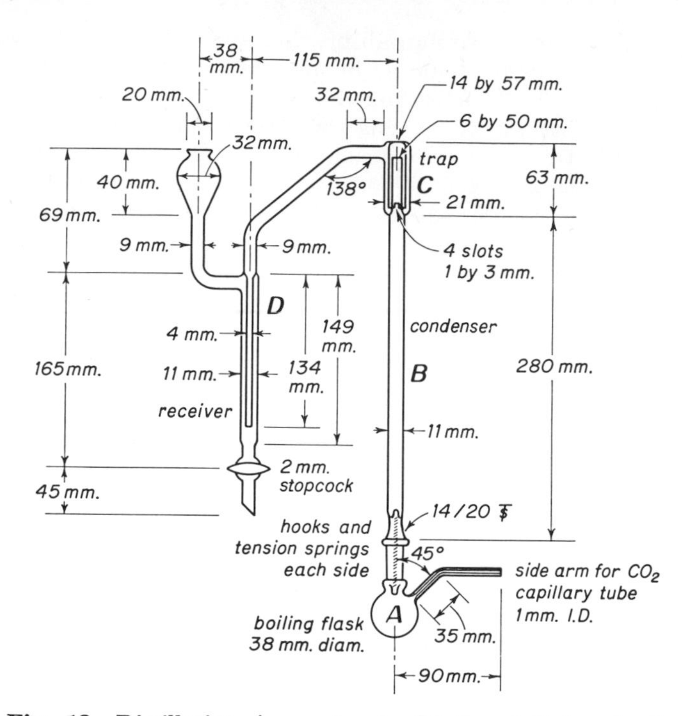

Fig. 13—Distillation Apparatus for Methoxyl Determination

acid absorption tube, *D*. The carbon dioxide is introduced from a device arranged to minimize pressure fluctuations and connected to the apparatus by a small capillary containing a small plug of cotton.

Reagents

Acetic Potassium Acetate. Dissolve 100 grams of potassium acetate in 1000 ml. of a mixture consisting of 900 ml. of glacial acetic acid and 100 ml. of acetic anhydride.

Bromine-Acetic Acid Solution. On the day of use, dissolve 5 ml. of bromine in 145 ml. of the *Acetic Potassium Acetate* solution.

Hydriodic Acid. Use special grade hydriodic acid suitable for alkoxyl determinations, or purify reagent grade as follows: Distil over red phosphorus in an all-glass apparatus, passing a slow stream of carbon dioxide through the apparatus until the distillation is terminated and the receiving flask has completely cooled. (*Caution: Use a safety shield and conduct the distillation in a hood.*) Collect the colorless, or almost colorless, constant-boiling acid distilling between 126° and 127°. Store the acid in a cool, dark place in small, brown, glass-stoppered bottles previously flushed with carbon dioxide, and finally sealed with paraffin.

Procedure. Fill trap *C* half-full with a suspension of about 60 mg. of red phosphorus in 100 ml. of water, introduced through the funnel on tube *D* and the side-arm that connects with the trap at *C*. Rinse tube *D* and the side-arm with water, collecting the rinsings in trap *C*, then charge absorption tube *D* with 7 ml. of *Bromine-Acetic Acid Solution.* Place the sample, accurately weighed in a tared gelatin capsule, in the boiling flask *A*, along with a few glass beads or boiling stones, then add 6 ml. of *Hydriodic Acid.* Connect the flask to the condenser, using a few drops of the acid to seal the junction, and begin passing the carbon dioxide through the apparatus at the rate of about 2 bubbles per second. Heat the flask in an oil bath at 150°, continue the reaction for 40 minutes, and drain the contents of absorption tube *D* into a 500-ml. Erlenmeyer flask containing 10 ml. of sodium acetate solution (1 in 4). Rinse tube *D* with water, collecting the rinsings in the flask, and dilute to about 125 ml. with water. Discharge the reddish brown color of bromine by adding formic acid dropwise, with swirling, then add 3 drops in excess. Usually a total of 12 to 15 drops of formic acid is required. Allow the flask to stand for 3 minutes, add 15 ml. of diluted sulfuric acid T.S. and 3 grams of potassium iodide, and titrate immediately with 0.1 *N* sodium thiosulfate, adding starch T.S. near the end-point. Perform a blank determination with the same quantities of the same reagents, including the gelatin capsule, and in the same manner, and make any necessary correction. Each ml. of 0.1 *N* sodium thiosulfate is equivalent to 0.517 mg. (517 mcg.) of methoxyl groups ($—OCH_3$).

NITROGEN DETERMINATION

(Kjeldahl Method)

Caution: Provide adequate ventilation in the laboratory, and do not permit accumulation of exposed mercury.

Note: All reagents should be nitrogen-free, where available, or otherwise very low in nitrogen content.

Method I

This method should be used unless otherwise directed in the individual monograph. It is not applicable to certain nitrogen-containing compounds that do not yield their entire nitrogen upon digestion with sulfuric acid.

A. Nitrites and Nitrates Absent. Unless otherwise directed, transfer from 700 mg. to 2.2 grams of the sample into a 500–800 ml. Kjeldahl digestion flask of hard, moderately thick, well-annealed glass, wrapping the sample, if solid or semisolid, in nitrogen-free filter paper to facilitate the transfer if desired. Add 700 mg. of mercuric oxide or 650 mg. of metallic mercury, 15 grams of powdered potassium sulfate or anhydrous sodium sulfate, and 25 ml. of 93–98 percent sulfuric acid. (If a sample weight greater than 2.2 grams is used, increase the sulfuric acid by 10 ml. for each additional gram of sample.) Place the flask in an inclined position, and heat gently until frothing ceases, adding a small amount of paraffin, if necessary, to reduce frothing. *Caution:* The digestion should be conducted in a fume hood, or the digestion apparatus should be equipped with a fume exhaust system. Boil briskly until the solution clears, and then continue boiling for 30 minutes longer (or for 2 hours for samples containing organic material). Cool, add about 200 ml. of water, mix, and then cool to below 25°. Add 25 ml. of sulfide or thiosulfate solution (40 grams of K_2S, or 40 grams of Na_2S, or 80 grams of $Na_2S_2O_3.5H_2O$ in 1000 ml. of water), and mix to precipitate the mercury. Add a few granules of zinc to prevent bumping, tilt the flask, and cautiously pour sodium hydroxide pellets, or a 2 in 5 sodium hydroxide solution, down the inside of the flask so that it forms a layer under the acid solution, using a sufficient amount (usually about 25 grams of solid NaOH) to make the mixture strongly alkaline. Immediately connect the flask to a distillation apparatus consisting of a Kjeldahl connecting bulb and a condenser, the delivery tube of which extends well beneath the surface of a measured excess of 0.5 *N* hydrochloric or sulfuric acid contained in a 500-ml. flask. Add from 5 to 7 drops of methyl red indicator (1 gram of methyl red in 200 ml. of alcohol) to the receiver flask. Rotate the Kjeldahl flask to mix its contents thoroughly, and then heat until all of the ammonia has distilled, collecting at least 150 ml. of distillate. Wash the tip of the delivery tube, collecting the washings in the receiving flask, and titrate the excess acid with 0.5 *N* sodium hydroxide. Perform a blank determination, substituting 2 grams of sucrose for the sample, and make any necessary correction (see page 2). Each ml. of 0.5 *N* acid consumed is equivalent to 7.003 mg. of nitrogen. (*Note:* If it is known that the substance to be determined has a low nitrogen content, 0.1 *N* acid and alkali may be used, in which case each ml. of 0.1 *N* acid consumed is equivalent to 1.401 mg. of nitrogen.)

B. Nitrites and Nitrates Present. (*Note:* This procedure is not applicable to liquids or to materials having a high chlorine to nitrate ratio.) Unless otherwise directed, transfer from 700 mg. to 2.2 grams of the sample into the Kjeldahl flask, and add 40 ml. of 93–98 percent sulfuric acid containing 2 grams of salicylic acid. Mix thoroughly by shaking, and then allow to stand for 30 minutes or more, with occasional shaking. Add 5 grams of $Na_2S_2O_3.5H_2O$, or 2 grams of zinc dust (as an impalpable powder, not granules or filings), shake, and allow to stand for 5 minutes. Heat over a low flame until frothing ceases, then remove the heat, add 700 mg. of mercuric oxide (or 650 mg. of metallic mercury) and 15 grams of powdered potassium sulfate (or anhydrous sodium sulfate), and boil briskly until the solution clears. Continue boiling for 30 minutes longer (or for 2 hours for samples containing organic material), and then continue as directed under *A*, beginning with "Cool, add about 200 ml. of water. . . ."

Method II (**Semimicro**)

Transfer an accurately weighed or measured quantity of the sample, equivalent to about 2 or 3 mg. of nitrogen, to the digestion flask of a semimicro Kjeldahl apparatus. Add 1 gram of a powdered mixture of potassium sulfate and cupric sulfate (10 to 1), using a fine jet of water to wash down any material adhering to the neck of the flask, and then pour 7 ml. of sulfuric acid down the inside wall of the flask to rinse it. Add cautiously down the inside of the flask 1 ml. of 30 percent hydrogen peroxide, swirling the flask during the addition. (*Caution: Do not add any peroxide during the digestion.*) Heat over a free flame or an electric heater until the solution has attained a clear blue color and the walls of the flask are free from carbonized material. Cautiously add 20 ml. of water, cool, then add through a funnel 30 ml. of sodium hydroxide solution (2 in 5), and rinse the funnel with 10 ml. of water. Connect the flask to a steam distillation apparatus, and immediately begin the distillation with steam. Collect the distillate in 15 ml. of boric acid solution (1 in 25) to which has been added 3 drops of methyl red-methylene blue T.S. and enough water to cover the end of the condensing tube. Continue passing the steam until 80 to 100 ml. of distillate has been collected, then remove the absorption flask, rinse the end of the condenser tube with a small quantity of water, and titrate with 0.01 *N* sulfuric acid. Each ml. of 0.01 *N* acid is equivalent to 140 mcg. of nitrogen.

When more than 2 or 3 mg. of nitrogen is present in the measured quantity of the substance is to be determined, 0.02 or 0.1 *N* sulfuric acid may be used in the titration if at least 15 ml. of titrant is required. If the total dry weight of the material taken is greater than 100 mg., increase proportionately the quantities of sulfuric acid and sodium hydroxide added before distillation.

OPTICAL ROTATION

Many chemicals in a pure state or in solution are optically active in the sense that they cause incident polarized light to emerge in a plane forming a measurable angle with the plane of the incident light. When this effect is large enough for precise measurement, it may serve as the basis for an assay or an identity test. In this connection, the optical rotation is expressed in degrees, as either *angular rotation* (observed) or *specific rotation* (calculated with reference to the specific concentration of 1 gram of solute in 1 ml. of solution, measured under stated conditions).

Specific rotation usually is expressed by the term $[\alpha]_x^t$, in which t represents, in degrees centigrade, the temperature at which the rotation is determined, and x represents the characteristic spectral line or wavelength of the light used. Spectral lines most frequently employed are the D line of sodium (doublet at 589.0 and 589.6 mμ) and the yellow-green line of mercury at 546.1 mμ. The specific gravity and the rotatory power vary appreciably with the temperature.

The accuracy and precision of optical rotatory measurements will be increased if they are carried out with due regard for the following general considerations.

The source of illumination should be supplemented by a filtering system capable of transmitting light of a sufficiently monochromatic nature. Precision polarimeters generally are designed to accommodate interchangeable disks to isolate the D line from sodium light or the 546.1 $m\mu$ line from the mercury spectrum. With polarimeters not thus designed, cells containing suitably colored liquids may be employed as filters [see "Technique of Organic Chemistry," A. Weissberger, Vol. I, Part II, 3rd ed. (1960), Interscience Publishers, Inc., New York, N.Y.].

Special attention should be paid to temperature control of the solution and of the polarimeter. Observations should be accurate and reproducible to the extent that differences between replicates, or between observed and true values of rotation (the latter value having been established by calibration of the polarimeter scale with suitable standards), calculated in terms of either *Specific rotation* or *Angular rotation*, whichever is appropriate, shall not exceed one-fourth of the range given in the individual monograph for the rotation of the article being tested. Generally, a polarimeter accurate to 0.05° of angular rotation, and capable of being read with the same precision, suffices for *Food Chemicals Codex* purposes; in some cases, a polarimeter accurate to 0.01°, or less, of angular rotation, and read with comparable precision, may be required.

Polarimeter tubes should be filled in such a way as to avoid creating or leaving air bubbles which interfere with the passage of the beam of light. Interference from bubbles is minimized with tubes in which the bore is expanded at one end. However, with tubes of uniform bore, such as semimicro- or micro-tubes, care is required for proper filling. At the time of filling, the tubes and the liquid or solution should be at a temperature not higher than that specified for the determination, to guard against the formation of a bubble upon cooling and contraction of the contents.

In closing tubes having removable end-plates fitted with gaskets and caps, the latter should be tightened only enough to ensure a leak-proof seal between the end-plate and the body of the tube. Excessive pressure on the end-plate may set up strains that result in interference with the measurements. In determining the specific rotation of a substance of low rotatory power, it is desirable to loosen the caps and tighten them again between successive readings in the measurement of both the rotation and the zero point. Differences arising from end-plate strain thus generally will be revealed and appropriate adjustments to eliminate the cause may be made.

Procedure

In the case of a solid, dissolve the substance in a suitable solvent, reserving a separate portion of the latter for a blank determination. Make at least five readings of the rotation of the solution, or of the substance itself if liquid, at 25° or the temperature specified in the individual monograph. Replace the solution with the reserved portion of the solvent (or, in the case of a liquid, use the empty tube), make the same number of readings, and use the average as the zero point value. Subtract the zero point value from the average observed rotation if the two figures are of the same sign, or add if opposite in sign, to obtain the corrected observed rotation.

Calculation. Calculate the specific rotation of a liquid substance, or of a solid in solution, by application of one of the following formulas: (I) For liquid substances, $[\alpha]_x^t = (a/ld)$; (II) For solutions of solids, $[\alpha]_x^t = (100a/lpd) = (100a/lc)$; in which a is the corrected observed rotation, in degrees, at temperature t; l is the length of the polarimeter tube in decimeters; d is the specific gravity of the liquid or solution at the temperature of observation; p is the concentration of the solution expressed as the number of grams of substance in 100 grams of solution; and c is the concentration of the solution expressed as the number of grams of substance in 100 ml. of solution. The concentrations p and c should be calculated on the dried or anhydrous basis, unless otherwise specified.

pH DETERMINATION

Colorimetric Method

Indicator Solutions. Prepare the indicator test solutions as directed for each solution required under *Test Solutions*, page 985.

Buffer Solutions. Prepare the necessary solutions as directed under *Standard Buffer Solutions*, page 983.

Procedure. In order to select a suitable indicator (see page 1004) for the determination of an unknown pH value, the approximate pH range of the solution being tested must first be found. This may be done by testing small portions of the solution with a few indicator solutions having different transition ranges. Thus, a solution which remains colorless after the addition of phenolphthalein T.S., but becomes yellow upon the addition of methyl orange T.S., has a pH between 8.0 and 4.4. Additional tests with methyl red (pH interval 4.2–6.2), bromothymol blue (6.0–7.6), or phenol red (6.8–8.2) will show which specific indicator should be used in the actual pH determination with a series of buffer solutions.

After the approximate pH range and a suitable indicator have been selected, place a measured volume, preferably 10 ml. of the liquid to be tested, in a suitable container such as a 16 × 150-mm. test tube, and to separate, similar containers add the same volume of two or more of the standard buffer solutions chosen from the pH range which the sample is expected to show. To each container add, in a volume equal to

1 to 2 percent of the volume of sample, the appropriate indicator T.S. Compare the color of the sample against the colors of the tubes containing the standard buffer solutions, and report the pH of the sample as being that of the buffer solution which its color most nearly matches. If the color of the sample falls approximately midway between the colors of two buffer solutions, a value obtained by interpolation of the nearest 0.1 pH unit may be reported as the pH of the sample. If desired, a suitable colorimeter may be used as an aid in comparing the solutions.

Potentiometric Method

Apparatus and Procedure. Use a suitable pH meter or other electronic instrument for measuring pH, calibrate the instrument against an appropriate *Standard Buffer Solution* (see page 983), or a similar solution available commercially, in accordance with the manufacturer's instructions, and determine the pH as directed by the manufacturer for the particular instrument being used.

Absolute pH values read from a pH meter are meaningful only shortly after the instrument has been calibrated. Readings obtained using buffered samples may not be assumed to represent the pH of the corresponding samples unless the readings can be reproduced after rinsing the electrode compartment with a portion of the sample or with distilled water, and unless two or more standard buffer solutions have been found to read correctly on the same instrument during the same period of operation. Readings obtained on unbuffered samples such as distilled water must be reproducible within the limits of accuracy for the instrument after checking it with a standard phthalate buffer and again after checking the instrument with a standard borate buffer. Readings obtained from "flow-type" electrodes may be used if comparable evidence of validity is obtained.

If a precision of greater than 0.1 pH unit is desired, adjust the temperature of the standard buffers, the glass and calomel electrodes, and the test solutions to within 2° of the same temperature, preferably at least 2 hours prior to making the measurement.

READILY CARBONIZABLE SUBSTANCES

Reagents

Sulfuric Acid T.S. Add a quantity of sulfuric acid of known concentration to sufficient water to adjust the final concentration to between 94.5 and 95.5 percent of H_2SO_4. Since the acid concentration may change upon standing or upon intermittent use, the concentration should be checked frequently and solutions assaying more than 95.5 percent or less than 94.5 percent discarded or adjusted by adding either diluted or fuming sulfuric acid, as required.

Cobaltous Chloride C.S. Dissolve about 65 grams of cobaltous chloride

($CoCl_2 . 6H_2O$) in enough of a mixture of 25 ml. of hydrochloric acid and 975 ml. of water to make 1000 ml. Pipet 5 ml. of this solution into a 250-ml. iodine flask, add 5 ml. hydrogen peroxide T.S. and 15 ml. of sodium hydroxide solution (1 in 5), boil for 10 minutes, cool, and add 2 grams of potassium iodide and 20 ml. of dilute sulfuric acid (1 in 4). When the precipitate has dissolved, titrate the liberated iodine with 0.1 *N* sodium thiosulfate. The titration is sensitive to air oxidation and should be blanketed with carbon dioxide. Each ml. of 0.1 *N* sodium thiosulfate is equivalent to 23.79 mg. of $CoCl_2 . 6H_2O$. Adjust the final volume of the solution by the addition of enough of the mixture of hydrochloric acid and water to make each ml. contain 59.5 mg. of $CoCl_2 . 6H_2O$.

Cupric Sulfate C.S. Dissolve about 65 grams of cupric sulfate ($CuSO_4 . 5H_2O$) in enough of a mixture of 25 ml. of hydrochloric acid and 975 ml. of water to make 1000 ml. Pipet 10 ml. of this solution into a 250-ml. iodine flask, add 40 ml. of water, 4 ml. of acetic acid, and 3 grams of potassium iodide, and titrate the liberated iodine with 0.1 *N* sodium thiosulfate, adding starch T.S. as the indicator. Each ml. of 0.1 *N* sodium thiosulfate is equivalent to 24.97 mg. of $CuSO_4 . 5H_2O$. Adjust the final volume of the solution by the addition of enough of the mixture of hydrochloric acid and water to make each ml. contain 62.4 mg. of $CuSO_4 . 5H_2O$.

Ferric Chloride C.S. Dissolve about 55 grams of ferric chloride ($FeCl_3 . 6H_2O$) in enough of a mixture of 25 ml. of hydrochloric acid and 975 ml. of water to make 1000 ml. Pipet 10 ml. of this solution into a 250-ml. iodine flask, add 15 ml. of water, 5 ml. of hydrochloric acid, and 3 grams of potassium iodide, and allow the mixture to stand for 15 minutes. Dilute with 100 ml. of water, and titrate the liberated iodine with 0.1 *N* sodium thiosulfate, adding starch T.S. as the indicator. Perform a blank determination with the same quantities of the same reagents and in the same manner and make any necessary correction. Each ml. of 0.1 *N* sodium thiosulfate is equivalent to 27.03 mg. of $FeCl_3 . 6H_2O$. Adjust the final volume of the solution by addition of the mixture of hydrochloric acid and water to make each ml. contain 45.0 mg. of $FeCl_3 . 6H_2O$.

Platinum-cobalt C.S. Transfer 1.246 grams of potassium chloroplatinate, K_2PtCl_6, and 1.00 gram of crystallized cobaltous chloride, $CoCl_2.6H_2O$, into a 1000-ml. volumetric flask, dissolve in about 200 ml. of water and 100 ml. of hydrochloric acid, dilute to volume with water, and mix. This solution has a color of 500 A.P.H.A. units. [*Note:* Use this solution only when specified in an individual monograph.]

Procedure

Unless otherwise directed, add the specified quantity of the substance, finely powdered if in solid form, in small portions to the comparison container, which is made of colorless glass resistant to the action of sulfuric acid and contains the specified volume of *Sulfuric Acid T.S.*

Stir the mixture with a glass rod until solution is complete, allow the solution to stand for 15 minutes, unless otherwise directed, and compare the color of the solution with that of the specified matching fluid in a comparison container which also is of colorless glass and has the same internal and cross-section dimensions, viewing the fluids transversely against a background of white porcelain or white glass.

When heat is directed in order to effect solution of the substance in the *Sulfuric Acid T.S.*, mix the sample and the acid in a test tube, heat as directed, cool, and transfer the solution to the comparison container for matching.

Matching Fluids[1]

Matching Fluid	Parts of Cobaltous Chloride C.S.	Parts of Ferric Chloride C.S.	Parts of Cupric Sulfate C.S.	Parts of Water
A	0.1	0.4	0.1	4.4
B	0.3	0.9	0.3	8.5
C	0.1	0.6	0.1	4.2
D	0.3	0.6	0.4	3.7
E	0.4	1.2	0.3	3.1
F	0.3	1.2	0.0	3.5
G	0.5	1.2	0.2	3.1
H	0.2	1.5	0.0	3.3
I	0.4	2.2	0.1	2.3
J	0.4	3.5	0.1	1.0
K	0.5	4.5	0.0	0.0
L	0.8	3.8	0.1	0.3
M	0.1	2.0	0.1	2.8
N	0.0	4.9	0.1	0.0
O	0.1	4.8	0.1	0.0
P	0.2	0.4	0.1	4.3
Q	0.2	0.3	0.1	4.4
R	0.3	0.4	0.2	4.1
S	0.2	0.1	0.0	4.7
T	0.5	0.5	0.4	3.6

[1] Solutions A–D, very light brownish yellow.
Solutions E–L, yellow through reddish yellow.
Solutions M–O, greenish yellow.
Solutions P–T, light pink.

Matching Fluids

For purposes of comparison, a series of twenty matching fluids, each designated by a letter of the alphabet, is provided, the composition of each being as indicated in the preceding table. To prepare the matching fluid specified, pipet the prescribed volumes of the colorimetric test solutions (C.S.) and water into one of the matching containers, and mix the solutions in the container.

REFRACTIVE INDEX

The refractive index of a transparent substance is the ratio of the velocity of light in air to its velocity in that material under like conditions. It is equal to the ratio of the sine of the angle of incidence made by a ray in air to the sine of the angle of refraction made by the ray in the material being tested. The refractive index values specified in this Codex are for the D line of sodium (589 mμ), unless otherwise specified. The determination should be made at the temperature specified in the individual monograph, or at 25° if no temperature is specified. This physical constant is used as a means for identification of, and detection of impurities in, volatile oils and other liquid substances. The Abbé refractometer, or other refractometers of equal or greater accuracy, may be employed at the discretion of the operator.

RESIDUE ON IGNITION

(Sulfated Ash)

Method I (**for solids**)

Transfer the quantity of the sample directed in the individual monograph to a tared 50- to 100-ml. platinum dish or other suitable container, and add sufficient diluted sulfuric acid T.S. to moisten the entire sample. Heat gently, using a hot plate, an Argand burner, or an infrared heat lamp, until the sample is dry and thoroughly charred, then continue heating until all of the sample has been volatilized or nearly all of the carbon has been oxidized, and cool. Moisten the residue with 0.1 ml. of sulfuric acid, and heat in the same manner until the remainder of the sample and any excess sulfuric acid have been volatilized. Finally ignite in a muffle furnace at 800° ± 25° for 15 minutes or longer, if necessary to complete ignition, cool in a desiccator, and weigh.

Method II (**for liquids**)

Unless otherwise directed, transfer the required weight of the sample to a suitable tared container, add 10 ml. of diluted sulfuric acid T.S., and mix thoroughly. Evaporate the sample completely by heating gently without boiling, and cool. Finally ignite in a muffle furnace at 800° ± 25° for 15 minutes, cool in a desiccator, and weigh.

ROSINS, ROSIN ESTERS, AND RELATED SUBSTANCES

ACID VALUE

The acid value is defined as the number of mg. of potassium hydroxide required to neutralize the free acids in 1 gram of the test substance.

Procedure. Unless otherwise directed in the individual monograph, transfer about 4 grams of the sample previously crushed into small lumps and accurately weighed, into a 250-ml. Erlenmeyer flask, and add 50–75 ml. of a 2:1 mixture of benzene-methanol, previously neutralized to phenolphthalein T.S. with sodium hydroxide. Dissolve the sample by shaking or heating gently, if necessary, then add about 0.5 ml. of phenolphthalein T.S., and titrate with 0.5 N alcoholic potassium hydroxide to the first pink color that persists for 30 seconds. Calculate the acid value by the formula $56.1V \times N/W$, in which V is the exact volume, in ml., and N is the exact normality, respectively, of the potassium hydroxide solution, and W is the weight of the sample, in grams.

SOFTENING POINT

Drop Method

The *drop softening point* is defined as that temperature at which a given weight of rosin or rosin derivative begins to drop from the bulb of a special thermometer mounted in a test tube that is immersed in a constant-temperature bath.

Apparatus. The apparatus illustrated in Fig. 14 consists of the components described in the following paragraphs.

Thermometer. A special total-immersion softening point thermometer,* covering the range from 0° to 250° and graduated in 1° divisions, should be employed. The bulb should be $5/8 \pm 1/32$ inch in length (16 mm.) and $1/4 \pm 1/64$ inch in diameter (6.4 mm.).

Heating Bath. Use an 800- or 1000-ml. beaker containing a suitable heating medium. For resins having a softening point below 80°, use water; for those having softening points above 80°, use glycerin, mineral oil, or other suitable vegetable oil, depending upon the temperature range required. The temperature of the heating medium must be maintained within ±1° of the temperature specified in the individual monograph. The medium must be stirred constantly during the test with a suitable mechanical stirrer to ensure uniform heating of the medium.

Test Tube. Use a standard glass 22 × 175-mm. test tube with rim, fitted with a cork stopper as shown in Fig. 14.

Sample Preparation. Place about 20 grams of the sample in a 50-ml. beaker, and heat in an oven, on a sand bath or hot plate, or in an oil bath until its softening point is exceeded, but not more than 25–30° above its softening point nor for any longer than necessary. Tare the softening point thermometer, and cautiously warm the bulb over a flame or hot plate until it registers 15–20° above the expected softening point of the sample. Immediately dip the thermometer bulb into the melted sample, withdraw, and rotate it so that a uniform film of the molten sample is deposited over the surface of the bulb, care being

* Available from The Taylor Instrument Co., 95 Ames Street, Rochester, N.Y. 14601.

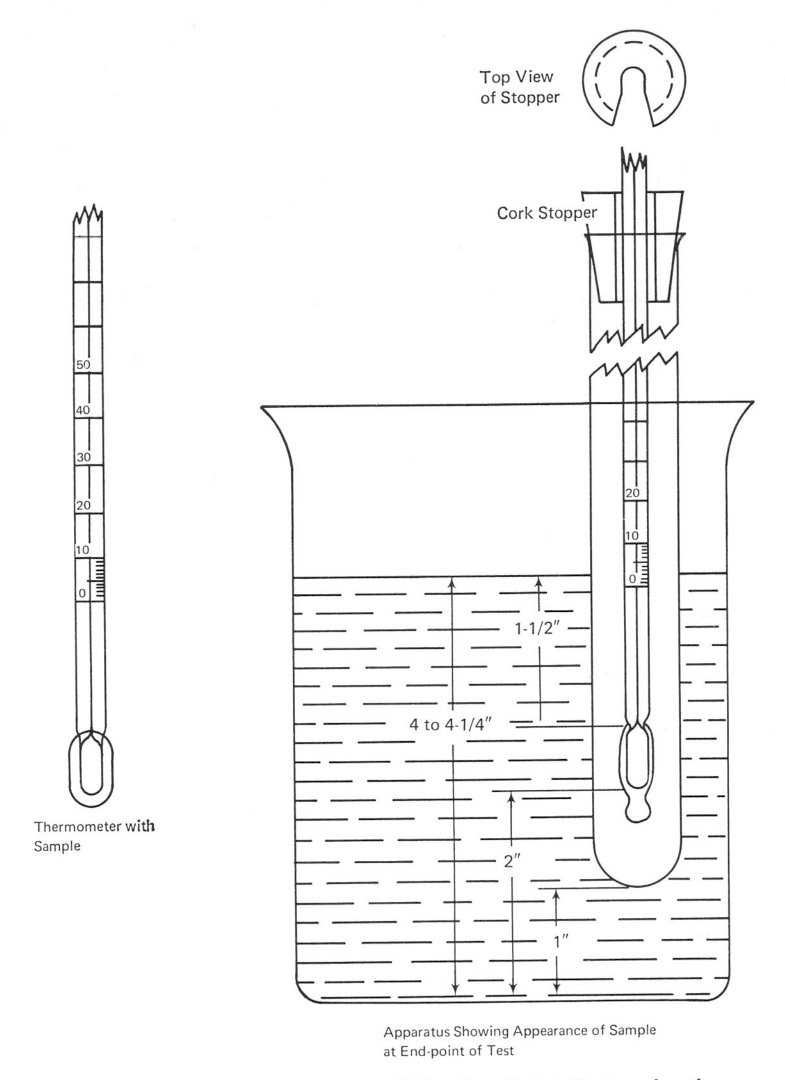

Fig. 14—Apparatus for Drop Softening Point Determination

taken not to extend the film higher than the top of the bulb. Quickly place the thermometer on a balance, and weigh. The weight of the sample on the thermometer bulb should be between 500 mg. and 550 mg. If the weight is low, again dip the bulb in the molten sample; if the weight is high, pull off some of the sample with the fingers. When the correct sample weight has been obtained, mold the sample uniformly around the bulb by rolling on the palm of the hand or between the fingers. The sample must be of uniform thickness over the bulb, and it must not extend up onto the thermometer stem (see

Fig. 14). (If the film of the sample is not uniform when cooled, it should be completely removed from the bulb and a new one applied. Do not reheat the film and try to remold.) Allow the film and thermometer to cool to room temperature, allowing about 15 minutes for cooling. (*Note:* If samples having high softening points crack or "check" on the thermometer bulb upon cooling to room temperature, prepare another sample film and cool only to about 50° below the expected softening point.)

Procedure. Fill the glass beaker to a depth of not less than 4 inches (100 mm.) nor more than 4.25 inches (108 mm.) with a suitable heating medium, support the beaker over a Bunsen burner, hot plate, or other suitable source of heat, and insert the bath stirrer and a bath temperature thermometer. Place the stirrer to one side so that the impeller clears the side of the beaker and is about 0.5 inch (13 mm.) above the bottom of the beaker. Start the stirrer, heat the bath to the temperature specified in the monograph, and maintain this temperature within ±1° throughout the test.

Insert the prepared sample thermometer in the test tube, supporting it with a notched cork stopper so that the lower end of the bulb is 1 inch (25 mm.) from the bottom of the test tube. Place the test tube in the bath so that the bottom of the thermometer bulb is 2 inches (51 mm.) from the bottom of the beaker; the top of the bulb should be about 1.5 inches (38 mm.) below the liquid level of the bath. Stir the bath in order to keep its temperature uniform throughout. Observe the sample thermometer, and record as the softening point the reading at which the elongated drop of sample on the end of the bulb first becomes constricted (see Fig. 14). Report the softening point to the nearest 0.5°. (*Precautions:* If the rosin crystallizes, thus making it difficult to obtain the correct softening point, prepare a new sample by heating the rosin rapidly, yet cautiously, over a flame to a temperature of 160–170° in order to destroy all crystal nuclei; then dip the thermometer bulb into the molten resin, remove momentarily, and rotate the thermometer to provide a uniform resin film on the bulb as it partially cools in air; dip the bulb in the melted sample repeatedly until the proper amount of resin has been deposited on the bulb. Results should not be reported if a crystal-free sample cannot be obtained.)

Ring-and-Ball Method

The *ring-and-ball softening point* is defined as the temperature at which a disk of the sample held within a horizontal ring is forced downward a distance of 1 inch (25.4 mm.) under the weight of a steel ball as the sample is heated at a prescribed rate in a water or glycerin bath.

Apparatus. The apparatus illustrated in Figs. 15 and 16 consists of the components described in the following paragraphs.

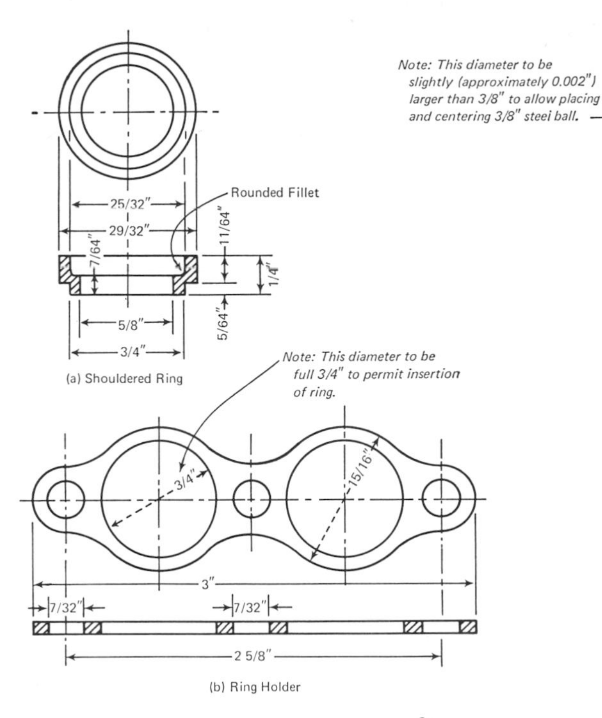

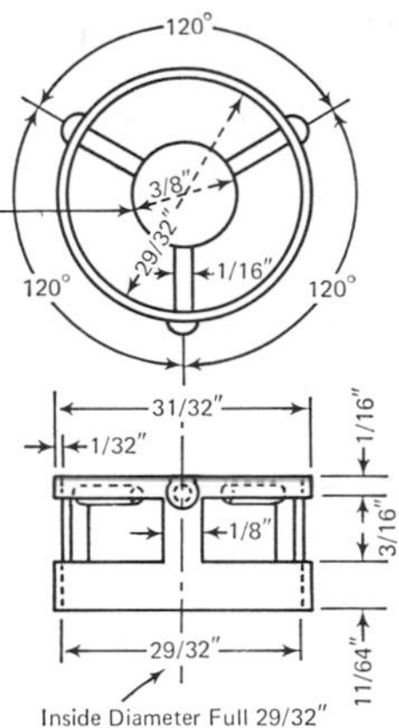

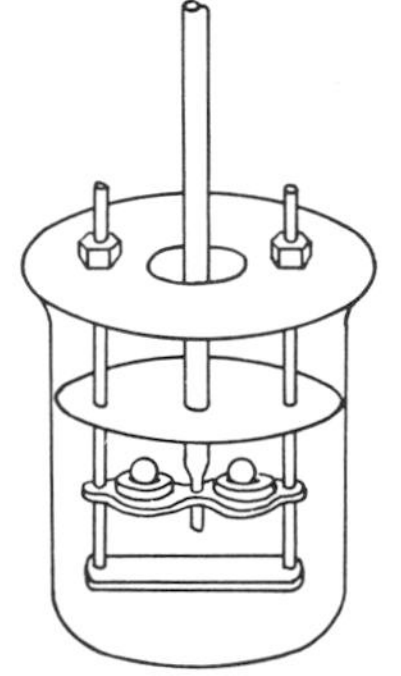

Fig. 15—Shouldered Ring, Ring Holder, Ball-Centering Guide, and Assembly of Apparatus Showing Two Rings

Ring. A brass-shouldered ring conforming to the dimensions shown in Fig. 15*a* should be used. If desired, the ring may be attached by brazing or other convenient manner to a brass wire of about 13 B & S gauge (0.06–0.08 inch, or 1.52–2.03 mm., in diameter) as shown in Fig. 16*a*.

Ball. A steel ball, ⅜ inch (9.53 mm.) in diameter, weighing between 3.45–3.55 grams, should be used.

Ball-Centering Guide. A guide for centering the ball, constructed of brass and having the general shape and dimensions illustrated in **Fig. 15c**, may be used if desired.

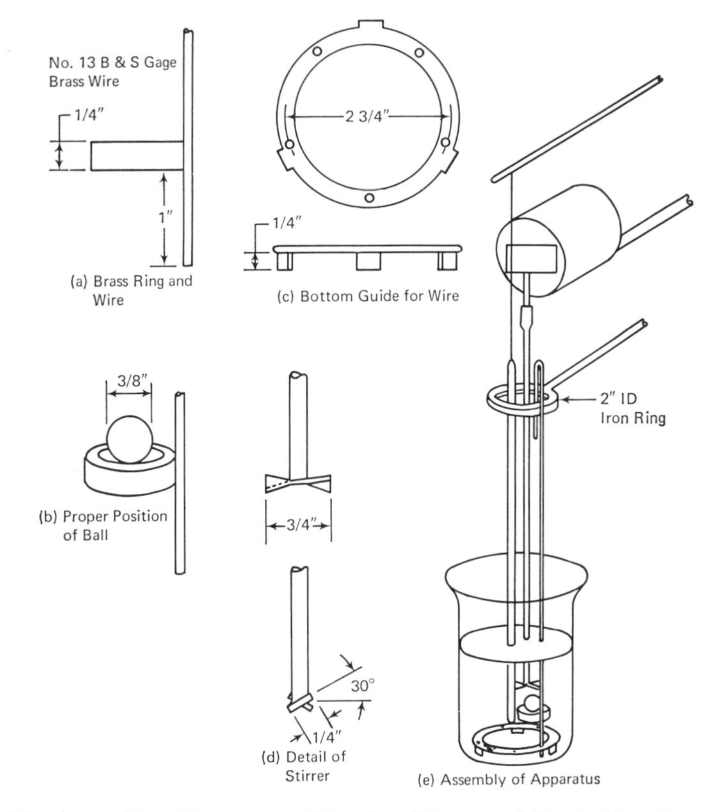

Fig. 16—Assembly of Apparatus Showing Stirrer and Single Shouldered Ring

Container. Use a heat-resistant glass vessel, such as an 800-ml. low-form Griffin beaker, not less than 3.34 inches (8.5 cm.) in diameter and not less than 5 inches (12.7 cm.) in depth from the bottom of the flare.

Support for Ring and Thermometer. Any convenient device for supporting the ring and thermometer may be used, provided that it meets the following requirements: (*a*) the ring shall be supported in a substantially horizontal position; (*b*) when using the apparatus shown in **Fig. 15***d*, the bottom of the ring shall be 1.0 inch (25.4 mm.) above the horizontal plate below it, the bottom surface of the horizontal plate shall be at least 0.5 inch (13 mm.) and not more than 0.75 inch (18 mm.) above the bottom of the container, and the depth of the liquid in the container shall be not less than 4.0 inches (10.2 cm.); (*c*) when using the apparatus shown in **Fig. 16***e*, the bottom of the ring shall be 1.0 inch (25.4 mm.) above the bottom of the container, with the bottom end of the rod resting on the bottom of the container, and the depth of the liquid in the container shall be not less than 4.0 inches (10.2 cm.), as shown in Figs. 16*a*, *b*, and *c;* and (*d*) in both assemblies, the thermometer shall be suspended so that the bottom of the bulb is level with the bottom of the ring and within 0.5 inch (13 mm.) but not touching the ring.

Thermometers. Depending upon the expected softening point of the sample, use either an A.S.T.M. 15C or 15F low-softening-point thermometer (−2° to 80° C.) or an A.S.T.M. 16C or 16F high-softening-point thermometer (30° to 200° C.), as described under *Thermometers*, page 966.

Stirrer. Use a suitable mechanical stirrer, rotating between 500 and 700 rpm, to ensure uniform heat distribution in the heating medium (see Fig. 16*d* for recommended dimensions).

Sample Preparation. Select a representative sample of the material under test consisting of freshly broken lumps free of oxidized surfaces. Scrape off the surface layer of samples received as lumps immediately before use, avoiding inclusion of finely divided material or dust. The amount of sample taken should be at least twice that necessary to fill the desired number of rings, but in no case less than 40 grams. Immediately melt the sample in a clean container, using an oven, hot plate, or sand or oil bath to prevent local overheating. Avoid incorporating air bubbles in the melting sample, which must not be heated above the temperature necessary to pour the material readily without inclusion of air bubbles. The time from the beginning of heating to the pouring of the sample shall not exceed 15 minutes. Immediately before filling the rings, preheat them to approximately the same temperature at which the sample is to be poured. While being filled, the rings should rest on an amalgamated brass plate. Pour the sample into the rings so as to leave an excess on cooling. Cool for at least 30 minutes, and then cut the excess material off cleanly with a slightly heated knife or spatula. Use a clean container and a fresh sample if the test is repeated.

Procedure

Materials Having Softening Points Above 80° C. Fill the glass vessel with glycerin to a depth of not less than 4.0 inches (10.2 cm). and not more than 4.25 inches (10.8 cm.). The starting temperature of the bath shall be 32° C. (90° F.), except for resins (including rosin) the glycerin should be cooled to not less than 45° C. (81° F.) below the anticipated softening point but in no case lower than 35° C. (95° F.). Position the axis of the stirrer shaft near the back wall of the container, with the blades clearing the wall and with the bottom of the blades 0.75 inch (18 mm.) above the top of the ring. Unless the ball-centering guide is used, make a slight indentation in the center of the sample by pressing the ball or a rounded rod, slightly heated for hard materials, into the sample at this point. Suspend the ring containing the sample in the glycerin bath so that the lower surface of the filled ring is 1.0 inch (25.4 mm.) above the surface of the lower horizontal plate (see **Fig.** 15*d*), which is at least 0.5 inch (13 mm.) and not more than 0.75 inch (18 mm.) above the bottom of the glass vessel, or 1.0 inch (25.4 mm.) above the bottom of the container (see **Fig.** 16*e*). Place the ball in the glycerin but not on the test specimen. Suspend an A.S.T.M. high-softening-point thermometer (16C or 16F) in the glycerin so that the bottom of its bulb is level with the bottom of the ring and within 0.5 inch (13 mm.) but not touching the ring. Maintain the initial temperature of the glycerin for 15 minutes, and then, using suitable forceps, place the ball in the center of the upper surface of the sample in the ring. Begin stirring, and continue stirring at 500–700 rpm until completion of the determination. Apply heat at such a rate that the temperature of the glycerin is raised 5° C. (or 10° F.) per minute, avoiding the effects of drafts by using shields if necessary. [*Note:* The rate of rise of the temperature shall be uniform and shall not be averaged over the test period. Reject all tests in which the rate of rise exceeds ±0.5° C. (or ±1° F.) for any minute period after the first three.] Record as the softening point the temperature of the thermometer at the instant the sample touches the lower horizontal plate (see **Fig.** 15*d*) or the bottom of the container (see **Fig.** 16*e*). Make no correction for the emergent stem of the thermometer.

Materials Having Softening Points of 80° C. or Below. Follow the above procedure, except use an A.S.T.M. low-softening-point thermometer (15C or 15F) and use freshly boiled water cooled to 5° C. (41° F.) as the heating medium. For resins (including rosins), use water cooled to not less than 45° C. (81° F.) below the anticipated softening point, but in no case lower than 5° C. (41° F.).

VISCOSITY

Unless otherwise directed in the individual monograph, transfer the sample into an 8-oz. wide-mouth glass jar, 10.8 cm. (4.25 inches) high and 5.7 cm. (2.75 inches) in inside diameter, equipped with a screw lid. Condition the sample in a water bath at 25° ± 0.2° for 2 hours (±5 minutes), taking care to prevent water from coming into contact with the sample. Insert a No. 4 spindle in a Brookfield Model RVF viscometer*, or equivalent, and move the jar into place under the spindle, adjusting the elevation of the jar so that the upper surface of the sample is in the center of the shaft indentation and the spindle is in the center of the jar. (*Note:* The viscometer must be kept level at all times during the test procedure.) Set the viscometer to rotate at 20 rpm, and allow the spindle to rotate until a constant dial reading is obtained. The viscosity, in poises, is the dial reading on the 100 scale, or the dial reading on the 500 scale divided by 5.

* Available from Brookfield Engineering Laboratories, Inc., Stoughton, Mass.

SELENIUM LIMIT TEST

Selenium Stock Solution. Transfer 120.0 mg. of metallic selenium (Se) into a 1000-ml. volumetric flask, add 100 ml. of dilute nitric acid (1 in 2), warm gently on a steam bath to effect solution, and dilute to volume with water. Transfer 5.0 ml. of this solution into a 200-ml. volumetric flask, dilute to volume with water, and mix. Each ml. of this solution contains 3 mcg. of selenium ion (Se).

Standard Solution. Just prior to use, transfer 20.0 ml. of *Selenium Stock Solution* (60 mcg. Se) into a 200 × 25-mm. test tube, add 20 ml. of hydrochloric acid, and mix.

Sample Solution. The solution obtained by treating the sample as directed in an individual monograph is used directly as the *Sample Solution* in the *Procedure* after transferring it into a 200 × 25-mm. test tube. If no directions are given in the monograph, prepare the sample as follows: Transfer 2.0 grams of the sample into a 250-ml. Erlenmeyer flask, and cautiously add 10 ml. of 30 percent hydrogen peroxide. After the initial reaction has subsided, add 6 ml. of 70 percent perchloric acid, heat slowly until white fumes of perchloric acid are copiously evolved, and continue heating gently for a few minutes to ensure decomposition of any excess peroxide. If the solution is brownish in color due to undecomposed organic matter, add a small portion of the peroxide solution and heat again to white perchloric acid fumes, repeating if necessary until decomposition of the organic matter is complete and a colorless solution is obtained. Cool, add 10 ml. of water, and filter into a 200 × 25-mm. test tube. Wash the filter with hot water until the filtrate measures 20 ml., add 20 ml. of hydrochloric acid, and mix.

Procedure. Place the test tubes containing the *Standard Solution* and the *Sample Solution* in a 45° water bath, and heat until the temperature of the solutions reaches 40°. To each tube add 400 mg. of ascorbic acid, stir until dissolved, and maintain at 40° for 30 minutes. Cool the solutions, dilute with water to 50 ml., and mix. Any pink color produced by the sample does not exceed that produced by the standard.

SIEVE ANALYSIS OF GRANULAR METAL POWDERS

(Based on A.S.T.M. Designation: B 214)

Apparatus

Sieves. Use a set of standard sieves, ranging from +80 mesh to −325 mesh, conforming to the specifications in A.S.T.M. Designation: E 11 (Sieves for Testing Purposes).

Sieve Shaker. Use a mechanically operated sieve shaker that imparts to the set of sieves a horizontal rotary motion of between 270 and 300 rotations per minute and a tapping action of between 140 and 160 taps per minute. The sieve shaker is fitted with a plug to receive the impact of the tapping device. The entire apparatus is mounted rigidly by bolting to a solid foundation, preferably of concrete. Preferably a time switch is provided to ensure accuracy of duration of the test.

Procedure. Assemble the sieves in consecutive order as to size of openings, with the coarsest sieve (+80 mesh) at the top, and place a solid collecting pan below the bottom sieve (−325 mesh). Place 100.0 grams of the test sample on the top sieve, and close the sieve with a solid cover. Securely fasten the assembly to the sieve shaker, and operate the shaker for 15 minutes. Remove the most coarse sieve from the nest, gently tap its contents to one side, and pour the contents onto a glazed paper. Using a soft brush, transfer onto the next finer sieve any material adhering to the bottom of the sieve and frame. Place the sieve just removed upside down on the paper containing the retained portion, and tap the sieve. Weigh the paper and its contents to the nearest 100 mg., and record the net weight of the fraction obtained. Repeat this process for each sieve in the nest and for the portion of the sample that has been collected in the bottom pan. Record the total of the fractions retained on the sieves as T and that portion collected in the pan as t. The combined total, S, of $T + t$ is not less than 99.0 grams. Add the weight $(100.0 - S)$ to the fraction t collected in the pan: the sum, $[t + (100.0 - S)]$, represents the portion of the sample that has passed through the −325 mesh sieve.

SOLIDIFICATION POINT

Scope

This method is designed to determine the solidification point of food grade chemicals having appreciable heats of fusion. It is applicable to chemicals having solidification points between −20° and +150°. Necessary modifications will be noted in individual monographs.

Definition

Solidification Point is an empirical constant defined as the temperature at which the liquid phase of a substance is in approximate equilibrium with a relatively small portion of the solid phase. It is measured by noting the maximum temperature reached during a controlled cooling cycle after the appearance of a solid phase.

Solidification point is distinguished from freezing point in that the latter term applies to the temperature of equilibrium between the solid and liquid state of pure compounds.

Some chemical compounds have two temperatures at which there may be a temperature equilibrium between the solid and liquid state depending upon the crystal form of the solid which is present.

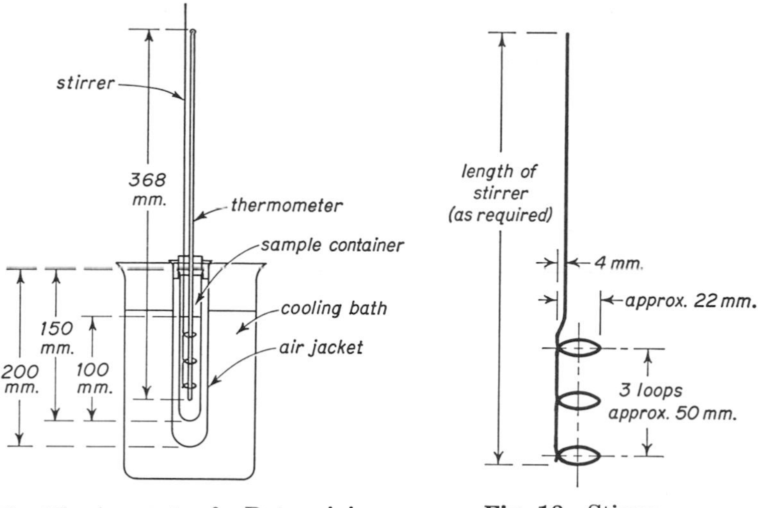

Fig. 17—Apparatus for Determining Solidification Point

Fig. 18—Stirrer

Apparatus

The apparatus illustrated in Fig. 17 consists of the components described in the following paragraphs.

Thermometer. A thermometer having a range not exceeding 30°, graduated in 0.1° divisions, and calibrated for 76 mm. immersion should be employed. A satisfactory series of thermometers, covering a range from −20° to +150°, is available as A.S.T.M.-E1 89C through 96C (see *Thermometers*, page 966). A thermometer should be so chosen that the solidification point is not obscured by the cork stopper of the sample container.

Sample Container. Use a standard glass 25 × 150-mm. test tube with lip, fitted with a two-hole cork stopper to hold the thermometer in place and to allow stirring with stirrer.

Air Jacket. For the air jacket use a standard glass 38 × 200-mm. test tube with lip, fitted with a cork or rubber stopper bored with a hole into which the sample container can easily be inserted up to the lip.

Cooling Bath. Use a 2-liter beaker or similar suitable container as a cooling bath. Fill it with an appropriate cooling medium such as glycerin, mineral oil, water, water and ice, or alcohol-dry ice.

Stirrer. The stirrer (Fig. 18) consists of a 1-mm. diameter (B and S gauge 18), corrosion-resistant wire bent into a series of three loops about 25 mm. apart. It should be made so that it will move freely in the space between the thermometer and the inner wall of the sample container. The shaft of the stirrer should be of a convenient length designed to pass loosely through a hole in the cork holding the thermometer. Stirring may be hand operated or mechanically activated at 20 to 30 strokes per minute.

Assembly. Assemble the apparatus in such a way that the cooling bath can be heated or cooled to control the desired temperature ranges. Clamp the air jacket so that it is held rigidly just below the lip and immerse it in the cooling bath to a depth of 160 mm.

Preparation of Sample

The solidification point is usually determined on chemicals as they are received. Some may be hygroscopic, however, and require special drying. Where this is necessary it will be noted in the monograph.

Products which are normally solid at room temperature must be carefully melted at a temperature about 10° above the expected solidification point. Care should be observed to avoid heating in such a way as to decompose or distil any portion of a sample.

Procedure

Adjust the temperature of the cooling bath to about 5° below the expected solidification point. Fit the thermometer and stirrer with a cork stopper so that the thermometer is centered and the bulb is about 20 mm. from the bottom of the sample container. Transfer a sufficient amount of the sample, previously melted if necessary, into the sample container to fill it to a depth of about 90 mm. when in the molten state. Place the thermometer and stirrer in the sample container and adjust the thermometer so that the immersion line will be at the surface of the liquid and the end of the bulb 20 ± 4 mm. from the bottom of the sample container. When the temperature of the sample is about 5° above the expected solidification point, place the assembled sample tube in the air jacket.

Allow the sample to cool while stirring, at the rate of 20 to 30 strokes per minute, in such a manner that the stirrer does not touch the thermometer. Stir the sample continuously during the remainder of the test.

The temperature at first will gradually fall, then become constant as crystallization starts and continues under equilibrium conditions, and finally will start to drop again. Some chemicals may supercool slightly below (0.5°) the solidification point; as crystallization begins the temperature will rise and remain constant as equilibrium conditions are established. Other products may cool more than 0.5° and cause deviation from the normal pattern of temperature change. If the temperature rise exceeds 0.5° after the initial crystallization begins, repeat the test and seed the melted compound with small crystals of the sample at 0.5° intervals as the temperature approaches the expected solidification point. Crystals for seeding may be obtained by freezing a small sample in a test tube directly in the cooling bath. It is preferable that seed of the stable phase be used from a previous determination.

Observe and record the temperature readings at regular intervals until the temperature rises from a minimum, due to supercooling, to a maximum and then finally drops. The maximum temperature reading is the solidification point. Readings 10 seconds apart should be taken in order to establish that the temperature is at the maximum level and continues until the drop in temperature is established.

SPECTROPHOTOMETRY, COLORIMETRY, AND TURBIDIMETRY

Absorption spectrophotometry is the measurement of the absorption, by substances, of electromagnetic radiation of definite and narrow wavelength range, approximating monochromatic radiation. *Colorimetry* is the commonly accepted term for the measurement of absorption of light, usually in the visible spectrum, which is not monochromatic but is restricted by the use of filters. *Turbidimetry* involves measurement of the degree of attenuation of a light beam incident on particles suspended in a medium, the measurement being made in the directly transmitted beam. In *atomic absorption spectrophotometry*, the amount of energy absorbed by the atoms of certain metallic elements is determined. A solution of the test sample is sprayed into a flame through which is passed radiation emitted by a lamp whose cathode is composed of the element being determined.

The wavelength range available for these measurements is roughly divided into the ultraviolet (185 to 380 mμ), the visible (380 to 780 mμ), the near-infrared (780 to 3000 mμ), and the infrared (3 to 40 μ).

Terminology. The following definitions and symbols are designated for Codex purposes.

Transmittance, T. The ratio of the radiant power transmitted by a sample to the radiant power incident on the sample.

Absorbance, A. The logarithm to the base 10 of the reciprocal of the transmittance. $A = \log_{10}(1/T)$.

Absorptivity, a. The absorbance divided by the product of the concentration of a substance (c, in grams per 1000 ml.) and the sample path length (b, in cm.). $a = A/bc$.

Apparatus. Many types of spectrophotometers are commercially available. Fundamentally, all types provide for passing essentially monochromatic radiant energy through a sample in suitable form and measuring the intensity of the fraction of the energy that is absorbed. The spectrophotometer comprises an energy source, a dispersing device with slits for selecting the wavelength band, a cell or holder for the sample, a detector of radiant energy, and associated amplifiers and measuring devices. Some instruments are manually operated, while others are equipped for automatic and continuous recording.

Atomic absorption determinations are usually made in the visible or near ultraviolet ranges. In addition to the usual spectrophotometric equipment, a hollow cathode lamp is required for each element to be determined, together with a burner, an aspirator, and a special detector.

The choice of a "suitable spectrophotometer," though essentially governed by the wavelength range in which the instrument must operate, depends upon a number of factors such as the composition and amount of available sample, the degree of accuracy, sensitivity, and specificity required, and the manner in which the sample is to be handled. No single type of instrument is capable of functioning at all wavelengths; consequently, instruments are available for use in the visible, in the visible and ultraviolet, in the visible and ultraviolet and near-infrared, and in the infrared regions of the spectrum.

Procedure

Spectrophotometry. Detailed instructions for operating spectrophotometers are supplied by the manufacturers. To achieve significant and valid results, the instruction manual or other publication should be followed closely on such matters as care, cleaning, and calibration of the instrument, and techniques of handling absorption cells, as well as instructions for operation.

Most Codex tests and assays employing spectrophotometry specify that a parallel comparison of the sample preparation be made against an appropriate reference standard. The expression "similarly measured," when referring to the reference standard preparations, indicates that the reference sample is to be prepared and observed in a manner identical for all practical purposes to that used for the test specimen. The absorptivity of the reference solution is calculated on the basis of the exact amount weighed out and, if a previously dried sample of the reference standard has not been used, on the anhydrous basis.

Comparisons of a sample with a reference standard are best made at a peak of spectral absorption for the compound concerned. Assays prescribing spectrophotometry give the commonly accepted wavelength for peak spectral absorption of the substance in question. It is known, however, that different spectrophotometers may show minor variation in the apparent wavelength of this peak. Good practice demands that use be made of the peak wavelength actually found in the individual instrument, rather than the wavelength given in the monograph, provided the two do not differ by more than 1 $m\mu$. If the difference is greater, re-calibration of the instrument may be indicated.

Colorimetry. Colorimetric assays usually call for comparing the absorbance produced by the *Assay preparation* with that produced simultaneously by a *Standard preparation* containing approximately an equal quantity of a reference standard. In some situations it is permissible to omit the use of a reference standard. This is true when colorimetric assays are made with routine frequency, and when a suitable standard curve is available, prepared with the respective Reference Standard, and when the substance assayed conforms to Beer's law within the range of about 75 to 125 percent of the final concentration used in the assay. Under these circumstances, the absorbance found in the assay may be interpolated on the standard curve, and the assay result calculated therefrom. Such standard curves should be confirmed frequently, and always when a new colorimeter and new lots of reagents are put into use.

In colorimetric assays that direct the preparation and use of a standard curve, it is permissible and preferable, when the assay is employed infrequently, not to use the standard curve but to make the comparison directly against a solution of the reference standard having a concentration approximately equal to that of the sample preparation.

Visual Color and Turbidity Comparisons. When color or turbidity comparison is directed, color-comparison tubes that are matched as closely as possible in internal diameter and in all other respects are to be used. The solutions to be compared are to be at the same temperature, preferably at room temperature.

For color comparisons, the tubes should usually be viewed downward against a white background, with the aid of a light source directed from beneath the bottoms of the tubes. If the colors to be compared are too dark to be viewed downward through the depth of the solutions, they should be viewed horizontally across the diameter of the tubes, with the aid of a light source directed from the back of the tubes. If the test procedure results in the formation of two layers, the designated layer must be viewed horizontally across the diameter of the tube.

For turbidity comparisons, the tubes should be viewed horizontally across the diameter of the tubes, with the aid of a light source directed against the sides of the tubes.

In conducting limit tests involving the comparison of colors or turbidities, suitable instruments may be used rather than the unaided eye.

STARCHES AND RELATED SUBSTANCES

ACETYL GROUPS

Transfer about 5 grams of the sample, accurately weighed, into a 250-ml. Erlenmeyer flask, suspend in 50 ml. of water, add a few drops of phenolphthalein T.S., and titrate with 0.1 N sodium hydroxide to a permanent pink end-point. Add 25.0 ml. of 0.45 N sodium hydroxide, stopper the flask, and shake vigorously for 30 minutes, preferably with a mechanical shaker. Remove the stopper, wash the stopper and sides of the flask with a few ml. of water, and titrate the excess alkali with 0.2 N hydrochloric acid to the disappearance of the pink color, recording the volume, in ml., of 0.2 N hydrochloric acid required as S. Perform a blank titration on 25.0 ml. of 0.45 N sodium hydroxide, and record the volume, in ml., of 0.2 N hydrochloric acid required as B. Calculate the percent of acetyl groups by the formula $(B - S) \times N \times 0.043 \times 100/W$, in which N is the exact normality of the hydrochloric acid solution, and W is the weight of the sample, in grams.

CRUDE FAT

Apparatus. The apparatus consists of a Butt-type extractor* (shown in the accompanying diagram), having a standard-taper 34/45 female joint at the upper end, to which is attached a Friedrichs- or Hopkins-type condenser, and a 24/40 male joint at the lower end, to which is attached a 125-ml. Erlenmeyer flask.

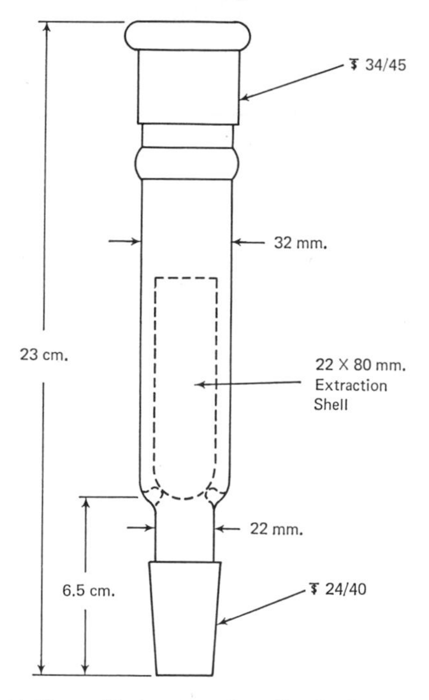

Fig. 19—Butt-Type Extractor for Crude Fat Determination

Procedure. Transfer about 10 grams of the sample, previously ground to 20-mesh or finer and accurately weighed, to a 15-cm. filter paper, roll the paper tightly around the sample, and place it in a suitable extraction shell. Plug the top of the shell with cotton previously extracted with carbon tetrachloride, and place the shell in the extractor. Attach the extractor to a dry 125-ml. Erlenmeyer flask containing about 50 ml. of carbon tetrachloride and to a water-cooled condenser, apply heat to the flask to produce 150–200 drops of condensed solvent per minute, and extract for 16 hours. Disconnect the flask, and filter the extract to remove any insoluble residue. Rinse the flask and filter with a few ml. of carbon tetrachloride, combine the washings and filtrate in a tared flask, and evaporate on a steam bath until no odor of solvent remains. Dry in a vacuum for 1 hour at 100°, cool in a desiccator, and weigh.

* Available from H. S. Martin & Co., Evanston, Ill.

MANGANESE

Manganese Detection Instrument. Use any suitable atomic absorption spectrophotometer (such as the Perkin-Elmer 290) equipped with a fast-response recorder or other readout device and capable of measuring the radiation absorbed by manganese atoms at the manganese resonance line of 2795 Å (199.4 setting on the Perkin-Elmer 290).

Standard Preparation. Transfer 1.000 gram of manganese metal powder into a 1000-ml. volumetric flask, dissolve by warming in a mixture of 10 ml. of water and 10 ml. of 0.5 *N* hydrochloric acid, cool, dilute to volume with water, and mix. Pipet 5.0 ml. of this solution into a 50-ml. volumetric flask, dilute to volume with water, and mix. Finally, pipet 5.0 ml. of this solution into a 1000-ml. volumetric flask, dilute to volume with water, and mix. The final solution contains 0.5 part per million of Mn.

Sample Preparation. Transfer 10.000 grams of the sample into a 200-ml. Kohlrausch volumetric flask, previously rinsed with 0.5 *N* hydrochloric acid, add 140 ml. of 0.5 *N* hydrochloric acid, and shake vigorously for 15 minutes, preferably with a mechanical shaker. Dilute to volume with 0.5 *N* hydrochloric acid, and shake. Centrifuge approximately 100 ml. of the sample mixture in a heavy-walled centrifuge tube at 2000 rpm for 5 minutes, and use the clear supernatant liquid in the following *Procedure.*

Procedure. Aspirate distilled water through the air-acetylene burner for 5 minutes, and obtain a base-line reading at 2795 Å, following the manufacturer's instructions for operating the atomic absorption spectrophotometer being used for the analysis. Aspirate a portion of the *Standard Preparation* (0.5 part per million Mn) in the same manner, note the reading, then aspirate a portion of the *Sample Preparation,* and note the reading. Prepare a standard curve by plotting a straight line between zero and the reading for the *Standard Preparation.* From the graph determine the parts per million of Mn in the *Sample Preparation,* and multiply this value by 20 to obtain the parts per million of Mn in the original sample taken for analysis.

PHOSPHORUS

Reagents

Ammonium Molybdate Solution (5 percent). Dissolve 50 grams of ammonium molybdate tetrahydrate, $(NH_4)_6Mo_7O_{24}\cdot 4H_2O$, in 900 ml. of warm water, cool to room temperature, dilute to 1000 ml. with water, and mix.

Ammonium Vanadate Solution (0.25 percent). Dissolve 2.5 grams of ammonium metavanadate, NH_4VO_3, in 600 ml. of boiling water, cool to 60–70°, and add 20 ml. of nitric acid. Cool to room temperature, dilute to 1000 ml. with water, and mix.

Zinc Acetate Solution (10 percent). Dissolve 120 grams of zinc acetate dihydrate, $Zn(C_2H_3O_2)_2 \cdot 2H_2O$, in 880 ml. of water, and filter through Whatman No. 2V or equivalent filter paper before use.

Nitric Acid Solution (29 percent). Add 300 ml. of nitric acid (sp. gr. 1.42) to 600 ml. of water, and mix.

Standard Phosphorus Solution (100 mcg. P in 1 ml.). Dissolve 438.7 mg. of monobasic potassium phosphate, KH_2PO_4, in water in a 1000-ml. volumetric flask, dilute to volume with water, and mix.

Standard Curve. Pipet 5.0, 10.0, and 15.0 ml. of the *Standard Phosphorus Solution* into separate 100-ml. volumetric flasks. To each of these flasks, and to a fourth blank flask, add in the order stated 10 ml. of *Nitric Acid Solution,* 10 ml. of *Ammonium Vanadate Solution,* and 10 ml. of *Ammonium Molybdate Solution,* mixing thoroughly after each addition. Dilute to volume with water, mix, and allow to stand for 10 minutes. Determine the absorbance of each standard solution in a 1-cm. cell at 460 mμ, with a suitable spectrophotometer, using the blank to set the instrument at zero. Prepare a standard curve by plotting the absorbance of each solution versus its concentration, in mg. P per 100 ml.

Sample Preparation. Transfer about 10 grams of the sample, accurately weighed, into a Vycor dish, and add 10 ml. of *Zinc Acetate Solution* in a fine stream, distributing the solution uniformly in the sample. Carefully evaporate to dryness on a hot plate, then increase the heat, and carbonize the sample on the hot plate or over a gas flame. Ignite in a muffle furnace at 550° until the ash is free from carbon (about 1 to 2 hours), and cool. Wet the ash with 15 ml. of water, and slowly wash down the sides of the dish with 5 ml. of *Nitric Acid Solution.* Heat to boiling, cool, and quantitatively transfer the mixture into a 200-ml. volumetric flask, rinsing the dish with three 20-ml. portions of water and adding the rinsings to the flask. Dilute to volume with water, and mix. Transfer an accurately measured aliquot (V, in ml.) of this solution, containing not more than 1.5 mg. of phosphorus, into a 100-ml. volumetric flask, and add 50 ml. of water to a second flask to serve as a blank. To each flask add in the order stated 10 ml. of *Nitric Acid Solution,* 10 ml. of *Ammonium Vanadate Solution,* and 10 ml. of *Ammonium Molybdate Solution,* mixing thoroughly after each addition. Dilute to volume with water, mix, and allow to stand for 10 minutes.

Procedure. Determine the absorbance of the *Sample Preparation* in a 1-cm. cell at 460 mμ, with a suitable spectrophotometer, using the blank to set the instrument at zero. From the *Standard Curve,* determine the mg. of phosphorus in the aliquot taken, recording this value as a. Calculate the amount, in parts per million, of phosphorus (P) in the original sample by the formula $(a \times 200 \times 1000)/(V \times W)$, in which W is the weight of the sample taken, in grams.

PROPYLENE CHLOROHYDRIN

Standard Preparation. Draw 25 μl. of mixed propylene chlorohydrin isomers (1-chloro-2-propanol containing 25 percent of 2-chloro-

1-propanol*) into a 50-μl. syringe, weigh the charged syringe, and inject its contents into a 200-ml. volumetric flask partially filled with diethyl ether. Reweigh the syringe, and record the weight of the chlorohydrins taken. Dilute the flask to volume with ether, and mix. Pipet 1.0 ml. of this solution into a 5-ml. volumetric flask, dilute to volume with ether, and mix. The final solution contains the equivalent of approximately 130 mcg. of mixed isomers in each 5 ml. The parent chlorohydrin solution is stable for at least 1 month if refrigerated. The dilute solutions should be prepared fresh on the day of use.

Sample Preparation. Transfer 20.0 grams of the sample into a dry 200-ml. pressure bottle, add 50 ml. of 2 *N* sulfuric acid, clamp the top in place, and mix by swirling. Place the bottle in a steam bath, and heat for 15–20 minutes or until the sample is translucent throughout. Cool to room temperature in the air (not in cold water or in ice), add 15 grams of anhydrous sodium sulfate, and shake until dissolved. Using 50 ml. of water, quantitatively transfer the contents of the bottle into a 250-ml. separator, and extract with four 50-ml. portions of diethyl ether, combining the extracts in a Kuderna-Danish concentrator.† Place the concentrator receiver in a water bath maintained at 55–60°, and concentrate the extract to a volume of 4 ml. Cool to room temperature, quantitatively transfer into a 5-ml. volumetric flask with the aid of small portions of ether, dilute to volume with ether, and mix.

Procedure. Inject 7.5 μl. of the *Standard Preparation* into a suitable gas chromatograph, such as the F&M Model 609, equipped with a single column flame ionization detector. The operating conditions of the apparatus may vary, depending upon the particular instrument used, but a suitable chromatogram is obtained with the F&M instrument having a 6.5-mm. × 3-m. stainless steel column packed with 7.5 percent Carbowax 20M on 100/120-mesh Gas Chrom Q, using a column temperature of 65°, detector and injection port temperature of 175°, and helium as the carrier gas flowing at a rate of 90 ml. per minute. Record the two peaks representing the isomers 1-chloro-2-propanol and 2-chloro-1-propanol. Similarly, inject 7.5 μl. of the *Sample Preparation*, and record the chromatogram.

Calculate the content of 1-chloro-2-propanol in the sample, in mcg. per gram, by the formula $(W \times a)/(w \times A)$, in which W is the weight, in mcg., of the mixed isomers in the 1.0-ml. aliquot used in making the final dilution of the *Standard Preparation;* a is the area of the peak produced by the *Sample Preparation;* w is the weight, in grams, of the sample used in preparing the *Sample Preparation;* and A is the area of the peak produced by the *Standard Preparation.* In a similar manner, calculate the content of 2-chloro-1-propanol in the sample, in mcg. per

* Available as P-1325 from Eastman Kodak Co., Rochester, N.Y. 14603.

† Available from Kontes Glass Co., Vineland, N.J.; concentrator, with 500-ml. flask, No. K-57000; size 4E concentrator tube, No. K-57005.

gram. Sum the amounts of the two isomers, and express the results as mcg. of total chlorohydrins per gram of sample.

SULFUR DIOXIDE

Apparatus. Use a Monier-Williams apparatus for the determination of sulfurous acid,* or construct the apparatus, using rubber stopper connections, as shown in the accompanying illustration. The assembly consists of a 1000-ml. two-neck round-bottom boiling flask to which a gas-inlet tube, a 60-ml. dropping funnel equipped with a 2-mm. bore stopcock, and a sloping Allihn reflux condenser are attached. A delivery tube connects the upper end of the condenser to the bottom of a 250-ml. Erlenmeyer receiving flask, which is followed by a Peligot tube.

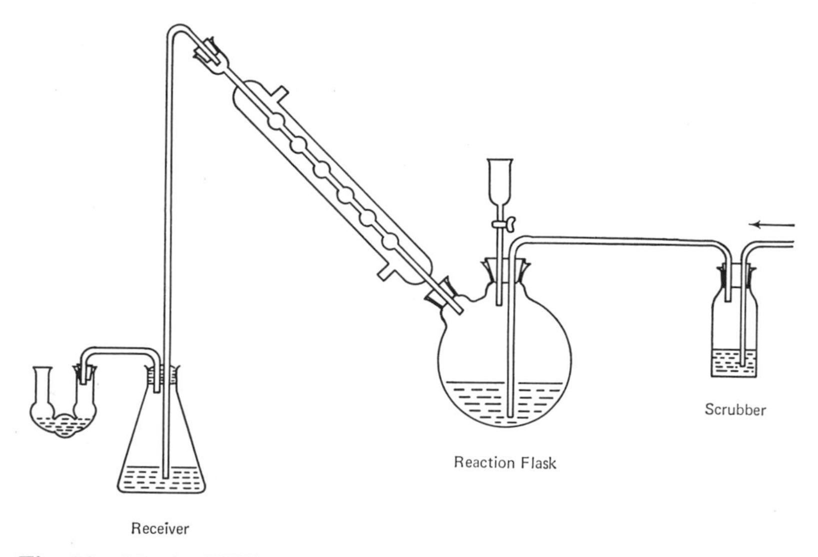

Fig. 20—Monier-Williams Apparatus for Determination of Sulfurous Acid

Procedure. Pass nitrogen from a cylinder through a 15 percent sodium carbonate scrubber solution to remove chlorine, thence into the gas-inlet tube of the boiling flask. Place 15 ml. of 3 percent hydrogen peroxide in the receiving flask and 5 ml. in the Peligot tube, connect the apparatus, and introduce into the boiling flask, by means of the dropping funnel, 300 ml. of water and 20 ml. of hydrochloric acid. Boil the contents of the flask for about 10 minutes in a current of nitrogen. Disperse about 100 grams of the sample, accurately weighed, in 300 ml. of recently boiled and cooled water, and transfer the slurry into the boiling flask through the dropping funnel, regulating the rate of sample addition and gas flow rate to prevent drawback of hydrogen peroxide, inclusion of air, or burning of the sample. Boil

* Available from Scientific Glass Apparatus Co., Bloomfield, N.J. 07003.

the mixture gently for 1 hour in a slow current of nitrogen, stopping the flow of water in the condenser just before the end of distillation. Immediately remove the delivery tube from the condenser when the portion of the tube just above the receiving tube becomes hot. Wash the contents of the delivery tube and the Peligot tube into the receiving flask, and titrate with 0.1 *N* sodium hydroxide, using bromophenol blue T.S. as indicator, recording the volume, in ml., required as *S*. Perform a blank determination, and record the volume of 0.1 *N* sodium hydroxide required as *B*. Calculate the percent of sulfur dioxide in the sample taken by the formula $(S - B) \times 0.0032 \times 100/W$, in which *W* is the weight of the sample, in grams.

SULFURIC ACID TABLE†

Be°	Sp. Gr.	Percent H_2SO_4	Be°	Sp. Gr.	Percent H_2SO_4
0	1.0000	0.00	36	1.3303	42.63
1	1.0069	1.02	37	1.3426	43.99
2	1.0140	2.08	38	1.3551	45.35
3	1.0211	3.13	39	1.3679	46.72
4	1.0284	4.21	40	1.3810	48.10
5	1.0357	5.28	41	1.3942	49.47
6	1.0432	6.37	42	1.4078	50.87
7	1.0507	7.45	43	1.4216	52.26
8	1.0584	8.55	44	1.4356	53.66
9	1.0662	9.66	45	1.4500	55.07
10	1.0741	10.77	46	1.4646	56.48
11	1.0821	11.89	47	1.4796	57.90
12	1.0902	13.01	48	1.4948	59.32
13	1.0985	14.13	49	1.5104	60.75
14	1.1069	15.25	50	1.5263	62.18
15	1.1154	16.38	51	1.5426	63.66
16	1.1240	17.53	52	1.5591	65.13
17	1.1328	18.71	53	1.5761	66.63
18	1.1417	19.89	54	1.5934	68.13
19	1.1508	21.07	55	1.6111	69.65
20	1.1600	22.25	56	1.6292	71.17
21	1.1694	23.43	57	1.6477	72.75
22	1.1789	24.61	58	1.6667	74.36
23	1.1885	25.81	59	1.6860	75.99
24	1.1983	27.03	60	1.7059	77.67
25	1.2083	28.28	61	1.7262	79.43
26	1.2185	29.53	62	1.7470	81.30
27	1.2288	30.79	63	1.7683	83.34
28	1.2393	32.05	64	1.7901	85.66
29	1.2500	33.33	64¼	1.7957	86.33
30	1.2609	34.63	64½	1.8012	87.04
31	1.2719	35.93	64¾	1.8068	87.81
32	1.2832	37.26	65	1.8125	88.65
33	1.2946	38.58	65¼	1.8182	89.55
34	1.3063	39.92	65½	1.8239	90.60
35	1.3182	41.27	66	1.8354	93.19

† Courtesy of the Manufacturing Chemists' Association.

Specific Gravity determinations were made at 60° F., compared with water at 60° F.

From the Specific Gravities, the corresponding degrees Baumé were calculated by the following formula:

$$\text{Baumé} = 145 - \frac{145}{\text{Sp. Gr.}}$$

Baumé Hydrometers for use with this table must be graduated by the above formula, which formula should always be printed on the scale. Acids stronger than 66° Bé should have their percentage compositions determined by chemical analysis.

THERMOMETERS

Thermometers suitable for *Food Chemicals Codex* use conform to the specifications of the American Society for Testing and Materials, A.S.T.M. Standards E 1, and are standardized in accordance with A.S.T.M. Method E 77.

The thermometers are of the mercury in glass type, and the column above the liquid is filled with nitrogen. They may be standardized for *total immersion* or for *partial immersion* and should be used as near as practicable under the same condition of immersion.

"Total immersion" means standardization with the thermometer immersed to the top of the mercury column, with the remainder of the stem and the upper expansion chamber exposed to the ambient temperature. "Partial immersion" means standardization with the thermometer immersed to the indicated immersion line etched on the front of the thermometer, with the remainder of the stem exposed to the ambient temperature. If used under any other condition of immersion, an emergent stem correction is necessary to obtain correct temperature readings.

Thermometer Specifications

A.S.T.M.	Range		Subdivisions		
No. E-1-	° C.	° F.	° C.	° F.	Immersion, mm.
		For General Use			
1 C	−20 to +150	. . .	1	. . .	76
1 F	. . .	0 to 302	. . .	2	76
2 C	−5 to +300	. . .	1	. . .	76
2 F	. . .	20 to 580	. . .	2	76
3 C	−5 to +400	. . .	1	. . .	76
3 F	. . .	20 to 760	. . .	2	76
		For Determination of Softening Point			
15 C	−2 to +80	. . .	0.2	. . .	Total
15 F	. . .	30 to 180	. . .	0.5	Total
16 C	30 to 200	. . .	0.5	. . .	Total
16 F	. . .	85 to 392	. . .	1	Total

Thermometer Specifications (*continued*)

A.S.T.M. No. E-1-	Range		Subdivisions		Immersion, mm.
	° C.	° F.	° C.	° F.	
	For Determination of Kinematic Viscosity				
44 F	...	66.5 to 71.5	...	0.1	Total
45 F	...	74.5 to 79.5	...	0.1	Total
28 F	...	97.5 to 102.5	...	0.1	Total
46 F	...	119.5 to 124.5	...	0.1	Total
29 F	...	127.5 to 132.5	...	0.1	Total
47 F	...	137.5 to 142.5	...	0.1	Total
48 F	...	177.5 to 182.5	...	0.1	Total
30 F	...	207.5 to 212.5	...	0.1	Total
	For Determination of Distillation Range				
37 C	−2 to +52	...	0.2	...	100
38 C	24 to 78	...	0.2	...	100
39 C	48 to 102	...	0.2	...	100
40 C	72 to 126	...	0.2	...	100
41 C	98 to 152	...	0.2	...	100
102 C	123 to 177	...	0.2	...	100
103 C	148 to 202	...	0.2	...	100
104 C	173 to 227	...	0.2	...	100
105 C	198 to 252	...	0.2	...	100
106 C	223 to 277	...	0.2	...	100
107 C	248 to 302	...	0.2	...	100
	For Determination of Solidification Range				
89 C	−20 to +10	...	0.1	...	76
90 C	0 to 30	...	0.1	...	76
91 C	20 to 50	...	0.1	...	76
92 C	40 to 70	...	0.1	...	76
93 C	60 to 90	...	0.1	...	76
94 C	80 to 110	...	0.1	...	76
95 C	100 to 130	...	0.1	...	76
96 C	120 to 150	...	0.1	...	76
100 C	145 to 205	...	0.2	...	76
101 C	195 to 305	...	0.5	...	76
	For Special Use				
14 C[1]	38 to 82	...	0.1	...	79
36 C[2]	−2 to +68	...	0.2	...	45
18 C[3]	34 to 42	...	0.1	...	Total
18 F[3]	...	94 to 108	...	0.2	Total
22 C[3]	95 to 103	...	0.1	...	Total
22 F[3]	...	204 to 218	...	0.2	Total
23 C[4]	18 to 28	...	0.2	...	90
24 C[4]	39 to 54	...	0.2	...	90

[1] For determination of melting range of Class III solids.
[2] For determining the titer of fatty acids.
[3] For determination of Saybolt viscosity.
[4] For determination of Engler viscosity.

In selecting a thermometer, careful consideration should be given to the conditions under which it is to be used. The preceding table lists several A.S.T.M. thermometers, together with their usual conditions of use, which may be required in Codex tests. Complete specifications for these thermometers are given in "A.S.T.M. Standards on Thermometers."

THIAMINE ASSAY

Special Solutions and Solvents

Potassium Chloride Solution. Dissolve 250 grams of potassium chloride, KCl, in sufficient water to make 1000 ml.

Acid Potassium Chloride Solution. Add 8.5 ml. of hydrochloric acid to 1000 ml. of *Potassium Chloride Solution.*

Potassium Ferricyanide Solution, 1 Percent. Dissolve 1 gram of potassium ferricyanide, $K_3Fe(CN)_6$, in sufficient water to make 100 ml. Prepare fresh on the day of use.

Oxidizing Reagent. Mix 4.0 ml. of *1 Percent Potassium Ferricyanide Solution* with sufficient 3.5 *N* sodium hydroxide to make 100 ml. Use this solution within 4 hours.

Quinine Sulfate Stock Solution. Dissolve 10 mg. of quinine sulfate, $(C_{20}H_{24}N_2O_2)_2 \cdot H_2SO_4 \cdot 2H_2O$, in sufficient 0.1 *N* sulfuric acid to make 1000 ml. Preserve this solution, protected from light, in a refrigerator.

Quinine Sulfate Standard Solution. Dilute 1 volume of *Quinine Sulfate Stock Solution* with 39 volumes of 0.1 *N* sulfuric acid. This solution fluoresces to approximately the same degree as the thiochrome obtained from 1 mcg. of thiamine hydrochloride and is used to correct the fluorophotometer at frequent intervals for variation in sensitivity from reading to reading within an assay. Prepare this solution fresh on the day of use.

Standard Thiamine Hydrochloride Stock Solution. Transfer about 25 mg. of U.S.P. Thiamine Hydrochloride Reference Standard, previously dried at 105° for 2 hours and accurately weighed, to a 1000-ml. volumetric flask, observing precautions to avoid absorption of moisture in weighing the dried Standard. Dissolve the weighed Standard in about 300 ml. of dilute alcohol solution (1 in 5) adjusted to a pH of 4.0 with diluted hydrochloric acid T.S., and add the acidified, dilute alcohol to volume. Store in a light-resistant container, in a refrigerator. Prepare this stock solution fresh each month.

Standard Preparation. Pipet a volume of *Standard Thiamine Hydrochloride Stock Solution,* equivalent to exactly 100 mcg. of U.S.P. Thiamine Hydrochloride Reference Standard, into a 100-ml. volumetric flask, and dilute with *Acid Potassium Chloride Solution* to volume. Dilute exactly 10 ml. of this solution with *Acid Potassium Chloride Solution* to 50.0 ml. Each ml. of the resulting *Standard Preparation* contains 0.2 mcg. of U.S.P. Thiamine Hydrochloride Reference Standard.

Assay Preparation. Prepare as directed in the individual monograph.

Procedure. Into each of four or more tubes (or other suitable vessels), of about 40-ml. capacity, pipet 5 ml. of *Standard Preparation.* To each of three of these tubes add rapidly (within 1 to 2 seconds), with mixing, 3 ml. of *Oxidizing Reagent,* and within 30 seconds add 20 ml. of isobutyl alcohol, then mix vigorously for 90 seconds by shaking the capped tubes manually, or by bubbling a stream of air through the mixture. Prepare a blank in the remaining tube of the standard by substituting for the *Oxidizing Reagent* an equal volume of 3.5 *N* sodium hydroxide and proceeding in the same manner.

Into each of four or more similar tubes pipet 5 ml. of the *Assay Preparation.* Treat these tubes in the same manner as directed for the tubes containing the *Standard Preparation.*

To each of the eight tubes add 2 ml. of absolute alcohol, swirl for a few seconds, allow the phases to separate, and decant or draw off about 10 ml. of the clear, supernatant isobutyl alcohol solution into standardized cuvettes, then measure the fluorescence in a suitable fluorophotometer. Use an input filter of narrow transmittance range with a maximum at about 365 mμ, and an output filter of narrow transmittance range with a maximum at about 435 mμ.

Calculation. The number of mcg. of $C_{12}H_{17}ClN_4OS.HCl$ in each 5 ml. of the *Assay Preparation* is given by the formula $(A - b)/(S - d)$, in which A is the average of the fluorophotometer readings of the portions of the *Assay Preparation* treated with *Oxidizing Reagent,* b is the fluorophotometer reading of the *Assay Preparation* blank, S is the average of the fluorophotometer readings of the portions of the *Standard Preparation* treated with *Oxidizing Reagent,* and d is the fluorophotometer reading of the *Standard Preparation* blank.

Calculate the quantity, in mg., of $C_{12}H_{17}ClN_4OS.HCl$ (thiamine hydrochloride) in the assay material on the basis of the aliquots taken. Where indicated, the quantity, in mg., of $C_{12}H_{17}N_5O_4S$ (thiamine mononitrate) may be calculated by multiplying the quantity of $C_{12}H_{17}ClN_4OS.HCl$ found by 0.9706.

TOCOPHEROLS

Weight-Unit Relationships

In accordance with the convention adopted by the National Formulary, label claims for Tocopherol products described in this Codex when expressed in International Units (I.U.) of Vitamin E are based on the following equivalents:

1 mg. *dl*-Alpha Tocopheryl Acetate	= 1 International Unit
1 mg. *dl*-Alpha Tocopherol	= 1.1 International Unit
1 mg. *d*-Alpha Tocopheryl Acetate	= 1.36 International Unit
1 mg. *d*-Alpha Tocopherol	= 1.49 International Unit
1 mg. *d*-Alpha Tocopheryl Acid Succinate	= 1.21 International Unit

VISCOSITY OF DIMETHYLPOLYSILOXANE

Apparatus. The Ubbelohde suspended level viscometer, shown in the accompanying diagram, is preferred for the determination of the viscosity of dimethylpolysiloxane.

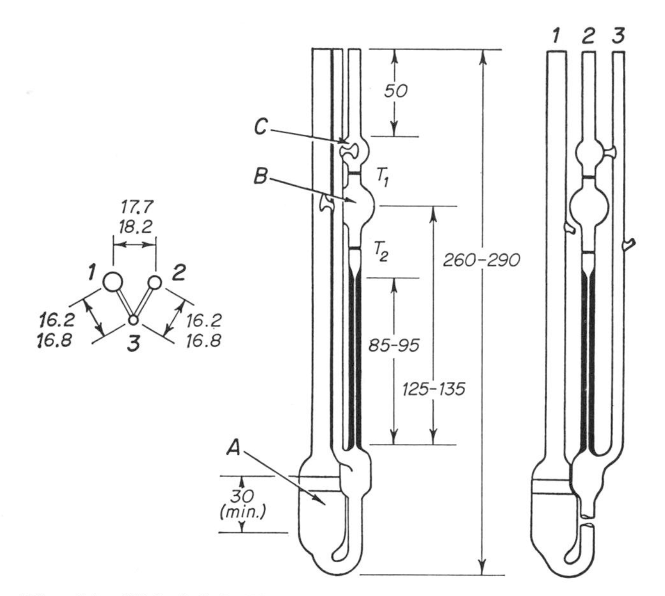

Fig. 21—Ubbelohde Viscometer for Dimethylpolysiloxane

(all dimensions are in millimeters)

For use in the range of 300 to 600 centistokes, a number 3 size viscometer, having a capillary diameter of 2.00 ± 0.04 mm., is required. The viscometer should be fitted with holders which satisfy the dimensional positions of the separate tubes as shown in the diagram, and which hold the viscometer vertical. Filling lines in bulb *A* indicate the minimum and maximum volumes of liquid to be used for convenient operation. The volume of bulb *B* is approximately 5 ml.

Calibration of the Viscometer. Determine the viscosity constant *C* for each viscometer by using an oil of known viscosity.* Charge the viscometer by tilting the instrument about 30 degrees from the vertical, with bulb *A* below the capillary, and then introduce enough of the sample into tube *1* to bring the level up to the lower filling line. The level should not be above the upper filling line when

* Oils of known viscosities, formerly supplied by the National Bureau of Standards, may be obtained from The Cannon Instrument Co., P.O. Box 812, State College, Pa. 16801. For determining the viscosity of dimethylpolysiloxane, choose an oil whose viscosity is as close as possible to that of the type of sample to be tested.

the viscometer is returned to the vertical position and the sample has drained from tube *1*. Charge the viscometer in such a manner that the U-tube at the bottom fills completely without trapping air.

After the viscometer has been in a constant-temperature bath (25° ± 0.2°) long enough for the sample to reach temperature equilibrium, place a finger over tube *3* and apply suction to tube *2* until the liquid reaches the center of bulb *C*. Remove suction from tube *2*, then remove the finger from tube *3* and place it over tube *2* until the sample drops away from the lower end of the capillary. Remove the finger from tube *2*, and measure the time, to the nearest 0.1 second, required for the meniscus to pass from the first timing mark (T_1) to the second (T_2). In order to obtain accurate results within a reasonable time, the apparatus should be adjusted to give an elapsed time of from 80 to 100 seconds.

Calculate the viscometer constant C by the equation $C = cs/t_1$, in which cs is the viscosity, in centistokes, and t_1 is the efflux time, in seconds, for the standard liquid.

Determination of the Viscosity of Dimethylpolysiloxane. Charge the viscometer with the sample in the same manner as described for the calibration procedure, determine the efflux time, t_2, and calculate the viscosity of the dimethylpolysiloxane by the formula $C \times t_2$.

VISCOSITY OF METHYLCELLULOSE

Apparatus. Viscometers used for the determination of the viscosity of methylcellulose and some related compounds are illustrated in Fig. 22 and consist of three parts: a large filling tube *A*, an orifice tube *B*, and an air vent to the reservoir *C*.

There are two basic types of viscometers, one for cellulose derivatives of a range between 1500 and 4000 centipoises, and the other for less viscous types. Each type of viscometer is modified slightly for the different viscosity types.

Calibration of the Viscometer. Determine the viscometer constant K for each viscometer by the use of an oil of known viscosity.* Place an excess of the liquid which is to be tested (adjusted to 20° ± 0.1°) in the filling tube *A* and transfer it to the orifice tube *B* by gentle suction, taking care to keep the liquid free from air bubbles by closing the air vent tube *C*. Adjust the column of liquid in tube *B* even with the top graduation line. Open both tubes *B* and *C* to permit the liquid to flow into the reservoir against atmospheric pressure. Failure to open air vent tube *C* before determining the viscosity will yield false values. Record the time in seconds for the liquid to flow from the upper mark to the lower mark in tube *B*.

* Oils of known viscosities, formerly supplied by the National Bureau of Standards, may be obtained from The Cannon Instrument Co., P.O. Box 812, State College, PA 16801. For determining the viscosity of methylcellulose, choose an oil whose viscosity is as close as possible to that of the type of sample to be tested.

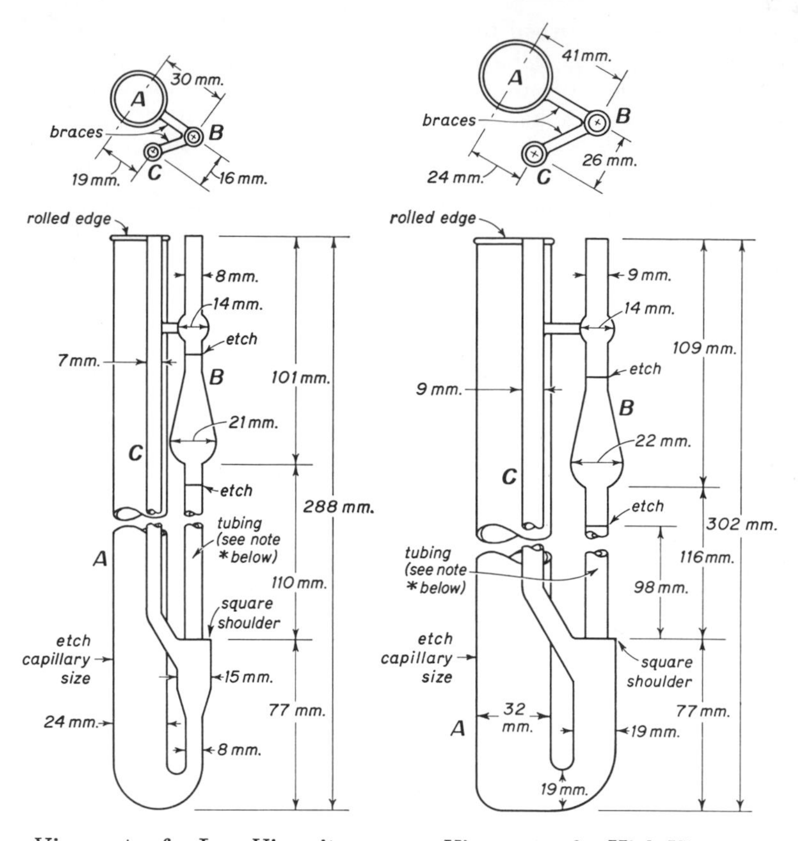

Viscometer for Low Viscosity

* Precision bore capillary tubing 1.5 mm. i.d. for 15 cps, 1.8 mm. i.d. for 25 cps, 2.4 mm. i.d. for 100 cps, and 3.2 mm. i.d. for 400 cps viscosities.

Viscometer for High Viscosity

* Precision bore capillary tubing 5.0 mm. i.d. for 1500 cps and 6.0 mm. i.d. for 4000 cps viscosities.

Fig. 22—Methylcellulose Viscometers

Calculate the viscometer constant K from the following equation:

$V = Kdt$, in which $V =$ the viscosity of the liquid in centipoises, $K =$ the viscometer constant, $d =$ the specific gravity of the liquid tested at 20°/20°, and $t =$ the time in seconds for the liquid to pass from the upper to the lower mark.

For the calibration, all values in the equation are known, or can be determined except K which must be solved. If a tube is repaired, it must be recalibrated to avoid obtaining significant changes in the value of K.

Determination of the Viscosity of Methylcellulose. Prepare a 2 percent solution of methylcellulose or other cellulose derivative, by weight, as directed in the monograph. Place the solution in the proper viscometer and determine the time required to flow from the upper mark to the lower mark in orifice tube *B*. Separately determine the specific gravity at 20°/20° and from the values of d and t thus determined, calculate V from the equation under *Calibration of the Viscometer.*

VISCOSITY OF SODIUM CARBOXYMETHYLCELLULOSE

Apparatus

Viscometer. A Model LVF Brookfield or equivalent type viscometer should be used for the determination of viscosity of aqueous solutions of sodium carboxymethylcellulose within the range of 25 to 10,000 centipoises at 25°. Instruments of this type are provided with spindles for use in determining the viscosity of different viscosity types of sodium carboxymethylcellulose. The spindles and speeds for determining viscosity within different ranges are tabulated below.

Viscometer Spindles Required for Given Speeds

Viscosity Range, Centipoises	Spindle No.	Speed, rpm	Scale	Factor
10 to 100	1	60	100	1
100 to 200	1	30	100	2
200 to 1000	2	30	100	10
1000 to 4000	3	30	100	40
4000 to 10000	4	30	100	200

Mechanical Stirrer. Use an agitator essentially as shown in Fig. 23 which can be attached to a variable speed motor capable of turning up to 1500 rpm. (*Note:* Stirrers equipped with 1 1/2-inch, 3-blade type, stainless steel propellers, such as A. H. Thomas Co. Catalog No. 9240-K, have also been found to be satisfactory.)

Sample Container. A glass jar about 6 1/2 inches deep having an outside diameter of approximately 2 3/4 inches at the bottom and 2 3/8 inches at the top.

Procedure. Transfer an accurately weighed sample, equivalent to 8 grams of sodium carboxymethylcellulose, on the dried basis, into a tared sample container and add, without stirring or shaking, sufficient water to make a total of 400 ± 0.1 grams of solution. Place the agitator in the mixture so that the blade is about halfway between the bottom of the jar and the surface of the liquid. Start the stirrer, and gradually increase the speed to as near 1500 rpm as the sample will permit without splashing or spilling. After 1 1/2 hours of stirring, adjust the stirrer speed so that agitation is continued at a rapid rate, but slowly enough to avoid beating air into the solution, and continue at this speed for another 30 minutes. Detach the agitator from the motor and transfer the sample container, without removing the agitator, to a constant-

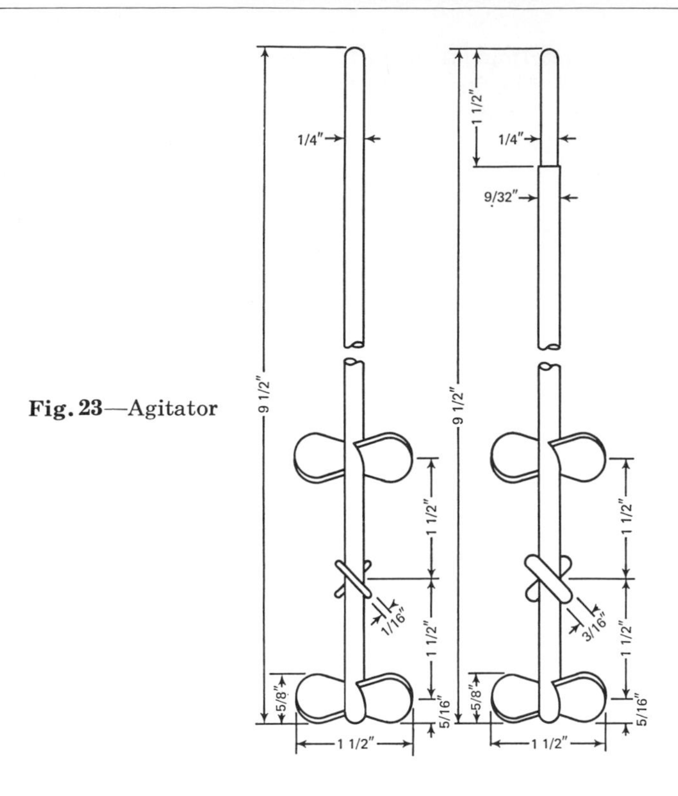

Fig. 23—Agitator

temperature water bath maintained at a temperature of 25° ± 0.2°. Check the temperature of the solution frequently and stir by hand with the agitator to insure that the test temperature of 25° is reached within a 30-minute time interval. Remove the container from the bath and measure the viscosity of the sample solution with a suitable viscometer, selecting the spindle and speed indicated in the Table. Allow the spindle to rotate until a constant reading is obtained. The viscosity in centipoises is calculated by multiplying the reading observed by the appropriate factor from the Table.

> *Note:* If the room temperature is considerably above or below 25°, the entire operation of stirring, standing, and measurement should be conducted with the sample suspended in the water bath.

VITAMIN A ASSAY

This procedure is provided for the determination of vitamin A intended for use as a food additive. It conforms to that which was adopted in 1956 for international use by the International Union of Pure and Applied Chemistry.

Complete the assay promptly and exercise care throughout the procedure to keep to a minimum exposure to atmospheric oxygen and other oxidizing agents and to actinic light, preferably by the use of an atmosphere of an inert gas and non-actinic glassware.

Special Reagents

Isopropyl Alcohol. Use reagent grade isopropyl alcohol. Redistil, if necessary, to meet the following requirements for spectral purity: When measured in a 1-cm. quartz cell against water it shows an absorbance not greater than 0.05 at 300 mμ and not greater than 0.01 between 320 mμ and 350 mμ.

Ether. Use freshly redistilled reagent grade ether, discarding the first and last 10 percent portions.

Procedure. Transfer into a saponification flask an accurately weighed portion of the sample containing the equivalent of not less than 0.15 mg. of retinol, but containing not more than 1 gram of fat. If in solid form, heat the portion taken in 10 ml. of water on a steam bath for about 10 minutes, crush the remaining solid with a blunt glass rod, and warm for about 5 minutes longer.

Add 30 ml. of alcohol if the sample is liquid or 23 ml. of alcohol and 7 ml. of glycerin if the sample is solid, followed by 3 ml. of potassium hydroxide solution (9 in 10). Reflux under an all-glass condenser for 30 minutes. Cool the solution, add 30 ml. of water, and transfer to a separator. Add 2 grams of finely powdered sodium sulfate. Extract by shaking for 2 minutes with one 150-ml. portion of ether, and, if an emulsion forms, with three additional 25-ml. portions of ether. Combine the ether extracts, if necessary, and wash by swirling gently with 50 ml. of water. Repeat the washing more vigorously with three additional 50-ml. portions of water. Transfer the washed ether extract to a 250-ml. volumetric flask, and add ether to volume.

Evaporate a 25.0 ml. portion of the ether extract to about 5 ml. *Without applying heat and with the aid of a stream of inert gas or vacuum,* continue the evaporation to about 3 ml. Dissolve the residue in sufficient isopropyl alcohol to give an expected concentration of the equivalent of 3 to 5 mcg. of retinol per ml. or such that it will give an absorbance in the range of 0.5 to 0.8 at 325 mμ. Determine the absorbances of the resulting solution at the wavelengths 310 mμ, 325 mμ, and 334 mμ, with a suitable spectrophotometer fitted with matched quartz cells.

Calculation. Calculate the retinol content as follows: Content (in mg.) $= 0.549A_{325}/LC$, in which A_{325} is the observed absorbance at 325 mμ, L is the length, in cm., of the absorption cell, and C is the amount of sample expressed as grams in each 100 ml. of the final isopropyl alcohol solution, provided that A_{325} has a value not less than $[A_{325}]/1.030$ and not more than $[A_{325}]/0.970$, where $[A_{325}]$ is the corrected absorbance at 325 mμ and is given by the equation $[A_{325}] = 6.815\,A_{325} - 2.555A_{310} - 4.260A_{334}$, in which A designates the absorbance at the wavelength indicated by the subscript.

Where $[A_{325}]$ has a value less than $A_{325}/1.030$, apply the following equation: Content (in mg.) $= 0.549[A_{325}]/LC$, in which the values are as defined herein.

Confidence Interval. The range of the limits of error, indicating the extent of discrepancy to be expected in the results of different laboratories at $P = 0.05$, is approximately ±8 percent.

VOLUMETRIC APPARATUS

Most of the volumetric apparatus available in the United States is calibrated at 20°, although the temperatures generally prevailing in laboratories more nearly approach 25°, which is the temperature specified generally for tests and assays. This discrepancy is inconsequential provided the room temperature is reasonably constant and the apparatus has been calibrated accurately prior to and under the conditions of its intended use.

Use. To attain the degree of precision required in many assays involving volumetric measurements and directing that a quantity be "accurately measured," the apparatus must be chosen and used with exceptional care. Where less than 10 ml. of titrant is to be measured, a 10-ml. buret or microburet generally is required.

The design of volumetric apparatus is an important factor in assuring accuracy. For example, the length of the graduated portions of graduated cylinders should be not less than five times the inside diameter, and the tips of burets should permit an outflow rate of not more than 0.5 ml. per second.

Volumetric Flasks

Designated Volume, ml. →	10	25	50	100	250	500	1000
Limit of error, ml.	0.02	0.03	0.05	0.08	0.12	0.15	0.30
Limit of error, %	0.20	0.12	0.10	0.08	0.05	0.03	0.03

Transfer Pipets

Designated Volume, ml. →	1	2	5	10	25	50	100
Limit of error, ml.	0.006	0.006	0.01	0.02	0.03	0.05	0.08
Limit of error, %	0.6	0.30	0.20	0.20	0.12	0.10	0.08

Burets

Designated Volume, ml. →	10 ("micro" type)	25	50
Subdivisions, ml.	0.02	0.10	0.10
Limit of error, ml.	0.02	0.03	0.05

Standards of Accuracy. The capacity tolerances for volumetric flasks, transfer pipets, and burets are those accepted by the National Bureau of Standards,* as indicated in the preceding tables.

The capacity tolerances for measuring (i.e., "graduated") pipets of up to and including 10-ml. capacity are somewhat larger than those for the corresponding sizes of transfer pipets, namely, 0.01, 0.02, and 0.03 ml. for the 2- , 5- , and 10-ml. sizes, respectively.

Transfer and measuring pipets calibrated "to deliver" should be drained in a vertical position and then touched against the wall of the receiving vessel to drain the tips. Volume readings on burets should be estimated to the nearest 0.01 ml. for 25- and 50-ml. burets, and to the nearest 0.005 ml. for 5- and 10-ml. burets. Pipets calibrated "to contain" may be called for in special cases, generally for measuring viscous fluids. In such cases, the pipet should be washed clean, after draining, and the washings added to the measured portion.

WATER DETERMINATION

Karl Fischer Titrimetric Method

Principle. The Karl Fischer Titrimetric Method for the determination of water is based upon the fact that iodine and sulfur dioxide react only in the presence of water and that the reaction between water and a solution of sulfur dioxide and iodine in pyridine and methanol is stoichiometric.

Apparatus. The titration with the Fischer Reagent may be conducted in any apparatus which provides for rigid exclusion of atmospheric moisture and determination of the end-point. The end-point of colorless solutions may often be detected visually by the change from a light brownish or canary yellow to an amber color. When the end-point is obscure, or for colored solutions, the electrometric method for its determination is indicated. The necessary apparatus, in this case, consists of a simple electrical circuit capable of passing 5 to 10 microamperes of direct current between a pair of platinum electrodes which are immersed in the solution to be titrated. At the beginning and throughout the titration, the flow of current does not persist beyond a few microamperes, but, at the end-point, a slight excess of the reagent increases the flow to between 50 and 150 microamperes for at least 30 seconds. Commercially available apparatus usually comprises a closed system in which the air is kept dry with suitable desiccants and usually consists of one or two automatic burets and a tightly closed titration vessel equipped with the two electrodes and a magnetic stirrer.

Preparation of the Fischer Reagent. To a mixture of 670 ml. of methanol and 170 ml. of pyridine contained in a flask, add 125 grams of iodine, immediately stopper the flask, and cool. Pass dry sulfur dioxide

* See "Testing of Glass Volumetric Apparatus," N.B.S. Circ. 602, April 1, 1959.

through 100 ml. of pyridine contained in a 250-ml. graduate and cooled in an ice bath, until the volume of the solution attains 200 ml. Slowly add this solution, with shaking, to the cooled iodine mixture, stopper immediately, and shake well until the iodine is dissolved. Transfer the combined solution to the apparatus, preferably an automatic buret protected from moisture with desiccants such as phosphorus pentoxide, anhydrous calcium chloride, or silica gel, and allow to stand for 24 hours or overnight before standardizing. Each ml. of this reagent when freshly prepared is equivalent to approximately 5 mg. of water. Since this solution deteriorates continuously, it should be standardized within 1 hour before use, or daily if in continuous use. Protect from light while in use and store bulk solutions in glass-stoppered containers and under refrigeration.

A stabilized Karl Fischer Reagent solution is commercially available which can be used satisfactorily instead of one prepared as directed herein.

Standardization of the Fischer Reagent (*Primary Method*). Place about 36 ml. of methanol in the titration vessel and titrate with the *Fischer Reagent* to the characteristic end-point. Quickly add about 200 mg. of sodium tartrate ($Na_2C_4H_4O_6 . 2H_2O$), accurately weighed, and again titrate, stirring vigorously, to the end-point. The water equivalence factor, F, in mg. of water per ml. of reagent, is represented by the formula 0.1566 W/V, in which W is the weight, in mg., of the sodium tartrate, and V is the volume, in ml., of the reagent consumed in the second titration.

Standardization of the Fischer Reagent (*Secondary Method*). After the water equivalence factor, F, has been determined in the primary standardization, the reagent may be standardized for each day's use against a water-methanol solution standardized as follows: Transfer quantitatively 2 ml. of water to a thoroughly dry 1000-ml. volumetric flask, dilute to volume with methanol, tightly stopper, and mix. Transfer quantitatively 25 ml. of the water-methanol solution to the titration vessel and titrate with the *Fischer Reagent* to the end-point. The water content of the water-methanol solution, in mg. per ml., is obtained by the formula $V'F/25$, in which V' is the volume of the *Fischer Reagent* required, and F is the water equivalence factor of the reagent as determined against sodium tartrate in the *Primary Standardization*.

Procedure. Unless otherwise directed, place about 25 ml. of methanol in the titration vessel and titrate with the *Fischer Reagent* to the end-point, disregarding the volume consumed. Quickly transfer to the titration vessel an accurately weighed or measured quantity of the sample, preferably containing 10 to 50 mg. of water, stir vigorously, and again titrate to the end-point. The water content of the sample, in mg., is obtained by multiplying the volume of the reagent used in titrating the sample by the equivalence factor, F, of the reagent.

Toluene Distillation Method

Principle. This method determines water by distillation of a sample with an immiscible solvent, usually toluene.

Apparatus. Glass distillation apparatus provided with 24/40 ground-glass connections, constructed and assembled in accordance with the specifications shown in the accompanying diagram, should be used. The components consist of a 500-ml. short neck, round bottom flask connected by means of a trap to a water-cooled condenser. The lower tip of the condenser should be about 7 mm. above the surface of

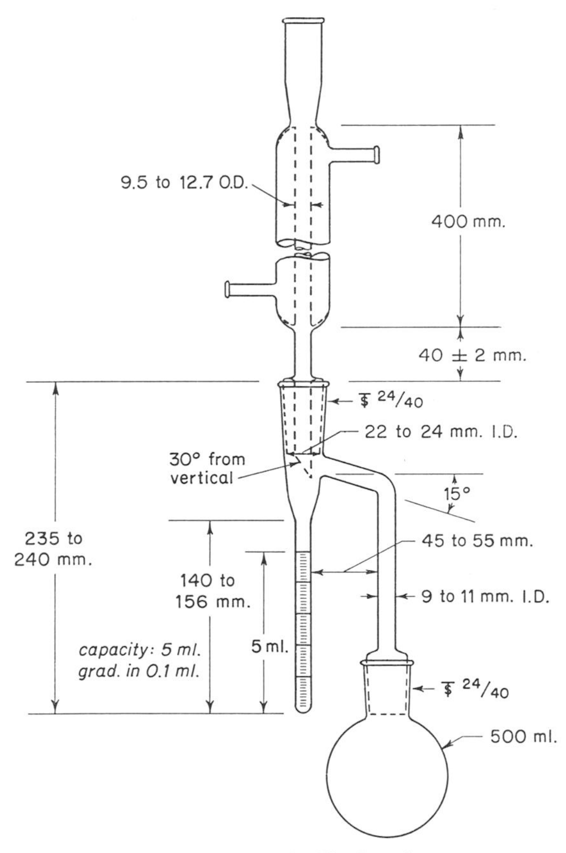

Fig. 24—Moisture Distillation Apparatus

the liquid in the trap after distillation conditions have been established (see *Procedure*).

The trap should be constructed of well-annealed glass, the receiving end of which is graduated to contain 5 ml. and subdivided into 0.1 ml. divisions with each 1-ml. line numbered from 5 ml. beginning at the top. Calibrate the receiver by adding 1 ml. of water, accurately measured, to 100 ml. of toluene contained in the distillation flask. Conduct the distillation and calculate the volume of water obtained as directed in the *Procedure*. To the cooled apparatus add another ml. of water and repeat the distillation. Continue in this manner until five 1-ml. portions of water have been added. The error at any indicated capacity should not exceed 0.05 ml.

The source of heat is either an oil bath or an electric heater provided with a suitable means of temperature control. The distillation may be better controlled by insulating the tube leading from the flask to the receiver with asbestos. It is also advantageous to protect the flask from drafts.

Clean the entire apparatus with potassium dichromate-sulfuric acid cleaning solution, rinse thoroughly, and dry completely before using.

Procedure. Place in the previously cleaned and dried flask a quantity of the substance, weighed accurately to the nearest 0.01 gram, which is expected to yield from 1.5 to 4 ml. of water. If the substance is of a paste-like consistency, weigh it in a boat of metal foil that will pass through the neck of the flask If the substance is likely to cause bumping, take suitable precautions to prevent it. Transfer about 200 ml. of A.C.S. reagent grade toluene into the flask and swirl to mix it with the sample. Assemble the apparatus, fill the receiver with toluene by pouring it through the condenser until it begins to overflow into the flask, and insert a loose cotton plug in the top of the condenser. Heat the flask so that the distillation rate will be about 100 drops per minute. When the greater part of the water has distilled, increase the distillation rate to about 200 drops per minute and continue until the water level increases no more than 0.1 ml. in 30 minutes, or until the distillation has proceeded for 2 hours, whichever is the first to occur. Discontinue the heating, dislodge any drops of water which may be adhering to the inside of the condenser tube or receiver with a copper or nichrome wire spiral, and wash down with about 5 ml. of toluene. Disconnect the receiver, immerse it in water at 25° for at least 15 minutes or until the toluene layer is clear, and then read the volume of water. Conduct a blank determination using the same volume of toluene as used when distilling the sample mixture, and make any necessary correction (see page 2).

WEIGHTS AND BALANCES

Codex tests and assays are designed for use with three types of analytical balances, known as macro- , semimicro- , and micro- .

Tolerances. The analytical weights meet the tolerances of the National Bureau of Standards for Class S if used without corrections, or meet the use tolerances for Class S-1 if used with corrections. Where quantities of 25 mg. or less are to be "weighed accurately," any applicable corrections for weights should be used.

Use. Where substances are to be "accurately weighed" in an assay or a test, the weighing is to be performed in such manner as to limit the error to 0.1 percent or less. For example, a quantity of 50 mg. is to be weighed to the nearest 0.05 mg.; a quantity of 0.1 gram is to be weighed to the nearest 0.1 mg.; and a quantity of 10 grams is to be weighed to the nearest 10 mg.

Solutions and Indicators

COLORIMETRIC SOLUTIONS (C.S.)

Colorimetric solutions are used in the preparation of colorimetric standards for certain chemicals, and for the carbonization tests with sulfuric acid that are specified in several monographs. Directions for the preparation of the primary colorimetric solutions and *Matching Fluids* are given under the test for *Readily Carbonizable Substances*, page 943. Store the solutions in suitably resistant, tight containers.

Comparison of colors as directed in the Codex tests preferably is made in matched color-comparison tubes or in a suitable colorimeter under conditions that insure that the colorimetric reference solution and that of the specimen under test are treated alike in all respects (see *Visual Color and Turbidity Comparisons*, page 959).

STANDARD BUFFER SOLUTIONS

Reagent Solutions. Previously dry the crystalline reagents, except the boric acid, at 110° to 120°, and use water that has been previously boiled and cooled in preparing the solutions. Store the prepared reagent solutions in chemically resistant glass or polyethylene bottles, and use within 3 months. Discard if molding is evident.

Potassium Chloride, 0.2 M. Dissolve 14.911 grams of potassium chloride, KCl, in sufficient water to make 1000.0 ml.

Potassium Biphthalate, 0.2 M. Dissolve 40.846 grams of potassium biphthalate, $KHC_6H_4(COO)_2$, in sufficient water to make 1000.0 ml.

Potassium Phosphate, Monobasic, 0.2 M. Dissolve 27.218 grams of monobasic potassium phosphate, KH_2PO_4, in sufficient water to make 1000.0 ml.

Boric Acid-Potassium Chloride, 0.2 M. Dissolve 12.366 grams of boric acid, H_3BO_3, and 14.911 grams of potassium chloride, KCl, in sufficient water to make 1000.0 ml.

Hydrochloric Acid, 0.2 M, and *Sodium Hydroxide, 0.2 M.* Prepare and standardize as directed under *Volumetric Solutions*, page 996.

Procedure. To prepare 200 ml. of a standard buffer solution having a pH within the range 1.2 to 10.0, place 50.0 ml. of the appropriate 0.2 *M* salt solution, prepared above, in a 200-ml. volumetric flask, add the volume of 0.2 *M* hydrochloric acid or a sodium hydroxide specified for the desired pH in the accompanying table, dilute to volume with water, and mix.

Composition of Standard Buffer Solutions*

Hydrochloric Acid Buffer		Acid Phthalate Buffer		Neutralized Phthalate Buffer	
To 50.0 ml. of 0.2 *M* KCl add the ml. of HCl specified		To 50.0 ml. of 0.2 *M* $KHC_6H_4(COO)_2$ add the ml. of HCl specified		To 50.0 ml. of 0.2 *M* $KHC_6H_4(COO)_2$ add the ml. of NaOH specified	
pH	0.2 *M* HCl, ml.	pH	0.2 *M* HCl, ml.	pH	0.2 *M* NaOH, ml.
1.2	85.0	2.2	49.5	4.2	3.0
1.3	67.2	2.4	42.2	4.4	6.6
1.4	53.2	2.6	35.4	4.6	11.1
1.5	41.4	2.8	28.9	4.8	16.5
1.6	32.4	3.0	22.3	5.0	22.6
1.7	26.0	3.2	15.7	5.2	28.8
1.8	20.4	3.4	10.4	5.4	34.1
1.9	16.2	3.6	6.3	5.6	38.8
2.0	13.0	3.8	2.9	5.8	42.3
2.1	10.2	4.0	0.1	—	—
2.2	7.8	—	—	—	—

Phosphate Buffer		Alkaline Borate Buffer	
To 50.0 ml. of 0.2 *M* KH_2PO_4 add the ml. of NaOH specified		To 50.0 ml. of 0.2 *M* H_3BO_3-KCl add the ml. of NaOH specified	
pH	0.2 *M* NaOH, ml.	pH	0.2 *M* NaOH, ml.
5.8	3.6	8.0	3.9
6.0	5.6	8.2	6.0
6.2	8.1	8.4	8.6
6.4	11.6	8.6	11.8
6.6	16.4	8.8	15.8
6.8	22.4	9.0	20.8
7.0	29.1	9.2	26.4
7.2	34.7	9.4	32.1
7.4	39.1	9.6	36.9
7.6	42.4	9.8	40.6
7.8	44.5	10.0	43.7
8.0	46.1	—	—

* Dilute all final solutions to 200.0 ml. (see *Procedure*). The standard pH values given in this table are considered to be reproducible to within ±0.02 of the pH unit specified at 25°.

STANDARD SOLUTIONS

FOR THE PREPARATION OF CONTROLS AND STANDARDS

The following solutions are used in tests for impurities which require the comparison of the color or turbidity produced in a solution of the test substance with that produced by a known amount of the impurity in a control. Directions for the preparation of other standard solutions are given in the monographs or under the general tests in which they are required.

Ammonium Standard Solution (10 mcg. NH_4 in 1 ml.). Dissolve 296.0 mg. of ammonium chloride, NH_4Cl, in sufficient water to make 100.0 ml., and mix. Transfer 10.0 ml. of this solution into a 1000-ml. volumetric flask, dilute to volume with water, and mix.

Barium Standard Solution (100 mcg. Ba in 1 ml.). Dissolve 177.9 mg. of barium chloride, $BaCl_2 . 2H_2O$, in water in a 1000-ml. volumetric flask, dilute to volume with water, and mix.

Iron Standard Solution (10 mcg. Fe in 1 ml.). Dissolve 702.2 mg. of ferrous ammonium sulfate, $Fe(NH_4)_2(SO_4)_2 . 6H_2O$, in 10 ml. of diluted sulfuric acid T.S. in a 100-ml. volumetric flask, dilute to volume with water, and mix. Transfer 10.0 ml. of this solution into a 1000-ml. volumetric flask, add 10 ml. of diluted sulfuric acid T.S., dilute to volume with water, and mix.

Magnesium Standard Solution (50 mcg. Mg in 1 ml.). Dissolve 50.0 mg. of magnesium metal, Mg, in 1 ml. of hydrochloric acid in a 1000-ml. volumetric flask, dilute to volume with water, and mix.

Phosphate Standard Solution (10 mcg. PO_4 in 1 ml.). Dissolve 143.3 mg. of monobasic potassium phosphate, KH_2PO_4, in water in a 100-ml. volumetric flask, dilute to volume with water, and mix. Transfer 10.0 ml. of this solution into a 1000-ml. volumetric flask, dilute to volume with water, and mix.

TEST SOLUTIONS (T.S.)

Certain of the following test solutions are intended for use as acid-base indicators in volumetric analyses. Such solutions should be so adjusted that when 0.15 ml. of the indicator solution is added to 25 ml. of carbon dioxide-free water, 0.25 ml. of 0.02 *N* acid or alkali, respectively, will produce the characteristic color change. Similar solutions are intended for use in pH measurement. Where no special directions for their preparation are given, the same solution is suitable for both purposes.

In general, the directive to prepare a solution "fresh" indicates that the solution is of limited stability and must be prepared on the day of use.

Acetic Acid T.S., Diluted. A solution containing about 6 percent (w/v) of CH_3COOH. Prepare by diluting 60.0 ml. of glacial acetic acid, or 166.6 ml. of 36 percent acetic acid (6 *N*), with sufficient water to make 1000 ml.

Alcohol, 70 Percent (at 15.56°). A 38.6:15 mixture (v/v) of 95 percent alcohol and water, having a specific gravity of 0.884 at 25°. To prepare 100 ml., dilute 73.7 ml. of alcohol to 100 ml. with water at 25°.

Alcohol, 80 Percent (at 15.56°). A 45.5:9.5 mixture (v/v) of

95 percent alcohol and water, having a specific gravity of 0.857 at 25°. To prepare 100 ml., dilute 84.3 ml. of alcohol to 100 ml. with water at 25°.

Alcohol, 90 Percent (at 15.56°). A 51:3 mixture (v/v) of 95 percent alcohol and water, having a specific gravity of 0.827 at 25°. To prepare 100 ml., dilute 94.8 ml. of alcohol to 100 ml. with water at 25°.

Alcohol, Aldehyde-free. Dissolve 2.5 grams of lead acetate in 5 ml. of water, add the solution to 1000 ml. of alcohol contained in a glass-stoppered bottle, and mix. Dissolve 5 grams of potassium hydroxide in 25 ml. of warm alcohol, cool, and add slowly, without stirring, to the alcoholic solution of lead acetate. Allow to stand for 1 hour, then shake the mixture vigorously, allow to stand overnight, decant the clear liquid, and recover the alcohol by distillation. Ethyl Alcohol F.C.C., Alcohol U.S.P., or U.S.S.D. #3A or #30 may be used. If the titration of a 250-ml. sample of the alcohol by *Hydroxylamine Hydrochloride Solution* (see page 894) does not exceed 0.25 ml. of 0.5 *N* alcoholic potassium hydroxide, the above treatment may be omitted.

Alcoholic Potassium Hydroxide T.S. *See Potassium Hydroxide T.S., Alcoholic.*

Alkaline Cupric Tartrate T.S. (*Fehling's Solution*). See *Cupric Tartrate T.S., Alkaline.*

Alkaline Mercuric-Potassium Iodide T.S. (*Nessler's Reagent*). See *Mercuric-Potassium Iodide T.S., Alkaline.*

Ammonia-Ammonium Chloride Buffer T.S. (approx. pH 10). Dissolve 67.5 grams of ammonium chloride, NH_4Cl, in water, add 570 ml. of ammonium hydroxide (28 percent), and dilute with water to 1000 ml.

Ammonia T.S. A solution containing between 9.5 and 10.5 percent of NH_3. Prepare by diluting 400 ml. of ammonium hydroxide (28 percent) with sufficient water to make 1000 ml.

Ammonia T.S., Stronger (*Ammonium Hydroxide, 28 percent, Stronger Ammonia Water*). A practically saturated solution of ammonia in water, containing between 28 and 30 percent of NH_3.

Ammoniacal Silver Nitrate T.S. Add ammonia T.S., dropwise, to a 1 in 20 solution of silver nitrate until the precipitate that first forms is almost, but not entirely, dissolved. Filter the solution, and store in a dark bottle.

Note: Ammoniacal silver nitrate T.S. forms explosive compounds on standing. Do not store this solution, but prepare a fresh quantity for each series of determinations. Neutralize the excess reagent and rinse all glassware with hydrochloric acid immediately after completing a test.

Ammonium Acetate T.S. Dissolve 10 grams of ammonium acetate, $NH_4C_2H_3O_2$, in sufficient water to make 100 ml.

Ammonium Carbonate T.S. Dissolve 20 grams of ammonium carbonate and 20 ml. of ammonia T.S. in sufficient water to make 100 ml.

Ammonium Chloride T.S. Dissolve 10.5 grams of ammonium chloride, NH_4Cl, in sufficient water to make 100 ml.

Ammonium Molybdate T.S. Dissolve 6.5 grams of finely powdered molybdic acid (85 percent) in a mixture of 14 ml. of water and 14.5 ml. of stronger ammonia T.S. Cool the solution, and add it slowly, with stirring, to a well-cooled mixture of 32 ml. of nitric acid and 40 ml. of water. Allow to stand for 48 hours, and filter through asbestos. This solution deteriorates upon standing and is unsuitable for use if, upon the addition of 2 ml. of sodium phosphate T.S. to 5 ml. of the solution, an abundant yellow precipitate does not form at once or after slight warming. Store it in the dark. If a precipitate forms during storage, use only the clear, supernatant solution.

Ammonium Oxalate T.S. Dissolve 3.5 grams of ammonium oxalate, $(NH_4)_2C_2O_4 . H_2O$, in sufficient water to make 100 ml.

Ammonium Sulfanilate T.S. To 2.5 grams of sulfanilic acid add 15 ml. of water and 3 ml. of ammonia T.S. and mix. Add, with stirring, more ammonia T.S., if necessary, until the acid dissolves, adjust the pH of the solution to about 4.5 with diluted hydrochloric acid T.S., using bromocresol green T.S. as an outside indicator, and dilute to 25 ml.

Ammonium Sulfide, T.S. Saturate ammonia T.S. with hydrogen sulfide, H_2S, and add two-thirds of its volume of ammonia T.S. Residue upon ignition: not more than 0.05 percent. The solution is not rendered turbid either by magnesium sulfate T.S. or by calcium chloride T.S. (*carbonate*). This solution is unsuitable for use if an abundant precipitate of sulfur is present. Store it in small, well-filled, dark amber-colored bottles, in a cold, dark place.

Ammonium Thiocyanate T.S. Dissolve 8 grams of ammonium thiocyanate, NH_4SCN, in sufficient water to make 100 ml.

Barium Chloride T.S. Dissolve 12 grams of barium chloride, $BaCl_2 . 2H_2O$, in sufficient water to make 100 ml.

Barium Diphenylamine Sulfonate T.S. Dissolve 300 mg. of *p*-diphenylamine sulfonic acid barium salt in 100 ml. of water.

Benedict's Qualitative Reagent. See *Cupric Citrate T.S., Alkaline.*

Benzidine T.S. Dissolve 50 mg. of benzidine in 10 ml. of glacial acetic acid, dilute to 100 ml. with water, and mix.

Bismuth Nitrate T.S. Reflux 5 grams of bismuth nitrate, $Bi(NO_3)_3 . 5H_2O$, with 7.5 ml. of nitric acid and 10 ml. of water until dissolved, cool, filter, and dilute to 250 ml. with water.

Bromine T.S. (*Bromine Water*). A saturated solution of bromine, prepared by agitating 2 to 3 ml. of bromine, Br_2, with 100 ml. of cold water in a glass-stoppered bottle, the stopper of which should be lubricated with petrolatum. Store it in a cold place, protected from light.

Bromocresol Blue T.S. Use *Bromocresol Green T.S.*

Bromocresol Green T.S. Dissolve 50 mg. of bromocresol green in 100 ml. of alcohol, and filter if necessary. For pH determinations, dissolve 50 mg. in 1.4 ml. of 0.05 *N* sodium hydroxide, and dilute with carbon dioxide-free water to 100 ml.

Bromocresol Purple T.S. Dissolve 250 mg. of bromocresol purple in 20 ml. of 0.05 *N* sodium hydroxide, and dilute with water to 250 ml.

Bromophenol Blue T.S. Dissolve 100 mg. of bromophenol blue in 100 ml. of dilute alcohol (1 in 2), and filter if necessary. For pH determinations, dissolve 100 mg. in 3.0 ml. of 0.05 *N* sodium hydroxide, and dilute with carbon dioxide-free water to 200 ml.

Bromothymol Blue T.S. Dissolve 100 mg. of bromothymol blue in 100 ml. of dilute alcohol (1 in 2), and filter if necessary. For pH determinations, dissolve 100 mg. in 3.2 ml. of 0.05 *N* sodium hydroxide, and dilute with carbon dioxide-free water to 200 ml.

Calcium Chloride T.S. Dissolve 7.5 grams of calcium chloride, $CaCl_2 . 2H_2O$, in sufficient water to make 100 ml.

Calcium Hydroxide T.S. A solution containing approximately 140 mg. of $Ca(OH)_2$ in each 100 ml. To prepare, add 3 grams of calcium hydroxide, $Ca(OH)_2$, to 1000 ml. of water, and agitate the mixture vigorously and repeatedly during 1 hour. Allow the excess calcium hydroxide to settle, and decant or draw off the clear, supernatant liquid.

Ceric Ammonium Nitrate T.S. Dissolve 6.25 grams of ceric ammonium nitrate, $(NH_4)_2Ce(NO_3)_6$, in 100 ml. of 0.25 *N* nitric acid. Prepare the solution fresh every third day.

Chlorine T.S. (*Chlorine Water*). A saturated solution of chlorine in water. Place the solution in small, completely filled, light-resistant containers. Chlorine T.S., even when kept from light and air, is apt to deteriorate. Store it in a cold, dark place. For full strength, prepare this solution fresh.

Chromotropic Acid T.S. Dissolve 50 mg. of chromotropic acid or its sodium salt in 100 ml. of 75 percent sulfuric acid (made by adding cautiously 75 ml. of 95–98 percent sulfuric acid to 33.3 ml. of water).

Cobaltous Chloride T.S. Dissolve 2 grams of cobaltous chloride, $CoCl_2.6H_2O$, in 1 ml. of hydrochloric acid and sufficient water to make 100 ml.

Cobalt-Uranyl Acetate T.S. Dissolve, with warming, 40 grams of uranyl acetate, $UO_2(C_2H_3O_2)_2.2H_2O$, in a mixture of 30 grams of glacial acetic acid and sufficient water to make 500 ml. Similarly, prepare a solution containing 200 grams of cobaltous acetate, $Co(C_2H_3O_2)_2 . 4H_2O$, in a mixture of 30 grams of glacial acetic acid and sufficient water to make 500 ml. Mix the two solutions while still warm, and cool to 20°. Maintain the temperature at 20° for about 2 hours to separate the excess salts from solution, and then filter through a dry filter.

Congo Red T.S. Dissolve 500 mg. of congo red in a mixture of 10 ml. of alcohol and 90 ml. of water.

Cresol Red T.S. Triturate 100 mg. of cresol red in a mortar with 26.2 ml. of 0.01 *N* sodium hydroxide until solution is complete, then dilute the solution with water to 250 ml.

Cresol Red-Thymol Blue T.S. Add 15 ml. of thymol blue T.S. to 5 ml. of cresol red T.S., and mix.

Crystal Violet T.S. Dissolve 100 mg. of crystal violet in 10 ml. of glacial acetic acid.

Cupric Citrate T.S., Alkaline (*Benedict's Qualitative Reagent*). With the aid of heat, dissolve 173 grams of sodium citrate, $C_6H_5Na_3O_7 \cdot 2H_2O$, and 117 grams of sodium carbonate, $Na_2CO_3 \cdot H_2O$, in about 100 ml. of water, and filter through paper, if necessary. In a separate container dissolve 17.3 grams of cupric sulfate, $CuSO_4 \cdot 5H_2O$, in about 100 ml. of water, and slowly add this solution, with constant stirring, to the first solution. Cool the mixture, dilute to 1000 ml., and mix.

Cupric Sulfate T.S. Dissolve 12.5 grams of cupric sulfate, $CuSO_4 \cdot 5H_2O$, in sufficient water to make 100 ml. and mix.

Cupric Tartrate T.S., Alkaline (*Fehling's Solution*). *The Copper Solution* (*A*). Dissolve 34.66 grams of carefully selected, small crystals of cupric sulfate, $CuSO_4 \cdot 5H_2O$, showing no trace of efflorescence or of adhering moisture, in sufficient water to make 500 ml. Store this solution in small, tight containers. *The Alkaline Tartrate Solution* (*B*). Dissolve 173 grams of crystallized potassium sodium tartrate, $KNaC_4H_4O_6 \cdot 4H_2O$, and 50 grams of sodium hydroxide, NaOH, in sufficient water to make 500 ml. Store this solution in small, alkali-resistant containers. For use, mix exactly equal volumes of solutions *A* and *B* at the time required.

Cyanogen Bromide T.S. Dissolve 5 grams of cyanogen bromide in water to make 50 ml. *Caution: Prepare this solution under a hood,* as cyanogen bromide volatilizes at room temperature and the vapor is highly irritating and *poisonous.*

Denigès' Reagent. See *Mercuric Sulfate T.S.*

2,7-Dihydroxynaphthalene T.S. Dissolve 100 mg. of 2,7-dihydroxynaphthalene in 1000 ml. of sulfuric acid and allow the solution to stand until the initial yellow color disappears. If the solution is very dark, discard it and prepare a new solution from a different supply of sulfuric acid. This solution is stable for approximately one month if stored in a dark bottle.

Diphenylamine T.S. Dissolve 1 gram of diphenylamine in 100 ml. of sulfuric acid. The solution should be colorless.

α,α'-Dipyridyl T.S. Dissolve 100 mg. of α,α'-dipyridyl, $C_{10}H_8N_2$, in 50 ml. of absolute alcohol.

Dithizone T.S. Dissolve 25.6 mg. of dithizone in 100 ml. of alcohol.

Eosin Y T.S. (adsorption indicator). Dissolve 50 mg. of eosin Y in 10 ml. of water.

Eriochrome Black T.S. Dissolve 200 mg. of eriochrome black T and 2 grams of hydroxylamine hydrochloride, $NH_2OH.HCl$, in sufficient methanol to make 50 ml., and filter. Store the solution in a light-resistant container and use within 2 weeks.

***p*-Ethoxychrysoidin T.S.** Dissolve 50 mg. of *p*-ethoxychrysoidin monohydrochloride in a mixture of 25 ml. of water and 25 ml. of alcohol, add 3 drops of hydrochloric acid, stir vigorously, and filter if necessary to obtain a clear solution.

Fehling's Solution. See *Cupric Tartrate T.S., Alkaline.*

Ferric Ammonium Sulfate T.S. Dissolve 8 grams of ferric ammonium sulfate, $FeNH_4(SO_4)_2.12H_2O$, in sufficient water to make 100 ml.

Ferric Chloride T.S. Dissolve 9 grams of ferric chloride, $FeCl_3.6H_2O$, in sufficient water to make 100 ml.

Ferric Chloride T.S., Alcoholic. Dissolve 100 mg. of ferric chloride, $FeCl_3.6H_2O$, in 50 ml. of absolute alcohol. Prepare this solution fresh.

Ferrous Sulfate T.S. Dissolve 8 grams of clear crystals of ferrous sulfate, $FeSO_4.7H_2O$, in about 100 ml. of recently boiled and thoroughly cooled water. Prepare this solution fresh.

Formaldehyde T.S. A solution containing approximately 37.0 percent (w/v) of HCHO. It may contain methanol to prevent polymerization.

Hydrochloric Acid T.S., Diluted. A solution containing 10 percent (w/v) of HCl. Prepare by diluting 236 ml. of hydrochloric acid (36 percent) with sufficient water to make 1000 ml.

Hydrogen Peroxide T.S. A solution containing between 2.5 and 3.5 grams of H_2O_2 in each 100 ml. It may contain suitable preservatives, totaling not more than 0.05 percent.

Hydrogen Sulfide T.S. A saturated solution of hydrogen sulfide made by passing H_2S into cold water. Store it in small, dark amber-colored bottles, filled nearly to the top. It is unsuitable unless it possesses a strong odor of H_2S, and unless it produces at once a copious precipitate of sulfur when added to an equal volume of ferric chloride T.S. Store in a cold, dark place.

Hydroxylamine Hydrochloride T.S. Dissolve 3.5 grams of hydroxylamine hydrochloride, $NH_2OH.HCl$, in 95 ml. of 60 percent alcohol, and add 0.5 ml. of bromophenol blue solution (1 in 1000) and 0.5 *N* alcoholic potassium hydroxide until a greenish tint develops in the solution. Then add sufficient 60 percent alcohol to make 100 ml.

8-Hydroxyquinoline T.S. Dissolve 5 grams of 8-hydroxyquinoline (oxine) in sufficient alcohol to make 100 ml.

Indigo Carmine T.S. (*Sodium Indigotindisulfonate T.S.*). Dissolve a quantity of sodium indigotindisulfonate, equivalent to 180 mg. of

$C_{16}H_8N_2O_2(SO_3Na)_2$, in sufficient water to make 100 ml. Use within 60 days.

Iodine T.S. Dissolve 14 grams of iodine, I_2, in a solution of 36 grams of potassium iodide, KI, in 100 ml. of water, add 3 drops of hydrochloric acid, dilute with water to 1000 ml., and mix.

Lead Acetate T.S. Dissolve 9.5 grams of clear, transparent crystals of lead acetate, $Pb(C_2H_3O_2)_2 . 3H_2O$, in sufficient recently boiled water to make 100 ml. Store in well-stoppered bottles.

Lead Subacetate T.S. Triturate 14 grams of lead monoxide, PbO, to a smooth paste with 10 ml. of water, and transfer the mixture to a bottle, using an additional 10 ml. of water for rinsing. Dissolve 22 grams of lead acetate, $Pb(C_2H_3O_2)_2 . 3H_2O$, in 70 ml. of water, and add the solution to the lead oxide mixture. Shake it vigorously for 5 minutes, then set it aside, shaking it frequently during 7 days. Finally filter, and add enough recently boiled water through the filter to make 100 ml.

Lead Subacetate T.S., Diluted. Dilute 3.25 ml. of lead subacetate T.S. with sufficient water, recently boiled and cooled, to make 100 ml. Store in small, well-filled, tight containers.

Litmus T.S. Digest 25 grams of powdered litmus with three successive 100-ml. portions of boiling alcohol, continuing each extraction for about 1 hour. Filter, wash with alcohol, and discard the alcohol filtrate. Macerate the residue with about 25 ml. of cold water for 4 hours, filter, and discard the filtrate. Finally digest the residue with 125 ml. of boiling water for 1 hour, cool, and filter.

Magnesia Mixture T.S. Dissolve 5.5 grams of magnesium chloride, $MgCl_2.6H_2O$, and 7 grams of ammonium chloride, NH_4Cl, in 65 ml. of water, add 35 ml. of ammonia T.S., set the mixture aside for a few days in a well-stoppered bottle, and filter. If the solution is not perfectly clear, filter it before using.

Magnesium Sulfate T.S. Dissolve 12 grams of crystals of magnesium sulfate, $MgSO_4 . 7H_2O$, selected for freedom from efflorescence, in water to make 100 ml.

Malachite Green T.S. Dissolve 1 gram of malachite green oxalate in 100 ml. of glacial acetic acid.

Mayer's Reagent. See *Mercuric-Potassium Iodide T.S.*

Mercuric Acetate T.S. Dissolve 6 grams of mercuric acetate, $Hg(C_2H_3O_2)_2$, in sufficient glacial acetic acid to make 100 ml. Store in tight containers, protected from direct sunlight.

Mercuric Chloride T.S. Dissolve 6.5 grams of mercuric chloride, $HgCl_2$, in water to make 100 ml.

Mercuric-Potassium Iodide T.S. (*Mayer's Reagent*). Dissolve 1.358 grams of mercuric chloride, $HgCl_2$, in 60 ml. of water. Dissolve 5 grams of potassium iodide, KI, in 10 ml. of water. Mix the two solutions, and add water to make 100 ml.

Mercuric-Potassium Iodide T.S., Alkaline (*Nessler's Reagent*). Dissolve 10 grams of potassium iodide, KI, in 10 ml. of water, and add slowly, with stirring, a saturated solution of mercuric chloride until a slight red precipitate remains undissolved. To this mixture add an ice-cold solution of 30 grams of potassium hydroxide, KOH, in 60 ml. of water, then add 1 ml. more of the saturated solution of mercuric chloride. Dilute with water to 200 ml. Allow the precipitate to settle, and draw off the clear liquid. A 2-ml. portion of this reagent, when added to 100 ml. of a 1 in 300,000 solution of ammonium chloride in ammonia-free water, produces at once a yellowish brown color.

Mercuric Sulfate T.S. (*Denigès' Reagent*). Mix 5 grams of yellow mercuric oxide, HgO, with 40 ml. of water, and while stirring slowly add 20 ml. of sulfuric acid, then add another 40 ml. of water, and stir until completely dissolved.

p-**Methylaminophenol Sulfate T.S.** Dissolve 2 grams of *p*-methylaminophenol sulfate, $(HO.C_6H_4.NHCH_3)_2.H_2SO_4$, in 100 ml. of water. To 10 ml. of this solution add 90 ml. of water and 20 grams of sodium bisulfite. Confirm the suitability of this solution by the following test: Add 1 ml. of the solution to each of four tubes containing 25 ml. of 0.5 *N* sulfuric acid and 1 ml. of ammonium molybdate T.S. Add 5 mcg. of phosphate (PO_4) to one tube, 10 mcg. to a second, and 20 mcg. to a third, using 0.5, 1.0, and 2.0 ml., respectively, of Phosphate Standard Solution, and allow to stand for 2 hours. The solutions in the three tubes should show readily perceptible differences in blue color corresponding to the relative amounts of phosphate added, and the one to which 5 mcg. of phosphate was added should be perceptibly bluer than the blank.

Methylene Blue T.S. Dissolve 125 mg. of methylene blue in 100 ml. of alcohol, and dilute with alcohol to 250 ml.

Methyl Orange T.S. Dissolve 100 mg. of methyl orange in 100 ml. of water, and filter if necessary.

Methyl Red T.S. Dissolve 100 mg. of methyl red in 100 ml. of alcohol, and filter if necessary. For pH determinations, dissolve 100 mg. in 7.4 ml. of 0.05 *N* sodium hydroxide, and dilute with carbon dioxide-free water to 200 ml.

Methyl Red-Methylene Blue T.S. Add 10 ml. of methyl red T.S. to 10 ml. of methylene blue T.S., and mix.

Methylrosaniline Chloride T.S. See *Crystal Violet T.S.*

Methyl Violet T.S. See *Crystal Violet T.S.*

Millon's Reagent. To 2 ml. of mercury in an Erlenmeyer flask add 20 ml. of nitric acid. Shake the flask under a hood to break up the mercury into small globules. After about 10 minutes add 35 ml. of water, and, if a precipitate or crystals appear, add sufficient dilute nitric acid (1 in 5, prepared from nitric acid from which the oxides have been removed by blowing air through it until it is colorless) to dissolve the separated solid. Add sodium hydroxide solution (1 in 10), dropwise, with thorough mixing, until the curdy precipitate that forms after the addition of each drop no longer redissolves but is dispersed to form a suspension. Add 5 ml. more of the dilute nitric acid, and mix well. Prepare this solution fresh.

Naphthol Green T.S. Dissolve 500 mg. of naphthol green B in water to make 1000 ml.

Nessler's Reagent. See *Mercuric-Potassium Iodide T.S., Alkaline.*

Neutral Red T.S. Dissolve 100 mg. of neutral red in 100 ml. of 50 percent alcohol.

Ninhydrin T.S. See *Triketohydrindene Hydrate T.S.*

Nitric Acid T.S., Diluted. A solution containing about 10 percent (w/v) of HNO_3. Prepare by diluting 105 ml. of nitric acid (70 percent) with water to make 1000 ml.

Orthophenanthroline T.S. Dissolve 150 mg. of orthophenanthroline, $C_{12}H_8N_2.H_2O$, in 10 ml. of a solution of ferrous sulfate, prepared by dissolving 1.48 grams of clear crystals of ferrous sulfate, $FeSO_4.7H_2O$, in 100 ml. of water. The ferrous sulfate solution must be prepared immediately before dissolving the orthophenanthroline. Store the solution in well-closed containers.

Oxalic Acid T.S. Dissolve 6.3 grams of oxalic acid, $H_2C_2O_4.2H_2O$, in water to make 100 ml.

Phenol Red T.S. (*Phenolsulfonphthalein T.S.*). Dissolve 100 mg. of phenolsulfonphthalein in 100 ml. of alcohol, and filter if necessary. For pH determinations, dissolve 100 mg. in 5.7 ml. of 0.05 *N* sodium hydroxide, and dilute with carbon dioxide-free water to 200 ml.

Phenolphthalein T.S. Dissolve 1 gram of phenolphthalein in 100 ml. of alcohol.

Phenolsulfonphthalein T.S. See *Phenol Red T.S.*

***p*-Phenylphenol T.S.** On the day of use, dissolve 750 mg. of *p*-phenylphenol in 50 ml. of sodium hydroxide T.S.

Phosphotungstic Acid T.S. Dissolve 1 gram of phosphotungstic acid, (approx. $24WO_3.2H_3PO_4.48H_2O$) in water to make 100 ml.

Picric Acid T.S. See *Trinitrophenol T.S.*

Potassium Acetate T.S. Dissolve 10 grams of potassium acetate, $KC_2H_3O_2$, in water to make 100 ml.

Potassium Chromate T.S. Dissolve 10 grams of potassium chromate, K_2CrO_4, in water to make 100 ml.

Potassium Dichromate T.S. Dissolve 7.5 grams of potassium dichromate, $K_2Cr_2O_7$, in water to make 100 ml.

Potassium Ferricyanide T.S. Dissolve 1 gram of potassium ferricyanide, $K_3Fe(CN)_6$, in 10 ml. of water. Prepare this solution fresh.

Potassium Ferrocyanide T.S. Dissolve 1 gram of potassium ferrocyanide, $K_4Fe(CN)_6.3H_2O$, in 10 ml. of water. Prepare this solution fresh.

Potassium Hydroxide T.S. Dissolve 6.5 grams of potassium hydroxide, KOH, in water to make 100 ml.

Potassium Hydroxide T.S., Alcoholic. Use 0.5 *N Alcoholic Potassium Hydroxide*, page 1000.

Potassium Iodide T.S. Dissolve 16.5 grams of potassium iodide, KI, in water to make 100 ml. Store in light-resistant containers.

Potassium Permanganate T.S. Use 0.1 *N Potassium Permanganate*, page 1000.

Potassium Sulfate T.S. Dissolve 1 gram of potassium sulfate, K_2SO_4, in sufficient water to make 100 ml.

Quinaldine Red T.S. Dissolve 100 mg. of quinaldine red in 100 ml. of glacial acetic acid.

Schiff's Reagent, Modified. Dissolve 200 mg. of rosaniline hydrochloride, $C_{20}H_{20}ClN_3$, in 120 ml. of hot water. Cool, add 2 grams of sodium bisulfite, $NaHSO_3$, followed by 2 ml. of hydrochloric acid, and dilute to 200 ml. with water. Store in a brown bottle at 15° or lower.

Silver Nitrate T.S. Use 0.1 *N Silver Nitrate*, page 1001.

Sodium Bisulfite T.S. Dissolve 10 grams of sodium bisulfite, $NaHSO_3$, in water to make 30 ml. Prepare this solution fresh.

Sodium Bitartrate T.S. Dissolve 1 gram of sodium bitartrate, $NaHC_4H_4O_6 . H_2O$, in water to make 10 ml. Prepare this solution fresh.

Sodium Borate T.S. Dissolve 2 grams of sodium borate, $Na_2B_4O_7 . 10H_2O$, in water to make 100 ml.

Sodium Carbonate T.S. Dissolve 10.6 grams of anhydrous sodium carbonate, Na_2CO_3, in water to make 100 ml.

Sodium Cobaltinitrite T.S. Dissolve 10 grams of sodium cobaltinitrite, $Na_3Co(NO_2)_6$, in water to make 50 ml., and filter if necessary.

Sodium Fluoride T.S. Dry about 500 mg. of sodium fluoride, NaF, at 200° for 4 hours. Weigh accurately 222 mg. of the dried sodium fluoride, and dissolve it in sufficient water to make exactly 100 ml. Transfer 10.0 ml. of this solution into a 1000-ml. volumetric flask, dilute to volume with water, and mix. Each ml. of this final solution corresponds to 10 mcg. of fluorine (F).

Sodium Hydroxide T.S. Dissolve 4.3 grams of sodium hydroxide, NaOH, in water to make 100 ml.

Sodium Indigotindisulfonate T.S. See *Indigo Carmine T.S.*

Sodium Nitroferricyanide T.S. Dissolve 1 gram of sodium nitroferricyanide, $Na_2Fe(NO)(CN)_5 . 2H_2O$, in water to make 20 ml. Prepare this solution fresh.

Sodium Phosphate T.S. Dissolve 12 grams of clear crystals of dibasic sodium phosphate, $Na_2HPO_4 . 7H_2O$, in water to make 100 ml.

Sodium Sulfide T.S. Dissolve 1 gram of sodium sulfide, $Na_2S . 9H_2O$, in water to make 10 ml. Prepare this solution fresh.

Sodium Thiosulfate T.S. Use 0.1 *N Sodium Thiosulfate*, page 1002.

Stannous Chloride T.S. Dissolve 40 grams of reagent grade stannous chloride dihydrate, $SnCl_2.2H_2O$, in 100 ml. of hydrochloric acid.

Starch T.S. Triturate 1 gram of arrowroot starch with 10 ml. of cold water, and pour slowly, with constant stirring, into 200 ml. of boiling water. Boil the mixture until a thin, translucent fluid is obtained. (Longer boiling than necessary renders the solution less sensitive.) Allow to settle, and use only the clear, supernatant liquid. Prepare this solution fresh.

Starch Iodide Paste T.S. Heat 100 ml. of water in a 250-ml. beaker to boiling, add a solution of 750 mg. of potassium iodide, KI, in 5 ml. of water, then add 2 grams of zinc chloride, $ZnCl_2$, dissolved in 10 ml. of water, and, while the solution is boiling, add with stirring a smooth suspension of 5 grams of potato starch in 30 ml. of cold water. Continue to boil for 2 minutes, then cool. Store in well-closed containers in a cool place. This mixture must show a definite blue streak when a glass rod dipped in a mixture of 1 ml. of 0.1 *M* sodium nitrite, 500 ml. of water, and 10 ml. of hydrochloric acid, is streaked on a smear of the paste.

Sulfanilic Acid T.S. Dissolve 800 mg. of sulfanilic acid, p-NH_2-$C_6H_4SO_3H.H_2O$, in 100 ml. of acetic acid. Store in tight containers.

Sulfuric Acid T.S. See page 942.

Sulfuric Acid T.S., Diluted. A solution containing 10 percent (w/v) of H_2SO_4. Prepare by cautiously adding 57 ml. of sulfuric acid (95–98 percent) or sulfuric acid T.S. to about 100 ml. of water, then cool to room temperature, and dilute with water to 1000 ml.

Tannic Acid T.S. Dissolve 1 gram of tannic acid (tannin) in 1 ml. of alcohol, and add water to make 10 ml. Prepare this solution fresh.

Thymol Blue T.S. Dissolve 100 mg. of thymol blue in 100 ml. of alcohol, and filter if necessary. For pH determinations, dissolve 100 mg. in 4.3 ml. of 0.05 *N* sodium hydroxide, and dilute with carbon dioxide-free water to 200 ml.

Thymolphthalein T.S. Dissolve 100 mg. of thymolphthalein in 100 ml. of alcohol, and filter if necessary.

Triketohydrindene Hydrate T.S. (*Ninhydrin T.S.*). Dissolve 200 mg. of triketohydrindene hydrate, $C_9H_4O_3.H_2O$, in water to make 100 ml. Prepare this solution fresh.

Trinitrophenol T.S. (*Picric Acid T.S.*). Dissolve the equivalent of 1 gram of anhydrous trinitrophenol in 100 ml. of hot water. Cool the solution, and filter if necessary.

Xylenol Orange T.S. Dissolve 100 mg. of xylenol orange in 100 ml. of alcohol.

VOLUMETRIC SOLUTIONS

Normal Solutions. A normal solution contains 1 gram equivalent weight of the solute per liter of solution. The normalities of solutions used in volumetric determinations are designated as 1 *N*, 0.1 *N*, 0.05 *N*, etc., in this Codex.

Molar Solutions. A molar solution contains 1 gram molecular weight of the solute per liter of solution. The molarities of such solutions are designated as 1 *M*, 0.1 *M*, 0.05 *M*, etc., in this Codex.

Preparation and Methods of Standardization

The details for the preparation and standardization of solutions used in several normalities are usually given only for the one most frequently required. Solutions of other normalities are prepared and standardized in the same general manner as described. Solutions of lower normalities may be prepared accurately by making an exact dilution of a stronger solution, but solutions prepared in this way should be restandardized before use.

Dilute solutions that are not stable, such as 0.01 *N* potassium permanganate and sodium thiosulfate, are preferably prepared by diluting exactly the higher normality with thoroughly boiled and cooled water on the same day they are to be used.

All volumetric solutions should be prepared, standardized, and used at the standard temperature of 25°, if practicable. When a titration must be carried out at a markedly different temperature, the volumetric solution should be standardized at that same temperature, or a suitable temperature correction should be made. Since the strength of a standard solution may change upon standing, the normality or molarity factor should be redetermined frequently.

Although the directions provide only one method of standardization, other methods of equal or greater accuracy may be used. For substances available as certified primary standards, or of comparable quality, the final standard solution may be prepared by weighing accurately a suitable quantity of the substance and dissolving it to produce a specific volume solution of known concentration. Hydrochloric and sulfuric acids may be standardized against a certified primary standard.

In volumetric assays described in this Codex, the number of milligrams of the test substance equivalent to 1 ml. of the primary volumetric solution is given. In general, these equivalents may be derived by simple calculation.

Ammonium Thiocyanate, 0.1 N (7.612 grams NH_4SCN per liter). Dissolve about 8 grams of ammonium thiocyanate, NH_4SCN, in 1000 ml. of water, and standardize by titrating the solution against 0.1 *N* silver nitrate as follows: Transfer about 30 ml. of 0.1 *N* silver nitrate, accurately measured, into a glass-stoppered flask. Dilute with 50 ml. of water, then add 2 ml. of ferric ammonium sulfate T.S. and

2 ml. of nitric acid, and titrate with the ammonium thiocyanate solution to the first appearance of a red-brown color. Calculate the normality, and, if desired, adjust the solution to exactly 0.1 *N*. If desired, 0.1 *N* ammonium thiocyanate may be replaced by 0.1 *N* potassium thiocyanate where the former is directed in various tests and assays.

Bromine, 0.1 N (7.990 grams Br per liter). Dissolve 3 grams of potassium bromate, $KBrO_3$, and 15 grams of potassium bromide, KBr, in sufficient water to make 1000 ml., and standardize the solution as follows: Transfer about 25 ml. of the solution, accurately measured, into a 500-ml. iodine flask, and dilute with 120 ml. of water. Add 5 ml. of hydrochloric acid, stopper the flask, and shake it gently. Then add 5 ml. of potassium iodide T.S., re-stopper, shake the mixture, allow it to stand for 5 minutes, and titrate the liberated iodine with 0.1 *N* sodium thiosulfate, adding starch T.S. near the end of the titration. Calculate the normality. Store this solution in dark amber-colored, glass-stoppered bottles.

Ceric Sulfate, 0.1 N [33.22 grams $Ce(SO_4)_2$ per liter]. Transfer 59 grams of ceric ammonium nitrate, $Ce(NO_3)_4 . 2NH_4NO_3 . 2H_2O$, to a beaker, add 31 ml. of sulfuric acid, mix, and cautiously add water, in 20-ml. portions, until solution is complete. Cover the beaker, let stand overnight, filter through a sintered-glass crucible of fine porosity, add water to make 1000 ml., and mix. Standardize the solution as follows: Weigh accurately 200 mg. of primary standard arsenic trioxide, As_2O_3, previously dried at 100° for 1 hour, and transfer to a 500-ml. Erlenmeyer flask. Wash down the inner walls of the flask with 25 ml. of sodium hydroxide solution (2 in 25), swirl to dissolve the sample, and when solution is complete add 100 ml. of water, and mix. Add 10 ml. of dilute sulfuric acid (1 in 3) and 2 drops each of orthophenanthroline T.S. and a solution of osmium tetroxide in 0.1 *N* sulfuric acid (1 in 400), and slowly titrate with the ceric sulfate solution until the pink color is changed to a very pale blue. Calculate the normality. Each 4.946 mg. of As_2O_3 is equivalent to 1 ml. of 0.1 *N* ceric sulfate.

Ceric Sulfate, 0.01 N for Tocopherol Assay [3.322 grams $Ce(SO_4)_2$ per liter]. Dissolve 4.2 grams of ceric sulfate, $Ce(SO_4)_2 . 4H_2O$, or 5.5 grams of the acid sulfate, $Ce(HSO_4)_4$, in about 500 ml. of water containing 28 ml. of sulfuric acid, and dilute to 1000 ml. Allow the solution to stand overnight, and filter. Standardize this solution daily as follows: Weigh accurately about 275 mg. of hydroquinone, $C_6H_6O_2$, dissolve it in sufficient 0.5 *N* alcoholic sulfuric acid to make 500.0 ml., and mix. To 25.0 ml. of this solution add 75 ml. of 0.5 *N* sulfuric acid, 20 ml. of water, and 2 drops of diphenylamine T.S. Titrate with the ceric sulfate solution at a rate of about 25 drops per 10 seconds until an end-point is reached which persists for 10 seconds. Perform a blank determination using 100 ml. of 0.5 *N* alcoholic sulfuric acid, 20 ml. of water, and 2 drops of diphenylamine T.S., and make any necessary correction. Calculate the normality of the ceric sulfate solution by the formula $0.05W/$

55.057V, in which W is the weight, in mg., of the hydroquinone sample taken, and V is the volume, in ml., of the ceric sulfate solution consumed in the titration.

Disodium Ethylenediaminetetraacetate, 0.05 M (16.81 grams $C_{10}H_{14}N_2Na_2O_8$ per liter). Dissolve 18.6 grams of disodium ethylenediaminetetraacetate, $C_{10}H_{14}N_2Na_2O_8 . 2H_2O$, in sufficient water to make 1000 ml., and standardize the solution as follows: Weigh accurately about 200 mg. of primary standard calcium carbonate, $CaCO_3$, transfer to a 400-ml. beaker, add 10 ml. of water, and swirl to form a slurry. Cover the beaker with a watch glass, and introduce 2 ml. of diluted hydrochloric acid T.S. from a pipet inserted between the lip of the beaker and the edge of the watch glass. Swirl the contents of the beaker to dissolve the calcium carbonate. Wash down the sides of the beaker, the outer surface of the pipet, and the watch glass, and dilute to about 100 ml. with water. While stirring, preferably with a magnetic stirrer, add about 30 ml. of the disodium ethylenediaminetetraacetate solution from a 50-ml. buret, then add 15 ml. of sodium hydroxide T.S. and 300 mg. of hydroxy naphthol blue indicator, and continue the titration to a blue end-point. Calculate the molarity by the formula $W/(100.09V)$, in which W is the weight, in mg., of $CaCO_3$ in the sample of calcium carbonate taken, and V is the volume, in ml., of disodium ethylenediaminetetraacetate solution consumed. Each 5.004 mg. of $CaCO_3$ is equivalent to 1 ml. of 0.05 *M* disodium ethylenediaminetetraacetate.

Hydrochloric Acid, 1 N (36.46 grams HCl per liter). Dilute 85 ml. of hydrochloric acid with water to make 1000 ml. and standardize the solution as follows: Accurately weigh about 1.5 grams of primary standard anhydrous sodium carbonate, Na_2CO_3, that has been heated at a temperature of about 270° for 1 hour. Dissolve it in 100 ml. of water and add 2 drops of methyl red T.S. Add the acid slowly from a buret, with constant stirring, until the solution becomes faintly pink. Heat the solution to boiling, and continue the titration until the faint pink color is no longer affected by continued boiling. Calculate the normality. Each 52.99 mg. of Na_2CO_3 is equivalent to 1 ml. of 1 *N* hydrochloric acid.

Hydroxylamine Hydrochloride, 0.5 N (35 grams $NH_2OH . HCl$ per liter). Dissolve 35 grams of hydroxylamine hydrochloride in 150 ml. of water and dilute to 1000 ml. with anhydrous methanol. To 500 ml. of this solution add 15 ml. of a 0.04 percent solution of bromophenol blue in alcohol and titrate with 0.5 *N* triethanolamine until the solution appears greenish blue by transmitted light. *Prepare this solution fresh before each series of analyses.*

Iodine, 0.1 N (12.69 grams I per liter). Dissolve about 14 grams of iodine, I, in a solution of 36 grams of potassium iodide, KI, in 100 ml. of water, add 3 drops of hydrochloric acid, dilute with water to 1000 ml., and standardize as follows: Weigh accurately about 150 mg. of primary standard arsenic trioxide, As_2O_3, previously dried at 100° for 1 hour, and dissolve it in 20 ml. of 1 *N* sodium hydroxide by warming if necessary. Dilute with 40 ml. of water, add 2 drops of methyl orange T.S., and follow with diluted hydrochloric acid T.S. until the yellow color is changed to pink. Then add 2 grams of sodium bicarbonate, $NaHCO_3$, dilute with 50 ml. of water, add 3 ml. of starch T.S., and slowly add the iodine solution from a buret until a permanent blue color is produced. Calculate the normality. Each 4.946 mg. of As_2O_3 is equivalent to 1 ml. of 0.1 *N* iodine. Store this solution in glass-stoppered bottles.

Oxalic Acid, 0.1 N (4.502 grams $H_2C_2O_4$ per liter). Dissolve 6.45 grams of oxalic acid, $H_2C_2O_4 . 2H_2O$, in sufficient water to make 100 ml. Standardize by titration against freshly standardized 0.1 *N* potassium permanganate as directed under *Potassium Permanganate, 0.1 N*. Store this solution in glass-stoppered bottles, protected from light.

Perchloric Acid, 0.1 N (10.046 grams $HClO_4$ per liter). Mix 8.5 ml. of perchloric acid (70%) with 500 ml. of glacial acetic acid and 30 ml. of acetic anhydride. Cool, and add glacial acetic acid to make 1000 ml. Allow the prepared solution to stand for 1 day, for the excess acetic anhydride to be combined, and determine the water content by the *Karl Fischer Titrimetric Method*, page 977. If the water content exceeds 0.05 percent, add more acetic anhydride, but if the solution contains no titratable water, add sufficient water to make the content between 0.02 and 0.05 percent of water. Allow to stand for 1 day, and again determine the water content by titration. Standardize the solution as follows: Weigh accurately about 700 mg. of primary standard potassium biphthalate, $KHC_6H_4(COO)_2$, previously dried at 105° for 3 hours, and dissolve it in 50 ml. of glacial acetic acid in a 250-ml. flask. Add 2 drops of crystal violet T.S., and titrate with the perchloric acid solution until the violet color changes to emerald-green. Deduct the volume of the perchloric acid consumed by 50 ml. of the glacial acetic acid, and calculate the normality. Each 20.42 mg. of $KHC_6H_4(COO)_2$ is equivalent to 1 ml. of 0.1 *N* perchloric acid.

Perchloric Acid, 0.1 N, in Dioxane. Mix 8.5 ml. of perchloric acid (70%) with sufficient dioxane, which has been especially purified by adsorption, to make 1000 ml. Standardize the solution as follows: Weigh accurately about 700 mg. of primary standard potassium biphthalate, $KHC_6H_4(COO)_2$, previously dried at 105° for 2 hours, and dissolve in 50 ml. of glacial acetic acid in a 250-ml. flask. Add 2 drops of crystal violet T.S., and titrate with the perchloric acid solution until the violet color changes to bluish green. Deduct the volume of the perchloric acid consumed by 50 ml. of the glacial acetic acid, and calculate the normality. Each 20.42 mg. of $KHC_6H_4(COO)_2$ is equivalent to 1 ml. of 0.1 *N* perchloric acid.

Potassium Acid Phthalate, 0.1 N [20.42 grams $KHC_6H_4(COO)_2$ per liter]. Dissolve 20.42 grams of primary standard potassium biphthalate, $KHC_6H_4(COO)_2$, in glacial acetic acid in a 1000-ml. volumetric flask, warming on a steam bath if necessary to effect solution and protecting the solution from contamination by moisture. Cool to room temperature, dilute to volume with glacial acetic acid, and mix.

Potassium Dichromate, 0.1 N (4.903 grams $K_2Cr_2O_7$ per liter). Dissolve about 5 grams of potassium dichromate, $K_2Cr_2O_7$, in 1000 ml. of water, transfer quantitatively 25 ml. of this solution to a 500-ml. glass-stoppered flask, add 2 grams of potassium iodide (free from iodate), KI, dilute with 200 ml. of water, add 5 ml. of hydrochloric acid, and mix. Allow to stand for 10 minutes in a dark place, and titrate the liberated iodine with 0.1 *N* sodium thiosulfate, adding starch T.S. as the end-point is approached. Correct for a blank run on the same quantities of the same reagents, and calculate the normality.

Potassium Hydroxide, 1 N (56.11 grams KOH per liter). Prepare and standardize 1 *N* potassium hydroxide by the procedure set forth for *Sodium Hydroxide*, 1 *N*, using 74 grams of the potassium hydroxide, KOH, to prepare the solution. Each 204.2 mg. of $KHC_6H_4(COO)_2$ is equivalent to 1 ml. of 1 *N* potassium hydroxide.

Potassium Hydroxide, 0.5 N, Alcoholic. Dissolve about 35 grams of potassium hydroxide, KOH, in 20 ml. of water, and add sufficient aldehyde-free alcohol to make 1000 ml. Allow the solution to stand in a tightly stoppered bottle for 24 hours. Then quickly decant the clear supernatant liquid into a suitable, tight container, and standardize as follows: Transfer quantitatively 25 ml. of 0.5 *N* hydrochloric acid into a flask, dilute with 50 ml. of water, add 2 drops of phenolphthalein T.S., and titrate with the alcoholic potassium hydroxide solution until a permanent, pale pink color is produced. Calculate the normality. Store this solution in tightly stoppered bottles protected from light.

Potassium Iodate, 0.05 M (10.70 grams KIO_3 per liter). Dissolve 10.700 grams of potassium iodate of primary standard quality, KIO_3, previously dried at 110° to constant weight, in sufficient water to make 1000.0 ml.

Potassium Permanganate, 0.1 N (3.161 grams $KMnO_4$ per liter). Dissolve about 3.3 grams of potassium permanganate, $KMnO_4$, in 1000 ml. of water in a flask, and boil the solution for about 15 minutes. Stopper the flask, allow it to stand for at least 2 days, and filter through asbestos. Standardize the solution as follows: Weigh accurately about 200 mg. of sodium oxalate of primary standard quality, $Na_2C_2O_4$, previously dried at 110° to constant weight, and dissolve it in 250 ml. of water. Add 7 ml. of sulfuric acid, heat to about 70°, and then slowly add the permanganate solution from a buret, with constant stirring, until a pale pink color which persists for 15 seconds is produced. The temperature at the conclusion of the titration should be not less than 60°. Calculate the normality. Each 6.700 mg. of $Na_2C_2O_4$ is equivalent to 1 ml. of 0.1 *N* potassium permanganate. Potassium permanganate is reduced on contact with organic substances such as rubber; therefore, the solution must be handled in apparatus entirely of glass or other suitably inert material. Store it in glass-stoppered, amber-colored bottles, and restandardize frequently.

Silver Nitrate, 0.1 N (16.99 grams $AgNO_3$ per liter). Dissolve about 17.5 grams of silver nitrate, $AgNO_3$, in 1000 ml. of water, and standardize the solution as follows: Dilute about 40 ml., accurately measured, of the silver nitrate solution with about 100 ml. of water, heat the solution, and add slowly, with continuous stirring, diluted hydrochloric acid T.S. until precipitation of the silver is complete. Boil the mixture cautiously for about 5 minutes, then allow it to stand in the dark until the precipitate has settled and the supernatant liquid has become clear. Transfer the precipitate completely to a tared filtering crucible, and wash it with small portions of water slightly acidified with nitric acid. Dry the precipitate at 110° to constant weight. Each 14.332 mg. of silver chloride obtained is equivalent to 1 ml. of 0.1 *N* silver nitrate. Protect the silver chloride from light as much as possible during the determination.

Sodium Arsenite, 0.05 N (3.248 grams $NaAsO_2$ per liter). Transfer 2.4725 grams of arsenic trioxide, which has been pulverized and dried at 100° to constant weight, to a 1000-ml. volumetric flask, dissolve it in 20 ml. of 1 *N* sodium hydroxide, and add 1 *N* sulfuric acid or 1 *N* hydrochloric acid until the solution is neutral or only slightly acid to litmus. Add 15 grams of sodium bicarbonate, dilute to volume with water, and mix.

Sodium Hydroxide, 1 N (40.00 grams NaOH per liter). Dissolve about 45 grams of sodium hydroxide, NaOH, in about 950 ml. of water, and add a freshly prepared saturated solution of barium hydroxide until no more precipitate forms. Shake the mixture thoroughly, and allow it to stand overnight in a stoppered bottle. Decant or filter the solution, and standardize the clear liquid as follows: Transfer about 5 grams of primary standard potassium biphthalate, $KHC_6H_4(COO)_2$, previously dried at 105° for 3 hours and accurately weighed, to a flask, and dissolve it in 75 ml. of carbon dioxide-free water. If the potassium biphthalate is in the form of large crystals, it should be crushed before drying. To the flask add 2 drops of phenolphthalein T.S., and titrate with the sodium hydroxide solution to a permanent pink color. Calculate the normality. Each 204.2 mg. of potassium biphthalate is equivalent to 1 ml. of 1 *N* sodium hydroxide.

> *Note:* Solutions of alkali hydroxides absorb carbon dioxide when exposed to air. They should therefore be stored in bottles with well-fitted, suitable stoppers, provided with a tube filled with a mixture of sodium hydroxide and lime so that air entering the container must pass through this tube, which will absorb the carbon dioxide. Standard solutions of sodium hydroxide should be restandardized frequently.

Sodium Methoxide, 0.1 N, in Pyridine (5.40 grams CH_3ONa per liter). Weigh 14 grams of freshly cut sodium metal, and cut into small cubes. Place about 0.5 ml. of anhydrous methanol in a round-bottom

250-ml. flask equipped with a ground-glass joint, add 1 cube of the sodium metal, and, when the reaction subsides, add the remaining sodium metal to the flask. Connect a water-cooled condenser to the flask, and slowly add 100 ml. of anhydrous methanol, in small portions, through the top of the condenser. Regulate the addition of the methanol so that the vapors are condensed and do not escape through the top of the condenser. After addition of the methanol is complete, connect a drying tube to the top of the condenser, and allow the solution to cool. Transfer 17.5 ml. of this solution (approximately 6 *N*) into a 1000-ml. volumetric flask containing 70 ml. of anhydrous methanol, and dilute to volume with freshly distilled pyridine. Store preferably in the reservoir of an automatic buret suitably protected from carbon dioxide and moisture. Standardize the solution as follows: Weigh accurately about 400 mg. of primary standard benzoic acid, transfer it into a 250-ml. wide-mouth Erlenmeyer flask, and dissolve it in 50 ml. of freshly distilled pyridine. Add a few drops of thymolphthalein T.S., and titrate immediately with the sodium methoxide solution to a blue end-point. During the titration, direct a gentle stream of nitrogen into the flask through a short piece of 6-mm. glass tubing fastened near the tip of the buret. Perform a blank determination (see page 2), correct for the volume of sodium methoxide solution consumed by the blank, and calculate the normality. Each 12.21 mg. of benzoic acid is equivalent to 1 ml. of 0.1 *N* sodium methoxide in pyridine.

Sodium Nitrite, 0.1 M (6.900 grams $NaNO_2$ per liter). Dissolve 7.5 grams of sodium nitrite, $NaNO_2$, in sufficient water to make 1000 ml., and standardize the solution as follows: Weigh accurately about 500 mg. of U.S.P. Sulfanilamide Reference Standard, previously dried at 105° for 3 hours, and transfer to a beaker or a casserole. Add 50 ml. of water and 5 ml. of hydrochloric acid, and stir well until dissolved. Cool to 15°, and add about 25 grams of crushed ice, then titrate slowly with the sodium nitrite solution, stirring vigorously, until a blue color is produced immediately when a glass rod dipped in the titrated solution is streaked on a smear of starch iodide paste T.S. When the titration is complete, the end-point should be reproducible after the mixture has been standing for 1 minute. Calculate the molarity. Each 17.22 mg. of sulfanilamide is equivalent to 1 ml. of 0.1 *M* sodium nitrite.

Sodium Thiosulfate, 0.1 N (15.81 grams $Na_2S_2O_3$ per liter). Dissolve about 26 grams of sodium thiosulfate, $Na_2S_2O_3.5H_2O$, and 200 mg. of sodium carbonate, Na_2CO_3, in 1000 ml. of recently boiled and cooled water. Standardize the solution as follows: Weigh accurately about 210 mg. of primary standard potassium dichromate, previously pulverized and dried at 120° for 4 hours, and dissolve in 100 ml. of water in a 500-ml. glass-stoppered flask. Swirl to dissolve the sample, remove the stopper, and quickly add 3 grams of potassium iodide, KI, and 5 ml. of hydrochloric acid. Stopper the flask, swirl to mix, and let stand in the dark for 10 minutes. Rinse the stopper and inner walls of the flask with water, and titrate the liberated iodine with the sodium thiosulfate solution until the solution is only faint yellow in color. Add starch T.S., and continue the titration to the discharge of the blue color. Calculate the normality.

Sulfuric Acid, 1 N (49.04 grams H_2SO_4 per liter). Add slowly, with stirring, 30 ml. of sulfuric acid to about 1020 ml. of water, allow to cool to 25°, and standardize by titration against primary standard sodium carbonate, Na_2CO_3, as directed under *Hydrochloric Acid,* 1 *N.* Each 52.99 mg. of Na_2CO_3 is equivalent to 1 ml. of 1 *N* sulfuric acid.

Sulfuric Acid, Alcoholic, 5 N (245.2 grams H_2SO_4 per liter). Add cautiously, with stirring, 139 ml. of sulfuric acid to a sufficient quantity of absolute alcohol to make 1000.0 ml.

Sulfuric Acid, Alcoholic, 0.5 N. Add cautiously, with stirring, 13.9 ml. of sulfuric acid to a sufficient quantity of absolute alcohol to make 1000.0 ml. Alternatively, this solution may be prepared by diluting 100.0 ml. of 5 *N* sulfuric acid with absolute alcohol to make 1000.0 ml.

Thorium Nitrate, 0.1 M [48.01 grams $Th(NO_3)_4$ per liter]. Weigh accurately 55.21 grams of thorium nitrate, $Th(NO_3)_4 \cdot 4H_2O$, dissolve it in water, dilute to 1000.0 ml., and mix. Standardize the solution as follows: Transfer 50.0 ml. into a 500-ml. volumetric flask, dilute to volume with water, and mix. Transfer 50.0 ml. of the diluted solution into a 400-ml. beaker, add 150 ml. of water and 5 ml. of hydrochloric acid, and heat to boiling. While stirring, add 25 ml. of a saturated solution of oxalic acid, then digest the mixture for 1 hour just below the boiling point and allow to stand overnight. Decant through Whatman No. 42, or equivalent, filter paper, and transfer the precipitate to the filter using about 100 ml. of a wash solution consisting of 70 ml. of the saturated oxalic acid solution, 430 ml. of water, and 5 ml. of hydrochloric acid. Transfer the precipitate and filter paper to a tared tall-form porcelain crucible, dry, char the paper, and ignite at 950° for 1.5 hours or to constant weight. Cool in a desiccator, weigh, and calculate the molarity of the solution by the formula $200W/264.04$, in which W is the weight, in grams, of thorium oxide obtained.

Triethanolamine, 0.5 N (74 grams $N(CH_2CH_2OH)_3$ per liter) Transfer 65 ml. (74 grams) of 98 percent triethanolamine into a 1000-ml. volumetric flask, dilute to volume with water, stopper the flask, and mix thoroughly.

Zinc Sulfate, 0.05 M (8.072 grams $ZnSO_4$ per liter). Dissolve about 15 grams of zinc sulfate, $ZnSO_4 \cdot 7H_2O$, in sufficient water to make 1000 ml., and standardize the solution as follows: Dilute about 35 ml., accurately measured, with 75 ml. of water, add 5 ml. of ammonia-ammonium chloride buffer T.S. and 0.1 ml. of eriochrome black T.S., and titrate with 0.05 *M* disodium ethylenediaminetetraacetate until the solution is deep blue in color. Calculate the molarity.

INDICATORS

The necessary solutions of indicators may be prepared as directed under *Test Solutions* (*T.S.*), page 985. The sodium salts of many indicators are commercially available and may be used interchangeably in water solutions with the alcohol solutions specified for the free indicators. It should be noted that two different methods for the preparation of an indicator solution are often given, depending on whether the test solution is to be used in volumetric analysis or in the determination of pH.

Useful pH indicators, listed in ascending order of the lower limit of their range, are: methyl yellow (pH 2.9–4.0), bromophenol blue (pH 3.0–4.6), bromocresol green (pH 4.0–5.4), methyl red (pH 4.2–6.2), bromocresol purple (pH 5.2–6.8), bromothymol blue (pH 6.0–7.6), phenol red (pH 6.8–8.2), thymol blue (pH 8.0–9.2), and thymolphthalein (pH 9.3–10.5).

Alphazurine 2G. Use a suitable grade.

Azo Violet [*4-(p-Nitrophenylazo) Resorcinol*]. A red powder, melting at about 193° with decomposition.

Bromocresol Blue. Use *Bromocresol Green.*

Bromocresol Green (*Bromocresol Blue; Tetrabromo-m-cresolsulfonphthalein*). A white or pale buff-colored powder, slightly soluble in water; soluble in alcohol and in solutions of alkali hydroxides. Transition interval: from pH 3.8 (yellow) to 5.4 (blue).

Bromocresol Purple (*Dibromo-o-cresolsulfonphthalein*). A white to pink, crystalline powder; insoluble in water; soluble in alcohol and in solutions of alkali hydroxides. Transition interval: from pH 5.2 (yellow) to 6.8 (purple).

Bromophenol Blue (*Tetrabromophenolsulfonphthalein*). Pinkish crystals, soluble in alcohol. Insoluble in water; soluble in solutions of alkali hydroxides. Transition interval: from pH 3.0 (yellow) to 4.6 (blue).

Bromothymol Blue (*Dibromothymolsulfonphthalein*). A rose-red powder. Insoluble in water; soluble in alcohol and in solutions of alkali hydroxides. Transition interval: from pH 6.0 (yellow) to 7.6 (blue).

Cresol Red (*o-Cresolsulfonphthalein*). A red-brown powder. Slightly soluble in water; soluble in alcohol and in dilute solutions of alkali hydroxides. Transition interval: from pH 7.2 (yellow) to 8.8 (blue).

Crystal Violet (*Hexamethyl-p-rosaniline Chloride*). Dark green crystals. Slightly soluble in water; sparingly soluble in alcohol and in glacial acetic acid. Its solutions are deep violet in color.

Sensitiveness. Dissolve 100 mg. in 100 ml. of glacial acetic acid, and mix. Pipet 1 ml. of the solution into a 100-ml. volumetric flask, and dilute with glacial acetic acid to volume. The solution is violet-blue in color and does not show a reddish tint. Pipet 20 ml. of the diluted solution into a beaker, and titrate with 0.1 *N* perchloric acid, adding the perchloric acid slowly from a microburet. Not more than 0.1 ml. of 0.1 *N* perchloric acid is required to produce an emerald-green color.

Dithizone (*Diphenylthiocarbazone*). A bluish black powder. Insoluble in water; soluble in alcohol, in chloroform, and in carbon tetrachloride, yielding intensely green solutions even in high dilutions.

Eriochrome Black T [*Sodium 1-(1-Hydroxy-2-naphthylazo)-5-nitro-2-naphthol-4-sulfonate*]. A brownish black powder having a faint metallic sheen. Soluble in alcohol, in methanol, and in hot water.

Sensitiveness. To 10 ml. of a 1 in 200,000 solution in a mixture of equal parts of methanol and water add sodium hydroxide solution (1 in 100) until the pH is 10. The solution is pure blue in color and free from cloudiness. Add 0.2 ml. of *Magnesium Standard Solution* (10 mcg. Mg ion). The color of the solution changes to red-violet, and with the continued addition of magnesium ion it becomes wine-red in color.

p-Ethoxychrysoidin Monohydrochloride [*4-(p-Ethoxyphenylazo)-m-phenylenediamine Monohydrochloride; 4′-Ethoxy-2,4-diaminoazobenzene Monohydrochloride*]. A reddish powder, insoluble in water. Transition interval: from pH 3.5 (red) to 5.5 (yellow).

Hydroxy Naphthol Blue. The disodium salt of 1-(2-naphtholazo-3,6-disulfonic acid)-2-naphthol-4-sulfonic acid deposited on crystals of sodium chloride. Small blue crystals, freely soluble in water. In the pH range between 12 and 13 its solution is reddish pink in the presence of calcium ion and deep blue in the presence of excess disodium ethylenediaminetetraacetate.

Suitability for calcium determinations. Dissolve 300 mg. in 100 ml. of water, add 10 ml. of sodium hydroxide T.S. and 1.0 ml. of calcium chloride solution (1 in 200), and dilute with water to 165 ml. The solution is reddish pink in color. Add 1.0 ml. of 0.05 *M* disodium ethylenediaminetetraacetate. The solution becomes deep blue in color.

Litmus. A blue powder, cubes, or pieces. Partly soluble in water and in alcohol. Transition interval: from approximately pH 4.5 (red) to 8 (blue). Litmus is unsuitable for determining the pH of solutions of carbonates or bicarbonates.

Methylene Blue [*3,7-Bis(dimethylamino)phenazathionium Chloride*]. Dark green crystals or a crystalline powder, having a bronze-like luster. Soluble in water and in chloroform; sparingly soluble in alcohol.

Methyl Orange (*Helianthin; Tropaeolin D; 4′-Dimethylaminoazobenzene-4-sodium Sulfonate*). An orange-yellow powder or crystalline scales. Slightly soluble in cold water; readily soluble in hot water; insoluble in alcohol. Transition interval: from pH 3.2 (pink) to 4.4 (yellow).

Methyl Red (*o-Carboxybenzeneazo-dimethylaniline Hydrochloride*). A dark red powder or violet crystals. Sparingly soluble in water; soluble in alcohol. Transition interval: from pH 4.2 (red) to 6.2 (yellow).

Methyl Red Sodium. The sodium salt of *o*-carboxybenzeneazo-dimethylaniline. An orange-brown powder. Freely soluble in cold water and in alcohol. Transition interval: from pH 4.2 (red) to 6.2 (yellow).

Methyl Yellow (*p-Dimethylaminoazobenzene*). Yellow crystals, melting between 114° and 117°. Insoluble in water; soluble in alcohol, in benzene, in chloroform, in ether, in dilute mineral acids, and in oils. Transition interval: from pH 2.9 (red) to 4.0 (yellow).

Murexide Indicator Preparation. Add 400 mg. of murexide to 40 grams of powdered potassium sulfate, K_2SO_4, and grind in a glass mortar to a homogeneous mixture. Alternatively, tablets containing 0.4 mg. of murexide admixed with potassium sulfate or potassium chloride, available commercially, may be used.

Naphthol Green B. The ferric salt of 6-sodium sulfo-1-isonitroso-1,2-naphthoquinone. A dark green powder, insoluble in water.

Neutral Red (*3-Amino-7-dimethylamino-2-methylphenazine Chloride*). A coarse, reddish to olive-green powder. Sparingly soluble in water and in alcohol. Transition interval: from pH 6.8 (red) to 8.0 (orange).

Phenol Red (*Phenolsulfonphthalein*). A bright to dark red crystalline powder, very slightly soluble in water; sparingly soluble in alcohol; soluble in solutions of alkali hydroxides. Transition interval: from pH 6.8 (yellow) to 8.2 (red).

Phenolphthalein. White or yellowish white crystals, practically insoluble in water; soluble in alcohol and in solutions of alkali hydroxides. Transition interval: from pH 8.0 (colorless) to 10.0 (red).

Quinaldine Red (*5-Dimethylamino-2-strylethylquinolinium Iodide*). A dark blue-black powder, melting at about 260° with decomposition. Sparingly soluble in water; freely soluble in alcohol. Transition interval: from pH 1.4 (colorless) to 3.2 (red).

Thymol Blue (*Thymolsulfonphthalein*). A dark, brownish green, crystalline powder. Slightly soluble in water; soluble in alcohol and in dilute alkali solutions. Acid transition interval: from pH 1.2 (red) to 2.8 (yellow). Alkaline transition interval: from pH 8.0 (yellow) to 9.2 (blue).

Thymolphthalein. A white to slightly yellow, crystalline powder. Insoluble in water; soluble in alcohol and in solutions of alkali hydroxides. Transition interval: from pH 9.3 (colorless) to 10.5 (blue).

Xylenol Orange [*3,3′-Bis-di*(*carboxymethyl*)*aminomethyl-o-cresolsulfonphthalein*]. An orange powder. Soluble in water and in alcohol. In acid solution it is colored lemon yellow, and its metal complexes intensely red. It gives a distinct end-point in the direct EDTA titration of metals such as bismuth, thorium, scandium, lead, zinc, lanthanum, cadmium, and mercury.

INDICATOR PAPERS AND TEST PAPERS

Indicator papers and test papers are strips of paper of suitable dimension and grade (usually Swedish O filter paper or other makes of like surface, quality, and ash) impregnated with a sufficiently stable indicator solution or reagent.

Treat strong, white filter paper with hydrochloric acid, and wash with water until the last washing shows no acid reaction to methyl red T.S. Then treat with ammonia T.S., wash again with water until the last washing is not alkaline toward phenolphthalein T.S., and dry thoroughly. Saturate the dry paper with the appropriate indicator solution prepared as directed below, and dry carefully by suspending from glass rods or other inert material in still air free from acid, alkali, and other fumes. Cut the paper into strips of convenient size, and store in well-closed containers, protected from light and moisture.

Indicator papers and test papers that are available commercially may be used, if desired.

Cupric Sulfate Test Paper. Use *Cupric Sulfate T.S.*

Lead Acetate Test Paper. Usually about 6 × 80 mm. in size. Use *Lead Acetate T.S.*, and dry the paper at 100°, avoiding contact with metal.

Litmus Paper, Blue. Usually about 6 × 50 mm. in size. It meets the requirements of the following tests.

Phosphate. Place 10 strips in 10 ml. of water to which have been added 1 ml. of nitric acid and 0.5 ml. of ammonia T.S. Allow to stand for 10 minutes, then decant the solution, warm, and add 5 ml. of ammonium molybdate T.S. Shake at about 40° for 5 minutes. No precipitate of phosphomolybdate is formed.

Residue on ignition. Ignite carefully 10 strips of the paper to constant weight. The weight of the residue corresponds to not more than 400 mcg. per strip of about 3 sq. cm.

Rosin acids, etc. Immerse a strip of the blue paper in a solution of 100 mg. of silver nitrate, $AgNO_3$, in 50 ml. of water. The color of the paper does not change in 30 seconds.

Sensitiveness. Drop a 10- to 12-mm. strip into 100 ml. of 0.0005 *N* hydrochloric acid contained in a beaker, and stir continuously. The color of the paper is changed within 45 seconds.

Litmus Paper, Red. Usually about 6 × 50 mm. in size. Red litmus meets the requirements for *Phosphate*, *Residue on ignition*, and *Rosin acids, etc.*, under *Litmus Paper, Blue.*

Sensitiveness. Drop a 10 × 12-mm. strip into 100 ml. of 0.0005 *N* sodium hydroxide contained in a beaker, and stir continuously. The color of the paper changes within 30 seconds.

Phenolphthalein Paper. Use a 1 in 1000 solution of phenolphthalein in dilute alcohol (1 in 2).

Starch Iodate Paper. Use a mixture of equal volume of *Starch T.S.* and potassium iodate solution (1 in 20).

Starch Iodide Paper. Use a solution of 500 mg. of potassium iodide, KI, in 100 ml. of freshly prepared *Starch T.S.*

Former and Current Titles of Food Chemicals Codex Monographs

First Edition Title	Second Edition Title
Aldehyde C-16	Ethyl Methyl Phenylglycidate
Allyl Ionone	Allyl α-Ionone
Amyl Butyrate	Isoamyl Butyrate
Amylcinnamaldehyde	α-Amylcinnamaldehyde
Amyl Formate	Isoamyl Formate
Butylated Hydroxyanisole	BHA
Butylated Hydroxytoluene	BHT
Calcium Disodium Ethylenediaminetetraacetate	Calcium Disodium EDTA
Capric Acid	Decanoic Acid
Caproic Acid	Hexanoic Acid
Caprylic Acid	Octanoic Acid
Caryophyllene	β-Caryophyllene
Cinnamon Oil	Cassia Oil
Dimethyl Octanol	3,7-Dimethyl-1-octanol
Dimethyl Silicone	Dimethylpolysiloxane
Disodium Ethylenediaminetetraacetate	Disodium EDTA
Ethyl Anisate	Ethyl *p*-Anisate
Ethyl Caproate	Ethyl Hexanoate
Ethyl Caprylate	Ethyl Octanoate
Ethyl Pelargonate	Ethyl Nonanoate
Hexyl Alcohol	Hexyl Alcohol (Natural)
Hexylcinnamaldehyde	α-Hexylcinnamaldehyde
Hydratropic Aldehyde	2-Phenylpropionaldehyde
Hydratropic Aldehyde Dimethyl Acetal	2-Phenylpropionaldehyde Dimethyl Acetal
Lauryl Alcohol	Lauryl Alcohol (Natural)
Methyl Acetophenone	4′-Methyl Acetophenone
Methylanisole	*p*-Methyl Anisole
Methylbenzyl Acetate	α-Methylbenzyl Acetate
Methylbenzyl Alcohol	α-Methylbenzyl Alcohol
Methyl Cinnamaldehyde	α-Methylcinnamaldehyde
Methyl Heptenone	6-Methyl-5-Hepten-2-one
Methyl Heptine Carbonate	Methyl 2-Octynoate
Myristica Oil	Nutmeg Oil
Nonalactone	γ-Nonalactone
Octanol	1-Octanol (Natural)
Peru Balsam Oil	Balsam Peru Oil
Phenyl Ethyl Acetate	Phenethyl Acetate
Phenyl Ethyl Alcohol	Phenethyl Alcohol
Phenyl Propyl Acetate	3-Phenylpropyl Acetate
Phenyl Propyl Alcohol	3-Phenyl-1-propanol
Phenyl Propyl Aldehyde	3-Phenylpropionaldehyde
Propyl Anisole	*p*-Propyl Anisole
Sage Oil, Clary	Clary Oil
Spike Oil	Spike Lavender Oil
DL-Tryptophane	DL-Tryptophan

Food Chemicals Codex Substances Classified by Functional Use in Foods

Acids, Acidifiers

Acetic Acid, Glacial
Citric Acid
Fumaric Acid
Glucono Delta-Lactone
Hydrochloric Acid
Lactic Acid
Malic Acid
Phosphoric Acid
Potassium Acid Tartrate
Sodium Bisulfate
Sulfuric Acid
Tartaric Acid

Alkalies

Ammonium Bicarbonate
Ammonium Hydroxide
Calcium Carbonate
Calcium Oxide
Magnesium Carbonate
Magnesium Hydroxide
Magnesium Oxide
Potassium Bicarbonate
Potassium Carbonate
Potassium Carbonate Solution
Potassium Hydroxide
Potassium Hydroxide Solution
Sodium Bicarbonate
Sodium Carbonate
Sodium Hydroxide
Sodium Hydroxide Solution
Sodium Sesquicarbonate

Anticaking Agents, Drying Agents

Calcium Phosphate, Tribasic
Calcium Silicate
Calcium Stearate
Cellulose, Microcrystalline
Kaolin
Magnesium Carbonate
Magnesium Hydroxide
Magnesium Silicate
Magnesium Stearate
Silicon Dioxide
Sodium Ferrocyanide
Sodium Silicoaluminate

Antimicrobial Agents

Benzoic Acid
Heptylparaben
Methylparaben
Propylparaben
Sodium Benzoate

Antioxidants

Ascorbic Acid
Ascorbyl Palmitate
BHA
BHT
Butylated Hydroxymethylphenol
Calcium Ascorbate
Dilauryl Thiodipropionate
Erythorbic Acid
Ethoxyquin
Gum Guaiac
Lecithin
Potassium Metabisulfite
Potassium Sulfite
Propyl Gallate
Sodium Ascorbate
Sodium Erythorbate
Sodium Metabisulfite
Sodium Sulfite
Sodium Thiosulfate
Stannous Chloride
dl-Alpha Tocopherol
Tocopherols Concentrate, Mixed

Binders, Fillers, Plasticizers

Calcium Stearate
Cellulose, Microcrystalline
Ethylcellulose
Food Starch, Modified
Glycerin
Lactylated Fatty Acid Esters of Glycerol and Propylene Glycol
Magnesium Stearate
Methylcellulose
Mineral Oil, White
Oleic Acid
Polyethylene Glycols

Bleaching, Oxidizing Agents

Acetone Peroxides
Benzoyl Peroxide

Calcium Peroxide
Hydrogen Peroxide

Bodying, Bulking Agents

Glycerin
Methylcellulose
PVP
Sodium Carboxymethylcellulose
Xanthan Gum

Buffers, Neutralizing Agents

Adipic Acid
Aluminum Ammonium Sulfate
Aluminum Potassium Sulfate
Aluminum Sodium Sulfate
Ammonium Carbonate
Ammonium Phosphate, Dibasic
Ammonium Phosphate, Monobasic
Calcium Citrate
Calcium Gluconate
Calcium Hydroxide
Calcium Lactate
Calcium Phosphate, Monobasic
Calcium Phosphate, Tribasic
Calcium Pyrophosphate
Magnesium Oxide
Potassium Acid Tartrate
Potassium Citrate
Potassium Phosphate, Dibasic
Potassium Phosphate, Monobasic
Sodium Acetate
Sodium Acetate, Anhydrous
Sodium Acid Pyrophosphate
Sodium Citrate
Sodium Phosphate, Dibasic
Sodium Phosphate, Monobasic
Sodium Phosphate, Tribasic
Sodium Potassium Tartrate
Sodium Pyrophosphate
Sodium Sesquicarbonate
Succinic Acid

Carriers, Disintegrating Agents, Dispersing Agents

Cellulose, Microcrystalline
Citric Acid
Magnesium Carbonate
Polyethylene Glycols
PVP

Chewing Gum Base Components

Butadiene-Styrene 75/25 Rubber
Butadiene-Styrene 50/50 Rubber
Candelilla Wax
Glycerol Ester of Partially Dimerized Rosin
Glycerol Ester of Partially Hydrogenated Wood Rosin
Glycerol Ester of Polymerized Rosin
Glycerol Ester of Tall Oil Rosin
Glycerol Ester of Wood Rosin
Isobutylene-Isoprene Copolymer
Lanolin, Anhydrous
Masticatory Substances, Natural
Methyl Ester of Rosin, Partially Hydrogenated
Pentaerythritol Ester of Partially Hydrogenated Wood Rosin
Pentaerythritol Ester of Wood Rosin
Polyethylene
Polyisobutylene
Polyvinyl Acetate
Rice Bran Wax
Terpene Resin, Natural
Terpene Resin, Synthetic

Clarifying Agents

PVP
Tannic Acid

Color Fixatives, Adjuncts; Color-retention Agents

Ferrous Gluconate
Magnesium Carbonate
Magnesium Chloride
Magnesium Hydroxide
Potassium Nitrate
Potassium Nitrite
Sodium Nitrate
Sodium Nitrite

Components in the Manufacture of Other Food-grade Additives

Decanoic Acid
Lauric Acid
Myristic Acid
Octanoic Acid
Oleic Acid
Palmitic Acid
Stearic Acid

Defoaming Agents

Decanoic Acid
Dimethylpolysiloxane
Lauric Acid
Mineral Oil, White
Myristic Acid
Octanoic Acid
Oleic Acid
Oxystearin
Palmitic Acid

Petrolatum
Silicon Dioxide
Sorbitan Monostearate
Stearic Acid

Dough Conditioners

Acetone Peroxides
Ammonium Chloride
Ammonium Phosphate, Dibasic
Ammonium Phosphate, Monobasic
Ammonium Sulfate
Calcium Bromate
Calcium Carbonate
Calcium Iodate
Calcium Lactate
Calcium Oxide
Calcium Peroxide
Calcium Phosphate, Dibasic
Calcium Phosphate, Monobasic
Calcium Stearoyl-2-lactylate
Calcium Sulfate
Ethoxylated Mono- and Diglycerides
Potassium Bromate
Potassium Iodate
Sodium Stearoyl-2-lactylate
Sodium Stearyl Fumarate

Emulsifiers; Foaming, Whipping Agents

Acacia
Acetylated Monoglycerides
Agar
Alginic Acid
Ammonium Alginate
Calcium Alginate
Calcium Stearate
Calcium Stearoyl-2-Lactylate
Carrageenan
Cholic Acid
Desoxycholic Acid
Diacetyl Tartaric Acid Esters of Mono- and Diglycerides
Dioctyl Sodium Sulfosuccinate
Guar Gum
Hydroxylated Lecithin
Hydroxypropyl Cellulose
Hydroxypropyl Methylcellulose
Karaya Gum
Lactated Mono-Diglycerides
Lactylated Fatty Acid Esters of Glycerol and Propylene Glycol
Lactylic Esters of Fatty Acids
Lecithin
Locust Bean Gum
Magnesium Stearate
Methylcellulose
Methyl Ethylcellulose
Mono- and Diglycerides
Pectin
Polyglycerol Esters of Fatty Acids
Polysorbate 20
Polysorbate 60
Polysorbate 65
Polysorbate 80
Potassium Alginate
Potassium Phosphate, Tribasic
Potassium Polymetaphosphate
Potassium Pyrophosphate
Propylene Glycol Alginate
Propylene Glycol Monostearate
Sodium Alginate
Sodium Aluminum Phosphate, Basic
Sodium Metaphosphate
Sodium Phosphate, Dibasic
Sodium Phosphate, Monobasic
Sodium Phosphate, Tribasic
Sodium Pyrophosphate
Sodium Stearoyl-2-lactylate
Sorbitan Monostearate
Succinylated Monoglycerides
Tragacanth
Xanthan Gum

Enzymes, Proteolytic

Papain
Pepsin

Firming Agents

Aluminum Potassium Sulfate
Aluminum Sodium Sulfate
Aluminum Sulfate
Aluminum Sulfate Solution
Calcium Carbonate
Calcium Chloride
Calcium Chloride, Anhydrous
Calcium Citrate
Calcium Gluconate
Calcium Hydroxide
Calcium Lactobionate
Calcium Phosphate, Monobasic
Calcium Sulfate
Magnesium Chloride

Flavor Enhancers, Intensifiers

Disodium Guanylate
Disodium Inosinate
Monoammonium L-Glutamate
Monopotassium L-Glutamate
Monosodium L-Glutamate
Sodium Chloride

Flavoring Adjuncts

Brominated Vegetable Oil

1,3-Butylene Glycol
Formic Acid
Poloxamer 331
Poloxamer 407
Polyethylene Glycols

Flavoring Agents

Acetaldehyde
Acetanisole
Acetic Acid, Glacial
Acetoin
Acetophenone
Allyl Cyclohexanepropionate
Allyl Hexanoate
Allyl α-Ionone
Allyl Isothiocyanate
Almond Oil, Bitter, FFPA
Ambrette Seed Oil
α-Amylcinnamaldehyde
Amyris Oil
Anethole
Angelica Root Oil
Angelica Seed Oil
Anise Oil
Anisole
Anisyl Acetate
Anisyl Alcohol
Balsam Peru Oil
Basil Oil
Bay Oil
Beeswax, White
Beeswax, Yellow
Benzaldehyde
Benzophenone
Benzyl Acetate
Benzyl Alcohol
Benzyl Benzoate
Benzyl Butyrate
Benzyl Cinnamate
Benzyl Phenylacetate
Benzyl Propionate
Benzyl Salicylate
Bergamot Oil, Expressed
Birch Tar Oil, Rectified
Bois De Rose Oil
Bornyl Acetate
2-Butanone
Butyl Acetate
Butyl Alcohol
Butyl Butyrate
Butyl Butyryllactate
Butyraldehyde
Butyric Acid
Cananga Oil
Caraway Oil
Cardamom Oil
Carrot Seed Oil
Carvacrol
d-Carvone
l-Carvone
β-Caryophyllene
Cascarilla Oil
Cassia Oil
Cedar Leaf Oil
Celery Seed Oil
Chamomile Oil, English
Chamomile Oil, German
Cinnamaldehyde
Cinnamic Acid
Cinnamon Bark Oil
Cinnamon Leaf Oil
Cinnamyl Acetate
Cinnamyl Alcohol
Cinnamyl Anthranilate
Cinnamyl Formate
Cinnamyl Isovalerate
Cinnamyl Propionate
Citral
Citric Acid
Citronellal
Citronellol
Citronellyl Acetate
Citronellyl Butyrate
Citronellyl Formate
Citronellyl Isobutyrate
Citronellyl Propionate
Clary Oil
Clove Leaf Oil
Clove Oil
Clove Stem Oil
Cognac Oil, Green
Copaiba Oil
Coriander Oil
Costus Root Oil
Cresyl Acetate
Cubeb Oil
Cumin Oil
Cyclamen Aldehyde
Δ-Decalactone
Decanal
1-Decanol (Natural)
Diacetyl
Diethyl Malonate
Diethyl Sebacate
Diethyl Succinate
Dill Seed Oil, European
Dill Seed Oil, Indian
Dillweed Oil, American
Dimethyl Anthranilate
Dimethyl Benzyl Carbinol
Dimethyl Benzyl Carbinyl Acetate
3,7-Dimethyl-1-octanol
Δ-Dodecalactone

Estragole
Ethyl Acetate
Ethyl Acetoacetate
Ethyl Acrylate
Ethyl *p*-Anisate
Ethyl Anthranilate
Ethyl Benzoate
2-Ethylbutyraldehyde
Ethyl Butyrate
2-Ethylbutyric Acid
Ethyl Cinnamate
Ethyl Formate
Ethyl Heptanoate
Ethyl Hexanoate
Ethyl Isovalerate
Ethyl Lactate
Ethyl Laurate
Ethyl Maltol
Ethyl Methyl Phenylglycidate
Ethyl Nonanoate
Ethyl Octanoate
Ethyl Oxyhydrate (So-called)
Ethyl Phenylacetate
Ethyl Phenylglycidate
Ethyl Propionate
Ethyl Salicylate
Ethyl Vanillin
Eucalyptus Oil
Eugenol
Fennel Oil
Fir Needle Oil, Canadian
Fir Needle Oil, Siberian
Fumaric Acid
Furfural
Garlic Oil
Geraniol
Geranium Oil, Algerian
Geranyl Acetate
Geranyl Benzoate
Geranyl Butyrate
Geranyl Formate
Geranyl Phenylacetate
Geranyl Propionate
Ginger Oil
Glutamic Acid Hydrochloride
Grapefruit Oil, Expressed
Heptanal
2-Heptanone
3-Heptanone
Heptyl Alcohol
Hexanoic Acid
Hexyl Alcohol (Natural)
α-Hexylcinnamaldehyde
Hops Oil
Hydroxycitronellal
Hydroxycitronellal Dimethyl Acetal
Indole
α-Ionone
β-Ionone
Isoamyl Acetate
Isoamyl Butyrate
Isoamyl Formate
Isoamyl Hexanoate
Isoamyl Isovalerate
Isoamyl Salicylate
Isobornyl Acetate
Isobutyl Acetate
Isobutyl Alcohol
Isobutyl Cinnamate
Isobutyl Phenylacetate
Isobutyl Salicylate
Isobutyraldehyde
Isobutyric Acid
Isoeugenol
Isoeugenyl Acetate
Isopropyl Acetate
Isopulegol
Isovaleric Acid
Juniper Berries Oil
Labdanum Oil
Laurel Leaf Oil
Lauryl Alcohol (Natural)
Lauryl Aldehyde
Lavandin Oil, Abrial
Lavender Oil
Lemongrass Oil
Lemon Oil
Lime Oil, Distilled
d-Limonene
Linaloe Wood Oil
Linalool
Linalyl Acetate
Linalyl Acetate, Synthetic
Linalyl Benzoate
Linalyl Isobutyrate
Linalyl Propionate
Lovage Oil
Mace Oil
Malic Acid
Maltol
Mandarin Oil, Expressed
Marjoram Oil, Spanish
Menthol
p-Methoxybenzaldehyde
4′-Methyl Acetophenone
p-Methyl Anisole
Methyl Anthranilate
Methyl Benzoate
α-Methylbenzyl Acetate
α-Methylbenzyl Alcohol
α-Methylcinnamaldehyde
Methyl Cinnamate
Methyl Cyclopentenolone

Methyl Eugenol
6-Methyl-5-hepten-2-one
Methyl Isoeugenol
Methyl β-Naphthyl Ketone
Methyl 2-Octynoate
4-Methyl-2-pentanone
Methyl Phenylacetate
Methyl Salicylate
2-Methylundecanal
Myrrh Oil
Nerol
Nerolidol
γ-Nonalactone
Nonanal
Nonyl Acetate
Nonyl Alcohol
Nutmeg Oil
Octanal
1-Octanol (Natural)
Octyl Acetate
Octyl Formate
Olibanum Oil
Onion Oil
Orange Oil
Orange Oil, Bitter
Origanum Oil, Spanish
Orris Root Oil
Palmarosa Oil
Parsley Seed Oil
Pennyroyal Oil
2,3-Pentanedione
2-Pentanone
Peppermint Oil
Pepper Oil, Black
Petitgrain Oil, Paraguay
α-Phellandrene
Phenethyl Acetate
Phenethyl Alcohol
Phenethyl Isobutyrate
Phenethyl Isovalerate
Phenethyl Phenylacetate
Phenethyl Salicylate
Phenoxyethyl Isobutyrate
Phenylacetaldehyde
Phenylacetaldehyde Dimethyl Acetal
Phenylacetic Acid
3-Phenyl-l-propanol
2-Phenylpropionaldehyde
3-Phenylpropionaldehyde
2-Phenylpropionaldehyde Dimethyl Acetal
3-Phenylpropyl Acetate
Pimenta Leaf Oil
Pimenta Oil
Pine Needle Oil, Dwarf
Pine Needle Oil, Scotch
Piperonal
Propenylguaethol
Propionaldehyde
p-Propyl Anisole
Quinine Hydrochloride
Quinine Sulfate
Rhodinol
Rhodinyl Acetate
Rhodinyl Formate
Rose Oil
Rosemary Oil
Rue Oil
Sage Oil, Dalmatian
Sage Oil, Spanish
Sandalwood, East Indian
Santalol
Santalyl Acetate
Savory Oil (Summer Variety)
Sodium Chloride
Spearmint Oil
Spike Lavender Oil
Tangerine Oil, Expressed
Tarragon Oil
Terpineol
Terpinyl Acetate
Terpinyl Propionate
Thyme Oil
p-Tolyl Isobutyrate
Tributyrin
γ-Undecalactone
Undecanal
10-Undecenal
Valeric Acid
Vanillin

Humectants, Moisture-retaining Agents

Glycerin
Potassium Polymetaphosphate
Propylene Glycol
Sorbitol
Sorbitol Solution
Triacetin

Leavening Agents

Ammonium Bicarbonate
Ammonium Phosphate, Dibasic
Ammonium Phosphate, Monobasic
Calcium Phosphate, Monobasic
Glucono Delta-Lactone
Potassium Bicarbonate
Sodium Acid Pyrophosphate
Sodium Aluminum Phosphate, Acidic
Sodium Bicarbonate

Lubricants; Antistick, Release Agents

Acetylated Monoglycerides
Castor Oil
Mineral Oil, White
Petrolatum
Polyethylene Glycols
Stearic Acid

Maturing Agents

Acetone Peroxides
Azodicarbonamide
Calcium Bromate
Calcium Iodate
Potassium Bromate
Potassium Iodate

Mold and Rope Inhibitors

Calcium Propionate
Propionic Acid
Sodium Diacetate
Sodium Propionate

Nonnutritive Sweeteners

Ammonium Saccharin
Calcium Saccharin
Saccharin
Sodium Saccharin

Nutrients, Dietary Supplements

DL-Alanine
L-Alanine
L-Arginine
L-Arginine Monohydrochloride
Ascorbic Acid
DL-Aspartic Acid
Biotin
Calcium Carbonate
Calcium Glycerophosphate
Calcium Oxide
Calcium Pantothenate
Calcium Pantothenate, Racemic
Calcium Pantothenate, Racemic-Calcium Chloride Complex
Calcium Phosphate, Dibasic
Calcium Phosphate, Monobasic
Calcium Phosphate, Tribasic
Calcium Pyrophosphate
Calcium Sulfate
β-Carotene
Choline Bitartrate
Choline Chloride
Copper Gluconate
L-Cysteine Monohydrochloride
L-Cystine
Dexpanthenol
Ferric Phosphate
Ferric Pyrophosphate
Ferrous Fumarate
Ferrous Gluconate
Ferrous Sulfate
Ferrous Sulfate, Dried
L-Glutamic Acid
Glycine
Inositol
Iron, Electrolytic
Iron, Reduced
DL-Isoleucine
L-Isoleucine
Kelp
DL-Leucine
L-Leucine
L-Lysine Monohydrochloride
Magnesium Phosphate, Dibasic
Magnesium Phosphate, Tribasic
Magnesium Sulfate
Manganese Chloride
Manganese Gluconate
Manganese Glycerophosphate
Manganese Hypophosphite
Manganese Sulfate
Mannitol
DL-Methionine
L-Methionine
Niacin
Niacinamide
Niacinamide Ascorbate
DL-Panthenol
DL-Phenylalanine
L-Phenylalanine
Potassium Chloride
Potassium Glycerophosphate
Potassium Iodide
L-Proline
Pyridoxine Hydrochloride
Riboflavin
Riboflavin 5′-Phosphate Sodium
DL-Serine
L-Serine
Sodium Ascorbate
Sodium Chloride
Sodium Ferric Pyrophosphate
Sodium Gluconate
Sodium Phosphate, Dibasic
Sodium Phosphate, Monobasic
Sodium Phosphate, Tribasic
Sodium Pyrophosphate
Thiamine Hydrochloride
Thiamine Mononitrate
L-Threonine
dl-Alpha Tocopherol
Tocopherols Concentrate, Mixed
d-Alpha Tocopheryl Acetate
dl-Alpha Tocopheryl Acetate

d-Alpha Tocopheryl Acetate Concentrate
d-Alpha Tocopheryl Acid Succinate
DL-Tryptophan
L-Tryptophan
L-Tyrosine
L-Valine
Vitamin A
Vitamin D_2
Vitamin D_3
Zinc Sulfate

Preservatives

Ascorbic Acid
Benzoic Acid
Calcium Disodium EDTA
Calcium Propionate
Dehydroacetic Acid
Disodium EDTA
Erythorbic Acid
Gum Guaiac
Heptylparaben
Hydrogen Peroxide
Methylparaben
Potassium Metabisulfite
Potassium Nitrate
Potassium Sorbate
Potassium Sulfite
Propionic Acid
Propylparaben
Sodium Benzoate
Sodium Bisulfite
Sodium Chloride
Sodium Dehydroacetate
Sodium Diacetate
Sodium Erythorbate
Sodium Metabisulfite
Sodium Propionate
Sodium Sulfite
Sorbic Acid

Salt Substitutes

L-Glutamic Acid
Glutamic Acid Hydrochloride
Monoammonium L-Glutamate
Monopotassium L-Glutamate
Potassium Chloride

Sequestrants

Calcium Acetate
Calcium Chloride
Calcium Chloride, Anhydrous
Calcium Citrate
Calcium Disodium EDTA
Calcium Gluconate
Calcium Phosphate, Monobasic
Calcium Sulfate
Citric Acid
Disodium EDTA
Glucono Delta-Lactone
Oxystearin
Phosphoric Acid
Potassium Citrate
Potassium Phosphate, Dibasic
Potassium Phosphate, Monobasic
Sodium Acid Pyrophosphate
Sodium Citrate
Sodium Diacetate
Sodium Gluconate
Sodium Metaphosphate
Sodium Potassium Tartrate
Sodium Tartrate
Sodium Thiosulfate
Sorbitol
Sorbitol Solution
Tartaric Acid
Triethyl Citrate

Solvents, Vehicles, Solubilizers

Acetone
Acetylated Monoglycerides
1,3-Butylene Glycol
Ethyl Alcohol
Ethylene Dichloride
Glycerin
Isopropyl Alcohol
Methyl Alcohol
Methylene Chloride
Propylene Glycol
Triacetin
Trichloroethylene

Stabilizers, Suspending Agents

Acacia
Agar
Alginic Acid
Ammonium Alginate
Brominated Vegetable Oil
Calcium Alginate
Calcium Stearoyl-2-lactylate
Carrageenan
Disodium EDTA
Food Starch, Modified
Glycerol Ester of Wood Rosin
Guar Gum
Hydroxypropyl Cellulose
Hydroxypropyl Methylcellulose
Karaya Gum
Lactated Mono-Diglycerides
Lactylated Fatty Acid Esters of Glycerol and Propylene Glycol
Locust Bean Gum
Methylcellulose
Methyl Ethylcellulose
Mono- and Diglycerides

Pectin
Poloxamer 331
Poloxamer 407
Polysorbate 20
Polysorbate 60
Polysorbate 65
Polysorbate 80
Potassium Alginate
Propylene Glycol Alginate
Propylene Glycol Monostearate
PVP
Sodium Alginate
Sodium Carboxymethylcellulose
Sodium Stearoyl-2-lactylate
Sorbitan Monostearate
Stearyl Monoglyceridyl Citrate
Tragacanth
Xanthan Gum

Surface-active Agents, Wetting Agents

Dioctyl Sodium Sulfosuccinate
Lactylated Fatty Acid Esters of Glycerol and Propylene Glycol
Lactylic Esters of Fatty Acids
Propylene Glycol
Sodium Lauryl Sulfate

Surface-finishing Agents, Glazes, Polishes Coating Agents, Film-formers

Acetylated Monoglycerides
Beeswax, White
Beeswax, Yellow
Carnauba Wax
Castor Oil
Ethylcellulose
Hydroxypropyl Cellulose
Methylcellulose
Mineral Oil, White
Petrolatum
Polyethylene Glycols
Rice Bran Wax

Texturizers, Texture-modifying Agents

Acetylated Monoglycerides
Mannitol
Potassium Pyrophosphate
Potassium Tripolyphosphate
Sodium Metaphosphate
Sodium Phosphate, Dibasic
Sodium Tripolyphosphate
Sorbitol
Sorbitol Solution

Thickeners, Gelling Agents

Acacia
Agar
Alginic Acid
Ammonium Alginate
Calcium Alginate
Carrageenan
Food Starch, Modified
Guar Gum
Hydroxypropyl Cellulose
Hydroxypropyl Methylcellulose
Karaya Gum
Locust Bean Gum
Methylcellulose
Pectin
Potassium Alginate
Potassium Chloride
Propylene Glycol Alginate
Sodium Alginate
Sodium Carboxymethylcellulose
Tragacanth
Xanthan Gum

Yeast Foods

Ammonium Chloride
Ammonium Phosphate, Dibasic
Ammonium Phosphate, Monobasic
Ammonium Sulfate
Calcium Carbonate
Calcium Lactate
Calcium Oxide
Calcium Phosphate, Dibasic
Calcium Phosphate, Monobasic
Calcium Sulfate
Potassium Chloride
Potassium Phosphate, Dibasic
Potassium Phosphate, Monobasic

Other Functional Uses, Miscellaneous

Caffeine (STIMULANT)
Carbon, Activated (DECOLORIZING AGENT)
β-Carotene (COLOR)
Diatomaceous Silica (FILTER AID)
Gibberellic Acid (ENZYME ACTIVATOR)
Hydrogen Peroxide (STARCH MODIFIER)
Hydroxylated Lecithin (CLOUDING AGENT)
Methyl Formate (FUMIGANT)
Mineral Oil, White (FERMENTATION AID)
Monoglyceride Citrate (ANTIOXIDANT ADJUNCT)
Oxystearin (CRYSTALLIZATION INHIBITOR)

Potassium Gibberellate (ENZYME ACTIVATOR)
Potassium Sulfate (WATER CORRECTIVE)
Sodium Methylate (CATALYST)
Sodium Sulfate (CARAMEL PRODUCTION)
Stannous Chloride (REDUCING AGENT)

Index

Monograph titles are shown in boldface type. Asterisk (*) indicates new monograph for Second Edition. Dagger (†) indicates monograph included in one of the supplements to the First Edition, but not in the bound volume.

A

PAGE

Abbreviation, F.C.C. 1
Abbreviations, Weights and Measures 6
Acacia **9**
Acetaldehyde **10**
Acetals 892
Acetanisole **12**
Acetic Acid, Glacial **13**
Acetic Acid T.S., Diluted . . . 985
Acetoacetic Ester 272
Acetoin **14**
2′-Acetonaphthone 529
***Acetone** **15**
Acetone Peroxides **17**
***Acetophenone** **18**
Acetylated Mono- and Diglycerides 19
Acetylated Monoglycerides **19**
Acetyl Groups 959
Acetyl Methyl Carbinol 14
3-Acetyl-6-methyl-1,2-pyran-2,4(3*H*)-dione 234
Acetyl Propionyl 587
Acetyl Value 902
Acidity, Volatile 916
Acid Value (Essential Oils and Related Substances) . . 893
Acid Value (Fats and Related Substances) 902
Acid Value (Rosins, Rosin Esters, and Related Substances) 945
Acknowledgments xii
Activated 7-Dehydrocholesterol 855
Added Substances 8
Adipic Acid **21**
Advisory Panel on the Food Chemicals Codex vii
Agar **22**
*DL-**Alanine** **24**
*L-**Alanine** **25**
Alcohol 277
Alcohol, Aldehyde-Free 986
Alcohol C-6 372
Alcohol C-8 560
Alcohol C-9 555
Alcohol C-10 233
Alcohol C-12 440
Alcohol, 70% 985
Alcohol, 80% 980
Alcohol, 90% 986
Alcoholic Potassium Hydroxide T.S. 986
Alcohols, Total 893
Aldehyde C-7 365
Aldehyde C-8 558
Aldehyde C-9 554
Aldehyde C-10 231
Aldehyde C-11 Undecyclic . . . 845
Aldehyde C-11 (Undecylenic) 846
Aldehyde C-12 441
Aldehyde C-16 295
Aldehyde C-12 MNA 536
Aldehyde C-14 Pure (so-called) 844
Aldehyde C-18 (So-called) . . . 553
Aldehydes 894
Aldehydes and Ketones—Hydroxylamine Method . . 894
Aldehydes and Ketones—Neutral Sulfite Method . . . 895
Algin 43, 117, 641, 721
Alginates Assay 863
Algin Derivative 680
Alginic Acid **26**
Alkaline Cupric Tartrate T.S. 986
Alkaline Mercuric-Potassium Iodide T.S. 986
Alkali Salts of Organic Acids Assay 864
Allspice Oil 618
p-Allylanisole 267
Allyl Caproate 29
Allyl Cyclohexanepropionate . **28**
Allyl-3-cyclohexanepropionate 28

4-Allyl-1,2-dimethoxy Benzene . . . 528
4-Allylguaiacol . . . 306
***Allyl Hexanoate** . . . **29**
Allyl Ionone . . . 30
Allyl α-Ionone . . . **30**
Allyl Isothiocyanate . . . **31**
4-Allyl-2-methoxyphenol . . . 306
Almond Oil, Bitter, FFPA **32**
Almond Oil, Bitter, Free From Prussic Acid . . . 32
Alphazurine 2G . . . 1004
Alternate Methods . . . 1
Aluminum Ammonium Sulfate . . . **34**
Aluminum Potassium Sulfate . . . **36**
Aluminum Sodium Sulfate . . . **37**
Aluminum Sulfate . . . **39**
***Aluminum Sulfate Solution** . . . **40**
Alum Liquor . . . 40
Ambrette Seed Liquid . . . 42
Ambrette Seed Oil . . . **42**
Aminoacetic acid . . . 359
L-1-Amino-4-guanidovaleric Acid . . . 63
L-1-Amino-4-guanidovaleric Acid Monohydrochloride . . 65
L-2-Amino-3-hydroxybutyric Acid . . . 818
DL-2-Amino-3-hydroxypropanoic Acid . . . 713
L-2-Amino-3-hydroxypropanoic Acid . . . 714
DL-α-Amino-3-indolepropionic Acid . . . 838
L-α-Amino-3-indolepropionic Acid . . . 841
L-2-Amino-3-mercaptopropanoic Acid Monohydrochloride . . . 226
L-2-Amino-3-methylbutyric Acid . . . 849
DL-2-Amino-4-(methylthio)-butyric Acid . . . 504
L-2-Amino-4-(methylthio)-butyric Acid . . . 506
DL-2-Amino-3-methylvaleric Acid . . . 414
L-2-Amino-3-methylvaleric Acid . . . 415
DL-2-Amino-4-methylvaleric Acid . . . 450
L-2-Amino-4-methylvaleric Acid . . . 451
L-2-Aminopentanedioic Acid. 347
2-Aminopentanedioic Acid Hydrochloride . . . 349
DL-α-Amino-β-phenylpropionic Acid . . . 607
L-α-Amino-β-phenylpropionic Acid . . . 610
DL-2-Aminopropanoic Acid . . 24
L-2-Aminopropanoic Acid . . . 25
DL-Aminosuccinic Acid . . . 69
Ammonia-Ammonium Chloride Buffer T.S. . . . 986
Ammoniacal Silver Nitrate T.S. . . . 986
Ammonia Solution, Strong . . 48
Ammonia Standard Solution 985
Ammonia T.S. . . . 986
Ammonia T.S., Stronger . . . 986
Ammonia Water, Stronger . . . 48
Ammonium Acetate T.S. . . . 986
Ammonium Alginate . . . **43**
Ammonium Alum . . . 34
Ammonium Bicarbonate. **44**
Ammonium Carbonate . . . **45**
Ammonium Carbonate T.S. . 986
Ammonium Chloride . . . **47**
Ammonium Chloride T.S. . . . 987
Ammonium Glutamate . . . 539
Ammonium Hydroxide . . . **48**
Ammonium Molybdate T.S. . 987
Ammonium Oxalate T.S. . . . 987
Ammonium Phosphate, Dibasic . . . **49**
Ammonium Phosphate, Monobasic . . . **50**
Ammonium Saccharin . . . **51**
Ammonium Sulfanilate T.S. . 987
Ammonium Sulfate . . . **52**
Ammonium Sulfide T.S. . . . 987
Ammonium Thiocyanate, 0.1 *N* . . . 996
Ammonium Thiocyanate T.S. 987
Amyl Acetate . . . 397
Amyl Butyrate . . . 398
Amylcinnamaldehyde . . . 53
α-Amylcinnamaldehyde . . **53**
Amyl Formate . . . 399
Amyl Hexanoate . . . 400
Amyl Isovalerianate . . . 401
Amyl Valerate . . . 401
Amyris Oil . . . **54**
Analytical Samples . . . 2
Anethole . . . **56**
Aneurine Hydrochloride . . . 815
Angelica Root Oil . . . **57**
Angelica Seed Oil . . . **58**
Angular Rotation . . . 939

p-Anisaldehyde 507
Anise Oil **59**
Anisic Alcohol 62
Anisic Aldehyde 507
Anisole **60**
Anisyl Acetate **61**
Anisyl Alcohol **62**
Apparatus 2
Apparatus and General Tests 861
D-Araboascorbic Acid 266
***L-Arginine** **63**
***L-Arginine Monohydrochloride** **65**
Arsenic Standard Solution 865
Arsenic Test 865
Ascorbic Acid **66**
L-Ascorbic Acid 66
Ascorbyl Palmitate **68**
Ash (Acid-Insoluble) 869
Ash (Total) 868
***DL-Aspartic Acid** **69**
Assays and Tests 2
Atomic Weights and Chemical Formulas 1
Azodicarbonamide **70**
Azo Violet 1004

B

Baking Soda 727
Balances and Weights 980
Balsam Fir Oil 322
Balsam Peru Oil **72**
Barium Chloride T.S. 987
Barium Diphenylamine Sulfonate T.S. 987
Barium Standard Solution 985
Basil Oil **73**
Bay Leaf Oil 438
Bay Oil **74**
Beeswax, White **75**
Beeswax, Yellow **77**
Benedict's Qualitative Reagent 987
Benzaldehyde **78**
Benzidine T.S. 987
1,2-Benzisothiazolin-3-one 1,1-Dioxide 706
1,2-Benzisothiazolin-3-one 1,1-Dioxide Ammonium Salt 51
1,2-Benzisothiazolin-3-one 1,1-Dioxide Calcium Salt 154
1,2-Benzisothiazolin-3-one 1,1-Dioxide Sodium Salt 764
Benzoic Acid **79**
Benzoic Aldehyde 78
Benzophenone **81**
o-Benzosulfimide 706
Benzoylbenzene 81
Benzoyl Peroxide **82**
Benzoyl Superoxide 82
Benzyl Acetate **83**
Benzyl Alcohol **84**
Benzyl Benzoate **85**
Benzyl Butyrate **86**
Benzyl *n*-Butyrate 86
Benzyl Cinnamate **87**
Benzyl Phenylacetate **88**
Benzyl Propionate **90**
Benzyl Salicylate **91**
***Bergamot Oil, Expressed** **92**
BHA **93**
BHT **95**
Biotin **96**
Birch Tar Oil, Rectified **97**
Bismuth Nitrate T.S. 987
Blank Tests 2
Bois De Rose Oil **98**
Bornyl Acetate **99**
Bound Styrene 873
Brominated Vegetable Oil **100**
Bromine, 0.1 *N* 997
Bromine T.S. 987
Bromocresol Blue 1004
Bromocresol Blue T.S. 987
Bromocresol Green 1004
Bromocresol Green T.S. 988
Bromocresol Purple 1004
Bromocresol Purple T.S. 988
Bromophenol Blue 1004
Bromophenol Blue T.S. 988
Bromothymol Blue 1004
Bromothymol Blue T.S. 988
Buffer Solutions, Preparation of 983
***Butadiene-Styrene 75/25 Rubber** **102**
***Butadiene-Styrene 50/50 Rubber** **103**
1,4-Butanedicarboxylic Acid 21
Butanedioic Acid 800
2,3-Butanedione 238
1-Butanol 107
†2-Butanone **104**
trans-Butenedioic Acid 331
†Butyl Acetate **106**
n-Butyl Acetate 106
†Butyl Alcohol **107**
Butyl Aldehyde 112
Butylated Hydroxyanisole 93
†Butylated Hydroxymethylphenol **108**
Butylated Hydroxytoluene 95

Butyl Butyrate 109
n-Butyl *n*-Butyrate 109
†**Butyl Butyryllactate** 110
2,6-Di-*tert*-butyl-*p*-cresol 95
1,3-Butylene Glycol 111
Butyl Rubber 407
†**Butyraldehyde** 112
Butyric Acid 114
Butyrin 835
Butyryllactic Acid, Butyl Ester 110

C

Caffeine 115
Calciferol 853
Calcium Acetate 116
Calcium Alginate 117
Calcium Ascorbate 119
Calcium Biphosphate 147
Calcium Bromate 120
Calcium Carbonate 121
Calcium Carbonate, Precipitated 121
Calcium Chloride 123
Calcium Chloride, Anhydrous 124
Calcium Chloride Double Salt of *dl*-Calcium Pantothenate 143
Calcium Chloride T.S. 988
Calcium Citrate 126
Calcium Cyclamate ix
Calcium Dioxide 144
Calcium Disodium Edetate 127
Calcium Disodium EDTA 127
Calcium Disodium Ethylenediaminetetraacetate 127
Calcium Disodium (Ethylenedinitrilo)tetraacetate 127
Calcium 4-(β,D-Galactosido)-D-gluconate 136
Calcium Gluconate 129
Calcium Glycerophosphate 130
Calcium Hydroxide 131
Calcium Hydroxide T.S. 988
Calcium Iodate 133
Calcium Lactate 134
Calcium Lactobionate 136
Calcium Oxide 138
Calcium Pantothenate 140
Calcium Pantothenate Assay 869
Calcium Pantothenate, Racemic 141
Calcium Pantothenate, Racemic-Calcium Chloride Complex 143
Calcium Peroxide 144
Calcium Phosphate, Acid 147
Calcium Phosphate, Dibasic 146
Calcium Phosphate, Monobasic 147
Calcium Phosphate, Precipitated 149
Calcium Phosphate, Tribasic 149
Calcium Propionate 151
Calcium Pyrophosphate 153
Calcium Saccharin 154
Calcium Silicate 156
Calcium Stearate 158
†**Calcium Stearoyl-2-lactylate** 160
Calcium Sulfate 163
Calcium Superoxide 144
Cananga Oil 164
***Candellila Wax** 165
Capraldehyde 231
Capric Acid 232
Caproic Acid 370
Capryl Alcohol 560
Caprylic Acid 558
Caprylic Aldehyde 558
Caraway Oil 166
***Carbon, Activated** 167
Carbon, Active 167
Carbon, Decolorizing 167
Cardamom Oil 169
Carnauba Wax 170
Carob Bean Gum 464
β-Carotene 171
Carrageenan 172
Carrot Seed Oil 174
Carvacrol 175
***d*-Carvone** 176
***l*-Carvone** 178
β-Caryophyllene 179
Cascarilla Oil 180
Cassia Oil 181
Castor Oil 182
Caustic Potash 652
Caustic Soda 743
Cedar Leaf Oil 183
Cedar Leaf Oil, White 183
Celery Seed Oil 184
Cellulose Gel 185
Cellulose Gum 731
†**Cellulose, Microcrystalline** 185
Ceric Ammonium Nitrate T.S. 988
Ceric Sulfate, 0.1 *N* 997
Ceric Sulfate, 0.01 *N* for Tocopherol Assay 997

Chamomile Oil, English . . **186**
Chamomile Oil, German. **187**
Chamomile Oil, Hungarian . . 187
Charcoal, Activated 167
Chemical Formulas and
Atomic Weights 1
Chewing Gum Base, General
Tests 873
Chloride Standard Solution . . 879
Chloride and Sulfate Limit
Tests 879
Chlorinated Compounds 895
Chlorine T.S. 988
Cholalic Acid 188
Cholecalciferol 855
Cholic Acid **188**
Choline Bitartrate **190**
Choline Chloride **191**
Chromatography 880
Chromatography, Adsorption 880
Chromatography, Ascending
Paper 883
Chromatography, Column . . . 880
Chromatography,
Descending Paper 882
Chromatography, Gas 886
Chromatography,
Ion-Exchange Resin 881
Chromatography, Paper 882
Chromatography, Partition . . 881
Chromatography, Use of
Reference Substances in
Identity Tests 883
Chromatography,
Thin-Layer 884
Chromotropic Acid T.S. 988
Cinene 456
Cineole, Percentage of 898
Cinnamal 192
Cinnamaldehyde **192**
†**Cinnamic Acid** **194**
Cinnamic Aldehyde 192
Cinnamon Bark Oil,
Ceylon **195**
Cinnamon Leaf Oil **196**
Cinnamon Oil 181
Cinnamyl Acetate **197**
Cinnamyl Alcohol **198**
Cinnamyl Anthranilate . . . **199**
Cinnamyl Formate **200**
Cinnamyl Isovalerate **201**
Cinnamyl Propionate **202**
Citral **202**
Citric Acid **204**
Citronellal **205**
Citronellol **206**
Citronellyl Acetate **208**
Citronellyl Butyrate **209**
Citronellyl Formate **210**
Citronellyl Isobutyrate . . . **211**
Citronellyl Propionate **212**
Citrus Oils, Ultraviolet
Absorbance of 900
Clary Oil **213**
Clary Sage Oil 213
Clove Leaf Oil **214**
Clove Oil **215**
Clove Stem Oil **217**
CMC 731
Coagulated or Concentrated
Latices of Vegetable Origin 501
Cobaltous Chloride C.S. 942
Cobaltous Chloride T.S. 988
Cobalt-Uranyl Acetate T.S. . 988
Codex Standards 1
Cognac Oil, Green **217**
Colorimetric Solutions (C.S.) 983
Colorimetry 957
Committee on Specifications viii
Committee on Specifications,
Advisors to viii
Congo Red T.S. 988
Constant Weight, Definition 3
Container, Definition 7
Container, Light-resistant . . . 7
Container, Tight 8
Container, Well-closed 8
Copaiba Oil **218**
Copper Gluconate **219**
Coriander Oil **220**
Costus Root Oil **221**
Cream of Tartar 639
Cresol Red 1004
Cresol Red T.S. 989
Cresol Red-Thymol Blue T.S. 989
Cresyl Acetate **222**
p-Cresyl Isobutyrate 831
p-Cresyl Methyl Ether 511
Crude Fat 960
Crystal Violet 1004
Crystal Violet T.S. 989
Cubeb Oil **223**
Cumin Oil **224**
Cupric Citrate T.S., Alkaline 989
Cupric Sulfate C.S. 943
Cupric Sulfate Test Paper . . . 1007
Cupric Sulfate T.S. 989
Cupric Tartrate T.S.,
Alkaline 989
Cyanogen Bromide T.S. 989
Cyclamen Aldehyde **225**
1,2,3,5/4,6-Cyclohexanehexol 389
Cyclohexylsulfamic Acid ix
*L-**Cysteine**

Monohydrochloride **226**
L-**Cystine** **227**

D

Δ-**Decalactone** **229**
Decanal **231**
Decanoic Acid **232**
*__1-Decanol__ (__Natural__) **233**
Decyl Alcohol 233
Dehydroacetic Acid **234**
Denigès' Regent 989
Deoxycholic Acid 235
Description in Monographs, Significance of 6
Desiccators and Desiccants . . 3
Desoxycholic Acid **235**
*__Dexpanthenol__ **236**
Dextro Calcium Pantothenate 140
Dextro-Carvone 176
Diacetyl **238**
†**Diacetyl Tartaric Acid Esters of Mono- and Diglycerides** **239**
2,6-Diaminohexanoic Acid Hydrochloride 467
Diammonium Phosphate 49
*__Diatomaceous Silica__ **242**
Dicalcium Phosphate 146
Dichloroethane 287
Dichloromethane 520
†**Diethyl Malonate** **243**
Diethylpyrocarbonate ix
†**Diethyl Sebacate** **244**
†**Diethyl Succinate** **245**
Dihydroanethole 678
13α,12α-Dihydroxycholanic Acid 235
2,7-Dihydroxynaphthalene T.S. 989
1,2-Dihydroxypropane 678
Dilauryl Thiodipropionate 246
Dill Herb Oil 250
Dill Oil 250
Dill Oil, Indian 248
Dill Seed Oil, East Indian . . . 248
Dill Seed Oil, European . . . **247**
Dill Seed Oil, Indian **248**
Dillweed Oil, American . . . **250**
Dimagnesium Phosphate 475
1,2-Dimethoxy-4-allylbenzene 524
Dimethyl Anthranilate . . . **251**
Dimethyl Benzyl Carbinol **252**
Dimethyl Benzyl Carbinyl Acetate **253**
Dimethyldiketone 238
Dimethylglyoxal 238
Dimethylketol 14
Dimethyl Ketone 15
cis-3,7-Dimethyl-2,6-octadien-1-al 202
trans-3,7-Dimethyl-2,6-octadien-1-al 202
cis-3,7-Dimethyl-2,6-octadien-1-ol 547
trans-3,7-Dimethyl-2,6-octadien-1-ol 335
3,7-Dimethyl-1,6-octadien-3-ol 458
3,7-Dimethyl-2,6-octadien-1-yl Acetate 337
3,7-Dimethyl-1,6-octadien-3-yl Acetate 459, 460
3,7-Dimethyl-2,6-octadien-1-yl Benzoate 338
3,7-Dimethyl-1,6-octadien-3-yl Benzoate 461
3,7-Dimethyl-2,6-octadien-1-yl Butyrate 339
3,7-Dimethyl-2,6-octadien-1-yl Formate 340
3,7-Dimethyl-2,6-octadien-3-yl Isobutyrate 462
3,7-Dimethyl-2,6-octadien-1-yl Phenylacetate 341
3,7-Dimethyl-2,6-octadien-1-yl Propionate 342
3,7-Dimethyl-2,6-octadien-3-yl Propionate 463
Dimethyl Octanol 254
3,7-Dimethyl-1-octanol . . . **254**
3,7-Dimethyl-6-octen-1-al . . . 205
3,7-Dimethyl-6-octen-1-ol . . . 206
3,7-Dimethyl-6-octen-1-yl Acetate 208
3,7-Dimethyl-6-octen-1-yl Butyrate 209
3,7-Dimethyl-6-octen-1-yl Formate 210
3,7-Dimethyl-6-octen-1-yl Isobutyrate 211
3,7-Dimethyl-6-octen-1-yl Propionate 212
α,α-Dimethylphenethyl Acetate 253
α,α-Dimethylphenethyl Alcohol 252
Dimethylpolysiloxane **255**
Dimethylpolysiloxane, Viscosity of 970
Dimethyl Silicone 255
Dioctyl Sodium Sulfosuccinate **256**

Diphenylamine T.S. 989
Diphenyl Ketone 81
Dipotassium Monophosphate 661
Dipotassium Phosphate 661
α,α′-Dipyridyl T.S. 989
Disodium Dihydrogen
Pyrophosphate 719
Disodium Edetate 258
Disodium EDTA **258**
Disodium Ethylenediamine-
tetraacetate 258
Disodium Ethylenediamine-
tetraacetate, 0.05 *M* 998
Disodium (Ethylenedinitrilo)-
tetraacetate 258
Disodium Guanosine-5′-
monophosphate 261
***Disodium Guanylate** **261**
***Disodium Inosinate** **263**
Disodium
Inosine-5′-monophosphate 263
Disodium Monohydrogen
Phosphate 756
Disodium Phosphate 756
Disodium Pyrophosphate . . . 719
Disodium Tartrate 777
Disodium *d*-Tartrate 777
Distillation Range 890
L-3,3′-Dithiobis(2-amino-
propanoic acid) 227
Dithizone 1005
Dithizone T.S. 989
Division of Biology and
Agriculture vii
Δ-Dodecalactone **265**
Dodecanal 441
Dodecanoic Acid 439
1-Dodecanol 440
Drop Method, Softening
Point 946
DSS 256

E

Enanthic Alcohol 368
Eosin Y T.S. 989
Epsom Salt 483
Ergocalciferol 853
Eriochrome Black T 1005
Eriochrome Black T.S. 990
Erythorbic Acid **266**
Essential Oils and Related
Substances, General Tests . 892
Ester Determination 896
Ester Determination (High
Boiling Solvent) 896
Esters 896
Ester Value 897
Estragole **267**
Estragon Oil 807
Ethanal 10
Ethanol 277
p-Ethoxychrysoidin
Monohydrochloride 1005
p-Ethoxychrysoidin T.S. 990
6-Ethoxy-1,2-dihydro-2,2,4-
trimethylquinoline 270
3-Ethoxy-4-hydroxybenzal-
dehyde 303
1-Ethoxy-2-hydroxy-4-pro-
penylbenzene 674
†**Ethoxylated Mono- and
Diglycerides** **268**
***Ethoxyquin** **270**
Ethyl Acetate **271**
†**Ethyl Acetoacetate** **272**
†**Ethyl Acrylate** **274**
***Ethyl Alcohol** **277**
Ethyl *o*-Aminobenzoate 280
Ethyl *p*-Anisate **279**
Ethyl Anthranilate **280**
Ethyl Benzoate **281**
Ethyl Butyl Ketone 367
†**2-Ethylbutyraldehyde** **282**
Ethyl Butyrate **283**
†**2-Ethylbutyric Acid** **283**
Ethyl Caproate 290
Ethyl Capronate 290
Ethyl Caprylate 297
Ethylcellulose **284**
Ethyl Cinnamate **286**
Ethyl Citrate 837
Ethyl Dodecanoate 293
trans-1,2-Ethylenedicar-
boxylic Acid 331
***Ethylene Dichloride** **287**
Ethylene Trichloride 836
Ethyl Formate **288**
Ethyl Heptanoate **289**
Ethyl Heptoate 289
Ethyl Hexanoate **290**
Ethyl 2-Hydroxypropionate . 292
Ethyl Isovalerate **291**
Ethyl Lactate **292**
Ethyl Laurate **293**
Ethyl Malonate 242
†**Ethyl Maltol** **294**
Ethyl *p*-Methoxybenzoate . . . 279
Ethyl 2-Methylbutyrate 291
**Ethyl Methyl
Phenylglycidate** **295**
Ethyl Nonanoate **296**
Ethyl Octanoate **297**
Ethyl Octoate 297

Ethyl 3-Oxobutanoate...... 272
†**Ethyl Oxyhydrate (So-Called)**............ **298**
Ethyl Pelargonate.......... 296
Ethyl Phenylacetate...... **300**
Ethyl Phenylglycidate.... **301**
Ethyl 3-Phenylpropenoate... 286
Ethyl Propionate......... **301**
Ethyl Salicylate.......... **302**
Ethyl Sebacate............ 244
Ethyl Succinate............ 245
Ethyl Vanillin............ **303**
Eucalyptus Oil........... **305**
Eugenic Acid.............. 306
Eugenol................. **306**
Eugenyl Methyl Ether...... 524

F

Fat, Crude................ 960
Fats and Related Substances, General Tests............ 902
Fatty Acids, Free.......... 903
Federal Statutes, Compliance with.................. iv
Fehling's Solution.......... 990
Fennel Oil.............. **307**
Ferric Ammonium Sulfate T.S.................... 990
Ferric Chloride C. S........ 943
Ferric Chloride T.S......... 990
Ferric Chloride T.S., Alcoholic............... 990
Ferric Orthophosphate...... 309
Ferric Phosphate........ **309**
Ferric Pyrophosphate..... **312**
Ferrous Fumarate........ **313**
Ferrous Gluconate....... **318**
Ferrous Sulfate.......... **320**
Ferrous Sulfate, Dried.... **321**
Ferrous Sulfate T.S......... 990
Fir Needle Oil, Canadian.. **322**
Fir Needle Oil, Siberian... **323**
Fluoride Limit Test........ 917
Food and Nutrition Board... vii
Food Protection Committee. vii
***Food Starch, Modified**.... **324**
Formaldehyde T.S.......... 990
Formic Acid............. **330**
Free Fatty Acids........... 903
Free Glycerin or Propylene Glycol.................. 904
Free Phenols.............. 899
Fumaric Acid............ **331**
Functional Use in Foods, Substances Classified by.. 1011
Functional Use in Foods, Significance of, Statements in Monographs.......... 7
2-Furaldehyde............. 333
†**Furfural**................ **333**

G

Gallic Acid, Propyl Ester.... 683
Garlic Oil............... **334**
Gaultheria Oil............. 534
General Provisions......... 1
General Specifications of Codex................. 6
General Tests and Apparatus 861
Geranial.................. 202
Geraniol................ **335**
Geranium Oil, Algerian... **336**
Geranium Oil, East Indian.. 570
Geranium Oil, Turkish...... 570
Geranyl Acetate.......... **337**
Geranyl Benzoate........ **338**
Geranyl Butyrate......... **339**
Geranyl Formate......... **340**
Geranyl Phenylacetate.... **341**
Geranyl Propionate....... **342**
***Gibberellic Acid**.......... **343**
Ginger Oil.............. **345**
D-Glucitol................ 786
Glucono Delta-Lactone... **346**
Gluside.................. 706
Glutamic acid............. 347
L-Glutamic Acid......... **347**
Glutamic Acid Hydrochloride.......... **349**
Glycerin................ **350**
Glycerin and Propylene Glycol, Free............. 904
Glycerol................. 350
***Glycerol Ester of Partially Dimerized Rosin**....... **354**
***Glycerol Ester of Partially Hydrogenated Wood Rosin**................ **355**
***Glycerol Ester of Polymerized Rosin**..... **356**
***Glycerol Ester of Tall Oil Rosin**................ **357**
***Glycerol Ester of Wood Rosin**................ **358**
Glyceryl Triacetate......... 833
Glyceryl Tributyrate....... 835
Glycine................ **359**
Glycocoll................. 359
Graham's Salt............ 749
Granular Metal Powders, Sieve Analysis of......... 953
Grapefruit Oil, Coldpressed.. 360

Grapefruit Oil, Expressed . **360**
Guaiac Resin . 363
Guar Gum . **361**
Gum Arabic . 9
***Gum Guaiac** . **363**

H

Hanus Method for Iodine . 905
Heavy Metals Test . 920
Heliotropine . 620
Heptaldehyde . 365
Heptanal . **365**
†2-Heptanone . **366**
†3-Heptanone . **367**
Heptyl Alcohol . **368**
n-Heptyl-*p*-hydroxybenzoate 369
†Heptylparaben . **369**
Hexadecanoic Acid . 571
2,4-Hexadienoic Acid . 783
2,4-Hexadienoic Acid, Potassium Salt . 668
cis-Hexahydro-2-oxo-1H-thieno[3,4]-imidazole-4-valeric Acid . 96
Hexanedioic Acid . 21
1,2,3,4,5,6-Hexanehexol . 786
Hexanoic Acid . **370**
1-Hexanol . 372
†Hexyl Alcohol (Natural) . **372**
α-Hexylcinnamaldehyde . **373**
Hops Oil . **374**
Hydrated Silica . 716
Hydratropic Aldehyde . 612
Hydratropic Aldehyde Dimethyl Acetal . 614
Hydrochloric Acid . **375**
Hydrochloric Acid, 1 *N* . 998
Hydrochloric Acid Table . 922
Hydrochloric Acid T.S., Diluted . 990
Hydrocinnamaldehyde . 613
Hydrocinnamyl Alcohol . 611
Hydrogen Peroxide . **377**
Hydrogen Peroxide T.S. . 990
Hydrogen Sulfide T.S. . 990
α-Hydro-*omega*-hydroxy-poly-(oxyethylene)poly-(oxypropylene)(51–57 moles)poly-(oxyethylene) Block Copolymer . 621
α-Hydro-*omega*-hydroxy-poly-(oxyethylene)poly-(oxypropylene)(63–71 moles)poly-(oxyethylene) Block Copolymer . 624
3-Hydroxy-2-butanone . 14
Hydroxycitronellal . **380**
Hydroxycitronellal Dimethyl Acetal . **381**
7-Hydroxy-3,7-dimethyl Octanal . 380
7-Hydroxy-3,7-dimethyl Octanal: acetal . 381
3-Hydroxy-2-ethyl-4-pyrone . 294
(2-Hydroxyethyl)trimethyl-ammonium Bitartrate . 190
(2-Hydroxyethyl)trimethyl-ammonium Chloride . 191
Hydroxylamine Hydrochloride, 0.5 *N* . 998
Hydroxylamine Hydrochloride T.S. . 990
†Hydroxylated Lecithin . **382**
Hydroxyl Value . 904
4-Hydroxy-3-methoxybenz-aldehyde . 850
4-Hydroxymethyl-2,6-di-*tert*-butylphenol . 108
5-Hydroxy-6-methyl-3,4-pyridinedimethanol Hydrochloride . 689
3-Hydroxy-2-methyl-4-pyrone . 487
Hydroxy Naphthol Blue . 1005
L-β-(*p*-Hydroxyphenyl)-alanine . 842
2-Hydroxypropionic Acid . 428
Hydroxypropoxyl Determination . 923
Hydroxypropyl Alginate . 680
Hydroxypropyl Cellulose . **385**
Hydroxypropyl Methylcellulose . **387**
8-Hydroxyquinoline T.S. . 990
Hydroxysuccinic Acid . 484

I

Identification—General Tests 925
Identification Tests in Monographs . 3
Indicator Papers and Test Papers . 1007
Indicators . 1004
Indicators, Quantity to Use . 3
Indigo Carmine T.S. . 990
Indole . **389**
Inositol . **389**
i-Inositol . 389
meso-Inositol . 389
Iodine, 0.1 *N* . 998
Iodine T.S. . 991
Iodine Value . 905

α-**Ionone** **391**
β-**Ionone** **392**
***Iron, Electrolytic** **393**
Iron Phosphate 309
Iron Pyrophosphate 312
Iron, Reduced **394**
Iron Standard Solution 985
***Isoamyl Acetate** **397**
Isoamyl Butyrate **398**
Isoamyl Caproate 400
Isoamyl Formate **399**
***Isoamyl Hexanoate** **400**
Isoamyl Isovalerate **401**
Isoamyl Salicylate **402**
Isobornyl Acetate **403**
Isobutanol 405
Isobutyl Acetate **404**
†**Isobutyl Alcohol** **405**
Isobutyl Cinnamate **406**
***Isobutylene-Isoprene Copolymer** **407**
Isobutyl Phenylacetate **408**
Isobutyl Salicylate **409**
†**Isobutyraldehyde** **410**
†**Isobutyric Acid** **411**
Isoeugenol **412**
Isoeugenyl Acetate **413**
Isoeugenyl Methyl Ether 528
*DL-**Isoleucine** **414**
*L-**Isoleucine** **415**
Isopropanol 417
†**Isopropyl Acetate** **416**
Isopropylacetic Acid 420
***Isopropyl Alcohol** **417**
Isopropylformic Acid 411
Isopulegol **419**
Isovaleric Acid **420**

J

Juniper Berries Oil **421**

K

***Kaolin** 421
Karaya Gum **423**
Karl Fischer Titrimetric Method for Water Determination 977
Kasal 724
***Kelp** **424**
Ketones and Aldehydes—Hydroxylamine Method 894
Ketones and Aldehydes—Neutral Sulfite Method 895
Kurrol's Salt 749

L

Labdanum Oil **426**
Labels, Indication of F.C.C. Grade on 1
Lactated Mono-Diglycerides **427**
Lactic Acid **428**
Lactic Acid, Butyl Ester, Butyrate 110
***Lactylated Fatty Acid Esters of Glycerol and Propylene Glycol** **430**
Lactylic Esters of Fatty Acids **433**
***Lanolin, Anhydrous** **437**
Laurel Leaf Oil **438**
Lauric Acid **439**
Lauryl Alcohol (**Natural**) **440**
Lauryl Aldehyde **441**
Lavandin Oil, Abrial **442**
Lavender Oil **443**
Lead Acetate Test Paper 1007
Lead Acetate T.S. 991
Lead Limit Test 929
Lead Subacetate T.S. 991
Lead Subacetate T.S., Diluted 991
Lecithin **444**
Legal Status of Codex xi
***Lemongrass Oil** **446**
Lemon Oil **448**
*DL-**Leucine** 450
L-**Leucine** **451**
Levo-Bornyl Acetate 99
Levo-Carvone 178
Lime 138
Lime Oil, Distilled **453**
Lime, Slaked 131
***Limestone, Ground** **454**
Limits of Impurities, General Policy on x
d*-Limonene** **456**
Linaloe Wood Oil **457**
Linalool **458**
Linalool Determination 897
Linalyl Acetate **459**
Linalyl Acetate, Synthetic **460**
Linalyl Benzoate **461**
Linalyl Isobutyrate **462**
Linalyl Propionate **463**
Litmus 1005
Litmus Paper, Blue 1007
Litmus Paper, Red 1007
Litmus T.S. 991
Locust Bean Gum **464**
Loss on Drying 931

Loss on Drying and Water . . 3
Lovage Oil . . . **466**
†**L-Lysine Monohydrochloride** . . . **467**

M

Mace Oil . . . **468**
Magnesia Mixture T.S. . . . 991
Magnesium Carbonate . . . **469**
†**Magnesium Chloride** . . . **471**
Magnesium Hydroxide . . . **472**
Magnesium Oxide . . . **473**
Magnesium Phosphate, Dibasic . . . **475**
Magnesium Phosphate, Tribasic . . . **477**
Magnesium Silicate . . . **479**
Magnesium Standard Solution . . . 985
Magnesium Stearate . . . **481**
Magnesium Sulfate . . . **483**
Magnesium Sulfate T.S. . . . 991
Malachite Green T.S. . . . 991
Malic Acid . . . **484**
DL-Malic Acid . . . 484
Malonic Ester . . . 243
Maltol . . . **487**
Mandarin Oil, Expressed . . **488**
Manganese . . . 961
***Manganese Chloride** . . . **489**
Manganese Gluconate . . . **491**
Manganese Glycerophosphate . . . **492**
Manganese Hypophosphite . . . **493**
Manganese Standard Preparation . . . 961
†**Manganese Sulfate** . . . **495**
Mannite . . . 496
Mannitol . . . **496**
D-Mannitol . . . 496
Marjoram Oil, Spanish . . . **499**
***Masticatory Substances, Natural** . . . **501**
Matching Fluids, Preparation of . . . 944
Mayer's Reagent . . . 991
Measures and Weights, Abbreviations . . . 6
Melting Range or Temperature . . . 931
p-Menth-8-en-3-ol . . . 419
d-*p*-Mentha-1,8-diene . . . 456
p-Mentha-1,5-diene . . . 597
3-*p*-Menthanol . . . 502
Menthen-1-ol-8 . . . 812
Menthen-1-yl-8 Acetate . . . 813
Menthen-1-yl-8 Propionate . . 814
Menthol . . . **502**
Mercuric Acetate T.S. . . . 991
Mercuric Chloride T.S. . . . 991
Mercuric-Potassium Iodide T.S. . . . 991
Mercuric-Potassium Iodide T.S., Alkaline . . . 992
Mercuric Sulfate T.S. . . . 992
Mercury Limit Test . . . 934
Methanol . . . 509
†**DL-Methionine** . . . **504**
†**L-Methionine** . . . **506**
Methods, Alternate . . . 1
p-Methoxyacetophenone . . . 12
****p*-Methoxybenzaldehyde** . . **507**
p-Methoxybenzyl Acetate . . . 61
p-Methoxybenzyl Alcohol . . . 62
Methoxyl Determination . . . 935
2-Methoxy-4-propenylphenol 412
2-Methoxy-4-propenyl Phenyl Acetate . . . 413
4′-Methyl Acetophenone . . **508**
Methylacetopyronone . . . 234
***Methyl Alcohol** . . . **509**
Methyl 2-Aminobenzoate . . . 512
p-Methylaminophenol Sulfate T.S. . . . 992
Methyl Amyl Ketone . . . 366
***p*-Methyl Anisole** . . . **511**
Methyl Anthranilate . . . **512**
Methyl Benzoate . . . **513**
α-Methylbenzyl Acetate . . **514**
α-Methylbenzyl Alcohol . . **515**
β-Methyl Butyl Acetate . . . 397
Methylcellulose . . . **516**
Methylcellulose, Viscosity of . 971
α-Methylcinnamaldehyde **517**
Methyl Cinnamate . . . **518**
Methyl *p*-Cresol . . . 511
3-Methylcyclopentane-1,2-dione . . . 519
†**Methyl Cyclopentenolone** . **519**
Methylene Blue . . . 1005
Methylene Blue T.S. . . . 992
***Methylene Chloride** . . . **520**
3,4-Methylenedioxybenzaldehyde . . . 620
***Methyl Ester of Rosin, Partially Hydrogenated** . **521**
Methylethylcellulose . . . **522**
Methyl Ethyl Ketone . . . 104
Methyl Eugenol . . . **524**
Methyl Formate . . . **525**
Methyl Glycol . . . 678
Methyl Heptenone . . . 527

6-Methyl-5-Hepten-2-one **527**
Methyl Heptine Carbonate. . 529
Methyl *p*-Hydroxybenzoate. 530
Methyl Isobutyl Ketone. 532
Methyl Isoeugenol. **528**
d-1-Methyl-4-isopropenyl-6-
cyclohexen-2-one. 176
l-1-Methyl-4-isopropenyl-6-
cyclohexen-2-one. 178
2-Methyl-3-(*p*-isopropyl-
phenyl)propionaldehyde. . . 225
Methyl *N*-Methyl
Anthranilate. 251
Methyl β-Naphthyl
Ketone. **529**
Methyl *n*-Nonyl
Acetaldehyde. 536
Methyl 2-Octynoate. **529**
Methyl Orange. 1005
Methyl Orange T.S. 992
Methylparaben. **530**
†**4-Methyl-2-pentanone**. . . . **532**
α-Methyl
Phenylacetaldehyde. 612
Methyl Phenylacetate. . . . **533**
Methyl Phenylcarbinol. 515
Methyl Phenylcarbinyl
Acetate. 514
Methylphenyl Ether. 60
Methyl Phenyl Ketone. 18
2-Methyl Propanoic Acid. . . 411
Methyl Propyl Ketone. 588
Methyl Red. 1006
Methyl Red-Methylene Blue
T.S. 992
Methyl Red Sodium. 1006
Methyl Red T.S. 992
Methylrosaniline Chloride
T.S. 992
Methyl Salicylate. **534**
Methyl *p*-Tolyl Ketone. 508
2-Methylundecanal. **536**
Methyl Violet T.S. 992
Methyl Yellow. 1006
Microbiological Criteria. 5
Millon's Reagent. 992
Mineral Oil, White. **536**
Modified Food Starch. 324
Molar Solutions. 996
Molecular Weight of
Polyethylene. 874
Polyisobutylene. 875
Polyvinyl Acetate. 875
Monoammonium Glutamate. 539
****Monoammonium**
L-Glutamate. **539**
Monoammonium Phosphate. 50
Monocalcium Phosphate. . . . 147
Mono- and Diglycerides. . . **538**
***Monoglyceride Citrate**. . . . **540**
1-Monoglycerides. 907
Monoglycerides, Total. 908
Monopotassium Glutamate. . 542
Monopotassium
L-Glutamate. **542**
Monopotassium Phosphate. . 663
Monosodium Dihydrogen
Phosphate. 757
Monosodium Glutamate. . . . 544
Monosodium L-Glutamate **544**
Monosodium Phosphate. 757
MPG. 542
MSG. 544
Murexide Indicator
Preparation. 1006
Mustard Oil, Volatile. 31
Myrcia Oil. 74
Myristic Acid. **545**
Myristica Oil. 556
Myrrh Oil. **546**

N

Naphthol Green B. 1006
Naphthol Green T.S. 993
Negligible, Definition. 3
Neral. 202
Nerol. **547**
Nerolidol. **548**
Nessler's Reagent. 993
Neutral Red. 1006
Neutral Red T.S. 993
Niacin. **549**
Niacinamide. **551**
Niacinamide Ascorbate. . . **552**
Nicotinamide. 551
Nicotinic Acid. 549
Ninhydrin T.S. 993
Nitre Cake. 728
Nitric Acid T.S., Diluted. . . . 993
Nitrogen Determination. . . . 937
Nomenclature. xi
γ-Nonalactone. **553**
Nonanal. **554**
1-Nonanol. 555
Nonyl Acetate. **555**
Nonyl Alcohol. **555**
Nonylcarbinol. 233
Nordihydroguaiaretic Acid. . ix
Normal Solutions. 996
Nutmeg Oil. **556**

O

Octadecanoic Acid 796
cis-9-Octadecenoic Acid 562
Octanal **558**
Octanoic Acid **558**
1-Octanol (Natural) **560**
Octyl Acetate **561**
Octyl Alcohol 560
Octyl Formate **561**
Odorless, Definition 3
Oil of Frankincense 564
Oil of Muscatel 213
Oil of Shaddock 360
Oleic Acid **562**
Olibanum Oil **564**
***Onion Oil** **564**
Optical Rotation 939
Orange Oil **565**
Orange Oil, Bitter **566**
Organization of the Food Chemicals Codex . . . . vii, ix
Origanum Oil, Spanish . . . **568**
Orris Root Oil **568**
Orthophenanthroline T.S. 993
Orthophosphoric Acid 616
Oxalic Acid, 0.1 *N* 999
Oxalic Acid T.S. 993
Oxyethylene Determination . 910
Oxystearin **569**

P

Pacific Kelp 424
Packaging and Storage Statements in Monographs 7
Palmarosa Oil **570**
Palmitic Acid **571**
Palmitoyl L-Ascorbic Acid . . . 68
Panthenol 236
***DL-Panthenol** **573**
D(+)-Pantothenyl Alcohol . . 236
DL-Pantothenyl Alcohol 573
Papain **574**
Parsley Seed Oil **576**
Peach Aldehyde 844
***Pectin** **577**
PEG 626
Pelargonic Aldehyde 554
Pennyroyal Oil **584**
Pennyroyal Oil, Imported . . . 584
***Pentaerythritol Ester of Partially Hydrogenated Wood Rosin** **585**
***Pentaerythritol Ester of Wood Rosin** **586**
†2,3-Pentanedione **587**
Pentanoic Acid 846
†2-Pentanone **588**
Pentapotassium Triphosphate 672
Pentasodium Triphosphate . . 780
Pentyl Hexanoate 400
Peppermint Oil **589**
Pepper Oil, Black **590**
†Pepsin **591**
Percentage of Cineole 898
Perchloric Acid, 0.1 *N* 999
Perchloric Acid, 0.1 *N*, in Dioxane 999
Petitgrain Oil, Paraguay . . **593**
Petrolatum **594**
Petrolatum, Liquid 536
Petrolatum, White 594
Petrolatum, Yellow 594
pH Determination 941
α-Phellandrene **597**
Phenethyl Acetate **598**
Phenethyl Alcohol **599**
α-Phenethyl Alcohol 515
Phenethyl Isobutyrate **600**
Phenethyl Isovalerate **601**
Phenethyl Phenylacetate . **602**
Phenethyl Salicylate **603**
Phenolphthalein 1006
Phenolphthalein Paper 1008
Phenolphthalein T.S. 993
Phenol Red 1006
Phenol Red T.S. 993
Phenols 898
Phenols, Free 899
Phenolsulfonphthalein T.S. . . 993
Phenoxyethyl Isobutyrate **603**
Phenylacetaldehyde **604**
Phenylacetaldehyde Dimethyl Acetal **605**
Phenylacetic Acid **606**
DL-Phenylalanine **607**
***L-Phenylalanine** **610**
Phenyl Carbinol 84
α-Phenyl Ethyl Acetate 514
2-Phenylethyl Acetate 598
2-Phenylethyl Alcohol 599
p-Phenylphenol T.S. 993
3-Phenyl-1-propanol **611**
3-Phenylpropenoic Acid 194
2-Phenylpropionaldehyde . **612**
3-Phenylpropionaldehyde . **613**
2-Phenylpropionaldehyde Dimethyl Acetal **614**
3-Phenylpropyl Acetate . . . **615**
Phenylpropyl Alcohol 611
Phenylpropyl Aldehyde 613
Phosphate Standard Solution 985

Phosphoric Acid **616**
Phosphorus 961
Phosphorus Standard
Solution 962
Phosphotungstic Acid T.S... 993
Picric Acid T.S............. 993
Pimenta Leaf Oil **616**
Pimenta Oil **618**
Pimento Leaf Oil........... 616
Pimento Oil............... 618
Pine Needle Oil...........323, 619
Pine Needle Oil, Dwarf.... **619**
Pine Needle Oil, Scotch... **619**
Piperonal............... **620**
Piperonyl Aldehyde......... 620
Platinum-Cobalt C.S........ 943
***Poloxamer 331**............ **621**
***Poloxamer 407**............ **624**
***Polyethylene**............. **625**
†**Polyethylene Glycols**..... **626**
Polyethylene, Molecular
Weight of............... 874
Polyglycerate 60........... 268
†**Polyglycerol Esters of
Fatty Acids**............ **629**
***Polyisobutylene**.......... **631**
Polyisobutylene, Molecular
Weight of............... 875
Poly[1-(2-oxo-1-pyrroli-
dinyl)ethylene].......... 686
Polyoxyethylene (20) Mono-
and Diglycerides of Fatty
Acids.................. 268
Polyoxyethylene (20)
Sorbitan Monolaurate.... 632
Polyoxyethylene (20)
Sorbitan Monooleate..... 637
Polyoxyethylene (20)
Sorbitan Monostearate.... 634
Polyoxyethylene (20)
Sorbitan Tristearate...... 635
Polysorbate 20........... **632**
Polysorbate 60........... **634**
Polysorbate 65........... **635**
Polysorbate 80........... **637**
***Polyvinyl Acetate**......... **638**
Polyvinyl Acetate, Molecular
Weight of............... 875
Polyvinylpyrrolidone....... 686
Potassium Acetate T.S...... 993
Potassium Acid Phthalate,
0.1 *N*.................. 999
Potassium Acid Tartrate.. **639**
Potassium Alginate....... **641**
Potassium Alum........... 36
Potassium Bicarbonate... **642**
Potassium Biphosphate..... 663
Potassium Bitartrate....... 639
Potassium Bromate...... **643**
Potassium Carbonate..... **644**
**Potassium Carbonate
Solution**............... **645**
Potassium Chloride...... **646**
Potassium Chromate T.S.... 993
Potassium Citrate........ **647**
Potassium Dichromate, 0.1 *N* 1000
Potassium Dichromate T.S.. 993
Potassium Dihydrogen
Phosphate............... 663
Potassium Ferricyanide T.S.. 993
Potassium Ferrocyanide T.S. 993
***Potassium Gibberellate**... **649**
Potassium Glutamate....... 542
**Potassium
Glycerophosphate**...... **651**
Potassium Hydroxide..... **652**
Potassium Hydroxide, 1 *N*.. 1000
Potassium Hydroxide, 0.5 *N*,
Alcoholic................ 1000
**Potassium Hydroxide
Solution**............... **653**
Potassium Hydroxide T.S.... 994
Potassium Hydroxide T.S.,
Alcoholic................ 994
Potassium Iodate......... **654**
Potassium Iodate, 0.05 *M*... 1000
Potassium Iodide......... **656**
Potassium Iodide T.S....... 994
Potassium Kurrol's Salt..... 666
Potassium Metabisulfite.. **657**
Potassium Metaphosphate.. 666
Potassium Nitrate........ **659**
Potassium Nitrite........ **660**
Potassium Permanganate, 0.1
N..................... 1000
Potassium Permanganate
T.S.................... 994
**Potassium Phosphate,
Dibasic**................ **661**
**Potassium Phosphate,
Monobasic**............. **663**
**Potassium Phosphate,
Tribasic**................ **664**
**Potassium
Polymetaphosphate**.... **666**
Potassium Pyrophosphate **667**
Potassium Pyrosulfite....... 657
Potassium Sorbate....... **668**
Potassium Sulfate........ **670**
Potassium Sulfate T.S....... 994
Potassium Sulfite........ **670**
Potassium Triphosphate.... 672
***Potassium
Tripolyphosphate**...... **672**

Povidone.................. 686
Preface, First Edition....... xiv
Preface, Second Edition..... ix
*L-**Proline**.................. **673**
1,2-Propanediol............. 678
2-Propanol................ 417
2-Propanone............... 15
p-Propenylanisole.......... 56
Propenylguaethol........ **674**
4-Propenyl Veratrole....... 528
†**Propionaldehyde**......... **675**
Propionic Acid........... **676**
***p*-Propyl Anisole**.......... **678**
Propylene Chlorohydrin..... 962
Propylene Glycol......... **678**
Propylene Glycol Alginate **680**
Propylene Glycol Ether of Methylcellulose.......... 387
Propylene Glycol and Glycerin, Free........... 904
Propylene Glycol Lactostearate............ 430
Propylene Glycol Monostearate.......... **682**
Propyl Gallate............ **683**
Propyl *p*-Hydroxybenzoate.. 685
Propylparaben........... **685**
***PVP**.................... **686**
3-Pyridinecarboxylic Acid... 549
Pyridoxine Hydrochloride. **689**
Pyromucic Aldehyde....... 333
L-2-Pyrrolidinecarboxylic Acid.................. 673

Q

Quinaldine Red............ 1006
Quinaldine Red T.S......... 994
Quinine Hydrochloride... **690**
Quinine Sulfate.......... **692**
Quinones................. 876

R

Racemic Pantothenyl Alcohol 573
Readily Carbonizable Substances.............. 942
Reagents, Specifications for.. 4
Reference Standards........ 4
Reference Substances, Use of, in Chromatographic Identity Tests........... 883
Refractive Index........... 945
Reichert-Meissl Value...... 912
Residual Styrene........... 877
Residue on Evaporation.... 899
Residue on Ignition........ 945
Rhodinol................. **694**
Rhodinyl Acetate......... **695**
Rhodinyl Formate........ **696**
Riboflavin............... **697**
Riboflavin 5′-Phosphate Ester Monosodium Salt... 699
***Riboflavin 5′-Phosphate Sodium**................ **699**
***Rice Bran Wax**........... **702**
Ring-and-Ball Method, Softening Point.......... 948
Rochelle Salt.............. 760
Rose Geranium Oil, Algerian. 336
Rosemary Oil............ **704**
Rose Oil................. **703**
Rosins, Rosin Esters, and Related Substances, General Tests............ 945
Rue Oil.................. **705**
Rum Ether (So-called)...... 298

S

Saccharin................ **706**
Sage Oil, Dalmatian...... **707**
Sage Oil, Spanish......... **709**
Samples, Analytical........ 2
Sample Solution for Arsenic Test (Chewing Gum Base). 877
Sample Solution for Lead Limit Test (Chewing Gum Base).................. 878
Sandalwood Oil, East Indian................ **709**
Santalol................. **710**
Santalyl Acetate.......... **711**
Saponification Value (Essential Oils and Related Substances)...... 896
Saponification Value (Fats and Related Substances).. 914
Savory Oil (**Summer Variety**)............... **712**
SBR 1027 Type Solid....... 102
SBR 1028 Type Solid....... 103
SBR 2000 Type Latex...... 103
SBR 2006 Type Latex...... 102
Schiff's Reagent, Modified... 994
Scope of Food Chemicals Codex................ ix
Selenium Limit Test........ 953
Selenium Standard Solution. 953
Selenium Stock Solution.... 953
*DL-**Serine**............... **713**
*L-**Serine**................. **714**
Sieve Analysis of Granular Metal Powders.......... 953

Significant Figures 4
Silica Aerogel 716
Silicon Dioxide **716**
Silver Nitrate, 0.1 *N* 1001
Silver Nitrate T.S. 994
Soap 915
Soda Alum 37
Sodium Acetate **717**
Sodium Acetate, Anhydrous **718**
Sodium Acid Pyrophosphate **719**
Sodium Acid Sulfate 728
Sodium Alginate **721**
Sodium Alum 37
Sodium Aluminosilicate 767
Sodium Aluminum Phosphate, Acidic **722**
†**Sodium Aluminum Phosphate, Basic** **724**
Sodium Arsenite, 0.05 *N* 1001
Sodium Ascorbate **725**
Sodium L-Ascorbate 725
Sodium Benzoate **726**
Sodium *o*-Benzosulfimide 764
Sodium Bicarbonate **727**
Sodium Biphosphate 757
Sodium Bisulfate **728**
Sodium Bisulfite **729**
Sodium Bisulfite T.S. 994
Sodium Bitartrate T.S. 994
Sodium Borate T.S. 994
Sodium Carbonate **730**
Sodium Carbonate T.S. 994
Sodium Carboxymethylcellulose **731**
Sodium Carboxymethylcellulose, Viscosity of 973
Sodium Chloride **734**
Sodium Citrate **736**
Sodium Cobaltinitrite T.S. . . 994
Sodium Cyclamate ix
Sodium Dehydroacetate . . **737**
Sodium Diacetate **738**
Sodium Erythorbate **739**
Sodium Ferric Pyrophosphate **740**
Sodium Ferrocyanide **741**
Sodium Fluoride T.S. 994
Sodium Gluconate **742**
Sodium Glutamate 544
Sodium 5′-Guanylate 261
Sodium Hexametaphosphate . 749
Sodium Hydrogen Diacetate . 738
Sodium Hydroxide **743**
Sodium Hydroxide Solution **744**
Sodium Hydroxide, 1 *N* 1001
Sodium Hydroxide T.S. 994
Sodium 3-(1-Hydroxyethylidene)-6-methyl-1,2-pyran-2,4(3*H*)-dione 737
Sodium Hyposulfite 778
Sodium Indigotindisulfonate T.S. . . 994
Sodium 5′-Inosinate 263
Sodium Iron Pyrophosphate . 740
Sodium Lauryl Sulfate **746**
Sodium Metabisulfite **748**
Sodium Metaphosphate . . **749**
Sodium Metaphosphate, Insoluble 749
Sodium Methoxide 751
Sodium Methoxide, 0.1 *N*, in Pyridine 1001
***Sodium Methylate** 751
Sodium Nitrate **754**
Sodium Nitrite **755**
Sodium Nitrite, 0.1 *M* 1002
Sodium Nitroferricyanide T.S. 994
Sodium Phosphate, Dibasic **756**
Sodium Phosphate, Glassy . . 749
Sodium Phosphate, Monobasic **757**
Sodium Phosphate, Tribasic **758**
Sodium Phosphate T.S. 994
Sodium Polyphosphate 749
Sodium Potassium Tartrate **760**
Sodium Propionate **761**
Sodium Pyrophosphate . . . **762**
Sodium Pyrosulfite 748
Sodium Saccharin **764**
Sodium Sesquicarbonate . **765**
Sodium Silicoaluminate . . **767**
***Sodium Stearoyl-2-Lactylate** **770**
Sodium Stearyl Fumarate **773**
***Sodium Sulfate** **775**
Sodium Sulfide T.S. 994
Sodium Sulfite **776**
Sodium Sulfite, Exsiccated . . 776
Sodium Tartrate **777**
Sodium Tetrametaphosphate 749
Sodium Tetraphosphate 749
Sodium Thiosulfate **778**
Sodium Thiosulfate, 0.1 *N* . . . 1002
Sodium Thiosulfate T.S. 994
Sodium Trimetaphosphate . . 749
Sodium Triphosphate 780
Sodium Tripolyphosphate **780**

Softening Point 946
Solidification Point 954
Solubility in Alcohol 899
Solubility, Descriptive Terms 6
Solubility Statements in Monographs, Significance of 6
Soluble Saccharin 764
Solutions, Colorimetric 983
Solutions, Defined 4
Solutions, Standard Buffer .. 983
Solutions, Test 985
Solutions, Volumetric 996
Sorbic Acid **783**
Sorbitan Monostearate ... **784**
D-Sorbite 786
Sorbitol **786**
D-Sorbitol 786
Sorbitol Solution **788**
D-Sorbol 786
Spearmint Oil **792**
Specifications, General, of Codex 6
Specifications, Sources of ix
Specific Gravity 5
Specific Gravity (Fats and Related Substances) 915
Specific Rotation 939
Spectrophotometry, Colorimetry, and Turbidimetry . 957
Spike Lavender Oil **793**
Standard Buffer Solutions ... 983
Standard Lead Solution 921
Standard Lead Solution, Diluted 929
Standards, Codex 1
Standard Solutions 984
***Stannous Chloride** **794**
Stannous Chloride T.S. 995
Starches and Related Substances, General Tests 959
Starch, Food, Modified 324
Starch Iodate Paper 1008
Starch Iodide Paper 1008
Starch Iodide Paste T.S. 995
Starch T.S. 995
Stearic Acid **796**
Stearyl Monoglyceridyl Citrate **797**
Sterculia Gum 423
Strawberry Aldehyde 295
Styrene, Bound 873
Styrene, Residual 877
Succinic Acid **800**
***Succinylated Monoglycerides** **800**
Sulfanilic Acid T.S. 995
Sulfate and Chloride Limit Test 879
Sulfate Standard Solution ... 879
Sulfur Dioxide 964
Sulfuric Acid **802**
Sulfuric Acid, 1 *N* 1003
Sulfuric Acid, Alcoholic, 5 *N*. 1003
Sulfuric Acid, Alcoholic, 0.5 *N* 1003
Sulfuric Acid Table 965
Sulfuric Acid T.S. 995
Sulfuric Acid T.S., Diluted .. 995
Summer Savory Oil 712
Sweet Orange Oil 565
Sweetwood Bark Oil 180

T

Tangerine Oil, Expressed .. **804**
Tannic Acid **806**
Tannic Acid T.S. 995
Tarragon Oil **807**
Tartaric Acid **808**
L(+)-Tartaric Acid 808
Temperatures for Test and Assay Procedures 5
***Terpene Resin, Natural** ... **809**
***Terpene Resin, Synthetic**. **811**
Terpineol **812**
Terpinyl Acetate **813**
Terpinyl Propionate **814**
Test Papers and Indicator Papers 1007
Test Solutions (T.S.) 5, 985
Tests and Assays 2
Tetradecanoic Acid 545
Tetrahydrogeraniol 254
Tetrapotassium Pyrophosphate 667
Tetrasodium Diphosphate ... 762
Tetrasodium Pyrophosphate. 762
Thermometers 966
Thiamine Assay 968
Thiamine Chloride 815
Thiamine Hydrochloride. **815**
Thiamine Mononitrate ... **816**
Thiamine Nitrate 816
Thorium Nitrate, 0.1 *M* 1003
***L-Threonine** **818**
Thuja Oil 183
Thyme Oil **819**
Thymol Blue 1006
Thymol Blue T.S. 995
Thymolphthalein 1006
Thymolphthalein T.S. 995
Time Limits, Defined 5
Title, Form of Codex 1

Titles of Codex Monographs, Former and Current...... 1009
dl-**Alpha Tocopherol**...... **820**
d-Alpha Tocopherol Acetate. 824
dl-Alpha Tocopherol Acetate. 825
d-Alpha Tocopherol Acetate Concentrate............ 826
d-Alpha Tocopherol Acid Succinate.............. 829
Tocopherols Concentrate, Mixed................. **821**
Tocopherols, Weight-Unit Relationships........... 969
d-**Alpha Tocopheryl Acetate**............... **824**
dl-**Alpha Tocopheryl Acetate**............... **825**
d-**Alpha Tocopheryl Acetate Concentrate**.... **826**
d-**Alpha Tocopheryl Acid Succinate**............. **829**
Tolerances, Purity......... 5
Toluene Distillation Method for Water Determination.. 979
α-Toluic Acid.............. 606
α-Toluic Aldehyde.......... 604
p-Tolyl Acetate............ 222
p-**Tolyl Isobutyrate**....... **831**
Total Alcohols............. 893
Total Monoglycerides....... 908
Total Unsaturation......... 878
Trace Impurities........... 5
Tragacanth............. **832**
Triacetin................ **833**
Tributyrin.............. **835**
Tricalcium Phosphate...... 149
***Trichloroethylene**........ **836**
Triethanolamine, 0.5 *N*..... 1003
Triethyl Citrate.......... **837**
3,7,12-Trihydroxycholanic Acid................... 188
Triketohydrindene Hydrate T.S.................... 995
Trimagnesium Phosphate... 477
4(2,6,6-Trimethyl-1-cyclohexenyl)-3-butene-2-one... 392
4(2,6,6-Trimethyl-2-cyclohexenyl)-3-butene-2-one... 391
3,7,11-Trimethyl-1,6,10-dodecatrien-3-ol.......... 548
1,3,7-Trimethylxanthine.... 115
Trinitrophenol T.S.......... 995
Triphosphate.............. 780
Tripotassium Phosphate.... 664
Trisodium Phosphate....... 758
DL-**Tryptophan**........... **838**
*L-**Tryptophan**........... **841**
Turbidimetry Comparisons.. 959
L-**Tyrosine**............... **842**

U

Ultraviolet Absorbance of Citrus Oils.............. 900
γ-**Undecalactone**......... **844**
Undecanal............... **845**
***10-Undecenal**............. **846**
Undecene-10-al............ 846
n-Undecyl Aldehyde........ 845
Unsaponifiable Matter...... 915
Unsaturation, Total........ 878

V

Vacuum, Defined.......... 6
†**Valeric Acid**.............. **846**
*L-**Valine**.................. **849**
Vanillin................... **850**
Viscosity (Rosins, Rosin Esters, and Related Substances)............. 952
Viscosity of Dimethylpolysiloxane..... 970
Viscosity of Methylcellulose. 971
Viscosity of Sodium Carboxymethylcellulose... 973
Visual Color Comparisons... 959
Vitamin A............... **851**
Vitamin A Assay........... 974
Vitamin B_1................ 815
Vitamin B_1 Hydrochloride... 815
Vitamin B_1 Mononitrate.... 816
Vitamin B_2................ 697
Vitamin B_6 Hydrochloride... 689
Vitamin C................ 66
Vitamin C Sodium......... 725
Vitamin $\mathbf{D_2}$.............. **853**
Vitamin $\mathbf{D_3}$.............. **855**
Volatile Acidity........... 916
Volatile Oil Content........ 901
Volumetric Apparatus...... 976
Volumetric Solutions....... 996

W

Water Determination, Karl Fischer Titrimetric Method................. 977
Water Determination, Toluene Distillation Method... 979
Water and Loss on Drying.. 3
Weights and Balances...... 980
Weights and Measures, Abbreviations........... 6

West Indian Sandalwood Oil. 54
White Wax. 75
Wijs Method for Iodine. 906
Wine Yeast Oil. 217
Wintergreen Oil. 534
Wool Fat. 437

X

*__Xanthan Gum__. **856**
Xylenol Orange. 1006
Xylenol Orange T.S. 995

Y

Yellow Prussiate of Soda. 740
Yellow Wax. 77

Z

*__Zinc Sulfate__. **859**
Zinc Sulfate, 0.05 *M*. 1003